Partially Ordered Systems

Partially Ordered Systems

For further volumes, go to
http://www.springer.com/series/1212

K.L. Ngai

Relaxation and Diffusion in Complex Systems

Springer

K.L. Ngai
CNR-IPCF Associate
Dipartimento di Fisica
Università di Pisa
Largo Bruno Pontecorvo 3, I-56127
Pisa, Italy
Formerly at Naval Research Laboratory
Washington, DC, USA
ngai@df.unipi.it

ISSN 0941-5114
ISBN 978-1-4419-7648-2 e-ISBN 978-1-4419-7649-9
DOI 10.1007/978-1-4419-7649-9
Springer New York Dordrecht Heidelberg London

Printed on acid-free paper

Springer is part of Springer Science+Business Media (www.springer.com)

Preface

Glass transition occurs on cooling or compressing a liquid when the characteristic time of molecular motions responsible for structural rearrangements becomes longer than the timescale of the experiment. As a result, structural relaxation toward equilibrium is arrested below some temperature, T_g, and the material is in the glassy state. This general phenomenon is found in condensed matters of diverse chemical composition and physical structure, and is of interest in research and applications in many disciplines. Considered as one of the many branches of study in physical sciences, glass transition is exceptional in the sense that the problem it presents has a very long history but still not solved till the present time. If one counts from the period of 1854 to 1866 when R. Kohlrausch and F. Kohlrausch in Göttingen, Germany published their scientific studies of relaxation in glassy materials, then the research field of relaxation and diffusion in glasses and liquids is by now about 160 years old. One may even argue that the field is more than two millennia old if starts counting from the days of the Babylonians recording their recipe of making glasses and studies of glasses on stone tablets. It is highly unusual that a problem in physics with such a long history dating back to the dawn of civilization has remained unsolved. Since the last century, many experimental techniques have been invented to macroscopically as well as microscopically examine the properties of glass-forming substances. The database has become so rich that the general and critical facts on glass transition have emerged and are ready for anyone to take into consideration in attempting to solve the problem. There have been many serious attempts to solve the problem throughout the past several decades. The activity is accelerated in recent years by the recognition of the challenge and the fundamental importance of the glass transition problem by many more scientists, Nobel laureates, and even *The New York Times*. Despite the current intensified efforts, it is a general consensus that solution of the problem is still at large, as reported in a 2008 editorial of *The New York Times*.

A modest decrease of temperature can change the viscosity or the structural relaxation time by many orders of magnitude to cause glass transition. The spectacular property germane to glass transition had lured many researchers to focus attention mainly on this experimental fact and to explain the phenomenon expediently by solely focusing on the change in thermodynamic quantities such as specific volume and configurational entropy. There is a general lack of recognition of the fact that at constant temperature, the intermolecular, interatomic, or interparticle interactions in the glassformer necessarily engender many-body effect in the dynamics of the basic relaxing or diffusing units. The interaction between basic units in glassformers is usually non-trivial and anharmonic, resulting in complex (chaotic) dynamics in the phase space spanned by the coordinates and momenta of the interacting units themselves. One of the major endeavors of this book is to collect multiple facts from experiments as convincing evidences for the presence of the many-body effects in the dynamics that cannot be ignored. To do this, a surrogate for many-body effects has to be used to bring out the evidences from experiments. Perhaps the simplest and the most direct choice of the surrogate is the degree of stretching of the structural relaxation time correlation function $\phi(t)$ beyond the exponential function for non-interacting systems. If the Kohlrausch's stretched exponential function $\phi(t) = \exp[-(t/\tau)^{1-n}]$ is used, then the fractional quantity n appearing in the stretch exponent is a natural choice for the surrogate. An equivalent one is the measured width $\mathcal{W}$ of the frequency dispersion of the susceptibility. Both n and $\mathcal{W}$ are easily accessible from experimental measurements, and they either govern or correlate with the relaxation time τ in its magnitude and its dependences on various parameters at a fixed temperature or on changing temperature. I submit these findings are indisputable evidences for the presence of many-body effects in relaxation and diffusion, which must be incorporated in conjunction with thermodynamic factors into any theory of glass transition. Otherwise, the theory is a failure because it certainly cannot explain the multiple experimental facts indicating that n or $\mathcal{W}$ governs or correlates with the relaxation time τ. Remarkably and surprisingly, in most if not all conventional or popular glass transition theories or models, n or $\mathcal{W}$ does not play any important role and is in some cases not considered at all. This state of affair is one of several reasons why the glass transition problem has not yet been solved.

Unfortunately, relaxation and diffusion in interacting systems, the origin of the many-body effects observed in the glass transition problem, is itself an unsolved problem in statistical mechanics. My encounter with this problem is an example of serendipity in scientific research. By accident, in 1978, I was obliged to attend a seminar on dielectric relaxation of liquids and glass. Prior to 1978, my research was limited to electronic properties using quantum mechanical methods, and I had absolutely no knowledge or appreciation of research in relaxation in classical systems, let alone the glass transition problem. The seminar got me exposed to the research field and stimulated my realization that these interacting systems and many-body effects must be important and necessarily be accounted for in the relaxation and diffusion processes. Since I had no previous knowledge, I was free to think about the problem without any preconceived notion or the influence by established schools of thought. My modest contribution to the problem published in 1979 (known nowadays as the coupling model) yielded predictions of the many-body effects in relaxation and diffusion of interacting systems. The central prediction from the coupling model (CM) equation is general and it spawns many different subsidiary or specific predictions when applied to various processes in different kinds of interacting systems. Despite the weak theoretical foundation and framework of the original CM, the many predictions it readily offers suggested to me that the best strategy for proceeding was to test these predictions with available experimental data at the outset. If the predictions fail, it would be timely for me to abandon the model and the sojourn in this new area, and return to my previous occupation. To my surprise, the initial tests of the predictions against experiments were successful. After that, I realized that the only way to really test the predictions was to get deeper into the various research fields. This decision forced me to change the course of my scientific career. The continued success of the applications of the CM predictions to increasing areas of research enticed me to explore more areas of application. This had the adverse effect of preventing me from bettering the theoretical foundation of the CM for many years, until more than 15 years later by improvements based on classical chaos, which connects back to the original model based on Gaussian orthogonal ensemble of random matrices, which is related to classical chaos via semiclassical quantization. The central prediction of the CM remains the same for more than 31 years. By no means can the CM be considered as a full solution of the problem of relaxation and diffusion in interacting systems. Notwithstanding, its constructs and predictions are in accord with numerous experiments. Until someone has a better model or theory that can do as much or more than the CM in relation with experiments, I hope I need not to apologize for its shortcomings

The application of the CM to the glass transition problem is natural because glass-forming liquids are just interacting systems belonging to a special class. The voluminous amount of experimental data accumulated over the years on glass-forming substances and systems presents great opportunity of testing the CM predictions. Also there are some occasions where predictions from the CM were made before experiments were carried out, and the results of the experiments were in accord with the forecast. Again the disclaimer must be made that the CM is not a full solution of the glass transition problem, although it accounts well for the many-body effects of the relaxation and diffusion processes, unrivaled by any extant theory of glass transition. One of the constructs in the CM is the primitive relaxation which initiates the evolution of many-body relaxation to reach the terminal steady state of structural relaxation responsible for glass transition. A remarkable achievement of the CM is the realization of the primitive relaxation in experiments by the secondary relaxation of a special kind involving the motion of the entire basic unit of the glass-former. There is a need to distinguish this class of secondary relaxation from the trivial ones. For this purpose, I have called it the Johari–Goldstein (JG) β-relaxation. Not only does this term distinguish the secondary relaxation of the special kind but it also serves to honor these two colleagues for their discovery of secondary relaxation in totally rigid small molecular glassformers, which necessarily involve the motion of the entire molecule. The relaxation time and the properties of the JG β-relaxation are analogous to that of the primitive relaxation of the CM. Like the primitive relaxation, the JG β-relaxation and the structural relaxation are inseparable as evidenced by many relations or connections between the two relaxations found by experiments in glassformers of many different kinds. Hence, the primitive relaxation or the JG β-relaxation plays a fundamental role in the dynamics of glassformers, and no glass transition theory is complete without considering it and explaining its connection with the structural relaxation. Moreover, there is abundant experimental evidence to show that the primitive relaxation or the JG β-relaxation time is not only sensitive to thermodynamic variables but also a function of exactly the same combination of thermodynamics variables as the primary (structural) relaxation time. From causality, these experimental facts indicate that the dependence of the structural relaxation on thermodynamic variables actually originates from that of the primitive relaxation or the JG β-relaxation, and is magnified in the former by the participation of many units than in the latter. This finding may revolutionize theoretical approaches solving the glass transition problem in the future. There are considerable interests in the early time dynamics manifested by molecules when they are mutually caged by the intermolecular potential. Experimental data also show that this caged dynamics regime is terminated when time becomes comparable to the primitive relaxation or the JG β-relaxation time. This is yet another important role played by the primitive relaxation or the JG β-relaxation time.

If the physics of relaxation and diffusion in interacting systems is general as anticipated by the CM, properties found in the glassformers should show up as analogues in other interacting systems that have nothing to do with glass transition. The final part of the book is devoted to the collection of the experimental measurements and computer simulations to show plenty of analogues of general properties of glassformers in diverse non-glass-forming interacting systems. By the way, this is the reason why in the title of the book I use the term complex systems to include other non-glass-forming interacting systems. The results combined support the existence of universal properties of relaxation and diffusion in interacting systems. It is a very important conclusion because of the following reasons. (1) It identifies a very fundamental problem in physics not pointed out by anyone before that is certainly worthwhile for the theoretical physics community to pursue and provide a solution. (2) The universal phenomena associated with this problem help experimentalists in many different areas of research in relaxation and diffusion in interacting systems (including the glass transition community) to understand their data, especially the anomalies, in the framework of many-body effects caused by interactions and plan more critical experiments. (3) It provides guidance for direction or strategy in application of the systems in technology and industry. (4) The general phenomenology found across different systems in various disciplines and described in the book should provide inspiration for all readers to explore further on their own in this new domain. Throughout the book, the CM is used successfully to rationalize or even quantitatively explain the universal features in the experimental data and to resolve the analogous anomalies found in different interacting systems. The success of the CM suggests it can be used as a basis for others to construct a more rigorous model or theory that is equally or even more successful in explaining experimental findings. Until the time when this more rigorous theory appears in the future, the CM is a useful tool to solve problems of relaxation and diffusion in interacting complex systems encountered in the future. The reader may be as surprised as I am that such a simple CM equation is applicable to so many different systems and problems. If this CM equation withstands the test of time and can be considered as truth at least to some degree, then the quote from Lord Byron, "For truth is always strange; stranger than fiction," seems fitting to describe the situation.

A major part of this book has been to collect the results of many experimental investigations cited in more than 1800 references. Only part of these is reflected in the few hundred figures selected. Without this large collection of experimental data, there would have been no way for me to bring out the universal behavior of relaxation and diffusion in interacting systems, and test the CM predictions. Therefore I salute all the experimentalists who have contributed to the advancement of our knowledge in this new exciting area.

In some sections of the book, connections of the basic physics are made to practical problems encountered in various technological areas and industries. Examples include the problem of improved solubility and bioavailability of pharmaceuticals, the stabilization of lyophilized biopharmaceuticals used in medical treatments, hydration of biomolecules for functioning in life science, fuel cell materials for the energy industry, and novel materials for improved performance in applications. I show that the basic features of dynamics of the systems in these other fields are isomorphic to glassformers, which is perhaps unsurprising because they are all interacting systems and the same physics of many-body relaxation and diffusion basically governs their behavior. I have demonstrated that the understanding of the relaxation and diffusion properties of prototypical glass-forming systems and substances gained has benefited some progress in these applied research fields important for the populace and the economy worldwide, and I expect it will bring more benefits in the future by more cross-field fertilization.

Considering many of the substances and systems discussed in this volume are so complex and widely different in chemical composition and physical structure, it is remarkable that universal properties of relaxation and diffusion exist and are shared by all of them. It seems to be the consequence of another unexplored fundamental physical law of Nature for irreversible processes in interacting systems. Only in the presence of a physical law can universal behavior be coming out from diverse condensed matter and systems. Capturing this physical law is a worthwhile scientific endeavor of the new century.

Acknowledgments

The research carried out by the author on the subject in the period from 1979 till 2010 would not be possible without the support, encouragement, and/or collaboration of many colleagues. I use this opportunity to express my heartfelt gratitude to them. Their names arranged in chronological order of my first association with each of them on subjects or events related to research discussed in this book are given as follows: Sidney Teitler, A. K. Jonscher, Carter T. White, P. W. Anderson, L. Lam, F.S. Liu, Donald J. Plazek, A. K. Rajagopal, R. W. Rendell, George B. Wright, Ulrich Strom, James Culbertson, Albert F. Yee, A. A. Jones, John Aklonis, G. B. McKenna, H. Jain, C. A. Angell, Gyan Johari, Satoru Mashimo, George Fytas, C. H. Wang, Ben Chu, Tadeusz Pakula, E.W. Fischer, G. Floudas, A. K. Rizos, T. V. Ramakrishnan, A. Schönhals, Otmar Kanert, Klaus Funke, John D. Ferry, T. P. Lodge, Roland Böhmer, A. Loidl, C. T. Moynihan, Ian Hodge, J. Colmenero, A. Arbe, U. Buchenau, R. Zorn, Ranko Richert, A. P. Sokolov, S. W. Martin, C. M. Roland, G. M. Borsuk, Peter Lunkenheimer, S.-L. Peng, K.-Y. Tsang, George Phillies, Bo Nyström, C. Xiao, Neville Greaves, Yening Wang, L. Magalas, S. Etienne, L. David, Carlos León, J. Santamaria, A. Rivera, T. Atake, M. Oguni, M. Hanaya, O. Yamamuro, C. L. Soles, J. H. Magill, Marian Paluch, M. Mierzwa, H. Z. Cummins, Riccardo Casalini, Junko Habasaki, Y. Hiwatari, F. Affouard, P. Bordat, Marc Descamps, Simone Capaccioli, Daniele Prevosto, Pierangelo Rolla, Mauro Lucchesi, Dino Leporini, K. Kessaire, Mario Beiner, A. Nogales, T. Ezquerra, M.-C. Bellissent-Funel, T. Odagaki, M. Shahin Thayyil, Naoki Shinyashiki, S. J. Rzoska, S. Pawlus, Li-Min Wang, and Giancarlo Ruocco. My special thanks to Sid Teitler for bringing me from MIT Lincoln Lab to NRL in 1971, to Pierangelo Rolla, Mauro Lucchesi, and Simone Capaccioli for providing me the opportunity of doing collaborative research with their group at the Università di Pisa in the past 5 years and into the future, and to S. Capaccioli, R. Casalini, J. Habasaki, C. León, M. Paluch, D. Plazek, D. Prevosto, A. K. Rajagopal, R. W. Rendell, and C.M. Roland for many years of fruitful collaborations. Finally, but not the least, I thank my wife, Linsen, and my daughters, Sianne, Seagan, and Serin, for their loving care throughout the years of my pursuit in scientific research.

Washington, District of Columbia K.L. Ngai

Contents

Chapter 1
Introduction to the Problems of Relaxation and Diffusion in Complex Systems

1.1 Historical Perspective

Relaxation and diffusion are physical and chemical processes that occur by various ways in condensed matter of all kinds (inorganic, organic, polymeric, biomolecular, colloidal, and metallic), in different states (crystalline, glassy, liquid, molten, and ceramic) and of different sizes or dimensions (from bulk to nanometer). The processes are often involved in the formation of natural substances and in the fabrication of synthetic materials. For the latter, we can go back to ancient history of the Babylonians making glasses out of sand. The recipe of making glasses, which is recorded in history, involves rapid cooling of the liquid material at a high temperature to a lower temperature to form the glass. The procedure necessarily requires change (i.e., relaxation) of the structure of the material from the thermal equilibrium state of the liquid to that of the glassy state in a time span determined by the cooling rate. In modern times, the fabrication of electronic devices requires the use of amorphous polymeric thin films for lithography and glassy oxides and nitrides as insulators. The quality of the devices depends on the structures of these glassy materials, which in turn depend on the relaxation processes in arriving at the structures. Pharmaceutical industry is interested to develop amorphous solid pharmaceuticals for oral delivery of water-insoluble compounds. This is because a pharmaceutical in the amorphous solid state has improved solubility and dissolves more rapidly than in the crystalline state. The stability of an amorphous pharmaceutical against slow crystallization is a major concern and it is also determined by relaxation and diffusion processes. For different purpose but for the same reason is the practice in food science and in the preservation of biomaterials where relaxation and diffusion are the controlling factors. In geoscience, the relaxation in georelevant melts including silicate melts is important for the estimation of magma properties, which are of great importance in volcanology and petrology. The many different applications of polymeric materials in materials engineering, aerospace, and everyday living are reflections of the tremendous range of relaxation (i.e., viscoelastic and rheological) properties that can be realized by manipulating the chemical structure. Conductivity relaxation and mobility of ions determine the feasibility of the ionic conducting materials for battery and fuel cells in energy source applications. The list

K.L. Ngai, *Relaxation and Diffusion in Complex Systems*, Partially Ordered Systems,
DOI 10.1007/978-1-4419-7649-9_1,

of the important roles played by relaxation and diffusion in diverse research fields and technological applications goes on and on.

The development of the study of relaxation and diffusion processes as a branch of physics possibly started with the experimental study by the Scottish botanist Robert Brown in 1827 [1]. He noticed under the lens of the microscope that pollen particles suspended in water jiggled about, and the motion of a particle follows a zigzag path. This phenomenon is now known as Brownian motion or diffusion. The next important work seems to be the measurements of electrical relaxation of alkali ions in the Leyden jar (a glass) in 1854 by R. Kohlrausch [2] and mechanical relaxation in the natural polymer, silk, in 1863 and 1866 by his son, F. Kohlrausch, both from Göttingen, Germany [3, 4]. They found that the decays of the relaxing quantities have time dependences given by

$$\phi_K(t) = \exp[-(t/\tau)^{1-n}], \text{ where } 0 < (1-n) \leq 1, \tag{1.1}$$

now named after Kohlrausch or technically described as the stretched exponential functions. R. Kohlrausch is a contemporary of C.F. Gauss and W.E. Weber at Göttingen. He also performed the important Weber–Kohlrausch experiment in 1856 to show that somehow a velocity (of light) enters into the electromagnetic picture. The experiment might have some influence on James Clerk Maxwell in the construction of his electromagnetic theory of light in the ensuing years from 1856 to 1868.

James Clerk Maxwell (1831–1879), a giant in modern physics, made an important contribution to the conceptual foundation of relaxation. Maxwell made use of the observation by others from measurements of glaciers that they move like liquid over long periods of time and introduced in 1867 the concept of "relaxation time" into physics, engineering, and glaciology. A material or a system such as a glacier acts like a solid at times shorter than the relaxation time but like a liquid at longer times. This contribution of Maxwell can be considered as the starting point in the development of the branch of study of relaxation now called rheology and viscoelasticity. Other contributions by Maxwell related to the concept of relaxation time include the Maxwell fluid [5], the Maxwell element consisting of a Hookean spring and dashpot in series still used by some modern textbooks to model mechanical relaxation, and the celebrated Maxwell relation

$$\eta = G_\infty \tau, \tag{1.2}$$

between the shear viscosity η and the structural relaxation time τ via the shear modulus G_∞, which is often used nowadays to estimate either τ or G_∞ of a liquid from the measured viscosity.

Another notable historic example of scientific study of relaxation and glasses is the observation of the effect of aging of silicate glass thermometer at room temperature over a period of 38.5 years, from April 1844 to December 1882, by James Prescott Joule [6]. The recorded data of Joule was recently revisited and discussed in relation to current concepts of dynamics in glassy materials [7].

The next notable development in relaxation and diffusion seems to be the theoretical solution of the Brownian motion problem by A. Einstein [8], which is one of his three important papers published in 1905, his magical year, the other two being the photoelectric effect and the special theory of relativity [9]. The particle is subjected to random forces due to frequent collisions with the solvent liquid molecules, making it to perform a random walk. Einstein showed that the mean-square displacement of the Brownian particle $\langle|\vec{r}(t) - \vec{r}(0)|^2\rangle$ is linearly proportional to time. In three dimensions

$$\langle|\vec{r}(t) - \vec{r}(0)|^2\rangle = 6D_t t, \tag{1.3}$$

where D_t is the translational diffusion coefficient. Furthermore, by considering simultaneously the driving force of diffusion due to osmotic pressure and the viscous drag force on a sphere moving in a fluid given by Stokes [10], Einstein derived the following expression for D_t through the Stokes–Einstein relation:

$$D_t = \frac{RT}{N_A}\left(\frac{1}{6\pi\eta r}\right), \tag{1.4}$$

where R is the gas constant, N_A is the Avogadro's number of molecules in 1 mol, and r is the radius of the spherical particle [11]. There is no doubt the solution of the Brownian diffusion in 1905 by Einstein is a great achievement in physics. This is because from the Brownian motion of large particles suspended in a liquid predicted by Einstein's statistical molecular theory of liquids, Perrin [12] was able to perform experiments to determine the Avogadro's number, and hence the dimension of the liquid molecules. Notwithstanding, the pollen particles studied by Brown or the narrowly dispersed particles measured by Perrin are very dilute suspensions in liquids. The particles themselves do not interact with each other and hence their (Brownian) motions are independent of each other. On the other hand, the relaxing or diffusing units in the condensed matters studied by Kohlrausch, such as repeat units of polymer chains or alkali ions in glasses, are densely packed together and there are non-trivial interactions between them. The motions of the interacting units cannot be independent of each other, at least from intuition. Therefore, none of the solutions of Brownian diffusion by Einstein in 1905, Smoluchowski in 1906 [13], and Langevin in 1908 [14] applies to the relaxation or the diffusion in these many-body interacting systems. In fact there is break down of the Stokes–Einstein relation (Eq. (1.4)) all to be shown in Chapters 2 and 3 of this book. The correlation function (to be defined later) of Brownian motion is the linear or monoexponential function [15]

$$\phi(t) = \exp -(t/\tau), \tag{1.5}$$

which obviously is far simpler than the stretched exponential correlation function of Kohlrausch given by Eq. (1.1). The exponential decay is also known as the Perrin's law with the relaxation time τ given by $3\eta V_H/kT$, where η is the solvent viscosity and V_H is the hydrodynamic volume of the particle [16].

Rapid developments of science and technology in the twentieth century and continued unabated till now have introduced a wide variety of new materials and systems, many of which share the common property of the materials studied by Kohlrausch, e.g., presence of interactions between the relaxing and the diffusing units. As shall be made clear later on in this chapter and the chapters to follow, relaxation or diffusion of many-body interacting systems is a long-standing but still unsolved problem in condensed matter physics and statistical mechanics. There is not even a universally accepted interpretation of the physical origin of the Kohlrausch correlation function found in the same material or the same class of materials, and not to mention many more dynamic properties of relaxation of diffusion discovered since the times of R. Kohlrausch. This situation persists up till the present time in spite of the voluminous amount of experimental data and well-established phenomenology accumulated in the past century that should help as clues pointing the way toward solution. Perhaps the world of science today would be different had Einstein, the recognized genius, been aware of the results of R. Kohlrausch and his son F. Kohlrausch at Göttingen since 1854 and solved their more difficult problem of Kohlrausch relaxation or diffusion in many-body interacting complex systems than the simpler problem of diffusion of non-interacting pollen particles in water of Robert Brown dated back earlier to 1827. Many-body interaction in condensed matter originating from forces between the relaxing or the diffusing species is omnipresent in condensed matter composed of densely packed atoms, ions, molecules, or particles. By "complex" in the term "many-body interacting complex systems," I mean additional complications such as randomness, disorder, heterogeneity, surface, interfaces, and the presence of more than one relaxing or diffusing species. Although many many-body interacting condensed matters of interest are also complex, one should not be distracted from the fact that the fundamental unsolved problem is relaxation and diffusion of many-body interacting systems without any additional complexity. The purpose is not served by constructing theory or model to deal with the complexity part and ignoring the many-body part of the problem. Unfortunately, this practice is often the case in recent years. As shown in other chapters of this book, many properties of the dynamics of relaxation and diffusion are shared by many-body interacting (complex or not) systems and materials of widely different chemical compositions and physical structures. These empirical facts indicate that the underlying physics of many-molecule relaxation dynamics is fundamental, making the problem most interesting.

An analogue of translational diffusion is rotational diffusion. In 1906, Einstein derived the following equation for the rotational diffusion coefficient D_r by making use of Stokes equation for the viscous drag on a sphere subjected to a torque [17]

$$D_r = \frac{RT}{N_A}\left(\frac{1}{8\pi\eta r^3}\right), \tag{1.6}$$

which has a stronger dependence on the radius than the translational diffusion coefficient D_t in Eq. (1.4). Closely related to this problem is rotational diffusion of a polar molecule that possesses a permanent dipole in a solvent considered by Debye [18–20]. Like Einstein's case, there is no interaction between the polar molecules in this related problem. Debye proposed and solved a model of reorientation of a particle through an appreciable angle after undergoing many collisions with the solvent particles. The model was applied for the consideration of rotational diffusion of polar molecules, which constituted part of his works on dielectric properties and relaxation of molecular systems that earned him the Nobel Prize in 1936. The result on the rotational diffusion coefficient is similar to that derived by Einstein and hence Eq. (1.6) is often referred to as the Stokes–Einstein–Debye relation. The solution of rotational diffusion of a polar molecule, now commonly referred to as the Debye model by workers in dielectric relaxation, has dipole–dipole time correlation function given by the linear exponential function $\exp -(t/\tau)$ like Brownian diffusion (Eq. (1.5)). Although the Debye model is an important step to understand dielectric relaxation, it does not apply to dielectric relaxation in complex interacting materials and systems, including most glass-forming substances. Actually, the Debye diffusion equation is a special case of the Smoluchowski equation of motion of the probability density function $P(\hat{u}^N, \vec{r}, t)$ for the orientations $\hat{u}^N$ and positions $\vec{r}^N$ of N interacting particles undergoing rotational and translational diffusion [13, 21, 22]:

$$\partial P(\hat{u}^N, \vec{r}^N; t)/\partial t = \hat{S} P(\hat{u}^N, \vec{r}^N; t), \tag{1.7}$$

where $\hat{S}$ is the Smoluchowski operator

$$\hat{S}(\cdots) = \sum_{i,j=1}^{N} \vec{\nabla}_i \overleftrightarrow{D}_{ij}(\hat{u}^N, \vec{r}^N) \cdot \left\{ \frac{1}{k_B T} [\vec{\nabla}_j \Phi(\vec{r}^N)](\cdots) + \vec{\nabla}_j(\cdots)] \right\}. \tag{1.8}$$

Here, $\vec{\nabla}_i$ is the gradient operator acting on the position and the orientation of particle i, and k_B is the Boltzmann's constant. $\overleftrightarrow{D}(\hat{u}^N, \vec{r}^N)$ is the time-dependent diffusion tensor. The term $\vec{\nabla}_j \Phi(\vec{r}^N)$ is the force exerted on particle j by the other N–1 particles through the total interaction potential $\Phi(\vec{r}^N)$, which is independent of $\hat{u}^N$ because it is assumed that there is no torque between particles. Solution of the Smoluchowski equation is a difficult and so far an unaccomplished task. This is precisely due to the presence of the many-body direct and hydrodynamic interactions.

In the absence of interaction, Φ vanishes and the diffusion tensor becomes diagonal. For non-spherical particles, the translational diffusion matrix elements are orientation dependent, and hence the dynamics of translation and rotation are coupled. For spherical particles, translation and rotation are not coupled, and the Smoluchowski equation of the N identical and non-interacting-particles reduces to an equation for a single molecule:

$$\partial P(\hat{u}, \vec{r}; t)/\partial t = D_t \vec{\nabla}^2 P(\hat{u}, \vec{r}; t) + D_r \vec{\nabla}^2 P(\hat{u}, \vec{r}; t), \tag{1.9}$$

where D_t and D_r are the translational and rotational Stokes–Einstein diffusion coefficients of a particle and $\vec{\nabla}^2 = \partial^2/\partial x^2 + \partial^2/\partial y^2 + \partial^2/\partial y^2$ is the Laplacian. Using spherical coordinates r, ϑ, and φ, the Laplacian can be decomposed into a radial part, $\vec{\nabla}_r^2 = r^{-2}(\partial/\partial r)r^2(\partial/\partial r)$, and an angular part $\vec{\nabla}_\Omega^2 = (r\sin^2\vartheta)^{-2}\{\sin\vartheta(\partial/\partial\vartheta)[\sin\vartheta(\partial/\partial\vartheta)] + \partial^2/\partial\phi^2\}$.

When there is only reorientation but no translation, this special case of the Smoluchowski equation has the following form:

$$\frac{\partial}{\partial t}P(\hat{u},t) = D_r\vec{\nabla}_\Omega^2 P(\hat{u},t), \tag{1.10}$$

which is the Debye equation, with $P(\hat{u},t)\mathrm{d}\hat{u}$ as the probability of finding a particle with orientation $\hat{u}$ at time t. This equation was solved [23] by using the spherical harmonics as eigenfunctions of $\vec{\nabla}_\Omega^2$. The mean of the product $\hat{u}(t)\hat{u}(0)$, denoted by $\langle\hat{u}(t)\hat{u}(0)\rangle$, is given by $\int P(\hat{u},t)\hat{u}(t)\hat{u}(0)\mathrm{d}\hat{u}$. It follows from Eq. (1.10) that the orientation time correlation function $\langle\hat{u}(t)\hat{u}(0)\rangle$ has the dependence:

$$\langle\hat{u}(t)\hat{u}(0)\rangle = \exp(-2D_r t). \tag{1.11}$$

This simple exponential time dependence from the Debye model is usually not observed in complex interacting systems and materials. Instead, the orientation time correlation functions measured by experiments are well approximated by the stretched exponential function of Kohlrausch (Eq. (1.1)). Attempts have been made in the past to patch up this problem by using a superposition of the exponential Debye functions with a distribution of D_r as the origin of the Kohlrausch function. However, as we shall see from later chapters, multiple experimental facts of relaxation and diffusion in complex interacting systems rule out this possibility. The failure of this generalization of the Debye model for relaxation and diffusion of complex interacting systems is another indication that many-body effects have to be taken into account seriously.

When there is pure translation without rotation, Eq. (1.9) is reduced to

$$\frac{\partial}{\partial t}P(\vec{r},t) = D_t\vec{\nabla}_r^2 P(\vec{r},t). \tag{1.12}$$

Introducing the Fourier components $P(\vec{k},t) = \int P(\vec{r},t)\exp(-i\vec{k}\vec{r})\mathrm{d}\vec{r}$, the Fourier transform of Eq. (1.12) is then given by

$$\frac{\partial}{\partial t}P(\vec{k},t) = -D_t k^2 P(\vec{k},t). \tag{1.13}$$

Assuming the particle is at the origin at $t = 0$, the solution of the last equation is

$$P(\vec{k},t) = \exp(-k^2 D_t t). \tag{1.14}$$

The solution $P(\vec{r}, t)$ of Eq. (1.12) is obtained from Eq. (1.14) after inverse Fourier transform and is given by

$$P(\vec{r}, t) = (4\pi D_t t)^{-3/2} \exp[-|\vec{r}|^2 / 4D_t t]. \tag{1.15}$$

The mean-square displacement $\langle |\vec{r}(t)|^2 \rangle$ is obtained by averaging over $P(\vec{r}, t)$ and the result is

$$\langle |\vec{r}(t)|^2 \rangle = 6D_t t. \tag{1.16}$$

The probability distribution function (Eq. (1.15)) is Gaussian and the mean-square displacement has linear time dependence for translation diffusion of non-interacting particles. These properties as well as others to be discussed in Chapter 2 are at variance with the observations by experiments in many different kinds of interacting many-body systems.

Glass-forming substances form an important class of complex interacting materials in the study of relaxation and diffusion. Upon decreasing temperature or increasing pressure, a non-crystallizing liquid will vitrify, that is, the structural relaxation becomes so slow and the relaxation time so long that the system cannot attain an equilibrium configuration in the available time and is transformed into the glassy state. This is the liquid vitrification problem and commonly referred to as the glass transition problem, although it is a kinetic phenomenon and there is no observed phase transition at the glass transition temperature T_g. It is an important problem because natural and synthetic glassformers are most abundant within the realm of condensed matter and occur in a great variety of chemical compositions that can be inorganic, metallic, organic, biomolecular, and polymeric. A popular scientific article published in 2008 in *The New York Times* entitled "The Nature of Glass Remains Anything but Clear" [24] has drawn the attention of the scientific community and the general public that glass transition is a challenging but worthwhile problem to be solved. It pays tribute to the fundamental nature and importance of the glass transition problem by stating: "Understanding glass would not just solve a long-standing fundamental (and arguably Nobel-worthy) problem and perhaps lead to better glasses. That knowledge might benefit drug makers, for instance. Certain drugs, if they could be made in a stable glass structure instead of a crystalline form, would dissolve more quickly, allowing them to be taken orally instead of being injected. The tools and techniques applied to glass might also provide headway on other problems, in material science, biology and other fields, that look at general properties that arise out of many disordered interactions."

This NY Times article cited Philip W. Anderson, Nobel Laureate, who wrote in 1995 [25(a)][1]: "The deepest and most interesting unsolved problem in solid state

[1] On a personal and historical note, the author was invited in 1982 by Prof. P.W. Anderson to give a talk at Princeton University on his paper published in *Comment Solids State Phys.* **9**, 127 (1979), Roman Smoluchowski was the editor (now a defunct journal). A year later, he was invited again by Prof. Anderson to participate and give a talk in the workshop on glass transition organized at

theory is probably the theory of the nature of glass and the glass transition." He added, "This could be the next breakthrough in the coming decade." Earlier in 1984 and together with others, Anderson himself published a model of hierarchically constrained dynamics for glassy relaxation having the Kohlrausch time dependence and the Vogel–Fulcher temperature dependence for its relaxation time [25(b)]. Recently, P.-G. de Gennes, another Nobel Laureate, tried his hand on the problem in 2002 [25(a)]. Since the 1980s, many scientists have attempted to solve the problem. Some examples are given in [26(b)–26(z)]. These are just chosen examples. The contributions from many other researchers including the present author will be cited later and discussed. In spite of these great research efforts by many including Anderson and coworkers and the promotion of the problem by him again in 1995, the 2008 NY Times article pointed out that the problem remains unsolved by the statement: "Thirteen years later, scientists still disagree, with some vehemence, about the nature of glass." This pessimistic but factual note sounded in the NY Times article on current state of affairs in research on glass transition is in my view caused by the many-body nature of relaxation and diffusion in glassformers, which is an unsolved problem and creates the impasse. Due to interactions between the structural units of the liquid, structural relaxation leading to the liquid–glass transition is necessarily a many-body relaxation problem. Since many-body relaxation of all complex interacting systems is an unsolved problem, naturally the glass transition problem remains a challenging problem in condensed matter physics and chemistry. There is still no generally accepted and totally satisfactory theory of the glass transition, a conclusion made also in the NY Times article.

Some practitioners in the research field of glass transition may disagree with this statement by citing the conventional theories of glass transition such as the free

---

the Aspens Center for Physics, Aspens, Colorado. This activity was followed by another workshop on glass transition at the Institute for Theoretical Physics, University of California, Santa Barbara organized by Prof. Anderson and Prof. E. Abrahams, and I was one of the invited speakers. For the record, I must admit that my talk at Santa Barbara did not impress many people in the audience, particularly the theoreticians, and I would call it a failure. Dr. George Wright at the Office of Naval Research was aware of the upsurge of interest in the glass transition problem by the condensed matter physics community, and he funded me to organize a workshop entitled *Relaxation in Complex Systems* in the campus of Virginia Tech. at Blacksburg, VA, in July 1983. The proceedings with the same title was published in 1984 and widely distributed free of charge by the Naval Research Laboratory. Prof. T.V. Ramakrishnan was also interested in the problem. He organized a workshop in Bangalore, India, in 1985, and I was invited as the keynote speaker. This activity was followed by the publication of the monogram *Non-Debye Relaxation in Condensed Matter*, T.V. Ramakrishnan and M. Raj Lakshmi editors, World Scientific, Singapore (1987). In this book I contributed a 170-page review. Anyway, all the activities mentioned in the above indicate the tremendous interest of the research community at that period in the 1980s. Theoreticians interested were mostly those involved in spin glasses who logically viewed the glass transition as the next important problem to attack. The outlook changes dramatically in 1987, after the discovery of high T_c superconductors by Bednorz and Müller in 1986. The excitement generated by the discovery of high T_c superconductors occupied the attention of many scientists especially the theoreticians, including possibly Prof. Anderson himself. It is gratifying to see Prof. Anderson returned in 1995 to remind the community of the importance of the glass transition problem. On behalf of all workers in the field, I thank Prof. Anderson for his unwavering support given to the research area of glass transition.

volume [27–29] and the configurational entropy models [30] or recent approaches by mode coupling theory [31–35], and potential energy landscape models [36–40]. These models and theories do have success in explaining or rationalizing some of the observed properties, but not all, especially those critical ones. None of them deals directly with the many-body relaxation problem. In this respect, the free volume and configurational entropy models are the best examples because the dynamics coming from many-body relaxation is not captured by considering macroscopic thermodynamic quantities alone in these models. Consequently, they cannot explain a number of experimentally observed general properties (to be presented in detail in chapters later) that are manifestly consequences of many-body relaxation.

The NY Times article made its conclusion by sampling the views principally from theoreticians who have considered or emphasized only a few familiar experimental facts and physical quantities all related to the structural α-relaxation but not those having direct connection to many-body relaxation. Several important and general experimental facts outside those considered by the theoreticians consulted have been discovered in recent years. Many of them are consequences of many-body relaxation. These experimental results have impact on theoretical understanding, and if taken into consideration, they should vastly improve the chance for someone to finally solve the glass transition problem. Unfortunately, these facts are not mentioned in the NY Times article and not taken into account by the theoreticians consulted either. These will be amply discussed in this book to heighten the awareness for those interested in the glass transition problem. Without incorporating these general experimental facts and providing explanations, any theory of glass transition would be incomplete and may even have contradicted some of them without knowing. Moreover, these additional experimental facts are found in "other problems, in material science, biology and other fields" [24]. In this regard, the NY Times article gave one specific example, which is the problem of improved solubility and bioavailability of drugs in pharmaceutical industry. We can supply more examples in this book that include carbohydrates for biopreservation and food science, hydration of biomolecules for functioning in life science, viscoelastic properties of polymers for materials engineering, ionic conductors for portable energy source, and fuel cell materials for the energy industry. We show, particularly in Chapter 3, the basic features of dynamics of these systems in other fields are isomorphic to glassformers, and hence solution of the glass transition problem can benefit progress in the other fields, and vice versa. Like glassformers, these are other classes of complex interacting materials in which many-body relaxation has to be dealt with before the dynamic and thermodynamic properties can be satisfactorily and fully understood. Examples include viscoelasticity and rheology of entangled polymer melts and semidilute polymer solutions; diffusion of concentrated colloidal particle suspensions; ionic conductivity in glassy, crystalline, and molten ionic conductors; and conformation fluctuations in biomolecules. Like glass transition, most theories, if not all, in these research fields have avoided the difficult task of confronting directly the many-body relaxation problem. The original reptation theory of entangled polymer melts starts by reptation of a chain within a static tube formed by other chains, which obviously becomes an exactly soluble one-chain relaxation problem [41, 42].

Reconciliation with the actual many-chain relaxation process is made in modifying the reptation motion of the chain by invoking additional mechanism such as "constraint release" due to motion of the confining chains forming the tube and "tube renewal or fluctuations" [43, 44]. It is unclear whether this way is adequate to solve the many linear-chain relaxation problem and dubious for viscous flow of multi-arm star polymers and H-polymers because all arms have to retract by reptation before flow can take place again by reptation of the transformed linear chain [45]. Many ionic conductors of interest in applications are glassy and there is no long-range order. Ion–ion interactions are important concern in glassy ionic conductors containing high concentration of mobile ions, and the many-ion relaxation dynamics cannot be ignored when considering ion conductivity relaxation. Nevertheless, perhaps due to the fact that many-ion relaxation is another difficult problem, theories of glassy ionic conductors have turned to emphasize other causes. Some considered disorder as the principal cause of the observed dynamics and proposed theories based on disorder. An example is the random energy barrier model [46]. Ignoring ion–ion interaction has the consequence of many generally observed properties of ion dynamics are left unexplained, which will be discussed in detail in following chapters. Another indication that many-body relaxation is common to all the systems mentioned is that similar or analogous dynamic properties have been found in them, despite very different systems and manifestations. Chapter 3 is dedicated to the universal properties. This is not expected from the disparate explanations given to the different systems such as reptation for entangled polymers, random energy barriers for ionics, and mode coupling, free volume, configurational entropy, or potential energy landscape models for glass-forming substances. Therefore, it is appropriate to generalize Anderson's statement for glass transition to other fields also involving many-body relaxation. The generalization of his statement could be like the following: "One of the deepest and most interesting unsolved problems in condensed matter physics and statistical mechanics theory is probably the theory of the relaxation and diffusion in many-body interacting complex systems and materials."

The above is a brief sketch of the past and present history of the development of relaxation and diffusion in science. The study has promulgated to many research fields and has the tendency to become interdisciplinary at modern times. The sketch is admittedly rough and incomplete but the purpose of this is to make the readers aware of the long history of the unsolved problem of relaxation and diffusion in complex interacting materials and systems. In spite of the fundamental importance and wide range of applications of the related physical phenomena, the solution of the problem is still at large. Nowadays, the research of materials with length scale just larger than the size of atoms and molecules (i.e., nanoscience and nanotechnology) is at the frontier of science and technology. Progress in some areas of nanoscience and nanotechnology requires good understanding of the relaxation and diffusion processes in materials of nanometer size. It is understandable that this requirement cannot be met at the present time because the same relaxation and diffusion processes are still not well explained in materials of macroscopic size. Similar remark is applicable to biomaterials because their structures are more complicated than the simpler model systems including molecular glassformers and yet their relaxations are still not fully understood.

In view of the absence of any viable and universally accepted theoretical solution, one can appreciate that the research in relaxation and diffusion in many-body interacting complex systems is experimentally driven for many decades. Abundant experimental data, facts, and phenomenology have accumulated in the past century and continued into the new century. New classes of materials and novel systems have entered into this research field everyday. Invented new experimental techniques have expanded immensely the time and frequency ranges as well as the temperature and pressure ranges in which the dynamics of relaxation and diffusion processes can be observed and measured, and thus important new results emerge every now and then. This situation is unlike other problems in condensed matter and statistical physics, where the expanse or the diversity of experimental data to be dealt with theoretically is considerably less, and ultimately the problems have been solved in much shorter time spans. Moreover, despite many well-learned theoretical methods in condensed matter physics and statistical mechanics available in the literature, so far anyone using them to develop conventional theories of relaxation and diffusion in interacting complex materials has failed to fully explain all the vital experimental facts and hence cannot be considered as a solution to the problem.

Therefore, due to these unusual circumstances of the problem, there is hardly any use for me to review the conventional theoretical approaches to relaxation and diffusion in a particular field such as glass transition. They can be found in many textbooks, but there the reader cannot find critical review of all vital experimental facts and the impact they have on theory. Instead, the emphasis of my presentation is on the important experimental facts generally found in the relaxation and diffusion of many-body interacting complex systems. The purpose is to make the readers become aware of these facts as completely as possible, convince them that there is fundamental physics at work, and any solution is no solution unless it can explain (or at least consistent with) all these general experimental facts. Relaxation and diffusion in glassformers deserve our utmost attention because of the wide range of systems and materials involved, and the plethora of experimental data and phenomenology present, and the glass transition has been highlighted by many eminent scientists as an important unresolved problem.

From the start, when I accidentally entered into this field 32 years ago in 1978, I have recognized many-body relaxation in interacting systems is the cornerstone to build a theory that can solve many problems. Although I have not succeeded to reach the lofty goal of solving the core problem, the fundamental concepts and predictions in the crude model I first introduced in 1979 remain viable [47, 48]. The model, improved in later years and now known in the literature as the coupling model (CM), has a general result that remains unchanged for the past 30 years. It spawns many different applications to various experimental investigations, and the predictions have resolved many challenging problems by explaining the anomalous properties generally found in various relaxation and diffusion processes in diverse complex systems with interactions. Some of these problems and their resolutions by the CM are presented for several purposes. One is to demonstrate that many-body relaxation is definitely the principal cause of the observed general and anomalous properties. Another is to provide clues to concepts needed for constructing a theory capable of solving the problem and to pave the way for others in coming up with a theory superior than the CM. Hopefully, this book will be the catalyst for someone else to obtain a fully rigorous and complete solution of the problem in not so distant future. The book is primarily written for researchers across several disciplines with widely different educational backgrounds. Therefore, minimal mathematical formalisms are introduced and used. Since most of the problems of interest are in the domain of classical mechanics, quantum mechanical description is not needed. The essentials of linear response theory necessary for macroscopic and microscopic descriptions of experimental data are given. For understanding the microscopic description in terms of correlation functions presented here, the reader only needs familiarity with some basic concepts of statistical mechanics. Tools required to represent the experimental data in physical and mathematical terms can be found in the text, and there is no need to seek help elsewhere unless the reader is interested in more details of their derivations from statistical mechanics. The book is written primarily for the research community. Nevertheless, if supplemented by other textbooks teaching the basics, it can be used in an advanced graduate course for introducing research opportunities to young investigators.

1.2 Relaxation and Diffusion

Within the terminology of statistical mechanics, "relaxation" means irreversible process in contradistinction to time-reversible process of Newtonian mechanics. It is the time-dependent change of a system or a sub-system from one physical or thermodynamic state to another, involving the dissipation of energy. Although the microscopic dynamics of the many-body interacting system is invariant under the reversal of time, nevertheless the system irreversibly moves toward equilibrium. Diffusion is a special case of relaxation process which occurs naturally without involving the change of any external input or parameter $\Im$. The external input can be mechanical stress and strain, electrical stress or electric field (voltage) and electrical strain or dielectric displacement (true charge) and the magnetic and chemical analogues, foreign agent such as humidity, or thermodynamic parameter (e.g., temperature and pressure). Relaxation process of different kinds, mechanical, electrical, chemical, and structural, is the time-dependent approach of the system from the initial stationary state to another as a consequence of the change of the external input or parameter. For example, if the external input is stress, then "relaxation" is observed by measuring strain. The simplest changes include sudden application of the external input at some time to the system initially stationary and kept constant thereafter or conversely, sudden removal of the external input after it has been applied for a sufficiently long time and the system already has become stationary.

In general, as for the definition of relaxation, there is no restriction on the size of the change. However, if the purpose of the experiment is to probe the intrinsic irreversible process of the system at equilibrium or even out of equilibrium, the change has to be sufficiently small such that none of the physical or chemical aspects of the system is altered by the applied external input and the change imposed. An alternative method to probe relaxation is the application of a small alternating external input $\Im=\Im_0 \cos(\omega t)$ with constant frequency ω, and relaxation of the system is observed as the linear steady-state response of a dynamic variable to the oscillations. These are linear response or linear relaxation, where the time or the frequency dependence of the relaxation process does not depend on the size of the perturbation. The time-dependent approach toward the final equilibrium state is usually observed on a parameter conjugate to the external input. If the external input is stress, then "relaxation" is observed on the strain. I put quotation marks on the word relaxation because the response of strain to change of stress is more appropriately referred to as creep or compliance. Also in this case if the stress applied at $t = 0$ is not constant but varies with time, the total strain at later time is a superposition of the increments or decrements of strains in response to the corresponding increments or decrements of stress in the course of time. This follows from the superposition principle of Boltzmann [49] proposed by him when analyzing creep compliance and stress relaxation properties of solids. The principle is valid only when the measurements involve small enough strain and stress. Similarly, for dielectric relaxation, the dielectric displacement at any instant due to time variation of the electric field is the sum of displacements in response to the increments in the electric field applied at earlier times.

Nowadays, many experimental techniques have been developed for the study of relaxation and diffusion. In addition to mechanical and dielectric methods, techniques built upon nuclear magnetic resonance, neutron scattering, and various electromagnetic and light-scattering techniques have emerged. The techniques combined can probe the dynamics of relaxation over an immense range of time from molecular vibration time to days. The techniques have different features and characters. They probe not only the dynamics through different dynamic variables such as density, enthalpy, shear stress or strain, electric and magnetic polarization, nuclear spin orientation, and mean-square and rotational angle displacements but also these variables at different length scales from the size of atoms and molecules to macroscopic dimension. The experimental results on the dynamics are presented in terms of time/frequency dependence of a quantity in response to the change of another quantity. The time/frequency response becomes independent of the magnitude of the change when the latter is sufficiently small, and this is the regime of linear response. Experimental results obtained by some techniques including dielectric and mechanical relaxation are given in terms of macroscopic variables (e.g., dielectric displacement D, electric field E, strain γ, and stress σ), and the dynamics are expressed by macroscopic relaxation and compliance functions (electric permittivity ε, shear compliance J, electric modulus M, and shear modulus G) of time or frequency. On the other hand, the dynamics measured by techniques such as neutron scattering and obtained by molecular dynamics simulations are given at the molecular level. If the experiment is performed in the linear response regime, time-dependent correlation functions of dynamic variables are most suitable quantities to represent the measurements and compare with microscopic theory. In systems obeying statistical mechanics, dissipation during relaxation is related to intrinsic fluctuations by the fluctuation–dissipation theorem, which involves the correlation function [15, 50, 51]. Hence, the correlation function gives all desirable information about the intrinsic spontaneous statistical fluctuations about the thermal equilibrium state of the system. Results from most experimental techniques will be discussed in this book but the author is mindful that the readers may have widely different backgrounds. The use of macroscopic response functions may not be familiar to those well versed in statistical mechanics working at the molecular level such as some of the condensed matter physicists, chemical physicists, and physical chemists. On the opposite side, those in the research of materials, polymer, food, pharmaceutical, and biomaterials may not have been exposed to the time correlation functions and their use to represent experimental data. Furthermore, in the following sections, both macroscopic response functions and time correlation functions will be used extensively to discuss experimental results. Therefore, a brief summary of both approaches to relaxation given here is appropriate.

1.2.1 Macroscopic Description of Dynamics: Time- and Frequency-Dependent Mechanical Properties

Viscoelastic behavior is a time-dependent mechanical response and is characterized by creep compliance and stress relaxation, or alternatively by dynamic mechanical measurements of modulus or compliance as a function of frequency. Since time is an additional variable to deformation and force, to obtain unique characterizing functions in these measurements, one of these other variables is usually held constant.

1.2.1.1 Shear Creep and Recovery

In a shear creep experiment a shearing stress σ_0 is created in a relaxed material at time $t = 0$ and held constant, while the resulting shear strain $\gamma(t)$ increases monotonically with t. Given a sufficiently long time of creep, the velocity of creep will decelerate to zero and $\gamma(t)$ attains an equilibrium limit if a viscoelastic solid is being measured. On the other hand, if the material is a viscoelastic liquid that eventually flows, the velocity of creep will decelerate and finally attains a finite constant value. Viscoelastic steady state is achieved, and $\gamma(t)$ increases indefinitely. The creep experiment has a second part when the stress is set to zero after a period of creeping. A portion or all of the strain accumulated during creeping is then recovered as a function of time for a viscoelastic liquid or solid, respectively. For a viscoelastic liquid, the portion that is permanently deformed and irrecoverable reflects the contribution of viscous flow to the total deformation accumulated during creep. Since a viscoelastic solid does not flow, all of its creep deformation is recoverable.

When the strains or the strain rates are sufficiently small, the creep response is linear. In this case, when the time-dependent strain is divided by the fixed stress, a unique creep compliance curve results, that is, at each time there is only one value for this ratio, which is the *compliance*, i.e., $\gamma(t)/\sigma_o \equiv J(t)$. The shear creep compliance function $J(t)$ (in units of cm^2/dyne or Pa^{-1}, 1 $Pa^{-1} = 10\ cm^2$dyne) obtained for an viscoelastic liquid can be written as the sum of several contributions:

$$J(t) = \gamma(t)/\sigma_o = J_\infty - [J_\infty - J_0]\Psi(t) + t/\eta \equiv J_r(t) + t/\eta, \tag{1.17}$$

where J_0 and J_∞ are effectively defined as the values of $J(t)$ when $t =$ and $t = \infty$ respectively. $J_d \equiv (J_\infty - J_0)$ is the delayed compliance; $\Psi(t)$ is the normalized function, which is equal to one when $t = 0$ and is zero when $t = \infty$; and η (in units of pascal second or poise) is the shear viscosity (10 poise = 1 Pa s.). The t/η term reflects the permanent viscous deformation. $J_r(t)$ is the recoverable shear compliance, which can be obtained from creep recovery measurements. For a viscoelastic solid, η is operationally infinite, since normally a molecular network is present to preclude any permanent deformation. The counterpart of Eq. (1.17) for a viscoelastic solid is

$$J(t) = \gamma(t)/\sigma_o = J_\infty - [J_\infty - J_0]\Psi(t). \tag{1.18}$$

For a viscoelastic liquid, $\lim_{t\to 0} J(t) = J_0$, $\lim_{t\to\infty} J(t) = t/\eta$, and $\lim_{t\to\infty} J_r(t) = J(t) - t/\eta = J_0 + J_d \equiv J_\infty$. The last limiting value J_∞ is sometimes replaced in notation by J_s which is called the steady-state recoverable shear compliance. It is the maximum recoverable strain per unit stress, which reflects the maximum configurational orientation achievable at the present stress.

For a viscoelastic solid, $J_r(t) = J(t)$, $\lim_{t\to 0} J(t) = J_0$, and $\lim_{t\to\infty} J(t) = J_0 + J_d \equiv J_\infty$. A different notation J_e instead of J_∞ is sometimes used to denote the equilibrium compliance of a solid.

The creep and recovery experiment is the only characterization of viscoelastic behavior that is readily comprehended since contributions to the creep compliance are *additive* in the strain as shown by Eq. (1.17). The molecular processes involved are simply short and long-range configurational orientations and viscous flow reflecting the permanent increasing separation of the centers of gravity of neighboring molecules.

In the linear viscoelastic regime, from the Boltzmann superposition principle and Eq. (1.18), the response in the shear strain $\gamma(t)$ to a series of step shear stresses $\Delta\sigma_j$ at times $t = 0, t_1, t_2, \ldots, t_j, \ldots t_N$, is given by

$$\gamma(t) = \sum_j^N J(t - t_j)\Delta\sigma_j = \int_0^t du\, J(t-u)\frac{\partial\sigma(u)}{\partial u}. \tag{1.19}$$

1.2.1.2 Shear Stress Relaxation

After a constant shear strain is created in a previously relaxed material at $t = 0$, the resulting shear stress decays with ensuing time to zero for a viscoelastic liquid and to a finite equilibrium value for a viscoelastic solid. The shear stress relaxation modulus (Pa or dynes/cm^2) is

$$G(t) = \sigma(t)/\gamma_0 = G_\mathrm{e} + [G_0 - G_\mathrm{e}]\phi(t), \tag{1.20}$$

where γ_o is the imposed fixed strain; $\phi(t)$ is the relaxation function decreasing from $\varphi(0) = 1$ at $t = 0$ to $\varphi(\infty) = 0$ at $t = \infty$; G_0 is the initial shear modulus at $t = 0$; and G_e is the equilibrium modulus, which is zero for a viscoelastic liquid and finite for a viscoelastic solid. Unlike strain (or compliance), the time-dependent stresses arising from different molecular mechanisms are *not additive*, and hence it is difficult if not impossible to isolate and characterize each one of them individually by $G(t)$. Nevertheless, it can be shown in general that $G(t)$ and $J(t)$ are related through a convolution integral equation:

$$\int_0^t G(t-u)J(u)\mathrm{d}u = t, \tag{1.21}$$

from which one function can be calculated from the other by a numerical procedure [52].

Again, in the linear viscoelastic regime, from the Boltzmann superposition principle and analogue of Eq. (1.19), the response in the shear stress $\sigma(t)$ to a series of step shear stresses $\Delta\gamma_j$ at times $t = 0,\ t_1,\ t_2,\ \ldots, t_j, \ldots, t_N$, is given by

$$\sigma(t) = \sum_j^N G(t-t_j)\Delta\gamma_j = \int_0^t \mathrm{d}u\, G(t-u)\frac{\partial\gamma(u)}{\partial u}. \tag{1.22}$$

1.2.1.3 Dynamic Shear Modulus

If the sinusoidal shear strain

$$\gamma(t) = \gamma_0 \sin(\omega t) \tag{1.23}$$

is imposed starting at $t = 0$, where γ_o is the strain amplitude and ω the angular frequency, Eq. (1.22) becomes

$$\sigma(t) = \omega\gamma_0 \int_0^t G(t-u)\cos(\omega u)\mathrm{d}u. \tag{1.24}$$

After changing the integration variable from u to $v = t - u$, using trigonometric formula for the cosine of the sum of two angles and letting t be large to obtain steady-state response

$$\sigma(t) = \gamma_0[G'(\omega)\sin(\omega t) + G''(\omega)\cos(\omega t)], \tag{1.25}$$

where

$$G'(\omega) = G_e + \omega \int_0^\infty [G(t) - G_e]\sin(\omega t)dt = G_e + \omega[G_0 - G_e] \int_0^\infty dt\, \varphi(t)\sin(\omega t), \tag{1.26}$$

$$G''(\omega) = \omega \int_0^\infty [G(t) - G_e]\cos(\omega t)dt = \omega[G_0 - G_e] \int_0^\infty dt\, \varphi(t)\cos(\omega t). \tag{1.27}$$

Again, G_e in Eqs. (1.26) and (1.27) is zero for a liquid but is the equilibrium modulus for a viscoelastic solid. For a liquid, it can shown that $\lim_{\omega\to 0} G'(\omega) = \omega^2\eta^2 J_\infty$ and $\lim_{\omega\to 0} G''(\omega) = \omega\eta$. For the solid, $\lim_{\omega\to 0} G'(\omega) = G_e$ and $\lim_{\omega\to 0} G''(\omega) = \omega \int_0^\infty [G(t) - G_e]dt$.

The results can be restated as follows. After the sinusoidal strain of constant frequency has been applied for a long time, a steady sinusoidal strain or stress results; there is a fixed phase angle between the input and the output. For the sinusoidal shear strain $\gamma(t) = \gamma_o \sin(\omega t)$, the stress σ will oscillate sinusoidally as

$$\sigma(t) = \sigma_o \sin(\omega t + \delta). \tag{1.28}$$

Since the stress always leads the strain, the phase angle δ between stress and strain is positive. On comparing the same result for $\sigma(t)$ given by Eqs. (1.25) and (1.28), we have $G(\omega) = (\sigma_o/\gamma_o)\cos\ \delta$ and $G''(\omega) = (\sigma_o/\gamma_o)\sin\ \delta$. The storage shear modulus G' is a measure of the elastic energy stored and recovered, and the loss shear modulus G'' is a measure of the energy dissipated as heat in the cyclic deformation. The ratio $G''(\omega)/G'(\omega)$ is $\tan\delta$, called the loss tangent. Although usually not written out explicitly, δ is a function of ω. The complex dynamic shear modulus $G^*(\omega)$ defined by $G'(\omega) + iG''(\omega)$ is given by

$$G^*(\omega) = (\sigma_o/\gamma_o)\exp(i\delta). \tag{1.29}$$

1.2.1.4 Dynamic Shear Compliance

The complementary case to the above is the steady sinusoidal strain

$$\gamma(t) = \gamma_o \sin(\omega t - \delta) \tag{1.30}$$

in response to the applied sinusoidal stress of constant frequency $\sigma(t) = \sigma_o \sin(\omega t)$. The minus sign in Eq. (1.30) reflects the strain lagging behind the stress. If the sinusoidal stress is applied starting at $t = 0$, with $J(t)$ from Eq. (1.18), the strain response for $t > 0$ is given by

$$\gamma(t) = \omega\sigma_0 \int_0^t J(t-u)\cos(\omega u)du. \tag{1.31}$$

Performing the same operations as in the derivation of Eqs. (1.26) and (1.27), we find

$$\gamma(t) = \sigma_o[J'(\omega)\sin(\omega t) - J''(\omega)\cos(\omega t)]. \tag{1.32}$$

Here, the storage (in phase) compliance $J'(\omega)$ and the loss (90° out of phase) compliance $J''(\omega)$ are related to the creep compliance $J(t)$ of a viscoelastic liquid by the one-sided Fourier transforms:

$$\begin{aligned} J'(\omega) &= J_\infty - \omega\int_0^\infty [J_\infty - J(t) + t/\eta]\sin(\omega t)\mathrm{d}t \\ &= J_\infty - \omega[J_\infty - J_0]\int_0^\infty \Psi(t)\sin(\omega t)\mathrm{d}t, \end{aligned} \tag{1.33}$$

$$\begin{aligned} J''(\omega) &= (1/\omega\eta) + \omega\int_0^\infty [J_\infty - J(t) + t/\eta]\cos(\omega t)\mathrm{d}t \\ &= (1/\omega\eta) + \omega[J_\infty - J_0]\int_0^\infty \Psi(t)\cos(\omega t)\mathrm{d}t. \end{aligned} \tag{1.34}$$

For a viscoelastic solid, the relations are given by

$$J'(\omega) = J_e - \omega\int_0^\infty [J_e - J(t)]\sin(\omega t)\mathrm{d}t = J_e - \omega[J_\infty - J_0]\int_0^\infty \Psi(t)\sin(\omega t)\mathrm{d}t, \tag{1.35}$$

$$J''(\omega) = \omega\int_0^\infty [J_e - J(t)]\cos(\omega t)\mathrm{d}t = \omega[J_e - J_0]\int_0^\infty \Psi(t)\cos(\omega t)\mathrm{d}t. \tag{1.36}$$

By comparing Eqs. (1.30) and (1.32) for the same result, it is clear that $J'(\omega) = (\gamma_o/\sigma_o)\cos\ \delta$ and $J''(\omega) = (\gamma_o/\sigma_o)$. The complex dynamics shear compliance $J^*(\omega)$ defined by $J'(\omega) - \mathrm{i}J''(\omega)$ is related to δ by the following equation:

$$J^*(\omega) = (\gamma_o/\sigma_o)\exp(-\mathrm{i}\delta), \tag{1.37}$$

and $J''(\omega)/J'(\omega) = \tan\delta$. Since δ appearing in Eqs. (1.28) and (1.30) has to be the same, it follows immediately from Eqs. (1.29) and (1.37) that

$$J^*(\omega) = 1/G^*(\omega). \tag{1.38}$$

For the liquid, the low-frequency limits are $\lim_{\omega\to 0} J'(\omega) = J_\infty$ and $\lim_{\omega\to 0} J''(\omega) = 1/\omega\eta$. For the solid, they are $\lim_{\omega\to 0} J'(\omega) = J_e$ and $\lim_{\omega\to 0} J''(\omega) = \omega\int_0^\infty [J_e - J(t)]\mathrm{d}t$.

1.2.1.5 Tensile (Bulk, Longitudinal) Compliance and Tensile (Bulk, Longitudinal) Modulus

Although the viscoelastic functions and their inter-relations have been given above for shear deformation, they apply to tensile and bulk (isotropic) deformations as well. The analogues of $J(t)$, $J'(\omega)$, $J''(\omega)$, $G(t)$, and $G''(\omega)$ for tensile deformation are the tensile compliance $D(t)$, the tensile storage compliance $D'(\omega)$, the tensile loss compliance $D''(\omega)$, the tensile or Young's modulus $E(t)$, the tensile storage modulus $E'(\omega)$, and the tensile loss modulus $E''(\omega)$, respectively. For bulk deformation the corresponding quantities are the bulk compliance $B(t)$, the bulk storage compliance $B'(\omega)$, the bulk loss compliance $B''(\omega)$, the bulk modulus $K(t)$, the bulk storage modulus $K'(\omega)$, and the bulk loss modulus $K''(\omega)$. All equations written above for shear hold for tensile (bulk) deformation after all quantities have been changed in connotation from shear to tensile (bulk). It can be shown that the three compliances and moduli are related by the following equations [29]:

$$D(t) = \frac{J(t)}{3} + \frac{B(t)}{9}, \tag{1.39}$$

$$E(t) = \left[\frac{1}{3G(t)} + \frac{1}{9K(t)}\right]^{-1}. \tag{1.40}$$

If the specimen is highly incompressible such that $K(t) \gg G(t)$, and $B(t) \ll J(t)$, we have the simpler relations:

$$E(t) = 3G(t), \tag{1.41}$$

$$D(t) = J(t)/3. \tag{1.42}$$

These relations enable one to relate the shear viscoelastic functions to their tensile counterparts. The Poisson ratio $\upsilon_{\text{Poisson}}$ is defined as the ratio of transverse contraction per unit dimensions of a bar of uniform cross section to its elongation per unit length, when subjected to a tensile stress. Through $\upsilon_{\text{Poisson}}$ we have the following relations between the tensile modulus E to the bulk modulus K and shear modulus G:

$$E = 3K(1 - 2\upsilon_{\text{Poisson}}) = 2(1 + \upsilon_{\text{Poisson}})G. \tag{1.43}$$

These relations give

$$\upsilon_{\text{Poisson}} = \frac{3K - 2G}{6K + 2G}. \tag{1.44}$$

K and G can be determined by either light scattering or the longitudinal and transverse velocities of ultrasonic waves [29]. When $E \approx K$ as in the case of very hard materials, $\upsilon_{\text{Poisson}} \approx 1/3$. On the other hand, when $K >> G$ as in the case of soft materials, $\upsilon_{\text{Poisson}}$ approaches its maximum value of ½.

The deformation that leads to the tensile modulus $E(t)$ is elongation in one direction accompanied by transverse contraction. Another mode of deformation is elongation in one direction but the transverse dimensions are constrained to be constant. If such a tensile strain $t = 0$ is applied in a previously relaxed material at $t = 0$, the stress can be calculated and it relaxes as a function of time according to $\sigma(t) = \gamma_0[K(t) + (4/3)G(t)] \equiv \gamma_0 M(t)$, where $M(t)$ is called the longitudinal modulus. The corresponding complex dynamic longitudinal modulus $M^*(\omega) = M'(\omega) + iM''(\omega)$ are obtained from $M(t)$ in the same manner as for shear modulus (see Eqs. (1.26) and (1.27)). $M^*(\omega)$ can be obtained from the measured velocity of propagation v and the absorption coefficient α for longitudinal wave propagation in ultrasonic attenuation experiment [29].

There are techniques that measure directly the response of materials under small dynamic bulk compression using hydrostatic pressure from 1 to 1000 atm [53]. Both the storage and loss bulk compliances $B'(\omega)$ and $B''(\omega)$ were obtained over a range of frequencies from 50 to 1000 Hz.

1.2.2 *Macroscopic Description of Dynamics: Time- and Frequency-Dependent Dielectric Properties*

1.2.2.1 Dielectric Permittivity

For dielectric relaxation of a material composed of polar molecules with permanent dipole moments, the electric field E and electric displacement D are the analogues of mechanical stress σ and strain γ, respectively. If the electric field E_o is applied to the material starting at time $t = 0$ and held constant, the resulting electric displacement $D(t)$ increases monotonically with t. In analogy to the creep compliance $J(t)$ for a viscoelastic solid given by Eq. (1.18), the time-dependent dielectric permittivity is defined by

$$\varepsilon_o \varepsilon(t) = D(t)/E_o, \tag{1.45}$$

where ε_o is the permittivity of vacuum. It increases from the initial value $\varepsilon_o \varepsilon_{t=0}$ to the long-term value $\varepsilon_o \varepsilon_{t=\infty}$, according to the time dependence given by

$$\varepsilon_o \varepsilon(t) = \varepsilon_o \varepsilon_{t=\infty} - \varepsilon_o(\varepsilon_{t=\infty} - \varepsilon_{t=0})\Phi(t) = \varepsilon_o \varepsilon_{t=0} + \varepsilon_o(\varepsilon_{t=\infty} - \varepsilon_{t=0})[1 - \Phi(t)], \tag{1.46}$$

where $\Phi(t)$ is equal to one when $t = 0$ and is zero when $t = \infty$. In dielectric experiments on a polar material, an oscillating sinusoidal electric field $E(t) = E_0 \sin(\omega t)$ produces an oscillating electric polarization P. After reaching steady state, it as well as the electric displacement lags behind the applied electric field by some phase angle δ:

$$D(t) = D_0 \sin(\omega t - \delta). \tag{1.47}$$

If the electric field is applied starting at $t = 0$, the electric displacement response for $t > 0$ is

$$D(t) = \omega E_0 \int_0^t \varepsilon_o \varepsilon(t - u)\cos(\omega u)du. \tag{1.48}$$

After a transformation of variable from u to $u' = t - u$, the use of the formula for the cosine of the sum of two angles, and passage to the steady state given by Eq. (1.48) by letting $t \to \infty$, we arrive at the following results:

$$D(t) = \varepsilon_o E_o[\varepsilon'(\omega)\sin(\omega t) - \varepsilon''(\omega)\cos(\omega t)], \tag{1.49}$$

$$\varepsilon'(\omega) = \varepsilon_{t=\infty} - \omega[\varepsilon_{t=\infty} - \varepsilon_{t=0}] \int_0^\infty dt\, \Phi(t)\sin \omega t, \tag{1.50}$$

$$\varepsilon''(\omega) = \omega[\varepsilon_{t=\infty} - \varepsilon_{t=0}] \int_0^\infty dt\, \Phi(t)\cos \omega t. \tag{1.51}$$

When dielectric relaxation measurements are carried out in the frequency domain, the notations are changed from $\varepsilon_{t=\infty}$ to ε_0, where now the suffix "0" indicates zero frequency, and $\varepsilon_{t=0}$ to ε_∞, where the suffix "∞" indicates high frequency. Using these conventional notations, the results given by Eqs. (1.50) and (1.51) can be combined into one equation given by

$$\frac{\varepsilon^*(\omega) - \varepsilon_\infty}{\varepsilon_0 - \varepsilon_\infty} = 1 - i\omega \int_0^\infty dt\, \Phi(t)e^{-i\omega t}, \tag{1.52}$$

for the complex relative permittivity

$$\varepsilon^*(\omega) = \varepsilon'(\omega) - i\varepsilon''(\omega). \tag{1.53}$$

1.2.2.2 Electric Modulus

Just like the relation between the complex mechanical modulus $G^*(\omega)$ and the mechanical compliance $J^*(\omega)$ given by Eq. (1.38), formally the electric modulus $M^*(\omega)$ can be defined as the reciprocal of the dielectric compliance or permittivity $\varepsilon^*(\omega)$:

$$M^*(\omega) \equiv M'(\omega) + i\,M''(\omega) = 1/\varepsilon^*(\omega). \tag{1.54}$$

$M^*(\omega)$ was first put to use to represent dielectric measurements of ionically conducting materials [54–56] because its imaginary part $M''(\omega)$, the electric loss modulus, exhibits a peak when plotted against log ω. The peak frequency readily gives a conductivity relaxation time of the ions, while the corresponding characteristic time of the same data represented in terms of $\varepsilon^*(\omega)$ cannot be determined that easily.

Experiments of either dipolar or ionically conducting materials in the frequency domain measure the complex quantities such as the voltage $V^*(\omega)$, the current $I^*(\omega)$, or the pair of capacitance $C(\omega)$ and the conductance $G(\omega)$. Whether these results are presented in terms of $\varepsilon^*(\omega)$ or $M^*(\omega)$ is a matter of choice. The same information, no more and no less, is given in either representation as well as the complex conductivity $\sigma^*(\omega) = \mathrm{i}\omega\varepsilon_0\varepsilon^*(\omega) = \mathrm{i}\omega\varepsilon_0/M^*(\omega)$, although the emphasis on the spectral features is different. It is true that the high-frequency dielectric constant ε_∞ enters into the electric modulus not simply as in $\varepsilon^*(\omega)$, and stray capacitance in experimental measurement can influence its value. However, this problem is avoided if measurements are carried out with care.

The situation is different for experiments performed in the time domain. This is because for polarization $P(t)$ as a time-dependent quantity, it depends on the experiment of whether $P = D - \varepsilon_0 E$ is proportional to $D(t)$ or to $E(t)$, i.e., to charge or to field. If a constant field E_0 is switched on starting at $t = 0$, then $\varepsilon(t)$ is given in terms of the electric displacement $D(t)$ by Eq. (1.45). Alternatively, if a constant charge or electric displacement D_0 is switched on at $t = 0$, then $M(t)$ is proportional to polarization and measured in terms of the time-dependent field $E(t)$ by

$$M(t) = E(t)/D_o. \tag{1.55}$$

An experimental technique designed for direct measurement of $M(t)$ was accomplished recently [57]. The fidelity of this time-domain modulus technique had been tested using polar glass-forming materials against the standard frequency-domain impedance method [57]. The time-domain $M(t)$ data obtained was Fourier transformed to $M^*(\omega)$, and the reciprocal $1/M^*(\omega)$ agrees quantitatively with $\varepsilon^*(\omega)$ measured on the same material by impedance method. Thus, $M(t)$ and $M^*(\omega)$ are experimentally accessible quantities like $\varepsilon(t)$ and $\varepsilon^*(\omega)$. Converting $M(t)$ and $M^*(\omega)$ to $\varepsilon(t)$ and $\varepsilon^*(\omega)$ and vice versa or to other representations such as the complex conductivity $\sigma^*(\omega) = \mathrm{i}\omega\varepsilon_0\varepsilon^*(\omega) = \mathrm{i}\omega\varepsilon_0/M^*(\omega)$ is a straightforward mathematical procedure without loss of any information. All attempts published in the literature to discredit the electric modulus are unwarranted as shown in a recent review [56] and further discussed later on.

Finally, $M^*(\omega)$ is related to the normalized electric field decay function $\Phi(t)$ by the one-sided Fourier transform:

$$M^*(\omega) = M' + iM'' = M_\infty\left[1 - \int_0^\infty dt \exp(-i\omega t)\left(-\frac{d\Phi}{dt}\right)\right]. \tag{1.56}$$

1.2.3 *Macroscopic Description: Spectroscopy Based on Other Variables*

1.2.3.1 Heat Capacity Spectroscopy

Mechanical and dielectric relaxations are the most commonly used experimental techniques to probe the dynamics through the time or the frequency dependence of the dynamic variables, stress and strain for mechanical, and polarization and electric field for dielectric. Other dynamic variables can also be used to probe the dynamics if experimental techniques were invented to harness them. Specific heat c_p related to enthalpy H by $c_p = (\partial H/\partial T)_P$ is one such example. If the material is kept at constant pressure, the change in enthalpy per unit volume h in response to the change in temperature by a small amount ΔT is given by the product $h = c_p\Delta T$. This increase in enthalpy in general is a function of time because of slow relaxation of some degrees of freedom. If ΔT is switched on in the material at time $t = 0$ and held constant, the resulting enthalpy $h(t)$ increases monotonically with t from the initial value $c_{p,t=0}$ to the long-term value $c_{p,t=\infty}$ according to the equation

$$h(t)/\Delta T = c_{p,t=\infty} - (c_{p,t=\infty} - c_{p,t=0})\Phi(t) = c_{p,t=0} + (c_{p,t=\infty} - c_{p,t=0})[1 - \Phi(t)]. \tag{1.57}$$

Contributions from all degrees of freedom that come to equilibrium quickly are included in $c_{p,t=0}$. The slow relaxation probed contributes to the increment $(c_{p,t=\infty} - c_{p,t=0})$ in reaching the long-term equilibrium specific heat $c_{p,t=\infty}$. One may recognize that this is the enthalpy analogue of Eq. (1.46) for dielectric relaxation. Here, $\Phi(t)$ describes the normalized time dependence and again it is equal to one when $t = 0$ and is zero when $t = \infty$. Following the same path in deriving the complex permittivity $\varepsilon^*(\omega)$ given by Eqs. (1.50), (15.1), (1.52), and (1.53), we have

$$\frac{c_p^*(\omega) - c_{p\infty}}{c_{p0} - c_{p\infty}} = 1 - \mathrm{i}\omega \int_0^\infty \mathrm{d}t\, \Phi(t)\mathrm{e}^{-\mathrm{i}\omega t}. \tag{1.58}$$

In going from Eq. (1.57) to Eq. (1.58), we have changed the notations from $c_{p,t=\infty}$ to c_{p0}, where now the suffix "0" indicates zero frequency, and $c_{p,t=0}$ to $c_{p\infty}$, where the suffix "∞" indicates high frequency. The experimental technique of heat capacity spectroscopy was developed by Birge and Nagel [58] to measure the dynamic compliance $c_p^*(\omega, T)$ caused by a small harmonic temperature perturbation. The measurement of $c_p^*(\omega, T)$ by heat capacity spectroscopy has been made and compared with dielectric and mechanical measurements to achieve a fuller understanding of the dynamics of polymeric glassformers [59, 60].

Less information on dynamics but much more often performed is the measurement of the specific heat c_p by differential scanning calorimetry (DSC) or differential thermal analysis (DTA). There is also calorimetric technique to measure the enthalpy H as a function of time toward the equilibrium enthalpy after a jump in temperature.

1.2.3.2 Spectroscopy Based on Other Macroscopic Dynamic Variables

There are other spectroscopies based on other macroscopic dynamic variables than those discussed above. An example is the dilatometric measurements of the specific volume V as a function of time after a jump in temperature or pressure [61, 62]. In passing, we mention some spectroscopies that are related to the others discussed above. These include solvation dynamics experiment [63], capacitive scanning dilatometry, frequency-dependent thermal expansion measurements [64], and various light-scattering techniques, which will be discussed later.

1.3 Molecular Description of Dynamics in the Linear Response Regime

Relaxation in materials has been described by time or frequency dependences of macroscopic dynamic variables in the previous section. Alternative descriptions of relaxation and diffusion start from the consideration of the motions of the atoms, ions, or molecules on the microscopic levels, and the results are expressed in terms of time correlation functions in the linear response regime. Linear response theory of irreversible processes (i.e., relaxation) was first developed by Kubo in 1957 [65], and since then the formalism of the theory is given amply in many textbooks [15, 50, 66–71] and reviews [72, 73]. Perhaps there is no need to reproduce the formalism here. However, some of the steps taken to derive the key results are either not transparent to those having no rigorous training in statistical mechanics or jumped over in these texts. Some texts give classical and quantum mechanical versions of the linear response theory. But, since all relaxation and diffusion processes discussed in this book involve classical mechanics, the unneeded quantum mechanical version tends to confuse the uninitiated readers. This book is deemed to be read by readers with widely different backgrounds. Therefore, it makes sense to rederive the key results step by step and exclusively in the context of classical statistical mechanics. This self-contained version given here has the advantage that the readers do not need to consult the advanced texts on statistical mechanics for the purpose of reading this book. The version of the mathematical formalism given below is by no means original, and much of the material is extracted from the references given above, particularly from [69]. Connections between the correlation function from the formalism to various spectroscopies will be made. These connections are useful and justify the presentation of the formalism here.

First, consider an isolated system consisting of a large number N of identical units each with coordinates q_i and conjugate momenta p_i (although not explicitly indicated as such, q_i and p_i are vectors in three dimensions). The system at any given time is completely described by the collection of $3N$ coordinates $q^N \equiv (q1,\ldots,q_N)$ and $3N$ momenta $p^N \equiv (p_1,\ldots,p_N)$ of the units, which define a phase point (q^N,p^N). The time dependence of q_i and p_i is governed by Hamiltonian's equations:

$$\dot{q}_i = \partial H/\partial p_i \tag{1.59}$$

and

$$\dot{p}_i = -\partial H/\partial q_i, \tag{1.60}$$

where H is the Hamiltonian of the unperturbed system. In classical statistical mechanics, an ensemble of systems identical to each other is considered, and the phase points of the ensemble are distributed according to a probability density function $f(q^N, p^N, t)$. Thus, $f(q^N, p^N, t)\mathrm{d}q^N\mathrm{d}p^N$ is the probability at time t that the system is in the dynamic state represented by a phase point within the infinitesimal neighborhood defined by the product $\mathrm{d}q^N\mathrm{d}p^N$. The time evolution of $f(q^N, p^N, t)$ is governed by the Liouville equation of classical mechanics involving also the Hamiltonian H given by

$$\partial f/\partial t = \{H,f\} = -\mathrm{i}Lf, \tag{1.61}$$

where $\{C, D\}$ is the Poisson bracket

$$\{C, D\} \equiv \sum_{i=1}^{N}\left(\frac{\partial C}{\partial q_i}\frac{\partial D}{\partial p_i} - \frac{\partial C}{\partial p_i}\frac{\partial D}{\partial q_i}\right). \tag{1.62}$$

Defined by the Liouville equation (1.61), naturally $\mathrm{i}L_0$ is called the Liouville operator and

$$-\mathrm{i}L = \sum_{i}^{N}\left(\frac{\partial H}{\partial q_i}\frac{\partial}{\partial p_i} - \frac{\partial H}{\partial p_i}\frac{\partial}{\partial q_i}\right). \tag{1.63}$$

The formal solution is $f(q^N, p^N, t) = \exp(-\mathrm{i}Lt)f(q^N, p^N, 0)$. From now on in this section, the curly brackets are exclusively used to designate the operation of the Poisson bracket.

The objective of linear response theory is to calculate the change in the statistical ensemble average of a dynamic variable B as a function of time caused by a weak time-dependent external field $K(t)$ conjugate to another dynamic variable $A(t)$, which can be the same as $B(t)$. The external field $K(t)$ is applied from the infinite past at $t = -\infty$ when the system is at thermal equilibrium. The Hamiltonian of the perturbed system is now given by

$$H = H_0 - AK(t), \tag{1.64}$$

where H_0 is the Hamiltonian of the unperturbed system. The time evolution of $f(q^N, p^N, t)$ is governed by the Liouville equation

$$\partial f/\partial t = \{H,f\} = \{H_0 - AK(t), f\} = -\mathrm{i}L_0 f - \{A, f\}K(t), \tag{1.65}$$

where $\mathrm{i}L_0 f \equiv \{H_0, f\}$. For linear response we need to consider only terms in Eq. (1.65) up to first order of the weak external field. Writing f as the sum

$$f(q^N, p^N, t) = f_{\mathrm{eq}}(q^N, p^N) + \Delta f(q^N, p^N, t), \tag{1.66}$$

where Δf accounts for the change up to first order of the weak external field, and neglecting the term of second order, $-\{A, \Delta f\}K(t)$, Eq. (1.64) becomes

$$\partial \Delta f(q^N, p^N, t) = -\mathrm{i}L_0 \Delta f(q^N, p^N, t) - \{A, f_{\mathrm{eq}}(q^N, p^N\}K(t). \tag{1.67}$$

Solution of Eq. (1.67) by integration gives

$$\Delta f(t) = \int_{-\infty}^{t} \exp[-\mathrm{i}(t-s)L_0]\{-A(q^N, p^N), f_{\mathrm{eq}}(q^N, p^N)\}K(s)\mathrm{d}s. \tag{1.68}$$

In solving the differential equation, the boundary condition is that at $t = -\infty$, f is the equilibrium distribution function f_{eq}. The observed quantity $\langle B \rangle$ of a dynamic variable $B(q^N, p^N)$ is calculated as averages over an ensemble of identical systems and is given by integrating the product $B(q^N, p^N) f(q^N, p^N, t)$ over phase space, i.e., $\mathrm{d}q^N \mathrm{d}p^N$. As a special case, B is the same as A.

The statistically averaged change in the dynamic variable B, $\langle \Delta B(t) \rangle$, resulting from the change in the distribution function Δf is given by

$$\begin{aligned}
\langle \Delta B(t) \rangle &= \int\int \mathrm{d}q^N \mathrm{d}p^N B(q^N, p^N) \Delta f(t) \\
&= -\int\int \mathrm{d}q^N \mathrm{d}p^N \int_{-\infty}^{t} B(q^N, p^N) \exp[-\mathrm{i}(t-s)L_0]\{A(q^N, p^N), f_{\mathrm{eq}}(q^N, p^N)\}K(s)\mathrm{d}s \\
&= \int\int \mathrm{d}q^N \mathrm{d}p^N \int_{-\infty}^{t} \{f_{\mathrm{eq}}(q^N, p^N), A(q^N, p^N)\} \exp[\mathrm{i}(t-s)L_0] B(q^N, p^N) K(s)\mathrm{d}s \\
&= \int\int \mathrm{d}q^N \mathrm{d}p^N \int_{-\infty}^{t} \{f_{\mathrm{eq}}(q^N, p^N), A(q^N, p^N)\} B(q^N_{t-s}, p^N_{t-s}) K(s)\mathrm{d}s \\
&= \int\int \mathrm{d}q^N \mathrm{d}p^N \int_{-\infty}^{t} f_{\mathrm{eq}}(q^N, p^N) \{A(q^N, p^N), B(q^N_{t-s}, p^N_{t-s}\} K(s)\mathrm{d}s \\
&\equiv \int_{-\infty}^{t} \langle \{A, B(t-s)\} \rangle_{\mathrm{eq}} K(s)\mathrm{d}s \\
&\equiv \int_{-\infty}^{t} \Psi_{BA}(t-s) K(s)\mathrm{d}s.
\end{aligned} \tag{1.69}$$

The third line above is obtained from the second by making the Liouville operator act on $B(q^N, p^N)$ instead of $\{A(q^N, p^N\},\ f_{\mathrm{eq}}(q^N, p^N)\}$. This is done by using the Hermitian property of the Liouville operator with respect to the inner product of $\{A(q^N, p^N), f_{\mathrm{eq}}(q^N, p^N)\}$, and $B(q^N, p^N)$. The inner product of two arbitrary functions $g(q^N, p^N)$ and $h(q^N, p^N)$ is defined by

$$\int\int \mathrm{d}q^N \mathrm{d}p^N g^*(q^N,p^N) L_0 h(q^N,p^N)$$
$$= \int\int \mathrm{d}q^N \mathrm{d}p^N \mathrm{i} g^*(q^N,p^N) \sum_i^N \left(\frac{\partial H_0}{\partial q_i}\frac{\partial}{\partial p_i} - \frac{\partial H_0}{\partial p_i}\frac{\partial}{\partial q_i} \right) h(q^N,p^N). \tag{1.70}$$

Using partial integration to transfer the operations $\partial/\partial p_i$ and $\partial/\partial q_i$ on $h(q^N,p^N)$ to those on $g^*(q^N,p^N)$, and if g and h vanish at the boundaries of the phase space, we have the Hermitian property:

$$\int\int \mathrm{d}q^N \mathrm{d}p^N g^*(L_0 h) = \int\int \mathrm{d}q^N \mathrm{d}p^N (L_0 g)^* h. \tag{1.71}$$

Generalizing this to the operator $\exp[-\mathrm{i}(t-s)L_0]$ and noting from Eq. (1.63) that $(L_0)^* = -L_0$, we arrive at line three.

The fourth line in Eq. (1.69) is obtained from the third by the relation

$$B(q_{t-s}^N, p_{t-s}^N) \equiv B(t-s) = \exp(\mathrm{i}L_0 t) B(q^N,p^N). \tag{1.72}$$

This comes from the fact that changes of the dynamic variable B originate from the changes of the phase point $(q^N,\ p^N)$ with time, and the time derivatives $\partial q_i/\partial t$ and $\partial p_i/\partial t$ are governed by the Hamilton's equations of motion (Eqs. (1.59) and (1.60)). Thus

$$\mathrm{d}B/\mathrm{d}t = \sum_i^N \left(\frac{\partial B}{\partial q_i}\frac{\partial q_i}{\partial t} + \frac{\partial B}{\partial p_i}\frac{\partial p_i}{\partial t} \right) = \sum_i^N \left(\frac{\partial B}{\partial q_i}\frac{\partial H_0}{\partial p_i} - \frac{\partial B}{\partial p_i}\frac{\partial H_0}{\partial q_i} \right) = \mathrm{i}L_0 B \tag{1.73}$$

and hence Eq. (1.72) is the formal solution of Eq. (1.73). The fifth line in Eq. (1.69) is obtained from the fourth again by partial integration to transfer the operations $\partial/\partial p_i$ and $\partial/\partial q_i$ on $f(q^N,p^N)$ as well as the Poisson bracket to those on $B(q_{t-s}^N, p_{t-s}^N)$. The subscript "eq" is supplied to remind us that the statistical average is taken by using the equilibrium distribution function $f_{\mathrm{eq}}(q^N,p^N)$.

The final result from Eq. (1.69)

$$\langle \Delta B(t) \rangle = \int_{-\infty}^{t} \Psi_{BA}(t-s) K(s) \mathrm{d}s \tag{1.74}$$

is rewritten as a superposition of responses to the time-dependent external field $K(s)\mathrm{d}s$, all governed by the response or after-effect function:

$$\Psi_{BA}(t-s) = \int\int \mathrm{d}q^N \mathrm{d}p^N f_{\mathrm{eq}}(q^N,p^N)\{A, B(t-s)\} = \langle \{A, B(t-s)\} \rangle. \tag{1.75}$$

Here and from now on, the angular brackets $\langle \cdots \rangle$ represent the operation of statistically averaging the quantity inside by integration over the phase space weighed by the equilibrium distribution function $f_{\text{eq}}(q^N, p^N)$. Thus, $\Psi_{BA}(t)$ is the *equilibrium* statistically expected value of the Poisson bracket of the dynamic variable A conjugate to the external field and B at time $(t - s)$. Another expression for $\Phi_{BA}(t)$ can be obtained from the fourth line in Eq. (1.69) by recognizing that $f_{\text{eq}} = C\exp(-H_0/k_{\text{B}}T)$, where C is a normalization constant and k_{B} is the Boltzmann's constant, and manipulating the Poisson bracket $\{f_{\text{eq}}, A\}$ therein as follows:

$$\begin{aligned}\{f_{\text{eq}}, A\} &= \sum_i^N \left(\frac{\partial f_{\text{eq}}}{\partial q_i}\frac{\partial A}{\partial p_i} - \frac{\partial f_{\text{eq}}}{\partial p_i}\frac{\partial A}{\partial q_i}\right) \\ &= (1/k_{\text{B}}T)\sum_i^N \left(\frac{\partial A}{\partial q_i}\frac{\partial H_0}{\partial p_i} - \frac{\partial A}{\partial p_i}\frac{\partial H_0}{\partial q_i}\right) f_{\text{eq}} \\ &= (1/k_{\text{B}}T)(\mathrm{i}L_0 A) f_{\text{eq}} = (1/k_{\text{B}}T)\dot{A} f_{\text{eq}}\end{aligned} \tag{1.76}$$

In the last line we have used the relation between $(\mathrm{i}L_0A)$ and $\dot{A}$, with the time derivative of A given by Eq. (1.73). Thus, an alternative expression for $\langle \Delta B(t)\rangle$ and $\Psi_{BA}(t)$ is

$$\langle \Delta B(t)\rangle = \left(\frac{1}{k_{\text{B}}T}\right)\int_{-\infty}^{t} \langle B(t-s)\dot{A}\rangle_{\text{eq}} K(s)\mathrm{d}s = -\left(\frac{1}{k_{\text{B}}T}\right)\int_{-\infty}^{t} \langle \dot{B}(t-s)A\rangle_{\text{eq}} K(s)\mathrm{d}s \tag{1.77}$$

and by comparing this with Eq. (1.74), the response function is

$$\Psi_{BA}(t) = (1/k_{\text{B}}T)\langle B(t)\dot{A}\rangle = -(1/k_{\text{B}}T)\langle \dot{B}(t)A\rangle. \tag{1.78}$$

In obtaining the last part of Eqs. (1.77) and (1.78), we have used the relation

$$\langle \dot{B}(t-s)A\rangle = -\langle B(t-s)\dot{A}\rangle. \tag{1.79}$$

This relation is a consequence of the fact that the correlation function $\langle B(t+t')A(t')\rangle_{\text{eq}}$ is independent of the choice of the time origin t', and that means

$$\frac{\mathrm{d}}{\mathrm{d}t'}\langle B(t+t')A(t')\rangle = \langle \dot{B}(t+t')A(t')\rangle + \langle B(t+t')\dot{A}(t')\rangle = 0 \tag{1.80}$$

and hence the relation (1.79) when we put $t' = 0$ and $t = t - s$.

If the change $\langle \Delta B(t)\rangle$ is induced by a time oscillating field $K(s) = K_0 f \exp(-i\omega s)$, Eqs. (1.77) and (1.79) give, after changing integration variable from s to t–s,

$$\begin{aligned}\langle \Delta B(t)\rangle &= -K_0\exp(\mathrm{i}\omega t)\int_0^{\infty} \langle \dot{B}(t')A\rangle_{\text{eq}}\,\mathrm{d}t' \\ &= K_0\exp(\mathrm{i}\omega t)\int_0^{\infty} \Psi_{BA}(t')\exp(-\mathrm{i}\omega t')\mathrm{d}t' \equiv \chi_{BA}(\omega)K_0\,\mathrm{e}^{\mathrm{i}\omega t},\end{aligned} \tag{1.81}$$

where $\chi_{BA}(\omega)$ is the complex dynamic susceptibility or dynamic response function:

$$\chi_{BA}(\omega) \equiv \chi'_{BA}(\omega) - \mathrm{i}\chi''_{BA}(\omega) = \int_0^\infty \Psi_{BA}(t)\exp(-\mathrm{i}\omega t)\mathrm{d}t. \tag{1.82}$$

All the above equations apply for the special case when B and A are the same dynamic variables, and Eqs. (1.78) and (1.82) become

$$\Psi_{AA}(t) = -(1/k_{\mathrm{B}}T)\langle\dot{A}(t)A\rangle \tag{1.78a}$$

and

$$\chi_{AA}(\omega) \equiv \chi'_{AA}(\omega) - \mathrm{i}\chi''_{AA}(\omega) = \int_0^\infty \Psi_{AA}(t)\exp(-\mathrm{i}\omega t)\mathrm{d}t, \tag{1.82a}$$

respectively.

1.3.1 Dielectric Relaxation

The linear response formalism given above will be applied to specific experimental measurements and the results compared with the macroscopic treatment. The example chosen is the relaxation of dipolar molecules in amorphous substances measured by dielectric relaxation spectroscopy. The application of linear response theory to treat this problem of an ensemble of N equivalent dipole molecules with permanent dipole moment $\boldsymbol{\mu}$ was first developed by Glarum [74] and Cole [75], and the earlier works were reviewed by Williams [76, 77]. The external field $K(t)$ is now a weak and uniform electric field $E_z(t)$ applied in the z-direction. The dynamic variable conjugate to the applied electric field is the dipole moment $\boldsymbol{M}(q^N) = \boldsymbol{\Sigma}_i\boldsymbol{\mu}_i$ and the perturbing Hamiltonian is $-[M_z(q^N)E_z(t)$, the interaction energy of the applied field with the moments of the dipoles. The quantity of interest in dielectric relaxation measurement is the average dipole moment in the field direction in response to the applied field, which is $\iint \mathrm{d}q^N\mathrm{d}p^N[M_z(q^N)]f(q^N,p^N,t) = \langle M_z(q^N)\rangle$. Thus, we can identify $A(q^N,p^N)$ with $M_z(q^N)$, $K(t)$, and $B(q^N,p^N)$ with $M_z(q^N)$. It follows from Eqs. (1.77) and (1.78)

$$\begin{aligned}\langle M_z(t)\rangle &= \int_{-\infty}^{t} \Psi_{MM}(t-s)E_z(s)\mathrm{d}s \\ &= -\frac{1}{kT}\int_{-\infty}^{t} E_z(s)\langle\dot{M}_z(t-s)M_z\rangle_{\mathrm{eq}}\mathrm{d}s \\ &= -\frac{\langle\vec{M}(0)\cdot\vec{M}(0)\rangle_{\mathrm{eq}}}{3kT}\int_{-\infty}^{t} E_z(s)\dot{\Phi}_\mu(t-s)\mathrm{d}s.\end{aligned} \tag{1.83}$$

The factor 1/3 is the result of replacing $\dot{M}_z(t-s)M_z$ by the scalar product of the corresponding vectors in the following expression for $\Phi_\mu(t)$, the dipole moment time correlation function:

$$\Phi_\mu(t) = \frac{\langle \vec{M}(t)\cdot\vec{M}(0)\rangle_{\text{eq}}}{\langle \vec{M}(0)\cdot\vec{M}(0)\rangle_{\text{eq}}}. \tag{1.84}$$

It is useful to write out explicitly here that

$$\Psi_{MM}(t) = -\dot{\Phi}_\mu(t). \tag{1.85}$$

(i) For an electric field E_o applied to the material starting at time $t=0$ and held constant thereafter, i.e., $E_z(t)=0$ for $t<0$ and $E_z(t)=E_0$ for $t\ge 0$, from Eq. (1.83) we have

$$\langle M_z(t)\rangle = -\left(\left\langle\vec{M}(0)\cdot\vec{M}(0)\right\rangle/3k_\text{B}T\right)E_0[1-\Phi_\mu(t)]. \tag{1.86}$$

(ii) For an oscillating applied electric field $E_z(t)=E_0\exp(\mathrm{i}\omega t)$, we have

$$\langle M_z(t)\rangle = (\langle\vec{M}(0)\cdot\vec{M}(0)\rangle/3k_\text{B}T)E_0\mathrm{e}^{\mathrm{i}\omega t}\left[1-\mathrm{i}\omega\int_0^\infty \mathrm{e}^{-\mathrm{i}\omega s}\Phi_\mu(s)\mathrm{d}s\right]. \tag{1.87}$$

To arrive at Eq. (1.87) from Eq. (1.83), we have performed partial integration to transfer the time derivative from $\Phi_\mu(t)$ to $\exp(\mathrm{i}\omega t)$ and used the boundary conditions that $\Phi_\mu(0)=1$ and $\Phi_\mu(\infty)=0$.

Connection can be made between the results from the current microscopic approach and the macroscopic approach in the previous section. In macroscopic description, the permanent dipoles contribute to polarization $P_z(t)$ in the z-direction of the small applied electric field $E_z(t)$. The dielectric displacement $D_z(t)$ can be written as

$$D_z(t) = \varepsilon_v\varepsilon_\infty E_z(t) + P_z(t) \tag{1.88}$$

and

$$P_z(t) = \varepsilon_v(\varepsilon_0-\varepsilon_\infty)\int_{-\infty}^t E_z(s)\psi(t-s)\mathrm{d}s, \tag{1.89}$$

where $\psi(t)$ is the normalized macroscopic response or after-effect function similar in spirit to $\Psi_{MM}(t)$ in Eq. (1.83). Consider an oscillating applied electric field $E_z(t) = E_0\cos(\omega t)$. Substituting it into Eq. (1.89), changing variable from s to $t'=t-s$, and using the trigonometric formula of the cosine of the sum of two angles, it follows from Eq. (1.88) that

$$D_z(t) = \varepsilon_v \varepsilon'(\omega) E_0 \cos(\omega t) + \varepsilon_v \varepsilon''(\omega) E_0 \sin(\omega t), \tag{1.90}$$

where

$$\varepsilon'(\omega) = \varepsilon_\infty + (\varepsilon_0 - \varepsilon_\infty) \int_0^\infty \cos(\omega t') \psi(t') \mathrm{d}t' \tag{1.91}$$

and

$$\varepsilon''(\omega) = (\varepsilon_0 - \varepsilon_\infty) \int_0^\infty \sin(\omega t') \psi(t') \mathrm{d}t'. \tag{1.92}$$

Equation (1.89) can be generalized to a complex $E_z(t) = E_0 \exp(\mathrm{i}\omega t)$ and the response is given by

$$D_z(t) = \varepsilon_v [\varepsilon'(\omega) - \mathrm{i}\varepsilon''(\omega)] E_0 \mathrm{e}^{\mathrm{i}\omega t} \equiv \varepsilon_v \varepsilon^*(\omega) E_z(t). \tag{1.93}$$

The complex dielectric constant $\varepsilon^*(\omega)$ as defined is related to the one-sided Fourier transform of $\psi(t)$ by

$$\varepsilon^*(\omega) = \varepsilon_\infty + (\varepsilon_0 - \varepsilon_\infty) \int_0^\infty \mathrm{e}^{-\mathrm{i}\omega t} \psi(t) \mathrm{d}t. \tag{1.94}$$

The connection to microscopic description is made by identifying polarization $P_{z(t)}$ with $\langle M_z(t) \rangle / V$, where V is the volume. Thus, by comparing Eq. (1.81) for $M_z(t)$ with Eq. (1.89) for $P_z(t)$, we have

$$\langle \vec{M}(0) \cdot \vec{M}(0) \rangle = 3 k_B T V \varepsilon_v (\varepsilon_0 - \varepsilon_\infty) \tag{1.95}$$

and

$$\psi(t) = -\dot{\Phi}(t), \tag{1.96}$$

showing the correspondence between the quantities in macroscopic and microscopic descriptions of dielectric relaxation. Furthermore, Eq. (1.94) leads us to the final result:

$$\frac{\varepsilon^*(\omega) - \varepsilon_\infty}{\varepsilon_0 - \varepsilon_\infty} = \int_0^\infty -\mathrm{e}^{-\mathrm{i}\omega t} \dot{\Phi}_\mu(t) \mathrm{d}t = 1 - \mathrm{i}\omega \int_0^\infty \frac{\langle \vec{M}(0) \cdot \vec{M}(t) \rangle}{\langle \vec{M}(0) \cdot \vec{M}(0) \rangle} \mathrm{e}^{-\mathrm{i}\omega t} \mathrm{d}t. \tag{1.97}$$

Thus, from this relation, $\langle \vec{M}(0) \cdot \vec{M}(t) \rangle$ calculated microscopically by molecular dynamics simulations [78–80] can be directly compared with $\varepsilon^*(\omega)$ obtained by macroscopic dielectric measurements. Also, from Eqs. (1.82) and (1.85), the quantity on the left-hand side of Eq. (1.97) is just the dynamic dielectric susceptibility $\chi_{MM}(\omega)$. Actually, in molecular dynamics simulation as well as real experiments, the statistical average of $A(0)B(t)$ such as $\vec{M}(0) \cdot \vec{M}(t)$ is obtained not as $\langle A(0) \cdot B(t) \rangle$ or $\langle \vec{M}(0) \cdot \vec{M}(t) \rangle$ by ensemble averaging as discussed in theory but by a time average over the dynamic history of a single system observed over a sufficiently long period of time τ. The time average is the integral

$$\langle A(0)B(t)\rangle = \lim_{\tau\to\infty}\frac{1}{\tau}\int_0^\tau A(s)B(s+t)\mathrm{d}s. \tag{1.98}$$

It is assumed that this time integral is the same as the ensemble average. When this assumption is valid, the system is called ergodic.

Closely related to the dipole moment time correlation function, $\Phi_\mu(t)$, is the time correlation of an arbitrary unit vector $\boldsymbol{u}(t)$ defined by

$$C_1(t) = <\boldsymbol{u}(t)\cdot\boldsymbol{u}(0)> . \tag{1.99}$$

The vector $\boldsymbol{u}(t)$ can represent the orientation of molecular dipole or the molecular bond. If the dynamic variable is a tensor $\sigma_{\alpha\beta} = (3u_\alpha u_\beta - \delta_{\alpha\beta})/2$, the correlation function

$$C_{\alpha\beta}(t) = \langle\sigma_{\alpha\beta}(t)\sigma_{\alpha\beta}(0)\rangle \tag{1.100}$$

describes the linear response of the change in the $\alpha\beta$ component of the tensorial quantity. In terms of the equilibrium statistical average of the first-order Legendre polynomial $P_1[\cos\theta(t)]$, $C_1(t)$ can be rewritten as

$$C_1(t) = \langle\cos\theta(t)\rangle \equiv \langle P_1[\cos\theta(t)]\rangle. \tag{1.101}$$

There is another experimentally accessible correlation function based on the second-order Legendre polynomial $P_2[\cos\theta(t)]$ given by

$$C_2(t) = \left\langle\frac{1}{2}(3\cos^2\theta(t) - 1)\right\rangle \equiv \langle P_2[\cos\theta(t)]\rangle. \tag{1.102}$$

In addition to the time correlation functions $C_1(t)$ and $C_2(t)$ of vectors, molecular dynamics simulators can also evaluate the autocorrelation function for the rotation of an angle ϑ in two different forms [79, 80]:

$$R_\vartheta(t) = \frac{\langle\cos\vartheta(t)\cos\vartheta(0)\rangle - \langle\cos\vartheta(0)\rangle^2}{\langle\cos^2\vartheta(0)\rangle - \langle\cos\vartheta(0)\rangle^2} \tag{1.103}$$

or

$$R_\vartheta(t) = \frac{\langle\vartheta(t)\vartheta(0)\rangle - \langle\vartheta(0)\rangle^2}{\langle\vartheta^2(0)\rangle - \langle\vartheta(0)\rangle^2}. \tag{1.104}$$

1.3.2 Light Scattering

Dynamic light scattering is a method for the study of static and dynamic properties of materials. The electric field of the incident monochromatic light at each one of N equivalent particle at $\vec{r}_j$ that scatters light is $E(\vec{r},t) = E_0 \exp(\mathrm{i}\vec{k}_I \cdot \vec{r}_j - \mathrm{i}\omega t)$, where $\vec{k}_\mathrm{I}$ is the incident wave vector and ω is the frequency. The electric field of the light scattered by the particle has wave vector $\vec{k}_\mathrm{F}$. The light-scattering experiments of interest have nearly the same initial and final frequencies (i.e., quasielastic scattering), and hence the wave vectors $\vec{k}_\mathrm{I}$ and $\vec{k}_\mathrm{F}$ of the incident and scattered light have nearly the same magnitude and differ only in direction. The scattering vector $\vec{q}$ is defined as the difference $\vec{k}_\mathrm{F} - \vec{k}_\mathrm{I}$. The scattered electric field detected is the sum of the electric fields scattered by the N particles given by

$$E_{\mathrm{scatt}}(\vec{q},t) = \xi E_0 \exp(-\mathrm{i}\omega t) \sum_{j=1}^{N} \exp[\mathrm{i}\vec{q} \cdot \vec{r}_j(t)], \tag{1.105}$$

where ξ, called the scattering length, determines the amplitude of the light scattered by each particle. The intensity of the scattered light at time t, given by the square of the absolute value of $E_{\mathrm{scatt}}(\vec{q},t)$, is

$$I(\vec{q},t) = (\xi E_0)^2 \sum_{j=1}^{N} \sum_{k=1}^{N} \exp[\mathrm{i}\vec{q} \cdot (\vec{r}_j(t) - \vec{r}_k(t))]. \tag{1.106}$$

Due to the motion of the particles, the scattered light intensity fluctuates in time, and the intensity–intensity time correlation function $\langle I(\vec{q},0) I(\vec{q},t)\rangle$ in return gives information on the dynamics of the particles whether simple or complex. In the literature, the intensity–intensity correlation function is denoted by $g^{(2)}(\vec{q},t)$. The scattered light obeys Gaussian statistics because $E_{\mathrm{scatt}}(\vec{q},t)$ is the sum over many statistically independent terms. From this, it can be shown that $g^{(2)}(\vec{q},t)$ is the sum of a constant B and the square of $g^{(1)}(\vec{q},t)$, the (electric) field correlation function of the scattered light is defined by

$$g^{(1)}(\vec{q},t) = (\xi E_0)^2 \left\langle \sum_{j,k=1}^{N} \exp[\mathrm{i}\vec{q} \cdot (\vec{r}_j(t) - \vec{r}_k(t))] \right\rangle. \tag{1.107}$$

Explicitly, we have

$$g^{(2)}(\vec{q},t) = B + [g^{(1)}(\vec{q},t)]^2. \tag{1.108}$$

The constant B is called the baseline. The scattering vector $q = (4\pi n/\lambda)\sin(\theta/2)$, where n is the refractive index of the system, θ is the scattering angle, and λ is the light wavelength.

In actual light-scattering experiment, $\langle I(\vec{q},0)I(\vec{q},t)\rangle$ is determined by time averaging instead of ensemble averaging as discussed before when the system studied is ergodic. The correspondent of Eq. (1.98) in this case is

$$\langle I(\vec{q},0)I(\vec{q},t)\rangle = \int_0^{\tau} \mathrm{d}s\, I(\vec{q},s)I(\vec{q},t+s), \tag{1.109}$$

with τ being a long time, the duration of the experiment.

The details of the methods and the information acquired can be found in books and review articles dedicated to this subject [81–87]. The method of light scattering with different techniques can cover a wide dynamic range from 10^5 up to 10^{-12} s. Fabry–Perot interferometry and Rayleigh–Brillouin spectra can reveal molecular motions and orientational fluctuations between 10^{-8} and 10^{-12} s. At longer timescales from 10^{-6} to 10^5 s, the molecular motions can be extracted from data obtained by the photon correlation spectroscopy. The scattering vector $\vec{q}$ determines the wavelength of the observed fluctuations, which selects the length scale of the motion to be characterized by the experiment. The range of length scale probed is enhanced by using X-ray in addition to conventional light sources. The particular motion detected by the light-scattering technique depends on the directions of the polarizations of the scattered radiation and the incident radiation. Provided that the incident and scattered light are polarized with the electric vector at right angles to the plane containing $\vec{k}_\mathrm{I}$ and $\vec{k}_\mathrm{F}$, VV scattering is observed. The so-called depolarized light scattering (DLS) or VH scattering, where V and H indicate a vertical and a horizontal direction of the polarization of the incident and scattered light relative to the scattering plane [83, 88], is related to fluctuations of the anisotropy of the polarizability tensor. Thus, re orientational motions of optically anisotropic molecules can be analyzed by using depolarized light scattering [89–91]. The field correlation function of depolarized light scattering is a product of the orientational and translational autocorrelation functions. The orientational autocorrelation function is the second-order Legendre polynomial of the orientation angle given before in Eq. (1.102). On the other hand, dielectric relaxation is related to the first-order Legendre polynomial.

It can be shown that the rotational correlation function obtained from infrared and Raman spectroscopies are $C_1(t)$ and $C_2(t)$ of unit vector along the symmetry axis of the molecule [70].

1.3.3 Nuclear Magnetic Resonance

Many methods and special techniques have been developed over the years for relaxation and diffusion measurements using nuclear magnetic resonance and related effects. As a result, there are so many different quantities and their correlation functions introduced that it would not be profitable to review them here. The reader can consult [92] for conventional techniques and [93, 94] for more modern techniques applied to relaxation and diffusion in complex interacting systems.

1.3.4 Neutron Scattering

The incoherent neutron scattering technique [95, 96] can probe the time evolution of spatial correlations of N identical entities, particles, or molecules. The latter is described microscopically by the autocorrelation function $G(\vec{r},t)$ of the density $\rho(\vec{r},t) = \sum_{i=1}^{N} \delta[\vec{r} - \vec{r}_i(t)]$:

$$G(\vec{r},t) = \langle \rho(\vec{r},t)\rho(\vec{0},0)\rangle/\rho = \frac{1}{N}\left\langle \sum_{i=1}^{N}\sum_{j=1}^{N} \delta[\vec{r} + \vec{r}_j(0) - \vec{r}_i(t)] \right\rangle. \quad (1.110)$$

The function $G(\vec{r},t)$, giving the space- and time-dependent distribution functions, was introduced by van Hove [97]. Thus, $G(\vec{r},t)\mathrm{d}\vec{r}$ is proportional to the probability of finding an entity i in the small neighborhood $\mathrm{d}\vec{r}$ of $\vec{r}$ at time t with the condition that there was an entity j at the origin at time $t = 0$. $G(\vec{r},t)$ can be written as the sum of two separate parts:

$$G(\vec{r},t) = G_s(\vec{r},t) + G_d(\vec{r},t). \quad (1.111)$$

Here $G_s(\vec{r},t)$, called the self-part, is the part of $G(\vec{r},t)$ when i and j are the same particle and is given by

$$G_s(\vec{r},t) = \frac{1}{N}\left\langle \sum_{i=1}^{N} \delta[\vec{r} + \vec{r}_i(0) - \vec{r}_i(t)] \right\rangle \quad (1.112)$$

and $G_s(\vec{r},0) = \delta(\vec{r})$ when $t = 0$. The physical interpretation of the self-part is that $G_s(\vec{r},t)\mathrm{d}\vec{r}$ is proportional to the probability of finding an entity i in the small neighborhood $\mathrm{d}\vec{r}$ of $\vec{r}$ at time t when previously it was at the origin at time $t = 0$. The space Fourier transform of $G(\vec{r},t)$ and $G_s(\vec{r},t)$ are called the coherent and incoherent intermediate scattering function $F(\vec{k},t)$ and $F_s(\vec{k},t)$, respectively, and are given by

$$F(\vec{k},t) = \int G(\vec{r},t)\exp(-\mathrm{i}\vec{k}\cdot\vec{r})\mathrm{d}\vec{r} = \frac{1}{N}\sum_i\sum_j \langle \exp\{\mathrm{i}\vec{k}\cdot[\vec{r}_i(0) - \vec{r}_j(t)]\}\rangle, \quad (1.113)$$

$$F_s(\vec{k},t) = \int G_s(\vec{r},t)\exp(-\mathrm{i}\vec{k}\cdot\vec{r})\mathrm{d}\vec{r} = \frac{1}{N}\sum_j \langle \exp\{\mathrm{i}\vec{k}\cdot[\vec{r}_j(0) - \vec{r}_j(t)]\}\rangle. \quad (1.114)$$

The distinct part $G_d(\vec{r},t)$ corresponding to i and j being different particles is given by

$$G_d(\vec{r},t) = \frac{1}{N}\left\langle \sum_{i\neq j}^{N} \sum \delta[\vec{r} + \vec{r}_j(0) - \vec{r}_i(t)] \right\rangle. \quad (1.115)$$

When $t = 0$, $G_d(\vec{r},0) = \rho g(\vec{r})$, where $g(\vec{r})$ is the static pair distribution function.

It turns out that the measurements of the inelastic scattering of thermal neutrons can be presented in terms of the Fourier transforms of $G(\vec{r},t)$ and $G_s(\vec{r},t)$. Consider incident neutrons with wave vector $\vec{k}_i$ and frequency ω_i are scattered by nuclei into a solid angle $d\Omega$ resulting in the scattered neutron to have wave vector $\vec{k}_f$ and frequency ω_f. The changes in wave vector and frequency are $\vec{k} = \vec{k}_i - \vec{k}_f$ and $\omega = \omega_f - \omega_I$, respectively, and the scattered intensity is the sum of an incoherent part and a coherent part, described by the differential cross sections [95]:

$$\frac{d^2\sigma}{d\Omega\, d\omega} = \frac{d^2\sigma_{inc}}{d\Omega\, d\omega} + \frac{d^2\sigma_{coh}}{d\Omega\, d\omega}. \tag{1.116}$$

The incoherent part is given by

$$\frac{d^2\sigma_{inc}}{d\Omega\, d\omega} = \left[(\langle b_j^2\rangle - \langle b_j\rangle^2)\left(\frac{k_f}{k_i}\right)N\right] S_s(\vec{k},\omega) \tag{1.117}$$

and the coherent part by

$$\frac{d^2\sigma}{d\Omega\, d\omega} = \langle b_j^2\rangle \left(\frac{k_f}{k_i}\right) NS(\vec{k},\omega). \tag{1.118}$$

Here b_j are the scattering lengths of the individual nuclei which may differ if they are different isotopes. $S_s(\vec{k},\omega)$ and $S(\vec{k},\omega)$ are the spectrum of the density autocorrelation functions and are called the "self-dynamic structure factor" and "dynamic structure factor," respectively. They in turn are given by the space and time Fourier transforms of the self-part and the entire van Hove functions $G_s(\vec{r},t)$ and $G(\vec{r},t)$, respectively:

$$S_s(\vec{k},\omega) = \frac{1}{2\pi}\int_{-\infty}^{\infty} dt\, \exp(i\omega t) \int G_s(\vec{r},t)\exp(-i\vec{k}\cdot\vec{r})d\vec{r} \tag{1.119}$$

and

$$S(\vec{k},\omega) = \frac{1}{2\pi}\int_{-\infty}^{\infty} dt\, \exp(i\omega t) \int G(\vec{r},t)\exp(-i\vec{k}\cdot\vec{r})d\vec{r}. \tag{1.120}$$

From these relations, it follows that

$$S_s(\vec{k},\omega) = \frac{1}{2\pi}\int_{-\infty}^{\infty} dt\, \exp(i\omega t)F_s(\vec{k},t), \tag{1.121}$$

$$S(\vec{k},\omega) = \frac{1}{2\pi}\int_{-\infty}^{\infty} dt\, \exp(i\omega t)F(\vec{k},t). \tag{1.122}$$

In other words, $S_s(\vec{k},\omega)$ and $S(\vec{k},\omega)$ are the frequency spectra of $F_s(\vec{k},t)$ and $F(\vec{k},t)$, respectively.

An alternative expression for the cross section $\mathrm{d}^2\sigma/\mathrm{d}\Omega\,\mathrm{d}\omega$ can be obtained when the density $\rho(\vec{r},t)$ is rewritten as $\rho(\vec{r},t) = \rho + \delta\rho(\vec{r},t)$, where $\delta\rho(\vec{r},t)$ is the fluctuations in density. Expressing $\delta\rho(\vec{r},t)$ as a Fourier transform in terms of its Fourier components $\delta\rho_{\vec{k}}(t) = \int \rho(\vec{r},t)\exp(-\mathrm{i}\vec{k}\vec{r})\mathrm{d}\vec{r}$ and putting this expression back into the equations above starting from Eq. (1.110), we have

$$S(\vec{k},\omega) = \frac{1}{2\pi N}\int_{-\infty}^{\infty} \langle \delta\rho_{\vec{k}}(t)\delta\rho_{-\vec{k}}(0)\rangle \mathrm{e}^{\mathrm{i}\omega t}\mathrm{d}t. \tag{1.123}$$

In arriving at Eq. (1.123), we have used the fact that the correlation function $\langle \delta\rho(\vec{r},t)\delta\rho(\vec{0},t)\rangle$ is a translational invariant which implies that the correlations between the Fourier components $\delta\rho_{\vec{k}'}(t)$ and $\delta\rho_{\vec{k}''}(0)$ are non-zero only when $\vec{k}' = -\vec{k}''$. Also the additive term $(2\pi)^3\rho\delta(\omega)\delta(\vec{k})$ is ignored because it conserves both energy and momentum and hence does not correspond to inelastic scattering and is unrelated to any relaxation or diffusion process.

For a system of N non-interacting particles, closer examination of the definition or interpretation of $G_s(\vec{r},t)$ indicates that it is the same as the Gaussian spatial distribution function $P(\vec{r},t)$ given by Eq. (1.15). Both $G_s(\vec{r},t)\mathrm{d}\vec{r}$ and $P(\vec{r},t)\mathrm{d}\vec{r}$ are probability of finding the particle in the small neighborhood $\mathrm{d}\vec{r}$ of $\vec{r}$ at time t, given the condition that it was previously at the origin at $t=0$. Thus, the solution for $P(\vec{r},t)$ given by Eq. (1.15) is shared by $G_s(\vec{r},t)$. Hence, the Fourier transform of $P(\vec{r},t)$, $P(\vec{k},t)$, is the same as the incoherent intermediate scattering function, $F_s(\vec{k},t)$, the Fourier transform of $G_s(\vec{r},t)$. It follows from Eq. (1.14), we have

$$F_s(\vec{k},t) = \exp(-k^2D_\mathrm{t}t) = P(\vec{k},t) \tag{1.124}$$

and the relaxation time given by $\tau = (D_\mathrm{t}k)^{-2}$ has the characteristic k^{-2} dependence.

For interacting systems, $G_s(\vec{r},t)$ usually deviates from the Gaussian distribution in r and $F_s(\vec{k},t)$ from the form of Eq. (1.124). One measures of the degree of deviation from the Gaussian form of $G_s(\vec{r},t)$ is the non-Gaussian parameter [98]:

$$\alpha_2(t) = \frac{3\langle r^4(t)\rangle}{5\langle r^2(t)\rangle^2} - 1, \tag{1.125}$$

which involves the moments of $G_s(\vec{r},t)$. If $G_s(\vec{r},t)$ is Gaussian, then $\alpha_2(t)$ is exactly zero.

1.3.5 The Green–Kubo Relation Between Transport Coefficients and Time Correlation Functions

In the above, we emphasize the time/frequency dependence of the dynamics of relaxation and diffusion and the use of time correlation functions to make connection with the microscopic origin. At long times or low frequencies, the linear response of the system reaches a steady state and it is then fully described by a time/frequency-independent transport coefficient. Examples include the diffusion constant, viscosity, electrical, and thermal conductivities, and they are important quantities of linear non-equilibrium statistical mechanics that can be directly measured in experiments. It turns out that there is relation between the transport coefficient and the time correlation function, first discovered by Green [99] and fully developed by Kubo into a formalism [100]. Simply stated, the Green–Kubo relation says that any transport coefficient can be obtained by the integral over all times of an appropriate correlation function.

A classical example is the self-diffusion constant D in -dimensional space, which defines the steady-state linear time dependence of the mean-square displacement at long times by

$$\lim_{t\to\infty} < |\vec{r}(t) - \vec{r}(0)|^2\rangle = 2\upsilon Dt. \tag{1.126}$$

From the fact that $\vec{r}(t) - \vec{r}(0)$ is given by integration of the velocity $\vec{v}(s)$ over time s from 0 to t, we have

$$< |\vec{r}(t) - \vec{r}(0)|^2\rangle = \int_0^t \mathrm{d}s \int_0^t \mathrm{d}s' \langle \vec{v}(s') \cdot \vec{v}(s)\rangle = \int_0^t \mathrm{d}s \int_0^t \mathrm{d}s' \langle \vec{v}(s'-s) \cdot \vec{v}(0)\rangle. \tag{1.127}$$

The last equality was obtained by making use of the time translation invariance of the system, which after a change of integration variable from s' to $t' = s' - s$ and further steps yields

$$\left\langle |\vec{r}(t) - \vec{r}(0)|^2 \right\rangle = 2t \int_0^t \mathrm{d}s(1 - s/t)\langle \vec{v}(s) \cdot \vec{v}(0)\rangle. \tag{1.128}$$

Identifying this expression in the limit of $t \to \infty$ with Eq. (1.124), we obtain the Green–Kubo relation between the self-diffusion coefficient and the integral of the velocity–velocity time correlation function over time:

$$D = \frac{1}{\upsilon} \int_0^\infty \mathrm{d}s \langle \vec{v}(s) \cdot \vec{v}(0)\rangle. \tag{1.129}$$

Similar relations have been derived for other transport coefficients. For example, the electrical conductivity σ of N identical carriers each with charge q is the time integral, from 0 to ∞, of the microscopic equilibrium autocorrelation of current J_z in the z-direction:

$$\sigma = \frac{q}{k_B TV} \int_0^\infty \mathrm{d}t \langle J_z(t) J_z(0)\rangle, \text{ with } J_z = \sum_{i=1}^{N} v_i^z. \tag{1.130}$$

The shear viscosity η is related to the time integral of autocorrelations $C_{\alpha\beta}(t)$ given by Eq. (1.100) of the off-diagonal components of the stress tensor $\sigma_{\alpha\beta} = (3u_\alpha u_\beta - \delta_{\alpha\beta})/2$, i.e.,

$$\eta = \frac{1}{k_B TV}\int_0^\infty dt\langle\sigma^{xy}(0)\sigma^{xy}(t)\rangle. \tag{1.131}$$

1.3.6 The Fluctuation–Dissipation Theorem

The Green–Kubo formulas relating the transport coefficient to time integral of autocorrelation function turn out to be a special case of a wider class of relations collectively known as the fluctuation–dissipation theorem. There are several forms of this theorem, but they all essentially state that the linear response to external fields of a thermodynamic equilibrium system is related to the equilibrium fluctuations. One form of the theorem is illustrated by Einstein's relation, which links the diffusion and friction coefficients in Brownian motion. Equation (1.14) for translational diffusion is the first statement ever made on the relation between fluctuation and dissipation. Another form of the fluctuation–dissipation theorem is the formula of Nyquist [100], relating the electrical resistance to voltage noise. In Nyquist formula, current flow in response to external voltage is dissipative and the dissipation is governed by the electrical resistance, while noise comes from thermal fluctuations of the charge carriers in the conductor at equilibrium.

The general form of the fluctuation–dissipation theorem was given by Callen and Welton [51] and by Kubo [15, 65, 101]. This general form of the theorem was given in the same mathematical structure as the time-dependent linear response formalism discussed in Section 1.2.2. Examples that involve an external electric field include the appearance of an electric polarization or an electric current due to linear response of permanent dipoles and mobile charged particles, respectively. Although the response is small if the external field is small, nonetheless the process is dissipative in energy. On the other hand, even in the absence of applied field, there are spontaneous random fluctuations from equilibrium occurring all the time in a system at thermodynamic equilibrium. The fluctuation of a dynamic variable $A(t)$ from its average value $\langle A\rangle$ is the difference $a(t) = [A(t) - \langle A\rangle]$. The fluctuations are characterized by $S_{AA}(t) = \langle a(t)a(0)\rangle$, the equilibrium time autocorrelation functions of the fluctuating parts. The spectral density or the power spectrum $S_{AA}(\omega)$ is obtained from $S_{AA}(t)$ by Fourier transform:

$$S_{AA}(\omega) = \frac{1}{2\pi}\int_{-\infty}^{\infty} dt\, S_{AA}(t)\exp(i\omega t). \tag{1.132}$$

$S_{AA}(\omega)$ is an important quantity because it is the spectra measured in experiments using various spectroscopic techniques. A case in point is the dynamic structure factors measured in neutron scattering $S_s(\vec{k},\omega)$ and $S(\vec{k},\omega)$, which are the spectral density of the density autocorrelation functions (see Eqs. (1.119), (1.20), (1.21), (1.22), and (1.123)).

It can be shown that the power spectrum $S_{AA}(\omega)$ is a real and even function of ω, and always positive. Furthermore, through some mathematical steps, $S_{AA}(\omega)$ from fluctuation consideration is related to the dynamic susceptibility $\chi_{AA}(\omega)$ of linear response (Eq. (1.82a)) by

$$S_{AA}(\omega) = \frac{k_B T}{\pi \omega} \mathrm{Im}[\chi_{AA}(\omega)], \tag{1.133}$$

where $\mathrm{Im}[\chi_{AA}(\omega)] \equiv \chi''(\omega)$ is the imaginary part of the complex $\chi_{AA}(\omega)$. This relation is one form of the fluctuation–dissipation theorem. Some authors define the power spectrum $S_{AA}(\omega)$ from $S_{AA}(t)$ by Fourier transform without the prefactor $1/2\pi$ in Eq. (1.132) and then $S_{AA}(\omega) = (2k_B T/\omega)\mathrm{Im}[\chi_{AA}(\omega)]$. From these relations, it is clear that an alternative method to study relaxation and diffusion is by making measurement of fluctuation or noise as implemented by others experimentally.

From the fluctuation and dissipation theorem, we can see once more that the correlation functions are effective in bridging microscopic processes in thermal equilibrium and the irreversible macroscopic response of many-body systems.

1.4 Obstacles of Progress in Finding a Solution

Although it takes almost three quarters of a century since the publication by R. Brown on Brownian diffusion [1], this simpler problem of diffusion in systems without many-body interactions was finally solved by Einstein and others [11–14]. However, the more complicated problem of relaxation and diffusion in systems with many-body interactions remained unsolved ever since the first recorded experimental observation by Kohlrausch in 1854 [2] and up to the present times of modern science for 156 years. This situation in scientific research is highly unusual because all important problems in chemistry, physics, and condensed matter physics have been solved in much shorter time spans. For example, superconductivity discovered by Kamerlingh-Onnes [102] was solved in few decades of time later [103]. What are the factors that impede progress in seeking for a solution of the problem? An answer to this question is given in this section by pointing out several obstacles encountered by anyone attempting to solve the problem.

1.4.1 An Unsolved Many-Body Problem

The foremost obstacle is the lack of any method to deal directly with the many-body relaxation dynamics originating from the presence of interactions in the complex systems of our interest. The methods of solutions by the giants in physics and chemistry in the past for the simpler problems in the absence of interaction cannot be used. Generalizations of the methods such as the introduction of memory function offer merely mathematical formalisms but not results that can be used to confront experimental data. Certainly the only approach to relaxation in interacting system that guarantees the essential physics of the dynamics will not be lost is one that starts from the Hamiltonian level with the interactions taken into account on a fundamental basis. This is an enormously difficult task because it is already difficult to treat irreversible process at a Hamiltonian level, and now on top of it many-body interactions have to be included into the theoretical treatment. This task is perhaps unavoidable if one demands a fundamental solution of the problems. Interactions in most complex materials such as van der Waals glass-forming liquids, polymers, vitreous ionic conductors, and dense colloidal particles involve potentials (e.g., dipolar, Lennard–Jones, screened Coulomb, and hard sphere, respectively) that are anharmonic. In classical mechanics, such anharmonic interactions cause complex non-linear Hamiltonian dynamics (*chaos*) in the phase space of the system. Soluble chaos problems usually consider a few bodies but not many bodies. These impediments compound the difficulty of constructing a theory for relaxation when irreversibility has to be incorporated. Therefore, a first principle rigorous theory for understanding relaxations of many-body interacting complex systems that can impact experiments may not be available for some times in the future. This situation means that molecular dynamics simulation that starts with realistic potential and force field is the only substitute and explains the large number of researchers now engaged in this line of work. Molecular dynamics simulations yield valuable information, nevertheless simulations are computer experiments and are no substitute for fundamental theory.

1.4.2 Plethora of Experimental Facts: Anomalies are the Real Guides to Solution

While the development of fundamental and viable theory stalls, experimental investigations of relaxation and diffusion of variety of complex materials flourished in the last century and continued unabated into the new millennium. Increasing number of new natural substances and synthetic materials and invention of new experimental techniques to probe relaxation and diffusion all contribute to the explosion of experimental activities. Examples are given here for molecular and polymeric liquids and glasses [29, 104–121], inorganic glassformers [122–124], metallic glasses [125–129], water and aqueous mixtures [130, 131], hydrated proteins and biomolecules [132–134], bioprotectants [135], pharmaceuticals [136, 137], carbohydrates [138, 139], ferrofluids [140], colloidal particle suspension [141–143], plastic crystals [144, 145], crystalline, glassy, and molten ionic conductors [146–149], fuel cell materials [149–153], room temperature ionic liquids [154–159], and more. By now, immense amount of data, general properties, and phenomenology of diverse nature and in many different classes of materials have accumulated. The experimental facts are certainly the key to gain understanding of relaxation in complex systems. However, the large volume and diversity (from different disciplines) of important and general experimental facts makes them not easily accessible to any researcher. For this reason alone, most theories and models constructed were aimed to explain a selected subset of the experimental facts. The danger of this practice is that the theories may have already contradicted the rest of the experimental facts not considered. In regard to this practice relevant to the glass transition problem, some advice for theorists on research strategy was given by P.W. Anderson [160], and it is highly appropriate. In an article with the title *Some Thoughtful Words (Not Mine) on Research Strategy for Theorists*, Anderson quoted some abbreviated passages from Francis Crick's autobiography [161]. Here I extract some of the words of Crick on research strategy as quoted by Anderson:

> "Theorists almost always become too fond of their own ideas.... It is difficult to believe that one's cherished theory, which works rather nicely, may be completely false. The basic trouble is that many quite different theories can go some way to explaining the facts. If elegance and simplicity are...dangerous guides, what constraints can be used as a guide through the jungle of possible theories?... The only useful constraints are contained in the experimental evidence." "It is thus not sufficient to have rough acquaintance with the evidence, but rather a deep and critical knowledge of many different types, since one never knows what type of facts is likely to give the game away ...".

These words are most relevant for research not only in glass transition but also in relaxation and diffusion in complex interacting systems because of the great amount and diversity of experimental data in the fields.

In the same article, Anderson himself offered his own admonition to theorists:

> They often seem to believe that a kind of "Miranda rule" determines what sort of evidence is admissible. Theorists discuss theory either in an experimental vacuum or else to experiments that were endorsed by some previous paper or produced by the most fashionable experimental methods, rather than searching out the anomalies, which are the real guide to the truth.

Anderson's remark was intended, as he said, for young theorists in condensed matter physics, but I submit that it is useful for theorists of all ages in the fields of relaxation and diffusion. Reproducible anomalous experimental facts are not explained by extant or conventional theories; otherwise they would not be called anomalies. To amplify Anderson's point, let us consider an example from superconductivity of metals, which is the isotope mass dependence $M^{-\kappa}$ with $\kappa \approx 1/2$ of the superconducting transition temperature and critical magnetic field [162]. At the time of discovery in 1950, the isotope mass effect could very well be called an anomaly because no previously proposed theory of superconductivity can explain it. Fröhlich [163] and Bardeen [164] immediately recognized the importance of this anomaly and the insight it provided. It turned out to be the critical experimental fact that led to the development of a satisfactory solution of the problem by Bardeen, Cooper, and Schrieffer 7 years later [103]. Fulfilling exactly the role mentioned by Anderson, the isotope mass anomaly was the real guide to the truth in superconductivity. It is ironic that a single anomaly had facilitated the solution of the superconductivity problem, whilst multiple anomalies found in the dynamics of glassformers to be discussed in Chapter 2 have not helped to solve the glass transition problem (not to mention analogous anomalies of relaxation and diffusion in complex interacting systems not involving glass transition discussed in Chapter 3).

Anderson also issued the following warning [160]:

> Theorists discuss theory either in an experimental vacuum or else to experiments that were endorsed by some previous paper or produced by the most fashionable experimental methods...

An excellent example that this warning is cogent can be found in the field of relaxations in complex systems. In the mid-1980s, when correlation functions found experimentally are the Kohlrausch's stretched exponential function, many different theories have been published with the sole purpose of showing how the stretched exponential (or related empirical functions) can be derived but without addressing any other experimental facts then or later. The deficiency of these theories as pointed out by Anderson was made explicit by a penetrating question raised by an astute experimentalist in a scientific conference after he had heard several such theories explaining the origin of the same stretched exponential. His question was basically how could he tell which theory was the right one to explain his experiment, since none of the theories gave additional prediction for him to verify or falsify the theory by further experiment. I may add the following to the admonition issued by Anderson. Due to the overabundance and diversity of important experimental data and phenomena in our field, it is understandable that theorists tend to limit themselves to consider a selected few. Sometimes the selected few, like Anderson said, "were endorsed by some previous paper" published by some influential authorities or eloquent individuals in the field. As we shall see in later sections, this occurs quite often in the major field of glass transition. Theories or models guided by a selected few experiments may invite eventual failure if some other equally important experimental facts not considered turn out to invalidate the theories. This unwanted situation actually happens often to theoretical efforts in the research on glass transition and related areas. It seems that theoreticians in the fields of glass transition and relaxation and diffusion in complex systems have no other choice but to take into considerations all important experimental facts without bias. This is an immense burden because of the plethora of diverse facts but cannot be avoided if one wants to have a truly satisfactory solution of the problem.

The difficulty in doing research in relaxation and diffusion in complex systems pointed out above may discourage many to seek a solution to the problem. However, the situation is not as dire as it seems at first sight. It turns out that some observed experimental facts are shared by different kinds of interacting complex materials and systems in various fields of relaxation and diffusion (see Chapters 2 and 3). Therefore, experimental facts from diverse materials and systems can be classified into groups, which reduce them to a manageable smaller number of general or "universal" properties. Moreover, these general properties are genuine anomalies. They should play the pivotal role of guiding us in seeking the solution. Some of them have been given in reviews [111, 114–117, 120, 166], and a more exhaustive review will be made in this book. Unfortunately, in the past years some of these anomalies have not received as much attention as they deserve. Conventional theories and models have not considered these anomalies and definitely not used them as guides, but some of them cannot be blamed entirely because a number of the anomalies were discovered after they were published. Nevertheless, there are still some general and yet anomalous properties that were discovered years before. Possibly because these anomalies are so difficult to explain, they were largely ignored or put aside in the developments of main stream theories. A good example can be given in the field of polymer viscoelasticity. The anomaly is the breakdown of thermorheological simplicity in viscoelasticity of polymers [29, 165–172] discovered 45 years ago and repeatedly confirmed to be a general property of amorphous polymers up to present days [165, 171, 173]. But these anomalies were not considered in conventional theories of polymer viscoelasticity over the past years since the first discovery was published in 1965. It is remarkable that these viscoelastic anomalies were ignored in textbooks on viscoelasticity except the one by John Ferry [29] and [173]. Leaving the viscoelastic anomalies untouched and unexplained, the conventional theories of viscoelasticity popular nowadays cannot claim to be on firm grounds. Inadvertently, this practice produces the ill effect of obscuring the importance of the anomalies for the newcomers in the field. Similar examples in other fields will be given in later sections.

1.4.3 An Interdisciplinary Research Area: Downside and Upside

1.4.3.1 The Downside

The research community in relaxation and diffusion come from different disciplines and they have diversely different educational and research backgrounds. They use different methods to approach a problem and apply different criteria to judge the importance of research results and what constitute a satisfactory solution. Physicists often employ physical principles and mathematical formalism to build theories, but details of the motions of the individuals in the many-body systems are not necessarily given. The theory is considered acceptable by other physicists if its results are consistent with experiments. The theory would be superior if it has falsifiable predictions [174] which can be checked by new experiments. However, some workers from other disciplines may consider this form of solution too abstract and not be able to appreciate it. There are also those who would not be satisfied with any solution unless it can describe in detail the motion of all individual units in the many-body interacting system. In fact, this is exactly what a prominent physical chemist told me years ago in a Gordon Research Conference. I can give an example to counter this belief. There is a powerful technique (confocal microscopy) and a system (suspension of colloidal particles) for which the motion of all particles at all times within a broad range is known from the experiment [141], and yet this knowledge does not permit immediate explanation of some properties observed on the same system by the same technique and other technique. An example is the anomalous scattering vector dependence of the relaxation time in dynamic light-scattering experiment [143]. This anomalous dependence will be discussed in detail in a later section.

Some researchers value more a theory that is simple enough that they can understand, even though the theory has no further predictions. The theory is accepted if it can explain a few experimental facts that they subjectively considered as important. Consistency with other experimental facts is not examined or required. No further demand is made that the theory has other verifiable or falsifiable predictions that can be further checked by experiments.

There are others approaches that start with a bold or unjustified assumption and then use some mathematical formalism to crank out the results. Since the starting assumption cannot be proven to be fundamentally correct, the work has accomplished nothing or at best it merely replaced one unsolved problem by another one. Nevertheless, some will champion the approach because the assumption, though not proven, is appealing to them.

The disparate views described above of what a solution should be are the downside of the interdisciplinary nature of the research in relaxation and diffusion in complex systems, even when the systems considered are restricted to those in which many-body interactions play an important role.

In the previous subsection, we have already seen from the admonition issued by Anderson to condensed matter physicists that they do not uniformly pay attention to anomalies. The lack of attention to anomalies is even more widespread in various disciplines having to deal with relaxation and diffusion of complex interacting systems, and most workers have not taken advantage of using anomalies as guides to the solutions of their problems.

Thus, it is difficult to find a solution from a particular approach that can satisfy all the workers from the different disciplines. The situation is different from most problems in physics and chemistry, which is usually the preoccupation of workers primarily within one discipline and with similar background. The discussions above should be helpful for readers to realize that no solution can do everything to satisfy all because of the widely different backgrounds of people involved in the interdisciplinary areas.

1.4.3.2 The Upside

The upside of the interdisciplinary nature of the research on relaxation and diffusion is the extremely broad ranges of materials (natural and synthetic) and phenomena that one can access and experience, which is unrivalled by any other problem that I can think of. Solution to the problem will have impact on many different fields. Though difficult and time consuming to master them all, the vast amount of data from different disciplines should provide sufficient clues to look for the fundamental pieces of physics and the correct conceptual basis to build a viable solution. This positive outlook is supported by the view of Anderson [160]: “I believe that in all its branches, physics is still an experimental science. Even in theoretical physics, most of the great advances have been conceptual rather than mathematical. The basic goal of physics is not mathematical elegance..., but learning the truth about the world around us,” which is definitely appropriate for this area of research. It is also supported by remarkable universal properties of relaxation and diffusion found in systems and materials of very different chemical and physical structures, which will be discussed in the next subsection. Plenty of experimental evidences to borne out the universality will be provided in Chapter 2 restricted to glass-forming materials and in Chapter 3 to other materials with no connection with the glass transition problem. In a lecture given in receiving a honorary degree from the University of Vigo, Vigo, Spain, in 1997, during the 3rd International Discussion Meeting on Relaxation in Complex Systems De Gennes emphasized that keen observation is the crux of making important scientific discovery. If one follows the advice of P.-G. De Gennes, keen observation of the universal properties will lead to advance.

Since the properties of relaxation and diffusion are universal in many different systems and disciplines, solution of one major problem in one discipline will bring benefit to other problems in other disciplines. An example is the glass transition problem in condensed matter physics, the solution of which will benefit the pharmaceutical industry to make amorphous drugs to enhance its bioavailability, the food industry to increase quality and shelf life, biomedical research to stabilize biomaterials and enhance their functions, the energy industry to make more efficient batteries for portable energy and fuel cells, and the materials engineering industry to fabricate better materials for specific applications. Some of these benefits were also pointed out in the *New York Times* article [24] and quoted before in Section 1.1.

1.5 Universal (Anomalous) Properties: The Outstanding Guides to Solution of the Problem

From the experimental data of different research areas presented in the next two chapters, the reader will see that some dynamic properties are shared by different relaxation and diffusion processes in chemically and physically dissimilar systems, but many-body interaction plays an important role in all of them. Naturally, the recognition of the existence of such similar properties by anyone requires knowledge (or at least to be informed) of the experimental data and phenomenology across many different subfields in diverse disciplines. Naturally, such a broad knowledge is not easy to acquire. To some extent, international meetings on relaxation in complex systems that brought together researchers from different disciplines had made more colleagues become aware of the remarkable similarities. Examples are the series of International Discussion Meeting on Relaxation in Complex Systems, Crete (1990), Alicante (1993), Vigo (1997), Crete (2001), Lille (2005), and Roma (2009), and the proceedings published in [109]. We are talking about not just dynamic properties of complex systems that are well known by the general public such as the Kohlrausch's stretched exponential time dependence of the correlation function but also the anomalous properties not explained by conventional or established theories/models in the different subfields and disciplines. Even if one has an explanation for a particular anomalous property in a relaxation or a diffusion phenomenon in one area of research, the explanation could be so specialized to the particular phenomenon that it is inapplicable to similar phenomena in other systems. For applicability to all similar phenomena, each explanation has to stem from a common fundamental principle. The latter has to be related to the many-body aspects of the relaxation or the diffusion process because the only common characteristic of the different systems is the many-body interaction between the basic units. Many researchers are working only in one area. They do not keep track of the developments in other fields and disciplines, and are not aware that the anomalies they see appear in other processes of totally different kinds of systems also. This lack of recognition of the universal nature of properties means lost opportunity for many workers to have another real guide in their quest for the ultimate truth.

Most importantly, the existence of universal properties in relaxation and diffusion in processes and systems of different kinds means that very fundamental physics is behind it. Thus, the effort of finding a central solution to the problem is very rewarding because the solution would likely be a basic law of physics. The effort is also most profitable because the solution applies to many different processes and systems in various disciplines. In view of its importance of the universal properties, it is worthwhile to demonstrate them clearly by examples. This is done in the Chapter 2 to follow where the similar dynamic properties observed from various relaxation and diffusion processes in diverse types of glassformers and related to the glass transition problem are summarized and compared. In Chapter 3, we bring in other complex systems which are not glass-forming or in which the interest is not on their glass-forming ability.

Detailed discussion of the properties in each research area and complex system is impossible because this would make this book much voluminous than planned. However, this can be done when coverage is confined to one area of research, which I chose to be the glass transition problem. This choice is justified by the extremely wide classes of materials and systems that are glassformers and the tremendous amount of experimental data obtained by different techniques covering 18 orders of magnitude in time/frequency, and the glass transition problem is currently touted as a challenging and worthwhile unsolved problem in condensed matter physics. In discussing the universal (anomalous) properties of many-body relaxation and diffusion in many different systems (including non-glassformers), it would be advantageous to accompany the discussion by a theory or a model that is deemed applicable to all of them. As far as the author knows, the only such model is the one proposed by the author himself, referred to in the literature as the coupling model (CM) [120, 175–196]. It is a far cry from being a full solution of the many-body relaxation problem. Notwithstanding, it has captured some essential elements of the physics, and most importantly it has predictions that compare favorably with experiments in the different systems and disciplines. The successes of the predictions have been collected together and discussed in several reviews [110, 111, 120] and will be further demonstrated throughout the following chapters. For this reason, the coupling model (CM) is discussed along with the experimental data, especially the anomalous ones that are most challenging to explain. The gist of the CM and its defining equations are introduced early on to facilitate the comparison of its predictions with experiments throughout the following chapters.

Chapter 2
Glass-Forming Substances and Systems

2.1 Current Status of the Glass Transition Problem

The glass transition or vitrification, which refers to the dramatic slowing down of kinetic processes, such as diffusion, viscous flow, and molecular reorientations, is a general phenomenon found in organic, inorganic, metallic, polymeric, colloidal, and biomolecular materials, as well as synthetic materials and systems. On decreasing temperature T or increasing pressure P, the structural relaxation time τ_α of supercooled liquids becomes increasingly long. Eventually the molecules cannot attain their dynamic equilibrium configurations on the timescale of observation, and vitrification commences. The science and technology of glass formation has a long history, starting from the first recorded recipe for glass that appeared a few millennia ago in Babylon. Nowadays, glasses are formed from many different kinds of materials. The use of glass is widespread, and the glass-making industry alone contributes significantly to the world economy.

A fundamental understanding of glass transition requires elucidation of the dynamics of the structural relaxation and identification of the factors that make τ_α to slow down drastically with decreasing temperature or increasing pressure. Surprisingly, in spite of the long history of the phenomenon and technological significance of glass, there is still no universally accepted view on the dynamics of the structural relaxation. There is no consensus on the factors governing the dramatic slowing down of the structural relaxation and related kinetic processes, such as viscous flow and diffusion. This situation is a testament of the complexity of the structural relaxation process in the precursor supercooled liquid. Development of a microscopic and quantitatively accurate theory of the glass transition that is generally applicable to real materials has become even more challenging with the improvement of experimental techniques and the introduction of new ones. These advances have led to the discovery of an increasing number of general properties of the dynamics of glass-forming materials spanning the range from picoseconds to years. Glass transition is still the subject of active research with many participants, and the theme of many international scientific conferences. Its study is no longer limited to the few traditional areas, but has branched out to electronics and optoelectronics, metallurgy, pharmaceuticals, food science, biomaterials, geoscience, volcanology, petrology, materials for energy source applications, and others. The

K.L. Ngai, *Relaxation and Diffusion in Complex Systems*, Partially Ordered Systems,
DOI 10.1007/978-1-4419-7649-9_2,

study of glass transition has recently received a boost from activities in nanoscience and technology, where the changes found in the dynamics of the structural relaxation and glass transition temperature present new challenges.

There are plenty of theoretical activities in the quest to understand the fundamentals of glass transition and to explain the multitude of experimental data and physical phenomena. It is worthwhile to reiterate the criterion of viability of a theory/model of glass transition laid out at the end of the last chapter. The criterion is that it has predictions that are consistent with *all* the observed properties, especially the anomalous ones. I do not accept the claim that the problem of glass transition is solved by any theory/model if only a selected few properties have been explained. Is this stringent requirement subjective or unnecessary? I do not think so, the reason being that because the other properties and anomalies not addressed are equally important and they may have already shown that the theory/model is either deficient or even untenable.

In the sections to follow, the unexpected but general experimental facts (i.e., anomalies) will be the subjects of detailed discussion. As shall be demonstrated, the root cause of the anomalies is the many-body effects in relaxation and diffusion in glass-forming substances and systems. Since many-body dynamics is either neglected altogether or not given an adequate treatment in most conventional theories and models, it is not surprising that they cannot explain the anomalies. New theory and model of glass transition that incorporate the many-body dynamics at the very start should have a better chance of being truly successful.

2.2 General Properties and Anomalies

An observed anomaly, or unusual behavior or property, is important only if it is general. Some anomalies are better known possibly because they have longer history or are more accessible from the literature. Some of the anomalies have become so familiar to workers in the field of glass transition that no one is surprised by their generality. There are still other anomalies that are not well known or discovered only recently. In considering anomalies, it is important to consider all of them because, a priori, no one can tell which anomaly is more pivotal than the other for the purpose of gaining a deeper understanding. It also allows us to see if the various anomalies are related or not. If they are related, the relations may allow us to extract the quintessential factor or parameter that governs all the anomalies. The quintessential parameter may reveal directly and unambiguously the fundamental physics behind glass transition that so far has escaped general attention. Finally, theories capturing the quintessential parameter may emerge that can explain all the anomalies and other experimental facts with the help of the quintessential parameter. As I shall demonstrate in this chapter, it turns out that this scenario can be realized.

Instead of keeping the readers in suspense, I hasten to point out the identity of the quintessential parameter. It is n appearing in the exponent of the Kohlrausch stretched exponential function, $\phi_K(t) = \exp[-(t/\tau)^{1-n}]$ where $(0 \leq n < 1)$, given before by Eq. (1.1). As we shall see, n not only governs all anomalies but also indicates the extent or degree of the many-body dynamics occurring in the structural α-relaxation of the glassformer. The proviso is that $\phi_K(t)$ describes well the time dependence of some correlation function, $\langle A(t)A \rangle$, of the structural relaxation. For example, when normalized, $\langle A(t)A \rangle$ can be $\Phi_\mu(t)$, the dipole moment time correlation function in Eq. (1.84), if the rotations of the dipoles are related to structural relaxation. In light scattering, it can be the field correlation function, $g^{(1)}(\vec{q}, t)$, given by Eq. (1.107) [81–91, 198–211]. If the structural relaxation is probed in the frequency domain and given in terms of susceptibility such as Eq. (1.88) for dielectric relaxation, then the one-sided Fourier transform of $\phi_K(t)$ describes well the data. Historically, Williams and Watts are the first to use the Fourier transform of $\phi_K(t)$ to fit the frequency dependence of isothermal dielectric relaxation data [212, 213], and hence the name Kohlrausch–Williams–Watts (KWW) is appropriate when referring to this practice of using the Fourier transform of the Kohlrausch function. The magnitude of n is a measure of the increase in the width of the dispersion of $\langle A(t)A \rangle$ over and above that of the linear exponential, $\exp(-t/\tau)$. A limiting case of the Kohlrausch function is when $n = 0$, which is the correlation function for the classical Brownian diffusion and Debye relaxation in simple systems without the effect of many-body interactions (see Chapter 1). According to Eqs. (1.78a) and (1.82a), the dynamic (frequency-dependent) susceptibility function $\chi_{AA}(\omega)$ is the one-sided Fourier transform of $\Psi_{AA}(t) = -(1/k_B T)\langle \dot{A}(t)A \rangle$. This, together with the fact that $\langle \dot{A}(t)A \rangle$ is the time derivative of $\langle A(t)A \rangle$, follows that $\chi_{AA}(\omega)$ is proportional to the one-sided Fourier transform of the time derivative of the Kohlrausch function. This connection to the Kohlrausch function applies also to the various dynamic response functions discussed in Chapter 1. For dielectric relaxation, $\varepsilon^*(\omega) = \varepsilon'(\omega) - i\varepsilon''(\omega)$, the relation is represented by Eqs. (1.52), (1.97), and (1.85). For shear modulus, $G^*(\omega) = G'(\omega) + iG''(\omega)$, the relation is given by Eqs. (1.26) and (1.27). For dynamics heat capacity, $c_p{}^*(\omega) = c_p{}'(\omega) - ic_p{}''(\omega)$, it is Eq. (1.58).

The Fourier transform of the Kohlrausch function [212, 213] with larger n is needed to fit the measured dynamic response in the frequency domain with broader dispersion. The imaginary part of the Fourier transform of the Kohlrausch function has the shape of a skew asymmetric peak. Naturally, the full-width at half-maximum of the peak can also be used as a measure of the width of the dispersion. There is an approximate relation between n and w, the full-width at half-maximum of the KWW loss peak normalized to that of the loss peak from the linear exponential, $\exp(-t/\tau)$, which is equal to 1.144 decade. It is given by [214]

$$n \approx 1.047(1 - w^{-1}). \tag{2.1}$$

From this relation between n and w, we can use the width of the structural relaxation dispersion as the quintessential parameter as an alternative to n.

The two claims made above that (i) the various observed anomalies are related and (ii) n is the quintessential parameter that governs or correlates with all the dynamic properties of the structural relaxation including the anomalous ones are quite extraordinary and not anticipated a priori. They need serious justification by experimental data. This task is carried out throughout the following subsections in this chapter, where each anomaly is discussed in detail and shown how it is related to or governed by the parameter n.

Although every worker has noticed that the dispersion of the structural α-relaxation of most glassformers is broader than that of the linear exponential, surprisingly they have not made use of the degree of non-exponentiality or n to get insight into the dynamics and explain the anomalies (except perhaps the author himself in the coupling model). Instead, emphasis is on other characteristics and parameters. These include the length scale of the structural α-relaxation and the temperature dependence of the α-relaxation time τ_α or viscosity η. The most popular one is the shape of the T_g-scaled temperature dependence of $\log \tau_\alpha$ or $\log \eta$, and the word "fragility" was coined to describe the degree of concaveness of the plot [215, 216]. It is remarkable that "fragility" has become the center of attention and one of the most discussed parameter in current glass transition research. Therefore, following non-exponentiality or n in the next subsection, length scale and "fragility" are reexamined in detail as well. Experimental results will be given to show that length scale is impractical and "fragility" is overly complex to be used as parameter or concept to gain insight into the dynamics. Thus, the non-exponentiality or n is not only practical but also essential to unravel the many anomalies exhibited by glassformers.

2.2.1 Non-exponential Time Correlation Function of the Structural α-Relaxation, $\exp[-(t/\tau_\alpha)^{1-n}]$, the Kohlrausch Stretched Exponential Function

In most glass-forming substances and systems, the slower structural α-relaxation usually does not decay linear exponentially as a function of time. The departure is well approximated in most cases by the Kohlrausch's stretched exponential expression [2, 3]:

$$\varphi(t) = \exp[-(t/\tau_\alpha)^{1-n_\alpha}]. \quad 0 \leq n_\alpha < 1. \tag{2.2}$$

There are other empirical functions given by Cole and Cole [217], Cole and Davidson [218], Havriliak and Negami [219], Dixon and Nagel [220], Bergman [221], Blochowicz [222], and others. Some of these may fit some raw experimental data better than the Kohlrausch function. However, one must be aware of the fact that some of these other forms have more parameters than the Kohlrausch function. Also, raw experimental data though principally coming from the α-relaxation may nevertheless have contributions from other relaxation processes such as secondary β-relaxation at higher frequencies and unwanted conductivity relaxation at lower frequencies from charged impurities. As a result, if present, these extraneous relaxation processes make it difficult to know the actual contribution from the α-relaxation. Thus, better fits to the raw data by these other functions do not necessarily mean that they are more cogent representation of the α-relaxation than the Kohlrausch (K) function. A good example is the Havriliak–Negami (HN) equation:

$$\varepsilon^*(\nu) = \Delta\varepsilon \frac{1}{\left(1 + (\mathrm{i}2\pi\nu\tau)^{1-\alpha}\right)^{\gamma}}, \tag{2.3}$$

where ν is the frequency, $\Delta\varepsilon$ the relaxation strength, and α and γ two shape parameters, both less than unity. The one extra exponent of the HN equation may fit some data of non-polymeric glassformers over a wider frequency range than the Fourier transform of the K-function. But, the lure to a "better" fit with the exponents α and γ chosen may make the limiting low-frequency dependence of $\varepsilon''(\nu)$, the imaginary part of $\varepsilon^*(\nu)$, not proportional to ν. This consequence violates the requirement that genuine structural α-relaxation must have $\varepsilon''(\nu)$ proportional to ν at low frequencies for non-polymeric glassformers. Some empirical function for the dispersion carries a specific interpretation of the α-relaxation. An example is the Dixon–Nagel function, which is now defunct because of violations by experiments found. The most probable or the average relaxation time determined from fits to the raw data will differ from one empirical form to another. However, in most cases the differences between these relaxation times are not large and their temperature dependences are very similar.

The Kohlrausch stretched exponential functions have some mathematical properties in probability theory that may make them more fundamental [223]. They are characteristic functions of the stable Lévy distributions, out of which the normal or Gaussian distribution is a special case. This property of the K-function has been exploited to make it more palatable to relaxation in complex systems [180] and Lévy flights [224, 225].

Any non-exponential time correlation function $\phi(t)$ including the K-function can always be rewritten formally as a sum of linear exponentials, $\phi(t) = \Sigma_i g_i \exp(-t/\tau_i)$, which may have led many in the past to interpret the dispersion of the structural α-relaxation as originating from superposition of exponential relaxation processes with different relaxation times τ_i and weighted by g_i. Macroscopic mechanical and dielectric measurements can neither support nor refute this interpretation. Thus, this easy interpretation of dispersion of the structural relaxation has often been conveniently used to rationalize the dispersion, thus rendering dispersion inconsequential. This interpretation is, however, not correct in the light of results by microscopic probes of the α-relaxation. Multidimensional NMR experiments and other techniques showed in poly(vinyl acetate) and other non-polymeric glassformers [226–233] that the structural relaxation is not a static superposition of independent exponential relaxation processes. Instead the non-exponential time dependence is due to the presence of rapidly and slowly moving molecular units exchanging their roles at a time of the order of τ. This complicated process is often referred to as dynamically heterogeneous and was anticipated conceptually in the light of the coupling model (CM) [234] in a year before the first publication of experimental evidence by Schmidt-Rohr and Spiess [226]. More recent multidimensional

NMR experiments by Sillescu et al. [235] have shown in the molecular glassformers studied that molecular reorientation in the heterogeneous α-relaxation occurs by relatively small jump angles, just like some secondary relaxations that have important connection with the α-relaxation to be discussed later in Section 2.3 on secondary relaxations.

The Kohlrausch exponent, $(1 - n_\alpha) \equiv \beta_\alpha$, usually tends to increase with temperature, although the amount of change varies from one glassformer to another as illustrated by examples [236–239]. Hence, objective comparison of n_α or β_α between different glassformers should be made at the same specified relaxation time, τ_α, or at their glass transition temperatures. Next, if one wants to discover any systematic trend that n_α may have, it is important to restrict the comparison to glassformers belonging to the same family. If not, then the widely different chemical bonds and physical structures of the glassformers examined may obscure the trend of n_α. Some examples of trends are given in the following paragraphs.

(i) Glycerol, threitol, xylitol, and sorbitol are hydrogen-bonded glassformers of the same family of polyhydric alcohols. The only difference between them is the number of carbon atoms, which is three for glycerol, four for threitol, five for xylitol, and six for sorbitol. The values of $n_\alpha(T_g)$ deduced from dielectric relaxation measurements [237–240] by the Kohlrausch–Williams–Watts (KWW) fits [195] systematically increase with the number of carbon atoms. The value of n_α is 0.29 for glycerol, 0.36 for threitol, 0.46 for xylitol, and 0.52 for sorbitol. Qualitatively, a molecule with more carbon atoms is expected to have larger intermolecular constraints to its motion by other molecules, simply due to the increased number of carbon atoms and intermolecular potentials. Thus, the observed trend suggests intermolecular constraint or coupling is responsible for it. Dielectric relaxation measurements were performed on propylene glycol and oligomers having different number N of repeat units (N=2, 3, and 69) [241]. The α-relaxation in all these glassformers has the Kohlrausch–Williams–Watts (KWW) form, with n_α which increases with increasing N. For propylene glycol ($N = 1$), $n_\alpha = 0.28$; dipropylene glycol (N=2), n_α=0.33; tri-propylene glycol (N=3), n_α=0.37; and poly(propylene glycol) (N=69 and molecular weight=4000 g/mol), n_α=0.37. As the number N of carbon atoms bonded together in the molecule increases, there is corresponding increase of intermolecular constraint or coupling. The observed trend of increase of n_α with N is attributed to the increase of intermolecular constraint or coupling. Similar increase of n_α with the molecular weight of the oligomers of diglycidylether of bisphenol A (DGEBA) can be seen by comparing the dielectric loss data of a sample with molecular weight of 380 g/mol [242] with that having molecular weight of 1750 g/mol [243]. Compared at the same value of $\tau_\alpha = 1$ s, the lower molecular weight DGEBA has n_α=0.47, smaller than the value of about 0.63 for the higher molecular weight sample.

(ii) Amorphous polymers of different chemical structures in general have different values of $n_\alpha(T_g)$. Correlation between $n_\alpha(T_g)$ and the capacity for intermolecular coupling can be inferred by comparing the chemical structures of closely related polymers [244]. Polyisobutylene (PIB) and polystyrene (PS) are two extensively studied amorphous polymers. The value of n_α is 0.45 for PIB [208–210] and in the range of 0.63–0.65 for PS from light scattering [85, 245, 246] and shear creep measurements [247, 248], and 0.67 from novel NMR technique [226]. Both have two carbon backbone atoms in the repeat unit. PIB has two methyl groups attached to one of the carbon atoms, but PS has one hydrogen atom and a bulky phenyl ring attached to the corresponding carbon. The chemical structure of PIB is compact and symmetric. By contrast, PS has a bulky and rigid phenyl ring attached to the backbone, making it less flexible than PIB as reflected by a much higher T_g. The differences in structure and flexibility suggest larger intermolecular constraint or coupling in PS than PIB, which again correlates with their difference in $n_\alpha(T_g)$. Aromatic backbone polymers such as bisphenol A-polycarbonate, polysulfone, poly(aryl ether ether ketone) [249–251] have bulky rings in the backbone and as expected they have much larger $n_\alpha(T_g)$ than polymers with carbon atoms in the backbone but without bulky rings attached. 1,2-Polybutadiene (1,2-PBD, $n_\alpha(T_g) = 0.60$) differs from 1,4-polybutadiene (1,4-PBD, $n_\alpha(T_g) = 0.50$) by having the unit with rigid carbon double bond sticking out of the polymer backbone instead of located on it. Isotactic poly(methylmethacrylate) (iPMMA) has the side chains all on one side while syndiotactic PMMA has them alternating on both sides. Intuitively we may expect the repeat units in syndiotactic PMMA to be intermingled to a higher degree, to experience higher constraints and hence have a larger n than in isotactic PMMA, as found experimentally by dielectric spectroscopic measurements [252].

Small molecular glassformers that are formed with rigid rings like 1,2-diphenylbenzene (OTP, $n_\alpha(T_g) = 0.50$) is expected to have higher intermolecular coupling than the more flexible linear chain molecules like 3-bromopentane (3BP, $n_\alpha(T_g) = 0.30$) [253], and in fact the former has larger $n_\alpha(T_g)$ than the latter. In all cases, increase of constraints between the relaxing units by changing the chemical structure invariably leads to a larger$n$, a smaller Kohlrausch exponent, or a broader dispersion. We shall see later that change in n correlates or governs the corresponding changes in dynamic properties of many classes of glassformers.

(iii) The motions of colloidal particles [141–143] and polymer-micronetwork colloidal particles [254] suspended at high concentrations in a liquid offer a clear example of the increase of n_α with increasing volume fraction ϕ of the particles or decreasing average distance between particles, and hence increasing particle–particle interaction. The particles are nearly monodisperse with a mean radius of about 10^2 nm. In contrast to molecular systems, the diffusion of the large colloidal particles occurs at macroscopic times longer than a μs. Dynamic light scattering was used to measure the normalized intensity autocorrelation function $g_2(\boldsymbol{q},t) =< I(\boldsymbol{q},0)I(\boldsymbol{q},t) > / < I(\boldsymbol{q},0) >^2$ from 10^{-6} to 10^3 s, where $\boldsymbol{q}$ is the scattering wave vector. As discussed in Section 1.3.2, the difference, $g_2(\boldsymbol{q},t) - 1$, is $< \rho(\boldsymbol{q},0)\rho(\boldsymbol{q},t) > / < \rho(\boldsymbol{q},0) >^2$ or the intermediate scattering function $f(\boldsymbol{q},t)$ for diffusion of the colloidal particles within the single scattering and the Gaussian approximation. At long times, $f(\boldsymbol{q},t)$ follows the time dependence of $\exp[-(t/\tau_s)^{1-n}]$. Information on dynamics of the particles is contained in $f(\boldsymbol{q},t)$. Although the colloidal suspensions are athermal systems and differ in several respects from glass-forming liquids, the light-scattering data readily show that the dispersion of $f(\boldsymbol{q},t)$ is broader in suspensions with higher volume fraction ϕ of the particles and stronger inter-particle interaction. In terms of the Kohlrausch function, the value of the exponent n is equal to 0.35 for $\phi = 0.465$ and increases monotonically with increasing ϕ [143].

Confocal microscopy was also used to directly observe three-dimensional dynamics of particles in colloidal suspensions [141]. The ensemble-averaged mean-squared displacement (MSD), $< \Delta r^2(t) >$, was determined as a function of time in the range from 10 to 10^5 s for different volume fractions ϕ in the supercooled liquid state from 0.46 to 0.57, and in the glassy states when $\phi = 0.60$ and 0.61. Since the displacement of a single particle is a Gaussian variable known to be reasonably accurate for colloidal suspensions in their equilibrium fluid states [255], the relation $f(\boldsymbol{q},t) = \exp[-2q^2 < \Delta r^2(t) > /6]$ holds. From this relation and Fig. 1, the increase of dispersion of $f(\boldsymbol{q},t)$ increasing with volume fraction follows from that of $< \Delta r^2(t) >$, all reflecting the increase of inter-particle interaction.

(iv) As already been mentioned, the comparison of n_α of glassformers with different kinds of chemical bonding and different basic structural units can lead to ambiguous conclusions. This situation is clear for someone who

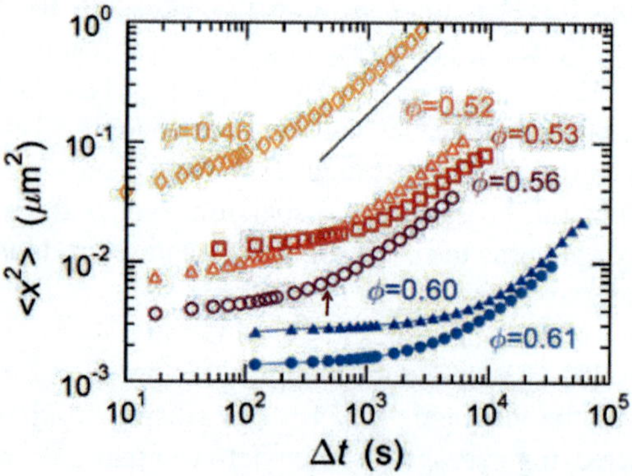

Fig. 1 One-dimensional mean-squared displacement $< x^2 >$ for several volume fractions ϕ of colloidal particles obtained by three-dimensional direct imaging by Weeks et al. [141]. *Open symbols* for $0.46 \leq \phi \leq 0.56$ are in the supercooled liquid state. The *vertical arrow* indicates 500 s, the time a typical particle takes to shift position and leave the cage for the case of $\phi = 0.56$. The *line* has slope =1.0. *Closed symbols* are glasses for $\phi = 0.60$ and 0.61. Reproduced from [141] by permission

approaches the problem by molecular dynamics simulation [256]. The first order of business is to construct a realistic potential for the glassformer. For 1,2-diphenylbenzene (*o*-terphenyl or OTP), it is the potential based on that of Lennard-Jones [257] proposed by Lewis and Wahnström [258], while for silica it is the very different BKS model potential [259] used by Horbach et al. [260] and by Sciortino et al. [261]. For sodium silicate melts the potential used [262] by Horbach et al. [263] is also different. These three different potentials yield different $n_\alpha(\tau_\alpha)$ at the same τ_α, comparable $n_\alpha(\tau_\alpha)$ for OTP and Na silicates, and small for silica. However, because these potentials are drastically different in quality and quantity, there is no sensible way to interpret the comparison of the observed values of $n_\alpha(\tau_\alpha)$.

The way may become clearer if the potentials are obtained by varying a parameter or parameters systematically, as has been done recently for the binary Lennard-Jones particles by controlled changes of the potential, $V(r)$, and performance of the molecular dynamics simulations by Bordat et al. [264]. Earlier, Kob and Andersen [265] had used the standard LJ potential

$$V(r) = \frac{E_0}{(q-p)}[p(r_0/r)^q - q(r_0/r)^p] \tag{2.4}$$

with q=12 and p=6 to simulate the binary Lennard-Jones system consisting of two kinds of particles A and B. The parameters r_0 and E_0 represent the position of the minimum of the well and its depth, respectively. These two parameters of the LJ potentials for interaction between particles A and A, A and B, and B and B are all different in the Kob–Andersen model, and these features of the model ensure the system does not crystallize to form a glass on cooling.

For the purpose of systematic investigation of the change of dynamics with controlled change of $V(r)$, two other models were constructed by changing only the exponents, q and p, of the LJ potential in Eq. (2.4) for the A–A interactions. They are ($q = 8,\ p = 5$) and ($q = 12,\ p = 11$) and these new LJ potentials are shown together with the standard ($q = 12,\ p = 6$) LJ potential in Fig. 2. The well–depth and the location of the minimum of $V(r)$ are the same for all three potentials, and the standard ($q = 12,\ p = 6$) LJ potentials of the Kob–Andersen model for the A–B and B–B interactions were kept, in order to retain the glass-forming ability of the Kob–Andersen model.

It can be seen by inspection of Fig. 2 that the ($q = 12,\ p = 11$) LJ potential is more harmonic than the classical ($q = 12,\ p = 6$) LJ potential, while the ($q = 8,\ p = 5$) LJ potential is a flat well and exceedingly anharmonic. Anharmonicity was not defined using the usual prefactor of the cubic term of a Taylor expansion of the potential close to its minimum. Rather, anharmonicity was defined as the width of the well at $V(r) = -0.5$ in order to take into account the global shape of the potential and its implication on the longer range interaction between the A particles allowed by each potential. The

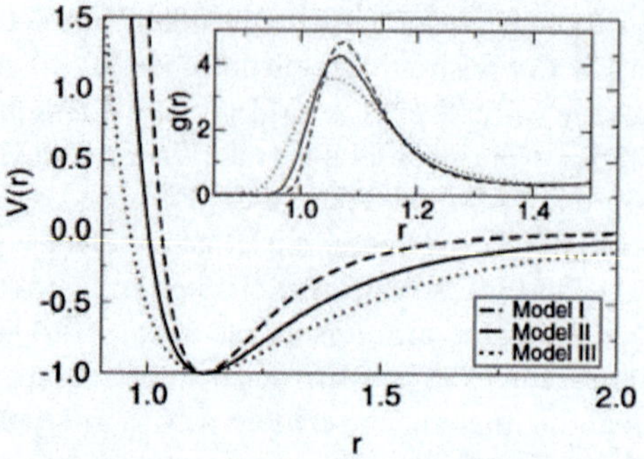

Fig. 2 Potential V(r) governing the A–A interaction of three binary LJ particles models. Model I ($q = 12,\ p = 11$): *dashed curve*, model II ($q = 12,\ p = 6$): *solid curve*, and model III ($q = 8,\ p = 5$): *dotted curve*. The *inset* shows the radial distribution function $g(r)$ of species A at $T = 1.02T_{\text{ref}}$ for the three models (see text for the definition of T_{ref}). Reproduced from [264] by permission

models using the ($q = 12,\ p = 11$), ($q = 12,\ p = 6$), and ($q = 8,\ p = 5$) potentials were referred to, in order of increasing anharmonicity, as models I, II and III, respectively.

Molecular dynamics simulation was carried out for the three models in the binary LJ system with a total of 1500 uncharged particles (1200 species A and 300 species B). The self-intermediate scattering function, $F_s(Q_0, t)$ (for definition, see Eq. (1.114)), for the A particles was calculated from the simulation data for $Q_0 = 2\pi/r_0$ and presented in Fig. 3. Here r_0 is the position of the maximum of the first peak of the static A–A pair correlation function, $g_{\text{AA}}(r)$, shown in Fig. 2.

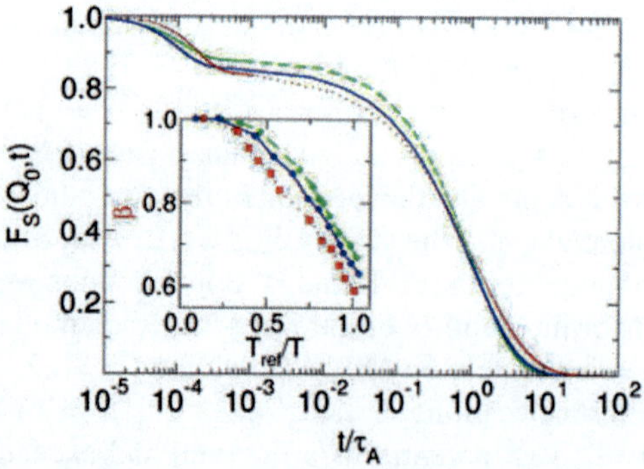

Fig. 3 Self-intermediate scattering function $F_s(Q_0, t)$ vs. scaled time t/τ_A. *Dashed, solid*, and *dotted lines* are for models I, II, and III, respectively. For all three models, $\tau_A(T_{\text{ref}}) = 46435.8$. The *inset* shows the stretched exponent $\beta = (1 - n)$ as a function of the scaled reciprocal temperature T_{ref}/T for the three models: (♦) model I, (•) model II, and (■) model III. For definitions of τ_A and β, see text. Reproduced from [264] by permission

At high temperatures, $F_s(Q_0, t)$ decays linear exponentially to zero with a characteristic time of about 0.45 in LJ units of time, which is close to crossover time $t_c \approx 1 - 2$ used as a fundamental time in the coupling model [190]. When temperature is lowered, the dynamics slows down dramatically and a two-step decay process appears in $F_s(Q_0, t)$. The decay in the first step is due to relaxation of particles A while confined by the fluctuating cages. The decay of $F_s(Q_0, t)$ in the second step at each of the lower temperatures T was fit to the Kohlrausch function, Eq. (2.2). From the fits, the relaxation times, τ_α, and the stretched exponents, $\beta \equiv (1-n_\alpha)$, of particles A were determined for a number of temperatures in models I, II, and III. A very long relaxation time from the simulation, $\tau_{\alpha,\mathrm{ref}}$ equal to 46435.8 (LJ units), was chosen to compare $n_\alpha(\tau_{\alpha,\mathrm{ref}})$ of the three models at the same $\tau_{\alpha,\mathrm{ref}}$. A more comprehensive comparison of n_α was made by first determining the temperature, T_{ref}, at which $\tau_\alpha(T_{\mathrm{ref}}) \equiv \tau_{\alpha,\mathrm{ref}} = 46435.8$ for each model, and plot $\beta \equiv (1 - n_\alpha(T))$ from all three models againstT_{ref}/T as shown in Fig. 3. The values of T_{ref} are 0.688, 0.431, and 0.263 (in LJ units) for models I, II, and III, respectively. It can be seen that with increasing anharmonicity (in going from models I to II and to III) not only does $\beta(\tau_{\alpha,\mathrm{ref}})$ decrease or $n_\alpha(\tau_{\alpha,\mathrm{ref}})$ increase at $T_{\mathrm{ref}}/T = 1$ and $\tau_\alpha(T_{\mathrm{ref}}) = \tau_{\alpha,\mathrm{ref}}$, but also $\beta(T_{\mathrm{ref}}/T)$ decreases or $n_\alpha(T_{\mathrm{ref}}/T)$ increases at the same T_{ref}/T for all T_{ref}/T.

In going from models I to III through II, the inset of Fig. 2 shows that the first peak of $g_{AA}(r)$ becomes broader and extends to smaller pair separation distance values, and hence the A-particles can come closer together and the interactions between them are enhanced. Since $n_\alpha(\tau_{\alpha,\mathrm{ref}})$ also increases in the same order of the models, the simulation results support the correlation between $n_\alpha(\tau_{\alpha,\mathrm{ref}})$ and the effects that inter-particle interaction have on slowing down the relaxation. We shall return in later sections to discuss several dynamic properties of the A-particles obtained by Bordat et al. from the three models that correlate with $n_\alpha(\tau_{\alpha,\mathrm{ref}})$. Taken together, these correlations strongly indicate that the non-exponentiality parameter n_α is an indicator of the strength of intermolecular coupling, and it controls the dynamic properties of relaxation in the binary LJ glassformers.

(v) In the same glassformer, different time correlation functions of the same dynamic variable or different dynamic variables relating to the structural relaxation may all be well described by Kohlrausch functions, but the parameters τ_α and n_α may be different. There are many examples of this behavior. Moynihan et al. [266] had reported the observation of such differences in using enthalpy, light scattering, and refractive index to monitor structural relaxation of the same glassformer, although for enthalpy relaxation the result was obtained not directly but through a model-dependent analysis. Directly obtained results of τ_α and n_α from isothermal photon correlation spectroscopy, dielectric relaxation, and mechanical relaxation experiments of several glassformers had found they were different [267–270]. Molecular dynamics simulations can easily obtain different correlation functions including $C_1(t)$ and $C_2(t)$ of the first- and second-order Legendre polynomials $P_1[\cos\theta(t)]$ and $P_2[\cos\theta(t)]$ of the rotation angle θ given by Eqs. (1.101) and (1.102), and the self- or incoherent intermediate scattering function, $F_s(\vec{k}, t)$, in Eq. (1.114). From $F_s(\vec{k}, t)$, we can obtain the mean-squared displacement $\langle|\vec{r}(t)|^2\rangle$, and the self-diffusion coefficient D_t from the Green–Kubo relation given in Section 1.3.5. The results of several simulations of polymeric [271–273] and non-polymeric [274, 275] glassformers all show different τ_α and n_α for the different correlation functions. Conventional theories usually have no explanation for these observations.

The differences in τ_α and n_α for different correlation functions can be attributed to intermolecular coupling. The origin is based on the reasoning that different dynamic variables i weigh the intermolecular coupling differently, and hence the slowing down (related to τ_α) and the stretching (related to n_α) of their correlation functions by the many-body dynamics are not the same. It is possible that the correlations $C_{\alpha\beta}(t)$ given by Eq. (1.100) of the off-diagonal components of the stress tensor $\sigma_{\alpha\beta} = (3u_\alpha u_\beta - \delta_{\alpha\beta})/2$ can be more susceptible to influence by intermolecular coupling than the correlation function $C_1(t)$ for the rotation of a certain part of a molecule, although both are related to the structural α-relaxation. Consequently, mechanical relaxation which measures $C_{\alpha\beta}(t)$ may be more stretched (a larger n_α) and have a longer τ_α than dielectric relaxation which comes from $C_1(t)$.

One obvious difference between center-of-mass diffusion and stress relaxation or rotation is that the former involves the correlation of the center of mass (one point), while rotation is determined by the time correlation of the transition dipole of the molecule, which is a vector, and stress relaxation involves a tensor quantity. The dipole vector and the stress tensor are defined by two or more points in space than self-diffusion. Hence, one can expect intermolecular constraints to have stronger slowing effect on rotation or stress relaxation than the translational motion of the center of mass [268]. This intuitive deduction can be justified on a firmer ground from the theoretical consideration of constraint entropy [181–183, 276] based on the Dirac constraint dynamics [277]. In this formulation, the dynamics under constraints is governed by the time-dependent constraint entropy, which is $S_{c,t}$ for translational diffusion and $S_{c,r}$ for rotation. In order that the constraint entropies $S_{c,t}$ and $S_{c,r}$ all lead to the Kohlrausch forms for the correlation functions, they necessarily have the time dependences given by $S_{c,t} = S_{max} - \kappa n_t \ln(\omega_c t)$ and $S_{c,r} = S_{max} - \kappa n_r \ln(\omega_c t)$, respectively. Here S_{max}, κ, and ω_c are constants. From the fact that there are more dynamic constraints imposed on motion when considering rotation of a vector or stress relaxation of a tensor than the translation of the center of mass, we have $S_{c,r} < S_{c,t}$, from which it follows that $n_t < n_r$. Similarly, we have $n_t < n_\eta$. Such rationalization is not sufficient to convincingly support the differences in τ_α and n_α for different correlation functions originating from intermolecular coupling in many-body relaxation. We need additional prediction or predictions of different nature from any theory/model and confirmation by experiment before this feature of the structural relaxation can be attributed to many-body relaxation. Such additional prediction has been provided by the coupling model to be presented and discussed in later sections.

(vi) Bonn and Kegel [278] studied the frequency dependence of the diffusion coefficient and viscosity of concentrated hard-sphere colloidal suspensions. On comparing the frequency dependence of $1/\eta(\omega)$ with that of $D(\omega)$ in Fig. 2 of Bonn and Kegel, it is clear that the frequency dispersion of $1/\eta(\omega)$ is broader than that of $D(\omega)$. Thus, for the colloidal hard spheres we have experimental evidence to support that, if $1/\eta(\omega)$ and $D(\omega)$ were fit to the Fourier transform of the Kohlrausch function, n_{D} would be smaller than n_η, just like molecular glassformers. Expressing all quantities in reduced units by Bonn and Kegel, the frequency-dependent Stokes–Einstein relation (1.48) takes the form of $\eta(\omega) = 1/D(\omega)$. In the low-frequency or long-time regime, significant deviations from this relation were observed. The values of the inverse viscosity are significantly smaller than the values of the diffusion coefficients, and the Stokes–Einstein relation is violated, which will be revisited in a section where the breakdown of Stokes–Einstein and Debye–Stokes–Einstein relations in various systems are discussed in detail and explained by the coupling model.

(vii) The differences in τ_α and n_α for different correlation functions for structural relaxation in glass-forming liquids and colloidal suspensions were found in entirely dissimilar processes in other systems. Although this subject belongs to Chapter 3, it is worthwhile to cite an example here. The example is the ionic motion in glassy ionic conductors, which can be probed by measurement of the frequency-dependent conductivity $\sigma(\omega)$ or by spin-lattice relaxation (SLR) rate of the diffusing ion nucleus, $T_1^{-1}(T)$, as a function of temperature at constant Larmor frequency in nuclear magnetic resonance. The frequency-dependent conductivity, $\sigma(\omega)$, is related to the Fourier transform of the current–current correlation function (see Eq. (1.130)) and is given by

$$\sigma(\omega) = (e/kTV)\int_0^\infty \langle J_z(0)J_z(t)\rangle \exp(\mathrm{i}\omega t)\mathrm{d}t, \tag{2.5}$$

where J_z is the sum of the z-component of the velocity of the ions, with total number equal to N and each carrying charge e. On the other hand, the SLR rate, $T_1^{-1}(T)$, is given in terms of another correlation function,

$$C_{\mathrm{SLR}}(t) = (1/N)\sum\nolimits_{i\neq j}\langle F_{ij}^{(q)}(t)F_{ij}^{(q)}(0)\rangle, \tag{2.6}$$

where $F_{ij}^{(q)}(t) = (q\sqrt{8\pi/15})Y_2^q(\Omega_{ij}(t))/r_{ij}^3$, Y_2^q is the spherical harmonics with q=1 or 2, r_{ij} is the distance between two ions i and j, and Ω_{ij} is the spherical coordinates of the vector $\boldsymbol{r}_{ij}$ [92]. The $(r_{ij})^{-3}$ factor in the SLR correlation function accentuates the contribution to the spin-lattice relaxation rate at shorter distances, where the ion–ion interaction is larger. Consequently, the effects of ion–ion interactions are stronger for SLR than $\sigma(\omega)$.

Experiments were able to determine from the $T_1^{-1}(T)$ data the Kohlrausch correlation function for SLR [279–286]

$$C_{SLR}(t) = \exp(-(t/\tau_{SLR}(T))^{1-n_{SLR}}), \tag{2.7a}$$

and, from the $\sigma(\omega)$ data, the Kohlrausch conductivity relaxation correlation function

$$C_{\sigma}(t) = \exp(-(t/\tau_{\sigma}(T))^{1-n_{\sigma}}). \tag{2.7b}$$

For several different glassy ionic conductors [147, 279–289] and a crystalline ionic conductor [290], the general experimental finding [291–295] is that

$$n_{SLR} > n_{\sigma}. \tag{2.8}$$

This general result supports once more that n (appearing in the exponent of the Kohlrausch correlation function) is a measure of the effect that many-body interaction has on relaxation. In the present case, they are n_{σ} and n_{SLR}, and their difference is determined by the stronger effect of inter-ion coupling on SLR than conductivity relaxation. Also from experiments, the SLR relaxation time, $\tau_{SLR}(T)$, is much longer than the conductivity relaxation time, $\tau_{\sigma}(T)$ and has a larger activation energy (see Chapter 3 for details). These accompanying experimental facts on the relation between $\tau_{SLR}(T)$ and $\tau_{\sigma}(T)$ turn out to be consequences of the relation between n_{σ} and n_{SLR} given by the inequality (2.8) in the context of the coupling model [291–294] and will be discussed in a later section devoted to ion dynamics.

Before we leave this section, some words of caution are in order for the readers when considering the width of the dispersion of the structural relaxation in certain situations. Let us start with the polymers in the family of poly(n-alkyl methacrylates) [296]. The members of the family have the same backbone but differ only in the length of the alkyl side chain measured here by the number C of carbons in the chain. The shortest is in poly(methyl methacrylate) with C=1 (PMMA). Higher members like poly(n-hexyl methacrylate) has C=6, poly(n-decyl methacrylate) has C=10, and poly(n-lauryl methacrylate) has C=12. In higher members of the poly(n-alkyl methacrylates) starting from C=4, the alkyl groups from different side chains aggregate in the melt and form self-assembled alkyl nanodomains of sizes in the range 0.5–2 nm, where a larger size corresponds to a larger C [297, 298]. Randomly arranged in space, the alkyl nanodomains create concentration fluctuations or heterogeneous spatial environments seen by the polymer backbone repeat units. The consequence for poly(n-alkyl methacrylates) with $C \geq 4$ is extra broadening of the dispersion of the structural relaxation beyond that caused by the many-body relaxation. This extraneous broadening may even modify the frequency dispersion to the extent that the observed result cannot be reconciled with the Fourier transform of any of the Kohlrausch functions. Similar examples are mixtures of two glassformers where concentration fluctuations additionally broaden the dispersion of the structural relaxation of either component. A more subtle case occurs in the glassformer *bis*-5-hydroxypentylphthalate that has two dipoles moving more or less independently of each other and contributing to the observed dielectric spectrum attributed to the structural relaxation [299]. In all the situations discussed above, if not recognized, the presence of the extraneous broadening may confound the connection of the width of dispersion to many-body relaxation.

The deviation from single exponential (referred to from now on as non-exponential) of structural relaxation function $\phi(t)$ was widely recognized since the beginning of research on glass transition. The prevalent rationalization of this property is by the superposition of independent and exponential relaxations with different relaxation times τ_i out of a distribution with weights g_i, i.e., $\phi(t) = \sum_i g_i \exp(-t/\tau_i)$. Some models attribute it to the physical presence of regions with different relaxation times, e.g., the environmental relaxation model of Simmons and Macedo [300], the density fluctuation model of Robertson [301], or the cooperatively rearranging region model of Donth [302]. This simple hypothesis was reasonable at earlier times when there was absence of microscopic investigations on the relaxation to understand the true nature of the structural relaxation and was widely used to rationalize the non-exponential property and to trivialize its fundamental importance unfortunately. It is still used by others for convenience at the present times as the basis of the Kohlrausch function [266, 303, 304]. One can argue any of the above correlations of the width of the dispersion or n with various changes is due to an increase in the width of the distribution of relaxation times τ_i. However, microscopic information on the structural relaxation now available by modern experimental techniques can be used to rule out this hypothesis. An example is the momentum transfer dependence in neutron and light scattering, which will be discussed later. Furthermore, the most probable τ and the width of the dispersion from the hypothesis, $\phi(t) = \sum_i g_i \exp(-t/\tau_i)$, are independent of each other. Experiments using variable combinations of pressure and temperature show that these two quantities are related, as will be shown in Section 2.2.4. Also worth pointing out is that some still write the correlation function as $\phi(t) = \sum_i g_i \exp(-t/\tau_i)$, while having accepted the fact that the structural relaxation is dynamically heterogeneous. The two may not be compatible.

Other models proposed an inherently non-exponential mechanism, e.g., the defect diffusion model of Glarum [305] and the fractal time version of it [306]. These models are not consistent with microscopic data either because all units relax uniformly with time and contradict the dynamic heterogeneous character found by experiments as pointed out in [235]; the subject will be further discussed in the next section.

2.2.1.1 Crossover of Correlation Function from $\exp(-t/\tau_0)$ to $\exp[-(t/\tau)^{1-n}]$ at t_c, a Temperature-Insensitive Time

Neutron Scattering

The first direct experimental evidence of the incoherent intermediate scattering function (i.e., the self-correlation function) associated with local segmental relaxation, $F_{rel}(Q,t)$, crossing over from $\exp(-t/\tau_0)$ to $\exp[-(t/\tau)^{1-n}]$ at a temperature-insensitive and scattering vector Q-independent time, t_c, came from quasielastic neutron scattering experiments in the polymer polyvinylchloride (PVC), in 1993 [307], and refinements later [308] by Colmenero et al. These authors modified a procedure of Kiebel et al. [309] who assumed that the intermediate scattering function $F(Q,t)$ measured for OTP is the product of a vibrational contribution $F_{\mathrm{vib}}(Q,t)$ and a relaxational contribution $F_{\mathrm{rel}}(Q,t)$. The former was obtained by scaling a spectrum measured at low T, where there is negligible relaxation to the temperature of interest by the ratio of Bose–Einstein and Debye–Waller factors. $F_{\mathrm{rel}}(Q,t)$ was then obtained by dividing $F(Q,t)$ by the scaled $F_{\mathrm{vib}}(Q,t)$. Actually, the data of OTP obtained by Kiebel et al. already indicate the presence of a faster relaxation preceding the Kohlrausch relaxation at about 1–2 ps, same as that found by Colmenero et al. The study of Colmenero et al. [307] was followed by similar measurements on polyisoprene and polybutadiene, by Zorn et al. [310], polyisobutylene [311], and more recently in other polymers, poly(ethylene oxide), either pure or in blends with poly(methyl methacrylate) by García Sakai et al. [312], and the same conclusion was made. The magnitude of t_c found is about 1–2 ps for all these polymers. As examples, the data of pure PEO and hPEO in mixtures with dPMMA are reproduced in Figs. 4 and 5 to show the crossover. The time dependence of the faster relaxation is consistent with $\exp(-t/\tau_0)$. This identification is particularly clear at high temperatures where $F(Q,t)$ decays principally via $\exp(-t/\tau_0)$ at times before t_c, and the rest after t_c by a Kohlrausch-like function, and τ_0 is of the order of picoseconds. In some of the neutron scattering studies [307, 308, 310, 312, 313, 314], τ_0 is found to have the signatures of independent motion, which include its Q^{-2} dependence of normal diffusion and its activation energy at higher temperatures being close to the conformation energy barrier of rotation of monomers in a single chain for polymers. At lower temperatures, the vibrational part is not the only other contribution present in the spectra. There is a nearly constant loss appearing in the frequency dependence of the susceptibility spectrum, which is the loss while molecules are confined within cages defined by the anharmonic intermolecular potential of other molecules or atomic units at times before escaping to execute relaxation. This complication makes it harder and sometimes impossible to detect the fast relaxation

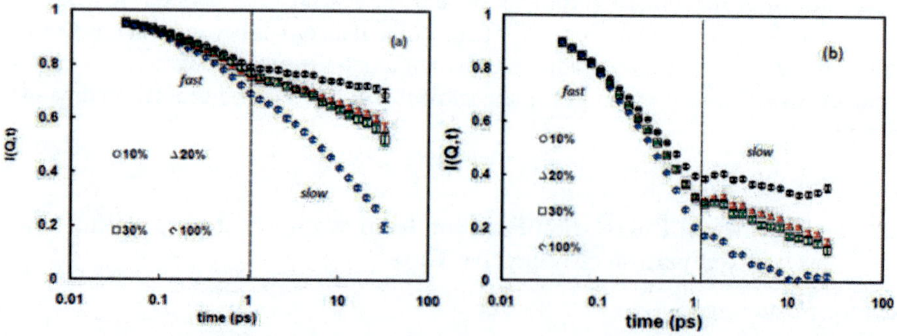

Fig. 4 Effect of composition on the self-intermediate scattering function of neat PEO and hPEO in blends with dPMMA at (**a**) Q=0.89 Å^{-1} and (**b**) Q=2.51 Å^{-1}. The temperatures of the measurements are 343 K (100% PEO), 345 K (10 and 30% PEO), and 348 K (20% PEO). Data of García Sakai et al. Reproduced from [312] by permission

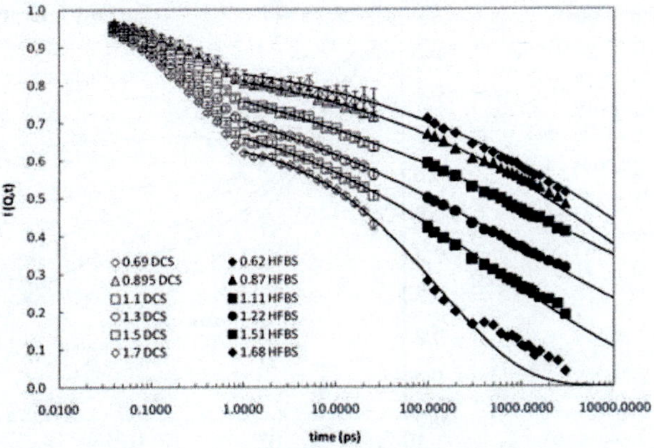

Fig. 5 Combined DCS and HFBS scattering decay curves for hPEO in dPMMA at 308 K. Seven spatial scales are shown. *Lines* represent a KWW fit with parameters falling within the error bars determined using the DCS data alone. The disk chopper time-of-flight spectrometer (DCS) was operated at an incident wavelength of 4.2 Å and an energy resolution of 81.6 μeV. A dynamic range of ± 20 μeV was used for the high-flux backscattering spectrometer (HFBS), with an energy resolution of 0.87 μeV. Reproduced from [312] by permission

at low temperatures where τ_0 is long. The subject of caged relaxation and nearly constant loss will be discussed later.

Molecular Dynamics Simulations

Furthermore, molecular dynamics simulations of different molecular systems with realistic Lennard-Jones intermolecular potentials invariably show that the correlation function has a faster and more rapidly decaying part at times shorter than $t_c \approx 1$ or 2 ps and crosses over to the slower part having stretched exponential (Kohlrausch) time dependence. Examples include molecular dynamics simulations of the dynamics of (i) *ortho*-terphenyl by Lewis and Wahnström [258, 315] and analysis of the data [187, 316, 317]; (ii) methanol [318, 319]; (iii) the rotational degrees of freedom in a supercooled system composed of rigid, diatomic molecules [320]; (iv) the standard glass-forming binary mixture of Lennard-Jones particles [321, 322] and variations with different anharmonicity of the interaction potential [264]; (v) orientational degrees of freedom in an equimolar mixture of ellipsoids of revolution and spheres [274]; (vi) biatomic molecular liquid [323]; (vii) local segmental relaxation of polyisoprene [324] and several polyolefins, namely poly(ethylene propylene), poly(ethylene butene), isotactic polypropylene, and head-to-head polypropylene [325]; and (viii) difluorotetrachloroethane ($CFCl_2$–$CFCl_2$) glassy crystal [326].

The results of the molecular dynamics simulation of OTP by Lewis and Wahnström are reproduced in Fig. 6. Shown are the incoherent part $F_s(q,t)$ of the intermediate scattering function as function of log(t) for the site–site correlations (corresponding to neutrons scattering off the centers of the different benzene rings). The change of concavity of $F_s(q,t)$ from concave to convex in a neighborhood near 1–2 ps can be seen at higher temperatures, 318 and 400 K, and can be

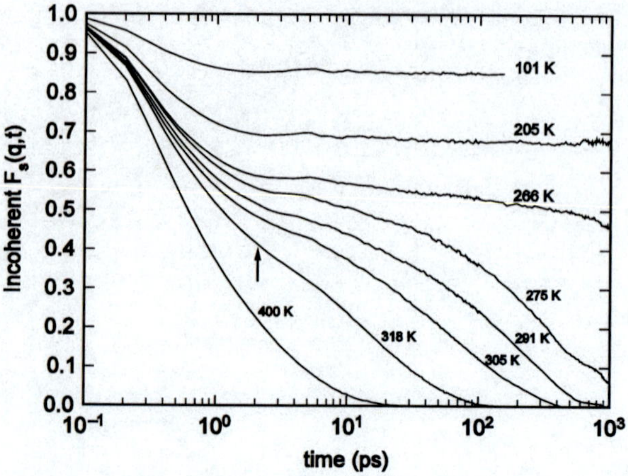

Fig. 6 Self-intermediate scattering function $F_s(q,t)$ vs. time of OTP for wave vector $q = 1.94$ Å^{-1} at various temperatures. Data of Lewis and Wahnström reproduced from [258, 315] by permission

taken as evidence of crossover. When temperature is lowered, the relaxation times of both the fast primitive (exponential) and the slow cooperative (Kohlrausch) processes become significantly longer, and caged dynamics appear before them. Over much of the time range, $F_s(q,t)$ is dominated by the motion of molecules confined within anharmonic cages, giving rise to the very gradual and plateau-like decay. This obscures the crossover at $t_c = 1 - 2$ ps, and it can no longer be easily discerned. Instead it looks more like a two-step decay of $F_s(q,t)$, which has led others [321, 322, 326] to analyze it using the idealized mode coupling theory [31, 32]. The above discussion applies to molecular dynamics simulation of binary mixture of Lennard-Jones particles. For this system, the crossover in the neighborhood of $t_c = 1$ (LJ unit) can be seen from the time dependence of the intermediate scattering function of the A particles in Fig. 20 of [321] and in Fig. 1 of [322].

Pakula and coworkers [327, 3283] introduced the microscopic simulation method with parallel algorithm which takes into account coincidences of elementary molecular movements resulting in local cooperative structural transformations. The efficient method, known as the dynamic lattice liquid (DLL) model, was used to obtain static and dynamic properties of different interacting systems on a lattice, which include simple liquids, linear polymers, multi-arm star polymers, and microgels. Here we show the results for the dynamics of solvent in polymer solution systems with various linear polymer chain lengths (N) in the range from $N = 2$ to N=32 and different concentrations of polymers (ϕ) in the range from $\phi = 0.05$ to $\phi = 0.95$. The mean-squared displacement of solvent beads $< r^2(t) >$ vs. time for the solutions with polymer chains of length N=32 at various polymer concentrations are shown in Fig. 7. The proportionality, $< r^2(t) >= D_S t$, at short times correspond to primitive diffusion with exponential correlation function. The crossover of $< r^2(t) >$ from linear t-dependence to fractional power law $< r^2(t) > \propto t^{1-n}$ in the neighborhood of t_c is clear for solutions with higher polymer concentrations due to enhanced interaction/constraint by the less mobile polymer. The combined $< r^2(t) > \propto t^{1-n}$ and the eventual return of $< r^2(t) >$ to another linear

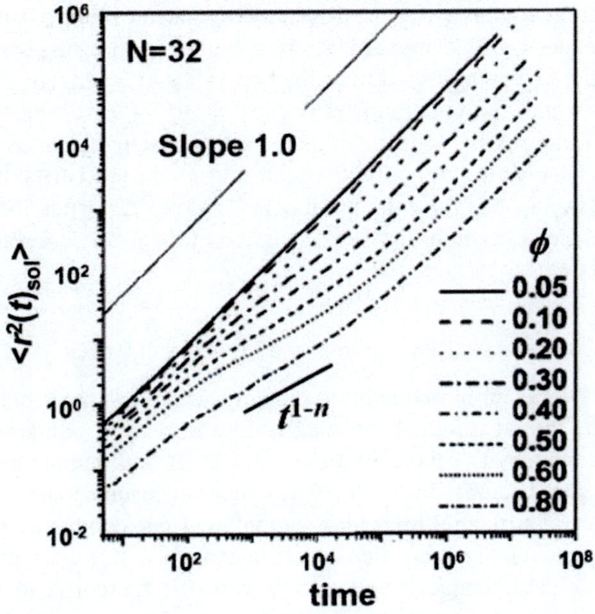

Fig. 7 The mean-squared displacement of solvent beads vs. time for the solution with chains of length N=32 for various polymer concentrations. Data from [328] are reconstructed and illustrated to show the sublinear t-dependence

t-dependence, $< r^2(t) > = D_L t$, can be derived from the Kohlrausch stretch exponential correlation function for diffusion [329].

The crossover from $< r^2(t) >= D_S t$ to $< r^2(t) > \propto t^{1-n}$ at $t_c \approx 1$ ps was found in the molecular dynamics simulation of several polyolefins. This can seen from Fig. 7 in [325], if only the data of $< r^2(t) >$ at times not much shorter than 1 ps are considered. The data of $< r^2(t) >$ at much shorter times may come from the ballistic motion.

Optical Kerr Effect Experiments

Optically heterodyne-detected optical Kerr effect (OHD-OKE) experiments can probe in a broad time range from sub-picoseconds to about 30 ns the orientational dynamics of glassformers such as salol [330], dibutylphthalate [331], benzophenone [332], and ionic organic liquids, *N*-propyl-3-methylpyridinium *bis*(trifluoromethylsulfonyl)imide (PMPIm) and 1-ethyl-3-methylimidazolium tosylate (EMImTOS) [333]. At times from sub-ps to a few ps, a faster and more rapid decay of the OHD-OKE signal was seen. Although occurring in a small range, this decay can be distinguished from the shortest timescale part of the OHD-OKE signal, which has oscillations arising from intramolecular modes [330]. It is represented in one paper [333] by a power law at^{-s} with s increasing with decreasing temperature from 2.3 to 4.3. This fast process was identified by the authors as the so-called fast β-process of mode coupling theory (MCT) [31], but the exponent s of the superlinear power law is much larger than allowed by the idealized MCT after having accounted for the fact that the OHD-OKE signal $F(t)$ measures the derivative of the correlation function. Instead, appearing only in a narrow range of time, it may as well be the fast $\exp(-t/\tau_0)$ decay particularly at higher temperatures and shorter τ_0, where it cannot be distinguished from the power law at^{-s} with large s appearing in the assumed form of the function $F(t) = (at^{-s} + pt^{-z} + dt^{b-1})\exp(-t/\tau_\alpha)$ used to fit the data. The term pt^{-z} with $z \approx 1$ is the time-domain equivalent of the nearly constant loss observed in susceptibility spectrum, and the term dt^{b-1} is the von Schweidler power law in MCT, and the product $dt^{b-1}\exp(-t/\tau_\alpha)$ may be considered to account for the α-relaxation. Further discussion of the nearly constant loss (NCL) will be given later.

Conductivity Relaxation of the Molten Salt $0.4Ca(NO_3)_2$–$0.6KNO_3$ (CKN)

CKN is an ion-containing material which has the structural relaxation time given by the measured shear relaxation time $< \tau_s >$ being nearly the same as the ionic conductivity relaxation time $< \tau_\sigma >$ at high temperatures where these relaxation times become short [55]. This condition ensures that one can study the fast dynamics of structural relaxation by making ionic conductivity relaxation measurements. Conductivity relaxation measurements of CKN were made by Cramer et al. [334, 335] at frequencies below and above 10^{11} Hz (corresponding to about 2 ps) and into the far infrared region, where the vibrational contribution is evident as absorption peaks. It was established phenomenologically in *glasses* [336] that the vibrational contribution to $\sigma(\omega)$ extends down to low frequencies with a ω^2-dependence. A similar frequency dependence of $\sigma(\omega)$ is found in CKN melt at high temperatures. These facts suggest the removal of the vibrational contribution from $\sigma(\omega)$ at lower frequencies by subtracting off the ω^2-dependent contribution extrapolated from the data at high frequencies. The difference at the highest temperature of measurements is solely coming from the conductivity relaxation contribution from ion diffusion, $\sigma_{ion}(\omega)$, and as shown in Fig. 8 taken from [335] it exhibits the crossover at some frequency between 10^{11} and 10^{12} Hz which corresponds to t_c of the order of 1 ps.

Dynamic Light Scattering in Colloidal Suspensions

Mention is made before in Section 2.2.1 (iii) of dynamic light-scattering measurements of the intermediate scattering function $f(\boldsymbol{q}, t) =< \rho(\boldsymbol{q}, 0)\rho(\boldsymbol{q}, t) > / < \rho(\boldsymbol{q}, 0) >^2$ for diffusion of the colloidal particles of PMMA with a mean radius of about 200 nm (size polydispersity $\sigma = 0.05$.) and volume fraction ϕ from 0.4 to 0.57 suspended in a liquid. The Brownian motions of colloidal particles suspended at higher concentrations in a liquid constitute another problem of relaxations in a many-body interacting system that bear some resemblance to the motions of atoms and molecules in a liquid. In contrast to molecular systems, the diffusion of the large colloidal particles occurs at macroscopic times longer than a μs and there is no vibrational contribution to complicate the analysis of the diffusional dynamics. This advantage of colloidal particles has enabled a direct observation of the crossover in dynamics proposed by the CM. For the colloidal suspension with $\phi = 0.465$ [143], Fig. 9 shows mean-squared displacement, $\Delta r^2(t)$, having the linear t-dependence

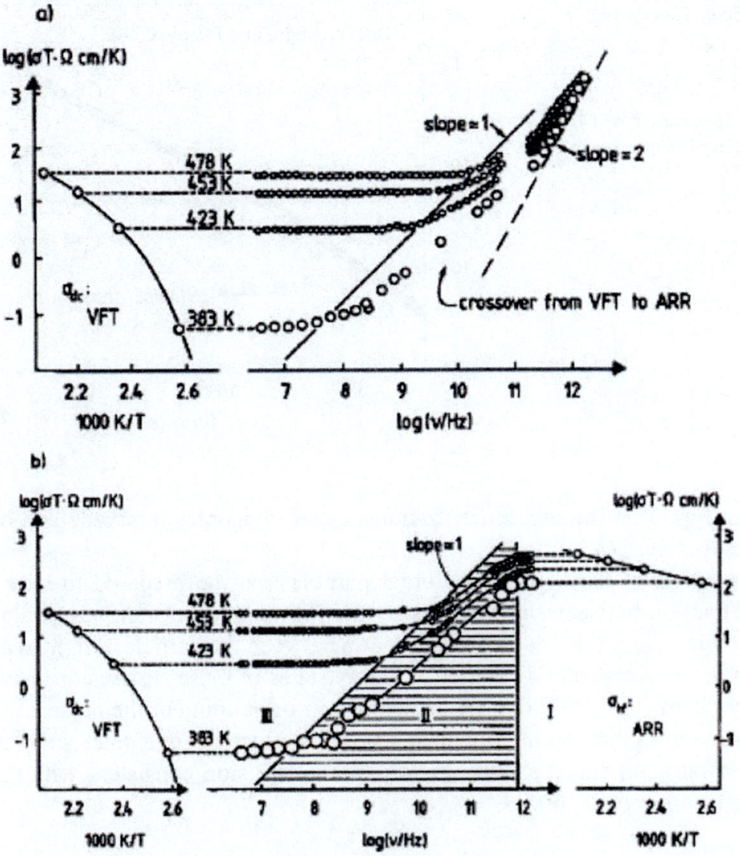

Fig. 8 (**a**) Conductivity spectra $\sigma(\nu)$ of 0.4Ca$(NO_3)_2$–0.6KNO_3 including the ν^{-2}-dependence of the low-frequency flank of the vibrational contribution (*on the right*), and the VFTH temperature dependence of σ_{dc} *on the left*. (**b**) Conductivity spectra from ions only $\sigma_{ion}(\nu)$ after the vibrational contribution has been subtracted off $\sigma(\nu)$, showing the crossover to the high-frequency (frequency-independent) ion conductivity σ_{hf}. *Symbol size* indicates experimental error. The temperature dependence of σ_{hf} is Arrhenius

for $t < 3$ ms, and changing to a fractional power law proportional to t^{1-n} for 6 ms $< t < 0.1$ s. This is consistent with the intermediate scattering function $f(\boldsymbol{q}, t)$ crossing over from exponential to the stretched exponential in the neighborhood of $t_c \approx 4$ ms because $f(\boldsymbol{q}, t) = \exp[-2q^2 < \Delta r^2(t) > /6]$. For $t < t_c \approx 4$ ms, when rewriting $f(\boldsymbol{q}, t)$ as $\exp(-t/\tau_f)$, it is clear that τ_f has the q^{-2}-dependence of normal Brownian motion. The return of $\Delta r^2(t)$ back to the linear t-dependence for $t > 0.2$ s is the natural consequence of diffusion even with the correlation function being a stretched exponential function [329]. The $\Delta r^2(t)$ data of colloidal particles with $\phi = 0.465$ are exact analogue of that of dynamic lattice liquid (DLL) shown

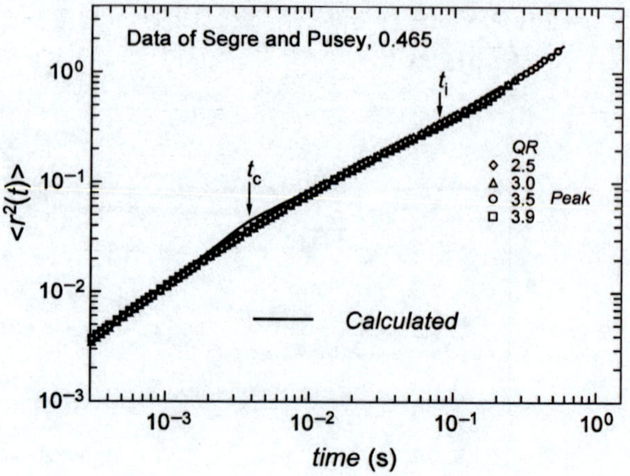

Fig. 9 Scaled scattering function $\ln[f(Q,t)/Q^2 D_S(Q)]$ for colloidal suspensions with volume fraction $\phi = 0.465$. *Symbols* are data from [143] replotted here. *Solid line* is the CM fit

before in Fig. 7. At higher concentrations, caged dynamics intervene and obscure the crossover near t_c.

Suspensions of hard-sphere colloidal particles can be prepared to have short-range attraction between the particles by the introduction of non-adsorbing linear polymer which induces a depletion attraction between the particles. Light-scattering study of the structure and dynamics of glassy states of these suspensions have been made by Pham et al. [337] to provide detailed information. For the present purpose, we only mention that the intermediate scattering function (dynamics structure factor) $f(\boldsymbol{q},t)$ at short times shows single particle diffusion consistent with the time dependence of $\exp(-t/\tau_0)$.

Dynamic Light Scattering in Aqueous Suspension of Laponite

Similar crossover of the intermediate scattering function from exponential to stretched exponential was found by dynamic light scattering in aqueous colloidal suspension of Laponite, a synthetic clay composed of discs with diameter=25 nm and thickness=1 nm that are not monodisperse [338–345]. At short times, the time dependence of $f(\boldsymbol{q},t)$ was found to be $\exp(-t/\tau_f)$, with τ_f having the q^{-2}-dependence normal Brownian motion [338], and thus it corresponds to the primitive diffusion in the CM. At longer times, $f(\boldsymbol{q},t)$ follows the time dependence of $\exp[-(t/\tau_s)^{1-n}]$. In practice, $f(\boldsymbol{q},t)$ was fitted by the sum $A\exp(-t/\tau_f) + (1-A)\exp[-(t/\tau_s)^{1-n}]$. Although this looks like the two-step decay of mode coupling theory (MCT), the faster relaxation is an exponential function rather than a power law as predicted by MCT.

2.2.1.2 Crossover of Temperature Dependence of Viscosity at High Temperatures

As mentioned in the previous subsection, at high temperatures where neutron scattering measurements of polymers observe the fast $\exp(-t/\tau_0)$ decay and Arrhenius temperature dependence of τ_0, the activation energy is close to the conformation energy barrier of the polymer [313, 314, 317]. Thus, the crossover to $\exp(-t/\tau_0)$ can also be detected by measuring either the viscosity η or the structural relaxation time as a function of temperature to reach high temperatures, where the structural relaxation time τ_η is of the order of picoseconds or shorter, and observing the crossover of their T-dependence from the VFTH law to the Arrhenius law at some high temperatures. If t_c is of the order of 2 ps, the crossover in T-dependence of η should occur near the temperature at which $\tau_\eta \approx 2\,\mathrm{ps}$ or η calculated from the Maxwell relation, $\eta = G_\infty \tau_\eta$, where $< \tau_\eta > \approx 2\,\mathrm{ps}$ and G_∞ is the high-frequency shear modulus typically of the order of 10^{10} dyne/cm^2 or slightly less. Figure 10 shows the shear viscosity data of *ortho*-terphenyl (OTP), *bis*(2-ethylhexyl)phthalate (DOP), polychlorinated biphenyl (Aroclor), and toluene in high-temperature and low-viscosity regime [346]. The corresponding open symbols stand for $\log(\tau/\mathrm{s})$ from Fabry–Perot interferometry (FPI) experiments. The horizontal dotted line

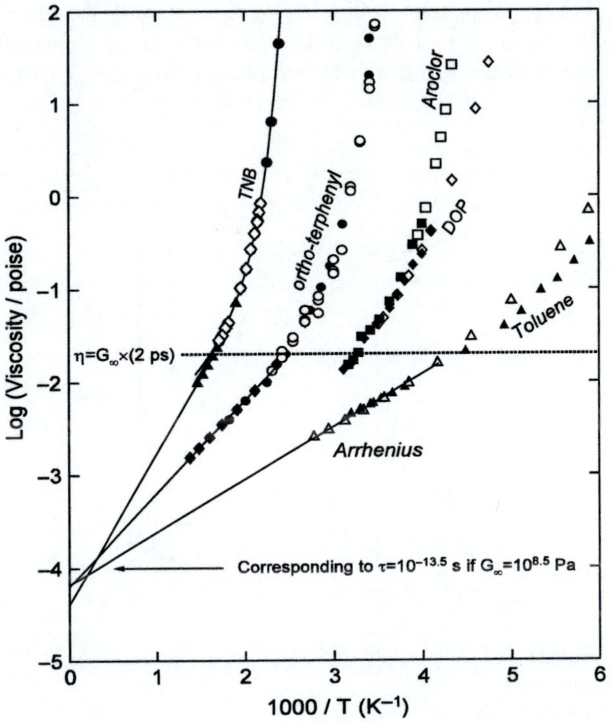

Fig. 10 The viscosity η of tri-naphthyl benzene (TNB), *ortho*-terphenyl, Aroclor, dioctalphthalate (DOP), and toluene at high temperatures to show the change to Arrhenius temperature dependence when the relaxation times are less than about 2 ps (*dotted horizontal line*). The *solid lines* extrapolating the apparent Arrhenius temperature dependences to infinite temperature indicate the prefactor τ_∞ has the physically reasonable value corresponding to angular attempt frequencies of the order of $10^{-13.5}$ s for independent rotation of the molecules at high temperatures

indicates the viscosity calculated from the Maxwell relation $\eta = G_\infty \tau_\eta$, with $< \tau_\eta > \approx 2$ ps and $G_\infty \approx 10^{10}$ dyne/cm^2. The value of G_∞ used here may be too high at these high temperatures where the glassformers are in the near fluid state. Evidence of the crossover at $t_c \sim 2$ ps can be gleaned from the facts that (1) $< \tau_\eta >$ assumes the Arrhenius temperature dependence $\tau_\infty \exp(H/RT)$ for time shorter than $< \tau_\eta > \sim 2$ ps, (2) τ_∞ is about $10^{-13.5}$ s, the reciprocal of which corresponds to a physically meaningful vibrational attempt frequency, and (3) H has the magnitude of a true barrier energy for a molecule of these van der Waals liquids.

In Fig. 11 and in the inset we show the viscosity, η, data of 1,3-*bis*(1-naphthyl)-5-(2-naphthyl)benzene (TNB) plotted against reciprocal temperature. The data are taken from three sets of measurements by Plazek, Magill, and Greet (see [347]) and shown by filled circles, open diamonds, and filled triangles separately. The two curves labeled VFTH(1) and VFTH(2) are the fits to the data for $T < T_B$ and for $T > T_B$, respectively, by the Vogel–Fulcher–Tammann–Hesse empirical form. Note that the VFTH(2) curve cannot fit the data for $T > T_A$, which has a true Arrhenius temperature dependence (not shown). The main figure is the Stickel plot of $[d \log \eta / d(1/T)]^{-1/2}$ against 1000/T using the viscosity data in the inset. The derivative, $[d \log \eta / d(1/T)]$, is approximated by the ratio of finite differences. The three straight lines are obtained from the Arrhenius fit and the two VFTH fits to the

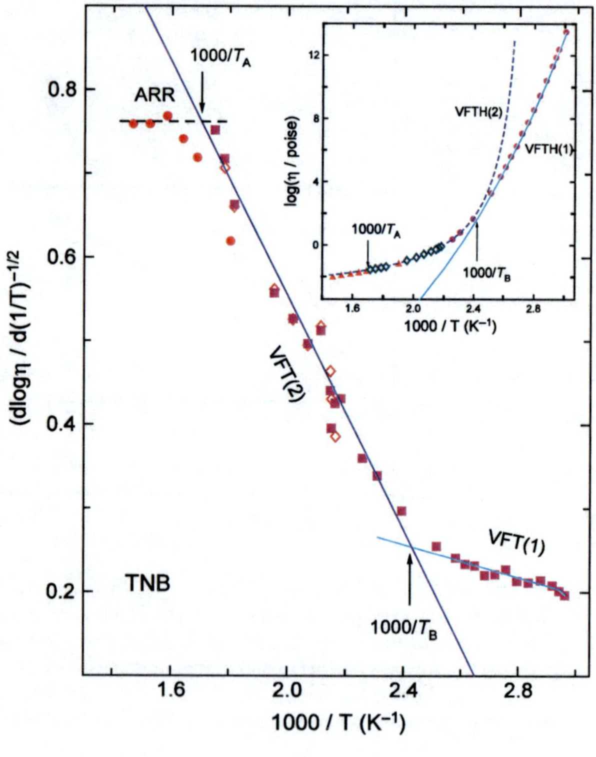

Fig. 11 The Stickel plot of $[d \log \eta / d(1/T)]^{-1/2}$ against 1000/T of the viscosity data of TNB. The viscosity data shown in the *inset* also show an Arrhenius temperature dependence for $T > T_A$, and one Vogel–Fulcher–Tammann–Hesse equation VFTH(2) for $T_A > T > T_B$, and another VFTH(1) for $T_B > T$. Data from [347] are replotted

viscosity data in the three separate regions shown in the inset. The same symbols are used for the viscosity data and the $[d \log \eta / d(1/T)]^{-1/2}$ calculated from the viscosity data sets by ratio of finite differences. The two crossover temperatures T_A and T_B are indicated in both the main figure and in the inset. VFTH fits to the viscosity data in the three separate regions are shown in the inset. Again at T_A, the viscosity is of the order of $10^{-1.7}$ poise, which corresponds to a shear relaxation time of the order of ps.

At temperatures higher than T_A and τ_α of the order of ps, extrapolations of dielectric relaxation data taken at lower frequencies suggest the dispersion narrows to the extent consistent with correlation function having the $\exp(-t/\tau_0)$ form [253]. The temperature dependence of τ_α has the Arrhenius form, $\tau_\infty \exp(H/RT)$, with the prefactor $\tau_\infty \sim 10^{-12.8}$ s like that of a vibrational time and H between 1 and 2 kcal/mol. These numbers are indicative of primitive relaxation.

Similar crossover to Arrhenius T-dependence of η at the same order of magnitude as TNB as well as dielectric relaxation time τ_α at temperatures higher than T_A were found in butylbenzene, propylbenzene, and salol by Hansen et al. [348]. Some of these data will be shown later in Section 2.2.5.1 where the change of T-dependence of τ_α will be discussed in detail.

2.2.1.3 A Relation Between Primitive Relaxation Time and Many-Body Relaxation Time Resulting from the Crossover at t_c (the Coupling Model)

The Original Model Based on Universal Statistics of Energy Levels

The crossover of the correlation function from $\exp(-t/\tau_0)$ to $\exp[-(t/\tau)^{1-n}]$ at a temperature-insensitive time t_c, the magnitude of which is determined by the strength of the interaction, has been amply shown by experiments and molecular dynamics simulations in the previous two subsections (see also [190]). Actually, this property was pointed out in the first version of the coupling model (CM) [175–179] on relaxation of interacting systems long before the experimental observations. In this very first version of the CM published in 1979 [175], the interacting system is semiclassically quantized and the energy levels distribution is described by Wigner's statistical theory [349, 350]. This theory of Wigner originated from his idea that the complex Hamiltonians of many-body interacting systems (in the original case by Wigner it is that of heavy atom nuclei such as uranium) could be approximated by a random Hamiltonian representing the probability distribution of individual Hamiltonians for the purpose of finding the energy levels. This idea was then further developed with advances in random matrix theory and statistics [350, 351]. For systems invariant under time reversal, it is given by the Gaussian orthogonal ensemble (GOE) in random matrix theory [350–353]. It makes sense to use GOE because it was shown in 1965 by Gorkov and Eliashberg [354] to apply to electronic energy levels of small (sounds like nanophysics) metallic particles, and variety of atomic, molecular, and nuclear systems [350–353]. In GOE, the distribution of level spacings E is given by the expression $P(E/D) = (\pi/2)(E/D)\exp[-(\pi/4)(E/D)^2]$ of

Wigner, where D is the average spacing. It has the characteristic linear dependence of $P(E) \propto E$ that can be traced back to be caused by energy level repulsions, and this dependence holds up to a high-energy cutoff at E_c. The CM considers a primitive or one-body relaxation with constant rate $W_0 \equiv (\tau_0)^{-1}$ and correlation function given by $\exp(-t/\tau_0)$ starting at short times. Due to interactions, this one-body relaxation cannot continue indefinitely, and many-body relaxation eventually takes over. It is the response of the level structure to the primitive relaxation that is used to account for the many-body relaxation which necessarily slows down its primitive relaxation rate W_0 [175]. Calculated by perturbation theory [175–179], the response coming from $P(E)$ with the linear dependence on E slows down the relaxation rate W_0 to have the time-dependent form of $W(t) = W_0(\omega_c t)^{-n}$ and the correlation function to have the Kohlrausch form given by Eq. (1.1) or Eq. (2.2), but only for times longer than $t_c \equiv (\omega_c)^{-1}$. This is because the linear dependence of the level spacing distribution no longer holds for $E > E_c$, which corresponds to times shorter than t_c, the reciprocal of the frequency, $\omega_c = E_c/\hbar$, where $\hbar$ is the Planck's constant. From this connection to E_c, it is clear that t_c decreases with strength of interaction, but is insensitive to temperature or pressure. The power, n, called the coupling parameter, is a fraction of unity which increases with the strength of interaction. Thus, the relaxation rate of interacting many-body systems is time dependent. It is the primitive relaxation rate

$$W(t) = W_0 \text{ for } t < t_c, \tag{2.9a}$$

and the many-body relaxation rate,

$$W(t) = W_0(\omega_c t)^{-n} \tag{2.9b}$$

takes over when $t > t_c$. This crossover from the primitive to the many-body relaxation rate does not occur necessarily sharply at t_c but rather smoothly in a neighborhood of t_c, so that the correlation function and its derivatives are continuous across t_c. The factor, $\exp[-(\pi/4)(E/D)^2]$, in $P(E)$ effects the transition between the two rates. Its width parameter D suggests that the width of the neighborhood is of the order of t_c itself, and hence narrow, if there is no other factor like polydispersity of relaxation units entering into the problem. Some of the experimental data discussed in Section 2.2.1.1 show that the crossover is quite sharp. In view of this and in the absence of reliable way to account quantitatively for the narrow crossover, the sharp crossover of the two relaxation rates at t_c is used to generate predictions. The correlation function $\phi(t)$ obtained from the CM rate equation, $\partial\phi(t)/\partial t = -W(t)\phi(t)$, by integration is given by

$$\phi(t) = \exp(-t/\tau_0) \quad \text{for } t < t_c, \tag{2.10}$$

where $\tau_0 \equiv 1/W_0$, and

$$\phi(t) = A\exp[-(t/\tau)^{1-n}] \quad \text{for } t > t_c, \tag{2.11}$$

where

$$\tau = \left[(1-n)\,(t_c)^{-n}\,\tau_0\right]^{\frac{1}{1-n}} \tag{2.12}$$

and

$$A = \exp\{[n/(1-n)](t_c/\tau_0)\}. \tag{2.13}$$

The average time is $\langle\tau\rangle = [\Gamma(1/\beta)/\beta]\tau$, where $\beta \equiv (1-n)$ and Γ is the gamma function.

When τ_0 is much longer than t_c, $A \to 1$, and requirement of continuity of $\phi(t)$ at t_c leads to the relation

$$\tau = \left[(t_c)^{-n}\,\tau_0\right]^{\frac{1}{1-n}}. \tag{2.14}$$

There is no difference between Eqs. (2.12) and (2.14) when the CM is used merely to predict the relation between the dependence of the many-body relaxation time τ with that of the primitive τ_0 on some variable U. For example if τ_0 has Arrhenius T-dependence with activation energy E_a, then both equations predict that the activation energy of τ is given by $E_a/(1-n)$. Quantitatively for the relation between τ_0 and τ, the two expressions make some difference particularly when n becomes larger. In that case, Eq. (2.12) should be used when τ_0 is not much longer than t_c for the sake of accuracy, as demonstrated in comparing prediction [355] with molecular dynamics simulation data [356] where n has unusually large value. When using the prediction to deduce quantitatively τ_0 from the experimentally observed τ with known values of n and t_c, Eqs. (2.12) and (2.14) lead, respectively, to $\tau_0 = (t_c)^n(\tau)^{1-n}/(1-n)$ and $\tau_0 = (t_c)^n(\tau)^{1-n}$. The difference is a factor 2 for $n = 0.5$.

The significance of the CM equation (2.12) or (2.14) is that it makes a connection between the many-body relaxation time τ usually endowed with anomalous properties and the primitive one-body relaxation, the properties of which are normal and known. Thus, the connection provides falsifiable explanation/predictions of the anomalous properties of τ from the known or familiar properties of τ_0. The connection is made via the Kohlrausch exponent n and the crossover time t_c. These two parameters of the many-body relaxation naturally are ultimately determined by the interaction and its strength. This is clearly so in the context of the CM because both quantities originate from interaction. Also this can be inferred from the dependence of the size of both quantities on the strength of interaction from many experimental examples given throughout this book, showing that n decreases and t_c becomes longer on weakening the interaction strength. Equation (2.12) or (2.14) coupled with the Kohlrausch function (2.11) of the CM spawns many predictions that can be tested by experiments and used to explain anomalous properties. Many such tests and applications are given in the sections to follow. Since it was derived for complex Hamiltonians in general, the predictions that form these equations should apply to relaxation and diffusion in interacting many-body systems of all kinds, and the dynamics of glass-forming systems in the glass transition problem is only a special case. Thus, ever since the inception of the CM in 1979, I have been the advocate of the existence of universal relaxation and diffusion properties of interacting many-body systems, and this belief has led to concurrent explorations of several fields using the three coupled equations (2.10), (2.11), and (2.12) or (2.14) as the tool [109–111, 113]. It must be borne in mind that these equations hold strictly for systems in which all relaxation/diffusing units are identical and monodisperse, and heterogeneity is not introduced by boundaries, randomness, mixing, etc. If extrinsic heterogeneity is present, these equations have to be modified by incorporating extraneous factors, and the test of applicability of the CM become less precise and direct, but nevertheless can be done.

The CM does not provide description of how many-body relaxation evolves with time in either space, configurational space, or phase space, except the exploitation of the crossover in some neighborhood of t_c. However, the CM has anticipated that relaxation in interacting many-body systems is dynamically heterogeneous, including the case of the structural α-relaxation in glass-forming liquids. The analogy to the heterogeneous process in the solution of the "dining philosophers problem" in computer science [357] was pointed out in the context of the CM [234] a year before the first experimental evidence of dynamic heterogeneity of structural α-relaxation was published [226]. In the CM, dynamic heterogeneity and Kohlrausch non-exponentiality are regarded as parallel consequences of the cooperative many-body molecular dynamics, but the former is not emphasized in the applications of the CM. Description of the motions as a function of time is best obtained by special experimental techniques like confocal microscopy for colloidal suspensions [141] or by molecular dynamics simulation and specially designed computer simulation method such as the dynamic lattice liquid model [327, 328]. Such description is worthwhile as well as pleasing to acquire, but being able to describe motions as a function of time does not necessarily mean that it can explain the anomalous properties of τ, while the CM equation can. It is this problem-solving capability of the CM equation that has attracted my attention and influenced my research activities ever since it was first derived in 1979 by the crude model. Before that, I had neither formal training nor experience in the various fields of relaxation and diffusion of classical systems including glass transition, polymer viscoelasticity, and ionically conducting materials. I thought the best way to enter these new fields without heavy investment in time and effort is to apply the CM predictions to some notable problems and see if that works. If the predictions were wrong, then it would be high time to abandon the CM and return to my principal occupation before 1979, which is research on electronic properties in condensed matter physics. It turned out that the initial applications of the CM predictions are not disappointing, and some are even surprisingly successful particularly in two areas, polymer diffusion/viscoelasticity [104, 172, 201, 202, 267, 358–361] and dynamics of ions in glassy ionic conductors [362]. Some of the early applications in the first half of the 1980s are summarized in a book chapter [111]. The initial successes lured me to educate myself in the sciences of the new fields and to look for more opportunities in applying the predictions to explain experimental results and established phenomenology. This effort has become my preoccupation and this comes at the expense of not being able to spend sufficient time in making substantial improvement in the theory for a long time. Nevertheless there are some developments in the early years based on related physics including stable Lévy distributions [180(a)], Paley–Wiener criterion for relaxation functions [180(b)], and classical constraint dynamics [181–184].

Tracing the Key Result of the CM, $W(t) = W_0(\omega_c t)^{-n}$, Back to R. Kohlrausch

In 1979 when I first published the CM [175], I was not aware of the works of R. Kohlrausch in 1854 [2] and his son F. Kohlrausch in 1863 and 1866 [3, 4], and not until 2 years later. This was acknowledged in reference 27 of [180(c)], where I cite the abstract of Struik from which I first found mention of the work of F. Kohlrausch's 1866 paper, and this led me thereafter to search for other works of Kohlrausch. One of the crucial results of the CM in my 1979 paper is the time-dependent relaxation rate, $W(t) = W_0(t/t_c)^{-n}$. Nowadays, to most workers in the field of relaxation, R. and F. Kohlrausch are known for their stretched exponential function. However, if one reads the 1854 paper by R. Kohlrausch, on page 198 he actually started with the time-dependent rate equation, $d(pQ_t - r_t)/dt = -bt^m(pQ_t - r_t)$, duplicated faithfully here. For the notations he used, m is a *negative* fraction of unity, b is a constant, and $(pQ_t - r_t)$ is the relaxing quantity he considered. Integrating the differential equation from $t = 0$ to $t = t$ and with $Q_t(t = 0) = Q_0$ and $r_t(t = 0) = 0$, he then arrived at $\log[(pQ_t - r_t)/pQ_0] = -[b/(m+1)]t^{m+1}$. This is just the same as $(pQ_t - r_t)/pQ_0 = \exp(-Bt^{m+1})$, where $B = b/(m+1)$. In other words, R. Kohlrausch derived the stretched exponential function from the time-dependent relaxation rate bt^m, with $-1 < m < 0$. Using his own data [2], he deduced $m = -0.5744$, and $b = 0.4289$. F. Kohlrausch [3, 4] wrote down a similar equation, $dx/dt = -g \cdot x/t^n$, before getting the stretched exponential function. Now it becomes clear that the primary result of the CM, $W(t) = W_0(t/t_c)^{-n}$, is exactly the same as time-dependent relaxation rate, bt^m, of R. Kohlrausch after putting $m = -n$ and $b = W_0(t_c)^n$. From the rate equation, $d\phi/dt = -W(t)\phi$, and $W(t)$ given by Eqs. (2.9a) and (2.9b), one can not only obtain the Kohlrausch function, $\exp[-(t/\tau)^{1-n}]$, but also the second relation $\tau = [(1-n)(t_c)^{-n}\tau_0]^{1/(1-n)}$, where $\tau_0 \equiv 1/W_0$. This second relation introduces new physics and immensely enhances the applications of the Kohlrausch function beyond an empirical function to fit data. In the CM, n is indicator of the extent of the many-body relaxation. As will be shown in many sections to follow, the second relation can explain the experimentally observed anomalous and universal properties of τ from the normal properties of the primitive relaxation time τ_0. These important advances cannot be achieved without using the time-dependent relaxation rate, $W(t) = W_0(t/t_c)^{-n}$, originally proposed in terms of bt^m with $-1 < m < 0$ by R. Kohlrausch phenomenologically. Therefore, when paying homage to Kohlrausch, I propose that it is more appropriate to cite Kohlrausch's fractional-power time-dependent relaxation rate, bt^m with $-1 < m < 0$, instead of his already famous stretched exponential function, or at least together with it. The function follows as a consequence of the rate, bt^m, but not vice versa. Moreover, there can be different interpretations of the stretched exponential function, such as the more trivial one of distribution of linear exponentials due to randomness, while the interpretation of the rate is more restrictive and possibly the consequence derivable from fundamental physics as discussed in the above and below.

Personally, I am flabbergasted to find in 1983 that one of the key results of the CM based on fundamental physics was already proposed by R. Kohlrausch back in 1854.

Coupling Model from Classical Chaos

In classical mechanics, it is well known that systems with anharmonic interactions universally exhibit classical non-linear dynamics (chaotic classical motion) [363]. In the 1980s, many publications show that classically chaotic systems when quantized have quantum energy spectra characterized by the universal statistics of random matrix eigenvalue ensembles [364–369]. Overwhelming experimental and numerical evidence supports this connection for systems as diverse as atomic nuclei, Rydberg atoms, and quantum billiards [350, 370]. The relation between classical non-linear dynamics and quantum energy spectra is natural because in the macroscopic size regime, laws of classical mechanics are expected to emerge in the semiclassical limit, $\hbar \to 0$. It suggests that the many-body relaxation stems from classical non-linear mechanics and from which one can find another way to derive the CM equations applicable to relaxation/diffusion in systems exhibiting classical chaos. The first attempt in this direction involves the use of the characteristic of classical chaos, which is the emergence of complexity on infinitely fine scales in classical phase space. The structure of the infinitely fine phase space was used as the basis of qualitative models to generate the Kohlrausch correlation function [184, 186]. Some residual order in phase space in terms of quasiperiodic motions (vague tori) was found to exist on a short timescale even in the chaotic regions of phase space for a large class of systems [371, 372]. The quasiperiodic motions are attributed to remnants of destroyed invariant tori in phase space This property ensures the relaxation/diffusion at short times is normal and its correlation function is given by the linear exponential of Eq. (2.10).

Starting from simple non-integrable Hamiltonian models that exhibit classical chaos, the results of the CM have been reproduced [186, 188, 189, 191, 192, 193]. The crux of the CM is the existence of a rather sharp crossover of the relaxation function from linear exponential to the Kohlrausch stretched exponential, and this has been found from the numerical solutions of the simple models. Moreover, stretching (or n) increases with increase in non-linearity or interaction strength of the Hamiltonian, consistent with the same expected heuristically from the CM on increasing the strength of coupling or constraints in glass-forming substances and other interacting systems.

Relaxation of Interacting Arrays of Phase-Coupled Oscillators

We [188, 192] start from an array of N oscillators. The phase $\varphi_i(t)$ of the ith oscillator, $1 \leq i \leq N$, is coupled non-linearly by a sine function to the phases of the other oscillators and obey the equation of motion

$$\frac{d}{dt}\varphi_i = -\frac{K}{N}\sum_{j=1}^{N}\sin(\varphi_j - \varphi_i), \tag{2.15}$$

with uniform interaction *K/N*. We are interested in the decay of the phase coherence *r*, which is the absolute value of the order parameter, $r\exp(i\psi)$, defined by

$$r = |\, re^{i\psi}\,| = |\frac{1}{N}\,\Sigma_j\, e^{i\,\varphi_j}\,|. \tag{2.16}$$

It has been shown that the decay of *r* for an array of phase-coupled oscillators is exponential [373], like the primitive relaxation of an isolated molecule in a solvent such as described by the Debye model. As an analogue of anharmonic potentials in interacting many-body systems, we consider now a number *M* (larger than one) of such arrays and couple these arrays non-linearly together, again by the sine function

$$\frac{K'}{MN}\,\Sigma_{\beta=1,\beta\neq\alpha}^{M}\,\Sigma_{j=1}^{N}\, sin(\,\phi_{j\beta} - \phi_{i\alpha}\,), \tag{2.17}$$

where K'/MN is the inter-array interaction strength. The new equation of motion of the phase of the *i*th oscillator in the αth array, $1 \le \alpha \le M$, is now given by

$$\phi'_{i\alpha} = \phi_{i\alpha} - \frac{K}{N}\,\Sigma_{j=1}^{N}\, sin(\,\phi_{j\alpha} - \phi_{i\alpha}\,) + \frac{K'}{MN}\,\Sigma_{\beta=1,\beta\neq\alpha}^{M}\,\Sigma_{j=1}^{N}\, sin(\,\phi_{j\beta} - \phi_{i\alpha}\,). \tag{2.18}$$

The problem is simplified to a map by picking the appropriate time steps and rescaling the time, so that time *t* is now discrete and incremented by 1 after each iteration of the map. The effect of the interactions between the *M* arrays on *r* of each array is studied by considering the new map. The interacting arrays mimic an assembly of interacting molecules in a small molecule liquid. In the absence of inter-array interaction (i.e., $K' = 0$), *r* of the non-interacting arrays of coupled oscillators decays exponentially to an incoherent state, i.e., $r = 0$, for $K > 0$, in analogy to the primitive Debye relaxation of isolated molecules in dilute solution. However, since the interacting arrays do not model translational or rotational motion, they cannot be identified exactly with the structural relaxation in a glass-forming liquid. A non-linear Hamiltonian that resembles more closely a glass-forming liquid has to be much more complicated than the interacting arrays, and likely its dynamics cannot be solved exactly. In spite of its limitations, the interacting arrays of oscillators model have the advantage that their dynamics can be obtained readily, as shown below, and is an analogue of the relaxation of interacting many-body systems.

With arrays of random initial oscillator phases, we iterate the map in Eq. (2.18) to obtain the evolutions of the coupled arrays numerically. The evolutions of three (i.e., M=3) interacting arrays, each of N=32 oscillators initially with random phases, are obtained by iteration of the map defined by Eq. (2.18). From the results we calculate the decay of the phase coherence *r(t)* for each array. First for a fixed *K* we calculate $r\,(t)$ for five values of $K' = 0$, 0.6, 0.8, 1.0, and 1.2. The results are shown in Fig. 12. The figure shows that with increasing inter-array coupling strength K', the

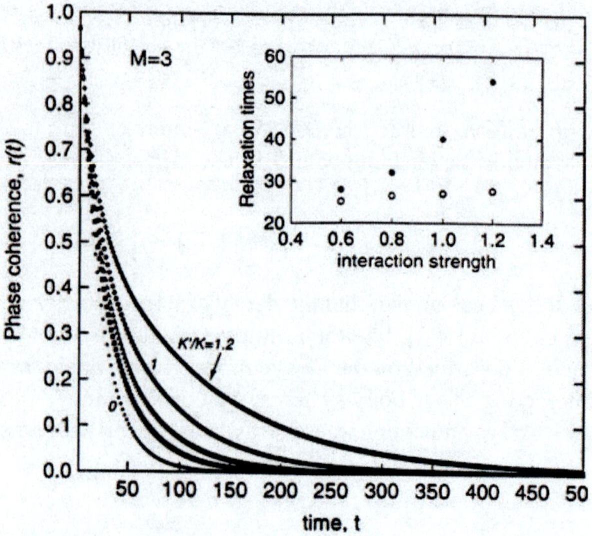

Fig. 12 Decay of $r(t)$ calculated for $M = 3,\ K = -0.03$, and $K'/K = 0, 0.6, 0.8, 1.0$, and 1.2. The *inset* shows τ_0 (O) and τ (•) as a function of K'/K. Reproduced from [188, 192] by permission

decay of $r(t)$ becomes slower and its time dependence more non-exponential. The accompanying Fig. 13 is replotting the same results as $\log_{10}\{-\log_e[r(t)]\}$ vs. $\log_{10} t$. At short times the slope is exactly 1, indicating that initially $r(t)$ is an exponential function of time, $\exp(-t/\tau_0)$, in analogy to the primitive relaxation of the CM, Eq. (2.10). However, the linear exponential decay does not hold at longer times. There exists a crossover time t_c after which the slope becomes less than 1, and $r(t)$ departs from the $\exp(-t/\tau_0)$ time dependence. Actually, for $t > t_c$, the slope varies slightly with time but the result is still reasonably well approximated by a straight line having a constant slope equal to $\beta \equiv (1-n) \leq 1$, indicating $r(t)$ has crossed over to assume the stretched exponential time dependence, Eq. (2.11), of the CM. Naturally, Eq. (2.12) of the CM is satisfied. The systematic decrease of the long time slope β (see inset in Fig. 13) with increase of K'/K supports the intuitively reasonable surmise that n, the coupling parameter in the CM, increases with the mutual interaction strength of the relaxing units in real materials. We observe that there is a slight increase of t_c with decrease in the interaction strength, which is also consistent with real materials.

The interacting arrays of oscillators is a prototype non-linear Hamiltonian. We expect the crossover property to be general and carried over to other Hamiltonians that model more realistically the non-linear interaction potentials in a molecular liquid. To some extent, the results can be considered as a justification of the physical principle behind the CM for molecular systems.

Structural Relaxation Properties of Glassformers Captured

Next we study the change of the decay of $r(t)$ on varying K at constant K'. As we shall see later in this section, K has the effect on $r(t)$ like temperature in relaxation

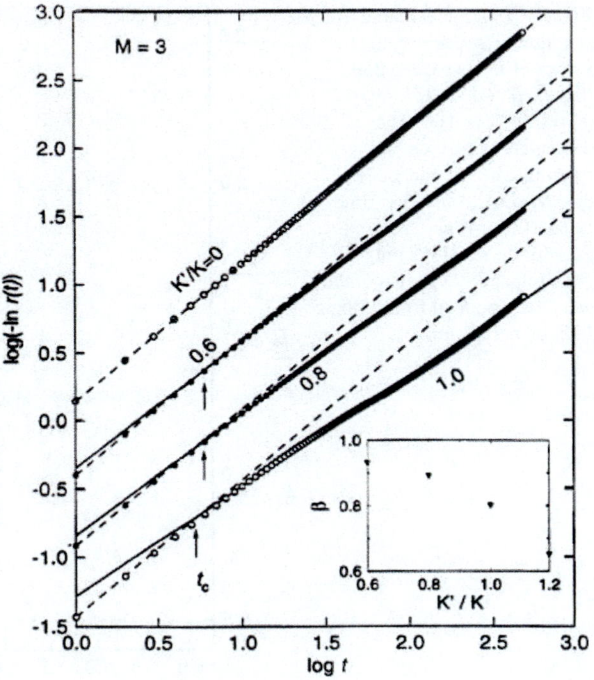

Fig. 13 Plot of $\log_{10}\{-\log_e[r(t)]\}$ vs. $\log_{10} t$ obtained numerically for $M = 3$, $K = -0.03$, and $K'/K = 0, 0.6, 0.8$, and 1.0. The *dashed lines* are the exponential fits and the *solid lines* are the stretched exponential fits. Curves for $K'/K = 0.0, 0.6$, and 0.8 are shifted vertically by multiples of 0.5 to avoid overlapping of results in the figure. The crossover time t_c is indicated by the *vertical arrows*. The calculated $r(t)$ conforms well to $\exp(-t/\tau_0)$ for $t < t_c$ and to $\exp[(-t/\tau)^{\beta}]$ for $t > t_c$. The *inset* shows that the stretch exponent β decreases with increasing inter-array interaction strength

of real materials. The results as a function of K at constant K' mimic remarkably well the characteristics of the temperature dependence of structural relaxation properties of glass-forming substances, including the temperature dependence of the stretch exponent β and the relaxation time τ of the Kohlrausch relaxation function, $\exp[-(t/\tau)^{\beta}]$. The results for increasing K' change progressively like corresponding quantities of structural relaxation in glassformers with increasing intermolecular coupling by changing the chemical structure (Fig. 14).

The values of τ and $(1–n)$ obtained as a function of K for several constant K' values equal to 0.030, 0.024, 0.018, 0.012, 0.006, and 0.000 are displayed in Fig. 15, which shows that τ decreases monotonically with K. Since the analogue of K' is the intermolecular interaction strength and K has the effect like temperature in glass-forming liquids, the results $r(t)$ are appropriately compared with the change of dielectric relaxation and dynamic light-scattering properties with temperature of the molecular liquids with different intermolecular interactions. For any K', on decreasing K, τ increases and the increase becomes increasingly more rapid. It is apparent also that the $\log_{10}\tau$ rises more rapidly with decreasing K for a larger K'. The model results for $K' = 0.030$ (•), 0.018 (♦), 0.012 (Δ), and 0.006 (▼) resemble the four glassformers, *ortho*-terphenyl (OTP), salol, glycerol, and propylene glycol (PG), respectively (see inset in Fig. 15).

For each K', a cross-plot of $(1 - n)$ vs. $\log_{10}\tau$ (i.e., at the same K) is made. The relations between $(1 - n)$ and $\log_{10}\tau$ for several values of K' are shown

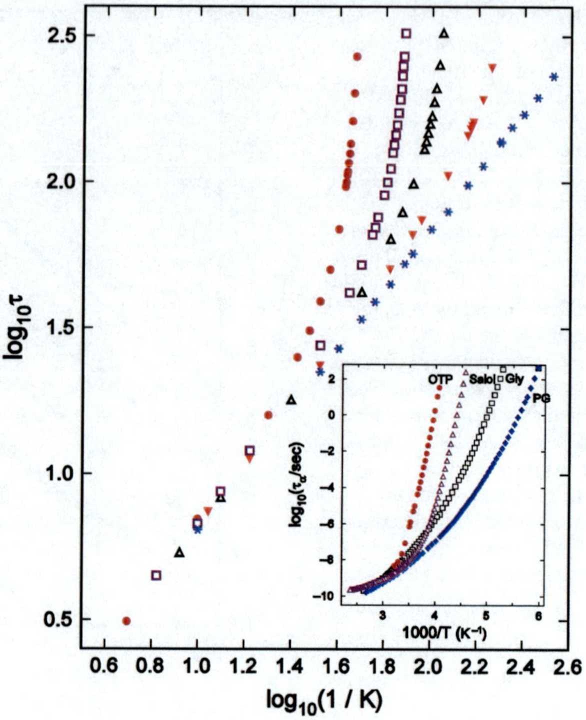

Fig. 14 Log τ calculated as a function of $1/K$ at various values of K': (•) K'=0.030, (□) 0.018, (Δ) 0.012, (▼) 0.006, and (∗) 0.00. The dependences of τ on $1/K$ resemble the dependences of the structural relaxation time τ_α on $1/T$ for four glassformers, OTP, salol, glycerol, and PG shown in the *inset*. Reproduced from [188, 192] by permission

in Fig. 15. From this figure it is observed that the dependences of $(1 - n)$ on $\log_{10}\tau$ are similar, i.e., all showing with decreasing $\log_{10}\tau$ monotonic increasing $(1 - n)$, which eventually reach the maximum value of 1. However, at the same value of $\log_{10}\tau$, $(1 - n)$ is larger for smaller K'. The limiting case $K' = 0$, corresponding to the absence of inter-array coupling, has $(1 - n) = 1$ (i.e., exponential relaxation) for any $\log_{10}\tau$ or K. These properties of the model are shown to be similar to that found in the structural or α-relaxation of small molecule glass-forming liquids with increasing temperature. For example, in the molecular liquids, OTP, salol, glycerol, and propylene glycol, the stretch exponent β_α of the α-relaxation correlation function, $\exp[-(t/\tau_\alpha)^{\beta_\alpha}]$, obtained by dielectric and light-scattering measurements increases toward unity as temperature increases and the effective α-relaxation time, τ_α, decreases. This is shown in the inset of Fig. 15. There is a correlation between $n_\alpha(T_g) \equiv [1 - \beta_\alpha(T_g)]$ and the T_g-scaled temperature dependence of τ_α. Here, T_g is the glass temperature at which $\tau_\alpha(T_g) = 10^2$ s. The temperature dependence of $\log_{10}\tau_\alpha$ of the four liquids, propylene glycol, glycerol, salol, and OTP, is mimicked by the K-dependence of $\log_{10}\tau$ for increasing K' (see Fig. 14). Thus the results obtained from our model at various fixed values of the inter-array interaction strength K' reproduce the pattern of changes in relaxation properties of "strong," intermediate, and "fragile" glass-forming liquids

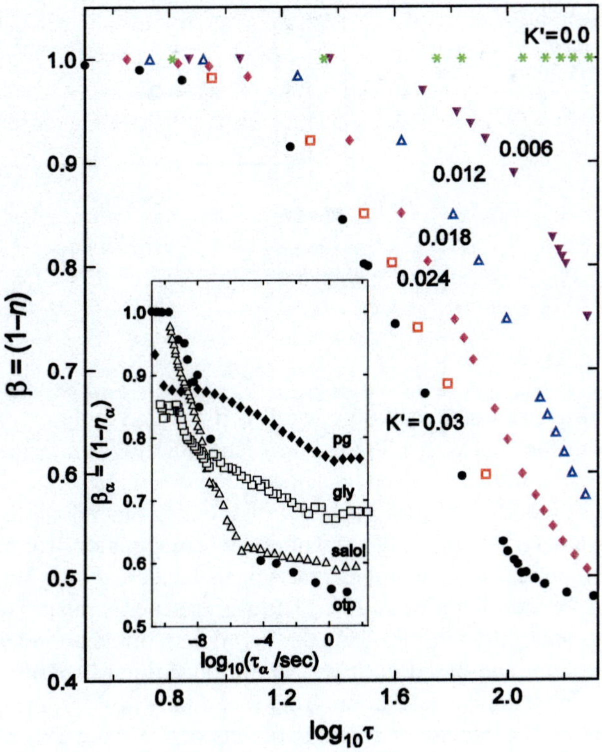

Fig. 15 A cross-plot of (1–n) vs. $\log_{10} \tau$ (i.e., for the same K). The results shown are for $K' = 0.030$ (•), 0.024 (□), 0.018 (♦), 0.012 (Δ), 0.006 (▼), and 0.00 (∗). Similar dependences of β_α on $\log_{10} \tau_\alpha$ are shown for OTP (•), salol (Δ), glycerol (□), and PG (♦) in the *inset*. Reproduced from [188, 192] by permission

in the Oldekop–Laughlin–Uhlmann–Angell plot [215, 216, 374, 375]. Since K' is the analogue of intermolecular interaction in glassformers, the results obtained here indicate that the relaxation properties of glassformers within the same class are determined principally by the intermolecular coupling.

The Dissipative Fermi Map in a Stadium: The Fermi-Stadium Map

In 1949, Fermi introduced a theory of cosmic radiation in which cosmic rays originated and accelerated in the interstellar space of the galaxy by collision against moving magnetic fields [376–378]. The model of Fermi was adapted to describe transient chaos by simplifying it to a ball bouncing in one-dimensional motion between a fixed and a moving wall [379]. The latter oscillating sinusoidally with time in the same direction, $x = a\cos(\omega t)$, and elastically imparts momentum to the ball dependent on its velocity u and without the wall changing its position in space. The ball suffers a fraction loss δ in velocity for every collision with the fixed

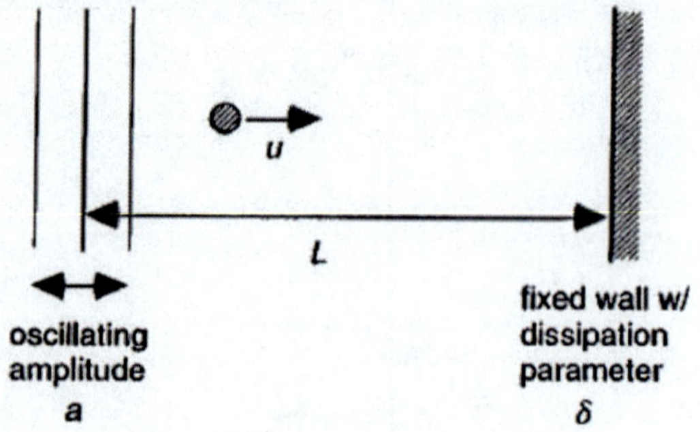

Fig. 16 One-dimensional dissipative Fermi map. Reproduced from [189] by permission

wall, and thus the Fermi map is dissipative. It is illustrated by Fig. 16, and L is the distance between the two walls. Tsang and Lieberman [377] considered the case where $L/16a >> 1$ and $\delta << 1$ and calculated the evolution of the distribution in velocities. They found that almost all initial distributions in velocities $f_0(u)$ evolve to the invariant one $f(u) \sim \exp(-2\delta u^2)$. Numerical computations for various values of $L/16a$ were reported to be in agreement with such calculations. The evolution is determined by a Fokker–Planck equation. The average energy relaxes to the equilibrium energy *exponentially* in time. Thus the Fermi map offers a model of primitive or normal relaxation, and based on it we can introduce non-linearity and see how chaos modifies the primitive relaxation. This was done [189] by introducing some curvature to the elastic boundary at the two ends of the fixed dissipative wall as shown in Fig. 17. The motion of the ball is now two dimensional. The composite system is actually the combination of the Fermi map system and half a stadium,

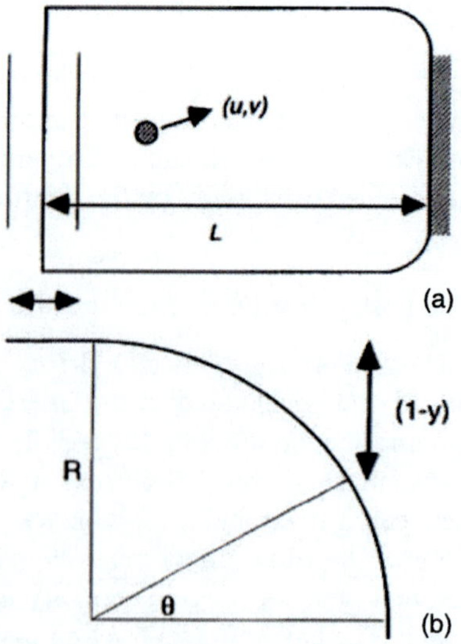

Fig. 17 (**a**) Two-dimensional Fermi-stadium map; (**b**) a curved corner of the added stadium portion. Reproduced from [189] by permission

appropriately called the Fermi-stadium map. The details of this map can be found in [189]. In the problem of a billiard ball idealized as a point reflected elastically from the boundary of a stadium without any loss of energy [368], it is well known that the motion of the ball is chaotic, following no regular pattern at all. Moreover, when the motion of the billiard in the stadium is quantized, the eigenvalue spacings have the Wigner distribution characteristic of a random Hamiltonian and GOE [364]. Hence, we can expect that chaos of the appropriate kind will appear in the Fermi-stadium map with a degree that is proportional to the curvature or the radius R. The relaxation of the average energy in the Fermi-stadium map, with parameters $\delta = 0.01$ and $M = L/16a = 100$, for various values of $R = 0.0, 0.02, 0.1$, and 0.5. The normalized difference Φ between the mean energy and the equilibrium energy is plotted against time t in Fig. 18 which shows the relaxation of Φ is slowed down when R is non-zero and the degree of slowing down increases with R. Note that for $R = 0$ (the dashed line at the bottom), the map is reduced to the Fermi map and there is no slowing down of the relaxation. At $t = 140$, the calculated solid curve, corresponding to $R = 0.02$, begins to significantly deviate from the Fermi curve. The next curve above corresponds to $R = 0.1$, and it begins to deviate significantly from the Fermi curve starting at $t = 120$. For $R = 0.5$ since the change from the Fermi map is substantial, we rescale the curve horizontally so that it coincides for a significant portion with the Fermi curve. The rescaling is shown by the arrow in Fig. 18. At $t = 85$, the curve corresponding to $R = 0.5$ begins to deviate significantly from the Fermi curve. Thus, there exists a crossover time t_c (the magnitude of which depends on R), before which the Fermi-stadium curve decays exponentially like the Fermi curve, but thereafter its decay is slower than the Fermi curve. The existence of such a crossover time from primitive to complex relaxation dynamics is thus present in the Fermi-stadium map, a result which can be expected to have general validity in classical chaotic systems as anticipated by the CM. The crossover is made more explicit in Fig. 19 where we plot $\log_{10}(-\log_{10}\Phi)$ against

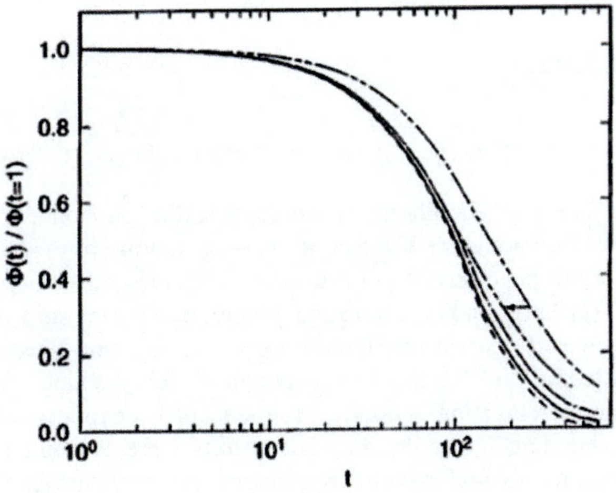

Fig. 18 Numerical result of the Fermi-stadium map with parameters $\delta = 0.01$ and $M = 100$, for various values of $R = 0.0, 0.02, 0.1$, and 0.5, from *left to right*. The normalized difference Φ between the mean energy and the equilibrium energy is plotted against time t. The *arrow* indicates a rescaling of the $R = 0.5$ curve to show a longer coincidence with the Fermi curve. Reproduced from [189] by permission

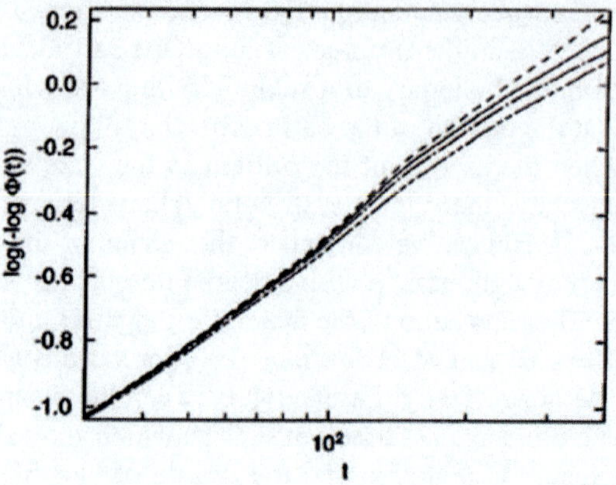

Fig. 19 $\mathrm{Log}_{10}(-\log_{10}\Phi)$ against $\log_{10} t$. The curves become nearly straight after a crossover time t_c, which depends on R, and can be fitted to *straight lines* of slopes $\beta = 1.0$, 0.86, 0.8, and 0.78, respectively, for $R = 0.0$, 0.02, 0.1, and 0.5 (*from left to right*). Reproduced from [189] by permission

$\log_{10} t$. After a crossover time t_c (which depends on R), the curves become nearly straight. The portion of the curves can be fitted to straight lines of slopes $\beta = 1.0$, 0.86, 0.8, and 0.78, respectively, for $R = 0.0$, 0.02, 0.1, and 0.5.

Similar Results from Related Problems

There are studies of relaxation and diffusion in other non-linear Hamiltonian systems which also show slow decay well described by the Kohlrausch stretched exponential. Here we cite a few examples. This was found by Pettini and Landolfi in 1990 who showed the time evolution toward equipartition of energy in non-linear Hamiltonian systems with a large number of degrees of freedom. The systems studied include the Fermi–Pasta–Ulam (FPU) β-model [380] and the lattice ϕ^4 model. The Hamiltonian of the former is given by

$$H(p,q) = \sum_{i=1}^{N} [p_i^2/2 + (q_{i+1} - q_i)^2/2 + \mu(q_{i+1} - q_i)^4] \tag{2.19}$$

and latter by

$$H(p,q) = \sum_{i=1}^{N} [p_i^2/2 + (q_{i+1} - q_i)^2/2 + mq_i^2/2 + \mu q_i^4/4]. \tag{2.20}$$

The two Hamiltonians would describe phonon dynamics in one-dimensional anharmonic lattice if the coordinates q_i are interpreted as deviations from the equilibrium positions of point masses. Tsironis and Aubry [381] and their coworkers [382] have studied relaxation properties of one- and two-dimensional non-linear lattices described by Hamiltonians with the non-linear potential given by the ϕ^4 potential and by the Morse potential. They found stretched exponential lattice energy relaxation in contrast to the standard exponential relaxation law of the corresponding linear system. The results were attributed to hierarchies of discrete breathers that relax with different time constants, leading to a hierarchy of relaxation timescales in the system. Discrete breathers are intrinsic localized and time periodic lattice solutions that exist in non-linear lattice systems. For more on breathers, see review by Flach and Gorbach [383].

Kalosakas et al. [384] investigated the diffusive motion of a charge interacting with the non-linear dynamics of a thermalized underlying lattice. Signatures of anomalous diffusive properties are found at relatively high temperatures, where highly non-linear excitations are present. A sublinear diffusion and a plateau appear before the standard long-time diffusion during the evolution of the mean-squared displacement with time, and a significant degree of heterogeneity is exhibited among individual trajectories. This behavior of the mean-squared displacement is similar to that found in other interacting systems. The results are in contrast with those of the linearized case, in the absence of breathers, indicating that they are the effects caused by non-linearity of the Hamiltonian.

Outlook

Hopefully, the results discussed above are sufficiently convincing and encouraging for others to pursue further and to take them to higher level and with mathematical rigor. As mentioned before, ever since the inception of the CM in 1979, the author has been preoccupied with applications of the CM predictions to experimental data in several fields. All theoretical efforts discussed above were made over short periods of time and with limited resource. Frankly, the theoretical skill of the author is overmatched by the difficulty of the problem of relaxation/diffusion in interacting many-body systems. Despite the meager theoretical support, the CM predictions have been remarkably successful in explaining experiments and phenomenology, as well as making predictions before experiments were performed and subsequently verified. These successes will be discussed in conjunction with the experimental findings throughout the rest of this book. I certainly agree that the problem is not solved until a rigorous physical-mathematical theory emerges that gives a full description of the motions at all times and also can explain the many properties, especially the anomalies found experimentally. For the latter, I submit that the three coupled equations (2.10), (2.11), and (2.12) have done a credible job, which the readers can judge for themselves from the contents of this book. The final solution of the problem we are waiting for has to match the multiple successes of these equations of the CM in applications to experimental data. On this connection between the final solution one is looking for in the future and the existing CM equations, it is perhaps appropriate to cite the philosopher, William of Occam, who wrote: "It is vain to do with more what can be done with less."

It is clear from the history of the CM given above that it has not attracted other theorists to contribute in making it a more complete theory, even though it has been applied successfully to many different fields. Thus, what is available in its theoretical development is limited, essentially that given by a one-man effort. In contrast, even in a subfield of relaxation and diffusion of complex systems, some theories have benefited from its development from participation and contribution of many able theorists. An example in polymer rheology is the reptation theory, and in glass transition it is the mode coupling theory. The materials in the sections in this chapter and the next will show how well the CM relations can explain the multitude of experimental data. The author hopes these will convince others that the CM has captured some fundamental physics of relaxation and diffusion in complex interacting systems in general, and it is a worthwhile starting point to build a rigorous theory.

2.2.2 *Length Scale and Dynamic Heterogeneous Nature of the Structural Relaxation*

When treating a relaxation process and particularly the structural relaxation in glass-forming liquids, naturally some length related to the motion is an important characteristic and has to be considered at some arbitrarily chosen value of τ_α, at any temperature, or at T_g.

2.2.2.1 Length Scale from the Free Volume Model

In the free volume model [385], the length is the size of the unoccupied (free) volume, which decreases with decrease in temperature, and can be measured by positronium annihilation lifetime spectroscopy (PALS) [386–390]. This length has no direct link to the length scale of dynamically heterogeneous dynamics or many-body relaxation dynamics because the free volume model does not have the concept of many-body relaxation. Nevertheless, correlation of the free volume measured by PALS with some outstanding characteristics of the observed dynamics, to be discussed in Section 2.2.5.1, was first found by Ngai et al. [387] in *ortho*-terphenyl and propylene carbonate, glycerol, and propylene glycol, and subsequently verified by others.

2.2.2.2 Length Scale from the Configuration Entropy Model

In the configurational entropy model of Adam and Gibbs [30], the size of the cooperative rearranging regions (CRR) offers a length scale of the structural relaxation. Kauzmann [391] pointed out that the extrapolated entropy of an equilibrium supercooled liquid may become less than that of a crystal and approach a negative value at 0 K. In order to avoid the paradox of Kauzmann, the entropy theory was developed by Gibbs and DiMarzio [392] originally for polymeric glassformers. The theory was extended to non-polymeric liquids by a phenomenological consideration by Adam and Gibbs (AG). They proposed that the rearrangements over energy barriers of molecular units must be cooperative involving altogether a number z^* of molecular units that necessarily increases with decreasing temperature. Several assumptions were made in the AG theory. The first one is that the transition of a CRR involves the z^* molecules simultaneously, and individually each molecule has to surmount the potential energy barriers, $\Delta\mu$, which is temperature independent. This assumption enabled AG to write the relaxation time, $\tau_\alpha(T)$, and the corresponding exponential relaxation correlation function as

$$\tau_\alpha(T) = A_{AG}\exp(z^*\Delta\mu/RT) \text{ and } \varphi(t) = \exp(-t/\tau_\alpha(T)), \quad (2.21)$$

where the preexponential term, A_{AG}, is a physically reasonable time that is largely determined by molecular vibrations of the liquid, and R is the gas constant. AG expressed z^* in terms of the molar configurational entropy $S_c(T)$ of the macroscopic sample as

$$z^*(T) = N_A s_c^*/S_c(T), \quad (2.22)$$

Here, s_c^* is the configurational entropy of a single molecule and N_A is Avogadro's number. The value of s_c^* is determined experimentally by extrapolating the $S_c(T)$ data to infinite temperature. Inserting this equation into the expression for the relaxation time, the result of the AG model is given by

$$\tau_\alpha(T) = A_{AG}\exp[\Delta\mu s_c^*/k_B T S_c(T)]. \quad (2.23)$$

This equation has four unknown quantities, A_{AG}, s_c^*, $\Delta\mu$, and S_c, and some of these have not been determined by experiment. In practice, the combination $s_c^*\Delta\mu/k_B$ in Eq. (2.23) is replaced by C_{AG}, which is treated as a temperature-independent constant of the glassformer, and the AG equation is rewritten as

$$\tau_\alpha(T) = A_{AG}\exp[C_{AG}/TS_c(T)]. \quad (2.24)$$

The AG equation was intended mainly to describe the super Arrhenius behavior of a liquid approaching T_g. Over some range of temperatures above T_g, the temperature dependence of η or τ_α of many liquids conforms to the Vogel–Fulcher–Tammann–Hesse (VFTH) equation [393–395]

$$\tau_\alpha = A_{VFTH}\exp\left[B/(T - T_0)\right]. \quad (2.25)$$

However, the preexponential factor, A_{VFTH}, is often found to be many orders of magnitudes shorter than the vibrational times [347, 348, 396–399], which is assumed for A_{AG}. This glaring discrepancy between A_{VFTH} and A_{AG} already has signaled the inadequacy of the AG theory. The critical entropy, $s_c{}^*$, in the AG equation supposedly given by $k_B\ln 2(= 0.9563 \times 10^{-23}$ J molecule^{-1}K^{-1}) is less than even the residual entropy per molecule in a glass at 0 K. Other limitations of the AG theory and the problems encountered in testing it against experimental data are best given in an article by Johari [400]. Since S_c is not known, the excess entropy of a liquid over that of its ordered crystal state, S_{exc}, is used instead of the required S_c in the AG equation. The validity of the AG equation can be tested by showing whether log η or log τ_α bears a linear relation with $(TS_{exc})^{-1}$ or not over a temperature range down to T_g. This procedure to examine the validity of the AG equation was used by Magill [396] in 1967 and by Laughlin and Uhlmann [375] in 1972. For the three liquids, including salol and *ortho*-terphenyl, whose viscosity they had carefully measured over a range of about 14 decades covering a wide temperature range down to T_g, they found the plots of log η against $(TS_{exc})^{-1}$ were non-linear and concluded that the AG equation is not valid for those liquids. The approximately linear relation between logarithm of the dielectric relaxation time τ_α and $(TS_{exc})^{-1}$ found at lower temperatures invariably breaks down at higher temperatures [398, 401]. In spite of the aforementioned problems of the AG model, the predicted linear relation between log η or log τ_α and $(TS_{exc})^{-1}$ has continued to be the main point used to test and support the validity of the AG theory [398]. Others found that the AG model extended to include pressure dependence is consistent with experimental data [399, 402], while an extended free volume model for the same purpose failed [403]. Even if in some cases the AG equation prediction is consistent with the experimentally observed temperature/pressure dependence of η or τ_α, the latter is only one among many notable properties and anomalies of glassformers. One must confront the AG theory with the other anomalous properties before any conclusion can be made on its ultimate validity. It is true that the AG theory does not have any immediate prediction other than the temperature or pressure dependence of η or τ_α. But this shortcoming of the AG theory does not mean we can use it as an excuse and accept the theory without reservation.

The relaxation process of a CRR envisaged by AG is so much simplified that it is unlikely to represent the actual many-body relaxation. Equation (2.22) for $z^*(T)$ indicates that it is determined exclusively by configurational entropy, a thermodynamic quantity. The dispersion of the structural relaxation in the AG model is a linear exponential. All these features of the AG model indicate that the length scale, $L_{AG}(T)$, associated with $z^*(T)$ is unrelated to the many-body relaxation dynamics of the structural relaxation.

Actually there is no way to test directly the entropy theory of AG simply because S_{conf} values are not available. Instead, studies which had tested the entropy theory and found it to be valid had used the excess entropy of a liquid over that of its ordered crystal state, S_{exc}, instead of the required S_c. This practice has assumed that the non-configurational part of a viscous liquid's entropy is equal to the entropy of its ordered crystal, S_{crys}, over the range of temperatures being studied. However, it was known as early as the Goldstein's analysis of entropy change in glasses below T_g that this is a poor approximation in many cases [404]. This study together with the amplification by Johari [405] has shown that the configurational part of the excess entropy S_{exc} may be as small as 40% of S_{exc} at T_g, and there is no obvious pattern for different kinds of liquids. This assumption together with other assumptions often used in applying the AG model predictions was proven incorrect by Johari in a series of papers [405–413].

The non-configurational part of a viscous liquid's entropy is the sum of the vibrational and anharmonic force contributions, S_{vib}, and any secondary relaxation contribution, and cannot be equated to S_{crys}. Thus, in general $S_c \neq S_{exc}$. However, if S_c were proportional to S_{exc} over the relevant temperature range, Eq. (2.24) is replaced by $\eta(T) = \eta_0 \exp[C'_{AG}/TS_{exc}(T)]$, with S_{exc} taking the place of S_{conf}, and the AG theory can still be tested directly against experiment. The assumption that S_c is proportional to S_{exc} was made [414] and the new equation used. This proportionality assumption has been questioned [409, 413] and shown quantitatively incorrect for 8 out of 10 standard glassformers [415]. Thus, in general S_c is not proportional to S_{exc}. Other tests of the relations based upon the Adam–Gibbs equation (2.23) were made by using newly obtained data of S_{exc} and heat capacity in some cases [416–419]. These studies have concluded that the AG entropy theory may not be valid.

Despite the difficulty in determining S_c by experiment, it can be estimated by subtracting S_{vib}, which is the sum of both harmonic phonons and anharmonic forces contributions, from the experimentally determined entropy. Yamamuro et al. [420] have calculated S_{vib} by using the lattice, the librational, and the internal modes in the Debye equation for C_V and then corrected the values for the volume change. Their estimate of S_c is more complete, but still approximate. The values of $z^*(T)$ were obtained by Eq. (2.22) with the approximate S_c calculated from the heat capacity measured by adiabatic calorimetry. $z^*(T)$ was determined from some high temperature down to below T_g for many glass-forming molecular liquids [420]. The glassformers include butyronitrile, isopropylbenzene, 1-propanol, 3-bromopentane, salol, 1-butene, ethylbenzene, 3-methylpentane, 2-methyltetrahydrofuran, propylene carbonate, 1-pentene, toluene, and *ortho*-terphenyl (Fig. 20). At $T = T_g$, the values of $z^*(T_g)$ fall within the range $3.5 < z^*(T_g) < 7.0$. Hence the length scale $L_{AG}(T_g)$ of AG from pure thermodynamic consideration is not large, and apparently significantly smaller than the length scale ξ_{het} determined from the heterogeneous nature of the dynamics by solid-state NMR experiments [226, 227, 421, 422] to be discussed in detail in a subsection to follow. The reported values of ξ_{het} [422] are $\xi_{het} = 3.7 \pm 1$ nm at $T = T_g + 9$ K for poly(vinyl acetate) (PVAc); $\xi_{het} = 2.2 - 2.9$ nm at $T = T_g + 9$ K for 1,2-diphenylbenzene (*ortho*-terphenyl); and $\xi_{het} = 1.3$ nm at

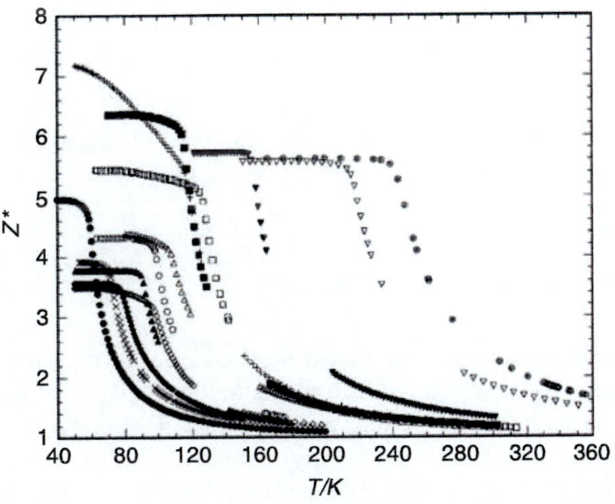

Fig. 20 Temperature dependence of the number of the molecules in CRR glass-forming molecular liquids. ○: butyronitrile, □: isopropylbenzene, ◇: 1-propanol, △: 3-bromopentane, ▽: salol, •: 1-butene, ■: ethylbenzene, ♦: 3-methylpentane, ▲: 2-methyltetrahydrofuran, ▼: propylene carbonate, ×: 1-pentene, +: toluene, ◎: *ortho*-terphenyl. Data taken from [420] and replotted

$T = T_g + 10$ K, $\xi_{het} = 1.1$ nm at $T = T_g + 14$ K, and $\xi_{het} = 1.0$ nm at $T = T_g + 18$ K for 1,2,3-propanetriol (glycerol). Since the length scale is expected to increase with decreasing temperature, the values of ξ_{het} at T_g would be even larger. Putting more faith in the length scale determined directly by NMR, the much smaller $L_{AG}(T_g)$ determined from S_c using the AG theory presents a problem.

The disparity can be explained as follows. $S_c(T)$ is a pure thermodynamic quantity and hence the AG theory at best can only account for the effect of configurational entropy on the relaxation rate, but not the effect from the heterogeneous many-body relaxation dynamics and its length scale, while the NMR measurement can. To explain the disparity, the consideration must include both effects. This was carried out by using the CM to account for the many-body dynamics as follows [401]. It is clear from the relation between τ_α and the primitive relaxation time τ_0 from Eq. (2.14) that the configurational entropy dependence of τ_α implies the same for τ_0 and vice versa. Chronologically, $S_c(T)$ actually first enters into the dynamics by making τ_0 to have the temperature dependence given by

$$\tau_0(T) = A_0 \exp[B_0/TS_c(T)]. \tag{2.26}$$

This dependence of τ_0 on configurational entropy can be rationalized by the same reasoning as in the original AG theory. From the CM relation (2.14), it follows that the dependence of $\tau_\alpha(T)$ on $S_c(T)$ is given by [401]

$$\begin{aligned}\tau_\alpha(T) = [t_c^{-n(T)}\tau_0(T)]^{1/(1-n(T))} &= \{t_c^{-n(T)} A_0 \exp[B_0/TS_c(T)]\}^{1/(1-n(T))} \\ &\propto \exp[B_0/T\hat{S}_c(T)],\end{aligned} \tag{2.27}$$

where

$$\hat{S}_c(T) = [1 - n(T)]S_c(T). \tag{2.28}$$

$\hat{S}_c(T)$ is smaller than $S_c(T)$, but $z^*(T)$ calculated from it by the CM-modified AG relation

$$z^*(T) = N_A s_c^*/\hat{S}_c(T) \tag{2.29}$$

would be larger than that from Eq. (2.22) of the original AG model by the factor $(1-n)^{-1}$, which helps to reconcile the disparity of length scales determined from the thermodynamic data and by NMR. Experimental evidence will be given later in Section 2.3.2, particularly Sections 2.3.2.19, 2.3.2.29, and 2.3.2.30 thereof, to support that τ_0 already depends on entropy and volume.

There is one problem in applying the original AG theory mentioned before which can be resolved by the modification discussed above. The problem is the deviation from the linear relation between $\log \tau_\alpha(T)$ and $1/[TS_c(T)]$ predicted by the Adam–Gibbs equation (2.24) and found in many glassformers [396–398, 401]. This problem does not appear in plotting $\log \tau_0(T)$ against $1/[TS_c(T)]$ [401], where $\tau_0(T)$ is calculated from the experimental data of $\tau_\alpha(T)$ and $n(T)$ by inverting the CM equation $\tau_\alpha(T) = [t_c^{-n(T)}\tau_0(T)]^{1/(1-n(T))}$. Figure 21 demonstrates this linear relation between $\log \tau_0(T)$ and $1/[TS_c(T)]$ by an example taken from salol [401]. While $\log \tau_\alpha(T)$ exhibits two very different linear dependences on $1/[TS_c(T)]$ separated by T_B, $\log \tau_0(T)$ is a linear function of $1/[TS_c(T)]$ in the whole temperature range within the error involved in calculating $\log \tau_0(T)$. The latter supports that the dependence of the dynamics on $S_c(T)$ starts at the primitive relaxation of the CM, and the dependence of its relaxation time $\tau_0(T)$ on $1/[TS_c(T)]$ may be described by Eq. (2.26). According to the CM, the structural relaxation time $\tau_\alpha(T)$ is determined by the slowing-down and stretching effects of many-body relaxation dynamics. Due to the more rapid increase of $n(T)$ with decreasing temperature on crossing the characteristic temperature T_B [423], the linear dependence of $\log \tau_0(T)$ on $1/[TS_c(T)]$ no longer holds for $\log \tau_\alpha(T)$.

Although for some of the glassformers shown in Fig. 20 $z^*(T_g)$ correlates with $n_\alpha(T_g)$, there are also some outright violations. For example, $z^*(T_g)$ of propylene carbonate and *ortho*-terphenyl are nearly the same, 5.49 and 5.46, respectively, but $n_\alpha(T_g)$ of propylene carbonate (= 0.27) is significantly smaller than *ortho*-terphenyl (=0.50) from dielectric relaxation measurements [237, 424]. The violations are not surprising if $n_\alpha(T_g)$ reflects many-body relaxation dynamics as suggested by the

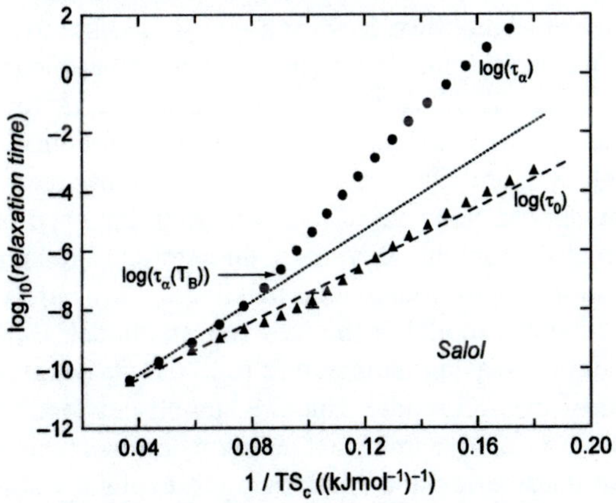

Fig. 21 Test of the Adam–Gibbs equation for the α-relaxation time $\tau_\alpha(T)$ of salol at temperatures above T_g. The data (*filled circles*) are from dielectric measurements. The *horizontal arrow* indicates the value of $\log(\tau_\alpha(T_B))$. The *filled triangles* are values of $\log \tau_0$ calculated from $\tau_\alpha(T)$ data according to the CM equation (2.14) with $n(T)$ obtained also from dielectric relaxation data. Replotted from data in [401]

evidences in the previous and later sections, while $z^*(T_g)$ or $L_{AG}(T_g)$ are constructs exclusively from thermodynamics.

Another attempt was made to calculate the values of $z^*(T_g)$, the size of the cooperatively rearranging region at T_g, for different liquids with the following assumptions [400, 413]. From the Vogel–Fulcher–Tammann–Hesse equation (2.25) for the dielectric relaxation time τ_α or for viscosity η, it is argued that $z^* = [1 - (T_0/T)]^{-1}$; the energy term, $\Delta\mu$, is equal to RB; and the pre-exponential term, τ_∞, of τ_α is equal to the pre-exponential term A_{VFTH}. For details see [400]. These $z^*(T_g)$ values of different liquids have been found to be in the range 2.4–14.3 [400]. The values $z^*(T_g)$ of some glassformers including propylene carbonate and *ortho*-terphenyl are about the same as those determined by Yamamuro et al. [420]. There are some that are difficult to accept. For example, tri-α-naphthylbenzene has the anomalously low value of $z^*(T_g) = 2.4$ while it is like *ortho*-terphenyl in many other respects of dynamic properties including the magnitude of $n_\alpha(T_g)$, which is about 0.50. For 1-butene, the value $z^*(T_g) = 10$ is much larger than the value of 4.3 obtained by Yamamuro et al.[420].

There are several experimental facts of glass transition that the Adam and Gibbs (AG) model either cannot explain or is contradicted. Some of these experimental facts have been given before [253, 401] and more will be given in the sections to follow. The problem of the AG model stems from its oversimplification of the many-body relaxation dynamics by the cooperative rearranging regions and their transition rate. At the time when Adam and Gibbs constructed their model more than 40 years ago, none of the experimental data that now give us better idea of the nature of the many-body relaxation dynamics were available to them. Considering this handicap, the AG model is a remarkable historic accomplishment, and Adam and Gibbs need not apologize for the shortcomings of their model. It is still much discussed, justifiably so because configurational entropy can be measured and must contribute in some manner to molecular mobility. However, one should not use the AG model in its original form or with minor modification to explain experimental data unless the fundamental problems of the 40-year-old theory have been rectified to incorporate many-body dynamics. Unfortunately, this practice still continues in the current literature of liquids, glasses, and glass transition. This current practice is indefensible because by now one should be aware of the overwhelming experimental evidences (the thrust of this treatise) showing that many-body effects are an important part of the structural relaxation and they cannot be handled by any purely thermodynamic formulation such as the AG model or the free volume model. Explaining the rapid increase of $\tau_\alpha(T)$ with decreasing temperature (i.e., the glass transition problem) is the focus and merit of the AG model. But, we should not forget that at any fixed temperature T, there are rich dynamics extending from microscopic times down to $\tau_\alpha(T)$ that are worth considering and challenging to explain. As we shall see later, the dynamics are fundamental and bear intriguing relations to the structural relaxation, which the AG model cannot handle. A solution of the glass transition problem is not complete unless it can explain these essential details of the dynamics at constant T. Put another way, if one has not fully understood the dynamics at a constant temperature, how can one have confidence to construct a theory designed principally to explain the temperature dependence of $\tau_\alpha(T)$?

2.2.2.3 Length Scale from the Thermodynamic Fluctuation Theory

Donth [302] basically accepted the notion of cooperatively rearranging regions (CRR) of Adam–Gibbs, but proposed a different method to calculate the size of cooperatively rearranging regions CRR of Adam and Gibbs, V_α, and the corresponding characteristic length $\xi_\alpha = (V_\alpha)^{1/3}$. The idea in Donth's theory was to use the relation between the mean square temperature fluctuation δT^2 of a CRR and N_a, the number of particles per CRR. The relation was originally proposed as $\delta T^2 = k_B T^2/N_a C$, where C is the heat capacity per particle and k_B is the Boltzmann constant [302]. In the modern version of the thermodynamic fluctuation theory [121, 425], the relevant quantities are given by

$$V_\alpha = \xi_\alpha^3 = k_B T^2[(1/C_V)^{\text{glass}} - (1/C_V)^{\text{liquid}}]/\rho\delta T^2 \quad \text{and} \quad N_\alpha = V_\alpha \rho/M_0, \tag{2.30}$$

where C_V is the specific heat at constant volume, M_0 is the molar mass of the fundamental unit (monomer in the case of polymer), ρ is the mass density, and $\delta T^2 \equiv (\delta T)^2$, where δT is taken as the width in temperature of the α-relaxation peak of $C''(\omega = \text{constant}, T)$, the imaginary part of the dynamics specific heat at constant frequency ω. The procedure to extract the relevant parameters from calorimetric data is given in detail in [425]. From the calorimetry data of many glassformers with different chemical and physical structures, the size of the CRR was determined by Eq. (2.30) and the values of ξ_α near T_g fall within the range of 1.0–3.6 nm. On close scrutiny, ξ_α does not seem to correlate with any other notable property such as non-exponentiality measured by the Kohlrausch exponent n, or the T_g-scaled temperature dependence of the α-relaxation time [112] quantified by the steepness or "fragility" index:

$$m = \left.\frac{\mathrm{d}\tau_\alpha(T_g/T)}{\mathrm{d}(T_g/T)}\right|_{T_g/T=1}. \tag{2.31}$$

For example, ξ_α of poly(methyl methacrylate) (PMMA) is 1.5 nm, which is significantly smaller than 2.9 nm, the ξ_α of glycerol [425, 296, 426], but both n and m of the former are much larger than the latter. Another example is the comparison of BMPC with OTP and polystyrene (PS). They all have the same $\xi_\alpha = 3.0$ nm, but the dielectric dispersion of BMPC is narrower than OTP and PS, and the T_g-scaled temperature dependence of the relaxation time of BMPC is much weaker than that of OTP and PS. The application of the theory to the family of poly(n-alkyl methacrylates) gave results on ξ_α that correlate with m. The higher members of the family with longer alkyl side chains have smaller ξ_α and correspondingly smaller m. However, the large width of the α-relaxation of higher members is contributed significantly by concentration fluctuations originating from internal plasticization by the long alkyl side chains. This contribution to the width conceptually has nothing to do with mean square temperature fluctuation of CRR, δT^2, but nevertheless is used in Eq. (2.30) to calculate ξ_α. Since only thermodynamic quantities are used to deduce ξ_α, the inconsistencies of the values obtained may be due to the likely scenario that many-body relaxation dynamics cannot be adequately described in terms of CRR, irrespective of whether ξ_α is calculated by the Adam–Gibbs method or by the Donth method. Support of this comes from dynamic properties (to be presented later) which cannot be explained by the thermodynamic fluctuation theory. In spite of the skepticism on the thermodynamic fluctuation theory, qualitatively its results including the onset of cooperativity at some temperature higher than T_g and the growing length scale with decreasing temperature are correct. Certainly, thermodynamics has a role in determining length scale. There is evidence of synergy between thermodynamics and many-body effects [253, 387], and this may explain why some of the results from Donth's theory are qualitatively correct.

2.2.2.4 Dynamic Heterogeneity and Its Length Scale

The Structural α-Relaxation Is Dynamically Heterogeneous

Anyone giving any thought to many-body relaxation dynamics in interacting systems (e.g., glass-forming substances) naturally would be led to the question: How do the fundamental units (molecules) move? It is not easy to answer this question for real systems and substances because it is nearly impossible to describe the movements of the large number of units, which is the Avogadro's number $N_{AV} = 6.022 \times 10^{23}$ molde^{-1} for molecular glassformers, altogether simultaneously and as a function of time. There was no experimental or computational technique that can do something toward this goal before 1991. Inability to answer this demanding but natural question may be the cause of lack of theoretical efforts to tackle directly the many-body relaxation problem. Asking anyone to describe the movements of a large number of interacting units in detail may be unrealistic, but some simpler questions on less specific properties of the motions in the many-body dynamics may have an answer. Simple questions coming to mind are the following. Are all units moving (relaxing) uniformly at the same time, and has each the same Kohlrausch time dependence? This scenario was referred to as homogeneous relaxation. If not, then necessarily there are some fast units which relax earlier and some slow units which relax later, and the structural relaxation is dynamically heterogeneous. But then, is there any interplay between the fast and the slow units? Glass-forming systems in general are amorphous, and the spatial and orientational relations between the structural units fluctuate. From this fact, intuition alone would be sufficient to tell us that it is highly unlikely that homogeneous relaxation can occur. In 1989, the approximate analogy of the CM's view on motion of molecules with the process in the solution of the "dining philosophers problem" in computer science [427] was pointed out [234] to gain insight into the dynamics of glassformers from the solution of the computer science problem. The dining philosophers problem is restated as follows. Imagine there are N philosophers sitting on chairs at the circumference of a round table and spend their entire time either eating or thinking (i.e., not eating). On the table, at the center is a bowl of noodles, but there is only one chopstick placed in between a neighboring pair of philosophers. To eat, any philosopher absolutely needs to have in possession two chopsticks at the same time. Each philosopher can only pick up the chopsticks on his immediate right or left. If a philosopher can pick up *both* chopsticks, he eats for a while. After a philosopher finishes eating, he puts down the chopsticks on the table and starts to think. However, at any time some philosophers may find either one or both chopsticks on his left and right in the possession of his neighbors. In that case, he keeps on thinking. The problem to solve is to find protocols for the philosophers to eat and think in such a manner that everyone eventually gets to eat. A symmetric and fully distributed solution to the dining philosophers problem was obtained by Lehmann and Rabin [428]. This solution guarantees with probability one that every philosopher gets to eat. The procedure of the solution can be described as follows [427]. When a philosopher wants to eat, he randomly chooses between the left and the right chopsticks with equal probability. Suppose his choice is the chopstick on his right. Now he waits for that chopstick to be available. When that chopstick is available, he picks it up. Next, he looks to his left. If the chopstick on the left is available, he picks it up, eats, and afterwards returns both chopsticks to their original places. If that chopstick is not available, he releases the first chopstick and starts the whole process over again. Demonstration of a simple Java solution to the dining philosophers problem can be found in the Internet. There are some "fast" philosophers who have eaten earlier and some "slow" philosophers who eat later, and they exchange roles later. The purpose of bringing up the solution of the dining philosophers problem [234] is not only to suggest the presence of fast and slow relaxing units and their interplay in structural relaxation in glassformers, but also to enunciate that heterogeneous dynamics is envisaged by the coupling model. The Kohlrausch function of the coupling model is a two-time correlation function. Like all the two-time and two-space point correlation functions discussed in Sections 1.2 and 1.3 above, the Kohlrausch correlation function provides no information on heterogeneous dynamics of glassformers because the slow and fast motions have been averaged out in the two-time correlation functions. Thus, there is no way to judge whether a model is advocating homogeneous or heterogeneously dynamics based on just the fact that the Kohlrausch function is invoked or employed. In particular, only averages over the dynamic heterogeneities are considered in the two-time correlation function in the usual coupling model equations or, by the same token, in the mode coupling equations. Besides, the CM abstains from giving a detailed description of the motions of the molecules, as is evident from its description given in Section 2.2.1.3.

It is unfortunate that in the 2000 review by Ediger [429], there is the following statement: "Ngai recently used the coupling model to provide an alternate, homogeneous explanation for experimental observations of enhanced translational diffusion (79a)." This statement effectively labels the coupling model as a model for homogeneous dynamics that contradicts experiments, in spite of my paper cited by him (reference 79a in Ediger's review, and [268] in the present treatise) explicitly making it clear that the dynamics in the coupling model is heterogeneous. The article on the analogy of the CM to the dining philosophers problem published in 1990 [234] was cited in [268] to make the point. The statement made by Ediger is plainly wrong. The amount of damage done by this statement on the CM although cannot be estimated but could be substantial because the 2000 review by Ediger has been cited already 680 times by others up to 2009.

There was a similar misunderstanding of the CM by Sillescu [430], but he corrected this mistaken view of the CM in a follow-up paper with others [431] by citing the CM exclusively by my 1990 paper on the dining philosophers problem [234].

The paper invoking the analogy to the dining philosophers problem [234] was published in 1990, a year before the seminal experimental work of Schmidt-Rohr and Spiess [226]. In stating this, the intent in no way is to take credit away from the contribution by Schmidt-Rohr and Spiess who had proven the dynamic are heterogeneous for the first time in 1991. Instead, the purpose is to let it be known that this aspect of the dynamics, considered before the actual experimental proof, is within the spirit of the coupling model.

The 4D-NMR experiment by Schmidt-Rohr and Spiess enabled them to examine multi-time correlation functions beyond the usual two-time correlation functions, which made possible to probe the complex dynamics of molecular reorientation. In the 4D-NMR experiment and additional experiments following it [227, 228, 235, 421, 422, 432, 435], one can select a subensemble of slow molecules that have not relaxed and subsequently monitor its relaxation on a different timescale. The heterogeneities are not static. Molecules slow at one time will exchange their dynamics with the fast ones to maintain ergodicity. In a temperature region slightly above T_g, the life-times of the heterogeneities, τ_{het}, are of the order of the α-relaxation time, τ_α, or the mean α-relaxation time, $< \tau_\alpha >$ [227, 431, 436, 437]. If the two-time correlation function is the Kohlrausch function (Eq. (2.22)), $< \tau_\alpha >$ is equal to $[\Gamma(1/(1-n))/(1-n)]\tau_\alpha$, where Γ stands for the gamma function.

Other experimental methods used to show the existence of dynamic heterogeneity in supercooled liquids include optical deep bleaching [438], solvation spectroscopy [231, 439], and polarization noise measurements [440]. The results from all these other methods confirmed that near the calorimetric glass transition temperature T_g the primary relaxation is dynamically heterogeneous. NMR and polarization noise measurements all found that the heterogeneity lifetime τ_{het} is of the order of τ_α in PVAc, PS, *ortho*-terphenyl, toluene, glycerol, and 2-methyl tetrahydrofuran (MTHF) [441] as shown in Fig. 22. However, optical deep bleaching [442] gave much larger values of the ratio τ_{het}/τ_α of about 10 at $T/T_g \approx 1.015$ that increase rapidly with decreasing T/T_g, reaching almost to about 1000 at $T/T_g \approx 1.0$ (see Fig. 22). Solvation experiments on MTHF [231, 439] concluded a long-lived heterogeneity, $\tau_{het}/\tau_\alpha \approx 10$, in the range of $T/T_g > 1.055$ at which NMR measurements on the same substance found $\tau_{het}/\tau_\alpha \approx 1$ (see Fig. 22). Since NMR measurement is more precise and applied to more glassformers, it is likely that the deduction of $\tau_{het}/\tau_\alpha \approx 10$ from solvation experiment has to be taken with a grain of salt. The glaring inconsistency of the lifetimes of the heterogeneities from optical deep bleaching and solvation experiments with NMR and polarization noise

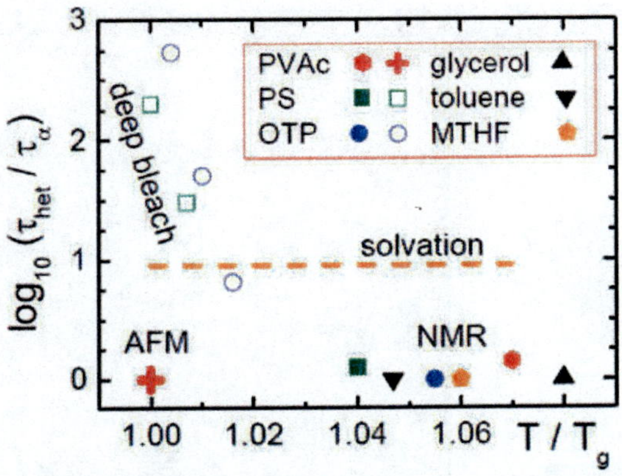

Fig. 22 Temperature-dependent ratio of typical heterogeneity lifetimes and of structural relaxation times, $\tau_{\rm het}/\tau_\alpha$, as deduced for various glassformers from NMR (*closed symbols*, the source of the data are given in [441]) and from polarization noise measurements (AFM, +, [440]), from optical deep bleaching (*open symbols*, [442]), and from solvation spectroscopy (*dashed line*, [39]). According to [12] the *dashed line* marks a lower bound. Data from [441] are replotted in *color*

measurements can be seen in Fig. 22. The deductions from optical deep bleaching is likely wrong, because if it were true then the ratio $\tau_{\rm het}/\tau_\alpha$ would increase by a factor of nearly 100 in the range $1.015 \geq T/T_{\rm g} \geq 1.0$, or about 1000 in the range $1.055 \geq T/T_{\rm g} \geq 1.0$ when NMR data are included. Such a large increase of $\tau_{\rm fhet}/\tau_\alpha$ would engender a corresponding large change in the dispersion of the structural α-relaxation. However, broadband dielectric relaxation of MTHF measured from 96.3 K (with $\tau_\alpha \approx 10^{-3}$ s down to 89.9 K (with $\tau_\alpha \approx 10^4$ s) shows little change of the α-relaxation spectrum of MTHF [441]. This observation can be used as evidence against the deduction of $\tau_{\rm het}/\tau_\alpha$ from optical deep bleaching experiments. There is another experimental fact that leads to the same conclusion. The study of tri-*m*-cresyl phosphate (TCP) in mixture with poly(methyl methacrylate) containing 50.1 wt% of TCP by one- and two-dimensional ^{31}P-NMR spectroscopy showed the TCP molecules reorient isotropically [443]. Two-dimensional spectra for TCP show the dynamics are heterogeneities and are transient in nature, i.e., $\tau_{\rm het}$ cannot be much longer than τ_α of TCP. This occurs for TCP even in a rigid polymer matrix at temperatures well below the glass transition of the plasticized PMMA.

Non-resonant hole burning spectroscopy is another technique that shows the existence of dynamic heterogeneity in supercooled liquids. This is a non-linear dielectric method which allows one to achieve a frequency-selective modification of the dielectric step-response function by the application of an intense electric field [444, 445].

Molecular dynamics simulations of the binary Lennard-Jones particles [321, 446, 447] and simulations of the dynamic lattice liquid [327, 328, 448] also show heterogeneous dynamics. So is the experimental technique of confocal microscopy which can monitor the motion of *all* the colloidal particles in suspension over several decades of time [141]. There are mobile, less mobile, and immobile particles. The

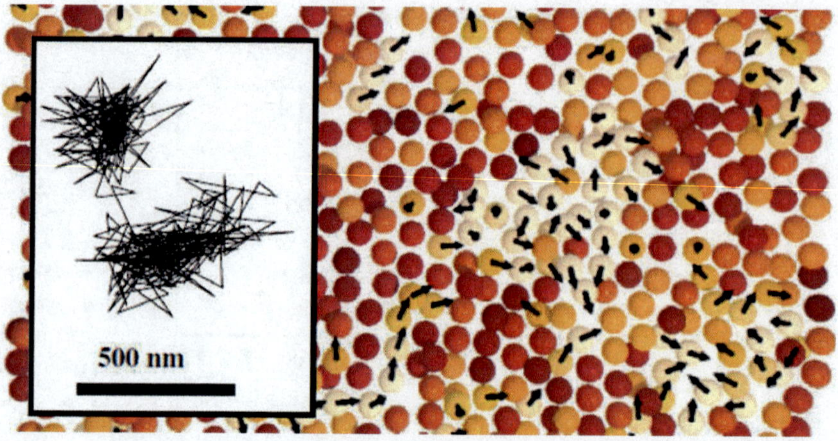

Fig. 23 A cut through a three-dimensional sample of colloidal suspension having $\phi = 0.52$, with *arrows* indicating the direction of motion for particles with displacements $\Delta r > 0.2\mu$m, at the cage rearranging time, $\Delta t^* = 600$ s, estimated by finding the maximum of the non-Gaussian parameter, $\alpha_2(\Delta t)$. The cut is 2.5 μm thick (~1 layer of particles). The *arrows* are all the same length in three dimensions, so shortened arrows indicate motion in or out of the picture. *Lighter colors* indicate particles with larger displacements. *Inset*: 120 min trajectory of a typical particle from this sample. It exhibits caged motion, with a sudden cage rearrangement which lasts ~600 s. Reproduced from [141(b)] by permission

mobile (or fast) particles were found to be strongly spatially correlated and exhibited large extended clusters. These features are illustrated in Fig. 23 and the result demonstrates that the structural relaxation in colloidal fluids occurs by means of cooperative particles motion.

All the above experimental studies have established as a fact that in general the structural relaxation is dynamically heterogeneous. However, one must not overestimate the degree of fundamental importance of dynamic heterogeneity. Like the dispersion of the correlation function (or the Kohlrausch fractional exponent) and other properties to be described later (such as the breakdown of Stokes–Einstein relation), dynamic heterogeneity is one of several notable consequences of the many-body relaxation. Each property is tantamount to just one glimpse into the multi-facets of the complex many-body dynamics, the details of which cannot be completely embodied by one property. All being *parallel* consequences, they are consistent with each other, but this does not mean that one can consider one of them more fundamental than the others and derive all the other properties from it. An example of this attempt (to be discussed in great detail in Section 2.2.5.6) is the use of dynamic heterogeneity (with additional assumption) to explain the breakdown of Stokes–Einstein relation [429, 449–454]. This explanation had been proven untenable by several experiments which showed its other consequence contradicts data [274, 275, 355, 455–458]. The situation has an analogy to the Indian fable often used in science to warn against jumping to conclusion based on limited or partial experimental facts. The fable describes several blind men each trying to identify an animal (which is an elephant) by his own touch. They touch different parts of the elephant and naturally they give different identifications of the animal but not as the elephant. Believing dynamic heterogeneity as most fundamental and identifying it as the cause of all other dynamic properties of glassformers is as wrong as one of the blind men who mistakenly identified the elephant as a mule by touching the tail of the elephant.

Controversy on Homogeneous or Heterogeneous Dynamics of the α-Relaxation of Polymers

In spite of the overwhelming evidences pointing to the dynamically heterogeneous nature of structural relaxation of glass-forming systems, there is still an opposite point of view from some workers studying dynamics using neutron scattering [459, 460] that might continue to influence others to do the same [461]. These authors [459, 460] professed that incoherent neutron scattering results agree well with a homogeneous scenario for the α-relaxation in glass-forming polymers, at least in the timescale covered by their neutron scattering measurements which is about a few nanoseconds. To them, homogeneous scenario is a process and I quote: "that all of the particles in the system relax identically but by an intrinsically non-exponential process." This drastic difference in opinion with others created a lasting controversy [462, 463] that continues on [235, 324, 464]. The homogeneous scenario was judged by others [235, 462] to be erroneous.

There are two sources of misunderstanding by the authors of [459,460] that may have led them to create the controversy. First, these authors defined heterogeneous dynamics specifically as when the correlation function is a "superposition of different simple exponential relaxations weighted by a broad distribution of relaxation times $g(\ln\tau)$," and they went further to state that "This picture is usually known as the 'heterogeneous' " [459, 460]. Their definition of heterogeneous dynamics in glass-forming liquids is at variance with that from other experiments because heterogeneities in supercooled liquids are transient in nature with lifetimes as found out by multidimensional NMR [227, 228, 235, 421, 422, 432, 435] and non-resonant dielectric hole-burning experiments [444, 445], and hence cannot be considered as a superposition of independent exponential relaxations with different relaxation times. At $T > T_g$, the dynamical heterogeneity of the α-relaxation probed by these experiments shows up in the manner that a dynamically distinguishable subensemble can be selected, and its return to the full equilibrium ensemble on a timescale of the order of the α-relaxation time τ_α has been subsequently monitored. Thus the supercooled liquid becomes ergodic when $t >> \tau_\alpha$.

Using their unconventional definition of heterogeneous dynamics, Arbe et al. [459, 460, 463] fitted the intermediate scattering function, $F_s(Q,t)$, observed by incoherent neutron scattering, by the Kohlrausch function

$$F_s(Q,t) = A\exp[-(t/\tau_\alpha(Q,T))^\beta], \tag{2.32}$$

where τ_α is the Q- and T-dependent relaxation time, $\beta < 1$ the stretching exponent, and $A = \exp(-\langle u^2 \rangle Q^2)$ the Lamb–Mössbauer factor to account for faster processes characterized by an effective mean-squared displacement $\langle u^2 \rangle$. Arbe et al. argued if a superposition of independent exponential relaxations were the origin of the Kohlrausch time dependence of $F_s(Q,t)$, then the Kohlrausch relaxation time τ_α would have the same Q^{-2}-dependence as each of the independent exponential relaxations. This was used by these authors to rule out heterogeneous relaxation (according to their definition as superposition of independent exponential relaxations) because from their own previous as well as subsequent incoherent neutron scattering studies of several polymers [307, 308, 310, 313, 314, 464] they found that the Q-dependence of τ_α, in the Q-range restricted to $Q < 1\text{Å}^{-1}$, can be approximately described by a power law determined by the stretching exponent β:

$$\tau_\alpha(Q) \propto Q^{-2/\beta}. \tag{2.33}$$

As polymers have β usually within the range 0.4–0.65, Eq. (2.33) predicts much stronger Q-dependences than the Q^{-2}-dependence of simple diffusion (see Eq. (1.14)). This way of ruling out heterogeneous dynamics is moot because the real heterogeneous dynamics in glass-forming liquids found by multidimensional NMR is *not* a superposition of independent exponential relaxations. Present are exchange processes between fast and slow particles or units. The question of heterogeneities or homogeneous dynamics cannot be settled by considering $F_s(Q,t)$ and the $Q^{-2/\beta}$-dependence of its correlation time. Moreover, dynamic light scattering in colloidal particles suspensions [143] also found the intermediate scattering function $F_s(Q,t)$ and relaxation time given by Eqs. (2.32) and (2.33), respectively, exactly the same as neutron scattering in polymers. The motions of all the particles seen by confocal microscopy [141] is without a doubt heterogeneous, and the results rule out homogeneous dynamics as would be expected by Arbe et al. from the $Q^{-2/\beta}$-dependence. Besides, even though the dynamics is spatially and dynamically heterogeneous, it is hard to believe that $F_s(Q,t)$ can be represented by a superposition of independent exponential relaxations according to the definition of Arbe et al.

Second, a special homogeneous scenario was arrived at by the authors of [464] by analyzing intermediate scattering function, $F_s(Q,t)$, of incoherent neutron scattering in glass-forming polymer melts. Results obtained were concluded by them to be only compatible with the homogeneous scenario. The arguments [460, 463] leading them to this conclusion go as follows. The t-dependence of $F_s(Q,t)$ given in Eq. (2.32) and the Q-dependence of τ_α therein given by Eq. (2.33) are results from neutron scattering experiments. Substituting Eq. (2.33) into Eq. (2.32), they [459, 464] obtained

$$F_s(Q,t) = A\exp[-Q^2\langle r^2(t)\rangle/6] = A\exp[-Q^2 D(T)t^\beta], \tag{2.34}$$

where $D(T)$ governs the T-dependence of τ_α. From this equation, they concluded that the increase of the mean-squared displacement associated with the α-relaxation is sublinear in time, i.e., $\langle r^2(T)\rangle \sim t^\beta$ with $\beta < 1$. This interpretation of neutron scattering data was used by them to support their homogeneous α-relaxation scenario. The controversy was created in several statements made by the authors in [459] that include the following: "In this Letter we show that incoherent neutron scattering results agree rather well with a homogeneous scenario for the α-relaxation in glass-forming polymers, at least in the time scale covered by neutron scattering techniques" and "From the Q dependent KWW relaxation times τ_W, it could be demonstrated that the apparent stretching of the α-relaxation function relates dominantly to sublinear diffusion and is not a result of heterogeneities in the material. Although this result was obtained at temperatures well above the glass transition, the almost temperature independent stretching exponents suggest that this behavior also prevails close to T_g." It is important to note that the authors did not give the origin of the sublinear diffusion, but nevertheless invoke it as the hallmark for homogeneous relaxation. This is not necessarily true. Counter examples can be found from the molecular dynamics simulation data of binary Lennard-Jones particles [321]. There is sublinear diffusion in the LJ liquid like that found in polymers by neutron scattering, but the dynamics are dynamically heterogeneous as illustrated by motions of the particles. Another counter example is the colloidal particles in suspension monitored by confocal microscopy, where there is sublinear diffusion but the motions of all particles are dynamically heterogeneous [141]. Hence, one cannot conclude, based on the sublinear diffusion alone, that the dynamics is homogeneous [463]. As will be shown in Chapter 3 on universal relaxation/diffusion properties of interacting systems, subdiffusion is found by neutron scattering in the center-of-mass diffusion of entangled linear polymer chains [465] and by molecular dynamic simulations in diffusion of ions in glass [149, 466–472]. In all these cited cases, the motions have been shown to be dynamically heterogeneous.

We have shown in the above that the controversy was created [459] first by the wrong definition of dynamic heterogeneities and second by inference [464] from the sublinear time dependence of the mean-squared displacement obtained by substituting the experimentally observed $Q^{-2/\beta}$-dependence of the correlation time into the intermediate scattering function $F_s(Q,t)$ having the Kohlrausch time dependence. The second step involves only experimental facts and no theory was introduced to support the claim that the resulting $\langle r^2(t)\rangle \sim t^\beta$ with $\beta < 1$ corresponds to homogeneous dynamics. Actually, there is a way to interpret neutron scattering data that avoids running into the controversy with others who have already shown the α-relaxation, having the Kohlrausch time dependence (Eq. (2.32)) for its correlation function, is heterogeneous by NMR and dielectric experiments [227, 228, 431, 444]. The way is to use the coupling model (CM), its concept, and predictions, and it confers the bonus of having the $Q^{-2/\beta}$-dependence of τ_α derived. As we recall, the CM has professed that the many-body relaxation dynamics is heterogeneous [234]. For α-relaxation having the Kohlrausch form $\varphi(t) = \exp[-(t/\tau_\alpha)^{1-n}]$, for its correlation function such as $F_s(Q,t)$ from neutron scattering given by Eq. (2.32), the exponent n is the indicator of many-body relaxation. The CM also has predicted a relation between τ_α and the primitive relaxation time τ_0 by the quantitative relation now rewritten to show the dependences on various variables U including Q:

$$\tau_\alpha(T,P,\cdot\cdot,U,\cdot\cdot,Q) = [t_c^{-n}\tau_0(T,P,\cdot\cdot,U,\cdot\cdot,Q)]^{1/(1-n)}. \quad (2.35)$$

This relation was the consequence of the crossover from independent relaxation to heterogeneous cooperative many-body relaxation at a temperature-insensitive time t_c predicted by the CM and actually observed at $t_c \approx 2$ ps by neutron scattering experiments [307, 308, 310, 312–314] and molecular dynamics simulations [258, 315, 316, 324–326]. From Eq. (2.35), dependences of τ_0 on any variable U, including temperature T, pressure P, and scattering vector Q shown explicitly, will all be modified by the power $1/(1-n)$, to stronger dependences. Since τ_0 of independent relaxation has the Q^{-2}-dependence, it follows from this equation that τ_α has the $Q^{-2/(1-n)}$-dependence (Eq. (2.33)) as observed by experiments. We shall see more of the utility and versatility of Eq. (2.35) in explaining other properties of the α-relaxation of glassformers as well as in other complex systems. We also hasten to add that the primitive relaxation time τ_0 calculated from the known values of t_c, n, and τ_α by the counterpart of Eq. (2.14), $\tau_0(T,P) = (t_c)^n[\tau_\alpha(T,P)]^{1-n}$, is in order of magnitude agreement with the relaxation time of an observed secondary relaxation with properties showing connection to the α-relaxation. This remarkable fact about the primitive relaxation time is the subject of extensive discussion in Section 2.3.

A new twist in the interpretation of $F_s(Q,t)$ obtained by neutron scattering and simulation was given later [324, 464, 473]. In these references, the Gaussian or non-Gaussian nature of the space distribution was used to characterize the dynamics. Gaussianity was either interpreted from the Q-dependence of $F_s(Q,t)$ of neutron scattering experiments or from the non-Gaussian parameter $\alpha_2(t)$ obtained by molecular dynamics simulation. As we may recall, $F_s(Q,t)$ is related to the self-part of the van Hove function $G_s(\vec{r},t)$ which give the space and time-dependent distribution function (see Eqs. (1.111)—(1.115)), and the non-Gaussian parameter $\alpha_2(t)$ is defined by Eq. (1.125). From the Q^2-term appearing explicitly in Eq. (2.34), the authors of [464, 473, 474] concluded that the self-part of the van Hove correlation function is a Gaussian function. For larger Q, $F_s(Q,t)$ no longer has the dependence on Q as appearing in Eq. (2.34), and this regime was considered to be non-Gaussian. These authors interpreted the data as a crossover from a Gaussian regime of sublinear diffusion at longer distances to a strongly non-Gaussian regime at short distances. Thus they reaffirmed the idea of identifying Kohlrausch time dependence of $F_s(Q,t)$ in the form of Eq. (2.34) as Gaussian and dynamically homogeneous behavior [324, 464]. They went on to say that their proposed crossover from Gaussian to non-Gaussian "... could be understood as a homogeneous to heterogeneous crossover of the incoherent dynamics involved in the α-relaxation." This understanding shows they maintain their view that $F_s(Q,t)$ having the Kohlrausch time dependence of $F_s(Q,t)$ in the form of Eq. (2.34) indicates homogeneous dynamics, contradicting the view of others and the CM interpretation. Nevertheless, they constructed an anomalous jump diffusion model with a distribution of jump lengths to explain the so-called Gaussian-to-non-Gaussian crossover. Actually, there is no need to invoke the new model to explain the crossover because the data are fully consistent with the transient nature and the finite life-time of dynamic heterogeneities of the α-relaxation. It was pointed out [468] that $\alpha_2(t)$ as a function of time and the relation between $\alpha_2(t)$ and $< r^2(t) >$ are similar in three different interacting systems, binary Lennard-Jones particles [321], the colloidal particles [141], and ions in oxide glasses [468]. In these systems, $\alpha_2(t)$ starts from zero at short times, increases with increasing time to reach a maximum, and finally goes to zero at long times. As T decreases (or as volume fraction increases in the case of colloidal particles), the position of the maximum $t_{\max}$ shifts toward longer times, and the height of the maximum $r^2(t)$ increases. For all temperatures or volume fractions, whichever is appropriate, the maximum occurs at time $t_{\max}$, which is after the plateau of $< r^2(t) >$ signifying caged dynamics [468, 475] has ended by primitive relaxations and near the time when the sublinear diffusion, $\langle r^2 \rangle \sim t^{\beta}$ with $\beta < 1$, starts (called t_{x2} in [468]). The decrease of $\alpha_2(t)$ down to zero at times longer than $t_{\max}$ is simply the consequence of a finite life-time, τ_{het}, of the heterogeneities. As time increases to approach τ_{het}, there are increasing exchanges between the fast and the slow units. Heterogeneity and $\alpha_2(t)$ decrease together. At even longer times when $t > \tau_{\text{het}}$, it is no longer possible to distinguish fast and slow particles; heterogeneity as well as $\alpha_2(t)$ vanishes. In other words, the systems becomes ergodic and the spatial distribution is Gaussian. The decrease of $\alpha_2(t)$ down to zero at times shorter than $t_{\max}$ is because the shorter is the time in the plateau regime, the more likely are the particles being caged.

The features described above can be found also in the $\alpha_2(t)$ and $\langle r^2 \rangle$ obtained for the polymer, polyisoprene, by molecular dynamics (MD) simulations (see Fig. 3 of [324]), although the terminal steady-state diffusion regime, $\langle r^2 \rangle \sim t$, is absent because the calculated $\langle r^2 \rangle$ is for a repeat unit of a long chain in a polymer, which does not translate to long distances and contribute to viscosity. The advantage that neutron scattering and MD simulation have on some other experimental techniques is the extra Q-dependence, which is evident from the form of intermediate scattering function in Eq. (2.34). From this expression, it is clear that a smaller Q means a longer relaxation time $\tau_{\alpha}(Q)$ for $F_s(Q, t)$. In discussing the data here, let us start at some Q_c such that $\tau_{\alpha}(Q_c) \geq \tau_{\text{het}}$. Under this condition and at times longer than $\tau_{\alpha}(Q_c)$, the exchange processes between slow and fast units are nearly complete, dynamic heterogeneity is no longer important, and the space distribution is hence Gaussian. On decreasing Q beyond Q_c, $\tau_{\alpha}(Q)$ increases further past $\tau_{\alpha}(Q_c)$; for times longer than $\tau_{\alpha}(Q)$, $\alpha_2(t \geq \tau_{\alpha}(Q))$ decreases down to near zero value (see Fig. 6 of [324]) and the spatial distribution is Gaussian. Also for the same times, $t > \tau_{\alpha}(Q)$, the system is ergodic. However, for times shorter than and not much longer than $\tau_{\alpha}(Q)$, the dynamics of the α-relaxation described by $F_s(Q, t)$ having the Kohlrausch time dependence in Eq. (2.34) is dynamically heterogeneous, and α_2 is non-zero. Therefore, in the region, $Q < Q_c$, the system is ergodic and the spatial distribution is Gaussian only for $t > \tau_{\alpha}(Q)$, but otherwise for $t \leq \tau_{\alpha}(Q)$ the dynamics of $F_s(Q, t)$ is heterogeneous, contradicting the opposite conclusion of homogeneous dynamics by authors of [324]. Like discussed before, this contradiction is artificial and can be avoided by using the CM, and the $Q^{-2/\beta}$-dependence of τ_{α} found for $Q < Q_c$

can be obtained via Eq. (2.35) without contradicting the heterogeneous dynamics of $F_s(Q,t)$.

The MD simulation at 363 K [324] found that Q_c is slightly smaller than Q_{max}, where Q_{max} is the value of Q at the first maximum of the static structure factor. Hence, when Q is increased to exceed Q_c, the length scale probed is small and $F_s(Q,t)$ is contributed by local and non-cooperative relaxation processes all having Q^{-2}-dependence for their relaxation times. This is supported by $\tau_\alpha(Q)$ becoming comparable and less than the onset time of the $\langle r^2\rangle \sim t^\beta$ with $\beta = 0.4$ which has the same order of magnitude (see Fig. 3 of [324]) as $\tau_\alpha(Q_c) \approx 30$ ps (see Fig. 6 of [324]). The observed crossover of the Q-dependence is thus explained. The maximum of α_2 is located at $t^* = 4$ ps. Its width at half-maximum encompasses $\tau_\alpha(Q)$ for all Q larger than Q_c reported [324], and hence the spatial distribution is non-Gaussian.

In the above, we have brought into consideration the relation of the lifetime of dynamic heterogeneity to $\tau_\alpha(Q)$, the very local length scales probed by $F_s(Q,t)$ when $Q > Q_c$, and the derivation of the $Q^{-2/\beta}$-dependence of τ_α for $Q < Q_c$ by the CM. These elements are necessary for interpreting neutron scattering and simulation data, and the crossover of the $Q^{-2/\beta}$-dependence of τ_α for $Q < Q_c$ to a weaker Q-dependence for $Q > Q_c$ can be explained without contradicting dynamically heterogeneous nature of the many-body α-relaxation for $Q < Q_c$. Alone, $F_s(Q,t)$ and its Q-dependences from neutron scattering and simulation cannot unravel the nature of the dynamics as attempted in [324]. Higher order space and time correlation functions than $F_s(Q,t)$ are needed for this purpose, and they have been given by multidimensional NMR measurements.

Length Scale of Dynamic Heterogeneity

As mentioned before, the length scale of the dynamic heterogeneity of several archetypal glassformers had been measured by an advanced solid-state NMR experiment [421]. This multidimensional NMR experiment of Tracht et al. makes possible selection of a slow subensemble which has distinct ^{13}C-NMR characteristics. Then, the magnetization of this subensemble is allowed to diffuse. After various times, ^{13}C spins are interrogated to determine if they are still the slow units. At sufficiently long times, the ^{13}C spins are found on sites whose dynamics are characteristic of the entire ensemble. From the time required to leave the slow subensemble, the heterogeneity length scale, ξ_{het}, can be calculated using an independent measurement of the spin diffusion coefficient. The most recent reported values of ξ_{het} [422] are $\xi_{het} = 3.7\pm1$ nm at $T = T_g+9$ K for poly(vinyl acetate) (PVAc); $\xi_{het} = 2.2-2.9$ nm at $T = T_g + 9$ K for 1,2-diphenylbenzene (*ortho*-terphenyl); and $\xi_{het} = 1.3$ nm at $T = T_g+10$ K, $\xi_{het} = 1.1$ nm at $T = T_g+14$ K, and $\xi_{het} = 1.0$ nm at $T = T_g+18$ K for 1,2,3-propanetriol (glycerol). The numbers of molecules per slow domain with length scale ξ_{het} are estimated to be 390 monomer units of PVAc, 76 molecules of OTP, and 10 molecules of glycerol. Due to the sophistication of the technique and the need of ^{13}C labeling of the material, no more measurements of ξ_{het} for other glassformers were made except for D-sorbitol [476]. At 275 K(T_g + 7 K), the value of ξ_{het} of sorbitol was reported to be 2.5 ± 1.2 nm.

Correlation of ξ_{het} with n

Although we have data of ξ_{het} for only four glassformers, they are worth further consideration since they were obtained directly from experiments. Simple reasoning in several possible ways can lead one to see correlation between ξ_{het} and the width of the dispersion of the structural relaxation. In [476] it was argued that, in a system with a smaller ξ_{het}, particles might move with a lesser spread of mobility and hence narrow distribution of relaxation rates than particles in a system with large ξ_{het}. Such argument must be right because the width of the dispersion and the size of ξ_{het} are both parallel consequences of the many-body relaxation dynamics, and hence the two quantities must be correlated. In fact, an attempt was made to test the correlation between n and ξ_{het} (actually the equivalent anti-correlation between the Kohlrausch exponent, $\beta_K \equiv 1-n$, and ξ_{het}) [476] with data of PVAc, OTP, sorbitol, and glycerol. The correlation between n and ξ_{het} is not perfect. The bad actor is sorbitol which has the smallest $\beta_K = 0.41$ or the largest $n = 0.59$ [476], but not the largest ξ_{het}. The problem of this test is that the values of β_K of PVAc, OTP, sorbitol, and glycerol were taken from different sources where different criteria were used to analyze the dispersions. The only objective way to carry out the test is to obtain β_K or n values for all four glassformers using dispersions all obtained by the same technique and the same criterion in fitting the data by the Kohlrausch functions. When fitting the frequency dependence of the α-loss peaks obtained by dielectric relaxation by the one-sided Fourier transform of the Kohlrausch function, one criterion used is the emphasis of good agreement with the loss data on the main peak especially the low-frequency side, if no significant conductivity contribution is present there or it has been removed if present. Deviations of the Kohlrausch fit to the data invariably occur at frequencies sufficiently high above the loss maximum. The deviations are considered natural in the coupling model interpretation of the evolution of dynamics with time as explained in [477] and practiced in [195] and other works. They come from processes of smaller length scales that transpire at shorter times before the dynamics evolves to the one with maximum length scale and correlation function given by the Kohlrausch function. This criterion had been applied uniformly to all four glassformers in previous works. The n value in decreasing order is 0.53–0.57 for PVAc [238, 267, 478], 0.52 for sorbitol [195], 0.50 for OTP [237], and 0.29 for glycerol [195]. These systematically and consistently determined values of n are well correlated with the ξ_{het} values of the four glassformers.

An attempt was made to correlate the steepness or the "fragility" index m with ξ_{het} for these four glassformers. No correlation between ξ_{het} and m was found. As will be discussed in greater details later, τ_α is determined not only by the effects of slowing down by many-body relaxation but also by thermodynamic factors including specific volume and entropy. The same can be said about m since it is the T_g-scaled temperature dependence of τ_α. Furthermore, glycerol and sorbitol belong to the same family of polyols, with hydrogen bonding; *ortho*-terphenyl has benzene rings and is a van der Waals liquid; and PVAc is a polymer. Differences in chemical bond and structure of these four glassformers cannot guarantee similar changes of specific volume and entropy with T_g-scaled temperature. All the different factors contributing to m will frustrate the relation between it and other more fundamental quantities such as ξ_{het} in the present case, and correlation cannot survive. We shall see later other examples of failure of correlations of m with other quantities including n. Nevertheless, if restricted to the same chemical family of glassformers to minimize the influence of extraneous factors, some correlations of m with other quantities can be restored [112, 479, 480]. Examples include restriction to carbon backbone polymers [479] and to Ge–As–Se glass-forming chalcogenides [480].

2.2.2.5 Length Scale from Relaxation Behavior of Nanophase-Separated Side-Chain Polymers

Nanophase separation of incompatible main and side-chain parts is a general phenomenon in amorphous side-chain polymers with long alkyl groups. This property was shown [297, 298] by X-ray scattering and relaxation spectroscopy data acquired from higher members of the poly(n-alkyl acrylates) (PnAA) and poly(n-alkyl methacrylates) (PnAMA) that alkyl groups aggregate in the melt and form self-assembled alkyl nanodomains with a typical size d of 0.5–2 nm. Such sizes have been deduced from Bragg spacings d_{II} obtained from prepeaks found by X-ray scattering.

This is shown in Fig. 24 together with a comparison with some data for other polymer series having alkyl groups, including poly(di-n-alkyl itaconates), hairy rod polyimides, and poly(alkylbenzimidazol-*alt*-thiophene). The data of Beiner and Huth reveal that the alkyl nanodomain size and the dynamics are mainly determined by the number, C, of alkyl carbon atoms per side chain and depends slightly on the microstructure of the main chain. A polyethylene-like glass transition driven by a process called α_{PE} within the alkyl nanodomains has been observed by dielectric, mechanical, and dynamic heat capacity spectroscopy [297, 298]. The results described for different series of side-chain polymers support that α_{PE} is a general phenomenon. The angular relaxation frequencies $1/\tau_{C,PE}$ of α_{PE} are shown as a function of $1000/T$ in Fig. 25 for three higher PnAA with C=4 (filled triangles), C=8 (filled diamonds), and C=10 (filled circles), together with the VFTH fits to $1/\tau_{8,PE}$ and $\tau_{10,PE}$ for C=8 and C=10, respectively, and an Arrhenius fit to $1/\tau_{4,PE}$ for C=4 [481]. The open symbols on the VFTH fits mark the temperatures at which $\tau_{8,PE}$ and $\tau_{10,PE}$ have reached 1 s. The nanodomain size d serves as an upper bound of the heterogeneity length scale, ξ_{het}, of α_{PE}. Hence the observed change of dynamics of α_{PE} with C provides the corresponding dependence of the dynamics on ξ_{het}. For example, increasingly weaker T-dependence of $1/\tau_{C,PE}$ with decreasing C or d is evident. The fragility $m(C)$ calculated from $1/\tau_{C,PE}$ with T_g defined by $\tau_{C,PE} = 1$ s for all $C \leq 10$ show correspondingly monotonic decrease. This trend can be seen from the values of m in the inset of Fig. 25. For C=4 with the small alkyl nanodomain size ≈ 0.5 nm, its α_{PE} processes have properties different from that of the higher PnAA and PnAMA members. The α_{PE} process in C=4 (PnBMA

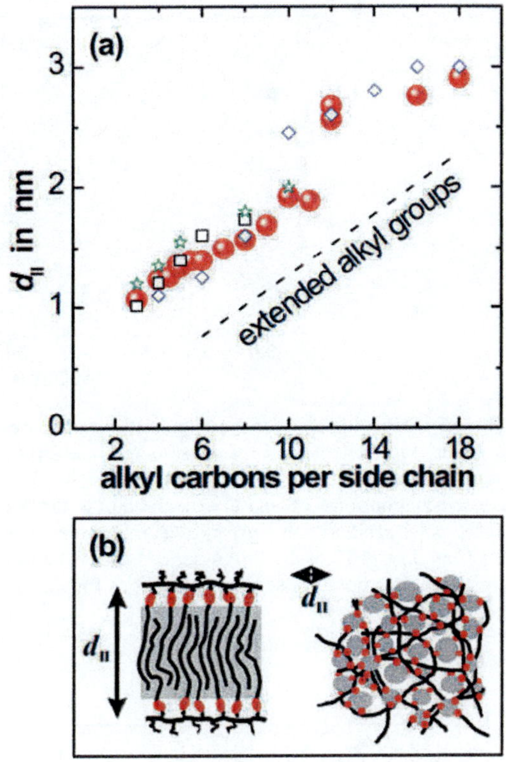

Fig. 24 (**a**) Equivalent Bragg spacings d_{II} as obtained from prepeaks for methacrylates (*circles*), acrylates (*squares*), itaconates (*stars*), and hairy rod polyimides (*diamonds*). The lengths of extended alkyl groups are given for comparison. (**b**) Schematic pictures for the situation in the nanophase-separated melt. A local 1D picture (*left*) and a global 3D view (*right*) are compared. The *light grey regions* are the alkyl nanodomains and the *small dark ellipses* represent the carboxyl groups. Reproduced from [297] by permission

or P*n*BA) has small intensity; its relaxation time, $\tau_{4,\mathrm{PE}}$, has nearly Arrhenius temperature dependence; and its steepness (fragility) index m is about 14, close to the minimum value of fragility. The Arrhenius fit to the $\tau_{4,\mathrm{PE}}$ data of C=4 in Fig. 25 is given by the expression $\tau_{4,\mathrm{PE}} \approx 10^{-13.8}\exp(40.3\ \mathrm{kJ}/RT)$. These features all seem to indicate that cooperativity is close to being totally removed in the α_{PE} process of C=4.

The α_{PE} of P*n*AMA exhibits the same property as shown for P*n*AA as shown in Fig. 26. The polyethylene-like glass transition in self-assembled alkyl nanodomains is perhaps the best case for experimental support of the correlation of fragility with length scale. Nothing is changed except for the size of the alkyl nanodomains. For C=4, the Arrhenius T-dependence of $1/\tau_{4,\mathrm{PE}}$ together with the near minimum value of m reached indicates the total absence of cooperativity when the alkyl nanodomain size is less than 0.5 nm. This estimate of the minimum size for cooperativity applies to the polyethylene-like dynamics and may not apply to other glassformers because of the difference in intermolecular interaction/constraint. The absence of cooperativity in α_{PE} of C=4 means it is no different from a secondary β-relaxation of the Johari–Goldstein (JG) kind [481]. In other words, the observed α_{PE} process of C=4 may appropriately be identified as the β_{PE} process. This indicates that

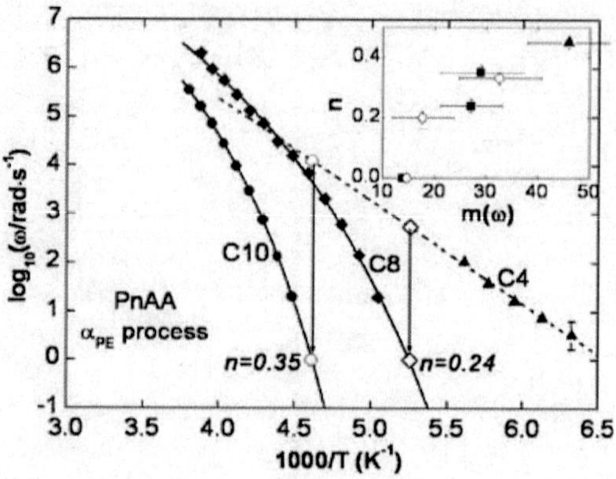

Fig. 25 Arrhenius plot for the angular relaxation frequencies, $\omega \equiv 1/\tau_{C,PE}$, of α_{PE} processes in higher P*n*AA with *C*=4 (*filled triangles*), *C*=8 (*filled diamonds*), and *C*=10 (*filled circles*) as obtained from dielectric spectroscopy [297]. The *lines* are fits to the Vogel–Fulcher–Tammann–Hesse equation for *C*=8 and *C*=10, and a fit to the Arrhenius equation for *C*=4. For explanation of the *other symbols* and *vertical arrows*, see text. The *inset* is the cross-plot of n_C against *m*(*C*) for *C*=4, 8, and 10 of the P*n*AA series (*squares*) and *C*=4, 7, and 10 of the P*n*AMA series (*circles*). The data point (*m*,*n*) from amorphous PE is included (*triangle*). Reproduced from [481] by permission

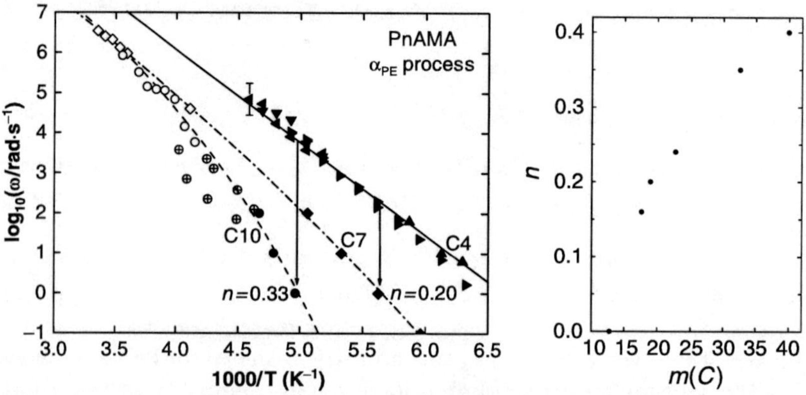

Fig. 26 (*Left*) Arrhenius plot for the relaxation times $\tau_{C,PE}$ of α_{PE} processes in higher P*n*AMA with *C*=4 (*triangles*), *C*=7 (*diamonds*), and *C*=10 (*circles*). Data from dielectric spectroscopy (*full symbols*), mechanical spectroscopy (*open symbols*), and heat capacity spectroscopy (⊕) are shown. The *lines* are fits to the Vogel–Fulcher–Tammann–Hesse equation for *C*=7 and *C*=10, and a fit to the Arrhenius equation for *C*=4. For explanation of the *vertical arrows*, see text. Plotted also are the JG β-relaxation times of bulk P*n*BMA (*C*=4) samples with different degrees of polymerization of 1000 (◀), 52 (▼), and 18 (▶). The figure on the right is the cross-plot of n_C against *m*(*C*) for various *C* values of the P*n*AA series and the P*n*AMA series. Reproduced from [481] by permission

for polyethylene-like dynamics, the length scale of the α-process must be larger than the small alkyl nanodomain size ≈ 0.5 nm of C=4 in order to have usual characteristics of the usual cooperative structural α-relaxation such as VFTH temperature dependence and heat capacity response at glass transition. For C>4, α_{PE} has these characteristics because the size of the nanodomain is larger than the minimum requirement of 0.5 nm. The higher glass transition temperature and larger m_C that the nanodomain of larger C has are a reflection of the increase of the length scale with increasing C up to the maximum value in bulk PE. The alkyl nanodomain size ≈ 1.5 nm of C=10 poly(n-decyl methacrylate) (PnDMA) could be used as a lower bound of the maximum length scale of the α-relaxation in PE.

The universality and fundamental importance of the JG secondary relaxation in glass-forming systems will be discussed in Section 2.3. Moreover, there is a relation between JG β-relaxation time and the α-relaxation time via the coupling or non-exponentiality parameter of the α-relaxation afforded by the CM. Applied here to $\tau_{C,JG}$ (or its approximant $\tau_{4,PE}$) and $\tau_{C,PE}$, the relation enables the coupling parameter n_C of the α_{PE} process to be calculated for $C > 4$ [481], and the results are plotted against m_C in the insets of Figs. 25 and 26. These plots serve to show the correlation between n_C and m_C, as well as the length scale of the α_{PE} process. This is because all three quantities increase monotonically with C.

In calculating n of the α_{PE} process for $C > 4$ in the above, we have assumed that the observed $\alpha_{PE} \equiv \beta_{PE}$ process of C=4 can be identified as the JG β-process for $C > 4$ and the universal existence of the JG β-process for all glassformers including alkyl nanodomains of $C > 4$. Although not resolved by normal cooling rates, the JG β-relaxation associated with α_{PE} for C>4 should exist if the JG β-relaxation in glass-forming systems is indeed universal. Also its relaxation time $\tau_{C,JG}$ should be nearly the same as $\tau_{4,PE}$ since α_{PE} for C=4 is non-cooperative and identifiable as the β_{PE} process [481]. These expectations were subsequently verified experimentally by Beiner and coworker by rapidly quenching C=10 poly(n-decyl methacrylate) (PnDMA) and C=7 poly(n-heptyl methacrylate) [482]. For rapidly quenched samples, sometimes one or two additional peaks were detected on the high-frequency flank of the α_{PE} process in the frequency range where the $\alpha_{PE} \equiv \beta_{PE}$ process is observed for the slowly cooled C=4 side-chain polymers with short alkyl groups and small alkyl nanodomains. Shown in the left panel of Fig. 27 is the case where a small peak appears on the high-frequency side of the α_{PE} process after rapid quenching (1st) the sample. This peak disappears after periods (2nd and 3rd) of physical aging slightly below T_g, the conventional glass transition temperature of the a process (not the α_{PE} process). The two peak maxima frequencies from dielectric measurements at different temperatures on quenched samples (1st sweeps) are shown in the right panel of Fig. 27. The frequencies of the two additional peaks in the quenched C=10 samples have Arrhenius-like temperature dependence. They are higher than the relaxation frequencies of the conventional structural relaxation process a and the polyethylene-like structural relaxation process α_{PE} (open symbols) in C=10 (PnDMA). Interestingly they are in good agreement with the frequencies of the $\alpha_{PE} \equiv \beta_{PE}$ process in C=4 (PnBMA) shown in Fig. 26 and represented in Fig. 27 (right panel) by the thick dashed-dotted line. The observation

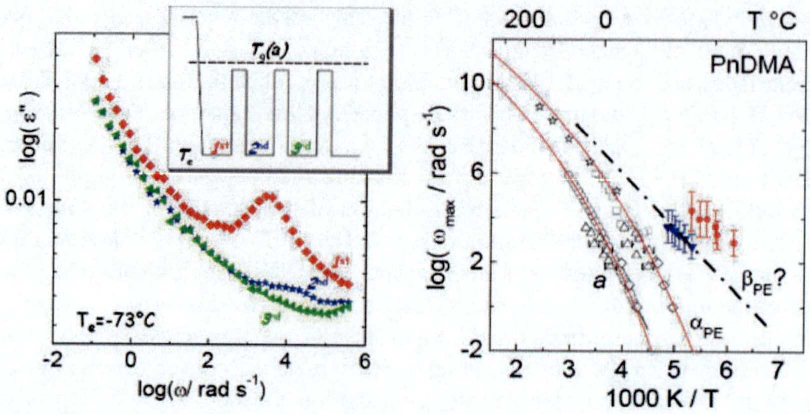

Fig. 27 (*Left*) Dielectric loss ε'' vs. logarithm of the angular frequency ω measured for C=10 poly(*n*-decyl methacrylate) (P*n*DMA) at –80°C after rapid quench without aging (1st sweep), and after physical aging of the quenched sample for 20 min (2nd sweep), and for 40 min (3rd sweep) at a temperature $T_e = -73°C$ slightly below the glass temperature [482]. (*Right*) Arrhenius plots of the two peak maxima frequencies from dielectric measurements on quenched samples (1st sweeps) are indicated by *closed triangles* and *circles*. The relaxation frequencies for the conventional glass transition *a* (*open symbols*) and for the polyethylene-like glass transition α_{PE} (*open symbols*) in C=10 (P*n*DMA) are shown for comparison. Data of *a* and α_{PE} for C=10 from dielectric spectroscopy (*triangles*), shear (*diamonds*), and heat capacity spectroscopy (*squares*) are included. The *thick dashed-dotted line* corresponds to the $\alpha_{PE} = \beta_{PE}$ process in C=4 (P*n*BMA) as shown in Fig. 26. Reproduced from [482] by permission

of the localized β-process in the alkyl nanodomains of C=10 after rapid quench is due to enhancement of its amplitude of motion by additional free volume introduced, which disappears during physical aging slightly below $T_g(\alpha)$. These effects of quenching and aging on the β-process are similar to that found for conventional bulk glassformers to be discussed in Section 2.3.

A recent experimental study of P*n*BMA and P*n*HMA [482(c)] has verified the conclusions made first in [481] and previously confirmed by experiment in [482(a), 482(b)]. Another confirmation is the study of alkyl nanodomains in poly(3-alkylthiophenes) by Pankaj and Beiner [297].

2.2.2.6 Length Scale from Nanoconfinement

In some models of glass transition, such as Adam–Gibbs [30], Donth [121], and several others such as [483], the cooperative structural relaxation dynamics is characterized by a temperature-dependent correlation length $\xi(T)$. The increase of $\xi(T)$ with decreasing temperature was used to explain the corresponding rapid increase of $\tau_\alpha(T)$ in the manner of the Vogel–Fulcher–Tammann–Hesse (VFTH). As we have seen above, near the glass transition temperature, ξ estimated from some glassformers is in the range of a few nanometers as expected [484]. Therefore, if one or more than one dimension is reduced to some nanometer size d, the increases of ξ on decreasing temperature will be arrested when $\xi(T_d)$ exceeds d. Consequently, the

VFTH temperature dependence of $\tau_\alpha(T)$ found at high temperatures $T > T_d$ cannot continue below T_d, where $\tau_\alpha(T)$ will have a milder temperature dependence. Such an effect was found in small molecule glassformers in nanometer spaces [485–506]. However, interpretation of the observed effect by the Adam–Gibbs model turns out to be problematic as will be discussed later in Section 2.2.5.8.

A comprehensive investigation of the effect on nanoconfinement was performed by Schönhals and coworkers [498–500] on poly(dimethyl siloxane) (PDMS, with $M_w = 1400$ g/mol, and $T_g = 151$ K) and poly(methyl phenyl siloxane) (PMPS, $M_w = 1000$ g/mol, and $T_g = 212$ K). Dielectric spectroscopy, temperature-modulated differential scanning calorimetry (TMDSC), and quasielastic neutron scattering were employed to investigate the molecular dynamics of poly(dimethyl siloxane) (PDMS) and poly(methyl phenyl siloxane) (PMPS) confined to random nanoporous glasses with mean pore dimensions of 2.5, 5.0, 7.5, and 20 nm to confine the two polymers. In all pores, the confined PDMS and PMPS had faster local segmental (structural α-) relaxation than the bulk state. The temperature dependence of τ_α from dielectric measurements still had the VFTH form for pore size of 20 and 7.5 nm (see Figs. 28 and 29).

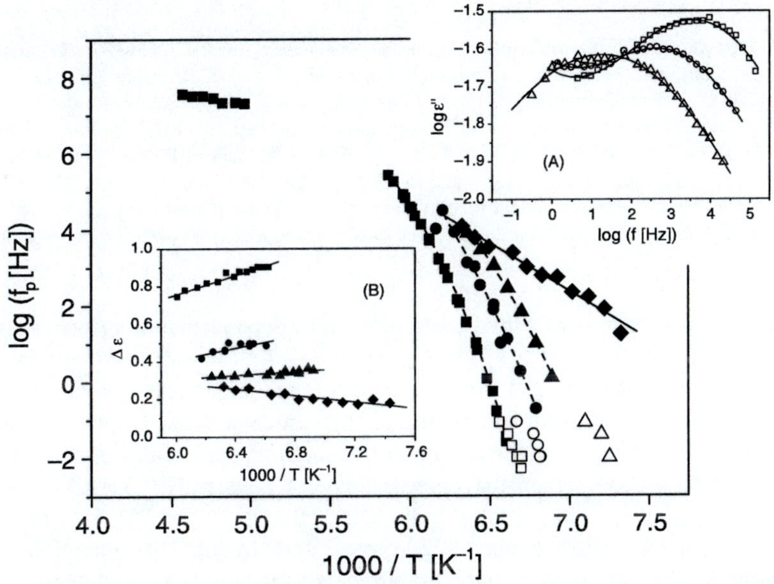

Fig. 28 Relaxation map for PDMS: (■) bulk, (•) 20 nm, (▲) 7.5 nm, (♦) 5 nm, from dielectric measurements. Corresponding *open symbols* are from TMDSC measurements. The *dashed lines* are fits of the VFT equation to the data. The *solid line* is a fit of the Arrhenius equation to the data. $E_a = 48.4$ kJ/mol, $\log(f_\infty[\text{Hz}]) = 20.1$. *Inset* (**A**) gives the dielectric loss vs. frequency for PDMS confined to nanoporous glasses with a pore size of 5 nm: (Δ) $T = 153.9$ K, (O) $T = 142.4$ K, (□) $T = 136.6$ K. The *lines* are fits of the HN-function including a conductivity contribution to the data. *Inset* (**B**) gives the uncorrected dielectric relaxation strength $\Delta\varepsilon$ vs. inverse temperature: (■) bulk, (•) 20 nm, (▲) 7.5 nm, (♦) 5 nm. The *lines* are linear regressions to the data. Reproduced from [498] by permission

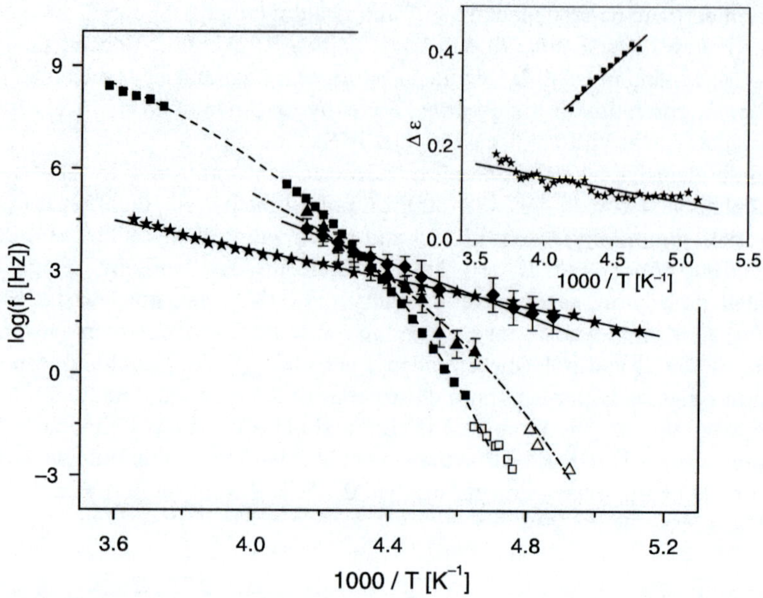

Fig. 29 Relaxation map of PMPS plotting the maximal loss f_p vs. $1/T$ of PMPS: (■) bulk, (▲) 7.5 nm, (◆) 5 nm, and (⋆) 2.5 nm are from dielectric measurements. Corresponding *open symbols* are from TMDSC measurements. The *dashed line* is fit by the VFTH equation to the bulk data. The *dashed-dotted line* is a fit by the VFTH equation to the data of PMPS confined in 7.5 nm pores. The *solid lines* are fits by the Arrhenius equation to the data of PMPS confined in 5 nm (E_a = 73.7 kJ/mol, log(f_∞[Hz] = 18.5) and in 2.5 nm (E_a = 39.8 kJ/mol, log(f_∞[Hz] = 11.9). The *inset* gives the uncorrected dielectric relaxation strength $\Delta\varepsilon$ vs. inverse temperature: (■) bulk, (⋆) 2.5 nm. The *lines* are linear regressions to the data. Reproduced from [499] by permission

However, at pore size of 5 nm, the temperature dependence of τ_α was changed to an Arrhenius dependence, $\tau_\alpha(T) = \tau_\infty \exp(E_a/K_B T)$. The activation energy E_a is 48.4 kJ/mol for PDMS and 73 kJ/mol for PMPS. The temperature dependence of τ_α of PMPS confined in 2.5 nm pores also is Arrhenius with E_a = 40 kJ/mol and $\tau_\infty = 10^{-12.7}$ s, like the sizes of the corresponding parameters of a local and independent relaxation. When the T-dependence of τ_α is still VFTH-like (bulk or confined in 20 and 7.5 nm pores), the dielectric strength $\Delta\varepsilon$ of PDMS and PMPS decreases with increasing temperature, typical for structural α-relaxation in all glassformers. The change in the temperature dependence of τ_α to Arrhenius dependence of PDMS when confined in 5.0 nm pores or PMPS in 5.0 and 2.5 nm pores is accompanied by a change in the temperature dependence of the dielectric relaxation strength: it increases with increasing temperature (see insets of Figs. 28 and 29). The changed temperature dependence of $\Delta\varepsilon$ is a characteristic for localized molecular motions.

Temperature-modulated differential scanning calorimetry (TMDSC) measurements also provide valuable information on dependence in pore size. The step in the specific heat ΔC_p at T_g, normalized to the weight of the confined polymer, decreases strongly with decreasing pore size. For PDMS and PMPS confined to pores with a

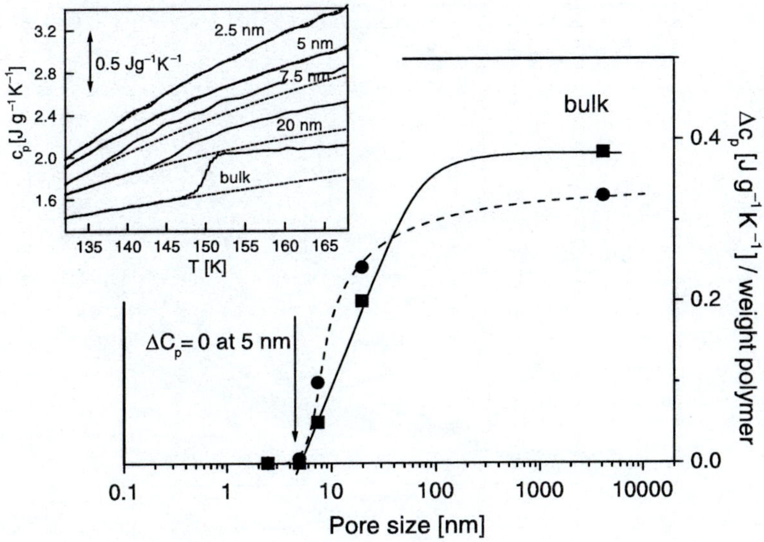

Fig. 30 ΔC_p normalized to the weight of the confined polymer vs. pore size: (■) PDMS, (•) PMPS. The *lines* are guides for the eyes. The *inset* gives the real part of the complex heat capacity obtained by TMDSC for PDMS at the labeled pore sizes. The temperature was modulated with a modulation time of 200 s and an amplitude of 0.5 K. The underlying heating rate was 0.25 K/min. The *dashed* lines are extrapolations of c_p below T_g to temperatures above T_g. Reproduced from [499] by permission

dimension of 5.0 and 2.5 nm, signature of glass transition was not observed from ΔC_p by the temperature-modulated differential scanning calorimetry measurements within the experimental uncertainty (see inset of Fig. 30 in the case of PDMS). These results provide independent evidence that a cooperative length scale is relevant for the glassy dynamics in PDMS and PMPS. From this trend of the calorimetry data, the cooperative length scale of about 5 nm was proposed, which is larger than other reported values. The discrepancy could be due to the presence of an immobilized boundary layer found for PMPS, which means that effective pore size is smaller than 5 nm.

Neutron scattering experiments were carried out only for PDMS [498] and PMPS [499]. The samples were protonated, and thus incoherent scattering dominates and the data are for the self-correlation function $F_s(\vec{k},t)$ defined in Eq. (1.114). Neutron backscattering instruments used has high energy resolution in the range of 1 μeV, corresponding to observation times of the order of $t \sim \hbar/E \sim 4$ ns. The elastically scattered intensity I_{el} measured as a function of temperature is normalized by the total scattering intensity I_0 measured below 2 K where all molecular motions are frozen. Assuming a Gaussian form for the elastically scattered intensities the mean-squared displacement $< r^2 >_{eff}$ is deduced from the relation

$$I_{el}/I_0 = \exp[-Q^2 < r^2 >_{eff} /3]. \tag{2.36}$$

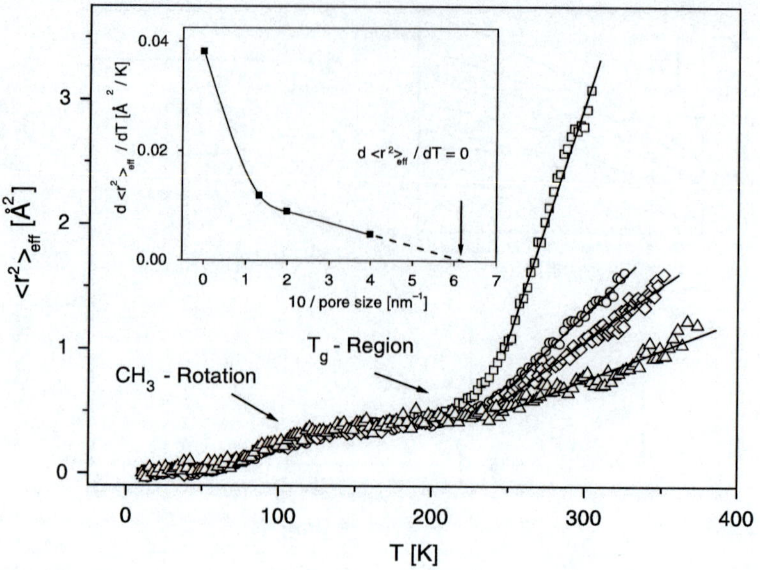

Fig. 31 Temperature dependence of $< r^2 >_{\text{eff}}$ of PMPS for the different pore sizes: *open squares*—bulk, *open circles*—7.5 nm, *open diamonds*—5 nm, *open triangles*—2.5 nm. The *inset* gives the slope d $< r^2 >_{\text{eff}}/dT$ for $T > T_{g,\text{bulk}}$ vs. inverse pore size. Reproduced from [500] by permission

The temperature dependence of $< r^2 >_{\text{eff}}$ for bulk PMPS and PMPS confined in 7.5, 5.0, and 2.5 nm pores are shown in Fig. 31. The PMPS sample used in neutron scattering studies has higher molecular weight and T_g than the one discussed above. The parameters are $M_w = 2600$ g/mol, polydispersity index $M_w/M_n = 1.20$, and $T_g = 223$ K. On increasing temperature, the first rise of $< r^2 >_{\text{eff}}$ in the vicinity of 100 K is due to the fast methyl group rotation (see Fig. 31). This is a localized and independent relaxation and therefore it is not much influenced by the confinement. After this step rise, $< r^2 >_{\text{eff}}$ exhibits a plateau of slow increase of $< r^2 >_{\text{eff}}$ with temperature, and the height of the plateau is not much changed by confinement. Shown before, the plateau reflects the caged molecule dynamics [115, 195, 507, 508], and this will be discussed in detail in Section 2.3. Since caged dynamics is insensitive to confinement, it follows the same for the plateau as observed. The plateau continues till near a temperature T_x, beyond which $< r^2 >_{\text{eff}}$ of bulk PMPS shows a steeper increase with temperature. For the unconfined PMPS, T_x is approximately 223 K, the T_g of bulk PMPS. This elbow shape of the temperature dependence of $< r^2 >_{\text{eff}}$ with $T_x \approx T_g$ is general and found in many bulk glass-formers of different types [114, 115, 509, 510]. Some examples of bulk glassformers including the classical case of Se by Buchenau and Zorn [509] are shown together with PDMS confined in 7.5 nm pores in Fig. 32. The same change was found for the PMPS confined in the pores (Fig. 31). The absolute value of $< r^2 >_{\text{eff}}$ and the slope of its change with increasing temperature characterized by the derivative

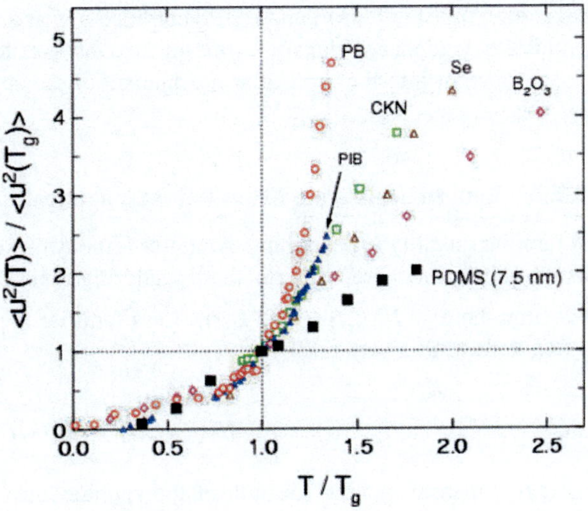

Fig. 32 Comparison of $< u^2(T) >$, normalized by its value at T_g for PDMS confined in 7.5 nm pores with other bulk glassformers

d$< r^2 >_{eff}$/dT for $T > T_x$ decrease significantly with decreasing pore size (Fig. 31). This trend is the same as that found when comparing different bulk glassformers, where glassformer with smaller n and fragility index m has smaller d $< r^2 >_{eff}$/dT for $T > T_g$, found empirically to hold, particularly when restricting to glassformers of the same class [115, 239]. This trend in bulk glassformers has been rationalized by the CM [507]. If we recall the dielectric and TMDSC data of nanoconfined PDMS and PMPS definitely show decrease of T_g and the fragility index m when confined in pore of smaller size. The change toward local and non-cooperative dynamics with Arrhenius T-dependence when decreasing pore size suggests decrease of cooperativity, and hence decrease of its index n. Thus, the trend of d $< r^2 >_{eff}$ /dT for $T > T_g$ observed in nanoconfined PMPS with decreasing pore size and concomitant decrease of m and n are analogous in all respects to that found in bulk glassformers.

Before we leave this section, some words of caution are needed. Although in this section the models of glass transition by Adam–Gibbs, Donth, and others were used to motivate the connection between VFTH temperature dependence of $\tau_\alpha(T)$ and increasing length scale with decreasing temperature, by no means the connection validates these models. The actual molecular dynamics in unconfined glassformers may be entirely different from that suggested by these models. More scrutiny of these models like that given before in Sections 2.2.2.2 and 2.2.2.3 and in later sections can reveal the deficiencies of these models. Therefore, one should not rush to make any judgment on the veracity of these models based only on their success in rationalizing the VFTH dependence of $\tau_\alpha(T)$ in the bulk, and the decrease of $\tau_\alpha(T)$ as well as the change toward Arrhenius temperature dependence by nanoconfinement. Later, we shall return to show that there is a connection between the structural relaxation times of PDMS and PMPS confined in 5 and 2.5 nm pores and the Johari–Goldstein secondary relaxation time in their bulk states. This connection offers a deeper insight into the molecular mechanism of the structural relaxation not given by these models.

2.2.2.7 Length Scale from Multi-point Dynamical Susceptibilities

A familiar quantity in describing dynamics is the two-point correlation function such as the self or incoherent intermediate scattering function given by Eq. (1.114), and rewritten here as $F_s(\vec{k},t) = \langle f_s(\vec{k},t)\rangle$. Less familiar is the four-point susceptibility, $\chi_4(t)$, defined as a variance by

$$\chi_4(t) = N[\langle f_s^2(\vec{k},t)\rangle - \langle f_s(k,t)\rangle^2] \tag{2.37}$$

It is a measure of the strength of the spontaneous fluctuations about the average dynamics. For density, energy, or orientation fluctuations generally denoted by $U(\vec{r},t)$, the four-point correlation function [511, 512] is defined by

$$G_4(\vec{r},t) = \langle U(0,0)U(0,t)U(\vec{r},0)U(\vec{r},t)\rangle - \langle U(0,0)U(0,t)\rangle\langle U(\vec{r},0)U(\vec{r},t)\rangle. \tag{2.38}$$

It was constructed to measure the correlation in space of the local two-time correlation functions, $\langle U(0,0)U(0,t)\rangle$, which is the density–density correlation when U is the density ρ, or the dipole moment–dipole moment time correlation functions when U is the dipole moment $\vec{M}$. Let us consider at point 0 that a decorrelation of the local density is occurring over time interval t. Then $G_4(\vec{r},t)$ measures the probability that a similar event has occurred at position $\vec{r}$ away within the same time interval t. In view of this, $G_4(\vec{r},t)$ is a measure of the heterogeneity and cooperativity of the dynamics. So is the associated four-point susceptibility $\chi_4(t)$ given by the integral over volume of $G_4(\vec{r},t)$, which is equal to the variance of the correlation function [513, 514]. The $\chi_4(t)$ for several model glassformers has been computed numerically, and all show $\chi_4(t)$ exhibits a maximum at the structural relaxation time τ_α [515–522]. The peak value, $\chi_4(\tau_\alpha)$, is found to increase as temperature decreases or density increases. From the relation between $\chi_4(\tau_\alpha)$ and $G_4(\vec{r},t)$, this indicates that the range of heterogeneity and cooperativity of the dynamics increases with increase of τ_α toward glass transition. Except for a factor of order unity, $\chi_4(\tau_\alpha)$ is a measure of a correlation volume, or a number of molecules $N_{corr,4}$ that are dynamically correlated over a time period of the order of τ_α. Although the maximum of $\chi_4(t)$ is located near $t = \tau_\alpha$, $N_{corr,4}$ is actually defined by the maximum value of $\chi_4(t)$ at some t near τ_α.

Unfortunately so far no one has been able to measure $\chi_4(t)$ in glass-forming liquids and polymers. Berthier et al. [523, 524] came to the rescue by proposing a method to obtain $N_{corr,4}$ from a three-point function, $\chi_T(t)$, related to the sensitivity of the two-time dynamics to external control parameters such as temperature or density. For temperature, $\chi_T(t)$ is the temperature derivative of the intermediate scattering function $F_s(\vec{k},t)$ or a local two-time correlation function, $C(t) = \langle U(0,0)U(0,t)\rangle$, i.e., $\chi_T(t) = \mathrm{d}C(t)/\mathrm{d}T$. By means of the fluctuation-dissipation theorem, Berthier et al. were able to relate $\chi_T(t)$ to $\chi_4(t)$, show $\chi_T(t)$ has a maximum at $t \sim \tau_\alpha$ like $\chi_4(t)$, and obtain the following expression for $N_{\mathrm{corr},T}$:

$$N_{\mathrm{corr},T}(T) = [k_B T^2/\Delta C_p(T)]\{\max_t \chi_T(t)\}^2, \tag{2.39}$$

where ΔC_p is the configurational heat capacity per molecule at constant pressure. $N_{\mathrm{corr,T}}$ is the number of molecules whose dynamics on the timescale of the α-relaxation at T is correlated to a local enthalpy fluctuation. Berthier et al. [523] initially suggested that $N_{\mathrm{corr,T}}$ from Eq. (2.39) is a lower bound for $N_{\mathrm{corr},4}$, but later showed [524, 525] that it is a very good estimate of $N_{\mathrm{corr},4}$ by numerical calculations of $\chi_T(t)$ and $\chi_4(t)$ from molecular dynamics simulations of binary LJ particles and silica. If $C(t)$ is well described by a stretched exponential function, $C(t) = \exp\{-[t/\tau_\alpha(T)]^{\beta(T)}\}$, $N_{\mathrm{corr,T}}$ is given by

$$N_{\mathrm{corr},T}(T) = [k_B \beta(T)^2/e^2 \Delta C_p(T)]\{\mathrm{d}\ln\tau_\alpha/\mathrm{d}\ln T\}^2. \tag{2.40}$$

Two other terms with one involving $\mathrm{d}\beta(T)/\mathrm{d}T$ have been neglected in Eq. (2.40) because in practice they only contribute 1% to the value of $N_{\mathrm{corr},4}$ [526].

Values of $N_{\mathrm{corr},T}(T)$ and its value at $T = T_g$ for many different glassformers were obtained from $\chi_T(t)$ in the manner as discussed above first by Berthier et al. [523] for a few archetypal glassformers including glycerol from dielectric relaxation data, colloidal hard spheres from dynamic light-scattering data, and binary LJ particles and silica [524] from molecular dynamics simulation. Additional values of $N_{\mathrm{corr},T}(T)$ of other small molecular glassformers and B_2O_3 are given from dielectric relaxation, light and neutron scattering, and optical Kerr effect experimental data by Dalle-Ferrier et al. [525]. Values of $N_{\mathrm{corr},T}(T)$ for even more glassformers of all kinds including amorphous polymers were obtained by Capaccioli et al. [526] principally from dielectric relaxation data. All these studies show, as a function of T, $N_{\mathrm{corr},T}(T)$ increases with decreasing temperature, or as a function of $\tau_\alpha, N_{\mathrm{corr},T}(\tau_\alpha)$ increases with increase of τ_α. If $N_{\mathrm{corr},T}(T)$ truly reflects $N_{\mathrm{corr},4}(T)$, these results do indicate the growth of some length scale of dynamic heterogeneity, cooperativity, or correlation in the motion on decreasing temperature. However, the quantitative details of the $N_{\mathrm{corr},4}(T)$ obtained are perplexing. After normalizing τ_α by a microscopic time τ_0 specific to different systems, Dalle-Ferrier et al. found the $N_{\mathrm{corr},4}(T)$ as a function of τ_α/τ_0 falls remarkably close to one another for many different glassformers (see Fig. 4 in [525]). Examples are results from molecular dynamics simulations of silica compared with binary LJ particles, propylene glycol compared with *ortho*-terphenyl, and B_2O_3 compared with decalin. In each of the three pairs, other characteristics of the dynamics, however, are very different. Notable are non-exponentiality (or n appearing in the fractional exponent of the

Kohlrausch correlation function) and fragility (or steepness index m). The result of $N_{corr,T}(T)$ that it differs little from one glassformer to another, including at glass transition (at the same τ_α/τ_0), poses a problem for many theories of glass transition that ascribe the strong temperature dependence of the dynamics *solely* coming from a length scale that grows as one approaches the glass transition, if the length scale is truly $N_{corr,T}$. This is obvious because in these theories the $N_{corr,T}$ results would say that many pairs of glassformers have the same dynamics, which is not borne out by the experimental facts on other properties shown immediately below.

Dependence of $N_{corr,T}$ on τ_α of more glassformers obtained by Capaccioli et al. [526] from experimental data shows there are differences for some. However, overall the results are still perplexing. For example, the strong glassformer GeO_2 has exponential time dependence (n=0) for the correlation function [112], Arrhenius T-dependence for τ_α, and small ΔC_p, and yet its $N_{corr,T} = 413$ is so much larger than $N_{corr,T} = 92$ of fragile glassformer tri-naphthyl benzene (TNB) which has non-exponential time dependence ($n \approx 0.44$) for the correlation function [455], VFTH T-dependence for τ_α, and much larger ΔC_p. Capaccioli et al. found at the glass transition temperature, $N_{corr,T}(T_g)$ and the configurational entropy per unit volume $S_c(T_g)$ are anticorrelated, as originally predicted by the Adam–Gibbs theory. Using the same data as tabulated in the paper by Capaccioli et al., we attempt to see if there is anticorrelation between $N_{corr,T}(T_g)$ and $\beta(T_g) \equiv [1 - n(T_g)]$ or correlation between $N_{corr,T}(T_g)$ and $m(T_g)$. The results shown in Figs. (33-1a) and (33-1b) indicate absence of any anticorrelation and correlation, respectively, even when restricted to glassformers in the same class.

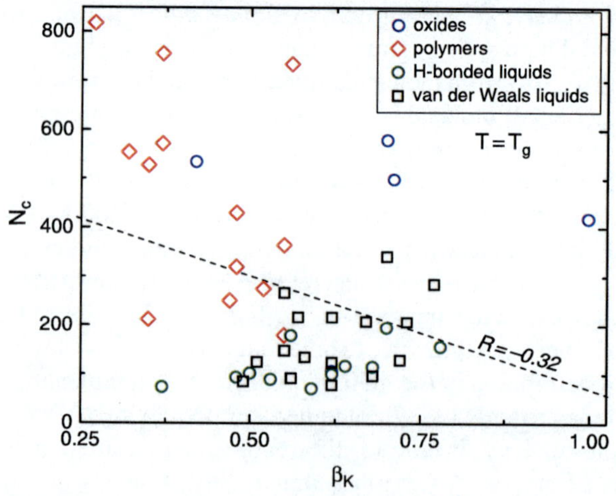

Fig. 33-1a Plot of N_c or $N_{corr,T}(T_g)$ obtained from the $\chi_T(t)$ approximation vs. Kohlrausch exponent at T_g and atmospheric pressure for 45 glassformers (data from [526]). Polymers (*diamonds*), oxide glassformers and selenium (*circles*), hydrogen-bonded materials (*triangles*), and van der Waals glassformers (*squares*). The absence of correlation is indicated by the small value of the Pearson linear correlation coefficient

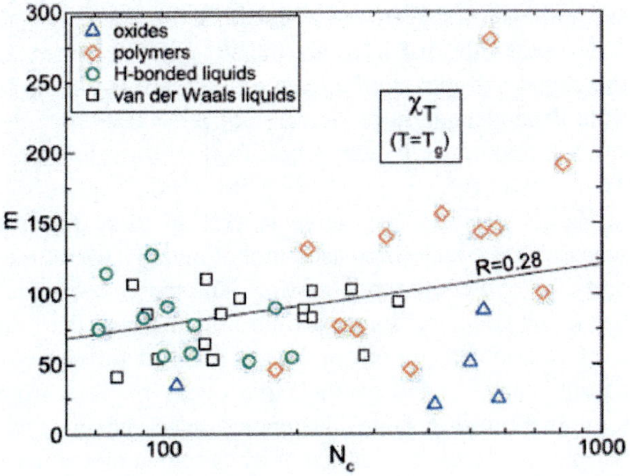

Fig. 33-1b Plot of fragility index m vs. N_c or $N_{corr,T}(T_g)$ corresponding to the same data in Fig. 33-1(a). Correlation coefficient is indicated

The lack of correlation $N_{corr,T}$ with other key indicators of dynamics especially non-exponentiality seriously undermines its relevance as a measure of the length scale of heterogeneous or cooperative dynamics. As mentioned before and later on in various other connections, in my view the heterogeneous dynamics and non-exponentiality are parallel consequences of the many-body structural relaxation dynamics. Therefore the length scale of the heterogeneous dynamics ξ_{het} and the non-exponentiality parameter n should correlate unless certain extraneous factor has entered in determining these two quantities. The problem encountered by $N_{corr,T}$ probably originates from Eq. (2.40) through the term τ_α. Not only many-body relaxation but also specific volume and entropy enter in determining τ_α, and thus $N_{corr,T}$ is not as directly and simply related to many-body relaxation as is the non-exponentiality parameter n. It is clear that n and τ_α are two essential parameters in $C(t) = \langle U(0,0)U(0,t)\rangle$ needed to characterize the structural relaxation. Furthermore, n and τ_α play independent roles. This fact is made evident by the invariance of n to different combinations of P and T which keep τ_α constant, a general experimental fact to be presented in Section 2.2.4 later. The fact that only τ_α enters in Eq. (2.40) but not n in determining $N_{corr,T}$ indicates not all aspects of many-body dynamics have been accounted for by $N_{corr,T}$, and this explains its lack of correlation with n or m. The essential and independent role played by n becomes transparent if dynamics are discussed in the framework of the CM. In this model, the primitive relaxation time τ_0 has already the pressure P/volume V and temperature T/entropy S dependences [196] (verified experimentally via the Johari–Goldstein relaxation, see Section 2.3). These dependences of τ_0 are magnified by the many-body dynamics leading to stronger corresponding dependences of τ_α, obtainable by the CM relation, $\tau_\alpha = [t_c^{-n}\tau_0]^{1/(1-n)}$, in which the role played by n is evident.

The shortcomings of $N_{corr,T}$ do not translate to $\chi_4(t)$. The latter has sound physical underpinning and provides important information on the dynamics despite not having been measured by experiment in glass-forming liquids and polymers so far. For the binary LJ particles system, $\chi_4(t)$ obtained numerically at $T = 0.42$

(LJ unit) shows $\chi_4(t) \sim t^4$ in the ballistic regime at very short times [527(a)]. Later, when the particles are caged, $\chi_4(t)$ becomes weakly dependent on t. After the caged particles dynamics regime, there is the onset of the fractional power law time dependence, $\chi_4(t) \sim t^{\mu}$, with $\mu = 0.73$, which starts from small value and rises to maximum located close to the relaxation time of the correlation function. This rise of $\chi_4(t) \sim t^{\mu}$ starting from small value is consistent with the existence of the uncorrelated primitive relaxation to be followed by many-body dynamics whose length scale increases monotonically with time as more and more particles participate in cooperative motion. The increase continues until the maximum number allowed by the inter-particle potentials of the system is reached at time near τ_α. Cessation of growth of length scale at times beyond times of the order of τ_α naturally is reflected by the rapid decay of $\chi_4(t)$ from the maximum down to its long time limiting value. The evolution of dynamics with time starts from the caged regime (1) which is terminated by the most elementary relaxation (2), which is the primitive relaxation in the context of the CM. Thereafter it is followed by cooperative motions with increasing participation of particles (3) and continues until the terminal regime at time of the order of the structural relaxation time (4) is reached. This description of evolution of dynamics proffered by $\chi_4(t)$ is consistent with that of the CM [120, 507, 508] as will become evident when dynamics at shorter times than τ_α are discussed in Section 2.3. The $\chi_4(t)$ and the corresponding dynamical correlation length $\xi_4(t)$ obtained from molecular dynamics simulations of binary LJ particles by Laĉević et al. [522] offer the same insight. $\xi_4(t)$ provides an estimate of the range of correlated particle motion. Like $\chi_4(t)$, $\xi_4(t)$ has a maximum as a function of time t. The value of the maximum of $\xi_4(t)$ is less than one particle diameter at higher temperature ($T = 0.94$ and 2) where for the coherent intermediate scattering function its relaxation time τ_α is short (of the order of a few LJ units of time) and its time dependence is close to exponential. The corresponding $\chi_4(t)$ is negligibly small, indicating particle motion is essentially uncorrelated. These results are consistent in all respects with the primitive relaxation of the CM, and it is observed because at these high temperatures τ_α is comparable to t_c, which is of the order of unity in LJ unit. On lowering temperatures, $\xi_4(t)$ also starts from small values and increases with time to reach maximum value. The latter increases with decreasing temperature, reaching a value of about nine particle diameters at $T = 0.6$.

We have seen from the analysis of experimental data for a broad range of real materials from Fig. 33-1(a), wherein values of N_c or $N_{corr,T}(T)$ or at $T = T_g$ were obtained from $\chi_T(t)$, that there is absence of any relationship between N_c and β_K. However, this does not happen for N_c calculated from simulations of the Kob–Andersen model of Lennard-Jones mixture with the 12-6 LJ potentials as a function of temperature T and density ρ [527(b)]. The non-exponentiality parameter β_K is obtained from the self-intermediate scattering function, and N_c from the maximum in $\chi_4(t)$ obtained as a function of t. The results for different state points (T and ρ) given in a plot of $N_c = \max\{\chi_4(t)\}$ in Fig. 33-1(c) show N_c does indeed correlate with β_K over a range of thermodynamic conditions. The good correspondence is in accord with the interpretation of increasing N_c or the degree of dynamic

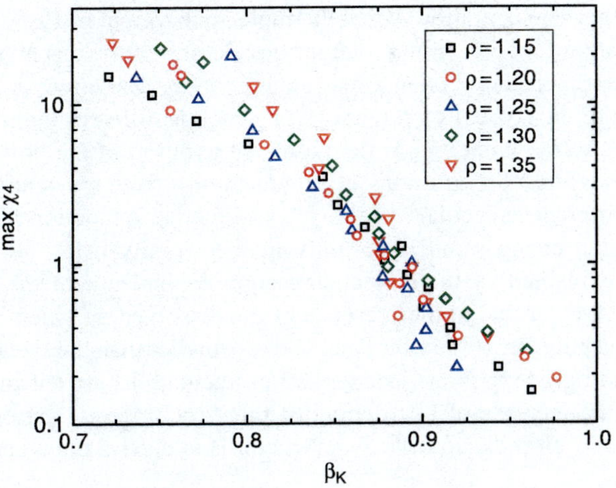

Fig. 33-1c N_c from $\chi_4(t)$ vs. Kohlrausch exponent for simulated Lennard-Jones particles. Each *symbol* represents a distinct state point. Increasing dynamic heterogeneity is associated with a broader distribution of relaxation times [527(b)]

heterogeneity is associated with an increasing breadth of the relaxation function or smaller β_K or larger n. The result here supports that both length scale of α-relaxation given by N_c and n are parallel consequences of the many-body α-relaxation.

2.2.2.8 Length Scale Is Not Practical to Use as Measure of Many-Body Dynamics

We have seen from the discussions above that the length scale of structural relaxation given by theories and models based on thermodynamic considerations and thermodynamic quantities is unreliable. Although the four-point susceptibility $\chi_4(t)$ is fundamentally sound as the quantity to extract a length scale, so far no one has been able to measure it for glass-forming liquids and polymers. The only dependable measurement of the heterogeneity length scale, ξ_{het}, is by multidimensional ^{13}C solid-state NMR technique. But this technique is elaborate and expensive to carry out and that is why so far it had only been carried out for four glassformers. Alternative experimental methods had been proposed and used to determine cooperative length scale or heterogeneity length scale. They all utilize some special effects that enable length scale to be deduced. References to the other techniques not covered in this section can be found in two reviews [228, 528].

One method used the effect of modification of the solvent by addition of a polymer to determine the length scale of the structural relaxation of the solvent, ξ [529]. The solvent used is polychlorinated biphenol (Aroclor) and the polymers are polyisoprene (PI) and 1,4-polybutadiene (PB). The characteristics of the Aroclor reorientation motion seen by photon correlation and dielectric relaxation spectroscopies in Aroclor solutions of high molecular weight PI and PB are distinctively different from those in Aroclor solutions of low molecular weight polymers. At constant polymer concentration, when the molecular weight of PI or PB is low, the photon correlation function shows a larger broadening of the dispersion of Aroclor relaxation and a more rapid change with temperature. The effect is broadening of the dispersion of the solvent by the polymer, which is different in the two regimes, $\xi > R_g$ and $\xi < R_g$, where R_g is the radius of gyration of the polymer. The dispersion was measured for solutions of polymers at a fixed concentration but with polymers of different molecular weights (R_g). Plotted as a function of R_g, the width of the dispersion changes rather abruptly in the vicinity of $R_g = 1.5$ nm. This property was explained by the molecular weight dependence of the concentration fluctuation of Aroclor due to the presence of the dissolved polymer. When the size of the dissolved polymer coil is comparable to or smaller than the size of the Aroclor cooperative rearranging regions, concentration fluctuations are enhanced, resulting in a concomitant increase of broadening of the Aroclor reorientational relaxation spectrum. This result indicates that ξ of Aroclor is about 1.5 nm when temperature is near T_g.

Another example worth further description is based on an effect seen by light scattering from glass-forming liquids. The light-scattering intensity vs. temperature curve exhibits a maximum during heating in the glass transition region from the glassy state to the liquid state [530, 531]. This effect does not afford a direct measurement of the heterogeneous dynamics. Nevertheless, with some assumptions and structural relaxation kinetic parameters taken from other experiments, Moynihan and Schroeder [531] were able to deduce the size of nanoscale inhomogeneities (density fluctuations) which relax at different rates. The predicted length scales obtained by way of the Adam–Gibbs model and given in terms of the cube root of volumes are 2.3, 2.7, and 4.6 nm for B_2O_3, glycerol, and PVAc, respectively. The corresponding values obtained by using the free volume model are larger and equal to 5.3, 5.4, and 8.4 nm for B_2O_3, glycerol, and PVAc, respectively.

Many experimental techniques discussed all have confirmed the existence of length scale that determines the number of molecules or particles involved in the structural relaxation. Thus, although length scale alone does not tell us the actual dynamics, it implies some form of many-body relaxation. However, most of the experimental techniques have given the length scales of a few glassformers, and it is uncertain the length scales of the same glassformer obtained by different techniques are compatible with each other.

Only calorimetry has determined the length scales of many more glassformers, but the results were obtained through some Adam–Gibbs model-dependent analyses of the thermodynamics data [420]. As we have discussed earlier, the results seem internally inconsistent and doubtful. Probably, this is because thermodynamic quantities alone cannot account for the many-body relaxation part of the dynamics. Besides, there is still disagreement on how to account for the configurational entropy, which is only one part of the supercooled liquid's entropy. Another part is the vibrational entropy. The importance of both parts of entropy and heat capacity C_p has been recognized since the Kauzmann's discussion of entropy of supercooled liquids [391]. On the one hand, Johari found the configurational entropy of a liquid cannot be assumed to be equal to, or even proportional to, the excess entropy of the liquid over its ordered crystal state, and this is compounded by the implausibility of Kauzmann's extrapolation [413]. On the other hand, Angell and Borick [414] argued in favor of the opposite that the vibrational part of C_p is the same as the C_p of the ordered crystal phase, or proportional to it, although subsequently this proposal seems to have been withdrawn [532]. However, the controversy continues. By analyzing the vibrational density of states in the low-frequency (Boson peak) region and combining it with a computer simulation, Angell and coworkers [532] deduced that there is a discontinuous increase in the vibrational C_p due to unfreezing of the structure of a glass on heating through T_g. They concluded that the vibrational part of a liquid's C_p is much greater than the value used in all earlier analyses of calorimetric data using the Adam–Gibbs theory [420], and questioned the results obtained by claiming that the behavior of the vibrational heat capacity was not properly taken into account. Later on, Johari found for three different glassformers that the change in the vibrational part at liquid–glass transformation is negligible [533]. A different conclusion was reached that there is no discontinuity in the vibrational parts of C_p on structural unfreezing in the T_g range, and hence the change in C_p at T_g is almost entirely due to change in the configurational part.

2.2.2.9 Why Fixation on the Length Scale of the α-Relaxation, and Disregard of the Width of the Dispersion?

The length scales of the α-relaxation determined from pure thermodynamic consideration, such as those using the Adam–Gibbs model or the Donth's model, show inconsistencies. This is unsurprising because the models have assumed that thermodynamics alone determine the α-relaxation and its relaxation time. The assumption can hardly be justified because the α-relaxation is a many-body process and its effects are not expected to be captured by thermodynamic variables including configuration entropy, heat capacity, free volume, and the like. The length scales determined from different methods hinged on thermodynamics often are inconsistent with each other, creating the situation of which one should use. For example, we can take some values of the length scale $\xi_\alpha(T_g)$ for some glassformers from Donth's model [425] and compare them with $N_{corr,T}(T_g)$, the number of molecules whose dynamics on the timescale of the α-relaxation at T_g is correlated to a local enthalpy fluctuation, of Bethier et al. [523–525]. For example, the values of $\xi_\alpha(T_g)$ for PMMA, OTP, and B_2O_3 are 1.5, 3.0, and 1.5 nm, respectively [425], and the corresponding values of $N_{corr,T}(T_g)$ are 572, 148, and 107, respectively [526]. Obviously the two sets of values for the length scale from the two sources are irreconcilable with each other.

The dynamic heterogeneous nature of the structural relaxation, like its non-exponentiality and other anomalous dynamics properties (to be discussed later), is a direct consequence of many-body relaxation. Hence the length scale of dynamic heterogeneity is a true measure of the extent of the many-body relaxation. Multidimensional NMR experiment is a microscopic technique that has been successfully applied to determine the length scale of dynamic heterogeneity [226–228]. The size of the length scale has been determined at temperatures near but still above

T_g, but unfortunately this has been done so far for a paucity of four glassformers within the last two decades. The four-point susceptibility $\chi_4(t)$ has the potential of providing the length scale, but experimental technique has yet to be developed to measure this quantity proficiently for a large number of glassformers in order to be useful. Thus, at least for the present time, it is not practical to use length scale from any source as the measure of the extent of many-body relaxation. A pragmatic choice is the width of the dispersion of the intermediate scattering function, $F_s(\vec{k}, t)$, or a local two-time correlation function, $C(t) = \langle U(0,0)U(0,t)\rangle$, pertaining to the structural relaxation. This is because these quantities are widely and easily accessible by a variety of spectroscopic techniques and are already available for many glassformers and polymers in the literature. If the Kohlrausch function or its Fourier transform describes well the time dependence or the frequency dispersion of the measurement, the exponent n can be used instead of the width of the dispersion. The correlation found between length scale of dynamic heterogeneity ξ_{het} from NMR and n also gives assurance that the two are equivalent in the usage as the indicator of the many-body nature of the structural relaxation, albeit ξ_{het} of only four glassformers is on hand. A longer ξ_{het} associated with a larger n is borne out by considering ξ_{het} for glycerol, *o*-terphenyl, and poly(vinyl acetate) obtained by multidimensional ^{13}C solid-state exchange NMR experiments by Reinsberg et al. [422], where ξ_{het} is 1.3, 2.2, and 3.7 nm for glycerol, *o*-terphenyl, and poly(vinyl acetate), respectively, all determined uniformly at $T = T_g + 9$ K. These values of ξ_{het} correlate with the corresponding values of 0.29 [195], 0.50 [237], and 0.55 [238] of n, all obtained by fitting dielectric measurements by the Fourier transform of the Kohlrausch function when $\tau_\alpha \approx 1$ s for the three glassformers. Description of the procedure used to fit the dielectric loss data can be found in [477]. An independent ^{13}C solid-state exchange NMR measurement of ξ_{het} for supercooled D-sorbitol at $T = T_g + 7$ K by Qiu and Ediger [476] yields 2.5 ± 1.2 nm. These authors quoted KWW parameter $\beta_{\text{KWW}} \equiv (1 - n)$ of sorbitol as well as glycerol, *o*-terphenyl, and poly(vinyl acetate) from publications of others. These values may not be used together fairly because they were obtained by various authors using different fitting procedures and criteria. For sorbitol, they took $\beta_{\text{KWW}} = 0.41$ or $n = 0.59$ (citing private communication from Richert) which is quite different from the value of $n = 0.52$ [195] obtained by the aforementioned procedure [477]. The large value of n=0.59 for sorbitol taken by Qiu and Ediger causes sorbitol not to conform to the correlation between $\xi_{\text{het}}$and n. On the other hand, when the fair value $n = 0.52$ is used together with $\xi_{\text{het}} = 2.5 \pm 1.2$ nm for sorbitol, the correlation between ξ_{het} and n holds for glycerol, *o*-terphenyl, sorbitol, and poly(vinyl acetate) as shown in Fig. 33-2.

The most convincing argument favoring the width of dispersion or n as the key parameter characterizing the many-body structural relaxation is the existence of correlations between n and many other properties of the structural relaxation to be discussed in Sections 2.2.3–2.2.5 to follow and the explanations by the CM. In contrast, such correlations with other properties do not exist between the length scale determined from pure thermodynamic consideration by using the Adam–Gibbs model or the Donth's model, or with $N_{\text{corr},T}(T_g)$. For $N_{\text{corr},T}(T_g)$, this conclusion

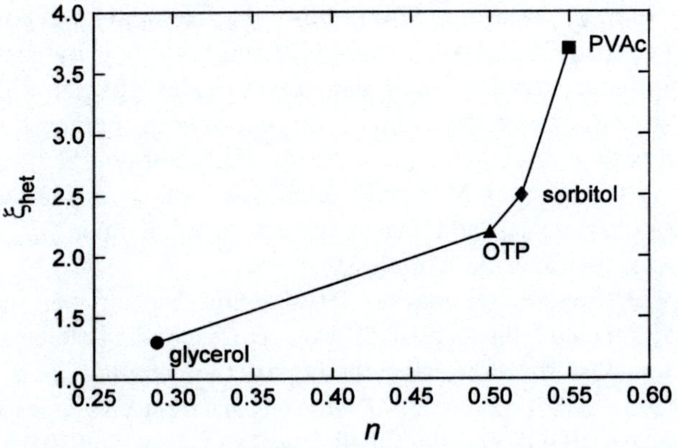

Fig. 33-2 Plot showing the correlation found between length-scale of dynamic heterogeneity ξ_{het} from NMR and n

comes from the absence of correlation between $N_{corr,T}(T_g)$ and n (see Fig. 33) while n correlates with the other properties. One such property is the breakdown of Stokes–Einstein (SE) and Debye–Stokes–Einstein (DSE) relations to be discussed in Section 2.2.5.6. Actually this property has attracted explanation from dynamical heterogeneous nature of the α-relaxation, and naturally $N_{corr,T}(T_g)$ is relevant and it should correlate with the size of the breakdown at T_g. The values of $N_{corr,T}(T_g)$ for silica and TNB are 579 and 92, respectively, and one would be led to expect larger degree of breakdown in silica than in TNB. Experimentally it is the opposite with major breakdown found in TNB and none in silica (see Section 2.2.5). Despite all that said above on the virtue of considering dispersion and its width or n, it is remarkable that main stream theories of glass transition did not take it seriously as a key parameter of the dynamics, and instead focused on the length scale. As far as I know, the CM is the only approach that emphasizes the dispersion and has its width or n determining dynamic properties.

2.2.3 T_g-Scaled Temperature Dependence of η or τ_α and the Steepness or "Fragility" Index

2.2.3.1 The T_g-Scaled Plot of η by Oldekop–Laughlin–Uhlmann–Angell

It is instructive to compare the same property of different glassformers under the same condition, such as the isobaric glass transition temperature T_g (or the isothermal glass transition pressure P_g) at which the structural relaxation time reaches a predetermined long time, say 10^2 s. Furthermore, if a quantity can be measured as a function of temperature T or pressure P, insight may be gained by comparing the variation of the quantity of different glassformers at the same corresponding-states variable such as the T_g-scaled temperature, T_g/T or T/T_g, or P_g-scaled pressure,

P_g/P or P/P_g. In 1957, Oldekop [374] seems to be the first one to compare in this way the liquid viscosity η of 10 different oxide and fluoride glassformers with and without the addition of alkali oxide or alkali earth oxides. His plot of log η against T_g/T shows a pattern of variation from one glassformer to another. It was assumed that the viscosity at $T = T_g$ is about the same for all his non-polymeric glassformers. This can be expected from the Maxwell's relation, $\eta = G_\infty \langle \tau_\alpha \rangle$, because the mean structural relaxation time $\langle \tau_\alpha \rangle$ and high-frequency shear modulus G_∞ at $T = T_g$ are about the same for the same class of glassformers.

Later in 1972, apparently unaware of the Oldekop's paper, Laughlin and Uhlmann [375] repeated the use of T_g as a corresponding-states parameter to correlate and compare the flow behavior of different non-polymeric liquids. By comparing log η in a plot against T_g/T of three different classes of materials, (i) SiO_2 and GeO_2, *(ii)* B_2O_3 and the alkali silicate glasses, and (iii) three organic small molecule liquids including *ortho*-terphenyl (OTP), salol, α-phenyl-*o*-cresol, and 1,3-*bis*(1-naphthyl)-5-(2-naphthyl)benzene (TNB), and molten salt $Ca(NO_3)_2$–KNO_3, Laughlin and Uhlmann found liquids within each class have nearly the same T_g-scaled temperature dependence. However, the T_g/T-dependences of log η of the three classes have very different curvatures. Laughlin and Uhlmann did not give a physical interpretation of the difference in the T_g/T-dependences they found for different glassformers.

Angell [215, 216] recognized the importance of the Laughlin–Uhlmann plot as a means to classify the transport properties of glassformers. He included more non-polymeric materials into the plot of Laughlin and Uhlmann and expanded the pattern found by them. Thus, it seems logical to refer to the plot of log η vs. T_g/T as the Oldekop–Laughlin–Uhlmann–Angell (OLUA) plot. But most workers called it the Angell plot, ostensibly giving all the credit to Angell for inventing this plot, which is arguably inappropriate. However, credit has to be given to Angell for recognizing the efficacy of this plot, and for coining the words "fragile" and "strong" to separate out those glassformers having, respectively, stronger and weaker T_g/T-dependence on log η. When using the phrase "strong and fragile classification of glassformers," it is appropriate to refer to Angell as the originator. However, it is not scientifically accurate to call the log η vs. T_g/T plot as the "Angell plot." This is a minor point, but there is no reason for anyone who values the importance of T_g-scaled plot to ignore the historical contributions of Oldekop, Laughlin, and Uhlmann.

The terms "fragile" and "strong" originated from an idea of Angell [215, 216] used to explain why some glassformers like 0.4 $Ca(NO_3)_2$–0.6 KNO_3, OTP, and propylene carbonate show rapid changes in η or relaxation time τ_α for the same change of scaled reciprocal temperature, T_g/T, at temperatures near $T = T_g$ in the OLUA plot. The key concept in his idea was the loss of the local structure (short range order) with increasing T above the glass transition. He suggested the structures of these glass-forming liquids change significantly with increase in temperature, and hence he labeled them "fragile" glassformers. On the other extreme in the OLUA plot are glassformers having smaller changes in η or τ_α for a given change of T_g/T near $T = T_g$. Their minimal slopes and curvatures in the OLUA plot mean that η or τ_α has Arrhenius or near Arrhenius temperature dependence. Angell attributed this behavior of these glassformers to their ability to maintain the short range order to higher temperatures and consequently there are relatively smaller changes in η or τ_α for a given change of T_g/T near $T = T_g$. Angell called them "strong" glassformers, naturally since in English "strong" has the opposite meaning of "fragile." In fact, "strong" glassformers like SiO_2 and GeO_2 are networks with strong covalent bonds, and the structure is not changed much by temperature. The class of glassformers lying between the "fragile" and "strong" classes in the OLUA plot was referred to as intermediate liquids. A similar pattern has been found for amorphous polymers by plotting log τ_α against T/T_g [479] where τ_α is the local segmental relaxation time. It was also found when plotting against T_g/T the α-relaxation time τ_α of the tertiary chalcogenide glassformers, $Ge_xAs_ySe_z$ [480], and in some combination of various glassformers [535]. The viscosity of polymers involves the motion of entire chains and its value depends sensitively on molecular weight and polydispersity, and the plot of log η against T_g/T should not be used. However, the pattern exhibited by the plot of log τ_α against T/T_g for polymers cannot be rationalized by Angell's interpretation of fragile–strong pattern given by him in early years because in polymers repeat units are bonded together and there is no obvious difference of the change of short range order from one polymer to another with change in temperature.

Nowadays, the words "strong" and "fragile" are often used to convey the curvature or location of either log η or log τ_α of the glassformer in the OLUA plot. However, one must bear in mind that the use of these words may have committed oneself to the interpretation by Angell of the source of the pattern either the one originally suggested [215, 216, 534] or his other alternative interpretations, one of which is based on landscape description. The landscapes may consist of a single megabasin for "strong" glassformers, whereas "fragile" ones have an abundance of well-separated megabasins [37]. These interpretations are qualitative and may not be accurate for some cases. There may be other factors including many-body relaxation dynamics which are neglected in these interpretations but do have effect on the observed T_g-scaled temperature dependence of η or τ_α. Therefore, it is prudent not to use the terms "fragile" and "strong" indiscriminately to describe the pattern seen in the OLUA plot for all different kinds of glassformers. Since the viscosity of a "fragile" glassformer is more sensitive to the temperature change than "strong" glassformer, it takes a shorter duration of time to shape an object from its melt and to anneal out the internal stresses from the finished object. Therefore, for a century, glass-makers have used the used the terms "long" and "short" instead of "strong" and "fragile," respectively [400]. But, the latter has been repeatedly used in the current literature; it is unlikely there will be any change in the practice. In fact, having said that, this author sometimes does conform to this practice in using the terms "fragile" and "fragility."

2.2.3.2 The Steepness or "Fragility" Index

One way to quantify the overall T_g/T-dependence of log η or log τ_α of a glassformer in the OLUA plot is by its slope m evaluated at $T_g/T = 1$ [480, 112]. That is,

$m = \mathrm{d}\log x/\mathrm{d}(T_g/T)$ evaluated at $T_g/T = 1$, where x is either η or τ_α. This quantity was mentioned before in Section 2.2.2.3 in connection with length scale of the structural relaxation. The terms "steepness index" and "fragility" are the notations commonly used nowadays for m. If τ_α has the VFTH dependence written in the form, $\tau_\alpha = \tau_\infty \exp[DT_0/(T - T_0)]$, then $m = DT_gT_0/[2.303(T_g - T_0)^2]$.

A different steepness index S was also introduced even a year earlier than m when considering viscoelastic data [479]. It is defined by $S = \mathrm{d}\log(a_T)/\mathrm{d}(T/T_g)$ evaluated at $T = T_g$, where a_T is the shift factor of the measured shear compliance of the local segmental relaxation. The difference between m and S comes from the scaled variable T_g/T used in the former and T/T_g in the latter. As soon as S for polymers had been introduced, examination was made for possible correlation between S and the parameter characterizing the non-exponential time dependence of the structural relaxation given by the exponent n of the Kohlrausch function also at $T = T_g$. This step taken was motivated by a previously published paper [535] which showed from the effect of many-body relaxation of the coupling model that a correlation can be expected between the T_g/T-dependence of η or τ_α and non-exponentiality of the structural relaxation of a glassformer.

The correlation was also seen between m and n in the family of the ternary chalcogenide glassformers, $Ge_aAs_bSe_c$, on varying the composition by changing the average coordination number, $<r>= 4a + 3b + 2c$, while keeping the ratio $y \equiv a/(a+b)$ constant and equal to 0.5 [480]. More glassformers were considered in the combined effort [112].

Glassformers come from many different classes distinguished by their chemical bondings and structures. Polymers are obviously different from non-polymeric materials because of bonds between repeat units in a long chain. Within polymers, there are the subclasses with carbon backbones or non-carbon backbones. Non-polymeric glassformers are further divided into liquids with hydrogen bonds, van der Waals liquids, plastic crystals, molten salts, inorganic glasses with covalent or non-covalent bonds, metallic glassformers, and others. Volume and entropy and their temperature dependences can be widely different in these different classes of glassformers and they may vary greatly in their influence on the temperature dependence of η or τ_α. Taking glassformers from different classes altogether at one time into consideration may suffer from the consequence of smearing out any correlation that m and n may have. Therefore, sharper or rigorous correlation can only be observed if the glassformers belong to the same class and are related in chemical structures. These aspects demonstrated already in the initial paper [112] will be further illustrated by other examples to be given later.

It is instructive to show the merit of the concept of fragility as well as pointing out possible perils encountered when interpreting it from the experimental data of glass-forming chalcogenides alone. For these covalently bonded network glassformers, Phillips introduced the constraint theory [536] to answer the fundamental question of why some of them, such as SiO_2 and the binary and ternary chalcogenide glasses As_2Se_3 and $Ge_aAs_bSe_c$, form glasses much more proficiently than others. He considered the strongest covalent bond stretching and bond bending forces between nearest neighbors to serve as Lagrangian (mechanical) constraints. The number of

Lagrangian bond-stretching constraints per atom with average coordination number $< r >$ is $< r > /2$, and of the bond-bending constraints is $2 < r > -3$. He suggested that glass-forming ability is optimal when the averaged total number of Lagrangian local-bonding constraints per atom, $5 < r > /2 - 3$, is equal to the number of degrees of freedom, which is 3 for three dimensional networks. Thus the optimal glass-forming composition corresponds to $< r >= 2.4$. Subsequently, Thorpe [537] reformulated the constraint theory as a problem of rigidity percolation. In random networks numerical calculations have shown that a floppy-to-rigid transition occurs when $< r >$ increases to a value quite close to the predicted mean-field value of 2.40. Experimental confirmations of the transition suggested by Phillips and Thorpe in various ways have been found in binary (Ge or Si)$_x$Se$_{1-x}$ and ternary Ge$_a$As$_b$Se$_c$ glasses [480, 538–541]. Of interest here is the finding in the same family of chalcogenides that the "fragility" exhibits a minimum as a function of $< r >$ with the minimum value occurring near $< r >= 2.4$, where the jump of heat capacity at T_g is small. More recently, measurement of the glass-forming $(Na_2O)_x(P_2O_5)_{1-x}$ liquids was made by photon correlation spectroscopy at temperatures near the glass transition for compositions extending from pure phosphorus pentoxide to the metaphosphate ($x = 0.5$) [542]. Pure P_2O_5 forms a three-dimensional network and has $< r >= 3$. On increasing Na_2O, $< r >$ decreases monotonically to 2. The "fragility" index m when plotted against $< r >$ also exhibits a very shallow minimum near $< r >= 2.4$.

Later on, Raman scattering and T-modulated differential scanning calorimetry (TMDSC) showed that in (Ge or Si)$_x$Se$_{1-x}$ actually there are two transitions, one near $r_c(1) = 2.40$ from a floppy to an (intermediate) unstressed rigid phase and another transition near $r_c(2) = 2.52$ from an (intermediate) unstressed rigid to a stressed rigid phase [543–545]. The intermediate phase lies within the compositional window defined by $r_c(1) < (< r >) < r_c(2)$ and has unusual properties compared with the other two phases. These include the fragility index m, the Arrhenius activation energy barrier of mechanical relaxation E, and the non-reversing heat flow term, ΔH_{nr}, obtained by MDSC. All these quantities exhibit a minimum in the intermediate phase. ΔH_{nr} provides a measure of the latent heat between the relaxed solid glass and its melt, and hence the degree of configurational difference between the glass and the liquid. In the intermediate phase, ΔH_{nr} is found not only to nearly vanish, but also not to age, in contrast to that found in the floppy and stressed rigid phases. The same was found in the ternary Ge$_a$As$_b$Se$_c$ chalcogenides which has $r_c(1) = 2.27$ and $r_c(2) = 2.42$ [546]. As $< r >$ increases across the floppy, intermediate, and the stressed rigid phases, fragility correlates well with the magnitude of ΔH_{nr}, and the implication it has on the structure and the degree of aging. As mentioned before for the Ge$_a$As$_b$Se$_c$ chalcogenides, if $y \equiv a/(a + b)$ is kept constant at 0.5, there is also a correlation between the fragility m and n at T_g on varying $< r >$ [480]. The values of $m \approx 80$ and $n \approx 0.58$ are largest at $< r >= 2.0$: both decrease monotonically on increasing $< r >$ to reach the minimum values of $m \approx 30$ and $n \approx 0.36$ when $< r >$ falls within the intermediate state regime defined by $r_c(1) = 2.27$ and $r_c(2) = 2.42$. However, when the ratio $y \equiv a/(a + b)$ is not kept constant at 0.5, for Ge$_a$As$_b$Se$_c$ having the same $< r >= 2.4$, m is the same, but $n \approx 0.67$ for $y = 0.2$ and $n = 0.52$ for $y = 1.0$. This breakdown of correlation between m and n was pointed out by Böhmer and Angell [480] themselves, which they attribute to the chemical effects near the binary edges of the ternary system. The admonition by these authors for the case of the $Ge_aAs_bSe_c$ chalcogenides is a special case of a more general breakdown of correlation between m and n when glassformers with widely different chemical structures are considered together.

2.2.3.3 Isobaric Fragility m_P Decreases with Increasing Pressure

The commonly quoted "fragility" and its index m are determined from the temperature dependence of η or τ_α at ambient pressure. Isobaric "fragility" at elevated pressure P and the index m_P had been obtained by relaxation measurements of many glassformers, small molecular [547–551] and polymeric [552–555]. Generally speaking, the isobaric m_P of van der Waals glassformers and polymers decreases with P although the rate of change varies greatly to the extent that for some such as poly(bisphenol A coepichlorohydrin) [552, 553], poly(vinyl acetate), and poly(ethylene acrylate) [555], m_P is independent of pressure. For examples from van der Waals liquids see Fig. 34a. Exceptions to this trend are found in strongly H-bonded materials such as glycerol [556].

2.2.3.4 The Isochoric "Fragility" m_V Is Significantly Less Than the Isobaric "Fragility" m_P

The isobaric "fragility" and its index m_P are obtained simply by plotting the temperature dependence of η or τ_α vs. T_g/T at each pressure. For the isochoric "fragility," one has to use *PVT* data in combination with the measured pressure and temperature dependence of η or τ_α to obtain these quantities at constant specific volume and then plot them against T_g/T. The isochoric "fragility" index m_V is then obtained as the slope of the plot log(η or τ_α) vs. T_g/T at $T = T_g$, in the same way as for m_P. In general for small molecular and polymeric glassformers, the values for m_V are all smaller than the corresponding isobaric m_P [555, 557]. For examples from van der Waals liquids see Fig. 34a. The dependence of the isobaric "fragility" on specific volume varies from no dependence for poly(vinyl acetate) and poly(ethyl acrylate), slight increase with specific volume for polystyrene, to significant increase with specific volume for poly(methyl acrylate) and poly(vinyl chloride). Interestingly, under isochoric conditions, the temperature dependence of the segmental relaxation time of poly(methyl acrylate) and poly(vinyl chloride) cannot be represented by a VFTH equation except over a very narrow temperature range. These results seem to suggest that although the VFTH equation is widely applicable to describe the temperature dependence of η or τ_α under isochoric condition, it may not have general validity under isochoric condition.

From the values of m_V obtained for molecular liquids and polymers [119, 558] one can observe that there is no correlation between m_V and the non-exponentiality parameter n even when restricted to the same chemical class of glassformers.

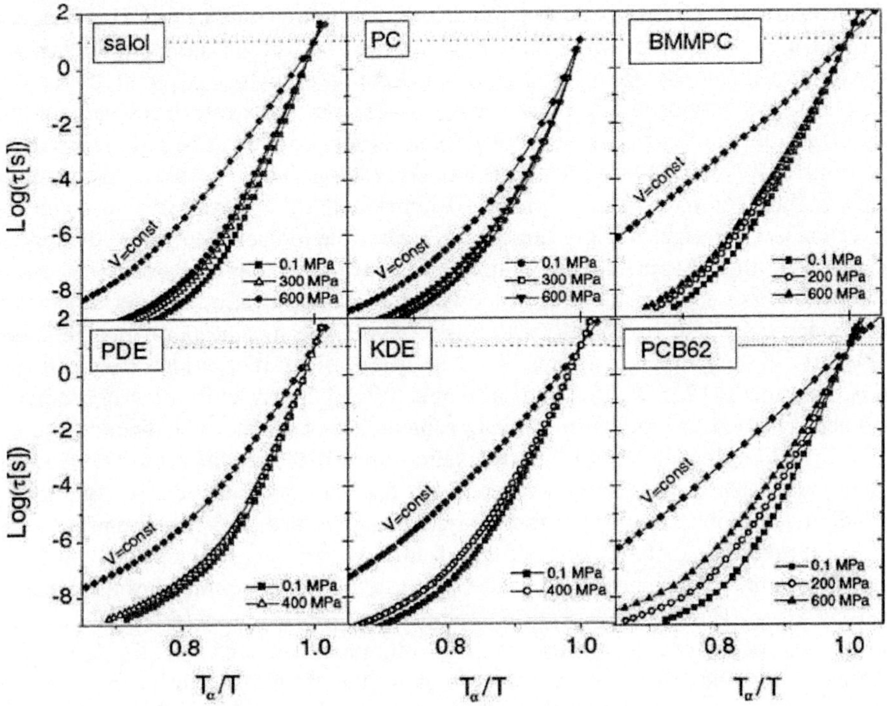

Fig. 34a Isobaric dielectric relaxation times for salol, PC, BMMPC, PDE, KDE, and PCB62 vs. T_α/T where $\tau(T_\alpha) = 10$ s. Isochors were calculated at the volume at which $\tau = 10$ s at atmospheric pressure; $V = 0.7907$ (salol), 0.7558 (PC), 0.9067 (BMMPC), 0.7297 (PDE), 0.7748 (KDE), and 0.6131 (PCB62) ml/g. Reproduced from [588(c)] by permission

2.2.3.5 Correlation Between Kinetic "Fragility" and Thermodynamic "Fragility"?

There is no doubt that volume and entropy do influence the mobility of basic structural unit of a liquid. On changing temperature/pressure, volume and entropy of the liquid change and in turn impart changes of η or τ_α and their temperature/pressure dependence. This certainly happens, independent of the presence of any addition determining factor of η or τ_α such as the many-body relaxation dynamics, which lies outside the realm of thermodynamic description. If an additional factor exists, then m is not a quantity that is determined solely by thermodynamics. Moreover, since the degree of the influence of volume and entropy on η or τ_α depends on the chemical and physical structure, even the thermodynamics contribution to m can vary greatly from one class of glassformers to another. Thus, the steepness or "fragility" index m is a complex or mixed parameter. It does not point exclusively to any single and fundamental factor determining glass transition, and hence its importance should not be overstated. When used as a central or key quantity to characterize or correlate glass transition dynamics of glassformers with widely different chemical and physical structures, it may lead to dubious or even wrong conclusions. This danger is acute at recent times because "fragility" has been so well promoted in the literature that by now it has become the center of attention of many workers in the field. Somehow, these workers think that "fragility" is the key to a full understanding of glass transition, and establishing any such correlation with it will be a step forward. This frame of mind has led to the pursuit of correlations between various properties and "fragility" or m (or other alternative parameters like $F_{1/2}$) exclusively.

The most serious push in this direction is the correlation between thermodynamic quantities or their T_g-scaled temperature dependence and "fragility" or m by Angell and coworkers with the intention of showing that glass transition is determined by thermodynamics and its change with temperature. The first proposal that related "fragility" to the jump in heat capacity C_p at T_g [37, 215, 216, 534] has been shown to be incorrect [417, 559–561] and subsequently withdrawn, even after an account had been taken of the number of rearrangeable units per mole of glassformer [562].

Still building on the notion that relaxation properties are related to thermodynamic quantities, Angell and co-workers put forth the proposition that fragilities could be determined from other thermodynamic quantities [563]. In one proposal, $1/S_{exc}$, the reciprocal of the excess of liquid entropy over that of the crystal, at different temperatures T_g/T above T_g is scaled by $1/S_{exc}(T_g)$, the reciprocal of the excess entropy at T_g. The initial paper [563] plotted $S_{exc}(T_g)/S_{exc}$ against T_g/T above T_g for three different glassformers, As_2Se_3, bromopentane, and $Ca(NO_3) \cdot 4H_2O$. The data show the same pattern as that found in the "fragility" plot of these three glassformers, and this result led the authors of [563] to conclude that thermodynamics, in particular entropy, determines "fragility." This different T_g/T-dependences of the scaled $S_{exc}(T_g)/S_{exc}$ called "thermodynamic fragilities" was probably motivated by the Adam–Gibbs equations (2.21) and (2.22), indicating that $\log(\tau_\alpha)$ is proportional to reciprocal of the configurational entropy, S_c, and the assumption that either $S_c = S_{exc}$ or S_c is proportional to S_{exc}. However, as we have discussed, neither of these two assumptions are valid. Nevertheless, the publication of [563] invited further the investigation of the correlation of the T_g/T-dependences of the scaled $S_{exc}(T_g)/S_{exc}$ (thermodynamic "fragilities") with their T_g/T-dependences of the $\log(\tau_\alpha)$ (kinetic "fragilities") [416]. The glassformers chosen are within the class of organic small molecular glassformers, where one can find plenty of the most accurately determined entropy data by adiabatic calorimetry with vibration contribution taken into account [420]. The results show a number of violations and cast doubt on the proposed correlation between "thermodynamic fragility" and "kinetic fragility" [416]. More glassformers of different classes were added by Martinez and Angell [564] to show the correlation. The inorganic glassformers have extended both "thermodynamic fragility" and "kinetic fragility" to much lower values than the organic small molecular glassformers. With all kinds of glassformers included, a rough correlation seems to exist between "thermodynamic fragilities" and "kinetic fragilities." However, when confined to the organic small molecular glassformers, their results (see Fig. 3 in [564]) do not show any correlation, consistent with the previous work [416] that show explicitly the various breakdowns of correlation. In another paper [562] that favors the correlation of m with thermodynamic fragility

[66], the thermodynamic fragility of glycerol ($m = 53$) was shown to be smaller than that of *cis*-decalin ($m = 147$) and hence the correlation. However, the thermodynamic fragility of $Ca(NO_3)_2 \cdot 4H_2O$ is still larger than that of *cis*-decalin, although the m value of the former was not given. The correlation would still break down unless m of $Ca(NO_3)_2 \cdot 4H_2O$ is larger than 147 of *cis*-decalin, which is unlikely because CKN has $m = 93$.

Here is an example of the difference in view that can occur between workers coming from different scientific backgrounds. While the breakdowns equated to anomalies are emphasized by physicists as clues to find the solution to the problem [25, 160, 161], they do not seem to be a matter of concern to chemists such as Martinez and Angell. Instead, they cited support from computer simulation studies of hard spheres [565], supercooled water [566], and binary Lennard-Jones particles [567]. Their belief in thermodynamic factors are sufficient to describe α-relaxation and glass transition is made clear by the statement "... it is now being suggested by molecular-dynamics (MD) studies that liquid diffusion is a process dominated by thermodynamic factors." It is risky to jump to this conclusion based on a correlation which has exceptions without taking into consideration other experimental aspects, particularly those pointing to the role played by the dispersion of the α-relaxation. There is no doubt that thermodynamic factors constitute the major cause for the rapid increase of τ_α and vitrification, but the role of other factors may not be so easily dismissed. The α-relaxation is definitely a many-body process, and most workers would not disagree. But, can all the effects of many-body dynamics be captured by thermodynamics alone, as some theories such as the Adam–Gibbs model and the free volume model suggest? For the sake of argument, let us not be so ambitious to solve the glass transition problem, and instead consider just the properties of the α-relaxation of a liquid either at constant T, P, or V, or at constant τ_α. As we shall see by many examples throughout this treatise, there are still several important and general properties of the α-relaxation that have no relation to thermodynamic factors but require explanation. These situations suggest thermodynamic factors alone cannot fully describe the α-relaxation, and some basic physics of many-body relaxation leading to the thermodynamic factors have yet to be found and implemented into any theory of glass transition before it can claim to have solved the problem. Undaunted by this possibility, Martinez and Angell sought to rationalize the correlation in terms of thermodynamic factors by the Adam–Gibbs equation. As has been discussed before in a previous section, the Adam–Gibbs theory shows explicitly that the reciprocal of the configurational entropy S_c governs $\log(\tau_\alpha)$. S_c is only a part of S_{exc}. In order to have the correlation as a consequence of the Adam–Gibbs equation, Martinez and Angell asserted that S_c and S_{exc} change proportionally with increase of temperature above T_g. This assertion was not supported by analysis of thermodynamic data by Johari [409, 413, 415].

For 11 amorphous polymers for which sufficient thermodynamic and relaxation data were available, it was found that their "thermodynamic fragilities" in the plot of $S_{exc}(T_g)/S_{exc}$ against T_g/T bear no relationship to their corresponding "kinetic fragilities" in the plot of $\log(\tau_\alpha)$ against T_g/T [417]. Though polymers have many repeat units bonded together to form long chains, their thermodynamic and dynamic properties related to glass transition are qualitatively no different from non-polymeric glassformers, and even selenium and some metaphosphate glassformers which have long chains. One has to reconcile with the non-existence of strong connection between "thermodynamic fragility" and "kinetic fragility" in polymers.

Establishing correlations between physical quantities is a worthwhile pursuit in a research problem such as glass transition that has no satisfactory solution. Yet even a perfect correlation is at best an indication of a possibility but cannot be used as a proof of any theory. If a correlation has exceptions or breakdowns, like in the present case of thermodynamics and kinetic "fragilities," they should not be ignored. They are anomalies which signal that some other piece of physics may be missing in the consideration of the correlation (see Section 1.4.2 for a discussion on the importance of anomalies as guides to solution of a problem). Now, thermodynamics and kinetic "fragilities" are just two out of more quantities that characterize the thermodynamics and dynamics of glassformers. Even if undisturbed by the imperfect correlation between thermodynamics and kinetic "fragilities," one should follow this up with the next question, which is whether other properties are also correlated with them. One outstanding property is the non-exponentiality of the structural relaxation characterized by n that appears in the exponent $(1 - n)$, of the Kohlrausch correlation function. We have mentioned before that if all kinds of glassformers are included in the consideration, there are many breakdowns of the correlation between m and n. Correlation is strong or perfect only when restricted to glassformer from the same family with related chemical and physical structures. Martinez and Angell defined a thermodynamics "fragility" index $F_{3/4}$ by the value of T_g/T at which $S_{exc}(T_g)/S_{exc}$ assumes the value of 0.75. The next question is: how good is the correlation between $F_{3/4}$ and n? By inspection of Fig. 3 of [564], we see $F_{3/4} \approx 0.44$ of B_2O_3 is smaller than $F_{3/4} \approx 0.76$ of 3-bromopentane and $F_{3/4} \approx 0.80$ of propylene carbonate (note that the scale of $F_{3/4}$ in the figure is different for values above 0.7). In the opposite direction, among the three, B_2O_3 has the largest $n \approx 0.40$ [112] compared with $n \approx 0.29$ for 3-bromopentane [253] and $n \approx 0.27$ for propylene carbonate [195]. Comparing $ZnCl_2$ with B_2O_3, both inorganic glassformers, $ZnCl_2$ has $F_{3/4} \approx 0.55$ larger than that of B_2O_3, but its $n \approx 0.27$ is smaller than that of B_2O_3. Selenium has $F_{3/4} \approx 0.65$ less than the $F_{3/4}$ values of all the small molecular glassformers considered by Martinez and Angell, and yet it has the largest $n \approx 0.58$ [480]. An example of such violation is given by comparing selenium with glycerol ($F_{3/4} \approx 0.70$, $n \approx 0.29$). More examples can be given to show that $F_{3/4}$ does not correlate with n. One may turn around and say that the above comparisons actually indicate the existence of an anticorrelation between $F_{3/4}$ and n, e.g., a small $F_{3/4}$ is associated with a large n. This hope is dashed by GeO_2 which has $F_{3/4} \approx 0.12$ and $n \approx 0.0$; both figures are the smallest. One may go to the extreme to declare that the time/frequency dispersion of the structural relaxation is not as important or fundamental a property as m and $F_{3/4}$, and hence correlation of the latter with n is not expected. However, this view needs serious justification. By considering all kinds of glassformers as done by Martinez and Angell, the upside is that it does give an impression that thermodynamics and kinetic "fragilities" correlate for a number of glassformers, but the downside is that it exposes the other problem.

The problem is no longer that serious if the glassformers chosen all belong to the same class as demonstrated for some small molecular glass-forming liquids [253]. For a selected few, it was shown there that $z^*(T_g)$ obtained from thermodynamic data correlates with the steepness index m and exponent n. Moreover, the temperature dependences above T_g of n, m, and z^* are quite similar. This observation may be the consequence of synergy of all the factors that determine n, m, and z^*. There is serious doubt on this correlation from the multiple violations of the proposed correlation found even within the same class of glassformers. The failure of extending the correlation between "thermodynamic fragility" and "kinetic fragility" to include the time/frequency dispersion of the structural relaxation or n is another indication that the correlation is not fundamental. According to the coupling model, thermodynamics is not the only factor that determines the properties of glassformers as envisaged in the Adam–Gibbs theory or other models. Many-body relaxation dynamics and their effects must be included before any good sense out of any correlation or lack of correlation between two quantities or parameters.

The effort of Angell and coworkers to establish a connection between the thermodynamic properties and dynamics and relaxation properties was motivated by the Adam–Gibbs model and the development of an energy landscape model and they used it to analyze data on various small molecular glassformers [37]. The central idea of the model for the structural and transport properties of a glassformer is that the manner by which they evolve with temperature is governed by the density of configurational states comprising the potential energy hypersurface. The topology of this energy landscape (i.e., the number of minima and the barrier heights between them) provides a measure of the steepness of the excitation profile for glass-forming liquids. According to the model, when heated through T_g, fragile liquids readily transition among many configurational states, giving rise to substantial changes in relaxation times and viscosities. In general, one may expect that the topographic features of the potential energy surface are possible sources of the apparent connection between thermodynamics and kinetics in supercooled liquids. However, this picture was not supported by the families of model "rugged landscape" potential energy functions using a modular basin approach by Stillinger and Debenedetti [568]. This study was intended to clarify the molecular level basis for the relationship between thermodynamic and kinetic properties of glassformers. Topographically, the families of model functions are sufficiently diverse such that pairs of model functions can be chosen to share exactly the same thermodynamic behavior (depth distribution of potential energy minima), but they differ drastically in those topographic attributes that control kinetic and relaxation behavior. Hence, within this landscape model, there is no correlation between "thermodynamic fragility" and "kinetic fragility." These authors did not question the possibility of validity of the correlation between thermodynamic and kinetic behaviors as suggested by the Adam–Gibbs equation and the empirical studies by Angell and coworkers. Instead, in order to bring the landscape model to produce the correlation, they suggested considering "an additional physical principle involving details of interparticle interactions, transcending the purely mathematical aspects of potential energy landscape topography." This suggestion interestingly originated from the authors who have some faith that the correlation between thermodynamic and kinetic behavior has been firmly established and is fundamental. The coupling model (CM) has often stressed besides thermodynamics the importance of including the effects of many-body relaxation dynamics due to intermolecular interactions in capturing the true nature of the structural relaxation. Apparently, many-body relaxation captured by the CM is the "additional physical principle involving details of interparticle interactions" that Stillinger and Debenedetti (in their own words) are looking for.

A different way of constructing the potential energy surface was given by Wales and Doye [569]. They found the model which exhibits higher kinetic fragility usually has a larger change in the heat capacity at the glass transition. However, this is not always the case. In some regions of parameter space, they found the correlations between dynamics and thermodynamics are not present. Thus, the conclusion of Wales and Doye are qualitatively similar to that of Stillinger and Debenedetti, namely the thermodynamic and kinetic properties are decoupled in the landscape models. There is another analysis of the relation between statistical properties of the landscape and fragility [570]. These authors cautioned that the knowledge of number, energy depth, and shape of the basins of the potential energy landscape may not be sufficient for predicting "fragility."

Sastry studied the relationship between kinetic and thermodynamic "fragilities," configurational entropy, and the potential energy landscape of binary LJ particles systems by molecular dynamics simulations [567]. From the simulation, the diffusion coefficient, D, and the configurational entropy, S_c, were obtained. All systems have the same number of A-type and B-type particles but differ in density. For each system, a kinetic "fragility" index, K_{VFT}, was obtained from the VFTH fit to the temperature dependence of D. It was found that K_{VFT} increases with density. The plot of $\log D$ against $(TS_c)^{-1}$ shows approximately a linear relation, in agreement with the Adam–Gibbs equation. From the scaled temperature dependence of TS_c, the slope, $K_{TI,AG}$, was defined and used as the quantitative index of thermodynamic "fragility." Another thermodynamic "fragility" index, $K_{PEL,AG}$, was defined and obtained from the potential energy landscape. $K_{TI,AG}$ is nearly the same as $K_{PEL,AG}$, and both are in agreement with kinetic "fragility" K_{VFT} for systems of different densities. The results led Sastry to conclude that the rate of increase of TS_c does indeed determine the kinetic "fragility," and there is a quantitative relationship between fragility and the energy landscape of the liquid. The fixation with "fragility," configurational entropy, and Adam–Gibbs equation by many including Sastry in [567] has somehow prevented them from looking into other equally important properties such as the time/frequency dispersion, and seeking more general interpretation by considering also the many-body dynamics. Sastry gives only the long-time steady-state diffusion coefficient, but not the time dependence of the diffusion coefficient and its change with density. This practice is like in the Adam–Gibbs theory where only η or τ_α is the quantity of interest, but not the dispersion. On decreasing density, on the average, the LJ particles are further apart. Intuitively, we can expect that entropy and volume become less sensitive to change in temperature. But, in addition there is concomitant reduction of inter-particle interaction and the effects of many-body relaxation on slowing down relaxation and diffusion that cannot be accounted for by the decrease of configurational entropy and specific volume. Thus, by decreasing density of the LJ particles, thermodynamics and many-body relaxation work in the same direction to reduce the scaled temperature dependence of D, as found by Sastry. But, attributing the change of the scaled temperature dependence of D (i.e., "fragility") exclusively to the configurational entropy is an oversimplification. The colloidal particles suspension systems with different volume fractions (densities) have been studied by confocal spectroscopy [141] and dynamic light scattering [143]. Both sets of data clearly show the dispersion of the time dependence of the mean-squared displacement or the diffusion coefficient decreases with decreasing volume fraction of the colloidal particles. This trend is expected to be there for the systems studied by Sastry, but unfortunately the data are not present. We recall here the molecular dynamics simulation by Bordat et al. [264] discussed in Section 2.2.1. They studied the change of dynamic properties of the binary LJ particles with change of the interaction potential. They showed the dispersion of the intermediate scattering function or n changes with the change in potential, caused by the change of the capacity of inter-particle coupling. The change in n correlates with the corresponding change in m, the kinetic "fragility," and the T_g-scaled temperature dependence of the non-ergodicity parameter determined by the vibrations at low temperatures in the glassy state observed experimentally by Scopigno et al. [571, 572]. Thus, from these other studies which have shown the important role played by the dispersion or n, it is premature for Sastry to interpret his data entirely in terms of thermodynamics and the Adam–Gibbs equation.

There is yet another interesting correlation suggested by Wang et al. [573, 574] which compares the "fragility" index m with a dimensionless combination of the heat capacity jump $\Delta C_p(T_g) = C_{p,\text{liquid}}(T_g) - C_{p,\text{glass}}(T_g)$ at T_g and the melting enthalpy H_m. This empirical relation, $m \approx m_{\text{cal}}$, found, where $m_{\text{cal}} = 56 T_g \Delta C_p(T_g)/H_m$, has no obvious connection to the Adam–Gibbs theory, but has subsequently been rationalized [575, 576] from the random first-order transition theory [577, 578]. For many glassformers of all kinds, there is good agreement between m obtained by experiment from the T_g/T-dependence of η or τ_α and the value of $56 T_g \Delta C_p(T_g)/H_m$ calculated from thermodynamic data, although are still a few exceptions. In spite of the correlation of m with m_{cal} found in many glassformers, there is no escape from facing the multiple breakdowns of correlation of these two parameters with n. This problem cannot be overcome at this time even with the help of the random first-order transition theory. However, the works by Wang et al. [573, 574] are important for further understanding of the connection between thermodynamics and dynamics.

The glass-forming metallic liquids are particularly intriguing. The non-exponential parameters n determined from the modulus measurements of these metallic liquids do not exhibit significant difference, while the kinetic fragility index m determined by mechanical modulus relaxation changes appreciably, and hence breakdown of the correlation between n and m [574(c)]. The same was found from extensive analyses of mechanical modulus relaxation data of molecular (mechanical) and electrical modulus data of ionic liquids [574(c)]. For the glass-forming metallic liquids, the analyses of enthalpy relaxation measurements of metallic glasses gave values of n_{cal} different from n, and the use of n_{cal} restored the correlation between n_{cal} and m. Nevertheless, one has to bear in mind that the determination of the non-exponential parameters n_{cal} from enthalpy relaxation is not direct, in contrast to n from mechanical modulus, but indirectly from fitting the data by the Tool–Narayanaswamy–Moynihan model [782–784(a)] using the program of Hodge [785, 786] that involves parameters in addition to n_{cal}.

2.2.3.6 Correlation of Kinetic "Fragility" with Other Quantities?

The kinetic "fragility" has been so well publicized that by now it is considered by most researchers to be a critical and fundamental property of glassformers in glass transition. Consequently, many other properties of glassformers have been compared with its measure, the "fragility" index m, with the purpose to find correlation. Correlation of a property with m if found would allow one to claim that the very property has a strong connection to the mechanism of structural relaxation and glass transition. We have seen before in this section several attempts to correlate some thermodynamic properties or quantities from experiments or theoretical models [565, 567, 570, 579] with m. Other properties invoked to demonstrate correlation with m include (i) the strength of the vibrational boson peak [580], (ii) the compressibility at T_g [572], (iii) the T_g-scaled temperature dependence of the non-ergodicity parameter determined by the vibrations at low temperatures in the glassy state [571, 572], (iv) T-dependence of the shear modulus [581–583], and (v) the Poisson's ratio in the glassy state [584]. The degree of success in showing the existence of a correlation with fragility varies from one property to another. The extreme happened in the case of the proposed correlation of m with Poisson's ratio. It was later found that the correlation of m with the ratio of elastic moduli breaks down even within the same family of glassformers including polymers and alkali borate glasses with different concentrations of alkali-metal atoms. In particular, it was found in polystyrene samples with different molecular weights that "fragility" changes in the direction opposite to the proposed correlation with Poisson's ratio [585]. Within the family of metallic glasses, subsequent addition of more experimental data has led to the final conclusion that the correlation is very poor, if present at all [586, 587]. The application of pressure also has led to the result that τ_α is a function of TV^γ where V is the specific volume, and γ is positive and varies over a wide range from near zero to about 9 depending on the glassformer [119]. By examining the values of γ in Table 2 and m_P or m_V in Table 3 in [119] of many glassformers, it is clear that there is no correlation between γ and m_P or m_V, even when restricted to the same class of glassformers. This is another indication that fragility is a complex and not fundamental parameter.

All the activities mentioned above rely on the hope that "fragility" is indeed the single most important concept of glass transition and key to the solution of the problem, or it is the sole parameter that controls other properties. These are false hopes that cannot be realized because "fragility" is a complex parameter, determined not only by the thermodynamic and mechanical properties but also by the dynamics of many-body relaxation.

2.2.3.7 Different Patterns of Change of *m* with the Molecular Weight *M* of Polymers

If the molecular weight of a polymer M exceeds some critical high molecular weight, M_c, polymer chains become entangled and T_g remains constant. For low molecular weight polymers with $M < M_c$, in general T_g increases with molecular weight without exception. However, in the same region with increasing M, "fragility" can increase [589–592] like in polystyrene, and remain the same as in poly(dimethylsiloxane) [593]. In polyisobutylene, "fragility" from dielectric relaxation measurements decreases with increasing M. On the other hand, "fragility" from dynamic mechanical measurements remains constant with increase in M [592]. These inexplicable widely different changes of "fragility" with M of different polymers, and in the same polymer obtained by different techniques, all indicate that "fragility" is neither simple nor fundamental.

2.2.3.8 Breakdown of Correlation Between *m* and *n*

Small Molecule Organic Glassformers

Even restricted to some glassformers within the class of small molecules made of carbon, oxygen, and hydrogen atoms, the diverse chemical structures available to them may bring in different volume and entropy contributions to change of τ_α to T_g-scaled temperature dependence, or m. As a result, there is no correlation of m with n. An example of bad actors is the much studied glassformer propylene carbonate (4-methyl-2-oxo-1,3-dioxolane), which has large m=90 which is larger than that of salol ($m = 73$) and *ortho*-terphenyl ($m = 84$), and much larger than that of glycerol ($m = 57$), and the methylated hydrocarbons including 2,4,6-trimethylheptane and 3-methylheptane with m values falling in the range from 46 to 63 [594, 595]. All these m values were obtained at $T = T_g$, where T_g is the temperature at which $\tau_\alpha = 100$ s. On the other hand, the n value of propylene carbonate from dielectric relaxation measurements is 0.27 [195, 424], which is significantly smaller than the dielectric values of salol ($n = 0.37$) [596], *ortho*-terphenyl ($n = 0.50$) [237], 2,4,6-trimethylheptane ($n = 0.44$), and 3-methylheptane ($n = 0.45$) [595]. The same situation applies to phenylphthalate dimethylether (PDE) and cresolphthalate dimethylether (CDE or KDE). The chemical structure of propylene carbonate has the oxygen-bonded unit not shared by the other glassformers. Plausibly, this unusual chemical structure of propylene carbonate with flexible oxygen bonds enhances the volume and entropy contribution to the T_g-scaled temperature dependence of τ_α. Support of this possibility comes from propylene carbonate having a larger $z^*(T_g)$, calculated from the excess entropy obtained by adiabatic calorimetry [597], than *ortho*-terphenyl. Corroborating piece of evidence comes from positronium annihilation spectroscopy that shows the unoccupied volume measured by the *ortho*-positronium lifetime, τ_3, has T_g-scaled temperature dependence as strong as that of *ortho*-terphenyl [387]. These two experimental results indicate the strong influence that volume and entropy may have on the T_g-scaled temperature dependence of the structural relaxation time of propylene carbonate because of its special chemical structure. On the other hand, comparing propylene carbonate with PDE and CDE that have similar oxygen-bonded units, the correlation between m and n works well for these three glassformers [597]. The lesson we learn is that correlation between m and n may hold well only for glassformers having similar chemical and physical structures. Consideration of glassformers with different chemical structures and physical interactions will not find any recognizable correlation found as early as 1990 by Torell et al. [598] and later by Schröter and Donth [599]. The latter considered metallic glass, inorganic glass, van der Waals, and hydrogen-bonded glassformers, and the absence of correlation between fragility and n is unsurprising. Schröter and Donth did not understand this point and instead used the lack of correlation against the CM. They cite the study of a series of polybutadienes of different microstructure by Zorn et al. [600], which are closely related glassformers and the correlation indeed works. Nevertheless, Schröter and Donth mentioned another (unrelated) prediction of the CM on the difference in temperature dependence between viscosity and local segmental relaxation, which Zorn et al. found to be qualitatively correct but not quantitatively. Actually the lack of quantitative agreement is due to the fact that Zorn et al. could not measure both the viscosity and the local segmental relaxation at the same temperatures close to T_g by dynamic mechanical method. They had to extrapolate the viscosity or terminal relaxation shift factor measured at higher temperatures to lower temperatures by using the VFTH dependence, introducing large uncertainties in the process. The data are insufficient to test the prediction quantitatively. The test can only be made with creep measurement of Plazek [165, 168–171, 360], who measured both terminal and local segmental relaxation near T_g. Schröter and Donth cited also the lack of correlation in three different substituted poly(*p*-phenylenes) by Connolly et al. [601], who mentioned the presence of internal plasticization in one sample and the influence of free volume inferred from macroscopic expansion coefficient. Schröter and Donth cited also the dynamic mechanical measurements by Santangelo and Roland on low and high molecular weight polystyrenes which show a variation in fragility but a constant shape of the relaxation curves at the dynamic glass transition [589], but they did not cite another publication by Rizos and Ngai [590] which explained this as due to broadening of the dispersion by chain ends in low molecular weight PS by providing measurement over a much broader frequency and temperature range.

For propylene carbonate PC, PDE, and CDE, however, more needs to be said about the narrow dielectric susceptibility dispersion $\varepsilon^*(\nu)$ of these highly polar glassformers with large ε_∞, the high-frequency dielectric constant. Light scattering [598, 211, 602] and mechanical shear modulus [599] measurements of these glassformers have found much broader dispersion. The Kohlrausch exponent $\beta_K = (1 - n)$ of PDE from light scattering is about 0.5–0.6, while from dielectric relaxation β_K lies between 0.7 and 0.84 in a similar temperature range. On the other hand, for less polar glassformers such as BMMPC and BMPC that have smaller

ε_∞, both the relaxation times and β_K obtained by dielectric susceptibility and by light scattering are the same. It was also found that after transforming the dielectric data of propylene carbonate and PDE from susceptibility, $\varepsilon^*(\omega)$, to electric modulus, $M^*(\omega)$, the frequency dispersion of $M^*(\omega)$ is as broad as seen by light scattering and mechanical relaxation [603, 604]. The larger width of $M^*(\omega)$ found in PC, PDE, and CDE having large ε_∞ could be due to the M'' α-loss peak overlapping with the unresolved secondary relaxation in these glassformers, and hence the broadening. These findings and implications need further investigation.

Plastic Crystalline Materials

Even restricted to glassformers having the same physical structure, m will not correlate with n if the glassformers have widely different chemical structures (and thus different unoccupied volume and entropy). An example can be given by the class of plastic crystalline materials. The plastic crystals refer to the class of molecular materials having the centers of mass of the molecules located on sites of a regular crystalline lattice but the molecules are dynamically disordered in their orientational degrees of freedom. The orientational relaxation dynamics of plastic crystals are similar in many respects to the relaxation dynamics of ordinary glassformers [144, 145, 605–612]. By sufficiently fast cooling, complete orientational ordering can be avoided and the dynamically disordered state can be supercooled. The orientational dynamics slow down on decreasing temperature and the relaxation time increases over many orders of magnitude. At temperatures below the orientational glass temperature, T_g, the relaxation time becomes so long that the orientational degrees of freedom are frozen and a glass-like orientationally disordered state is reached. Six plastic crystalline materials having widely different chemical structures were studied by Brand et al. [145] by dielectric relaxation measurements over a broad frequency range. They are 1-cyanoadamantane (CNA), adamantanone (AON), pentachloronitrobenzene (PCNB), cyclo-hexanol (CHEX), ethanol [145], and cyclo-octanol (COCT) [613]. The "fragility" index m was determined with T_g defined as the temperature at which the dielectric α-relaxation time reaches 100 s. In increasing order, the m values are 17 for PCNB, 19 for AON (phase II), 33 for COCT, 38 for ethanol, and 48 for CHEX. On the other hand, the n values are 0.5 for PCNB, 0.53 for AON (phase II), 0.25 for COCT, 0.25 for ethanol, and 0.38 for CHEX. On comparing the values of m and n in the five plastic crystals, it is clear there is no correlation. For example, PCNB has the smallest m but almost the largest n. The chemical and physical structures of these plastic crystals are quite different, and these differences may be the cause of the dissimilar dependences of their τ_α on temperature.

Neat Glassformer and When Mixed with Another Glassformer

One way to demonstrate that volume and/or entropy enter into the determination of "fragility" is by comparing the dynamics of a neat glassformer A with that of the same molecules A when mixed with another glassformer B. The packing of the A molecules, and hence volume and entropy, can be quite different in the two situations, resulting in breakdown of the correlation between n and the "fragility" index m of the same molecules A. Such breakdown was found in *tert*-butylpyridine (TBP) or quinaldine (Qn) in their pure states and in mixtures with trimer of styrene (tristyrene or 3Styr). TBP and Qn are polar rigid molecules with large dipole moments, while 3Styr has negligible dipole moment and a much higher T_g. Consequently, the dielectric relaxation experiment [614] only probes the dynamics and determines the quantities m and n of TBP or Qn in the mixtures. In binary mixtures, concentration fluctuations are present, which cause additional broadening of the dispersion of the α-relaxation and the true value of n cannot be determined by fitting the experimental complex dielectric constant $\varepsilon^*(\omega)$ in Eq. (1.94) with $\psi(t)$ given by the Kohlrausch function. There is a way to deduce the true value of n from the data by a relation of the ratio of τ_α/τ_β, where τ_β is the secondary relaxation time. This method will be discussed later in Section 2.3. Anticipating the results, on increasing the concentration of 3Styr in the binary mixtures, it was found that m of either TBP or Qn decreases almost monotonically whilst n increases monotonically. As an example, for the TBP/3Styr mixtures, m (n) assumes the maximum (minimum) value of 101 (0.36) for neat (100%) TNB, and decreases (increases) nearly monotonically with increasing 3Styr concentration down to 79 (0.63) at 5% TBP. This spectacular breakdown of correlation between m and n is likely due to the change in packing of the rigid TBP or Qn molecules in the mixtures with increasing presence of the more flexible 3Styr molecules. Consequently, the influence of volume and entropy on the dynamics of TBP or Qn is much reduced, and hence also the "fragility." On the other hand, when present, the much slower 3Styr molecules slow down the dynamics of TBP or Qn and stretch its α-relaxation over a longer period of time, thus increasing n. This increase of n has the potential of increasing the fragility of TBP or Qn, but it is opposed by the stronger decrease from the diminishing sensitivity of τ_α to volume and entropy, resulting in the observed decrease of m. This explanation is qualitative. Nevertheless, the complex nature of "fragility" is brought out by the experimental results.

2.2.3.9 Restoration of Correlation Between *m* and *n* When Restricted to the Same Family

The pursuit of correlations is in part due to the lack of a reliable theory of glass transition. The correlation of n with m discussed above was one of the very first of such attempts in 1987 [535], in 1991 [479], and in 1993 [112], and we have seen that there are many instances of breakdown of the correlation if glassformers of all kinds are considered all at one time. The change of thermodynamics with temperature, which is one of the factors that determine m, is not expected to be significantly different if the glassformers considered belong to the same family with similar chemical and physical structures. Examples include the carbon backbone polymers [244, 479] and the supercooled chalcogenide systems composed of different amounts of Ge, As, and Se [480]. These systems studied in these previous works show the existence of correlation between m and n.

The same chemical composition of the repeat units of a polymer but having the pendant group arranged in different ways in repeat units (i.e., different tacticity) is perhaps the best case for testing the correlation between m and n. The most complete study of the dependence of dynamics on tacticity is on poly(methyl methacrylate) (PMMA) and poly(ethyl methacrylate) (PEMA). For the same molecular weight, syndiotactic and atactic PMMA and PEMA have higher T_g and smaller fragility index m than their isotactic homologues, i-PMMA and i-PEMA. The i-PMMA and i-PEMA are more ordered because the pendant groups are all on the same side as opposed to more random arrangements in the syndiotactic and atatic polymers. This suggests at least intuitively weaker intermolecular coupling and smaller coupling parameter n in i-PMMA and i-PMMA. Experimental support of this was given in [252], establishing the correlation between m and n for the polymers differing only in tacticity.

Another example is the family of hydrogen-bonded polyalcohols. Glycerol, threitol, xylitol, and sorbitol belong to this family and they differ mainly in the number of carbon atoms on the backbone which increases from 3 to 6 in the order given. The values of (m,n) for them are glycerol ($m = 57$, $n = 0.29$), threitol ($m = 79$, $n = 0.36$), xylitol ($m = 94$, $n = 0.46$), and sorbitol ($m = 128$, $n = 0.52$). The m values are from [239, 615], and n values are from [195].

Another example can be drawn from the family of epoxy resins [616]. The members of this family have the same basic chemical structure. These include *t*-butylphenylglycidylether (*t*BPGE), diglycidylether of bisphenol F (DFDGE), phenylglycidylether (PGE), the monoepoxy *o*-cresylglycidylether (*o*CGE), the diepoxies DGEBA and PPGE, the triepoxy TPMGE. The good correlation between m and n is shown in Fig. 34b.

The glass-forming $(Na_2O)_x(P_2O_5)_{1-x}$ liquids with compositions extending from pure phosphorus pentoxide to the metaphosphate ($x = 0.5$) have been studied by dynamic light scattering [542]. Correlation between m and n was found. This is

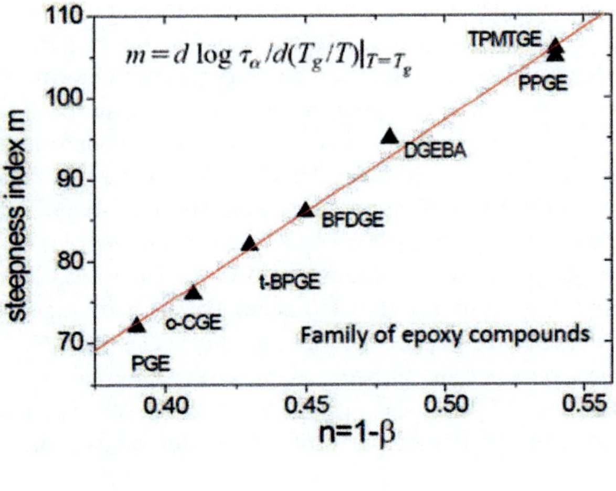

Fig. 34b Plot of steepness index evaluated at T_g vs. $n \equiv (1 - \beta)$ where β is the Kohlrausch exponent for epoxy systems. The *continuous line* represents a linear fit of the data. Reproduced from [616] by permission.

no surprise because these glassformers belong to the same family. The value of n has not been determined at T_g and its value should be taken from the Kohlrausch function fitting the dynamic light-scattering data with τ_α equal to some fixed but sufficiently large value say 1 s. This is because in samples with higher x, the Na ion conductivity relaxation overlaps the structural relaxation at shorter τ_α and broadens the α-dispersion.

We have previously described the molecular dynamics simulations of the family of binary Lennard-Jones (LJ) systems consisting of two kinds of particles A and B, but with different LJ potentials [264]. The general form of the LJ potential is given by Eq. (2.68). The three LJ potentials come from three different combinations of q and p : ($q = 12$, $p = 6$), ($q = 8$, $p = 5$), and ($q = 12$, $p = 11$). From the molecular dynamics simulation data of $\tau_\alpha(T)$, m for each LJ system is defined as the T_{ref}/T-dependence of τ_α where T_{ref} is the temperature at which $\tau_\alpha(T_{ref}) = 46{,}435.8$ (LJ units of time), which is a very long time in simulations. It was found that m correlates well with n for these three binary Lennard-Jones systems. Therefore, when restricted to glassformers that are closely related members in the same family, "fragility" and its index m correlate well with n. This shows again that n does not fail in correlation with other quantities when the latter are unambiguous.

2.2.3.10 Colloidal Suspension of Soft Spherical Particles: Proving Non-exponentiality (n) and Fragility (m) Are Parallel Consequences of Inter-particle Interaction

An enlightening experimental fact indicating that many-body α-relaxation from intermolecular interaction is one of the determining factors of fragility in ordinary glassformers can be inferred from a report by Weitz and collaborators [617a] and Mattsson et al. [617b]. They studied the dynamics of colloidal suspensions of soft spherical particles, specifically closely packed microgels. The softness of the particles fabricated can be adjusted, ranging from very soft, intermediate soft, to the conventional hard spheres (with PMMA core, studied before in [617c]). For hard-sphere suspensions, the particle concentration is quantified by the volume fraction, $\phi = NV_p$, where N is the number density of particles and V_p is their volume. However, because the microgel particles are deformable, their volume is not fixed and ϕ is no longer a good measure of concentration. Instead, $\zeta = NV_0$ was used, where $V_0 = 4\pi(R_0)^3/3$ is the volume of an undeformed particle of radius R_0 measured in dilute suspension. The dynamics of each of these colloidal suspensions was measured by confocal microscopy. The time-dependent correlation function was fitted to the Kohlrausch correlation function, $\exp[-(t/\tau_\alpha)^\beta]$, for all the colloidal suspensions at different volume fractions ζ. Their relaxation times τ_α as a function of ζ are compared in Fig. 34c, left panel. For each microgel sample, they also determined the frequency-dependent shear moduli, $G'(\omega)$ and $G''(\omega)$, by oscillatory rheology. They obtained the viscoelastic τ_α as the timescale corresponding to the frequency for which $G'(\omega) = G''(\omega)$. Shown by plus symbols in the figure, the viscoelastic τ_α has the same dependence on ζ as τ_α observed by light scattering.

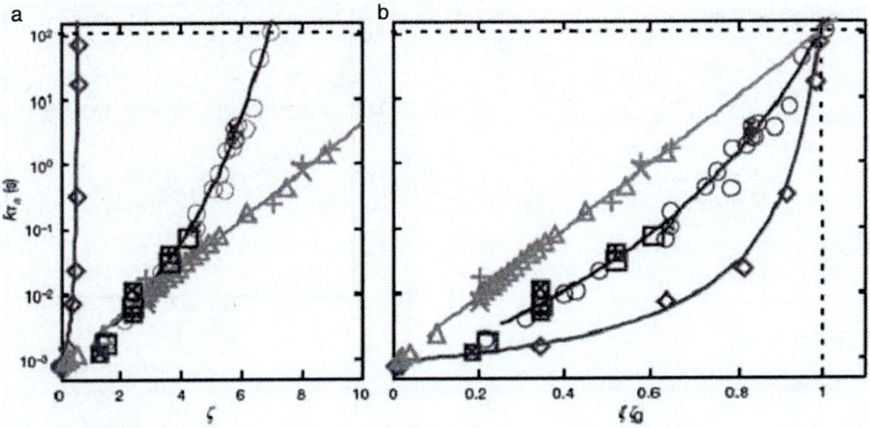

Fig. 34c (*Left*) Plot of $k\tau_\alpha$ vs. ζ for stiff (*diamonds*, $R_0 = 95$ nm), intermediate (*empty circles*, $R_0 = 92$ nm), and soft (*triangles*, $R_0 = 80$ nm) microgels, where k is chosen to collapse the data onto those of the intermediate sample at low ζ values. Data for a second intermediate sample (*empty squares*, $R_0 = 168$ nm) scale onto those of the first for $\zeta > \zeta^*$, as expected. Rescaled shear viscosities (intermediate: *crosses in circles*, $R_0 = 92$ nm, and *crosses in squares*, $R_0 = 168$ nm; soft: *crosses*, $R_0 = 80$ nm) and rheological structural relaxation times (intermediate: *pluses in circles*, $R_0 = 92$ nm, and *pluses in squares*, $R_0 = 168$ nm; soft: *pluses*, $R_0 = 80$ nm). (*Right*) Same as (*Left*), with ζ normalized by $\zeta_g = \zeta(\tau_\alpha = 100\text{ s})$. Reproduced from [617b] by permission

Defining ζ_g as the value of ζ at which $\tau_\alpha = 100$ s uniformly for all suspensions, the data of log τ_α are replotted against ζ/ζ_g. This plot shown in the right panel of Fig. 34c is the analogue of the OLUA plot of log τ vs. T_g/T for molecular glassformers. A pattern emerges from this plot for the colloidal suspensions. The softer the spherical particles, the weaker the dependence of log τ_α on ζ/ζ_g, and smaller the steepness index m_ζ defined by $\mathrm{d}\log\tau/\mathrm{d}(\zeta/\zeta_g)$ evaluated at $\zeta/\zeta_g = 1$. This pattern reproduces the OLUA plot for ordinary glassformers.

Mattsson et al. also compared the stretch exponent $\beta \equiv (1-n)$ of the Kohlrausch correlation function, $\exp[-(t/\tau_\alpha)^\beta]$ (Fig. 34d). They found, for ζ greater than some critical ζ^*, β is larger or n is smaller in colloidal suspension of softer particles, with $\beta \approx 0.8$ or $n \approx 0.2$ for the softest spheres, and $\beta \approx 0.6$ or $n \approx 0.4$ for the intermediate soft spheres. The photon correlation spectroscopy data of Bartsch et al. on the hard spheres show monotonic decrease of β with increasing ζ or τ_α. β is about 0.6 when $\tau_\alpha \approx 10^{-1}$ s and decreases to $\beta \approx 0.4$ or $n = 0.6$ when $\tau_\alpha \approx 10^2$ s or $\zeta = \zeta_g$. Since softer spheres have weaker dependence of log τ_α on ζ/ζ_g, and smaller steepness or fragility index m_ζ, therefore n correlates with m_ζ, which is equivalent to the correlation between n and m for ordinary glassformers. In the present case of athermal colloidal particles, thermodynamic factors do not enter. Thus, both n and m_ϕ are controlled solely by the softness of the particles, or more exactly the strength of the particle–particle interaction. Naturally softer the particles, weaker the inter-particle interaction, and smaller the degree of many-particle cooperative dynamics. Thus

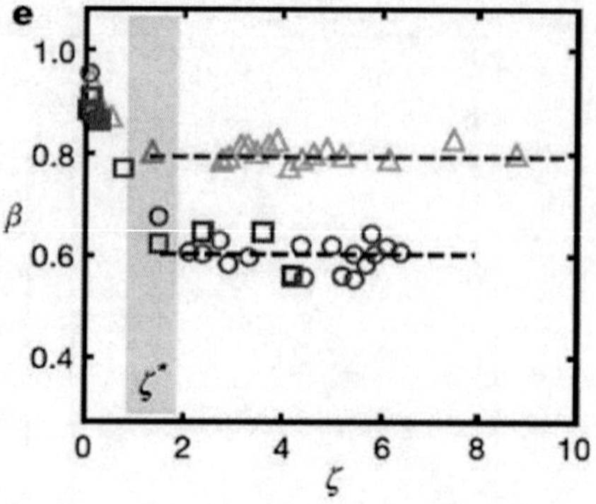

Fig. 34d The stretching exponent β remains constant for values of ζ greater than ζ^* (the range of ζ^* is indicated by the *shaded region*) for both intermediate (*circles*, $R_0 = 92$ nm; *squares*, $R_0 = 168$ nm) and soft (*triangles*, $R_0 = 80$ nm) microgels. Reproduced from [617b] by permission

the results of Weitz et al. nicely elucidate two important points. First, n and m_ζ or "fragility" are *parallel* consequences of solely the many-particle dynamics. Second, the correlation between them is necessarily true because no other factor enters into the problem, unlike the case of molecular glassformers where thermodynamic factors also enter. Having identified many-particle dynamics to be the sole source in determining n and m_ζ in colloidal suspensions, it is rational to presume that many-molecule dynamics are important also for molecular glassformers in determining their non-exponentiality (n) and "fragility" (m).

Although I have emphasized the correlation (anticorrelation) of n (β) with fragility m_ζ in the colloidal systems, this was mentioned only in passing by Mattsson et al. Moreover, we have identified the trends observed in both n and fragility are caused by the change in inter-particle interaction in varying the softness of the microgel. Thus the correlation between n and fragility is a natural consequence On the other hand, Mattsson et al. put their emphasis on fragility and went on to show fragility is dictated by elastic properties on the scale of individual colloidal particles. Their findings are interesting and instructive. However, elastic property is just one of those properties that are derivable from the inter-particle interaction. So is fragility, because no doubt τ_α is governed by the inter-particle potential. More properties of these colloidal systems other than elasticity may also correlate with fragility. I speculate that another one is dynamic heterogeneity. If this is true and proven in the future by experiment or simulation, then one may be led by the importance of elasticity argued by Mattsson et al. to think that dynamic heterogeneity is also dictated by elastic properties. This seems questionable. In my view, all properties mentioned including fragility, elastic properties, dynamic heterogeneity, and non-exponentiality are consequences of the many-particles dynamics, the intensity of which is dictated by the inter-particle potential. Therefore I argue that it is more fundamental to relate fragility, elasticity, and dynamic heterogeneity to the simplest and most direct indicator of the many-particle dynamics; a candidate is the degree of non-exponentiality or $n \equiv (1 - \beta)$. Softer the microgel particles, weaker the inter-particle interaction, and lesser the degree of anomaly in these properties. Naturally fragility, elasticity, dynamic heterogeneity, and non-exponentiality of colloidal suspensions, all derivable from the inter-particle potential, correlate well with each other. The correlations are even better than molecular glassformers because the colloidal suspensions are athermal systems and thermodynamic factors including volume and entropy no longer enter in conjunction with many-particle potential to determine fragility. The reader can find in this chapter many examples of breakdown of correlation of fragility molecular glassformers with other properties because the temperature dependence of τ_α is governed by volume and entropy in addition to the inter-particle potential. Study of the glass transition problem by the soft colloidal systems has the advantage that thermodynamic factors are absent, hence not only simplifying the problem but also exposing the importance of the many-particle dynamics governed by the inter-particle interaction. It is because of the simplification and revelation that I agree with Mattsson et al. in saying "Colloidal suspensions may thus provide new insight into glass formation in molecular systems" [617b].

Concluding Remarks on "Fragility" and the "Fragility" Index *m*

The concept of "fragility" or its index m or $F_{1/2}$ stems totally from the dependence of τ_α on T (after scaling T by T_{ref} or T_g). For it to be considered as a fundamental quantity in glass transition, the assumption has to be made that no other factor determines τ_α except the scaled temperature and associated thermodynamic quantities. This is a bold assumption which is unjustifiable because the effects of many-body α-relaxation dynamics on τ_α cannot be completely described by scaled temperature. We have seen that m_P changes with pressure P and m_V changes with specific volume V. Also m_P does not correlate faithfully with other observed dynamic and thermodynamic properties when different kinds of glassformers are considered altogether, or even within the same class of glassformers. The cause of these shortcomings of "fragility" is that it is a complex quantity because τ_α depends not only on T/T_{ref} but also on other factors. These include P for isobaric condition and V for isochoric condition, as well as the effects coming from the many-body nature of the α-relaxation. The latter can vary greatly with the intermolecular potentials in different glassformers. From these multiple dependences of τ_α, "fragility" and its index m or $F_{1/2}$ cannot faithfully correlate with other observed dynamic and thermodynamic properties when glassformers of different chemical bonding and physical structures are considered together.

While "fragility" and its indices m_P and m_V can vary a great deal with pressure and specific volumes in the same glassformer and for the same $\tau_\alpha(T_{\mathrm{ref}})$ or $\tau_\alpha(T_g)$, the time/frequency dispersion (or n) of the α-relaxation remains the same independent of the different possible combinations of P, V, and T leading to the same $\tau_\alpha(T_{\mathrm{ref}})$ or $\tau_\alpha(T_g)$ [120, 588, 618]. This remarkable property of the time/frequency dispersion (or n), applicable not only at constant $\tau_\alpha(T_{\mathrm{ref}})$ or $\tau_\alpha(T_g)$ but also at any constant τ_α value, is the subject of discussion of the following section. This remarkable link between τ_α and n, independent of P, V, and T, demonstrates that the dispersion of the α-relaxation or its non-exponential parameter n is a more faithful indicator of a fundamental mechanism behind the α-relaxation of glassformers than "fragility" or the indices m_P and m_V. A fundamental mechanism is the many-body relaxation. This is indicated by the strong correlations found between the dispersion (or n) and other properties of τ_α presented in previous sections and more to come in the next section. Thus, while n is solely determined by many-body relaxation dynamics, m_P and m_V depend on thermodynamic variables as well as many-body dynamics. The combined dependences make "fragility" a complex quantity not as useful as n in solving the problem of glass transition.

To illustrate once more that fragility has the combination dependences, a comparison of its value for hyper-branched polystyrene (HBPS) with that for linear PS was made by mechanical relaxation data [619]. While the linear PS is entangled, the HBPS is not, even though their glass transition temperatures are nearly the same. The lack of intermolecular chain entanglements in HBPS suggests the segmental motion for HBPS is less cooperative than that of the linear PS. The density of HBPS was measured, and it is comparable to or less than that of linear PS. Both factors may contribute in causing the fragility index of HBPS to be lower than that of the linear PS as was observed experimentally. Evidence for both factors in action also comes from the results of the study of hyper-branched poly(ether ketone) and its linear analogue [620]. The study reported that hyper-branched poly(ether ketone) was more fragile than the corresponding linear analogue opposite to the trend found in polystyrene [619], but here the hyper-branched polymer was denser than the linear polymer. These two experimental results suggest both density and cooperative many-body dynamics determine the fragility index.

2.2.4 *Invariance of the α-Dispersion to Various Combinations of* T *and* P *While Keeping* τ_α *Constant*

Studies of changes of relaxation dynamics of glassformers with temperature at ambient pressure (0.1 MPa) have traditionally been the way to study glass transition. This practice is due largely to experimental convenience in varying temperature. Most of the experimental data of structural α-relaxation in the literature are ambient pressure measurements carried out at different temperatures. From the experimental data, it is well established that the shape of the α-dispersion (or the exponent n in Eq. (1) if fitted by the Kohlrausch function) can vary from one glassformer to another, when compared at T_g or some other reference value of τ_α [112, 244]. Many experimental studies have shown also that for a given material, often the width of the α-relaxation dispersion monotonically increases with decreasing temperature or increasing τ_α [238, 423, 621, 622], although the rate of change varies. Less common than temperature studies at ambient pressure are experiments employing hydrostatic pressure, notwithstanding the fact that some pioneering dielectric measurements at applied elevated pressures were carried out half a century ago by Robert Cole and coworkers [623] and followed by others in the earlier days to cite a few [76, 624–632].

Recently, the technique of applying high pressure in broadband dielectric relaxation measurements has been greatly improved. Now, pressure from 0.1 MPa up to several GPa can be routinely employed as an experimental variable in various spectroscopies including light scattering, neutron scattering, and broadband dielectric spectroscopy to probe the dynamics of glassformers and other substances. Only some representative publications are cited here in [633–679]. Consequently, experimental measurements can be made at different combinations of P and T over wide ranges of both variables. The specific volume V corresponding to any combination of P and T can be inferred from P–V–T measurements made separately. Elevated pressure increases τ_α, but the increase can be compensated by raising temperature. Hence, various combinations of P and T can be chosen for which the α-loss peak frequency ν_α and the corresponding relaxation time τ_α are the same.

If measurements under this condition are carried out, then an interesting question is whether the shape of the frequency dispersion of the α-relaxation changes or not? Sufficiently large amount of experimental data of this kind have accumulated till 2005 so that the answer can be given. An important experimental fact has emerged from the combined pressure and temperature studies of many different glassformers. At any chosen constant value of the structural relaxation time τ_α or frequency ν_α, the dispersion of the structural α-relaxation is invariant. Some examples of data prior to 2005 found in [618] are shown in Figs. 35 and 36. There are instances in which the heights of the dielectric α-loss peak, ε''_{max}, of some of the P and T combinations are not exactly the same. Nevertheless, after the measured dielectric loss $\varepsilon''(\nu)$ has been normalized by ε''_{max}, the frequency dispersions of $\varepsilon''(\nu)/\varepsilon''_{max}$ is invariant for any chosen ν_α. Thus, generally at any fixed value of τ_α, the time/frequency dependence of the relaxation function (or n) is invariant to changes in thermodynamic condition (temperature, pressure, and volume). In other words, temperature–pressure superpositioning holds for the dispersion of the structural α-relaxation at constant τ_α. Lack of superposition may occur at frequencies high compared with ν_α. Such deviations at high frequencies or shorter times are attributed to the contribution to dielectric loss from resolved or unresolved faster secondary relaxation, whose dielectric relaxation strength does not necessarily have the same P and T dependences as the α-relaxation. The superpositions in Figs. 35 and 36 are shown as $\log \varepsilon''(\nu)$ vs. $\log \nu$. Had the superpositions been shown as $\varepsilon''(\nu)$ vs. $\log \nu$, the small deviations from perfect superposition at high frequencies seen in the $\log \varepsilon''(\nu)$ vs. $\log \nu$ superpositions will be invisible.

Since the general property is remarkable and important, some examples of dielectric experimental data from different kinds of glassformers shown or not shown before are presented here. In the paper of 2005 [618], the glassformers considered include both molecular liquids (Fig. 35) and amorphous polymers (Fig. 36) with diverse chemical structures. All show the property of temperature–pressure superpositioning of the dispersion of the structural α-relaxation at constant τ_α. In some cases, experimental data of the same glassformer at several different dielectric relaxation times are presented to demonstrate that P–T superpositioning holds for any choice of constant τ_α. In many of them, the width of the α-relaxation increases with increasing τ_α. Therefore, one cannot trivialize the observation by saying that it is due to the dispersion being independent of temperature and pressure.

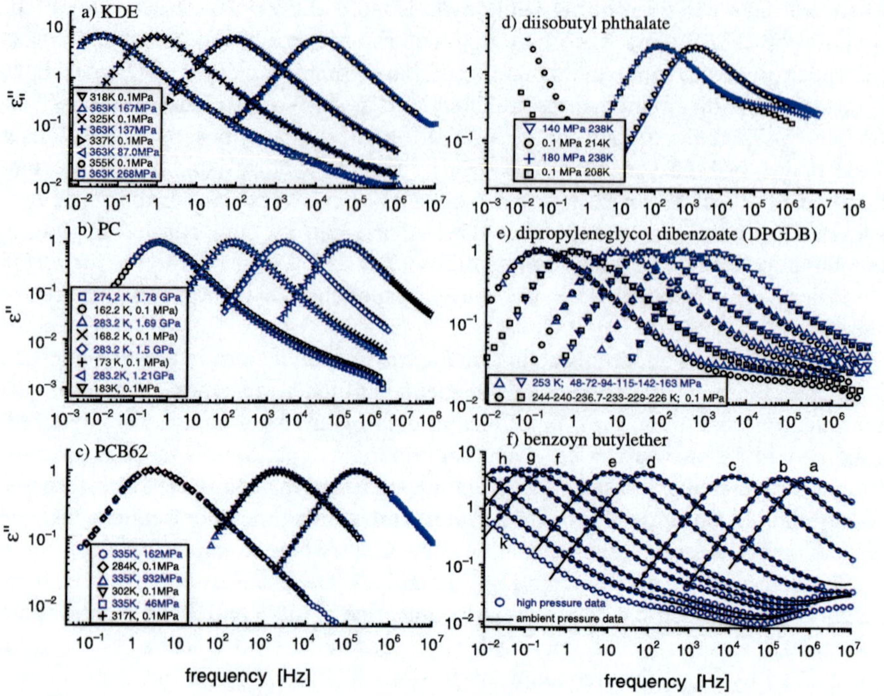

Fig. 35 Dielectric loss data at various combinations of temperature and pressure as indicated to demonstrate the invariance of the dispersion of the α-relaxation at constant α-loss peak frequency ν_α or equivalently at constant α-relaxation time τ_α for (**a**) cresolphthalein-dimethylether (KDE), (**b**) propylene carbonate (PC) (loss normalized to the value of the maximum of the α-loss peak), (**c**) polychlorinated biphenyl (PCB62), (**d**) diisobutyl phthalate (DiBP). (**e**) Dielectric loss of dipropyleneglycol dibenzoate (DPGDB). Loss normalized to the value of the maximum of the α-loss peak. The dc conductivity contribution has been subtracted. *Red triangles* are isothermal measurements at T=253 K and P=48, 72, 94, 115, 142, 163 MPa (*from right to left*). *Black symbols* are isobaric measurements done at P=0.1 MPa and T= 244, 240, 236.7, 233, 229, 226 K (*from right to left*). The spectrum at T=226 K has been shifted along X-axis by multiplying frequency by a factor 1.3. (**f**) Dielectric loss of benzoyn isobutylether (BIBE) at different T and P. The dc conductivity contribution has been subtracted. Spectra obtained at higher P are normalized to the value of the maximum of the loss peak obtained at the same frequency at atmospheric pressure. *From right to left*: *Black lines* are atmospheric pressure data at T= 271 (a), 263 (b), 253 (c), 240 (d), 236 (e), 230 (f), 228 (g), 226 (h), 223 (i), 220.5 (j), 218 (k) K. *Symbols* are high-pressure data: T=278.5 K and P=32 (a), 65 (b), 118 (c), 204 (d), 225 (e), 320 (h), 370 (j), 396 (k) MPa, T=288.2 K and P=350 (f), 370 (g), 423 (i), 450 (j) MPa, T=298 K and P=330 (d), 467 (h) MPa. Figures redrawn from original data in each case

2.2.4.1 Molecular Glassformers

Numerous molecular glass-forming liquids have narrow dispersions of the α-relaxation and an excess wing on the high-frequency flank, but otherwise no other resolved secondary relaxation in their dielectric spectra. There are experimental results [424, 508, 657, 658, 679–684] indicating that the excess

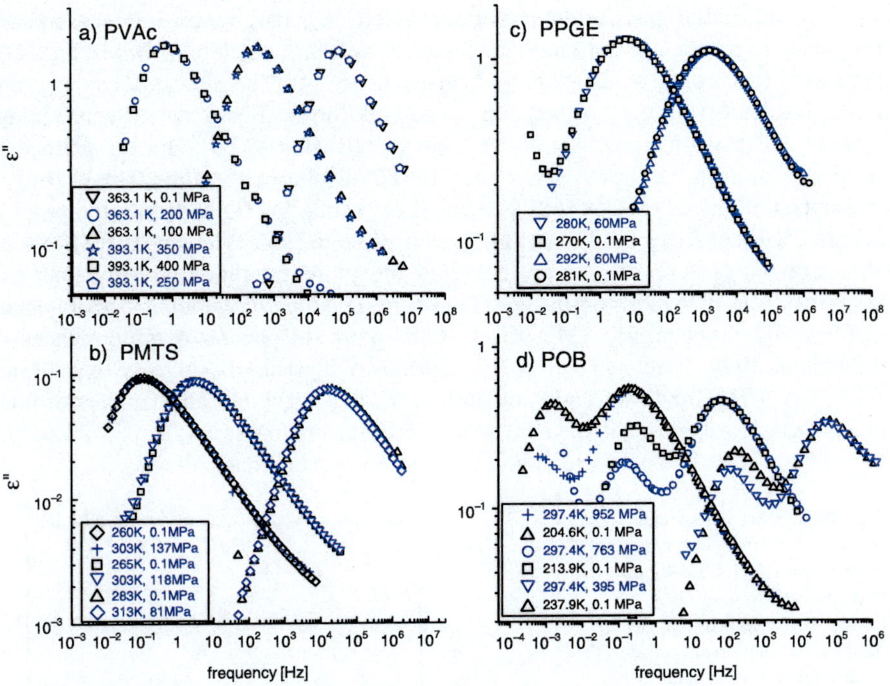

Fig. 36 Dielectric loss data at various combinations of temperature and pressure as indicated to demonstrate the invariance of the dispersion of the α-relaxation at constant α-loss peak frequency ν_α or equivalently at constant α-relaxation time τ_α, for (**a**) poly(vinyl acetate) (PVAc); (**b**) poly(methyltolylsiloxane) (PMTS); (**c**) poly(phenylglycidylether)-*co*-formaldehyde (PPGE); (**d**) poly(oxybutylene) (POB). In all cases spectra obtained at higher *P* are normalized to the value of the maximum of the loss peak obtained at the same frequency at atmospheric pressure. Figures redrawn from original data in each case

wing is an unresolved Johari–Goldstein secondary relaxation. More detailed discussion of these experimental results will be given in a later section on Johari–Goldstein secondary relaxation. The glassformers of this type that exhibit *P–T* superpositioning of the α-relaxation holding for any choice of constant τ_α are cresolphthalein-dimethylether (KDE), phenylphthalein-dimethylether (PDE), propylene carbonate (PC), and polychlorinated biphenyl (PCB62) included in Fig. 35. In each of these figures, data are used to show that this property holds for more than one value of τ_α. Shown elsewhere [120] and not be duplicated here are the same behavior found in phenylphthalein-dimethylether (PDE), phenyl salicylate (salol), 3,3′,4,4′-benzophenonetetracarboxylic dianhydride (BPTCDaH), 1,1′-*bis* (*p*-methoxyphenyl)cyclohexane (BMPC), 1,1′-di(4-methoxy-5-methylphenyl) cyclohexane (BMMPC), diethyl phthalate (DEP), and di-isooctal phthalate (DiOP). More small molecular glassformers obeying *P–T* superpositioning at constant τ_α have been found since the publication of [120, 618].

Some molecular glassformers have a resolved secondary relaxation whose peak frequency is practically pressure independent; these are not Johari–Goldstein (JG) processes (according to the definition given in [685]). The slower JG relaxation is not resolved from the α-relaxation in the equilibrium liquid state. Such liquids include 1,1′-*bis*(*p*-methoxyphenyl)cyclohexane (BMPC) [659], diethyl phthalate (DEP) [660], di-*n*-butyl phthalate (DBP) [661], diisobutyl phthalate (DiBP) [661], di-isooctal phthalate (DiOP) [684], decahydroisoquinoline (DHIQ) [667], dipropyleneglycol dibenzoate (DPGDB) [669], and benzoin isobutylether (BIBE) [687]. Also are the epoxy compounds including the more familiar diglycidylether of bisphenol A (EPON828) [640], shown in Fig. 37, and less common 4,4′-methylene *bis*(*N,N*-diglycidylaniline) (MBDGA) [688, 689], bisphenol A propoxylate(1-PO/phenol)diglycidylether) (1PODGE) [690], *N,N*-diglycidyl-4-glycidyloxyaniline (DGGOA) [691], and *N,N*-diglycidylaniline (DGA) [691]. For all members of this class of glassformers, the dispersion associated with any fixed value of τ_α is invariant to change of thermodynamic condition (*T* and *P* combinations).

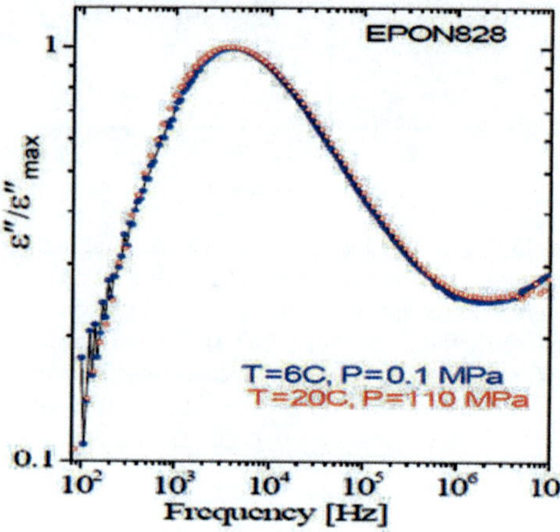

Fig. 37 Dielectric loss data of EPON828, diglycidylether of bisphenol A (DGEBA), at two combinations of temperature and pressure as indicated to demonstrate the invariance of the dispersion of the α-relaxation at constant α-loss peak frequency. Open symbols (P = 110 MPa), closed symbols (0.1 MPa). Data from [676] replotted here

Earlier dielectric studies under elevated pressure [632] had found temperature–pressure superpositioning at constant τ_α in a few molecular glassformers including *ortho*-terphenyl (OTP), di(2-ethylhexyl) phthalate, tricresyl phosphate, polyphenyl ether, and refined naphthenic mineral oil, albeit the temperature and pressure ranges are not as wide as achieved in the more recent measurements.

We add in Fig. 38 to show the *P*–*T* superposition of the dielectric loss data of liquid triphenyl phosphite [676] at ambient and elevated pressure of 500 MPa, and also in Figs. 39 and 40 here the same for diphenyl-vinylene carbonate (DPVC) with $T_g = 251$ K [697] for more than one purpose. The figures show (1) superpositioning in the $\log \varepsilon''(\nu)$ vs. $\log \nu$ plot as well as in the $\varepsilon''(\nu)$ vs. $\log \nu$ plot of not only the main α-loss peak but also the excess wing at high frequencies; (2) the broadening of the dispersion on increasing τ_α; (3) the loss data plotted linearly against $\log \nu$ that are very well fitted by the one-sided Fourier transform of the Kohlrausch function

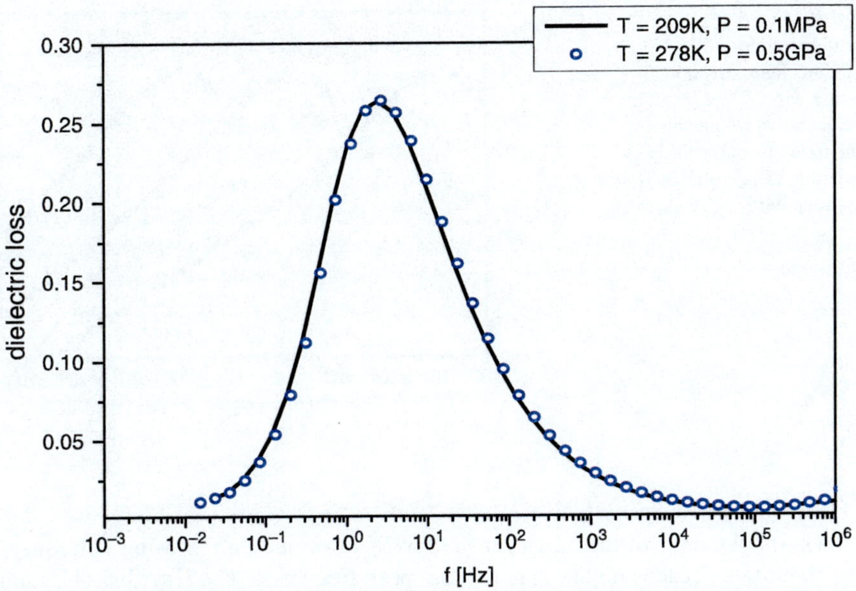

Fig. 38 *T–P* superposition of dielectric loss data of liquid triphenyl phosphite at ambient pressure (*line*) and elevated pressure of 500 MPa (*circles*). Data from [24] replotted

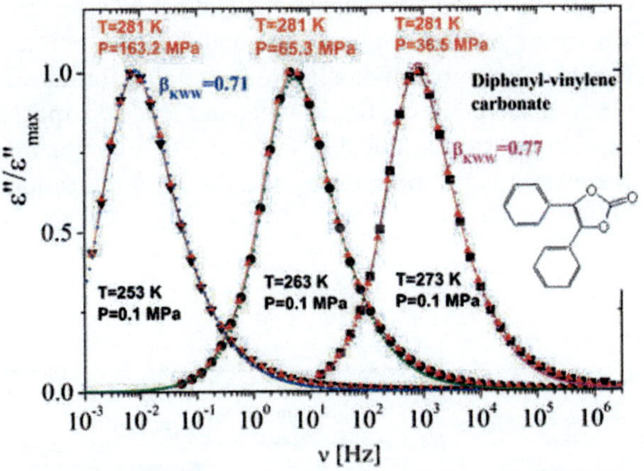

Fig. 39 Normalized loss spectra of DPVC. *Symbols* represent experimental data (atmospheric and high-pressure data) showing same frequency dispersion if the loss peak frequency is the same, and broadening of the dispersion with decreasing peak frequency. *Dotted lines* are fits to the data by the Fourier transform of the KWW function ($\beta_{KWW} = 0.77$, 0.73, and 0.71 from *right to left*). Discrepancies between the KWW fit and experimental data only occur at high frequencies. Data from [697] are replotted here

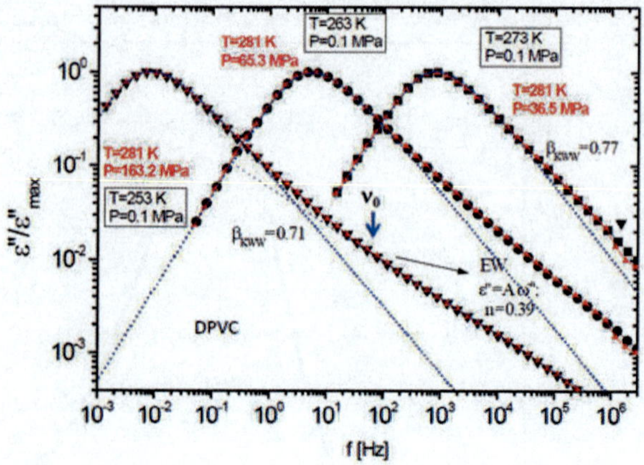

Fig. 40 Log–log plot of the same spectra of DPVC in previous figure to show the excess wing. The *arrows* are the primitive frequencies ν_0 calculated from the CM equation. Data from [697] are replotted here

(lines). Small excess of the data over the KWW fit occurs only starting at frequencies about two decades higher than the loss peak frequency. The Havriliak–Negami function gives a better fit to the high-frequency part of the data but at the expense of introducing one more fitting parameter that has to be interpreted than the Kohlrausch function.

The dielectric loss spectra of benzoyn isobutylether (BIBE) shown before in Fig. 35f has two resolved secondary relaxation. The peak frequency of the slower one shifts to lower values on elevating pressure and is the JG β-relaxation, but not the faster one. This is an example of small molecular glassformer in which the JG β-relaxation has been resolved in the liquid state, and *P*–*T* superpositioning at constant τ_α holds for both the α- and the JG β-relaxation. A complement to Fig. 35f to show *P*–*T* superpositioning of both the α- and the JG β-relaxation was given by Capaccioli et al. [697] in Fig. 41.

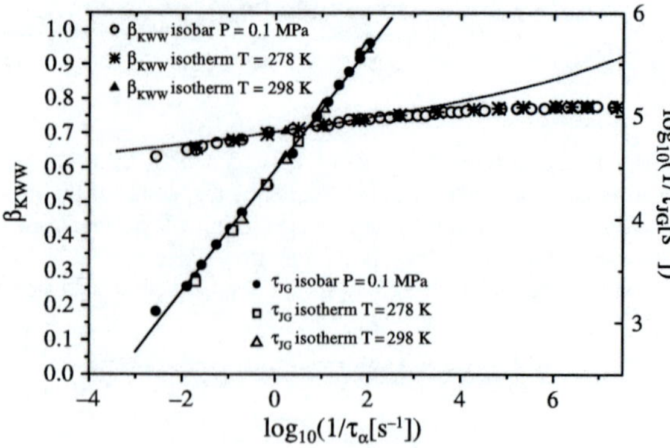

Fig. 41 KWW parameter for the α-relaxation (*left y*-axis) and logarithm of relaxation time, τ_{JG} of the JG relaxation (*right y*-axis) of BIBE plotted vs. logarithm of α-relaxation time for different *T* and *P*. *Symbols* are from experimental data of BIBE. *Continuous and dotted lines* are fitting functions explained in the text of [697]. The figure is reconstructed from the data in [697]

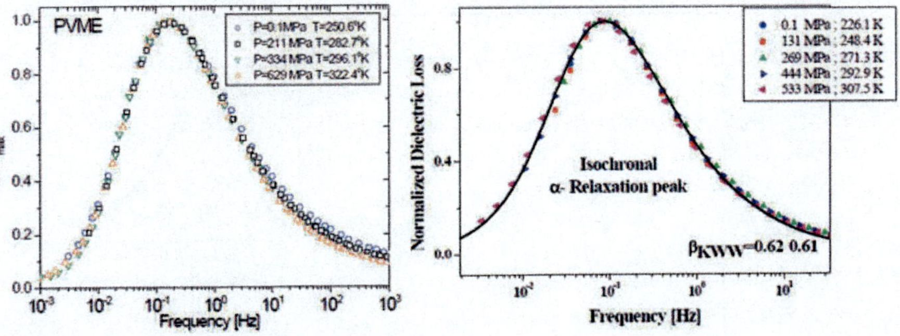

Fig. 42 (*Left*) Superposed dielectric loss spectra of poly(vinylmethylether) (PVME) with weight average molecular weight of 99,000 Da and a polydispersity=2.1 as a function of pressure, for various temperatures such that the relaxation times are almost equal (frequencies were shifted by less than 30%). Data from [618] are replotted here. (*Right*) Superposition of normalized dielectric loss data of DPGDB at various combinations of P and T as indicated. All have the same loss peak frequency. Data from [678] are replotted here

As a last example, the left panel of Fig. 42 shows that P–T superpositioning at constant τ_α also holds for dipropyleneglycol dibenzoate (DPGDB) [678]. We shall see in a later section that P–T superpositioning for both the α- and the JG β-relaxation of DPGDB holds at T_g.

2.2.4.2 Amorphous Polymers

Dielectric relaxation measurements under pressure have been carried out on several amorphous polymers, and for all cases studied the dispersion of the local segmental relaxation (i.e., the structural α-relaxation) conforms to temperature–pressure superpositioning at constant τ_α. These polymers include poly(vinylmethylether) (PVME) [652], poly(vinyl acetate) (PVAc) [631], poly(ethylene-*co*-vinyl acetate) (EVA, having 70 wt% vinyl acetate) [698]. polymethylphenylsiloxane (PMPS) [549], poly(methyltolylsiloxane) (PMTS) [549], 1,2-polybutadiene (1,2-PBD, also referred to as polyvinylethylene, PVE) [699], poly(phenylglycidylether)-*co*-formaldehyde (PPGE) [553], 1,4-polyisoprene (PI) [700], poly(propylene glycol) (molecular weight: 4000 Da) PPG4000 [701], poly(oxybutylene) POB [668], and poly(isobutyl vinylether) (PiBVE) [703]. Constant dispersions at a fixed value of τ_α independent of thermodynamic conditions (T and P) are shown for more than one τ_α in Fig. 36 for PVAc, PMTS, PPGE, and POB. Note that for POB there is a dielectrically active normal mode at lower frequencies. The normal mode originating from global chain motion has entirely different character than the local structural α-relaxation, different P and T dependences, and hence its dispersion is not expected to obey T–P superpositioning at constant τ_α. The dielectric data of some more common polymers showing T–P superposition of the dispersion of the α-relaxation at fixed τ_α are presented in Fig. 42 (right panel) for PVME, Fig. 43 for PVE, Fig. 44 for PPG400, and Fig. 45 for the less common PiBVE.

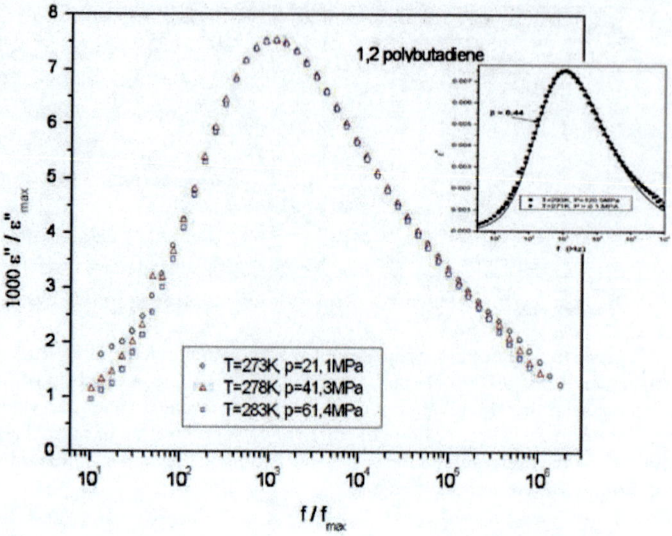

Fig. 43 Dielectric loss data of 1,2-polybutadiene (1,2-PBD) with molecular weight of 3000 Da (about 56 monomer units) at various combinations of temperature and pressure as indicated to demonstrate the invariance of the dispersion of the α-relaxation at constant α-loss peak frequency ν_α or equivalently at constant α-relaxation time τ_α. Data from [699] redrawn here

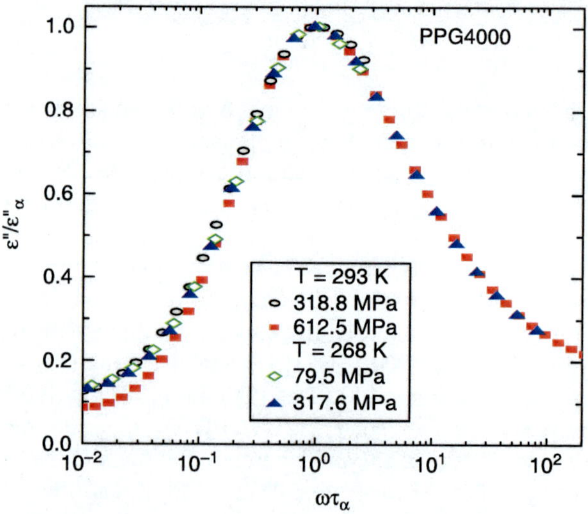

Fig. 44 Dielectric loss data of poly(propylene glycol) (PPG4000, molecular weight: 4000 Da) at various combinations of temperature and pressure as indicated to demonstrate the invariance of the dispersion of the α-relaxation at constant α-loss peak frequency ν_α or equivalently at constant α-relaxation time τ_α. Data of Casalini and Roland replotted here

As illustrated in some of the figures of small molecular and polymeric glass-formers, all the α-loss peaks are well fit by the one-sided Fourier transform of the Kohlrausch function (Eq. (1.1)) over the main part of the dispersion. Thus, the experimental fact of constant dispersion at constant τ_α can be restated as the invariance

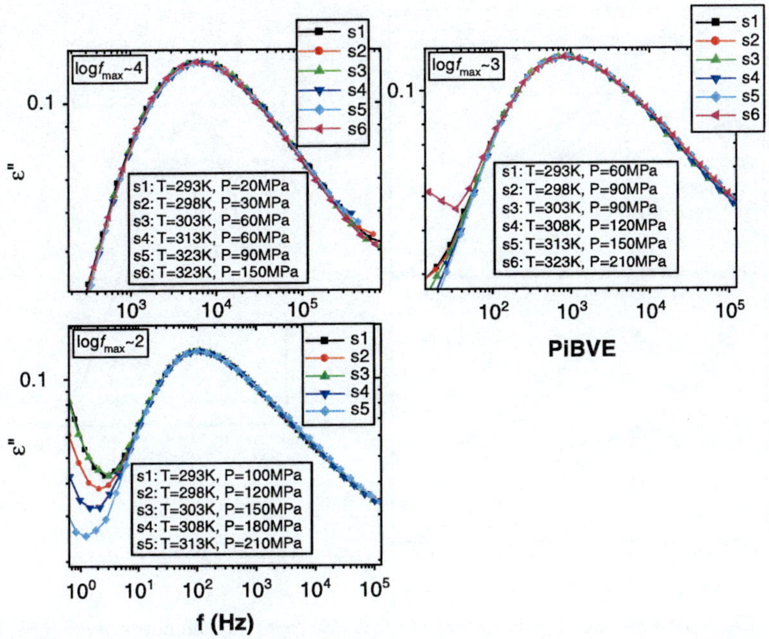

Fig. 45 *P–T* superposition of dielectric loss data of PiBVE at three $\log f_\alpha$. Reproduced from [703] by permission

of the fractional exponent n (or $\beta_{KWW} = 1 - n$) at constant τ_α. In other words, τ_α and n are co-invariants of changing thermodynamic condition (T and P). So is the full-width at half-maximum of the dielectric loss peak, since after normalized to that of Debye loss peak, its value w is uniquely related to n by $n = 1.047$ $(1 - w^{-1})$ [214].

Hydrogen-bonded networks or clusters, if present, are modified at elevated pressure and temperature, changing the structure of the glassformer in the process. This occurs, for example, in glycerol, threitol [556], dimer (2PG) and trimer (3PG) of propylene glycol [657, 658], and *m*-fluoroaniline [671]. These hydrogen-bonded glassformers do not obey temperature–pressure superpositioning of the α-relaxation at constant τ_α because the material itself is changed with different combinations of T and P. Such behavior is shown for glycerol and threitol in Fig. 46, 2PG in Fig. 46, and *m*-fluoroaniline in Fig. 47. In higher members of the polyols, such as xylitol and sorbitol, the departure from *T–P* superpositioning at constant τ_α is smaller compared with the lower member glycerol. This is shown for sorbitol in Fig. 48, left panel [659(b)]. Also this trend can be seen when comparing PPG4000 (see Fig. 44) with dimer and trimer of PG.

Isobaric fragility index m_P generally decreases with increasing pressure [588(c)] for non-hydrogen-bonded glassformers. On the other hand, for hydrogen-bonded glassformers, m_P increase with pressure is found in lower molecular weight propylene glycols [652(b)], heptapropylene glycol (7PG) [636], PPG400, PPG4000

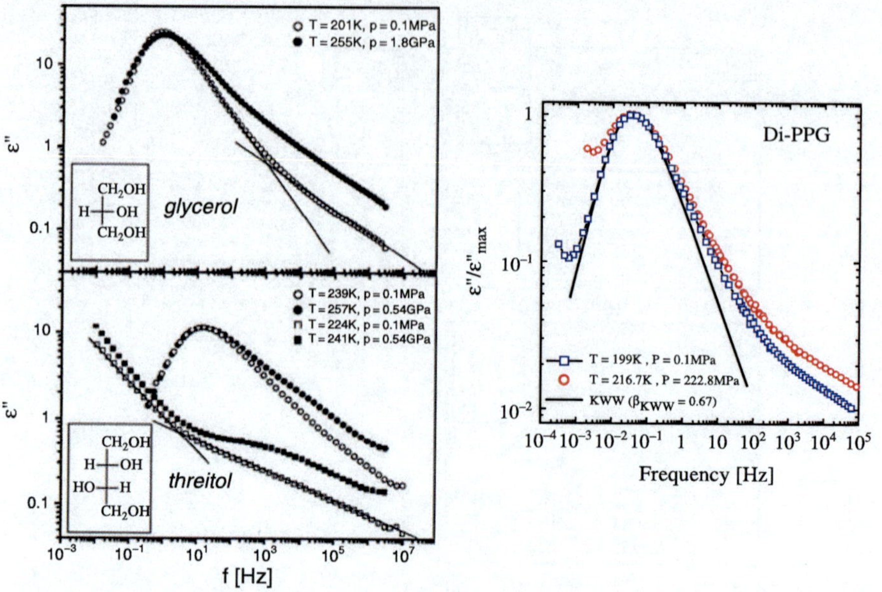

Fig. 46 Dielectric loss data of glycerol and threitol [659] (*left side*) and dimer of PG [658] (*right side*) at various combinations of temperature and pressure as indicated to demonstrate the departure of invariance of the dispersion of the α-relaxation at constant loss peak frequency ν_α [556]. Data from the references cited have been replotted in all the figures

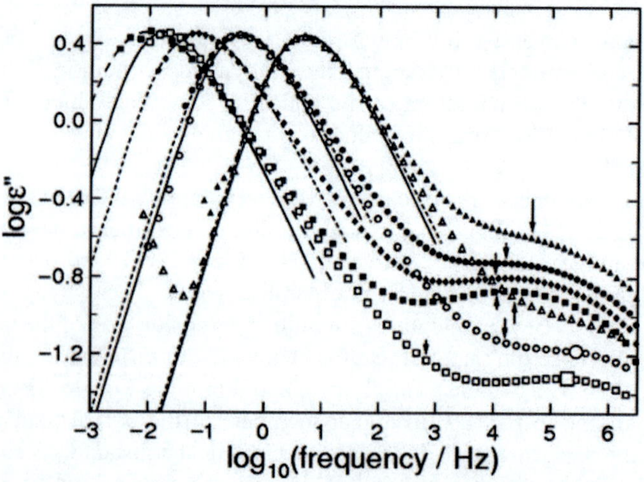

Fig. 47 Dielectric loss spectrum of *m*-FA at 279 K and 1.69 GPa (■), 1.60 GPa (♦), 1.52 GPa data (•), and 1.4 GPa data (▲). Dielectric loss spectrum of *m*-FA at ambient pressure and 174 K (□), 177 K (○), and 180 K (Δ). The *dashed lines* are fits to the data at 279 K and under GPa pressures by the one-sided Fourier transform of the KWW function. The *solid lines* are similar fits to the ambient pressure data. The dispersion of the α-loss peak is always wider at higher temperature and pressure for the same loss peak frequency ν_α. The *vertical arrows* indicate the calculated primitive relaxation frequencies, ν_0, for all the data sets and will be discussed in Section 2.3. Reproduced from [671] by permission

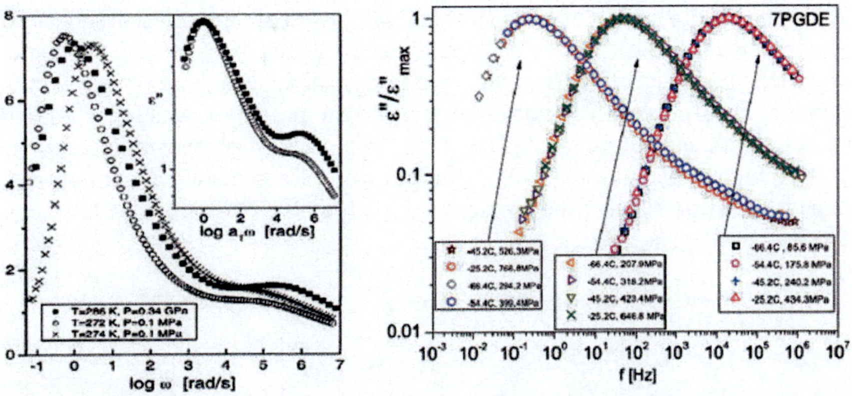

Fig. 48 (*Left*) The dielectric loss of sorbitol (■) at elevated pressure (340 MPa) at 286.4 K and at atmospheric pressure at (○) 272 and (×) 274 K. The *inset* shows the superposition obtained by shifting the 272 K spectrum by 0.8 decades. The conductivity contribution, assumed proportional to $1/\omega$, has been subtracted from the spectra. Reproduced from [659(b)] by permission. (*Right*) Invariance of the α dispersion at constant α peak frequency upon removal of the hydrogen bonding. Loss spectra of 7PGDE at various combinations of *T* and *P* (as indicated). The data on the high-frequency flank also superpose. Reproduced from [652(c)] by permission

[652(d)], and in other hydrogen-bonded materials [633(b)]. The increase of m_P with increasing pressure is accompanied by the broadening of the α-dispersion in accord with the correlation between *m* and *n*. The broadening of the α-dispersion suggests an increase of the intermolecular coupling parameter *n*, and this explains the increase of m_P with increasing pressure of hydrogen-bonded glassformers by using the CM. While hydrogen-bonded glassformers such as 7PG do not strictly obey *P*, *T* superpositioning at constant τ_α, its non-hydrogen-bonded analogue, heptapropylene glycol dimethyl ether (7PGDE), does. For the 7PGDE, the terminal hydroxyl groups are exchanged for methoxy groups (O-CH_3) in 7PG to create an analogous structure lacking hydrogen bonding. In fact, 7PGDE obeys *P*, *T* superpositioning at constant τ_α (see Fig. 48, right panel), and its m_P decreases with increasing pressure [652(c)].

2.2.4.3 Ionic Liquids

Room-temperature ionic liquids (IL) are made of oppositely charged molecules that are liquids at or near room temperature. For environmental considerations, they are favorable compared with the highly volatile organic solvents currently used in many chemical processes because of their low vapor pressures, thermal stability, high electric conductivity, and low viscosity [704, 705]. The ILs are also glass-forming substances.

Dielectric relaxation measurements were reported on the ionic liquid 1-butyl-1-methylpyrrolidinium *bis*[oxalato]borate (BMP-BOB), over wide temperature (123–300 K) and pressure ranges (0.1–500 MPa) [706]. The measured complex dielectric susceptibility $\varepsilon^*(\nu)$ was presented in terms of the electric modulus $M^*(\nu) = 1/\varepsilon^*(\nu)$

[707]. The imaginary part of $M^*(\nu)$, $M''(\nu)$, exhibits a maximum. The frequency dispersion of the loss peak and the peak frequencies characterize the ionic conductivity relaxation. These quantities have features that are similar to those observed in conventional glassformers. The relaxation time of the primary relaxation τ_σ strongly increases with applied pressure. The $M''(\nu)$ loss peak broadens with increasing τ_σ on lowering temperature or by elevating pressure. Remarkably, at constant τ_α, its shape is the same whether the pressure is 0.1 MPa or 500 MPa, as demonstrated in Fig. 49.

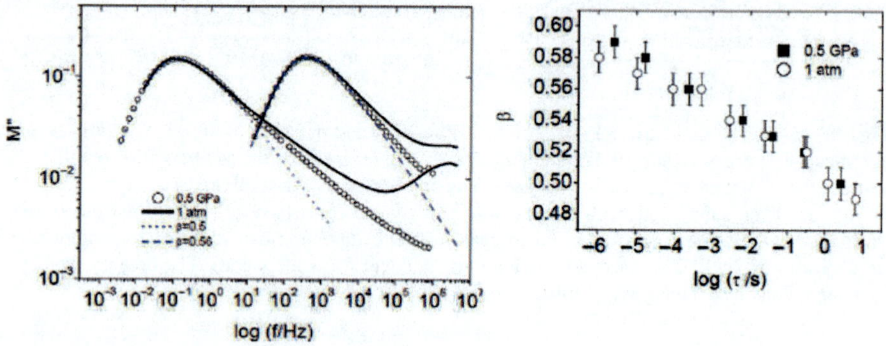

Fig. 49 (*Left*) Electric modulus relaxation spectra (M'') of the ionic liquid BMP-BOB at ambient pressure and 231 and 245 K are plotted as *solid lines*. High-pressure M'' data (0.5 GPa) at the temperatures that yield relaxation times similar to those of the ambient pressure data, 283 and 308 K, are included in the figure as *squares*. Data at 0.5 GPa data are slightly shifted in frequency to match perfectly the atmospheric peak frequencies. *Long and short dashed lines* are fits to a Kohlrausch relaxation function with $\beta \equiv (1-n) = 0.56$ and 0.50, respectively. (*Right*) The figure shows the co-invariance of β and the relaxation time at different temperatures and at atmospheric pressure and at elevated pressure of 0.5 GPa. Data from [706] are replotted here

The invariance of the shape of the $M''(\nu)$ loss peak to P and T combinations that maintain τ_α constant was also found in another room temperature ionic liquid, 1-hexyl-3-methylimidazolium chloride (or bromide), by Mierzwa et al. [702] (see Fig. 50, left panel).

2.2.4.4 Pharmaceutical and Saccharides

An example from the pharmaceuticals, indomethacin used for treatment of inflammation like gout, is chosen to show invariance of the α-dispersion to variations of P and T at constant τ_α in Fig. 50 (right panel). The data come from [397].

Glucose, fructose, galactose, sorbose, and ribose are monosaccharides and members of the sugar family and they are hydrogen-bonded glass-forming organic substances. They have important applications in food science, medicine, and biology. For this reason, the molecular relaxation dynamics of these sugars has been investigated in their liquid and glassy states by various techniques including dielectric relaxation and enthalpy relaxation [708]. The α-loss peaks of ribose measured

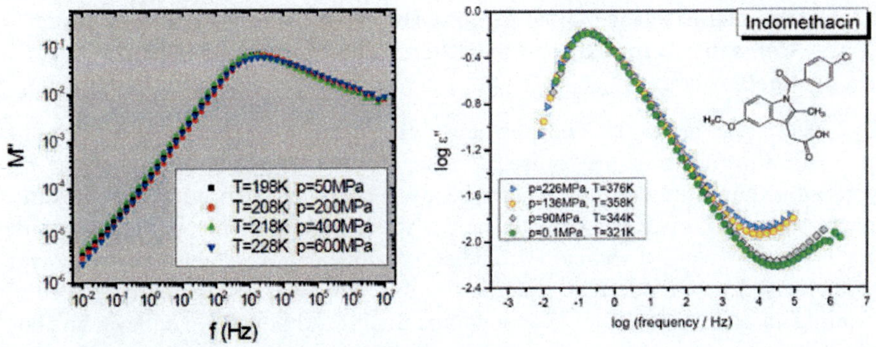

Fig. 50 (*Left*) Invariance of the shape of the electric modulus $M''(\nu)$ loss peak to P and T combinations that maintain τ_α constant in another room temperature ionic liquid, 1-hexyl-3-methylimidazolium bromide. Courtesy of M. Mierzwa, the data will be published elsewhere in the future as [702]. (*Right*) Master dielectric loss $\varepsilon''(\nu)$ curve of the pharmaceutical, indomethacin, constructed from the loss spectra obtained at different combinations of T and P. All loss spectra were chosen to have almost the same τ_α. Data from [397] are replotted here

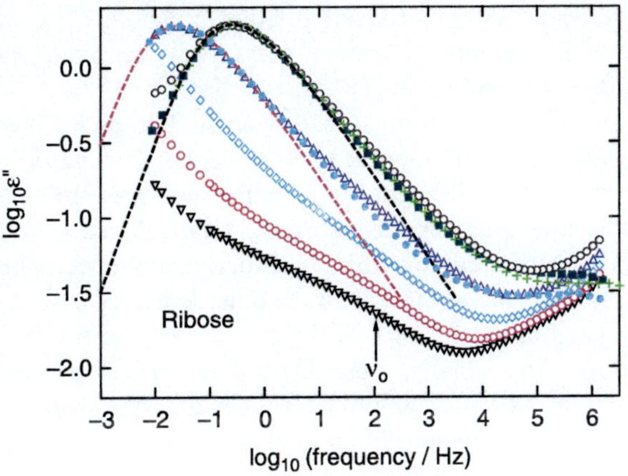

Fig. 51 Ribose. *Open symbols* are loss data taken at 500 MPa at T=297.8, 292.9, 288.1, 283.2, and 278.2 K (*from right to left*). The *closed squares*, *crosses* (+), and *closed circles* are loss data taken at ambient pressure (0.1 MPa) at T=269, 267, and 265 K and shifted horizontally by –0.25, –0.275, and +0.205 decades, respectively. The *dashed lines* are fits by the one-sided Fourier transform of the Kohlrausch function with n=0.45. The loss data taken at the elevated pressure and higher temperatures superpose well with that taken at ambient pressure and lower temperatures. To be discussed later in Section 2.3, the data (*inverted triangles*) taken below T_g show the presence of a secondary relaxation and its location in frequency is consistent with the primitive relaxation frequency ν_0 of the CM. Reproduced from [708] by permission

isothermally at ambient pressure (0.1 MPa) and at 500 MPa and various temperatures are shown in Fig. 51. The α-loss peak shape at the elevated pressure is almost the same as that at ambient pressure provided temperature is raised to maintain constant α-loss peak frequency [708].

2.2.4.5 Invariance of the α-Dispersion to Different T and P Combinations at Constant τ_α Investigated by Other Techniques than Dielectric Spectroscopy

The spectra for molecular and polymeric glassformers shown above to demonstrate the invariance of the α-dispersion to changes of T and P at constant τ_α, were all acquired by dielectric spectroscopy. This reflects the utility of dielectric spectroscopy in investigating broadband dynamics under pressure, particularly with recent developments of techniques. Therefore, an extensive database of broadband dielectric spectra of different materials in different T–P conditions is available, enabling an assessment of the T–P superpositioning. Although we believe the phenomena to be quite general, there is a paucity of data from other experimental techniques. However, some results are available, as described below.

For polymeric systems, only a few photon correlation spectroscopy (PCS) studies have been carried out with applied pressure and these were done more than 20 years ago. Within the experimental resolution, the shape of the α-relaxation was found to be essentially invariant to temperature and pressure at fixed τ_α, and time–temperature–pressure (t–T–P) superposition was valid. The investigated systems were poly(ethylacrylate) [709, 710], poly(methylacrylate) [711], and polystyrene [712]. A recent PCS experiment done on poly(propylene oxide) also found that t–T–P superposition was applicable [713].

More small molecular glassformers had been studied under pressure by more techniques than PCS. For *ortho*-terphenyl (OTP), the Kohlrausch parameters for the α-relaxation by PCS at different temperatures and pressures [714, 715] have been reported. The stretching parameters β_{KWW} vs. α-relaxation time τ_α for different T and P fall on a single curve within the experimental uncertainty, with β_{KWW} decreasing slightly with increasing τ_α. On the other hand, a shape invariance of the α-relaxation has been observed for OTP by specific heat spectroscopy under elevated pressure [716]. Additionally, experiments done on OTP by neutron scattering at different T and P revealed a negligible dependence of the shape of the structural relaxation, while the static structure factor yielded a master curve only for isochronal conditions, i.e., for the same relaxation time τ_α [717].

Salol is another system where the invariance of the shape of the α-relaxation dispersion was reported under pressure. Recent PCS experiments [718, 719] revealed that the correlation functions acquired at different pressures up to 180 MPa and at room temperature superposed. The stretching parameter β_{KWW} was 0.68, in agreement with the PCS measurements done at ambient pressure [720, 721].

Very accurate PCS measurements under different T and P conditions were carried out on different molecular glass-forming systems by Patkowski and coworkers, including epoxy oligomers [722–724], and the van der Waals liquid PDE [725, 726], BMPC [727], and BMMPC [728]. In most of the systems investigated, master curves are obtained for $\beta_{\mathrm{KWW}}(T,P)$ of the α-relaxation plotted vs. $\tau_\alpha(T,P)$, with the value decreasing (broader dispersion) as the dynamics slows (longer $\tau_\alpha(T,P)$). These are additional evidences indicating that the dispersion and the relaxation time are strongly correlated with each other and the relation remains unchanged with changes in the combinations of T and P.

From the above discussion, we can conclude the result that the α-dispersion is invariant to T and P at constant τ_α is generally found in different kinds of glassformers and by different experimental techniques.

2.2.4.6 The α-Dispersion of a Component in Binary Polymer Blends Is Invariant to T and P When τ_α Is Constant

Dielectric spectroscopic study of the component dynamics in the miscible 50/50 blend of poly(vinylmethylether) (PVME) and poly(2-chlorosytrene) (P2CS) was reported [729]. For the PVME component (which has the more intense loss peak due to its higher polarity), the shape of the segmental relaxation loss peak depends only on the relaxation time τ_α, and is otherwise independent of thermodynamic conditions, i.e., different P and T combinations. An example is shown in Fig. 52a. The same result was obtained before for the PVME component in the miscible blend of PVME with polystyrene (PS) [730]. This property of the component dynamics of miscible polymer blends is in accord with the general behavior of neat materials discussed in the preceding subsections. Several models have been proposed to address the component dynamics of polymer blends [731–733]. None of these models except that based on the coupling model [734–739] consider the dispersion of the α-relaxation of a component in the blend (or in its neat state), and naturally

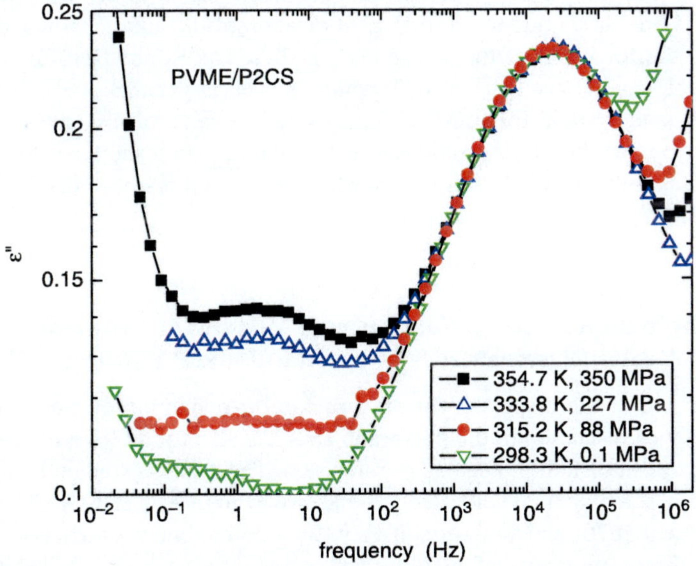

Fig. 52a Comparison of loss spectra for the 50 PVME/50 P2CS blend at combinations of T and P such that τ_α for the PVME component is essentially constant. To superpose the peaks, the frequencies for $P = 0.1$ and 227 MPa have been multiplied by 1.2 and 1.4, respectively, while the ordinate values for $P = 88$ and 227 MPa have been multiplied by 1.33 and 1.3, respectively. Data from [729] are replotted here

these models cannot address the observation of the invariance of the dispersion of the segmental relaxation of the component PVME for different T and P at constant relaxation time. Besides, these models are based on thermodynamic considerations and if considered the dispersion will change on changing T and P. On the other hand, the CM for component dynamics of polymer blends and mixtures of glassformers is based on considering the dispersion of the segmental relaxation of component in each environment i created by possible composition or concentration fluctuations. As discussed in detail in published works, the segmental relaxation in environment i with relaxation time $\tau_{\alpha i}$ has its own coupling parameter n_i, and the corresponding Kohlrausch function with stretch exponent, $(1 - n_i)$, determines the dispersion. In the same manner as shown for neat glassformers in the preceding section, the CM equation, $\tau_{\alpha i} = [t_c^{-n_i} \tau_0]^{1/(1-n_i)}$, ensures co-invariance of $\tau_{\alpha i}$ and n_i (or the shape of loss contributed by i) for different combinations of T and P. The observed segmental relaxation of the component is composed of contributions from all i, and hence its overall shape depends only on the most probable relaxation time $\hat{\tau}_\alpha$ and is otherwise independent of thermodynamic conditions (T and P combinations), as observed for PVME in two different blends mentioned above

That the generality of the invariance of the α-dispersion of a component in binary polymer blends to variations of T and P when τ_α is kept constant is further borne out by the dielectric loss data taken at ambient and elevated pressures of neat poly(cyclohexyl methacrylate) (PCHMA) and its 50/50 blend with poly(α-methylstyrene) (PαMS) [740], and presented in Fig. 52b. Superposition of α-loss peak is achieved in both neat PCHMA and the blend. As an aside, the width of the neat PCHMA is broad due to internal plasticization of the alkyl chain. This is evidenced by the forced fit of the α-loss peak at T_g to the Kohlrausch function with $\beta_{KWW} \equiv (1 - n) = 0.41$ or n=0.59, which is even larger than n=0.51 for atactic PMMA, and n=0.40 for PEMA. Thus, using the value of n=0.59 to calculate the primitive relaxation time would seriously obtain a much shorter value than the experimental value, a conclusion also reached by the authors of [740]. The correct value of n for PCHMA seems to be 0.40.

2.2.4.7 The α-Dispersion of a Component in Mixtures of Two Small Molecular Glassformers Is Invariant to T and P When τ_α Is Constant

The invariance of the dispersion of the structural α-relaxation of a component in polymer blends discussed in the paragraph above is also found for a component in binary mixtures of small molecular glassformers. The systems studied by dielectric spectroscopy at ambient and elevated pressure are a mixture of 25 wt% of 2-picoline with tri-styrene [676] and a mixture of 10 wt% of quinaldine with tri-styrene [677]. Since the dipole moments of picoline and quinaldine are much larger than tri-styrene, the observed spectra are contributed effectively by the motions of picoline or quinaldine. Both picoline and quinaldine are rigid molecules without any internal degree of freedom. In each mixture, the shape of the α-loss peak of picoline

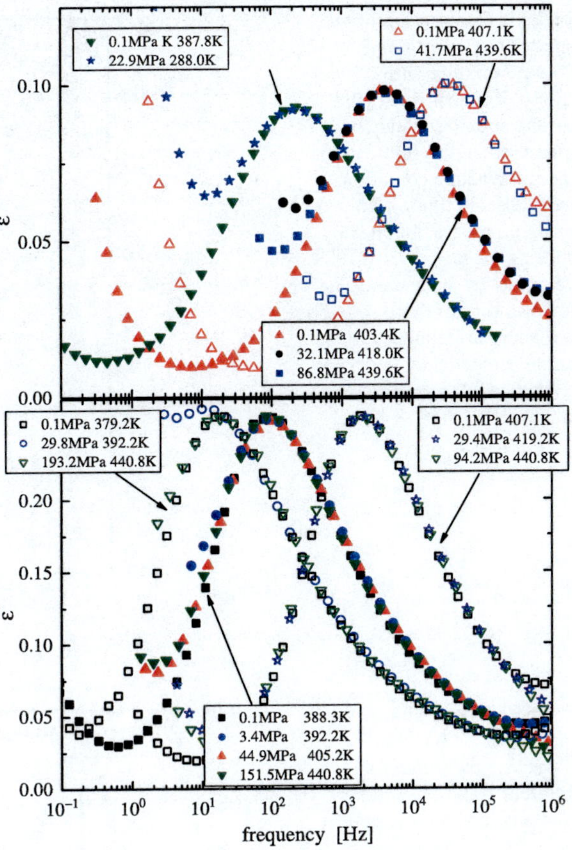

Fig. 52b Local segmental relaxation peaks for various T and P such that τ_α is essentially constant: (*bottom*) neat PCHMA, (*top*) PCHMA/PαMS 50/50 blend. Spectra were shifted slightly to superpose the peak maxima. The rise in the dielectric loss toward low frequency, especially prominent in the blend spectra, is due to dc conductivity. Data from [740] are replotted here

or quinaldine is invariant to T and P when τ_α is held constant, just like PVME in blends with P2CS and PS. The invariance of the shape was found to hold for more than one constant τ_α value. Examples of superposed spectra from the two mixtures are shown in Figs. 53 and 54. In Fig. 55, the α-relaxation of the polar rigid molecule benzonitrile with 10 wt% in mixture with polystyrene with molecular weight of 370 shows perfect superposition for several combinations of P and T [741].

Moreover, the data show that not only the dispersion of the α-loss peak, but also the JG β-relaxation time τ_{JG} ($\sim$ the primitive relaxation time τ_0 of the CM, see Section 2.3) of picoline or quinaldine in the mixtures is invariant to different combinations of T and P while keeping τ_α constant, like neat glassformers DPGDB and BIBE (see Fig. 35e and f). This means that τ_α, τ_0 (or τ_{JG}), and the dispersion (or n) of picoline or quinaldine in the mixtures are co-invariants to changes in T and P, exactly as prescribed by the CM equation, $\tau_\alpha = [(t_c)^{-n}\tau_0]^{1/(1-n)}$.

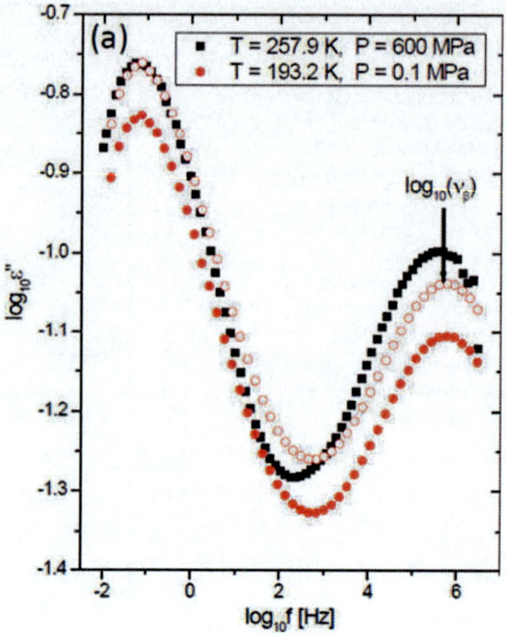

Fig. 53 Comparison of two spectra of 2-picoline in 25 wt% mixture with tri-styrene having the same α-relaxation frequency but measured at different thermodynamic conditions (see labels in figure). The *open circles* are obtained by shifting the data at 0.1 MPa vertically by a constant. The *arrow* indicates that JG β-relaxation frequency remains practically unchanged. Data from [676] are replotted here

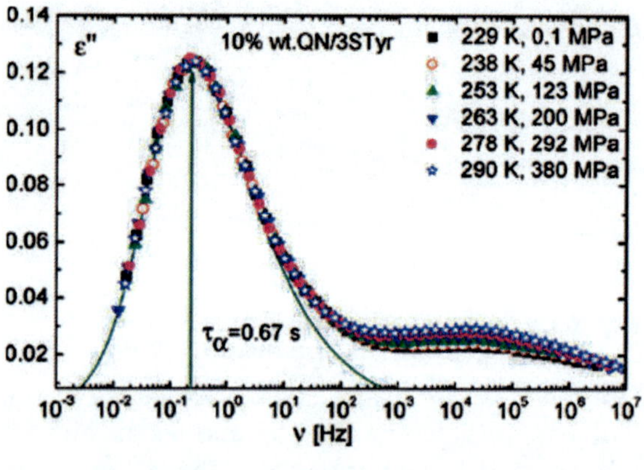

Fig. 54 Superposition of loss spectra for 10% QN in tri-styrene measured for different T and P combinations but the same $\tau_\alpha = 0.67$ s. The *line* is a Fourier-transformed Kohlrausch function fit with $(1 - n) = 0.5$. Reproduced from [677] by permission

2.2.4.8 Impact on Theory by *T–P* Superpositioning of the α-Dispersion at Constant τ_α

We now discuss the impact of this general property on theories and models of the glass transition. The primary concern of most published theories seems to explain the temperature and pressure dependences of the structural relaxation time τ_α. The dispersion (n or β_{KWW}) of the structural relaxation is either not addressed, or else considered separately with additional input not involved in the main part of the theory. For example, the original free volume models [26, 29] and the Adam–Gibbs

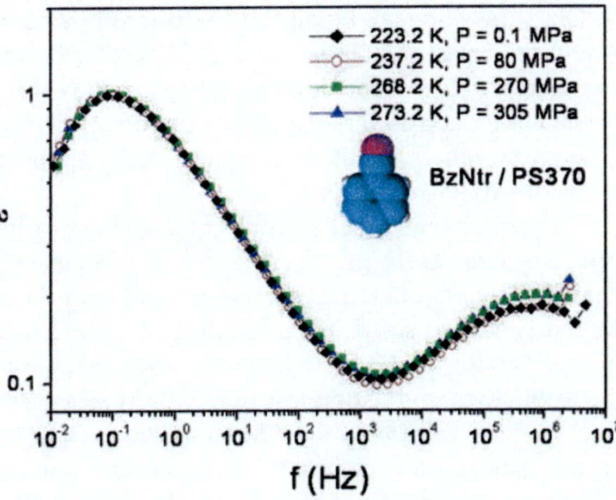

Fig. 55 Superposition of loss spectra for 10% BzNtr in PS370 measured for different T and P combinations but the same τ_α. Reproduced from [741] by permission

model [30] treat the variation of relaxation times with T and P but do not predict the dispersion of the α-relaxation or the distribution of molecular relaxation rates. Additional inputs such as some specific fluctuation or distribution of some parameter were introduced to generate a distribution of relaxation times consistent with the empirical KWW time correlation function, as have been done by others in some extended versions of these theories. It is not difficult for any model to find combinations of T and P such that the predicted $\tau_\alpha(T,P)$ is constant. However, it is unlikely that the same combinations will also keep the predicted dispersion or $n(T,P)$ constant. There is change in volume and also in entropy S for different combinations of T and P as can be expected from the thermodynamic relation, $(\partial S/\partial P)_T = (\partial V/\partial T)_P$ [742]. For one glassformer it may be possible to introduce additional assumptions to force both $\tau_\alpha(T,P)$ and $n(T,P)$ to be simultaneously constant. However, this would not be a worthwhile undertaking since $\tau_\alpha(T,P)$ and $n(T,P)$ are simultaneously constant for many glassformers, with different physical and chemical structures and broadly different sensitivities to temperature and density. Thus, the experimental observations (i.e., simultaneous constancy of $\tau_\alpha(T,P)$ and $n(T,P)$) impose severe constraint on any theoretical effort to understand the glass transition phenomena in toto, past and present. Theories and models, in which the structural relaxation time does not define or govern the dispersion of the structural relaxation, cannot explain the simultaneous constancy of $\tau_\alpha(T,P)$ and the dispersion of the α-relaxation or $\beta_{\mathrm{KWW}}(T,P)$. Most theories and models of glass transition fall into this category. Revision is required to bring them back to consistency with this general experiment fact.

Also can be seen from Figs. 53 to 55 is, in addition to the dispersion of the α-relaxation, the JG β-relaxation time τ_{JG} (or τ_0) of the CM is invariant to different combinations of T and P while keeping τ_α constant, not only in mixtures but also in neat glassformers. This apparently general property of the dynamics of glass-forming systems poses another challenge for formulating a perfect theory of glass transition.

Glass transition is a broad and interdisciplinary field of research. Due to the large amount of experimental data and multiple phenomena, it is possible that even this spectacular *t–T–P* superposition property of the α-relaxation may not be recognized or valued by some workers. If this important experimental fact is not taken into account by others in future development, I see no hope that the glass transition problem will be genuinely solved.

The experimental fact of constant dispersion of the α-relaxation at constant τ_α can be restated as the invariance of the fractional exponent or the Kohlrausch exponent, $(1 - n)$, to different combinations of T and P that hold τ_α constant. Stated this way, it is easy to see that the coupling model (CM) is consistent with this experimental fact. In the CM, the heterogeneous many-molecule dynamics gives rise to the Kohlrausch correlation function with its relaxation time τ_α directly linked to its fractional exponent n by the CM equation, $\tau_\alpha = [t_c^{-n}\tau_0]^{1/(1-n)}$. This CM equation clearly indicates that constant n is a necessary condition in order for τ_α to hold for different combinations of T and P. In Section 2.3.2.1 we present experimental data showing that not only n but also τ_0 remains constant to variations in combinations of T and P that hold τ_α constant. Anticipating this result and the fact that t_c is insensitive to T and P, the CM equation proves that constant n or α-dispersion is both necessary and sufficient condition to have τ_α maintained constant for different combinations of T and P. Not only is τ_α uniquely defined by the dispersion and vice versa on changing thermodynamic conditions as shown herein, but also many properties of τ_α are governed by the dispersion or the fractional exponent n seen by other examples to be discussed in the sections to follow.

2.2.5 Other Structural Relaxation Properties Either Governed by or Correlated with the Dispersion of the α-Relaxation

In Section 2.2.1, we have shown that the width of the dispersion of the structural α-relaxation (or n appearing in the exponent of the Kohlrausch function) systematically increases with the number of intermolecular constraints and the strength of intermolecular interaction. The trend indicates that the dispersion reflects the many-body nature of the α-relaxation. If the latter plays a role in determining the dynamic and thermodynamic properties, then the dispersion or n can be expected to either govern general properties of the α-relaxation or at least correlate with them. In Section 2.2.2, we have seen the correlation between n and ξ_{het}, the length scale of the heterogeneous dynamics of the α-relaxation measured by multidimensional NMR. In Section 2.2.3, it is pointed out that "fragility" and its index m also correlate to n so long as the glassformers are restricted to the same family with similar chemical and physical structures. The restriction helps to avoid interference from thermodynamic factors on which m also depends. The strong connection between τ_α and n is shown in Section 2.2.4 by the co-invariance of the two quantities to various combinations of T and P at constant $\tau_\alpha(T,P)$. These previously discussed results are indicators that the dispersion or n is a key parameter in understanding the dynamics of glassformers. In this section, we review other general properties of the α-relaxation that lead to the same conclusion. Explanation of a selected one or a small subset of these properties is possible without ever invoking the dispersion or n as demonstrated by some theoretical attempts. However, these explanations have the shortcomings of ignoring the empirically established correlation of the properties with the dispersion or n, and leaving the other equally important properties unexplained. Naturally, simultaneous explanation of *all* the properties by a theory based on the dispersion or n will not have the shortcomings, as achieved in the coupling model.

2.2.5.1 Failure of a Single Vogel–Fulcher–Tammann–Hesse (VFTH) Expression to Describe the Temperature Dependence of $\tau_\alpha(T)$

The temperature dependence of τ_α or the viscosity η of the majority of glass-forming liquids is non-Arrhenian and the VFTH expression,

$$\tau_\alpha(T) \text{ or } \eta(T) = A\exp[B/(T - T_0)], \tag{2.41}$$

is often used to fit the experimental data. More than 30 years ago, Plazek, Magill, and Greet [743–745] measured the viscosity of 1,3-*bis*(1-naphthyl)-5-(2-naphthyl)benzene (TNB) over a wide range of almost 16 decades. They showed that a single VFTH expression cannot fit the data over such a wide range. A satisfactory account can only be obtained by an Arrhenius expression at high temperatures above T_A with a crossover to a VFTH expression denoted here by $(\text{VFTH})_1$ below T_A. However, VFTH_1 is no longer adequate when temperature falls below $T_B = 412$ K. A second VFTH equation, VFTH_2, has to be used to describe η for $T_B > T > T_g$. The complete measurements of the viscosity η spanning almost 15 decades together with the VFTH_1 and VFTH_2 fits to the data are shown in the inset of Fig. 11.

It is much easier to spot the failure of a single VFTH expression to fit temperature dependence of τ_α or η over a wide range by replotting the data as the derivative of $\log\tau_\alpha$ with respect to reciprocal temperature, $[\mathrm{d}\log\tau_\alpha/\mathrm{d}(1/T)]^{-1/2}$, vs. $1/T$ [236, 347, 348, 621]. In this plot, VFTH temperature dependence for τ_α or η is transformed to a linear dependence of $[\mathrm{d}\log\tau_\alpha/\mathrm{d}(1/T)]^{-1/2}$ on $1/T$ with non-zero slope equal to $(B/2.303)^{-1/2}T_0$, and is given by

$$\left[\frac{\mathrm{d}\log\tau_\alpha}{\mathrm{d}(1/T)}\right]^{-1/2} = (B/2.303)^{-1/2}\left(1 - \frac{T_0}{T}\right). \tag{2.42}$$

For Arrhenius temperature dependence, $A\exp(E_a/RT)$ is transformed to

$$\left[\frac{\mathrm{d}\log\tau_\alpha}{\mathrm{d}(1/T)}\right]^{-1/2} = (E_a/2.303R)^{-1/2}, \tag{2.43}$$

which is a constant or a line with zero slope in the plot. The inadequacy of any single expression, Arrhenius or VFTH, to fit the data over the whole temperature range becomes obvious. The change of temperature dependence at T_A and T_B can be seen by inspection. The requirement for obtaining $d \log \tau_\alpha/\mathrm{d}(1/T)$ with high accuracy demands measurement of τ_α or η at many temperatures with spacings that are sufficiently small. This requirement is not always met in the data of τ_α or η reported in the literature. Broadband dielectric relaxation is the technique to obtain τ_α meeting the requirement. Therefore, most of the data showing the need of an Arrhenius expression at high temperatures and two VFTH expressions separately at lower temperatures were obtained by dielectric relaxation and a few by viscosity measurements. The high-temperature Arrhenius dependence of τ_α is not usually uncovered by dielectric measurements because it occurs when ν_α is larger than 1 GHz which is not easy to make dielectric relaxation measurement. Consequently, the crossover to Arrhenius dependence of viscosity was only seen in several small molecule glassformers [236, 237, 347, 348, 746], and in a few polymers [242, 361].

The viscosity data of TNB are historic in first revealing the presence of three distinctly different temperature regimes. The complete data satisfy the requirement for the derivative analysis. Indeed, the data when replotted as $[\mathrm{d}\ \log \eta/\mathrm{d}(1/T)]^{-1/2}$ vs. 1000/T in Fig. 11 show the presence of three regimes. The points in the figure are obtained from the experimental viscosity data by using the ratio of differences as approximation of $\mathrm{d}\ \log \eta/\mathrm{d}(1/T)$. The scattering of points about the straight lines is largely due to an insufficient number of data points to represent the derivative by the ratio of finite differences. The two straight lines with non-zero slopes and labeled by VFT(1) and VFT(2) correspond to the two VFTH expressions found originally by Plazek and Magill [743–745]. The two lines intersect at $T_B = 412$ K, the temperature below which $VFTH_1$ fits the temperature dependence of η, and above which $VFTH_2$ holds instead. The dashed horizontal line in the figure indicates that the T-dependence of the viscosity data is Arrhenius at high temperatures. The intersection of horizontal line with the steep full line determines $T_A = 588$ K, the temperature near which another change of temperature dependence of η from $VFTH_2$ to the Arrhenius dependence occurs. The same behavior as TNB was found in salol, butyl-benzene, and propyl-benzene by Hansen et al. [348]. The data and analysis of propyl-benzene are shown in Fig. 56.

Many more small molecular glassformers [236, 347, 348, 621] and polymers including PVAc [236, 238], PPGE, and DGEBA [242] show the crossover of T-dependence of either τ_α or η at T_B. Here I add the data of two inorganic glassformers to demonstrate generality. The viscosity of boron trioxide (B_2O_3) also has two crossovers, one at T_A from Arrhenius to $VFTH_2$, and at T_B from $VFTH_2$ to $VFTH_1$, which turns out to be almost Arrhenius as shown in Fig. 57. The dielectric and mechanical relaxation times combined of the molten salt 0.4 $Ca(NO_3)_2$–0.6 KNO_3 (CKN) also show the crossover from $(VFTH)_2$ to $VFTH_1$ at T_B in the inset of Fig. 58 [748]. Included in the main part of the figure are the secondary relaxation times of CKN determined by mechanical spectroscopy deep in the glassy state [749].

By now, the crossover of the temperature dependence of τ_α or η from $VFTH_1$ to $VFTH_2$ at some T_B has been found in so many glass-forming liquids that it is

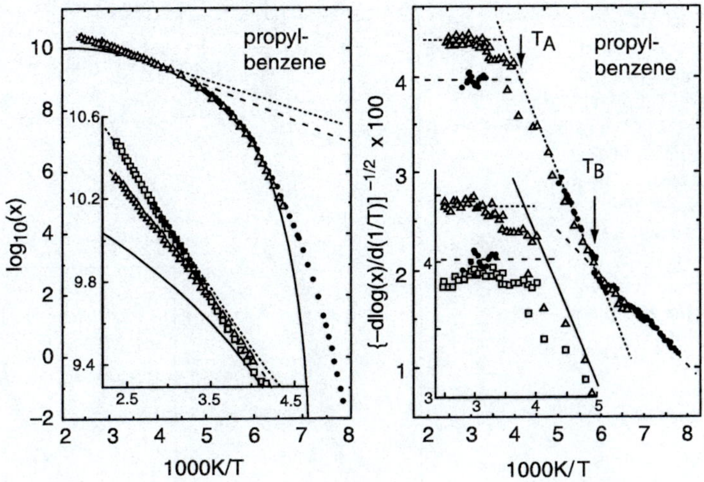

Fig. 56 *Left panel*: Temperature dependence of the dielectric peak frequency (•, $x = f_{\max}/\mathrm{Hz}$) and of the inverse viscosity (Δ, $x = \kappa \cdot \eta^{-1}$ Poise, $\log \kappa = 7.70$) for propylbenzene. The *solid line* is a VFTH fit to the $T_B \leq T \leq T_A$ data, the *dashed line* is an Arrhenius fit to the $T > T_A$ dielectric data. The *dotted line* is an Arrhenius fit to the $T > T_A$ viscosity data, $\log_{10}(\kappa \cdot \eta^{-1}\mathrm{Poise}) = 11.92 - 620\ \mathrm{K}/T$. The *inset* shows the Arrhenius regime on enlarged scales and including the data for $x \propto T \cdot \eta^{-1}$ (□) in addition to $x \propto \eta^{-1}(\Delta)$. *Right panel*: Temperature dependence of the dielectric peak frequency and of the viscosity for propylbenzene, plotted as $[\mathrm{d}\log x/\mathrm{d}(1/T)]^{-1/2}$ vs. 1/*T*. The data, fits, and *symbols* correspond to those in the *left panel*. The indicated characteristic temperatures are $T_A(\eta) = 264$ K, $T_A(D) = 240$ K, and $T_B = 170$ K. The *inset* shows the Arrhenius regime on enlarged scales and including the data for $x \propto \kappa \cdot \eta^{-1}$ (□) in addition to $x \propto \eta^{-1}$ (Δ). Reproduced from [348] by permission

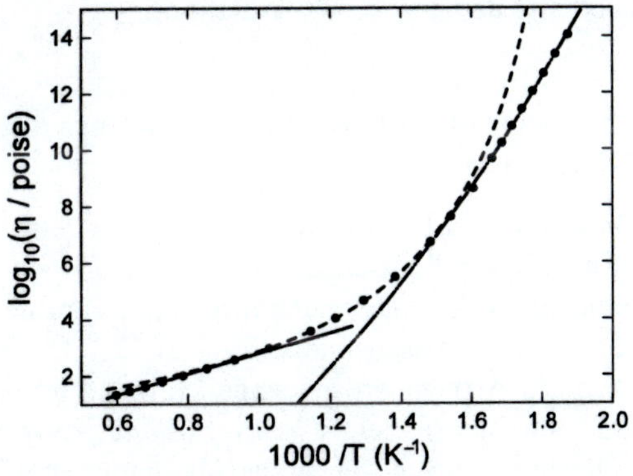

Fig. 57 Viscosity data of anhydrous B_2O_3 showing Arrhenius *T*-dependence at high temperatures (*solid line*) changing to $VFTH_2$ dependence at intermediate temperatures (*dashed line*), and changing to $VFTH_1$ which is practically Arrhenius (*dashed-dotted line*) at low temperatures (*dashed-dotted line*). Data taken from Ph.D. thesis of K. Bermatz (University of Pittsburgh 2004)

a general property or phenomenon of glass-forming liquids. Fewer cases show the crossover at T_A because this usually occurs when τ_α is much shorter than nanosecond and data of this sort are not common. From the results one can observe several correlations between the parameters characterizing the phenomena and *n* (or the width of the dispersion of the α-relaxation) as a function of temperature [237, 238,

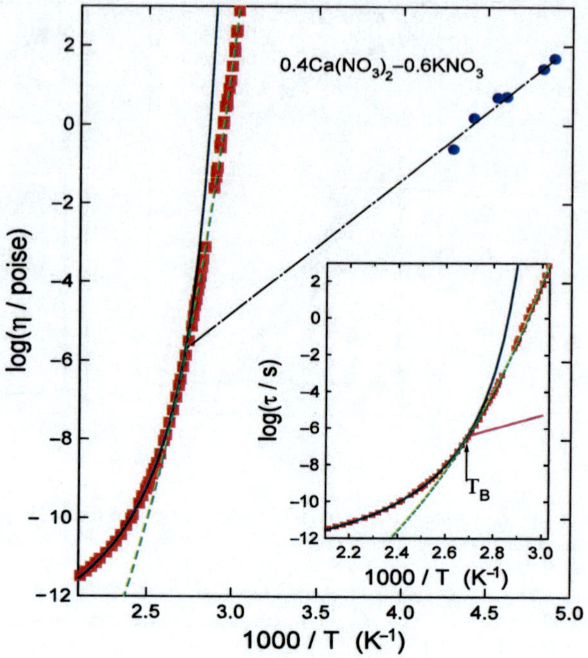

Fig. 58 Dielectric (*squares*) and mechanical (*circles*) relaxation time τ_α of CKN data from [748] showing the change of dependence from VFTH$_2$ to VFTH$_1$ at T_B clearly in the *inset*. The main figure shows also the secondary relaxation time from mechanical spectroscopy deep in the glassy state (*large circles*). Data from [749]

253, 192, 115(a)]. They are discussed in greater detail below than in previous publications. Theories claiming to have the temperature dependence of τ_α or η explained need also address these correlations.

Correlation of Crossover in T-Dependence of τ_α or η with Corresponding *T*-Dependence of *n*

Dielectric and light-scattering data of α-relaxation [192, 236, 237, 253, 347, 348] of mostly small molecular glassformers when analyzed show that for $T > T_A$, n is either unity (i.e., exponential relaxation) or nearly equal to unity (see Fig. 59 for a collection of glassformers, inset of Fig. 15, and Fig. 1A in [237]), and relaxation time has Arrhenius *T*-dependence (see Fig. 60 and Fig. 14 inset). These features are shown more clearly in Fig. 61 for propylene carbonate alone. Both behaviors indicate absence of intermolecular cooperativity of the motions when $T > T_A$. The relaxation times of four liquids have Arrhenius dependence $T > T_A$ as shown before in Fig. 10. The prefactor $10^{-13.5}$ has the physically reasonable value corresponding to angular attempt frequencies of the order of $10^{-13.5}$ s for independent rotation of the molecules at high temperatures.

In a table of [750], relaxation times at the onset of Arrhenius behavior for several liquids and one polymer are given. Averaging the 10 values yields $\log(\tau(T_A)/\mathrm{s}) = -10.5 \pm 0.4$; that is, the relaxation time at T_A is approximately the same for all the

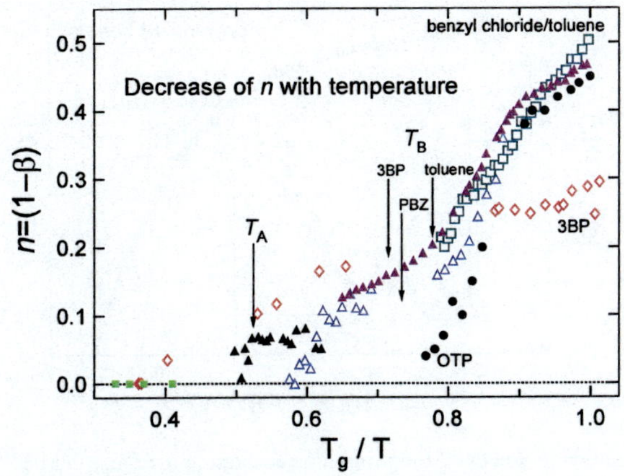

Fig. 59 *n(T)* obtained from the exponent (1–*n*) of the Kohlrausch function used to fit dispersion of dielectric relaxation data plotted as function of T_g/T for 25% benzyl chloride/75% toluene (*open squares*), 3BP (◇), propylbenzene (Δ), and salol (▲). Shown also are *n(T)* of *ortho*-terphenyl (OTP) (*closed squares*) and toluene (*closed circles*) obtained from dynamic light-scattering data. The normalized temperatures T_g/T_B indicated by *vertical arrows* are for propylbenzene (PBZ) and 3-bromopentane (3BP) and toluene. The normalized temperature T_g/T_A is nearly the same for all liquids and is indicated by a *single vertical arrow*. Data assembled in [253] are replotted

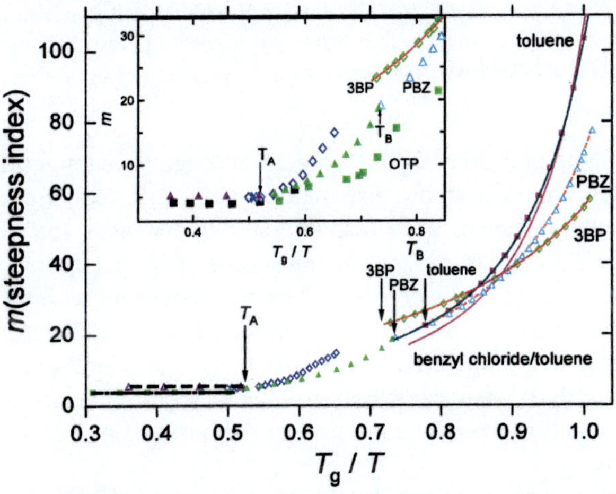

Fig. 60 The steepness index, $m = [\mathrm{d}\log\tau_\alpha/\mathrm{d}(T_g/T)]$, calculated from experimental τ_α as a function of T_g/T for toluene (■), propylbenzene (PBZ) (Δ,▲), 3BP (◇), and 25% benzyl chloride/75% toluene (*solid curve*). The normalized temperatures T_g/T_B defined in the text are indicated by *vertical arrows* for propylbenzene (PBZ) and 3-bromopentane (3BP) and toluene. T_B of toluene is not available and instead T_β is used. The normalized temperature T_g/T_A is nearly the same for all liquids and is indicated by a *single vertical arrow*. The viscosity data of toluene are used to calculate *m* at high temperatures. The *inset* enlarges the region around T_g/T_A and T_g/T_B to exhibit the constant *m* for $T > T_A$ for 3BP (◇), PBZ (Δ,▲), toluene (□), and OTP (•). Data collected in [253] are replotted

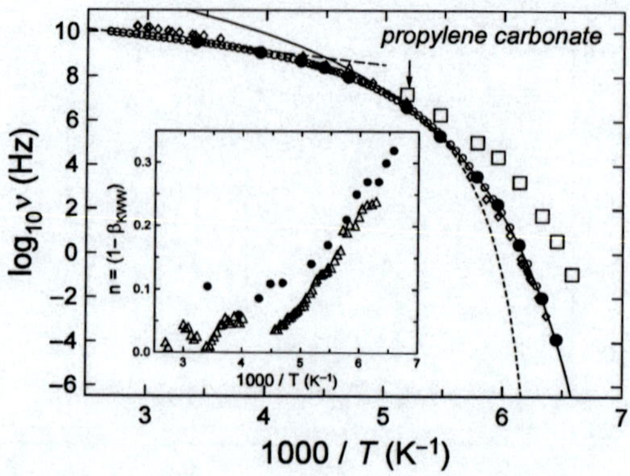

Fig. 61 Temperature dependences of the propylene carbonate α-relaxation peak frequency (*filled circles*) and the β-relaxation peak frequency (*open squares*) from the Cole–Cole curves from [424]. Shown also are the α-relaxation peak frequency (*open circles*) from the dielectric data from [236, 621]. The *open diamonds* are logarithm of the reciprocal of the viscosity data from [236, 621] shifted by a constant to match the dielectric data. The *solid and dashed lines* are, respectively, the VFTH law for the α-relaxation peak frequency for the lower temperature regime (T<192 K) and for the intermediate higher temperature regime of (192 K<T<265 K), with T_B=192 K as indicated. The *dashed straight line* is the Arrhenius fit to the high-temperature data of [236, 621], with T_A indicated. The *inset* shows the temperature dependence of the coupling parameter, n, or equivalently $(1-\beta_{KWW})$, used in the KWW function to fit the dielectric dispersion (*filled circles*). The *open triangles* are $(1-\beta_{KWW})$ calculated from w, the full-width at half-maximum of the dielectric loss peak normalized to that of an ideal Debye loss peak, given in [236, 621] by using Eq. (2.1). Data collected in [424] are replotted here

materials. This value of ~30 ps is three orders of magnitude longer than the vibrational frequencies, naturally so because the relevant length scale for liquid motions is the intermolecular distance, rather than interatomic distances. Interestingly, $\tau(T_A)$ is within about one decade from 2 ps, the onset time of cooperative motions according to the coupling model. This correspondence is consistent with the idea that the loss of Arrhenius behavior upon cooling to temperatures below T_A is due to development of intermolecular cooperativity. The narrow spread of $\tau(T_A)$ for the different liquids indicates only a weak sensitivity to chemical structure, from which a reasonable inference is that the relaxation time at the onset of Arrhenius behavior is a material constant.

There is an increase of n as temperature is decreased from T_A down to T_B. The increase is slight such that at $T = T_B$, $n(T_B)$ is still a fraction of $n(T_g)$. A more rapid increase of n is evident as temperature falls below T_B until it reaches $n(T_g)$ at $T = T_g$ [751]. This feature can be seen in Figs. 59 and 61. Thus, T_B is the temperature at which $\mathrm{d}n(T)/\mathrm{d}T$ exhibits a step-like change, suggesting that the latter is the cause of the change of temperature dependence of τ_α from VFTH$_2$ to VFTH$_1$ at T_B. Data of $\tau_\alpha(T)$ and $n(T)$ of several other small molecular glassformers [423] and poly(vinyl acetate) [238] all show the step-like increase of $\mathrm{d}n(T)/\mathrm{d}T$ when crossing T_B from above to below. Hence, generally at T_B, the change of T-dependence of $\tau_\alpha(T)$ or η is correlated with that of $\mathrm{d}n(T)/\mathrm{d}T$.

The Disparity Between $(\mathrm{VFTH})_1$ and $(\mathrm{VFTH})_2$ Correlates with $n(T_g)$

Moreover, one can observe [115, 237, 253] that glassformer having larger $n(T_g)$ or smaller Kohlrausch exponent $\beta(T_g)$ has larger difference between VFTH_2 and VFTH_1. Difference between VFTH_1 and VFTH_2 can be gauged by the difference between the values of τ_α given by these two functions, at any chosen value of the scaled temperature T_B/T. VFTH_1 and VFTH_2 can be extrapolated from their valid ranges to a common temperature range of $T_A > T > T_g$. Generally it is true that glassformer having larger $n(T_g)$ at $T = T_g$ exhibits a larger difference between the two values of $\tau_\alpha(T_B/T_{ref})$ at a common reference temperature, T_{ref}, obtained by the extrapolated VFTH_1 and VFTH_2. This trend is shown in Fig. 62 by comparing the ratio as a function of T_B/T for three glassformers in descending order of n: OTP ($n(T_g) \approx 0.5$, $T_g = 244$, $T_B = 290$ K), glycerol ($n(T_g) \approx 0.30$, $T_g = 190$ K, $T_B = 285$ K), and propylene glycol ($n(T_g) \approx 0.25$, $T_g = 167$ K). The one with larger $n(T_g)$ shows larger difference between $\mathrm{VFTH}_1(T_{ref})$ and $\mathrm{VFTH}_2(T_{ref})$. This trend suggests that the difference between $(\mathrm{VFTH})_1$ and $(\mathrm{VFTH})_2$ would vanish for glassformer having $n(T_g) = 0$, and a single VFTH is sufficient. GeO_2 can be considered to be such an example, which has n=0 and Arrhenius T-dependence if the latter is taken as a special case of VFTH dependence. This trend is another way to see that $n(T_g)$ governs the change in T-dependence of τ_α when crossing T_B.

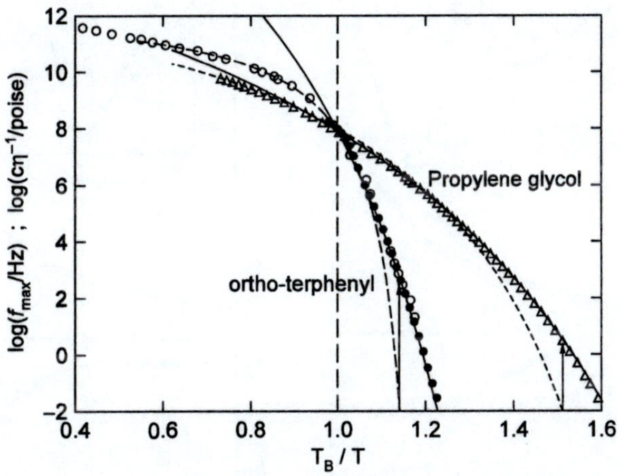

Fig. 62 Frequency of the maximum of the dielectric loss, f_{max}, of OTP and propylene glycol compared on a plot against T_B/T, where T_B is the crossover temperature from VFTH_2 (*dashed curve*) to another VFTH_1 (*solid curve*). OTP (*black circles*), propylene glycol (*open triangles*). Reciprocal of the viscosity data of OTP scaled by a constant log c=8.77 (*open circles*). At the highest temperature, the reciprocal viscosity of OTP has a true Arrhenius temperature dependence. The *two vertical arrows* indicate that there is a larger difference between $\mathrm{VFTH}_1(T_{ex})$ and $\mathrm{VFTH}_2(T_{ex})$ for OTP than propylene glycol

The CM Explanation

From the experimental facts discussed above, it is clear that the change of T-dependence of $\tau_\alpha(T)$ or η from $VFTH_2$ to $VFTH_1$ at T_B is correlated with that of $dn(T)dT$. Since n is the indicator of many-body dynamics, the cause of the crossover can be restated as due to the onset at T_B of more rapid increase of the extent of many-body dynamics with falling temperature toward T_g. This explanation by the CM of the crossover from $VFTH_2$ to $VFTH_1$ was tested [423] against experimental data of $\tau_\alpha(T)$ and $n(T)$ of several small molecular glassformers by calculating the primitive relaxation time $\tau_0(T)$ from these data by the CM equation $\tau_\alpha(T) = [t_c^{-n(T)}\tau_0(T)]^{1/(1-n(T))}$. All calculated $\tau_0(T)$ are well described by a single VFTH law and no crossover appears as shown by the example of OTP in Fig. 63, thus indicating that $n(T)$ governs the T-dependence of $\tau_\alpha(T)$ or η. More such examples can be found in [423]. These general results support the CM explanation of the crossover as T-dependence of τ_α from $VFTH_2$ to $VFTH_1$ at T_B is due to corresponding changes in the T-dependence of n.

Correlation with Structural Changes Supporting the CM Explanation

Wide Angle X-Ray Scattering

Studies of the change in structure of two low-molecular weight glass-forming liquids, propylene carbonate and salol, with temperature were carried out by using

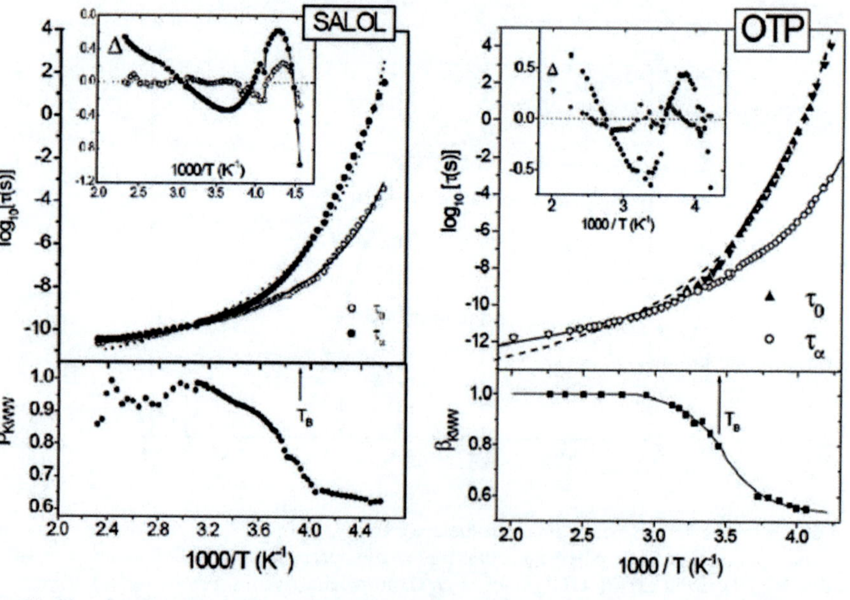

Fig. 63 The experimental $\tau_\alpha(T)$ and the calculated $\tau_0(T)$ in *upper panel*, together with $\beta_{KWW} \equiv (1-n(T))$ (*lower panel*) of salol (*left*) and OTP (*right*), plotted as a function of inverse temperature. The *lines* in the *upper panel* are the best fit to VFTH. The location of $1000/T_B$ is indicated by the *vertical arrow*. The *inset* shows the differences between the experimental points and the best fit to VFTH. Reproduced from [423] by permission

wide angle X-ray scattering (WAXS) and molecular dynamics simulation [752]. The experiment measured the WAXS intensities of propylene carbonate and salol as a function of the magnitude of the scattering vector q, and at different temperatures. For PC, at temperatures $T > T_A$(290 K), an almost symmetric peak as amorphous halo is centered at about $q = 1.38$ Å^{-1}. There is no indication for a shoulder on either the right or left side of the maximum. However, below T_A on decreasing T, intensity from the q-region of the amorphous halo around 1.4 Å^{-1} is transferred to higher qs around 2 Å^{-1}. The pronounced changes in the diffraction curves with decreasing T indicate development of new peaks with different temperature dependences, which correspond to next neighbor and next neighbor distances. Consequently, there is increasing orientational correlation of the molecules with decreasing temperature below T_A. The intensity of the main peak around 1.4 Å^{-1} also sees more rapid drop with decreasing temperature when crossing $T_A = 290$ K and $T_B = 200$ K. These structural changes observed by WAXS support increased spatial correlations of molecules and hence enhanced many-body dynamics and increase in dispersion or n. Thus, the step-like increase of $|dn(T)/dT|$ observed when crossing T_A and T_B on decreasing temperature can be understood, and the CM explanation of the changes in T-dependence of τ_α on crossing T_A and T_B is made more evident by the structural changes observed by WAXS.

Structure of atatic poly(ethylmethacrylate) ($T_g = 339$ K) was studied by WAXS [753]. The diffraction WAXS curves show two peaks. The peak at the highest q value corresponds to an equivalent "Bragg" spacing of $d = 0.5$ nm. It is assigned to the mean van der Waals distance of not chemically bonded atoms. The q value of the peak position as a function of temperature exhibits an abrupt change of slope at some temperature $T_c = 390$ K which coincides with $T_{\alpha\beta}$, the merging/splitting temperature of the α and β relaxations of atactic PEMA. Since the β relaxation is local and non-cooperative, when merging with it, the α relaxation will tend to have the same character, and has small n at $T_c = 390$ K. Dielectric [252, 296] and NMR [754, 755] have shown that n of atactic PEMA is a significant fraction of unity (~0.40) at T_g. From this and above, one may infer that the observed changes of structure of atactic PEMA at T_c has something to do with the corresponding change in dynamics due to the increase of n with decreasing temperature toward T_g.

Positron Annihilation Lifetime Spectroscopy

Positron annihilation lifetime spectroscopy (PALS) is a technique to probe the unoccupied volume, or so-called "free volume," of glassformers. In organic glasses, the *ortho*-positronium (o-Ps) bound state of a positron has a strong tendency to localize in heterogeneous regions of low electron density. If the heterogeneity is assumed to be a spherical cavity, the o-Ps lifetime τ_3 can be related to an average cavity or nanopore radius R. Over the range of τ_3 values usually measured, the relation between τ_3 and R is nearly linear. The variations of τ_3 with temperature of propylene carbonate (PC), *ortho*-terphenyl (OTP), glycerol, and propylene glycol (PG) from [387] are shown in Fig. 64. The thermal variations of τ_3 are qualitatively similar in all four small molecule glassformers. At low temperatures in the glassy

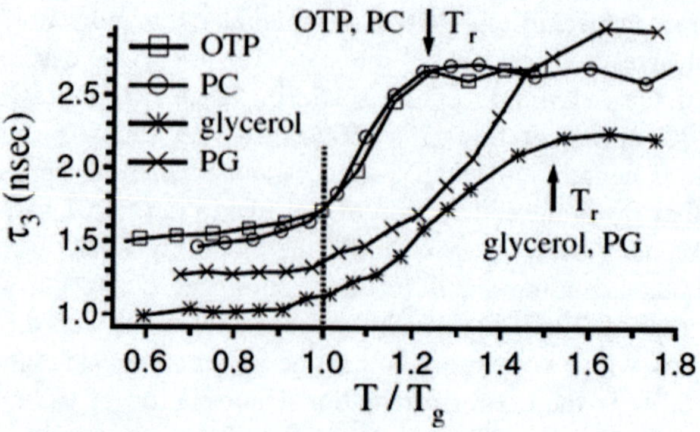

Fig. 64 τ_3 as a function of the T_g-scaled temperature for OTP, PC, PG, and glycerol. A crossover in the thermal dependence of τ_3 is observed both at T_g and an upper temperature T_r, indicated by the *vertical arrows*. Typical standard uncertainties in τ_3 are ± 0.025 ns. T_r is nearly the same as T_B for all four glassformers. Reproduced from [387] by permission

state, τ_3 exhibits a weak temperature dependence which changes to a much stronger dependence after crossing T_g. This change reflects the glass–liquid transition. On further increasing T, a second crossover of the temperature dependence of τ_3 occurs at T_r beyond which τ_3 becomes slowly varying and assumes a plateau-like value. This high-temperature crossover occurs at $T_r/T_g \approx 1.2$ for OTP and PC and ≈ 1.5 for PG and glycerol. Remarkably, T_r is nearly the same as T_B for all four glassformers. The observed more rapid decrease of τ_3 or nanopore radius R with falling temperature past T_r or T_B suggests the molecules are closer together, and hence the more rapid increase of the intermolecular coupling parameter n below T_B as observed via the change in T-dependence of the Kohlrausch stretch exponent $\beta(T)$ [423, 424] or the full-width at half-height of the dielectric loss peak [236].

From Configurational Entropy $S_c(T)$

Another thermodynamic quantity that has been extensively measured is the molar configurational entropy $S_c(T)$. In Section 2.2.2.2, we discussed the Adam–Gibbs model which gives the number of molecules inside the cooperative rearranging region z^* from $S_c(T)$ via $z^*(T) = N_A s_c^*/S_c(T)$ (given before by Eq. (2.22)). Although the Adam–Gibbs model is inadequate as discussed before in Section 2.2.2.2 [401], the $z^*(T)$ calculated as a function of temperature from $S_c(T)$ shows a more rapid increase with decreasing temperature after crossing T_B for many small molecular glassformers, some of them as shown in Fig. 65. This general result suggests the onset at $T = T_B$ of more rapid increase in the number of molecules participating in the α-relaxation (and hence also n) with decreasing temperature. This inferred behavior of n is the cause of the change of temperature dependence of τ_α from VFTH$_2$ to VFTH$_1$ at T_B. Worth mentioning here is the plot of $\log \tau_\alpha(T)$ against $1/[TS_c(T)]$. A linear relation between $\log \tau_\alpha(T)$ and $1/[TS_c(T)]$ is predicted by the Adam–Gibbs relation, $\tau_\alpha(T) = A_{AG} \exp[\Delta\mu s_c^*/k_B T S_c(T)]$ (given before by

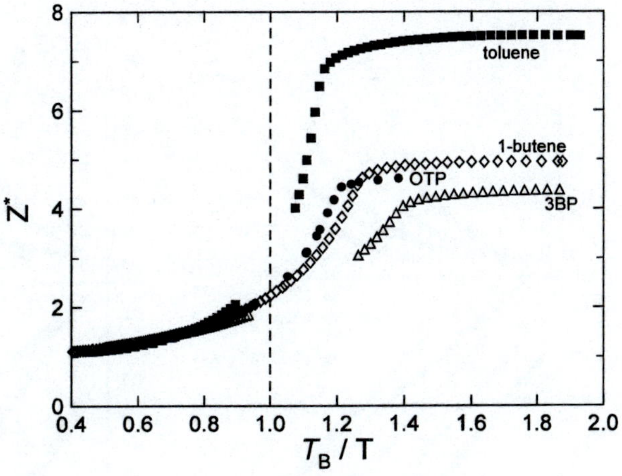

Fig. 65 T_B-scaled temperature dependence of z^*, the number of molecules in the group considered together in motion because of the entropy crisis. For toluene T_B is not available, and instead T_β is used

Eq. (2.23)). In reality, there are two separate linear relations, a weaker one for $\tau_\alpha < \tau_\alpha(T_B)$ and a stronger one for $\tau_\alpha > \tau_\alpha(T_B)$. The first such observation of the departure from Adam–Gibbs prediction was based on viscosity data by Magill [396], to be followed by Richert and Angell [398] and the author [401]. Since τ_α is determined by both configurational entropy $S_c(T)$ and many-body effects on dynamics, it is unsurprising that linear dependence of τ_α on $1/[TS_c(T)]$ cannot hold. On the other hand, many-body effects is absent in the primitive relaxation time τ_0, and the relation $\tau_0 \propto 1/[TS_c(T)]$ holds. Like the change of T-dependence of τ_α at T_B, the appearance of two different linear dependences between $\log \tau_\alpha(T)$ and $1/[TS_c(T)]$, one for $T < T_B$ and another for $T > T_B$, is caused by the onset of more rapid increase of n when T falls below T_B [401].

Crossovers at Elevated Pressures

(a) ***At*** $\boldsymbol{T_B}$: The crossover from VFTH_1 to VFTH_2 was observed isobarically at ambient pressure but also at elevated pressures. The crossover temperature T_B generally increases with applied pressure P, but remarkably the value of τ_α or the viscosity at the crossover, $\tau_\alpha(T_B)$, is the *same* for a given glassformer, independent of pressure or under the condition of constant volume V [653, 756]. Four examples, PC, KDE, BMMPC, and salol, are shown in Figs. 66 and 67. The reader may recall all these glassformers have the same dispersion or n at constant τ_α, independent of T and P, a general property discussed before in Section 2.2.4. Hence, the dispersion or n is invariant at all crossover temperatures T_B and crossover relaxation time τ_B independent of P or at constant V. The value of $n(T_B)$ as well as the increase of $dn(T)/dT$ when crossing T_B, from above to below it, is also independent of P.

(a′) **At** $\boldsymbol{T_A}$: The same is found of the crossover from Arrhenius T-dependence to the $(\text{VFTH})_1$ at T_A for salol [750]. For glassformers that have data above T_A, the value of $\tau_\alpha(T_A)$ falls within the range of 5–150 ps, and for the majority

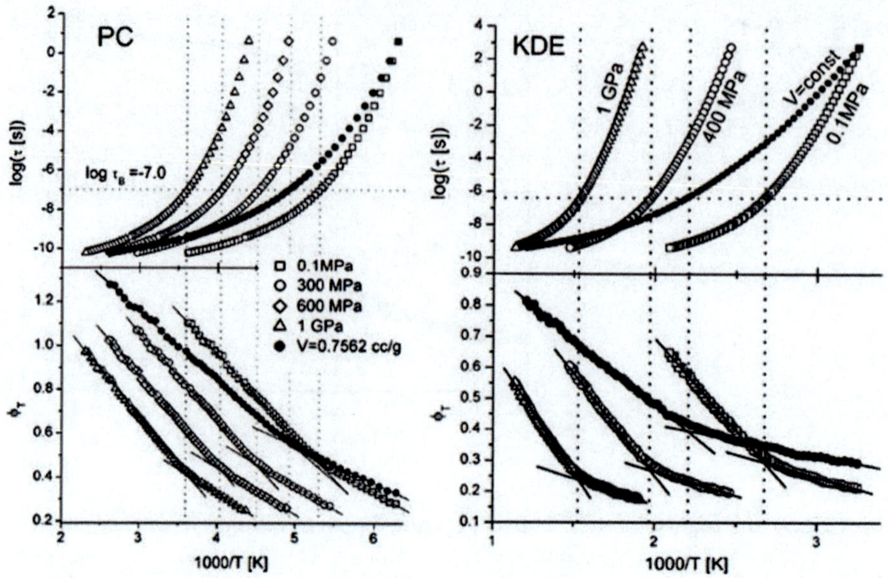

Fig. 66 *Left upper panel*: Dielectric relaxation time for PC (experimental data for 0.1 MPa; other isobars and isochoric curve were calculated). The *horizontal dotted line* indicates the average of $\log(\tau_{\mathrm{B}})$ for the different curves. *Left lower panel*: Stickel function for PC, with low and high T linear fits, done over the range $-6.14 < \log_{10}(\tau/\mathrm{s}) < 0.63$ and $-10.21 < \log_{10}(\tau/\mathrm{s}) < -8.03$, respectively. The *vertical dotted lines* in both panels represent the dynamic crossover. *Right upper panel*: Dielectric relaxation time for KDE (experimental data for 0.1 MPa; other isobars and the isochoric curve at $V = 0.7709$ ml/g were calculated). *Dotted line* indicates the average of $\log(\tau_{\mathrm{B}}) = -6.35$ for the different curves. *Right lower panel*: Stickel function for KDE, with low and high T linear fits, done over the range $-4.62 < \log_{10}(\tau/\mathrm{s}) < 2.72$ and $-9.4 < \log_{10}(\tau/\mathrm{s}) < -7.28$, respectively. *Vertical dotted lines* indicate the dynamic crossover. Reproduced from [653, 756, 757] by permission

it is of the order of 100 ps [750]. The relaxation times $\tau_{\alpha}(T)$ for salol under different pressures are presented in Fig. 68-1. The plot includes experimental data at ambient pressure and those calculated τ_{α} for two higher pressures using the scaling law, $\tau = f(TV^{\gamma})$, in which f represents a function, V is the specific volume, and γ is a material constant specifically for salol. The deviation from the Arrhenius slopes at high temperature occurs at a constant value of τ_{α}= 95 ps, as indicated by the horizontal dashed line. Thus, the relaxation time at the crossover is the *same* independent of pressure. From the previously established co-invariance of τ_{α} and the shape (or n) of the α-relaxation, this implies that the value of $n(T_{\mathrm{A}})$ as well as the increase of $\mathrm{d}n(T)/\mathrm{d}T$ when crossing T_{A}, from above to below it, is also independent of P. These quantities are function of τ_{α}.

The property seems general because the same behavior was found in propylene carbonate (PC) from dielectric relaxation data and *ortho*-terphenyl (OTP) from viscosity data. These behaviors are shown in Figs. 68-2 and 68-3.

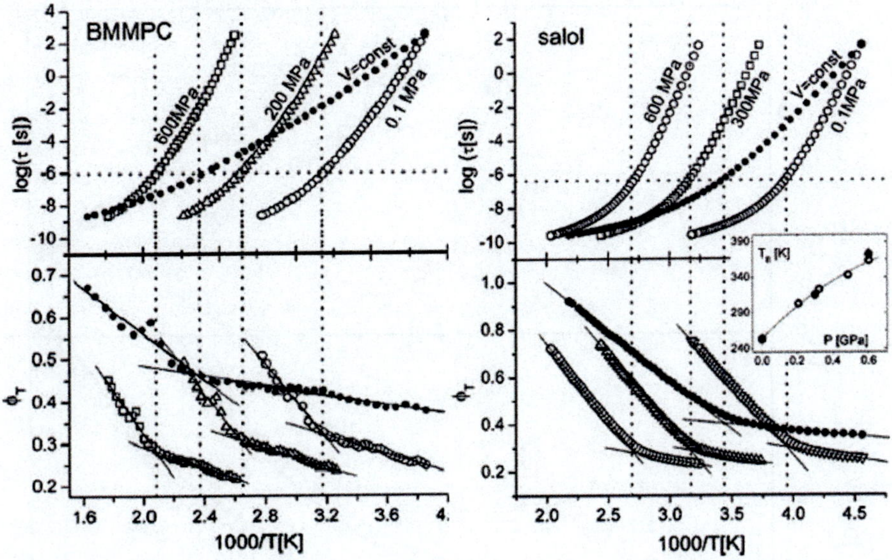

Fig. 67 *Left upper panel*: Dielectric relaxation time for BMMPC experimental data for 0.1 MPa, other isobars at 200 and 600 MPa and the isochore at $V = 0.9032$ ml/g were calculated. *Dotted line* indicates the average of $\log(\tau_{\mathrm{B}}) = -6.1$ for the different curves. *Left lower panel*: Stickel function for BMMPC, with low and high T linear fits, done over the range $-4.68 < \log_{10}(\tau/\mathrm{s}) < 3.85$ and $-8.55 < \log_{10}(\tau/\mathrm{s}) < -6.4$, respectively. *Vertical dotted lines* indicate the dynamic crossover. *Right upper panel*: Dielectric relaxation times for salol experimental data for 0.1 MPa, calculated curves for P=0.3 and 1 GPa, and $V = 0.7896$ ml/g. *Right lower panel*: Stickel functions for salol, with linear fits over the respective ranges $-4.58 < \log_{10}(\tau/\mathrm{s}) < 1.6$ and $-9.59 < \log_{10}(\tau/\mathrm{s}) < -7.44$. Reproduced from [653, 756, 757] by permission

(b) At some chosen constant temperature T, increase of τ_α or η can be made by applying pressure. It is customary to use the empirical VFTH-like equation [627],

$$\tau_\alpha(P) \text{ or } \eta(P) = A_P \exp[D_P P/(P_0 - P)], \tag{2.44}$$

to fit the P-dependence of experimental data. The corresponding function, $\varphi_P \equiv [\mathrm{d}\log\tau_\alpha/\mathrm{d}P]^{-1/2}$ or $[\mathrm{d}\log\eta/\mathrm{d}P]^{-1/2}$, transforms Eq. (2.44) to become the linear dependence $\phi_P = a_P - b_P P$. Hence, deviation of experimental data from the fit using Eq. (2.44) can be easily detected. For several small molecular glassformers, crossover from one VFTH-like P-dependence to another was found in τ_α and η at some pressure P_{B}. Examples from τ_α data are presented for PDE in Fig. 69, and two samples of poly(chlorinated biphenols), PCB62 and PCB42, are shown in Fig. 70 [653], and viscosity data of salol in Fig. 71 [547]. The same is found from the viscosity data of OTP (not shown here). It is remarkable that the crossover values of $\tau_\alpha(P_{\mathrm{B}})$ or $\eta(P_{\mathrm{B}})$ under isothermal condition for different choices of constant T and the crossover values $\tau_\alpha(T_{\mathrm{B}})$ or

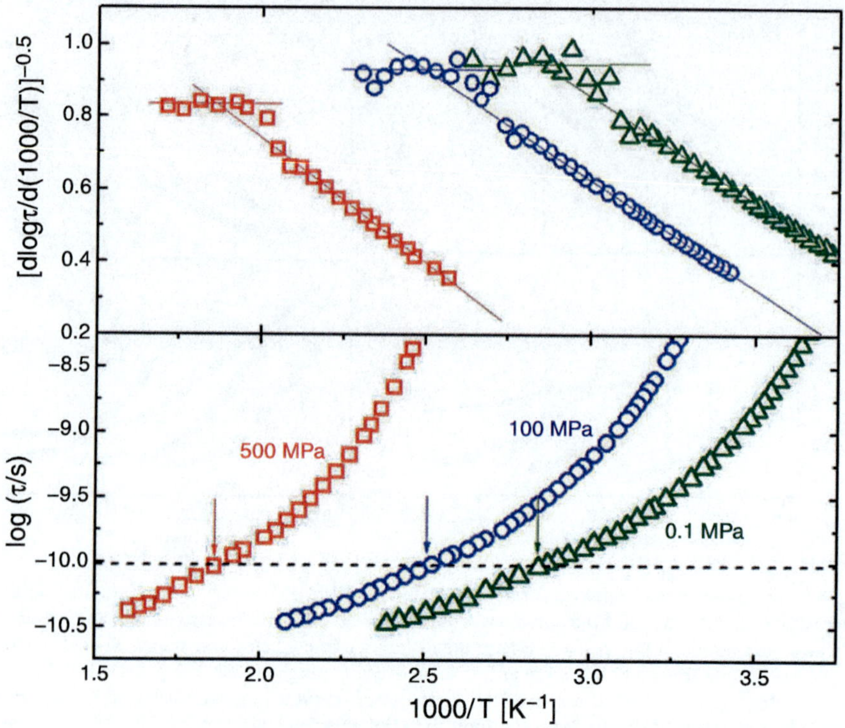

Fig. 68-1 Relaxation times for salol, including experimental data for ambient pressure and calculated τ for two higher pressures using the scaling law $\tau = f(TV^{\gamma})$, in which f represents a function, V is the specific volume, and γ is a material constant (see section on this topic). The deviation from the Arrhenius slopes at high temperature occurs at a constant value of $\tau = 95$ ps, as indicated by the *horizontal dashed line*. The *vertical arrows* indicate the crossover occurs at the same τ. Reproduced from [750] by permission

$\eta(T_B)$ at isobaric condition are the same (see Figs. 69 and 70). Again, from the general property of the same α-dispersion or n for the same τ_α independent of T and P, it follows that at the various crossovers the values of $n(P_B)$ or $n(T_B)$ are the same. In other words, $n(P_B)$ or $n(T_B)$ is a material constant associated with a characteristic of the P or T dependence of τ_α, i.e., its change when crossing P_B or T_B. Since this characteristic of τ_α is invariant to change in thermodynamic conditions, we can further say that this characteristic of τ_α is not caused by P, T, and V, and instead by the many-body dynamics manifested by the α-dispersion, $n(P_B)$ or $n(T_B)$, as well as the change in the P or T dependence of n when crossing P_B or T_B. This is clear evidence of contribution from many-body dynamics to P or T dependence of τ_α besides volume and entropy.

(c) Although encountered before, it is worth to re-emphasize that the crossover from $VFTH_1$ to $VFTH_2$ also was found generally under constant volume (isochoric) condition for PCB62, PDE, KDE, PC, BMMPC, and salol [757]. This has been shown in Figs. 66 and 67 where the isochoric curve,

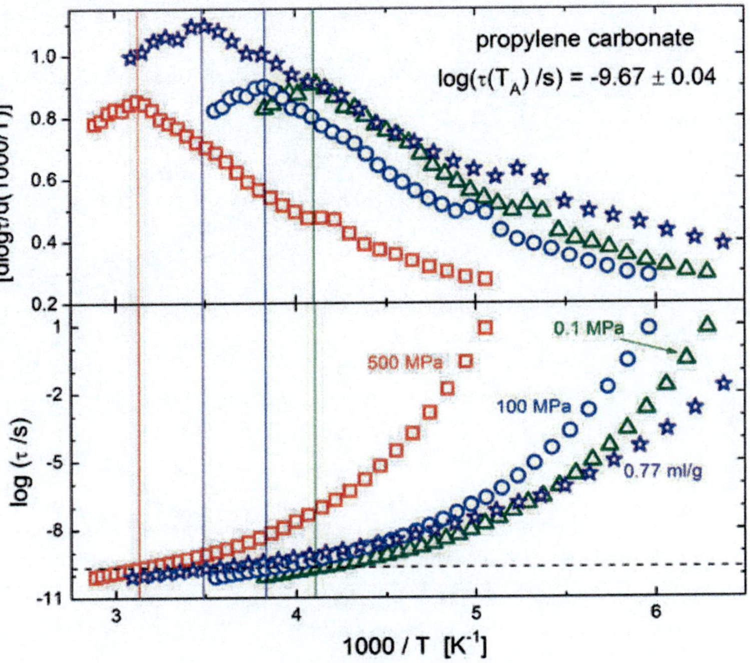

Fig. 68-2 Caption same as previous figure, now for PC. The *vertical lines* indicate T_A at various P, and the *horizontal line* indicates a constant value of τ at the crossovers

$[d\log\tau_\alpha/d(1/T)]^{-1/2}$, was obtained by calculation at some constant specific volume. From the crossover, the crossover temperatures T_B and relaxation times $\tau_\alpha(T_B)$ were obtained. Remarkably, $\tau_\alpha(T_B)$ under isochoric condition is the same as $\tau_\alpha(T_B)$ under isobaric condition, and $\tau_\alpha(P_B)$ under isothermal condition.

(d) Before we discuss further other properties, items (*a*)–(*c*) in the above can be summarized as a remarkable experimental fact: the two crossover parameters $\tau_\alpha(T_B)$ and $n(T_B)$ or $\tau_\alpha(P_B)$ and $n(P_B)$ are constant independent of the thermodynamic condition (constant P on varying T, constant T on varying P, or constant V on varying T). This can be readily shown to be an immediate consequence of the co-invariance of $\tau_\alpha(P,V,T)$ and $n(P,V,T)$ to all thermodynamic conditions discussed in Section 2.2.4.4, and explained by the CM. The invariance of $n(T_B)$ and $n(P_B)$ to thermodynamic condition is to be contrasted with the decrease of the isobaric fragility index m_P with increasing P and densification found in many glassformers [119]. Also the isochoric fragility index m_V at constant volume V is less than m_P. This result may be expected since for m_V only temperature affects τ_α, while for m_P both volume and temperature contribute to change of τ_α. The widely different values of m_V and m_P obtainable for one and the same glassformer are sufficient evidence to show that "fragility" is not as fundamental as the time/frequency dispersion (or n) of the α-relaxation.

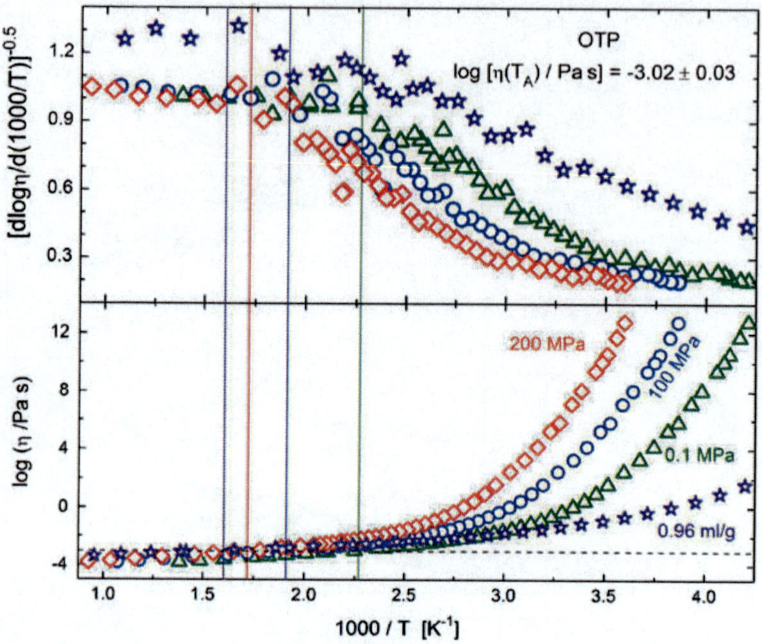

Fig. 68-3 Caption same as previous figure, now for OTP, except here the data are from viscosity. The *vertical lines* indicate T_A at various P, and the *horizontal line* indicates a constant value of viscosity at the crossovers

An Important Deduction from the Invariance of $\tau_\alpha(T_B)$ and $n(T_B)$, or $\tau_\alpha(P_B)$ and $n(P_B)$, to P, T, V

From the above subsections, on the change of T-dependence of τ_α at T_B under constant P (isobaric) or constant V (isochoric), and the change of P-dependence of τ_α at P_B under constant T (isothermal), the experimental facts presented are telling us that the general characteristics of τ_α are independent of P, T, V. The only invariant other than $\tau_\alpha(T_B)$ or $\tau_\alpha(P_B)$ itself is $n(T_B)$ or $n(P_B)$. From this, we can conclude that the P- and T-dependences of τ_α are determined by not only thermodynamic factors including volume and entropy, but also many-body dynamics exemplified by n.

Other Facts and Fiction on the Crossover of Temperature Dependence of τ_α

Correlated Changes of Behaviors of τ_α and the Dielectric Relaxation Strength $\Delta\varepsilon$ at T_B

The change of the temperature dependence of the dynamics causing the crossover of the T-dependence of τ_α from VFTH_1 to VFTH_2 at T_B is expected to have an effect of the relaxation strength. This is because both the relaxation strength $\Delta\varepsilon$ and

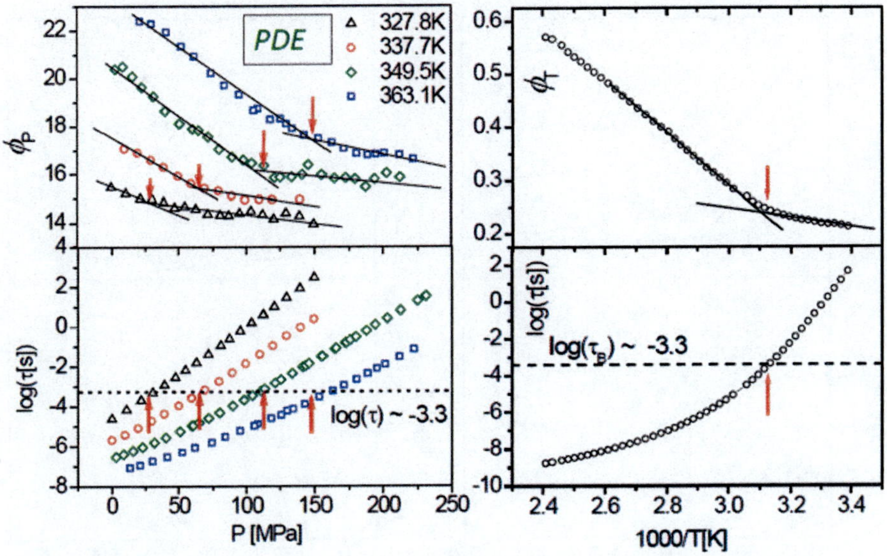

Fig. 69 Dielectric relaxation time data for PDE from [653]. *Left upper panel*: Derivative function ϕ_P vs. pressure, calculated for isotherms at temperature indicated. *Left lower panel*: log τ vs. pressure for T=327.8 (△), 337.7 (○), 349.5 (◇), and 363.1 K (□). *Right upper panel*: Derivative function ϕ_T vs. inverse temperature at atmospheric pressure. *Right lower panel*: log τ vs. inverse temperature from Stickel et al. [621]. Note the same crossover time τ_B from change of T-dependence at ambient pressure and change of P-dependence at constant temperature. Data from [653] are replotted here

the relaxation time τ_α are characteristics of the dynamics, and has led Schönhals [758] to look for correlated changes of behaviors of τ_α and the dielectric relaxation strength $\Delta\varepsilon$ at T_B by broadband dielectric measurements. The correlated changes indeed exist as shown in Figs. 72 and 73, where $\Delta\varepsilon$ is plotted against log f_p for dibutyl phthalate, salol, propylene carbonate, propylene glycol, dipropylene glycol, and poly(propylene glycol). Here f_p is the dielectric α-loss peak frequency and is approximately equal to $1/(2\pi\tau_\alpha)$. These plots indicate two different frequency regions of dynamics separated by f_B. The two relations when extrapolated intersect at a crossover frequency f_B. It turns out that for all the glassformers f_B is nearly the same as $1/[2\pi\tau_\alpha(T_B)]$. Hence $\Delta\varepsilon$ shows also a change of T-dependence at T_B.

Similar results were found also for the polymer, poly(vinyl acetate) (PVAc), first by Ngai and Roland [238] and reaffirmed later by Tyagi et al. [759]. These are shown altogether in Fig. 74. The change of the dependence of $\Delta\varepsilon$ on log f_p also occurs at the temperature T_B.

This phenomenon can also be explained in terms of the CM by the more rapid increase of n with decreasing temperature (frequency) after crossing $T_B(f_B)$, because n reflects the extent or length scale of the many-body relaxation and in turn the magnitude of $\Delta\varepsilon$. The temperature dependence of $\Delta\varepsilon$ is roughly proportional to $1/T$ at temperature much higher than T_B consistent with the Kirkwood–Fröhlich theory

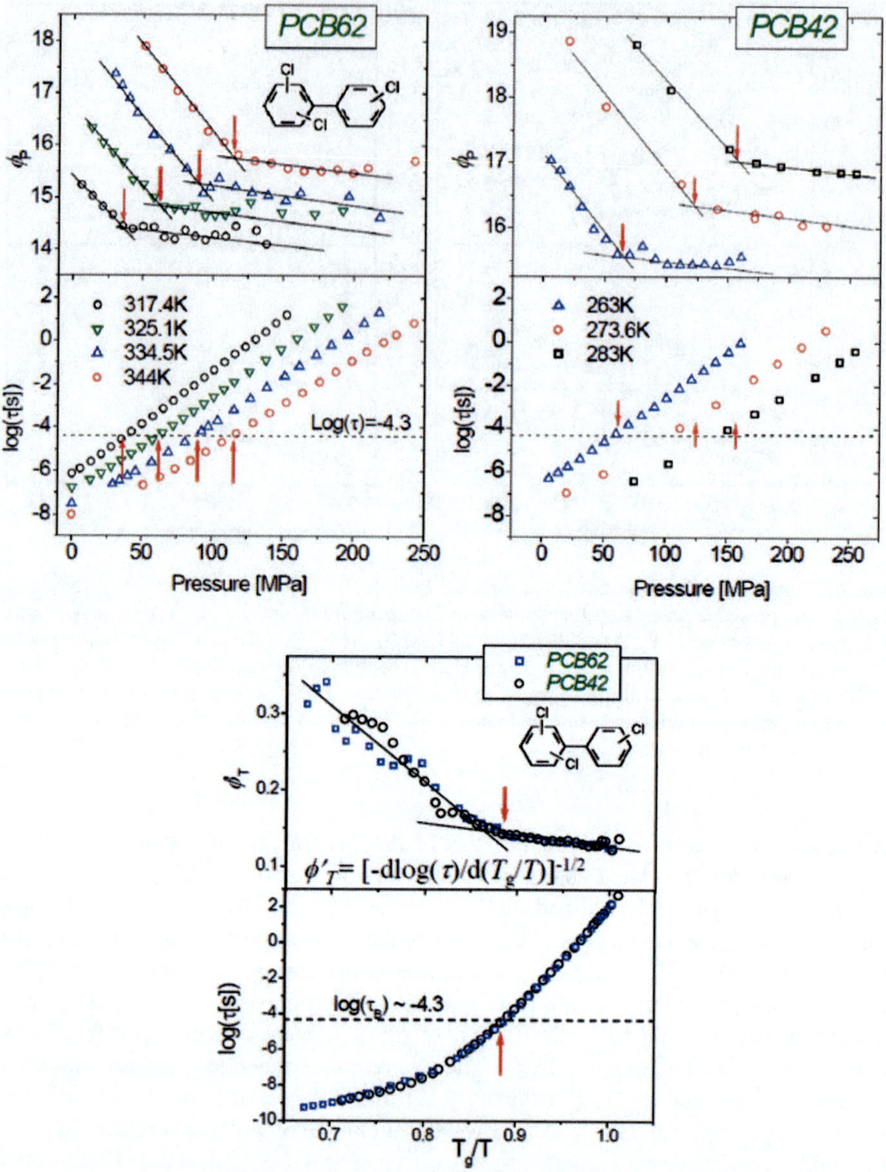

Fig. 70 *Above*: Dielectric relaxation time data for PCB62 (*upper left*) and PCB42 (*upper right*). *Upper panel*: Derivative function ϕ_P vs. pressure, calculated for isotherms at temperatures indicated. Lower panel: log τ vs. pressure. *Right side*: *Lower panel*: Derivative function ϕ_T vs. inverse temperature at atmospheric pressure vs. inverse temperature normalized to T_g for PCB62 and PCB42. Note the same crossover time τ_B with $\log(\tau_B) = -4.3$ is obtained from change of *P*-dependence at various constant temperatures and from change of *T*-dependence at ambient pressure. Data supplied by authors of [653] are replotted in color in all the figures here

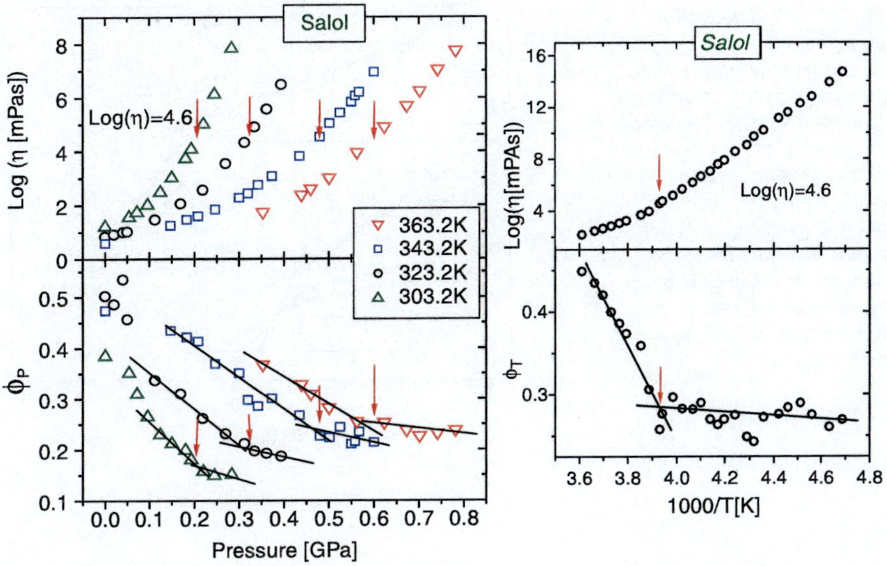

Fig. 71 *Left upper panel*: Crossover of pressure dependence of viscosity of salol at constant temperatures (reproduced from [547(b)] by permission). *Left lower panel*: Derivative function ϕ_P vs. pressure, calculated for isotherms at temperatures indicated. *Right upper panel*: Crossover of T-dependence of viscosity of salol at ambient pressure (reproduced from [547(b)] by permission). Value of η at the crossover is independent of pressure and temperature. *Right lower panel*: Derivative function ϕ_T

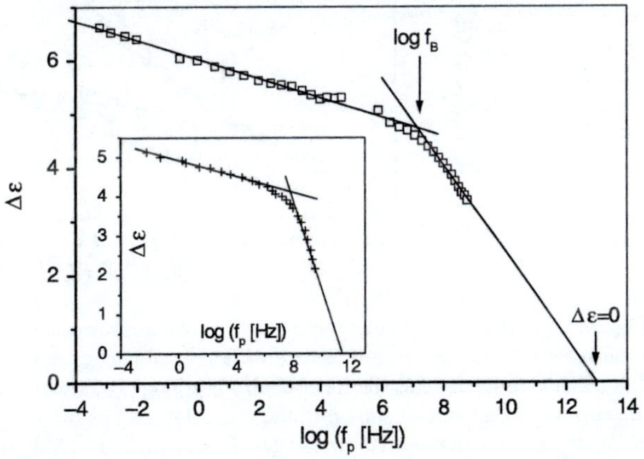

Fig. 72 $\Delta\varepsilon$ vs. $\log f_p$ for dibutyl phthalate (□). The *inset* shows the same for salol (+). The *lines* are linear regressions to the different branches of the data. Reproduced from [758] by permission

based on the assumption of non-interacting isolated dipoles, which is consistent with the small values of n. However, this dependence does not continue when temperature is lowered to approach T_B and fall below T_B. Thus, the observed change of T-dependence of $\Delta\varepsilon$ is another indication of the increase of coupling and cooperativity (or n) of the α-relaxation with falling temperature after crossing T_B. The

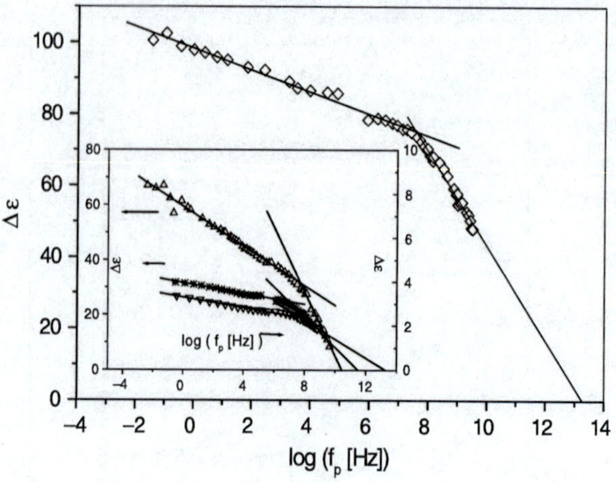

Fig. 73 $\Delta\varepsilon$ vs. $\log f_{\mathrm{p}}$ for propylene carbonate (◇). The *inset* shows the same for the propylene glycols of different molecular weights: from *top* to *bottom*, (Δ) propylene glycol, (∗) dipropylene glycol (*right scale*), (∇) poly(propylene glycol) (*left scale*). The *lines* are linear regressions to the different branches of the data. Reproduced from [758] by permission

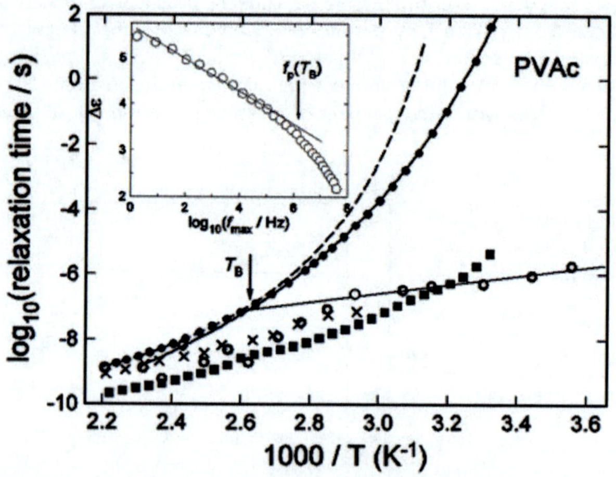

Fig. 74 The *inset* is a plot of the dielectric relaxation strength of the α-relaxation, $\Delta\varepsilon$, of PVAc against $\log_{10}(f_{\max})$ showing a change of temperature dependence of $\Delta\varepsilon$ when crossing T_{B} [236, 238]. The *arrow* indicates the location of $\log_{10}[f_{\max}(T_{\mathrm{B}})]$, which is determined by the fits of τ_{α} by the two VFTH equations shown in the main figure. The main figure is the plot of $\log_{10}(\tau_{\alpha}/\mathrm{s})$ of PVAc against $1000/T$. *Filled circles* are data from [236, 238], and the *two curves* through some of the data points are the relaxation times that correspond to the two VFTH equations for the frequencies. *Open circles* are τ_{β}s obtained by assuming the β-relaxation and the α-relaxation make additive contributions to the dielectric spectra [238, 760]. The *crosses* are τ_{β}s obtained by assuming the dielectric spectra is a convolution of the β-relaxation and the α-relaxation [238, 760]. The *straight line* drawn through the open circles at low temperatures intersect $\log_{10}(\tau_{\alpha}/\mathrm{s})$ at a temperature close to T_{B}. The *filled squares* are the primitive relaxation times, τ_0, calculated by the CM equation with the experimental values of n [238], τ_{α}, and $t_{\mathrm{c}} = 2$ ps

reader may find it surprising that the change of $\Delta\varepsilon$ with f_p in the $f_p > f_B$ regime is more rapid than in the $f_p > f_B$ regime. The cause of this is the slow variation of f_p with temperature in the range $T > T_B$ where τ_α is described by VFTH$_2$, and by contrast f_p varies more rapidly with temperature in the range $T < T_B$ where τ_α is described by VFTH$_1$. Thus, the same increment of f_p corresponds to a larger increment in temperature, and hence causes a larger decrement of $\Delta\varepsilon$ in the $f_p > f_B$ regime than in the $f_p < f_B$ regime.

Analogy to Relaxations of Different Length Scales from Quasielastic Neutron Scattering

Relaxations of different spatial scales can be probed by quasielastic neutron scattering experiments at various scattering vectors Q. The reciprocal of Q gives the spatial scale of the relaxation. Incoherent neutron scattering experiment follows the motion of protons within distances of size Q^{-1}. The first maximum of the static structure factor, $S(Q)$, located at $Q = Q_{max}$, gives a measure of the intramolecular spatial distance through $(Q_{max})^{-1}$. Data of the self intermediate scattering function $F_s(Q, t)$ obtained at $Q^{-1} < (Q_{max})^{-1}$, corresponding to small spatial scales, show its relaxation time τ_α has Arrhenius temperature dependence. For larger spatial scales data taken at $Q^{-1} > (Q_{max})^{-1}$, deviation from Arrhenius behavior of τ_α to become VFTH-like is observed in PMMA [761].

A similar result was observed for pure polybutadiene [762], although the neutron data for polybutadiene are for collective motion. These observations of the change of temperature dependence with spatial scale across $(Q_{max})^{-1}$ by neutron scattering provide support for the explanation we have given for the change of temperature dependence of τ_α across T_A or T_B obtained by dielectric relaxation. The reason is given as follows. The length, $(Q_{max})^{-1}$, corresponds to the interchain spacing. Therefore, data taken at $Q^{-1} < (Q_{max})^{-1}$ come from motions of protons within individual polymer chains. The motions probed are of local and intramolecular nature, without the participation of other chains, and hence many-body relaxation index n is small or even zero. On the other hand for low Q data such that $Q^{-1} > (Q_{max})^{-1}$, the data give information of motions in space containing protons in other chains. Many-body dynamics are involved and the index n is not zero. Thus, the change of temperature dependence of τ_α when spatial scale is changed in crossing $(Q_{max})^{-1}$ is associated with the change in n, in analogy to the change of the temperature dependence of τ_α when crossing T_A or T_B.

It is worthwhile to caution the reader that even local segmental relaxation of an isolated polymer chain has correlation function which is more dispersive than a linear exponential function due to connectivity of repeat units along the chain [763, 764]. This means that the intermediate scattering function obtained by neutron scattering at $Q^{-1} < (Q_{max})^{-1}$ when fitted to the Kohlrausch function has the stretch exponent β significantly less than 1, even though the many-body relaxation index n is small or even zero. There is support of the above interpretation from the Q-dependence of τ_α. As discussed before in Section 2.2.2.4, τ_α changes its Q-dependence when crossing Q_{max}, from $Q^{-2/(1-n)}$ for $Q^{-1} > (Q_{max})^{-1}$ to

Q^{-2} for $Q^{-1} < (Q_{max})^{-1}$. The $Q^{-2/(1-n)}$-dependence of τ_α for $Q^{-1} > (Q_{max})^{-1}$ is the consequence of cooperative and heterogeneous many-body relaxation taking place at longer length scales with involvement of repeat units of other chains, and can be derived from the coupling model as shown in Section 2.2.2.4. On the other hand, Q^{-2}-dependence of τ_α for $Q^{-1} < (Q_{max})^{-1}$ reflects local and non-cooperative motions of protons within individual polymer chains, which naturally have the Q^{-2}-dependence of simple rotation and diffusion. More discussion of the Q-dependence of τ_α will be given in Section 2.2.5.2 to follow.

Comments on Works of Others

Before closing this subsection, a remark is appropriate here to dispel an incorrect notion circulated in the literature that the magnitude of the crossover time, $\tau_\alpha(T_B)$, of all glassformers is located within the narrow range of $10^{-6.5} - 10^{-7.5}$s [765]. The following facts on $\tau_\alpha(T_B)$ are sufficient for this purpose. At $T = T_B$, PDE has the longest relaxation times with $\tau_\alpha(T_B) = 10^{-3.6}$ s ($10^{-3.3}$ s in Fig. 69), and propanol [348(b)] has $\tau_\alpha(T_B) = 10^{-3.7}$ s. Figure 70 shows that PCB62 has $\tau_\alpha(T_B) = 10^{-4.3}$ s. From Figs. 66 and 67, BMMPC has $\tau_\alpha(T_B) = 10^{-6.1}$ s, KDE and salol have $\tau_\alpha(T_B) = 10^{-6.3}$ s, and PC has $\tau_\alpha(T_B) = 10^{-7.0}$ s, and polymethylacrylate has $10^{-10.3}$ s [766]. The much studied epoxy resin, diglycidylether of bisphenol A (DGEBA), has T_B=275 K and $\tau_\alpha(T_B) = 10^{-4.3}$ s [242]. From these results, it is clear that $\tau_\alpha(T_B)$ varies over a wide range (seven orders of magnitude from $10^{-3.3}$ to $10^{-10.3}$ s) when all glassformers are considered. Hence, the results invalidate the claim that $\tau_\alpha(T_B)$ has "universal" or "magic" value lying within the narrow range of $10^{-6.5} - 10^{-7.5}$ s [765]. Connection was made of the crossover temperature T_B to the critical temperature T_c of mode coupling theory [765]. This proposal is also refuted by PDE, propanol, PCB62 having $\tau_\alpha(T_B)$ ranging from $10^{-3.3}$ to $10^{-4.3}$ s. This is because it is inconceivable that such a long or macroscopic relaxation time can be identified as relaxation time at T_c of mode coupling theory.

On the other hand, lest the reader forgets, the smaller values of n at temperatures higher than T_B and relatively more rapid increase of n with decreasing temperature (frequency) after crossing $T_B(f_B)$ are always found in all glassformers despite the large variations in their values of $\tau_\alpha(T_B)$. This general property of n associating with the crossover of T-dependence of τ_α at T_B is the basis of the explanation of the latter by the coupling model [423]. As mentioned before, $\tau_\alpha(T_B)$ at the crossover of T-dependence is independent of pressure or $\tau_\alpha(P_B)$ at the crossover of P-dependence is independent of temperature. Hence, we have the invariance of $\tau_\alpha(T_B)$ and $\tau_\alpha(P_B)$ to thermodynamic conditions. The experimental results demonstrate that the dynamic crossover phenomenon arises not at some critical temperature or volume, rather it is governed by τ_α and the corresponding value of n. The last conclusion follows from the co-invariance of τ_α and n to changing thermodynamic conditions, consistent with the prediction of the CM. The invariance of $\tau_\alpha(T_B)$ and $\tau_\alpha(P_B)$ to thermodynamic conditions further supports that the α-relaxation is governed by many-body relaxation through the surrogate n, since the phenomenon is independent of thermodynamic conditions.

There are recent theoretical [767] and experimental studies [768] that reported a change in the temperature dependence of the *equilibrium* structural relaxation time at temperatures near and below the conventional T_g. The VFTH temperature dependence valid for $T > T_g$ is changed to a milder Arrhenius temperature dependence as T approaches and falls below T_g. In one experiment, glassy samples of bisphenol A polycarbonate (BPA-PC) were aged into equilibrium at temperatures up to 17°C below $T_g \approx 140$°C [768]. Mechanical stress relaxation measurements were made to obtain the temperature dependence of the *equilibrium* viscoelastic response of BPA-PC for temperatures extending from 3°C above to 17°C below T_g. A gradual transition from VFTH temperature dependence to Arrhenius temperature dependence below T_g was concluded from the shift of the mechanical torque time response with T. There is no doubt that the measurements are accurate. But, before the conclusion is fully accepted, the following possibility has to be excluded. Deep below T_g, the measured torque did not decay more than half a decade. The small decay could possibly be contributed in part by secondary relaxation, and this contribution could be responsible for the equilibrium dynamics being faster than an extrapolation of the VFTH dependence at higher temperatures would suggest. Nevertheless, there is support of the conclusion from study of the rotational and translational diffusion of probe molecules (rubrene and tetracene) in the same BPA-PC sample aged to equilibrium [769]. The temperature dependence of probe motions in equilibrium PC was studied from 149 to 124°C, and a change in the temperature dependence of probe molecule motion near 134°C was reported. Below this temperature, the activation energies of the translational diffusion coefficient and the rotational correlation time decreases by factors about 3.

2.2.5.2 The $Q^{-2/(1-n)}$-Dependence of τ_α

Quasielastic neutron scattering experiments and molecular dynamics simulations on polymeric and non-polymeric glassformers are capable of investigating the dependence of τ_α on the scattering vector Q. These techniques give an important and measurable dependence of τ_α on another variable that is worth consideration in constructing theory of glass transition. As discussed before, for non-interacting particles, the incoherent intermediate scattering function obtained by incoherent neutron scattering or simulation has the form given by $F_s(Q,t) = \exp(-Q^2 D_t t)$ (see Eq. (1.124)). Rewritten as $\exp[-t/\tau_\alpha(Q)]$, it follows from the last equation that $\tau_\alpha(Q) = Q^{-2}/D_t$. From studies by neutron scattering and molecular dynamics simulations, this Q^{-2}-dependence of τ_α was observed for $Q > Q_{max}$ or at spatial scales shorter than intermolecular distance or interchain spacing in the case of polymers [324, 474, 761, 762]. Here, Q_{max} is the location of the first maximum of the static structure factor, $S(Q)$. This result was explained before in the subsection above. At short spatial scales corresponding to $Q > Q_{max}$, many-body participation is suppressed and the local relaxation probed is effectively as if in non-interacting systems and hence the Q^{-2}-dependence of τ_α.

For $Q < Q_{max}$ or longer spatial scales, the Q-dependence of τ_α instead was found by quasielastic neutron scattering experiments and molecular dynamics simulations to be given by $Q^{-2/(1-n)}$ for several polymers, PVC, PIB, PBD, PVE, PIP, PPO, PDMS, PH, PE, PVME, PMMA [307, 310, 313, 314, 324, 325, 459, 461, 474, 762, 770–773], PMMA in blends with PEO [761], and a few small molecular glassformers, OTP [316], glycerol [774], and methanol [318]. Here, n is the exponent appearing in Eq. (1.1) for the Kohlrausch function that was used to fit the time dependence of the intermediate scattering function, $F_s(Q,t)$, measured by incoherent neutron scattering or obtained by molecular dynamics simulations. An interesting case is the "strong" glassformer, methanol. Molecular dynamics simulations [318] found n assumes nearly zero value at high temperatures and τ_α has Q-dependence approaching Q^{-2}, as pointed out in [319]. *Hence the observed Q-dependence of τ_α is governed by n or the breadth of the α-dispersion.*

An exception was found in poly(vinyl acetate) (PVAc) [759]. Neutron scattering data of $F_s(Q,t)$ were fitted to the Kohlrausch function with $n \approx 0.50$. From this, the dependence of τ_α expected from $Q^{-2/(1-n)}$ in the low Q-region would be Q^{-4}. However, the experimentally observed dependence is $Q^{-2.52}$, which indicates that $n = 0.21$. The discrepancy found in PVAc but not in the other polymers could be due to the presence of flexible side chain in PVAc. The PVAc sample studied was fully protonated and thus the motions of both the backbone and the side chain contribute to $F_s(Q,t)$, broadening its dispersion and resulting in a larger n than the actual value. Interestingly, the dispersions of the α-relaxation from dielectric relaxation measurements are significantly narrower than $F_s(Q,t)$ in the same temperature range. The values of n obtained by fitting the dielectric data in the temperature range from 400 to 460 K where neutron scattering measurements were performed [759] fall between 0.30 and 0.25 (see Fig. 4 in [238]). These values are not far from the value $n = 0.21$ required to explain the $Q^{-2.52}$-dependence of τ_α. A similar explanation was offered in [759]. I have examined the neutron scattering data of the room temperature ionic liquid, BMIM-PF_6, obtained by Triolo et al. in 2003 [311]. They found $\tau_\alpha(Q) = \tau_\alpha^0 Q^{-\nu}$, with $\nu = 2.5$. The stretch exponent $\beta = (1–n)$ is not constant but the average value is about 0.75. If this value is taken, the predicted exponent ν has the value of $(2/0.75) = 2.67$, which is not far from the experimental value.

A superposition of $F_s(Q,t) = \exp(-Q^2 D_t t) \equiv \exp(-t/Q^{-2}D_t^{-1})$ with a distribution of D_t to make the resulting intermediate scattering function having the Kohlrausch form of Eq. (1.1) will not be able to explain the observed $Q^{-2/(1-n)}$-dependence of τ_α. As mentioned before in Section 2.2.2.4, the resulting Kohlrausch relaxation time still has the Q^{-2}-dependence [460]. This is another indication that the observed $Q^{-2/(1-n)}$-dependence of τ_α is a consequence of many-body relaxation and the exponent n of the Kohlrausch function is its index.

The observed $Q^{-2/(1-n)}$-dependence of the α-relaxation time of glass-forming liquids is also found in other systems including suspensions of colloidal particles [143], semidilute polymer solutions [198, 776, 777], associating polymer solutions [778, 779], and polymer cluster solutions [780]. This rather universal dependence of relaxation time on the scattering vector is the exemplification of many-body relaxations that are common to these systems all having mutual interaction between the basic units. These together with more common properties of these interacting systems will be further discussed later on in Chapter 3.

The anomalous $Q^{-2/(1-n)}$-dependence of τ_α is derivable from the CM equation (2.35) simply by substituting the known Q^{-2}-dependence of τ_0 into it.

2.2.5.3 Non-linear Enthalpy Relaxation of Glassformers Near and Below T_g

The properties of the α-relaxation near and below T_g are important as well as interesting. In the neighborhood of the glass transformation region, τ_α is dependent on thermal history, including cooling and heating rates, aging protocol, and pressure. These have been challenging research problems of practical importance. The two essential features of structural relaxation near and below T_g were brought out by experiments [62, 266, 303]. They are (1) the structural relaxation time depends not only on temperature T, but also on the instantaneous structure (non-linearity) and (2) the time dependence of the structural relaxation process is not a simple exponential function (non-exponentiality). Non-linearity is demonstrated by the asymmetry of relaxation in structural recovery following positive or negative departures from equilibrium. Non-exponentiality is shown by the memory effect, in which relaxation from some initial state depends on how that state was reached. These effects were found in the pioneering experimental studies of borosilicate glass by Ritland [781] and of poly(vinyl acetate) (PVAc) by Kovacs and coworkers [62, 303], and in more glassformers by Moynihan and coworkers [266]. There are two models that have taken into account non-linearity and non-exponentiality and they are essentially equivalent. The one formulated by Moynihan and coworkers [266] for enthalpy H relaxation is based on the constructs of Tool [782] and Narayanaswamy [783] and is now known as the TNM model. The other is the KAHR model developed by Kovacs and coworkers for volume relaxation [303]. For reviews see [173, 304].

In the TNM model, non-linearity is taken into account by modifying the linear differential equation, $\mathrm{d}(H - H_e)/\mathrm{d}t = -(H - H_e)/\tau$, describing relaxation of enthalpy H toward equilibrium value H_e. This is handled by making τ dependent not only on T but also on H. Actually the model is formulated on the evolution of the fictive temperature T_f, instead of H. T_f can be defined as the instantaneous contribution of the structural relaxation process to H expressed in temperature units. For example, following quenching of an equilibrium liquid at temperature T_0 down to T_1, the enthalpy $H(t)$ relaxes from the initial value H_0 toward the equilibrium enthalpy H_{e1} at temperature T_1. Correspondingly $T_f(T)$ varies from T_0 to T_1 in parallel with the changes in $H(t)$. Departure from equilibrium is measured by $T_f - T$. The progress of structural relaxation with time described by the normalized relaxation function, $\phi(t) \equiv [H(t) - H_{el}]/(H_0 - H_{el})$, is now replaced by $[T_f(T) - T_1]/(T_0 - T_1)$. Non-exponentiality is implemented by replacing $\phi(t) = \exp(-t/\tau)$ by the Kohlrausch relaxation function,

$$\varphi(t) = \exp\{-\int_0^t [\mathrm{d}t'/\tau(T, T_f)]^\beta\}. \tag{2.45}$$

In the TNM model, the following expression for $\tau(T, T_f)$ was proposed:

$$\tau(T, T_f) = \tau_0 \exp\left[\frac{x\Delta h}{RT} + \frac{(1-x)\Delta h}{RT_f}\right]. \tag{2.46}$$

Here $x(0 \leq x \leq 1)$ is the non-linear parameter, τ_0 a pre-exponential factor, Δh an activation enthalpy, and R the ideal gas constant and all of them are taken to be fitting parameters. Moynihan has warned that the TNM model cannot take care of temperature or structural dependence of β, if present.

Another way of introducing non-linearity is by modifying [784–786] the Adam–Gibbs equation for the relaxation time of an equilibrium liquid, $\tau = \tau_0 \exp[C_{AG}/TS_c(T)]$, by replacing $S_c(T)$ therein by $S_c(T_f)$. The resulting equation is $\tau(T, T_f) = \tau_0 \exp[C/TS_c(T_f)]$. Hodge [786] used the approximate hyperbolic form for the temperature dependence of the configurational heat capacity $\Delta C_p(T) \propto 1/T$ to obtain $S_c(T_f)$ and the following Hodge–Scherer expression for $\tau(T, T_f)$:

$$\tau(T, T_f) = A \exp\left[\frac{B}{T(1 - T_2/T_f)}\right]. \tag{2.47}$$

In the specific case when equilibrium condition is satisfied, $T_f = T$ and this expression is reduced to the VFTH T-dependence.

(i) *Correlation Between n and Degree of Non-linearity (i.e., Smaller x) and* Δh: The parameters of Eqs. (2.46) and (2.47) were shown to be related to one another by appropriate differentiation with respect to temperature in the equilibrium and glassy states [785]. The results are $x \approx (1 - T_2/T_f')$ and $\Delta h/R \approx B/(1 - T_2/T_f')^2 \approx B/x^2$, where T_f' is the value of T_f in the glassy state. These relations have been confirmed for polymers [785] by fitting the same experimental heat capacity data using either expressions for $\tau(T, T_f)$ and obtaining comparable best fits. Structural relaxation in response to any thermal history, simple or complicated, can now be calculated by the TNM model using the Kohlrausch relaxation function given by Eq. (2.45) and $\tau(T, T_f)$ by Eq. (2.46) or (2.47). The Kohlrausch exponent β in Eq. (2.45) obtained from analysis of non-linear enthalpic relaxation was found to be generally weak functions of thermal history, and usually in agreement with the exponent $(1-n)$ in Eq. (1.1) obtained by dielectric relaxation and other techniques in the linear response region above T_g [785, 786].

The important results for our present purpose is the finding by Hodge [785, 786] of the strong correlation between the parameters x, Δh, and β. The individual correlations between x and β and correlation between Δh and $(1-\beta) \equiv n$ are shown in Figs. 75 and 76, respectively. In other words, larger n (smaller β) corresponds to higher degree of non-linearity (i.e., smaller x) and larger Δh. Thus, n also governs the parameters of non-linear enthalpy relaxation near and below T_g.

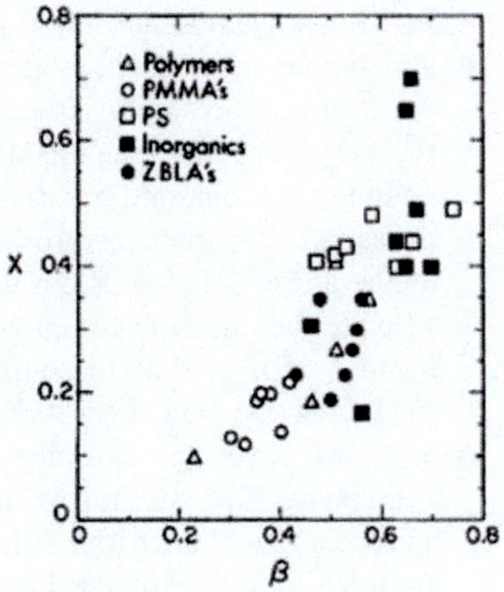

Fig. 75 Correlation between x and β for the glassformers considered in [786] by Hodge. Reproduced from [786] by permission

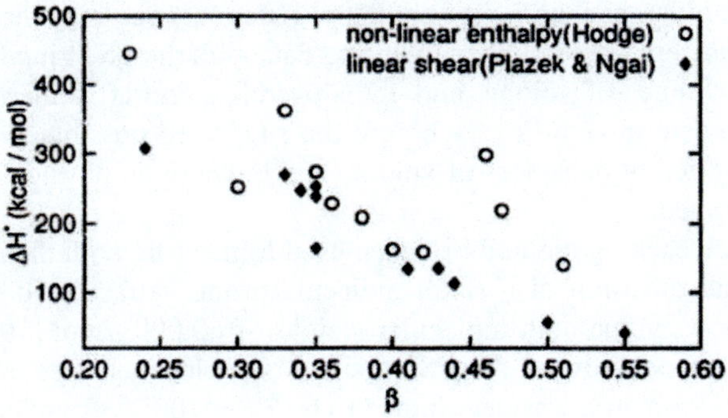

Fig. 76 Correlation between Δh and $(1 - \beta) \equiv n$ for the materials considered in [786], which include polymeric, small molecular, and inorganic glassformers (○). The correlation is better if restricted to glassformers of the same class. Shown also is similar correlation between Δh and $(1 - \beta) \equiv n$ for polymers only obtained by linear shear compliance measurements at T_g

(ii) *Correlation Between n and Equilibration Time* t_{eq} *for Different Quantities in Physical Aging*: A glass is not in a state of thermodynamic equilibrium. Consequently, its physical and mechanical properties change with time as the material attempts to reach equilibrium. In the literature, this phenomenon is referred to as structural recovery or physical aging [788]. Studies of physical aging often find that enthalpy, volume, dielectric relaxation, and creep compliance (or stress relaxation) reach equilibrium at widely different times. For examples from a number of polymeric glassformers, see [789].

In BPA-polycarbonate there is evidence that volume relaxation occurs more slowly than mechanical relaxation [790].

A specific example is taken here from the study of volume and enthalpy recovery of polystyrene [791]. The sample has $T_g = 94.9°C$ determined by the volume vs. temperature curve obtained at a cooling rate of 0.28°C/min. After the sample was quenched from equilibrium at a temperature above T_g to various temperatures T below about 94°C, the study shows that the times required for volume and enthalpy to reach equilibrium, t_∞, are different. It takes a longer t_∞ or t_{eq} for volume to equilibrate than enthalpy. Deeper the quench (lower T), longer are both equilibration times and larger their difference becomes. Since we have seen from the above that n governs non-linear enthalpy relaxation below T_g, a possible explanation is that the dynamics of enthalpy, volume, and shear compliance (or modulus) not necessarily have the same n, although all are related to the structural relaxation of molecules, as shown before in [267]. Analysis of the data using the TNM model does show that both the parameters x and $\beta \equiv (1 - n)$ are smaller, and Δh is larger for volume than enthalpy. Hence, from the correlations between x and β and correlation between Δh and $(1 - \beta) \equiv n$ shown before in Figs. 75 and 76, it can be concluded that n_V of volume is larger than n_H of enthalpy. However, one must be mindful that these parameters were obtained by fitting data with the TNM model. The actual n_V and n_H values of volume and enthalpy relaxation at isothermal condition have to be determined directly before the suggested possible role of n in the observed different behaviors of volume and enthalpy in physical aging can be fully confirmed.

A photobleaching method has been used to measure both the rotational and translational diffusion of a small molecule probe, tetracene, during physical aging in polystyrene with molecular weight of 60,000 g/mol [792]. The glass transition temperature of the PS/tetracene sample, measured by differential scanning calorimetry, was determined to be $T_g = 100°C$. Rotational relaxation time τ_c and translational diffusion coefficient D_t increase during isothermal aging after quenching the sample from above T_g into the glass. Physical aging affects rotational relaxation and translational diffusion differently. For shallow quenches down to $T = 98.6$ and 97.1°C, the rotation time and translation coefficient reach equilibrium at about the same time. However, for a deeper quench to $T = 95.6°C$, translational diffusion reaches equilibrium before rotational relaxation (see Fig. 77), and t_∞ or t_{eq} of rotational relaxation is thus longer than that of translational diffusion (see Fig. 78). The temperature dependence of t_{eq} is different for rotational relaxation and translational diffusion, with rotation exhibiting stronger temperature dependence than translation. The difference in t_{eq} values and T-dependence for rotational relaxation and translational diffusion in physical aging is one demonstration of the breakdown of Debye–Stokes–Einstein relation in the glassy state. Breakdown of the relation is found also in the equilibrium state above T_g, which is the subject of detailed discussion in the subsection to follow.

It will become clear later from discussion in a later section that a likely cause of the breakdown of the Debye–Stokes–Einstein relation above T_g is the larger n

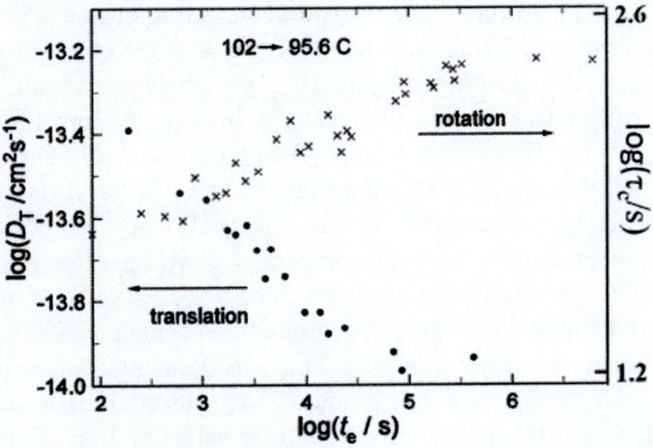

Fig. 77 Comparison of rotational and translational diffusion measurements of tetracene during isothermal physical aging at 97.1°C. Rotational correlation times τ_c correspond to the right-hand axis while the translational diffusion coefficient D_t corresponds to the left. The *solid line* located in the *upper left corner* of the figure marks the equilibrium value of D_t at 102°C before the temperature quench. The *line* in the *lower left corner* corresponds to the equilibrium value of τ_c at 102°C before quenching into the glass. The *arrows* guide the eye from the initial equilibrium dynamics to the corresponding aging curves. Data from [792] are digitized and replotted

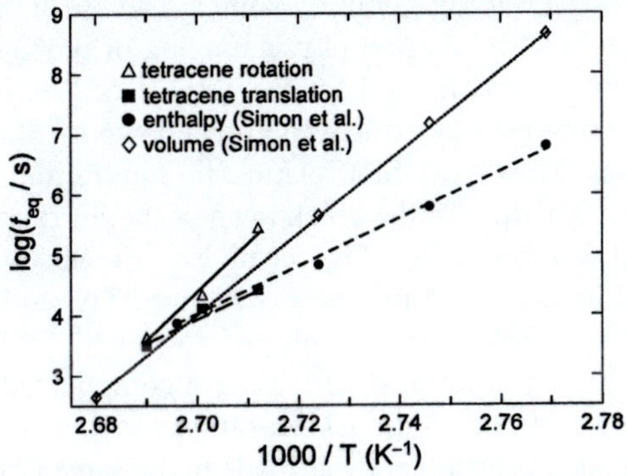

Fig. 78 Comparison of equilibration times t_{eq} for various observables in polystyrene. The times to reach equilibrium for probe rotation and translation show different temperature dependences. Compared to macroscopic observables, translation and enthalpy exhibit similar equilibration times, while the t_∞ values for probe rotation and for volume show similar temperature dependence. Volume and enthalpy data from [791], probe rotation and translation from [792]. All data from these references are replotted here in this figure

value for rotational relaxation than for translational diffusion. Figure 78 shows that translation diffusion and enthalpy exhibit similar equilibration times, while t_{eq} for probe rotation and for volume show similar and stronger temperature dependence. From this, and also recalling that we have previously deduced that $n_V > n_H$, we infer that a larger n value is associated with rotational than with translational diffusion. The latter is the possible cause for the difference in the aging behavior of rotational relaxation from that of translational diffusion. Later on in Section 2.2.5.6, we show that this is indeed the cause by applying the coupling model to address physical aging to derive the result that relaxation property having a larger n has a longer t_{eq}. This explanation is to be contrasted with that given by Thurau and Ediger [792] based on spatial heterogeneities in the segmental dynamics of PS being responsible for the differences in aging behavior of tetracene translational and rotational diffusion. What they offered is actually not an explanation but a transcription of the observed difference in the way these two macroscopic observables age to the Kohlrausch exponent $\beta \equiv (1 - n)$ parameter used to characterize the dispersion (non-exponentiality) of the rotational correlation function. This was done by using an empirical correlation between (1–β) and the product $D_t\tau_c$ (normalized to the ideal value when the Debye–Stokes–Einstein relation is valid) at $T = T_g$ established by data of various glassformers [451]. The observed change of the normalized $D_t\tau_c$ during physical aging was converted to change in β or the dispersion, which in turn was reinterpreted as due to regions of different mobility age at different rates in the context of the spatially heterogeneous model. However, this change of β had not been directly observed in the experiment because the error bars for β are large. Thus, this is not a direct explanation of why probe translation and rotation age differently, but is at best a demonstration that the observation can be consistent with spatial heterogeneous dynamics. The use of the spatial heterogeneous dynamics model to interpret physical aging of probe rotation relaxation and translation diffusion is natural for the authors of [792] because the same model had been employed to explain the breakdown of Stokes–Einstein (SE) and Debye–Stokes–Einstein (DSE) relation in equilibrium liquids [450–452, 793–796]. The explanation for the breakdown of the SE relation turns out to be invalid because it was contradicted by recent experiments [454–458, 274, 275, 797], which will be discussed in the next section. The collapse of the explanation by spatial heterogeneous dynamics of a simpler problem in equilibrium liquids does not help it in rationalizing a more complicated problem (i.e., the difference in aging of probe rotational relaxation and translational diffusion) by using the same model. A statement was made in the same [792]: "Current models of physical aging, such as the TNM and KAHR models, assume spatially homogeneous dynamics during aging, and fail to explain what causes different observables to age differently. Our results suggest that this homogeneous assumption is incorrect, and that heterogeneous dynamics are important for the description and prediction of physical aging deep in the glassy state." This charge against the TNM and KAHR models in the statement is unwarranted. The TNM model and its other forms [785, 786] do not address the microscopic nature of the structural relaxation, and are used principally to consider macroscopic quantities such as enthalpy and volume. In these models, the use of a Kohlrausch function merely describes the time dependence of the macroscopic structural relaxation, which is obtained after averaging over the microscopic and dynamically heterogeneous processes. Therefore, these models make no assumption that the dynamics are necessarily homogeneous. Moynihan, one of the architects of the TNM model, is an advocate of heterogeneous dynamics as evidenced by his paper in [531]. Moreover, he had shown in the 1975 paper [266] that various relaxing properties including enthalpy, volume, stress, strain, and refractive index behave differently in the glass transition region, and have different TNM model parameters Δh, β, and x. Probe rotation and translational diffusion are merely two additional relaxing properties and their difference in aging would be explained again by the TNM model from their different parameters Δh, β, and x. Thus, the criticism of the TNM model, "fail to explain what causes different observables to age differently," is not correct. In the TNM model's explanation of probe rotation and translation diffusion to age differently, the difference of the values of β for these two observables turns out to be main point in the coupling model explanation of the breakdown of the SE and the DSE relations in Section 2.2.5.6 [268, 275, 797].

2.2.5.4 Correlation Between *n* and Aging Time

Suspension of Colloidal Particles (Radius=1.18 μm, Packing Fraction=0.62)

Courtland and Weeks [798] used confocal microscopy to directly monitor the dynamics of aging in colloidal glasses prepared from colloidal suspensions of PMMA particles of radius=1.18 μm at high densities (packing fraction ϕ larger than $\phi_g = 0.58$), after stirring the sample. They followed the motion of several thousand colloidal particles and observed that their motion significantly slowed as the sample aged but continued to occur on all times scales. The aging is both spatially and temporally heterogeneous. For their aging samples and different aging times t_w, they identified the locations of mobile particles and showed that they are generally grouped into large clusters for all values of t_w. The dependence of the spatial clustering of mobile particles on t_w was obtained by observing the clusters and measuring their average size for each t_w. To their surprise, overall the cluster sizes do not increase with t_w. Definitely, for the colloidal glass to age, the particles must move to rearrange themselves, but Weeks and coworkers were not able to find any characteristics of the rearrangements which changes as the sample ages. Although from confocal microscopy aging is both spatially and temporally heterogeneous, the physical mechanism that ultimately causes the dynamics to slow down as their sample ages is not clear. In another effort [799], the static structure was analyzed in terms of tetrahedral packing. It was found that none of the geometrical quantities associated with the tetrahedra changes with age. Nevertheless, Courtland and Weeks did point out that, with increasing t_w, the changing character of the mean-squared displacement (MSD) seems to be the most significant dynamical changes in aging. The changes of the aging MSD, $< r^2(\Delta t) >$, are shown in Fig. 79 for a colloidal glass at packing fraction $\phi = 0.62$ [797]. The three sets of data represent three different ages of the sample, $t_w = 0\,\tau_{diff}$, $t_w = 100\,\tau_{diff}$, and $t_w = 300\,\tau_{diff}$, where $\tau_{diff} = 11$ s. The dotted line has a slope of 1 and represents diffusive behavior at much longer times than can be seen by experiment in this glassy sample. For each sample of different ages, at shorter lag times the MSD varies very slowly with

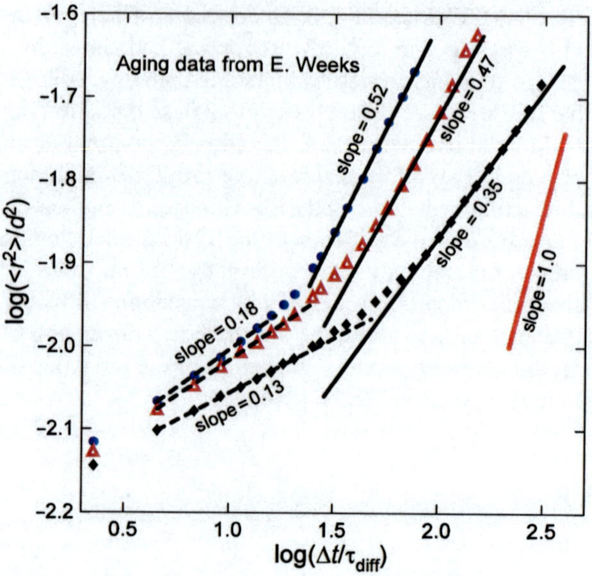

Fig. 79 Aging mean-squared displacement for a colloidal glass with ϕ=0.62. The three *curves* represent three different ages of the sample. The three sets of data represent three different ages of the sample. From *top to bottom*, $t_w = 0\,\tau_{\text{diff}}$ (*closed triangles*), $t_w = 100\,\tau_{\text{diff}}$ (*open squares*), and $t_w = 300\,\tau_{\text{diff}}$ (*closed circles*) where $\tau_{\text{diff}} = 11$ s. The *dotted line* has a slope of 1 and represents steady-state diffusion regime, not seen in this glassy sample. The data of $t_w = 100\,\tau_{\text{diff}}$ at shorter times are from the particles when they are still caged. The *dashed line* drawn through them has slope=0.13. Data from [798] supplied digitally by Eric Weeks and replotted here

time essentially on a plateau, because particles are locally confined or caged by their neighbors. It was interpreted as loss coming from cage decay [141, 797, 800, 801] or equivalently onset of particles leaving the cages [195, 196, 468, 802]. The dashed lines in Fig. 79 [797] drawn with small slopes ranging from 0.18 for $t_w = 0\,\tau_{\text{diff}}$, 0.16 for $t_w = 100\,\tau_{\text{diff}}$, and 0.13 for $t_w = 300\,\tau_{\text{diff}}$ merely serve to illustrate the slow increase of MSD, and its actual time dependence may not be strictly a power law. At larger lag times, Δt, each MSD curve shows a more rapid rise shown as a power law $(\Delta t)^{1-n}$ as found before in the liquid states [141]. It can be seen that (1–n) is a function of t_w, decreasing from 0.52 for $t_w = 0\,\tau_{\text{diff}}$ to 0.47 for $t_w = 100\,\tau_{\text{diff}}$ and 0.35 for $t_w = 300\,\tau_{\text{diff}}$. The change of the slope of increase of the MSD indicates the crossover from the shorter time regime of caged dynamics to many-body diffusion dynamics.

If the displacement r of a single particle is approximated as a Gaussian variable (the Gaussian approximation is known to be valid for colloidal suspensions in their equilibrium fluid states [255]), the self intermediate scattering function $F_{\text{self}}(Q,t)$ can be written in terms of the MSD by $F_{\text{self}}(Q,t) = \exp(-Q^2 < r^2(\Delta t) >)$. With power law $(\Delta t)^n$ for $< r^2(\Delta t) >$, $F_{\text{self}}(Q,t)$ assumes the Kohlrausch form (Eq. (1.1)), and hence n can be taken as a measure of the many-particle relaxation dynamics in the aged colloidal glass. The increase in n with t_w is due to the non-trivial motions of the particles observed at all timescales of aging, which somehow enhance the many-particle relaxation. As mentioned before, Courtland and Weeks did point out themselves that the changing character of the MSD is the most significant dynamical changes in aging. However, this pertinent observation by itself does not lead to an explanation of the principal effect of aging, i.e., the timescale for motions needed for a further structural change increases with the age of the sample or t_w. A theory or model is needed to rationalize this observation. This task is fulfilled expediently by the coupling model which does relate the increase of $n(t_w)$ to increase of relaxation time on aging, as shown in [797], and to be discussed further later.

Aqueous Colloidal Suspension of Laponite (Discs with Diameter=25 nm, Thickness=1 nm)

The aging dynamics of aqueous colloidal suspensions of a synthetic clay, Laponite, from fraction of 1 wt% to a few wt% was investigated by several groups [338–345] using dynamic light scattering which measured the normalized intensity autocorrelation function from 10^{-6} to 10^3 s in time $g_2(\boldsymbol{q}, t, t_w) =< I(\boldsymbol{q}, t_w)I(\boldsymbol{q}, t + t_w) > / < I(\boldsymbol{q}, t_w) >^2$, where t_w is aging or waiting time, t the delay time, and $\boldsymbol{q}$ the scattering vector. This system and the aging behavior are typical of soft glass-forming materials. As discussed in Section 1.3.2, $g_2(\boldsymbol{q}, t, t_w) - 1 =< \rho(\boldsymbol{q}, t_w)\rho(\boldsymbol{q}, t + t_w) > / < \rho(\boldsymbol{q}, t_w) >^2$ is the intermediate scattering function $F(\boldsymbol{q}, t, t_w)$ for diffusion of the colloidal discs within the Gaussian approximation. At short times, the time dependence of $F(\boldsymbol{q}, t, t_w)$ was found to have the $\exp(-t/\tau_f)$, with τ_f having the q^{-2}-dependence normal Brownian motion [338], and thus it corresponds to the primitive diffusion in the CM. The same q^{-2}-dependence at short times was found by dynamic light scattering in colloidal suspensions of PMMA particles [143]. At longer times, $F(\boldsymbol{q}, t, t_w)$ follows the time dependence of $\exp[-(t/\tau_s)^{1-n}]$. In practice, $F(\boldsymbol{q}, t, t_w)$ was fitted by the sum $A\exp(-t/\tau_f) + (1 - A)\exp[-(t/\tau_s)^{1-n}]$. Although this looks like the two-step decay of mode coupling theory (MCT), the faster relaxation is an exponential function rather than a power law as predicted by MCT. On increasing t_w, there is a large increase of τ_s accompanied by increase of n, indicating that the two quantities of the structural relaxation of the suspension are correlated. This correlation is expected by the CM. A preliminary test of the CM used the experimental data from dynamic light-scattering studies on a charged colloidal suspension of Laponite by Zulian et al. [344]. These authors derived a general relation from the CM which relates τ_s, τ_f, n, and the crossover time t_c from primitive diffusion to many-body Kohlrausch diffusion. Good agreement of the relation with experimental data was found for very low concentration samples. The discrepancies with the model with experiments may be due to the fact that the experimental parameters used for this test have been obtained from the fit of the intensity autocorrelation function [344] which is not the same as that prescribed by the CM. Moreover the presence of polydispersity in the system should be taken into account but was not.

Aging of a Very Rapidly Cooled High Molecular Weight Polystyrene

The two colloidal suspension systems discussed above have the advantage that the increase of dispersion or n with aging can be monitored experimentally. Ordinary organic, inorganic, and polymeric glassformers may show the same behavior on aging, but experimental verification is much more difficult because the relaxation time τ_α at T_g is already very long and increases with aging. There is no dependable experimental technique to observe the change of dispersion on aging. Moreover, the reductions of specific or free volume and entropy on aging play perhaps a major role in the increase of the τ_α of ordinary glassformers. If present in some ordinary glassformers, the increase of n on aging plays only a supplementary role in the increase of τ_α. Significant change of dispersion on aging can only be seen if the glassformer is very rapidly cooled. This will ensure that the structure of the very rapidly quenched glass is far from equilibrium, and there is a chance to detect the change of dispersion after aging for some time. Most aging experiments have not met this condition, and change or no change of dispersion cannot be ascertained. Apparently the favorable condition to observe change in dispersion was met in the study of aging of a high molecular weight (3.8×10^6) and monodisperse polystyrene sample after it had been very rapidly cooled from 100°C to 95°C and to 90°C [803]. Measurements of the creep compliance $J(t)$ were made after the quenched sample had been aged for times up to 1 day. The data of $J(t)$ after aging at 95°C for a sequence of increasing times are shown in Fig. 80. The $J(t)$ measured within the range $10^{-10} < J(t) < 2.5 \times 10^{-10}\text{cm}^2/\text{dyne}$ are contributed entirely by the local

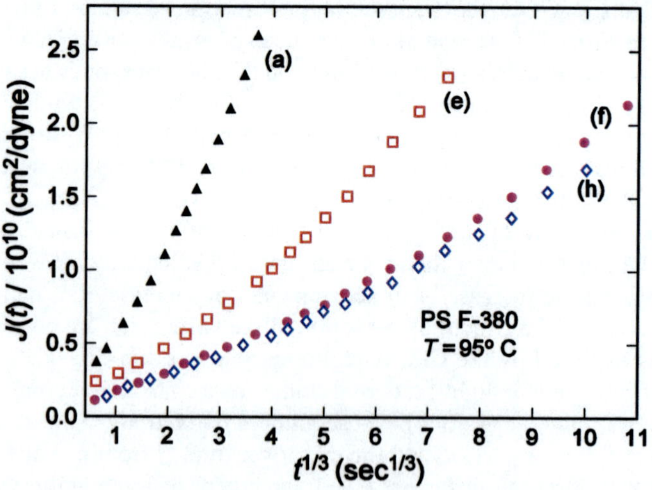

Fig. 80 Selected creep-compliance, $J(t)$, curves for a high molecular weight (3.8×10^6) polystyrene obtained at 95°C following rapid cooling from 100°C plotted against the cube root of the time of creep, $t^{1/3}$. Creep measurements were started (*a*) 5, (*e*) 90, (*f*) 1140, and (*h*) 2530 min after temperature equilibrium is obtained. Reproduced from [803] by permission

segmental (structural) relaxation of polystyrene [248]. The plot of $J(t)$ against $t^{1/3}$ follows the practice of Plazek and O'Rourke [168] for the local segmental relaxation of polystyrene at equilibrium, which is a straight line. By inspection of the creep curves measured after aging for $t_a = 5$, 90, 1140, and 2530 min, it is clear that departure from $t^{1/3}$-dependence is most significant at $t_a = 5$ min, and becomes less for longer t_a. Obviously the width of the dispersion increases with aging. The data were fitted by the approximate expression of $J(t) = J_g + bt^{1-n}$ that is consistent with the Kohlrausch function, $\exp[-(t/\tau)^{1-n}]$. The results show n=0.55, 0.60, 0.61, and 0.64 after aging for $t_a = 5$, 90, 1140, and 2530 min, respectively, showing that the increase of n is correlated with the increase of τ on aging.

2.2.5.5 The Effect of Shear on the Non-equilibrium Structural Dynamics of an Aging Colloidal Suspension of Laponite

Interesting effects were found in the non-linear rheological behavior resulting from the interplay of the microscopic dynamics of aging and shear flow in aqueous colloidal suspensions of Laponite. The viscosity as well as the structural relaxation time τ_s decreases strongly with the shear rate $\dot{\gamma}$ (the velocity gradient) applied to the system [804]. Aging in the absence of shear causes concomitant increases of τ_s and dispersion or n of the intermediate scattering function, $F(\boldsymbol{q}, t, t_w)$, as discussed in the previous section. Di Leonardo et al. [343] used dynamic light scattering to probe the change of $F(\boldsymbol{q}, t, t_w)$ of an aging 3 wt% Laponite suspension by shear at different levels of $\dot{\gamma}$. They found the shear flow influences significantly the aging dynamics as soon as structural relaxation enters the timescale set by the inverse shear rate, $\dot{\gamma}^{-1}$, as shown in Fig. 81. Aging is strongly reduced in this shear-dominated regime when $\tau_s > \dot{\gamma}^{-1}$, which can be seen in the figure by the decrease of the average slow relaxation time $< \tau_s >$, and the decrease increases with the shear rate $\dot{\gamma}$. Remarkably, the change of $< \tau_s >$ and its dependence on $\dot{\gamma}$ are the same as that of n shown in Fig. 81 by its counterpart $\beta \equiv (1 - n)$. Di Leonardo et al. also studied the effect of sudden increase of $\dot{\gamma}$ at some t_w. Rejuvenation was observed leading to faster $< \tau_s >$ and narrower dispersion (i.e., larger β or smaller n). This effect is shown by the crosses in Fig. 81, which are $< \tau_s >$ in the upper panel and β in the lower panel of two rejuvenated samples obtained by the two subsequent shear rate jumps: $67\,\mathrm{s}^{-1} \rightarrow 223\,\mathrm{s}^{-1}$ and $223\,\mathrm{s}^{-1} \rightarrow 446\,\mathrm{s}^{-1}$. Thus, the effect of shear on aging of Laponite suspension alters both $< \tau_s >$ and n, but the changes of the two quantities are always correlated.

Later on in Section 2.2.5.9, correlated changes of τ_α and the dispersion or n of the structural (local segmental) relaxation were found by non-linear deformation of "hard" glassy polymers, in the same manner as shown here for the "soft" aqueous suspension of Laponite. Despite many orders of magnitude of difference in mechanical modulus of the two systems, the phenomenon is similar because the underlying physics of many-body relaxation is the same.

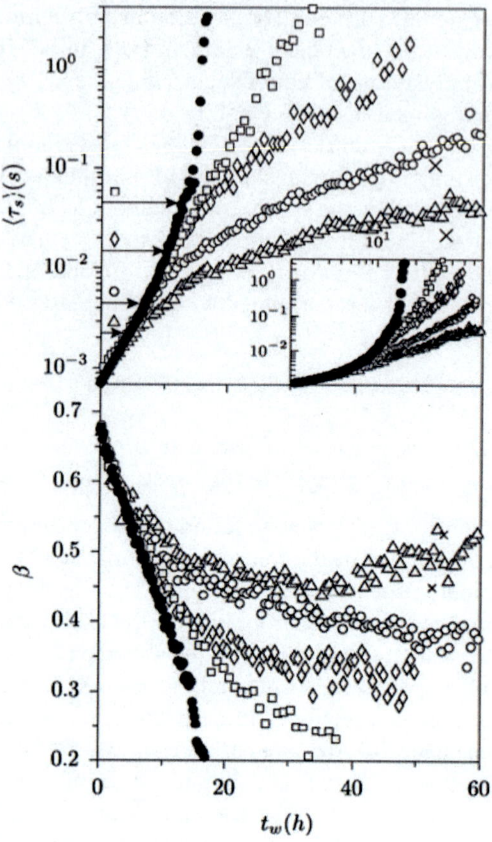

Fig. 81 Average slow relaxation time $<\tau_s>$ and stretching exponent $\beta \equiv (1-n)$ as a function of waiting time t_w during aging under different shear rates $\dot{\gamma}$: (Δ) 446, (O) 223, (◊) 67, (□) 22 s^{-1}. *Solid symbols* (•) refer to aging without shear. The uncertainties on the reported fitted values are comparable to *symbol size*. *Arrows* in top frame indicate the $\dot{\gamma}^{-1}$ values corresponding to each curve. *Inset* in top frame shows the same data in a double-logarithmic scale. For the *crosses*, see text. Reproduced from [343] by permission

2.2.5.6 Breakdown of the Stokes–Einstein Equation and the Debye–Stokes–Einstein Relation

The Stokes–Einstein Equation

When presenting the historical development of the field in Chapter 1, it has been made clear that the classic Stokes–Einstein (SE) equation (1.48) and the Debye equation (1.48) were originally designed for application to translational and rotational diffusion, respectively, in simple systems in which the diffusing and relaxing units are not interacting with each other. Thus, breakdown of these relations in glassformers and other materials is unsurprising because in these systems invariably non-trivial mutual interactions between the molecular units are present. The SE relation given before by Eqs. (1.48), with r_s, the spherical radius of the molecule, is

$$D = kT/6\pi\eta r_s. \tag{2.48}$$

It is not an easy task to measure self-diffusion coefficient D and viscosity in any glassformer over an extended common temperature range down to near T_g. As a result, the breakdown of the SE in any glassformers was not discovered experimentally until 1966 in *tris*-naphthylbenzene (TNB) and 1,2-diphenylbenzene (OTP) [347, 744, 793–797]. In these studies, the transport-dominated crystal growth rates $G'(T)$ of TNB and OTP were deduced from the measured crystal growth rate $G(T)$. Since $G'(T)$ is proportional to the self-diffusion coefficient $D(T)$, with the proper choice of a reference temperature T_s, the temperature dependence of the ratio, $D(T)/D(T_s)$, of TNB can be obtained from $G'(T)/G'(T_s)$. The viscosity of TNB ranging over 15 orders of magnitude had been measured [743, 744]. The temperature dependence of the normalized reciprocal viscosity $\eta(T_s)/\eta(T)$ was compared with $G'(T)/G'(T_s)$. It was found [805, 806] that the temperature dependence for $\eta(T_s)/\eta(T)$ is significantly stronger than that of $G'(T)/G'(T_s) = D(T)/D(T_s)$ as shown in Fig. 82-1.

Another way to show this disparity is by plotting $\log[G'(T)/G'(T_s)]$ against log η in the main Fig. 82-2 (left panel). On comparing the temperature dependence of $\eta(T_s)/\eta(T)$ and $G'(T)/G'(T_s)$, there is good agreement at high temperatures. The departure at the highest temperature may arise because the growth rate G is increasingly dominated by nucleation growth, as well as the larger uncertainty that accompanies the analysis and the values of G'. Departure is evident below $T_B \approx 413$ K. Incidentally this temperature is the same as the temperature $T_B = 412$ K at which occurs the crossover of temperature dependence of τ_α of TNB from VFTH$_1$ to VFTH$_2$, as shown before in Fig. 56. The departure increases as the temperature is lowered toward $T_g(= 342\text{ K})$. For $T_B > T > T_g + 26$ K, the temperature dependence of $G'(T)/G'(T_s) = D(T)/D(T_s)$ is approximately the same as $\eta(T)^{-0.74}$. The phenomenon when expressed as

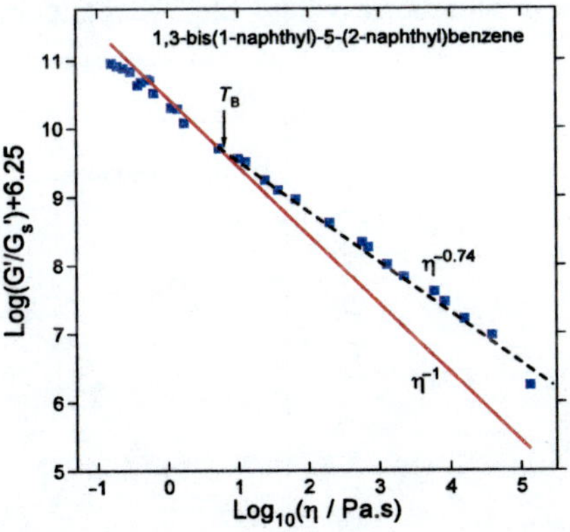

Fig. 82-1 The transport-dominated crystal growth rate $\log(G'/G'_s)$ of TNB plotted against $\log\eta$ (■). The subscript s indicates the values of G' and η at the reference temperature $T_s = 373.15$ K. The *solid* and *dashed lines* have slopes given by η^{-1} and $\eta^{-0.74}$ respectively. $T_B = 140°$C. The difference in the temperature dependences of G/G_s and η/η_s is shown in another manner in the next figure. Data from [347] are reconstructed in the new figure here

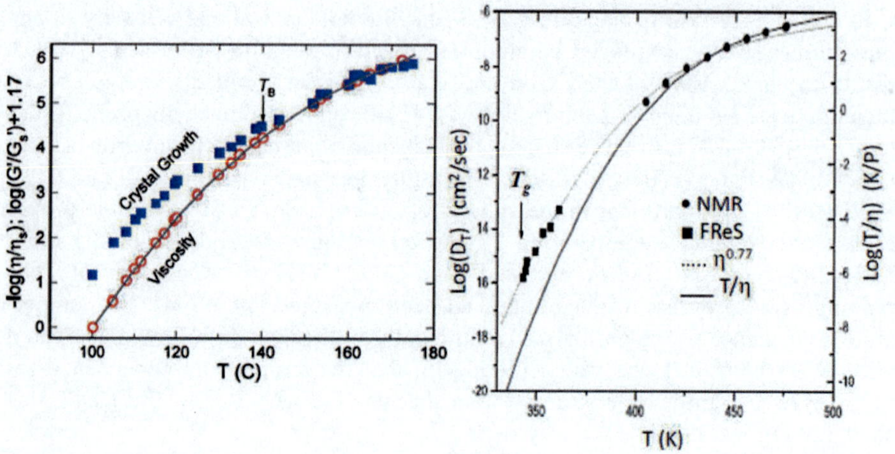

Fig. 82-2 (*Left*) Same data as Fig. 82-1 now plotted against temperature. (*Right*) Comparison of the temperature dependence of translational diffusion with the viscosity. D_T for TNB determined from FReS and NMR is given on the *left axis*. Temperature over viscosity is given by the *solid line*, with scale on the right-hand axis, shifted in order to permit overlap of the high-temperature values of the viscosity and D_T values. The *dashed line* is $\eta^{-0.77}$, vertically shifted. Data from [453] reproduced by permission

$$D \propto \eta^{-\xi} \text{ or } D \propto T/\eta^{\xi}, \text{ where } \xi < 1, \tag{2.49}$$

is sometimes referred to as fractional SE relation [64, 65, 105].

Similar results of breakdown of SE equation in 1,2-diphenylbenzene (OTP) were obtained by comparing $G'(T)/G'(T_s)$ with viscosity data as shown in Fig. 83. For OTP ($T_g = 243$ K), the breakdown occurs at temperatures below $T_B = 290$ K, which is again near the temperature where the T-dependence of τ_α crosses over

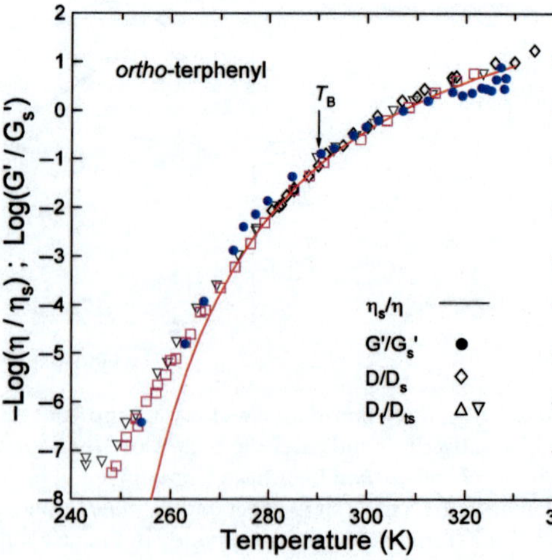

Fig. 83 Temperature dependence of the reduced viscosity $\log(\eta/\eta_s)$ (*solid curve*), transport-dominated crystal growth rate $\log(G'/G'_s)$(•), self-diffusion coefficient D/D_s(◇), and tracer diffusion coefficient D_t/D_{ts} (△ and ▽) for OTP. The subscript s indicates the value of the quantities at the reference temperature of 307 K. Data from [347] are replotted here

from $VFTH_1$ to $VFTH_2$ (see Fig. 63). The increasing weaker T-dependence of $G'(T)/G'(T_s)$ compared with $\eta(T_s)/\eta(T)$ with decreasing temperature from T_B down to about 10 K above T_g can be seen in Fig. 83. For OTP, the fractional SE relation was found to be $G'(T)/G'(T_s) = D(T)/D(T_s) \propto [\eta(T)]^{-0.74}$.

Disparity in magnitude and T-dependence of relaxation times of different dynamic variables was found and explained by the CM in 1988 [267]. Thus, if the breakdown of SE relation is simply put as disparity between translational diffusion and viscosity, then SE is just a special case of a more general phenomenon.

Starting from 1992, there were renewed interest in breakdown of the SE and the DSE equations in glassformers. Measurements by an NMR technique (proton nuclear magnetic resonance stimulated echoes in a static magnetic field gradient) of the self-diffusion coefficient, D, of small molecular glass-forming liquid such as 1,2-diphenylbenzene (OTP) were reported [808, 809]. Starting at high temperatures, it was found that D follows the temperature dependence $1/\eta$ of the Stokes–Einstein relation. But, at a certain temperature $T_B \approx 290$ K, approximately equal to 1.2 times the glass temperature $T_g = 244$ K, D shows a weaker temperature dependence than does $1/\eta$. However, this NMR technique is unable to measure D values smaller than 10^{-14} m^2/s and at temperatures significantly below $T_B \approx 290$ K to see larger enhancement of translational self-diffusion. At temperatures below T_B, only diffusion coefficients, D_t, of the tracer molecules similar in size to OTP were obtained by forced Rayleigh scattering technique. The normalized self-diffusion and tracer diffusion data D/D_s and D_t/D_{ts} from [808] and [809] are included in Fig. 83 for comparison with the transport-dominated crystal growth rate and viscosity. There is quantitative agreement between $G'(T)/G'(T_s)$ and D/D_s as well as D_t/D_{ts}, and they all have a weaker temperature dependence than $\eta(T_s)/\eta(T)$.

In recent years, direct measurements of self-diffusion of TNB were reported by forward recoil spectrometry to measure the concentration profiles of deuterio and protio TNB following annealing-induced diffusion in a vapor-deposited bilayer [453]. For OTP, self-diffusion coefficients were obtained by isothermally desorbing thin film bilayers of deuterio and protio o-terphenyl in a vacuum chamber [454]. The directly measured self-diffusion coefficients D of TNB at temperatures from T_g+18 K down to $T_g(= 342$ K) shown in Fig. 82 (right panel) extend the range of the data of $G'(T)/G'(T_s) = D(T)/D(T_s)$. The much weaker temperature dependence of D than T/η becomes clearer. When the forward recoil spectrometry data of D are combined with NMR measurements at higher temperatures, the relation $D(T) \propto \eta(T)^{-0.77}$ is obtained for TNB, which is about the same as $D(T) \propto \eta(T)^{-0.74}$ obtained from using the data of $G'(T)/G'(T_s) = D(T)/D(T_s)$ [347, 805–807]. For OTP, the later work by Mapes et al. [454] also shows self-diffusion coefficients they determined are in agreement with that deduced from crystal growth in o-terphenyl by Magill and Li [807], which is controlled by self-diffusion from 253 K to the melting point. Mapes et al. also include the crystal growth rate data below 253 K by Oguni and coworkers [810, 811]. These data by Oguni should not be included by Mapes et al. because at temperatures below 253 K a different crystal growth mechanism takes over, and the observed crystal growth rate is no longer proportional to self-diffusion. The mechanism is the so-called "homogeneous nucleation" and was attributed by Oguni and coworkers to the secondary relaxation of OTP, giving rise to a new crystal shape. More discussion of this can be found in Section 2.3.2.22.

The breakdown of the Stokes–Einstein relation was observed also in the supercooled liquid state of a metallic glass, $Zr_{46.7}Ti_{8.3}Cu_{7.5}Ni_{10}Be_{27.5}$, as revealed by a comparison of beryllium diffusion data with viscosity data [812]. Translation diffusion coefficient of Be is increased by 1–2 orders of magnitude and has a weaker T-dependence when compared to the Stokes–Einstein prediction. The similarity of the effect to that found in molecular glassformers was considered as an indication of cooperative diffusion mechanism in the supercooled liquid state of $Zr_{46.7}Ti_{8.3}Cu_{7.5}Ni_{10}Be_{27.5}$. Breakdown was found also in room temperature ionic liquid by Ito and Richert [158] and explained by the CM [159].

The Debye–Stokes–Einstein Relation

The Debye rotational diffusion equation [18–20] was originally derived for time-dependent orientation of non-interacting spherical Brownian particles (see Section 1.1). It was assumed that the Brownian particle undergoes many collisions with the host solvent molecules before it reorients through an appreciable angle. The Debye model is governed by Eq. (1.10), which can be solved by using the spherical harmonics, $Y_{lm}(\hat{u})$, as eigenfunctions of $\vec{\nabla}_{\Omega}^{2}$ in this equation. The orientation time correlation function $\langle \hat{u}(t)\hat{u}(0)\rangle$ is an exponential function of time given by Eq. (1.11). The orientational correlation function, $C_{\mathrm{r}}(t) = 4\pi \langle Y_{lm}^{*}(\hat{u}(0))Y_{lm}(\hat{u}(t))\rangle$, can be rewritten in terms of the ensemble average of the Legendre polynomials $P_l(\cos\theta(t))$ for any order l as follows:

$$C_l(t) = (2l+1)\langle P_l(\cos\theta(t))\rangle. \tag{2.50}$$

Here $\cos\theta(t)$ is the cosine of the angle, $\theta(t) = \hat{u}(t)\hat{u}(0)$, through which a particle rotates in time t. As shown before in Section 1.3.1, dielectric relaxation measurements pick out the response corresponding to $l = 1$ and give $C_1(t) = 3\langle P_1(\cos\theta(t))\rangle$, where $P_1(\cos\theta(t)) = \cos\theta(t)$. On the other hand, measurements using photon correlation spectroscopy, holographic fluorescence recovery after photobleaching (holographic FRAP) technique [813], and some NMR techniques pick out the responses corresponding to $l = 2$, and yield $C_2(t) = 5\langle P_2(\cos\theta(t))\rangle$, where $P_2(\cos\theta(t)) = (5/2)[3\cos^2(\theta(t)) - 1]$. In the Debye model, for all l, $C_{\mathrm{r}}(t)$ is a linear exponential function of time for all times and is given by

$$C_l(t) = \exp[-l(l+1)D_{\mathrm{r}}t] \tag{2.51}$$

A typical relaxation time τ_{c} for $l = 2$ is

$$D_{\mathrm{r}} \equiv \frac{1}{6\tau_{\mathrm{c}}} = \frac{kT}{8\pi\eta r_{\mathrm{s}}^{3}}, \tag{2.52}$$

where r_{s} is the radius of the spherical particle.

For a probe molecule in viscous glassformer, its rotational motion is not expected to have exponential correlation function as in the Brownian spherical particle in the Debye model. This is obvious in the case that the probe is identical to the host, where dielectric ($l = 1$) and photon correlation spectroscopic ($l = 2$) measurements of neat glassformers have shown that the correlation function has the Kohlrausch stretched exponential time dependence,

$$C_1(t) \text{or} C_2(t) = \exp[-(t/\tau_c)^{1-n_r}]. \tag{2.53}$$

Nevertheless, the result of the Debye model can be generalized by replacing the Debye τ_c in Eq. (2.52) by the Kohlrausch rotational correlation time τ_c of the probe molecule. D_r, defined in the same way as in the Debye equation, is the average rotational diffusion coefficient. This generalized Debye relation seems to hold in neat OTP where the measured D_r followed the temperature dependence practically at all temperatures above T_g where the viscosity η changes by 12 orders of magnitude [808, 809].

On combining this Debye relation with the SE relation for the probe molecule, $D_t = kT/6\pi\eta r_s$, we obtain the so-called Debye–Stokes–Einstein relation

$$D_t\tau_c = 2r_s^2/9. \tag{2.54}$$

That is, if both the SE and the Debye relations hold for the probe, the product $D_t\tau_c$ should be equal to $2r_s^2/9$. Henceforth, this ideal value of the product is designated by $(D_t\tau_c)_{SE,DSE}$. Measurements of probe translational diffusion and rotational diffusion were made in a variety of probes of low concentrations in many different host glassformers principally by the holographic fluorescence recovery after photobleaching (holographic FRAP) technique [457, 813] and forced Rayleigh scattering technique for translational diffusion, and rotational FRAP technique for probe rotation [449–452, 793–796, 814, 815], and other techniques [816–819]. In many systems it was found that the product $D_t\tau_c$ is a function of temperature, increases with decreasing temperature toward T_g of the host, and can reach values much larger than $(D_t\tau_c)_{SE,DSE}$. To quantify translational enhancement, the product $D_t\tau_c$ is normalized to $(D_t\tau_c)_{SE,DSE}$. Some examples are shown in Fig. 84. This breakdown of the Debye–Stokes–Einstein (DSE) relation is due to enhancement of probe translational diffusion in comparison with rotational diffusion. The ratio $D_t\tau_c/(D_t\tau_c)_{SE,DSE}$ evaluated at $T = T_g$ is a measure of the degree of the breakdown of the SE and DSE relations for various combinations of probes and host glassformers. In OTP the enhancement of tetracene translation measured by $\log[D_t\tau_c/(D_t\tau_c)_{SE,DSE}]$ is about 2.5 decades near T_g. Different behaviors are shown by larger probe molecules such as rubrene in OTP where the enhancement of translational diffusion becomes negligible as evidenced by D_t, T/η, and $< \tau_c >$ all having about the same temperature dependence, and $\log[D_t\tau_c/(D_t\tau_c)_{SE,DSE}]$ is small for all temperatures down to T_g (see Fig. 84). The study of the rotational motion and translational diffusion of probe molecules was extended [449–452, 793–796, 814, 815] to other hosts including several polymers. In all cases, the time dependence of the probe rotational

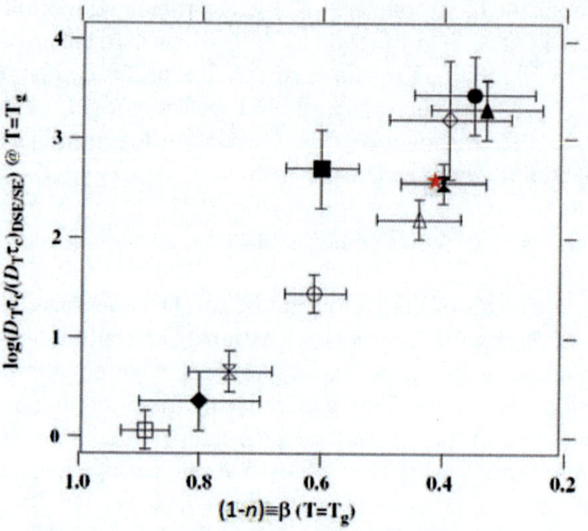

Fig. 84 The figure shows the correlation of enhanced translation $\log[D_t\tau_c/(D_t\tau_c)_{SE,DSE}]$ at T_g with $n_r(T_g) \equiv [1-\beta(T_g)]$, where $\beta(T_g)$ is the Kohlrausch exponent for the probe rotation in four matrices, OTP, TNB, polystyrene (PS), and polysulfone (PSF) at $T = T_g$. The probes are tetracene, rubrene, anthracene, and BPEA. The *symbols* represent PS/tetracene (*closed circle*), PS/rubrene (*open circle*), PSF/tetracene (*closed triangle*), PSF/rubrene (*open triangle*), OTP/tetracene (*closed square*), OTP/rubrene (*open square*), OTP/anthracene (*open diamond*), OTP/BPEA (*closed diamond*), TNB/tetracene (*closed hourglass*), TNB/rubrene (*open hourglass*). Data taken from [449, 451, 452, 795], and replotted here together with the *lone star*, which represents the datum of rubrene/sucrose benzoate [457]

time correlation functions $F_r(t)$ was found to be well described by the Kohlrausch function of Eq. (2.53). At $T = T_g$, a strong correlation was observed between the ratio $D_t\tau_c/(D_t\tau_c)_{SE,DSE}$ and $n_r(T_g) \equiv [1 - \beta(T_g)]$. This is equivalent to an anti-correlation between $\beta(T_g) \equiv (1 - n_r(T_g))$ and $D_t\tau_c/(D_t\tau_c)_{SE,DSE}$, which is actually shown in Fig. 84. In other words, at T_g, more enhanced probe translation compared with probe rotation (or the degree of breakdown of SE and DSE relations) is found for probe having rotation correlation functions which are more dispersive or larger n_r.

Figure 84 shows the correlation using practically all the data from different probes in several matrices. Restricted to the same probe, e.g., rubrene, the increase of the ratio $D_t\tau_c/(D_t\tau_c)_{SE,DSE}$ with increasing $n_r(T_g)$ or decreasing $\beta(T_g)$ is also clear. For example, $D_t\tau_c/(D_t\tau_c)_{SE,DSE}$ increases with decreasing values of $\beta(T_g)$ for rubrene rotation at T_g from 0.86 (in polyisobutylene), 0.89 (in OTP), 0.74 (TNB), 0.61 (in polystyrene), 0.44 (in polysulfone), and 0.42 (in BPA-polycarbonate). When the probe is identical to the host, probe diffusion becomes self-diffusion in a neat glassformer. By extrapolating the results of probe/host systems to neat glass glassformers, the breakdown of SE and DSE relations in neat glassformers is expected, and the correlation between $D_t\tau_c/(D_t\tau_c)_{SE,DSE}$ and $n_r(T_g)$ should still hold.

Explanation of the Breakdown of SE-DSE Relations by the Coupling Model

The empirical correlation found between the size of the deviation from the SE-DSE relation and $n_r(T_g)$ shown in Fig. 84 is an indication that many-body relaxation dynamics are responsible for the occurrence of such a general phenomena, and the explanation can be obtained from the coupling model (CM). Such a CM explanation was indeed proposed in 1999 [268]. In the framework of the CM, the explanation is based on the fact that different dynamic variables μ weigh the intermolecular coupling differently and have different coupling parameters (i.e., different degrees of intermolecular cooperativity), n_μ, which enter into the stretch exponents of their correlation functions, $\langle\mu(0)\mu(t)\rangle/\langle\mu^2(0)\rangle = \exp[-(t/\tau_\mu)^{1-n_\mu}]$, represented in the Kohlrausch form. This together with CM equations,

$$\tau_\mu(T) = [t_c^{-n_\mu}\tau_{0\mu}(T)]^{1/(1-n_\mu)}, \tag{2.55}$$

immediately leads to differences between the τ_μs as well as their temperature dependences. A larger n_μ for the observable μ will bestow a stronger temperature dependence for the relaxation time τ_μ. This is because the primitive relaxation times of all observables $\tau_{0\mu}$, uninfluenced by many-body relaxation dynamics, should have one and the same temperature dependence.

Let us here consider the case of *tris*-naphthylbenzene (TNB). Although the value of n_η for viscosity is known from shear mechanical measurement [743, 744] or light-scattering data ($n_{LS} = 0.45$) [820], n_d for dielectric relaxation [455] and n_{NMR} for NMR [821] are 0.50, n_D for self-diffusion is not yet available from experiment. Nevertheless, theoretical arguments based on comparing constraints have been given before to show that the coupling parameter n_r for rotational relaxation obtained by dielectric relaxation (first-order Legendre polynomial) or light scattering (second-order Legendre polynomial) and the coupling parameter n_η for shear viscosity are all larger than the coupling parameter for center-of-mass diffusion n_D [182, 183, 268], and the explanation of the breakdown of the SE and DSE relations follows from this relation. There is support of the inequality, $n_r > n_D$, from the results of molecular dynamics simulation of relaxation in the van der Waals liquid *o*-terphenyl by Wahnström and Lewis [315]. They found the orientational correlation function shows somewhat more stretching ($\beta_r \equiv 1 - n_r = 0.69$ (for $q = 1.94\,\text{Å}^{-1}$)) compared with $\beta_D \equiv 1 - n_D = 0.82$ for the center-of-mass motion. There is also similar support from experimental measurements of a room temperature ionic liquid [158, 159].

This CM explanation holds independent of whether n_r and n_η are temperature dependent or independent. We demonstrate this for TNB by taking $n_r = 0.50$ of dielectric relaxation to be temperature independent from below $T_g = 342 - 417$ K as found experimentally by Richert et al. [455]. The recoverable shear creep compliance, $J_r(t)$, data near and above T_g of TNB in [182, 183] have been fitted by the function $J_r(t) = J_g + (J_e - J_g)\{1 - \exp[-(t/\tau_{\alpha\eta})^{1-n_\eta}]\}$, where J_g and J_e are the glassy and the steady-state compliances, respectively, and n_η is also close to 0.50

[275, 797]. Richert et al. [455] gave the dielectric α-relaxation time by the Vogel–Fulcher–Tammann–Hesse equation, $\log[\tau_r(T)] = -18.05 + 1620/(T - 246)$. With these values of $\tau_r(T)$ and t_c known to be equal to about 2 ps, Eq. (2.55) is used to calculate the primitive dielectric relaxation times $\tau_{0r}(T)$. We assume the primitive self-diffusion relaxation time $\tau_{0D}(T)$ is the same as $\tau_{0r}(T)$, and calculate $\tau_D(T)$ by using a n_D to have a smaller value than $n_r = 0.50$ as suggested by the CM explanation [268]. In Fig. 85 we present the results of $\tau_{0r}(T)$ calculated by Eq. (2.55) for $\mu \equiv r$ with the *temperature-independent* value of $n_r = 0.50$, and $\tau_{\alpha d}(T)$ also calculated by the same equation for $\mu \equiv D$ with constant $n_D = 0.37$, and $t_c = 2$ ps for both. Before making quantitative comparisons, from $n_r > n_D$ and $\tau_{0D}(T)$ having the same T-dependence as $\tau_{0r}(T)$, readily Eq. (2.55) leads to the conclusion that $\tau_{\alpha r}(T)$ has a stronger T-dependence than $\tau_D(T)$. As a result, the separation between $\tau_r(T)$ and $\tau_D(T)$ increases with decreasing temperature even if n_r is constant, as observed experimentally for TNB [455], OTP [456], and sucrose benzoate [457]. We hasten to mention here that in contrast to the CM explanation of the breakdown of the SE-DSE relations, the explanation based on spatial heterogeneous dynamics by Ediger and coworkers [429, 450, 453, 454] runs into problem for TNB, OTP, and sucrose benzoate. This is because validity of the latter explanation requires the dispersion of the rotation correlation function to broaden or β to decrease with decrease in temperature. More discussion of this will be given later.

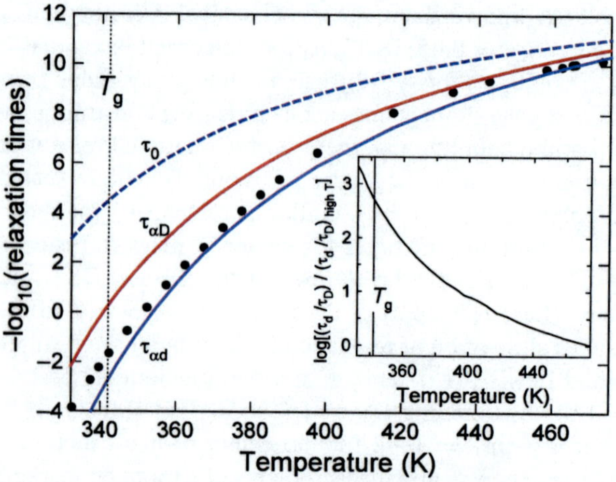

Fig. 85 Plots of various relaxation times of TNB against temperature. $\tau_{\alpha d}(T)$ is from the Vogel–Fulcher–Tammann–Hesse fit to dielectric data (*continuous line*), $\tau_{0d}(T) = \tau_{0D}(T) \equiv \tau_0(T)$ is calculated with constant $n_d = 0.50$ (*dashed line*), and $\tau_{\alpha d}(T)$ calculated with constant $n_d = 0.37$ (*dashed-dotted line*). Plotted is also T/η (*closed circles*), after a vertical shift of 7.1 decades upward has been applied. The results from the CM reproduce the decoupling of self-diffusion from dielectric relaxation, which increases with decreasing temperature even though n_d is constant. In the *inset*, the ratio $\tau_{\alpha d}(T)/\tau_{\alpha d}(T)$ normalized to its value at the high temperature of 475 K, $(\tau_{\alpha d}/\tau_{\alpha d})_{T=475\,\mathrm{K}} = 1.5$, is shown as a function of temperature. Data from various sources described in [275] are replotted here

The ratio, $\tau_r(T)/\tau_D(T)$, is a way to show the breakdown of the DSE relations. At the high temperature of 475 K, the ratio $(\tau_r/\tau_D)_{T=475\,K}$ is nearly unity. After normalizing it by this high temperature value of 1.5, the monotonic increase of the ratio $\tau_r(T)/\tau_D(T)$ with decreasing temperature toward T_g is shown in the inset of Fig. 85. The ratio, $\tau_r(T)/\tau_D(T)$, is equivalent to the product $D(T)\eta(T)$ often used to show the breakdown of the SE relation. It exhibits monotonic increase as temperature is lowered toward T_g in the same manner as $D(T)\eta(T)/(D\eta)_{\mathrm{highT}}$. The results in Fig. 85 are *qualitatively* the same for other choices of n_D less than $n_r = 0.50$. The particular choice of $n_D = 0.37$ was made in Fig. 85 because the decoupling of $\tau_D(T)$ from $\tau_r(T)$ becomes *quantitatively* approximately the same as that deduced from a comparison of the temperature dependences of the measured $D(T)$ [453, 454] and the measured $\tau_r(T)$ [455] of TNB. Next, we show the breakdown of the SE relation by comparing in Fig. 85 the temperature dependence of $\tau_D(T)$ with that of T/η. The shear viscosity data were taken from the works of Plazek and Magill [805, 806]. For comparison, the actual values of T/η have been shifted vertically by 7.1 decades (shown by closed circles) such that its value is about the same as $\tau_r(T)$ and $\tau_D(T)$. It is clear by inspection of Fig. 85 that T/η decreases more rapidly with falling temperature than $\tau_D(T)$. Since $\tau_D(T)$ is inversely proportional to D, from this we have demonstrated the breakdown of the SE relation from the CM in those cases where n_r or n_η is constant.

Explanation of the Correlation of Size of Breakdown with $n_r(T_g)$

Bainbridge and Ediger [451] found the existence of a correlation between the ratio $D_t\tau_c/(D_t\tau_c)_{\mathrm{SE,DSE}}$ at $T = T_g$ and $n_r(T_g) \equiv [1 - \beta(T_g)]$ shown in Fig. 84 by compiling data from different combinations of probes and matrices. From the CM explanation given above, it is plausible that probe with larger $n_r(T_g)$ can lead to larger $D_t\tau_c/(D_t\tau_c)_{\mathrm{SE,DSE}}$ at $T = T_g$, if the ratio $n_D(T_g)/n_r(T_g)$ is maintained constant. Although plausible, this assumption cannot be verified until $n_D(T_g)$ has been determined by measurement of the diffusion correlation function, which has not been achieved so far for the probes and matrices shown in Fig. 84 (only the long-term diffusion has been measured). The correlation certainly has drawn attention, but a deeper question that follows is what causes the variation of $n_r(T_g)$ of different probes and host combinations?

This question has been answered by the CM on adapting its previous application to mixtures to the case of probe dynamics [268, 736, 737], which after all is the special case of binary mixtures when concentration of one component becomes small. The component dynamics of mixtures is the subject of detailed discussion in Section 2.2.5.7. A probe in a host matrix cannot rotate with large angles without participation of motions of the host molecules. In other words, it is still involved in a many-body relaxation unless the host molecules are much more mobile than the probe. In that case, the motion of the probe occurs in the hydrodynamic regime of the host. There is absence of many-body effects and $n_r(T_g) = 0$, analogous to the Debye relaxation of a molecule in a solvent. If the probe is more mobile than the host molecules, its motion is necessarily stretched to longer times by the host molecules.

A practical measure of the difference in mobility of the probe and the host is the ratio $\tau_r(T)/\tau_\alpha(T)$ at $T = T_g$, where $\tau_\alpha(T)$ is the structural α-relaxation time of the host. It is obvious that longer is $\tau_\alpha(T)$ compared with $\tau_r(T)$, the more the rotational correlation function of the probe will be stretched out to longer times or smaller is its Kohlrausch exponent $\beta(T_g)$ and larger coupling parameter $n_r(T_g)$ of the CM. This deduction following the precept of the CM suggests $\beta(T_g)$ of probe in matrix is controlled by $\tau_r(T_g)/\tau_\alpha(T_g)$, and hence there should be a correlation between these two quantities. Indeed, such a correlation was found from experimental data [268] and is shown here in Fig. 86 (left panel). Since $\tau_r(T_g)/\tau_\alpha(T_g)$ is the control parameter of the cause, it should correlate with the consequence which is $D_t\tau_c/(D_t\tau_c)_{SE,DSE}$ at $T = T_g$. This correlation was also found from the experimental data [268] and shown in the right panel of Fig. 86. Thus, the correlation between $D_t\tau_c/(D_t\tau_c)_{SE,DSE}$ at $T = T_g$ and $n_r(T_g) \equiv [1 - \beta(T_g)]$ found empirically by Bainbridge and Ediger [451] shown in Fig. 84 is the result from combining the two more fundamental correlations in the left and right panels of Fig. 86. All said above is further supported by the correlation between $n_r(T_g)$ and the fragility or steepness index m obtained from the T-dependence of $\tau_c(T)$ shown in [268].

No Breakdown of SE Relation in Silica

For Stokes–Einstein relation between self-diffusion and viscosity, the best example to show the correlation of the size of its breakdown with the coupling parameter of

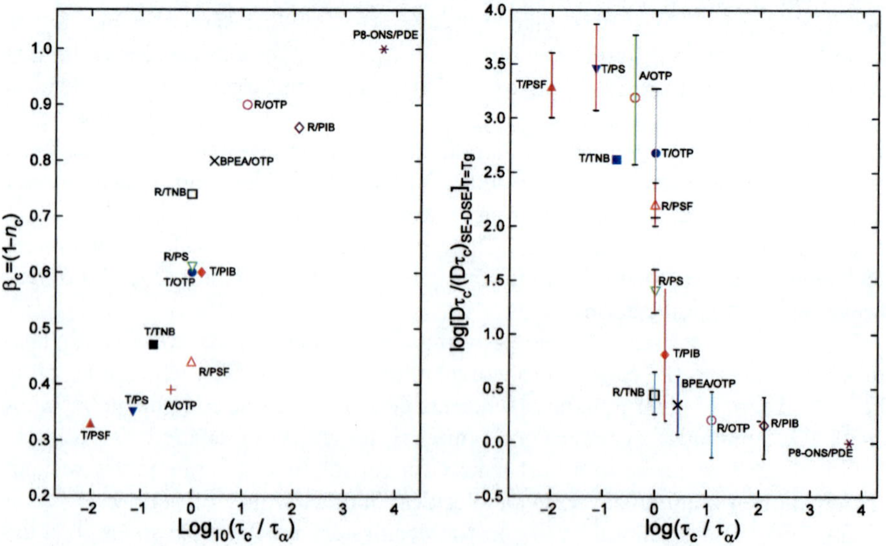

Fig. 86 *Left panel*: Correlation of the KWW β parameter for probe rotation at T_g in various hosts with the extent of probe–host coupling measured by the ratio $\tau_c(Tg)/\tau_\alpha(Tg)$. *Right panel*: Correlation of enhanced translation $\log[D_t\tau_c/(D_t\tau_c)_{SE,DSE}]$ at T_g of the probes in various hosts with the extent of probe–host coupling measured by the ratio $\tau_c(Tg)/\tau_\alpha(Tg)$. For the source of data and more explanation of the plots see [268]. Data from [268] are replotted in both figures here

viscosity, n_η, is afforded by the data of silica that has $T_g = 1451$ K and $T_m = 2007$ K [822]. In the temperature range $1.1T_g < T < T_m$, the diffusion coefficient deduced from viscosity is nearly the same as the self-diffusion coefficient, and thus SE relaxation is obeyed by silica. This is expected by the CM because many-body dynamics are practically absent in silica as evidenced by $n_\eta \approx 0.1$ from a molecular dynamics simulation [823], and suggested from its small value of the steepness index m together with the correlation between n and m. Hence, both coupling parameters being small, any difference between n_D for self-diffusion and n_η for viscosity is correspondingly small. The correlation is clear when comparing no breakdown of the SE relation in silica having $n_\eta \approx 0.1$ with the prominent breakdown found in o-terphenyl and TNB, all of which have larger n_η. These data were part of a larger collection showing a correlation between degree of breakdown of SE relation and fragility by Ediger et al. [824]. These authors rationalize the correlation by spatially heterogeneous dynamics and the argument that more fragile liquid is more heterogeneous. There is a flaw in their rationalization. As we have discussed before, spatially heterogeneous dynamics cannot explain breakdown of SE relation in the most well-studied cases including TNB, OTP, and sucrose benzoate. Nevertheless, since fragility roughly correlates with non-exponentiality or n particularly when restricted to the same class of liquids, their empirical observation is consistent with the correlation of degree of breakdown of the SE relation with n expected by the CM.

The CM Was Mistaken as a Homogeneous Relaxation Theory

The advances in this field up to the year 2000 were reviewed by Ediger [429]. It mentioned the explanation of the breakdown of SE and DSE relations given by the CM in 1999 [268]. Unfortunately, this review mistakenly labeled the coupling model (CM) as a homogeneous relaxation model, even though it was explicitly stated in the very same 1999 paper [268] that the dynamics envisaged in the CM is dynamically heterogeneous, and a previous work explicitly demonstrating this [234] was referenced therein. It is hard to imagine that many-body relaxation in glass-forming liquids can be homogeneous, and the CM has constantly emphasized the importance of many-body aspect of the α-relaxation. This misrepresentation of my work is unfortunate because the review by Ediger is highly cited by others. This mistake by Ediger may have caused some colleagues including himself to ignore the CM explanation till now. As it will become clear from the discussions below, the explanation by spatially heterogeneous dynamics is contradicted by more recent experiments, while the CM explanation remains a viable explanation [275], which will be revisited later in a separate section devoted to the many different applications of the CM.

Explanations for the Breakdown of the SE and DSE Relations Based on Spatially Heterogeneous Dynamics

The enhancements of translational diffusion compared with viscosity and probe rotational diffusion (i.e., breakdown of the SE and DSE relations) are so general a phenomenon that it attracts attention. Several attempts to explain it are all based on spatially heterogeneous dynamics in one form or the other [450, 825, 826]. For example, in one model it is assumed that in some regions of the liquid both rotation and translation are orders of magnitude faster than the corresponding dynamics in other regions. There is experimental evidence for dynamic heterogeneity as discussed before in Section 2.2.2, although it is far from clear that the assumed spatial heterogeneity in the model is consistent with the observed dynamic heterogeneity. It was argued in such a heterogeneous system, translation and rotation experiments average over the heterogeneity differently. The orientation correlation function is a superposition of the orientation relaxation functions for the different regions of the sample. Since the average rotational correlation time $< \tau_c >$ is the integral of the correlation function, it weighs regions of slower mobility to a much greater extent. By contrast, molecules tend to diffuse in paths through regions of faster dynamics. The self or the probe translational diffusion coefficient is determined effectively by regions of higher mobility, and it is enhanced over the rotational diffusion. Spatially heterogeneous dynamics engenders a distribution of relaxation times and is consistent with non-exponential rotational correlation function. Therefore, this explanation implies that a direct correlation should exist between enhanced translational diffusion and the non-exponentiality of the rotational correlation function, which is in fact observed (see Fig. 84). However, there are problems of this explanation as discussed below.

Problems of Explanations Based on Spatial Heterogeneous Dynamics

The models based on spatial heterogeneous dynamics sound reasonable and have been considered for a long while to provide the correct explanation of the breakdown of SE and DSE relations until recent experiments [455–457] and simulation [274, 275] have found contradiction to their premise. The contradictions invalidate the explanation. The problems of these models can be traced to the fact that all dynamic properties originate from the many-body relaxation, and heterogeneous dynamics and non-exponentiality are just two among many properties (including the breakdown of SE and DSE relations) that are parallel consequences of the many-body relaxation. Naturally, these parallel consequences are consistent with each other, and correlations between them must exist, such as that between $D_t\tau_c/(D_t\tau_c)_{SE,DSE}$ and the degree of non-exponentiality or n_r shown in Fig. 84. However, it is not possible to fully rationalize or derive one property from another. This is simply because, being parallel consequences, neither property retains the full attributes of the source, which is the many-body relaxation. Only by going back to the source (and with the help of a theory that can predict how many-body relaxation governs relaxation time and transport coefficients) can each of the parallel consequences be fully explained. An analogy is the attempt to completely solve the glass transition problem by considering only thermodynamics and thermodynamic variables, which obviously cannot succeed because many-body relaxation and its consequences are left out.

In the following, the contradictions with experiments as well as conceptual difficulties encountered by the models based on spatial heterogeneous dynamics (particularly the broadening of the distribution of local relaxation times as the temperature is lowered required by the models) are presented and discussed. Since spatial heterogeneous dynamics is an appealing and popular notion and has been highly promoted to be a fundamental, considerable number of points is given below to show that it is not a viable explanation. Again, the reason is that it is merely one of the consequences of many-body dynamics. Understandably, one consequence may not be able to explain another parallel consequence such as the breakdown of SE and DSE relations.

[1] *Why viscous flow is not enhanced compared with local orientation like translational diffusion*? Like translation diffusion, viscous flow also could emphasize regions of higher mobility. If considered, by the same argument for enhancement of translational diffusion in the spatial heterogeneous dynamics model, one would expect also enhancement of viscous flow compared with local orientational relaxation. But this is not observed in neat OTP [808, 809] because the rotational diffusion coefficient, D_r, and viscosity η are found to obey the DSE equation practically at all temperatures above T_g where the viscosity η changes by 12 orders of magnitude. This is explained by the coupling parameters n_r and n_η having the same value as found in TNB.

[2] *The distribution of rotational relaxation times does not broaden as the temperature is lowered toward* T_g: In order to explain translational motion having weaker temperature dependence than rotational motion, models based on spatially heterogeneous dynamics require the distribution of local relaxation times to broaden as the temperature is lowered toward T_g. Other models proposed to explain the same effect [827, 828] are also associated with a strong temperature dependence of the relaxation time distribution. This basic premise of the models is contradicted by several measurements of the orientational correlation function $F_r(t)$ of TNB by several techniques. The exponents β of the Kohlrausch function used to fit the time/frequency dispersion of various data are quite consistent with each other and all indicate a temperature-independent β or distribution of relaxation times. Dielectric relaxation measurements of TNB in the range of T/T_g from 1.01 to 1.22, where $D\tau_c$ changes significantly, show no change in the shape of the loss spectra. The dielectric frequency dispersions at all these temperatures are well described by the one-sided Fourier transform of the Kohlrausch function (see Eq. (1.97)) with β equal to 0.50. The orientation correlation function, $C_1(t) = 3\langle P_1(\cos\theta(t))\rangle$, measured by dielectric relaxation is thus consistent with the Kohlrausch function with constant $\beta = 0.50$ [455]. The orientational correlation function, $C_2(t) = 5\langle P_2(\cos\theta(t))\rangle$, of TNB obtained by photon correlation spectroscopy (see Eq. (1.108)) as the field correlation function of the scattered light, $g^{(1)}(\vec{q}, t)$, was reported to have $\beta = 0.55 \pm 0.01$ for T/T_g ranging from 1.02 to 1.12 [820]. The orientation correlation functions obtained by NMR measurements, also $F_{2,r}(t)$, are consistent with a constant $\beta = 0.5$ over the range of T/T_g from 1.01 to 1.09 [821].

These results indicate that the rotational FRAP experiments, which also measure $F_{2,r}(t)$, has a constant β. The recoverable creep compliance data $J_r(t)$ of TNB taken at different temperatures obey time–temperature superposition [743, 805], which is evidence of insensitivity of the mechanical dispersion to change in temperature.

A similar situation holds for *o*-terphenyl and sucrose benzoate. Dielectric relaxation measurements on *o*-terphenyl [456] and sucrose benzoate [457] also found very little or no change in the frequency dispersion in the temperature range where $D\tau_c$ changes significantly. The spectra of *o*-terphenyl [829] correspond to a constant value of $\beta = 0.50 \pm 0.04$ in the range of T/T_g from 1.02 to 1.17 with $T_g = 242.3$ K [237]. The values of β measured directly from photon correlation spectroscopy [205] are roughly consistent with the dielectric relaxation measurements, with a change from 0.55 to 0.62 over a range of T/T_g from 1.01 to 1.12. NMR studies of OTP using stimulated-echo methods found that β varies from 0.30 to 0.64 as the temperature increases from 252 to 264 K with a plateau near $\beta = 0.5$ in the intermediate range. [809]. Another study using deuteron NMR also reported temperature dependence of β but in narrow relaxation time range of $10^{-3} < \tau_c < 1$ s [830]. As stated in the case of OTP by Richert in [456]: "It remains unclear whether the deviations from the present dielectric results are a matter of experimental limitations of the stimulated-echo methods or the result of the different natures of the correlation functions probed by the NMR and the dielectric techniques."

As a reminder of the results demonstrated in Fig. 85, the CM can explain the breakdown of the SE-DSE relations independent of whether the distribution of rotational relaxation times broadens or does not broaden as the temperature is lowered toward T_g.

[3] *Self-diffusion and viscosity of entangled polymers are still decoupled despite the spatial heterogeneities having been averaged out*: It was argued by Urakawa et al. [831] from solid-state NMR measurements that the typical size of a slow region of spatial heterogeneous dynamics in polymeric and non-polymeric glassformers near T_g is 1–4 nm [422]. The average end-to-end distance of the polystyrene (PS) chains with low molecular weight of 2000 g/mol is 3 nm. If the heterogeneity size in polystyrene is comparable to other glass-forming systems measured by NMR, it was concluded that the 2000 g/mol PS molecules experience an average environment, which eliminates decoupling of their translation diffusion from viscosity due to the posited spatial heterogeneities. Actually, the physics of mechanical relaxation and shear viscosity η is more complicated than envisaged by Urakawa et al. as pointed out by Roland et al. [832]. The zero shear viscosity η is related to mechanical retardation time τ and the steady-state shear compliance J_s by $\eta = \tau/J_s$, the analogue of the Maxwell relation, $\eta = G_\infty\tau$, where G_∞ is the short time shear modulus, and τ is the relaxation time. For low molecular weight polymers such as the 2000 g/mol PS, J_s strongly decreases with decreasing temperature, and hence the temperature dependence of η does not directly reflect the behavior of the global chain modes. In fact, $D\eta/T$ increases with decreasing T reflecting a stronger T-dependence of

viscosity than self-diffusion coefficient D. However, for τ deduced from the product ηJ_s, the product $D\tau$ is independent of T [832]. If the Rouse model is a valid description of the chain modes, this last result is no surprise. This is because self-diffusion and shear relaxation, although contributed by different Rouse modes, have the same friction factor like all Rouse modes. The findings that $D\tau$ is independent of T is thus a trivial consequence of chain dynamics of unentangled polymers, and cannot be used by Urakawa et al. to support spatial heterogeneous dynamics as the cause of breakdown of Stokes–Einstein relation.

This misuse led Urakawa et al. further to expect that elimination of decoupling of translation diffusion and viscosity by spatial averaging would be even more complete in high molecular weight entangled PS because of larger size of the macromolecule. This expected effect, stated explicitly by Urakawa et al., is contradicted by experimental data of high molecular weight entangled polyethylene [833, 834] and hydrogenated polybutadiene [835, 836]. The viscosity and self-diffusion coefficient exhibit different temperature dependences. The Arrhenius activation energy for the viscosity is significantly larger than that for self-diffusion [359, 837]. These molecules are much larger in size than 2000 g/mol PS, and so the putative averaging of spatial heterogeneities is complete. Furthermore, the decoupling of η and D in these polymers was observed at high temperatures above T_g, where spatial heterogeneous dynamic of the local segmental relaxation is not an issue. Thus, the explanation based on spatially heterogeneous dynamics is not operative in this regime, and yet decoupling of η and D is still present. This suggests that decoupling of η and D originates for a more fundamental reason than spatially heterogeneous dynamics. On the other hand, this decoupling of η and D of entangled polymers was explained by an approach directly linked to many-body entangled chain relaxation dynamics [359, 837], and explained in the context of the coupling model. The explanation, although is for entangled chain dynamics, is closely related to the explanation [268] of enhancement of the diffusion of small probes in matrices or the self-diffusion of low molecular weight glassformers (e.g., TNB and *o*-terphenyl). More on the decoupling of η and D of entangled polymers is given in Chapter 3.

[4] *Decoupling between translational diffusion and rotational diffusion (or viscous flow) is a special case of a more general phenomenon*: The SE and DSE relations are noteworthy because they are important and historical steps in the development of modern physics. Therefore, the discovery of breakdown of these relations in glassformers has attracted special attention among researchers on the problem of glass transition. As discussed in Section 1.3, self and probe translation diffusion coefficients, D and D_t, are given via the long time limit of either $< |\vec{r}(t) - \vec{r}(0)|^2 >$ or $F_s(\vec{k}, t)$, the incoherent intermediate scattering function (see Eqs. (1.126) and (1.124)), or from the integral of the velocity–velocity correlation function $\langle v_x(0) \cdot v_x(t) \rangle$ over time (see Eq. (1.129)). The rotational diffusion counterparts of these functions are $C_1(t)$ for dielectric relaxation measurements with relaxation time τ_{diel}, and $C_2(t)$ for rotational FRAP, light scattering, and NMR measurements with rotational correlation time τ_c.

The correlation function for viscous deformation is $C_{\alpha\beta}(t) = \langle\sigma_{\alpha\beta}(t)\sigma_{\alpha\beta}(0)\rangle$, where $\sigma_{\alpha\beta}$ is the off-diagonal component of the stress tensor, and the viscosity η is given by integrating $C_{\alpha\beta}(t)$ over time (see Eq. (1.100)). In the ensemble average performed to obtain any non-exponential time-dependent correlation function such as $C_1(t)$, $C_2(t)$, and $C_{\alpha\beta}(t)$ from measurements, the details of heterogeneous dynamics are lost. But this does not mean that homogeneous dynamics is implied when only the non-exponentiality of the correlation functions is considered. This is because many-body relaxation is the source of both heterogeneous dynamics and non-exponentiality.

Seen as evidence of breakdown of SE and DSE relations, the difference between the temperature dependences of D^{-1} and η or $(D_t)^{-1}$ and τ_c originates from the difference between $F_s(\vec{k},t)$ and $C_{\alpha\beta}(t)$ or $F_s(\vec{k},t)$ and $C_2(t)$, respectively. Similarly, any observation of difference in temperature dependence between τ_{diel}, τ_c, and η also originates from the corresponding difference between $C_1(t)$, $C_2(t)$, and $C_{\alpha\beta}(t)$, and hence the observation should be viewed as equally important as the breakdown of the SE and DSE relations.

In going back to examine and compare $F_s(\vec{k},t)$, $C_1(t)$, $C_2(t)$, and $C_{\alpha\beta}(t)$, not only the differences in their relaxation times or the transport coefficients are recaptured but also the possible differences in their time dependences or dispersions. The latter provide at least additional information, which may be critical for the construction of a general and valid explanation. This is because time dependence is the direct consequence of many-body relaxation, and the differences in the time dependences of $F_s(\vec{k},t)$, $C_1(t)$, $C_2(t)$, and $C_{\alpha\beta}(t)$ reflect the dissimilar effects that many-body relaxation has on the various observables. If the Kohlrausch functions, $\exp[-(t/\tau_\mu)^{1-n_{\mu t}}]$, are valid representations of the time dependences for all these correlation functions, the profitable steps are (1) to see any difference between the exponents n_t, n_1, n_2, and n_η for $F_s(\vec{k},t)$, $C_1(t)$, $C_2(t)$, and $C_{\alpha\beta}(t)$, respectively, and (2) to examine possible correlation between n_t, n_1, n_2, and n_η and the temperature dependences of the corresponding relaxation times, τ_t, τ_1, τ_2, and τ_η. One may wonder why these steps were not taken in the past when considering the breakdown of SE and DSE relations in the models cited above. The reason is because at that time only D_t was obtained by the holographic FRAP technique [796] from the observed $F_s(\vec{k},t)$ that has exponential time dependence of Fickian diffusion. In this technique the characteristic length scale for diffusion is the holographic grating period, a distance equal to few μm, although non-Fickian diffusion would be expected for spatially heterogeneous dynamics with a broad enough distribution of relaxation times [769, 815]. Indeed, non-Fickian diffusion was later found in the translation of tetracene in polystyrene [769] and in BPA-polycarbonate [815] by holographic FRAP. The decay of $F_s(\vec{k},t)$ was non-exponential and in the manner of the Kohlrausch function, $\exp[-(t/\tau_t)^{1-n_t}]$. For the translation of tetracene in BPA-polycarbonate across a grating with a spacing $d = 1.5\,\mu\text{m}$, values of n_t equal to 0.84, 0.81, and 0.77 were reported at 129, 126, and 124°C, respectively. Due to the large distance of the spacing compared with the size of the tracer molecule, these results may not be the same as the microscopic

$F_s(\vec{k}, t)$. Experimental technique has not yet been invented or reported to measure $\exp[-(t/\tau_t)^{1-n_t}]$ directly for distances of the order of the size of the tracer molecule, and to determine the value of n_t experimentally. So far, this has only been achieved by molecular dynamics simulations, which will be discussed in detail later.

(a) *From experimental data of polymers*: As early as in 1988, attention was paid to the large differences between the temperature dependences of the dielectric relaxation time τ_1 of $C_1(t)$, the correlation time τ_2 of $C_2(t)$ from depolarized light scattering, and mechanical relaxation time τ_{mech} of $C_{\alpha\beta}(t)$ from shear compliance measurements, all for the local segmental relaxation in the same amorphous polymer [267]. There, comparisons of the exponents $n_{1,r}$, $n_{2,r}$, and n_η for $C_1(t)$, $C_2(t)$, and $C_{\alpha\beta}(t)$, respectively, were made for each of several polymeric glassformers, and the corresponding differences in the temperature dependences of the relaxation times, τ_1, τ_2, and τ_{mech}, explained quantitatively by the CM equation (2.55). Such differences in non-exponentiality of other physical quantities in non-polymeric glassformers near the glass transition region have been pointed out also by Moynihan and coworkers [266]. The comparison made was indeed profitable because the following correlation was found. Larger n_μ, or smaller Kohlrausch exponent $(1 - n_\mu)$ is accompanied by a stronger temperature dependence of τ_μ, a general consequence of the CM. This correlation signals that the decouplings of the different observables originate from many-body relaxation. The relaxation time τ_μ of a relaxing observable having longer time and stronger temperature dependence is because the process involves a higher degree of many-body dynamics and is stretched out to longer times (i.e., larger n_μ in the exponent of the Kohlrausch function).

(b) *From a molecular dynamics simulation of polyethylene*: Molecular dynamics simulation of local segmental relaxation in polyethylene have obtained the correlation functions of the time correlation functions, $M_1(t) =< \mathbf{u}(0) \cdot \mathbf{u}(t) >$ and $M_2(t) = (1/2) < 3[\mathbf{u}(0) \cdot \mathbf{u}(t)]^2 - 1 >$, where $\mathbf{u}$ is a unit vector embedded in a certain direction [79, 838]. Important properties out of the simulations were extracted and discussed in [272]. One can recognize that $M_2(t)$ is the analogue of the correlation functions measured by NMR, time-resolved optical spectroscopy, and dynamic light scattering. On the other hand, $M_1(t)$ is the dielectric relaxation correlation function. Both $M_i(t)$, $i = 1, 2$, were well fitted by $M_i(t) = \exp[-(t/\tau_i)^{1-n_i}]$. For $\mathbf{u}$ along the chain axis, τ_1/τ_2 is much larger than unity. For example, in the system consisting of chains of length 500 at 300 K, $(\tau_1/\tau_2) \approx 10$, $n_1 = 0.60$, and $n_2 = 0.50$ [79, 838]. The ratio τ_1/τ_2 for vector along the chain axis can reach values over 100 at lower temperatures. Such extremely large value of τ_1/τ_2 cannot be reconciled with either the Debye rotational diffusion model which has $\tau_1/\tau_2 = 3$ [839] or large angle jump motion which has $\tau_1/\tau_2 = 1$ [840]. Moreover, the temperature dependences of τ_1 and τ_2 are different with τ_1 having a stronger dependence, and consequently the ratio τ_1/τ_2 increases with decreasing temperature.

The aforementioned deviations of the behavior of τ_1 and τ_2 from the Debye model are correlated with the difference between n_1 and n_2, and in particular that $n_1 > n_2$. There is an intuitive way to understand why n_1 is larger than n_2 for **u** along the chain axis in terms of many-body relaxation due to intermolecular interactions. $M_1(t)$ is related to $\cos\theta(t)$ while $M_2(t)$ is related to $\cos^2\theta(t)$. From the monotonic decrease of $\cos\theta(t)$ with increasing $\theta(t)$, it is clear that $M_1(t)$ places more importance on the contributions from larger polar angles $\theta(t)$ than $M_2(t)$. Larger $\theta(t)$ corresponds to closer encounters between the repeat units on neighboring chains and hence stronger intermolecular interactions, more intense many-body relaxation, and larger n_1.

The simulation found that the local segmental relaxation depends strongly on the direction of the vector **u**. The correlation functions $M_1(t) =< \mathbf{c}(0)\cdot\mathbf{c}(t) >$ and $M_1(t) =< \mathbf{b}(0)\cdot\mathbf{b}(t) >$, obtained, respectively, for vectors **c** along the chain axis and the out-of-plane vector **b**, were fitted to the Kohlrausch functions with relaxation times $\tau_{1\mathbf{c}}$ and $\tau_{1\mathbf{b}}$, and exponents $n_{1\mathbf{c}}$ and $n_{2\mathbf{b}}$. It was found that $\tau_{1\mathbf{c}}$ is the longest and at some temperature $\tau_{1\mathbf{c}}$ is longer than $\tau_{1\mathbf{b}}$ by about three orders of magnitude. This finding that $\tau_{1\mathbf{c}}$ is much longer than $\tau_{1\mathbf{b}}$ is correlated with $n_{1\mathbf{c}}$ being larger than $n_{2\mathbf{b}}$. This can be seen from the smaller stretch exponent $(1-n_{1\mathbf{c}})$ than $(1-n_{2\mathbf{b}})$, as can be seen by inspection of Fig. 5 of [79]. The observed strong dependence of the local segmental relaxation of polyethylene on the direction of the vector **u** out from the same correlation function $M_1(t)$ obviously cannot be explained by the spatial heterogeneous dynamics model.

The mean-squared displacement $< r^2(t) >$ has a time dependence of $t^{0.65}$ [273], which in the Gaussian approximation means the intermediate scattering function has the time dependence of $F(t) = \exp[-(t/\tau_t)^{1-n_t}] = \exp[-(t/\tau_t)^{0.65}]$. Thus, $n_t = 0.35$ which is smaller than $n_1 = 0.60$ and $n_2 = 0.50$. The result of this molecular dynamics simulation verifies that n_t for translational diffusion is in fact the smallest among other coupling parameters.

All the anomalies found by the simulations have been explained by the CM in [272].

(c) *From the comparison of dielectric relaxation with viscosity*: Dielectric relaxation and shear viscosity data have been compared [819] for several supercooled liquids including OTP for which translational diffusion was found to decouple from viscous flow and from rotational diffusion. Like the behavior of $D\eta/T$ showing breakdown of the SE relation, the quantity $\eta/T\,\tau_{\mathrm{diel}}$ increases on decreasing temperature to approach T_g. Here τ_{diel} is the dielectric relaxation time of the glassformer. Also found is that τ_{diel} is shorter than rotational correlation time τ_c obtained from ^{2}H-NMR in deuterated OTP by about one order of magnitude. This difference between τ_{diel} and τ_c has been seen also in supercooled toluene [841]. In some other liquids like salol, the increase of $\eta/T\tau_{\text{diel}}$ with decreasing temperature down to T_g is even larger, reaching to values of about one and a half decades [819]. Thus, not only translational diffusion but also dielectric relaxation is enhanced compared with rotational diffusion and viscosity, and this observation cannot be explained by the spatial heterogeneous dynamics model. The same result was found from D, η, and τ_{diel} measurements in the system of 200 ppm rubrene in sucrose benzoate by Rajian et al. [457]. The findings of Chang and Sillescu [819] show that decoupling of different dynamic variables is a general phenomenon not restricted to just translational diffusion and rotational diffusion or viscosity. The latter attract more attention because it is the breakdown of the better known laws of Stokes–Einstein and Debye–Stokes–Einstein.

The generality of the phenomenon is no surprise if the reader recalls the discussion in Section 2.2.1 (v) that different dynamic variables such as D, η, and τ_{diel} are related to different correlation functions, and have different stretch exponents of the Kohlrausch function or different coupling parameters in the coupling model (CM).

(d) *From the comparison of dynamic light scattering with viscosity and dielectric relaxation*: The combined studies of diglycidylether of bisphenol A (DGEBA) by dynamic light scattering (DLS), mechanical relaxation [458], and dielectric spectroscopy (DS) [842] gave the opportunity to compare these different dynamic processes and examine the validity of the DSE relation. The mechanical measurements gave the viscosity and the average shear relaxation times $< \tau_\eta >$. The DSE relation was verified for dielectric relaxation over the entire temperature range, i.e., $< \tau_{\text{diel}} > \approx < \tau_\eta >$ and $\tau_{\text{diel}} \propto \eta$. It was found that $< \tau_{\text{DLS}} >$ agrees with $< \tau_{\text{diel}} >$ in a limited temperature range above T_g but beyond that at higher temperatures they differ by an amount that increases with increasing temperature. The fractional power relations, $\tau_{\text{DLS}} \propto \eta^{0.89}$ and $\tau_{\text{DLS}} \propto (\tau_{\text{diel}})^{0.89}$, hold over seven decades (see Fig. 87).

This decoupling occurs at *high* temperatures in contrast to the breakdown of the SE-DSE relations of other glass-forming liquids, which occurs at *low* temperatures near T_g due to an enhancement of translational diffusion over rotational diffusion or shear stress relaxation rate. Nevertheless, the similarity between this decoupling and the fractional SE and DSE relations (see Eq. (2.49)) suggests that enhanced translation is a special case of a more general phenomenon. DLS (by depolarized Rayleigh scattering), dielectric relaxation, and shear stress relaxation all probe the rotational dynamics of the molecules, which cannot be explained by theories that explain enhancement of translational diffusion from spatial heterogeneity of the α-relaxation. Although $\tau_{\text{DLS}} \propto \eta^{0.89}$ and $\tau_{\text{DLS}} \propto (\tau_{\text{diel}})^{0.89}$ show that τ_{DLS} has a weaker temperature dependence than η and τ_{diel}, like D_t compared with η and τ_c, $< \tau_{\text{DLS}} >$ is actually *longer* than $< \tau_{\text{diel}} >$ and $< \tau_\eta >$. Here there is a *reduction* of DLS relaxation rate compared with viscous flow rate, which is to be contrasted with an *enhancement* of

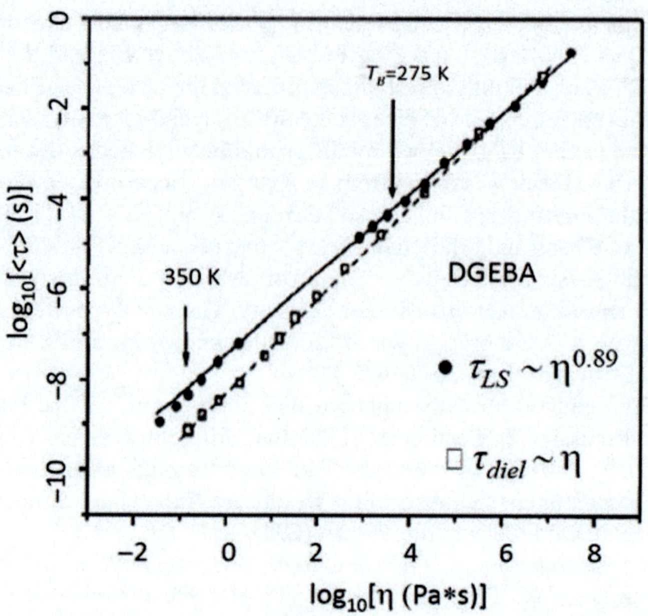

Fig. 87 Plot of relaxation times of dynamic light scattering (*closed circles*) and dielectric relaxation (*open squares*) against viscosity of DGEBA. Data from [842] and [843]

translational diffusion. This seems to contradict the spatial heterogeneous dynamics models, which lead to only enhancement of local relaxation over flow or viscosity. The authors of [842] also concluded that theories proposed to explain enhanced translation over rotational diffusion based on fluidized and correlated domains cannot explain their data. They argued that application of these models would lead to an untenable situation that different parts of the same molecule belong to different domains, i.e., epoxy rings probed by dielectric relaxation in correlated domain and aromatic rings probed by DLS in fluidized domain. The comparison of the true exponents n_{DLS} and n_{diel} for DLS and dielectric relaxation cannot be carried out by using the Havriliak–Negami parameters, $(1-\alpha_1)$ and β_1 that describe the frequency dependence of isothermal dielectric data with that of DLS by depolarized light scattering in the high-temperature range where $\tau_{\mathrm{DLS}} \propto (\tau_{\mathrm{diel}})^{0.89}$ was observed. This is because the secondary relaxation of DGEBA appears prominently in the dielectric spectrum and broadens the dispersion, due to the large dipole moment of the epoxy group (see Fig. 4 of [842]). The actual width of the α-relaxation dispersion is narrower than indicated by the product of $(1-\alpha_1)$ and β_1. On the other hand, secondary relaxation makes no contribution visible in the susceptibility spectrum from DLS.

A quantitative explanation of the decoupling of DLS from viscosity and dielectric relaxation has been given by the CM [270].

(e) *From molecular dynamics simulation of an equimolar mixture of Gay-Berne ellipsoids of revolution and Lennard-Jones spheres*: Chakrabarti and Bagchi (CB) [274], performed molecular dynamics simulation of an equimolar mixture of interacting Gay-Berne ellipsoids of revolution and Lennard-Jones spheres along an isochore at a series of temperatures down to the deeply supercooled state. This work is a natural extension of the simulations of the binary mixtures of Lennard-Jones spheres [265], and allow for the study of rotational relaxation. For this system with orientational degrees of freedom, CB obtains for the ellipsoids of revolution the self intermediate scattering function $F_s(k_{max}, t)$, and the first and second rank single particle orientational time correlation functions, $C_1(t)$ and $C_2(t)$. The latter two are given by $C_l(t) = \langle \sum_i P_l(\vec{e}_i(t) \cdot \vec{e}_i(0)) \rangle / \langle \sum_i P_l(\vec{e}_i(0) \cdot \vec{e}_i(0)) \rangle$. Here $\vec{e}_i$ is the unit vector along the long axis of the ellipsoid of revolution i, P_l is the lth order Legendre polynomial, and the angular brackets stand for ensemble averaging. All three functions have time dependences well described by the Kohlrausch form at all temperatures. The Kohlrausch correlation time and exponent are designated by $\tau_D(T)$ and $\beta_D(T) \equiv (1 - n_D(T))$, respectively, for $F_s(k_{max}, t)$; $\tau_1(T)$ and $\beta_1(T) \equiv (1 - n_1(T))$, respectively, for $C_1(t)$; and $\tau_2(T)$ and $\beta_2(T) \equiv (1 - n_2(T))$ respectively, for $C_2(t)$. The translational diffusion coefficient D_t was obtained from $F_s(k_{max}, t)$ in the usual way. At high temperatures ($T > 1.0$ in LJ units), the ratio $\tau_1(T)/\tau_2(T)$ has a value close to 3 as predicted by the Debye model, and $F_s(k_{max}, t)$, $C_1(t)$, and $C_2(t)$ have exponential or nearly exponential time dependence. On decreasing T starting at $T \approx 1.0$, $\tau_1(T)/\tau_2(T)$ decreases monotonically until it reaches a value about unity at $T \approx 0.5$, meanwhile the time dependences of all three functions deviate increasingly from exponential as temperature falls. Hence, the departure from the Debye model is correlated with non-exponentiality. But this cannot be explained by the spatial heterogeneous model because both $C_1(t)$ and $C_2(t)$ are local process, for the same reason as in the cases discussed above in (*b*) for polyethylene simulation and (*d*) for DGEBA experiments. However, this is consistent with the emergence and progressive increase of many-body relaxation with falling temperature, as can be seen from the temperature dependences of $n_D(T)$, $n_1(T)$, and $n_2(T)$ in Fig. 88. It can be seen from the figure that all of them are very close to zero at high temperatures and increase as temperature falls.

At high temperatures, $0.7 < T \leq 2.0$, $n_D(T)$ and $n_2(T)$ are both smaller and nearly the same, and the product $D_t(T)\tau_2(T)$ is nearly independent of temperature as predicted by the DSE relation. However, as temperature is falling below $T = 0.7$, $n_D(T)$ and $n_2(T)$ increase rapidly and $n_D(T)$ becomes increasingly less than $n_2(T)$ (see Fig. 88 left panel). Simultaneously, the product $D_t(T)\tau_2(T)$ increases rapidly from its nearly constant value at high temperature by a factor of about 2

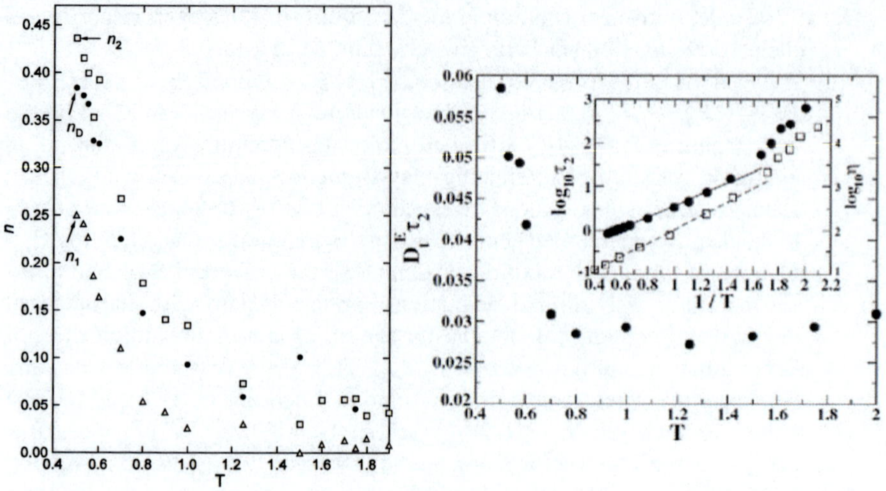

Fig. 88 *Left panel*: The values of $n_{\mathrm{D}}(T)$, $n_1(T)$, and $n_2(T)$ obtained by Chakrabarti and Bagchi [274] replotted. Reproduced from [275] by permission. *Right panel*: The product of the translational diffusion coefficient D_{t}^{E} and the second-rank rotational correlation time τ_2 for the ellipsoids of revolution as a function of temperature. The *inset* shows the inverse temperature dependence of the logarithm of τ_2 (*filled circles*). On a different scale (appearing on the *right* of the inset) shown is the inverse temperature dependence of the logarithm of the shear viscosity (*squares*). The *solid and dashed lines* are the respective Arrhenius fits to data over a restricted temperature range. Reproduced from [274] by permission

from $T = 0.7$ down to $T = 0.5$ (see Fig. 88 right panel), i.e., $[D_t\tau_2(T = 0.5)/(D_t\tau_2)_{\mathrm{high}T}] \approx 2$. Thus, enhancement of translation over rotation described by $C_2(t)$ is observed as found by FRAP and NMR experiments for probes in glassformers (compare Fig. 88 with Fig. 85). The advantage of the simulation over experiments is availability of $n_{\mathrm{D}}(T)$, which enables the conclusion that the departure from the DSE relation is correlated with $n_2(T)$ being larger than $n_{\mathrm{D}}(T)$, and support the alternative explanation of the breakdown of the DSE relation originating from $n_{\mathrm{D}}(T)$ becoming smaller than $n_2(T)$ [268]. CB also obtained the viscosity η from their simulations, and they found the fractional power law dependence $D_{\mathrm{t}} \propto \eta^{-\xi}$, with $\xi = 0.75$ for the ellipsoids of revolution. The results of CB restated so far in this paragraph are consistent with the spatial heterogeneous model. However, the observed much weaker temperature dependence behavior of $D_{\mathrm{t}}(T)/\tau_2(T)$ contradicts the explanation for the decoupling between rotational and translational diffusion in terms of dynamical heterogeneity. Since the ratio $\tau_1(T)/\tau_2(T)$ decreases by a factor of 3 when temperature falls down to $T = 0.5$ while $D_{\mathrm{t}}(T)\tau_2(T)$ increases by only a factor of 2 (see Fig. 88), one may even be led to conclude that $[D_{\mathrm{t}}\tau_1(T = 0.5)/(D_{\mathrm{t}}\tau_1)_{\mathrm{highT}}] \approx 0.7$. Either one can say that $D_t(T)\tau_1(T)$ exhibits no breakdown of the DSE relation or there is a breakdown of the relation but in the opposite direction of $D_t(T)\tau_2(T)$. It is worthwhile to point out that $n_1(T)$, $n_2(T)$, and $n_D(T)$ all have different values at lower temperatures, and $n_1(T)$ is less than $n_D(T)$ (see Fig. 88) opposite to the relation between $n_2(T)$ and $n_D(T)$. Therefore, $[D_t\tau_1(T = 0.5)/(D_t\tau_1)_{\mathrm{highT}}] \approx 0.7$ as found is also consistent with the CM because $n_1(T) < n_D(T)$.

CB found the ratio $\tau_1(T)/\tau_2(T)$ has a value close to 3 at high temperatures when both $n_1(T)$ and $n_2(T)$ are small as expected for the ratio of the first to second rank rotational correlation times of the Debye model involving small steps in orientational motion. The ratio starts to decrease monotonically at $T \approx 1.0$ until it reaches a value nearly unity at low temperatures of $T \approx 0.5$. Values of $\tau_1(T)/\tau_2(T)$ larger than 3 was observed in molecular dynamics simulations of a model polymer [79] and discussed in Section (b) before. The deviation from the normal value of 3 is in opposite direction to that found by CB in their system, but interestingly in the polymer case, $n_1(T)$ is larger than $n_2(T)$. For example, in one polymer chain of length 500 at 300 K, $\beta_1 \equiv (1 - n_1) = 0.399$ and $\tau_1 = 94.7$ ps, while $\beta_2 \equiv (1 - n_2) = 0.498$ and $\tau_2 = 9.79$ ps. The polymer result that τ_1/τ_2 is more than three times larger than 3 was explained by the CM by the fact that n_1 is larger than $n_2(T)$ [272]. The CM can also explain the result when τ_1/τ_2 is less than 3 in the system simulated by CB because here n_1 is smaller than n_2 at lower temperatures.

(f) *From other systems*: Before closing discussion on this subject, it is worth mentioning that this general phenomenon is found not only for structural relaxation in molecular glass-forming systems, but also for different correlations functions of other systems. Here we cite two examples, which have been discussed in Section 2.2.1 for a different purpose. One is the motion of ions in crystalline and glassy ionic conductors. Different experimental measurements using different techniques (e.g., NMR spin-lattice relaxation vs. conductivity relaxation) have obtained different relaxation times, activation energies [279–290] as well as different stretch exponents in the Kohlrausch correlation functions. For spin-lattice relaxation, its Kohlrausch correlation function is more non-exponential (larger n_{SLR} than n_σ), its relaxation time τ_{SLR} is longer and has higher activation energy E_{SLR} than the corresponding quantities τ_σ and E_σ of conductivity relaxation. The difference between τ_{SLR} and τ_σ as well as that between their activation energies E_{SLR} and E_σ have been explained quantitatively by the CM equation [291–294].

The other is the study by Bonn and Kegel [278] of the frequency dependence of the diffusion coefficient and viscosity of concentrated hard-sphere colloidal suspensions. Expressing all quantities in reduced units, their frequency-dependent Stokes–Einstein relation was given by $\eta(\omega) = 1/D(\omega)$. In the low-frequency or long-time regime, significant deviations from this relation were observed. The values of the inverse viscosity are significantly smaller than the values of the diffusion coefficients, and the SE relation is violated. Moreover, on comparing the frequency dependence of $1/\eta(\omega)$ with that of $D(\omega)$ in Fig. 2 of the paper by Bonn and Kegel [278], it is clear that the frequency dispersion of $1/\eta(\omega)$ is broader than that of $D(\omega)$. Thus, for the colloidal hard spheres we have experimental evidence to support that n_D is smaller than n_η, like in the MD simulation discussed above in (e). Although the metallic glass, $Zr_{46.7}Ti_{8.3}Cu_{7.5}Ni_{10}Be_{27.5}$, is different, its viscosity and the diffusion coefficient of beryllium are analogues of η and D of colloidal suspensions, and similar breakdown of the SE relation was observed in the supercooled liquid state [812a]. The explanation of the breakdown of SE relation in colloidal suspensions, by n_D being smaller than n_η, may apply to the case of metallic glass.

(g) *Inference from radiotracer diffusivities in a $Pd_{43}Cu_{27}Ni_{10}P_{20}$ melt [812(b)]*: Radiotracer diffusivities of all components in a $Pd_{43}Cu_{27}Ni_{10}P_{20}$ melt were measured over extended temperature range from near $T_g = 582$ K and up to nearly 1000 K for some components [812(b)]. Two important observations were made by the authors, Bartsch et al. First, for Pd, the Stokes–Einstein relation holds in the whole range investigated encompassing more than 14 orders of magnitude, while the SE relation breaks down for the smaller components. Second, a large decoupling of more than 4 orders of magnitude is observed between the diffusivity of the slowest Pd and of the smaller components (Cu, Ni, P, and added tracer amount of Cr), at the glass transition temperature T_g, but the decouplings decrease on increasing temperature and vanishes at T_c or $T_B \approx 710$ K. In commenting on the breakdown of SE relation and explanation of it in the literature, Bartsch et al. mentioned only that based on dynamical heterogeneities. In their own words, the explanation is "just a consequence of the effect of the broadening of the distribution of mobilities and the different ways transport and relaxation sample that distribution...." Apparently they are not aware of the contradiction of this dynamical heterogeneities' explanation with invariance of the α-dispersion in the entire temperature range from T_B down to T_g in TNB, OTP, and sucrose benzoate from shear mechanical [743, 744], light scattering [820], dielectric relaxation [455, 457], and NMR [821] measurements.

Interestingly, their observation of no breakdown of SE relation over at least 14 orders of magnitude for the slow Pd component offers another opportunity to show that the explanation of the breakdown of SE relation cannot be based on dynamical heterogeneities. The slowest Pd component is the determining factor for viscosity, and its motion has to be still heterogeneous. Homogeneous motion of Pd atoms in response to mechanical shear would be inconceivable although one cannot prove that it is impossible without direct experimental evidence. Bartsch et al. also stated that "... viscous flow requires rearrangement of the Pd subsystem," and rearrangement cannot be homogeneous relaxation. It is also difficult to reconcile homogeneous relaxation with the stretched exponential time dependence of the α-relaxation observed by neutron scattering in the same system [812(c)] and in mechanical relaxation [957]. If indeed the dynamics of Pd is heterogeneous, from the belief of the connection between the breakdown of the SE relation and dynamic heterogeneities [812(d), 449–454] it follows that breakdown of the SE relation for the Pd component has to occur. However, Bartsch et al. reported that it does not occur for the Pd component over 14 orders of magnitude. I was told that this report had led some colleague, who firmly and correctly believes dynamic heterogeneity is universal and hence necessarily causes the breakdown of the SE relation even for Pd, to express his doubts on the experimental data of Pd from Bartsch et al. There should be no doubt on the verity of the experimental data obtained meticulously by Bartsch et al. who are experts of using radiotracer technique to measure diffusivity. Therefore, observation of Pd obeying the SE law by Bartsch et al. contradicts the necessary consequence of the heterogeneous dynamics explanation of the breakdown of the SE relation. This contradiction can be considered as yet another evidence showing that the heterogeneous dynamics explanation of the breakdown of the SE relation is untenable in general.

In the $Pd_{43}Cu_{27}Ni_{10}P_{20}$ multicomponent system, the predominant metallic bonds have no fixed directions and hence no permanent relations among the components. Thus, the diffusion of each component has its own coupling parameter for the correlation function represented by Eq. (2.2.55) with μ now used to label the components. The coupling parameters of the smaller components are smaller than the largest Pd majority component naturally because of lesser inter-atomic constraints. From the CM relation it follows that the diffusion coefficients of the smaller components have weaker T-dependence than that of Pd. Hence this difference between the T-dependence of the diffusivity of Pd and the diffusivities of smaller components explains their decoupling which starts near T_c or $T_B \approx 710$ K and reaches more than 4 orders of magnitude at the glass transition temperature $T_g = 582$ K. After having explained the observed large decoupling in diffusivity of all the smaller components from Pd, it is easy to rationalize by the CM why the SE relation holds for Pd. From the much higher mobility of the smaller components than Pd inferred from their diffusivities, it is clear that viscosity of $Pd_{43}Cu_{27}Ni_{10}P_{20}$ is principally determined by the least mobile Pd. Since atomic Pd is in effect a point particle, we do not expect any significant difference in the time dependence as well as the correlation time of its correlation functions for viscosity or diffusion. Thus, there is no difference in the coupling parameters of viscosity and diffusion of Pd, and hence no breakdown of the SE relation from the CM explanation. In contrast, OTP, TNB, and sucrose benzoate molecules have chemical structures extending into space. The correlation function of the center of mass diffusion will differ from the correlation functions for rotation and viscosity, and hence breakdown of the SE relation in these molecular glassformers.

It is worthwhile to compare this CM explanation of the decoupling of diffusivities of the smaller components from Pd with that offered by Bartsch et al. They proposed that Pd atoms form a slow subsystem in the supercooled melt inside which the smaller elements carry out fast diffusion. This proposal is not much more than a paraphrase of the observed decoupling. This restatement of the fact, however, comes short of addressing the questions of why the diffusivity of Pd has a stronger T-dependence than the smaller components and why the slow Pd subsystem obeys the SE relation.

(h) *Concluding remarks*: The cases (a)–(g) given above have amply shown in general that different correlation functions of structural relaxation in glass-forming systems have different relaxation times as well as their dependences on temperature. Therefore the breakdown of the SE and the DSE relations are just special cases of this general phenomenon, when the consideration is limited to translational diffusion, rotational diffusion, and shear viscosity. The spatial heterogeneous dynamics model is exclusively constructed to explain the breakdown of SE and DSE relations, but even for this purpose the model is contradicted by some experimental data and simulations discussed above. The other examples of this general phenomenon not involving translation diffusion absolutely cannot be explained by the spatial heterogeneous dynamics model. To me, it is irrational to ignore that this is a general phenomenon requiring an all-purpose explanation, and instead holding on the spatial heterogeneous dynamics explanation for the breakdown of SE and DSE relations.

2.2.5.7 Changes Effected by Mixing with Another Glassformer

The properties of the structural α-relaxation of a glassformer A are modified in a miscible binary mixture by the presence of another glassformer B with large difference in mobility gauged by the difference in their T_gs (for a review and references to the literature, see [736]). On a fundamental level, the modifications are caused by several factors including the different intermolecular potential and chemical structures, possible reinforcement or mitigation of steric constraints, and change in occupied volume and entropy introduced by the partial replacement of A by B in the mixture. The frequency dispersion of susceptibility or time correlation function of the α-relaxation of component A in the mixture is modified by these factors. In the following, we discuss experimental data that show evidence that the change in dynamics of a component in mixtures is governed at least in part by the change of intermolecular coupling due to the presence of the other component. In those cases where broadening of dispersion by concentration fluctuations is unimportant, the component dynamics is governed by or correlated with the dispersion of the α-relaxation or n of the same component in the mixture.

Component A Is More Mobile Than Component B

If molecules B are less mobile than molecules A in the mixture, the α-relaxation of component A would be slowed down and its correlation function stretched to longer times by component B further than the α-relaxation of pure glassformer A. This heuristic description suggests increasing stretching of the correlation function or broadening of the frequency dispersion of the α-relaxation of component A on increasing the concentration of component B. In the context of the coupling model (CM), this effect is due to increasing coupling parameter n_A by increasing intermolecular interactions/constraints imposed by the slower B molecules, which leads to increased stretching of the Kohlrausch correlation function, $\phi_{\alpha A} = \exp[-(t/\tau_{\alpha A})^{1-n_A}]$, by the decrease of the stretch exponent $(1 - n_A)$ [739]. However, this is not the only source of broadening. Even if the two components are miscible, presence of concentration fluctuations also contributes to broadening. In the CM description, the concentration fluctuations engender a distribution of environments $\{i\}$ and coupling parameters, $\{n_{Ai}\}$. The observed correlation function $\phi_{\alpha A}$ is given by the sum $\phi_{\alpha A}(t) = \sum_i p_i \phi_{Ai}(t)$, where p_i is the normalized probability of environment i for the distribution, $\varphi_{Ai}(t) = \exp[-(t/\tau_{Ai})^{1-n_{Ai}}]$, and

$$\tau_{Ai} = [t_c^{-n_{Ai}} \tau_{0A}]^{1/(1-n_{Ai})}. \tag{2.56}$$

Here the primitive relaxation time τ_{0A} is assumed the same for all i, but the possibility of a distribution of it cannot be excluded. The most probable relaxation time $\hat{\tau}_{\alpha A}$ of A in the mixture can be obtained from Eq. (2.56) using the most probable $\hat{n}_A$ in the distribution $\{n_{Ai}\}$:

$$\hat{\tau}_{\alpha A} = [t_c^{-\hat{n}_A} \tau_{0A}]^{1/(1-\hat{n}_A)}. \tag{2.57}$$

By incorporating these distributions, a CM theory of component dynamics was proposed [739]. For earlier works, see [245, 734–736]. Nevertheless, for the uninitiated public, the broadening by concentration fluctuations is so palatable that it is difficult to ascertain from the observed broadening of dispersion of component A that it is at least partly caused by increased intermolecular coupling or n_{Ai}. There is help from results of studies over a wide composition range down to dilute concentration of A [672, 674, 677], where heterogeneous broadening by concentration fluctuations should diminish at low ϕ_A, the concentration of component A, and vanish at dilute and certainly in the limit of probe concentrations when $\phi_A \to 0$. Dielectric measurements have been made on the α-loss peak of picoline, quinaldine, or *tert*-butylpyridine serving as the low T_g component A at 5 wt% low concentration in mixtures with oligomers of styrene (B) having much higher T_g. The fact that the α-loss peak of the A component in these low concentration mixtures being much broader than that of pure A with the same peak frequency is evidence for the broadening being due to increase of intermolecular coupling or n_A. The steepness or "fragility" index m_A determined from the temperature dependence of $\hat{\tau}_{\alpha A}$ of some (with exception) of these mixtures shows monotonic decrease on decreasing ϕ_A. The correlation between m and n should be reliable since the system is the same except for variation of concentration. Hence, the observed monotonic increase of m_A supports the corresponding increase of n_A.

Another example comes from the study of the much more mobile poly(ethylene oxide) (PEO) in mixtures with poly(methyl methacrylate) (PMMA). The miscible polymer blends PEO/PMMA contain 10, 20, and 30% hydrogenated PEO in deuterated PMMA. Quasielastic neutron scattering in combination with deuterium labeling enabled determination of the characteristic relaxation times of the PEO component A over spatial scales from 3 to 10 Å [312]. The intermediate scattering function $F_s(Q,t)$ of pure PEO and hPEO in the blends can be found in Figs. 4 and 5. On decreasing ϕ_A, the data show parallel increases of $\tau_{\alpha A}$ and n_A of the Kohlrausch function, $\exp[-(t/\tau_{\alpha A})^{1-n_A}]$, employed to fit $F_s(Q,t)$. Concentration fluctuation is expected to decrease on decreasing ϕ_A from 30% down to 10%. Therefore the observed increase of n_A must be due to increase in intermolecular coupling and is the cause of the concomitant increase of $\tau_{\alpha A}$.

When concentration fluctuations are present to contribute broadening of the isothermal dielectric loss peak, the broadening is more emphatic on the low-frequency side, making it asymmetric and skew toward the low-frequency side [245, 845]. It is interesting to point out that the dielectric peak of neat A is not only narrower than in the mixture but also has the opposite skew asymmetry. The one-sided Fourier transform of the Kohlrausch function, the Cole–Davidson and the HN functions, which give a reasonably good fit to the loss peak of most neat glass-forming liquids, are asymmetric but all skew toward high frequencies. Thus, the opposite skew asymmetry can be taken as indicator of concentration fluctuations. The loss peak becomes progressively broader and more skew asymmetric toward low frequencies as concentration of B is increased. This is observed in PVME [845] or *bis*(2-ethylhexyl) phthalate [245] acting as A in mixtures with polystyrene acting as B. These properties can be derived from the CM [245, 734–736, 739] from a normal symmetric distribution of $\{n_{Ai}\}$ with the magnitude of n_{Ai} increasing with concentration of B to reflect increased stretching of the correlation functions $\varphi_{Ai}(t)$ and slowing down of A. The relation between τ_{Ai} and n_{Ai} in Eq. (2.56) is non-linear, and larger shift of τ_{Ai} with change in temperature of τ_{0A} occurs for larger n_{Ai}. These two properties of the CM account for the asymmetric broadening of the component A loss peak skewing toward low frequencies, and the change of its shape of the dispersion with temperature (i.e., breakdown of frequency–temperature superposition of spectra). These features are commonly experimentally observed in mixtures where concentration fluctuations are non-negligible.

The reader may wonder why the dielectric spectra of these mixtures are not shown here. The reason is the existence of yet another experimental fact coming from the correlation between n_A and the ratio, $\tau_{\alpha A}/\tau_{\beta A}$ or $\tau_{\alpha A}/\tau_{0A}$, where $\tau_{\beta A}$ is the relaxation time of the so-called Johari–Goldstein secondary relaxation of A.

Increase of $\tau_{\alpha A}/\tau_{\beta A}$ or $\tau_{\alpha A}/\tau_{0A}$ on decreasing ϕ_A was indeed observed in many mixtures and polymer blends, and thus the correlation gives another support of corresponding monotonic increase of n_A. This empirical correlation between n_A and $\tau_{\alpha A}/\tau_{\beta A}$ is a natural consequence of the CM, and is the subject of detailed discussion in the next section. For this reason, we defer presentation of the data showing changes of the dynamics of the α-relaxation discussed here to Section 2.3 in conjunction with the presentation of the correlation between n_A and $\tau_{\alpha A}/\tau_{\beta A}$ or $\tau_{\alpha A}/\tau_{0A}$.

Probe Dynamics ($\phi_A \to 0$)

At very low concentrations, ϕ_A, the component A is reduced to the role of a probe. Each probe molecule experiences the same environment, which eliminates complications from concentration fluctuations. Although not necessary, it is certainly desirable for a model of component dynamics of mixtures to have valid predictions for this limiting case. Since the crux of other models of component dynamics of mixtures [731, 732, 846, 847] is concentration fluctuations or self concentration [733], they cannot address probe molecule dynamics. The CM description is unique in having incorporated, in addition to concentration fluctuations, the essential physics of intermolecular coupling and its change on mixing, which are applicable at all concentrations. At probe concentration of A (i.e., $\phi_A \to 0$), concentration fluctuation vanishes and self concentration has no meaning, leaving only intermolecular coupling as the sole controlling factor of the dynamics of the probe A molecule. Thus, for probe molecules A dispersed in host B, a *single* coupling parameter n_A characterizes the degree of intermolecular coupling/constraint imposed on the rotation or translation of the probe. The magnitude of n_A may depend on the chemical structures of the probe and the host, but more importantly on the mobility difference between A and B. A practical measure of the latter is the ratio, τ_c/τ_α, where τ_c is the rotation correlation time of probe A in the host B and τ_α is the α-relaxation time of the host B. According to the CM, the larger the ratio of τ_α to τ_c, the larger is the intermolecular coupling and stretching of the correlation function of the probe to longer times, and hence a larger n_A for the probe molecule [268]. In the absence of concentration fluctuations, the predicted correlation function of the probe rotation is just the Kohlrausch function, $\exp[-(t/\tau_c)^{1-n_A}]$. Indeed, this correlation function is found [268] from experimental data of probe rotation experiments [793–796], as mentioned before in Section 2.2.5.6 on the breakdown of DSE relation. The coupling parameter n_A can be determined directly from the measured probe rotational correlation function, and the prediction of its increase with the ratio, τ_α/τ_c, can be falsified or verified. The data collectively presented in Fig. 86 show that n_A increases in the predicted way with increasing τ_α/τ_c. These results from probe dynamics experiments provide the strongest support for an important physics for component dynamics in mixtures and polymer blends incorporated into the CM, which is the role played by intermolecular coupling and its change on varying the composition.

Component A Is Less Mobile Than Component B

The situation here is opposite to that discussed above. In this opposite case, the more mobile B molecules of the mixture mitigate the intermolecular coupling/constraints of molecular motions in the relaxation originating from the component A. Hence the CM expects a decrease of $\{n_{Ai}\}$, and the decrease contributes narrowing of the dispersion, which counters the broadening by concentration fluctuations. If the α-relaxation originating from A in the mixture is observed not broader than in pure A, it may be considered as direct evidence of the narrowing effect by decrease of $\{n_{Ai}\}$ and the average $\hat{n}_A$. An indirect evidence is the decrease of steepness or "fragility" index m_A determined from the T_{gA}-scaled temperature dependence of $\hat{\tau}_{\alpha A}$ on decreasing ϕ_A, from which the decrease of $\hat{n}_A$ follows from the correlation between m_A and $\hat{n}_A$. These effects were seen experimentally in some mixtures to be discussed below. In some of the following examples, the frequency dispersion of the loss peak of A in the mixture does not show broadening that is skew toward the low-frequency side, which can be taken as evidence that concentration fluctuations are not important.

Effect of Low T_g Diluent (B) on Local Segmental Relaxation of Polymer (A)

A good example can be taken from the dielectric study of the local segmental relaxation of poly(vinyl chloride) (PVC) diluted with tetrahydrofuran (THF) by Adachi and Ishida [848]. T_g from differential thermal analysis was found to agree with the dielectric one, defined as the temperature at which the dielectric relaxation time of the α-relaxation of the PVC component reaches 100 s, and it is 349, 275, 195, 166 K, respectively, for the four concentrated solutions with 100, 84, 51, and 38 wt% of PVC studied. The presence of the low-frequency conductivity relaxation in dielectric response hampers the normal procedure of determining the coupling parameter $\hat{n}_A$ by fitting the loss peak by the Fourier transform of the Kohlrausch function. Notwithstanding, Adachi and Ishida resorted to an analysis of the values of $2\varepsilon''(\max)/\Delta\varepsilon$ for the α-relaxation to determine its dispersion for each solution. Here $\varepsilon''(\max)$ denotes the maximum value of the dielectric loss and $\Delta\varepsilon = (\varepsilon_0 - \varepsilon_\infty)$ is the relaxation strength determined from a Cole–Cole plot. As is well known, the ratio $2\varepsilon''(\max)/\Delta\varepsilon$ is unity if the relaxation is a Debye process having $\exp(-t/\tau)$ as the correlation function, and the ratio decreases monotonically with increasing width of the dispersion. Adachi and Ishida found the ratio decreases, and hence the width of the dispersion increases, with increasing PVC concentration. By evaluating $2\varepsilon''(\max)/\Delta\varepsilon$ from the $\varepsilon^*(\nu)$ obtained from one-sided Fourier transforms of the Kohlrausch functions and comparing with the experimental values of Adachi and Ishida, the $\hat{n}_A$ of the four concentrated PVC solutions were deduced [849], and the results show increasing reduction of $\hat{n}_A$ of PVC on increasing concentration of the diluent THF. Data of α-relaxation of PVC at dilute concentrations down to weight fraction of 0.059 shows τ_α has Arrhenius T-dependence with activation energy of about 5 kcal/mol corresponding to the energy barrier of local conformational transition, suggesting that intermolecular coupling is drastically reduced at dilute concentrations. The many-body nature of the α-relaxation is removed, and conformational energy barrier of PVC of about 5 kcal/mol governs the relaxation rate.

0.1 Molar Fraction of Bromoethylbenzene in Ethylbenzene

Pure ethylbenzene (EBz) has T_g 20 degrees lower than bromoethylbenzene (BrEBz). By dielectric relaxation measurements, the loss peak of 0.1 molar fraction of polar BrEBz in mixture with the apolar host EBz is slightly narrower than that of pure BrEBz [850].

50% Poly(methyl methacrylate) Plasticized by tri-m-Cresyl Phosphate

Study by 2H-NMR of the dynamics of poly(methyl methacrylate) (PMMA, $T_g = 381$ K) plasticized with tri-*m*-cresyl phosphate (TCP, $T_g = 205$ K) at a TCP concentration of 50% (w/w) [851] found the dispersion of the segmental α-relaxation of the slow PMMA remains essentially unchanged upon addition of the plasticizer TCP, despite a large decrease in the correlation time. The result indicates that concentration fluctuations are not dominant, otherwise the dispersion of PMMA would be much broader, and skew toward low frequencies.

It is also found that faster TCP molecules reorient isotropically even in a rigid polymer matrix at temperatures well below the glass transition of the plasticized PMMA, and the dynamics of TCP is heterogeneous but transient in nature. The orientational correlation functions of TCP change from Kohlrausch decay with $n_B \approx 0.40$ for pure TCP to extremely broad and quasi-logarithmic decay in the mixture. Since concentration fluctuations are reciprocal for both components and not overwhelming for the PMMA component, this dramatic broadening of the dispersion of TCP cannot be due solely to concentration fluctuations. The increase of intermolecular coupling and $\{n_{Bi}\}$ of TCP by the presence of the much slower PMMA seems to be crucial for the observed effect.

2% Polystyrene in Poly(vinylmethylether)

The segmental α-dynamics of pure backbone-deuterated polystyrenes with varying molecular weights (1.7–67 kg/mol, $331 \leq T_g \leq 376$ K) as well as 2% of it in blends with poly(vinylmethylether) (PVME, $M_n = 48.5$ kg/mol, $T_g = 250$ K) was measured by 2H-NMR spin-lattice relaxation at 15 and 77 MHz [852]. For all molecular weights, the relaxation time of PS becomes shorter by several orders of magnitude when mixed with PVME, but the stretch exponent β of the Kohlrausch correlation function of 2% PS in PVME within errors is essentially the same as that of pure PS, indicating the width of the dispersion is effectively unchanged, and broadening by concentration fluctuations is countered by narrowing due to reduction of intermolecular coupling by the presence of the fast PVME host.

Poly(methyl methacrylate) in Blends with Poly(ethylene oxide)

There is no reliable dispersion data of the α-relaxation of the PMMA component in blends with poly(ethylene oxide) (PEO, $T_g = 214$ K) to compare with pure PMMA. Nevertheless, the ratio, $\tau_{\alpha A}/\tau_{\beta A}$, was observed to decrease on addition of PEO. This will be shown in Section 2.3. Therefore the correlation between n_A and the ratio, $\tau_{\alpha A}/\tau_{\beta A}$ (also to be established in Section 2.3), can be used to deduce that $\{n_{Ai}\}$ of PMMA is decreased by the presence of the more mobile PEO.

Sorbitol in Mixtures with Glycerol

The same situation as in the previous paragraph was found for sorbitol in mixtures with the more mobile glycerol [853], and the same conclusion can be reached by the correlation between n_A and the ratio, $\tau_{\alpha A}/\tau_{\beta A}$ [737]. In many other mixtures including those discussed, the steepness index m_A of the slower component decreases on increasing the concentration of the faster component A, and hence the correlation between n and m can be used to deduce the corresponding decrease of $\{n_{Ai}\}$.

Molecular Dynamics Simulations

Bedrov and Smith [80] performed molecular dynamics simulations of model miscible polymer blends consisting of chemically realistic 1,4-polybutadiene (CR-PBD) as the slow component (higher T_g) and PBD chains with reduced dihedral barriers as the fast component (LB-PBD) with lower T_g. The simulation data also provide information on the slower CR-PBD component in the blend. The α-relaxation time of the slower CR-PBD component becomes shorter, and interestingly its frequency dispersion becomes *narrower* when compared with neat CR-PBD. At low LB-PBD content in the blend, broadening of the α-relaxation by concentration fluctuations is not that important, and hence the narrowing caused by reduction of the coupling parameter of CR-PBD dominates, and it shows up in the simulation data of Bedrov and Smith. Consistency of this feature of the simulation result with prediction of the CM on relaxation on the slower component was pointed out in [854].

Anomalous Component Dynamics in Polymer Blends

If the intention is to solve the fundamental problem of many-body relaxation and its specialization to glass transition, anomalies found are useful guides to the right approach and help to differentiate theories. This is in accord with the opinion offered by Crick [161] and Andersen [160] and discussed before in Section 1.4.2. Several anomalies of component dynamics found in blend and mixtures are highlighted here for this purpose as well as to show different attitude of others who either knowingly or unknowingly ignore these anomalies.

The Unusual Component Dynamics in Poly(ethylene oxide)/Poly(methyl methacrylate) Blends as Probed by Deuterium NMR

One example is $\tau_{seg}(\phi, T)$ or τ_{0A} of poly(ethylene oxide) (PEO, $T_g = 214$ K) in blends with poly(methyl methacrylate) (PMMA, $T_g = 391$ K), x% PEO–(1–x)% PMMA, for $x = 0.5$, 3, 6, 10, 20, 30, and neat PEO by deuteron NMR [855]. It was found and shown in the left panel of Fig. 89 that τ_{seg} of the PEO component

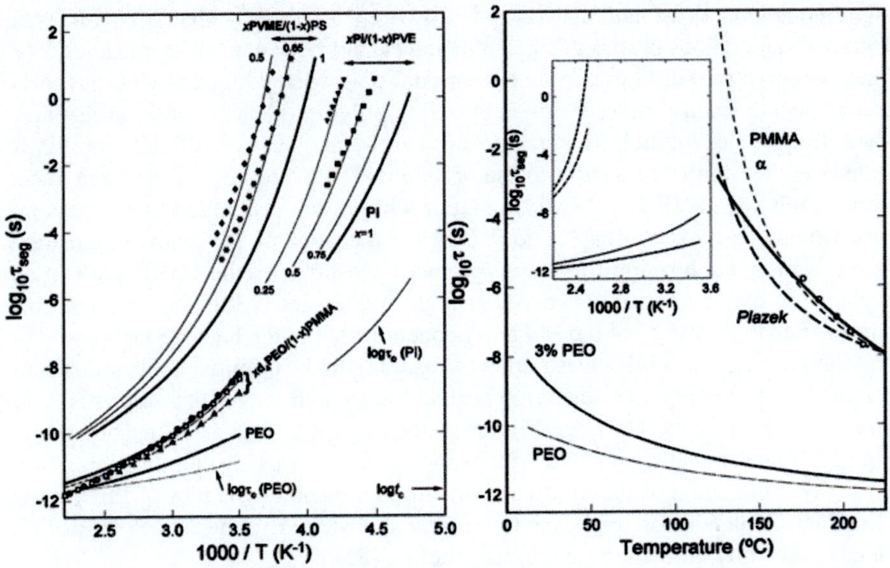

Fig. 89 *Left panel*: The segmental relaxation times for PEO neat (*bold solid line*) and in blends with PMMA (*dashed lines*) containing 3–30% PEO (*from top to bottom*) taken from the VFTH fits of [855]. The most probable relaxation times are also shown, as calculated from the CM equation for the lower (O, $\hat{n} = 0.76$), mid-range (□, $\hat{n} = 0.75$), and higher concentrations of PEO (Δ, $\hat{n} = 0.715$), respectively. Also shown is the independent relaxation time for PEO (*dotted line*, using $n = 0.5$), which lies close to the characteristic time, $t_c = 2$ ps. Similar results are shown for PI neat (*bold solid line*) and in blends with PVE containing 25–75% PI (*labeled thinner lines*), together with the calculated τ_{seg} for 25% (▼) and 50% (■) compositions. Note the independent relaxation time for PI (*dotted line*) is four or more decades longer than t_c. Similar results are included for PVME neat (*bold solid line*) and blended with PS (*dashed lines*), along with the calculated curves for 55% (♦) and 65% (∗) PVME compositions. For PVME, τ_0 (not shown because it overlaps the PEO data) is also much larger than t_c in the temperature range where we calculate τ_{seg}. Reproduced from [738] with permission. *Right panel*: Comparison between the temperature dependences of the segmental relaxation time for (1) the PEO homopolymer (*thin dashed-dotted line*), and (2) 3% of PEO in PMMA (*thick solid line*) with that of (3) the dielectric segmental relaxation time of PMMA (*thin dashed line*), (4) the terminal relaxation time of PMMA from Zawada et al. shifted downward by 9.1 decades (*open circles*), (5) the shift factor of the softening dispersion of PMMA given by shear creep measurements shifted vertically downward by 0.82 decade to coincide with the dielectric segmental relaxation time at the highest temperature (*thick dashed line*), and (6) the monomeric friction factor of the PEO tracers in PMMA matrix from Haley and Lodge (*continuous thick line*). The *inset* shows the same data of (1), (2), (3), and (5) plotted against reciprocal temperature. Reproduced from [876] with permission

is nearly composition-independent over the entire composition range from 0.5 to 30% PEO, and is retarded by *less than one order of magnitude* than τ_{seg} of pure PEO over a wide temperature range extending to well below the glass transition of the PMMA matrix, where the segmental relaxation times of PMMA are about 12 orders of magnitude greater than the fast PEO relaxation times. Similar behavior of τ_{seg} of polyisoprene (PI) in blends with poly(vinylethylene) (PVE) [856] and PEO in blends with poly(vinyl acetate) (PVAc) [857] was found by the same technique. Quasielastic neutron scattering [858] performed on a 50% PI/50% PVE blend at high frequencies also found that τ_{seg} of PI is barely affected by blending with PVE. Thus, this anomalous property is general, and cannot be explained by most models. These observations at high frequencies are unusual when compared with the component segmental dynamics of other miscible polymer blends measured at much lower frequencies by dielectric spectroscopy or by another kind of deuteron NMR technique [738, 859]. Shown in the same figure are data of τ_{seg} of the component PI in blends with PVE and PVME in blends with PS taken at much lower frequencies, which clearly show for PI and PVME large change of τ_{seg} with composition of the blends, conforming to normal behavior. So far only the CM has given an explanation of the anomaly observed at high frequencies [738]. The cause of this unusual behavior of τ_{seg} of the PEO component is due to the high frequencies (31–76 MHz) used in the NMR measurements, resulting in the primitive relaxation times τ_{0A} of the PEO component that are short and not much longer than the crossover time $t_c \approx 2$ ps of the CM (see Fig. 89). Consequently, τ_{seg} or $\hat{\tau}_A$ calculated by Eq. (2.57) does not change much with change of $\hat{n}_A$ on varying composition [738]. This CM explanation of the unusual short-time segmental dynamics of PEO in the PEO/PMMA blends also applies to the similar short-time dynamics of PI in PI/PVE blends probed by quasielastic neutron scattering [858].

More recently, Haley and Lodge [861] measured the tracer diffusion coefficient of 0.3% unentangled PEO ($M = 1000$ g/mol) in a matrix of PMMA ($M = 10^4$ g/mol) over a temperature range from 125 to 220°C by forced Rayleigh scattering. Also measured are the dynamic viscosities of blends of two different high molecular weight PEO tracers ($M = 440,000$ and 900,000 g/mol) in the same PMMA matrix at temperatures ranging from 160 to 220°C. The monomeric friction factors determined by diffusion and rheology were in good agreement. Surprisingly to them, the monomeric friction factors of the PEO tracer diffusion in the PMMA matrix are much greater than that of the segmental dynamics of PEO ($M_w = 1.25 \times 10^5$ g/mol) at concentrations of 0.5 and 3% in a matrix of PMMA ($M_w = 1.06 \times 10^5$ g/mol) measured by Lutz et al. [855] using NMR at high frequencies from 31 to 76 MHz. The difference is about two orders of magnitude at 220°C and increases with decreasing temperature to more than six orders of magnitude at 125°C, as can be seen in the right panel of Fig. 89. Hence, in the PMMA matrix, the PEO tracer terminal relaxation time is a much stronger function of temperature than the corresponding PEO segmental relaxation time obtained by high-frequency NMR. The much stronger temperature dependence of the PEO chain diffusion relaxation time than PEO segmental dynamics is not unique to PEO tracer in PMMA matrix. Lutz et al. [855] also found that in the 20% PEO/80% PMMA blend by comparing the terminal relaxation time of PEO in this blend [861] to its segmental dynamics obtained by high-frequency NMR in a common temperature region. The temperature dependence of terminal relaxation time τ_{terminal} of PEO in the 20% PEO/80% PMMA blend is much stronger than τ_{seg} of segmental dynamics.

The same effect was found in another blend of 2% PEO with PVAc. The segmental τ_{seg} of PEO in this blend with PVAc measured by NMR at high frequencies of 15.6 and 76.7 MHz shows weaker temperature dependence than the terminal τ_{terminal} of PEO in the same blend [857].

Conventional theories of polymer viscoelasticity have the same monomeric friction factors and temperature dependence for segmental and global dynamics [29]. The same holds for polymer blend theories that follow this traditional approach including that of [733, 862]. Hence, the observed anomalous behavior of PEO in PMMA matrices either defies explanation by these polymer blend models. The malady of these models is that they are extensions of conventional theories of viscoelasticity of neat polymers which have not taken into account the many-body relaxation dynamics of local segmental motion, and cannot explain even the properties of segmental relaxation of neat polymers [165–168, 170–173]. In fact, in neat polymer, the temperature dependences of τ_{terminal} (whether unentangled or entangled) and τ_{seg} (obtained by low-frequency mechanical or dielectric relaxation) are different, and this so-called "breakdown of thermorheological simplicity" problem [29, 165, 171, 172, 201–203, 360, 837, 863–875] cannot be resolved by conventional viscoelasticity theory including the reptation model. Through the coupling parameter n_α which accounts for many-body relaxation dynamics of the local segmental motion, the breakdown of thermorheological simplicity has been explained by the CM [201–203, 360, 837, 868–872].

It is interesting to note that data of τ_{terminal} and τ_{seg} of neat polymers taken at low temperatures/longer times show τ_{seg} usually has a stronger dependence than τ_{terminal} [29, 165–168, 170, 171, 837, 863]. This relation between the temperature dependences of τ_{terminal} and τ_{seg} of pure polymers is opposite to that for PEO in blends with PMMA or PVAc found at high temperatures/short times (see Fig. 89, right panel). The opposite directions of the two anomalies found in pure polymer and in blends make explanation of both from a single theory even more challenging, but has been achieved by the CM as demonstrated next. The CM resolved the anomaly of PEO in blends with PMMA or PVAc by combining the explanation of the segmental relaxation of the PEO component in PEO/PMMA blends at high frequencies [738] and the consideration of the data of diffusion of tracer PEO in PMMA given before in [275] and as follows. The diffusion of a tracer PEO chain in PMMA matrix involves global motion which is different from the local segmental relaxation in nature and length scale of the dynamics. The tracer PEO chain cannot diffuse without some motion of the matrix PMMA chains. Thus, the temperature dependence of the diffusion coefficient or the monomeric friction coefficient of the PEO tracer chain is determined also by the motion of the matrix PMMA chains. The length scale of the PMMA motion necessary for the PEO tracer to diffuse depends on the molecular weight of the PEO tracer. In Fig. 89 (right panel) we reproduce for high molecular weight PMMA its dielectric segmental relaxation time, the shift factor of the Rouse dynamics in the softening dispersion from 114 to 189°C, and the shift factor of the entangled terminal dispersion from 176 to 212°C. All these PMMA shift factors have much stronger temperature dependence than the segmental relaxation time of 0.5 and 3% PEO in PMMA matrix measured by high-frequency NMR [855]. For the low molecular weight unentangled PEO tracer used by Haley and Lodge [861], the shift factors of the PEO tracer diffusion coefficient may be closer to that of the Rouse dynamics of the PMMA matrix given by Plazek and coworkers. It could be accidental, but nevertheless remarkable, that this Rouse dynamics shift factors of PMMA changes by about the same orders of magnitude as the PEO tracer diffusion coefficient over the same temperature range as shown in Fig. 89. Exact agreement is not expected between the temperature dependence of the PEO tracer diffusion coefficient and any of the shift factors of pure PMMA shown in Fig. 89. Thus, the data of diffusion of tracer PEO in PMMA [861] and hence the difference in the temperature dependence over a common temperature range from 125 to 220°C of its monomeric friction factor from that of τ_{seg} of PEO (measured at high frequencies) are explained [876]. This explanation was adopted and used by Zhao et al. [857] to explain their data of 2% PEO blended with PVAc.

The results of PEO in PEO/PMMA blends discussed immediately above were not found for either the polyisoprene (PI) or the poly(vinylethylene) (PVE) component in PI/PVE blends where Haley et al. found that the segmental and terminal chain dynamics exhibited nearly the same dependences on temperature and composition [877]. Since τ_{seg} of PI in the PI/PVE blends was obtained also at high frequencies by the same NMR method [856] as that of PEO in PEO/PMMA blends, the dramatically different result poses another challenge and requires explanation. The segmental relaxation times of PI and PVE in several blends including those of neat PVE and PI measured by Min et al. [856] are reproduced in Fig. 90. To compare the temperature dependence of the same segmental and terminal dynamics of PI in 30% PI/70% PVE blend (lower solid squares in Fig. 90) with those of PVE in the same blend, the segmental relaxation times of PI (lower solid squares) were shifted vertically so that the shifted PI data (upper solid squares) coincide with the PVE data at high temperatures. By inspection, it is clear that the segmental and terminal dynamics of PI and PVE have nearly the same dependence on temperature throughout the range of measurements. Similar conclusion can be drawn for the other compositions. Therefore, in the PI/PVE blends, the diffusing PI chain sees all chains including PVE with dynamics having about the same temperature dependence at all length scales from segmental to terminal. Naturally, the global chain relaxation time of PI has about the same temperature dependence as its segmental relaxation time in the PI/PVE blends. The same argument holds for the PVE component. Hence, for each component, the ratio of the global chain relaxation time to the segmental relaxation time is independent of both temperature and composition as observed in experiment [877].

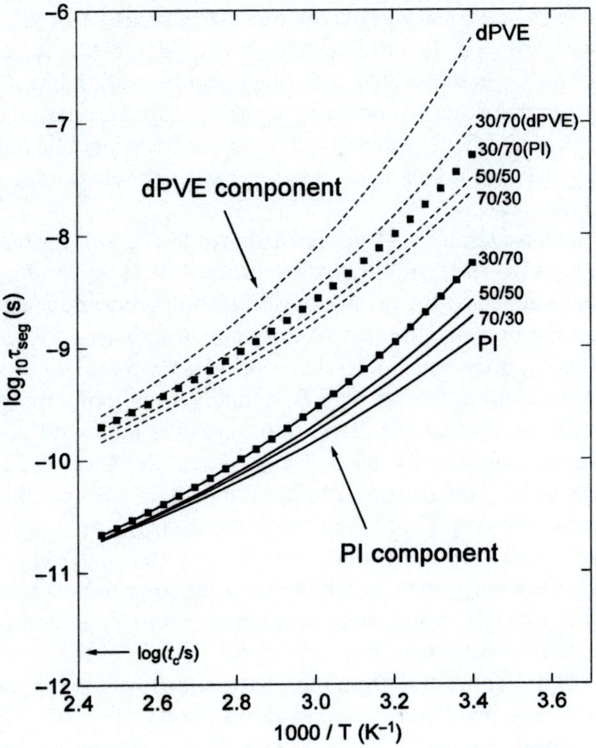

Fig. 90 Segmental correlation times for both components in PI/PVE blends at various compositions. The data of the PI component in the 30% PI/70% PVE blend are shown by the *line with solid squares* on top of it. These data are shifted vertically to compare with the temperature dependence of the segmental relaxation time of the PVE component in the same blend. For details, see text and figures in [856, 877]

Anomalous Component Dynamics in Mixtures

Anomalous component dynamics also were found in some solutions of polymers by small molecular glassformers. Dielectric and dynamical mechanical measurements have been performed on mixtures of the small molecule glass-forming liquid 1,1-*bis*(*p*-methoxyphenyl) cyclohexane (BMC) with poly(methylphenylsiloxane) (PMPS) [878, 879]. In these studies, some of the neat high molecular weight PMPS samples have higher T_gs than that of the neat solvent BMC, but the relaxation times of BMC in the mixtures of 10% BMC/90% PMPS are shorter than in neat BMC. Density measurements were also made for the neat solvent and the polymer solution. It was found that the density is increased with addition of PMPS, and this change should increase the relaxation times of BMC. The same anomaly was found in 1,2-polybutadiene (1,2-PB or also called poly(vinylethylene) PVE) dissolved in Aroclor (A1248), a polychlorinated biphenyl [90, 880(a)]. The 1,2-PB used had a molecular weight of 4300 and T_g of –25°C, which is 19 degrees higher than that of A1248. These data showed the reorientational relaxation time of A1248 becomes increasingly shorter on adding 1,2-PB, notwithstanding the fact that the polymer has a higher T_g. The same anomaly was also seen in a polymer blend, epoxidized natural rubber with polychloroprene [881]. Models based on a distribution of local T_gs due to concentration fluctuations like that of [733] all use the Fox-Flory equation, $1/T_g(\phi) = \phi/T_{gA} + (1-\phi)/T_{gB}$, or a modification to account for self-concentration of polymer, to generate the component relaxation times in one way or the other. The glass transition temperature of the mixture and the relaxation times of the components predicted by these models, therefore, cannot lie outside the bounds determined by the neat components. Hence, these models do not explain the aforementioned anomalies.

However, the CM has been shown to account for the aforementioned anomalies [90, 878–882]. In the context of the CM, $\tau_{A\alpha}$ (or T_{Ag}) of neat glassformer A is determined by the many-body relaxation process characterized by n_A but it is based on the primitive relaxation time τ_{A0}. Another neat glassformer B has its own $\tau_{B\alpha}$ (or T_{Bg}), n_B, and τ_{B0}. When considering the many-body relaxation of molecules A in a mixture of A with B, naturally it is more appropriate and realistic to start with the comparison of τ_{A0} and τ_{B0} rather than with $\tau_{A\alpha}$ and $\tau_{B\alpha}$ (or T_{Ag} and T_{Bg}). Let us consider the case of $\tau_{A\alpha} \ll \tau_{B\alpha}$ (or $T_{Ag} < T_{Bg}$), an example is for A to be BMC, and B to be PMPS. But, due to the fact that n_B is larger than n_A, we have from Eq. (2.55) the reverse relation, $\tau_{B0} \ll \tau_{A0}$ (see figure in [879]), i.e., the primitive motions of molecules B are faster than molecules A in the mixture A/B. Consequently, the many-body relaxation probed via component A in the mixture is faster than in neat A, and the relaxation time $\hat{\tau}_A$ of component A in the mixture A/B is to be shorter than $\tau_{A\alpha}$ of pure A.

The explanation from the CM given above also explained a similar but different anomaly involving the cooperative γ-relaxation of polymers related to bisphenol A polycarbonate (BPA-PC) [882–887]. Addition of the diluent *tris*(2-ethyl-hexyl) phosphate (TOP) has opposite effects on the γ-relaxation times of bisphenol A polycarbonate (BPA-PC) and of tetramethyl-bisphenol A polycarbonate (TMBPA-PC) [888–890]. Dynamic mechanical [888] and dielectric relaxation [889] measurements showed that addition of TOP to TMBPA-PC increases the relaxation rate of the γ-process; however, the opposite effect was found when TOP was added to BPA-PC, the γ-process became slower. In the jargon of polymer physics, the anomaly is restated as antiplasticizing γ-relaxation of BPA-PC but plasticizing of γ-relaxation of TMBPA-PC by TOP. NMR experimental data on the same mixtures confirmed these findings [890]. Positronium annihilation lifetime spectroscopy (PALS) found in BPA-PC/TOP blends that the total hole volume fraction is insensitive to the addition of TOP up to 10 wt% but increases at higher TOP concentration which should make the γ-relaxation faster, contrary to the antiplasticization observed. In the case of TMBPA-PC/TOP blends, the total hole volume fraction decreases monotonically with TOP concentration up to 15% and thereafter increases slightly from 15 to 20% which should make the γ-relaxation slower, contrary to plasticization observed. The CM successfully explained the opposite effect of TOP by comparing the primitive relaxation time of TOP with the primitive γ-relaxation times of BPA-PC/TOP and TMBPA-PC, which is $\tau_{0,\gamma\text{BPA-PC}} \ll \tau_{0,\text{TOP}} \ll \tau_{0,\gamma\text{TMBPA-PC}}$ (see Fig. 91).

Wang et al. [880(b)] found another anomaly in the dynamics of the higher T_g component in a binary mixture of associating glass-forming liquids: propylene glycol (PG) with 2-ethylhexylamine (EHA). Here PG is the higher T_g component

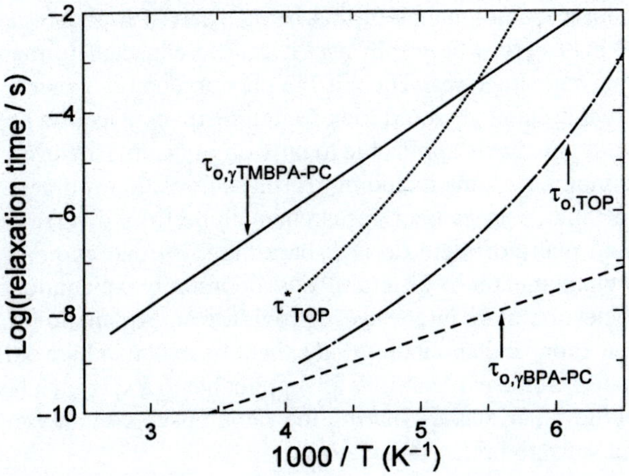

Fig. 91 The dotted curve represents the Vogel–Fulcher fit to the effective relaxation time, τ^*_{TOP}, of pure TOP obtained from dielectric measurement. The corresponding primitive relaxation time of TOP, $\tau_{0,TOP}$, is shown as *dashed-dotted curve*. Shown also are the primitive relaxation times $\tau_{0,\gamma BPA\text{-}PC}$ (*dashed line*) and $\tau_{0,\gamma TMBPA\text{-}PC}$ (*solid line*) calculated from the mechanical γ-relaxation times

and it shows changes of dynamics on increasing the concentration of the lower T_g component EHA very different or even opposite in trend from those of the higher T_g component found in other associating and non-associating binary mixtures. For example, the fragility index m increases and the coupling parameter n of the PG component increases on adding EHA, while T_g of the PG component does not change up to 0.5 molar fraction of EHA. >From the features presented in the broadband dielectric spectra of the PG component at different compositions, the anomalies have been identified [880(c)] to originate from the enhanced intermolecular coupling in the primary relaxation of the PG component in the presence of the EHA component. The enhancement can be traced to the fact that EHA has larger intermolecular cooperativity or coupling parameter than PG, despite the former having a lower T_g than the latter. The enhancement increases on increasing the concentration of the EHA component in the mixture, as evidenced by the appearance of the JG β-relaxation of the PG component when the molar fraction of EHA has been increased to 0.75. In contrast neat PG does not have a resolved JG β-relaxation [424]. The anomalous dynamics of PG in the mixtures is challenging for all theories and models of glass transition to explain, but it can be rationalized by the CM [880(c)].

Illustrating the Different Attitudes Toward Anomalies and Generalizations

Trained as a physicist, I share the view of Crick [161] and Anderson [160] that anomalies are indispensable guides to find the correct and fundamental explanation of physical phenomena. That is the reason why I have taken every opportunity to discuss the observed anomalies in this book. Also, if a phenomenon is fundamental and originates from basic physics, it can be expected to manifest itself in different systems and situations. Thus, if the phenomenon is general and observed in widely different systems, I would look for a holistic explanation applicable to all systems instead of one that is applicable to only one system. However, in the field of relaxation in complex systems including the glass transition problem, dynamics of polymer and soft matter in general, researchers come from diverse disciplines and backgrounds, and many of them do not share the view that anomalies are important clues, and phenomenon in general needs a common explanation coming from basic physics. They are more interested in providing an explanation for the norm of observed phenomenon, and anomalies that cannot be explained are either ignored or laid on the wayside. Some explanations are applicable to a specific class of systems, and inapplicable to other classes sharing the same observed general phenomenon, but this does not seem to bother them.

An example from component dynamics of mixtures is the Lodge–Mcleish (LM) model designed specifically for polymer solutions and blends to be described as follows. A polymer component has more repeat units of its own kind nearby due to chain connectivity. This is correctly pointed out by Lodge and McLeish (LM) [733]. They went on to suggest the local composition is altered from the average value ϕ to an effective concentration, $\phi_{\text{eff}} = \phi_{\text{self}} + (1 - \phi_{\text{self}})\phi$, where ϕ_{self} accounts for the excess of segments due to the chain connectivity. The modified Fox equation $T_g(\phi) = [\phi_{\text{eff}}/T_{g1} + (1 - \phi_{\text{eff}})/T_{g2}]^{-1}$ is then used to obtain $T_g(\phi)$ of the component. The segmental relaxation time in the blend, $\tau_{\text{seg}}(\phi)$, is calculated by modifying the Vogel–Fulcher–Tammann–Hesse (VFTH) equation obtained for the component when it is in its neat state, $\tau_{\text{seg}} = \tau_\infty \exp[B/(T - T_0)]$, by only replacing T_0 by $T_0(\phi)$. Here $T_0(\phi) = T_0 + [T_g(\phi) - T_g]$. In other words, the difference between $T_0(\phi)$ of component in the blend and T_0 of the neat polymer is exactly the same as the difference between $T_g(\phi)$ and T_g of the neat polymer. The remaining VFTH parameters, B and τ_∞, are assumed not to change. Thus, in the LM model, the component segmental relaxation time is given by $\tau_{\text{seg}}(\phi, T) = \tau_\infty \exp[B/(T - T_0(\phi))]$. Lodge and McLeish initially suggested that ϕ_{self} is to be calculated from the Kuhn length of the component. However, the ϕ_{self} calculated with the Kuhn length does not work in many test cases and it has to be treated as an adjustable parameter to yield agreement with experimental results [855, 877, 891, 892]. Even letting ϕ_{self} be an adjustable parameter, there are a number of cases where the observed $\tau_{\text{seg}}(\phi, T)$ of components cannot be explained [80, 729].

Bedrov and Smith [80] performed molecular dynamics simulations of model miscible polymer blends consisting of chemically realistic 1,4-polybutadiene (CR-PBD) as the slow component (higher T_g) and PBD chains with reduced dihedral barriers as the fast component (LB-PBD) with lower T_g. They found that only $\phi_{\text{self}} = 0.0$ provides a better description of the α-relaxation times for the LB-PBD component in the 10% LB-PBD blend with CR-PBD, indicating that, within the LM formulation, the segmental relaxation of the fast component is determined entirely by the bulk composition of the blend. This is contrary to the spirit of the LM model which assumes that the local, dynamically relevant environment for a segment differs in composition from the bulk blend due to chain connectivity. The situation of the 50% LB-PBD blend is worse; Bedrov and Smith found even the best with $\phi_{\mathrm{self}} = 0.0$; the Lodge–McLeish model prediction still underestimates the slowing down of the α-relaxation of the LB-PBD component.

Roland et al. [729] applied their data of the PVME/P2CS blends and found that the modified Fox equation gives $\phi_{\mathrm{self}} = 0.437$ for the P2CS *less than* the actual concentration. The result contradicts the enrichment due to the intramolecularly bonded neighboring segments.

Absence of chain connectivity in non-polymeric glassformers means inapplicability of the LM model to mixtures of these glassformers, and yet the essential properties of component dynamics (including those not addressed by the LM model) are similar to that in polymer blends. The anomalies found in polymer solutions such as BMC/PMPS [878, 879], and Aroclor/PVE [90, 880], and in polymer blends of PEO/PMMA and PEO/PVAc [855, 857] cannot be explained either. These limitations and deficiencies of the LM model are sufficient for me not to take it seriously as the solution of the component dynamics even restricted to polymers. However, the LM is the center of attention of almost all recent publications on mixtures involving polymers. It is exclusively the model used currently by others to compare with experimental data, although the comparison is limited to just the temperature dependence of $\tau_{\mathrm{seg}}(T)$. It does not address the other properties of the component dynamics including various changes of dispersion discussed before. The only link of the LM model to the glass transition problem is modification of the empirical Fox equation, and therefore it offers no help to understand the more fundamental problem of glass transition of pure polymers. Despite ad hoc assumptions (e.g., the VFTH parameters, B and τ_∞, of the polymer in the blend is same as those of the neat polymer, and the formula for $T_0(\phi)$) being made, the LM would be useful for purpose of polymer engineering in predicting $\tau_{\mathrm{seg}}(T)$ and T_g of a polymer component in blends and solutions if ϕ_{self} is pre-determinable. Unfortunately, in reality, ϕ_{self} turns out to be a floating parameter to fit data, and its value has been found to be unphysical in some cases.

2.2.5.8 Decrease of Relaxation Time by Nanoconfinement

Large decrease of τ_α causing reduction of T_g when the liquid is confined in nanometer size glass pores was observed in OTP and other glassformers [485–505, 893–897], and first discussed in connection with length scale in Section 2.2.2.6. The effect is more pronounced in pores that are smaller in size. Care must be exercised to eliminate formation of chemical bonds of the liquid molecules with the glass walls by chemical treatment of the latter. Other techniques of confining a liquid have been used. In one study of ethylene glycol, a countable few molecules confined in zeolite, silicalite, and sodalite show large decrease of τ_α [489].

This general effect, mentioned before in Section 2.2.2.6, is a challenge for theory of glass transition to explain. Let us take for example the Adam–Gibbs model discussed in Section 2.2.2.2, and its expression, $\tau_\alpha(T) = \tau_\infty \exp(z^* \Delta\mu/kT)$, where z^* is the number of molecules in the cooperatively rearranging region (CRR). On lowering the temperature of a liquid confined in pore, z^* increases. At a temperature T_p when $z^*(T_p)$ matches the size of the pores others have argued that z^* stops growing and remains constant. Hence, for $T < T_p$, in the pore τ_α assumes the Arrhenius temperature dependence, $\tau_\alpha(T) = \tau_\infty \exp(z^*(T_p)\Delta\mu/kT)$, which is shorter than τ_α in the unconfined liquid at the same temperature. Dielectric and light-scattering data of liquids confined in pores and zeolites show evidence that the temperature dependence of τ_α becomes more like Arrhenius. However, the departure of τ_α from the bulk values in these confined liquids, including salol [490], OTP [492–494], and PDMS (see Fig. 29), starts at temperatures where the bulk τ_α is of the order of 10^{-6} s or shorter. For OTP and salol, the onset temperature T_p is close to their respective T_B, defined in Section 2.2.5.1. As discussed therein, at T_B the size of z^* calculated by Eq. (2.22) from calorimetric data is about 2 [420], which is much smaller than the number of molecules that can be accommodated in the pores larger than 2.5 nm used in experiments. This consequence from the AG theory contradicts the assumption that $z^*(T_p)$ has already filled up the pore at the onset temperature T_p. Therefore, the explanation of the reduction of τ_α by nanoconfinement based on AG theory has contradiction, and hence it is untenable. [898, 899].

The finding that the AG theory cannot explain the effects of confinement is unsurprising in view of the difficulty that it has in fully rationalizing even just relaxation data of bulk liquids [253, 401] as discussed before in Section 2.2.2.2, and in encountering other problems [405]. The deficiency of the AG theory is based entirely on thermodynamic consideration of the configurational entropy $S_c(T)$, and hence it neglects the effect of the many-body relaxation dynamics manifested by n of the Kohlrausch correlation function. The AG theory is solely based on a thermodynamic quantity, the configurational entropy $S_c(T)$. Thus, even though the proposed transition rate has the connotation of molecules rearranging cooperatively, it cannot capture the complexities of molecular motion, i.e., the many-body dynamics. Certainly, molecular relaxation time depends on entropy and volume, and this dependence also applies to the primitive relaxation time τ_0 of the CM. Much later in Section 2.3.2 and particularly Sections 2.3.2.29 and 2.3.2.30 therein, experimental evidences will be presented to show that τ_0 already depends on entropy and volume. Hence, we have $\tau_0(T) = A_0 \exp[B_0/TS_c(T)]$, on retaining AG's idea on the importance of configurational entropy. The primitive relaxation function is given by $\exp(-t/\tau_0(T))$. But according to the CM this primitive α-relaxation mode cannot continue indefinitely. After t_c, the intermolecular interactions slow down the averaged relaxation process and the averaged correlation function assumes the stretched exponential form $\phi(t) = \exp[-(t/\tau_\alpha(T))^{1-n}]$, and the relation

$$\tau_\alpha(T) = [t_c^{-n}\tau_0(T)]^{1/(1-n)} = \{t_c^{-n}A_0 \exp[B_0/TS_c(T)]\}^{1/(1-n)} \tag{2.58}$$

between $\tau_\alpha(T)$ and $\tau_0(T)$ follows. The results of this modified AG theory by the CM have the benefit of being consistent with the experimental data of bulk [401] as well as nanoconfined glassformers [494, 899], while the original AG theory does not. This modified theory is referred to here as the CM/AG theory. The cooperative length scale, ξ, of the structural α-relaxation obtained from NMR [422] is consistent with the length scale of cooperative *many-body* dynamics in the CM/AG theory, but it should not be identified with the length scale associated with z^* of the AG theory. This is because z^* for OTP at T_g deduced from thermodynamic data is only about 4 [420] which is much smaller than the number of molecules inside the volume with length ξ at T_g deduced by NMR. However, confinement introduces a new length scale, which will alter the relaxation of the liquid when it is comparable to or smaller than ξ. The degree of intermolecular coupling will be reduced, and if intermolecular coupling is one of the causes of slowing down the relaxation, then it is not surprising to see a decrease of the structural relaxation time by confinement. The cause of the decrease comes from the small pore size limiting the number of neighboring molecules participating in the cooperative motions. The effect is stronger for molecules located closer to the surface, naturally because part of the molecules within ξ is absent. The molecular dynamics inside the pore depends on the location and is therefore spatially heterogeneous. We do not exclude here the possibility that the density of the liquid in the pores in some experiments was indeed lower than in the bulk [493], which contributes another cause of the reduction in intermolecular coupling. Regardless of how the reduction of intermolecular coupling comes about, in the coupling model it is modeled by a decrease of the coupling parameter, $n_d(T)$, of the glassformer in the pore of size, d, from the bulk value $n(T)$ at any temperature. Application of the CM/AG theory shows consistency with the observation of reduction of τ_α from the bulk values in confined OTP starts at T_B where the bulk τ_α is of the order of 10^{-6} s or shorter, as demonstrated in [494(a)] and shown in Fig. 92. The fact that nanoconfined glassformers has much weaker T_g-scaled reciprocal temperature dependence of $\log[\tau_\alpha(T)]$ than the bulk is another support for the decrease of n.

Correlation Between the Magnitude of the Change and n

The explanation of the reduction of τ_α and T_g by nanoconfinement by the CM/AG theory has the consequence that the size of the effect will be smaller for glassformers having smaller n. This result holds irrespective of whether the effect is due to reduction in intermolecular coupling or lower density of the liquid in the pore. From dielectric measurements of bulk glassformers, $\beta(T_g) \equiv 1 - n(T_g)$ is equal to 0.48 for OTP, 0.56 for salol, 0.70 for propylene glycol, and about 0.75 for *N*-methyl-ε-caprolactam (NMEC). One way to compare the finite-size effect on different glassformers is by the decrease in glass transition temperature, ΔT_g, measured by calorimetry. In 5 nm glass pores, ΔT_g is equal to 12.6 K [492], 11.0 K [490], 8 K [488(c)], and 6 K [488(c)], respectively, for OTP, salol, propylene glycol, and NMEC. ΔT_g decreases with increasing $n(T_g)$ of the bulk glassformer as depicted in Fig. 93. The inset shows the difference, $\log\,\tau_\alpha(T_g) - \log\,\tau_0(T_g)$, calculated from the CM equation, $\tau_\alpha = [t_c^{-n}\tau_0]^{1/(1-n)}$, and plotted against $1-n(T_g)$. By calorimetry, Jackson and McKenna observed that the shift of the glass transition temperature in the pores of the same size is much larger for OTP than butyl alcohol. Also broadening of T_g is measured in OTP confined in 8.5 nm pores, but not observed in the

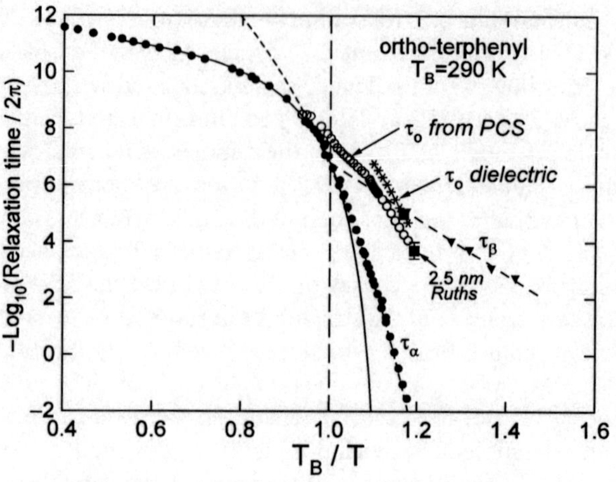

Fig. 92 The α-relaxation frequencies from light scattering (•) and dielectric relaxation (same as PCS but not shown), β-relaxation frequencies (▼ from dielectric relaxation), the calculated primitive α-relaxation frequencies from PCS (◯) and from dielectric (∗) plotted against $1000/T$ for OTP. The *full* and the *dashed-dotted curves* are the two VFTH dependencies of the α-relaxation frequency. The *dashed line* is an Arrhenius extrapolation of the β-relaxation frequencies determined below T_g. The α-relaxation frequencies of OTP in 2.5 nm pores obtained by light scattering (■) at two temperatures are in good agreement with the calculated primitive frequencies. Reproduced from [494(a)] by permission

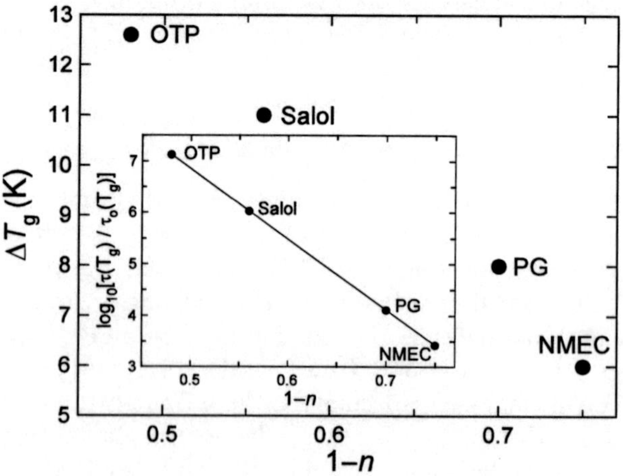

Fig. 93 Experimentally observed decrease of glass transition temperature by calorimetry of four glassformers, OTP, salol, propylene glycol (PG), and NMEC confined in 5 nm pores from their bulk values, ΔT_g, plotted against the Kohlrausch exponent, $1 - n(T_g)$ of the bulk glassformer at T_g. The *inset* shows plot of the calculated difference, $\log \tau_\alpha(T_g) - \log \tau_0(T_g)$, against $1 - n(T_g)$

case of butyl alcohol confined in the same size pores. These observations are in accord with the correlation shown in Fig. 93 because it is expected that $n(T_g)$ of OTP (1,2-diphenylbenzene) composed of three carbon rings is larger than that of butyl alcohol.

Explanation by Decrease of n in the Framework of the CM

When there is no physical interaction or chemical bonding at the interface to slow down the structural relaxation of the nanoconfinement glassformer, there is decrease of intermolecular coupling due to the more mobile surface layers and the finite size effect on the inner layers. The averaged coupling parameter n is expected to decrease monotonically with decreasing size, and it follows from the CM equation, $\tau_\alpha = [t_c^{-n}\tau_0]^{1/(1-n)}$, that τ_α decreases correspondingly as observed. In the limit of extreme confinement when the size becomes smaller than the length scale of the α-relaxation, $n \to 0$ and $\tau_\alpha \to \tau_0$. The value of τ_0 calculated from parameters of the bulk glassformer can be used to check if it agrees with the experimental value of τ_α under the condition of extreme nanoconfinement. The test is positive for OTP [494] and for poly(dimethyl siloxane) [494] confined in silanized glass pores with average diameter of 2 nm, which is comparable to the length scale of the α-relaxation of OTP determined by NMR at temperatures about 10 degrees above T_g. Also good agreement is found [900] for the much reduced α-relaxation time of 1.5–2.0 nm thin films of poly(methylphenylsiloxane) [495]. More discussion of the data of nanoconfinement will be given in Section 2.3.

2.2.5.9 Breakdown of Thermorheological Simplicity of Relaxation Mechanisms of Different Time/Length Scales, and Viscoelastic Anomalies of Polymer: Degree Depends on *n*

Unlike non-polymeric glassformers, the structural α-relaxation of amorphous polymers is not the terminal relaxation responsible for flow and diffusion. Because of chain connectivity, there are motions of longer length scale than that of the structural α-relaxation if the number of repeat units of the chains is sufficiently large. These chain modes give rise to viscoelastic properties unique to polymers, and those with longer wavelengths contribute more to viscous flow and self-diffusion [29]. The structural α-relaxation in polymers is often referred to as segmental relaxation, or more accurately as local segmental relaxation because the number of repeat units within a chain involved in cooperative motion with those in other chains is smaller than that of a segment usually defined in the polymer physics literature. The structural α-relaxation of amorphous polymers has all the properties as non-polymeric glassformers. This is clear since we have made no distinction between polymers and non-polymers in the previous discussions of properties. Bonded interaction along the chain enhances intermolecular constraint, resulting in the larger coupling parameter n, more stretched Kohlrausch correlation function, and broader frequency dispersion of the local segmental relaxation than the monomeric or oligomeric analogues. An example can be taken from the comparison of polystyrene with tri-styrene having only three repeat units [844], or PMMA with its oligomers.

The extent of intermolecular coupling in polymer depends on the molecular structure of the repeat unit. Polymers with chemical structures that have a larger capacity for intermolecular coupling tend to have a larger n appearing in the fractional exponent (1–n) of the Kohlrausch correlation function of the structural α-relaxation (Eq. (1.1)), in accord with the physical meaning attached to n by the CM. When different amorphous polymers are considered, this dependence causes large variations of the degree of non-exponentiality or n, from which different properties of the local segmental relaxation time τ_α originate and are observed, similar to non-polymeric glassformers [110–112, 244, 479]. For example, polystyrene (PS) has a bulkier phenyl ring attached to a carbon in the main chain while polyisobutylene (PIB) has a more compact and simpler backbone. It is expected that intermolecular coupling or n of PS is larger than that of PIB. Experimentally, n(PS) = 0.64 [479] and n(PS) = 0.45 [208–210], and PS has various dependences of τ_α on physical quantities that are different from PIB, such as stronger T_g/T-dependence of τ_α of PS than PIB.

Polymer chains have relaxation processes of longer length scales than the local segmental motion. These chain modes give rise to broader viscoelastic spectrum with properties not shared by small molecular glassformers. Unique to the structural α-relaxation of polymers are the relation it has with the chain modes and the viscoelastic properties they determine. The relation turns out to be governed by n, and is exhibited as viscoelastic anomalies [165–173, 901]. The viscoelastic anomalies or aberrations from the paradigm of viscoelastic behavior of polymers are general. These viscoelastic anomalies are important problems in the viscoelastic properties of polymers and the explanation of them is necessary for a complete understanding of viscoelasticity of polymers. Also, they have repercussion on theory of glass transition because they occur as the glass transition temperature is approached from above, and are caused by the structural α-relaxation. A glass transition theory is incomplete if does not have explanation of these anomalies, and it is suspect of missing an important physics of the structural relaxation. *Therefore, the materials presented in the following are not digression from discussion on important aspects of glass transition. On the contrary, these are facts that any proponent of glass transition theory should address, otherwise the theory cannot be considered applicable to amorphous polymers.*

Most of the viscoelastic anomalies discussed here were first found by Plazek and coworkers. The results clearly show general breakdown of thermorheological simplicity in amorphous polymers [87, 165–173, 206, 209, 210, 360, 832, 837, 869–879, 902–907]. Plazek's observations have been repeatedly confirmed by others using similar and various other techniques [202, 203, 864–868, 874, 908–912]. Elevated pressure was introduced in light scattering [199–201] and dielectric studies [638, 668, 700, 913] to show the local segmental relaxation and chain relaxation have difference pressure and temperature dependences. Plazek's seminal works were recognized by his peers in awarding him the Bingham Medal from the American Physical Society. His Bingham Medal Lecture can be found in [171]. Traditionally in the field of polymer viscoelastic properties, it is generally believed that molecular retardation and relaxation mechanisms including the structural α-relaxation and all chain modes have the same temperature dependence for their retardation/relaxation times [29, 87, 165, 173, 901]. Based on the notion that the monomeric friction coefficient $\zeta_0(T)$ governs the T-dependence of retardation/relaxation times of all mechanisms, this ideal property is called thermorheological simplicity. In reality and as found experimentally, the relaxation times of the local segmental relaxation, the unentangled chain modes, and the entangled chain modes (in the case of high molecular weight entangled polymer) have their own dependence on temperature, which can differ greatly at temperatures that are not far above T_g. Thus the viscoelastic spectrum, including the glass–rubber softening dispersion (contributed by the local segmental mode, the sub-Rouse mode [906, 914], and the Rouse modes) [208–210], the rubbery plateau, and the terminal dispersion zone (contributed by the entangled chain modes in the case of high molecular weight polymers), is not thermorheologically simple either. The anomalies are caused by the local segmental (structural) relaxation from its many-body nature, which is further supported by showing that the extent of anomalies increases with non-exponentiality (or n) of the structural relaxation [870–872]. Also the anomalies tend to go away when the polymer is diluted by a solvent to reduce the intermolecular coupling and hence n [169, 247, 907]. As far as I know, the only theoretical explanation of all these anomalies is from the CM [173, 837]. This is perhaps unsurprising because the CM has captured the many-body nature of the structural relaxation due to intermolecular interaction/coupling. It is a pleasure for me to mention that Don Plazek shared the same belief. He called this "crowding coupling."

In the subsections to follow we outline some of the apparent anomalies of polymer dynamics in different viscoelastic zones and indicate how they are resolved. The discussion will be abbreviated, but details as well as other experimental facts can be found in the references given.

The Glass–Rubber Softening Dispersion of High Molecular Weight Polymers

For a high molecular weight polymer in which the chains are sufficiently long for them to entangle, the viscoelastic response measured in shear compliance $J(t)$ that rises from the glassy compliance $J_g \approx 10^{-9}\,\text{Pa}^{-1}$ (glassy modulus $G_g \approx 10^9$ Pa) and ends at the beginning of the rubbery entanglement plateau with $J_N \approx 10^{-6}\,\text{Pa}^{-1}$ ($G_N \approx 10^6$ Pa) is referred to as the glass–rubber softening dispersion (or transition zone). For ease in identifying the various viscoelastic modes in the softening dispersion from mechanical creep compliance $J(t)$ or shear modulus data, the choice is an amorphous polymer with wide glass–rubber softening dispersion such as polyisobutylene (PIB). Isothermal data of high molecular weight PIB obtained by mechanical measurements with a spectral range over eight decades [906, 915] and additional photon correlation measurements [208–210] have found that there are three distinct viscoelastic mechanisms in the glass–rubber softening dispersion. The Rouse modes and the shorter sub-Rouse modes found by mechanical spectroscopy are shown in Fig. 94 by two peaks in the loss tangent as a function of frequency.

Dielectric spectroscopy studies reveal unusual shape of segmental relaxation and excess contribution on the low-frequency side of the α-loss peak in PIB, polyisoprene (PIP), and head-to-head polypropylene (hhPP) (see [906(a)]). This was

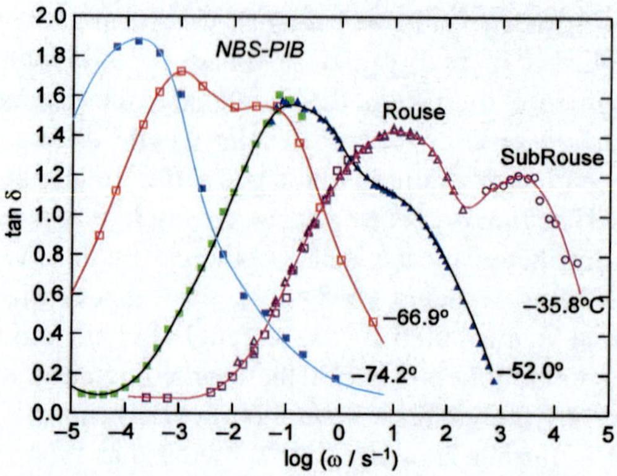

Fig. 94 The tan δ as a function of actual frequencies at several temperatures for NBS-PIB. The data were obtained by using several instruments spanning the frequency range as shown in the abscissa. The high frequency data at –35.8°C (*open circles*) are from [915]. The rest of the data were obtained by a combination of creep compliance and dynamic modulus measurements. Data from [906] are replotted here and the sub-Rouse and Rouse modes are indicated

identified to be the contribution of the sub-Rouse mode to the dielectric loss spectra of these polymers. The location of the sub-Rouse modes of PIB from dielectric spectroscopy is in accord with the previous studies by mechanical and by photon correlation spectroscopy.

Theoretical considerations [209] have helped to identify these three mechanisms to originate separately from the local segmental (α) mode, the sub-Rouse (sR) modes, and the Rouse (R) modes. The temperature dependences of these mechanisms given, respectively, by the shift factors, $a_{T,\alpha}$, $a_{T,\mathrm{sR}}$, and $a_{T,\mathrm{R}}$, determined over a common temperature range are found to be all different (see Fig. 95). This results in the failure of time–temperature superpositioning of mechanical compliance and relaxation data in the softening dispersion. Actually this fact was shown first by creep compliance measurements in polystyrene (PS) in the softening dispersion [167] long before sub-Rouse modes were clearly resolved in polyisobutylene. It has been confirmed in polystyrene by dynamic modulus measurements [864], in PS and tetramethyl polycarbonate (TMPC) [165], and in polybutadiene [865]. Most recently, this effect was found by internal friction measurements in PMMA [906(b)], chlorinated butyl rubber [906(c)], PVAc [906(d)], and PVAc blended with PEO [906(e)]. These works show the sub-Rouse modes are intermolecularly coupled. Change of temperature dependence at nearly the same T_B is found not only for the local segmental relaxation time but also for the relaxation time of the sub-Rouse modes in PVAc in Refs. [906(d)] and [906(e)] and in PMMA in a paper to be published by same research group. The explanation for the crossover of T-dependence of the structural relaxation time of small molecular glassformers [423] also applies to the local segmental relaxation the sub-Rouse modes.

Shift factor, a_T, obtained by time–temperature superpositioning of viscoelastic data taken in the softening transition over a limited experimental window is a combination of the three individual shifts factors, $a_{T,\alpha}$, $a_{T,\mathrm{sR}}$, and $a_{T,\mathrm{R}}$. Consequently,

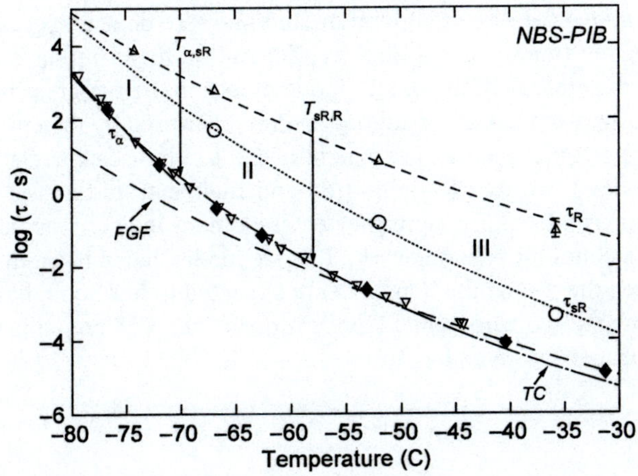

Fig. 95 The Rouse relaxation time τ_R (*open triangles*) and the sub-Rouse relaxation time τ_{sR} (*open circles*) of NBS-PIB for several temperatures obtained from the low-frequency and high-frequency tanδ peaks in Fig. 94. The *curves* that interpolate the data points are the WLF fits. Also shown are the shift factors of Tobolsky and Catsiff (TC) [916] from their stress relaxation data (*inverted open triangles*), and of Plazek et al. [906] from their creep data (*filled diamonds*). The *two vertical arrows* partition the temperature into three regions, I, II, and III, in which the viscoelastic response is, respectively, mainly contributed by the local segmental motion ($J_g < J(t) < 10^{-9.5}$ cm^2/dyne)), the sub-Rouse modes ($10^{-9.5} < J(t) < 10^{-9.5}$ cm^2/dyne), and the Rouse modes ($10^{-8} < J(t) < J_{\text{plateau}}$ cm^2/dyne). The *dashed-dotted line* through the inverted *open triangles* is according to the WLF equation given by TC. The *dashed line* is the WLF equation given by Fitzgerald, Grandine and Ferry (FGF) [915]. The *thick solid line*, passing close by the mechanical data points (*inverted open triangles and the filled diamonds*) in the lowest temperature region I corresponding to local segmental motion, is calculated. Reproduced from [209] by permission

the Williams–Landel–Ferry (WLF) equation or the equivalent VFTH equation that fits the T-dependence of a_T, obtained by shifting and superpositioning of modulus or compliance data to obtain a master curve for the entire softening dispersion, does not describe the temperature dependence of any one of the three viscoelastic mechanisms [209]. The local segmental mode contributes to compliances ranging from the glassy compliance, $J_g \approx 10^{-9}$ Pa^{-1}, up to $10^{-8.5}$ Pa^{-1}. The sub-Rouse modes contribute in the compliance range, $10^{-8.5} \leq J(t) \leq 10^{-7}$ Pa^{-1}. The Rouse modes consisting of a sequence of Gaussian submolecules and generalized to undiluted polymers [29] account for the compliances in the range of 10^{-7} Pa^{-1} $\leq J(t) \leq J_{\text{plateau}}$, where J_{plateau} is the plateau compliance. The magnitudes of the bounds given here are only rough estimates from data of polystyrene, and may change somewhat for other polymers.

The different T-dependences of $a_{T,\alpha}$, $a_{T,\text{sR}}$, and $a_{T,\text{R}}$ have been explained by the CM principally from the many-body dynamics of the local segmental (α) mode. In the CM, the independent (uncoupled) relaxation times, $\tau_{0\alpha}$, $\tau_{0\text{sR}}$, and $\tau_{0\text{R}}$, of the α, sR, and R modes all are governed by the same primitive friction factor, $\zeta_0(T)$, and hence they have the same temperature and pressure dependence. However, the effective relaxation times, τ_α, τ_{sR}, and τ_R, depend on the different T-dependences because of the differences of the α, sR, and R modes in their degrees of many-body participation. The α-mode has the highest degree measured by the coupling parameter n_α appearing in the fractional exponent of the Kohlrausch correlation function. The Rouse modes from its own definition and their entropic nature are not subjected to any intermolecular coupling or entanglement interactions, and have zero values for their coupling parameter n_R. The sR-modes having length scales intermediate between the α- and the R-modes are expected to have coupling parameters n_{sR} interpolating n_α and zero value. Applications of the CM equation lead to three different T-dependences given by

$$\tau_\alpha \propto (\zeta_0(T))^{1/(1-n_\alpha)}, \tag{2.59}$$

$$\tau_{sR} \propto (\zeta_0(T))^{1/(1-n_{sR})}, \tag{2.60}$$

$$\tau_R \propto \zeta_0(T). \tag{2.61}$$

The width of the softening dispersion is determined by the separation between the logarithm of the shortest τ_α and the longest τ_R. It is evident from Eqs. (2.59) and (2.61) that τ_α shifts more to longer times on decreasing temperature than τ_R, causing the α-relaxation to move closer to the Rouse modes. Consequently, a polymer with larger n has a smaller difference between log τ_α and log τ_R or a narrower softening dispersion. This explains why the softening dispersion is broad for a polymer with a smaller $n = 0.45$ such as PIB but narrow for a polymer with a larger $n = 0.64$ such as PS [165, 173]. Here the widths of the softening dispersions of high molecular weight PIB and PS are compared by the widths of the first peaks of their corresponding retardation spectra L obtained from $J_r(t)$ by the relation

$$J(t) = J_r(t) + t/\eta = \int_{-\infty}^{\infty} L(\lambda)(1 - e^{-t/\lambda}) d\ln\lambda + t/\eta \tag{2.62}$$

As shown in Fig. 96 in a log–log plot of L against the retardation time τ/a_T, the first peak of L is significantly narrower for PS than PIB. As discussed before in Section 2.2.5.7, intermolecular coupling and hence n of local segmental relaxation of PS are reduced when PS is dissolved in a solvent that has much lower T_g such as tri-cresylphosphate (TCP). From the decreasing in n upon dilution, it follows from Eq. (2.59) that the softening dispersion of the dissolved PS becomes broader [169]. The softening dispersion of the 25% PS solution in TCP [169] shown in Fig. 96 indeed is broader than PS and not too different from PIB. The loss tangent of the 25% PS solution in the softening dispersion shown in the inset is also similar in width as that of PIB, but the sub-Rouse peak has not yet been resolved. A better match with PIB is expected for solutions with lower PS concentration than 25%. In fact, the isothermal tanδ data of 17% PS solution in TCP shown in Fig. 97 have both the sub-Rouse and the Rouse peaks matching the same in bulk PIB.

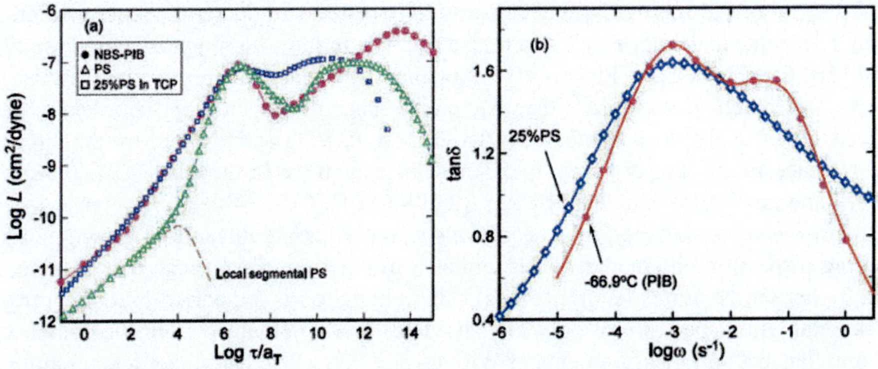

Fig. 96 (**a**) Comparison of the retardation spectra L of a high molecular weight PS (*open triangles*), a 25% PS solution in TCP (*open squares*), and PIB (*filled circles*). The shift factors are arranged such that the maximum of the first peak occurs at the same reduced frequency for all three samples. Downward vertical shifts of 0.869 and 1.39 of $\log_{10}L$ have been applied, respectively, to PS and 25% PS solution to make all data have about the same height at the first maximum. The disparity in width of the softening dispersion of bulk PS and PIB is clear. The small peak near the bottom (*dashed line*) is the contribution to L from the local segmental motion in bulk PS. (**b**) The figure shows isothermal data of tanδ of PIB in the softening region at –66.9°C, and tanδ of the 25% PS solution in TCP obtained from a reduced recoverable compliance curve after applying time–temperature superposition to the limited isothermal data. Data from [907] are replotted in these two figures

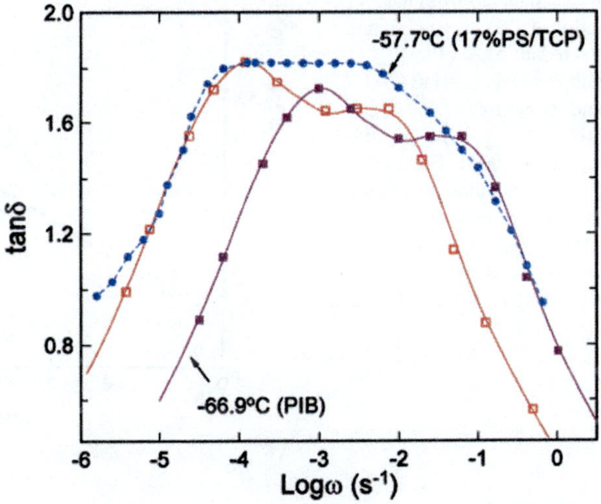

Fig. 97 Comparison of isothermal tanδ of the 17% PS solution in TCP at –57.7°C (*filled circles*) and isothermal tanδ of PIB at –66.9°C (*filled squares*) in the softening region. The *open squares* are tanδ of PIB at –66.9°C after the first peak is shifted horizontally and scaled vertically to match the position and height of the first peak of the 17% PS solution. The *lines* connecting the data points of each set are drawn to guide the eye. Data are from [907] and replotted here

The T_g-scaled temperature dependence of the shift factor of the softening dispersion of the 17% PS solution in TCP also resembles that of bulk PIB [907]. Thus several characteristic properties of the softening dispersion of PS, which differ greatly from PIB, are made the same as PIB by reducing the intermolecular coupling (n of the local segmental relaxation) in PS with the addition of a diluent.

The failure of thermorheological simplicity or time–temperature superpositioning of mechanical compliance and relaxation data in the softening dispersion found in PIB, PS, TMPC, and PBD is also explained by τ_α increasing more rapidly than τ_{sR}, and much more rapidly than τ_R on decreasing temperature. The degree of breakdown is higher in PS than in PIB [165, 870, 871] as indicated by the larger difference in the temperature dependences of τ_α and τ_R in the case of PS than in PIB, and this is also explained by Eqs. (2.59)–(2.61).

For the *same* polymer, these equations also predict significantly narrower softening dispersion obtained at lower temperatures (longer times) than that obtained at higher temperatures (shorter times). This prediction is supported by comparing the retardation spectrum of poly(methacrylate) (PMA) obtained from the complex compliance $J^*(\omega)$ measurements of William and Ferry [917] at higher temperatures (higher frequencies, $10 < \omega/2\pi < 6 \times 10^3$ Hz) with that obtained by Plazek et al. [903] from their $J(t)$ data at lower temperatures (longer times, $1 < t < 10^6$ s) data. The retardation spectra L of the softening dispersion shown in Fig. 98 are significantly broader at lower temperature as predicted.

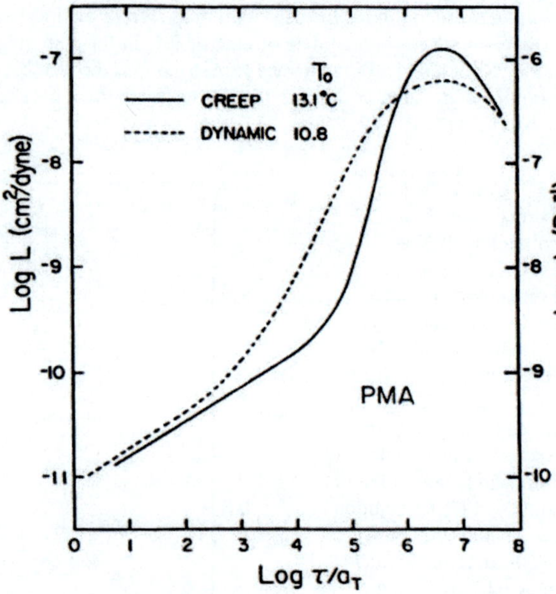

Fig. 98 Logarithm of the retardation spectrum L of poly(methylacrylate) (PMA) as a function of the logarithm of the reduced retardation time τ/a_T. The *solid curve* was calculated from the reduced $J_r(t)$ curve obtained from data taken at lower temperatures (14.4–34.7°C) and longer times ($10^\circ < t < 10^5$ s) and shifted to 13.1°C [903]. The *dashed line* was calculated from the dynamic compliances of [917] obtained at higher temperatures and frequencies; T_0 was chosen to be 10.8°C

The fact that τ_{sR} has a stronger temperature dependence than τ_R (Fig. 95) indicates that the sub-Rouse modes have some degree of intermolecular coupling, which is reasonable because they have length scales intermediate between the local segmental mode (with intermolecular coupling) and the Rouse modes (without intermolecular coupling). Neutron scattering can probe modes of different length scales, L, by different momentum transfers, Q, where $L = Q^{-1}$. Thus when Q^{-1} falls below the length scale of the smallest Gaussian submolecule, Rouse dynamics will give way to the slower sub-Rouse dynamics. This was observed [918, 919] in polyisobutylene by neutron spin-echo measurement which found significant slowing down of the relaxation compared to the Rouse model predictions at $Q^{-1} < 6.7$ Å. The slowing down of modes with $Q^{-1} < 6.7$ Å was interpreted [918, 919] not as intermolecular coupling of the sub-Rouse modes like here [209], but by the introduction of an additional dissipative mechanism (internal viscosity) into a Rouse-like single-chain theory of polymer melt dynamics. From results of simulations, Krushev et al. [920] gave reasons that this is a many-particle effect (i.e., intermolecular coupling consistent with our sub-Rouse modes interpretation [209]), which cannot be captured by an effective single particle theory explanation as suggested in [918, 919].

According to Eqs. (2.59) and (2.61), the T-dependences of the shift factor of the α-relaxation, $a_{T,\alpha}$, and that of the Rouse relaxations, $a_{T,\mathrm{R}}$, are related by

$$(1 - n_\alpha)\log(a_{T,\alpha}) = (1 - n_R)\log(a_{T,\mathrm{R}}), \tag{2.63}$$

where n_R should be equal to zero unless there is extraneous coupling between the Rouse modes such as caused by hydrogen bonding [868, 921]. Nicolai and Floudas (NF) [912] observed a weaker T-dependence of $a_{T,\mathrm{R}}$ than that of $a_{T,\alpha}$ in dielectric and viscoelastic relaxation of poly(oxypropylene) (POP) diols and triols of low molecular weights in the range 260–11,000 g/mol, as expected because n_α is larger than n_R. NF found diminishing difference between the T-dependences of $a_{T,\mathrm{R}}$ and $a_{T,\alpha}$ on increasing temperature above T_g. This behavior is not new, and was found before dielectrically in low molecular weight poly(propylene glycol) [868, 922] and polyisoprene [910, 911]. The relaxation frequencies ν_α and ν_R of the hydrogen-bonded polypropylene glycol with molecular weights of 1000, 2000, and 4000 are shown in Fig. 99 together with ν_α obtained from $a_{T,\alpha}$, which is calculated by

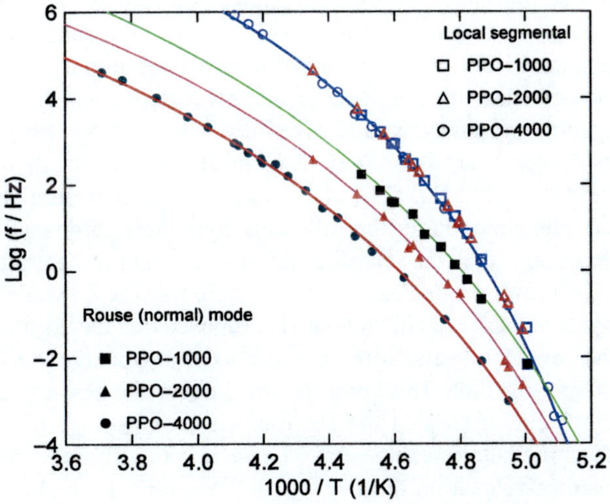

Fig. 99 Frequency at the dielectric loss maximum plotted vs. inverse temperature for the normal mode process and the segmental relaxation. The *points* in this figure are the experimental data points, and the *solid lines* are from the VFTH equations of the segmental relaxation and the normal mode of PPG1000, 2000, and 4000. The *other lines* are frequency of segmental relaxation calculated by using Eq. (2.63). Data from [868] are replotted here

Eq. (2.63) from $a_{T,\mathrm{R}}$ and the values of n_α and n_R determined from the spectra near T_g. Both n_α and n_R are assumed to be T-independent. The calculated ν_α agrees well with experiment at lower temperatures but overestimate the observed value at higher temperatures. The discrepancy at higher temperatures is likely due to failure of the assumption that n_α is constant. Many low molecular weight glass-formers show decrease of n_α when τ_α becomes shorter on increasing temperature. Nevertheless, from the diminishing difference between the T-dependences of $a_{T,\mathrm{R}}$ and $a_{T,\alpha}$ they observed in POP on increasing temperature above T_g, NF concluded this behavior violates Eq. (2.63), because they found the ratio, β_s/β_n equivalent to $(1-n_\alpha)/(1-n_\mathrm{R})$ here, to be constant. Before the conclusion of NF to be taken seriously, one has to recognize several problems in their experimental data that should not allow them to jump to their conclusion. First, all their samples are polydisperse. They only mentioned that the polydispersity index is less than 1.2 for most samples but without giving real figures. Polydispersity will broaden the Rouse modes, making the value β_n they obtained from the data not accurately reflecting $(1 - n_\mathrm{R})$. This is consistent with the fact that the normal mode loss peak reported by NF is significantly broader than that coming from the Rouse modes. Possibly, besides polydispersity, hydrogen bonding may contribute to the broadening of the Rouse modes like found in oligomers of PPG. Second, only spectra of one sample D4200 at two temperatures, 218 and 173 K, were shown, which leads the reader to suspect the spectra of other samples not shown are of the same quality or worse. By inspection of the loss spectra in Fig. 1 of NF, it can be seen that the width of the segmental relaxation at 173 K is much broader than that at 218 K, and hence the corresponding values of β_s at these two temperatures are very different. Moreover, the loss peak of the normal mode at 218 K is so weak that one wonders how reliable the value of β_n determined by NF is. Third, at higher temperatures, the α-relaxation merges with the slower one of the two secondary relaxations, and this may contribute additional broadening of the α-relaxation and make the observed β_s smaller than $(1 - n_\alpha)$ in Eq. (2.63). Despite the uncertainties and irregularities pointed out in the above, NF offered the blanket statement: "... the experimental finding that β_s/β_n is constant over the whole temperature range. In addition, the ratio β_s/β_n obtained from DS increases with increasing molar mass while the data in Figure 4 do not show a significant molar mass dependence." Until NF have shown their spectra and analysis, and given the values of β_s and β_n together with their error estimates, there is no way to judge the validity of their statement and conclusion they drew that Eq. (2.63) is violated. NF also offered an alternative explanation of the phenomenon by the following hypothesis proposed by Schönhals et al. [909]. Assuming that the length scale of the dynamic heterogeneity of the α-relaxation, ξ_het, grows on decreasing temperature toward T_g, when ξ_het matches the radius of gyration of the chain, Schönhals proposed that the normal modes relax together with the segmental relaxation, and the local segmental and the Rouse relaxation times merge together. This proposal does not make sense because within its framework the friction factor of the chain mode is an average over the friction factors of the dynamic heterogeneities, and hence it is no different from the friction factor of the average relaxation time, $< \tau_\alpha >$. Therefore, τ_R and $< \tau_\alpha >$ shift in parallel with temperature when approaching T_g, opposite to the observation of them having different shift factors. Furthermore, the normal modes are still Rouse-like modes because its relaxation time is observed to have the M^2 or stronger molecular weight dependence. Hence, the local segmental and the Rouse relaxation are different processes and cannot merge together to become one and the same.

Temperature Dependence of Terminal Relaxation or Viscosity of High Molecular Weight Polymers Compared with that of the Local Segmental Relaxation

Precision measurements of creep compliance $J(t)$ of polystyrene [166, 170, 171], poly(vinyl acetate) [169(b)], and atatic polypropylene [202, 875, 902] by Plazek and others have shown the temperature dependence of viscosity is weaker than the local segmental relaxation. This is demonstrated for entangled polystyrene with nearly uniform molecular weight 46,900 g/mol in Fig. 100 by a plot of creep compliance measurements shown as $J_p(t) \equiv J(t)(T\rho/T_0\rho_0)$ measured at different temperatures but each shifted by the shift factor determined solely by the viscosity, $a_{T\eta} = (\eta_0 T_0 \rho_0)/(\eta T \rho)$, which has been measured down to about 97°C, the glass transition temperature. The shift factors $a_{T\eta}$ naturally superposes well the data in the terminal zone because they have the same T-dependence as with viscosity, but they fail to superpose the data at shorter reduced times in the softening dispersion including the local segmental relaxation. From the direction of the deviation from superpositioning, it can be deduced that $a_{T,\alpha} \equiv \tau_\alpha(T)/\tau_\alpha(T_g)$, (or τ_α) has stronger T-dependence than $a_{T,\eta}$ (or the terminal relaxation time $\tau_\eta = \eta J_e$, since the plateau compliance J_e is nearly independent of T for entangled polymers).

The local segmental relaxation time τ_α of polystyrene was measured as a function of temperature by two-dimensional exchange NMR up to long times exceeding 100 s [923, 924] together with deuteron spin-lattice relaxation measurements of τ

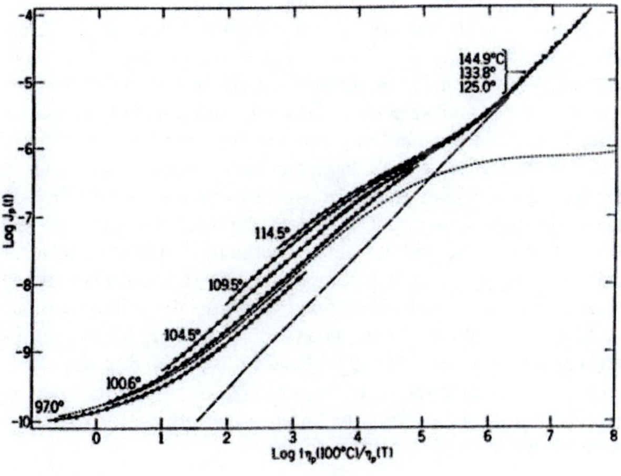

Fig. 100 Creep compliance measurements at several temperatures indicated in the figure on a polystyrene with molecular weight 46,900, reduced to 100°C with shift factors calculated from steady-state viscosity. Subscript p denotes multiplication by $T\rho/T_0\rho_0$. Reproduced from [170] by permission

between 10^{-6} and 10^{-7} s. The actual data, not shown, is well fitted by the WLF equation, $\log[\tau_\alpha(T)/\tau_\alpha(T_g)] = -C_1(T-T_g)/(C_2+T-T_g)$, where $T_g = 373\,\text{K}$, $\tau_\alpha(T_g) = 100\,\text{s}$, $C_1 = 16.35$, and $C_2 = 52.5$ K. The solid line in Fig. 101 is $\tau_\alpha(T)$ from the WLF fit and is drawn to span the same range as the measured $\tau_\alpha(T)$, and it truly represents the local segmental relaxation time. We compare the temperature dependence of $\tau_\alpha(T)$ with that of the shift factor, $a_{T,S}$, from time–temperature superposition of recoverable creep compliance, $J_r(t)$, curves in the glass–rubber softening region of another high molecular weight polystyrene described earlier (Fig. 100).

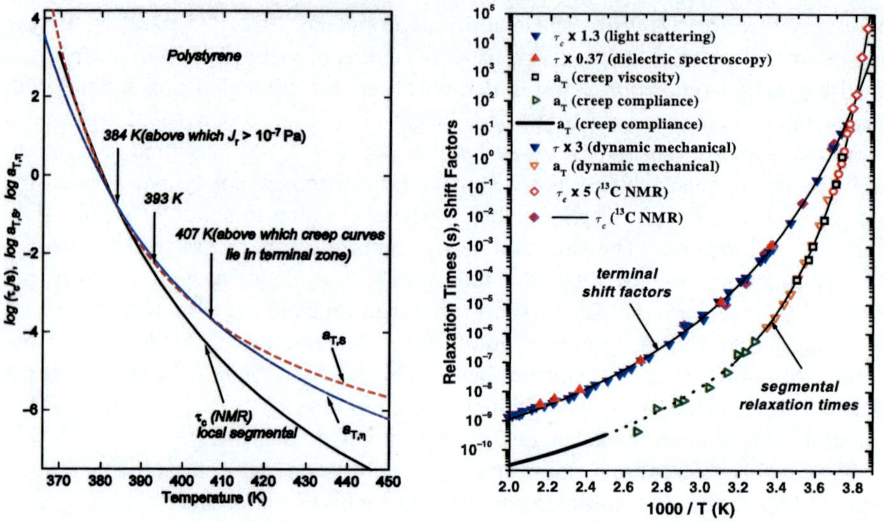

Fig. 101 *Left panel for polystyrene*: The local segmental relaxation correlation time, τ_c, of high molecular weight polystyrene as a function of temperature obtained by two-dimensional exchange NMR [923, 924] up to long times exceeding 100 s, and compared with the shift factor, $a_{T,S}$, from time–temperature superposition of recoverable creep compliance, $J_r(t)$, curves of a entangled polystyrene with $M = 46,900$ g/mol (PS-A25) in the glass–rubber softening region [170]. Only the fits to the data of τ_c (*solid line*) and $a_{T,S}$ (*short dashed line*) by the WLF equation are given. The viscosity shift factor, $a_{T,\eta}$, is shown also (*short-dashed-long dashed line*). The NMR τ_c clearly has stronger temperature dependence than the viscosity in the *entire* temperature range. There is also good agreement between the temperature dependence of τ_c and $a_{T,S}$ at temperatures below 384 K where $a_{T,S}$ becomes sequentially the shift factor of first the sub-Rouse modes and second the local segmental modes as temperature is decreased toward T_g. *Right panel for atatic polypropylene*: Atactic polypropylene segmental relaxation times obtained from dynamic mechanical spectroscopy (O), dynamic light scattering (□), dielectric relaxation (∇), and ^{13}C NMR (- - - -) (Δ), along with terminal time–temperature shift factors for aPP from dynamic mechanical spectroscopy (▲), creep compliance (♦), (◇), and viscosity (▼). The *hollow symbols* represent local segmental modes, while *filled symbols* refer to the terminal relaxation. The legend indicates vertical shifts applied to the relaxation time data. The symbol τ_c indicates the time integral of the segmental portion of the relevant time correlation function, while for the frequency domain techniques the symbol τ indicates $(2\pi f_{max})^{-1}$. All other data are shift factors, vertically scaled for superpositioning. The *thin, solid curves* are best fits to the VFTH equation [202, 875, 902]. Reproduced from [875] with permission

Comparison is also made with the viscosity shift factor, $a_{T,\eta} = (\eta_0 T_0 \rho_0)/(\eta T \rho)$ from [168, 170], in Fig. 101 (left panel). Again only the WLF fits to $a_{T,\mathrm{S}}$ and $a_{T,\eta}$ are shown in Fig. 101 (left panel), and the range of $a_{T,\eta}$ shown by the line drawn corresponds to that of the actual measurements. We have already known from the failure of time–temperature superpositioning in Fig. 100 that $a_{T,\mathrm{S}}$ and $a_{T,\eta}$ from creep compliance measurements do not have the same temperature dependence. By inspection of Fig. 101 (left panel), the NMR τ_α has the same temperature dependence as $a_{T,\mathrm{S}}$ in the temperature range below 384 K. This agreement supports the fact that the part of $a_{T,\mathrm{S}}$ for $T < 384$ K belongs to the local segmental relaxation [248] and possibly some more local sub-Rouse modes since NMR probes only local motion. In this lower temperature range, the $a_{T,\mathrm{S}}$ is determined principally from measurements of $J_r(t)$ that are less than 10^{-7} Pa^{-1} [168, 248].

Above 384 K, τ_α starts to exhibit stronger temperature dependence than $a_{T,\mathrm{S}}$. Between approximately 384 and 407 K, $a_{T,\mathrm{S}}$ is determined by curves with $J_r(t)$ larger than 10^{-7} Pa^{-1} and consists of Rouse modes and possibly some shorter time modes in the plateau [165]. Thus the NMR data provide another proof that the local segmental relaxation time has stronger temperature dependence than the Rouse modes shown before by other kinds of measurements [165, 169(b), 202, 875, 902]. Above approximately 407 K, the creep compliance data are contributed entirely by the terminal viscoelastic mechanism, which has exactly the same temperature dependence as $a_{T,\eta}$. It is interesting to observe that the extrapolation of $a_{T,\mathrm{S}}$ to this high-temperature regime shows a different (weaker) temperature dependence than the actual shift factor for the viscosity. More important is that the NMR local segmental relaxation time τ_α or its shift factor $a_{T,\alpha}$ clearly has stronger temperature dependence than the viscosity shift factor $a_{T,\eta}$ in the *entire* temperature range shown in Fig. 101 (left panel).

The same conclusion was reached for atactic polypropylene [202, 875, 902] by using photon correlation spectroscopy (PCS) instead of NMR to measure $\tau_\alpha(T)$ and compare its shift factor $a_{T,\alpha}$ with $a_{T,\mathrm{S}}$ and $a_{T,\eta}$ from shear creep and stress relaxation measurement. The results are shown in Fig. 101 (right panel).

The disparity between $a_{T,\alpha}$ and $a_{T,\eta}$ becomes larger as T_g is approached from above, but conventional glass transition theories offer no explanation, indicating possible negligence of an important piece of physics in these theories. An explanation was given by the CM [173, 360, 837, 904, 925]. This explanation recognizes that the terminal relaxation leading to viscous flow of entangled polymer chains is also a many-body problem. The interchain interactions come from the entanglements. The primitive chain relaxation are the Rouse modes, some of which with wavelength smaller than the entanglement distance survive and appear at the end of the softening dispersion of $J(t)$ and before the onset of the plateau. Since the CM is generally applicable to relaxation and diffusion in interacting systems, it follows that the many-chain relaxation time τ_η and the primitive Rouse relaxation time τ_R are related by

$$\tau_\eta = [\hat{t}_c^{-n_\eta} \tau_R]^{1/(1-n_\eta)}, \tag{2.64}$$

where $\hat{t}_c$ is the crossover time for entanglement coupling, which turns out to be of the order of nanoseconds, and $n_\eta \approx 0.40$. These figures will be substantiated by experimental data later in Section 3.2.3.2. On combining Eq. (2.64) with Eq. (2.61), the T-dependence of τ_η is given by

$$\tau_\eta \propto (\zeta_0(T))^{1/(1-n_\eta)}. \tag{2.65}$$

On comparing this T-dependence of τ_η with that of τ_α in Eq. (2.59), it is obvious that τ_α has a stronger T-dependence than τ_η if $n_\alpha > n_\eta$ as in the case of PS ($n_\alpha = 0.64$), aPP ($n_\alpha = 0.60$), and PVAc ($n_\alpha = 0.57$) [112, 165, 479].

On the other hand, if n_α is not much larger than n_η in some polymers, then the difference in T-dependence between τ_η and τ_α becomes insignificant. This case was demonstrated experimentally by Plazek et al. [870] in entangled polyisobutylene (PIB), which has a smaller $n_\alpha(T_g) = 0.45$ [208–210] comparable to $n_\eta \approx 0.40$

Strong Temperature Dependence of J_S in Low Molecular Weight Polymers

It is well known [29] that the steady-state recoverable compliance, J_S, of unentangled polymers obtained at temperatures sufficiently high above the glass temperature is remarkably well described by the modified Rouse model prediction: $J_S = 0.40(M/\rho RT)$, where M is the molecular weight and ρ the density. However, as the temperature is decreased toward T_g, J_S is found to decrease dramatically [168] (see left panel of Fig. 102a). The effect is large and measurable at low molecular weights. For example, in PS with $M = 3400$, J_S drops by 30 times from the

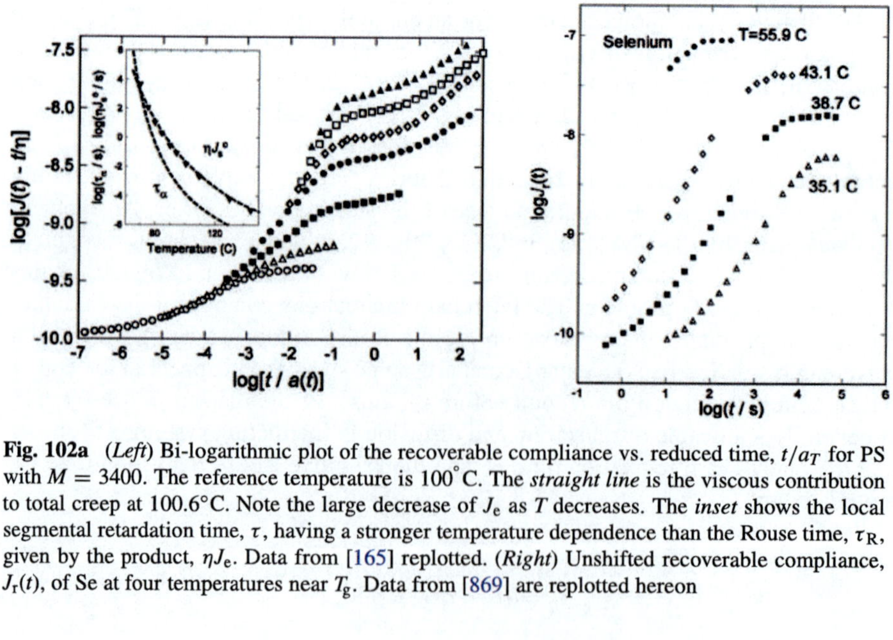

Fig. 102a (*Left*) Bi-logarithmic plot of the recoverable compliance vs. reduced time, t/a_T for PS with $M = 3400$. The reference temperature is 100°C. The *straight line* is the viscous contribution to total creep at 100.6°C. Note the large decrease of J_e as T decreases. The *inset* shows the local segmental retardation time, τ, having a stronger temperature dependence than the Rouse time, τ_R, given by the product, ηJ_e. Data from [165] replotted. (*Right*) Unshifted recoverable compliance, $J_r(t)$, of Se at four temperatures near T_g. Data from [869] are replotted hereon

Rouse value within 30°C as T_g is approached and reached [165, 168]. The result in PS was confirmed by others [866]. This effect is found in other polymers including polypropylene glycol (PPG) [867], selenium [869], and poly(methylphenylsiloxane) [206] and therefore is considered to be general. For a review see [165]. The same behavior is found for selenium [869], the inorganic and natural polymer, which is shown in the right panel of Fig. 102a.

Such a dramatic viscoelastic effect first discovered a long time ago [168] has only been rationalized by the CM as due to encroachment of the local segmental relaxation toward the chain modes [206, 871, 869]. Again, the fundamental cause of the effect lies in τ_α of the local segmental relaxation for unentangled low molecular weight polymers having a much stronger T-dependence than the relaxation times of the sub-Rouse and Rouse modes, τ_{sR} and τ_R, according to Eqs. (2.59)–(2.61) and the fact that n_α is larger than n_{sR} and n_R. This is shown in Fig. 102 by experimental data from PS. In the inset of this figure, the relaxation time of the chain modes is given in terms of the product ηJ_S where η is the viscosity. The local segmental mode encroaches upon the chain modes on the timescale as T is lowered, and the observed reduction of J_S follows as a consequence as explained in [871].

Figure 102b (left panel) shows the large drop of J_S with decrease in temperature when approaching T_g in a low molecular weight (5000 Da) poly(methylphenylsiloxane) [206]. The right panel shows τ_α encroaching the chain mode relaxation time ηJ_S on decreasing temperature because the former has

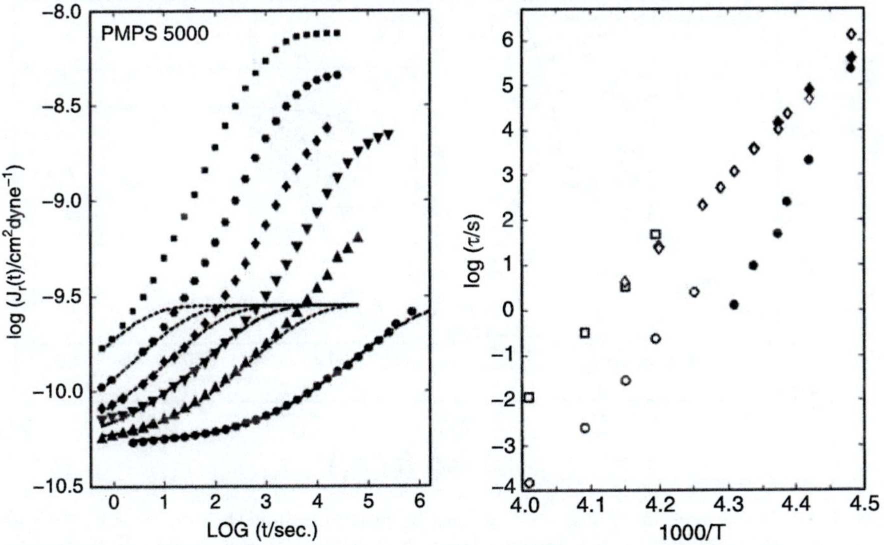

Fig. 102b (*Left*) Unshifted recoverable compliance, $J_r(t)$, of PMPS-5000. The *lines* are Kohlrausch fits to the data. (*Right*) Arrhenius plot of various relaxation times. *Filled circles* stand for α-relaxation time determined by the Kohlrausch fit in the left panel. *Filled diamonds* are the relaxation times of the chain modes. *Open diamonds* are log(ηJ_e) shifted upward by 1.25 decades. *Open circles and open squares* denote the α-relaxation and chain relaxation times, respectively. For more details see [206]. Reproduced from [206] by permission

stronger temperature dependence. Photon correlation spectroscopy measurement on the same sample was able to detect the chain mode, and its relaxation times determined are in agreement with ηJ_S from creep data.

Dielectric relaxation data of the local segmental relaxation and chain normal modes in polypropylene glycols [868, 909] and low molecular weight polyisoprene [910] have confirmed that the τ_α of the local segmental relaxation for unentangled low molecular weight polymers has a much stronger T-dependence than the relaxation times of the normal mode, which should represent the Rouse modes and the sub-Rouse when the molecular weight is low. This has been shown for PPG1000, 2000, and 4000 in Fig. 99. For more experimental data, see the review [165].

Effect Reduced on Decreasing Intermolecular Coupling

Intermolecular coupling between repeat units of polystyrene together with the coupling parameter n_α certainly will be reduced when diluting the polymer with a solvent with a much lower T_g such as tri-*m*-tolyl phosphate [247]. The effects observed in bulk polystyrene will be weakened by addition of tri-*m*-tolyl phosphate. This expected change indeed has been observed [247], and is shown in Fig. 103 in a plot of the retardation spectra $L(\log(\tau/a_T))$. A large reduction of L of the bulk PS with $M = 3400$ on decreasing temperature down to T_g can be seen from the

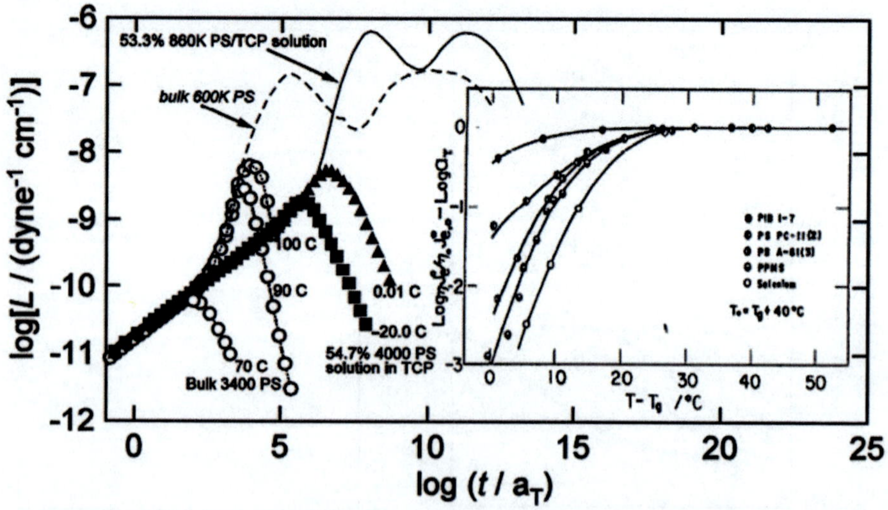

Fig. 103 Logarithmic plot of $L(t/a_T)$ as a function of reduced time for: (**a**) the bulk PS with $M = 3400$ (PS-4000) at seven temperatures, 100.6, 89.9, and 70°C (*open circles*); (**b**) the bulk PS with $M = 600,000$ (*dashed curve*); (**c**) the 54.7% solution of PS-4000 in *m*-TCP at two temperatures, 0.01 (*closed triangles*) and –20.0°C (*closed squares*), and (**d**) the 53.3% solution of PS with $M = 860,000$ (*solid curve*). (*Inset*) Showing the weaker T-dependence of the chain modes shift factor, $\frac{\eta(T)J_e(T)}{\eta(T_0)J_e(T_0)}$, than the shift factor of the structural α-relaxation time, $\alpha_T = \frac{\tau_\alpha(T)}{\tau_\alpha(T_0)}$ for five low molecular weight polymers, PIB (*closed circles on top*), two samples of PS, PPMS (or PMPS), and Se. Data from [247] are replotted here

figure, while the reduction is much smaller in the 54.7% solution of PS-4000 in *m*-TCP. The reduction of n_α on dilution has the effect of making lesser the difference in the temperature dependence between the structural (local segmental) relaxation and the sub-Rouse/Rouse modes (see Eqs. (2.59), (2.60), and (2.61)). This change explains the smaller decrease of J_e or the reduction of L as T decreases toward T_g, as explained in more detail in [247].

As mentioned before, PIB has a smaller $n_\alpha(T_g) = 0.45$ than polystyrene with $n_\alpha(T_g) = 0.64$ [208–210]. Again the same effects are expected to be weaker in low molecular weight PIB than in low molecular weight PS, which indeed has been observed [871], as shown in Fig. 103. Because Se and PMPS do not have carbon backbone like PIB and PS, comparison of the effect between the polymers of different class is not straightforward.

2.2.5.10 Non-linear Deformation of Amorphous Polymers

Polymer glass under mechanical stress/strain changes structure and at sufficient strain level, it yields and undergoes plastic flow [926, 927]. The first model to explain this phenomenon is by Eyring [928] who proposed in general for all glasses that the energy barriers for molecular motion are lowered by external loading and which effectively transforms a glass into a viscous liquid. Since then, various modifications of Eyring's approach have appeared [929–933]. There is, however, a different approach to explain the phenomenon that emphasizes the reduction of intermolecular coupling and the cooperativity parameter n by the change of structure with applied strain in the context of the coupling model (CM) [934–936]. When proposed in the 1980s, this approach was not obviously the right approach to the problem, but by now it has become self-evident in view of the fact that n governs the structural relaxation in so many ways as shown before in the various subsections. In contrast, the original and the modified Eyring models do not consider the dispersion of the structural relaxation and make no prediction of its change by the strain.

Verification by experiment of the basic premise of the CM approach to explain the increase of molecular mobility by mechanical deformation of glassy polymers was first made more than 22 years ago [936]. In this experimental study of the effects of non-linear strain, an initial large strain is applied in the form of a ramp, and then held constant. The holding time t_s must be sufficiently long such that the structure has approached essentially a constant state or at least it is changing very slowly. The specimen is then probed with a perturbation, i.e., an additional small strain step. It is the response to this small strain that is considered as representative of the relaxation behavior of the structure at that particular non-linear strain magnitude. The magnitude of the small strain must be small so as not to cause further change of the structure. It was found through experimentation on the polymer bisphenol A polycarbonate (BPA-PC) that as long as the small strain is kept below 0.5%, the relaxation modulus is independent of the magnitude of the small strain. This strain independence ensures that the small strain does not further change the structure. To determine the time t_s necessary for the structure to reach a constant state, the stress relaxation response to a small strain of 0.2% or less was measured after the initial large strain step (up to 5.1%) beginning at different times t_i. When t_i exceeds some time, the small strain response becomes constant to within the experimental error, then t_s is considered to have been reached. For example, with 4% applied strain at 35°C, no further change of the small strain response was found after $t_i = 10^5$ s, consistent also with the volumetric measurements. This constant small strain response was well fitted by the Kohlrausch function (Eq. (1.1)), and the exponent n and the relaxation time τ are obtained as functions of the non-linear strain. The spectrum narrows (or n decreases) and the effective relaxation time τ becomes shorter on increasing strain. The non-linear strains increasing up to 5.1% have the effect of reducing τ by many orders of magnitude accompanied by decrease of n quite drastically from about 0.9 for the undeformed BPA-PC to nearly 0.65 at applied strain of 5.1% (see Fig. 104). The data, when extrapolated to yield strain (about 6%), gives an n value of 0.6, which is smaller than the value of $n = 0.65$ at T_g of undeformed BPA-PC. It is worthwhile to point out that the temperature at which the small strain–stress relaxation measurements were made is about 100 degrees below T_g. For the undeformed BPA-PC, the stress relaxation measured is contributed by the secondary relaxations rather than the local segmental relaxation. Only at sufficiently high applied non-linear strain, when τ is not unreasonably long, are the parameters n and τ identifiable with those of the local segmental motion. Thus, this 1988 mechanical experiment [936] is the first to show that the enhanced molecular mobility of deformed BPA-PC is strongly correlated with a narrowing of the dispersion of the local segmental relaxation, a feature predicted in the CM explanation of the phenomenon. Furthermore, from the CM relation rewritten as $\log \tau = [n/(1-n)](-\log t_c) + \log \tau_0/(1-n)$, it is clear that a smaller n does produce a shorter τ as observed experimentally.

Criticism was raised by McKenna and Zapas [937, 938] against the technique used. The point made by these authors is that apparent aging behavior can be observed in a concentrated polymer solution subjected to non-linear strains. Furthermore, the BKZ theory [939] describes the non-linear behavior of the solution very well. Therefore, as their argument goes, the type of behavior observed cannot serve as evidence for structural changes since polymer liquids cannot undergo physical aging. Although the latter point is certainly correct, the implicit assumption that polymer liquids cannot undergo "structural" changes at large strains is probably not. One can certainly imagine flow-induced orientation of the chain molecules which can eventually lead to disentanglement. In fact, their data can be interpreted as strain-induced structural change and recovery: a plausible molecular interpretation of phenomena they choose to describe as "non-linear effects." As far as possible structural changes are concerned, whether or not the BKZ theory can describe the non-linear behavior of a polymer solution is irrelevant, since the theory does not take into account explicitly changes in structure. Finally, the volume relaxation results demonstrate that, insofar as one can assume that structural changes can be reflected by volumetric measurements, strain-induced structural changes have certainly occurred. It is important to note that the results of McKenna and Zapas fail to show a strain sensitivity of the parameter n after long hold times, whereas the

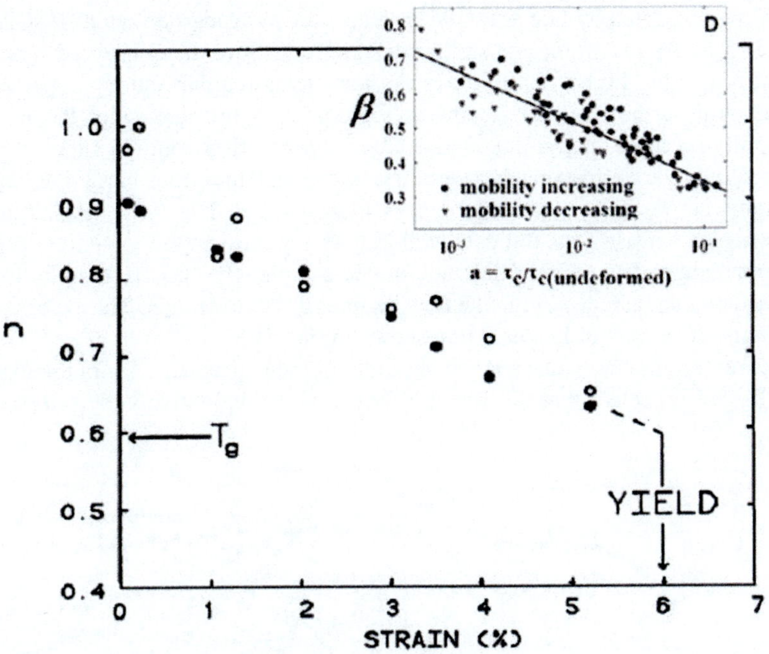

Fig. 104 Plot of n vs. non-linear strain of BPA-PC [936]. The derivative technique was used to obtain an initial value of n at each strain level (*open circles*). Curve fitting of entire response to the small strain was used to obtain n (*closed circles*). There are good agreements between the values of n obtained by the two methods. The data, when extrapolated to yield strain (about 6%), gives an n value of 0.6, which is smaller than the value of n at T_g. *Inset* is taken from [941] to show essentially the same result is obtained for lightly cross-linked PMMA. There is correlation between the Kohlrausch exponent, $\beta \equiv (1 - n)$, and enhanced mobility measured by the ratio of the relaxation time of deformed polymer to that of the undeformed polymer (x-axis) for three different trials with stresses in the range of 15.5–16.0 MPa. Data were acquired as the mobility was increasing (•) and decreasing (▾). The *solid line* is a fit to the data. Reproduced from [936, 941] by permission

results obtained by the technique described above do. There is another support of the verity of the results from the technique. Small strain (0.1% and 0.2% extension) measurement was made on a specimen taken from the necked region after BPA-PC was elongated to the point of yielding [940]. The drawing was done at a mean strain rate of 10^{-5} s^{-1} to an extension ratio of 1.70. This independent measurement of the necked specimen is free from any of the possible problem suggested by McKenna and Zapas. The data of the elongated necked sample show shorter relaxation time and narrower dispersion than that of the undeformed sample. When the two sets of relaxation data are fitted to the Kohlrausch function,$E(t) = E_{\text{U}} \exp[-(t/\tau)^{1-n}]$, we find for the undeformed sample $E_{\text{U}} = 2.43$ GPa, $n = 0.86$, and $\tau = 7 \times 10^{10}$ s, and for the stretched specimen $E_{\text{U}} = 3.54$ GPa, $n = 0.77$, and $\tau = 4 \times 10^{7}$ s.

In 2009, twenty-one years after the mechanical probe study by Yee et al. [936], optical measurements of stress-induced molecular mobility in a lightly cross-linked poly(methyl methacrylate) (PMMA) glass by the anisotropy decay function $r(t)$ of a probe was reported by Lee et al. [941]. Their study was carried out at 19 K below $T_g = 395$ K. As the strain and strain rates are increased, they observed faster $r(t)$ decays, indicating higher mobility. In the low-stress regime, the Eyring model is capable to describe the changes in the segmental mobility, but not the increase of mobility by more than a factor of 1000 as deformation continues into the flow regime. When the strain exceeds a certain level to reach the strain-hardening regime, the strain rate decreases, and $r(t)$ decays more slowly. The Kohlrausch function $r(t) = r_0 \exp[-(t/\tau)^{\beta}]$ fits the data well in both regimes. In the low-stress regime, β was unchanged from that of the undeformed sample ($\beta = 0.32$). As mobility was increased by increasing strain in the flow regime, β became larger. The experimental procedure and results of Lee et al. are shown in Fig. 105.

Apparently this observation is the highlight of the experiment. From the identity, $\beta \equiv (1 - n)$, it is clear that this result of Lee et al. is qualitatively the same as that

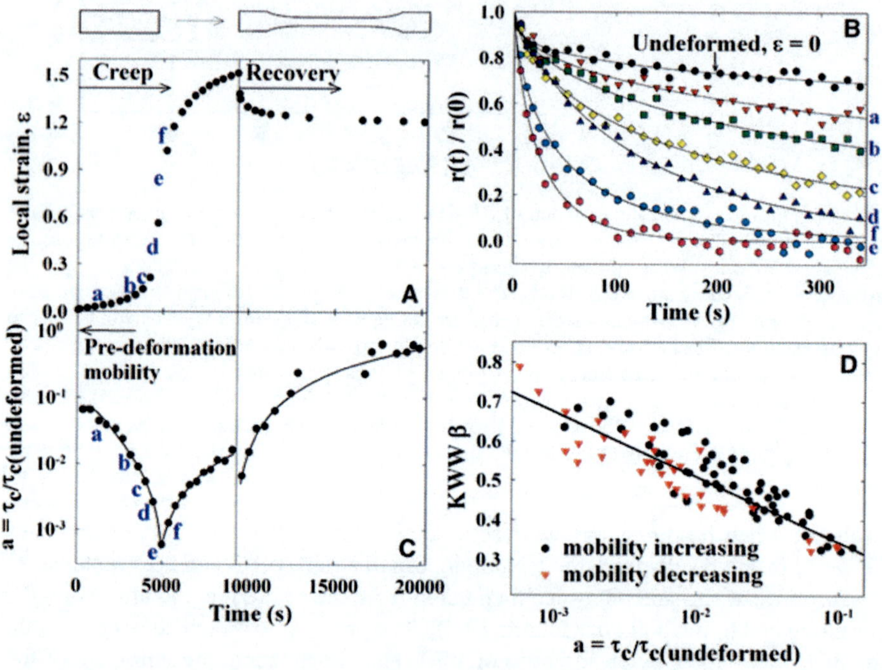

Fig. 105 Simultaneous local measurements of strain and mobility in PMMA glass. (**A**) Strain during creep experiments at 375.7 K with an engineering stress of 16.0 MPa, followed by recovery. The initial and final shapes of the sample are shown. (**B**) Normalized anisotropy decays obtained during the creep experiment at times indicated by letters *a* to *f*. As the strain rate increased, higher mobility (faster anisotropy decay) was observed. The *solid lines* are KWW fits to the data. (**C**) Mobility shift factor during creep and recovery. Mobility increased by up to a factor of 1000. The *solid lines* are guides to the eye. (**D**) Correlation between the KWW β parameter and mobility for three different trials with stresses in the range of 15.5–16.0 MPa. Data were acquired as the mobility was both increasing (•) and decreasing (▼). The *solid line* is a fit to the data. Reproduced from [941] by permission

of Yee et al. although the study of the former is at temperature much closer to T_g than the latter. It is therefore justified to say that the experiment of Lee et al. confirms the assumption of decrease of intermolecular coupling and n to explain the stress-induced increase of molecular mobility in the framework of the CM [934–936], although the CM was not given credit for this by Lee et al. None of the other models discussed by them predict this strong anticorrelation between increase of molecular mobility and width of dispersion or n. Lee et al. offered no explanation except basically reinterpreting their results by the statements: "This result indicates a narrowing of the distribution of relaxation times that reflects the dynamics in slow regions accelerating more than in mobile regions" and ".... in addition to lowering the barriers for molecular motion, deformation apparently modifies spatially heterogeneous dynamics in the glass." The appeal to spatially heterogeneous dynamics is not unreasonable because it correlates with the width of the dispersion or n. But, as we have seen before in the subsection on the breakdown of Stokes–Einstein relation, spatially heterogeneous dynamics fails in some cases to explain the phenomenon because it is just one of the parallel consequences of many-body relaxation.

The experimental results [934, 941] presented here cannot be explained by simple free volume concepts because Sternstein have observed similar shifts in relaxation behavior by shear deformation [942], where any volume change is but a second-order effect, and in compression, [943] where the volume change is actually negative. In fact, based on their studies on the effects of dilatative vs. deviatoric stresses, Sternstein [942] and Questad [943] have concluded that it is the deviatoric (distortional) component that is primarily responsible for the shifts in relaxation behavior. It is apparent that simple free-volume concepts or simple modifications of these concepts will not explain the effects of deviatoric stress on shifts in relaxation behavior. Volume recovery and physical aging measurements on quenched BPA-PC with or without mechanical deformation were made by Pixa and coworkers [944]. They found that volume recovery after cold drawing was faster than in the absence of deformation. On the other hand, the mass density of undeformed BPA-PC is lower than the cold drawn sample after elongated past the yield point. Thus the faster volume recovery is at odds with a decrease in specific volume after cold drawing. Pixa et al. [944] rationalized this apparent contradiction by assuming first that the volume decrease due to chain orientation affects only the occupied volume, and second that the free volume is increased by cold drawing. These are bold assumptions, which cannot be easily verified given the difficulty of separating free volume from occupied volume.

The decrease of the width of the α-dispersion or n in the exponent $(1 - n)$ of the Kohlrausch function by mechanical deformation, a critical experimental fact as well as anticipated by the coupling model, was developed into an approach to non-linear viscoelasticity [935]. The relaxation function predicted by the coupling model is used as a time-dependent kernel in a constitutive equation for a description of non-linear viscoelasticity. The model incorporates the strain history dependence and allows for the evolution of material structure and parameters n and τ. Simulations are presented showing how the model reproduces the essential features of constant strain rate, strain-rate change, and strain-step experiments. Experiments are described that can partially characterize the coupling model parameters and lead to quantitative predictions of the model.

Before closing this subsection, the reader may recall the discussion in Section 2.2.5.5 of the non-linear rheological behavior in aqueous colloidal suspensions of Laponite. When the shear rate $\dot{\gamma}$ is larger than the average structural relaxation rate, $< \tau_s >^{-1}$, Di Leonardo et al. observed [343] concomitant decrease of $< \tau_s >$ and the width of the dispersion or n. Both the decrease of $< \tau_s >$ and the decrease of n increase with the shear rate $\dot{\gamma}$. This observed non-linear rheological behavior of the *soft* glass-forming Laponite system is similar to that of *hard* glassy polymers effected by non-linear deformation, and consistent with the basic premise of the CM that n governs the relaxation time.

2.3 A Fundamentally Important Class of Secondary Relaxations

2.3.1 Background

The discussion in Sections 2.1 and 2.2 are almost exclusively on the structural α-relaxation. The emphasis on the structural α-relaxation in glass transition is understandable because it is the process responsible for vitrification, i.e., structural arrest on cooling when its relaxation time τ_α becomes too long for the glassformer to attain equilibrium. The many properties of the α-relaxation presented all indicate that it is the many-body relaxation process (the degree of which is best represented by the fractional exponent n in the Kohlrausch correlation function in Eq. (1.1)) that determines τ_α and its properties, in addition to the obvious dependence of τ_α on thermodynamic variables T, P, S, and V. To solve the glass transition problem, many-body relaxation must be incorporated into the theory in a manner that captures all its function. One way to accomplish this goal is first finding out how it is generated from simpler or more elementary processes at earlier times. The simpler processes of interest are those that either bear some relation or connection to the α-relaxation or show some correlation with the α-relaxation in observed properties. An obvious candidate is the so-called secondary relaxation found in many glassformers that have transpired at times long before the primary α-relaxation makes its appearance. Traditionally, secondary relaxations are considered unimportant for glass transition. This is because in the early days the secondary relaxations found were usually identified with local motion involving either only a part of the molecule or some internal (intramolecular) degree of freedom [106], in contrast to the α-relaxation which involves the motion of the entire molecule and cooperatively (intermolecularly) with other molecules.

Such a view of all secondary relaxations was challenged by the discovery of the presence of a secondary relaxation in totally structurally rigid glassformers having no internal degree of freedom for motion. The original example is the totally rigid and polar chlorobenzene molecules in a mixture with non-polar *cis*-decalin by dielectric relaxation measurements [105, 945]. A secondary β-relaxation originating from chlorobenzene was observed together with the α-relaxation also coming from the dipole moments of the chlorobenzene molecules in the mixture. This result removes the myth that all secondary relaxations are trivially intramolecular in origin. It also suggests that secondary relaxation may be a universal feature in all glassformers, and had led Johari and coworkers and others to look for secondary relaxations in different kinds of glassformers. Examples include hydrogen-bonded glassformers, alcohols [945, 946], plastic crystals [945, 947], glucose [948], epoxies [949], polymers [950], molten salt, $0.4Ca(NO_3)_2$–$0.6KNO_3$ (CKN) [951], and inorganic phosphate-silicate glass ($0.243Na_2O$–$0.269CaO$–$0.026P_2O_5$–$0.462SiO_2$) [952]. However, it must be mentioned that some of the secondary relaxations found by Johari and coworkers in these references do not belong to the special class that has fundamental importance in glass transition. Also, secondary relaxations in polymers with or without pendant molecular units on the backbone have been found by dielectric, mechanical, and NMR spectroscopies even long before Johari and Goldstein [106], but many of them do not belong to the special class. The special class was only recently identified and characterized by properties [865, 953], and will be the subject of extensive discussions in the following subsections. Particularly worth mentioning in discovery of secondary relaxations belonging to the special class are metallic glasses by Chen and coworkers [954, 955] and others following him [956–960]. These efforts raised awareness of the common occurrence of secondary relaxation and stimulated interpretations of its origin, but most if not all come short of asserting of any connection it has with the α-relaxation or any important role it plays in glass transition. The interpretation of secondary relaxations, given before [105, 945, 961–965] and maintained till recent years [966], is some local rotational and/or translational diffusion, in loosely packed isolated regions in a mechanically rigid structure of a *glass*. The loosely packed isolated regions or "interstices" between the clusters were referred to as "islands of mobility" [945], a term also used by Goldstein in 1969 [36], or "defects," and the size of such regions was assumed to decrease with temperature. The specific reference to *glass* and kinetic freezing-in of density fluctuation to produce regions of low density is perhaps due to the fact that most often observation of the secondary relaxations were made in the glassy states, particularly in earlier years when dielectric and mechanical spectroscopies were limited to measurements over narrow bands of frequency. Although it was noted by Johari [945] that the secondary relaxation is also a "property of the liquid state above T_g," the "islands of mobility" or "interstices" interpretation of secondary relaxation and the emphasis on the glassy state may have the adverse effect of dispelling the fundamental connection it may have with the α-relaxation in the liquid state.

In the potential energy landscape model proposed by Goldstein [36] or the inherent structure model [39, 967], the state of a glass is represented by a deep energy minimum with a corrugated bottom representing a number of shallow energy minima of different depths. Each shallow minimum corresponds to one configuration of the secondary relaxation. No connection between the secondary relaxation and the α-relaxation seems to be present in these models for the glass or liquid when generalized to above T_g. A recent view of Goldstein on this subject can be found in his publication in J. Non-Cryst. Solids of papers presented in the 6th IDMRCS held in Rome, Italy in 2009 [964].

Williams and Watts in 1971 [76, 968] proposed a different view of the secondary relaxation that it corresponds to faster, small angle (thermally activated) reorientational motion of *all* molecules. After the small angle reorientations have occurred, the same molecules undergo a large angle reorientation in a co-operative manner identified with the α-relaxation. The total correlation function $\Phi(t)$ is composed of that for the α-process $\Phi_\alpha(t)$ and that for the secondary β-process $\Phi_\beta(t)$ according to

$$\Phi(t) = [(1 - S(T))\Phi_\beta(t) + S(T)]\Phi_\alpha(t). \quad (2.66)$$

Here $(1-S(T))$ is the fraction of the correlation relaxed by the β-relaxation $\Phi_\beta(t)$, and the remnant is independently and completely relaxed by the α-relaxation. The secondary relaxation is manifestly *independent* of the α-relaxation in Eq. (2.66), and is called the Williams ansatz by others. So is the model of Donth and coworkers which assumes simple superposition of the two processes, and simple additivity of the frequency-dependent susceptibility [969]. The Williams ansatz and the simple additivity assumption are currently being used by researchers to analyze experimental data. In the "island of mobility" model, the secondary β- and the primary α-relaxations occur in spatially separate regions, and therefore simple additivity of the contributions from the two processes is apparently the rule for this model.

The works cited above make secondary relaxation more popular and intriguing, but they all indicate that it is independent of the α-relaxation and hence implying that it plays no role in glass transition. In the past few decades, secondary relaxations have been found in many more glassformers. Broadband dielectric spectroscopy and NMR techniques have helped to observe secondary relaxation in the liquid state of some glassformers, and in many instances more than one secondary relaxation were found in the same glassformer. However, little effort has been made to distinguish different kinds of secondary relaxations and classify them according to differences in properties until recently [685]. This lack of effort did not help the cause of finding out if some specific class of secondary relaxation may be important for glass transition. The first clear indication that a specific class of secondary relaxations does bear relation to α-relaxation seems to be works in 1998 [251, 953, 970], which demonstrate a strong correlation between n and the ratio τ_α/τ_β (or the separation distance, $\log \tau_\alpha - \log \tau_\beta$) at a predetermined value of τ_α. This correlation first established by examining existing data from many different glassformers is repeatedly confirmed by more experimental data coming out after 1998. Here τ_β is the relaxation time of the secondary β-relaxation, and n is the quantity appearing in the Kohlrausch correlation function (Eq. (1.1)) which measures the degree of non-exponentiality of the α-relaxation. Glassformers with larger n have larger ratio of τ_α/τ_β for the same

τ_α. The search and discovery for this correlation was motivated by a similar relation between n and the ratio, τ_α/τ_0, where τ_0 is the primitive relaxation time in the coupling model (CM) [251, 953, 195]. The primitive relaxation entails the motion of all parts of the molecule, and thus anyone of the secondary relaxations chosen to establish the correlation is of this nature, and naturally it is usually the slowest secondary relaxation of the glassformer when more than one is present. The secondary relaxation found in totally rigid glassformer by Johari and Goldstein necessarily involves motion of the entire molecule. It has obvious connection with the primitive relaxation and it should conform to the correlation, as has been shown [672, 676, 677]. The polyalcohols, glycerol, threitol, xylitol, and sorbitol, obey the correlation [239, 195], but they are flexible glassformers with internal degrees of freedom and hydrogen bonding. Nevertheless, deuteron NMR found that deuterons at all positions in sorbitol show the same spin-lattice relaxation, indicating that the whole molecule contributes to the secondary relaxation [239]. For other glassformers in general, it is important to distinguish secondary relaxations that conform to the correlation and have properties like that of the primitive relaxation. For these secondary relaxations, I have been calling them the *Johari–Goldstein β-relaxations*. This nomenclature serves me to kill two birds with one stone. First, a term is needed to distinguish secondary relaxations that resemble the primitive relaxations, and like the primitive relaxations, they have properties mimicking the α-relaxation. Second, it honors Johari and Goldstein for what I consider as an important discovery of secondary relaxation in totally rigid glassformers. It is for the same intents and purposes that the term *Johari–Goldstein β-relaxation* (JG β-relaxation) or simply JG relaxation is used throughout this book by me. However, it must be made clear that neither Johari nor Goldstein had distinguished and separated out secondary relaxations belonging to the special class from other unimportant ones, like I did starting in 1998 [251]. Therefore, their points of view or model interpretations are not necessarily the same as mine, especially on those I call JG β-relaxation.

The term, JG β-relaxation, is used merely to distinguish secondary relaxation within a class that bears symbiotic relation with the α-relaxation. It does not give any detail about the nature of the molecular motion involved in the secondary relaxation belonging to this class, which can vary greatly from one system to another. There is a hidden danger when using this term. Someone who has knowledge of the nature of the JG β-relaxation in some family of glass-forming systems, say amorphous polymers, may object the use of the same term for secondary relaxation in other systems, such as metallic glassformers or water, having very different nature. JG β-relaxation in polymer only involves restricted rotation of the bonded repeat unit, while the JG β-relaxation in metallic glass can translate, and in water can rotate and translate. This objection has been issued in the recent past based on past experience in the JG β-relaxation of polybutadiene to say that the observed secondary relaxation of water cannot be the JG β-relaxation. This confusion can be avoided if one bears in mind that JG β-relaxations in different systems can differ in characteristics. The use of the non-specific term, primitive relaxation, instead of JG β-relaxation for the secondary relaxation belonging to the special class has the advantage of eliminating the unnecessary confusion and objection.

Objection to the use of the term, JG β-relaxation, even for my purpose was raised by some practitioners of dielectric spectroscopy, especially by Graham Williams at the 2007 International Dielectric Society Conference in Lyon, France. The arguments he gave was that secondary relaxations, including the kind I emphasize, have been found in polymers (see [106]) long before the works of Johari and Goldstein. Williams maintained his objection although I explained to him that the term is merely used to distinguish a special class of secondary relaxation that is the true precursor of the structural α-relaxation analogous to the primitive relaxation of the coupling model (see Section 2.2.1.3).

Other workers have used the same term, Johari–Goldstein β-relaxations, for a different reason. The mode coupling theory (MCT) has a fast relaxation process in the GHz regime which is also called the β-process [31]. This fast process is not thermally activated and is not a true secondary β-relaxation in the usual sense. Nevertheless, the similar terminology, β-process, used by MCT has caused confusion with the true secondary β-relaxation. To distinguish the true secondary β-relaxation from the fast β-process of the MCT, the names Johari–Goldstein has been appended to it by others, calling it the Johari–Goldstein β-relaxation [682, 971, 972]. This renaming of secondary relaxation is understandably motivated by the desire by others also to recognize the contribution of Johari–Goldstein in the field of secondary relaxations. However, in this practice by others, any observed secondary relaxation would be called Johari–Goldstein relaxation without differentiating their properties as was done in many papers. An example is the secondary relaxation showing up as the only resolved dielectric loss peak at ambient pressure in dipropylene glycol (2PG), tri-propylene glycol (3PG) [657, 658], diethyl phthalate (DEP) [660], and dibutyl phthalate (DBP) [661]. On applying pressure, the loss peak does not shift to lower frequencies, unlike the α-loss peak. Thus, the resolved secondary relaxation in 2PG, 3PG, DEP, and DBP is not the JG β-relaxation because it is unlike the α-relaxation at least as far as pressure dependence is concerned. Based on the chemical structure of DEP and DBP, the resolved relaxation originates from the rotation of the alkyl group about the flexile oxygen bond and only involves a part of the molecule, and this is another reason why it is not the JG relaxation. As will be discussed later, the JG relaxation in these glassformers does exist but is not resolved, but become resolved on elevating pressure in the case of 2PG and 3PG [657, 658], and by mixing with a higher T_g glassformer in the case of DEP and DBP [660, 661]. It is interesting to note that the authors of [657] inadvertently called the resolved secondary relaxation at ambient pressure as the JG β-relaxation even though its relaxation time is pressure independent, and should be called the non-JG relaxation. This misrepresentation is understandable because [657] was published in 1993, a year before the paper on classification of secondary relaxations into JG and non-JG appeared [685].

On the other hand, secondary relaxations in some other glassformers such as poly(phenylglycidylether) (PPGE) from [673, 973], dipropyleneglycol dibenzoate (DPGDB) [669, 678], BIBE [669, 974], PMMA with three repeat units [975], picoline [676], or quinaldine [677, 976] in mixtures with tristyrene [974] have been observed to shift to lower frequencies on elevating pressure, as well as in other glassformers to be named later. Calling all of them Johari–Goldstein β-relaxations would obscure the difference in this and other important properties of secondary relaxation.

On applying sufficient high pressure on dipropylene glycol and tripropylene glycol, it turns out that a new secondary relaxation emerges in the isothermal dielectric loss spectra, which shifts to lower frequencies with increasing pressure. Also in other glassformers including DPGDB [678], PPGE [673, 973], DHIQ [686, 977], there are two resolved secondary relaxations, with one sensitive to pressure and the other not. It would be a disservice to call both secondary relaxations by the same name of Johari–Goldstein relaxation.

JG relaxation seems to be a universal feature of glassformers of all kinds. Its presence is clearly evident in the isothermal or isochrono relaxation spectra in many glassformers but may require special procedures to detect it in other glassformers. It is commonly found in small molecule glassformers and amorphous polymers, and also in wider classes of materials. In addition to CKN, phosphate-silicate glass, plastic crystals, metallic glasses mentioned before, JG relaxation was found in pharmaceuticals such as indomethacin, fananserine, aspirin, and ibuprofen, monosaccharide, disaccharides, polysaccharides, room temperature ionic liquids, aqueous mixtures, and hydrated proteins. All these cases will be discussed separately later.

In the subsections to follow, account is given of the interesting properties of the JG β-relaxation that show it is well connected to the α-relaxation, and hence it has fundamental importance and must be included when considering the dynamics leading to glass transition.

2.3.2 The Important Class of Secondary Relaxations That Are Well Connected to the Primary α-Relaxation: The Johari–Goldstein β-Relaxations

Since glass transition is manifested by the structural α-relaxation, secondary relaxation is important for glass transition consideration if and only if it bears some connection to the α-relaxation. If we know for sure that the secondary relaxation involves the motion of all parts of the molecule, then it is likely connected to the α-relaxation by virtue of the fact that α-relaxation involves more molecules but each molecule entirely. Also, if the α-relaxation is the product of the cooperative dynamics originating from the JG or primitive relaxation, it is natural to expect that the properties of the JG relaxation will mimic that of the structural α-relaxation. This connection between the two relaxations is not restricted to totally rigid small molecule glassformers. For example, the repeat units of the polymers, polyisoprene [978], 1,4-polybutadiene [762, 979, 980], and 1,2-polybutadienes [981] and poly(vinyl chloride) have all atoms in the main chain and no side group. Therefore, the secondary relaxation observed in these polymers by experiment and by simulations [80, 356] involves motion of all atoms of the repeat unit. Even for some polymers with side chains such as poly(n-methyl methacrylates) and poly (n-ethyl methacrylates), the secondary relaxation involves not only 180° flip of the $-O-(C\equiv O)-$plane of the side group between two potential energy minima but also the rocking motion of the part of the repeat unit on the main chain, the amplitude of which increases with temperature [754, 755]. Hence essentially all parts of the repeat unit are involved in the motion. Deuteron NMR studies have shown that all parts of the sorbitol molecule contribute equally to the dynamics of the β-relaxation [615].

If in general the properties of the secondary relaxation bear some relation to that of the α-relaxation, then it is intuitively obvious that the former must involve the motion of the entire molecule. The connection of the secondary relaxation to the α-relaxation we are looking for can be any of the following: (1) similarity in properties, (2) correlation between the two relaxations or relation of parameters characterizing them, and (3) modification of one relaxation leading to change in the other. Those secondary relaxations that exhibit one or more such connections belong to an important class, which we call the Johari–Goldstein (JG) β-relaxations. When specialized, each of these general criteria spawns many possible specific connections between the two relaxations that can be checked by experiments. The experimental data that show the various connections and support the "universal" presence of the JG β-relaxation are given item by item below.

2.3.2.1 Correlation Between the Ratio τ_α/τ_{JG} and n at a Predetermined Value of τ_α

As mentioned in the background material, the thought of possible existence of a class of JG secondary relaxations that are well connected to the α-relaxation comes from this very attribute of the primitive relaxation of the CM [251, 953], and the fact that secondary relaxations like the primitive relaxation are usually local and non-cooperative processes. The primitive relaxation involves the motion of the entire molecule and is the initiator of the structural α-relaxation. Thus, if a class of secondary relaxation important for glass transition (i.e., the JG relaxations) exists, it should have similar properties as the primitive relaxation, and its relaxation time, τ_{JG}, is approximately the same as the primitive relaxation time, τ_0, of the CM. In other words, the approximate equality, $\tau_{JG} \approx \tau_0$, is expected. Although the CM has been discussed before in some detail in Section 2.2.1.3, it is worthwhile to introduce once more here its signature equation,

$$\tau_\alpha = [t_c^{-n}\tau_0]^{1/(1-n)}, \tag{2.67}$$

which links τ_α and τ_0. Here n referred to as the coupling parameter also is the same quantity appearing in the Kohlrausch correlation function for the α-relaxation in Eq. (1.1). It is the indicator of non-exponentiality or the extent of the many-body α-relaxation dynamics. t_c is the onset time of many-body dynamics, the magnitude of which is entirely determined by the interaction potential. It should be approximately the same for glassformers belonging to the same family, and is insensitive to T and P. For carbon-based small molecule and polymeric glassformers, neutron scattering experiments and molecular dynamics simulations have indicated that t_c is about 2×10^{-12} s, as discussed before in Section 2.2.1. Equation (2.67) when rewritten as

$$\tau_0/\tau_\alpha = (\tau_\alpha/t_c)^{-n}, \text{ or } (\log\tau_\alpha - \log\tau_0) = n(\log\tau_\alpha - \log t_c) \tag{2.68}$$

clearly shows, at a fixed τ_α, the ratio τ_0/τ_α decreases or the difference ($\log\tau_\alpha - \log\tau_0$) increases monotonically with increasing n since $(\tau_\alpha/t_c) >> 1$ is usually satisfied in most experimental investigation. Hence, a correlation exists between τ_0 and n for any chosen fixed value of τ_α. This result together with the hunch that $\tau_{JG} \approx \tau_0$ led to the approximate relations

$$\tau_{JG}/\tau_\alpha \approx (\tau_\alpha/t_c)^{-n}, \text{ or } (\log\tau_\alpha - \log\tau_{JG}) \approx n(\log\tau_\alpha - \log t_c). \tag{2.69}$$

These approximate relations encouraged me to search for the possible correlation between τ_{JG} and n by examining published experimental data of many carbon-based small molecule glassformers and amorphous polymers at ambient pressure. A strong correlation was found and shown amply by tables and figures in [195, 251, 953]. Instead of duplicating these older results, new results are presented here. This is a *cross* correlation between the quantity τ_{JG} coming from the secondary relaxation and n (or the width of the dispersion) which is a characteristic of the primary α-relaxation. This *cross* correlation evidently indicates the existence of a non-trivial connection between the JG relaxation and the α-relaxation. Before 1998, criteria for JG relaxation are virtually non-existent for glassformers in general except in the obvious cases of totally rigid molecular glassformers and polymers with repeat unit devoid of side group. Fortunately, most of the glassformers used to show the correlation turn out to be genuine JG relaxations as judged by criteria established thereafter.

The most assuring test of such a cross correlation is best carried out on glassformers from the same family. An example is the polyalcohols, sorbitol, xylitol, threitol, and glycerol [615]. The coupling parameter n decreases with decreasing molecular weight [195], and indeed the separation between the α- and the JG β-relaxation measured by the ratio τ_{JG}/τ_α at constant τ_α increases in this order. Here I show the correlation obtained by Capaccioli et al. from mainly van der Waals glassformers in Fig. 106 (right panel).

Another example is the family of poly(n-alkyl methacrylates) [953(b), 969, 982]. Higher member of the family with longer side chains has smaller n because of increase of internal plasticization by more mobile methylene units in the side chain. Again the same correlation of τ_β with n at constant τ_α is found.

One can change the coupling parameter of poly(n-butylmethacrylate) (PnBMA) by incorporating styrene (S) into it to form a series of P(nBMA-*stat*-S) random copolymers with different styrene contents. Styrene is less mobile than butylmethacrylate [970(a)]. Consequently, the random copolymer with larger styrene content has stronger intermolecular coupling for the motion of the butylmethacrylate

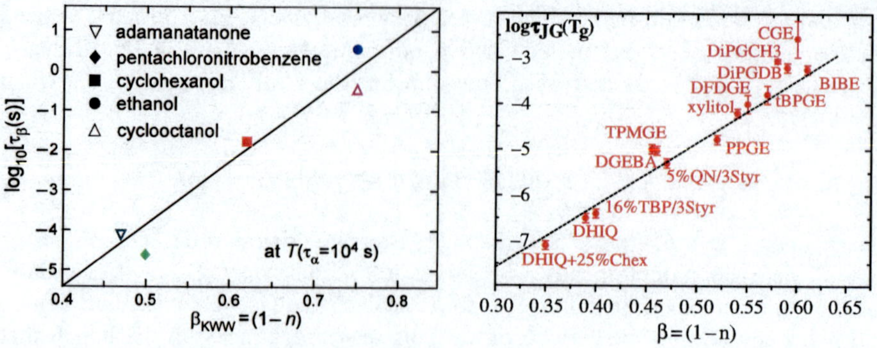

Fig. 106 (*Right*) JG relaxation time evaluated at T_g vs. coupling parameter n, where T_g is uniformly defined as the temperature at which $\tau_\alpha = 10^2$ s. The *dotted line* is calculated according to Eq. (2.69). Reproduced from [616] by permission. (*Left*) β-relaxation time vs. α-relaxation width-parameter β_{KWW} for five plastic-crystalline materials. Both parameters were taken at T_g, defined as the temperature where $\tau_\alpha = 10^4$ s. The *line* shows the prediction of the coupling model according to Eq. (2.69). Data from [145] redrawn

units probed by dielectric relaxation, resulting in increasing separation between the α- and the JG β-relaxation measured by the difference ($\log\,\tau_\alpha - \log\,\tau_{JG}$) [970(a)]. This is in fact observed experimentally [970(b)]. Actually, this effect is general and has been found in the α- and the JG β-relaxation of a component (such as picoline) when mixed with a less mobile (higher T_g) component (such as tri-styrene). The difference ($\log\,\tau_\alpha - \log\,\tau_{JG}$) increases with increasing concentration of the less mobile component [843, 844]. These data from mixtures will be discussed later in Section 2.3.2.10.

Plastic Crystals

For the family of plastic crystals [145], not only τ_β correlates with n at constant τ_α, but also τ_β is in good agreement with the primitive relaxation time τ_0 calculated from Eq. (2.67). The results are shown in the left panel of Fig. 106, where the label of the abscissa, β_{KWW}, is the same as $(1 - n)$.

More recently Bauer et al. [145] reported dielectric relaxation study of the plastic crystalline mixture of 60% succinonitrile and 40% glutaronitrile (60SN-40GN), the plastic phase of which can be easily supercooled. They found evidence of a slower secondary β-relaxation lying in between the α-relaxation and a pronounced faster secondary γ-relaxation. Bauer et al. found good agreement between the β-relaxation time and the primitive relaxation time calculated by the CM relation, indicating that the β-relaxation is the Johari–Goldstein of 60SN-40GN. The evidence for the presence of the β-relaxation in 60SN-40GN from the isothermal dielectric spectra is not strong because it is not directly resolved. The spectra of 60SN-40GN are similar to those of other glassformers including dipropylene glycol and tripropylene glycol [657, 658], diethyl phthalate [660], di-n-butyl phthalate (DBP) [661] and its isomer, diisobutyl phthalate (DiBP) [661], dioctal phthalate [684], m-fluoroaniline [671], and benzophenone [508, 1121]. Like 60SN-40GN, the isothermal dielectric spectra of all these glassformers are dominated by the α- and the resolved γ-relaxation, and the β-relaxation is not resolved, albeit its location indicated by the primitive relaxation frequency calculated by the CM relation in relation to the α- and the γ-loss peaks is similar to that found by Bauer et al. in 60SN-40GN. The reader can verify the similarity by comparing the spectra of 60SN-40GN with those of the others. Nevertheless, by various means including elevated pressure [657, 658, 671], physical aging [684], and mixing with another glassformer with higher T_g [508], the JG β-relaxation has been fully resolved in the other glassformers, and its existence is not in doubt. Moreover, the γ-relaxation does not shift its position on elevating pressure and hence it bears no relation to the α-relation [508, 657, 658, 671]. The proofs of existence of the JG β-relaxation in the other glassformers having similar dielectric spectra as 60SN-40GN should boost confidence in the JG β-relaxation found by Bauer et al. in 60SN-40GN.

Inorganic Glasses

It is difficult to obtain relaxation spectra of inorganic glassformers in the first place because of their high glass transition temperature, and limited number of available experimental techniques that can be used. Thus, the relation between the α- and the JG β-relaxation times and the correlation of τ_β with n at constant τ_α of inorganic glasses have not been checked except for a few exceptions [251] such as $2Ca(NO_3)_2$–$3KNO_3$ [951] by shear modulus, and B_2O_3 by calorimetry [983, 984]. Dynamic modulus measurements of the phosphate-silicate ($0.243Na_2O$–$0.269CaO$–$0.026P_2O_5$–$0.462SiO_2$) [952] and various oxide glasses [985] do not give sufficient information for the present purpose of examining the correlation.

However, recently Yu and coworkers [986, 987] were able to examine the relation between the α- and the JG β-relaxation times and its variation in a number of inorganic glasses of different compositions. The method is enthalpy relaxation by DSC scan of the rapidly quenched glass after annealing at temperature well below T_g apparently first used by Chen and coworkers to find the JG β-relaxation in polymers (polystyrene) [988], inorganic glasses (B_2O_3) [989], and metallic glasses, $Pd_{48}Ni_{32}P_{20}$ and $(Fe,Co,Ni)_{75}Si_{10}B_{15}$ [954, 955, 990, 991]. Annealing of the quenched glass well below T_g relaxes the enthalpy, particularly the part associated with the JG β-process. On reheating the annealed glass during the DSC scan, the lost enthalpy is recovered and detected by an endothermic sub-T_g peak, in the specific heat C_p spectrum that precedes the larger endothermic peak associated with the α-process and glass transition. Thus, the sub-T_g endothermic peak originates from the JG β-relaxation. Increase of annealing time leads to increase of the intensity of the sub-T_g peak and shifts it to higher temperatures. Chen and coworkers emphasized their observed relaxation at temperatures well below T_g by calling it sub-sub-T_g relaxation. They also pointed out that the sub-sub-T_g relaxations of polystyrene, B_2O_3, and metallic glasses are qualitatively similar in properties despite great differences in structure and bonding.

Bershtein and Egorov [984] applied the method for polymers, and obtained the activation enthalpy, E, of the process responsible for the sub-T_g peak from the change of the peak temperature T_p, with heating rate, q, by the relation $E = -R[\mathrm{d}\ln q/\mathrm{d}(1/T_p)]$. Identification of the sub-T_g peak with the JG β-relaxation is confirmed by good agreement with values obtained by dielectric and mechanical measurements for several polymers. The same was found in pharmaceutical substances, poly(vinylpyrrolidone) (PVP), indomethacin (IM), and ursodeoxycholic acid (UDA) [992]. The same equation was also used by Chen and coworkers in their study of inorganic and metallic glasses, and the value of the activation enthalpies obtained is significantly smaller than that of the α-relaxation, and it belongs to the JG β-relaxation.

Yue and coworkers [986, 987, 994] applied the method to hyperquenched inorganic glasses obtained by cooling from the melt at very large cooling rate of about 10^6 K/s. The glasses studied include GeO_2 ($m = 17.5$), NIST710a ($m = 44.6$), E-glass ($m = 50.2$), Basalt ($m = 62.5$), 45CaO–55SiO_2 ($m = 90.4$), and 55CaO–45SiO_2 ($m = 141.2$) in order of increasing "fragility" index m obtained from viscosity. In one experiment, the excess specific heat $C_{p,\mathrm{exc}}$ spectrum was obtained by subtracting the specific heat C_p spectrum of a slow cooled glass from that of the hyperquenched glass of the same inorganic material. The exothermic peak in the $C_{p,\mathrm{exc}}$ spectrum as a function of temperature is asymmetric and its width increases with increasing m (see Fig. 107-1). Two peaks seen in the broader spectrum of the more "fragile" glass are coming from the α- and the JG β-relaxations. The

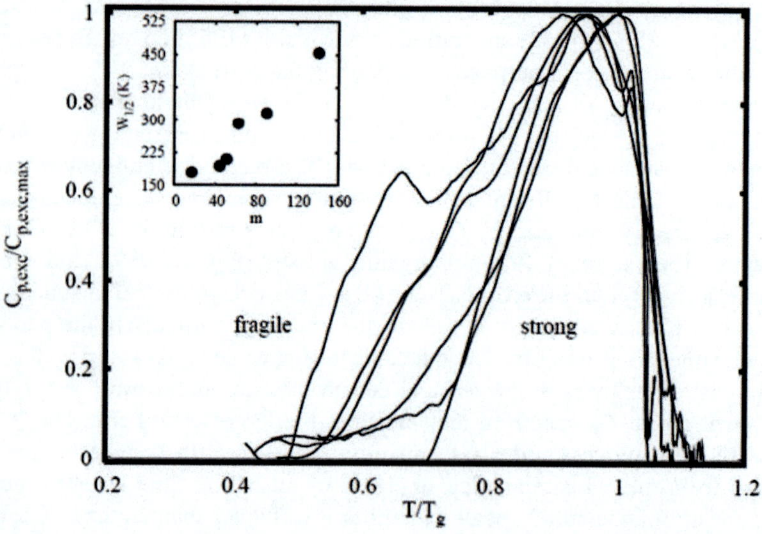

Fig. 107-1 The $C_{p,\mathrm{exc}}/C_{p,\mathrm{exc,max}}$ ratio as a function of the T_g-normalized temperature, T/T_g, for all samples ranging from fragile to strong, where $C_{p,\mathrm{exc,max}}$ is the peak value, i.e., the maximum excess heat capacity of the $C_{p,\mathrm{exc}}$ spectrum. The *inset* shows the dependence of the half-widths ($W_{1/2}$) on fragility. $W_{1/2}$ is determined as the width of the original spectra at a value of $0.5C_{p,\mathrm{exc,max}}$, and dimensionless. Reproduced from [986, 987] by permission

peak at lower temperature is less intense and is associated with the JG β-relaxation The separation in temperature between the two peaks decreases with decreasing m to the point that the lower temperature peak can no longer be resolved and appears as a shoulder in the $C_{p,\mathrm{exc}}$ spectrum of GeO_2 ($m = 17.5$) and NIST710a ($m = 44.6$). From the results, it is clear that the separation between the α- and the JG β-relaxations correlates with m. The non-exponentiality parameters n of all the glasses studied are unavailable. If indeed n correlates with m for the glasses studied, the results of Yu and coworkers provide another example of correlation of n with the separation between the α- and the JG β-relaxations in inorganic glasses.

GeO_2 has the smallest m and also n is nearly 0 from the exponential time dependence found by volume relaxation measurement [993]. More detailed study of the C_p spectrum of hyperquenched GeO_2 was made by Hu and Yue [994]. From the data they find τ_{JG} has Arrhenius temperature dependence with activation energy $E = 155.7$ kJ/mol and is equal to $23.5RT_g$ typical for JG β-relaxations [971, 995, 996], and $\tau_{JG}(T_g)$ is equal to 10 s. This long $\tau_{JG}(T_g)$ of GeO_2 together with the nearly zero n value is in accord with the empirical linear relationship between log $\tau_{JG}(T_g)$ and n established before by considering many glassformers [251, 997], and the prediction of the extended CM [195].

The fact that Yue and coworkers can resolve the JG β-relaxation in GeO_2 is primarily due to hyperquenching, which increases the relaxation strength of the JG β-relaxation relative to the α-relaxation. This is a technique to reveal the JG β-relaxation used by others in organic glassformers such as *ortho*-terphenyl having much lower T_g than inorganic glassformers to be discussed later. The rationale for the technique is the larger amplitude of motion for the JG β-relaxation in hyperquenched glasses with larger specific volume and entropy. An analogue can be found in the isothermal dielectric spectra of organic glassformers in the equilibrium liquid state. At temperatures high above T_g, where the JG β-relaxation is merging with the α-relaxation, the relaxation strength of the JG β-relaxation can become comparable to the α-relaxation in some glassformers.

Amorphous thin films of chalcogenide glasses of $Ge_1Sb_4Te_7$, $Ge_1Sb_2Te_4$ and $Ge_2Sb_2Te_5$ have found the JG β-relaxation [994], with $\tau_{JG} \approx \tau_0$.

Pearl Necklace Model of Polymer

Molecular dynamics (MD) simulations of polybutadiene (PBD) chains with reduced dihedral barriers (LB-PBD) by Bedrov and Smith [80, 356(a), 356(b)] shows the presence of both the α- and JG β-relaxation well separated from each other within the timescales of MD simulations. Nevertheless, the structure of LB-PBD chains contains multiple intramolecular mismatches (different atom sizes, bond lengths, bend angles, and dihedral potentials), and it is not clear which entity is involved in the relaxation, and whether the existence of the JG β-relaxation requires specific molecular motion in the polymer. To clarify this point, Bedrov and Smith [356(c)] have recently made MD simulations of the simplest polymer devoid of molecular structure, where the polymer chains were represented by a simple bead-necklace model to see if they could find a clearly resolved JG process. The polymer chains were represented by a collection of identical beads interacting via Lennard-Jones

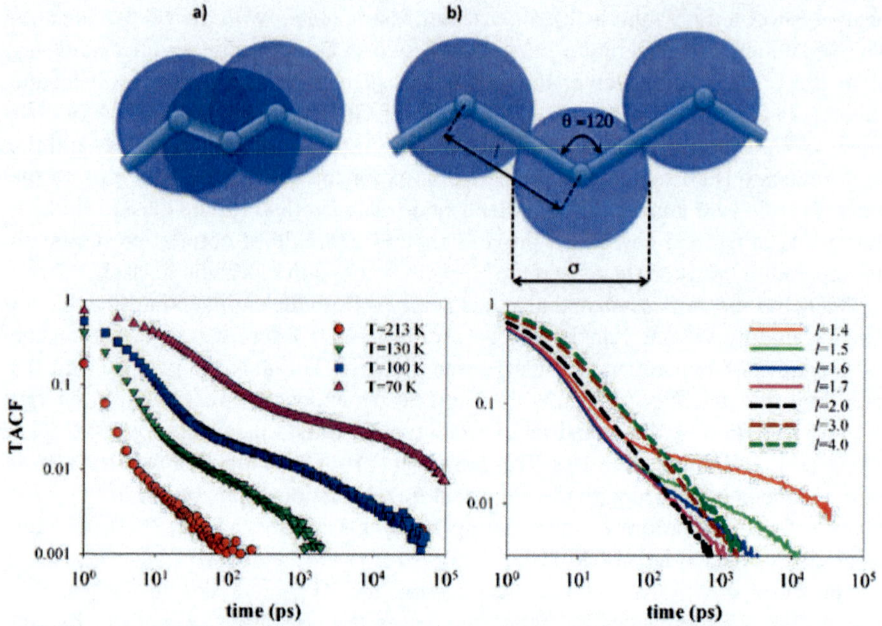

Fig. 107-2 (*Upper*) Schematic representation of bead-necklace model used in this work (**a**) $\sigma/l = 2.85$, (**b**) $\alpha/l = 1.0$. (*Lower left*) Torsional autocorrelation function (TACF) of bead-necklace model with bond length $l = 1.5$ Å obtained at several temperatures. (*Lower right*) TACF of bead-necklace model with bond length l ranging from 4.0 to 1.4 Å obtained at 100 K. Courtesy of D. Bedrov and G. Smith

potential with $\sigma = 4.0$ Å and $\varepsilon = 0.0936$ kcal/mol. The beads were connected by rigid (constrained) bonds of length l. Bends were also kept rigid at 120° to minimize interference from short-time dynamics by the high-frequency vibrational motions (see upper panel of Fig. 107-2). No dihedral potentials were applied. The system contained 100 chains, each comprised of 76 beads. Segmental dynamics/relaxations have been characterized by torsional autocorrelation function (TACF).

Bedrov and Smith [356(c)] studied the dynamics of melts comprised of chains with l ranging from 1.4 to 4.0 Å. The TACF for melts with $l = 1.5$ Å obtained at several temperatures in the lower left panel of Fig. 107-2 shows the two processes. The slower α-relaxation shifts to longer times much more than the faster β-relaxation as temperature decreases, resulting in increased separation between these processes. Thus, the β-relaxation exists even in the bead-necklace model which has no internal motional degree of freedom. This result for polymers by Bedrov and Smith is as fundamentally significant as the discovery of secondary relaxation in totally rigid molecules by Johari and Goldstein.

The influence of the extent of bead overlap or interaction on the dynamics can be deduced by comparing the TACFs of melts with different l ranging from 1.4 to 4.0 Å. For l=4.0 Å the bond length is equal to the bead diameter resulting in tangent bead-necklace chains, while for $l = 1.4$ Å there is considerable overlap between bonded beads. In the lower right panel of Fig. 107-2 the TACFs from MD simulations at atmospheric pressure and 100 K for bead-necklace chains with different bond lengths l are compared. On decreasing l or increasing overlap between bonded beads, several changes can be observed. First, for the α-relaxation, its time dependence becomes more stretched (larger n) and its relaxation time becomes longer. Second, the separation between the α- and the JG β-relaxation increases. Hence, the results of bead-necklace model from Bedrov and Smith proves once more the correlation between n and the separation between the α- and the JG β-relaxation. This figure clearly shows that for melts with $l > 2.0$ Å the time dependence of the α-relaxation approaches exponential (i.e., n is small), showing no clear separation between the α- and the JG β-relaxation, and hence the latter cannot be resolved. Fourier transform of the TACF may show an excess wing on the high-frequency flank of the α-loss peak. This serves as an analogue of unresolved JG β-relaxation in molecular glassformers caused by small n. In the bead-necklace model, this is due to small overlap between bonded beads. Bedrov and Smith also study the dynamics of the bead-necklace model at elevated pressures. The results of these studies will be discussed in Section 2.3.2.8 below together with other results [997–999].

Pharmaceuticals. The JG β-relaxation has been found in pharmaceuticals including indomethacin [992(b), 1000(a),(b)], acetyl salicylic acid (aspirin) [1003], fananserine ($C_{23}H_{24}FN_3O_2S$) [1000(c)], and ibuprofen ((2*RS*)-2[4-(2-methylpropyl)phenyl]propanoic acid, $C_{13}H_{18}O_2$) [1004]. The data will be shown later in another subsection.

2.3.2.2 Good Correspondence Between τ_{JG} and the Primitive Relaxation Time τ_0 at Ambient Pressure

In Section 2.3.2.1, JG relaxations of different glassformers when examined collectively reveal the cross correlation between τ_{JG} and n. Here we consider the glassformers individually. Since the publication of the 1998 paper on relation between τ_{JG} and τ_0 [251], a number of papers have shown for many glassformers that there is good correspondence between τ_{JG} and τ_0 at various temperatures and pressures, i.e., $\tau_{JG} \approx \tau_0$. Before we proceed further to show examples, it is worthwhile to comment that approximate equality between τ_{JG} andτ_0 can only be expected. This is because in practice τ_{JG} is usually obtained from a characteristic time of the experimentally observed spectrum but it is not entirely clear that it should be exactly τ_0 because the exact evolution of many-body relaxation with time and the resulting frequency spectrum after the onset at τ_0 is not yet known. If the technique used is dielectric spectroscopy and the JG relaxation is observed as a loss peak, it is common practice to model the JG secondary relaxation by a Cole–Cole distribution of relaxation times. To obtain τ_{JG}, the Cole–Cole distribution is added onto the α-relaxation represented by another empirical function, and τ_{JG} is determined by fitting the spectrum. The William–Watts ansatz (Eq. (2.66)) is another method of choice. Both methods are invalid because they have assumed that the secondary and the primary relaxations are unrelated processes, which is at odds with the cross correlation found and other relations to be described below [120, 195, 618, 685, 997], and the NMR experiment which has proven directly that they are intertwined [998, 999]. The results of this NMR experiment are important and shall be discussed in a subsection to follow. Therefore, the τ_{JG} obtained by these commonly used methods of analyzing spectra may not reflect its actual value. This was done even by me in the past in several studies that are collaborations with others. Examples are the analyses of dielectric loss spectra of the plastic crystal cyclohexanol [145], glycerol, propylene carbonate, and propylene glycol [424], where the Cole–Cole distribution is added onto the α-relaxation modeled by the one-sided Fourier transform of the Kohlrausch function or the other empirical functions. Figure 108 shows this practice for cyclohexanol, where the vertical arrow indicates the location of the primitive frequency, $\nu_0 \equiv 1/2\pi\tau_0$, calculated with $n \approx 0.38$ from the Kohlrausch fit to the α-loss peak. It can be seen that the peak frequency of the Cole–Cole fit underestimates ν_0, although the two have the same order of magnitude. For glycerol and propylene carbonate at temperatures near T_g, the Cole–Cole distribution is so broad that a significant part of its contributions occurs at frequencies lower than the α-loss peak frequency, contradicting the JG β-relaxation is the precursor of the α-relaxation, a fact which will be amply substantiated by experimental data to be discussed in the many subsections to follow. Ideas of how the relaxation dynamics proceed with time can be gained from the results of real-time evolution of dynamics of colloidal particles seen by confocal microscopy [141, 800] and molecular dynamics simulations of Li ions in Li metasilicate glass [802]. In order of increasing time, the relaxation spectrum is contributed first by relaxation of caged molecules exhibited as nearly constant loss in the frequency-dependent susceptibility spectrum, which is terminated when the local primitive relaxation makes its appearance. This is followed by continuous buildup of many-body relaxation with participation of increasing number of molecules with time until reaching the terminal relaxation with the maximum possible number of molecules (or length scale) and the correlation function follows the time dependence of the Kohlrausch function. Although this qualitative description of evolution of many-body relaxation with time deduced from the experimental and simulation data is reasonable, a quantitative theoretical description of the evolution with time has yet to be constructed by anyone. Thus, the experimentally deduced τ_{JG} from the loss peak frequency may not be exactly the same as τ_0, and thus only the approximate relation, $\tau_{JG} \approx \tau_0$, can be expected in the absence of a viable theory to quantitatively account for the evolution of dynamics by the author or anyone else.

The tests of good correspondence between τ_{JG} and τ_0 at various temperatures, i.e., $\tau_{JG} \approx \tau_0$, are mostly carried out at temperatures above T_g where τ_{JG}, τ_α, and n can be determined from the spectra and τ_0 calculated by

$$\tau_0 = (t_c)^n (\tau_\alpha)^{1-n}, \tag{2.70}$$

the equivalent of Eq. (2.67). There are many such examples in the past 10 years which will be introduced step by step in subsections to follow. Some examples from different classes of glassformers have been given before [139, 145, 195, 252,

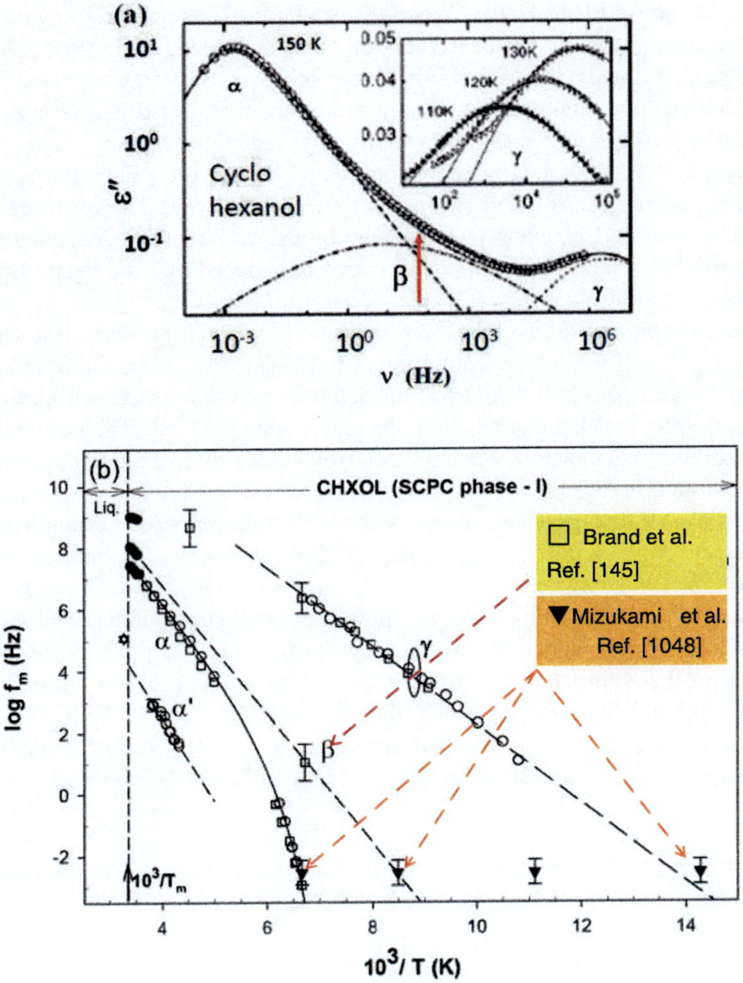

Fig. 108a (**a**) Dielectric loss spectrum of cyclo-hexanol at 150 K. The *solid line* is a fit with the sum of a HN function (*dashed line*) and two CC functions (*dash-dotted and dotted lines*). The *vertical arrow* indicates the location of the primitive relaxation frequency, ν_0, calculated by the CM equation with the value of n determined by fitting the α-loss peak by the Fourier transform of the Kohlrausch function. The *inset* shows the region of the γ-relaxation at low temperatures. The *lines* are fits with the CC function. Reproduced from [145] by permission. (**b**) Complete Arrhenius diagram depicting the α′, α, β, and γ-processes found in plastic phase (I) of cyclo-hexanol (CHXOL) obtained by authors of [681(a)]. Also included are the data of others for the purpose of comparison. A *vertical line* is drawn at the melting temperature T_m. Also shown by *filled inverted triangles* are the adiabatic calorimetry data for the several processes reported by Mizukami et al. [1010(b)] which approximately correspond to enthalpy relaxation of $\log f_m(\text{Hz}) = 2.45 \pm 0.45$. Reproduced from [681(a)] by permission

569(a), 660, 669–680, 684, 703, 978, 997, 1000–1010]. Reasonably good agreement between τ_{JG} and τ_0 has been found in many different glassformers wherever the resolved secondary relaxation is truly the JG relaxation. Hence it is sufficient here to demonstrate this by a few chosen examples in widely different families of glassformers to illustrate the general validity of the relation $\tau_{JG} \approx \tau_0$.

Plastic crystals. The example of cyclohexanol can be seen in Fig. 108. The spectra shown in part (a) are taken from [145] by Brand et al. The location of the primitive relaxation frequency ν_0 calculated by the CM equation indicated by the arrow in the figure is in good agreement with that deduced by Brand et al. from the spectrum.

The relaxation map of cyclohexanol presented as Part (b) of Fig. 108 showing all processes found is adapted from a figure in [681(a)] by Singh and Murthy. The most important aspect in this figure is the detection of the JG β-relaxation by adiabatic calorimetry by Mizukami et al. in the glassy state [681(c)], which supports the identification of this process by Brand et al. from dielectric relaxation data above T_g as demonstrated in Fig. 108.

van der Waals liquids. There are many such examples. Here one example chosen from dielectric loss data of ethylbenzene [1002] is shown here in the left panel of Fig. 109.

On the right panel of Fig. 109, we show the good correspondence of τ_{JG} and τ_0 calculated from τ_α and n (determined by fitting the α-loss peak by the Fourier transform of the Kohlrausch function) in the relaxation map of two glassformers, dipropyleneglycol dibenzoate (DiPGDB) and DiPG-CH_3 [616, 678]. These two glassformers have different molecular structures and widely different glass transition temperatures ($T_g = 220$ K for DiPGDB and 137 K for DiPG-CH_3). DiPGDB

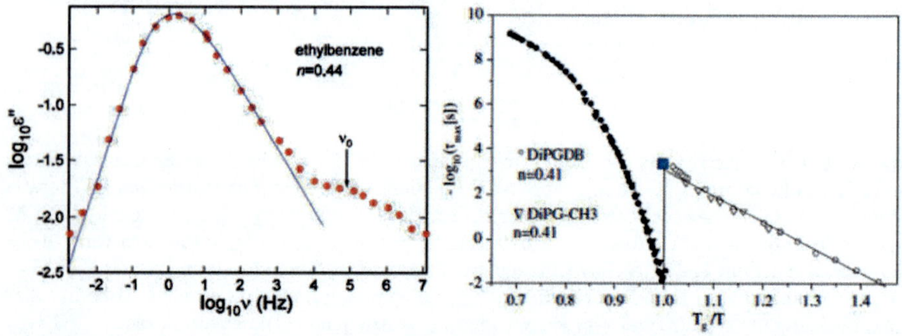

Fig. 109 (*Left*) Dielectric loss data of ethylbenzene. Data from [1002] replotted. (*Right*) Relaxation map representing the structural (*closed symbols*) and secondary JG (*open symbols*) relaxation time in two glassformers having the same n parameter. The temperature scale is scaled for the value of T_g (220 K for DiPGDB and 137 K for DiPG-CH_3). *Circles* refer to DiPGDB data and *triangles* to DiPG-CH_3. At the glass transition the distance of the timescales of the two processes is the same for both glassformers. The *filled large square* is the value of $-\log \tau_0$ calculated by the CM equation at $\tau_\alpha = 100$ s for both glassformers. Reproduced from [616, 678] by permission

is dominated by the presence of the bulky benzoate group, which is replaced by the small methyl group to form DiPG-CH3. Here the dielectric T_g is defined by $\tau_\alpha(T_g) = 100$ s. By scaling temperature by T_g, the τ_{JG} and τ_α data of the two glassformers can be presented together. Interestingly and coincidentally the two glassformers have the same value of n for the same τ_α, and in particular $n(T_g) = 0.41$ at $\tau_\alpha = 100$ s. The lone filled square in the right panel of Fig. 109 is the value of –log τ_0 calculated by the CM equation at $\tau_\alpha = 100$ s for both glassformers. There is good agreement between τ_{JG} and τ_0 when $\tau_\alpha = 100$ s. One can see the relaxation map of τ_{JG} and τ_α of the two glassformers are isomorphic after scaling T by T_g, and obviously the two glassformers have the same fragility. As a bonus, the example presented in the figure illustrates that $n(T)$ controls the entire dynamics irrespective of the difference in chemical structure and T_g.

Amorphous polymers. For the class of glassformers that have resolved JG relaxation in the equilibrium liquid state, I chose the amorphous polymer 1,4-polyisoprene (PI) [978]. Figure 110 presents the isothermal dielectric data of PI taken as several temperatures above the dielectric $T_g = 207.5$ K defined by $\tau_\alpha(T_g) = 100$ s. The fits of the asymmetric α-loss peaks at 216.0, 211.2, 208.2, and 204.2 K by the one-sided Fourier transform of the Kohlrausch function yield $n = 0.53$ independent of temperature (as shown for 211.2 and 208.2 K by *dashed lines*

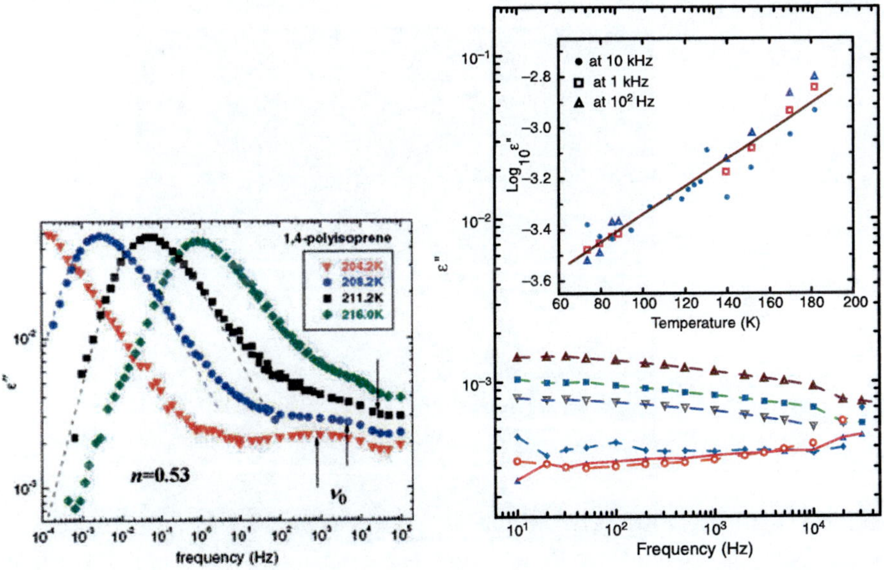

Fig. 110 (*Left*) Dielectric loss data of 1,4-polyisoprene at four temperatures showing the presence of the JG β-relaxation. The *dashed lines* are fits by the Fourier transform of the Kohlrausch function with n=0.53. The *vertical arrows* indicate the primitive relaxation frequency ν_0 calculated. (*Right*) The *inset* shows the nearly constant loss measured in PI at representative temperatures below T_g. *From top to bottom*: 169.7, 151.7, 139.7, 85.5, 79.5, and 73.5 K. The main figure shows the nearly constant loss as a function of temperature at 0.10, 1, and 10 kHz. The *line* is a least-squares fit to all data points by the function $A\exp(T/T_0)$, where A = 0.456, and $T_0 = 79.8$ K

in the figure). With $n = 0.53$, τ_0 is calculated using Eq. (2.67) and $t_c = 2 \times 10^{-12}$ s from the τ_α at 212.2 K, 208.2 K, and 204.2 K. The frequency $\nu_0 = 1/(2\pi\tau_0)$ for each temperature is indicated in the figure by a vertical arrow. For $T = 212.2$ and 208.2 K, the calculated ν_0 is located on the broad shoulder of the JG β-relaxation. For $T = 204.2$ K, ν_0 is near the secondary relaxation peak maximum. The typical VFTH T-dependence of τ_α above T_g and the Arrhenius T-dependence of τ_{JG} near and below T_g are shown in Fig. 110. The agreement between τ_0 and τ_β or τ_{JG} at 204.2 K is further displayed in Fig. 110 by near coincidence of the calculated τ_0 (*the lone open circle*) and τ_{JG} (*closed circles*) at this temperature. The good correspondence between τ_0 and τ_β or τ_{JG} is verified in 1,4-polyisoprene as well as in 1,4-polybutadiene (see Fig. 2 in [979]).

Room temperature ionic liquids: 1-hexyl-3-methylimidazolium chloride (or bromide). The room temperature ionic liquid, 1-hexyl-3-methylimidazolium chloride (or bromide), was discussed before in Section 2.2.4.3 and shown in Fig. 50 for superpositioning of the electric loss modulus $M''(\nu)$ data for different combinations of P and T at fixed ν_α [747(a)]. Here in Fig. 111-1, we present some sample spectra of the electric modulus $M''(\nu)$ and $M'(\nu)$, which show the JG β-relaxation and a much faster γ-relaxation at temperatures above and below T_g. The fit of the one-sided

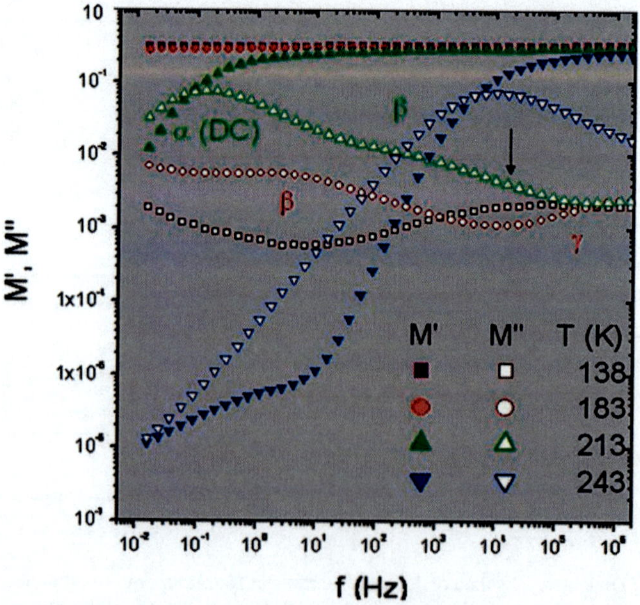

Fig. 111-1 Electric loss modulus of the room temperature ionic liquid, 1-hexyl-3-methylimidazolium chloride, showing the presence of the JG β-relaxation at two temperatures 183 and 213 K together with the γ-relaxation at higher frequency. The *vertical arrow* indicates the primitive relaxation frequency ν_0 calculated for $T = 213$ K. Courtesy of M. Mierzwa. To be published as [747(a)]

Fourier transform of the Kohlrausch function to the loss peak $M''(\nu)$ taking into consideration the broadening on the high-frequency side by the β-loss yields $n = 0.44$ at temperatures near T_g. The ν_0 calculated via Eq. (2.67) and $t_c = 2 \times 10^{-12}$ s from the ν_α at 213.0 K is about 10^4 Hz. Its location, indicated by the arrow in Fig. 111, is within the frequency range where $M''(\nu)$ shows an excess loss indicating the presence of the JG β-relaxation, and thus ν_0 is in order-of-magnitude agreement with ν_{JG}.

Room Temperature Ionic Liquids: 1-Butyl-3-methylimidazolium bis(trifluoromethylsulfonyl)imide. The appearance of the JG β-relaxation in another ionic liquid, 1-butyl-3-methylimidazolium *bis*(trifluoromethylsulfonyl)imide (BMIM-BMSF), was found by Rivera and Rössler [747(b)]. These authors identified it as the JG β-relaxation based on the facts that it is the slower of two secondary relaxations (see Fig. 111-2, upper left panel at 130 K), it merges with the structural relaxation,

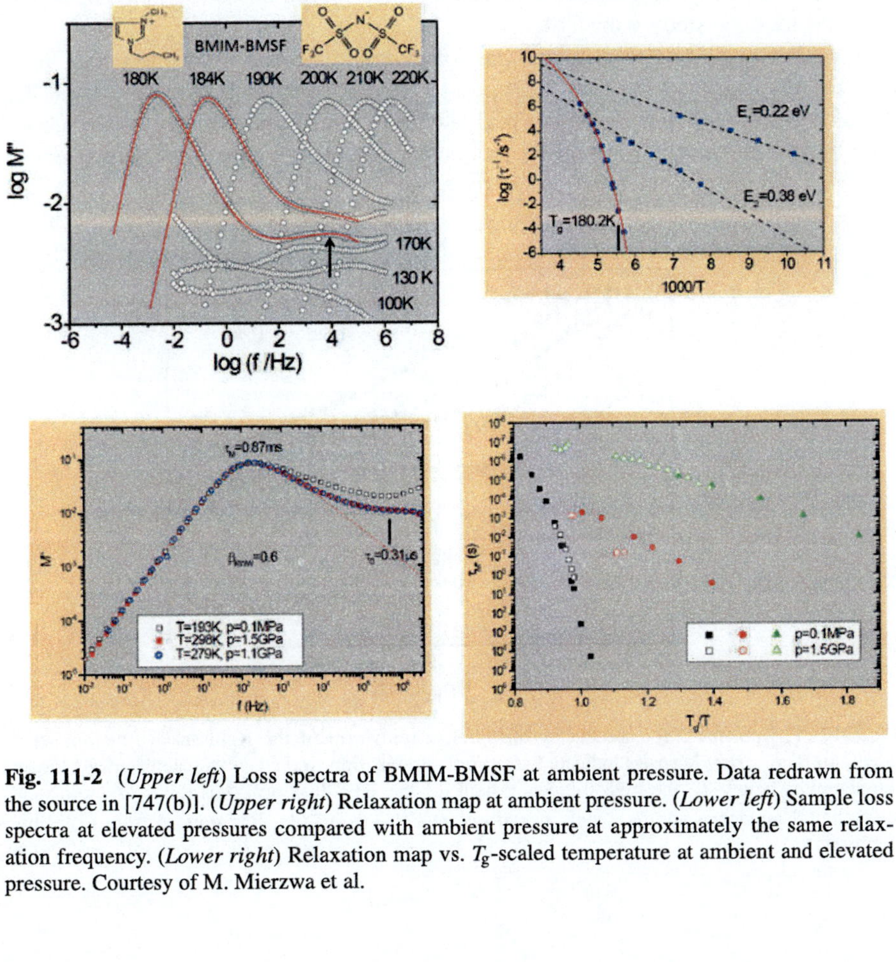

Fig. 111-2 (*Upper left*) Loss spectra of BMIM-BMSF at ambient pressure. Data redrawn from the source in [747(b)]. (*Upper right*) Relaxation map at ambient pressure. (*Lower left*) Sample loss spectra at elevated pressures compared with ambient pressure at approximately the same relaxation frequency. (*Lower right*) Relaxation map vs. T_g-scaled temperature at ambient and elevated pressure. Courtesy of M. Mierzwa et al.

and its relaxation time has Arrhenius activation energy of about $24k_BT_g$ and pre-exponential factor of $10^{-14} - 10^{-15}$ s (see Fig. 111-2, upper right panel). Their dielectric loss data taken at ambient pressure are redrawn and shown in the left panel of Fig. 111-2. The best fit to the loss peak at 180 K by the Fourier transform of the Kohlrausch function yields $n \approx 0.50$. The location of ν_0 calculated via Eq. (2.67) with $n \approx 0.50$ and $t_c = 2 \times 10^{-12}$ s for this temperature is indicated by the vertical arrow in the figure. The good agreement between ν_0 and the frequency of the resolved loss peak provides further confirmation that the slower secondary relaxation is the JG β-relaxation of BMIM-BMSF.

Mierzwa et al. [747(c)] measured the dielectric response of BMIM-BMSF at elevated pressures up to 1.5 GPa. At the same α-loss peak frequency, sample spectra at 1.1 and 1.5 GPa are superposable but have narrower width than the spectrum at ambient pressure (see lower left panel of Fig. 111-2). There is again agreement between the calculated ν_0 with ν_{JG} at the elevated pressures. The relaxation maps at 0.1 MPa and 1.5 GPa are plotted together vs. T_g-scaled reciprocal temperature in the lower right panel. On elevating pressure from 0.1 MPa to 1.5 GPa, T_g is increased from 180 to 276 K. Notwithstanding, the relation between τ_{JG} and τ_α remains about the same as suggested by the CM.

Polyalcohols. The dielectric loss data of xylitol and sorbitol are chosen to illustrate the relation $\tau_{JG} \approx \tau_0$ (Figs. 112 and 113). The xylitol figure shows the broad JG β-loss peak merges with the α-loss peak at higher temperatures [195]. The remarkable breadth of the JG β-loss peak, particularly on the high-frequency side, seems to be due to the presence of the nearly constant loss with $\varepsilon''(\nu) \propto \nu^{-0.11}$. The

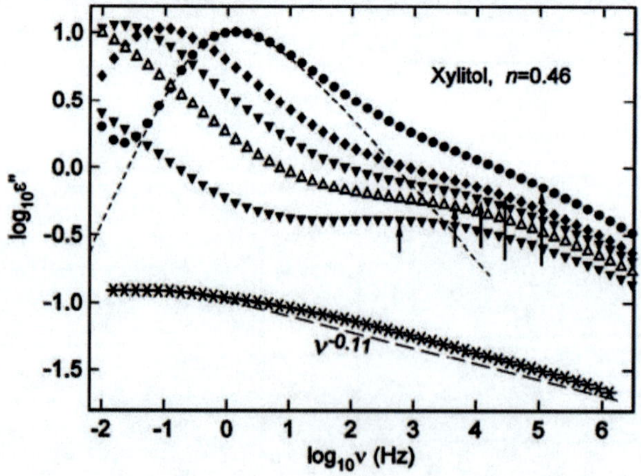

Fig. 112 The *dashed lines* are fits by the Fourier transform of the Kohlrausch function with $n = 0.46$. The *vertical arrows* indicate the primitive relaxation frequency ν_0 calculated. At lower temperatures below T_g, the high-frequency flank of the JG β-loss peak has the nearly constant frequency dependence approximately described by $\varepsilon''(\nu) \propto \nu^{-0.11}$. Reproduced from [195] by permission

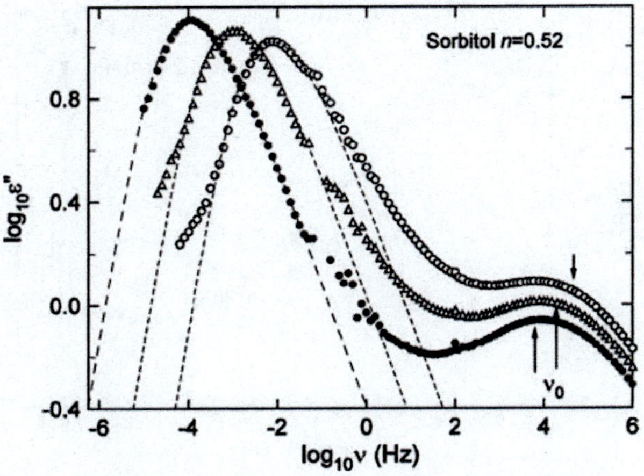

Fig. 113 Dielectric loss spectra of D-sorbitol at three temperatures: 264, 266, and 268 K (data from [240] replotted). The *dashed lines* are fits by the Fourier transform of the Kohlrausch function with $n = 0.52$. The *vertical arrows* indicate the primitive relaxation frequency ν_0 calculated from the coupling model equation

sorbitol figure shows good fit of the Fourier transform of the Kohlrausch function to increasingly higher frequencies as the α-relaxation is further separated from the JG β-relaxation at lower temperatures. The vertical arrow pointing toward the data indicates the location of the primitive relaxation frequency, ν_0, calculated by the coupling model (CM) equation. There is good agreement between ν_0 and the JG β-relaxation peak frequency.

Secondary Alcohols. Dielectric relaxation studies of glassformers of this kind have a long history. Many dielectric studies have found that the slowest relaxation is a Debye process describable by a single exponential time correlation function and has extremely large dielectric strength. This starts as early as in 1951 by Davidson and Cole [218, 1005] and continued to present times by others [348(b), 961, 1006–1012]. The dielectric loss spectrum of the secondary alcohol, 5-methyl-2-hexanol [1008], shown in Fig. 114 serves to illustrate its presence together with two faster but much less prominent relaxations. The slower one of the two is the α-relaxation, and the faster one is the JG β-relaxation.

The fit of the α-loss peak by the Fourier transform of the Kohlrausch function with $n = 0.46$ is shown by the line in the figure. The vertical arrow pointing toward the data taken indicates the location of the primitive relaxation frequency, ν_0, calculated by the coupling model (CM) equation. There is good agreement between ν_0 and the JG β-relaxation peak frequency.

Monohydroxyl Alcohols. The temperature dependences of the three processes are typical and are shown in Fig. 115 for another monohydroxyl alcohol, 1-propanol [348(b)]. The feature of the Debye process is unusual for supercooled liquids including even some other hydrogen-bonded glassformers. More recent studies have found

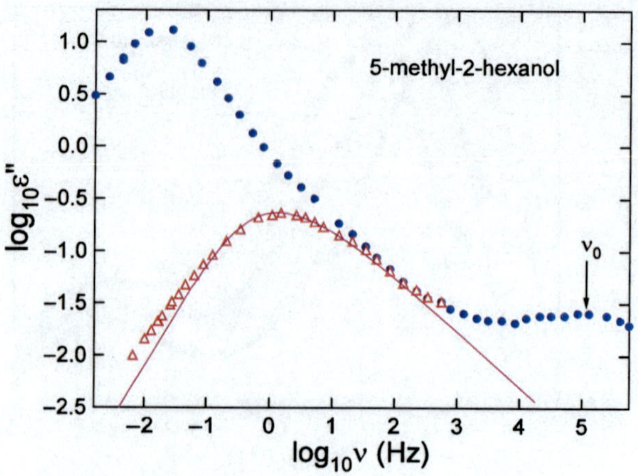

Fig. 114 Isothermal dielectric loss data of 5-methyl-2-hexanol at $T = 158.2$ K from [1008] replotted here showing resolved JG β-relaxation in the supercooled liquid state above T_g. KWW fits with $n = 0.46$ to the α-relaxation peak are shown by the line. The *vertical arrow* pointing toward the data indicates the location of the independent relaxation frequency, ν_0, calculated by the CM

studies that this dielectric Debye peak has no equivalent in mechanical or calorimetric experiments and suggest it is not the structural relaxation [348(b), 1010, 1011]. Although the exact process responsible for the Debye process is unknown, it is clear that it arises from the dielectric polarization fluctuations of the hydrogen-bonded network. One proof of this is by replacement of a H atom by a phenyl group in a 1-propanol molecule to form 1-phenyl-1-propanol. The replacement creates steric hindrance and prevents intermolecular H bonding, and it was found that the Debye-type relaxation process which occurs in 1-propanol vanishes in 1-phenyl-1-propanol [348(b)]. Present still are the structural α-relaxation and the JG β-relaxation. By subtracting off the contribution of the Debye process to the loss spectrum, the authors of [1008] and [348(b)] obtained the loss spectra of the α-relaxation in 5-methyl-2-hexanol and 1-propanol, and fits by the Fourier transform of the Kohlrausch function requires value of n equal to 0.46 and 0.40, respectively. The calculated primitive frequencies, ν_0, are in accord with that of the JG β-relaxation as indicated by the position of the arrow in Fig. 114, and by the lone closed circle in the right panel of Fig. 115 where the relaxation map of all three processes are presented.

Carbohydrates. The carbohydrates are an important class of glassformers having important applications in food science, medicine, and biology. It includes the monosaccharides such as glucose, fructose, galactose, sorbose, and ribose [1012(a), (e)]; the disaccharides such as sucrose, maltose, trehalose, lactulose, and leucrose [1012(b), (c)]; and polysaccharides such as cellulose, pullulan, and dextran [1012(d)]. They all have JG β-relaxation and will be further discussed in another subsection to follow. Here we show the example from fructose in Fig. 116.

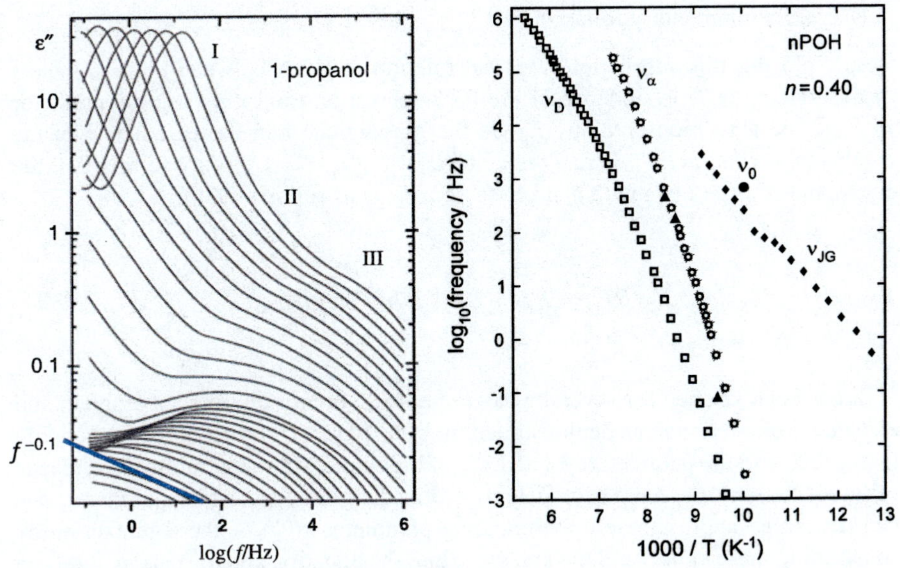

Fig. 115 (*Left*) Experimental dielectric loss results for nPOH in the temperature range 120.8 K > T > 64.8 K in steps of 2 K and in the order from *upper to lower curves*. Four distinct contributions, I, II, and III relaxations and the NCL with $\varepsilon'' \propto f^{-0.12}$, are observed. Peak I resembles a Debye-type process. Replotted from data of [348(b)]. (*Right*) Temperature dependence of the peak frequencies $\nu_{\max}$ for nPOH. The dielectric results ($\nu_{\max}$/Hz) for peak I (□), peak II (O), and peak III (♦). The light-scattering data ($\nu_{\max}$/Hz) are shown by (▲). $T_\beta \approx 138$ K marks the bifurcation temperature if the Arrhenius T-dependence of peak III above T_g is extrapolated to meet peak II. Data from [348(b)] and replotted. The *lone closed circle* (•) is the primitive relaxation frequency ν_0 calculated from the CM with $n = 0.40$

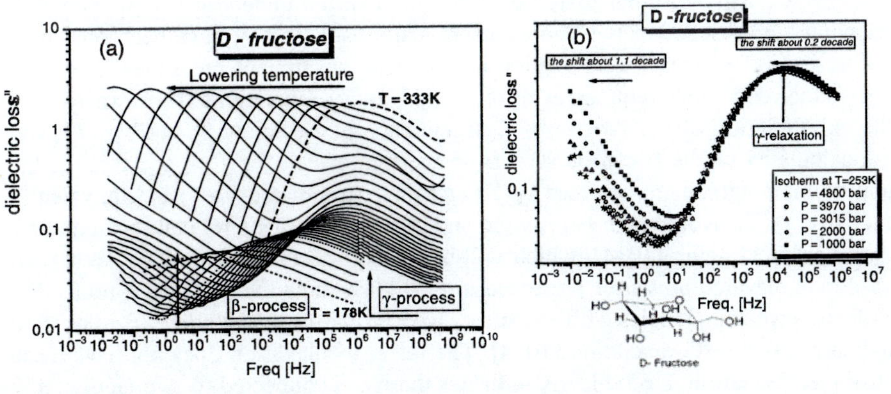

Fig. 116 (**a**) The dielectric loss spectra of fructose at ambient pressure showing the presence of two secondary relaxations, after dc conductivity has been subtracted. The slower β-relaxation is the JG process, but not the faster γ-relaxation, suggested in (**b**) by the shift of the β-relaxation with pressure but not the γ-relaxation. Data from [1012(a)] are replotted here

A Useful Supplementary Relation

Before closing this subsection, a useful relation is given between the parameters of the Arrhenius T-dependence of the JG β-relaxation time τ_{JG} (i.e., the prefactor τ_∞ and the activation enthalpy E_β) in the glassy state and the parameters of the α-relaxation (i.e., $n(T_g)$ where T_g is defined by $\tau_\alpha(T_g) = 100$ s). The relation, the equivalent of Eq. (2.67) with t_c taken to be 2 ps, is as follows [996]:

$$E_\beta / RT_g = 2.303(2 - 13.7n - \log_{10}\tau_\infty). \tag{2.71}$$

It has been verified for several glassformers and binary mixtures at ambient and elevated pressures such as demonstrated in [996, 1013]. If the secondary relaxation is truly JG, and the parameters E_β and τ_∞ of the Arrhenius temperature dependence of its relaxation time have been determined in the glassy state by experiment, one can determine approximately the coupling parameter $n(T_g)$ of the α-relaxation by substituting these values and the known T_g into the equation above. This method can be handy in case the dispersion of the α-relaxation is either unavailable or cannot be measured.

Many glassformers have a genuine JG relaxation but at ambient pressure it can only be resolved at temperatures below T_g and τ_{JG} determined, but not τ_α because the α-loss peak is outside the experimental frequency window. The Arrhenius T-dependence of τ_{JG} is then used to extrapolate to higher temperatures to obtain its values $\tau_{JG}(T)$ at temperature T near T_g, where $\tau_\alpha(T)$ and $n(T)$ can be determined by fitting the observed α-relaxation to the Fourier transform of the Kohlrausch function, and whence $\tau_0(T)$ calculated, and the relation $\tau_{JG}(T) \approx \tau_0(T)$ verified for T near T_g. An example among many is taken from dielectric measurements of dipropyleneglycol dibenzoate (DPGDB) [1014]; the dielectric loss spectra of it showing the presence of two secondary relaxations, β and γ, have been presented before in Fig. 35 (e). The relaxation map of data at ambient pressure (Fig. 117a) shows the relaxation frequencies of the α-, β-, and γ-relaxations from experiment, the extrapolation of the Arrhenius T-dependence of ν_β below T_g to $T \geq T_g$, and the calculated $\nu_0(T)$. It can be seen from the figure that, for $T \geq T_g$, there is the approximate agreement between $\nu_0(T)$ and the extrapolated $\nu_\beta(T)$ or equivalently $\tau_\beta(T) \approx \tau_0(T)$. Not shown here is the significant shift of its relaxation frequency ν_β on physical aging [1014], which will be further discussed later. The isothermal spectra at elevated pressures presented in Fig. 117b show the resolved β-relaxation shifts to lower frequencies with pressure. These properties of the slower β-relaxation indicate it is the JG relaxation [1014]. The faster γ-relaxation does not have these attributes. Equation (2.67) clearly indicates that τ_0 is connected to τ_α quantitatively and qualitatively in dependence on any variable (see Eq. (2.35)). This property of τ_0 in conjunction with the general experimental result, $\tau_{JG} \approx \tau_0$, leads immediately to the conclusion that the JG relaxation is well connected to the α-relaxation and instrumental in effecting glass transition.

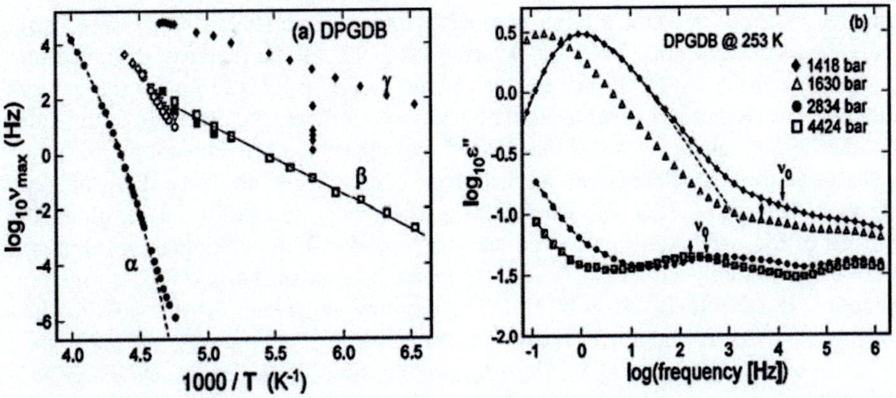

Fig. 117 (**a**) Relaxation map, frequency vs. 1000/*T*, of the α-relaxation (*filled circles*), β-relaxation (*open and filled squares*), and γ-relaxation (*filled triangles*). The *open diamonds* are ν_0 calculated for 228.15, 225.15, and 223.15 K from the parameters, ν_α and *n*, used to fit the α-relaxation by the KWW function. *Open circles* and *open inverted triangles* are ν_0 calculated for the lower temperatures [1014]. The *filled diamonds* indicate the decrease of the β-relaxation frequency, ν_β, with time on annealing a sample after it was rapidly quenched from 300 to 173.15 K. (**b**) Isothermal dielectric loss vs. frequency at 253 K at different levels of elevated pressure as indicated and presented to show sensitivity of the slower β-relaxation to pressure but not the faster γ-relaxation. The *line* is the fit to the α-loss peak of data taken at 1418 bar by the KWW function with $n = 0.38$. The *arrows* indicate the location of ν_0 calculated by the CM equation. Data from [1014] replotted here

Fallacious Attempts to Calculate τ_0 and Compare with τ_{JG}: Use and Misuse of the CM

There are glassformers in which the JG relaxation and the α-relaxation are not well separated in frequency and the dielectric strength of the former is not small compared with the latter. Consequently, the unresolved JG relaxation significantly broadens the α-relaxation. An example is the dielectric relaxation data of poly(3-methylbenzyl methacrylate) (P3MM), poly(3-fluorobenzyl methacrylate) (P3FM), poly(3-chlorobenzyl methacrylate) (P3CM), and poly(2,4-difluorobenzyl methacrylate) (P24FM) reported by Dominguez-Espinosa et al. [1015]. The isothermal loss data show a broad primary α-loss peak and a prominent and well-resolved secondary relaxation labeled γ by the authors and identified to originate from a local rotation of modified phenyl end group in the side chain. There is no feature in any of the isothermal spectra to indicate the presence of a slower Johari–Goldstein (JG) β-relaxation. Nevertheless, they fit the isothermal loss data by adding up the individual contributions from the α, β, and γ relaxations and the conductivity contribution $(\sigma(\omega)/\varepsilon_0\omega)^s$. The frequency dependence of the α-relaxation is assumed to be given by the Havriliak–Negami (HN) function, and the β- and γ-relaxations are modeled by either the Fuoss–Kirkwood function or the Cole–Cole function. These arbitrarily chosen empirical functions, their relaxation strengths, and the term $(\sigma(\omega)/\varepsilon_0\omega)^s$ used by D-E et al. to fit the isothermal dielectric loss data altogether involve assumptions and a large number of fitting parameters, which necessarily have large uncertainties but none was reported. This is particularly disconcerting in the determination of the parameters of the α- and β-relaxations by the authors because the isothermal spectra show no feature to indicate the presence of the latter on the high-frequency side of the former. Moreover, the JG relaxation and the α-relaxation are not independent processes and it is incorrect to fit the dielectric loss data as the sum of their losses especially when they are close to each other. The widths of the α-relaxation reported are very broad at lower temperatures (but still above T_g) as indicated by the small exponent, β_{KWW}, in the KWW α-correlation function, $\phi_\alpha(t) = \exp[-(t/\tau_\alpha)^{\beta_{KWW}}]$. The reported values are $\beta_{KWW} \approx 0.24$ when $f_{max} \approx 1/2\pi\tau_\alpha$ is about 10^2 Hz for P3FM and P3CM, and $\beta_{KWW} \approx 0.27$ when f_{max} is about 1 Hz for P24FM. Here f_{max} is the α-loss peak frequency. These values of β_{KWW} are smaller than practically all amorphous polymers that have been reported including non-aromatic and aromatic hydrocarbon backbone polymers that have much higher T_g than P3FM, P3CM, and P24FM. According to the anti-correlation [112, 479] between "fragility" and β_{KWW} established for amorphous polymers, the exceptionally small β_{KWW} of P3FM, P3CM, and P24FM would imply that these polymers are extremely fragile, even more fragile than polystyrene and PMMA. However, this is not the case. The fragility index m of these glassformers such as P24FM calculated from the VFTH fit is 80.1, which is much smaller than polystyrene and PMMA [112, 479]. Nevertheless, these authors used their highly dubious values of β_{KWW} to test the prediction by the coupling model (CM) on the relation between τ_α and τ_β. It is unsurprising that they found calculated primitive relaxation time τ_0 much shorter than τ_{JG} from their fit and the CM prediction fails. This conclusion is not valid, and it creates unnecessary confusion in the literature.

Another example of the danger of deducing τ_{JG} by the CM using complex data is the study of ethylcyclohexane (ECH) by Mandanici et al. in [1016(a)]. The mechanical and dielectric data obtained are of the highest quality for such low loss material, and the results obtained are very instructive for showing that a secondary relaxation observed has no relation to the α-relaxation and it is totally insensitive to glass transition. This is particularly clear from its persistence in the liquid state with relaxation times longer than the α-relaxation, and maintaining its Arrhenius T-dependence. Needless to say, this result of Mandanici et al. is an important demonstration that some prominently observed secondary relaxations have no fundamental importance in glass transition. Nevertheless, the presence of this secondary relaxation created complication in the analysis of the spectra and led the authors in this first [1016(a)] on ECH to give a doubtful figure to the primitive relaxation time that gives a false impression that the prediction of the CM is violated. The ethylcyclohexane plastic crystal has very low dielectric loss and data were obtained by use of a highly sensitive spectrometer that has a narrow frequency range of 50 Hz–20 kHz less than three decades wide. A conformation transition of intramolecular nature contributes dielectric loss additively to the α-relaxation with comparable magnitude, and is not the JG process. Due this complication by the conformational transition and the narrow frequency window, there is no escape from large uncertainty in determining the true exponent β_{KWW} of the Kohlrausch correlation function and coupling parameter n of the α-relaxation, and using it to calculate τ_0. Nevertheless, the authors of [1016] fitted the broad loss peaks by the Havriliak–Negami equation, and the fit parameters at 112.2 K were converted to yield $\beta_{KWW} = 0.32$ [1016(a)]. From this and the CM equation, they conclude that the primitive frequency and hence the JG peak frequency is at $10^{8.4}$ Hz, much faster than any observed dielectric secondary relaxation (see Fig. 4 of [1016]). This exercise should not be carried out in the first place because there is no reliable method to account for the broadening of the α-loss peak by the overlapping conformational transition at 112.2 K. The results of related plastic crystals such as cyclohexanol ($\beta_{KWW} = 0.62$) by Brand et al. [145] and cyanocyclohexane ($\beta_{KWW} = 0.70$) [1017, see Section 2.3.2.16] do not have such a small $\beta_{KWW} = 0.32$. In a follow-up report [1018(a)] by the authors of [1016] on their study of ethylcyclohexane, horizontal and vertical shifts of isothermal data were applied to compile highly questionable master curves for the α-relaxation. The revised values of $\beta_{KWW} = 0.53$ and the calculated primitive frequencies given [1018(a)] are several decades lower than that calculated with $\beta_{KWW} = 0.32$. Again, the broadening on both sides of the observed α-loss peak by the conformational transition cannot be ignored and the revised $\beta_{KWW} = 0.53$ may or may not still be different from the actual value. I apologize to the authors of [1016] for making the comment on their high-quality experimental work, but this has to be done to avoid others getting the wrong idea about the CM from Fig. 4 of [1016]. All said so far concerning [1016, 1018(a)] is the large uncertainty of the calculated primitive frequency and hence the JG relaxation frequency from the inability to obtain the actual β_{KWW} of the α-relaxation. The important question actually is whether the JG relaxation exists or not in ECH? The answer has been given by the latest experimental work of Mandanici and Cutroni [1018(b)] by measuring ultrasonic attenuation in the MHz range as a function of temperature. They found from the isochronal spectra evidence of another secondary relaxation that is faster than and tends to merge with the α-relaxation, and is likely the JG relaxation of ECH. However, the actual locations of the JG relaxation frequencies in ECH are still uncertain because fits to the isochronal data were required for their deduction.

Another similar misuse of CM prediction is the paper on dielectric relaxation of *bis*-5-hydroxypentylphthalate (BHPP) by Maślanka et al. [1019]. The molecular structure of BHPP is like dihexyl phthalate but having appended to it two hydroxyl end groups, resulting in two separate dipoles of comparable dielectric strength contributing to the α-relaxation. There is excess broadening of the α-loss peak by the overlapping contributions of two independent dipole moments in BHPP. Due to this complication, the actual coupling parameter is smaller than that reflected by the width of the α-loss peak. This is shown in [299] where actually good agreement of the data of BHPP with the CM was demonstrated after the overlapping contributions of two independent dipole moments had been taken into consideration.

The examples given above may help to deter further use of ambivalent or ambiguous experimental data to test the coupling model prediction. Another lesson can be learned from the studies of relaxation dynamics of decahydroisoquinoline (DHIQ). The first broadband dielectric study of DHIQ at ambient pressure was made by Richert and coworkers [455]. They observed a prominent secondary relaxation with unusually high amplitude relative to the α-relaxation process, which they readily identified as the JG β-relaxation process. This is natural because it is the only secondary relaxation observed by them and DHIQ has a compact molecular structure, similar to that of decalin. DHIQ is an interesting glassformer because it has very high steepness ("fragility") index $m = 156$ and very low value of $\beta_{\mathrm{KWW}} = 0.35$ (estimate from the HN fit and not accurate) according to Richert et al., and $m = 168$ and $\beta_{\mathrm{KWW}} = 0.40$ (obtained directly from the KWW fit) according to a later study by Paluch et al. [667]. The magnitudes of both quantities are unusually extreme for small molecular glassformers. Shown here instead in Fig. 118 are the isothermal spectra measured by Paluch et al. at ambient and elevated pressure of 0.5 GPa. The data at 0.1 MPa are similar to those obtained by Richert et al. The primitive frequency ν_{JG} calculated by the CM and indicated by the vertical arrows in Fig. 118 (left side) with $n \equiv (1 - \beta_{\mathrm{KWW}}) = 0.60$ [667] is nearly three orders of magnitude

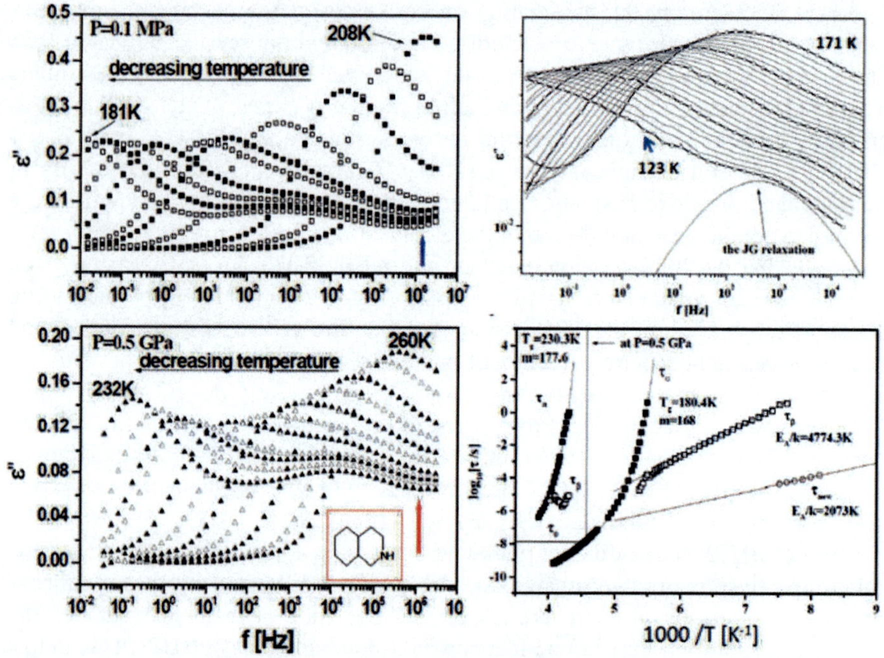

Fig. 118 On the *left side* are dielectric loss curves obtained for DHIQ on cooling at $P = 0.1$ MPa from 208 to 181 K (*upper panel*) and P=0.5 GPa from 260 to 232 K (*lower panel*). On the *right side*, the *upper panel* shows the temperature evolution of the secondary relaxation process from 171 to 123 K with step of 2 K. *Dotted line* represents the contribution at $T = 123$ K of the *faster* JG-process, which emerged from the non-JG β-process with decreasing temperature. The *lower panel* is the relaxation map showing isobaric α-relaxation times at 0.1 MPa and 0.5 GPa (*closed squares*), along with the corresponding β-relaxation (*open squares*) and the JG-relaxation (*open circles*) times at ambient pressure. Predictions for JG-relaxation times above T_g according to CM are plotted as *open stars*. *Rectangular inset* shows the high-pressure data ($P = 0.5$ GPa). Data from [667] are replotted in all the figures

higher than the peak frequency of the observed secondary relaxation assumed to be the JG β-relaxation by Richert et al. If their assumption were correct, then the DHIQ data would be an exception to the CM prediction. It turns out this is not an exception as proven by measurements carried out at isobaric and isothermal conditions up to 1.75 GPa by Paluch et al. [667]. The data show that the peak frequency of the observed secondary β-relaxation is insensitive to pressure, uncorrelated to the pressure-sensitive α-relaxation, and hence it is not the JG relaxation according to one of the criteria [685]. It is intra-molecular in origin, and possibly due to some conformational interconversion mechanisms contributing to dipole fluctuations.

A new and faster secondary relaxation of DHIQ was found by dielectric measurements made in the glassy state at temperatures much lower than done before [667]. This is shown in Fig. 118 on the right side in the upper panel. The relaxation map of all three processes at 0.1 MPa and 0.5 GPa is presented in Fig. 118 on the right side in the lower panel. There the Arrhenius T-dependence of the relaxation time τ_{new} of the newly found JG relaxation in the glassy state has been extrapolated to higher temperatures up to 180 K. The extrapolated values of τ_{new} are in good agreement with the primitive relaxation time τ_0 calculated by the CM with $n = 0.60$ [667] as demonstrated in the relaxation map. The agreement between τ_{new} and τ_0 supports the *faster* secondary relaxation is the JG relaxation of DHIQ. There is further support of this conclusion from dielectric relaxation measurements of the mixture of 25% w/w cyclohexane in DHIQ [977]. Cyclohexane is non-polar and its presence modifies the relaxation dynamics of pure decahydroisoquinoline. Two secondary relaxation processes of DHIQ were found below the glass transition temperature, as in pure DHIQ. The relaxation time of the slower secondary relaxation in the mixture is practically the same as in pure DHIQ, supporting the intramolecular character of non-JG relaxation. The relaxation time of the faster secondary relaxation of DHIQ in the mixture becomes faster and has a higher activation energy of 27 kJ/mol compared with 16.3 kJ/mol in pure DHIQ. There is again good agreement between the relaxation time of the faster secondary relaxation and τ_0 calculated by the CM, which is used to identify it as the genuine secondary JG relaxation of DHIQ in the mixture. From the discussion above, one can learn the lesson from DHIQ that it is dangerous to identify any observed secondary relaxation as the JG relaxation without further characterization of its properties.

2.3.2.3 Excess Loss over the Kohlrausch Fit of the α-Relaxation, or the Excess Wing

Many but not all glassformers with narrow dielectric α-relaxation loss peaks do not have a resolved β-relaxation at temperatures above and below T_g. Neither a peak nor a shoulder appears in the loss spectrum on the high-frequency side of the α-relaxation. However, the high-frequency part of the α-loss peak of $\varepsilon''(\nu)$ cannot be fully described by a single power law with a negative fractional exponent as required by any of the empirical functions such as Cole–Davidson (CD), Havriliak–Negami (HN), and the Fourier transform of the Kohlrausch function. In the description of the α-relaxation by the Kohlrausch function written as Eq. (1.1), $\varepsilon''(\nu) \propto \nu^{-(1-n)}$.

At sufficiently high frequencies and extended over some range, a second negative fractional power-law regime appears with an exponent having a smaller absolute value. This high-frequency tail is now commonly referred to as the "excess wing." This is the most general definition of the excess wing irrespective of whether the glassformer has a resolved secondary relaxation or not, and if resolved, whether it is the genuine JG β-relaxation or not. According to this general definition, the excess wing always exists in broadband dielectric relaxation spectroscopic measurements of most if not all molecular and polymeric glassformers, but the nature of the excess wing can be different dependent on the situation as discussed below.

Glassformers Have No Resolved Secondary Relaxation, JG or Non-JG

The presence of the excess wing in glassformers having no resolved JG β-relaxation or non-JG secondary relaxation was known long time ago [124, 218]. This feature can be seen for propylene carbonate in Fig. 35b and more clearly in Fig. 119, and for propylene glycol in Fig. 120 from broadband dielectric measurements on propylene carbonate by Lunkenheimer and coworkers [424, 682, 972]. These glassformers do not have a resolved JG β-relaxation. Starting from the works of Nagel and coworkers [220, 1021], with the introduction of broadband dielectric spectroscopy in recent years, the excess wing has been found in many more glassformers [241, 424, 682, 971, 972, 1021–1026]. Moreover, Nagel and coworkers [220] found a scaling procedure which makes the entire dielectric spectra, including the excess wing, measured at different temperatures fall onto a master curve. The early success of this scaling procedure [144, 220, 1021] generated some excitement, although later glassformers were found where the scaling failed [971], and no theoretical justification for the scaling procedure has been given so far. Nevertheless, the scaling approach gives the impression that the excess wing is an integral part of the α-relaxation, and β-relaxation is simply absent in these glassformers. This is one view of the excess wing, which has become defunct.

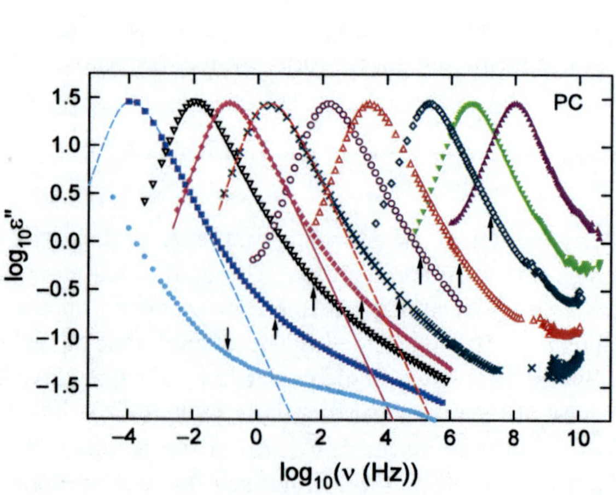

Fig. 119 Frequency-dependent dielectric loss of PC at various temperatures [424, 682, 972]. The curves at $T < T_g$ were measured after suitable aging times to ensure thermodynamic equilibrium was reached. The *vertical arrows* indicate the primitive relaxation frequency of the CM, ν_0, at several temperatures. Data from [424] are replotted here

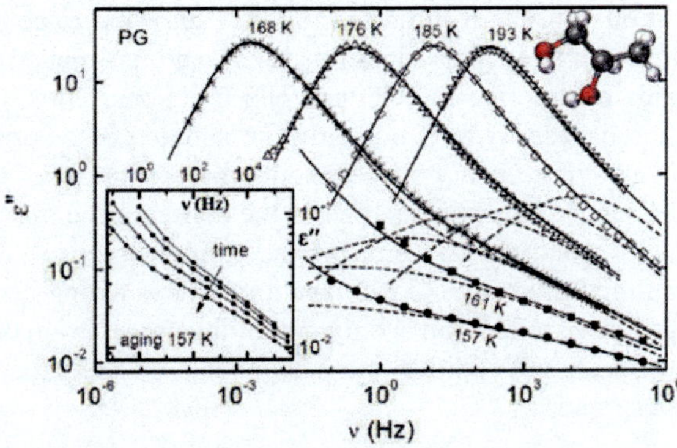

Fig. 120 Frequency-dependent dielectric loss of PG for various temperatures (to maintain readability, only data from a selection of temperatures are shown). The *curve at 161 K* was measured in thermodynamic equilibrium after aging for 53 min. The *157 K curve* was measured after aging for 6.5 days, which was not sufficient to reach thermodynamic equilibrium. The *solid lines* are fits of the α-peak and excess-wing region with the sum of a KWW function and a CC function, with τ_β fixed at the primitive relaxation time of the coupling model τ_0 at all temperatures. The broad peaks at the bottom (*dashed lines*) show the CC-curves for the β-relaxation used to explain the data in the excess wing region. The *inset* shows the dielectric loss at 157 K for selected aging times (from *top to bottom* $\log_{10}[t(S)] = 2.5, 3.5, 4.5, 5.75$). Data from [424] are replotted here

However, again for glassformers that show no resolved JG β-relaxation, there is another view coming from the empirical correlation between the ratio τ_α/τ_{JG} and n, discussed in Section 2.3.2.1 above. It suggests that glassformers with narrow α-dispersion (or small n) like propylene carbonate, propylene glycol, and glycerol will have smaller value of the ratio τ_α/τ_{JG}. If so, the JG relaxation is not well separated from the more intense α-relaxation, being hidden under and seen as the excess wing [237, 241, 613, 900, 1027–1030]. Discussed in Section 2.3.2.2, the relation $\tau_{JG} \approx \tau_0$ found valid for resolved JG relaxation in glassformers with larger n also supports this interpretation of the excess wing because from Eq. (2.67), smaller n leads to smaller ratio τ_α/τ_0. Other reasons for the possibility that part of the excess wing is the unresolved universal JG β-relaxation were given in several publications [237, 241, 1031–1034]. One reason is that the calculated primitive frequency $\nu_0 = (1/2\pi\tau_0)$ by the CM relation falls on the excess wing. Another is the shift of the excess wing with temperature or pressure is less than the α-loss peak (see for an example in Fig. 119).

According to the CM, the primitive relaxation is just the local relaxation that signals the onset of the many-body relaxation which develops with time by increasing participation of more molecules or length scale. At times sufficiently longer than τ_0 or at frequencies much lower than ν_0, the observed spectrum is contributed by these evolving many-body relaxation processes. It is not until some time (denoted by t_{x2} in sections presented later), when the maximum length scale allowed by the intermolecular interactions has been reached, will the loss spectrum follow the power law, $\varepsilon''(\nu) \propto \nu^{-(1-n)}$ or the time correlation follow the Kohlrausch function. It is interesting to point out that the Kohlrausch function turns out to be the characteristic function of stable probability distributions [223], and this makes the Kohlrausch function more fundamental than the CD and HN functions. Thus, from the above CM description, the excess wing is not only the submerged or unresolved JG β-relaxation but also evolving many-body relaxation before it reaches the steady state of maximum length scale. This conclusion has the consequence that one should not fit the dielectric loss spectrum as a sum of the Kohlrausch (or CD) contribution from a Cole–Cole function for the JG β-relaxation. This is wrong because the JG β-relaxation or the primitive relaxation is only part of the processes in the excess wing, and also the processes are not additive and independent of each other. Nevertheless, the practice is widespread

Glassformers Having the JG β-Relaxation Resolved

The excess wing exists also in the dielectric loss spectra of many glassformers that show a resolved genuine JG β-relaxation. Examples in Section 2.3.2.2 include polyisoprene in Fig. 110, sorbitol in Fig. 112, and DiPGDB in Fig. 117. The CM interpretation of the excess wing in these cases is no different from the above and that is part of the evolution of the many-body relaxation toward the Kohlrausch steady state with maximum length scale. The only difference is that the JG β-relaxation or the primitive relaxation is not a part of the excess wing.

Glassformers Having a Resolved Faster Non-JG Secondary Relaxation But No Resolved JG β-Relaxation

Except for the possible distortion by the resolved faster non-JG secondary relaxation in the loss spectrum, the excess wing still represents the unresolved JG β-relaxation and the evolving many-body relaxation processes following it.

There are experimental proofs that *part* of the excess wing is an unresolved β-relaxation submerged under the much stronger α-peak by applying a procedure to the glassformers and detecting a structure in the excess wing, such as a shoulder or even a second peak in the loss spectra, that is identifiable as a relaxation process. One such procedure is physical aging of the glassformers fallen out of thermodynamic equilibrium at temperature below T_g [682, 1033, 1034]. The results from Lunkenheimer and coworkers are shown in Fig. 121 by the dielectric loss spectra of glycerol ($T_g = 185$ K), propylene carbonate (PC, $T_g = 159$ K), and propylene glycol (PG, $T_g = 168$ K), each at a temperature somewhat below their own T_g for different aging times after reaching this temperature.

The α-loss peaks are located at very low frequencies and only their high-frequency flanks represented by the steeper increase of ε'' toward low frequencies can be seen. The shift of the high-frequency flank to lower frequencies with longer aging time reflects the increase of τ_α due to densification and/or decrease in entropy of the glassformer on physical aging. Simultaneously the excess wing shifts to lower frequencies. For all three glassformers, the typical power law,

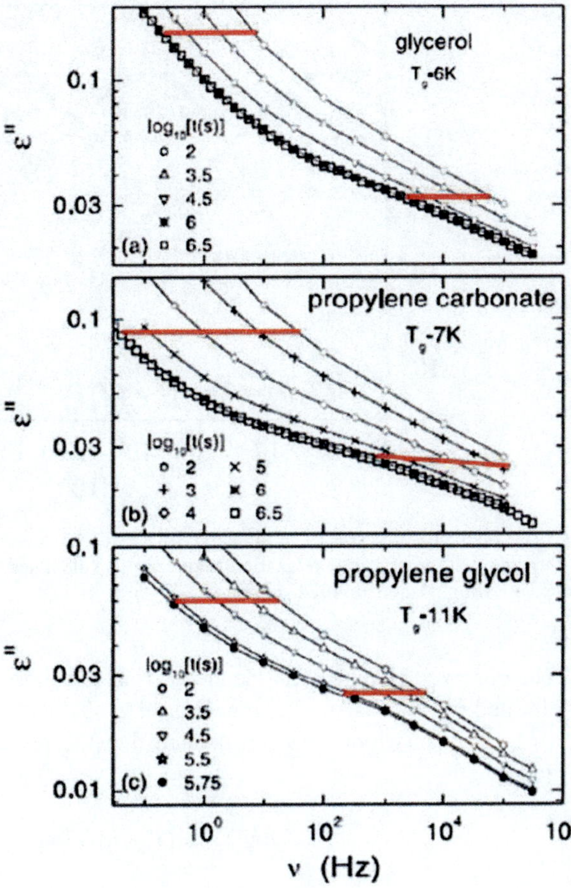

Fig. 121 Aging changes the excess wing to a shoulder, which is evidence of the presence of the JG β-relaxation in glycerol, propylene carbonate, and propylene glycol. Data replotted from [682, 1033, 1034]

characteristic of an excess wing, shows up for short aging times. After the maximum aging times of up to 5 weeks for glycerol and PC and 6.5 days for PG, the excess wing has developed into a shoulder (see Fig. 121). The same behavior of the excess wing on physical aging found in three different glassformers and also in phenyl salicylate (salol) [1035] strongly supports the identification of the excess wing as the unresolved β-relaxation. The excess wing/unresolved JG β-relaxation of all of these glassformers shifts to lower frequencies on aging shown here in Fig. 121, and also on elevating pressure for glycerol [556], PC [654], salol [655], cresolphthalein-dimethylether (KDE) [597], phenylphthalein-dimethylether (PDE) [1036], and polychlorinated biphenyl (Aroclor 1242) [596]. These are two properties of the excess wing/unresolved JG β-relaxation that mimic that of the α-relaxation, and hence it is the JG secondary relaxation according to the classification of secondary relaxation [685]. This identification as the JG relaxation is further substantiated by the calculated ν_0 falling within the excess wing (see Fig. 119 for propylene carbonate, and Fig. 122 for diphenylvinylene carbonate (DPVC) [508]).

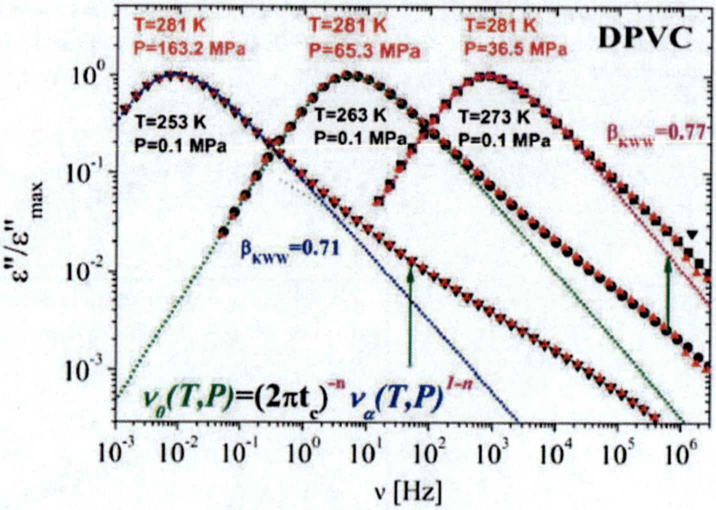

Fig. 122 Dielectric loss spectra of diphenylvinylene carbonate. A repeat of Fig. 40 for the purpose of showing that the primitive relaxation frequency ν_0 calculated by the CM equation is located at the excess wing. Data from [974] replotted

The observed change of τ_β deduced from the shift of the excess wing by aging or elevated pressure is smaller than that of τ_α. This is expected from the CM equation (2.67) from which one can deduce that the relation between the changes is given by

$$\Delta \log \tau_\beta = (1 - n)\Delta \log \tau_\alpha, \tag{2.72}$$

assuming n remains constant and $\tau_\beta \approx \tau_0$.

2.3.2.4 Excess Wing (Unresolved JG β-Relaxation) Eclipsed by the γ-Relaxation

The glassformers with narrow α-dispersion discussed in the previous paragraph all have only the excess wing/unresolved β-relaxation but no other resolved secondary relaxation. There are also glassformers with narrow α-relaxation dielectric loss peak (smaller n) that show no significant excess wing but a resolved secondary relaxation in the equilibrium *liquid* state above T_g. Examples include the hydrogen-bonded glassformers di- and tri-propylene glycol (2PG and 3PG) and polypropylene glycol of molecular weight M_w = 400 g/mol (PPG400) [657, 658, 1038, 1039], and van der Waals liquids, diethyl phthalate (DEP) [660], dibutyl phthalate (DBP) [661], diisobutyl phthalate (DiBP), di-n-octyl phthalate (DOP) [684], and isoeugenol [1041]. The epoxy resins diglycidylether of bisphenol A (DGEBA) and poly(phenylglycidylether) (PPGE) behave in the same way above T_g, but the JG β-relaxation can be resolved in the glassy state [242, 973], as also for the cases of PPG400 [1039], PPG1025, and PPG4000 [1040]. This secondary relaxation is not a JG β-relaxation (hence referred to as the γ-relaxation) because its relaxation time τ_γ does not change with applied pressure in both the liquid and glassy state [657, 658, 660, 661, 664, 673, 973, 1038–1040] and to physical aging in the glassy state [1042]. Also, τ_γ is much shorter than the τ_0 calculated by Eq. (2.67).

The high-frequency flank of the α-loss peak for the most part shows a single fractional power dependence of $\varepsilon''(\nu)$, ν^{-g}, with no sign of a noticeable excess wing coming out of it at higher frequencies. This situation is illustrated here by several examples taken from DGEBA in Fig. 123, PPGE in Fig. 124, and DPG, PPG1025, and PPG4000 in Fig. 125.

The excess wing/unresolved JG relaxation could be present but not seen because it is eclipsed by the more intense γ-loss peak which dovetails the ν^{-g} dependence. This possibility can be verified in the resolving excess wing/unresolved JG relaxation by increasing the separation of the γ-relaxation from the α-relaxation. The latter can be accomplished in several ways including application of pressure in the case of 2PG and 3PG [657, 658], aging for DOP [684] and isoeugenol [1041], and simply cooling to sufficiently low temperatures in the glassy state for PPG400, PPG1025, PPG4000 [1039, 1040], DGEBA, and PPGE [242, 973]. These procedures make the γ-relaxation further separated from the α-relaxation and the excess wing/JG relaxation was found in between the two relaxations (see the examples in Figs. 124 and 125). In the cases of 2PG, 3PG, PPG400, PPG1025, and PPG4000, the excess wing is further transformed to a β-loss peak, which is the sought-after JG β-relaxation because its relaxation time τ_β is pressure sensitive [657, 658] and in good agreement with the τ_0 calculated. The τ_β determined from the resolved JG

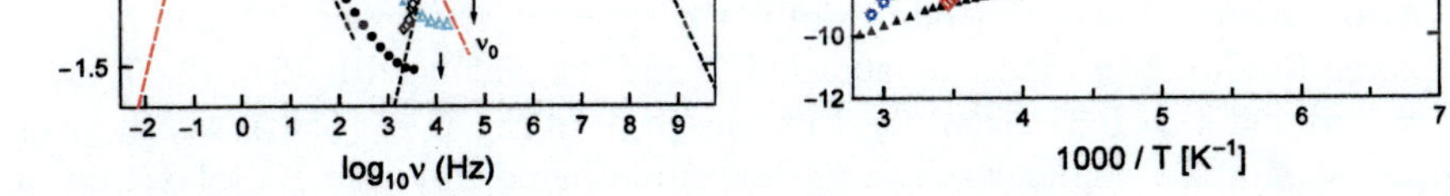

Fig. 123 (*Left*) Dielectric loss spectra at ambient pressure of DGEBA. The *vertical arrows* indicate the location of the calculated primitive frequency ν_0 calculated with $n = 0.47$. The *dashed lines* are fit to the α-relaxation by the Fourier transform of the Kohlrausch function with $n = 0.47$. (*Right*) Relaxation map of DGEBA. The *open diamonds* are the primitive relaxation times τ_0 calculated with $n = 0.47$ from the Kohlrausch function fit. The *open squares* are the relaxation times τ_{JG} of the resolved JG relaxation in the glassy state. Reproduced from [1043] by permission. Good agreement between τ_0 and τ_{JG} near T_g

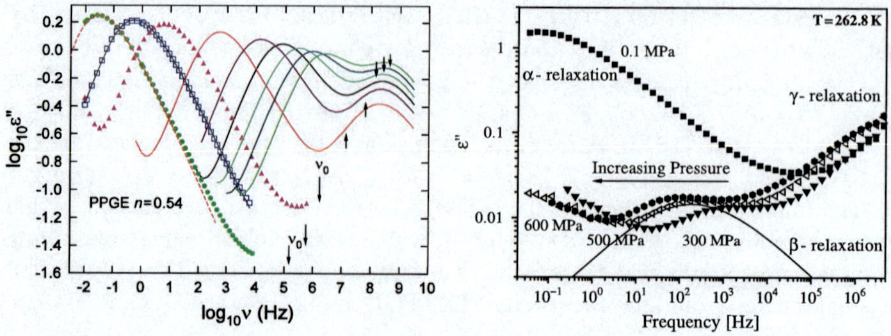

Fig. 124 (*Left*) Dielectric loss spectra at ambient pressure of PPGE. The *vertical arrows* indicate the location of the primitive frequency ν_0 calculated with $n = 0.54$. The *dashed lines* are fit to the α-relaxation by the Fourier transform of the Kohlrausch function with $n = 0.54$. (*Right*) Loss spectra (*symbols*) of PPGE measured under isothermal conditions (263 K) at different values of applied pressure as indicated in the figure. All spectra are measured in the glassy state. The *continuous line* is the contribution of the JG β-process at 500 MPa. Reproduced from [1043] and [973] by permission

relaxation in the glassy state seems to meet the calculated τ_0 above T_g when both are extrapolated as shown in Fig. 123 (right side).

More examples are given below from other glassformers to show that the excess wing is not seen above T_g because it is eclipsed by the γ-relaxation, but actually either it or the JG β-relaxation is present. In Fig. 126 (left panel in the inset) are the dielectric loss spectra of 3PG at $T = 220.5$ K and applied pressures increasing from 33.4 to 591.3 MPa. The excess wing seen at ambient pressure and intermediate pressures is transformed to a resolved secondary β-relaxation at higher pressures. This is the JG relaxation of 3PG because it shifts to lower frequencies with pressure and there is good correspondence between τ_β and τ_0. On the other hand, the maximum of the resolved secondary γ-relaxation at high frequencies is virtually pressure independent. The difference in pressure dependence of the β- and the γ-relaxations is clearly from the plot of relaxation time against pressure at $T = 218.4$ and 245.2 K in Fig. 126.

Also shown in Fig. 126 (right side) is the approximate agreement between the observed β-relaxation time τ_β (labeled JG) and τ_0 calculated by Eq. (2.67) with $t_c = 2$ ps and $n = 0.37$ (from the Kohlrausch fit in Fig. 5 of [658]), which gives further support for the excess wing/β-relaxation being truly the JG relaxation of 3PG. In addition, aging 2PG for 12 h at $T = 216.7$ K and $P = 510$ MPa further shifts the excess wing to lower frequencies [658].

Parallel to 2PG and 3PG in behavior is di-octalphthalate (DOP) [684]. Figure 127 shows in DOP no evidence of the excess wing at ambient pressure in the liquid state (left panel), but the excess wing emerges upon aging glassy DOP for sufficient lengths of time (middle panel).

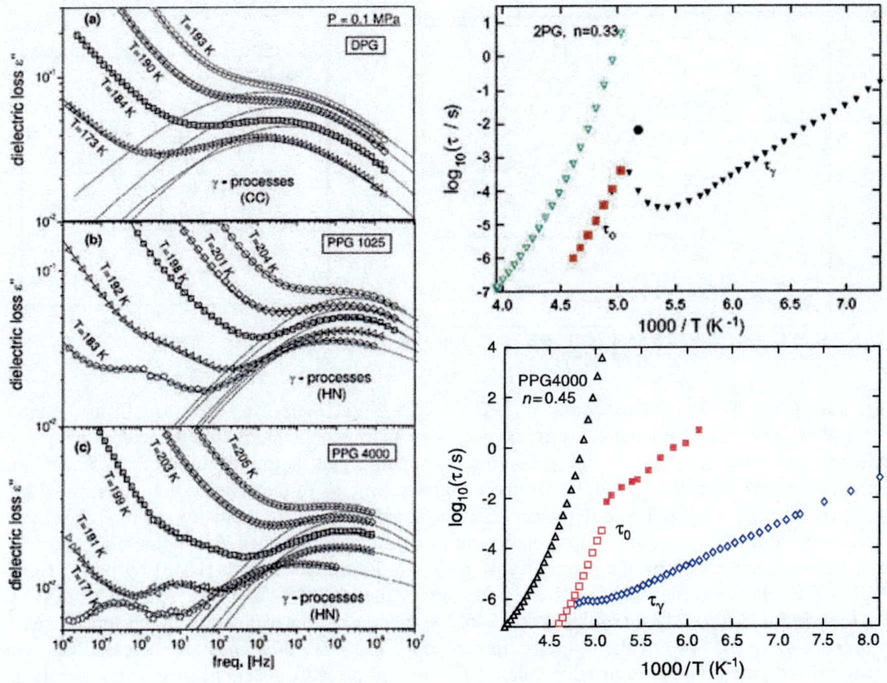

Fig. 125 (*Left*) Selected dielectric loss spectra of the γ-relaxation near the glass transition at ambient pressure and at different temperatures for DPG (**a**), PPG1025 (**b**), and PPG4000 (**c**). The *dotted lines* represent the fits of the γ processes. The JG β-relaxation is resolved in the glassy state of PPG1025 and PPG4000, but not in DPG. Data from [1039] and [1040] are redrawn here. (*Right*) Relaxation maps of DPG and PPG4000. The *upper figure* is for DPG, where the *filled squares* and the *lone filled circle* are τ_0 calculated by the CM equation with $n = 0.33$ from the Kohlrausch function fit of the α-relaxation. The *lower figure* is for PPG4000 where the *filled squares* are τ_{JG} determined from the resolved JG relaxation below T_g, while the *open squares* are τ_0 calculated by the CM equation with $n = 0.45$ from the Kohlrausch function fit of the α-relaxation. Reproduced from [1043] by permission

A high-frequency γ-relaxation is present in isoeugenol [1041] but, unlike 3PG, its intensity is much weaker that it only can be detected deep in the glassy state. It is not detected in eugenol. The excess wing is present (see Fig. 128 left side) and the spectra resembles the glycerol and PC which definitely have no detectable γ-relaxation. On cooling isoeugenol to temperatures 199 K and below (not shown) or aging at 209 K (see Fig. 128 in the middle), the excess wing is transformed to a shoulder revealing a slower β-relaxation. The shift of the shoulder to lower frequency along with the high-frequency flank of the α-relaxation on increasing aging time (but not the γ-relaxation) is one evidence for identifying it as the JG relaxation. The relaxation map of isoeugenol is presented in Fig. 128 (right). Two VFTH laws are needed to fit the T-dependence of τ_α and the crossing temperature

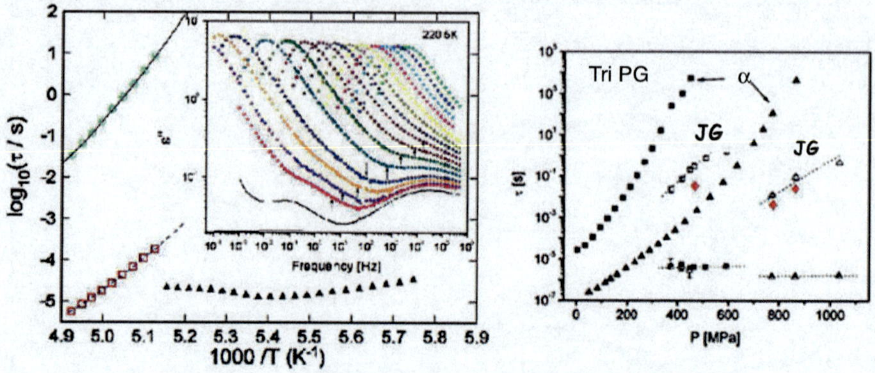

Fig. 126 (*Left*) Dielectric loss data of 3PG at 22.5 K on *increasing pressure* from 33.4 to 591.3 MPa [657, 658] showing the γ-relaxation is pressure independent, and the excess wing/JG relaxation becomes resolved at high pressures. The main figure is the relaxation map of 3PG at *ambient* pressure. The *filled triangles* are the relaxation time τ_γ of the γ-relaxation determined by fitting the spectra [1039]. The *open squares* are τ_0 calculated above T_g with $n = 0.37$ from the Kohlrausch fit of the α-loss peak. Encroachment of τ_{JG} toward τ_γ when T is increased toward T_g from below is suggested by the proximity of τ_0 to τ_γ. Reproduced from [1043] by permission. (*Right*) The relaxation times as function of pressure of the α-, JG β-, and γ-relaxations of 3PG at two temperatures, $T = 218.4$ (*squares*) and 245.2 K (*triangles*), determined from the spectra such as that shown in the *inset* of the figure on the *left side*. The *filled diamonds* are the calculated τ_0 at several pressures. The pressure dependence of τ_{JG} and pressure independence of τ_γ are made clear in this plot

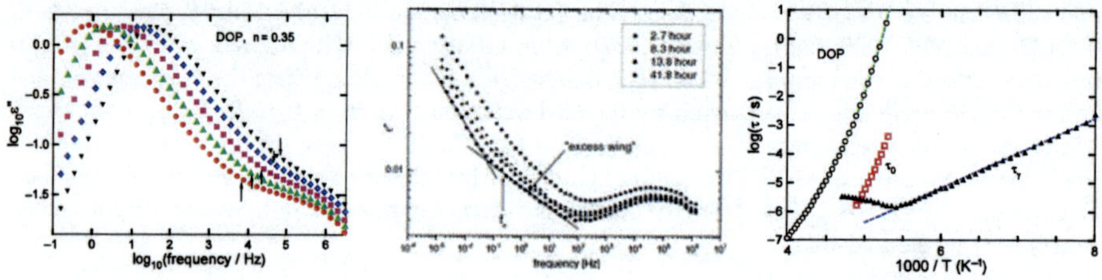

Fig. 127 (*Left*) Dielectric loss spectra at atmospheric pressure and five temperatures of DOP from 183 to 191 K. (*Middle*) Physical aging of DOP at $T = 177$ K. The dielectric loss spectra were measured every 5.6 h. Four spectra collected after 19.4, 25, 30.6, and 36.2 h are not shown in the figure. The *vertical arrow* indicates the calculated frequency of the primitive relaxation, which should be approximately the frequency of the JG relaxation. (*Right*) Relaxation map of α- and γ-relaxation times of DOP at ambient pressure. The *open squares* are τ_0 calculated above T_g with $n = 0.35$ from the Kohlrausch fit of the α-loss peak. Encroachment of τ_{JG} toward τ_γ when T is increased past T_g from below is suggested by the proximity of τ_0 to τ_γ [684]. Reproduced from [1043] by permission

T_B is indicated. The primitive relaxation times, τ_0 (stars), calculated by the CM are in good agreement with τ_{JG}.

The loss spectra of isoeugenol at $T = 263$ K for various pressures increasing from 0.1 to 440 MPa are presented in Fig. 129 (right). It shows no resolved JG

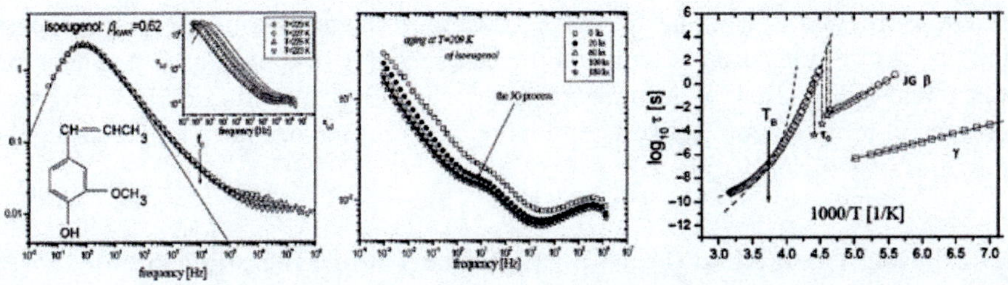

Fig. 128 (*Left*) Superimposed dielectric spectra of isoeugenol taken at ambient pressure ($P = 1$ atm), at four selected temperatures. In the *inset* are the same spectra before superposition. Included in the figure is the KWW fit (*solid line*) with $\beta_{KWW} = 0.62$. The *arrow* indicates the frequency $f_0 = 1/(2\pi\tau_0) \sim 10^4$ Hz calculated from the CM relation at $T = 229$ K. (*Middle*) Physical aging of isoeugenol ($T_g = 220$ K) at $T = 209$ K. Dielectric loss curves were measured every 20 ks. The spectra collected after 40, 80, 120, 140, and 160 ks are not shown in the figure. (*Right*) Relaxation map of isoeugenol at ambient pressure showing the relaxation times of the JG relaxation time resolved below T_g (O), the γ-relaxation (*open squares*), and the α-relaxation (*open diamonds*). The two VFTH laws needed to fit the T-dependence of τ_α and their crossing temperature T_B also are presented. The primitive relaxation times, τ_0 (*stars*), calculated by the CM are in good agreement with τ_{JG}. Data from [1041] are replotted here

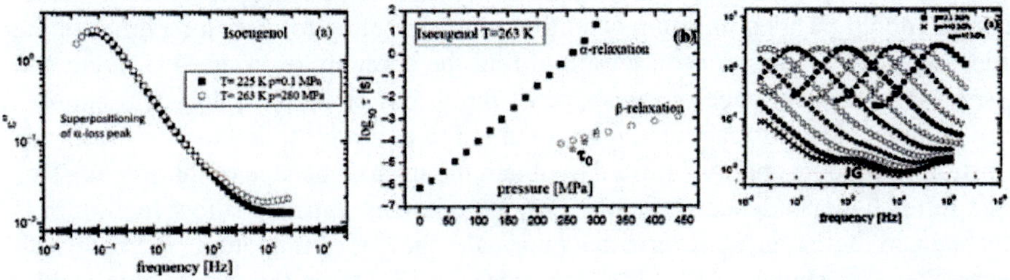

Fig. 129 (*Right*) Dielectric loss spectra of isoeugenol at $T = 263$ K and pressures increasing from 0.1 to 440 MPa. (*Middle*) Relaxation times of the JG relaxation and the α-relaxation as functions of pressure at 263 K. The primitive relaxation times, τ_0 (*stars*), calculated by the CM are in good agreement with τ_{JG}. (*Left*) Dielectric loss of isoeugenol at $T = 225$ K, $P = 0.1$ MPa (*filled squares*) and $T = 263$ K, $P = 280$ MPa (*open circles*), showing superposition of the two α-loss peaks having the same peak frequency. Data from [1041] are replotted in all figures here

relaxation in the liquid state and the development of a shoulder as P_g is approached, indicative of the presence of the β-relaxation The shoulder shifts to lower frequencies on increasing pressure which is another evidence that the β-relaxation is the JG relaxation. The relaxation map at 263 K as a function of pressure (Fig. 129 middle) shows approximate agreement between the observed β-relaxation time τ_β and τ_0 calculated by Eq. (2.67) with $t_c = 2$ ps and $n = 0.38$ (from the Kohlrausch fit, see Fig. 128 left). This is another confirmation of JG relaxation. Finally, on the left panel of Fig. 129, the superposition of α-loss peak taken at 225 K and ambient pressure superpose well with that taken at 263 K and 280 MPa, like many other glassformers. Combining the results of Figs. 128 and 129, it can be said that several properties of isoeugenol conform to the general trends established by many other glassformers.

The isothermal $\varepsilon''(\nu)$ spectra of DPGDB at ambient pressure show no excess wing but the γ-relaxation at temperatures above T_g [1014], but the JG β-relaxation was found below T_g. On elevating pressure, the JG β-relaxation of DPGDB was resolved at temperatures above T_g and its relaxation time becomes longer at higher pressures (see Fig. 117).

2.3.2.5 Encroachment of the JG β-Relaxation Toward the γ-Relaxation: The Cause of the Purported Observation of Anomalous *T*-Dependence of τ_γ

Attempts were made to obtain τ_γ as a function of temperature near and above T_g from dielectric loss spectra of glassformers showing no excess wing, but a non-JG γ-relaxation discussed in the previous subsection, first by Olsen [1031] and by Dyre and Olsen [1044], followed by others [684, 1038–1040]. The isothermal $\varepsilon''(\nu)$ data were fitted by the sum of the contributions from the α- and γ-relaxations with assumed frequency dependences for both. This practice was started by Dyre and Olsen on 3PG [1031, 1044], and near T_g they reported that τ_γ *decreases* on *cooling* from the equilibrium liquid state down to the glassy state which was confirmed by others in other glassformers [684, 1038–1040]. The γ-relaxation frequency dispersion was obtained at equilibrium condition after annealing for very long times of the order of 100 h. In the procedure used to find the anomaly of τ_γ in 3PG, Dyre and Olsen assumed invariance of the shape of the γ-loss peak; the α-loss was supposedly eliminated by subtracting a term assumed to be proportional to $\nu^{-1/2}$ with the constant of proportionality adjusted to obtain the invariant shape of the γ peak. This unexpected *T*-dependence of τ_γ on cooling continues until a shallow minimum is reached and thereafter τ_γ reverts to its normal behavior of monotonic increase with decreasing *T* as shown in Fig. 126 (left side) for 3PG from the results obtained by Casalini et al. [657], and for 2PG and oligomers of PG (see Fig. 125) by Grzybowska et al. [1039]. Again by fitting $\varepsilon''(\nu)$ data, such anomaly was reported in 2PG, 3PG, PPG400, PPG1025, PPG4000 [1039, 1049], and DOP [684]. A change of temperature dependence of τ_γ on cooling toward T_g, although no minimum and further above T_g, was observed in the epoxy resins DGEBA and PPGE [242, 616], and attention of this related phenomenon was drawn in [1043]. To explain this peculiar behavior of 3PG, Dyre and Olsen [1044] have utilized an asymmetric double-well potential to model the γ-relaxation and with special assumptions they constructed the so-called "minimal model" that can successfully explain the behavior of 3PG as well as PPG400 and DOP [684]. The model involves several parameters and special assumptions about them. Such a model does not seem likely to be the explanation of an effect observed in glassformers with different chemical structures and origins of the γ-relaxation.

In the previous subsection, we have shown that the JG β-relaxation exists in all the glassformers where the anomalous *T*-dependence of τ_γ was reported. Although not resolved near and above T_g, it lies in between the α- and the γ-relaxations. Its contribution to $\varepsilon''(\nu)$ could have caused the anomaly but none of the papers reporting the anomaly have taken the JG β-relaxation into consideration when analyzing the data in the temperature region where the anomaly occurs. This possibility is substantiated by the proximity of γ-loss peak frequency ν_γ (from the fits) and ν_{JG} estimated by the calculated primitive frequency ν_0 of the CM as shown in Fig. 123 (right side) for DGEBA, in Fig. 125 (right side) for 2PG, PPG1025, and PPG4000, in Fig. 126 (left side) for 3PG, and in Fig. 127 (right side) for DOP. To show the generality of the phenomenon, the observed minimum of τ_γ near T_g of PPG400 is presented in Fig. 130 (right side), and the change in T-dependence of τ_γ like that found in DGEBA is shown for PPGE in Fig. 131 (right side) taken from [242, 973], and ibuprofen ((2*RS*)-2[4-(2-methylpropyl)phenyl]propanoic acid, $C_{13}H_{18}O_2$) taken from [1004]. Ibuprofen has also a resolved JG β-relaxation. Again the measured ν_{JG} of ibuprofen as well as the calculated ν_0 is observed to encroach toward ν_γ on increasing T past T_g as shown in Fig. 131 (left side).

With the location of ν_0 or approximately ν_{JG} indicated by the vertical arrow for each isothermal $\varepsilon''(\nu)$ data of 3PG (Fig. 126), DOP (Fig. 127), DGEBA (Fig. 123), and PPGE (Fig. 124), or collectively by the relaxation maps particularly in the case of PPG400 (Fig. 130), ibuprofen (Fig. 131) [1004], 2PG and PPG4000 (Fig. 125), the encroachment of ν_0 or ν_{JG} toward ν_γ on increasing T is evident. These all indicate connection between the encroachment of ν_0 toward ν_γ and the origin of the anomaly in T-dependence of ν_γ.

Not only the proximity of ν_γ and ν_{JG} but also a more fundamental reason may have contributed to the purported observation of the anomaly. The JG β-relaxation involves essentially the motion of the entire molecule. On the other hand, the γ-relaxation involves the motion of a part of the molecule. This part of the molecule eventually also participates in the slower JG β-relaxation. Deep in the

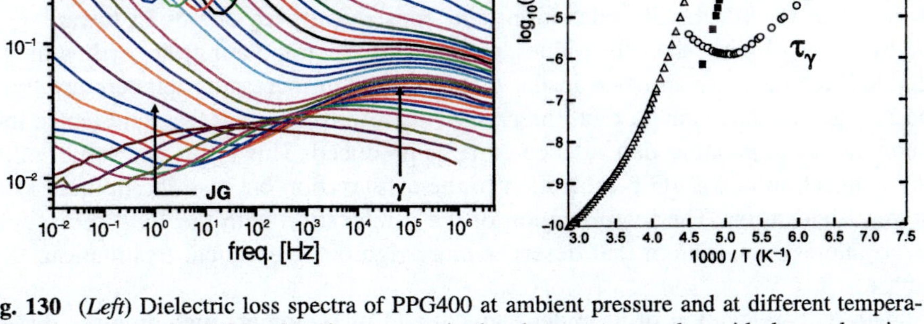

Fig. 130 (*Left*) Dielectric loss spectra of PPG400 at ambient pressure and at different temperatures. The excess wing/JG β-relaxation appears in the glassy state together with the γ relaxation. (*Right*) Relaxation maps of PPG400. The *filled squares* are τ_0 calculated by the CM equation with $n = 0.38$ from the Kohlrausch function fit of the α-relaxation. Reproduced from [1043] by permission

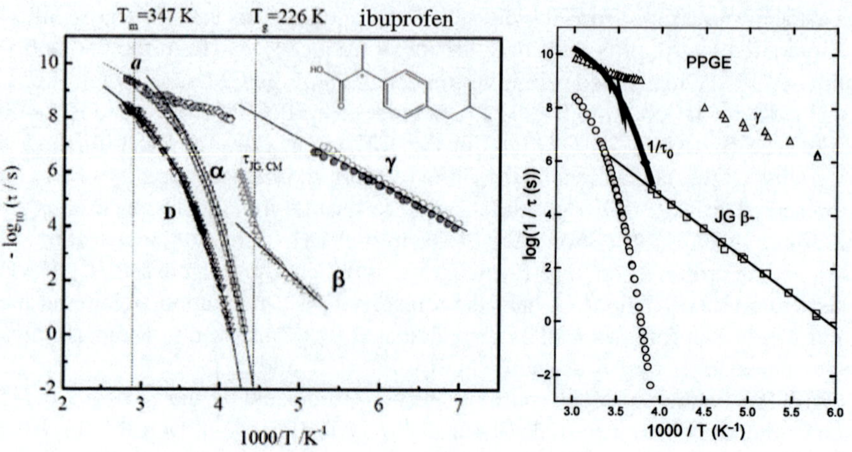

Fig. 131 (*Left*) Relaxation map of ibuprofen, which will be further discussed in a later section. Data from [1004] replotted. The observed JG β-relaxation time τ_{JG} (*open triangles*) is nearly the same as the calculated primitive relaxation time τ_0 (*stars*, labeled by $\tau_{JG,CM}$ in the figure). There is change in *T*-dependence of τ_γ, and the data suggest it is due to τ_{JG} encroaching τ_γ as temperature is raised. (*Right*) Relaxation map of PPGE reproduced from [1043] by permission. The *lone closed diamond* and the *thick line* are the primitive relaxation times τ_0 calculated with $n = 0.54$ from the Kohlrausch function fit of the α-loss peak near T_g, and assumed constant. The *open squares* are the relaxation times τ_{JG} of the resolved JG relaxation in the glassy state [242, 1043]. There is good agreement between τ_0 and τ_{JG} near T_g. The change in *T*-dependence of τ_γ occurs when τ_{JG} encroaches τ_γ as temperature is raised

glassy state where τ_γ is much shorter than τ_{JG}, the γ-relaxation is not influenced by or effectively decoupled from the JG β-relaxation, and τ_γ has the usual Arrhenius temperature dependence. However, near T_g and especially above T_g, τ_{JG} becomes comparable to τ_γ; the γ-relaxation and the JG relaxation are no longer independent of each other because that part of the molecule responsible for the γ-relaxation has to relax in concert with the rest to execute the JG relaxation. This means that the two Langevin equations describing the two relaxations are coupled. The γ-relaxation is hybridized with the JG relaxation, and consequently τ_γ is shifted toward τ_{JG} and becomes longer than the value determined by extrapolating the Arrhenius *T*-dependence of τ_γ established in the glassy state. On increasing temperature, τ_{JG} and τ_γ get closer to each other, the shift of τ_γ toward τ_{JG} is larger, and hence the anomalous temperature dependence of τ_γ is produced. This effect shows not only the connection of the JG β-relaxation to the α-relaxation but also the influence on some γ-relaxation. The hybridization of the γ-relaxation with the JG β-relaxation is an interesting problem that deserves more rigorous theoretical treatment in the future.

The JG β-relaxation of propylene glycol and its oligomers including the dimer (2PG), trimer (3PG), and PPG400 has weak dielectric strength and either not resolved or hard to resolve at ambient pressure (see Fig. 125 for the weakly resolved

JG β-relaxation of PPG400). But, by elevating pressure on 2PG and 3PG, Casalini and Roland [657, 658] were able to see JG β-relaxation first as an excess wing in the equilibrium state and finally resolve it as a loss peak in the glassy state, as shown before in Fig. 126. These materials obtained from Aldrich Chemical Company were dried by using molecular sieves and, immediately prior to all measurements, were maintained for 1 h at 125°C in a nitrogen atmosphere. The procedure is to remove water in the sample as much as possible. The remaining traces of water, each hydrogen bonded to a small fraction of the 2PG or 3PG molecules, change only somewhat the entire rotation/translation of the entire molecules, and hence the relaxation time is not much different from that of the JG β-relaxation of the neat 2PG or 3PG molecules. The enhanced dipole moment of the modified 2PG or 3PG molecules due to the water attached makes larger contribution to dielectric loss, albeit at frequencies not too different from the JG β-relaxation of the neat 2PG or 3PG molecules. In the dried sample, Casalini and Roland did not resolve the JG β-relaxation of the 2PG or 3PG molecules at ambient pressure, suggesting that the water-modified 2PG or 3PG molecules make no important contribution to the relevant secondary relaxation in their dried sample. Therefore, the JG β-relaxation resolved by them at elevated pressure in the same dried sample is clear indication that it comes at least in part from the 2PG or 3PG molecules. The shift of this relaxation to lower frequencies with pressure is indication that it is the true JG β-relaxation related to the α-relaxation. On the other hand, the prominent γ-relaxation does not change position with pressure and its intramolecular origin is clear. In a very recent paper, Gainaru et al. [1047(c)] showed after heating the sample up to 150 or 200°C the β-loss peak of a PPG sample with $N = 314$ repeat units that has lost intensity significantly. This observation is perfectly reasonable in view of the discussion above that water enhances the dielectric strength of the β-relaxation, but based on their result alone they concluded the β-relaxation originates from water. By this conclusion, Gainaru et al. cast doubt on the previously established interpretation as the true JG β-relaxation of PPG oligomers with its dielectric strength enhanced by very small amount of water that may still remain. Unfortunately, in reaching their conclusion, Gainaru et al. did not cite the works of Casalini and Roland. The procedure of drying the sample used by Casalini and Roland is similar to that of Gainaru et al. According to the reasoning of Gainaru et al., the β-relaxation can no longer be resolved in the dried 2PG and 3PG samples, and yet on elevating pressure the β-relaxation of the dried samples appears prominently in the loss spectra (see inset of Fig. 126 (left)). Another argument given by Gainaru et al. to support their conclusion is the similarity of the activation energy of the β-process in PPG oligomers ≈ 6400 K to that found in many water-containing systems. These other aqueous systems cited all have much high water content than in any of the PPG oligomer samples studied by all workers. Hence this comparison is not appropriate. Besides, the water secondary relaxation times in the aqueous mixtures cited are several decades faster than the β-relaxation time observed in the PPG oligomers. Gainaru et al. paid special attention to the γ-relaxation by pointing out that its activation energy E_γ is close to $20T_g$ and thus similar to that usually observed for secondary relaxation processes in molecular glass formers. I do not appreciate the significance of this statement because [739], not cited by Gainaru et al., has shown there is large spread in values of the ratio of the activation energy E_a to T_g, E_a/T_g, for all kinds of secondary relaxations, intermolecular (JG) or intramolecular (non-JG) [739]. Gainaru et al. seem not aware of the experimental fact that the γ-relaxation time does not change with pressure [657, 658], and hence not an important secondary relaxation in connection with glass transition. They mentioned that the γ-relaxation could be described by the "minimal model" of Dyre and Olsen [1044] for the anomalous T-dependence of its relaxation time, but neglected to mention the more general and appropriate explanation by the hybridization with JG β-relaxation in [1043]. If only the paper of Gainaru et al. were consulted, one could get the idea that the JG β-relaxation in PPG oligomers is non-existent and the γ-relaxation is the only native secondary relaxation. However, if other experimental facts of the β- and γ-processes are considered altogether, then the idea is not supported.

In the procedure used by Dyre and Olsen [1044] to find the anomalous T-dependence of τ_γ of 3PG, they assumed the $\varepsilon''(\nu)$ spectrum is contributed by the α- and the γ-relaxations, and the high-frequency flank of the α-loss peak is well represented by a term $\propto \nu^{-1/2}$. The contribution of the α-relaxation was supposedly entirely eliminated by subtracting the term $A\nu^{-1/2}$ from the $\varepsilon''(\nu)$ data, and the remnant was the γ-loss peak. I have criticized this procedure principally because they ignored the presence of the JG β-relaxation though not resolved. The assumed $A\nu^{-1/2}$ frequency dependence of the α-loss peak initially was based on the empirical observations from $\varepsilon''(\nu)$ data of a few glassformers [1045]. Recently, $\varepsilon''(\nu)$ data of more small molecular organic glassformers were examined by Nielsen et al. [1046]. Although they found more glassformers to have $\varepsilon''(\nu) \propto \nu^{-g}$ with $g = 1/2$, many other glassformers have g departing from ½. The authors admitted that "No attempts were made to subtract possible contributions from beta processes." They excluded for consideration the polymeric glassformers for no good reason. Except for the presence of chain modes at lower frequencies than the α-relaxation, the glass transition dynamics of polymers is the same in every respect as non-polymeric organic glassformers. Except for a few polymers that have dipole moment along the chain backbone, the slower chain modes do not contribute to $\varepsilon''(\nu)$. Even if the chain modes contribute, the contribution would not affect the high-frequency flank of the α-loss peak. Nielsen et al. offer no explanation of their own on why more glassformers have $g = 1/2$ except quoting obsolete models such as defect diffusion in one dimension of Glarum [74, 305] and others, which have paucity of predictions and cannot address the plentiful experimental facts about glass transition of modern days. Nielsen et al. make connection of the glassformers all having $g = 1/2$ with their fragilities, pointing out that the latter vary over a wide range. This observation is no surprise because, as discussed before, "fragility" is a composite quantity dependent on both thermodynamics and many-body relaxation (or n), and correlation with any other property can break down. Only by restricting to glassformers with similar or related chemical structures, it may be possible to find the correlation between "fragility" and n. Not specified, the $g = 1/2$ glassformers of Nielsen et al. may come from diversely different chemical classes with widely different sensitivity to thermodynamic variables. Moreover, the observations of $g = 1/2$ in some of their glassformers may be those having narrower width of α-loss peak (or n) and the $g = 1/2$ is partly due to the presence of unresolved JG β-relaxation on the high-frequency flank. In fact, 2PG, 3PG, PPG400, eugenol, isoeugenol, and DOP discussed before all have $g \approx 1/2$ and all of them have narrower α-dispersion with $n \leq 0.37$. On the other hand, for glassformers having broader α-dispersions and larger n like DGEBA ($n = 0.47$) and PPGE ($n = 0.54$), the entire α-loss peak including its high-frequency flank is well fitted by the one-sided Fourier transform of the Kohlrausch function (see Figs. 123 and 124). The latter has rigorously the ν^{-g}-dependence, with $g = (1 - n)$, which is not far from $g = 1/2$ for DGEBA and PPGE. The JG β-relaxation does not interfere because it is located at higher frequencies outside the ν^{-g}-dependence. This is supported by the location of the primitive ν_0 in the figures for $\varepsilon''(\nu)$ data taken at T near T_g, the same condition used by Nielsen et al. in examining their data on organic liquids. Thus, the ν^{-g}-dependence with $g \approx 1/2$ of some glassformers comes entirely from the Kohlrausch α-relaxation with $n \approx 0.5$ without the influence of the JG β-relaxation, while for some other glassformers having smaller n (including 2PG, 3PG, PPG400, isoeugenol, and DOP) it comes from the omnipresence of the unresolved JG β-relaxation nearby and possibly also the additive contribution from the γ-relaxation if also present. As will be shown in Section 2.3.2.16 later for a different emphasis, the much studied dialkyl phthalates and benzophenone, as well as the pharmaceutical, ibuprofen, also have value of g near ½ but definitely an unresolved JG β-relaxation is present and located in between the α- and the γ-relaxations. This assertion for the dialkyl phthalates and benzophenone has been proven by experiment in mixtures with another glassformer having much higher T_g. For ibupropen, the JG β-relaxation has been resolved above and below T_g.

Also we recall the dielectric loss data of DPVC plotted linearly against log ν shown before in Section 2.2.4. Figure 39 demonstrates that the Fourier transform of the Kohlrausch functions with $\beta_{KWW} \equiv (1-n) = 0.71$ and 0.77 at two temperatures fits the data very well except at frequencies about two decades higher than the loss peak frequency. In a plot of the same data as $\log \varepsilon''(\nu)$ vs. log ν in Fig. 122, the high-frequency flanks of these Kohlrausch fits have slopes –0.71 and –0.77 far from –0.50. The excess wing of DPVC has steeper slope but it depends on temperature or τ_α.

Without taking into account and making connection to other known experimental facts, the promotion of the loose phenomenology of minimum slope $-g$ having a distribution centered near –0.5 by Nielsen et al. does not serve any useful purpose, and instead it distracts or muddles the real and important issues of glass transition.

The frequencies where Nielsen et al. found $\varepsilon''(\nu) \propto \nu^{-g}$ with $g = 1/2$ data are high above the loss peak frequency. The mechanical counterpart of the dielectric susceptibility like the shear modulus $G^*(\nu)$ at these high frequencies would contribute very little to the viscosity, an important characteristic of the α-relaxation. From this we can say for the region of the spectrum having $\varepsilon''(\nu) \propto \nu^{-g}$ with g in the neighborhood of 1/2 that it is an unimportant feature as far as the essential properties derivable from the α-relaxation are concerned. Furthermore, Fragidakis and Roland (to be published) have found from a number of dielectric loss data of small molecular glassformers not considered by Nielsen et al. that the minimum slope deviates significantly from $g = 1/2$.

2.3.2.6 Removing the Confusion Caused by the Interpretation of the Excess Wing (EW) of Others

Ample evidences have been given above and elsewhere to show that for glass-formers that have no resolved JG β-relaxation, part of the excess wing (EW) is the submerged JG relaxation. We have seen in previous sections that the excess wing in several glassformers including PG, glycerol, PC, 3PG, and isoeugenol is transformed by pressure or physical aging into a well-resolved JG relaxation. The other part of the EW at lower frequencies than the unresolved JG β-relaxation is the evolving many-body relaxation processes before reaching the terminal Kohlrausch α-relaxation with the maximum cooperative length scale allowed by the intermolecular interaction. Both parts are connected to the structural α-relaxation, and their relaxation times shift with pressure. Let us call EW of this kind, type $\mathcal{A}$. The same is true for the EW in glassformers that shows resolved non-JG relaxation but no JG β-relaxation, and the EW also has the two parts because the JG β-relaxation and the development of the many-body relaxation following it are universal. Let us call glassformers and the associated EW of this kind, type $\mathcal{B}$. However, in glassformers that have a resolved JG β-relaxation, the EW is another part of the many-body relaxation process that evolves with time. Let us call these glassformers and the EW they have of this kind, type $\mathcal{C}$. The interpretation of the EW and the classification of glassformers follow from the CM, putting emphasis on the universal presence and fundamental importance of the JG or the primitive relaxation. The difference in the nature of the EW in glassformers of the three types, $\mathcal{A}$, $\mathcal{B}$, and $\mathcal{C}$, discussed before in Section 2.3.2.3 are reiterated here, because if not taken into account it can lead to confusion in interpreting the EW. This happens to the authors of [1025, 1047(a)] who believe that the excess wing (EW) is a distinct process of the same nature present in all glassformers irrespective of them having a resolved secondary relaxation or not. In the 2003 paper [1025], one justification given by these authors is the statement: "In dielectric experiments under pressure, the excess wing and JG process behave quite differently." This was based on the comparison of the excess wings of cresolphthalein-dimethylether (KDE) and phenolphthalein-dimethylether (PDE) and the resolved secondary relaxation of *bis*(methoxyphenyl)cyclohexane (BMPC) in the 2002 paper by Hensel-Bielowka and Paluch [1037]. There it was shown that the excess wings of KDE and PDE have strong P-dependence, but none for the secondary relaxation of BMPC. This belief was reaffirmed by the authors of [1047(b)] in their statement: ". . . . as already suggested on the basis of high-pressure experiments we will argue that the excess wing and the β relaxation cannot be treated on the same footing," and again citing Hensel-Bielowka and Paluch [1037].

Hensel-Bielowka and Paluch mistakenly identified the resolved secondary relaxation of BMPC as the JG process in their paper, and this has led the authors of [1025, 1047(a), 1047(b)] to do the same. Apparently, they are aware of the fact that since the publication of criteria to identify JG β-relaxation in 2004 [685], and from other experimental evidences [120, 618, 685], Paluch had changed his view of [1037] and instead he now attributes the excess wing or part of it in KDE and PDE as unresolved JG relaxation with pressure dependence, and the secondary relaxation in BMPC as a non-JG intramolecular process from its independence on pressure and thermo-history [1042]. Also, the secondary relaxation in BMPC (as well as in PDE) shown by deuterium NMR measurements from Meier and coworkers [1049, 1050] is due to the phenyl ring flip. It has intramolecular origin, and is not the JG relaxation according to the criteria of [685]. Therefore, in the statement by the authors of [1025] quoted above they are comparing type $\mathcal{A}$ excess wing of KDE, part of which is the JG β-relaxation, with the non-JG secondary relaxation of BMPC (which is of type $\mathcal{B}$). Of course they behave differently under pressure or aging, and therefore the statements made by the authors of [1025, 1047(a), 1047(b)] become irrelevant. This unnecessary confusion is due to the authors of [1025] not differentiating excess wings appearing in glassformers of different types, $\mathcal{A}$, $\mathcal{B}$, and $\mathcal{C}$, and having different characters. Certainly, they were misled by the 2002 paper by Hensel-Bielowka and Paluch [1037]. However, the confusion of these authors arises partly from them not paying attention to the universal presence of the JG β-relaxation, and not distinguishing it from the intramolecular non-JG secondary relaxation. The last problem is especially clear in [1407(b)], where among the glassformers considered, DHIQ, *m*-fluoroaniline (*m*-FA), and DGEBA, the resolved secondary relaxation above T_g have been proven to be not JG β-relaxation.

For glassformers that have a resolved secondary relaxation such as trimethyl phosphate in [1025] and triphenyl phosphate and DGEBA, the authors of [1047(a)] suggested their universal excess wing is present in between the α-relaxation and the secondary relaxation. They made the following statement: "On a closer examination of figure 11(b) of DGEBA, at low temperatures ($\leq$ 252 K) an excess wing seems to emerge from underneath the well-separated β-peak." The feature in the spectrum claimed by these authors as evidence of the excess wing is the same as that shown for DGEBA at the lowest temperature here in Fig. 123. The reader may recall that the genuine JG β-relaxation in DGEBA was found as resolved peak in the glass state and also near T_g [242]. It lies in between the α- and the prominent γ-relaxation (called *the well-separated β-peak* in [1047(a)]), and its relaxation time τ_{JG} corresponds well to τ_0 as can been seen in Fig. 123 (right), and also in [1051]. Thus, Fig. 123, or Fig. 11b of [1047], indicates that the excess wing claimed by the authors of [1047(a)] partly arises from the unresolved JG β-relaxation of DGEBA in the liquid state. On increasing temperatures above T_g, all three relaxations of DGEBA become less separated from each other, and the JG relaxation with the least dielectric strength cannot be resolved. From previous studies of the dielectric loss of DGEBA, it is known from the chemical structure that the prominent and fast γ-relaxation of DGEBA comes from the significant dipole moment of the epoxide group and not the entire molecule, and hence it is not the JG relaxation. Reacting EPON828 with cross-linking agents has the epoxide groups consumed; the γ-relaxation is suppressed and the JG β-relaxation strengthened, making the presence of the JG β-relaxation absolutely clear as shown in Fig. 132-1 after 5 h of reaction time [1051].

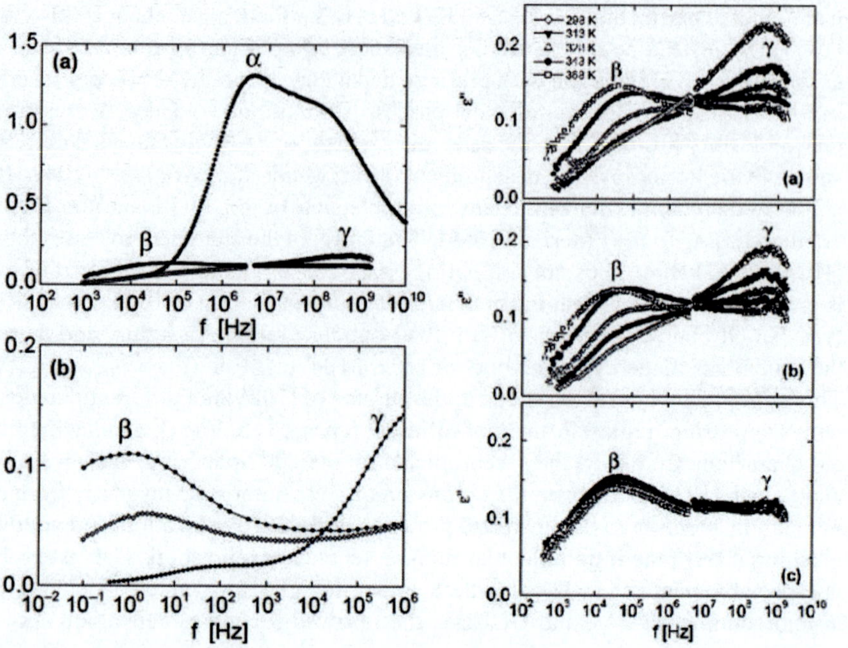

Fig. 132-1 (*Left*) Dielectric loss as obtained from isothermal measurements at (**a**) 297 K and (**b**) 203 K on unpolymerized EPON828 (★) and on EPON828/EDA mixtures after polymerization for 24 h at $T_{\mathrm{Poly}} = 297$ K (◊) and at $T_{\mathrm{Poly}} = 343$ K (•), showing suppression of the γ-relaxation and strengthening of the JG β-relaxation as the epoxide group is consumed by reaction. (*Right*) The dielectric loss spectrum of the DGEBA-EDA measured at 297 ± 1 K after polymerization at different temperatures, as indicated, for different times: (**a**) polymerized for 5 h, (**b**) polymerized for 24 h, and (**c**) polymerized for a further period of 5 h at 368 K, after polymerizing for 24 h at the indicated temperatures. Increase in T_{poly} is more effective in reducing the loss of the γ-relaxation and strengthening of the JG β-relaxation. From unpublished data of Mario Beiner

The authors of [1047(a)] seem unaware of these facts showing the JG relaxation does exist in DGEBA. Instead, they suggested that the feature in the region separating the α- and the γ-relaxations is evidence of the presence of their universal excess wing. The loss spectra of two other epoxies, the glass-forming poly[(phenylglycidylether)-*co*-formaldehyde] (PPGE) ([242, 973], see also Fig. 124 herein) and tri-epoxy triphenylolmethane triglycidylether (TPMTGE) [1052], are similar to DGEBA in every respect, including the presence of the JG β-relaxation in between the α- and the γ-relaxations, and the same feature as in DGEBA suggested as the excess wing by the authors of [1047]. The relaxation map presents the relaxation times of all three relaxations, and there is good agreement of the JG relaxation time τ_{JG} with the calculated τ_0 for PPGE as shown in Fig. 131 (right), and also for TPMTGE (not shown).

Contrasting the Interpretation of the Dynamics with That of the CM

The interpretation of the susceptibility spectra of the authors of [1025, 1047(a), 1047(b), 1053, 1054] emphasizes the universal presence of the EW $\propto \nu^{-\gamma}$ with $\gamma \approx 0.2$ of the same nature irrespective of whether there is resolved secondary relaxation or not. The contrast with the interpretation of the CM is brought out by comparing the two different ways data of the dielectric loss spectrum of 4-*tert*-butylpyridine (4-TBP) at 161 K [1047(b)] shown in Fig. 132-2 are analyzed. The authors of [1047(b)] assume their EW $= A\nu^{-\gamma}$ with $\gamma \approx 0.2$ is an additive contribution to the loss spectrum, and A has the magnitude that places it in the position as shown by the dashed line in Fig. 132-2. After subtraction, they obtain the symmetrical β-peak, and the high-frequency flank of the α-peak has the power law $\propto \nu^{-0.6}$. The result is three additive contributions. On the other hand, in the CM interpretation, the shoulder shown in the loss data is already an indication of the presence of the JG β-relaxation. This is confirmed by the primitive relaxation frequency, ν_0, locating at the shoulder as indicated by the vertical arrow in Fig. 132-2. Here, ν_0 has been calculated with ν_α and $n \approx 0.40$ from the Kohlrausch fit to the α-loss peak, which also has the power law $\propto \nu^{-0.6}$ on the high-frequency side. The excess of the observed loss $\varepsilon''(\nu)$ over the power law $\propto \nu^{-0.6}$ is considered as coming from the contributions of the evolving many-body relaxation processes starting from ν_0 and continuing down to the lower frequency, ν_{x2}, at which the maximum length scale of the α-relaxation is reached and the time correlation function assumes the

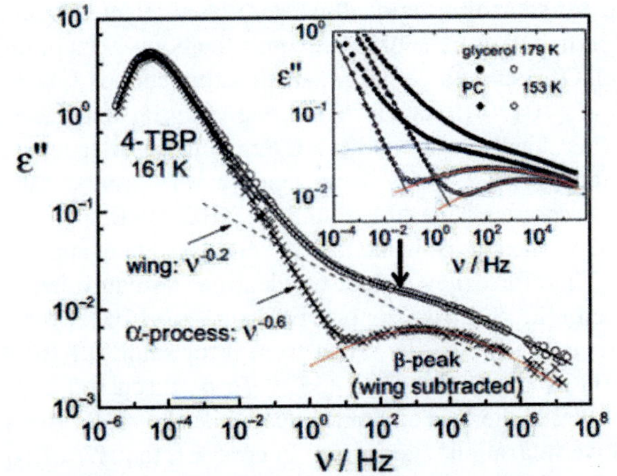

Fig. 132-2 *Circles*: The dielectric loss spectrum of 4-TBP as measured at 161 K under equilibrium conditions. *Crosses*: The same data but with the excess wing contribution $\propto \nu^{-0.2}$ (*dashed line*) subtracted on a phenomenological basis so that a symmetric β-peak results (*solid line*). The *dash-dotted line* highlights the power law on the high-frequency side of the α-peak. *Inset*: a similar subtraction analysis was performed for published data of glycerol and PC [682(a)]. The *solid lines* show the resulting β-peaks with those previously obtained by Schneider et al. [682(a)] (*dotted lines*). Reproduced from [1047(b)] by permission. The *two vertical arrows* indicate the location of the primitive relaxation frequencies, ν_0, calculated from ν_α, and $n = 0.40$ and $n = 0.44$ by the CM equation

Kohlrausch form of Eq. (1.1) with $n = 0.40$. None of the processes are additive, a major difference from the interpretation of [1047(b)]. Neither the CM nor anyone else has offered a quantitative account of the evolution of the many-body dynamics from the primitive relaxation to the Kohlrausch relaxation. That is an unsolved problem of many-body relaxation in statistical mechanics of complex interacting systems.

The authors of [1047(b)] also perform a similar subtraction of the EW $= A\nu^{-0.2}$ from the published dielectric loss data of glycerol at 179 K and propylene carbonate at 153 K from Schneider et al. [682(a)] and the resulting symmetric β-loss peaks are shown in the inset of Fig. 132-2. The β-peak frequencies are several decades higher than determined by Schneider et al. without subtraction of the EW $= A\nu^{-0.2}$, but still assuming the β- and α-loss peaks are additive contributions to $\varepsilon''(\nu)$. As can be seen in Fig. 134 to be presented later, the calculated ν_0 of glycerol at 179 K is about 10^{-1} Hz, while the peak frequency of the symmetric β-loss peak in the inset is at about 10^4 Hz, a difference of five decades. This very large separation between the α-relaxation and the β-relaxation resulting from the procedure of [1047(b)] violates the correlation of the ratio τ_α/τ_β with the number of carbons in the series of sorbitol, xylitol, threitol, and glycerol [239], and with n or the width of the dispersion of the α-relaxation in general [251].

EW $= A\nu^{-\gamma}$ with $\gamma \approx 0.2$ Is Not Equivalent to the IPL Observed by OHD-OKE

There is yet another confusion introduced by the authors of [1025, 1047(a), 1047(b), 1053, 1054] by identifying the EW discussed above with the intermediate power law (IPL) found by short-time optically heterodyne-detected optical Kerr effect (OHD-OKE) experiments from sub-picoseconds to about 30 ns in the orientational dynamics of salol, OTP, dibutyl phthalate, and benzophenone [330–332]. For example, the authors of [1047(b)] stated: "Here, we mention that the excess wing is equivalent[39] to what has been termed the intermediate power-law on the basis of optical Kerr effect studies.[21, 40, 47] There, again a very similar exponent close to $\gamma \approx 0.2$ has been reported." The IPL found by OHD-OKE appears at time *before* the merged α- and β-relaxations, while the EW discussed here appears at times *after* the β-relaxation. Therefore, they cannot be identified as equivalent. Moreover, the exponent γ (called c in [332]) does not have value ≈ 0.20. Fayer and coworkers reported that γ or c varies over the range from 0 for salol, 0.1 for BPM, 0.13 for BZP, 0.15 for OTP, and 0.21 for DBP [332]. To be discussed in detail later, the IPL from OHD-OKE is the loss at short times while the molecules are caged, and before the primitive relaxation takes place. In contrast, the EW of type either $\mathcal{A}$ or $\mathcal{B}$ is comprised of the JG β-relaxation and the subsequent development of the many-body dynamics, and the EW of type $\mathcal{C}$ is only made of the latter part. The two have to be distinguished.

Excess Wing Should Not Be Confused with Nearly Constant Loss

In the publications of the authors of [1047(a), 1047(b), 1053, 1054] on the dynamic susceptibility from dielectric and optical Kerr effect (OKE) data of glassformers without resolved secondary relaxation, they reaffirmed their belief that the excess wing (EW) is an invariant feature for a range of temperatures and has the "intermediate power law" (IPL) frequency dependence, $A\nu^{-\gamma}$, with temperature-independent γ. To fit dielectric loss data above T_g of five different glassformers including glycerol, PC, and PG, they fixed $\gamma \approx 0.2$ and the magnitude of the excess wing A to be a temperature-dependent parameter [1053, 1054]. This IPL is convoluted with the Cole–Davidson function for the α-relaxation with spectral shape that is temperature independent, and the results fit the data well. However, the fits were made [1053, 1054] to the data of glycerol at 195 K and above (where $\tau_\alpha < 1$ s), but not data at lower temperatures in the equilibrium liquid state closer to T_g and in the glassy state such as for glycerol and PC in Figs. 3 and 1, respectively, of [971], and in [424, 1023]. On decreasing T below 195, the IPL (excess wing) extends over increasing number of decades of frequency with γ decreasing down to value as small as 0.1 or less, and is more appropriately called the nearly constant loss (NCL). Certainly the fits for data at 195 K and above with fixed $\gamma \approx 0.2$ [1047(a), 1053, 1054] cannot work for the NCL at these lower temperatures. Therefore, the NCL should be considered as a feature of the dynamics distinctly different from IPL or excess wing of the authors of [1047(a), 1053, 1054]. To demonstrate this, isothermal dielectric loss data of *N*-methyl-ε-caprolactam (NMEC) [508] are presented in Fig. 133 together with *tert*-butylpyridine (TBP) [675] to show three distinct features: the α-loss peak

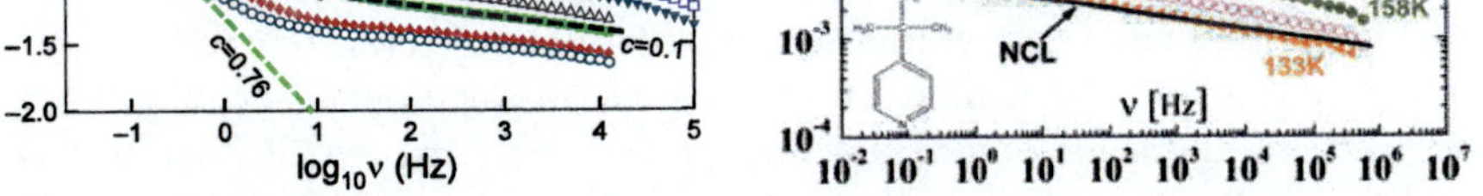

Fig. 133 (*Left*) Dielectric loss spectra of NMEC at temperatures above and slightly below T_g (from *right to left*: 186, 184, 182, 180, 178, 176, 174, 172, 170, 168, 164, 162 K) redrawn from the data supplied by Ranko Richert to show the following three distinct relaxation mechanisms. (1) The α-loss peaks and the fits (*red and pale blue dashed lines*) at two temperatures by the Fourier transform of the KWW function, $\varphi(t) = \exp[-(t/\tau_\alpha)^{1-n}]$, with $n = 0.24$. (2) The excess loss $\varepsilon''(\nu) \sim \nu^{-c}$ with $c = 0.29$ (*long dashed line*) at 178 K, and with $c = 0.47$ (*solid line*) at 168 K. The very different slopes of the excess loss at the same levels at the two temperatures rule out frequency-temperature superposition of the loss data of the excess loss in this region. (3) The emergence of the NCL, $\varepsilon''(\nu) \sim \nu^{-c}$ with $c \approx 0.1$ (*black short dashed line*), with lower intensity at lower temperatures. The *green dashed line* with the label $c \approx 0.76$ shows the frequency dependence of the high-frequency flank of the KWW fit to the (unobserved) loss peak at 168 K. The *vertical arrows* indicate the relaxation times of the unresolved Johari–Goldstein β-relaxation suggested by τ_0 of the CM [508]. (*Right*) Similarly for *tert*-butylpyridine, redrawn from the data supplied by D. Prevosto

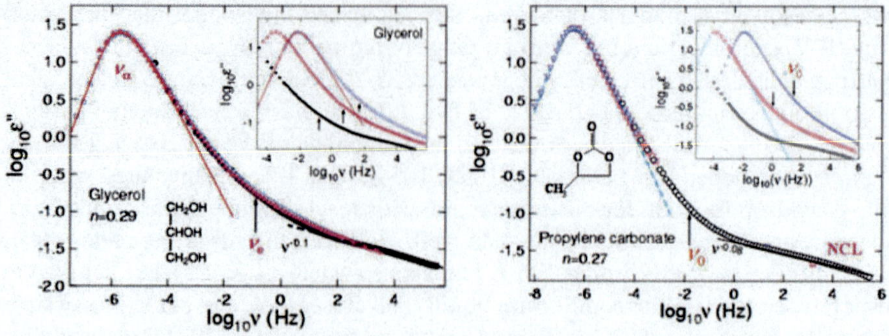

Fig. 134 (*Left*) The *inset* shows the dielectric loss data of glycerol at three temperatures: 179, 185, and 190 K. The data at 179 K were obtained after aging to achieve thermodynamic equilibrium. The curve is fitted to the α-relaxation peak by the one-sided Fourier transform of the Kohlrausch function with $n = 0.29$. Each *vertical arrow* pointing toward certain data taken at some temperature indicates the location of the independent relaxation frequency, ν_0, calculated for that temperature. The main figure shows the master curve obtained by shifting the data at the two higher temperatures to superpose on the data at the lowest temperature 179 K. The *vertical arrow* indicates the location of the independent relaxation frequency, ν_0, calculated. The *dashed line* with frequency dependence, $\nu^{-0.1}$, indicates the slow variation of ε'' at high frequencies or the NCL regime. (*Right*) The *inset* shows the dielectric loss data of propylene carbonate at three temperatures: 152, 155, and 158 K. The data at 152 K were obtained after aging to achieve thermodynamic equilibrium. The curve is fitted to the α-relaxation peak by the one-sided Fourier transform of the Kohlrausch function with $n = 0.27$. Each *vertical arrow* pointing toward certain data taken at some temperature indicates the location of the independent relaxation frequency, ν_0, calculated for that temperature. The main figure shows the master curve obtained by shifting the data at the two higher temperatures to superpose on the data at the lowest temperature 152 K. The *vertical arrow* indicates the location of the independent relaxation frequency, ν_0, calculated. The *dashed line* with frequency dependence, $\nu^{-0.08}$, indicates the slow variation of ε'' at high frequencies or the NCL regime. Reproduced from [195] by permission

fitted by the one-sided Fourier transform of the Kohlrausch function with $n = 0.24$, the excess loss over the Kohlrausch fit in an intermediate frequency range (excess wing, according to the terminology commonly used), and the NCL $\propto \nu^{-y}$ with $\gamma \approx 0.1$ at higher frequencies. The frequencies ν_{JG} of the unresolved JG relaxation, as suggested by the calculated primitive frequencies ν_0, are indicated by the vertical arrows. All ν_{JG} or ν_0 are located within the excess wings and therefore suggest the excess wing is indeed the unresolved JG relaxation. Isothermal dielectric loss data of glycerol and propylene carbonate in Fig. 134 show the same features as NMEC. These results are warnings that one should not confuse the IPL having fixed $\gamma \approx 0.2$ (excess wing of [1047(a), 1053, 1054]) with the NCL. They are distinctly different features of the dynamics.

2.3.2.7 Digression on NCL

Such a loss $A\nu^{-\gamma}$ with small γ and extending over many decades (six decades for glycerol as well as PC, and can be even longer had measurement been made at higher

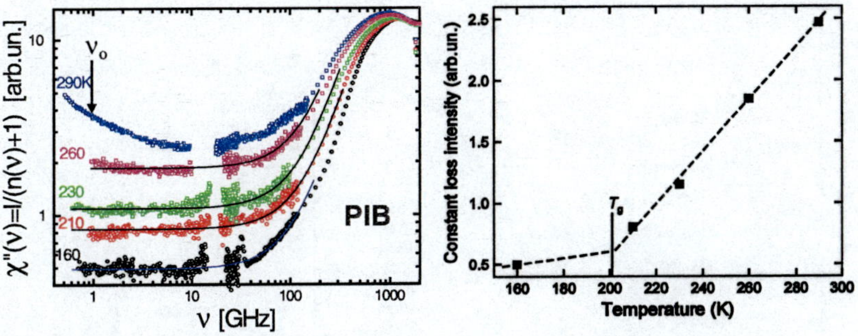

Fig. 135 (*Left*) The susceptibility from quasielastic light-scattering data of PIB from [1062] at five temperatures replotted. The existence of the NCL at temperatures below and above $T_g = 201$ K is evident. The location of the estimated primitive frequency ν_0 of about 1 GHz at 260 K is indicated by the *arrow*. (*Right*) Plot of the temperature dependence of the NCL intensity suggests that there is a break near T_g of PIB

frequencies) is also found long time ago in ionically conducting glasses, crystals, and molten salts from experiments [115(b), 124, 148, 149, 195, 507, 1055–1061] and from molecular dynamics simulations [802, 1058]. It is appropriately called the nearly constant loss (NCL) because of the small value of γ, and has been interpreted as relaxation due to anharmonicity of the potential while the ions are all still caged [802, 1057]. In molecular dynamics simulations or experiments, it is manifested as very slow rise of mean-squared displacement $< r^2(t) >$ over many decades of time at low enough temperatures in glassformers including binary LJ particles [321, 1058], OTP from the intermediate scattering function (see Fig. 6), and silica [823], colloidal particles (see Figs. 1 and 79), and ionic conductors [802, 1058]. Quasielastic neutron scattering experiments have captured it in many glassformers as well, not by its characteristic frequency dependence but by its weak temperature dependence [507].

Dynamic light-scattering experiments performed on the amorphous polymers, polyisobutylene (PIB) [1062] and polyisoprene (PI) [149, 507], have found the NCL in the GHz frequency range as shown in Figs. 135 and 136. For PIB, the imaginary part of the susceptibility $\chi''(\nu)$ shows a virtually frequency-independent NCL (i.e., nearly zero γ) starting from 160 K, which is 40 degrees below $T_g \approx 200$ K, and up to 260 K which is $1.3T_g$.

For PI ($T_g \approx 210$ K), $\chi''(\nu)$ from light scattering shows in Fig. 136 (Left) the presence of the NCL with $\gamma = 0.15$ from 153 to 240 K in the frequency range from 1 to 40 GHz [1063]. Dielectric loss $\varepsilon''(\nu)$ data of PI taken at temperatures below T_g and shown in Fig. 110 also found NCL at much lower frequencies from 10 to 10^4 Hz. Although dielectric data of the NCL cannot be obtained at higher frequencies at these temperatures, the dielectric NCL from $\varepsilon''(\nu)$ at 153 K or about in Fig. 110 if continued to higher frequencies is likely the NCL from $\chi''(\nu)$ by light-scattering NCL at the same temperature in Fig. 136. This result indicates that the IPL found by OKE [330–333] and light scattering [1047, 1053, 1054] at high frequencies and

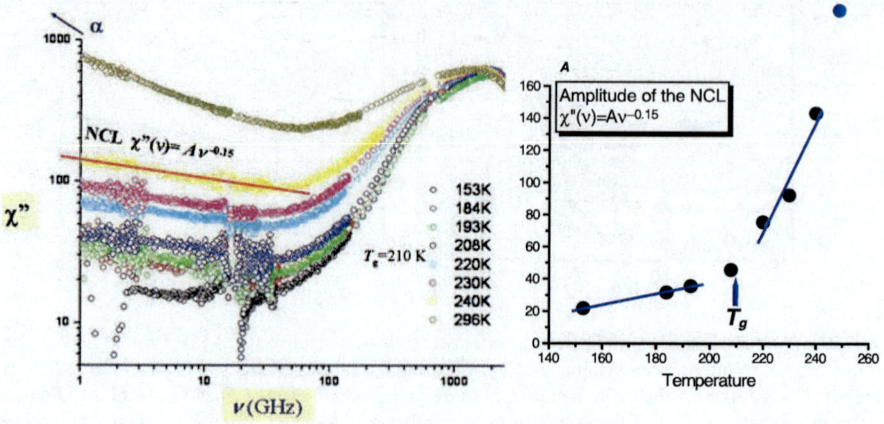

Fig. 136 (*Left*) The susceptibility from quasielastic light-scattering data of polyisoprene (PI) from [1063] at temperatures indicated. The existence of the NCL at temperatures below and above $T_g = 201$ K is evident. (*Right*) Plot of the temperature dependence of the NCL intensity suggests that there is a break near T_g of PI

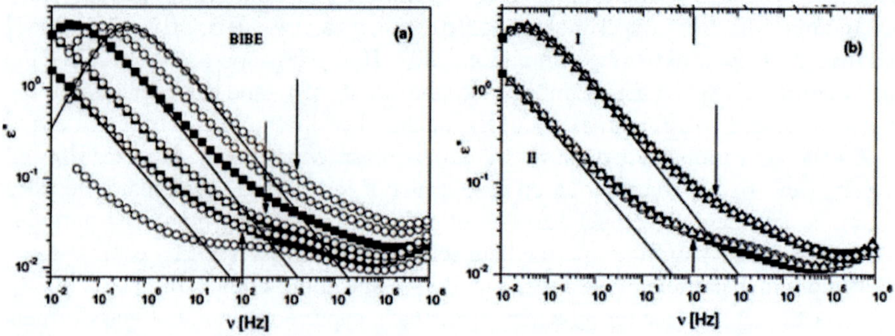

Fig. 137 Dielectric loss of BIBE vs. frequency at different pressures and temperatures. (**a**) *Open circles*: $T = 288.2$ K and $P = 3502$, 3699, 4234, 4507, 5165 bar (*from right to left*); *solid squares* $P = 1$ bar and $T = 226.1$, 223.0, 220.5 K (*from right to left*); *solid lines* are KWW fitting curves for α-process (coupling parameter $n = 0.33 - 0.38$). *Arrows* indicate the frequency location of JG process according to the CM predictions from the primitive frequency. (**b**) Comparison of different spectra with the structural peak in position I or II. I-peak includes: $T = 278.4$ K and $P = 3202$ bar (*open triangles*), $T = 298$ K and $P = 4666$ bar (*open diamonds*), $T = 226.1$ K and $P = 1$ bar (*solid squares*). II-peak includes: $T = 278.4$ K and $P = 3699$ bar (*open triangles*), $T = 288.2$ K and $P = 4507$ bar (*open circles*), $T = 220.5$ K and $P = 1$ bar (*solid squares*). Reproduced from [669] by permission

temperature above T_g and the NCL seen by dielectric relaxation at lower frequencies and near or below T_g is one and the same process.

Seen before in the dielectric spectra (Fig. 110, left panel), PI has a JG β-relaxation which shows up in about the same frequency range as the NCL but at temperatures above T_g (see Fig. 110, right panel) Thus, at the same temperature, this NCL of PI is on the high-frequency side of the JG β-relaxation. For glassformers having a resolved secondary relaxation, we recall that authors of [1047, 1053, 1054] believe their universal excess wing is present at intermediate frequencies between the α-relaxation and the secondary relaxation. One of the examples they used to demonstrate this belief is the dielectric loss data of DGEBA [1047]. It can be verified by inspection; the dielectric loss data of polyisoprene (PI) shown before in Fig. 110 have the same feature of the DGEBA data used by authors of [1047, 1053, 1054] to arrive at their conclusion. Thus, if one follows their logic, one would be led to conclude that there is an excess wing lying in between the α- and the JG β-relaxations of PI. However, the excess loss between the α-loss peak described by the Kohlrausch fit and the JG β-relaxation in the data of PI at 204.2 K in Fig. 110 is due to the contribution from the developing many-body relaxation with increasing length scale. Actually the excess wing or NCL has been found but is on the high-frequency side of the JG β-relaxations.

2.3.2.8 τ_{JG} Like τ_α Is Pressure Dependent, and Co-invariance of n and τ_α/τ_{JG} at Constant τ_α

The α-relaxation time τ_α increases with pressure at constant temperature. The sensitivity of τ_α to pressure, or the density-dependence of τ_α, varies depending on the chemical structure of the glassformer. The JG relaxation time τ_{JG} also increases with applied pressure, although less so than τ_α. The pressure dependence of τ_{JG} is important because it not only mimics the α-relaxation but also shows that τ_{JG} is sensitive to change in specific volume as well as to temperature, the latter is well known. On the other hand, secondary relaxations that are not of the JG kind (involving intramolecular degrees of freedom) usually have lesser or no pressure dependence. Several examples are given here to illustrate the different pressure dependences of the two kinds of secondary relaxations. Thermodynamic based theories of glass transition exclusively consider the dependences of volume and entropy of the α-relaxation but not those of the secondary relaxation not to say the JG relaxation, and this is a potential problem for these theories. In the following we discuss the pressure dependence of τ_{JG} by examples taken from glassformers in different situations.

Neat Glassformers: Resolved JG β-Relaxation Above T_g

Benzoin isobutylether (BIBE) and dipropyleneglycol dibenzoate (DPGDB) discussed in Section 2.2.4.1 and shown before in Fig. 35e and f are examples. There are two secondary relaxations in both cases, and the slower one shifts with pressure and is the JG relaxation. In contrast, the faster γ-relaxation loss peak does not seem to shift at all with pressure. For BIBE, it can be deduced from the loss data of BIBE with the longest τ_α in Fig. 35 f. The slower one shown as a bump in the loss spectrum is the JG relaxation. Although the structure indicating the JG relaxation is not as well resolved at shorter τ_α, its shift to lower frequencies with increasing pressure is an indication that τ_{JG} increases on elevating pressure. Furthermore, concurrent invariance of the α- and the JG β-loss peaks to different combinations of T

and P shown in Fig. 35e and f is another interesting feature. It indicates that at constant τ_α, for different combinations of P and T, not only the frequency dispersion (or n) of the α-relaxation but also the ratio, τ_α/τ_{JG}, is invariant. This co-invariance of the frequency dispersion (or n) of the α-relaxation and the ratio τ_α/τ_{JG} is another sign of the existence of some profound connection of the JG β-relaxation to the α-relaxation. As we shall see below this co-invariance property is general. Since this is a very important dynamic property, it is worthwhile to show once more the data of BIBE [669] more clearly in Fig. 137.

For DPGDB, Fig. 117 given before in Section 2.3.2.2 shows more explicitly than Fig. 37e that the slower JG secondary relaxation in DPGDB shifts to lower frequencies with increasing pressure, but not the faster γ-relaxation [1014]. For each temperature, there is good correspondence between the peak frequency of the JG relaxation ν_{JG} and the primitive frequency ν_0 calculated by Eq. (2.67). The co-invariance of the frequency dispersion (or n) of the α-relaxation and the ratio τ_α/τ_{JG} in the liquid state has been shown in Fig. 35e. There are more data of τ_{JG} in the glassy state from the work by Prevosto and coworkers [678]. These are presented in Fig. 138 as a function of T at various constant pressures P_{con}, and as a function of P at various constant temperatures T_{con} together with τ_α and τ_γ. The τ_{JG} data shown are all for $T < T_g$ or $P > P_g$, because the loss peak of the JG relaxation of DPGDB is best resolved in the glassy state over wider range of T or P. Here for all P_{con} and T_{con}, the reference temperature and pressure T_{ref} and P_{ref} are defined by $\nu_\alpha(T_{ref}, P_{con}) = 10^{-2}$Hz and $\nu_\alpha(P_{ref}, T_{con}) = 10^{-2}$Hz, respectively. It can be

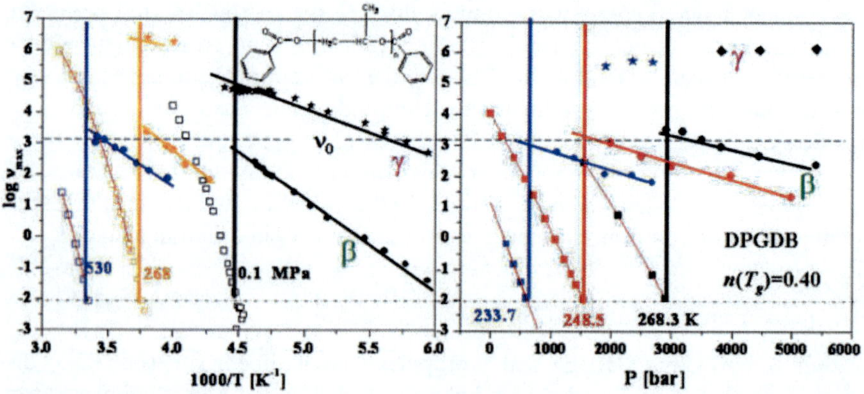

Fig. 138 (*Left*) Relaxation frequencies, ν_α, ν_β, and ν_γ, of the α-, β-, and γ-relaxations in DPGDB as functions of reciprocal temperature at three different constant pressures, 0.1, 268, and 530 MPa. (*Right*) Relaxation frequencies, ν_α, ν_β, and ν_γ, of the α-, β-, and γ-relaxations as functions of pressure at three different constant temperatures, 248.3, 268.5, and 233.7 K. In both panels, the *lower horizontal dotted line* indicates $\nu_\alpha = 10^{-2}$ Hz. The results show ν_β/ν_α is invariant to different combinations of P and T at constant value of $\nu_\alpha = 10^{-2}$ Hz. The *upper horizontal line* indicates the primitive frequency ν_0 calculated with $n = 0.40$ of the Kohlrausch function that fits the α-relaxation, and it shows $\nu_0 \approx \nu_\beta$, independent of T and P. Reproduced from [678] by permission

seen by inspection of Fig. 138 that the JG relaxation frequencies, $\nu_{JG}(T, P_{con})$ and $\nu_{JG}(P, T_{con})$, have approximately the same value at T_{ref} and P_{ref}, respectively. Thus, at T_{ref} or P_{ref} where $\nu_\alpha = 10^{-2}$ Hz, the ratio $(\nu_{JG}/\nu_\alpha) \equiv (\tau_\alpha/\tau_{JG})$ is invariant. Together with the previously established invariance of the frequency dispersion (or n) of the α-relaxation of DPGDB at constant τ_α or ν_α to T and P combinations, this new result again leads us to the property of co-invariance of n and (τ_α/τ_{JG}) at $\nu_\alpha = 10^{-2}$ Hz. The primitive frequency ν_0 has been calculated by Eq. (2.67) at T_{ref} or P_{ref} with $n = 0.40$ of the Kohlrausch function that fits the α-relaxation and its location in Fig. 138 indicated by the horizontal line. It is in approximate agreement with the values ν_{JG} would have at T_{ref} or P_{ref}.

Pressure dependence of τ_{JG} together with co-invariance of n and τ_α/τ_{JG} at constant τ_α is found in two hydrogen-terminated poly(methylmethacrylate) (PMMA) with degrees of polymerization equal to 3 and 4, i.e., trimer and tetramer with molecular weights of 302 and 402 g/mol, respectively [1064]. The left panel of Fig. 139 shows the normalized dielectric loss data of the trimer at three different combinations of P and T all having α-loss peaks with the same peak frequency and frequency

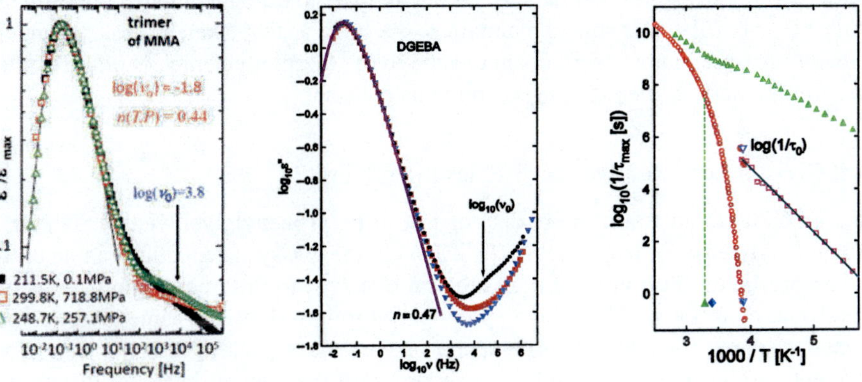

Fig. 139 (*Left*) Dielectric loss data of MMA trimer at three combinations of P and T at the same α-loss peak frequency. The line is the KWW fit with $n = 0.44$. The *arrow* indicates the location of the primitive relaxation frequency, ν_0. Data from [1064] replotted. (*Middle*) The frequency dependence of the dielectric loss, $\varepsilon''(\nu)$, of DGEBA (EPON828) taken at three selected combinations of temperature and pressure, $T = -16.0$°C, $P = 0.1$ MPa (*open triangles*); $T = 20.19$°C, $P = 246.5$ MPa (*open squares*); and $T = 30.19$°C, $P = 345.5$ MPa (*closed circles*), are shown in the main part of the figure. The loss peaks of the three sets of data have practically the same peak frequency, ν_{max}, and nearly the same frequency dependence. The *dashed line* is the fit by the imaginary part of the one-sided Fourier transform of the Kohlrausch function with $n = 0.47$. The location of the calculated ν_0 is indicated by the *arrow*. (*Right*) *open circles, squares*, and *triangles* are for the α-, β-, and γ-relaxation times reported in [242] at ambient pressure, and the *two closed circles* are for τ_0 calculated for two temperatures from τ_α. The *open inverted triangle* is the τ_0 calculated from the α-relaxation time at $T = -16.0$°C, $P = 0.1$ MPa in the main figure. The *closed diamond* and *triangle* are for $\log(1/\tau_\alpha)$ at $T = 20.19$°C, $P = 246.5$ MPa and at $T = 30.19$°C, $P = 345.5$ MPa, respectively. Data from [676] are replotted here

dependence in the liquid state. The latter is well fitted by the Kohlrausch function with $n = 0.44$ (solid line). The shoulder on the high-frequency side is the JG relaxation as evidenced by the good agreement of its location by the primitive frequency ν_0 calculated by Eq. (2.67) with $\nu_\alpha \approx 10^{-1.8}$ Hz and $n = 0.44$ (vertical arrow). Since the location of the shoulder does not change, τ_α/τ_{JG} is invariant to different combinations of P and T at constant τ_α. The shape of the JG relaxation changes because there is no guarantee that its dielectric strength in relation to that of the α-relaxation remains the same.

The situation is similar in the case of DGEBA (EPON828). Dielectric study of DGEBA at elevated pressure of 345.5 MPa and $T = 30.19$°C was able to resolve a shoulder in the loss spectrum that can be identified with the JG β-relaxation in the liquid state (see the middle panel of Fig. 139). The frequency where the shoulder appears is near ν_0 calculated by Eq. (2.67) of the CM. The corresponding τ_0 is shown in the right panel by the inverted triangle, and it suggests that τ_β has a stronger T-dependence above T_g than its Arrhenius dependence below T_g. The data of liquid DGEBA support that τ_β remains constant to changes of the combinations of P and T when τ_α is held constant. This is a remarkable result because τ_α is eight decades shorter at 30.19°C than at –16.0°C when both are at ambient pressure of 0.1 MPa (see vertical dashed line in inset). Although applied pressure of 341.1 MPa at 30.19°C is able to maintain τ_α the same as that taken at –16.0°C under ambient pressure, there is significant change in the thermodynamic condition of the glassformer, and yet the ratio, τ_β/τ_α, remains constant.

Neat Glassformers: Resolved JG β-Relaxation Below T_g

Like DPGPG, there are lot more data of τ_{JG} of poly(phenylglycidylether) (PPGE) and diglycidylether of bisphenol A (DGEBA) in the glassy state, which can be used to support the co-invariance of n and the ratio τ_α/τ_{JG} to different combinations of P and T at constant τ_α. These two epoxy glassformers show the same behavior as discussed above for DPGDB. The shape of the α-loss peak is also the same for different combinations of P and T, in accord with the general behavior discussed before in Section 2.2.4. The fit by the Kohlrausch function gives $n = 0.54$ for PPGE and $n = 0.47$ [678] for DGEBA [616].

The data from the work of Prevosto and coworkers are shown in Fig. 140 for PPGE and Fig. 141 for DGEBA. These figures are similar to those of DPGDB, and the descriptions given above apply. The conclusions are the same, i.e., (i) τ_{JG} like τ_α is pressure dependent, (ii) n and (τ_α/τ_{JG}) are co-invariant at a chosen fixed value of τ_α, and (iii) the calculated primitive τ_0 at T_{ref} or P_{ref} is close to the projected τ_{JG} (only determined at $T < T_g$ or $P > P_g$) at the same T and P. However, the faster γ-relaxation is invariant to P above and below T_g. Later on, in another subsection, it will be shown that the Arrhenius T-dependence or P-dependence of τ_{JG} established in the glassy state no longer holds and changes to assume a stronger T-dependence in the equilibrium liquid state. Thus, the extrapolation of the Arrhenius T-dependence τ_{JG} in the glassy state to T_{ref} would tend to overestimate the true value of $\tau_{JG}(T_{ref})$.

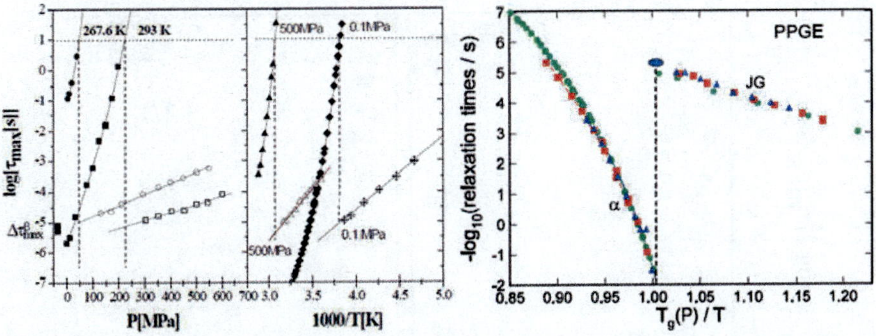

Fig. 140 (*Left panel*) Plot of $\log[\tau_\alpha(P)/\mathrm{s}]$ and $\log[\tau_{JG}(P)/\mathrm{s}]$ of PPGE against P at two temperatures, 267.6 and 293 K. P_g is defined by $\log[\tau_\alpha(P_g)/\mathrm{s}] = 1$. (*Middle panel*) Plot of $\log[\tau_\alpha(T)/\mathrm{s}]$ and $\log[\tau_{JG}(T)/\mathrm{s}]$ against $1000/T$ at two constant pressures, 0.1 and 500 MPa. T_g is defined by $\log[\tau_\alpha(T_g)/\mathrm{s}] = 1$. Extrapolation of the linear P dependence of $\log[\tau_{JG}(P)/\mathrm{s}]$ to P_g in *left panel* or the linear $1000/T$ dependence of $\log[\tau_{JG}(T)/\mathrm{s}]$ to T_g in *middle panel* all give values within the spread indicated by $\Delta\tau^{\beta}_{max}$ and the length of the *filled black rectangle* located on the *y*-axis of the *left panel*. (*Right panel*) The same data in the middle panel with the addition of the data taken at $P = 240$ MPa replotted as functions of the scaled reciprocal temperature $T_g(P)/T$. Here $T_g(P)$ is defined by $\log[\tau_\alpha(T_g)/\mathrm{s}] = 1.42$ for all three constant pressures, $P = 500$ MPa (*triangles*, $T_g = 323.3$ K), 240 MPa (*squares*, $T_g = 294.1$ K), 0.1 MPa (*circles*, $T_g = 260.4$ K). There is total collapse of both sets of the τ_α and τ_{JG} data, indicating the correlation between the two processes. The *oval symbol* is $-\log(\tau_0)$, where τ_0 is the primitive relaxation time calculated at $\log(\tau_\alpha/\mathrm{s}) = 2$, and $n = 0.54$ from fit of the α-loss peak by the Kohlrausch function. Reproduced from [678] by permission

As the last example from unpublished data of D. Prevosto, the relaxation frequencies ν_α and ν_{JG} of poly(vinyl acetate) (PVAc) with molecular mass $M_w = 167$ kg/mol and $T_g = 310$ K shown as functions of pressure P at two temperatures, 323.2 and 343.3 K, in the left panel of Fig. 141a are compared with ν_α and ν_{JG} as functions of $1000/T$ at ambient pressure of 0.1 MPa. Again, at the glass transition pressure P_g or temperature T_g defined by $\nu_\alpha = 0.1$ Hz, $\nu_{JG}(P_g)$ and $\nu_{JG}(T_g)$ are nearly the same for all three cases (as indicated by the narrow rectangular box in the figure). Also shown is the logarithm of the primitive relaxation frequency, ν_0, calculated by the CM equation (2.68) with $n = 0.58$ obtained from fitting the α-loss peak by the Fourier transform of the Kohlrausch function and $\nu_\alpha = 0.1$ Hz. The value $\log(\nu_0) = 5.9$ is close to $\nu_{JG}(P_g)$ and $\nu_{JG}(T_g)$, and the good agreement demonstrates the applicability of the CM.

Unresolved JG Relaxations

From the discussions given in Sections 2.3.2.3 and 2.3.2.4 above for glassformers having narrow dispersion (smaller n), the ratio τ_α/τ_{JG} is not large enough to resolve the weaker JG β-relaxation. The unresolved JG relaxation is responsible for the excess of the dielectric loss over the Kohlrausch fit of the α-loss peak, irrespective of whether the excess wing shows up prominently or not, and with or without a

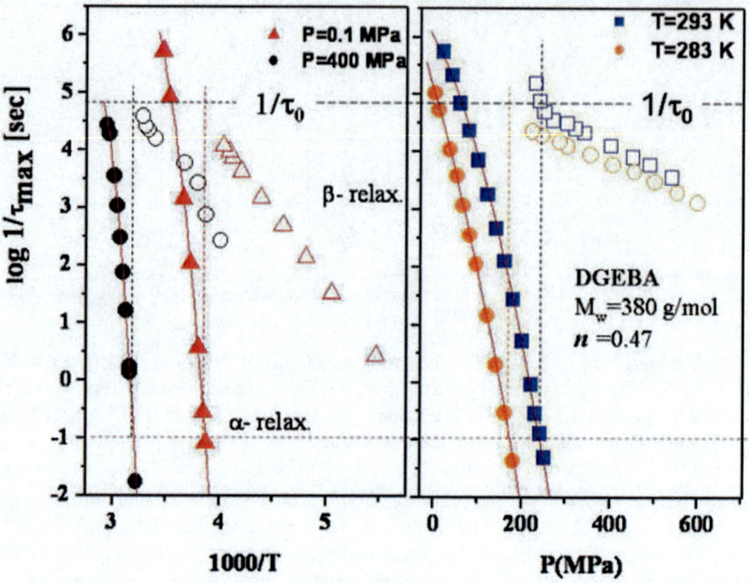

Fig. 141 (*Right panel*) Plot of $\log[\tau_\alpha(P)/\mathrm{s}]$ and $\log[\tau_{JG}(P)/\mathrm{s}]$ of DGEBA against P at two temperatures, 283 and 293 K. P_g is defined by $\log[\tau_\alpha(P_g)/\mathrm{s}] = 1$ (*lower dotted line*). (*Left panel*) Plot of $\log[\tau_\alpha(T)/\mathrm{s}]$ and $\log[\tau_{JG}(T)/\mathrm{s}]$ against 1000/T at two constant pressures, 0.1 and 400 MPa. T_g is defined by $\log[\tau_\alpha(T_g)/\mathrm{s}] = 1$ (*lower dotted line*). Extrapolation of the linear P dependence of $\log[\tau_{JG}(P)/\mathrm{s}]$ to P_g in *right panel* or the linear 1000/T dependence of $\log[\tau_{JG}(T)/\mathrm{s}]$ to T_g in *left panel* all give nearly the same value, which is close to the primitive relaxation time $\log(1/\tau_0)$ calculated at $\log(\tau_\alpha) = -1$, and $n = 0.47$ from fit of the α-loss peak by the Kohlrausch function (*upper dotted line*). Reproduced from [616] by permission

γ-relaxation at higher frequencies. For pressure dependence of τ_{JG} alone, let us recall the dielectric loss data of 3PG measured at 220.5 K and increasing pressures in Fig. 126. At lower P when $\nu_\alpha \geq 10^{-2}$ Hz there is only excess of $\varepsilon''(\nu)$ over the Kohlrausch fit with $n = 0.37$. At higher P and $\nu_\alpha < 10^{-2}$ Hz, the excess wing appears and is transformed to the resolved JG relaxation at higher pressures. Thus the excess, the excess wing, and the unresolved JG relaxation are one and the same process and they all shift to lower frequencies with pressure P. However, the α-relaxation of hydrogen-bonded glassformers including 2PG, glycerol, threitol, sorbitol (see Figs. 46 and 48a) [659], and *m*-fluoroaniline (*m*-FA) (see Fig. 47) [671] changes frequency dispersion on elevating T and P because of disruption of the hydrogen bondings, and the $\varepsilon''(\nu)$ obtained at different combinations of T and P at constant τ_α do not superpose. They are not the glassformers of choice to test the co-invariance of n and τ_α/τ_{JG} at constant τ_α.

Glassformers having narrow dispersion (smaller n) also have advantage to show both the pressure dependence of τ_{JG} and the co-invariance property of n and (τ_α/τ_{JG}) at constant τ_α. Since τ_α/τ_{JG} is not large in these glassformers, both the

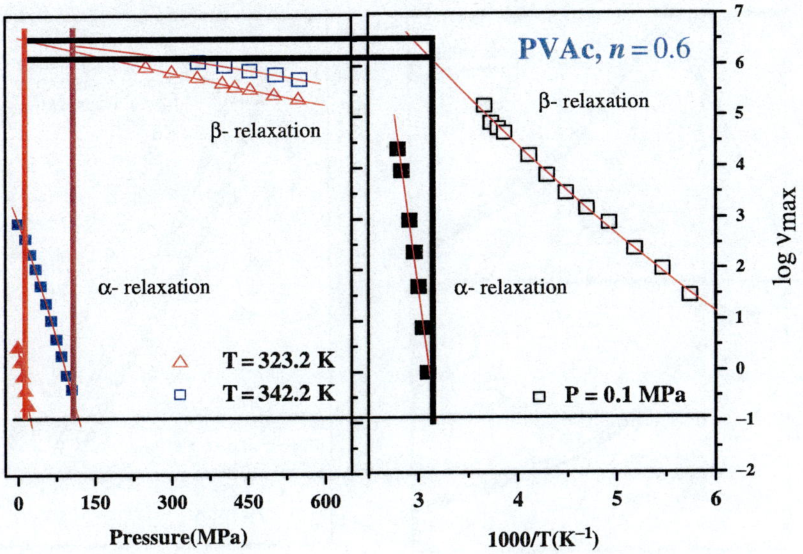

Fig. 141a (*Left panel*) Plot of $\log[\nu_\alpha(P)/\mathrm{s}]$ and $\log[\nu_{JG}(P)/\mathrm{s}]$ of high molecular weight PVAc against P at two temperatures, 323.2 and 342.2 K. P_g is defined by $\log[\nu_\alpha(P)/\mathrm{s}] = -1$ (*lower line*). (*Right panel*) Plot of $\log[\nu_\alpha(T)/\mathrm{s}]$ and $\log[\nu_{JG}(T)/\mathrm{s}]$ against $1000/T$ at ambient pressures of 0.1 MPa. T_g is defined by $\log[\tau_\alpha(T_g)/\mathrm{s}] = -1$ (*lower line*). Extrapolation of the linear P dependence of $\log[\nu_{JG}(P)]$ to P_g in *left panel* or the linear $1000/T$ dependence of $\log[\nu_{JG}(T)]$ to T_g in *right panel* all give nearly the same value, which is close to the primitive relaxation frequency $\log(\nu_0)$ calculated at $\log(\nu_\alpha) = -1$, and $n = 0.58$ from fit of the α-loss peak by the Kohlrausch function. Courtesy of D. Prevosto

α- and the JG β-processes in the liquid state can be observed in a single spectrum taken within the spectral range of the instrument, particularly those having no γ-relaxation in the dielectric spectrum. Demonstrated before for many glass-formers of this kind is the invariance to different combinations of P and T at constant τ_α of the entire spectrum encompassing the α-loss peak and the excess contributed by the unresolved JG β-process. Examples shown before in Section 2.2.4 by spectra include KDE, PC, PCB62, and DiBP (see Fig. 35), TPP (see Fig. 38), DPVC (see Figs. 39 and 122), 7PGDE (see Fig. 48b), and ribose (see Fig. 51). The same behavior was found in phenylphthalein-dimethylether (PDE), phenyl salicylate (salol), 3,3′,4,4′-benzophenonetetracarboxylic dianhydride (BPTCDaH), 1,1′-*bis*(*p*-methoxyphenyl)cyclohexane (BMPC), 1,1′-di(4-methoxy-5-methylphenyl)cyclohexane (BMMPC), diethyl phthalate (DEP), and di-isooctal phthalate (DiOP), and the data presented in [120]. More small molecular glass-formers obeying P–T superpositioning at constant τ_α have been found since the publication of [120, 618]. The data of propylene carbonate and 3,3′,4,4′-benzophenonetetracarboxylic dianhydride (BPTCDaH) are presented once more in Fig. 142, and salol and BMMPC in Fig. 143 to support the general property.

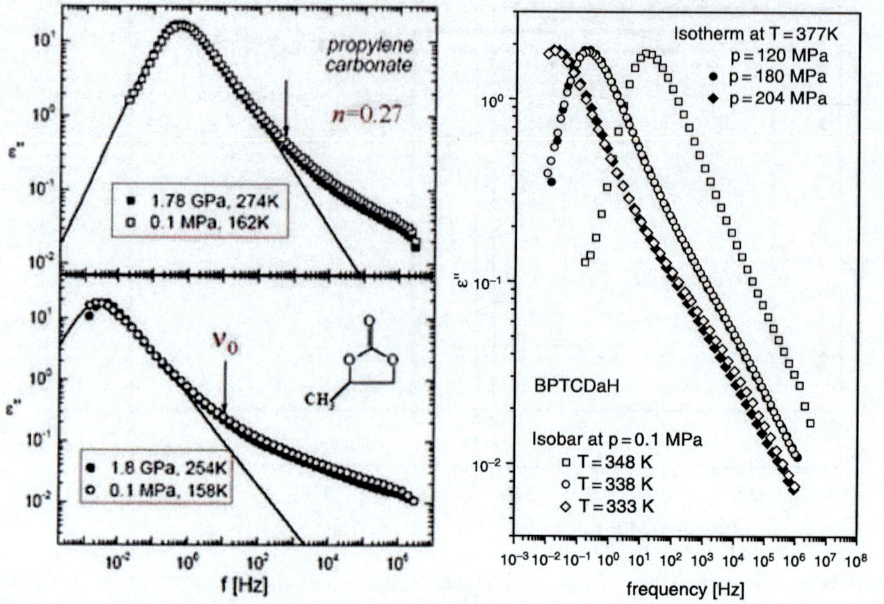

Fig. 142 (*Right*) Dielectric loss data of 3,3′,4,4′-benzophenonetetracarboxylic dianhydride (BPTCDaH) at various combinations of temperature and pressure as indicated to demonstrate the invariance of the dispersions of both the α-relaxation and the excess wing/unresolved JG β-relaxation at constant α-loss peak frequency ν_α or equivalently at constant α-relaxation time τ_α. (*Left*) The same for PC. Reproduced from [120] by permission

From these examples, we have seen that the α-loss peak fitted by the Fourier transform of the Kohlrausch function and the excess of the measured $\varepsilon''(\nu)$ over it on the high-frequency side obtained for different combinations of T and P at any fixed value of τ_α are invariant. The primitive frequencies ν_0 calculated by Eq. (2.67) located within the excess contribution, as shown by the vertical arrows in some of these figures, are an indication that the excess is the unresolved JG relaxation. The width of the α-loss peak increases and n of the Kohlrausch function increases with increasing τ_α. Nevertheless, for any constant τ_α, the frequency dependence of $\varepsilon''(\nu)$ obtained at different combinations of T and P are invariant over the entire frequency range shown covering the α-relaxation and the excess wing (i.e., the unresolved JG β-relaxation). This result implies that τ_{JG} increases with pressure along with τ_α although at different rates. At any τ_α, n as well as the ratio τ_α/τ_{JG} remains constant, for otherwise the entire shape of $\varepsilon''(\nu)$ will change with the combinations of P and T. At temperatures above $T_g(P)$ corresponding to any pressure P, the T-dependence of τ_{JG} (as indicated by τ_0 calculated from τ_α) will be non-Arrhenius and possibly well described by a Vogel–Fulcher–Tammann–Hesse equation, $(\text{VFTH})_{JG}$, albeit its T-dependence is weaker than that of the VFTH dependence of the α-relaxation, $(\text{VFTH})_\alpha$.

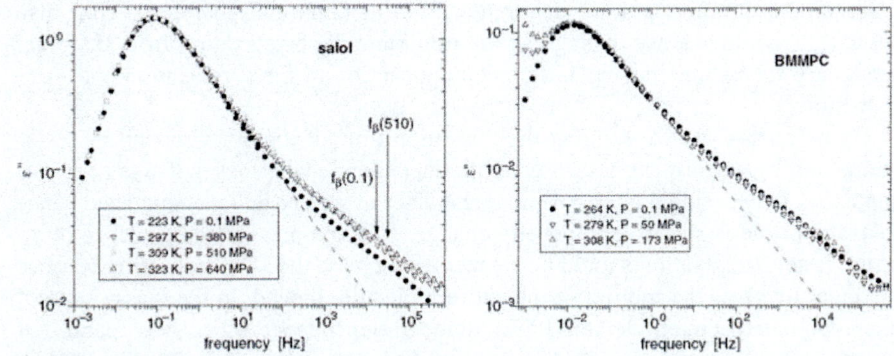

Fig. 143 (*Left*) Dielectric loss data of phenyl salicylate (salol) at various combinations of temperature and pressure as indicated to demonstrate the invariance of the dispersion of the α-relaxation at constant α-loss peak frequency ν_α or equivalently at constant α-relaxation time τ_α. The *dashed line* represents extrapolation of the power law for frequencies just past the maximum. (*Right*) Dielectric loss data of 1,1′-di(4-methoxy-5-methylphenyl)cyclohexane (BMMPC) at various combinations of temperature and pressure as indicated to demonstrate the invariance of the dispersion of the α-relaxation together with the excess wing/unresolved JG β-relaxation at constant α-loss peak frequency ν_α or equivalently at constant α-relaxation time τ_α. The *dashed line* is the imaginary part of the one-sided Fourier transform of the KWW function with $\beta_{KWW} \equiv (1-n) = 0.55$. The logarithmic ordinate scale makes evident the presence of an excess wing at higher frequencies. Reproduced from [120] by permission

Some narrow α-relaxation molecular glassformers have a resolved secondary relaxation whose peak frequency is practically pressure independent; these are not Johari–Goldstein (JG) processes (according to the criteria given in [685]). The slower JG relaxation is not resolved from the α-relaxation in the equilibrium liquid state, and appears as either excess or excess wing. Notwithstanding the presence of the γ-relaxation, the measured $\varepsilon''(\nu)$ at different combinations of T and P are invariant over the entire frequency range shown covering the α-relaxation and the excess wing (i.e., the unresolved JG β-relaxation) for any constant τ_α. As discussed before in Section 2.2.4.1 and some figures therein, such liquids include 1,1′-*bis*(*p*-methoxyphenyl)cyclohexane (BMPC), diethyl phthalate (DEP), di-*n*-butyl phthalate (DBP), diisobutyl phthalate (DiBP), di-isooctal phthalate (DiOP), decahydroisoquinoline (DHIQ), dipropyleneglycol dibenzoate (DPGDB), benzoin isobutylether (BIBE), the epoxy compounds including diglycidylether of bisphenol A (EPON828), 4,4′-methylene *bis*(*N,N*-diglycidylaniline) (MBDGA), bisphenol A propoxylate(1PO/phenol)diglycidylether) (1PODGE), *N,N*-diglycidyl-4-glycidyloxyaniline (DGGOA), and *N,N*-diglycidylaniline (DGA) [120].

From the discussions above, the pressure dependence of τ_{JG} and the co-invariance of n and τ_α/τ_{JG} to variations of P and T at constant τ_α or τ_{JG} are general properties of glassformers. More examples of these properties are found in the dynamics of a component of binary mixtures of glassformers to be given in the subsection to follow. These properties reveal that, like the α-relaxation, the JG β-relaxation is sensitive to changes in thermodynamic variables including pressure, volume, and entropy, and not just temperature as commonly believed. They also serve as ways to sharply distinguish the important JG β-relaxation from the other secondary relaxations having little or no connection to the α-relaxation and glass transition.

By no means can we say that the relaxation time of the other secondary relaxations in all glassformers has no pressure dependence whatsoever. This is because some may have residual intermolecular character or will be bulky enough and hence relaxation time is sensitive to density change. But, the pressure dependence of τ_γ of the faster γ-relaxation should be weaker than τ_{JG} of the JG β-relaxation because the latter involves the motion of the entire molecule. Indeed, in the glassy state of phenolphthalein dimethylether (PDE), some P-dependence of τ_γ was found, but to a much lesser extent in 1,1′-*bis*(4-methoxyphenyl) cyclohexane (BMPC) [1065]. Unfortunately, the experiment on PDE cannot obtain the P-dependence of τ_{JG} of PDE and compare with that of τ_γ to check the expected difference.

Well-Resolved JG Relaxation of a Bead-Necklace Model of Polymer

In Section 2.3.3.1, molecular dynamics simulation of the dynamics of a bead-necklace model for polymer by Bedrov and Smith [356(c)] was discussed and the torsional autocorrelation function (TACF) obtained at ambient pressure presented in Fig. 107-2. The results at ambient pressure have shown the existence of the JG β-relaxation with properties like molecular glassformers and real polymers, including the correlation between n and the separation of the JG β-relaxation from the α-relaxation. Simulations of the dynamics at elevated pressures also have been made by Bedrov and Smith. In the left panel of Fig. 144 the TACF for melt with $l = 1.5$ Å ($\sigma/l = 2.7$) at $T = 130$ K and $P = 1$ bar is compared with that at $T = 213$ K and $P = 5000$ bar, and also at $T = 60$ K and $P = 1$ bar with that at

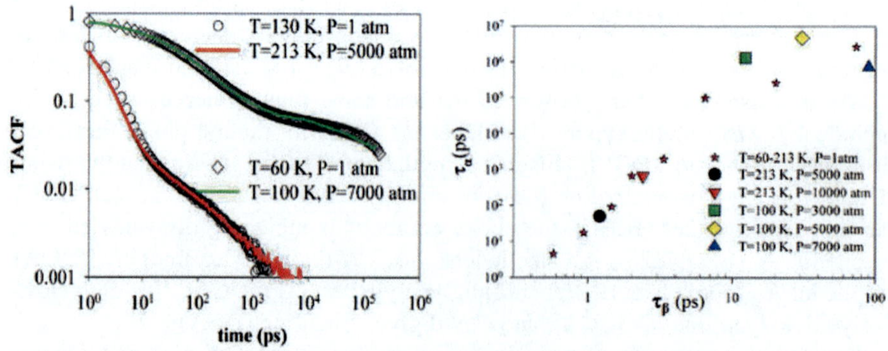

Fig. 144 (*Left*) Comparison of torsional autocorrelation function (TACF) of bead-necklace model for polymer with bond length $l = 1.5$ Å ($\sigma/l = 2.7$) obtained at $T = 130$ K and $P = 1$ bar with that at $T = 213$ K and $P = 5000$ bar, and at $T = 60$ K and $P = 1$ bar with that at $T = 100$ K and $P = 7000$ bar. (*Right*) Correlation between relaxation times for α- and β-processes as obtained from simulations of melts with $l = 1.5$ Å ($\sigma/l = 2.7$) at various thermodynamic conditions. Courtesy of D. Bedrov and G. Smith [356(c)]

$T = 100$ K and $P = 7000$ bar. The two TACFs superpose onto each other, showing the ratio τ_α/τ_{JG} and n are co-invariants to changes in the combinations of P and T that keep either τ_{JG} or τ_α constant. This result further supports the identification in Section 2.3.3.1 of the secondary relaxation observed in the simple bead-necklace polymer model as the JG process. The right panel shows the correlation between relaxation times for α- and β-processes as obtained from simulations of melts with $l = 1.5$ Å ($\sigma/l = 2.7$) holding at more thermodynamic conditions. According to Bedrov and Smith, the scattering in data seen for data points with large τ_α is likely due to uncertainty of τ_α extracted from their analysis of the data.

Well-Resolved JG Relaxation of a Component in Mixtures of van der Waals Liquids

As shown before in Section 2.2.4.5, the α-dispersion of a component in binary polymer blend mixtures of van der Waals liquids also is invariant to T and P variations when τ_α is constant. Here we shall show, for the same component, the JG β-relaxation shifts with pressure and the ratio τ_α/τ_{JG} remains the same on varying T and P while keeping τ_α constant.

2-Picoline in Tri-styrene

The experiment is dielectric relaxation of guest rigid polar molecules such as 2-picoline in mixtures with apolar host molecules such as tri-styrene. Only the motions of the guest molecules are observed in the dielectric spectra because the dipole moment of the host molecules is so small that their motions make no contribution. Since 2-picoline is a rigid molecule, its secondary relaxation is of definitely the JG β-relaxation. Temperature and applied pressure were varied over wide ranges in the experiment. Elevated pressure slows down the α-relaxation and increases its relaxation time τ_α, but the increase can be compensated by raising temperature. Naturally, widely different combinations of P and T can be found to have the same α-relaxation time τ_α, although there are significant variations in density. Dielectric relaxation measurements at 0.1 MPa of 50 wt% or less of 2-picoline in mixtures with tri-styrene (3S) had shown a well-resolved JG β-relaxation of 2-picoline at frequencies much higher than the α-relaxation that involves motion of 2-picoline [843, 844].

Subsequently, isothermal measurements of the same mixture with 25 wt% of 2-picoline at 0.1 and 600 MPa were made [676]. The loss spectra of 2-picoline in the mixture at 600 MPa are shown in Fig. 145 for some representative temperatures. In the inset, the loss data at –20.7°C are shown with P increasing from 600 to 1100 MPa. At these temperatures and pressures, the picoline component remains in the equilibrium liquid state. The secondary relaxation shifts to lower frequencies with increasing pressure, which is an indication that it is the JG β-relaxation. At –15.3°C and $P = 608$ MPa, both the JG β-relaxation and the α-relaxation loss peaks can be seen within the experimental frequency window. We have taken this spectrum and compare with the spectrum taken at $P = 0.1$ MPa and $T = -80$°C

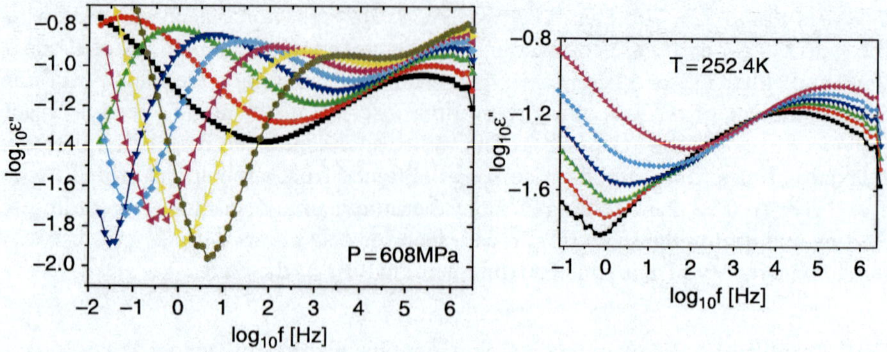

Fig. 145 (*left*) Loss data of 2-picoline in mixture of 25 wt% with tri-styrene at 608 MPa. The temperatures of measurement are −20.7, −15.3, −10.2, −5.2, −0.2, 4.95, 10.1, and 15.03°C (*from left to right*). (*Right*) The loss data at $T = 252.4$ K with $P = 1100$, 1000, 900, 800, 700, and 600 MPa (*from left to right*) show the secondary relaxation shifting to lower frequencies with pressure even in the glassy state. Data from [676] are replotted here

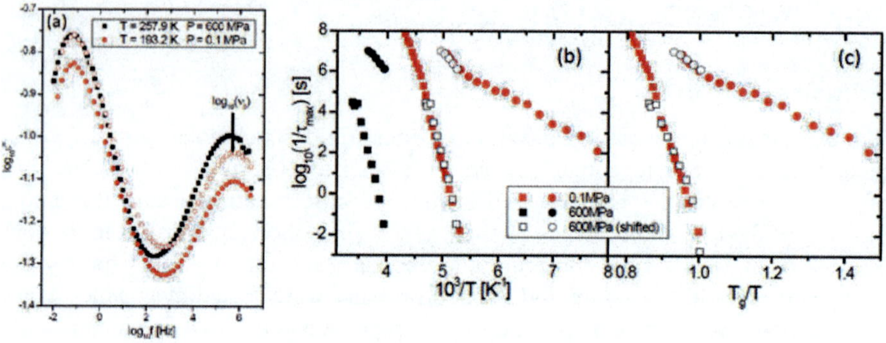

Fig. 146 (**a**) Comparison of two spectra of 2-picoline in mixture with tri-styrene with the same α-relaxation frequency but measured at different thermodynamic conditions (see labels in figure). The *open circles* are obtained by shifting the data at 0.1 MPa vertically by a constant. The *arrow* indicates that JG β-relaxation frequency remains practically unchanged. (**b**) Relaxation map of τ_α and τ_β of 2-picoline in mixture with tri-styrene. *Circles* are for 608 MPa, and *squares* are for 0.1 MPa. *Closed and open triangles* are the data of 608 MPa after a horizontal shift of 1.3 decades to the right has been made. (**c**) Same data in (**b**) plotted against T_g/T. Data from [676] are replotted in all figures here

that has the same α-relaxation loss peak frequency $\nu_\alpha \equiv 1/(2\pi\tau_\alpha)$. The comparison is made in Fig. 146 (left panel) directly and after the height of the α-loss peak at 0.1 MPa is scaled up by a multiplicative factor to coincide with that at 600 MPa (open circles). By inspection of the figure, it can be seen that the JG loss peak frequency,ν_β, is practically the same for the two different combinations of P and T. The most probable τ_α and τ_β (obtained from the frequencies of the maxima of the α- and β-loss peaks at various temperatures for $P = 0.1$ MPa and for $P = 600$ MPa) when plotted against reciprocal temperature are presented in Fig. 146 (right panel).

More data of τ_α and τ_β at other temperatures and pressures than those in Fig. 146 (left panel) are shown in the relaxation map of Fig. 146 (right panel). The invariance of the relation between τ_α and τ_β to change in the P and T combination for each τ_α is demonstrated by horizontally shifting the data of τ_α and τ_β at 600 MPa by the same amount of 1.3 along the 1000/T axis in Fig. 146 (right panel). This shift maintains the ratio τ_β/τ_α at 600 MPa constant. The shifted τ_α (closed triangles) and τ_β at 608 MPa (open triangles) overlap the corresponding τ_α (open squares) and τ_β (closed squares) data at 0.1 MPa. In this way, the fixed relation between τ_α and τ_β for different combinations of P and T at constant τ_α is demonstrated expediently for a number of values of τ_α. Additionally, in Fig. 146(c), the same data of $\tau_\alpha(T,\ P)$ and $\tau_\beta(T,\ P)$ at constant $P = 0.1$ MPa and $P = 600$ MPa are plotted as a function of $T_g(P)/T$, where the reference temperature $T_g(P)$ is defined by $\tau_\alpha(T_g(P), P) = 100$ s for the two constant pressures. This is another way to demonstrate the invariance of the relation of the ratio τ_β/τ_α for any τ_α, and additionally it shows that the $T_g(P)$-scaled temperature dependences of both τ_α andτ_β, which may be slightly dependent on P.

Quinaldine and Tert-butylpyridine in Tri-styrene

Two other mixtures of van der Waals liquids studied in another laboratory show exactly the same properties as the 25% picoline/3S mixture. The polar rigid molecule in the two mixtures are quinaldine (QN) ($T_g = 180$ K) and *tert*-butylpyridine (TBP) ($T_g = 164$ K), and the apolar host is again tri-styrene ($T_g = 234$ K). The α- and JG β-relaxations of QN or TBP were found in the dielectric spectra [976] (see Fig. 147, left panel). For mixtures with concentrations below 60% of TBP, and 40% of QN, the JG relaxation appears as a well-resolved peak in the dielectric loss spectrum. Therefore, the most probable α- and JG-loss peak frequencies ν_α and ν_{JG} and hence the corresponding τ_α and τ_{JG} were determined directly and unequivocally by the locations of the maximum loss frequencies of the two peaks. Several features worth pointing out are (1) the α-loss peak of QN or TBP is much broader (and hence n larger) in the mixture than in the pure material, (2) there is good correspondence between the calculated frequency τ_0 of the primitive process and the most probable relaxation time τ_{JG} of the resolved JG β-relaxation in the mixture (see Fig. 147, right panel), and (3) there is a clear change of T-dependence of τ_{JG} when crossing T_g of the same component, QN or TBP, in the mixture (see Fig. 147, right panel). These features will be revisited later on.

The measurements of the dynamics of QN and TBP in the mixtures were extended to elevated pressures (see Fig. 148, left panel) [677]. The remarkable interdependence of τ_{JG} and τ_α is shown in Fig. 148 (right panel) by comparing the spectra of 10% QN in tri-styrene taken at the six different combinations of P and T that have $\tau_\alpha = 0.67$ s [677].

The maximum loss of the α-relaxation differs in these six sets so slightly that no effort is made to adjust them to have exactly the same height. It can be seen by inspection that the frequency at the maximum loss of the JG relaxation, ν_{JG}, and hence τ_{JG} is almost unchanged. The height of the JG loss peak is enhanced with

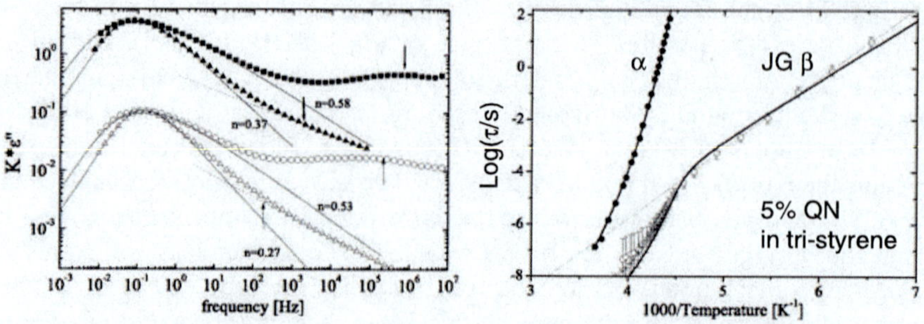

Fig. 147 (*Left*) Dielectric loss vs. frequency for spectra with the same loss maximum frequency: TBP/tri-styrene (16% wt.) at 218 K (*solid squares*), pure TBP at 167 K (*solid triangles*), QN/tri-styrene (5% wt.) at 232 K (*open circles*), pure QN at 185 K (*open triangles*). The vertical axis has been shifted by appropriate factors K in order to clearly show the spectra of the two polar molecules. The *continuous lines* are fits to the α-loss peak by the KWW function (coupling parameter n is shown). *Dotted lines* are from fitting a power law to the excess wing. *Vertical arrows* indicate the locations of the calculated frequency ν_0 of the primitive process. (*Right*) Relaxation map of QN/tri-styrene (5% wt.). Logarithm of the experimentally determined relaxation times vs. $1000/T$ for the α-process (*solid circles*) and β-process (*open circles*). *Continuous lines* represent the VFTH equation for α-relaxation and simulation data for β-relaxation obtained by numerical solution of predictions of the Hodge and CM models (n is fixed at 0.53). *Dotted line* is the Arrhenius fit of β-relaxation obtained only from data below T_g. *Asterisks* are the values of calculated frequency τ_0 of the primitive process, assuming n or the Kohlrausch exponent $(1\text{–}n)$ is constant up to the shortest τ_α. Usually n decreases slightly at shorter τ_α and this may explain the discrepancies at higher temperatures. Reproduced from [976] by permission

increasing P and T, showing that the dielectric strength of JG relaxation changes differently with P and T than that of the α-relaxation. The invariance of the frequency dispersion of the α-relaxation for different combinations of P and T that have $\tau_\alpha = 0.67$ s is also shown in the figure. Thus, the ratio τ_α/τ_{JG} and n are co-invariants to changes in the combinations of P and T that keep either τ_{JG} or τ_α constant. The same results were obtained for other choices of τ_α. Invariance of the ratio τ_α/τ_{JG} for $\log(\tau_\alpha) = -0.17$ (i.e., $\tau_\alpha = 0.67$ s) is shown alternatively by the relaxation map presented in Fig. 149, and in a different style in Fig. 4 in [677], where τ_α and τ_{JG} are plotted against $T_g(P)/T$ for two different choices of constant P, and against P for four different choices of constant T. One can see by inspection that approximately $\log(\tau_{JG}) = -5.46$ (i.e., $\tau_{JG} = 3.5\ \mu$s) for all cases.

The TBP component in mixtures with tri-styrene shows exactly the same behavior as QN in tri-styrene, but the data are not reproduced here.

Benzonitrile (10 wt%) in PS380

The dynamics of the polar cyanobenzene or benzonitrile of 10 wt% in mixture with oligomer of PS with molecular weight equal to 380 and $T_g = 232$ K was obtained most recently in dielectric measurement at ambient and elevated pressures by Shahin Thayyil and Capaccioli [1067]. The loss data at ambient pressure and

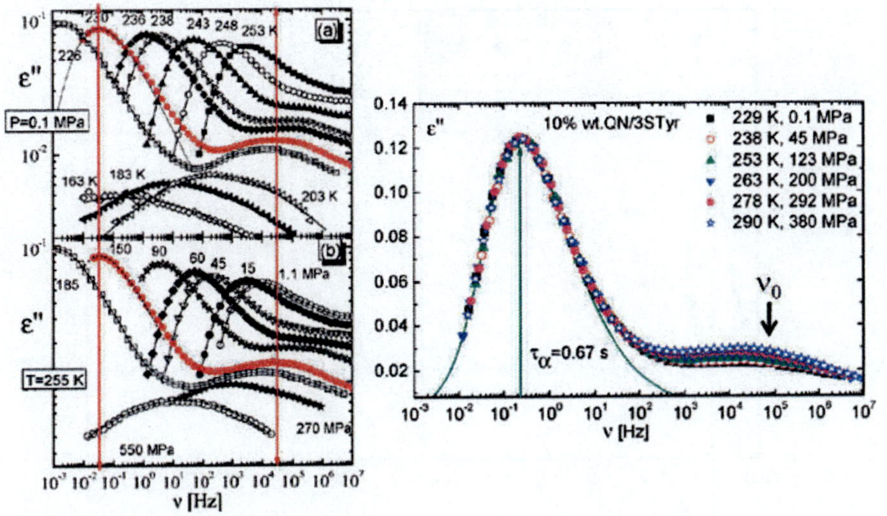

Fig. 148 (*Left*) Log–log plot of selected dielectric loss spectra of the mixture QN/3Styr (5% wt.) vs. frequency at atmospheric pressure at different temperatures (*upper panel*) and at high pressures at $T = 255$ K (*lower panel*). The low-frequency part of some spectra has been omitted for the sake of clearness. *Symbols* are experimental data, *continuous lines* are fitting curves, *thin dotted lines* are KWW function curves with n=0.50. *Vertical arrows* indicate the locations of the calculated frequency τ_0 of the primitive process. (*Right*) Superposition of loss spectra for 10% QN in tristyrene measured for different T and P combinations but the same $\tau_\alpha = 0.67$ s. The *line* is a Fourier-transformed Kohlrausch function fit with $n = 0.50$, and *arrow* indicates the location of the primitive relaxation frequency ν_0 calculated by the CM relation. Reproduced from [677] by permission

various temperatures as well as at elevated pressure and constant temperature of 268.2 K presented in Fig. 150 show the temperature and pressure dependence of the α-relaxation and the resolved JG β-relaxation of benzonitrile in the mixture. The spectra from measurement for four different combinations of P and T with the same α-loss peak frequency are compared. Not only the frequency dispersion of the α-loss peak but also the ratio τ_α/τ_{JG} is the same, just like that found in other mixtures. Also ν_0 calculated is in agreement with experiment (see Fig. 151).

Concentration Fluctuations

Except at low concentration of one of the components, concentration or composition fluctuations in binary mixtures are unavoidably present. They introduce a distribution of environments for the polar molecules and corresponding distributions of coupling parameters $\{n_i\}$ and the α-relaxation times $\{\tau_{\alpha i}\}$. One consequence is an additional broadening of the α-loss peak than the Kohlrausch function with the most probable $\hat{n}$ and $\hat{\tau}_\alpha$ would indicate. In the CM description of component dynamics in mixtures [736, 737], for each i there exists the relation between $\tau_{\alpha i}$ and $\tau_{\alpha 0}$ given by $\tau_{\alpha i}(T,P) = [t_c^{-n_i}\tau_0(T,P)]^{1/(1-n_i)}$ and $\tau_{JGi} \approx \tau_0$, analogous to Eq. (2.69).

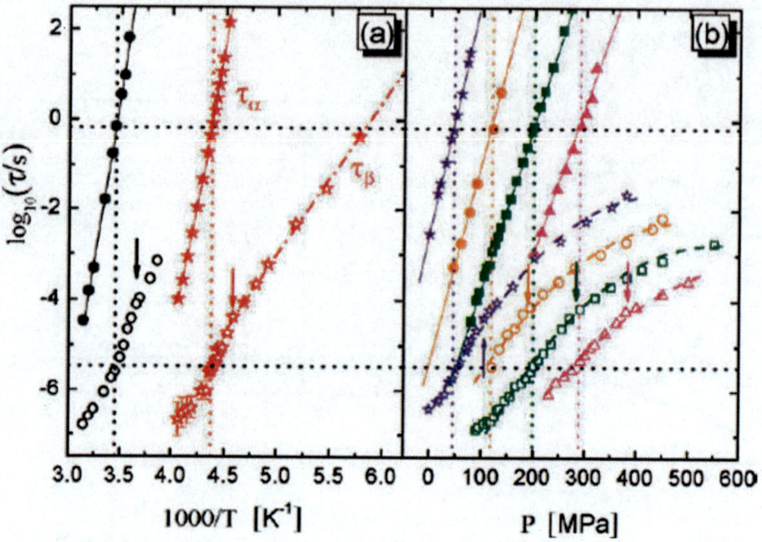

Fig. 149 Relaxation map for the mixture 10% QN in tri-styrene. (**a**) Isobaric data: 0.1 MPa (*stars*), 380 MPa (*circles*). (**b**) Isothermal data: 238 K (*stars*), 253 K (*circles*), 263 K (*squares*), 278 K (*triangles*). *Closed and open symbols* are for τ_α and τ_{JG}, respectively. *Continuous lines* through the τ_α data represent VFTH fits in (**a**) and linear fits in (**b**). The *dashed-dotted line* through τ_{JG} in (**a**) is an Arrhenius fit to the ambient pressure data at $T < T_g$, and the *dashed curves* in (**b**) are guides for eyes in following its *P*-dependence. *Upper and lower horizontal dotted lines* indicate $\log(\tau_\alpha) = -0.17$ (i.e., $\tau_\alpha = 0.67$ s) and $\log(\tau_{JG}) = -5.46$ ($\tau_{JG} = 3.5$ μs); *vertical dotted lines* mark the corresponding *x*-axis values. *Vertical arrows* correspond to the values of the *x*-axis for which $\log(\tau_\alpha) = 3$ and mark the change of dynamics for τ_{JG}. Reproduced from [677] by permission

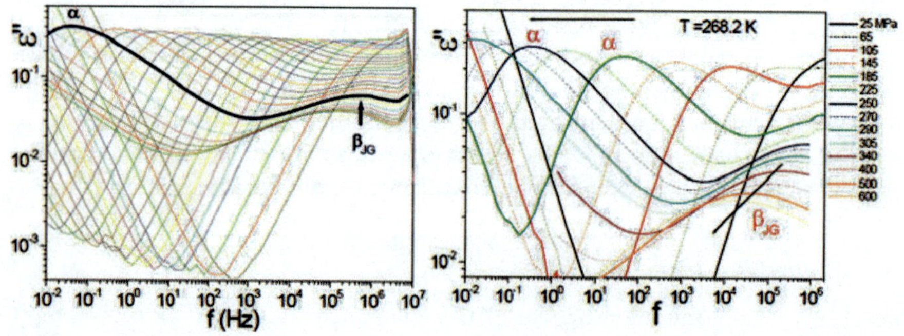

Fig. 150 (*Left*) Dielectric loss spectra of (10 wt%) benzonitrile in mixture with PS380 at ambient pressure. (*Right*) Dielectric loss data of (10 wt%) benzonitrile in mixture with PS380 at 268.2 K and various pressures indicated. The shift of the JG β-relaxation with pressure can be seen. Reproduced from [1067] by permission

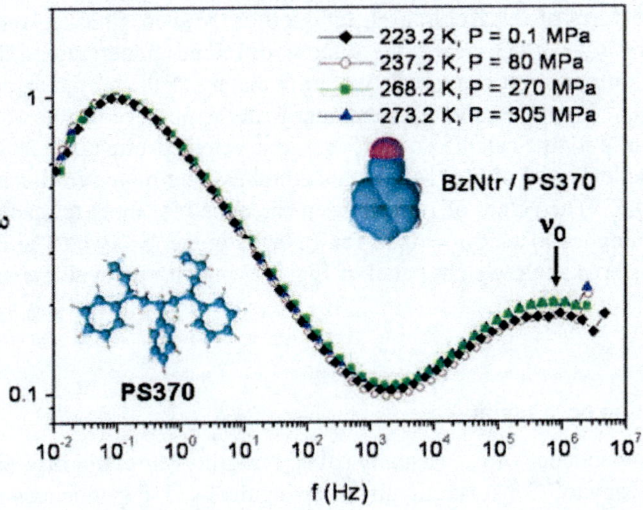

Fig. 151 Near perfect superposition of the spectra of (10 wt%) benzonitrile (BzNtr) in mixture with PS380 taken at four different combinations of P and T. Fitting the α-loss by the Fourier transform of the Kohlrausch function gives $n = 0.57$. With this value of n, the primitive frequency ν_0 calculated from the α-loss peak frequency by the CM is indicated by the *arrow*. Reproduced from [1067] by permission

In particular, for the most probable $\hat{n}$ and $\hat{\tau}_\alpha$, there are the relations given by

$$\hat{\tau}_{\alpha i}(T,P) = [t_c^{-\hat{n}}\tau_0(T,P)]^{1/(1-\hat{n})} \tag{2.73}$$

and

$$\hat{\tau}_\alpha(T,P)/\tau_{JG}(T,P) = [\hat{\tau}_\alpha(T,P)/t_c]^{\hat{n}}. \tag{2.74}$$

This last relation readily explains the invariance of the ratio $\hat{\tau}_\alpha(T,P)/\tau_{JG}(T,P)$ at constant $\hat{\tau}_{\alpha i}(T,P)$ or equivalently the invariance of the ratio of the observed α- and JG β-loss peak frequencies of the polar component observed experimentally.

Since the dispersion of the α-relaxation of 2-picoline is additionally broadened by concentration fluctuations in the mixture, $\hat{n}$ of 2-picoline or QN would be smaller than the value, n_{fit}, from the forced fit of the data by the Kohlrausch function. Therefore, the value of $\tau_{JG} \approx \tau_0$ calculated from $\tau_{JG} \approx \tau_0(T,P) = (t_c^{n_{\text{fit}}})[\hat{\tau}_\alpha(T,P)]^{1-n_{\text{fit}}}$ would be shorter than that actually predicted by the CM, which is $\tau_{JG} \approx \tau_0(T,P) = (t_c^{\hat{n}})[\hat{\tau}_\alpha(T,P)]^{1-\hat{n}}$. This has been verified for the data of 25% picoline/3S mixture. The forced Kohlrausch fit requires $n_{\text{fit}} = 0.65$ which yields $\nu_{JG} \approx 4.6 \times 10^6$ Hz, while the observed $\nu_{JG} \approx 4.9 \times 10^5$ Hz is a decade smaller.

At the lower concentration of 5% QN in 3S, concentration fluctuations and the excess broadening of the α-loss peak they cause are much reduced. The fit by the

Fourier transform of the Kohlrausch function to the α-loss peak gives a coupling parameter of $n_{\mathrm{fit}} = 0.53$, nearly the same at different temperatures. The exponent n_{fit} in the Kohlrausch fit can be taken approximately as $\hat{n}$, and the calculated value of ν_0 from n_{fit} would be in good agreement with the observed value of ν_{JG}. This is indeed the case for the calculated ν_0 for some selected spectra ($T = 238$ and 230 K) shown in the left panel of Fig. 148, with ν_α directly determined from the maximum, and $n_{\mathrm{fit}} = 0.53$. The values of τ_0 have been calculated for more temperatures above T_g by assuming a constant $n = 0.53$ for QN/tri-styrene (5% wt.) The results of τ_0 shown by asterisks in the right panel of Fig. 147 match very well the experimental data for τ_{JG} (open circles).

An Implication on "Fragility"

The T_g/T-dependence of τ_α is usually called "fragility" or α-fragility. Generalizing this terminology to JG β-relaxation by referring the T_g/T-dependences of τ_{JG} as JG β-fragility, the result discussed in Sections "Unresolved JG Relaxations" and "Well-Resolved JG Relaxation of a Component in Mixtures of van der Waals Liquids" can be restated as follows. The α-fragility and the JG β-fragility are related at any pressure P. If the "fragility" or steepness index of the JG relaxation, $m_{\mathrm{JG}}(P)$, at P is defined by $m_{\mathrm{JG}}(P) = \mathrm{d}(\log_{10}\tau_{\mathrm{JG}}(T,P))/\mathrm{d}(T_g(P)/T)|_{T=T_g(P)}$ in analogy to the fragility index m of the α-relaxation, then the two indices are approximately related by $m = m_{\mathrm{JG}}/(1-n)$ in the case of neat glassformers and $m = m_{\mathrm{JG}}/(1-\hat{n})$ for a component in binary mixtures. This relation holds for all P because n or $\hat{n}$ is independent of P as long as τ_α at $T_g(P)$ is the same for all P. Anyone who entertains "fragility" as an important concept in glass transition may consider this relation and its implication that the α-fragility is actually a derived quantity from the more basic JG β-fragility after accounting for the slowing of the α-relaxation by the cooperative many-body dynamics (represented by n or $\hat{n}$).

Pressure Dependence of Primitive Relaxation Time from Neutron Scattering

Before closing this subsection on pressure dependence of the JG relaxation time, mention is made of the observation by incoherent neutron scattering experiments on polybutadiene (PB) and polyisobutylene (PIB) [461, 474]. As discussed before in Sections 2.2.2.4 and 2.2.5.2, the relaxation times $\tau(Q_c)$ determined from the incoherent intermediate scattering function $F_s(Q,t)$ show crossover from $Q^{-2/\beta}$-dependence for $Q < Q_c$ to Q^{-2}-dependence for $Q > Q_c$, where Q_c is of the order of 1 Å^{-1}. At the small length scales corresponding to $Q > Q_c$, the $F_s(Q,t)$ is contributed by local and independent relaxation, with the primitive relaxation of the CM as the natural candidate. Frick et al. [461] found in PB that its relaxation time $\tau(Q_c)$ for $Q > Q_c$ increases with pressure, and hence the neutron data further support the sensitivity of the primitive relaxation time or the JG relaxation time to pressure.

2.3.2.9 From Causality: Dependence of τ_α on T, P, V, and S Originates from That of τ_0 (or τ_{JG})

The observed pressure dependence of τ_{JG} and the co-invariance of n and τ_α/τ_{JG} at constant τ_α is a remarkable and general property that deserves attention from any serious model or theory of glass transition. It is predicted by Eq. (2.69) of the CM when rewritten in the form

$$\tau_\alpha(T,P)/\tau_{JG}(T,P) = [\tau_\alpha(T,P)/t_c]^n, \tag{2.75}$$

which clearly demonstrates that n and τ_α/τ_{JG} are co-invariant if τ_α is kept constant. The pressure dependence comes from the CM relation between τ_α and $\tau_0 \approx \tau_{JG}$ given by Eq. (2.69). When rewritten as

$$\tau_{JG}(T,P) \approx \tau_0(T,P) = [\tau_\alpha(T,P)]^{1-n} t_c^n, \tag{2.76}$$

it shows that the pressure dependence of τ_α implies the pressure dependence of τ_{JG} or vice versa. The deep implication of this symbiotic relation between τ_α and τ_{JG} (or τ_0) should not be overlooked. Also from it, an important point in the study of glass transition deduction follows: not only τ_α but also τ_{JG} (or τ_0) depends on volume V and entropy S, and the conjugate variables of P and T. Later on in Section 2.3.2.29 we shall see that both τ_α and τ_{JG} (or τ_0) are functions of the same product variable TV^γ, where γ is a material constant, which is another way to show the dependence of τ_{JG} (or τ_0) on T and V. In terms of dependence on configurational entropy $S_c(T)$ and the Adam–Gibbs model, it has been shown that the linear dependence of the logarithm of the relaxation time on $1/TS_c(T)$ is better obeyed by $\log(\tau_0)$ than $\log(\tau_\alpha)$ [401].

The above suggests that the well-known dependence of τ_α on T, P, V, and S may have originated from that of τ_{JG} (or τ_0). Actually this follows from *causality* in classical physics, which requires the cause should always precede its effect. Since the JG β-relaxation (or the primitive relaxation) always precedes the α-relaxation in time, the dependence of τ_{JG} (or τ_0) on T, P, V, and S has to be the cause of the dependence of τ_α on T, P, V, and S, which is just the effect. If recognized by others, this standpoint may change their entire outlook in the research on glass transition. Amusingly, from this standpoint, it becomes more appropriate to say that the JG β-relaxation is the *primary* relaxation, and the structural α-relaxation is the *secondary* relaxation.

2.3.2.10 Systematic Increase of the Ratio τ_α/τ_{JG} (or τ_α/τ_0) of a Glassformer A (by Increase of n) on Mixing with Increasing Concentration of a Less Mobile Glassformer B

We have seen in Section 2.3.2.1 above from experimental data of many different glassformers that the ratio τ_α/τ_{JG} increases with n at a predetermined value of τ_α. This correlation between τ_α/τ_{JG} and n was explained by the coupling model

(CM) as due to increase of intermolecular coupling or the breadth of the dispersion of the α-relaxation, or n appearing in the fractional exponent (1–n) of the Kohlrausch correlation function in Eq. (1.1). This can be seen from the prediction $(\log \tau_\alpha - \log \tau_{JG}) \approx n(\log \tau_\alpha + 11.7)$, which follows from Eqs. (2.75) and (2.76) together with $t_c \approx 2\,\text{ps}$ for molecular glassformers, and the expected increase of n of the faster component with increasing presence of the slower component. The correlation is strong, but it is established by compiling data from different glassformers. The correlation would be more cogent if it can be demonstrated to exist for the same JG β-relaxation of the same glassformer A by changing intermolecular coupling and the parameter n. This can be accomplished in dielectric spectroscopy by mixing a dipolar glassformer A with an apolar glassformer B that has widely different mobility or T_g. The theoretical basis of this method was given by coupling model [739].

Mixtures of Two van der Waals Glassformers

If component B has lower mobility or higher T_g than component A, presence of the B molecules in the mixture will enhance the intermolecular constraints and slow down the α-relaxation involving the component A. Consequently the correlation function of component A is stretched to longer times and equivalently the frequency dispersion is broadened. In the context of the coupling model, the effect corresponds to an increase in the coupling parameter n appearing in the exponent of the Kohlrausch stretched exponential function in Eq. (1.1). If the B molecules are apolar like oligomers of styrene, and the A molecules are highly polar such as picoline [843, 844], *tert*-butylpyridine (TBP), and quinaldine (QN) [672–674], the dielectric spectra show only the relaxations that the polar A molecules take part in, and that include the α- and the JG β-relaxations of component A. Pure picoline, TBP, and QN all have narrow α-loss peak (small n), and show the unresolved JG β-relaxation as an excess wing. Upon addition of tri-styrene, the JG β-relaxation is further separated from the α-relaxation, has higher relaxation strength, and becomes increasingly well resolved. This trend of increase of the ratio τ_α/τ_{JG} of component A for a predetermined value of τ_α with increases of concentration of the higher T_g component B (tri-styrene) is shown in Fig. 152 for picoline taken from [843, 844]. Shown on the left panel of this figure are the spectra of pure picoline (100%) and picoline in five mixtures with 75 down to 25% picoline. The spectra all having roughly the same loss peak frequency of 10^{-2} Hz are shifted vertically for the sake of seeing the emergence of the JG β-relaxation at high frequencies on increasing the tri-styrene content. The relaxation map on the right panel also shows, at T_g, $\log(\tau_\beta)$ is shorter in mixture with higher tri-styrene content, and also the change in T-dependence of τ_β when crossing T_g.

Other examples include quinaldine (QN) or *tert*-butylpyridine (TBP) [672–674] in tri-styrene, which again is less mobile component. The data of TBP in tri-styrene [672(b)] are reproduced in Fig. 153 in two ways, as dependence on temperature at a fixed frequency (left panel) and as dependence on frequency at fixed α-loss peak frequency (right panel).

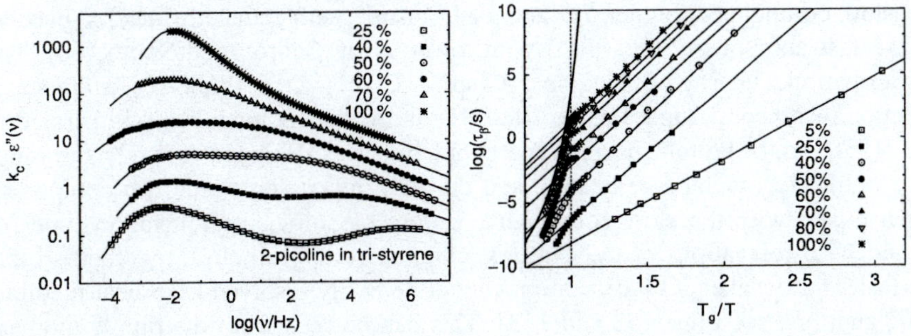

Fig. 152 (*Left*) The dielectric loss of 2-picoline in tri-styrene mixtures with similar α-relaxation times. For clarity each spectrum is shifted by a concentration-dependent prefactor k_c. *Solid lines* are interpolations using appropriate model functions together with the Williams–Watts approach [843, 844]. (*Right*) The relaxation times of the secondary process of picoline in picoline/tri-styrene mixtures. *Solid lines* stand for Arrhenius law. For comparison: the α-relaxation times of the 40% picoline in tri-styrene mixture is shown by the *bold line*. Reproduced from [843] by permission

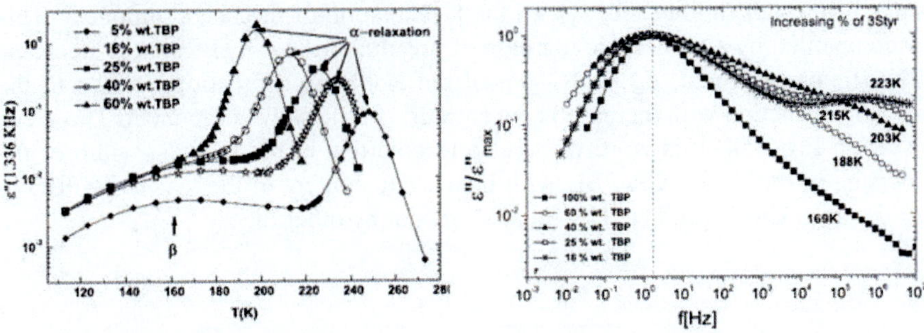

Fig. 153 (*Left*) Temperature dependence of dielectric loss of five different mixtures of TBP in tri-styrene at a fixed frequency of 1.366 KHz, showing increasing separation (in temperature) of the JG β-relaxation from the α-relaxation with increasing content of tri-styrene in the mixture. (*Right*) Log–log plot of normalized dielectric loss spectra of pure TBP and mixtures of TBP in tri-styrene at different T but similar value of f_{max}, showing increasing separation in frequency of the two relaxations. Reproduced from [672, 673] by permission

Although the α-loss peak continuously broadens with addition of tri-styrene, only at low concentration of either component is the exponent β_{KWW} in the Kohlrausch function used to fit it truly reflects the coupling parameter n in the mixture. At other compositions of the mixture, the presence of concentration fluctuations further broadens the loss peak in addition to the increase of intermolecular coupling.

Molecular Dynamics Simulations of Model Miscible Polymer Blends

Another example is given by molecular dynamics simulations of model miscible polymer blends consisting of chemically realistic 1,4-polybutadiene (CR-PBD) as the slow component (higher T_g) and PBD chains with reduced dihedral barriers as the fast component (LB-PBD) with lower T_g by Bedrov and Smith [80]. The conformational energy of the C(sp^2)–C(sp^3)—C(sp^3)–C(sp^2) (alkyl) dihedrals as a function of dihedral angle for the CR-PBD and LB-PBD models are illustrated in Fig. 155-1 (right bottom panel). The simulation was designed to study the influence on the segmental α-relaxation and the JG β-relaxation of the fast component when mixed with the slow component. Before blending, the relaxation times of α- and JG β-relaxations of the neat fast component are too close together, so that the faster but weaker JG β-relaxation cannot be easily resolved (like neat picoline, TBP, and QN, see Figs. 152 and 153). This can be seen from the dipole moment autocorrelation function (DACF) of the pure LB-PBD melt in Fig. 154 or from the normalized dielectric loss calculated by Fourier transform of the DACF (Fig. 155-1, left panel).

However, with the addition of the slow CR-PBD component, they found a monotonic increase in the separation between α- and the JG β-relaxations of the fast component, whereby the latter becomes resolved. The increased separation between the two is due to the strong increase of the α-relaxation time, $\tau_{\alpha f}$, but a much smaller change of the relaxation time, $\tau_{\beta f}$, of the β-relaxation of the fast component. This is accompanied by a monotonic increase of breadth of the α-dispersion or decrease of the stretch exponent, $\beta_\alpha \equiv 1 - n_f$, of the Kohlrausch function used to fit the correlation function with increased concentration of the slow component. These can be seen in Fig. 154. It is worthwhile also to note the decrease in the width of the dispersion of the slower CR-PBD with increase in content of the faster LB-PBD in the mixtures. This is predicted by the CM but not by other theory.

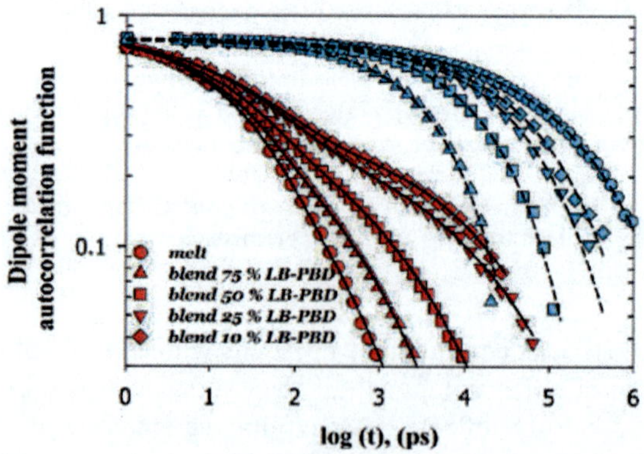

Fig. 154 DACF as a function of time for LB-PBD (*open symbols*) and CR-PBD (*filled symbols*) components in the LB-PBD/CR-PBD blends and pure melts at 198 K. Note the increasing separation between the α- and JG β-relaxations of the LB-PBD with increasing CR-PBD content in the mixtures. Also note the decrease in the width of dispersion of the slower CR-PBD with increase of the faster LB-PBD content in the mixtures. Data from [356(a)] reconstructed and kindly provided by D. Bedrov

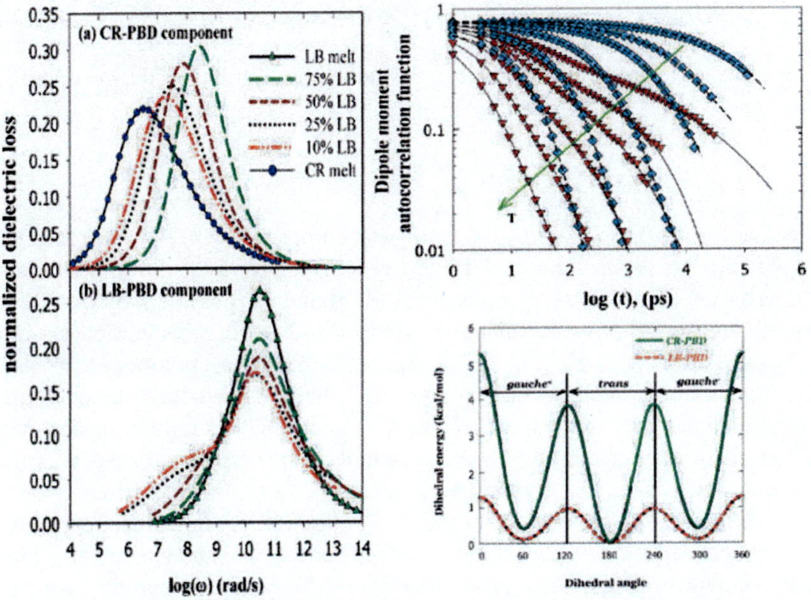

Fig. 155-1 (*Left*) The normalized dielectric loss of (**a**) CR-PBD component and (**b**) LB-PBD component, at 198 K for several blend compositions and pure melts of the components. All curves are normalized such that the total dielectric loss is unity. (*Right, bottom*) The conformational energy of the C(sp^2)–C(sp^3)–C(sp^3)–C(sp^2) (alkyl) dihedrals as a function of dihedral angle for the CR-PBD and LB-PBD models. Also shown is the division of alkyl dihedral angles into *gauche*$^+$, *trans*, and *gauche*$^-$ conformational states. (*Right, top*) The DACF as a function of time for LB-PBD (*open symbols*) and CR-PBD (*filled symbols*) components in the 10/90 LB-PBD/CR-PBD blends at 198, 222, 240, 273, 323, and 400 K. *Arrow* indicates the direction of increasing temperature. Uncertainties are smaller than the *symbol size*. All figures are reconstructed from data in [356(a)], and kindly supplied by D. Bedrov

These results from simulations obtained by Bedrov and Smith are fully compatible with the experimental data of mixtures of van der Waals liquids presented in the previous subsection and the explanation by the CM applies here as well as pointed out in [854]. The decrease of β_α and the concomitant increase of the separation between $\tau_{\alpha f}$ and $\tau_{\beta f}$ of the faster LB-PBD component with increase in the concentration of the slower CR-PBD are consistent with the prediction of the CM as due to increase of the coupling parameter $\hat{n}_f$. On the other hand, the narrowing of the dispersion of the slower CR-PBD on increasing the content of the faster LB-PBD is evidence of reduction of the coupling parameter $\hat{n}_s$ of CR-PBD as expected by the CM, which readily explains the observed concomitant decrease of the α-relaxation time of CR-PBD, $\tau_{\alpha s}$, (see Fig. 154) by the CM equation (2.67).

Bedrov and Smith [356(a)] tested the CM quantitatively by using their parameters obtained from their fits of the dipole moment autocorrelation function (DACF) of the fast LB-PBD component, assuming that

$$\mathrm{DACF}_{(f)}(t) = A_\beta \exp[-(t/\tau_{\beta f})^{\beta_{\beta f}}] + A_\alpha \exp[-(t/\tau_{\alpha f})^{\beta_{\alpha f}}]. \tag{2.77}$$

From the relaxation times, $\tau_{\alpha f}$ and $\tau_{\beta f}$, obtained by simulation, they used the CM relations,

$$\tau_{\alpha f} = [t_c^{-\hat{n}_f} \tau_{0f}]^{1/(1-\hat{n}_f)} \approx [t_c^{-\hat{n}_f} \tau_{\beta f}]^{1/(1-\hat{n}_f)}, \tag{2.78}$$

to calculate the most probable $\hat{\beta}_{\alpha f} \equiv 1 - \hat{n}_f$ and compare it with $\beta_{\alpha f}$ from their fit to the DACF. The crossover time t_c of the CM is not available from the simulation, and 2 ps is assumed for its value as experimentally found for polymers by quasielastic neutron scattering. The $\hat{\beta}_{\alpha f}$ calculated by Bedrov and Smith are significantly larger than $\beta_{\alpha f}$ for blends with 10, 25, 50, 75, and 100% of the fast component as shown in Fig. 6 of [356(a)], which would suggest that the CM prediction is not in quantitative agreement with molecular dynamics simulation. This failure of the CM for simulation data may cast doubt on the quantitative success of the CM in explaining the dynamics of the fast component in dielectric relaxation experiments on real mixtures. However, as pointed out before in Section 2.2.1.3 in connection with the difference between the two CM equations (2.12) and (2.14) and specifically for the present problem in [355], there is an important difference between the two cases. In dielectric experiment, usually τ_{0f} is much longer than $t_c = 2$ ps, and in this case the relation, $\tau_{\alpha f} = [t_c^{-\hat{n}_f} \tau_{0f}]^{1/(1-\hat{n}_f)} \approx [t_c^{-\hat{n}_f} \tau_{\beta f}]^{1/(1-\hat{n}_f)}$, obtained by using the continuity of the correlation function across t_c, is applicable. On the other hand, $\tau_{\beta f}$ from simulation (and hence also τ_{0f}) is only a little more than a decade longer than $t_c = 2$ ps. In this case, the relation,

$$\tau_{\alpha f} = [(1-\hat{n}_f) t_c^{-\hat{n}_f} \tau_{0f}]^{1/(1-\hat{n}_f)} \approx [(1-\hat{n}_f) t_c^{-\hat{n}_f} \tau_{\beta f}]^{1/(1-\hat{n}_f)}, \tag{2.79}$$

obtained by using the continuity of relaxation rate across t_c, is the right choice [355]. Thus, these equations of the CM should be used to test the simulation data against the CM as shown in Fig. 156 for the pure LB-PBD and 25% LB-PBD blend. Normalized dielectric losses from the α- and β-relaxations of the fast LB-PBD in the pure melt and in the 25/75 LB-PBD/CR-PBD blend from simulation at 198 K are shown (same as in Fig. 8 of Bedrov and Smith in [366(a)]). The one-sided Fourier transform of the Kohlrausch function, $\exp[-(t/\tau_{\alpha f})^{\beta_{\alpha f}}]$, that fits well the frequency dependence of α-relaxation loss peak has parameters $\{\tau_{\alpha f} = 1.34$ ns, $\beta_{\alpha f} = 0.22\}$ for the 25/75 LB-PBD/CR-PBD blend and $\{\tau_{\alpha f} = 134.9$ ps, $\beta_{\alpha f} = 0.44\}$ for the pure LB-PBD melt. From these parameters, we calculate $\tau_{\beta f}$ from the appropriate relation,

$$\tau_{\beta f} = \tau_0 = [(1-\hat{n}_f)^{-1} (t_c)^{\hat{n}_f} (\tau_{\alpha f})^{1-\hat{n}_f}], \tag{2.80}$$

where $t_c = 2$ ps and $(1 - \hat{n}_f) = \beta_{\alpha f}$. The last relation is justified by the negligible concentration fluctuation found in these blends by Bedrov and Smith [1068]. The calculated value of $\tau_{\beta f}$, shown as the reciprocal $\omega_{\beta f} \equiv 1/\tau_{\beta f}$ in Fig. 155-2 by the

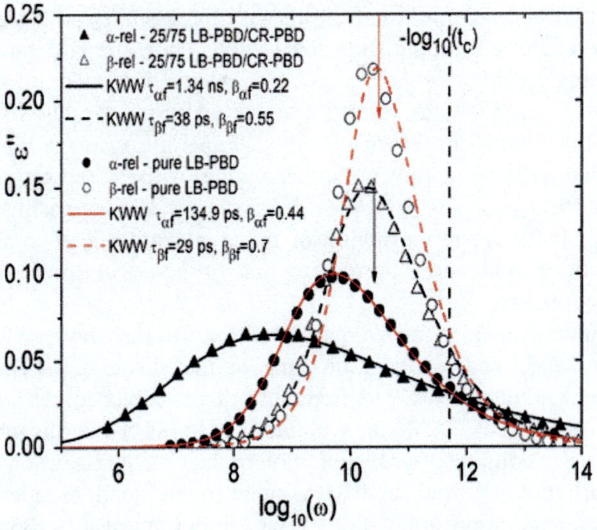

Fig. 155-2 Normalized dielectric loss contributions due to α- and β-relaxation processes of LB-PBD in the pure melt (*closed and open circles*, respectively) and the 25/75 LB-PBD/CR-PBD blend (*closed and open triangles*, respectively) at 198 K obtained by Bedrov and Smith, [356(a)]. The *lines* are fits by one-sided Fourier transforms of Kohlrausch functions. The *vertical arrows* indicate the angular frequencies, $\omega_{\beta f} \equiv 1/\tau_{\beta f}$, where $\tau_{\beta f} = \tau_{0f}$ and τ_{0f} is calculated by the appropriate CM equation (2.80) when τ_{0f} is not long compared with $t_c = 2$ps. Reproduced from [355] by permission

location of the vertical arrow, is in good agreement with the angular frequency of the LB-PBD β-relaxation loss peak for the 25/75 LB-PBD/CR-PBD blend and the pure LB-PBD melt. Similar good agreements have been found for the other blend compositions within the uncertainties or errors involved in deducing the parameter

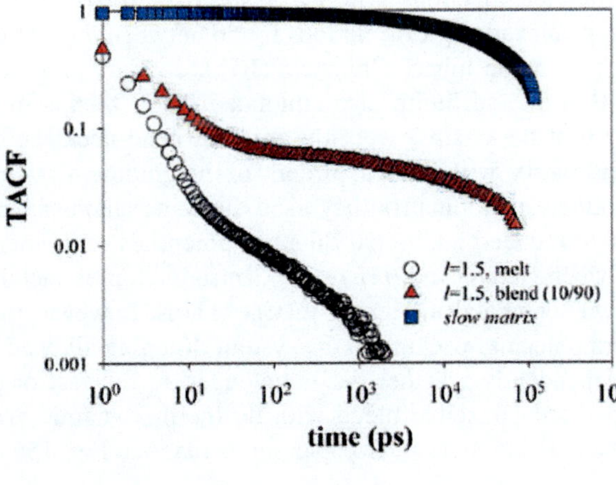

Fig. 156 TACF of pure melt and components of the blend (see text for definitions) for system A with $l = 1.5$ Å ($\sigma/l = 2.7$) at atmospheric pressure and $T = 130$ K. Courtesy of D. Bedrov and G. Smith [356(c)]

values from the data by Bedrov and Smith with the assumption that the α- and β-relaxations are additive in composing the DACF. Thus, the CM fares reasonably well in explaining quantitatively the simulation results presented in [356(a)].

Returning to Fig. 155-1, the right bottom panel shows the DACFs for the LB-PBD and CR-PBD components for different temperatures in the blend with 10% LB-PBD. The interesting feature is the narrowing of the α-relaxation of both the DACFs of CR-PBD and LB-PBD with decreasing τ_α on increasing temperature. In the case of LB-PBD, the narrowing of the α-relaxation is accompanied by its merging with the β-relaxation, similar to that observed in some molecular and polymeric glassformers.

The simulation data also provide information on the slower CR-PBD component in the blend. The α-relaxation time of the slower CR-PBD component becomes shorter, and interestingly its frequency dispersion becomes *narrower* when compared with neat CR-PBD. We have pointed out in [854] that this interesting simulation result is also consistent with prediction of the CM of relaxation on the slower component. Particularly at low LB-PBD content in the blend, broadening of the α-relaxation by concentration fluctuations is not important at all [1068], and hence the narrowing is caused by reduction of the coupling parameter of CR-PBD, and it shows up in the simulation data of Bedrov and Smith. In fact, reduction of n of a slower (higher T_g) component when blended with a faster (lower T_g) component had been seen before for sorbitol or xylitol in mixtures with water [1069] and sorbitol mixed with glycerol [737] and explained by using the CM.

Molecular Dynamics Simulations of Bead-Necklace Model of Miscible Polymer Blends

In Sections 2.3.2.1 and 2.3.2.8, we have introduced the extensive simulations of polymer melts using a simple bead-necklace polymer model by Bedrov and Smith. Polymer chains with different bond length l were investigated allowing to study the influence of bead diameter and bond length ratio σ/l (i.e., degree of the bonded beads overlap) on polymer segmental relaxation characterized by the torsional autocorrelation functions (TACF). Systems with $\sigma/l < 2.0$ showed well-separated α- and β-relaxations. Correlations found between the two indicate that the β-relaxation is the genuine Johari–Goldstein (JG) secondary process.

Bedrov and Smith also simulated 10/90 blends in which minority component (10% of the chains) were the original bead-necklace chains and 90% were polymer matrix with several orders of magnitude slower segmental relaxations. To obtained such a matrix they used the same chains as in their bead-necklace model but introduced additional dihedral potentials with energy barriers of 2.0 kcal/mol between *gauche* and *trans* states. Introduction of such barriers significantly slowed down segmental motion of matrix chains; however, it had no effect on inter- and intramolecular packing in the system. Keeping all bead–bead interactions the same, independently whether beads belonged to the fast or slow component, they have a perfectly miscible blend with no thermodynamic driving forces for segregation between components. Comparison is made in Fig. 156 of the TACFs obtained from simulations of pure melts of systems with $\sigma/l = 2.7$ and TACF of these chains blended with significantly slower matrix. TACFs for the pure slow matrix are also shown in the figure. Figure 156 shows that for $\sigma/l = 2.7$ ($T = 130$ K) blending with slow matrix results in slight changes of the β-relaxation and significant changes of the α-relaxation. The corresponding relaxation times were $\tau_{JG} = 1.5$ ps and $\tau_\alpha = 92$ ps in the melt and $\tau_\beta = 2.7$ ps and $\tau_\alpha = 368,000$ ps in the blend, showing increase of the ratio τ_α/τ_{JG} (or τ_α/τ_0) of the melt upon blending. It is clear by inspection of the TACFs in the figure that the TACF of the melt becomes more non-exponential in time dependence corresponding to an increase of n. Thus, here is another example of the increase of the ratio τ_α/τ_{JG} of a component on mixing with increasing concentration of a less mobile host due to increase of the coupling parameter n.

PVME Component in the Polymer Blends: x % PVME–(1–x)% PS

For polymer blends, an example we can consider is the α- and the JG β-relaxation of the component PVME in blends with PS that has a higher T_g [1070], where the unresolved JG relaxation of neat PVME becomes resolved with the addition of sufficient amount of PS (it is worthwhile to mention that the two resolved secondary relaxations of neat PVME, located below 120 K at 1 Hz, are not the JG relaxation, instead they originate from intramolecular degrees of freedom). The situation is like the α- and the JG β-relaxation of picoline or quinaldine in tri-styrene, where the JG relaxation of picoline and quinaldine is unresolved in the pure state but becomes resolved with addition of tri-styrene as discussed previously. The appearance of the JG relaxation of PVME in the isochronal spectra is particularly clear when more than 50% of PS is present in the blends. For example the data of 70% PS in Fig. 3 of Lorthioir et al. [1070(a)] is shown more clearly by the colored Fig. 2 of [1070(b)] where both the α- and the JG β-relaxations of the PVME component are evident in the isochronal loss spectrum at 1 Hz. Lorthioir et al. [1070(a)] gave in their Fig. 1b the temperature of about 280 K at which the α-relaxation frequency is 1 Hz, while from their Fig. 7 we can see that the JG β-relaxation frequency is 1 Hz at about 225 K. The α-relaxation time increases much more than the JG relaxation time of the PVME with increasing PS content. Thus, the separation between the α-relaxation and the JG relaxation of PVME, measured by ($\log \tau_\alpha - \log \tau_{JG}$) at constant τ_α, increases with increasing concentration of PS in the PVME/PS blends. Concomitantly, the dielectric strength of the JG relaxation grows at the expense of the α-relaxation, and in blends with 80% PS or higher, the JG relaxation dominates the α-relaxation in the dielectric spectrum. This is how we interpret the spectra of 80% PS blend based on that found in 50–70% PS blends, where both the α- and JG β-relaxations are resolved dielectrically, and the trend of dielectric strengths on increasing PS concentration. The trend of increasing relaxation strength of JG β-relaxation time relative to the α-relaxation of the faster component on increasing concentration of the slower component has been seen before in mixtures of van der Waals liquids (see Fig. 152 for 2-picoline in tri-styrene and Fig. 153 for TBP in tri-styrene). Both the α- and JG β-relaxations of 2-picoline, TBP, and QN are present and they are distinctly different even at 95% of tri-styrene. This also can be seen in Fig. 148 for 5% of QN in tri-styrene and in the relaxation map for 10% of QN in tri-styrene in Fig. 149 where the α-relaxation of the QN component has the VFTH *T*-dependence with decreasing temperature all the way down to vitrification.

However, Lorthioir et al. [1070(a)] interpreted their data of the 80 and 90% PS blends differently. They interpret the observed loss peak of the 80 and 90% PS blends at temperatures below the calorimetric glass transition temperature of the blend as the *α-relaxation* of the PVME component, which is changed from its normal cooperative nature with VFTH *T*-dependence to local relaxation with Arrhenius *T*-dependence as observed. This change to local motion of the *α-relaxation* of the PVME component was rationalized by "confinement" of the PVME by the immobile PS component when temperature falls below the T_g of the blend, and a confinement length of about 1 nm was suggested.

Actually, the calorimetric glass transition of the 80% PS blend is 330 K, nearly the same temperature as the location of the shoulder observed in the isochronal dielectric loss of the PVME component at 1 Hz contributed by its α-relaxation (Fig. 3 of [1070(a)]). For the 70% PS blends, the calorimetric glass temperature is about 300 K, which is to be compared with 285 K, the temperature of the broad isochronal α-loss peak of PVME at 1 Hz. Therefore PS relaxes at not much slower rate than PVME in the 80% PS blend, and it is questionable that Lorthioir et al. can consider PS as frozen matrix to "confine" the PVME. Rather, the observed relaxation of PVME in 80% PS blend shown in the isothermal spectra (Fig. 5 of [1070(a)]), the relaxation time of which has Arrhenius temperature dependence (filled diamonds in Fig. 6 of [1070(a)]), should be identified as the JG relaxation of PVME in the blend. In fact, the temperature at which this loss peak frequency in the 80% PS blend equals 1 Hz is about 220 K, about the same as that of the JG α-loss peak of PVME found in the 70 and 50% PS blends. More critique of the confinement interpretation of Lorthioir et al. can be found in [1071].

Another way to refute the "confinement" interpretation by Lorthioir et al. is to use the data of mixtures of either picoline or QN with tri-styrene (see Figs. 147, 149, and 152). In the mixture with 5 wt% of QN, the α-relaxation of QN is about seven orders of magnitude longer than the JG β-relaxation at temperature T_g for which the α-relaxation time of QN is about 100 s (see Fig. 147). In such mixture with dilute concentration of QN, the α-relaxation of tri-styrene has to be even much slower than the α-relaxation of QN. However, both the α-relaxation (with VFTH *T*-dependence) and the JG β-relaxation (with Arrhenius *T*-dependence below T_g of QN and a stronger *T*-dependence above T_g) in the mixture are present and detected. This contradicts the "confinement" interpretation which does not expect that the two processes together are present at T_g and above. More examples of finding both the α- and the JG β-relaxation of a polar component in mixture with a non-polar component having much higher T_g are given in Section 2.3.2.16, indicating this is a general property. The simulations of the LB-PBD/CR-PBD blends by Bedrov and Smith also show the presence of both the α- and the JG β-relaxation of the LB-PBD above T_g.

PEMA Component in the Polymer Blends: x % PEMA–(1–x)% PVPh

Study was made of the α- and JG β-relaxations of neat poly(ethyl methacrylate) (PEMA, $T_g = 74$ C), and the changes when PEMA was blended with poly(4-vinylphenol) (PVPh, $T_g = 171$ C) [1072]. Neat PEMA has a resolved JG relaxation, which is continued to be observed and practically unchanged in the blend. Since T_g of the PEMA component in the blend is shifted to higher temperature by the slower poly(4-vinylphenol) component, ($\log\ \tau_\alpha(T_g) - \log\ \tau_{JG}(T_g)$) of the PEMA component increases at constant $\tau_\alpha(T_g)$.

PEO Component in the Polymer Blends: x % PEO–(1–x)% PMMA

An interesting example is the blends of poly(ethylene oxide) (PEO, $T_g = 214$ K) with poly(methyl methacrylate) (PMMA, $T_g = 391$ K), the pure components of which have very large difference in T_g of 177 K. Several interesting observations of the dynamics of neat PEO and PEO in the blends have been reported [312(a), 855, 860, 1072–1074], which challenge any theory to explain. In these investigations, the PEO component dynamics of the amorphous x% PEO/(1$-x$)% PMMA blend have been widely studied: (1) by deuteron NMR for $x = 0.5, 3, 6, 10, 20, 30$, and neat PEO [855]; (2) by dielectric relaxation and DSC for $x = 10, 15, 20, 30$, and neat PEO [1073, 1074]; and (3) by quasielastic neutron scattering for $x = 10, 20, 30$, and neat PEO [312(a)]. It is worthwhile to mention that the neutron scattering data of Sakai et al. [312(a)] have verified the crossover of the intermediate scattering function of neat PEO as well as the component PEO in blends with PMMA from $\exp(-t/\tau_0)$ to $\exp[-(t/\tau)^{1-n}]$ at a temperature-insensitive and scattering vector Q-independent time of $t_c \approx 1$ ps, which is the premise of the CM leading to its predictions (see Section 2.2.1.1). The data may be considered as microscopic justification of applying the CM to the PEO component dynamics in the x% PEO–(1–x)% PMMA blends and, by extrapolation, to other blends and mixtures. Figure 157 shows the relaxation times of neat PEO and neat PMMA, together with the two components in the 20% PEO/80% PMMA blend mostly by dielectric measurements by Jin et al. [1073, 1074], except the high-frequency data of PEO obtained by deuterium NMR [855] as indicated. The lone star is used to locate the lower T_g at 243 K of two T_gs observed in the 30% PEO/70% PMMA blend by DCS. The figure is an alteration of Fig. 4 in [1074] of Jin et al. with some of the labels of the relaxation processes of PEO changed for expediency and to be compatible with the interpretation of one of the secondary relaxation as the JG β-relaxation of PEO given here.

First, neat PEO has four relaxation processes labeled by α, β, γ′, and γ [1073, 1074] in order of increasing frequency at the same temperature. Because neat PEO is semicrystalline, the slowest process traditionally called the α-relaxation is a local process associated with the crystalline phase. The next called the β-process (β_{PEO}) has been identified as the cooperative segmental motions in the amorphous region because it is absent in the dielectric spectrum of a PEO single crystal mat (PEO molecular weight $> 10^5$), while the fastest γ-process remains. The latter fact indicates that the γ-process has nothing to do with glass transition in the amorphous

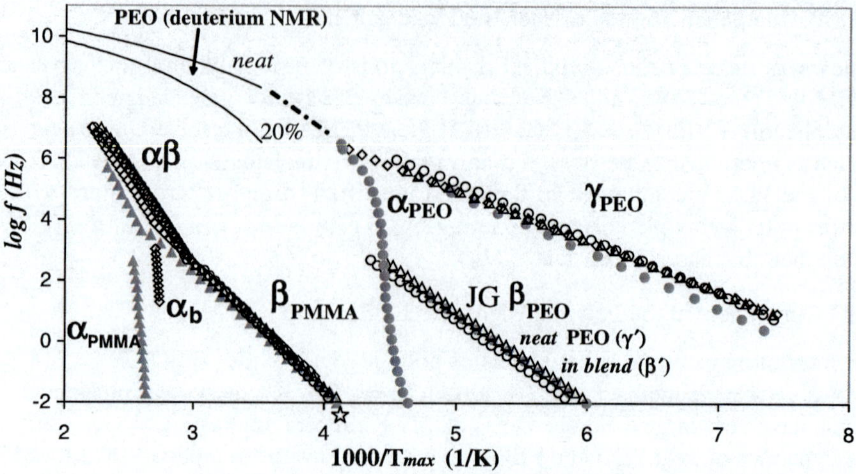

Fig. 157 Plot of data of loss peak frequency vs. T^{-1} obtained by Jin et al. [1074] from dielectric relaxation measurements of aged amorphous x% PEO/(1–x)% PMMA blends compared with those of the neat components. NMR data of neat PEO and $x = 20$ blend from Lutz et al. [855] have also been added to the plot for comparison purposes. *Symbols*: (1) neat components (*filled symbols*): *triangles*, PMMA; *circles*, PEO. (2) Aged blends (*unfilled symbols*): *diamonds*, $x = 10$ (M90); *circles*, $x = 15$ (M85); *triangles*, $x = 20$ (M80). *Solid lines* are data for the PEO segmental relaxation from Lutz et al. *From top to bottom*: neat PEO, PMMA/PEO blend having 20% PEO. The *dashed line* is used to guide the eyes for data of neat PEO. Calculated value of n at T_g defined as log ν_α/Hz = −2 is 0.31 according to this figure. It is larger of about 0.37 from the plot of Jin et al. The *lone star* near the x-axis and located at position corresponding to $T = 243$ K is used to indicate the lower T_g near −30°C of two T_gs observed in the DSC thermogram of the 30% PEO/70% PMMA (M70) blend, which is interpreted at that of the PEO component. The corresponding T_g of PEO in the 20% PEO/80% PMMA blend should be even higher. Reproduced from [1074] by permission

region of PEO and is not the JG relaxation. Since here the interest is on the amorphous PEO, I have in Fig. 157 replaced β by α_{PEO} to indicate the relaxation is the cooperative α-relaxation in the amorphous regions, and γ′ by JG β_{PEO} because, as discussed later, I have reason to identify it as such instead of the other possibilities discussed by Jin et al. [1073, 1074]. This interpretation is consistent with the possibility pointed out by Jin et al. [1073] that the γ′-process is similar to the "fast process" (i.e., the primitive relaxation) observed in nanoconfined glass-forming systems [494(a)]. They also offered argument to support that the γ′-process can be explained by the primitive relaxation of the CM. Here I support this by calculating the coupling parameter n of neat PEO on substituting the relaxation times of α_{PEO} (closed circles) and JG β_{PEO} (closed circles) shown in Fig. 157 into Eq. (2.69), and assume $t_c = 2$ ps as usual. The calculated value of n obtained from the data of [1073, 1074] are $n = 0.37$ and $n = 0.31$, respectively. Both values are reasonable for flexible and low T_g polymer such as PEO, as compared with $n = 0.45$ for polyisobutylene [872], and other polymers with the flexible C–O bond on the backbone [244]. The predicted $Q^{-2/(1-n)}$-dependence of the α-relaxation time for n within the range $0.31 \leq n \leq 0.37$ is also consistent with that obtained by neutron scattering [312(a)].

For the dynamics of PEO in the amorphous x% PEO/(1–x)% PMMA blends with $x = 10, 15$, and 20, dielectric relaxation can resolve its two secondary processes, called β' (open symbols) and γ (open symbols) by Jin et al. [1074]. The α_{PEO}-relaxation was not resolved in these blends. Nevertheless, it was found in the 30% PEO/70% PMMA ($x = 30$) blend by thermal analysis, where two T_gs were observed in the DSC thermogram. The lone star located at position corresponding to $T = 243$ K in Fig. 157 is used to indicate the T_g near -30°C (the lower of two T_gs observed) of the PEO component in the $x = 30$ blend. This is to be compared with $T_g \approx 213$ K of neat PEO. The corresponding T_g of PEO in the $x = 10, 15$, and 20 blend should be even higher. Thus, as can be seen by inspection of Fig. 157, the dielectric loss of the α_{PEO}-relaxation falls inside the domain of the known prominent loss from the β_{PMMA}-relaxation of the PMMA component (open symbols) [106] and cannot be resolved dielectrically. The α_{PEO}-relaxation has a better chance to be seen in heat capacity measurement by DSC because the contribution of secondary relaxation such as the β_{PMMA}-relaxation to heat capacity is weak [1075]. Most interesting is that the f_{max}–T locations of the β'-relaxation of PEO in the blends are essentially the same as those of the JG β_{PEO} (or γ') relaxation of neat *semicrystalline* PEO. This coincidence can be considered as evidence that the β'-relaxation is the JG β_{PEO} of the PEO component in the blends. The T_g of the PEO component in the $x = 30$ blend is already 30 K higher than neat PEO, and the difference should increase monotonically for blends on decreasing x in the order of 20, 15, and 10. From this and the invariance of the JG β_{PEO}-relaxation time, the conclusion that $(\log \tau_\alpha(T_g) - \log \tau_{JG}(T_g))$ of the PEO component increases with increasing concentration of PMMA at constant $\tau_\alpha(T_g)$ can be made with consideration including the neat PEO.

It is important to remind the reader once again that DSC and dielectric relaxation revealed, respectively, the concurrent presence of the α_{PEO}-relaxation and the JG β_{PEO} both from the PEO component at $T_g \approx 243$ K, which is well below $T_g \approx 320$ K of the PMMA component in the $x = 30$ blend. Deuterium NMR data of Lutz et al. [855] also show the presence of the α_{PEO}-relaxation of the PEO component with VFTH-like temperature dependence for its relaxation time in blends with $3 \leq x \leq 30$ at temperatures down to 285 K, which is way below the T_g of 391, 383, and 356 K for blends with $x = 3, 6$, and 20, respectively.

If the reader recalls that according to Lorthioir et al. [1070], the α-relaxation of the PEO component in all blends with $3 \leq x \leq 30$ would be confined by the deeply frozen PMMA matrix, and it should be converted to a local relaxation having Arrhenius dependence. This assertion is obviously contradicted by the combined data of Jin et al. and Lutz et al.

Additional Problems of Dynamics of PEO in x % PEO–(1–x)% PMMA Blends

Although the subsection is intended to discuss the increase of $(\log \tau_{\alpha f}(T) - \log \tau_{JGf}(T))$ of the faster component on increasing the concentration $(1 - x)$, of the slower component of the polymer blend, the change of the $\tau_{\alpha f}(T)$ itself is also part of the consideration. In most mixtures, $\tau_{\alpha f}(T)$ increases with increasing $(1-x)$, and the increase becomes significantly larger at lower temperatures and for larger difference between the mobility of the two components gauged by the difference in their T_gs. The segmental relaxation time, $\tau_{\alpha f}(T)$, of PEO in amorphous x% PEO/(1–x)% PMMA blends probed by deuterium NMR at high frequencies from 31 to 76 MHz by Lutz et al. [855] found that $\tau_{\alpha f}(T)$ of deuterated PEO is nearly independent of composition in blends ranging from 0.5 to 30% PEO, and is not much different from that of neat PEO. For example, $\tau_{\alpha f}(T)$ of 0.5% PEO chains dissolved at near tracer compositions in a PMMA matrix is retarded by less than one order of magnitude when compared with neat PEO samples. These behaviors of $\tau_{\alpha f}(T)$ holds for a wide range of temperatures extending down to 285 K, well below the T_g of 391, 383, and 356 K for blends with $x = 3, 6$, and 20, respectively, where the segmental relaxation times of PMMA are about 12 orders of magnitude greater than $\tau_{\alpha f}(T)$. These observations at high frequencies are unusual when compared with the component segmental dynamics of other miscible polymer blends measured at much lower frequencies by dielectric spectroscopy or by another kind of deuteron NMR technique at much lower frequencies [738, 859]. The results have been presented before in Fig. 89 and discussed in Section (c) of Section 2.2.5.7 where the emphasis is solely on anomalous properties of the α-relaxation without mentioning the connection to the JG β-relaxation. It is worthwhile for the reader to revisit the discussion of the anomalies of the α-relaxation of the PEO component in the blends because it comes together with the problem presented by JG β-relaxation of the PEO component discussed here.

It was concluded by Lutz et al. that their observations of PEO component dynamics at high frequencies from 31 to 76 MHz cannot be described by current models including the ones by Lodge and coworkers [733, 862]. The only explanation of this unusual behavior or anomaly comes from the CM [738]. The primitive relaxation times $\tau_0(T)$ of the PEO component calculated by Eq. (2.68) from $\tau_{\alpha f}(T)$ obtained by NMR at 31–76 MHz is very short and not much longer than the crossover time t_c at all temperatures $\geq$ 285 K. The predictions of the CM on the PEO dynamics in the PEO/PMMA blends from 0.5 to 30% PEO are consistent with the experimental findings that $\tau_{\alpha f}(T)$ of the PEO component is nearly composition-independent over the entire blend composition range, and is retarded by less than one order of magnitude when compared with $\tau_\alpha(T)$ of pure PEO.

2.3.2.11 Increase of the Ratio τ_α/τ_{JG} (or τ_α/τ_0) on Polymerizing or Cross-Linking a Glassformer (by Increase of n)

Although not a miscible mixture, cross-linking agents added to the model epoxy resins DGEBA (EPON828) will react with the host to form higher molecular weight polymer or cross-linked network depending on the functionality of the agents [1076–1080]. From the discussion given in Section 2.3.2.6 above, reacting the epoxy resins DGEBA (EPON828) with cross-linking agents have the epoxide groups of EPON828 consumed, the γ-relaxation originating from them suppressed, and the presence of the JG β-relaxation resolved as demonstrated in Fig. 132 for different lengths of reaction time. More the chemical bonds formed to either polymerize EPON828 or cross-link it to form a network, more enhanced are the intermolecular couplings and constraints imposed on the local segmental motion (structural α-relaxation) and hence its coupling parameter n. Concomitant with the increase of coupling, the ratio τ_α/τ_{JG} at a fixed value of τ_α has been observed to increase [1051] and explained by the CM, in analogy to the same effect found in the lower T_g component of mixtures.

Another example can be taken from another study [1081, 1082]. It was reported that τ_{JG} does not change on cross-linking PMMA, which has been rationalized as due to the local nature of the JG β-relaxation. On the other hand, τ_α increases significantly with the density of cross-links. Hence $[\log \tau_\alpha(T_g) - \log \tau_{JG}(T_g)]$ increases with cross-link density and this confirms the expected increase of n from the CM equation (2.69). Further support comes from the increase of the steepness index m and the width of the dispersion of the α-relaxation obtained by dynamic mechanical relaxation and creep compliance measurements [1081].

2.3.2.12 Systematic Decrease of the Ratio τ_α/τ_{JG} (or τ_α/τ_0) of a Glassformer A (by Decrease of n) on Mixing with Increasing Concentration of a More Mobile Glassformer B

It is also interesting to find the change of the ratio τ_α/τ_{JG} of glassformer A when mixed with glassformer B that has higher mobility and lower T_g than A. This complementary case to Section 2.3.2.10 can be realized experimentally by choosing the apolar component B to have higher mobility and lower T_g compared with the polar component A probed by dielectric spectroscopy. Presence of the more mobile molecules B causes reduction of intermolecular coupling/constraints imposed on the cooperative α-relaxation of the polar and slower A molecules, resulting in a decrease of the coupling parameter n of component A with increasing concentration of the faster component B. The situation here is the opposite of that discussed in Section 2.3.2.10, and the decrease of the ratio τ_α/τ_{JG} at constant τ_α is expected by the CM. The following are some examples.

Concentrated Solutions of Poly(vinyl chloride) (PVC) in Tetrahydrofuran (THF)

Discussed before in Section 2.2.5.7 is the effect of the low T_g diluent THF (B) on local segmental relaxation of the polymer PVC (A). The dielectric data of Adachi and Ishida [848] and analysis of the data [849] have shown monotonic decrease of $\hat{n}_A$ of PVC on increasing concentration of the diluent THF. The β-relaxation of PVC found by Adachi et al. is the JG relaxation because PVC has no side group. In terms of the loss peak temperatures, their isochronal data at 1 kHz indeed show decrease of the separation between the α- and JG β-relaxations with increasing concentration of THF.

PMMA Component in the Polymer Blends: x% PEO–(1–x)% PMMA

The amorphous x% PEO/(1–x)% PMMA blends already discussed in Section 2.3.2.10 offer a test of this expected change of dynamics of the slower PMMA component by the faster PEO component, although the latter is not apolar. The calorimetry and dielectric data of Jin et al. [1073, 1074], indeed show that $(\log \tau_\alpha(T_g) - \log \tau_{JG}(T_g))$ of the PMMA component at constant $\tau_\alpha(T_g)$ decreases with increasing concentration x of PEO continuously from $x = 0$ to $x = 30$. This conclusion is justified by the fact on increasing x of the PEO component that the T_g of the PMMA component decreases monotonically while its $\tau_{JG}(T)$ remains practically unchanged, as shown in Fig. 157. For example, $\log(f/\text{Hz})$ of the JG β-relaxation at $T = T_g \approx 320$ K of the PMMA component in the $x = 30$ blend [1073, 1074] is 2, corresponding to $\tau_{JG}(T_g) = 1.6 \times 10^{-3}$ s, which is about one and a half decades longer than log $\tau_{JG}(T_g)$ of neat PMMA.

PMMA Component in Concentrated Solutions of PMMA in DOP

Photon correlation spectroscopy and dielectric spectroscopy were employed to study the dynamics of the local segmental relaxation and JG secondary relaxation at temperatures above and below T_g of bulk PMMA with weight average molecular weight of 7×10^3 and of the PMMA component in concentrated solutions of PMMA/*bis*(2-ethylhexy1 phthalate) (PMMA/DOP) [246]. As determined by differential scanning calorimetry (DSC) at a heating rate of 10 K/min, the T_g of the bulk PMMA is 351 K. The T_g of the mixtures with concentration of PMMA equal to 0.9 and 0.8 g/mL are 313 and 266 K, respectively. The relaxation map in Fig. 158 shows $(\log \tau_\alpha(T_g) - \log \tau_{JG}(T_g))$ of the PMMA component at constant $\tau_\alpha(T_g)$ decreases with decreasing concentration of PMMA.

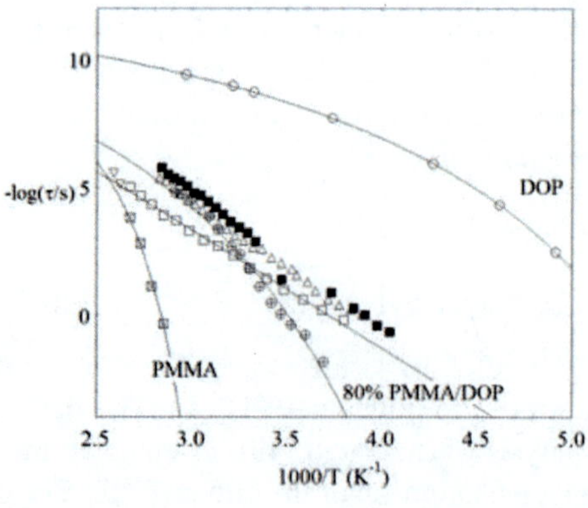

Fig. 158 Arrhenius plot of the relaxation times of the α- and β-relaxations in bulk (open squares with and without cross inside respectively) and plasticized PMMA with $C_{PMMA} = 0.8$ g/mL (circle with + inside and open triangles respectively). Data from Ref. [246] replotted. For data of $C_{PMMA} = 0.9$ g/mL, see figure in this reference

Bromoethylbenzene (BrEBz) Component in Host Ethylbenzene (EB)

Another test of the prediction was made by dielectric relaxation measurements of 0.1 molar fraction of dipolar bromoethylbenzene (BrEBz) with $T_g = 135.6$ K in the apolar host, ethylbenzene (EB), with $T_g = 115$ K that is 20 degrees lower, and also in another apolar host, tristyrene (3S), with $T_g = 232$ K which is 96.4 degrees higher [850]. The results are shown in Fig. 159 by the isochronal (at 1 kHz) and isothermal dielectric loss spectra of pure BrEBZ, and BrEBZ in mixtures with ethylbenzene and tri-styrene.

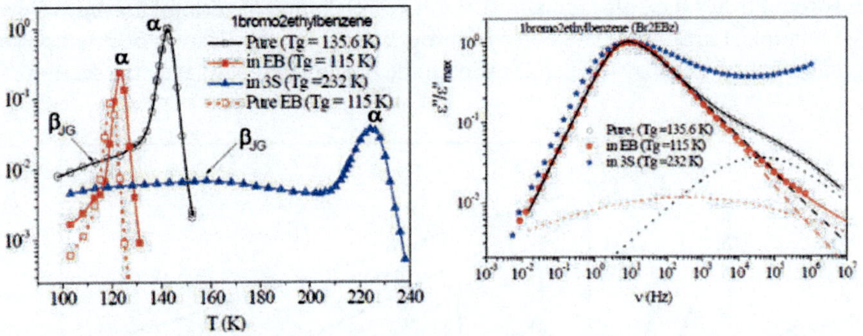

Fig. 159 (*Left*) Logarithm of dielectric loss at 1 kHz vs. temperature for BrEBz: pure (*open circles*); mixed with EB (*full squares*); mixed with 3S (*full triangles*). Data for pure EB (solvent) are shown as *open squares*. *Lines* are guides for the eye. (*Right*) Log–log plot of dielectric loss, normalized to the peak maximum, vs. frequency for pure BrEBz at 144.15 K (*open circles*), for the mixture BrEBz/EB at 125.15 K (*full circles*) and for the mixture BrEBz/3S at 230.15 K (*full stars*). *Lines* are for fitting functions; *solid lines* represent the overall fits, *dashed lines* the Havriliak–Negami functions for α-relaxation and *dotted lines* the Cole–Cole functions for β-relaxation. Data from [850] are replotted here

The JG relaxation is already resolved in pure BrEBz both isochronally and isothermally. In the mixture with 3S, the JG loss peak of BrEBz becomes more prominent compared with the case of neat BrEBz, and is further separated from the BrEBz α-loss peak, similar to cases discussed in Section 2.3.2.10. On the other hand, in the mixture with EB having a lower T_g, the α-loss peak of BrEBz is narrower than that of pure BrEBz on the high-frequency flank, and an excess wing (unresolved JG) appears instead of the shoulder (resolved JG). The former is consistent with a decrease of n of pure BrEBz after taking into consideration the spatially heterogeneous broadening by composition fluctuations in the mixture that it has to encounter, and the fact that the T_g of EB is only 20 K lower than BrEBz. The latter is indication of the decrease in separation between the α- and JG relaxations of BrEBz when mixed with EB.

Sorbitol in Mixtures with Glycerol

Sorbitol has $T_g = 268$ K compared with $T_g = 191$ K for glycerol. The α- and the JG β-relaxation times of neat sorbitol and amorphous mixtures of x% sorbitol/(1–x)%

glycerol for $x = 82, 67$, and 54 have been characterized by dielectric relaxation measurements for frequencies between 10 mHz and 10 MHz [853]. As shown in [737], both the dielectric T_g and the steepness or "fragility" index m of the slower sorbitol component decreases with increasing concentration of the faster glycerol, indicating corresponding decrease of the average coupling parameter n of sorbitol. As expected and shown in Fig. 160, the separation between the α- and the JG β-relaxations of the sorbitol component, given by $(\log \tau_\alpha(T_{ref}) - \log \tau_{JG}(T_{ref}))$ at constant $\tau_\alpha(T_{ref})$, decreases with increasing concentration x of glycerol continuously from $x = 100$ to $x = 54$. It is interesting to note that while addition of glycerol makes the α-relaxation of sorbitol faster (i.e., plasticization), the JG β-relaxation of sorbitol becomes slower (i.e., antiplasticization). The latter has implication on stabilization of dried protein in sugar-based glass by the β-relaxation to be discussed much later in a section.

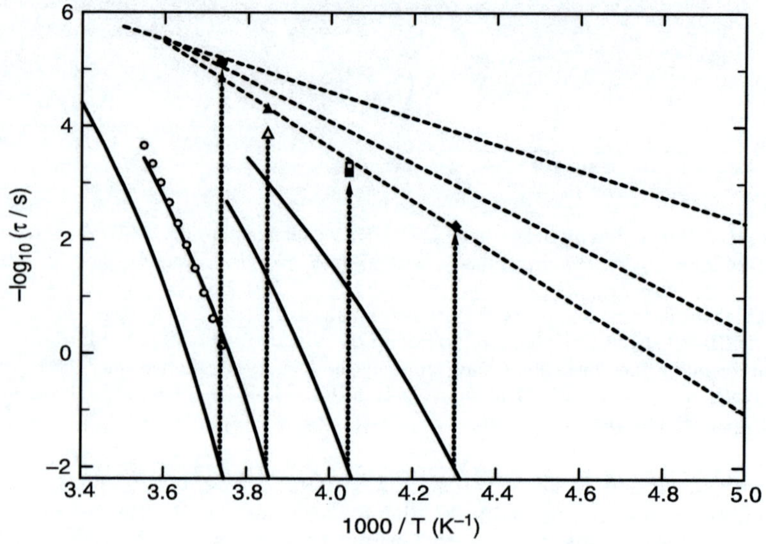

Fig. 160 The thick lines are the VFTH fits to the experimentally determined α-relaxation times of neat sorbitol and sorbitol in the x% sorbitol/(1–x)% glycerol mixtures with $x = 0.82$, 0.67, and 0.54 (*from left to right*). The Arrhenius fits to the JG relaxation times (τ_{JG}) of neat sorbitol and the x% sorbitol/(1–x)% glycerol mixtures for $x = 0.82$ and 0.67 are shown by the *three dashed lines* (*from top to bottom*). The *vertical arrows* show the $(\log \tau_\alpha(T_{ref}) - \log \tau_{JG}(T_{ref}))$ at constant $\tau_\alpha(T_{ref})$, decreases with increasing concentration x of glycerol continuously from $x = 100$ to $x = 54$. Reproduced from [737] by permission

2.3.2.13 Systematic Increase of the Ratio τ_α/τ_{JG} (or τ_α/τ_0) on Increasing the Molecular Weight of Polymers, Constancy of τ_{JG} or τ_0

Decreasing the molecular weight of a polymer below its entanglement molecular weight usually is accompanied by a decrease in T_g. The change is due to the increasing concentration of chain ends which, having higher mobility than the inner bonded repeat units, act like internal plasticization, and reduce the coupling parameter n of the polymer. The poly(n-ethyl methacrylate) (PEMA) offers a way to test this prediction by examining the change of [$\log \tau_\alpha(T_g) - \log \tau_{JG}(T_g)$] on decreasing molecular weight [252]. Shown in Fig. 161 is the relaxation map of τ_α and τ_{JG} of four PEMA samples obtained dielectrically by Reissig et al. [1083]. The samples all have nearly the same 78% of syndiodiads but different average molecular weights M of 150, 21, 7.1, and 1.6 kg/mol with the calorimetric T_gs equal to 74, 55, 42, and 10°C, respectively. Both τ_α and its fragility increases with M, but τ_{JG} remains practically the same. It is clear that [$\log \tau_\alpha(T_g) - \log \tau_{JG}(T_g)$] decreases with decreasing M, and the coupling parameters $n(T_g)$ deduced from the CM equation (2.69) are 0.37, 0.31, and 0.24 for PEMAs with molecular weights of 150, 7.1, and 1.6 kg/mol, respectively. The deduced monotonic decrease of $n(T_g)$ is in accord with expectation. Also included in Fig. 161 are $\tau_\alpha(*)$ of i-PEMA with $M_w = 47.2$ kg/mol, which has much lower T_g and fragility because it has exceptional high isotactic triads of 92% than i-PEMA samples [252]. The data will be used for the discussion in the next subsection.

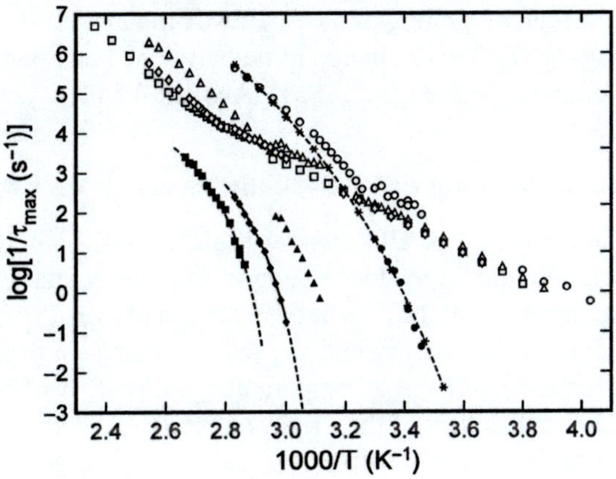

Fig. 161 Arrhenius plot of τ_α and τ_{JG} for PEMA samples with nearly the same 78% of syndiodiads but different molar weights (*squares*, 150 kg/mol; *diamonds*, 21 kg/mol; *triangles*, 7.1 kg/mol; *circles*, 1.6 kg/mol). Original data from [1083]. Shown also by *stars* are τ_α of i-PEMA with 92% isotactic triads and $M_w = 47.2$ kg/mol, and $M_w/M_n = 1.17$. Reproduced from [252] by permission

2.3.2.14 Changing the Ratio τ_α/τ_{JG} (or τ_α/τ_0) by Change in Tacticity of Polymers, Constancy of τ_{JG} or τ_0

In Section 2.2.3.9, we have already discussed change of structure of poly(methyl methacrylate) (PMMA) and poly(ethyl methacrylate) (PEMA) without changing the chemical composition of the repeat units by changing its tacticity. For example isotactic PEMA (i-PEMA) have all pendant groups on one side as opposed to more random arrangements in syndiotactic PEMA (s-PEMA) and atactic PEMA (a-PEMA). The α-relaxation dynamics is dependent on the tacticity of PMMA as well as of PEMA. Heuristically, the more ordered structure of i-PEMA suggests weaker intermolecular coupling and smaller coupling parameter n in i-PEMA than s-PEMA and a-PEMA. On the other hand, the characteristics of the local JG relaxation are expected not to be sensitive to change in tacticity. This is supported by the CODEX NMR results [1084] that τ_{JG} of i-PEMA and high molecular weight s-PEMA are the same, and τ_{JG} of i-PEMA coincides with that of atactic PEMA at temperatures where both are in the glassy states.

From Fig. 161 it can be seen that τ_α of the low molecular weight PEMA (M_w = 1.6 kg/mol) with 80% syndiodiads (•) is nearly the same as the τ_α of i-PEMA with a much higher molecular weight M_w = 47.2 kg/mol and 92% isotactic triads. Shown also is the τ_{JG} of the former, which can be assumed to be the same as τ_{JG} of the latter because of the CODEX NMR results. The indication is that τ_α and τ_{JG} are the same either by lowering the molecular weight of syndiotactic PEMA to 1.6 kg/mol or by changing tacticity to 92% isotactic triads, both have the same effects of lowering intermolecular coupling.

On comparing in Fig. 161 the relation between τ_α and τ_{JG} of s-PEMA having M_w = 21 kg/mol (triangles) with that of the i-PEMA having 92% isotactic triads and comparable M_w = 47.2 kg/mol (*), it is clear that $[\log \tau_\alpha(T_g) - \log \tau_{JG}(T_g)]$ is much larger for s-PEMA than i-PEMA. This is due to the expected larger intermolecular coupling or n of s-PEMA than i-PEMA. The same trend of $[\log \tau_\alpha(T_g) - \log \tau_{JG}(T_g)]$ with change in tacticity has been demonstrated to hold for PMMA by comparing the data of s-PMMA with i-PMMA [252].

2.3.2.15 Change of *T*-Dependence of τ_{JG} on Crossing T_g

Generally, τ_β of all secondary relaxations, both JG and non-JG, has an Arrhenius temperature dependence deep in the glassy state, i.e., $\tau_{JG} = \tau_\infty \exp(E_a/RT)$ with constant τ_∞ and E_a. Usually, near and above T_g, τ_{JG} and τ_α are not far apart and it is not possible to determine τ_{JG} without using some procedure. The actual temperature dependence of τ_{JG} at temperatures above T_g is of central importance to any theoretical explanation of the origin of the JG relaxation. Evidences have been presented from the co-invariance of τ_{JG} and τ_α to changes in combinations of P and T while maintaining either τ_{JG} or τ_α constant in Sections 2.3.2.8 and 2.3.3.9 that the relaxation strength and time of the JG relaxation are dependent on density ρ, entropy S, and fictive temperature T_f. More evidence will be presented in other subsections to follow. Since the T-dependences of these three quantities characterize the structure change after crossing T_g, naturally one would expect the Arrhenius T-dependence of τ_{JG} and T-dependence of the JG relaxation strength also change when crossing T_g.

Nevertheless, some researchers hold the belief that τ_{JG} has the same Arrhenius temperature dependence in the liquid state as that of the glassy state and the JG relaxation merges with the α-relaxation at T_β determined by the intersection of the extrapolated Arrhenius T-dependence of τ_{JG} with the VFTH T-dependence of τ_α. Different procedure of determining T-dependence of τ_{JG} near and above T_g were proposed and used in practice. One commonly used procedure assumes the complex dielectric susceptibilities of the β-relaxation, $\varepsilon_\beta^*(\nu)$, and the α-relaxation, $\varepsilon_\alpha^*(\nu)$, are additive, and they can be described by empirical functions such as that of Havriliak–Negami (HN) for both, or HN or Cole–Davidson for the α-relaxation, and Cole–Cole for the β-relaxation. The best fit to the spectrum by the sum, $\varepsilon_\beta^*(\nu)+\varepsilon_\alpha^*(\nu)$, is used to determine τ_{JG}. An example is [1085]. More sophisticated procedure used by Donth and coworkers [969, 1086] still assumes that the spectral densities of the α- and β-relaxations are additive in some temperature range above T_g.

In another procedure proposed by Williams [76, 77], the total correlation function including the α- and the β-relaxations is also additive, $\varphi(t) = f_\alpha \varphi_\alpha(t) + (1 - f_\alpha)\varphi_{\beta,\text{eff}}(t)$, except for the assumption that $\varphi_{\beta,\text{eff}}(t) = \varphi_\alpha \varphi_\beta(t)$. At the time Williams proposed this ansatz, he did not have the benefit of recent experimental facts to distinguish β-process that are JG β-relaxation from non-JG secondary relaxation. The procedure was applied by others to resolve JG and non-JG secondary relaxations without distinguishing them, as for example those secondary relaxations considered in Figs. 1a–f of [760]. These two procedures involve parameters in fitting data. In some cases, the results show that τ_β has stronger T-dependence in the liquid state than the Arrhenius T-dependence in the glassy state would predict if continued above T_g. An example of this result is the application of the Williams procedure to the JG relaxation of 2-picoline in mixtures with tri-styrene by Blochowicz [843] mentioned in Section 2.3.2.8, and shown in the right panel of Fig. 152. However, the same procedure of Williams used by others [1087] in sorbitol concluded that τ_β has the same Arrhenius T-dependence above and below T_g. The contradictory conclusions indicate that the procedure cannot be relied upon to ascertain the true T-dependence of τ_β above T_g. Finally, in all the fitting procedures described above including that of Donth and of Williams, the α- and β-relaxations are assumed to be independent of each other. Users of these fitting procedures must be aware that this assumption is incorrect and the procedures invalid when the β-relaxation is of the JG kind. The interdependence of the α- and JG β-relaxations was proven by the many correlations between the two processes and also by new spin-lattice relaxation weighted stimulated-echo experiments of Böhmer and coworkers on several glass-formers [998, 999]. The experimental data show a correlation between the spectral densities of the α- and the JG β-relaxations, and hence they are not independent processes.

Sorbitol and Xylitol Under High Pressure

The best situation to find out unequivocally the true T-dependence of τ_{JG} is when the JG β-relaxation is resolved and well separated from the α-relaxation at T_g and temperatures above, such that its relaxation time can be determined directly from the frequency at the maximum of the β-loss peak without using any procedure. Examples of such cases are sorbitol and xylitol at elevated pressure [1088]. At ambient pressure the JG β-relaxation is not that well separated albeit resolved at temperature near and above T_g in sorbitol and xylitol, and τ_{JG} can only be determined by employing some fitting procedure [1085]. Although the results indicate there is change of T-dependence from the Arrhenius one below T_g to a stronger one above T_g as shown by the symbols corresponding to 0.1 MPa in Fig. 162 by Fujima et al. [1085], skeptics may still question the conclusion. However, the two

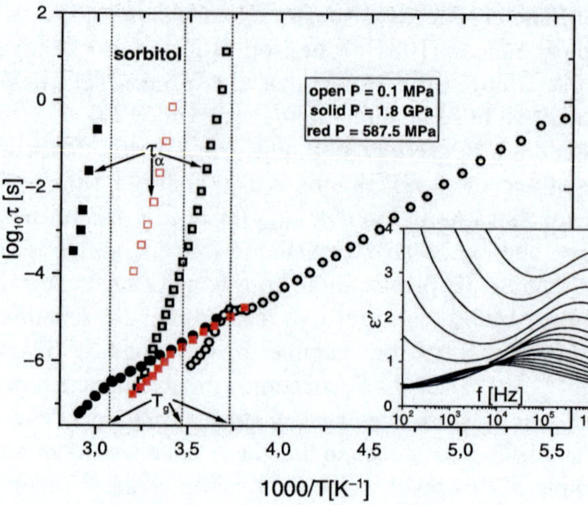

Fig. 162 Isobaric α-relaxation times at 0.1 MPa (*open square*), 0.59 MPa (*half-filled square*), and 1.8 GPa pressure (*filled squares*), along with the corresponding β-relaxation times at 0.1 MPa (*open circles*), 0.59 MPa (*half-filled circles*), and 1.8 GPa (*filled circles*). The slope of log τ_β is independent of pressure, although it differs markedly for low vs. high temperatures. The *inset* shows the JG peak in the dielectric loss at $P = 1.8$ GPa for temperatures from 273 to 343 K, in five increments (*bottom to top*). The α-peak is too low in frequency to appear within the measured frequency range. Reproduced from [1088] by permission

relaxations become more separated at elevated pressures and associated higher temperatures. This is due to the breaking of hydrogen bondings at higher P and T, resulting in increase in the coupling parameter n and hence shorter τ_{JG} for the same τ_α at T_g at ambient pressure or at elevated pressure according to the CM equation (2.69). At elevated pressure, the β-loss peak of sorbitol and xylitol is fully resolved and τ_{JG} has been determined directly without the use of any fitting procedure. The results unequivocally show the change of T-dependence from the Arrhenius one below T_g to a stronger one above T_g (see Fig. 162).

Low Concentration of 2-Picoline or Quinaldine in Tri-styrene

The peak of the JG β-relaxation of 5 wt% of QN in mixture with tri-styrene shown before in Fig. 152 (right panel), and 25% 2-picoline in tri-styrene in Fig. 153 (right panel) in the previous subsection is well separated from the α-loss peak at temperatures over a considerable range above T_g. For 5 wt% of QN in mixture with tri-styrene, τ_{JG} is shorter than τ_α by seven decades when $\tau_\alpha = 100$ s. The resolved JG β-relaxation of QN allows a direct determination of the peak frequency ν_{JG} and hence τ_{JG} without the need of using any procedure for fitting or deconvolution. The fact that there is a change of τ_{JG} from the Arrhenius T-dependence below T_g to a stronger T-dependence above T_g is beyond any doubt. Similar results are found for τ_{JG} in dielectric measurements at high pressures of 400 MPa in 17.2% chlorobenzene in decalin [641].

Copolyesters of Poly(ethylene terephthalate) and Poly(ethylene isophthalate)

The dielectric loss spectra of a series of copolyesters of poly(ethylene terephthalate) and poly(ethylene isophthalate) measured by Sanz et al. in a broad frequency range (10^{-1}–10^{9}Hz) show the secondary relaxation is the JG β-process from the good agreement of its relaxation time τ_β with the primitive relaxation time τ_0 calculated by Eq. (2.68) [1089]. The *T*-dependence of τ_β exhibits the change from the Arrhenius law at temperatures below T_g to a stronger *T*-dependence above T_g in these aromatic backbone polymers as shown in the left panel of Fig. 163. The same is found for the secondary relaxation in a series of amorphous aromatic polyesters, poly(ethylene terephthalate) PET (shown in the right panel of Fig. 163), poly(ethylene naphthalene dicarboxylate) PEN, poly(ethylene isophthalate) PEI, and poly(butylene isophthalate) PBI as found by Nogales et al. [1090].

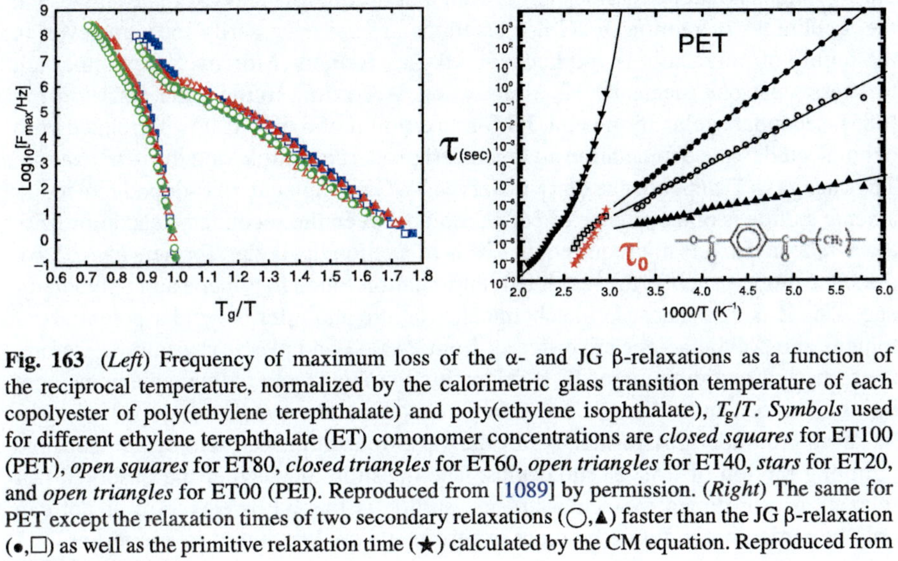

Fig. 163 (*Left*) Frequency of maximum loss of the α- and JG β-relaxations as a function of the reciprocal temperature, normalized by the calorimetric glass transition temperature of each copolyester of poly(ethylene terephthalate) and poly(ethylene isophthalate), T_g/T. *Symbols* used for different ethylene terephthalate (ET) comonomer concentrations are *closed squares* for ET100 (PET), *open squares* for ET80, *closed triangles* for ET60, *open triangles* for ET40, *stars* for ET20, and *open triangles* for ET00 (PEI). Reproduced from [1089] by permission. (*Right*) The same for PET except the relaxation times of two secondary relaxations (○,▲) faster than the JG β-relaxation (•,□) as well as the primitive relaxation time (★) calculated by the CM equation. Reproduced from [1090] by permission

P-Doped Molecular Arsenic Sulfide

Another system showing the same change of *T*-dependence across T_g of a secondary relaxation much faster than the α-relaxation is a P-doped molecular arsenic sulfide glass by Gjersing et al. [1091]. Doping with P results in the replacement of apical As of the As_4S_3 molecule by P. The secondary relaxation was observed by ^{31}P NMR spectroscopy to be an isotropic tumbling motion of the constituent molecules PAs_3S_3. At T_g, this fast secondary relaxation has relaxation time of about 10^{-5} s. The glassformer has a rather high steepness or "fragility" index ($m \approx 102$). The molecular dynamics found were recognized by the authors of [1091] to be similar to that of plastic crystals where the molecules reorient while their positions remain fixed at the lattice sites of the structure [145, 1092], and in fact, high-temperature polymorphs of As_4S_3 and P_4S_3 are plastic crystals. Since JG β-relaxation is found in many plastic crystals, it is likely the observed isotropic tumbling of the constituent molecules PAs_3S_3 is the JG β-relaxation of the P-doped molecular arsenic sulfide glassformer. This possibility was not taken on by Gjersing et al. [1091], citing that the isotropic rotational tumbling observed is different from the restricted reorientation of the JG β-relaxation observed in toluene and polybutadiene. This conclusion of Gjersing is based on the bold and unsubstantiated assumption that the motion of JG β-relaxation in all glassformers has to be exactly the same as toluene and polybutadiene irrespective of chemical structure and bonding. Second, there is a similar ^{2}H-NMR study of JG β-relaxation of neat polyalcohols, including glycerol, threitol, xylitol, and sorbitol [239]. Representing the spatially restricted motion of the JG β-relaxation in a cone, the semi-angle **Θ** of the cone for sorbitol deep in the glassy state is about three times larger than that of threitol. This example shows that the amplitudes of motion of JG β-relaxations are not necessarily the same even in the family of polyalcohols and hence in all glassformers. Moreover, it is important to stress what one means by JG β-relaxation. According to the criteria set forth in [685], secondary relaxation is the JG β-relaxation if and only if it is correlated with or connected to the α-relaxation and manifest properties mimicking the α-relaxation. The change in T-dependence of τ_β observed by Gjersing et al. in P-doped molecular arsenic sulfide is one evidence of connection between the secondary relaxation with the α-relaxation. It is not required that JG β-relaxation in all glassformers has to have the same motion such as the restricted reorientation found in toluene and polybutadiene. This is natural because the chemical structure and intermolecular potentials of toluene and polybutadiene are different from PAs_3S_3 and plastic crystals. Therefore, the restricted reorientation of JG β-relaxation in toluene and polybutadiene may not be used to conclude that the secondary relaxation observed in P-doped molecular arsenic sulfide glassformer is not the JG β-relaxation because of the isotropic tumbling motion of PAs_3S_3 molecules. The question of whether the observed fast relaxation in P-doped molecular arsenic sulfide is the JG β-relaxation or not can only be settled in the future by more experimental exploration of its properties.

The Ionic Liquid, 1-Hexyl-3-methylimidazolium Bromide

JG β-relaxation was found in room temperature ionic liquid, 1-hexyl-3-methylimidazolium bromide, by Mierzwa et al. [702] and discussed before in Sections 2.2.4.3 and 2.3.2.2. Change of τ_{JG} from the Arrhenius T-dependence below T_g to a stronger T-dependence above T_g was observed and shown in Fig. 164-1.

Water in Aqueous Mixtures

The same is found for the JG β-relaxation of the water component in aqueous mixtures of many different hydrophilic glassformers [1093] and water in the hydration shell of the hydrated proteins [1094]. Since water, aqueous mixtures, and hydrated proteins belong to important areas of research, the data will be presented in Sections

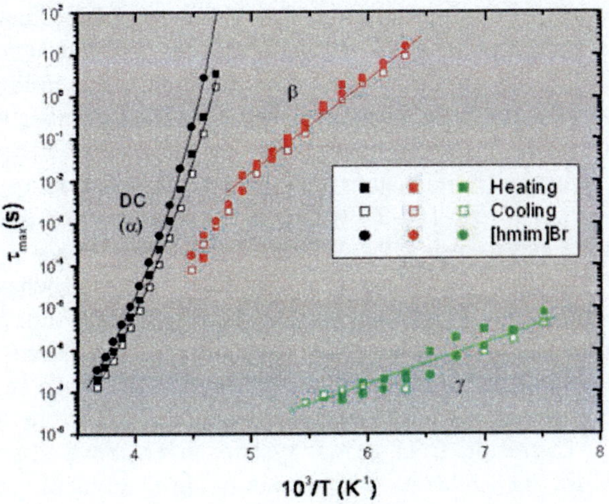

Fig. 164-1 Relaxation map of α-relaxation, JG β-relaxation, and γ-relaxation of the room temperature ionic liquid, 1-hexyl-3-methylimidazolium bromide. Reproduced from [702] by permission

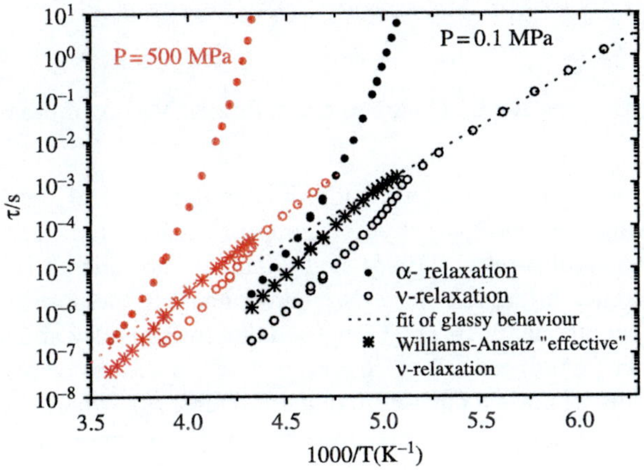

Fig. 164-2 Semilog plot of relaxation times vs. reciprocal T of 26% wt. fraction of water in PPG400. *Closed and open circles* indicate α- and ν-processes, respectively. *Asterisks* are the relaxation times of the "effective" relaxation obtained from the convolution procedure indicated by the Williams ansatz. *Black and red symbols* indicate isobaric scan done at $P = 0.1$ and 500 MPa, respectively. *Dotted lines* are Arrhenius fits to the data in the glassy state

2.3.2.27 and 2.3.2.28. In aqueous mixture the water-specific JG β-relaxation is sometimes called the ν-relaxation. Here in Fig. 164-2 are shown the α-relaxation and the β-relaxation times vs. reciprocal T of 26% wt. fraction of water in PPG400 at ambient and elevated pressures. The data offer Capaccioli et al. [1071(b)] an opportunity to test the effect of the convolution of the α- and β-relaxations according to the Williams ansatz against experiment. The same procedure used by others to obtain the results from the Williams ansatz and described in [1087(b), 1087(c)] was adopted by Capaccioli et al. [1071(b)]. First the Arrhenius T-dependence of the ν-relaxation time τ_ν in the glassy state was determined, then this dependence was extrapolated to higher temperatures up to where it merges with the α-relaxation. Following Eq. (2.66), the "effective" relaxation time for the convoluted of α- and ν-relaxation was calculated, and examined whether it can account for the experimentally observed change in T-dependence of τ_ν when crossing T_g or not. The "effective" ν-relaxation times from this exercise using the Williams ansatz are shown in Fig. 164-2. It is clear by inspection of the figure that the results from the analysis using the Williams ansatz bring only a much milder change of the "effective" ν-relaxation times from the Arrhenius behavior, and the change occurs at temperature significantly higher than T_g. These results are both in disagreement with experimental observation, and therefore the Williams ansatz is not able to explain the large deviation from the Arrhenius dependence of τ_ν after crossing T_g which is experimentally observed at ambient and elevated pressures.

The results from various systems given above all show change of T-dependence of the JG β-relaxation time on crossing T_g mimicking that of the α-relaxation time. This is another indication of the close connection between the JG β-relaxation and the α-relaxation.

Problem of the Explanation by Confinement in Frozen Matrix of the Higher T_g Component

Mentioned before, a different interpretation of the dynamics of the faster component in binary mixtures was given by Lorthioir et al. [1070, 1071], which has been elaborated by Cangiolosi et al. [1095]. These authors considered binary mixtures with large difference in the T_gs of the two components in the pure states, and rich in the higher T_g component (h). They proposed that the minority and lower T_g component (l) will be "confined" in the "frozen matrix" of the majority component at temperature below its glass temperature in the mixture, T_{gh}, where the α-relaxation time of the majority component, $\tau_{\alpha h}$, is very long. The consequence of confinement according to them is that, at temperatures below T_{gh}, the *α-relaxation* of the minority component is forced to execute localized motion within the frozen matrix of the majority component, and the temperature dependence of its relaxation time becomes Arrhenius. This interpretation violates the observations of the α-relaxation of the lower T_g component (l) having VFTH temperature dependence for its relaxation time $\tau_{\alpha l}$ at temperatures below T_{gh} as well as above T_{gh} in several miscible polymer blends and mixtures of small molecular glassformer with polymer.

As discussed in Section 2.3.2.10 and in [1096], the α-relaxation of the PEO component in all blends x% PEO–(1–x)% PMMA with $3 \le x \le 20$ is such an example. Deuterium NMR study of blends of deuterated PEO with PMMA [855] found that the segmental α-relaxation of PEO in any of the blends with 3, 6, 10, and 20% PEO is many orders of magnitude faster than the dynamics of PMMA at the T_{gh} of PMMA. The T-dependence of the segmental α-relaxation time of PEO, $\tau_{\alpha l}$, is not Arrhenius (see Fig. 157). For the 3% PEO blend, $T_{gh} = 391$ K, and the α-relaxation time of PEO is found by deuteron NMR down to 285 K, more than 100 degrees

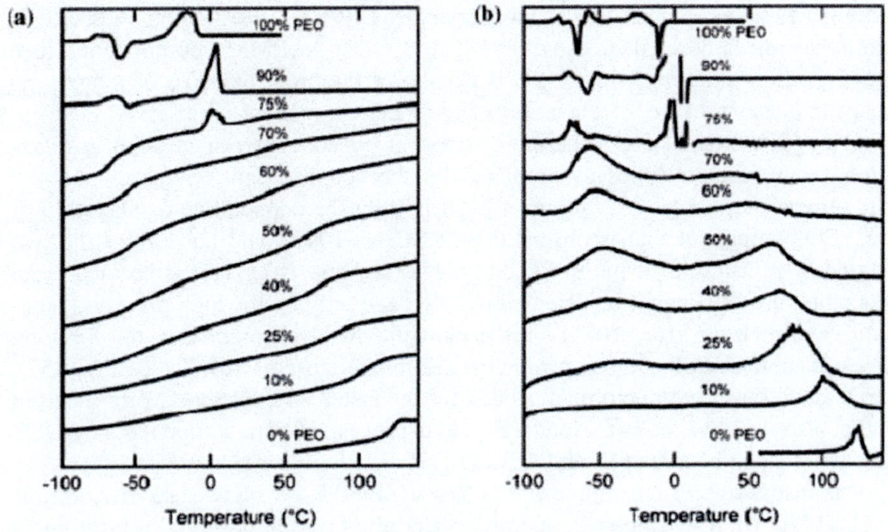

Fig. 165 DSC traces for PEO, PMMA, and the indicated blends containing 70, 60, 50, 40, and 25% PEO, two distinct glass transitions are clearly evident: (**a**) heat flow obtained at 10°C/min, and (**b**) corresponding derivative curves. The two transitions have relative amplitudes that vary as the composition of the associated component, that is the lower T_g can be assigned to the PEO and the upper T_g to the PMMA. Reproduced from [1097] by permission

below T_{gh}. The observation contradicts Lorthioir et al. that the 3% PEO will be confined by the frozen PMMA, and the segmental motion of PEO becomes local and Arrhenius (Fig. 165).

Miscible blends of polyisoprene (PI) and poly(4-*tert*-butylstyrene) (P4tBS) were measured by Zhao et al. [1097] using conventional and temperature-modulated differential scanning calorimetry (DSC) over a broad composition range. Before mixing, the two components have extraordinarily large difference in their glass transition temperatures T_g (~215 K). Nevertheless, two distinct calorimetric glass transitions of the two components were observed in blends of intermediate compositions (25–50% PI). The observation of α-relaxation of the PI component in the 25% PI blend by DSC at 100–1000 s time scale is another experimental fact indicating that the α-relaxation of the PI component has not been transformed to local relaxation at short times as would be predicted by the mechanism of "confinement in the frozen matrix of P4tBS" proposed by Lorthioir et al. [1070].

Differential scanning calorimetry measurement was made by Lodge et al. [1097] of blends of a PMMA ($M_n = 10{,}000$ g/mol) with a low molar mass PEO ($M_n = 300$ g/mol), as opposed to the high molecular weight PEO in the experiments of Lutz et al. and Jin et al. Lodge et al. found blends with 25% PEO has two broad but distinctly different glass transitions in the range $-64 < T < -8$°C with an average of -36°C for the PEO component, and $55 < T < 106$°C with an average of 81°C for the PMMA component. Thus, the cooperative α-relaxation originating from the PEO component was still observed by DSC at $T_{gl} = -36°C = 237.2$ K, more than one hundred degrees below $T_{gh} = 81°C$ of the PMMA component. Since the relaxation time probed by DSC is usually of the order of 100 s or longer, the result indicates that the α-relaxation of the PEO component is observed down to such long times at T_{gl}, and therefore it had not been transformed to local relaxation at any shorter times as predicted by the "confinement by frozen matrix" mechanism proposed by Lorthioir et al. It is truly the cooperative α-relaxation of the PEO component that is observed by DSC in [1097], and not any of the two secondary relaxations of the PEO component (see Fig. 157). This is because local relaxation usually cannot be detected by DSC and only by the high precision adiabatic calorimetry [681(c), 1075]. Furthermore, the JG β-relaxation time reaches long times of the order of 100 s at a much lower temperature of 167 K (see Fig. 157), and a much lower temperature is needed for the faster γ-relaxation. Worth mention is that also Arrese-Igor et al. found T_g and α-process of both components by DSC and dielectric in blends of PI and PtBS [1097].

It is instructive to reexamine the 20% PVME/80% PS blend data of Lorthioir et al. [1070(a)]. The emphasis of Lorthioir et al. is on the observed relaxation of the PVME component in this mixture called α′ at temperature below the calorimetric glass transition temperature of the blend (see inset of Fig. 166), where its relaxation time $\tau_{\alpha'}$ has Arrhenius T-dependence. The data of $\tau_{\alpha'}$ are shown as function of 1000/T by filled diamonds in Fig. 166 adapted from Fig. 7 in [1070(a)], and indeed $\tau_{\alpha'}$ has Arrhenius T-dependence in the range $3.44 < (1000/T) < 4.5$ or $290 > T > 222$ K. I hasten to point out that the α′-relaxation in 80% PS blend is not only observed below about 290 K where $\tau_{\alpha'}$ is Arrhenius, but a continuation of it is observed at temperature higher than 290 K, where the T-dependence of the relaxation time is stronger than the Arrhenius dependence of $\tau_{\alpha'}$ below about 290 K. The change in T-dependence is made clear with the help of the two lines with different slopes intersecting at $(1000/T) = 3.44$, as well as by the data (filled diamonds) present in the region $(1000/T) < 3.44$ in Fig. 166. The isothermal spectra in Fig. 167 actually demonstrate the loss peak of the α′ relaxation is seen not only below 290 K but also above 290 K. The prominent loss peak is located at 10^6 Hz when $T = 328$ K (see Fig. 167). After conversion to relaxation time, this data point at $1000/T = 1000/(328) = 3.05$ K^{-1} is represented by a closed diamond enclosed by a circle in the lower left corner in Fig. 166. For $T = 308$ K, the corresponding relaxation time from the loss peak in Fig. 167 actually has been represented by the last filled diamond at $(1000/T) = 3.25$ in Fig. 166.

Lorthioir et al. interpret the change of the T-dependence of $\tau_{\alpha'}$ near $(1000/T) = 3.44$ or approximately $T = 290$ K in the 20% PVME/80% PS blend as due to change of the *α-relaxation* of PVME from cooperative to local relaxation due to confinement by the frozen PS matrix. Now we show this interpretation by them is contradicted by their own data.

First, according to the inset in Fig. 166, the calorimetric glass transition temperature T_g (DSC) of the blend is ≈330 K (see inset), and its location in Fig. 166 on the 1000/T axis is indicated by the vertical line on the extreme left and labeled on top by "τ_α(PS) 80% PS." It can be seen by inspection of Fig. 166, the change of

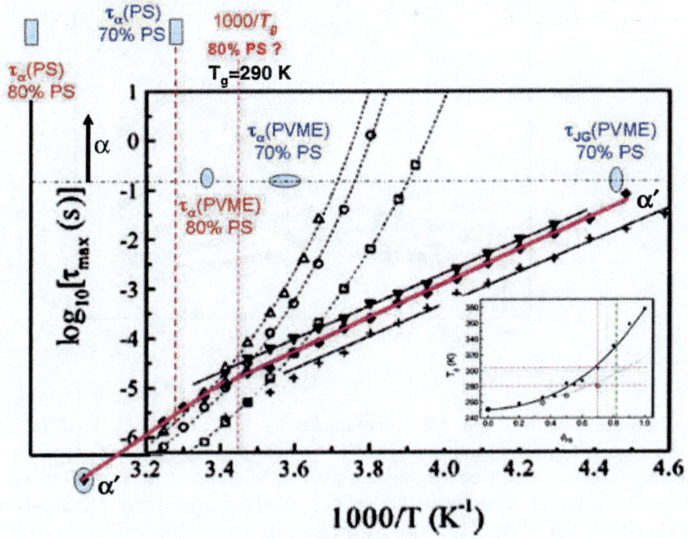

Fig. 166 Relaxation map of the α-relaxation time τ_α for neat PVME (*open squares*) and PS-PVME blends with wt% of PS, ϕ_{PS}, equal to 35 (*open circles*), 50 (*open triangles*) with VFTH temperature dependence. These data are irrelevant for the present purpose. The others with Arrhenius T-dependence are for PS-PVME blends with ϕ_{PS} equal to 70 (*closed inverted triangles*), 80 (*filled diamonds*), and 90 (*crosses*). The *solid lines* are fits of the experimental data of $\tau_{\alpha'}$ with an Arrhenius law. All in *black* adopted from a figure in [1070(a)]. Added are relaxation times determined from the isochronal spectrum at 1 Hz presented in [1070(a)]: τ_α (*horizontal oval*) and τ_{JG} (*vertical oval*) of the PVME component in the blend with 70 wt% of PS, and also τ_α (*vertical oval*) in the blend with 80 wt% of PS. The value of $T_g \approx 305$ and 330 K determined by DSC (see *inset*) suggests that τ_α of the PS component in the blend with 70 and 80 wt% of PS are located approximately within the two closed rectangles having $(1000/T_g) \approx 3.28$ and 3.0 (indicated by the *dashed and solid vertical lines*), respectively. The data point in the *lower left corner* and represented by a *closed diamond enclosed by a circle* is τ_{JG} of PVME in mixture with 80 wt% PS obtained from isothermal loss spectrum at 328 K shown in Fig. 167. The *two magenta lines* demonstrate the change of temperature dependence of τ_{JG} at $1000/T \approx 3.44$ corresponding to a temperature of 291 K that is more like T_g of the PVME component but not the $T_g \approx 330$ K of the PS component which is much higher. Obviously there is no relation between the temperature at which the crossover of T-dependence of τ_{JG} of PVME is found and T_g of the majority PS component. The *inset* presents the ϕ_{PS} dependence of the calorimetric glass transition temperature T_g(DSC) of the blends by *closed circles*. The *upper horizontal dashed lines* indicate that T_g(DSC) ≈ 305 and 330 K for the blend with $\phi_{PS} = 0.70$ and 0.80, respectively. The *open circles* are for the dielectric glass transition temperature T_g (1 Hz) determined from the temperature of the α-loss peak maximum of PVME component in the isochronal spectrum obtained for all blends at 1 Hz. The *lower dashed line* indicates that T_g (1 Hz) ≈ 280 K

T-dependence of $\tau_{\alpha'}$ in the 20% PVME/80% PS blend occurs at near 290 K far below T_g (DSC) ≈ 330 K of the blend. Therefore, there is no relation between the onset temperature of the "frozen PS component matrix" and the crossover temperature of $\tau_{\alpha'}$, at variance with the interpretation of Lorthioir et al.

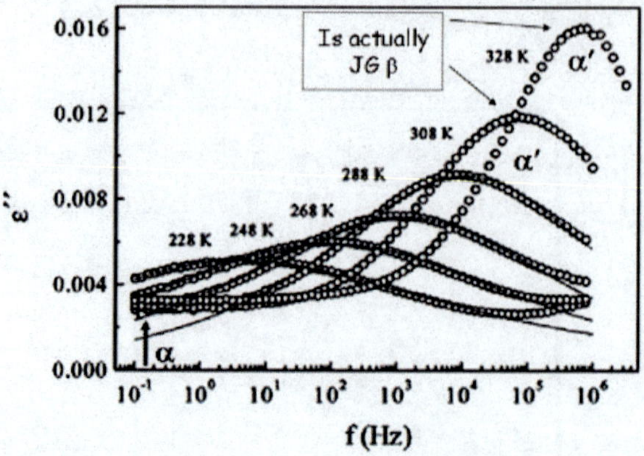

Fig. 167 Isothermal dielectric loss spectra obtained by Lorthioir et al. [1070(a)] on a 20% PVME/80% PS blend at different temperatures. The *arrow* indicates possible location of the frequency of the very broad α-relaxation loss peak of the PVME component. The tall loss peaks called the α′-process by Lorthioir et al. are actually the JG β-relaxation (see text). The *solid lines* are fits of the loss peak. Reproduced from [1070(a)] by permission

Second, from the isothermal spectra at 308 and 328 K in Fig. 167, Lorthioir et al. themselves have concluded that the broad loss at lower frequencies is contributed by the α-relaxation of the PVME component (distinctly different from the α′ relaxation). In order to make no mistake about this, here I quote the remarks made by Lorthioir et al. [1070(a)] on the isothermal spectra in their Fig. 4 which is Fig. 167 here: "For temperatures around 280 K, the a peak enters the considered spectral window, as can be seen in Fig. 4. This may be the origin of the flat signal observed in the low-frequency part of the dielectric loss $\varepsilon''(\omega)$ for temperatures higher than 288 K. This assumption is consistent with the slight increase of the level of this flat signal between 288 and 348 K. Besides, at 348 K, a broad shoulder is observed in the low-frequency side of the α' process we are interested in: it probably corresponds to the α-peak which is indeed known to be particularly broad in the high PS concentration regime. Thus, above 288 K, the $\varepsilon''(\omega)$ spectra also include a contribution from the α peak …." Although the α-loss peak frequency at 308 and 328 K cannot be determined from the broad spectrum in Fig. 167, any reasonable estimate at 328 K will put it lower than the vertical arrow in the figure. The corresponding α-relaxation time at 328 K will be in the neighborhood of 1 s, as suggested by the span of the vertical arrow drawn just above the label of the *y*-axis in Fig. 166. Now we can see from either Fig. 166 or Fig. 167 that two processes, α′ and α, are present at the same temperature such as 308 and 328 K. At both temperatures, the relaxation times of the α′ and α processes differ by many orders of magnitude. This finding invalidates the explanation by Lorthioir et al. because there is only one α-process at any temperature.

While the data contradict the explanation of the crossover in *T*-dependence by Lorthioir et al. based on the α-relaxation of the PVME component, there is no problem if the α′-process is identified with the JG β-process of the PVME component, and $\tau_{\alpha'}$ becomes τ_{JG}, the JG β-relaxation time of the PVME component. The α- and JG β-relaxations of the PVME component can coexist as found in the experiment. Due to the very broad α-loss peak of the PVME component in either the isochronal or the isothermal spectra, no exact value of T_{gl} of the PVME component can be given. In the inset of Fig. 166, the open circles are for the dielectric glass transition temperature T_g (at 1 Hz) determined from the temperature of the α-loss peak maximum of PVME component in the isochronal spectrum obtained for all blends at 1 Hz. The lower dashed line indicates that T_g (1 Hz) $\approx$ 280 K for blend with $\phi_{PS} = 0.70$. From this and the isothermal spectra at 288 K in Fig. 167, one can estimate that 290 K is possibly a candidate for T_{gl} of the 20% PVME/80% PS blend. The location of this $1000/T_{gl}$ is indicated by the vertical dotted line labeled "$T_g = 290$ K." The picture in Fig. 166 is now complete. It shows the observed change of T-dependence of $\tau_{\alpha'}$ when crossing 290 K is the same as found in the mixtures of picoline, *tert*-butylpyridine, or quinaldine with tri-styrene, as well as in some neat glassformers shown before, and the explanation follows from the correlation between the α-relaxation and the JG β-relaxation or the primitive relaxation of the coupling model.

Lesson from the Previous Study of the Mixture of Toluene with Polystyrene

Calorimetric measurements found signature of glass transition of the heat capacity at T_{gl} of the low T_g component (l) distinctly lower than another one at T_{gh} in several mixtures, indicating that the α-relaxation of l component exists up to long times of order of 10^2 or 10^3 s, and is not converted to local α′-process by confinement in the "frozen matrix" of the higher T_g component (h). We have already seen one example in the 25% PEO–75% PMMA blend in Fig. 165. More examples are given in [1071]. Here we mention just one case.

The ideal system to test the interpretation is mixtures of toluene with polystyrene. This is because the difference between the T_gs of the two components is much larger than toluene/PCB54, the system studied recently by Cangiolosi et al. [1095]. The mixture toluene/PS has the advantage that PS does not contribute to the dielectric relaxation spectrum because of small dipole moment in contrast to PCB54 in the toluene/PCB54 system. The toluene/PS system was investigated as early as 1975 by Adachi et al. [1098] by dielectric, thermal analysis, and NMR, and they found both the primary α-relaxations of toluene and PS together with some local relaxations. This work was followed up by others on the same toluene/PS systems by dielectric and depolarized Rayleigh scattering experiments [1099]. In the latter experiment, the primary α-relaxations of PS and toluene were found over a frequency range of 10 decades, and their relaxation times all have the usual VFTH temperature dependence of cooperative α-relaxations. The most recent investigation reported on the toluene/PS mixtures uses the sensitive adiabatic calorimetry method in combination with dielectric relaxation by Taniguchi et al. [1100]. The high molecular weight PS has $T_g = 373$ K, and the difference between T_g of PS and toluene is 256 K. This enormous difference in T_g, together with the absence of domination by the PS contribution in dielectric spectra, is an ideal condition to test any interpretation. The measured heat capacity C_p on toluene/PS mixtures containing 20, 30, 40, 50, and 70 wt% PS using an adiabatic calorimeter indicates that the C_p vs. temperature T curve exhibits a double-sigmoidal shape [1100]. The T_gs corresponding to the two sigmoids are denoted here by T_{gh} and T_{gl} with $T_{gh}>T_{gl}$. The same was found in other systems [169(a), 1098]. For the mixture with 70 wt% PS, $T_{gh} = 230$ K is associated with the segmental α-relaxation of the PS component and $T_{gl} = 145$ K is related to the α-relaxation of the toluene component. The difference between the two T_gs decreases with decreasing wt% of PS as expected. For example, the mixture with 30 wt% PS has $T_{gh} = 153$ K and $T_{gl} = 130$ K. The dielectric measurements found three relaxation processes. The slowest and intermediate processes are identified as the segmental α-relaxation of the PS component and the rotational α-relaxation of toluene, respectively. The dielectric T_gs of these two processes, defined as the temperatures at which their dielectric relaxation times attain 10^3 s, agree well with the calorimetric T_{gh} and T_{gl}. From the isochronal loss spectra at 1 kHz of the mixtures, the temperature at the maximum loss of the fastest relaxation is about 110 K and insensitive to composition of the mixture. Converting 1 kHz to time, we find that its relaxation time is 1.6×10^{-4} s at 110 K, which falls on top of the JG β-relaxation times of neat toluene, indicating that the fastest relaxation in the toluene/PS mixtures is the JG β-relaxation of the toluene component. From the calorimetric and dielectric data of the toluene/PS mixtures, Taniguchi et al. found the cooperative rotational α-relaxation of toluene (with VFTH dependence for its relaxation time) and a much faster local relaxation also originating from toluene. This result contradicts the confinement interpretation of Lorthioir et al. [1070]. This is best elucidated by the mixture with 70 wt% PS. Taniguchi et al. [1100] still find the cooperative rotational α-relaxation of toluene at $T_{gl} = 145$ K, which is 85 K below the $T_{gh} = 230$ K of the PS component. However, according to Lorthioir et al., the PS chains should be frozen below $T_{gh} = 230$ K to confine the toluene and preempt cooperative motion involving the toluene, resulting in the local toluene motion at temperatures below $T_{gh} = 230$ K.

Another binary mixture system that supports our interpretation and refutes that of Lorthioir et al. [1070(a)] and Cangiolosi et al. [1095] is the study by Svanberg et al. of propylene carbonate (PC) dissolved in PMMA polymer matrix of a gel consisting of 100% PC down to 50% PC [683]. The difference in T_gs of the two components is about 240 K. The confinement in the gel is of a fractal type with randomly interpenetrating and percolating phases of PMMA and PC. Thus, it is even more appropriate to say that PC is confined in this system than PVME is confined in frozen PS by Lorthioir et al. and toluene is confined in frozen PCB54 by Cangiolosi et al. Nevertheless, both the primary α-relaxation and the JG β-relaxation of PC were observed. In neat PC, the JG β-relaxation is not resolved and shows up as an excess wing like in neat picoline, *tert*-butylpyridine, or quinaldine. On increasing PMMA content, the α-relaxation of PC is slowed down while the excess wing continuously transforms toward a resolved JG β-relaxation of PC. The effects are similar to that seen in mixtures of picoline, *tert*-butylpyridine, or quinaldine with tristyrene. Dielectric relaxations of MTHF mixed with 3S show α-process with VFTH law (also by DSC), and α′, β, and γ processes all with Arrhenius law [1070(c)]. α′

is suggested to come from some MTHF by confinement, but this is not exactly the scenario of Lorthioir et al. because α- and β-relaxations coexist.

Mixtures of Toluene with Polychlorinated Biphenyl (PCB54)

From the discussion given above, it is clear that had Lorthioir et al. accepted the overwhelming evidence pointing to universal presence of the JG β-relaxation although not resolved in neat PVME, and it becomes resolved when mixed with PS with much lower mobility (like picoline, *tert*-butylpyridine, or quinaldine when mixed with tri-styrene), they would have no need to propose the superfluous concept of confinement in frozen matrix and invent the untenable α′-relaxation. However, their proposal is maintained in a recent study of mixtures of toluene with polychlorinated biphenyl (PCB54) with 0, 20, 35, 50, 70, and 100 wt% of PCB54 by Cangialosi et al. [1095]. Although the $T_g = 246$ K of pure PCB54 is much higher than the $T_g = 117$ K of toluene, the toluene/PCB54 mixtures are not equivalent in properties to PVME/PS blends or mixtures of tri-styrene with picoline, *tert*-butylpyridine, or quinaldine. For one, the higher T_g component PCB54 has much larger dielectric strength than toluene, dominates the loss spectra, and makes it impossible to see the α-relaxation of the toluene component even in mixtures with 20 wt% PCB54. Another difference is that the lower T_g component toluene already has a well-resolved secondary relaxation in isothermal dielectric loss spectra [841, 971], which is the JG β-relaxation because toluene is a rigid molecule and also its relaxation time is in agreement with the calculated primitive relaxation time of the coupling model [671]. These undesirable properties should make the toluene/PCB54 mixtures the wrong system to test either the interpretation of Lorthioir et al. or ours. The proposal of Lorthioir et al. would expect a local toluene relaxation, the analogue of the α′-relaxation in PVME/PS blends, with relaxation time having Arrhenius temperature present at temperature below T_{gh} of PCB54 component in the mixture. But this is hard to verify because toluene already has its JG β-relaxation with the same characteristics. In fact, the local relaxation found in 20, 30, and 50 wt% PCB54 have relaxation times indistinguishable from that of the JG β-relaxation of neat toluene (see Fig. 168). Our interpretation expects both the α-relaxation and the JG β-relaxation of the toluene component to be present. But this cannot be checked because the α-relaxation of toluene if present would be eclipsed by the dominant PCB54 loss peak in the spectrum. Shahin Thayyil and Capaccioli et al. [1101] have recently made dielectric measurements of a more ideal system of 10% toluene/90% OTP mixtures because OTP gives negligible contribution to the loss spectrum. They find both the α- and the JG β-relaxation of the toluene component and the change of T-dependence of the JG β-relaxation time at T_g of the toluene component, exactly the same as like picoline, *tert*-butylpyridine, or quinaldine when mixed with tri-styrene (see Figs. 147, 149, and 152).

Despite the toluene/PCB54 mixtures are less than ideal systems, it is worthwhile to compare the data from Cangialosi et al. [1095] with the calorimetric and dielectric data of toluene/PS mixtures from Taniguchi et al. [1100]. This comparison is made in Fig. 168. There the glass transition temperatures T_g at which

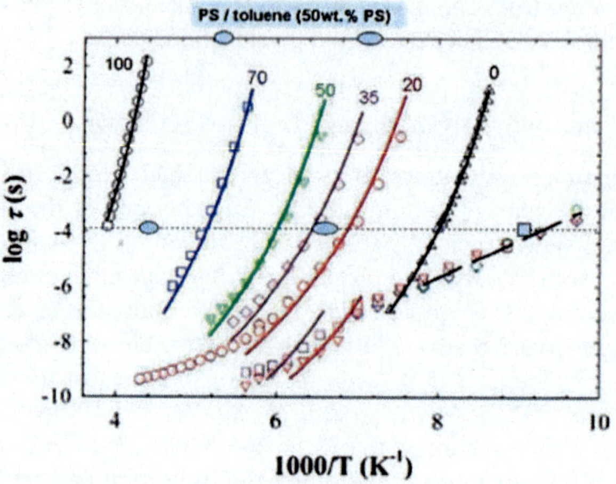

Fig. 168 Comparison of PCB54/toluene data of Cangiolosi et al. with PS/toluene mixture with 50 wt% PS data of Taniguchi et al. The glass transition temperatures T_g at which the calorimetric and dielectric α-relaxation times τ_α of PS and toluene in the 50 wt% PS mixture with toluene attain the value of about 1500 s are indicated by the *blue ovals* on the upper *x*-axis. Isochronal dielectric loss data at 1 kHz (equivalent to relaxation time of = 1.59×10^{-4} s) show the presence of three peaks coming from *left to right* the α-relaxation of PS component, the α-relaxation of toluene component, and the JG β-relaxation of toluene component. The temperatures at the peak maxima are shown on the *dashed line* at constant relaxation time of 1.59×10^{-4} s by *two filled ovals* for the α-relaxations of PS component and the toluene component, and the *filled square* for the JG β-relaxation of the toluene component. The JG β-relaxation of neat toluene (*thick dashed straight line*). The PCB54/toluene data of Cangiolosi et al. represented by *symbols* and *lines* are the same as in the original article by these authors. PCB54 wt%: 100% (*circles*), 70 (*squares*), 50 (*inverted triangles*), 35 (*diamonds*), 20 (*circles*), 0 (*triangles*). Data from [1095] redrawn and illustration added

the calorimetric and dielectric α-relaxation times τ_α of the PS and the toluene components in the PS/toluene mixture with 50 wt% PS attain the value of about 1500 s (the timescale of calorimetric measurement) are indicated by the filled oval symbols on the upper *x*-axis. The widths of the symbols represent the differences between the calorimetric and dielectric T_g. Isochronal dielectric loss data at 1 kHz (equivalent to relaxation time of = 1.59×10^{-4} s, dotted line in the middle) show the presence of three peaks coming from the α-relaxations of the PS component and the toluene component, and the JG β-relaxation of the toluene component. The temperatures at the isochronal peak maxima are shown on the dashed line at the constant relaxation time of 1.59×10^{-4} s by two filled ovals for the α-relaxations of the PS component and the toluene component, and the larger filled square for the JG β-relaxation of the toluene component. The PCB54/toluene data of Cangialosi et al. represented by symbols and lines are the same as in the original article by these authors. They are for mixtures of toluene with polychlorinated biphenyl (PCB54) with 0, 20, 35, 50, 70, and 100 wt% of PCB54. I have added the JG β-relaxation time of neat toluene from [841, 971]. By inspection of Fig. 168, it can be seen that there is good agreement of the JG β-relaxation time of both neat toluene and toluene in the PS/toluene mixture with 50 wt% PS with the relaxation time of the α′process found by Cangialosi et al. in their toluene/PCB54 mixtures. The agreement shows once again the α′ process in toluene/PCB54 mixtures is the JG β-relaxation of the toluene component in these mixtures.

Why So Much Discussion on the Interpretation by Lorthioir et al.?

The reader may wonder why so much discussion has been given in the above to refute the explanation of the faster component dynamics in mixtures offered by Lorthioir et al. and Cangialosi et al. There are two reasons for this. The first one is the true interpretation of the dynamics, showing the α- and the JG β-relaxation of the fast component l are connected, is very important for the glass transition problem. The connections include (1) the coexistence of the α- and the JG β-relaxation of the fast component l at all temperatures down to $T_g(l)$ of the fast component where the α-relaxation time, $\tau_\alpha(T_g(l))$, is a long time of the order of 10^2 s and (2) the JG β-relaxation time, $\tau_{JG}(T)$, changes its T-dependence when crossing $T_g(l)$, mimicking the well-known behavior of $\tau_\alpha(T)$. Therefore, we should not let the wrong interpretation of Lorthioir et al. confuse the facts and obscure this important connection between the α- and the JG β-relaxation. The second reason is that the interpretation of Lorthioir et al. or similar ideas has been widely applied to nanoconfined water [1102–1105], aqueous mixtures [1106–1108], and hydrated biomolecules [1109–1112]. These are more complicated systems and interpretation of the dynamics is more difficult than the mixtures discussed above. Continued use of the interpretation of Lorthioir et al. which has been proven wrong in simpler systems, to these more complicated cases will bear no fruit and only retard real progress. Therefore it is necessary to document in detail the experimental data that prove the interpretation of Lorthioir et al. untenable.

Reminder of Similar Behavior of JG β-Relaxation in Neat Glassformer

For glassformers that have an unresolved JG relaxation, which appears as an excess wing, the temperature dependence of τ_{JG} above T_g can be inferred from the shift of the excess wing with temperature. Since τ_α is non-Arrhenius above T_g and the excess wing is on the high frequency side of the α-peak, the shift of the excess wing with temperature will also be non-Arrhenius, albeit less than that of τ_α. This non-Arrhenius behavior of the excess wing for $T > T_g$ has been shown for glycerol, propylene carbonate, propylene glycol [424], and KDE [597]. Thus, for all JG relaxations, hidden or resolved, τ_{JG} has different temperature dependences above and below T_g.

The implication of the above discussion is that a change in T-dependence of τ_{JG} above T_g is a general feature in glass-forming liquids. The temperature dependence of τ_{JG} above T_g is much stronger than the dependence below T_g. Theoretical support of this behavior comes from the CM equation (2.67), $\tau_{JG}(T) \approx \tau_0(T) = [\tau_\alpha(T)]^{1-n} t_c^n$. This equation indicates the temperature dependence of τ_{JG} has Arrhenius temperature dependence below T_g because there τ_α is Arrhenius and n

is temperature independent. Above T_g, again from the CM equation, τ_{JG} necessarily has a stronger temperature dependence than the Arrhenius dependence as a consequence of the even stronger VFTH temperature dependence of τ_α. This is the case whether n decrease with increasing temperature or not.

2.3.2.16 Doubt on the Universal Presence of the JG β-Relaxation? Glassformers Only Showing a Non-JG Secondary Relaxation

If the JG β-relaxation is truly of fundamental importance, a prerequisite is its presence in all glassformers. This requirement is fulfilled in many molecular and polymeric glassformers as well as in some exotic glassformers such as plastic crystals [145, 681], inorganic oxide glasses [952, 985, 987], the molten salt $0.4Ca(NO_3)_2$–$0.6KNO_3$ (CKN) [951], and metallic glasses, $Pd_{77.5}Cu_6Si_{16.5}$ [1113], $La_{55}Al_{25}Ni_{20}$ [955], $Pd_{43}Ni_{10}Cu_{27}P_{20}$ [956], $Zr_{65}Al_{7.5}Cu_{27.5}$ with fragility index $m = 36.4-38.4$ [957, 960, 1114], $Zr_{46.75}Ti_{8.25}Cu_{7.5}Ni_{10}Be_{27.5}$ (Vit4), $Pd_{40}Ni_{10}Cu_{30}P_{20}$, and others [959, 1115], $Pd_{77}Cu_6Si_{17}$ with $m = 52.8 - 77.0$ [960]. We show this by the 1 Hz isochronal shear modulus data of $Pd_{43}Ni_{10}Cu_{27}P_{20}$ [956] in Fig. 169.

The isochronal loss shear modulus data of $Zr_{65}Al_{7.5}Cu_{27.5}$ [957] show the JG β-relaxation appearing in the form of an excess wing, as well as the presence of the nearly constant loss at lower temperatures as pointed out in [1114] (see Fig. 170).

Despite the common occurrence of the JG β-relaxation, there are glassformers in which none of the commonly used experimental methods has been able to produce evidence of the existence of the JG β-relaxation. Ironically, these cases are found in some small molecular and organic glassformers, a class of materials that have usually shown the presence of the JG β-relaxation by dielectric relaxation spectroscopy. This conundrum raises serious doubt in skeptics on the claim that the JG β-relaxation is universal and plays an important role in glass transition. In some of these cases, application of special procedure such as aging of the glassformer in the glassy state [424, 682, 684, 1014] or elevating the pressure [657, 669, 671, 708] were able to find the JG β-relaxation by resolving it from the higher intensity

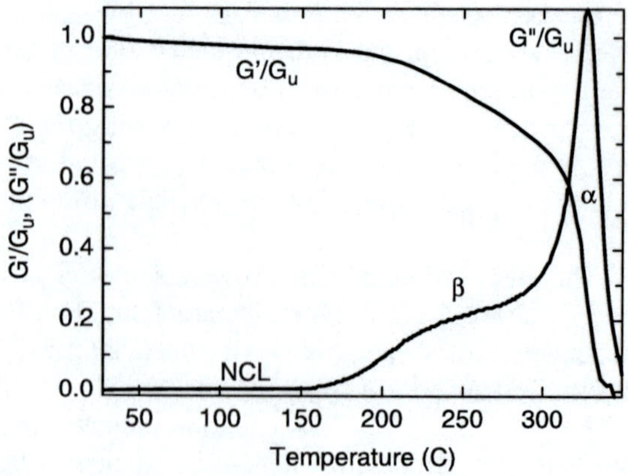

Fig. 169 Storage modulus G' and loss modulus G'' vs. temperature of $Pd_{43}Ni_{10}Cu_{27}P_{20}$ at a frequency of 1 Hz and at a heating rate of 1 K/min, showing the α- and β-relaxations and the low-temperature NCL (redrawn from data published in [956]). The G'' data have been scaled up by a factor of 4.8

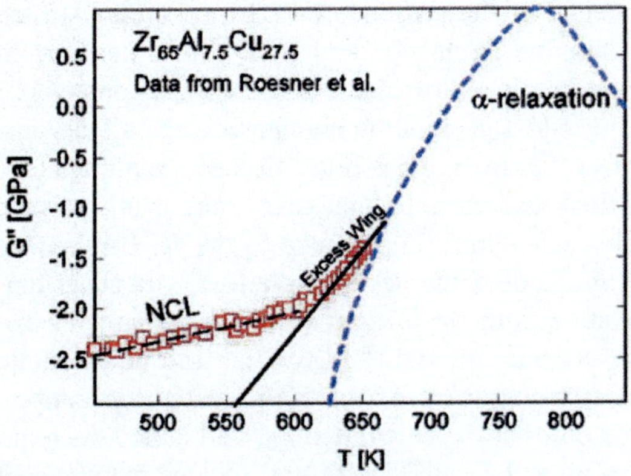

Fig. 170 The loss modulus $G''(T)$ of $Zr_{65}Al_{7.5}Cu_{27.5}$ measured by Rösner et al. (*open squares*), and the calculated the isochronal shear loss modulus contributed by the dominant α-relaxation, $G''_{\alpha}(T)$ (*dashed line*), at 5.4 kHz. The JG relaxation is not resolved as a peak and instead appears as an excess wing. The most probable temperature of the JG relaxation is estimated to lie between $T_{JG} = 647$ K and $T_{JG} = 635$ K. Reproduced from [1114] by permission

α-relaxation appearing at lower frequencies, and sometimes also from the high-frequency non-JG secondary relaxation. However, there are some glassformers for which none of these procedures have produced evidence for the presence of the JG β-relaxation. These glassformers include the lower members of the dialkyl phthalates, viz., dimethyl phthalate (DMP), diethyl phthalate (DEP) [660], and dibutyl phthalate (DBP) [661] or diisobutyl phthalate (DiBP). A secondary γ-relaxation is present in all these phthalates, but it is not the JG β-relaxation we are looking for because its dynamics bear no correlation with the α-relaxation. For example, its relaxation time is insensitive to applied pressure, in stark contrast to the α-relaxation time, τ_{α}. Based on the chemical structure of the phthalates, this γ-relaxation is related to the rotation of the alkyl group about the flexile oxygen bond and only involves a part of the molecule. These phthalates are good glassformers and extensively used in industry for various purposes. Their dynamic properties over broad frequency/time range have been studied by several experimental techniques including optical Kerr effect [332], dynamic light scattering [1116], dielectric relaxation measurements at ambient and elevated pressures [558, 660, 661, 1117], and inelastic X-ray scattering [1118]. The results of these studies have impact on theoretical models of glass transition [34, 35, 1119].

Another glassformer with similar situation as the phthalates is benzophenone (BZP) or diphenylketone, H_5C_6–CO–C_6H_5. It was studied by optical Kerr effect [332], dynamic light scattering [1047], and broadband dielectric spectroscopy [1120, 1121]. In fact, although BZP shows a well-resolved secondary relaxation, there is some evidence that such process could be of intramolecular nature [1122, 1123]. Moreover, its relaxation time near T_g is much shorter than the calculated primitive relaxation time τ_0 of the CM [508].

The lack of evidence of the presence of JG β-relaxation in some well-studied glassformers including the phthalates and BZP can undermine its importance in glass transition. In a log–log plot of dielectric loss against frequency, the high-frequency flank of the α-loss peak of the phthalates and BZP has maximum absolute value of the slope near 0.5. If the JG β-relaxation of the phthalates and BZP indeed were non-existent, then these exceptional cases may even invite the interpretation of α-relaxation which gives little importance to the JG β-relaxation [1045, 1046]. As we have seen from some examples given before, whenever the JG β-relaxation is resolved, its relaxation time τ_{JG} is approximately the same as the primitive relaxation time τ_0 of the coupling model (CM), which can be calculated by Eq. (2.68) from τ_α and the stretch exponent of the Kohlrausch correlation function of the α-relaxation. It turns out that the calculated τ_0, and hence the expected τ_{JG}, of the phthalates (see Fig. 4 in [661]) and BZP [508] is located in between the α- and the γ-relaxations. Thus, it is possible that the JG β-relaxation cannot be resolved in these glassformers because it is sandwiched between the slower and much more intense α-relaxation on one side and the well-resolved faster non-JG γ-relaxation on the other side. To resolve the JG β-relaxation we must increase the separation between the α- and the γ-relaxations. One way to do this is by mixing the phthalates and BZP with a host that have a much higher glass transition temperature T_g as demonstrated before for other polar rigid molecules including picoline, QN, and TBP in mixtures with tri-styrene, other oligomers of styrene, or *ortho*-terphenyl (OTP) discussed in a previous subsection. However, there is a difference of the present case from the previously discussed rigid molecules, which has no γ-relaxation. Nevertheless, the presence of the less mobile host molecules slows down the α-relaxation of the phthalate molecules in the mixture more than in the neat phthalate. On the other hand, the local and intramolecular γ-relaxation is not sensitive to the presence of the host molecules. Hence, the increased separation between τ_α and τ_γ of the phthalate in the mixture gives a better chance for the supposedly universal JG β-relaxation to be resolved. This is the rationale of the method to uncover JG β-relaxation, which is particularly effective in using dielectric spectroscopy when the host molecules are apolar like oligomers of styrene or OTP, and the dynamics of the polar phthalates or BZP are seen exclusively in the observed dielectric spectra. In the recent past, the method was applied to the phthalates, BZP, and other glassformers by mixing with low molecular weight polystyrene, and demonstrated that the JG β-relaxation exists in DMP, DEP, DBP, DiBP, and BZP by broadband isothermal dielectric relaxation of all these glassformers in the pure state as well as in mixtures with polystyrene and OTP. Results of some selected glassformers are presented in the following subsections.

Butylacetate (BuAc)

The first demonstration of the method was given for a lesser known dipolar glassformer, butylacetate (BuAc, $T_g = 125.7$ K) [1124]. Figure 171 shows the temperature dependence of the isochronal (i.e., obtained at 1 kHz) dielectric loss ε'' data of neat BuAc, 10 wt% BuAc in isopropylbenzene (IPB, $T_g = 128.5$ K), and 10 wt%

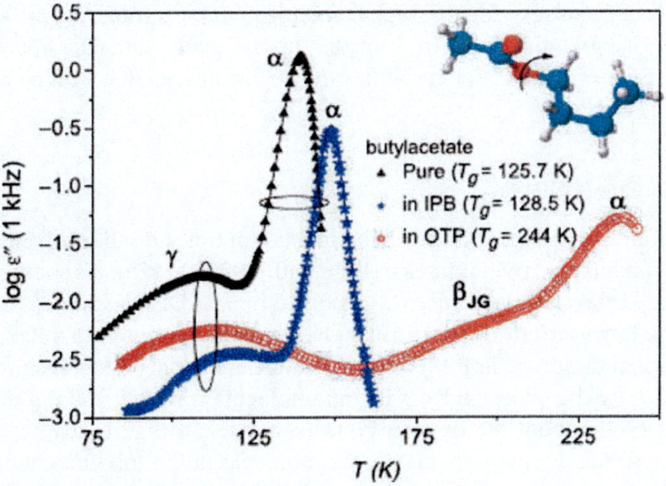

Fig. 171 Logarithm of dielectric losses at 1 kHz vs. temperature for butylacetate: pure (*triangles*), mixed with IPB (*stars*), mixed with OTP (*open circles*). *Lines* are guides for eyes. *Inset* shows the molecular structure and the internal rotation that gives rise to the γ-relaxation, which is insensitive to mixing with OTP. Data from [1124] replotted together with structure of butylacetate

BuAc in *o*-terphenyl (OTP, $T_g = 244$ K). Shown in the same figure is the chemical structure of BuAc together with the rotation of the –OOC–CH_3 group around its oxygen bond. This internal motion is responsible for the γ-relaxation of neat BuAc seen as a maximum of ε'' at $T = 109$ K (Fig. 171). The T_g of IPB is only 3 degrees higher than that of BuAc, and the mixing shifts the isochronal α-loss peak (at 1 kHz) from about 139.5 K of neat BuAc to 149 K of BuAc in IPB. The shift of less than 10 degrees is too small to significantly increase the separation between τ_α and τ_γ of the BuAc in this mixture to resolve the JG β-relaxation. A much higher T_g of the host such as OTP is required toward accomplishing the goal. In fact, isochronal loss spectra at 1 kHz of BuAc in OTP (Fig. 171) show that the α-loss peak has been shifted from about 139.5 K for neat BuAc to 245 K for BuAc in OTP. On the other hand, the shift of the γ-loss peak is small as expected for local and intramolecular secondary relaxation. It is clear by inspection of the figure that a new secondary relaxation appears in the isochronal spectrum of BuAc in OTP concomitant with the large separation between τ_α and τ_γ of the BuAc in OTP.

The Dialkyl Phthalates

The same method was applied more recently to prove the existence of the JG β-relaxation in the dialkyl phthalate. In these experiments, the motions of the polar dialkyl phthalates (DaP) including dimethyl phthalate (DMP, $T_g = 193$ K), diethyl phthalate (DEP, $T_g = 188$ K), and diisobutyl phthalates (DiBP, $T_g = 191$ K) were studied by dielectric spectroscopy in 10 wt% mixtures with PS of molecular weight equal to 800 Da (PS800). For DMP this is equivalent to 0.095 mole fraction of

DMP in PS800, and less for the higher members. These mixtures all satisfy the requirement of large difference in T_g of the dipolar guest phthalate and the apolar host PS800 ($T_g = 282$ K) for resolving the so far unseen JG β-relaxation of the phthalates [1125].

Diethyl Phthalate (DEP)

The relaxation map in Fig. 172 shows the relaxation times of all resolved processes in neat DEP, and in 10 wt% mixtures of DEP with PS800 [1125]. As a reminder, only one secondary relaxation labeled γ was resolved in neat DEP at ambient or elevated pressure. The large shift to significantly higher temperatures of the α-relaxation time τ_α of DEP when dissolved in PS800 is evident. There is only a small shift of the γ-loss peak, as it can be expected from its intramolecular origin involving the rotation of the alkyl group about the oxygen bond (see inset in Fig. 172 left panel), and therefore the different environment for the molecule in the mixture cannot change the energy barrier for the internal motion. In the mixture a newly resolved secondary (β-) relaxation appears at temperatures intermediate between the α- and the γ-loss peaks in some selected isothermal spectra of 10 wt% of DEP in PS800 presented in the right panel of Fig. 172. All three relaxations, α, β, and γ, originate from the DEP component in the mixture. The relaxation frequencies, ν_α, ν_β, and ν_γ, of all three relaxations of DEP in the mixture as well as ν_α and ν_γ of neat DEP are shown as a function of temperature in Fig. 172. It can be seen that ν_α is practically the same in

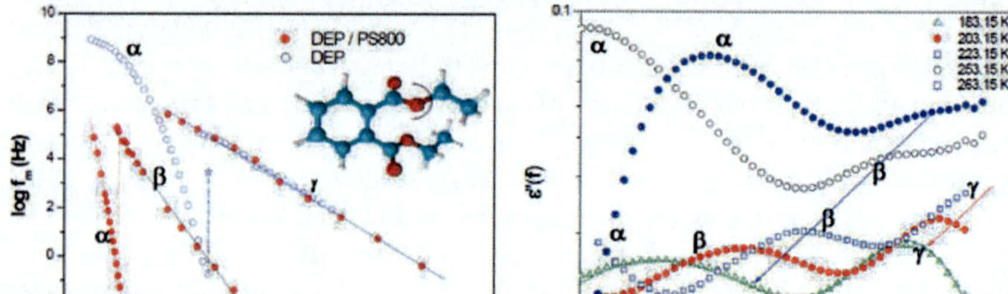

Fig. 172 (*Left*) Relaxation map of neat DEP (*open circles*) and 10 wt% of DEP in PS800 (*filled circles*). Only one secondary relaxation of DEP, namely γ, is resolved in neat DEP at ambient or at elevated pressure. This γ-relaxation originates from the rotation of the alkyl group about the flexile oxygen bond as illustrated by the picture in the *inset*. In the mixture with PS800, DEP has two resolved secondary relaxations, β and γ, as can be seen from the isothermal spectra shown in the *right panel*. The relaxation frequency of the γ-relaxation of DEP in the mixture is nearly the same as of neat DEP. The slower β-relaxation of DEP in the mixture is the JG β-relaxation that we are looking for. The *lone star* is the primitive frequency of neat DEP calculated from the α-relaxation frequency at the lowest measurement temperature and *n* obtained by fitting the α-loss peak by the Kohlrausch function. Its proximity to the γ-relaxation frequency explains why in neat DEP the β-relaxation cannot be resolved from the more intense γ-relaxation. Reproduced from [1125] by permission

the pure state and in the mixture. The new secondary relaxation found comes from motion of the DEP molecule, and since it is slower than the γ-relaxation, it is likely the candidate for the JG β-relaxation. The situation is similar to the mixtures with tri-styrene with other polar molecules including picoline, QN, and TBP discussed in Section 2.3.2.8 except no γ-relaxation is present in these rigid molecules. From the pressure dependence of relaxation time shown before in Figs. 146, 148, and 149, the secondary relaxation of picoline and QN found in the mixtures are definitely their JG β-relaxation. By inference, the slower β-relaxation of DEP found when mixed with PS800 is its JG β-relaxation. In the left panel of Fig. 172, the lone star indicates the location of the JG β-relaxation frequency ν_β of neat DEP estimated by the primitive frequency ν_0 calculated by Eq. (2.68) with $n = 0.36$ from [660] and the lowest ν_α data obtained experimentally. The separation between log ν_α and log ν_β indicated by the dashed-dotted line in neat DEP is shorter by about two decades compared with that found for DEP in the mixture for the same ν_α. There is increase of intermolecular coupling of DEP (or n) when executing the α-relaxation with the slower host PS800 molecules, and the increase of separation between log ν_α and log ν_β follows as a consequence of Eq. (2.69), which in turn confers the bonus of having the JG β-relaxation resolved in support of its universal presence.

If the found β-relaxation is truly the JG β-relaxation of DEP that we are looking for, it can be confirmed by applying pressure to the mixture of DEP with PS800 and observing it shift to lower frequencies, in contrast to the pressure insensitivity of the γ-relaxation seen before in neat DEP [660]. Confirmation is obtained by performing dielectric measurements under elevated pressures on the mixture of 10 wt% of DEP in PS800 [1125]. The loss spectra shown in the left panel of Fig. 173 confirm the pressure dependence of ν_β. The measurements confer a bonus. As shown in the right panel of Fig. 173, there is a change of pressure dependence of ν_β near the

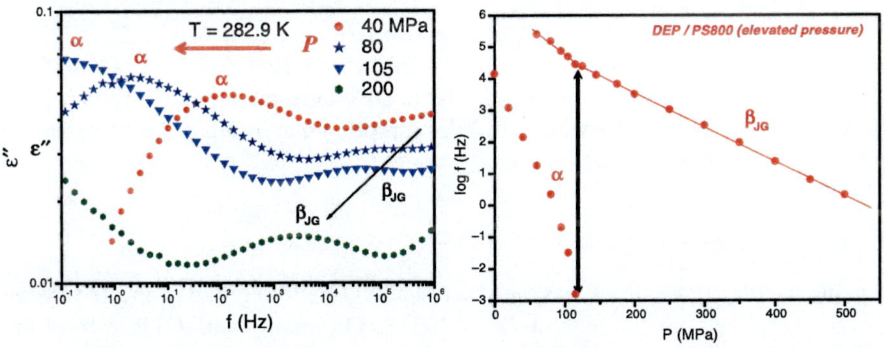

Fig. 173 (*Left*) Loss spectra of DEP component in 10 wt% DEP mixture with PS800 at four different elevated pressures and constant temperature of 282.9 K The shift of the β-relaxation of DEP to lower frequencies along with the α-relaxation is clear. (*Right*) Relaxation map of ν_α and ν_β as function of pressure at constant $T = 282.9$ K. There is a change to a stronger P-dependence of ν_β near the vitrification pressure (of the DEP component) at which ν_α corresponds to the α-relaxation time $\tau_\alpha = 100$ s as indicated by the *vertical arrow*. Reproduced from [1125] by permission

vitrification pressure P_g of the DEP component at which $\nu_\alpha(P_g)$ corresponds to the relaxation time $\tau_\alpha(P_g) = 100$ s. This is another evidence of the connection between the β-relaxation and the α-relaxation, characteristic of the JG β-relaxation.

There is yet another remarkable feature of the JG β-relaxation of DEP in the mixture that is worth pointing out by comparing its relaxation time $\nu_\beta(P_g)$ at the vitrification pressure P_g with $\nu_\beta(T_g)$ at the vitrification temperature T_g of the DEP component. Both P_g and T_g are defined by the same α-relaxation frequency, i.e., $\nu_\alpha(P_g) = \nu_\alpha(T_g)$, and both correspond to α-relaxation time $\tau_\alpha = 100$ s as indicated by the horizontal dotted lines near the bottom in both panels of Fig. 174. Guided by the vertical arrows, it can be seen by inspection that $\nu_\beta(P_g)$ is nearly the same as $\nu_\beta(T_g)$. Therefore, we have invariance of ν_β to combinations of P and T while maintaining ν_α constant, another remarkable characteristic of JG β-relaxation.

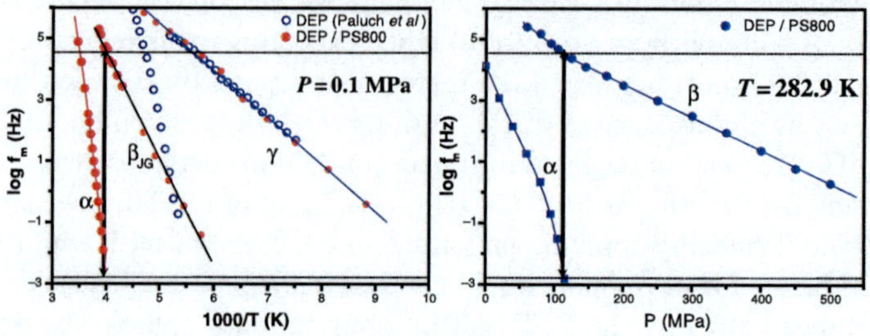

Fig. 174 (*Left*) Relaxation map of DEP in 10 wt% DEP mixture with PS800 as a function of reciprocal temperature at ambient pressure. (*Right*) As a function of pressure at constant $T = 282.9$ K. The *lower horizontal dotted lines* indicate α-relaxation time $\tau_\alpha = 100$ s. The *vertical arrows* indicate the location of either the vitrification temperature T_g or pressure P_g where $\tau_\alpha = 100$ s. The *upper horizontal lines* in both panels correspond to the same frequency. The constructions show that $\nu_\beta(P_g)$ is nearly the same as $\nu_\beta(T_g)$. Reproduced from [1125] by permission

Measurements similar to those described for DEP and 10 wt% of DEP in PS800 have been performed for DBP, DiBP, and DOP in their 10 wt% mixtures with OTP and PS800. The results obtained are similar to DEP in resolving the JG β-relaxation of these phthalates in the mixtures [1125], supporting that results are general for the phthalates, which will not be presented here.

Dimethyl Phthalate (DMP)

Among the phthalates, DMP has the shortest alkyl (methyl) group connected by the oxygen bond and has the highest $T_g = 193$ K. On mixing with OTP to form the 5% DMP/95% OTP mixture, the β-relaxation shows up in the loss data at ambient pressure (see left panel of upper level of Fig. 175). The relaxation maps of pure DMP and component DMP in the mixture are present in the right panel of Fig. 175. Dielectric measurements at elevated pressure have been made on a mixture of 90% DMP with PS2460, polystyrene with molecular mass of 2460 Dalton (lower level of Fig. 175). The β-relaxation is resolved in this DMP-rich mixture, and it shifts to

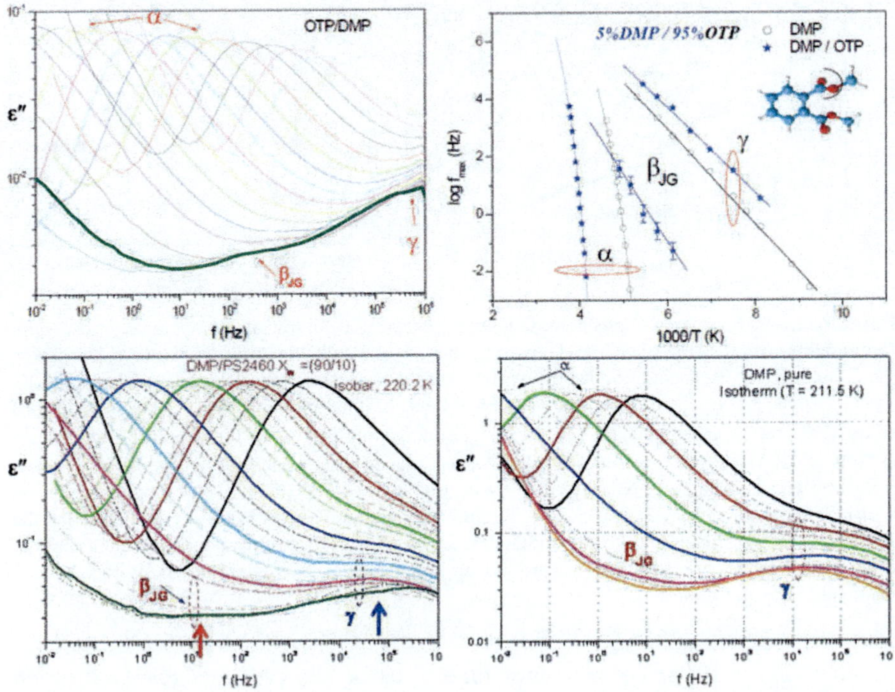

Fig. 175 *Upper level (left)* Isothermal loss spectra of DMP in 5% DMP/95% OTP mixture at ambient pressure. The slower β-relaxation is resolved at lower temperatures. (*Right*) Relaxation map of neat DMP and the DMP component in 5% DMP/95% OTP mixture. Only one secondary relaxation of neat DMP has been resolved at all temperatures. *Lower level (left)* Dielectric loss spectra of DMP at constant $T = 220.2$ K and various pressures in 90% DMP/10% PS2460, a mixture rich in DMP. PS2460 is polystyrene with molecular mass of 2460 Da. From *left to right*, the *solid lines* are data taken at $P = 0.5$, 0.62, 0.75, 112.5, 150, 190, and 300 MPa. The β-relaxation is resolved when $P = 300$ MPa (*bottom solid line*). (*Right*) Loss spectra of pure DMP at constant $T = 211.5$ K and various pressures. *From right to left*, the *solid lines* are data taken at $P = 60$, 85, 115, 150, 249.4, and 348.4 MPa. The β-relaxation is barely resolved at the highest pressure. Courtesy of M. Shahin Thayyil and S. Capaccioli

lower frequencies on increasing pressure. It is barely resolved in pure DMP at high pressure.

Benzophenone (BZP)

The dielectric loss data of BZP obtained over a wide range of temperatures and 12 decades of frequencies by Pardo et al. [1120, 1121] are reproduced in the left panel of Fig. 176. Within the wide ranges of temperature and frequency covered, only one secondary γ-relaxation is found together with the α-relaxation. The α-loss peak is narrow and near T_g it is well fitted by the Fourier transform of the Kohlrausch function with exponent $(1 - n) = 0.70$. Here T_g is defined as the temperature at

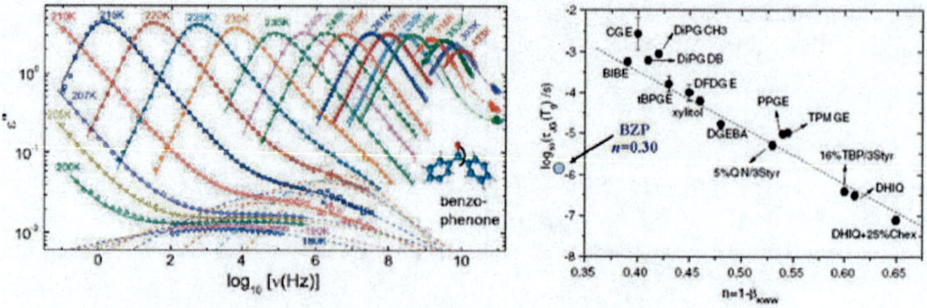

Fig. 176 (*Left*) Frequency-dependent dielectric loss of BZP for various temperatures. The *solid lines* are fits with the sum of a Cole–Davidson function and a CC function. The *dashed lines* show the CC part of the fits accounting for the observed secondary γ-relaxation. From Pardo et al. [1120, 1121]. (*Right*) Fitting the α-loss peak by the Kohlrausch function yields $n = 0.30$ near T_g defined as the temperature at which $\tau_\alpha(T_g) = 10^2$ s. The relaxation time τ_γ of the observed γ-relaxation at T_g is about 1.4×10^{-6} s. The *large filled circle* placed just outside the *y*-axis of the figure is the point with coordinates given by ($n = 0.30$, $\log\ \tau_\gamma = -5.8$). The *arrow* helps the reader to find it. This point falls way outside the relation between $\log\ \tau_{JG}(T_g)$ and n established for the JG β-relaxation of many glassformers including those shown. From this alone, it is clear that the observed secondary γ-relaxation by Pardo et al. is not the JG β-relaxation of BZP. Reproduced from [1121] by permission

which $\tau_\alpha(T_g) = 10^2$ s. The relaxation time τ_γ of the observed secondary relaxation at T_g is about 1.4×10^{-6} s. In the right panel of Fig. 176, the data from BZP are represented by the large filled circle with coordinates ($n = 0.30$, $\log\ \tau_\gamma = -5.8$) in a plot of the logarithm of the secondary relaxation time against n. It falls way outside the empirical correlation between $\log\ \tau_{JG}(T_g)$ and n established by the JG β-relaxation of many glassformers including those shown in Fig. 105. The line is the relation between the primitive relaxation time τ_0 and n. There is large deviation of $\tau_\gamma(T_g)$ and n of BZP from either that expected from the empirical correlation of $\tau_{JG}(T_g)$ with n or τ_0 with n. This indicates that the observed secondary γ-relaxation by Pardo et al. is not the JG β-relaxation of BZP.

To find the JG β-relaxation of BZP, it is dissolved in PS800 at 0.123 mole fraction. The right panel of Fig. 177 shows the spectra of BZP in the mixture at two temperatures in the glassy state. A new β-relaxation appear on the low-frequency flank of the faster γ-relaxation. The relaxation map in the left panel of Fig. 177 shows the new β-relaxation located in between the α- and the γ-relaxation, which has larger activation energy than the γ-relaxation. It is likely the JG β-relaxation of BZP. The relaxation times τ_γ of BZP in the mixture of 0.123 mole fraction of BZP in PS800 are within experimental errors not too different from that of neat BZP determined by Pardo et al. [1120, 1121] and by Walker and coworkers [1126].

The Plastic Crystals

Some liquids, usually consisting of globular or spherically symmetric molecules, can be vitrified on cooling to form plastic crystals, which are translationally ordered

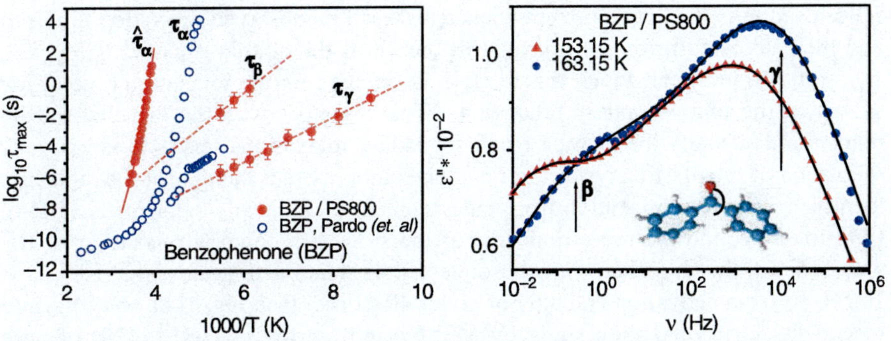

Fig. 177 (*Right*) The loss spectra of BZP in the mixture with PS800 show the presence of a slower secondary relaxation on the low-frequency flank of the γ-relaxation seen before in pure BZP. (*Left*) Relaxation map of the α- and γ-relaxations of neat BZP, and of the α-, β-, and γ-relaxations of BZP in the mixture with PS800. Courtesy of M. Shahin Thayyil and S. Capaccioli

on a lattice but are orientationally disordered. On cooling, the orientational motion slows down. When cooled past some temperature T_g the orientational relaxation time becomes too long for the structure to reach equilibrium, resulting in a orientational glassy state. Despite lacking translational degree of freedom, the dynamics of reorientation of molecules on lattice sites of plastic crystals resembles in many ways ordinary glass-forming liquids possessing both rotational and translational degrees of freedom. These include the presence of the primary α-relaxation and the secondary relaxation with typical behavior, and signature of vitrification exemplified by thermodynamic properties on crossing T_g, all with properties similar to ordinary glassformers [144, 145, 681, 947, 1016, 1048, 1127–1129]. The orientational relaxation dynamics at high temperature and short relaxation times from molecular dynamics simulations and neutron scattering performed on difluorotetrachloroethane ($CFCl_2$–$CFCl_2$) plastic crystal is like that found in ordinary small molecular glassformers [326]. The simulation used the Buckingham potential $\phi(r) = A\exp(-\rho \cdot r) - C/r^6$ for interaction between molecules. The similar properties include the self-angular correlation functions C_l decaying all the way to zero exponentially at high temperatures, a crossover from exponential to stretched exponential at t_c of about 0.6 ps is seen at some intermediate temperatures. At sufficiently lower temperatures, C_l changes from exponential decay to a plateau-like decay before the decay to zero at long time, which is well fitted by a stretched exponential.

If the JG relaxation is universal, it should be present in plastic crystals and should have similar properties and relations to the α-relaxation as regular glassformers. An earlier broadband dielectric relaxation study by Brand et al. [145] have identified and found the JG relaxation in five plastic crystals, adamantanone, ethanol, cyclo-hexanol, cyclo-octanol, and pentachloronitrobenzene (see Fig. 106). In some of these plastic crystals, more than one secondary relaxation is present. One of the criteria Brand et al. and Bauer et al. [145] used to identify which one is the true

JG relaxation is the good correspondence between its observed relaxation time τ_{JG} and the calculated primitive relaxation time τ_0 of the coupling model (CM), i.e., $\tau_{JG} \approx \tau_0$. Brand et al. found this relation is satisfied by the relaxation time of the slower of the two secondary relaxations. This, together with the fact that the JG relaxation is usually the slowest of all secondary relaxations serves to identify JG relaxation of the plastic crystals. For cyclohexanol, there is support of the dielectric identification of JG relaxation from measurement by adiabatic calorimetry [1048]. Calorimetry found the relaxation time of the slowest secondary relaxation as 10^3 s at 116 K. This together with the dielectric relaxation time at about 10^{-2} s at 150 K gives an activation enthalpy of about 40 kJ/mol (see Fig. 108a and b). More recent dielectric relaxation study by Murthy and coworkers [681, 1129] of pure plastic crystals and supercooled orientationally disordered binary solid solutions have revived the interest of identifying the JG relaxation in this type of glass-formers. In all three different solid solutions of cyclohexanol (with neopentylglycol (CHXOL–NPGOL) [681(a)], cyclohexanol–cycloheptanol (CHXOL–CHPOL), and neopentanol-neopentylglycol (NPOL–NPGOL) [681(b)] they found two sub-T_g processes, β and γ. The slower β-process has relaxation time in good agreement with the primitive relaxation time of the CM, and this led Murthy and coworkers to identify it with the JG β-process. These results are not so surprising because in the pure state one of the components already shows the presence of the JG β-process.

On the other hand, JG β-relaxation has not been resolved and found in pure plastic crystals, isocyanocyclohexane (ICNCH), cyanocyclohexane (CNCH), and 1-cyanoadamantane (CNADM). These polar plastic crystals all have the main dipole moment situated in their side group $-C \equiv N$ or $-N \equiv C$. As shown by Singh and Murthy [1129], isothermal and isochronal spectra of these plastic crystals can resolve only one secondary relaxation (let us call it the γ-relaxation from now on) accompanying the α-relaxation. For ICNCH, the dielectric loss spectra presented in the left panel of Fig. 178 show no trace of a slower secondary relaxation in between the α- and the γ-relaxation. The α-loss peak is rather narrow, and the fit by the Fourier transform of the Kohlrausch function yields coupling parameter $n = 0.30$. The relaxation time of this resolved γ-relaxation is much shorter than the primitive relaxation time calculated by Eq. (2.68) with this value of n, and therefore it is likely not the JG β-relaxation. The relaxation frequencies f_α and f_γ of the α- and the γ-relaxation of pure ICNCH are plotted as functions of reciprocal temperature and shown by open squares in Fig. 179, left panel. Although in [1129], Singh and Murthy have mistakenly identified it as the universal JG β-relaxation, they no longer hold this belief as professed in papers published afterward [681(a), (b)]. Other than this unimportant mistake, Singh and Murthy made important contribution by their dielectric and calorimetric measurements of these three polar molecules dispersed in less mobile and effectively non-polar *ortho*-terphenyl (OTP). A new secondary relaxation appears in the mixtures and it is the JG β-relaxation of ICNCH, CNCH, and CNADM ($C_{10}H_{15}CN$), which has not been seen before. The results are presented below.

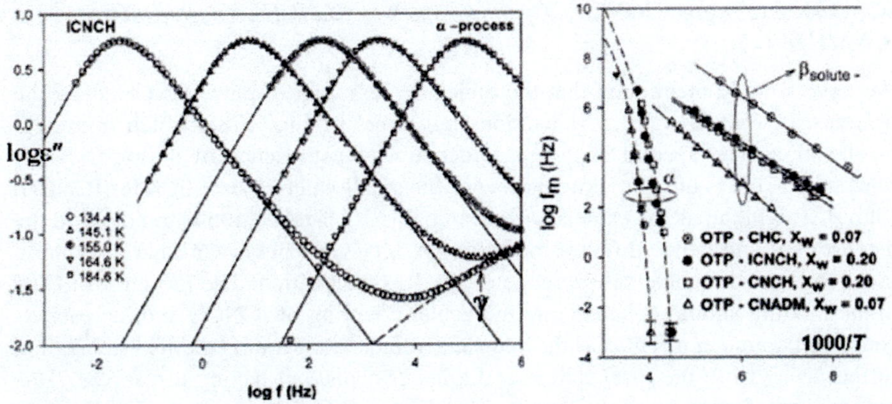

Fig. 178 (*Left*) Double logarithmic plot of ε'' vs. frequency at different temperatures showing the α- and the γ-relaxation of pure ICNCH. (*Right*) Arrhenius plot of the α- and the β-loss peak frequencies of the polar solutes, ICNCH, CNCH, CNADM, and CHC, in the solvent OTP. Also shown are the approximate low f_α values of the solutes in the mixtures corresponding to the enthalpy relaxation at T_g estimated from the DSC curves (for a heating rate of $10°\ \mathrm{min}^{-1}$). The *dashed lines* along the α-process are guide to the eye and the *straight lines* along the β-process is the Arrhenius fit. CHC has not been discussed in the text. Data from [1129] are replotted here in both figures

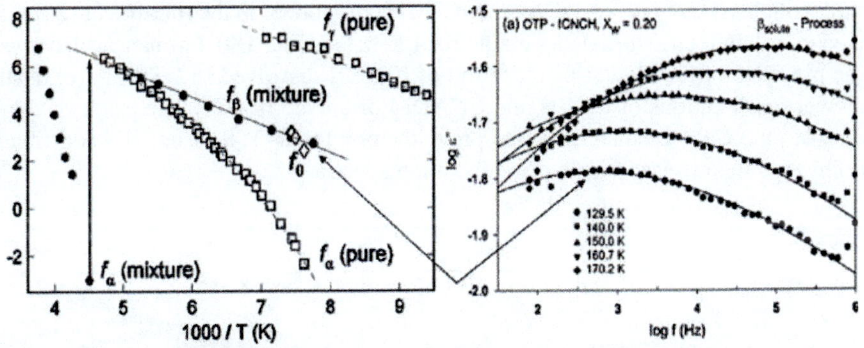

Fig. 179 (*Left*) Arrhenius plot of the logarithm of the α- and the γ-loss peak frequencies of pure ICNCH (shown by *open squares*), and the logarithm of the α- and the β-loss peak frequencies of ICNCH in the mixture of 20 wt% of ICNCH in OTP (shown by *filled circles*). Also shown is the approximate low f_α value of ICNCH in the mixture corresponding to the enthalpy relaxation at T_g estimated from the DSC curve (for a heating rate of $10°\ \mathrm{min}^{-1}$). The *vertical arrow* indicates the large separation between f_α and f_β at the DSC $T_g = 237$ K of ICNCH in the mixture. The three *open diamonds* are the primitive frequencies f_0 of pure ICNCH calculated from f_α (*open circles*) with $n = 0.30$. (*Right*) Isothermal β-loss peak at several temperatures. The two *arrows* indicate the same loss peak frequency f_β at 120.5 K in the *left and right panels*. Data from [1129] are replotted together with additional illustrations

Isocyanocyclohexane (ICNCH), Cyanocyclohexane (CNCH), 1-Cyanoadamantane (CNADM)

We have already mentioned that the dielectric loss data of pure ICNCH show the presence of only the fast γ-relaxation (left panel of Fig. 178), which is not JG β-relaxation. The spectra show no evidence of the presence of a slower JG β-relaxation. This is unsurprising in view of the small value of $n = 0.30$ for ICNCH. Like BZP, which has the same small n value, the JG β-relaxation is too close to the α-relaxation and hence difficult to be resolved. Nevertheless, work has to done in order to show the universal presence of the JG β-relaxation. The presence of OTP in the mixture should enhance intermolecular coupling of ICNCH and increase its coupling parameter n. This has the beneficial effect of separating the JG β-relaxation further away from the α-relaxation and a better chance of having it resolved. This scenario is in fact realized in 20 wt% of ICNCH in OTP. The right panel of Fig. 179 shows a new β-relaxation slower than the γ-relaxation appearing in the loss spectra of the polar ICNCH component in the mixture of 20 wt% of ICNCH in OTP. The relaxation frequencies f_α and f_β of the α- and the β-relaxation of ICNCH dissolved in OTP are shown as functions of reciprocal temperature by filled circles in the left panel of Fig. 179. They are also shown in the right panel of Fig. 178 together with similar data of CNCH and CNADM dissolved in OTP to be discussed later. There, all data were obtained from dielectric measurements except for one data point of f_α with the lowest value. It was obtained by calorimetric measurement of the mixture, which gives $T_g = 237$ K for the ICNCH component in the mixture [1129]. The isochronal dielectric loss data obtained at 1 kHz (see Fig. 180, left panel) also reveal the presence of the slower β-relaxation of ICNCH dissolved in OTP, as well as the slower β-relaxations of CNCH and CNADM dissolved in OTP. The isochronal loss data of pure OTP are also shown to prove the point that OTP essentially contributes nothing to the observed dielectric relaxation spectra.

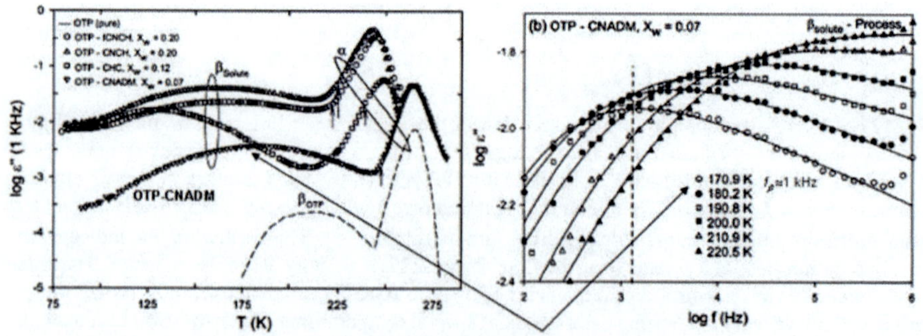

Fig. 180 (*Left*) Isochronal loss data of the solutes, ICNCH, CNCH, CNADM, and CHC, dissolved in OTP, as well as that of pure OTP. (*Right*) Isothermal β-loss spectra of 7% CNADM dissolved in OTP. The *vertical dashed line* indicates the loss peak frequency f_β at 170.9 K is slightly larger than 1 kHz. The two *arrows* indicate the isochronal β-loss peak temperature in the *left panel* is nearly the same as 170.9 K. Data from [1129] are replotted together with additional illustrations

When dissolved in OTP, f_α of ICNCH are shifted to higher temperatures, have steeper reciprocal temperature dependence, and hence large fragility index m. These changes of f_α are expected from the increase in intermolecular coupling of the guest ICNCH molecules in less mobile OTP host. The vertical arrow and its sizeable length are used to show that f_α and f_β are now well separated in the mixture, and hence the JG β-relaxation can be resolved.

In the left panel of Fig. 179, the three open diamonds are the primitive frequencies f_0 of pure ICNCH calculated from the α-relaxation frequency f_α (open circles) at the same temperatures with $n = 0.30$ obtained from the fit of the α-loss peak of pure ICNCH by the Kohlrausch function. These values of f_0 of pure ICNCH are nearly the same as the relaxation frequencies f_β of the directly observed JG β-relaxation of ICNCH dissolved in OTP (filled circles). This is perhaps no accident because in many cases, the JG β-relaxation frequency of a glassformer is not changed much when mixed with another glassformer.

There is yet a faster relaxation of ICNCH in the mixture besides the JG β-relaxation. It appears as a weak shoulder of ε'' in the isochronal spectrum at 1 kHz starting at about 100 K and extended to lower temperatures, as shown in the left panel of Fig. 180. The location in temperature of this shoulder of $\varepsilon''(f = 1\ \text{kHz})$ for ICNCH in the mixture is close to that of the isochronal γ-loss peak found in $\varepsilon''(f = 1\ \text{kHz})$ for pure ICNCH (see Fig. 1b in [1129]). The weaker intensity of the former compared with the latter is understandable because there is only 20 wt% of ICNCH in the mixture. This observation means that the faster γ-relaxation of pure ICNCH survives in the mixture with OTP, and its relaxation time is practically unchanged. Although not shown in the relaxation map (Fig. 179, left panel), it coexists with the slower JG β-relaxation of ICNCH in the mixture.

We have given in the previous paragraphs detailed discussion, analysis, and interpretation of the data of pure ICNCH and ICNCH dissolved in OTP. The plastic crystal CNCH is closely related to ICNCH in molecular structure, and one might expect similar behavior in relaxation dynamics. In fact this is the case for both pure CNCH and CNCH dissolved in OTP (see Fig. 178, right panel, and Fig. 180 left panel). The data of CNCH are isomorphic to that of ICNCH, and all discussions of ICNCH in the previous subsection apply to CNCH here. The isochronal spectrum of CNCH in mixture with OTP at 1 kHz exhibits a shoulder starting at below 100 K (see Fig. 7 in [1129]). Just like ICNCH, this indicates that the faster and resolved secondary relaxation of pure CNCH is still present when mixed with OTP, and it has about the same frequency.

The results of CNADM from the study of Singh and Murthy are also similar to ICNCH and CNCH. The resolved JG β-relaxations of 7% CNADM dissolved in OTP are shown by isothermal and isochronal data in the right and left panels of Fig. 180. The two arrows indicate good agreement of the JG β-relaxation frequency at 170.9 K and its temperature at 1 kHz.

Breakdown of Correlation Between Kinetic Fragility and Thermodynamic Fragility

Singh and Murthy calculated the thermodynamic fragility index m_{th} of pure CNCH, ICNCH, and CNADM from their own DSC data using the formula given by Wang and Angell [573]. They found m_{th} = 49.5, 40.1, and 20.7 for CNCH, ICNCH, and CNADM, respectively. When compared with the kinetic fragility index m of CNCH, ICNCH, and CNADM, the trend in kinetic fragility index is not the same as that of the thermodynamic fragility index, and the trend is even reversed. Here is another example of the thermodynamic fragility not correlating with the kinetic fragility. This is no surprise because not only thermodynamics but also many-body relaxation dynamics determined τ_α and hence the kinetic fragility index m.

Ethanol

Ethanol can be prepared in either a supercooled liquid or a plastic phase as shown first in the heat capacity measurements [1130]. Dielectric measurements in structurally disordered ethanol [621, 1131–1133] were compared to that of the plastic crystalline phase of ethanol by several groups [145, 612, 1134, 1135]. Brand et al. [145] obtained spectra of ethanol in its supercooled plastic phase. At higher temperatures the α-relaxation is significantly slower in the plastic phase than the supercooled liquid, but approaches values close to each other at lower temperatures to show identical glass temperatures of 99 K for both disordered phases. Since the T-variation of τ_α is weaker in the plastic crystalline phase than in the supercooled liquid, the fragility index m of the former is smaller than the latter. However, the α-peak is significantly broader for the plastic crystalline state than the supercooled liquid at lower temperatures near T_g, as shown in Fig. 181 at 96 K where the loss peaks of the two phases having the same peak frequency, f_α, are fitted by the Cole–Davidson functions.

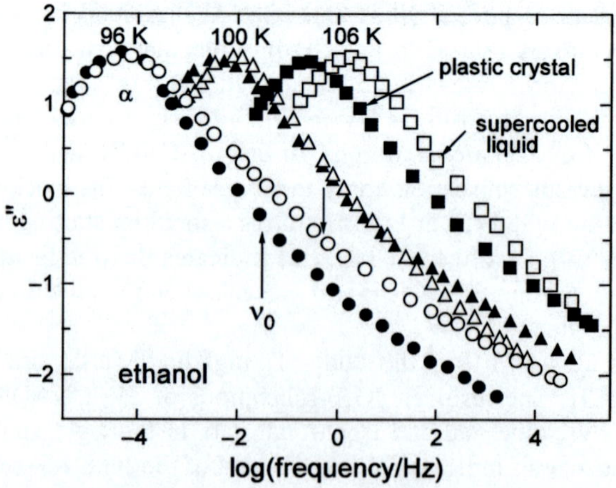

Fig. 181 Comparison of the dielectric loss spectra of structurally (*open symbols*) and orientationally (*closed symbols*) disordered ethanol. The *solid lines* are fits with the sum of a Cole–Davidson (for α-relaxation) and a Cole–Cole (for β-relaxation) function. For 96 K, the two functions used in the fit are shown by *dashed and dotted lines*. The *arrows* indicate the primitive frequencies calculated for the two cases using their Kohlrausch exponents (fits not shown). Data from [145] are replotted

Hence, the correlation between fragility index m and non-exponentiality index n breaks down even for the same compound but in different phases, indicating once more that fragility is dependent on both thermodynamics and many-body relaxation. Figure 181 also shows the JG β-relaxations of the two phases obtained by Brand et al. by additive superposition of the Cole–Davidson fit to the α-relaxation and a Cole–Cole function for the JG relaxation with peak frequency f_{JG}. Also shown is the primitive frequency f_0 calculated by Eq. (2.68) with the n values obtained from the Kohlrausch fits to the α-loss peaks (not shown). By inspection of the locations of f_{JG} or f_0 in Fig. 181, it can be seen that the correlation of $\log(f_{JG}/f_\alpha)$ with n, or $\log(f_0/f_\alpha)$ with n, continues to hold. This is because the relation between f_{JG} and f_α is determined solely by many-body relaxation exemplified by n.

2.3.2.17 Change of T-Dependence of Relaxation Strength $\Delta\varepsilon_{JG}$ on Crossing T_g

The β-relaxations in 12.1 and 16.6 mol% chlorobenzene–decalin mixture [945, 971, 1136], 71%/29% mixture of pyridine and toluene [1138], polybutadiene, toluene [971], 5-methyl-2-hexanol [1137], and D-sorbitol [1031, 1139] are all of the JG kind. The relaxation strength, $\Delta\varepsilon_{JG}$, of the JG relaxation in all these glassformers is found to change on heating through the glass transition temperature in a similar manner as the changes observed in the enthalpy H, entropy S, and volume V. Results from studies on 16.6 mol% chlorobenzene–decalin mixture and D-sorbitol are shown in Fig. 182, and polybutadiene and toluene in Fig. 183 (left panel). The derivative of $\Delta\varepsilon_{JG}$ with respect to temperature, $d\Delta\varepsilon_{JG}/dT$, is positive and increases from relatively low values below T_g to higher values above T_g. This is the same behavior observed for the specific heat C_p and the thermal expansion coefficient,

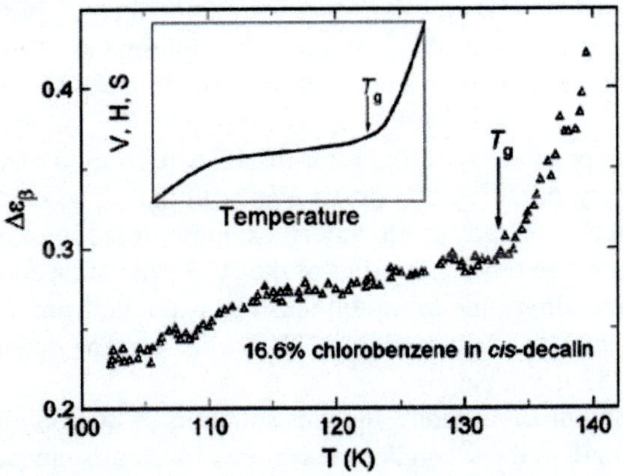

Fig. 182 Plot of $\Delta\varepsilon_{JG}$ against T of 16.6 mol% chlorobenzene–decalin mixture. Note the change in slope when crossing T_g. The decrease in the slope in the 115–125 K range is due to spontaneous structural relaxation in the glassy state. *Inset* is a schematic plot of H, S, and V of a glass heated through its T_g. Data from [1136] are digitized and replotted. Similar result for D-sorbitol [1139] is not shown here

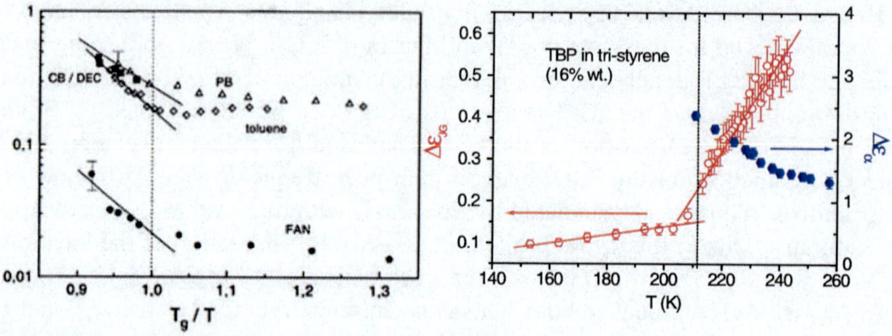

Fig. 183 (*Left*) The relative relaxation strength of the JG β-process plotted against T_g/T : 12.1% chlorobenzene in *cis*-decalin, as measured by Johari (CB/DEC, *full squares*) together with polybutadiene (PB, *open triangles*), and toluene (*open diamonds*). Shown also is relative relaxation strength of the resolved secondary relaxation of *m*-fluoroaniline (FAN, *full circles*), which originates from the hydrogen bondings and is not the JG β-process. *Solid lines* are drawn to guide the eye to note the strong increase above T_g. Reproduced from [971] by permission. (*Right*) Dielectric strengths $\Delta\varepsilon_{JG}$ and $\Delta\varepsilon_\alpha$ of the mixture of 16 wt% *tert*-butylpyridine (TBP) in tri-styrene. *Closed symbols* are for $\Delta\varepsilon_\alpha$ of the α-relaxation (*right axis*) and open symbols for $\Delta\varepsilon_\alpha$ of the JG β-relaxation (*left axis*). *Dotted lines* are linear fits of $\Delta\varepsilon_\alpha$ below and above T_g. *Dashed vertical line* indicates $T = 211.5$ K [where $\log(\tau_\alpha/\mathrm{s}) = 3$] and is near the temperature at which occurs the crossover of temperature dependence of $\Delta\varepsilon_{JG}$ with an elbow shape. Reproduced from [1093] by permission

which are proportional to the derivatives dH/dT and dV/dT, respectively. The rotation angle and/or translational displacement for the motions underlying the JG relaxation, and hence $\Delta\varepsilon_{JG}$, likely depends on the specific volume and the entropy. Thus, it is expected that the rate of change of $\Delta\varepsilon_{JG}$ with temperature should be similar to that of these thermodynamic quantities. For the same reason, the angle of rotation or $\Delta\varepsilon_{JG}$ is expected to depend on the density, entropy, and thermal history of the glass, a denser glass having a smaller $\Delta\varepsilon_{JG}$ as found by experiments. These general properties of $\Delta\varepsilon_{JG}$ indicate once more a connection between the JG β-relaxation and the α-relaxation.

The same behavior of $\Delta\varepsilon_{JG}$ is found for the JG β-relaxation of the more mobile component of binary mixtures of van der Waals liquids. The case of 16.6 mol% chlorobenzene–decalin mixture is already an example. It is found in the other mixtures amply discussed on other properties of the JG β-relaxation such as quinaldine, picoline, and *tert*-butylpyridine in tri-styrene. The example from measurements on the mixture of 16 wt% *tert*-butylpyridine (TBP) in tri-styrene taken from [1093] is presented in Fig. 183 (right panel).

Water is one component of many aqueous solutions of hydrophilic glassformers. The dielectric strength of the secondary relaxation of water in aqueous mixtures also shows change of T-dependence on crossing T_g of the mixture. The data from studies on mixtures of 35 wt% of water with oligomers of ethylene glycol, 3EG, 5EG, and PEG600, and mixtures of water with fructose with wt% of fructose ranging from 70.0 to 94.6 are shown in Figs. 184 and 185, respectively.

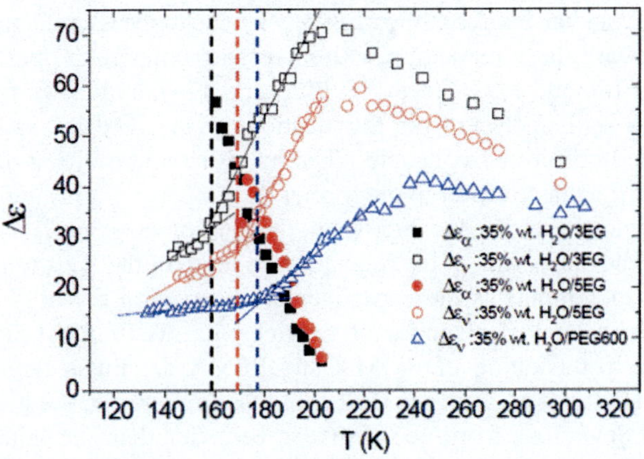

Fig. 184 Dielectric strengths $\Delta\varepsilon_{JG}$ and $\Delta\varepsilon_\alpha$ of the mixture of 35 wt% of water with 3EG, 5EG, and PEG600. *Full symbols* are for $\Delta\varepsilon_\alpha$ of the α-relaxation and *open symbols* for $\Delta\varepsilon_{JG}$ of the JG β-relaxation of water in the mixtures. *Full lines* are linear fits of $\Delta\varepsilon_{JG}$ below and above T_g. Each *dashed vertical line* indicates the temperature T_g near which the elbow-shape crossover of temperature dependence of $\Delta\varepsilon_{JG}$ occurs. Here T_g is defined as the temperature at which $\log[\tau_\alpha(T_g)/\mathrm{s}] = 3$. Reproduced from [1093] by permission

Fig. 185 Plots of the relaxation strength of the α-process (*closed symbols*) and JG β-process (*open symbols*) against the reciprocal temperature scaled by $1/T_g$ of the fructose–water mixtures with various fructose concentrations of 94.6, *circles*; 88.2, *squares*; 86.5, *diamonds*; 77.8, *triangles*; 74.0, *inverted triangles*; and 70.0 wt%, *left-directed triangles*. Reproduced from [1140] by permission

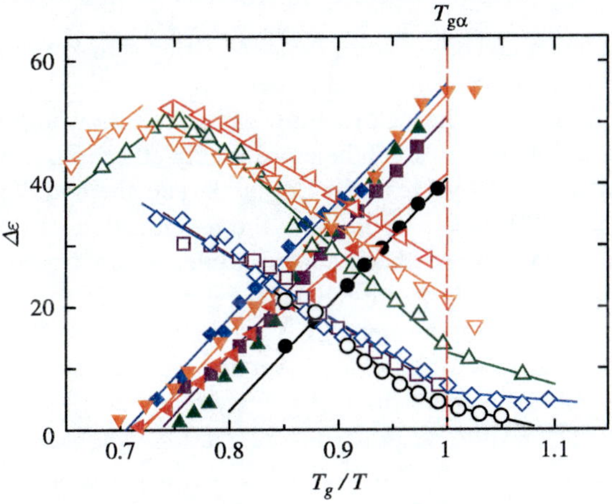

It should be noted that similar changes are observed also for some secondary relaxations that are not JG β-relaxations. The secondary relaxation of *m*-fluoroaniline shown in Fig. 183 (left panel) is not the JG β-relaxation as demonstrated in [671], and yet its dielectric strength changes its *T*-dependence on crossing T_g. Other examples include the γ-relaxation in the epoxies, DGEBA and PPGE [242, 842], and the γ-relaxation in tetramethyl BPA-polycarbonate (TMPC) and 50% TMPC/50% PS blends [1141]. These secondary γ-relaxations are also sensitive to the change in the dependence of density on temperature when crossing T_g. For

DGEBA, PPGE, and *m*-fluoroaniline; this may be due to the encroachment of the JG β-relaxation toward the γ-relaxation with increasing temperature past T_g discussed before for DGEBA and PPGE (see Figs. 123 and 124), and for *m*-fluoroaniline it can be seen in Fig. 1 of [671] from the locations of the JG β-relaxation suggested by the calculated primitive frequencies. The hybridization of the γ-relaxation with the JG β-relaxation may transfer to the former the property of the latter, which is the change of *T*-dependence of relaxation strength $\Delta\varepsilon_{JG}$ on crossing T_g.

In line with the discussion of the relaxation strength of the JG β-relaxation in this section, worth mentioning is the experimental observation that it also determines the non-linear mechanical response of polymer glasses [938(b)]. The relaxation strengths of JG β-relaxation of PMMA and PEMA are much larger than BPA-polycarbonate (PC) and polysulfone (PSF), and the former group of glassy polymers exhibits larger deviations from neo-Hookean behavior than the latter group. The larger relaxation strength of the JG β-relaxation of PMMA and PEMA originates from motion of the side group together with rotation of the main chain. In contrast, the PC and PSF have no side groups, and their JG β-relaxations only involve motion of the backbone.

2.3.2.18 Correlation of JG β-Relaxation with α-Relaxation: Evidence from Spin-Lattice Relaxation Weighted Stimulated-Echo Spectroscopy

Böhmer and coworkers [998, 999] using spin-lattice relaxation weighted stimulated-echo spectroscopy found evidence for a correlation of the α- and the JG β-relaxation times above the calorimetric glass transition temperature of *ortho*-terphenyl, D-sorbitol, and cresolphthalein-dimethylether (CDE or KDE when kresol in German is replaced by cresol in English). The methodology was justified by Geil et al. in [1142]. It was found that the α-relaxation can be modified by suppressing the contributions of those subensembles characterized by relatively slow secondary relaxation, which are the JG β-relaxation in these glassformers. For CDE, its JG β-relaxation is located at the excess wing seen by dielectric spectroscopy. An earlier deuteron NMR experiment also gave indication of a possible correlation of the α- and JG β-relaxations of polystyrene [226(b)].

The results of spin-lattice relaxation weighted stimulated-echo spectroscopy are important additional evidence for the existence of correlation of the α- and JG β-relaxations. Nevertheless, it was pointed out by Nowaczyk et al. [999] that the experiment only demonstrates the existence of the correlations but does not address the question of whether one process triggers the other. On the other hand, according to the coupling model (CM), the primitive relaxation or the faster JG β-relaxation is the precursor that "triggers" the α-relaxation.

Suppressing or delaying the JG β-relaxation or the primitive relaxation will certainly affect the α-relaxation, by retarding it further. JG β-relaxation at the interface of ultrathin film of PMMA supported on silica is suppressed by hydrogen bonding of the ester group on the side chain of PMMA and the hydroxyl groups on the silica surface [1143]. The suppression of JG relaxation of PMMA results in increase of the α-relaxation time and the corresponding increase of T_g was observed. A detailed description of this effect is given in Section 2.3.2.21.

2.3.2.19 JG β-Relaxation in the Glassy State, like the α-Relaxation, Is Sensitive to Thermodynamic (*T*,*P*) Path, Thermal History, and Annealing

Sensitivity to (*T*,*P*) Path

Glasses with different density and entropy can be obtained from the same glass-former by different thermodynamic paths involving pressure and temperature. Starting at some temperature T_i above T_g at ambient pressure, one path (A) is first elevating the pressure to P_f till the liquid is transformed to a glass and then cooling the glass down to a final temperature T_f. A different path (B) is to cool the liquid under ambient pressure from T_i to T_f at which the liquid is transformed to a glass, and then elevate the pressure up to P_f. Thus, the initial and final temperatures and pressures are the same, but the two glasses have different densities and entropies. Glass formed by path A is denser than that by path B because pressure has a larger effect on the liquid than the glass. The spectra of the resolved secondary relaxation measured for the two paths are shown in Fig. 186 for four glasses, poly(phenylglycidylether) (PPGE), PDE, BMPC, and PPG400. Only in the case of PPGE, the secondary relaxation is the JG β-relaxation. The interesting observation on poly(phenylglycidylether) (PPGE) [1042] is that the JG relaxation frequency ν_{JG} in Hz is lower by a factor of about 5, and the dielectric strength of the JG relaxation is *decreased* by about 10% in the denser glass obtained by path A than by

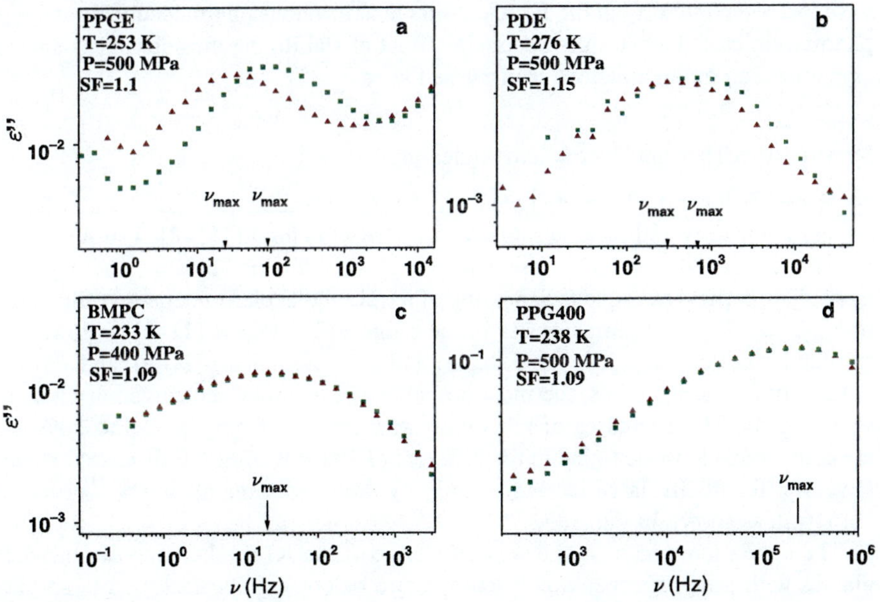

Fig. 186 Loss spectra of the secondary relaxation measured after vitrification along path A (*triangles*) and B (*squares*) for (**a**) PPGE, (**b**) PDE, (**c**) BMPC, and (**d**) PPG400. The spectra measured after vitrification along path A are vertically shifted by shift factor (SF) to obtain the same value of maxima loss of those measured after vitrification along path B. Reproduced from [1042] by permission

path B. The resolved non-JG secondary relaxations in phenolphthalein (PDE) show dependence of its relaxation time on the thermodynamic path, but the change is significantly smaller than the JG β-relaxation in PPGE [1146] (see Fig. 185b). The non-JG secondary relaxation in BMPC and PPG400 is the same independent of the path (see Fig. 185c and d)

A similar observation as PPGE is made on the JG β-relaxation of *bis*-5-hydroxypentylphthalate (BHPP) [1144]. The observed change of τ_{JG} and strength on varying the (*T*,*P*) path to vitrification indicates that the JG relaxation depends sensitively on the structure of the final glassy state, just like τ_α, although the latter is usually inaccessible deep in the glassy state by experiment.

Another example is cooling the polymer 1,2-polybutadiene (1,2-PBD) at high pressure of 350 MPa, followed by release of the pressure below T_g, which produces a polymer glass expected to be denser than that obtained by cooling under ambient pressure to the same temperature of 242.8 K. The ν_{JG} (τ_{JG}) of the pressure-densified 1,2-PBD glass is lower (longer) by about a factor of 3. However, the dielectric strength of the JG relaxation is *increased* by about 10% [1145(a)], in opposite direction to that found in PPGE. In diisobutyl phthalate (DiBP) the γ-process is almost totally pressure independent and is not the JG process, but it reveals a non-negligible dependence on the thermodynamic path [1147]. Hence, some non-JG secondary relaxations can have some property similar to JG processes, but not all the properties.

The technique of applying high pressure on cooling to vitrify a glass can be used to suppress the mobility of the JG relaxation in amorphous pharmaceuticals and bio-pharmaceuticals. This has the beneficial effect of stabilizing the substances against crystallization or degradation, as discussed later.

Sensitivity to Thermal History and Annealing

Much earlier Johari had already noted that the JG relaxation in the glass was affected by thermal history and cooling rate used to vitrify the liquid [1148]. The JG dielectric strength $\Delta\varepsilon_{JG}$ generally is less in denser glasses obtained by slow cooling or by aging. The earliest example is annealing of a rigid molecular toluene–pyridine mixture ($T_g = 127$ K) at temperatures in the range 112–123.4 K [1148], followed by similar studies on poly(propylene oxide) and poly(carbonate) [1149], and PMMA [1150]. In all these studies, the most probable τ_{JG} is either unchanged or changed very slightly. The decreases of relaxation strength vary from about 6 to 10% and are comparable to those found in the different (*T*,*P*) paths of PPGE discussed above. However, for PPGE, large increase of τ_{JG} by more rapid quenching was observed [1151(a)], as shown in Fig. 187.

There are more recent studies of the change of the JG β-relaxation of quenched glasses with time on annealing at temperature below T_g. One study on D-sorbitol glass by Power et al. [1152] is by dielectric relaxation measurements in real time during annealing at 221.1 K, which is 47 K below its T_g of 268 K. As the glass structurally relaxes during annealing up to 16 ks, the relaxation strength $\Delta\varepsilon_{JG}$ decreases from 4.1 to 3.6, the relaxation peak initially at 48 Hz shifts to a higher frequency

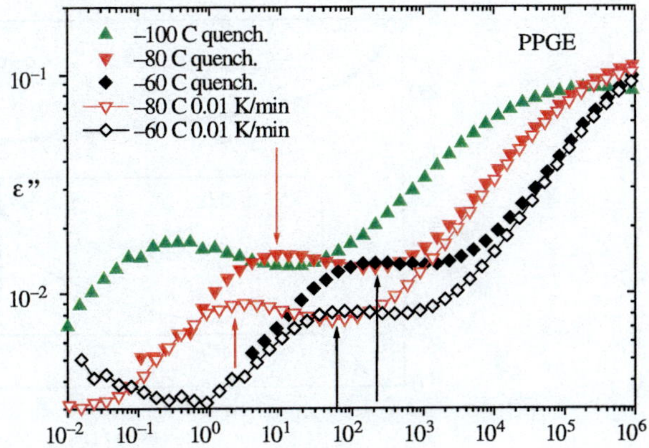

Fig. 187 Sensitive dependence of τ_{JG} in PPGE on cooling rate. The *arrows* indicate τ_{JG} of PPGE observed under the different conditions. Reproduced from [1151(a)] by permission

up to 70 Hz (corresponding to a decrease in τ_{JG}), and the relaxation spectra become narrower as indicated by the increase from 0.252 to 0.254 of the exponent α_{JG} of the Cole–Cole function that measures the breadth of the JG relaxation. Although the changes are systematic, they are small and hence interpretation can be tricky. Nevertheless, Power et al. proposed that shift of the relaxation peak to a higher frequency and narrowing of the relaxation spectra occur when local, loosely packed regions of molecules in the glass structure collapse non-uniformly and the relaxation time of some of the molecules in the collapsed state becomes too long to contribute to the JG-relaxation spectra. Consequently, the half-width of the spectra decreases, and the relaxation peak shifts to a higher frequency. Similar to sorbitol, slight decrease of dielectric strength and 12% decrease in τ_{JG} were also found after aging 1,2-PBD (T_g = 272.5 K) for 163 hours at 240.9 K by Casalini and Roland [1145(a)]. It is worth pointing out that on annealing glasses, the decrease in τ_{JG} of sorbitol and 1,2-PBD is opposite to the increase in τ_{JG} of PMMA (T_g = 383 K) at 247 K. Thus, although the JG β-relaxation is sensitive to thermal history and annealing, the changes of a denser glass depend on the glassformer. Another example is the aforementioned *increase* of the $\Delta\varepsilon_{JG}$ of the denser 1,2-PBD glass obtained by the pressure-densification path compared to the glass vitrified at atmospheric pressure by cooling [1145(a)]. Moura Ramos and coworkers also found the slowest secondary relaxation of 4,4′-methylene *bis*(*N*,*N*-diglycidylaniline) and of ethyl cellulose changes significantly with aging [1151(b), (c)], and identified them as the JG β-relaxation.

Detection of JG β-Relaxation by Rapid and Deep Quenching

OTP

From the above, the fact that rapidly quenched glass has larger $\Delta\varepsilon_{JG}$ can be used advantageously to detect JG β-relaxation in glassformers in experiment, which otherwise is not easy or possible. Such a case is dielectric relaxation measurement of

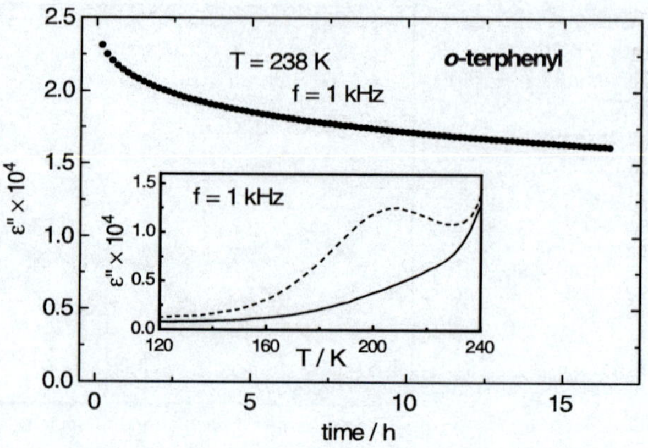

Fig. 188 Isothermal annealing experiment for the initially quenched *o*-terphenyl glass at $T = 238$ K. The annealing effect is monitored in terms of the time-dependent dielectric loss $\varepsilon''(t)$ measured at a fixed frequency of $f = 1$ kHz. The *inset* shows $\varepsilon''(T)$ for $f = 1$ kHz for quenched *o*-terphenyl (*dashed line*) and after annealing at $T = 234$ K for 24 h (*solid line*), which has suppressed the JG loss peak around $T = 205$ K. Reproduced from [829] by permission

o-terphenyl (OTP) that has small dipole moment. The JG β-relaxation was detected in the quenched glassy state of *o*-terphenyl ($T_g = 246$ K) over a wide temperature range from T_g down to 180 K by Wagner and Richert [829]. On annealing the quenched glass, the JG dielectric strength $\Delta\varepsilon_{JG}$ was observed to decay as shown in Fig. 188, and as result the isochronal JG loss peak at 1 kHz of the quenched glass disappears (see inset).

Indomethacin

The same technique was used by Carpentier et al. in their study of the pharmaceutical, indomethacin ($T_g = 316$ K) [1000(b)]. After a rapid and deep quench below T_g, they were able to resolve a slower JG β-relaxation as well as a faster γ-relaxation in the isochronal loss spectra (see Fig. 189, left panel). The JG β-relaxation appears as a shoulder in the loss spectra in the liquid state as suggested by the calculated primitive frequency ν_0 (see Fig. 189, right panel). The γ-relaxation is not the JG relaxation because its relaxation time at T_g is 2–3 orders of magnitude shorter than the calculated primitive relaxation It is also present in a crystalline phase of indomethacin and is attributed to rotation of the chlorobenzyl end group.

The JG β-relaxation of amorphous indomethacin was also detected using differential scanning calorimetry (DSC) by Vyazovkin and Dranca. It appears as a small endothermic feature when reheating quenched sample after annealing in the temperature region from 253 to 278 K [992(b)]. The same was applied to poly(vinylpyrrolidone) (PVP) and ursodeoxycholic acid (UDA), and the JG β-relaxation of these pharmaceuticals is found in the glassy state. This method, enthalpy relaxation by differential scanning calorimetry (DSC) scan of the rapidly

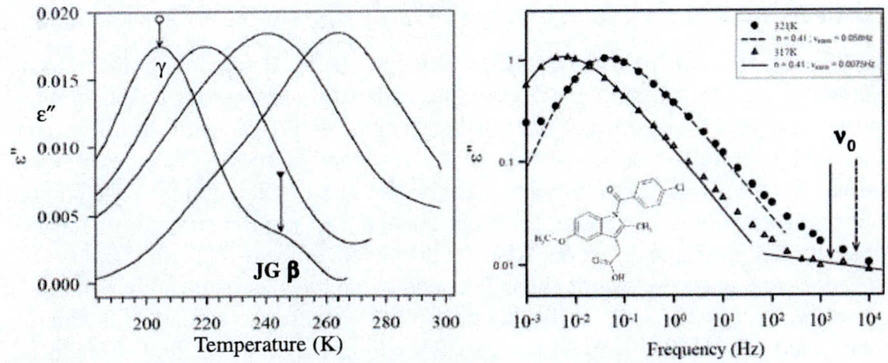

Fig. 189 (*Left*) Isochronal dielectric loss of glassy indomethacin after a rapid and deep quench below T_g measured while scanning temperature upon heating at the rate of 1 K/min. Data are shown for frequencies 10, 100 Hz and 1 KHz and 10 KHz. At 10 Hz, the *vertical arrow with circle on its tip* points to the temperature of the maximum loss of the fast γ-relaxation of intramolecular in origin, while the other arrow points to the location of the resolved Johari–Goldstein process. (*Right*) Dielectric loss of indomethacin at 317 and 321 K. The *lines* are KWW fits to the α-relaxation peaks with $n = 0.41$. The *arrows* indicate the locations of the primitive relaxation frequencies calculated from the CM equation. Reproduced from [1000(b)] by permission

quenched glass after annealing at temperature well below T_g, of detecting JG β-relaxation has been discussed before in Section 2.3.2.1.

Dipropyleneglycol Dibenzoate

In a detailed study of glass of dipropyleneglycol dibenzoate ($T_g = 220$ K) obtained by rapid cooling from 300 K, on aging the glass at $T_a = 173.15$ K, the dielectric strength decreases by about 10% as usual [1014]. But τ_{JG} shows an *increase* by more than a factor of 10 after 7272 s, not seen in aging of the other glasses. This will be further discussed in the next section.

Cross-Linking 1,2-PBD and DGEBA

Casalini and Roland [1145(b)] also studied the effect of chemical cross-linking of 1,2-PBD on the JG relaxation. They compare the imaginary part of the permittivity at atmospheric pressure and $T = 224$ K for the neat sample and cross-linked sample by cooling both from the equilibrium state at the same rate (2 K/min). They found that ν_{JG} (τ_{JG}) of the cross-linked sample is about a factor of 3 times higher (shorter) than the neat sample. In other words, cross-linking makes the JG relaxation significantly faster than for the linear 1,2-PBD. This trend found in cross-linking 1,2-PBD is opposite to that found in cross-linking or polymerizing DGEBA, where the JG relaxation is found to become slower (see Figs. 1, 3, 4, and 9 in [1051] and references therein, and also Fig. 5 in [1153]).

Possible Factors

Glasses obtained with different thermodynamic paths and cooling rates have different degrees of structural heterogeneities and distributions of local density fluctuations (as seen by small and intermediate angle X-ray scattering by Song and Roe [1154]), as well as different average density ρ and entropy S. Thus, more than one parameter are needed to fully characterize the glassy state, and their difference from one glassy state to another may not change τ_{JG} and the strength of the JG β-relaxation in the same direction. Also the dispersion of the α-relaxation and the α-relaxation time may or may not change depending on the glass state, although there is no spectroscopy that can verify this possibility because the α-relaxation cannot be observed. These may explain the variations in changes of τ_{JG} and strength of the JG relaxation in glasses having different thermal history, or vitrified by different thermodynamic and chemical paths.

The lack of correlation of the changes with density change may lead one to jump to the conclusion that the dynamics of the JG relaxation bears no relation to the density or entropy. This conclusion must be tempered by the fact that this occurs only in the glassy state and the variations causing the lack of correlation are small. Lest one forgets, a considerable amount of evidences in the equilibrium liquid state and across T_g given in many subsections have shown that the dynamics of the JG relaxation like the α-relaxation is dependent on density ρ and entropy S. Therefore, one should not be distracted from this important property of the JG relaxation in the equilibrium liquid state by the minor complications encountered in the glassy state. Let us explore possible explanations. Denoting all the parameters that characterize the glassy state collectively by λ, the dependence of τ_{JG} on λ are related to that of τ_α by the CM equations (2.67)–(2.69):

$$\tau_{JG}(\lambda) \approx \tau_0(\lambda) = [\tau_\alpha(\lambda)]^{1-n(\lambda)} t_c^{n(\lambda)} \text{ or } \tau_\alpha(\lambda) \approx [t_c^{-n(\lambda)} \tau_{JG}(\lambda)]^{1/(1-n(\lambda))}. \quad (2.81)$$

Here we have taken into consideration possible dependence of n on λ. If indeed n is not the same for different glasses, it can affect the corresponding observed changes of τ_{JG}. Since $\tau_\alpha(\lambda)$ and $\tau_\alpha(\lambda')$ are unknown, the possibility that $n(\lambda)$ and $n(\lambda')$ of glassy states λ and λ' with different thermal histories are different cannot be ruled out, and Eq. (2.81) is satisfied without any significant difference between $\tau_{JG}(\lambda)$ and $\tau_{JG}(\lambda')$, or change in $\tau_{JG}(\lambda)$ in the simpler case of physical aging.

An analogy of this scenario can be given by the comparison of the pressure dependences of τ_{JG} and τ_α for sorbitol at constant temperature [659(b)] as shown in Fig. 190. The dielectric α-loss peak of sorbitol was observed to broaden on elevating the pressure and temperature, indicating increase of intermolecular coupling or n (see left panel of Fig. 190). The increase is due to the breaking of hydrogen bonds in sorbitol at higher temperature associated with elevated pressure, as in the case of *m*-fluoroaniline [671]. This observed increase of $n(P)$ with increasing P at constant T results in a weak P-dependence of τ_{JG} (see right panel of Fig. 190), a result which

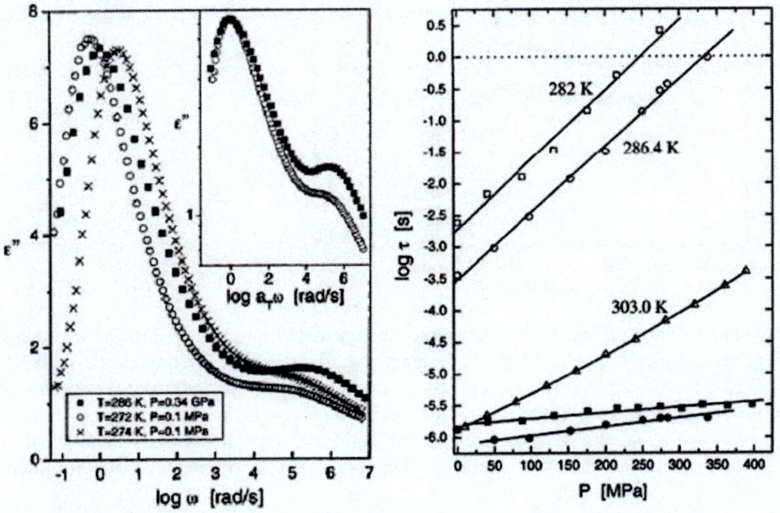

Fig. 190 (*Left*) The dielectric loss of sorbitol (■) at elevated pressure (340 MPa) at 286.4 K and at atmospheric pressure at (O) 272 and (×) 274 K. The *right panel* shows the superposition obtained by shifting the 272 K spectrum by 0.8 decades, and broadening of the loss peak at the elevated pressure/temperature. The conductivity contribution, assumed proportional to $1/\omega$, has been subtracted from the spectra. (*Right*) The structural (*open symbols*) and secondary (*filled symbols*) relaxation times for sorbitol as a function of pressure at several temperatures: (*squares*) 282.0; (*circles*) 286.4, and (*triangles*) 303.0 K. Data from [659(b)] are replotted in all figures here

follows from the equation, $\tau_{JG}(P) \approx \tau_0(P) = [\tau_\alpha(P)]^{1-n(P)} t_c^{n(P)}$. In fact, an approximately linear increase of $n(P)$ with increasing P can quantitatively relate the two experimentally observed relaxation times, $\tau_{JG}(P)$ and $\tau_\alpha(P)$, throughout the range of pressure.

2.3.2.20 Increase of τ_β on Aging in Some Glassformers

The glassy state is thermodynamically a non-equilibrium state of frozen-in disorder. Left at a temperature T_e below T_g, the glass spontaneously densifies with the passage of time at a rate that is higher, the higher is T_e. This occurs by irreversible structural relaxation toward equilibrium liquid state. In this process called aging, all thermodynamic and kinetic properties of the glass change with time. The change of the disordered structure by this spontaneous process is described as a decrease of the fictive temperature, T_f, of the glass [782, 266]. Here we are interested in the change of the JG β-relaxation with aging time t_e. On aging a glass of dipropyleneglycol dibenzoate (DPGDB) ($T_g = 220$ K) formed by rapid quenching from the liquid state at 300 K to the glass state at 173.15 K, it was found [678] that τ_{JG} increases with aging time, t_a, as shown in Fig. 191 (see also Fig. 117 in Section 2.3.2.2). Similar results were obtained for several other glassformers [1151]. These results indicate that τ_{JG} is sensitive to the structure of the glass characterized by the

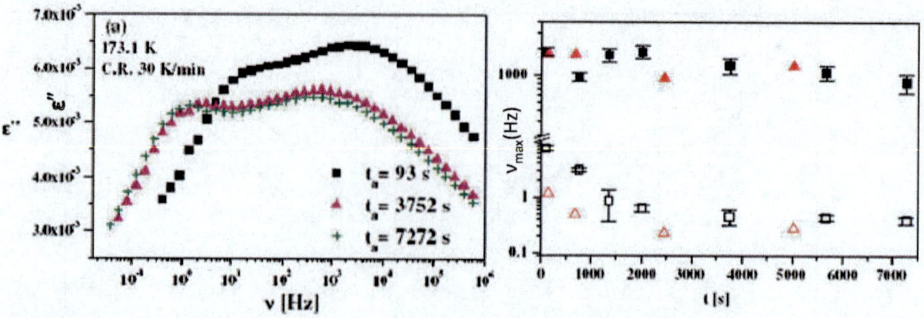

Fig. 191 (*Left*) Dielectric loss spectra of two secondary relaxations of DPGDB measured at different times during annealing at 173.1 K after a quench at about 30 K/min (the rate through T_g) from 323 K. (*Right*) Maximum loss frequency ν_{max} of the JG β-(*open symbols*) and the γ-(*closed symbols*) processes at 173.15 K, plotted as a function of the elapsed time after thermal stabilization. *Squares* refer to the spectra measured after vitrification at quench rate of 30 K/min and *triangles* to spectra measured after vitrification at quench rate of 9 K/min. Reproduced from [678] by permission

fictive temperature T_f, just like the well-known dependence of τ_α on T_f [266]. This dependence of τ_{JG} on T_f follows from the CM equations (2.67)–(2.69) rewritten as $\tau_{JG}(T_f) \approx \tau_0(T_f) = [\tau_\alpha(T_f)]^{1-n} t_c^n$. The increase of τ_{JG} on aging a glass follows from this relation and the well-known increase of τ_α with t_a. The increase in τ_{JG} with aging mimics the behavior of τ_α, as illustrated in Fig. 117 and explained in [1014]. Although the change in τ_{JG} is less than that of τ_α, the effect on the JG β-relaxation is much greater than on the faster, non-JG γ-relaxation. Similar results have been found for resolved JG β-relaxation in xylitol by Lunkenheimer et al. [1155] and sorbitol by Olsen [1031].

The JG β-relaxations of glycerol, propylene carbonate (PC), and propylene glycol (PG) are not resolved and appear in the loss spectrum as an excess over the Kohlrausch fit and an excess wing [424, 972]. On aging, the excess wing is converted to a shoulder, which is evidence for the resolved JG β-relaxation. The fact that these excess losses shift to lower frequencies on physical aging at constant temperature below T_g as shown in Fig. 192-1 for PC and PG [1156] is indication that the JG relaxation is sensitive to densification or loss of entropy on aging (see also Fig. 121 in Section 2.3.2.3).

Dipropylene glycol (2PG) [658], dibutyl phthalate (DBP), and isoeugenol [684] each has a resolved secondary γ-relaxation. The JG relaxation lying in between the α- and γ-relaxations is not resolved. Aging further separated the JG relaxation from the α- and γ-relaxations enabling the excess losses to become visible and the shift of the excess wing to lower frequencies with aging time evident. The results are shown for 2PG in Fig. 192-2 (left), and DBP in Fig. 193 (right).

As mentioned before in the previous section, despite the number of glass-formers cited above that have τ_{JG} shifted to longer times with aging, there are glassformers which have τ_{JG} insensitive to aging or even exhibiting a slight shift

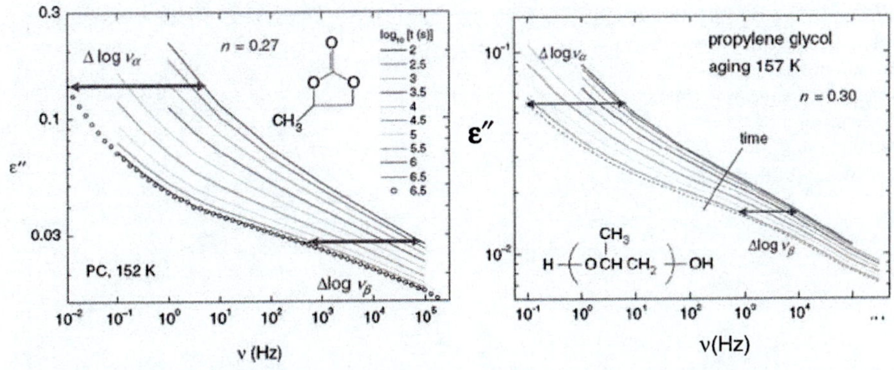

Fig. 192-1 (*Left*) Shifts of the α-relaxation and the excess wing of propylene carbonate at 152 K after aging for the periods of time as indicated. (*Right*) Shifts of the α-relaxation and the excess wing of propylene glycol at 157 K after aging for the periods of 100 s to nearly 10^6 s

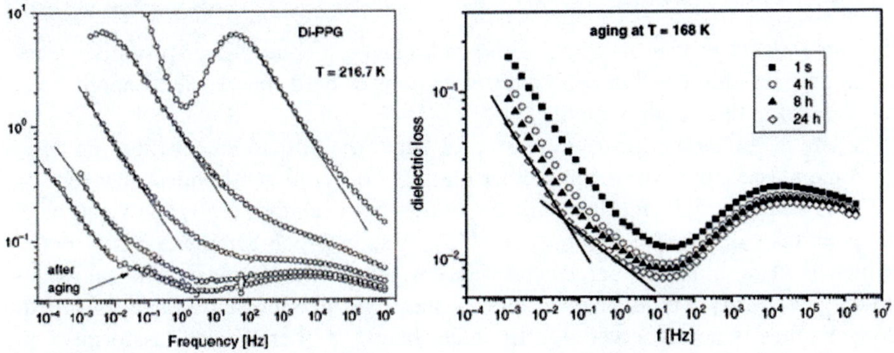

Fig. 192-2 (*Left*) Dielectric loss for PPG dimer at pressures (*from right to left*) of 67.6, 248.7, 335.7, 520, and 510 MPa. The last one was measured after 12 h of aging. There is a pressure-independent secondary peak at $\sim 10^4$ Hz, which exhibits a negligible response to aging. Reproduced from [658] by permission. (*Right*) Dielectric loss of dibutyl phthalate (DBP) at ambient pressures measured after aging for different periods of time as indicated. The excess wing becomes evident after aging for a sufficiently long time. Reproduced from [684] by permission

of the most probable value to shorter times. The examples are the rigid molecular toluene–pyridine mixture [1148], sorbitol [1031, 1152] poly(propylene oxide), poly(carbonate) [1149], PMMA [1150], and 1,2-polybutadiene or PVE [1157], some of which have already been discussed in the previous Section 2.3.2.19. Again, for the same reason given before in Section 2.3.2.19, if aging in these glassformers causes an increase in $n(t_e)$ with increasing t_e, as a consequence of the CM relation $\tau_{JG}(t_e) \approx \tau_0(t_e) = [\tau_\alpha(t_e)]^{1-n(t_e)} t_c^{n(t_e)}$, there will be no significant increase of τ_{JG}. This scenario is plausible for these glassformers all having larger intermolecular coupling or n, and even slight densification on aging can cause increase of n and a large increase of τ_α. Such increase of n was found on aging of a very rapidly quenched polystyrene discussed before in Section 2.2.5.4, and shown in Fig. 80.

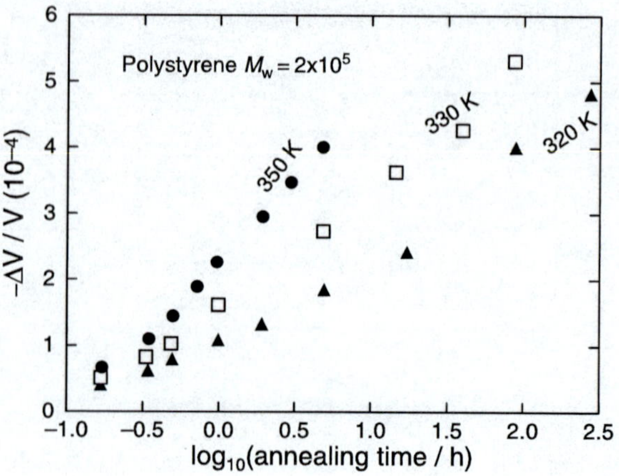

Fig. 193 Isothermal volume relaxation of a high molecular weight polystyrene as function of annealing time at three different annealing temperatures. Data of Chen and Wang [988] replotted

Sizeable increase of n on aging with small change of the fast or primitive relaxation time was observed in colloidal suspensions of hard spheres and Laponite also considered in that same section.

There is yet another possible reason for why τ_{JG} shifts to longer times on aging in some glasses, but not in some other glasses. For hydrogen-bonded glassformers such as sorbitol, τ_α is not that sensitive to density as evidenced by the small value of $\gamma = 0.13$ in its TV^γ-dependence [119]. Also, 1,2-polybutadiene has $\gamma = 1.9$ which is also small compared with van der Waals glassformers. The sensitivity of τ_{JG} to density would be even less in these two glasses, and hence its shift to longer times is not observed. On the other hand, van der Waals glassformers are much more sensitive to density as can be deduced from the larger values of γ they have. For example, $\gamma = 8.5$ for 1,1′-di(4-methoxy-5-methylphenyl)cyclohexane (BMMPC), $\gamma = 7$ for 1′-*bis*(*p*-methoxyphenyl)cyclohexane (BMPC), $\gamma = 4.5$ for cresolphthalein-dimethylether (KDE), $\gamma = 6.2$ for *o*-terphenyl/*o*-phenylphenol (2/1) mixture, $\gamma = 5.2$ for salol, $\gamma = 4.5$ for phenolphthalein-dimethylether (PDE), and $\gamma = 3.7$ for propylene carbonate. For BMMPC, BMPC, KDE, PDE, and PC, we know that the excess wing (which is the unresolved JG relaxation) shifts to lower frequencies on aging in concert with the α-relaxation.

2.3.2.21 JG Relaxation Responsible for Structural Change Deep in the Glassy State by Aging

Throughout the history of scientific study, it is well known that glasses or amorphous solids in general are not in thermodynamic equilibrium at temperatures below glass transition [6, 61, 62, 788, 1158–1162]. The specific volume, enthalpy, and entropy of a glass are all in excess of the equilibrium state. Slow relaxation toward equilibrium can be observed by relaxation of volume and enthalpy by experiment at temperatures that can be significantly below T_g. This kind of aging is called physical aging [788] to distinguish it from aging caused by change in chemical structure with time.

A historic example is the observation of the effect of aging of silicate glasses thermometer at room temperature over a period of 38.5 years, from April 1844 to December 1882 by James Prescott Joule [6], which was recently discussed by Nemilov and Johari [7]. Aging of the silicate glass caused change of the thermometer scale which in turn was measured by the shift in the "zero-point temperature." The T_g of silicate glasses, typically in the range of 680–900 K, is much higher than room temperature, and the α-relaxation time τ_α would be much longer than a century. However, structural change of the silicate glass was found by Joule from the change of zero-point temperature of about 8°F over 38.5 years. Nemilov [123, 1163] fitted the time dependence of the zero-point temperature data of Joules by the expression, $13.58 - 9.56 \times [\exp(-t/13.50)]$, where the unit of time is year. Since τ_α is much longer than 38.5 years, the observed change by Joule cannot be effected by the structural α-relaxation. However, τ_{JG} is much shorter and some local changes of the glass structure can occur through the JG β-relaxation. The mechanism for this observed spontaneous relaxation of glass at room temperature is attributed to the JG β-relaxation which causes local regions in the network and in turn the Si–O–Si bond angles in the immediate surroundings to change [7].

In many applications of glasses and particularly amorphous polymers, physical aging is an important concern because it results in time dependences of important properties such as increases in density, modulus, and brittleness, and decreases in permeability, impact strength, fracture energy, and deformation [788, 1164–1167]. Physical aging is measured by various rates. For volume changes, it is defined by $r_V = -V^{-1}[\mathrm{d}V/\mathrm{d}\log t_e]$, where t_e is the aging time. If aging is monitored by the shift to longer times by a factor a of mechanical response such as shear creep compliance, the aging rate μ is defined by $\mu = -\mathrm{d}\log a/\mathrm{d}\log t_e$. Struik [788] reported for amorphous polymers that μ is constant over wide ranges of t_e, has value near unity at temperatures slightly below T_g, decreases with decreasing temperature, and is still more than 0.5 down to some temperature, $T_{1/2}$, appreciably below T_g. For PMMA, PS, and PVC, $T_{1/2}$ is about 100 K below T_g. For aromatic backbone polymers BPA-PC and polysulfone, their $T_{1/2}$s are even much further down below T_g, which for BPA-PC may be due to the low temperature and intermolecular coupled γ-relaxation involving the π-flips of the phenyl rings on the main chain [358, 883–890], in addition to the JG β-relaxation. Therefore, the appreciable aging rate observed deep in the glassy state at or near $T_{1/2}$ has to originate from JG β-relaxation of these polymers. In fact Struik stated as one of his main conclusions that the temperature range in which aging occurs is not restricted to a narrow region below T_g but instead extends from T_g down to the secondary relaxation transition temperature T_β, which is −50°C for PVC ($T_g = 71$°C), and is −100°C for BPA - PC ($T_g = 144$°C). Aging disappears at temperatures below T_β, and this is another indication that the secondary relaxations including the JG β-relaxation can effect aging. These results of aging had led Struik to make statements "... secondary motion appears to be prerequisite for segmental motion ..." and "A general explanation for this behavior might be that the motions of segments as a whole (α process) are strongly facilitated by internal rearrangements (β processes)." This qualitative deduction from aging studies is consistent with the strong connection between the α-relaxation and the JG β-relaxation described in all the subsections above and to follow. However, in rationalizing the observation that the aging range falls between T_g and T_β, Struik suggested that aging does not affect secondary relaxations. This is not generally true. We have seen the resolved JG β-relaxation of small molecular glassformers such as DiPGDB as well as the unresolved one of dipropylene glycol, DBP, DOP, and isoeugenol, shifts to lower frequencies on aging. Recent and precise measurements of dynamic shear modulus of PMMA and PS showed changes of the JG β-relaxation with aging time [1162].

Supplementary evidence of JG β-relaxation being responsible for aging deep in the glassy state is given by results of the aging rate study of PMMA in supported ultrathin film on silica substrate at 305 K (i.e., 88 K below T_g of bulk PMMA) [1143], and in nanocomposites of silica (10–15 nm diameter) nanoparticles in PMMA at 296 K [1168]. Observed in both cases is the vastly reduced aging rate compared with bulk PMMA. The molecular origin of the suppressed physical aging is due to the strong hydrogen bonding interactions between the ester groups of PMMA and the hydroxyl groups on silica. As discussed before in earlier sections, the JG β-relaxation of PMMA originates from the motion of the ester group accompanied by rotation of the main chain. The interfacial hydrogen bonding of the ester groups arrests the JG β-relaxation. This explains the large reduction in aging rate since it is the molecular mobility of the JG β-relaxation which is responsible for physical aging of PMMA deep in the glassy state. This explanation is supported by the observation of large reduction in the dielectric strength of the JG β-process of PMMA in silica–PMMA nanocomposites due to the interfacial hydrogen bonds [1168].

If the temperature T_e of aging is much lower than T_g such that τ_α is so long that it cannot be instrumental in aging, for sure it is the JG β-relaxation which is responsible for the observed aging. If the aging temperature T_e is increased and becomes sufficiently close to T_g such that τ_α is not much longer than t_e, not only the JG β-relaxation but also the α-relaxation may become active in effecting aging. Nevertheless, the JG β-relaxation is still the initiator of aging, and τ_α of the α-relaxation in the glassy state is governed by τ_{JG} according to Eq. (2.69). Thus, the JG β-relaxation plays a role in aging even in this special case.

The conclusion, JG β-relaxation is responsible for physical aging of PMMA deep in the glassy state, has additional support from different results coming from aging experiments of polystyrene (PS) films. When aged deep in the glassy state of 20 nm thick PS film on silica [1143(b)] or 0.4 vol% silica–PS nanocomposites or 0.4 vol% silica–PS nanocomposites [1169], arrested or retarded aging is not found. PS contains no atoms that will form hydrogen bonds with hydroxyl groups and hence no reduction in the strength of the JG β-process of PS and no suppression of aging.

Another difference between 20 nm thick PS film and PMMA film, both supported on silica, should be pointed out. As mentioned before, attractive PMMA–silica interactions enhance the T_g of the ultrathin PMMA film to 407 K compared with 393 K of bulk PMMA. On the other hand, data by Ellison et al. [1170] and Keddie et al. [1171] indicate that a 20 nm thick PS film has T_g that is lower than that of the bulk, $T_{g,bulk}$, by 14–17 K. There are two factors determining T_g of ultrathin films supported on silica. The free-surface effect reduces intermolecular interaction (or n), increases molecular mobility, and hence lowers T_g. The other is the substrate effect from hydrogen-bond interactions between the polymer and the hydroxyl groups on the silica surface, which increases T_g and is present in PMMA but not in PS. Drawing from what we have learned from aging studies, let us restate the substrate effect more precisely as originating from the reduction in strength or suppression of the JG β-relaxation by interfacial hydrogen bonding of the ester groups of PMMA with the hydroxyl groups on the silica surface. With this more precise restatement of the substrate effect, the increase of T_g of the ultrathin PMMA is due to suppression of the JG β-relaxation. In the CM, the JG β-relaxation is the initiator of the α-relaxation. Also NMR experiment has shown that the α-relaxation is modified when part of the JG β-relaxation is suppressed [998, 999]. Thus, the substrate effect of suppressing the JG β-relaxation necessarily also suppresses the α-relaxation in the ultrathin PMMA film at the interface. The absence of suppression of JG β-relaxation of ultrathin PS film naturally explains the decrease of T_g by the free-surface effect acting alone.

Actually studies of physical aging or annealing of quenched glasses at temperatures well below T_g had been made starting as early as the 1980s and continued to present days in various glassformers. The glassformers studied include polymers [988, 984], an inorganic glass (B_2O_3) [989], metallic glasses ($Pd_{48}Ni_{32}P_{20}$, $(Fe,Co,Ni)_{75}Si_{10}B_{15}$) [954, 990, 991, 958, 960(b)], pharmaceuticals [992, 1000(b)], and inorganic oxide glasses [986, 987]. Aging the quenched glass well below T_g relaxes the enthalpy and volume, particularly the part associated with the JG β-process. On reheating the aged glass, the lost enthalpy is recovered and detected by an endothermic peak associated with the JG β-relaxation, called sub-T_g peak, in the specific heat C_p spectrum that precedes the larger endothermic peak associated with the α-process and glass transition. All studies cited above indicate that JG β-relaxation is responsible for physical aging of deeply and rapidly quenched glasses. In the study by Chen and Wang [988], they found sub-T_g structural relaxation from specific heat change and volume contraction in a nearly monodisperse high molecular weight polystyrene ($M_w = 2 \times 10^5$, $T_g = 382$ K by DSC at 10 K/min) by annealing at temperatures ranging from 32 to 62 K below T_g over times up to 90 h for volume as shown in Fig. 193. This sub-T_g structural relaxation is likely associated with the JG relaxation of polystyrene because the α-relaxation time is much longer than 90 h, and the samples recover the initial heat content and volume without reheating through T_g.

Hachenberg et al. [960(b)] studied physical aging of deeply and rapidly quenched (at the rate of $\approx 10^5$ K/s) metallic glasses, $Pd_{77}Cu_6Si_{17}$ ($m = 52.8 - 77.0$) and $Zr_{65}Al_{7.5}Cu_{27.5}$ ($m = 36.4 - 38.4$), by dynamic mechanical measurements. A significant increase in the mechanical modulus was observed on reheating the aged glasses at temperature significantly below T_g, and the increase is dependent on heating rate. This observation supports that the physical aging of deeply and rapidly quenched metallic glasses is promoted by the JG β-relaxation. As in some molecular glasses, the relaxation strength of the JG β-process is considerably decreased by aging [829, 1148].

2.3.2.22 JG β-Relaxation Governs the Rate of Crystal Nucleation, the Initial Process of Crystallization

Conventional Glassformers

At temperatures below and near T_g, an amorphous glass also crystallizes but the process is different from crystallization in the liquid state. Crystallization is preceded by nucleation and can start only after the formation of stable nuclei that reach some critical size [1172]. Studies of organic molecular glasses have shown that localized motions may lead to nucleation and crystallization of the glassy state over a long period of time. According to Oguni and coworkers [810, 811, 1173, 1174], in the rigid structure of the glass it is the Johari–Goldstein relaxation that leads to nucleation or crystal embryo formation, and then crystallization proceeds by the coalescence of crystal nuclei formed homogeneously in the vicinity of crystal surface into the crystalline phase. This new process, called the homogeneous nucleation based (HNB) crystallization, was found by optical spectroscopy and calorimetry in typical glassformers including *o*-terphenyl, salol, toluene, triphenylethylene, and benzophenone at temperature below and around their glass-transition temperatures. The homogeneous nucleation process is thus the precursor of HNB crystallization. The suggestion that the local JG β-relaxation is responsible for crystal nucleation is consistent with the proposal by Turnbull and Fisher [1175] that crystal nucleation proceeds via the attachment or release of one molecule at a time on the surface of a crystal embryo. It was confirmed by order of magnitude agreement of the calculated and observed crystal growth rate in *o*-terphenyl and toluene by using the experimental values of τ_{JG} for the relaxation time of molecular rearrangement by Oguni and coworkers. The crystal growth rates calculated with τ_α are many orders of magnitude too small compared with the observed values. The evidences are pointing to the JG β-relaxation in governing the crystal nucleation process. Although, like Oguni and coworkers [1174, 1175], I have attributed the JG β-relaxation to be responsible for enhanced nucleation and glassy crystal growth near and below T_g, I am not committed to their view that the crystal growth is necessarily the coalescence of homogeneous crystal nuclei, not individual molecules, onto existing crystal surfaces [1173]. I should mention also the entirely different view of Tanaka [1176] that enhanced crystallization in the glass results from increase in mobility of the particles surrounding the crystal provided by the free volume created from the stress around a crystal growing in the glass.

Recently, Sun et al. [1177(a)] studied the growth of crystals in the glass in a number of polymorphs from the fragile liquid of 5-methyl-2-[(2-nitrophenyl)amino]-3-thiophenecarbonitrile with $m = 71$, called ROY for its red, orange, and yellow crystal polymorphs. Some of the polymorphs show crystal growth in the glass. From the observed properties and kinetics of the growth, Sun et al. did not favor the explanations either by Oguni and coworkers, or by Tanaka. Instead, they suggested that "it is solid-state transformation enabled by local molecular motions native to the glassy state." Their term "local molecular motions" is no different from the primitive relaxation of the coupling model which involves local motion of the entire molecule, and hence the conclusion of Sun et al. [1177(a)] is no different from my own on crystallization in the glass. It is true that the primitive and the JG β-relaxations are not identical, but they are related because the former is part of the latter.

Notwithstanding the same conclusion reached by Sun et al. and myself, confusion could be created by an accompanying paper by Sun et al. [1177(b)] in which they presented dielectric relaxation data of ROY and showed no secondary relaxation peak was observed, even in glasses made by fast quenching. This is no surprise because the width of the α-loss peak of ROY is narrow and characterized by $\beta_{KWW} = 0.73$, or $n = 0.27$, and hence the separation of the JG β-relaxation from the α-relaxation is small and the former cannot be resolved from the latter even by rapid quenching. Not able to resolve and observe the JG β-relaxation in ROY by dielectric spectroscopy does not mean it is not present, but this was used by Sun et al. to rule out the role of JG β-relaxation in crystal growth in the glass. They cited also the case of ortho-terphenyl (OTP), which has weak dipole moment and is the main reason why its JG β-relaxation only can be barely observed by dielectric relaxation after rapid quenching and can disappear on annealing [829]. This fact of the dielectric JG β-relaxation of OTP not being observed while there is crystal growth in the glass has led them [1177(b), 1177(c)] to conclude this secondary relaxation of OTP is irrelevant to the growth, and suggested that other secondary processes perhaps detectable with other techniques could be associated with the growth. I do not see any secondary relaxation of OTP other than the motion of the entire OTP molecule (i.e., the JG β-relaxation) can contribute to crystal growth in the glass. Another technique to observe secondary relaxation in OTP is by dynamic light scattering by confining OTP in 2.5 nm silanized glass pores [492–494]. The large optical anisotropy of OTP makes possible detection of relaxation process even when confined. Under confinement to 2.5 nm pores, the cooperativity is reduced by reduction in either length scale or density, and the primitive or JG β-relaxation of OTP has been found. The relaxation time of OTP confined in 2.5 nm glass pores is about the same as that found by dielectric relaxation [494]. When the JG β-relaxation is not resolved in glassformers with small n like in ROY, it is usually located at the high-frequency flank of the α-loss peak and shows up as an excess wing in the plot of log ε'' against log(frequency). From the published log ε'' of ROY, the shift of the excess wing with temperature has temperature dependence which turns out to be parallel to that of the crystal growth rate in the glassy state $(\tau_u)_g$ at temperatures less than 250 K [1177(d)], which is another evidence of the JG β-relaxation being the rate-determining factor of crystal growth. The same is true for OTP as can be seen by comparing the temperature dependence of $(\tau_u)_g$ with that of the JG β-relaxation found by Wagner and Richert [829].

Sun et al. [1177(a), 1177(b)] also studied crystallization of ROY above T_g, the mechanism of which is controlled by the self-diffusion of the ROY molecules. Like that found previously for TNB and OTP [347, 453, 454, 805, 807], breakdown of Stokes–Einstein relation (see Section 2.2.5(E)) can be deduced for ROY polymorphs not showing fast crystal growth in the glassy state from the fractional power law relation between the growth rate u and structural relaxation time τ_α or with viscosity η given by $u \propto [(\tau_u)]^{-\xi}$ and $u \propto \eta^{-\xi}$, where $\xi \approx 0.7$. The α-relaxation process in ROY obeys time–temperature superposition for τ_α over the range from 100 s to 100 ns, with the constant width of the α-loss peak characterized by a constant $\beta_{KWW} = 0.73$. Thus, as explained in Section 2.2.5.6, the breakdown of Stokes–Einstein relation in ROY polymorphs is another example demonstrating the failure of the explanation by spatial heterogeneous dynamics [450, 454–457] discussed in Section 2.2.5.6. The viable explanation for ROY is also given by the CM based on the smaller coupling parameter n_D for diffusion than n_α for rotation or n_η for viscosity [275], analogous to that illustrated in Fig. 85 for TNB. The only difference is ROY having smaller $n_\alpha = 0.27$ compared to $n_\alpha \approx 0.50$ of OTP and TNB.

For OTP, the time $(\tau_u)_D$ required for the crystal to grow one molecular layer for diffusion-controlled growth above T_g is compared with the corresponding $(\tau_u)_g$ for the "diffusionless" crystal growth in the glassy state. When extrapolated to T_g, $(\tau_u)_D$ is more than 4 orders of magnitude longer than $(\tau_u)_g$. This difference reflects the disparity in the mechanism of the crystallization at temperatures above and below T_g. Above T_g, it is diffusion and n_D for diffusion-controlled growth is non-zero albeit smaller than n_α or n_η, like the demonstration given for TNB in Fig. 85. On the other hand, it is the "local" or primitive relaxation for growth in the glass. The difference of more than 4 orders of magnitude between $(\tau_u)_D$ and $(\tau_u)_g$ at T_g can be rationalized by n_D having some non-zero value. Sun et al. also found the crystal growth in the glass of ROY slows slightly over time, as has been reported for glassy OTP. In 10 h the crystal growth rate u at 248 K (T_g–11 K) decreases by 13% while the glass τ_α increases by 23 times. This slight slowing down of u is readily explained by the small effect of physical aging on the JG β-relaxation time, which is known to increase much less than that found on τ_α.

Metallic Glasses

Crystallization of the metallic glass, $Pd_{42.5}Ni_{7.5}Cu_{30}P_{20}$, by ultrasonic annealing at temperature below T_g was also found by Ichitsubo et al. to be caused by the accumulation of atomic jumps associated with the local JG β-relaxation (with relaxation frequencies ~MHz range at temperature slightly below $T_g \approx 300$ C) being stochastically resonant with the ultrasonic vibrations, which eventually leads to crystallization [958]. Under ultrasonic vibrations at 0.35 MHz, the metallic glass was found to be fully crystallized within only 18 h at 290°C. On the other hand, without ultrasonic vibrations at this temperature, crystallization was negligible when the samples were annealed for 75 h.

Ichitsubo et al. further proved that indeed the JG β-relaxation is responsible for the accelerated crystallization at 290°C under ultrasonic vibrations at 0.35 MHz by successfully detecting the increase of internal friction, Q^{-1}, at MHz frequencies (see left panel of Fig. 194) at temperatures where the atomic motions are associated with the JG β-relaxation. The latter was supported by the independent internal friction measurements of a similar Pd-based metallic glass by Pelletier et al. [956] obtained at lower frequencies and extended to MHz frequencies by Ichitsubo et al. by using the activation energy ($E = 1$ eV) and prefactor ($\tau_\infty = 10^{-14}$ s) of the

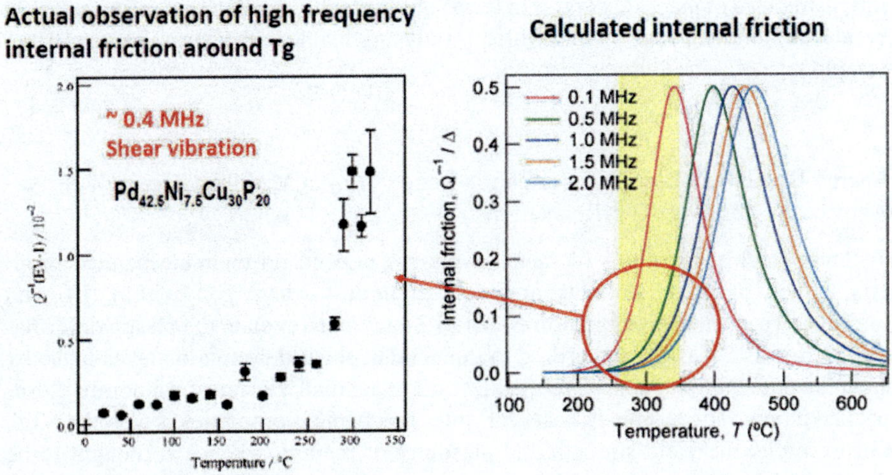

Fig. 194 (*Left*) Temperature dependence of high-frequency (0.4 MHz) internal friction Q^{-1} as a function of temperature measured for $Pd_{40}Ni_{40}P_{20}$ metallic glass ($T_g \approx 300°C$). Shear vibration modes were used for the internal friction measurements, and the rate of temperature scan was about 1°C/min. The increase of the internal friction means that the periodic ultrasonic strain is resonant with atomic motion. (*Right*) Q^{-1}–T curves of the JG β-relaxation in Pd-based metallic glass in the MHz range calculated by Ichitsubo et al. using the activation energy and prefactor of the JG β-relaxation obtained by Pelletier et al. [956] at lower frequencies also from internal friction measurement for a Pd–Ni–Cu–P glass by Pelletier et al. The *circle* and *shaded region* indicate the JG β-relaxation is responsible for the atomic motion in resonance with the periodic ultrasonic strain in the *left panel*. Figure courtesy of Ichitsubo

JG β-relaxation frequency determined by Pelletier et al. [956]. This is illustrated in the right panel of Fig. 194. Crystallization at such a low temperature (below T_g) is related to the JG β-relaxation, the atomic motions of which are stochastically resonant with the ultrasonic vibrations. Thus, the JG β-relaxation is proven to be responsible for the observed ultrasound-induced crystallization of bulk metallic glasses. Ichitsubo also observed crystallization of $Zr_{70}Ni_{30}$, $Pd_{40}Ni_{40}P_{20}$, and ZrAlNiCu at temperature below T_g that is attributed also to the action of the JG β-relaxation.

Caution on Interpretation of Microstructure of Metallic Glasses by Ichitsubo et al.

Ichitsubo et al. [958] identified the microstructure of the metallic glass, $Pd_{42.5}Ni_{7.5}Cu_{30}P_{20}$, before crystallization with the microstructure of the partially crystallized metallic glass that has isolated as amorphous islands of the size of ~5 nm surrounded by crystallized regions. This identification is not justified and likely to be incorrect, because the microstructure of the partially crystallized sample obviously is not the same as the original metallic glass. The reader should be aware of this problem of the interpretation by Ichitsubo et al. on the nanoscale microstructure of *metallic glasses* in terms of the weakly bonded region and strongly bonded region with size ~5 nm, which is only applicable to the partially crystallized sample.

Energy Barrier of Shear Transformation Zone (STZ) in Metallic Glasses, $W_{STZ} \approx$ Activation Energy of JG β-Relaxation

Technological applications of metallic glasses depend on their mechanical property, namely plasticity, ductility, and yield strength. It was Argon [959(a), (b)] who introduced a shear transformation zone (STZ) model to explain the plastic deformation of metallic glasses. According to this model, shear deformation takes place by spontaneous and cooperative reorganization of a small cluster of randomly close-packed atoms. The atomistic basis for other mechanical properties is also the STZ. For example, the yield strength and plasticity of metallic glasses is thought to be determined by the shear transformation zones (STZs) [959(c)–(h)]. According to the cooperative shearing model (CSM) developed by Johnson and Samwer [959(g)], the total potential energy barrier for an STZ is $W_{STZ} = (8/\pi^2)G\gamma_C^2\varsigma\Omega$, where Ω is the actual volume of STZ defined by the plastic core, $\zeta \sim 3$ is a correction factor arising from matrix confinement of an STZ, $\gamma_C = 0.0267$, and G is the shear modulus. The expression for W_{STZ} has been simplified by Wang and coworkers [959(i)] to $W_{STZ} \approx 0.39GV_m$, where V_m is the molar volume of the STZ. The STZ volumes of bulk metallic glasses (BMGs) were determined by means of nano-indentation experiment [959(j)]. The measured STZ volumes of 200–700 atoms are in good agreement with those predicted by cooperative shearing model of Johnson and Samwer [959(g)] and MD simulations [959(k)]. The results demonstrate that the plastic yielding of metallic glasses occurs upon cooperative shearing of STZs. The ductility of bulk metallic glasses was found to correlate with their STZ volumes [959(j)]. From the measured values of G and V_m of many BMGs, Wang and coworkers [959(i)] have obtained the potential energy barrier of STZ, W_{STZ}, and compared with the corresponding activation energy, E_{JG}, of the primitive or JG β-relaxation. Remarkably Wang and coworkers found the two quantities to be almost equal, i.e., $W_{STZ} \approx E_{JG}$ [959(i)]. This finding indicates that the flow resistance of STZs and the primitive or JG β-relaxation have common microstructural origin. The fact that the JG β-relaxation is instrumental in the dynamics of the STZ may not be surprising because, at temperatures way below T_g of metallic glasses, the mobility of atoms can only be initiated by the JG β-relaxation. The experimental data from [959(i)] demonstrated all that said above are shown in Fig. 194-1 from a to d.

Figure 194-1a illustrates typical isochronal mechanical data of metallic glasses exhibiting the JG β-relaxation and how their activation energies are determined. Figure 194-1b shows the relation between E_{JG} or E_β and RT_g. Although there is a rough correlation, there is no strictly linear relation between the two as demonstrated before for non-metallic glassformers in [996]. Figure 194-1c depicts the nearly linear relation between E_β and W_{STZ}.

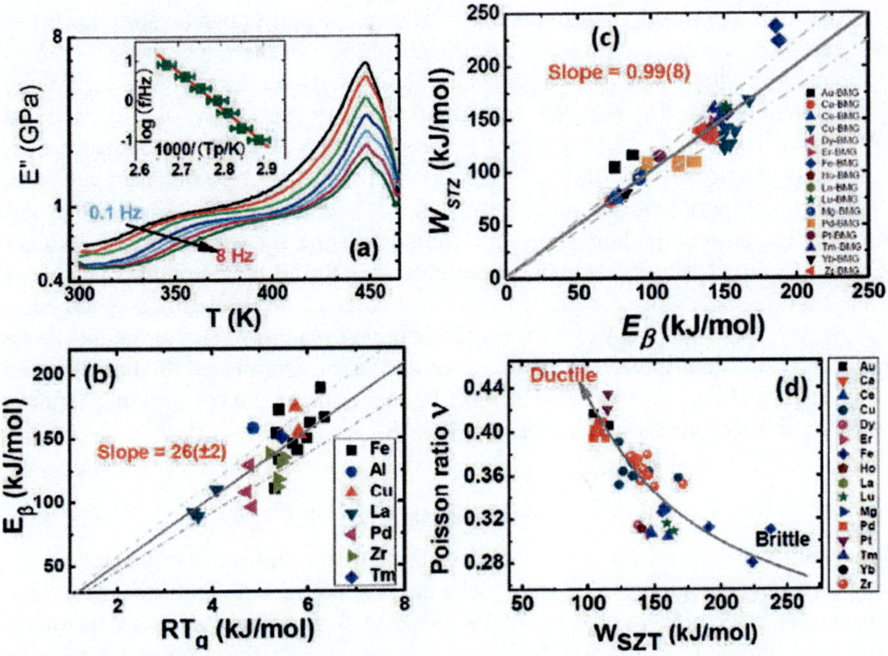

Fig. 194-1 (**a**) The temperature dependence of the loss modulus E'' for a $La_{55}Al_{15}Ni_{10}Cu_{10}Co_{10}$ metallic glass, measured at frequencies f (*from top to bottom*, indicated by the *arrow*) of 0.1, 0.2, 0.5, 1, 2, 4, and 8 Hz, at the heating rate of 3 K/min. The *inset* plots log f vs. $1000/T_p$, where T_p is peak temperature of the β-relaxation hump. (**b**) Activation energy of β-relaxation E_β plotted against RT_g for metallic glasses. The *solid line* is a least-squares linear fit. (**c**) Relationship between activation energy of β-relaxation E_β and energy barriers of STZs, W_{STZ}. (**d**) Relationship between W_{STZ} and Poisson ratio ν for metallic glasses. *Solid curve with arrow* is drawn to guide the eye. MGs based on La, Ca, Mg, Ce, and Yb not included in this plot for reason explained in [959(i)]. All figures redrawn from the data of [959(i)] are kindly provided by W.H. Wang

The Poisson ratio ν when plotted against $W_{STZ} \approx E_{JG}$ for many Au-, Pt-, Zr-, Cu-, Fe-, and some late rare-earth-based metallic glasses in Fig. 194-1d demonstrates yet another correlation between these quantities. This result has important implication for applications because the Poisson ratio is an indicator of ductility. Thus, the JG β-relaxation like the STZs is fundamental in the consideration of plasticity, yield stress, and ductility of metallic glasses. Lower the value of $W_{STZ} \approx E_{JG}$, better the plasticity and ductility of the BMG. Also, larger the Poisson ratio of the BMG, better its plasticity [959(l)]. The characteristics of the primitive or JG β-relaxation is fundamental for understanding the Poisson ratio criterion for plasticity of BMGs [959(i)]. The important findings by Wang and coworkers, $W_{STZ} \approx E_{JG}$ and the primitive or JG β-relaxation is the microstructural origin of STZ, will make the primitive relaxation or JG β-relaxation of metallic glasses a center of attention in improving their mechanical properties.

Although E_{JG} or E_{β} is an important parameter of the JG β-relaxation for STZs and resulting mechanical properties of BMGs in view of the relation $W_{STZ} \approx E_{JG}$, another relevant parameter is the actual values of its relaxation time τ_{JG} in the glassy state. Faster τ_{JG} together with smaller E_{JG} in the glassy state is expected to act in concert to facilitate thermal events driven by STZs and to enhance ductility and plasticity. Thus the value of $\tau_{JG}(T_g)$ is another relevant parameter. As found from experiment and predicted by the CM [251], larger the intermolecular coupling parameter n, shorter $\tau_{JG}(T_g)$. Moreover from the ancillary CM relation, $E_{JG} = E_0 = (1 - n)E_{\alpha}$ in connection with Fig. 200 to be introduced later, it is clear that E_{JG} is smaller for glassformers with a larger n if they all have comparable E_{α}, the activation energy of the α-relaxation in the glassy state. From the above considerations, intermolecular coupling (or n) and the established relation between (τ_{JG}, E_{JG}) and (τ_{α}, E_{α}) via the CM may be the fundamental for the consideration of STZs and mechanical properties of BMG.

Similar Correlation of Mechanical Properties of Glassy Polymers to β-Relaxation

It is worthwhile to follow the authors of [959(i)] to point out the general recognition in polymer glasses that the mechanical properties are determined to some extent by the secondary relaxation [883–888, 934–936, 959(n)–959(s)]. The transition from ductile to brittle was found in many polymeric glassformers to occur at the characteristic temperature of the secondary relaxations [959(n)–959(q)]. Impact toughness, yield strength, and failure modes were also often correlated with secondary relaxations [959(n)], and polymers with strong secondary relaxations often possess good ductility and vice versa [959(n), 959(p)]. Again these correlations between mechanical properties of glassy polymers with secondary relaxation may ultimately originate from the established relation between (τ_{JG}, E_{JG}) and (τ_{α}, E_{α}) via the intermolecular coupling (or n) predicted by the CM.

Pharmaceuticals

Recently, there is interest in pharmaceutical industry to develop amorphous solid pharmaceuticals. This is because the amorphous state of a pharmaceutical has various advantages over the crystalline state such as improved solubility and bioavailability [136, 1178, 1179]. However, the non-equilibrium glassy state is thermodynamically unstable, and tends to revert to the crystalline form on storage as found in the much studied anti-inflammation drug, indomethacin. Slow crystallization of indomethacin was reported at temperatures as low as about 25°C below T_g [1180–1184]. The tendency toward crystallization results in poor product performance over passage of time, which is a major concern. When stored at temperature sufficiently below T_g and the structural α-relaxation is completely frozen, the origin of nucleation and crystallization instabilities of amorphous pharmaceuticals might be attributed to the molecular motions that can still exist below T_g [1183, 1185–1187]. By analogy to nucleation and crystallization of organic molecular glasses studied by Oguni and coworkers, the molecular motion below

T_g is likely the JG β-relaxation. In fact, recently JG β-relaxation has been found in indomethacin [992, 1000(b)], and other pharmaceuticals including acetyl salicylic acid (aspirin) [1188, 1189], fananserine ($C_{23}H_{24}FN_3O_2S$) [1000(c)], and (2*RS*)-2[4-(2-methylpropyl)phenyl]propanoic acid ($C_{13}H_{18}O_2$), commonly known as ibuprofen [1004].

Occurrence of nucleation in glassy indomethacin was proven experimentally by detecting crystallization and melting in DCS curves on heating samples annealed for 10 days at 5°C (38 degrees below T_g). The nuclei formed during annealing set off the observed crystallization when the annealed samples are heated [992]. There is also evidence of homogenous nucleation occurring somewhere between the glassy state and the onset of the α-relaxation in ibuprofen. Nucleation is instrumental for cold crystallization observed when heating from the completely amorphous state to the temperature of crystallization near and above T_g [1004]. Nucleation is likely caused by the JG β-relaxation of indomethacin and ibuprofen. This is reasonable because at temperatures much below T_g, the kinetics are governed by the primitive or JG β-process. Additional evidence that the JG β-relaxation is responsible for the nucleation preceding crystallization in indomethacin came from the comparison of the critical size of the nucleus formed by nucleation, r^*, and the length scale, ξ, of the cooperative α-relaxation. The fact that $r^* = 0.7 - 0.9$ nm is much smaller than $\xi = 3.4$ nm for indomethacin rules out the α-relaxation and suggests the more local JG β-relaxation [992(b)]. Similarly, Surana et al. [992(c)] reported increased nucleation upon annealing samples of amorphous trehalose below the glass transition temperature.

From the results of the studies mentioned above, deterioration of amorphous pharmaceuticals by nucleation and crystallization can be avoided by storage at temperatures where the JG β-relaxation time, τ_{JG}, becomes sufficiently very long. For any predetermined storage time t_S, the temperature T_S may be determined by the condition $\tau_{JG}(T_S) \approx t_S$, with the help of the Arrhenius T-dependence of $\tau_{JG}(T)$ in the glassy state.

Less direct evidence of the crystallization kinetics being coupled with the JG β-relaxation and not the α-relaxation comes from the investigation of the isothermal crystallization kinetics of a drug called SSR at different temperatures above T_g by Alie et al. [1186]. The temperature dependence of the crystallization time is much weaker than τ_α and is similar to that of a secondary relaxation found. Although the JG β-relaxation is responsible for initiating the crystallization, it may be hidden and is not the observed and much faster secondary relaxation. This is because the observed secondary relaxation has relaxation time of the order of 10^{-7} s at T_g, which is much too short compared with τ_{JG} expected from the narrow isothermal dielectric α-loss peak (after conductivity has been subtracted) with full-width at half-maximum less than two decades, and also the low value of $m = 54.7$. Thus, the JG β-relaxation may not have been resolved but still be active in crystallization.

Pikal and coworkers measured the crystallization onset time θ_c by polarized light microscopy and the dielectric α-relaxation time τ_α above T_g of amorphous pharmaceuticals. They found linear relation between $\ln(\theta_c)$ and $\ln(\tau_\alpha)^\beta$ in sucrose [1190], and in indomethacin, flopropione, and felodipine [1191(a) and (b)] given by

$$\ln(\theta_c) = (M/\beta)\ln(\tau_\alpha)^\beta + C, \tag{2.82}$$

where $\beta \equiv (1-n)$ is the stretch exponent of the Kohlrausch correlation function of the α-relaxation, and M and C are constants. This relation predicts well the order of magnitude of the crystallization onset time θ_c at (pharmaceutically relevant) temperatures below T_g, from $(\tau_\alpha)^\beta$ measured at the same temperatures by DSC for sucrose, indomethacin, and flopropione. Thus, Eq. (2.82) may have practical applications. Interestingly, this equation is equivalent to $\ln(\theta_c) = (M'/\beta)\ln(\tau_0) + C'$, or $\ln(\theta_c) = (M'/\beta)\ln(\tau_{JG}) + C'$. The equivalence follows from the CM equation (2.68) when rewritten as $(\tau_\alpha)^\beta = t_c^{-n}\tau_0 \approx t_c^{-n}\tau_{JG}$ and recognizing that t_c is a constant, $\beta \equiv (1-n)$, and $\tau_0 \approx \tau_{JG}$. Thus, the correlation of θ_c with $(\tau_\alpha)^\beta$ found by Pikal and coworkers can be restated as correlation between θ_c and τ_0 or τ_{JG}, suggesting that the primitive or JG β-relaxation also is responsible for the nucleation and crystallization of the pharmaceuticals studied. The crystal growth rate of glassy phenobarbital and solid dispersions of phenobarbital with PVP and L-proline also were found to correlate with $(\tau_\alpha)^\beta$ [1191(c)]. The authors of [1191(c)] called $(\tau_\alpha)^\beta$ explicitly the structural relaxation time, but more appropriately it is the primitive or JG β-relaxation time except for the factor t_c^{-n}. Deep in the glassy state, τ_α is much longer than $(\tau_\alpha)^\beta$.

Bhugra et al. [1190(a)] reported for spray-dried sucrose at T_g that $(\tau_\alpha)^\beta = 0.0069$ with $\beta = 0.61$, the unit of τ_α being hour. The pharmaceuticals considered are all small molecular glassformers, and hence t_c can be taken approximately equal to 2 ps or 5.6×10^{-16} h. The value of $\tau_{JG}(T_g)$ for sucrose calculated from $\tau_{JG} \approx (\tau_\alpha)^{0.61}(5.6\times10^{-16})^{0.39}$ is $10^{-8.1}$ h or $10^{-4.6}$ s, which is within an order of magnitude the same as that of $\tau_{JG}(T_g)$ determined by dielectric spectroscopy (see Fig. 206 in Section 2.3.3.26 on carbohydrates). For indomethacin, Bhugra fitted the dielectric relaxation data with the Cole–Davidson function having the fractional exponent $\beta_{CD} = 0.41$ and gave $(\tau_\alpha)^\beta = 0.1$ at T_g where again the unit of τ_α is hour. It is known that β_{CD} is smaller than β of the Kohlrausch function used to fit the same data [85(c)]. In fact, the value $\beta = 0.59$ was given by Carpentier et al. from their fits to the dielectric loss data of indomethacin [1000(b)]. The value of $\tau_{JG}(T_g)$ calculated from $\tau_{JG} \approx (\tau_\alpha)^{0.59}(5.6\times10^{-16})^{0.41}$ is $10^{-7.3}$ h or $10^{-3.7}$ s, which is in good agreement with that deduced from the study of Carpentier et al. The good agreement found in sucrose and indomethacin further substantiates other evidences of the role of the primitive or JG β-relaxation in nucleation and crystallization of pharmaceuticals.

Bhattacharya and Suryanarayanan [1192] have recently reviewed the literature that recognized that local molecular mobility originating from the secondary relaxation is responsible for the physical as well as chemical instability of amorphous pharmaceuticals including crystallization. The review emphasizes the significance of local motions in amorphous pharmaceuticals, primarily the Johari–Goldstein relaxations, and the utility of the CM in relating the JG β-relaxation to global mobility effected by the α-relaxation. These authors have also reviewed the influence of additives including water on the local motions in an amorphous matrix, as in molecular dispersions.

2.3.2.23 JG β-Relaxation in Pharmaceuticals

Having just discussed the importance of the JG β-relaxation in crystallization of amorphous pharmaceuticals, it is worthwhile to give more details on the nature of JG β-relaxation in some examples. The objective here is to show the general presence of the JG β-relaxation in pharmaceuticals, and the same concepts used to describe the properties of JG β-relaxation in non-pharmaceutical supercooled liquids and glasses apply directly to pharmaceuticals. These demonstrations are deemed beneficial for researchers in pharmaceutical sciences by the transfer of knowledge from basic research of glass transition to industrial applications.

Aspirin

Acetyl salicylic acid ($HCOO–C_6H_4–COOCH_3$), commonly known as aspirin, has been studied by calorimetry [1193, 1194], optical heterodyne-detected optical Kerr effect [1195], deuteron NMR [1196], and dielectric relaxation [1189]. It has T_g = 243 K. The dielectric spectra measured in the frequency range $10^{-1} < \nu < 10^6$Hz show the presence of a well-resolved secondary relaxation at temperatures below T_g from 230 K down to 130 K. The T-dependence of its relaxation time is Arrhenius and given by $\tau_\beta(T) = 3 \times 10^{-15} \exp(4900/T)$ s, and is shown in Fig. 195 together

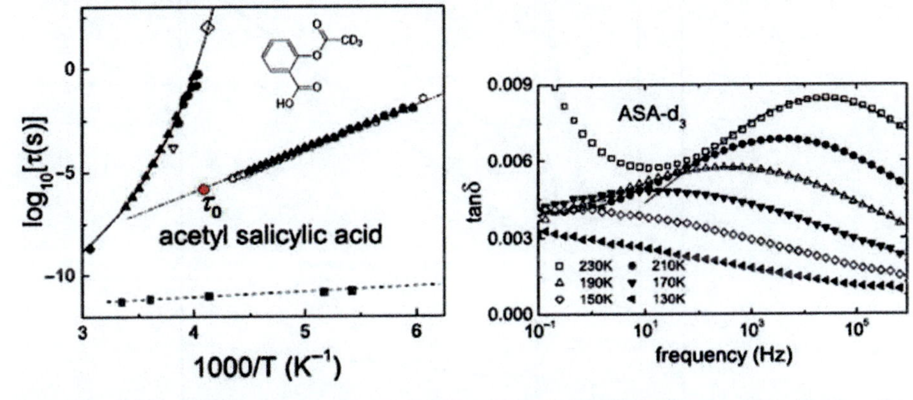

Fig. 195 (*Left*) Arrhenius plot of the correlation times of methyl deuterated acetyl salicylic acid (ASA) (*closed symbols*) and of fully protonated ASA (*open symbols*). The correlation times were obtained from dielectric measurements at 1 kHz (∇, from G.P. Johari) and in a broader range (▲ and ○ from [1188]), from differential scanning calorimetry (DSC) at 10 K/min (◇ from [1193]), from stimulated-echo NMR (• from [1189]) and from T_1 experiments (◆ from [1188]). The *dash-dotted line* represents an Arrhenius law and the *solid line* a Vogel–Fulcher law. The DSC results from [1194] depended somewhat on the heating rate which was varied between 20 and 150 K/min. Here shown are those data which approach the dielectric ones closest. All results described so far were obtained for supercooled or glassy ASA. The *closed squares* refer to CD_3 group reorientation times in crystalline ASA as reported in [1197] with the *dashed line* representing an energy barrier of 510 K. The *lone red filled circle* is the primitive relaxation time calculated from the CM equation (see text). (*Right*) Frequency-dependent dielectric loss factor of ASA. The upturn seen at 230 K and low frequencies is due to the high-frequency flank of the α-relaxation. The *lines* for $T = 70$ K are fits using the Cole–Cole expression. Reproduced from [1188] by permission

with $\tau_\alpha(T)$ obtained by dielectric, calorimetric, stimulated-echo NMR, and spin-lattice relaxation measurements and fitted to the VFTH dependence. Using the VFTH dependence, a steepness index $m(T_g) = 79 \pm 5$ was determined at $\tau_\alpha(T_g) = 100\,\text{s}$. The closed squares refer to methyl CD_3 group reorientation times in crystalline acetyl salicylic acid with the dashed line as fit to the Arrhenius dependence with an energy barrier of 510 K. The Kohlrausch correlation function of the α-relaxation was determined by the stimulated-echo amplitude measurements for supercooled partially deuterated aspirin ($HCOO–C_6H_4–COOCD_3$) as a function of the mixing time at several temperatures from about 258 K down to 247 K. The Kohlrausch exponents $\beta \equiv (1 - n)$ determined exhibit mild linear decrease with decreasing temperature dependence and was used to obtain its value at T_g. The result is $\beta(T_g) = 0.42$ or $n(T_g) = 0.58$. Inserting the values of $n(T_g) = 0.58$, $t_c = 2\,\text{ps}$, and $\tau_\alpha(T_g)$ from the VFTH fit into the CM equation $\tau_0(T_g) = [\tau_\alpha(T_g)]^{1-n(T_g)} t_c^{n(T_g)}$, the calculated $\tau_0(T_g)$ shown as the lone solid red circle in Fig. 196 is in excellent

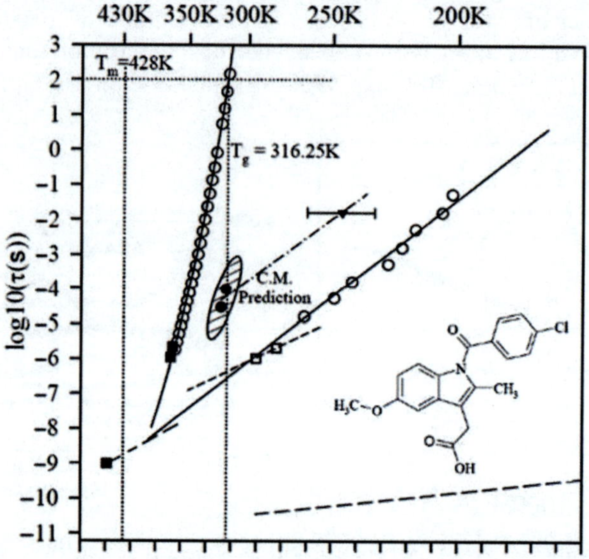

Fig. 196a Relaxation map of indomethacin. The *open circles* are relaxation time determined by dielectric measurements: isothermally above T_g and fitted with the Havriliak–Negami function; isochronally below T_g and determined from the temperature of the maximum loss factor below T_g. The *solid lines* represent the VFTH fit to the *T*-dependence of the α-process, and the Arrhenius fit to *T*-dependence of the γ-process. The *two closed circles* correspond to the primitive relaxation times predicted by the CM by using Eq. (2.68), the low-temperature point (*the closed inverted triangle*) is obtained from isochronal dielectric loss factor of the deeply quenched liquid. The *dashed-dotted line* has activation enthalpy of 56 kJ/mol. The *full squares* correspond to the ^{1}H-NMR T_{1z} (short times) and $T_{1\rho}$ (long time) minima of the amorphous phase, the *open squares* to the $T_{1\rho}$ minima measured in the crystalline γ-phase. Reproduced from [1000(b)] by permission

agreement with the experimental value of $\tau_{JG}(T_g)$ obtained by extrapolating the Arrhenius dependence to T_g. This good agreement between $\tau_0(T_g)$ and $\tau_{JG}(T_g)$ was found by Nath et al. [1189], and is indication that the observed secondary relaxation of aspirin is the JG β-relaxation.

Fast dynamics of aspirin was studied by optical heterodyne-detected optical Kerr effect (OHD-OKE) experiment by Cang et al. [1195]. The OHD-OKE signal $S(t)$ of aspirin is qualitatively no different from that of salol (phenyl salicylate) [330(a)], *ortho*-terphenyl [330(b)], and dibutyl phthalate (DBP) [331]. An intermediate power law (IPL) having time dependence t^{-1+c} with $c \approx 0.2$ in the time region from about 2 to 20 ps was observed in aspirin at 312 K. At 284 and 276 K, the IPL with $c \approx 0$ was observed immediately after a highly damped vibration with a period of 2 ps. When c becomes small at these lower temperatures, the t^{-1+c} decay of $S(t)$ corresponds to the near constant loss (NCL) spectrum in the susceptibility $\chi''(\nu)$ discussed before in Section 2.3.2.6. Thus, even the dynamics of aspirin at short times when molecules are caged is similar to that of ordinary glassformers.

Ibuprofen

Correia and coworkers investigated the structure and dynamics in the supercooled liquid and glassy states of another common drug (2*RS*)-2[4-(2-methylpropyl)phenyl]propanoic acid ($C_{13}H_{18}O_2$), commonly known as ibuprofen, which is a widely used non-steroidal pharmaceutical having analgesic, antipyretic, and anti-inflammatory properties [1004]. Two secondary relaxations, β and γ, are found together with the structural α-relaxation by broadband dielectric relaxation spectroscopy (DRS). The slower β-relaxation was observed at temperatures above and below T_g, and the T-dependence of its relaxation time $\tau_\beta(T)$ determined has been shown before in Fig. 131 in Section 2.3.2.5 together with $\tau_\alpha(T)$ and $\tau_\gamma(T)$. There, the discussion is on the encroachment of the JG β-relaxation toward the γ-relaxation near T_g, which causes the anomalous T-dependence of τ_γ. Here the focus is on the properties of the β-relaxation and its relation to the α-relaxation.

The temperature dependence of $\tau_\alpha(T)$ is well described by the VFTH law, $\text{VFTH}(T) = 1.2 \times 10^{-14} \exp\left[1426/(T - 187)\right]$, which shows that $\tau_\alpha(T)$ reaches $\tau_\alpha = 100$ s at $T = 226$ K, which can be taken as T_g. From the VFTH dependence, the steepness index $m(T_g) = 93$ was calculated showing that ibuprofen is a "fragile" glassformer. The slowest relaxation labeled D in Fig. 131 is associated with the hydrogen-bonding network, similar to that found in n-propanol and other hydrogen-bonded glassformers. The isothermal dielectric loss spectra at temperatures near and above T_g are fitted well by the one-sided Fourier transform of the Kohlrausch function with $\beta \equiv (1 - n) = 0.52$. Inserting this value of $n(T) = 0.48$, $t_c = 2$ ps, and $\tau_\alpha(T)$ from the VFTH fit into the CM equation $\tau_0(T) = [\tau_\alpha(T)]^{1-n} t_c^n$, the calculated $\tau_0(T)$ in a narrow range of temperatures near and above T_g are shown as stars (and labeled as $\tau_{JG,CM}$) in Fig. 131). The calculated $\tau_0(T)$ are in good agreement with the experimental value of $\tau_\beta(T)$ shown as open triangles in the figure, indicating that the β-relaxation is the JG β-relaxation of ibuprofen. Another indication of the same comes from change of the temperature dependence of $\tau_\beta(T)$ of ibuprofen from the

Arrhenius dependence below T_g to a manifestly stronger T-dependence above T_g, like that found in other glassformers. This temperature dependence of $\tau_\beta(T)$ above T_g albeit is weaker than that of VFTH(T) for $\tau_\alpha(T)$.

Change of temperature dependence of $\tau_\alpha(T)$ from one VFTH function to another and accompanying changes of temperature dependence of the dielectric strength of the α-relaxation were found in ibuprofen at $T_B = 265$ K [1004]. As shown before, these features are commonly found in non-pharmaceutical small molecular glassformers as well as in amorphous polymers such as poly(vinyl acetate). Thus, in many respects, the dynamic properties of ibuprofen are no different from ordinary glassformers.

Indomethacin

Indomethacin, $C_{19}H_{16}ClNO_4$ (1-(*p*-chlorobenzoyl)-5-methoxy-2-methylindole-3-acetic acid), is another anti-inflammation pharmaceutical that is commonly prescribed for patient suffering from gout. Carpentier and coworkers [1000(b)] found evidence of the presence of the JG β-relaxation in indomethacin by isothermal dielectric measurements above T_g, and in the glassy state obtained by a rapid and deep quench in the isochronal spectra (see Fig. 189). The $T_g = 315$ K was determined by DCS scan at the rate of 2 K/min. Good fits to the isothermal dielectric α-loss peak at 317 and 321 K were obtained by the Fourier transform of the Kohlrausch function with $n = 0.41$ (Eq. (2.1)). There is excess loss appearing on the high-frequency side of the α-loss peak at frequencies consistent with the location of the primitive relaxation frequency ν_0 calculated from the CM equation with $n = 0.41$. This coincidence suggests that the excess loss is due to the JG β-relaxation. The isochronal spectrum at 10 Hz as a function of temperature exhibits a loss peak at 245 K. This result together with the primitive relaxation times τ_0 calculated from the experimental values τ_α and $n = 0.41$ at two temperatures near T_g are shown by two filled circles in the relaxation map (Fig. 196a). Interpolation of these data points by a straight line is the predicted Arrhenius T-dependence of the JG β-relaxation time, $\tau_{JG} = (9.5 \times 10^{-17}\,\text{s})\exp(56\,\text{kJ/mol}/RT)$, of indomethacin. The faster γ-relaxation (open circles and open squares) is not the JG relaxation because it is also present in a crystalline phase of indomethacin and is attributed to rotation of the chlorobenzyl end group.

A secondary β-relaxation of amorphous indomethacin was also detected as a small endothermic peak by Vyazovkin and Dranca in differential scanning calorimetry (DSC) when reheating quenched sample after annealing in the temperature region from 253 to 278 K [992(b)]. Shift of the peak temperature with the heating rate allows the activation energy of β-relaxation to be estimated. The activation energy of the β-process obtained by this method is 57 kJ/mol, in good agreement with 56 kJ/mol determined by the combination of the dielectric τ_β and the calculated τ_0 shown in Fig. 196a by the dashed-dotted line, and reaffirms its identify as the JG β-relaxation. Another confirmation comes from the thermally stimulated depolarization current study by Correia et al. [1198].

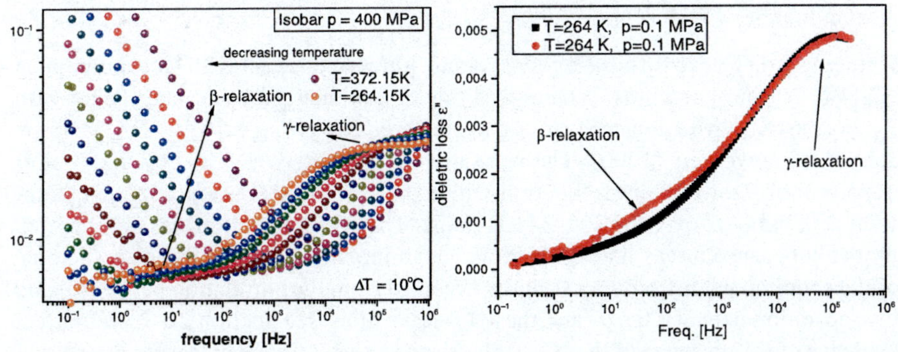

Fig. 196b (*Left*) Dielectric loss spectra of indomethacin all obtained at $P = 400$ MPa and temperatures in the range $372.15 \geq T \geq 264.15$ K. (*Right*) Comparison of dielectric loss spectra at $P = 0.1$ MPa and $T = 264$ K of two different glassy indomethacin obtained from the liquid state by different pressure and temperature paths. One path is by simply supercooling the melt at ambient pressure (*filled black squares*). The other path is first compressing the sample at $T = 310$ K at pressure of 400 MPa, then decreasing the temperature of sample down to 264 K while maintaining the pressure at 400 MPa, and finally reducing the pressure back to $P = 0.1$ MPa. Reproduced from [1199] by permission

Broadband dielectric measurements of indomethacin under high pressure of 400 MPa by Z. Wojnarowska et al. [1199] managed to find the JG β-relaxation appearing as a broad flat loss in the glassy state of indomethacin located at frequencies below the resolved γ-relaxation. This is shown in the left panel of Fig. 196b. Forced fitting of the flat loss was attempted in [1199] to determine the JG β-relaxation under elevated pressure of 400 MPa. However, one must be beware of large uncertainties accompanying these reported values.

Study of the dependence of the secondary relaxations of indomethacin on the thermodynamic paths carried out in [1199] is helpful to show the difference between the unresolved JG β-relaxation and the resolved γ-relaxation. One path is by simply supercooling the melt at ambient pressure (filled black squares) down to 264 K. The other path is first compressing the sample at $T = 310$ K at pressure of 400 MPa, then decreasing the temperature of sample down to 264 K while maintaining the pressure at 400 MPa, and finally reducing the pressure back to $P = 0.1$ MPa. It can seen in the right panel of Fig. 196b that the JG β-relaxation is sensitive to the difference in the thermodynamic paths but not the resolved γ-relaxation.

Poly(vinylpyrrolidone) (PVP) and Ursodeoxycholic Acid (UDA)

The DSC method used by Vyazovkin and Dranca also found the JG β-relaxation in poly(vinylpyrrolidone) (PVP) and ursodeoxycholic acid (UDA). The activation enthalpies are 68 and 67 kJ/mol for PVP and UDA, respectively, and are supported by dynamic mechanical measurements [992(a)].

Acetaminophen (Tylenol)

Isothermal dielectric spectra of glassy and ultraviscous states of acetaminophen ($C_8H_9NO_2$, and known as Tylenol and others), an analgesic and antipyretic with $T_g = 296$ K, have been measured over the frequency range 10 Hz–0.4 MHz by Johari and coworkers [1200]. The relaxation spectra are well fitted by the Fourier transform of the Kohlrausch function with exponent, $\beta \equiv (1 - n)$, remaining constant at 0.79±0.02 over the 305–341 K range. The JG β-relaxation was not reported probably because it was not resolved in acetaminophen because of the small value of the coupling parameter, $n = 0.21$, as expected from the correlation between n and the separation between the α- and the JG β-relaxation (see Section 2.3.2.1). Indirect evidence of the presence of the JG β-relaxation in acetaminophen was deduced from the temperature dependence of ε_∞, the limiting high-frequency permittivity of the α-process.

Dielectric spectroscopy was used to study supercooled liquid and glassy mixtures of acetaminophen and nifedipine (calcium channel blocker) with compositions ranging from pure acetaminophen (ACE) to pure nifedipine (NIF) by El Gorsey and Böhmer [1201]. The glass transition temperatures defined by $\tau_\alpha(T_g) = 100$ s vary linearly with composition, indicating good miscibility. The steepness index $m(T_g)$ characterizing the α-relaxation as well as the dispersion width of this process was almost independent of composition. Acetaminophen has $m(T_g) = 90$, and $\beta \equiv (1 - n) = 0.75$, which is consistent with the value 0.79±0.02 given by Johari et al. [1200]. The large $m(T_g) = 90$ and yet the small $n = 0.21 - 0.25$ of acetaminophen and nifedipine is another example of cases in which "fragility" does not correlate with non-exponentiality such as propylene carbonate (for other examples see Section 2.2.3). A well-resolved low-temperature γ-relaxation was found and likely originates from the side group. In fact, the ratio of its activation enthalpy E_γ to RT_g is 11.7 much smaller than the usual value of JG β-relaxation. Neither the isothermal nor the isochronal dielectric spectra resolve the JG β-relaxation, which is expected in view of the small n. Nevertheless, its presence was suggested by the wing on the low-temperature side of the α-loss peak in the isochronal spectra at 1.6 Hz (see Fig. 197). The wing of $\varepsilon''(T)$ at 1.6 Hz is similar to that found in the loss modulus $G''(T)$ at 5.4 kHz of metallic glass shown before in Fig. 170, except the absence of the γ-relaxation in the metallic glass. Assuming the ratio of the activation enthalpy E_{JG} of the unresolved JG β-relaxation of the mixtures of acetaminophen and nifedipine to RT_g is in the neighborhood of 24 as found in many glassformers, and assuming the prefactor τ_∞ of $\tau_{JG}(T)$ is 10^{-14} s, the location of the JG relaxation on the isochronal spectrum measured at $\nu = 1.6$ Hz was calculated. Its location on the wing as indicated by an arrow in Fig. 197 supports the interpretation given above.

Complex heat capacity, $C_p^* = C_p' - iC_p''$, where C_p' and C_p'' are the real and imaginary components, of ultraviscous melt and glassy acetaminophen was measured by dynamic heat capacity spectroscopy [1202]. The Kohlrausch exponent β from the C_p^* measurements is 0.65 for acetaminophen. It is somewhat less than the value of 0.79±0.02 obtained from dielectric relaxation studies, but the corresponding value

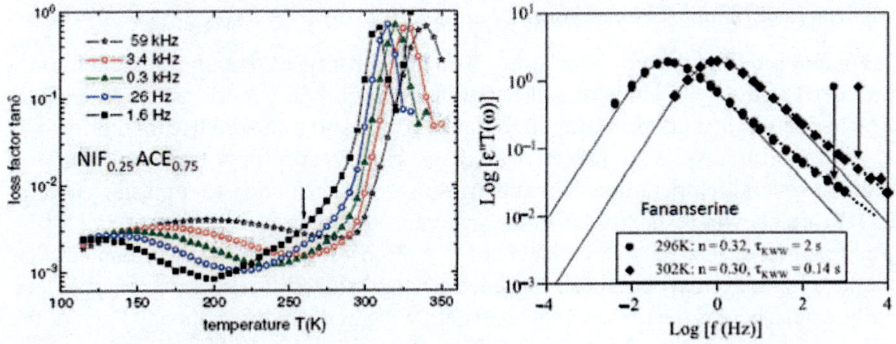

Fig. 197 (*Left*) Loss factor of $NIF_{0.25}ACE_{0.75}$ from El Gorsey and Böhmer [1201]. The *lines* are drawn to guide the eye. The *arrow* marks the JG loss peak temperature at $\nu = 1.6$ Hz (*filled squares*) expected on the basis of the estimate made in the text. Reproduced from [1201] by permission. (*Right*) The dielectric loss factor of fananserine at 296 and 302 K. The *lines* are KWW fits to the α-relaxation peaks with $n = 0.30$ (302 K) and $n = 0.32$ (296 K). *Vertical arrows* indicate the location of the corresponding primitive relaxation frequencies f_0 calculated from the CM equation and the parameters τ_α and n of the α-relaxation. Reproduced from [1000(c)] by permission

of $n = 0.35$ is still small to resolve the JG β-relaxation in the C_p^* spectra. This difference was explained by the fact that the configurational changes that contribute to the entropy and therefore to the C_p involve all types of molecular arrangement in a melt, including translational diffusion without rotation, and the breaking and reforming of hydrogen bonds (particularly relevant to acetaminophen having the –OH, and –NH- groups), all of which contribute to structural fluctuations. In contrast, dielectric relaxation involves only those modes of motions that change the dipole vector. The greater multiplicity of the modes involved in calorimetric relaxation was used to rationalize why β is less for C_p measurements than for dielectric relaxation.

Fananserine

Fananserine ($C_{23}H_{24}FN_3O_2S$) is a pharmaceutical having anxiolytic properties. The dynamics of amorphous fananserine was studied by dielectric relaxation, complex heat capacity, and NMR measurements by Carpentier et al. [1000(c)]. Its calorimetric T_g at the scanning rate of 5 K/min is 292 K, which is about the same as 290 K, the dielectric T_g arbitrarily defined as the temperature at the dielectric $\tau_\alpha = 100$ s. From the VFTH T-dependence of τ_α, the steepness index value of $m(T_g) = 77$ was determined. The Kohlrausch fit to the isothermal dielectric loss spectrum at 296 K gave $n = 0.32$, which is small and similar to that of acetaminophen (0.25 from dielectric, and 0.35 from calorimetry). Like acetaminophen, the consequence of small n is that the JG β-relaxation of fananserine was not resolved and showed up as an excess wing on the high-frequency flank of the α-loss peak. The location of the excess wing is consistent with the calculated primitive frequency ν_0. These are shown in the right panel of Fig. 197.

Amorphous Tramadol Monohydrate and Its Salt

Tramadol is a drug used to treat severe pain and most types of neuralgia and usually produced as the hydrochloride salt in two forms: solid dosage forms and injectable. The molecular dynamics of tramadol monohydrate and tramadol hydrochloride was studied by Kaminski et al. [1203]. The dielectric loss spectra of tramadol monohydrate at several temperatures and superpositions for reduction to a master curve at 251 K are shown, respectively, in the inset and the main part of Fig. 198a [1203]. The features of the frequency dispersion are almost the same as propylene carbonate (see Fig. 35b) except there is absence of the resolved γ-relaxation in propylene carbonate. There is the presence of an excess wing at the high-frequency flank of the α-loss peak, which is well fitted by the Fourier transform of the Kohlrausch function with $\beta_{KWW} = (1 - n) = 0.70$. The location of the primitive relaxation frequency at 251 K calculated by the CM equation indicated by the arrow in the figure lies within the domain of the excess wing, and suggests the excess wing is the unresolved JG β-relaxation. The steepness or fragility index m is 76. All these characteristics of the dynamics of tramadol monohydrate are similar to propylene carbonate.

Changing tramadol monohydrate to its hydrochloride salt results in significant increase in glass transition temperature. Large conductivity contribution to dielectric loss in tramadol hydrochloride makes it impossible to characterize the α-relaxation by the data represented by $\varepsilon''(f)$. Kaminski et al. resorted to analyzing the data represented by the complex electric modulus $M^*(f) \equiv 1/\varepsilon^*(f)$. The electric loss modulus, $M''(f)$, at 333 K of tramadol hydrochloride in the liquid state shows the presence of the JG β-relaxation by the shoulder instead of the excess wing in the monohydrate

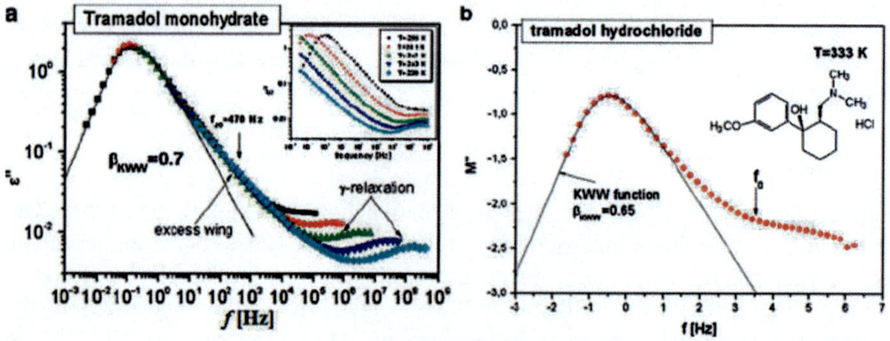

Fig. 198 (**a**) Dielectric loss spectra of tramadol monohydrate and five indicated temperatures after shifting in frequency to superimpose the high-frequency side of α-relaxation. The *inset* shows the same spectra before superposition. The *solid line* is a KWW function with $\beta_{KWW} = 0.70$. The *arrow* denotes the position of the primitive relaxation predicted by the CM equation at 251 K. (**b**) Frequency dependence of the imaginary part M'' of the electric modulus for tramadol hydrochloride at $T = 333$ K. The *solid line* is a KWW function with $\beta_{KWW} = 0.65$. The *arrow* denotes the position of the primitive relaxation predicted by the CM equation at 333 K. The structure of tramadol monohydrate is the same as tramadol hydrochloride shown except HCl is replaced by H_2O. Reproduced from [1203] by permission

(see Fig. 198b). The α-loss modulus peak is fitted by the Fourier transform of the Kohlrausch function with $\beta_{KWW} = (1 - n) = 0.65$. The calculated primitive relaxation frequency is consistent with the probable frequency of the barely resolved JG β-relaxation.

As Kaminski et al. pointed out, the fact that the JG β-relaxation in the monohydrate is closer to the α-process than in the hydrochloride salt may explain why the monohydrate is much more stable against crystallization than the hydrochloride. This means that tramadol monohydrate has less local mobility from its JG β-relaxation than tramadol hydrochloride.

Telmisartan

Telmisartan is a drug commonly prescribed to patients with high blood pressure. The dielectric loss data of Telmisartan obtained at ambient pressure by Adrjanowicz et al. [1204(a)] are shown in Fig. 199-1. It has an extremely broad β-process, which can be resolved at the lowest temperature 403.15 K but still above T_g (left panel). Below T_g, a faster γ-relaxation (right panel of Fig. 199) appears together with the β-process. The full-width at half-maximum of the β-process in the glassy state is estimated to be about eight decades in frequency. The α-loss peak at 403.15 K can be fitted by the Fourier transform of the Kohlrausch function with $\beta_{KWW} = 0.61$. The calculated primitive frequency $\nu_0 \approx 10^3$ Hz is indicated by the location of the thick vertical arrow consistent with the probable relaxation time of the flat β-process, and thus the latter is the JG β-relaxation of telmisartan.

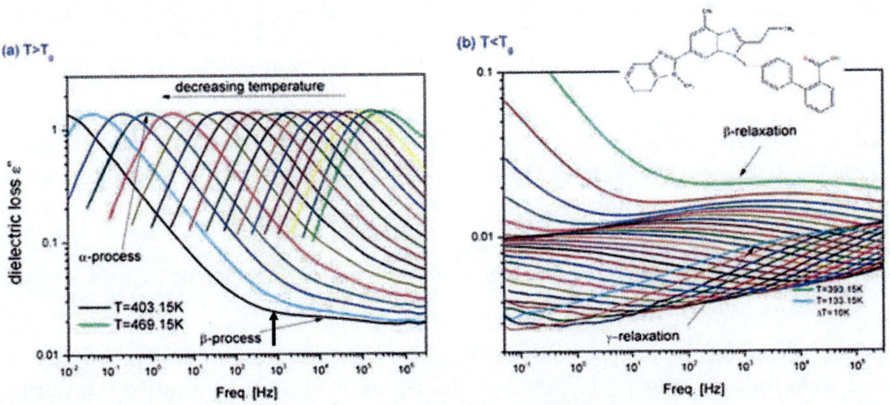

Fig. 199-1 (*Left*) Dielectric loss spectra of telmisartan obtained on cooling at ambient pressure and temperatures above the glass transition temperature between temperatures 469.15 and 403.15 K, after subtracting the conductivity. The *thick arrow* indicates the primitive frequency $\nu_0 \approx 10^3$ Hz calculated for 403.15 K. (*Right*) Dielectric loss data in the glassy state. The chemical structure of TEL is shown in the *inset*. Reproduced from [1204(a)] by permission

Celecoxib

Another pharmaceutical, celecoxib, was studied by dielectric relaxation, and the JG β-relaxation was found together with a faster intramolecular γ-relaxation [12049(b)]. Some data are shown in Fig. 199-2. Good agreement was found between the calculated τ_0 and τ_{JG} near T_g, and this suggests the change of T-dependence of τ_{JG} when crossing T_g like in other neat glassformers and mixtures. See caption of Fig. 199-2 for more details.

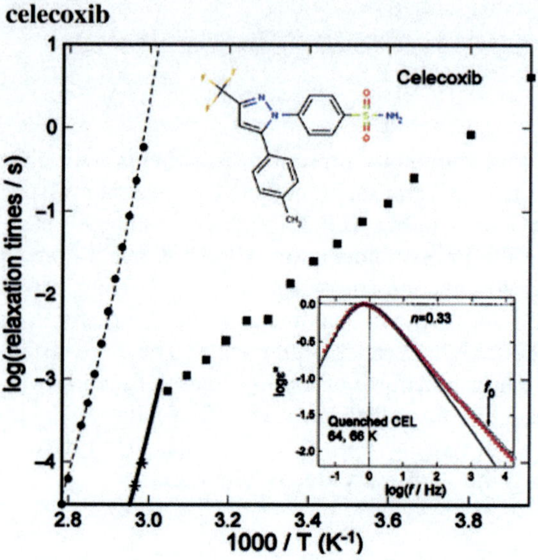

Fig. 199-2 The *inset* shows the loss spectrum of amorphous celecoxib at 64 and 66°C (after shifting to superpose with the data of 64°C) fit by the Fourier transform of the Kohlrausch function with $n = 0.33$. The *arrow* indicates the calculated primitive f_0. The main figure shows the α-relaxation time τ_α (•) and the VFTH fit (*dashed line*), the JG β-relaxation time τ_{JG} (■) with Arrhenius T-dependence, and the calculated primitive relaxation time τ_0 from the VFTH fit to τ_α (*thick line*). The *two stars* are τ_{JG} obtained by fitting the spectra at the two temperatures by the Cole–Cole representation of the JG β-relaxation. Note the near agreement between the calculated τ_0 and τ_{JG} near T_g, and the suggested change of T-dependence of τ_{JG} when crossing T_g

2.3.2.24 Relation Between the Arrhenius Activation Energies of τ_α and τ_{JG} in the Glassy State

As demonstrated in the previous sections, at temperatures above T_g, the JG β-relaxation time τ_{JG} has been shown to correspond well to the primitive relaxation time τ_0, and both are related to the structural α-relaxation by Eqs. (2.68) and (2.69). This equation should continue to hold at temperatures below T_g. However, testing this relation in the glassy state is more difficult because of either the scarcity or the unspecified thermal history of the data on the α-relaxation time τ_α. Full and reliable information on the characteristics of the structural relaxation can be acquired only when the structure is at equilibrium and such condition is rarely satisfied at temperatures well below T_g. In fact, glassy systems are non-ergodic and their properties can depend on aging time and thermal history. However, for glasses in isostructural state with a constant fictive temperature T_f, τ_α as well as τ_{JG} and τ_0 should have Arrhenius T-dependences with activation enthalpies E_α, E_{JG}, and E_0, respectively. The CM equations (2.68) and (2.69) in this case lead us to the relation

$$(1-n)E_\alpha = E_0 = E_{JG} \tag{2.83}$$

between the activation enthalpies. This relation was predicted by another model of relaxation by Perez and coworkers in glassformers [1205–1208], although it differs from the CM in quantitative details. When checking this prediction, it is important to bear in mind that both τ_α and τ_{JG} used to determine E_α and E_{JG} are derived from experiment under isostructural condition with T_f being held constant. Wrong conclusion could result from data of τ_α and τ_β obtained in glassy states having different T_f, a consequence of different cooling rates, annealing temperatures. and waiting times. In fact, it has been demonstrated in physical aging experiments [1209] that the structural states induced in three different types of thermal history (isothermal, isochronal, and isostructural) are very different. Thus, to verify the relation (2.83), it is important to know the thermal history for fulfillment of isostructural condition. This was done for dipropyleneglycol dibenzoate (DPGDB) by a procedure explained in [1014]. Knowing E_α of the isostructural state of DPGDB and $n = 0.43$, the product $(1-n)E_\alpha$ has the value of 48.4 kJ/mol. On the other hand, by fitting the experimental JG β-relaxation times in the lower temperature region to an Arrhenius equation, its activation energy E_{JG} has the value of 48.0±0.6 kJ/mol. There is good agreement between $(1-n)E_\alpha$ and E_{JG}.

Unfortunately, most of the papers in literature concerning JG relaxations do not report any information about thermal history and one cannot be sure if the relaxation times were derived isostructurally. The simple relation (Eq. (2.83)) can be tested for other systems where genuine JG relaxation has been reported in the literature, although the thermal history followed to obtain the glass is not given for all of them. The results from the work of Capaccioli et al [669] are shown in Fig. 200, where the product $(1-n)E_\alpha$ is plotted against the experimentally measured activation energy E_{JG} of the JG β-relaxation. A linear regression of data yields an angular coefficient of 0.99±0.01, meaning a remarkable agreement between experiments and model predictions, in spite of possible errors due to unknown thermal history of the glassy state. The data reported in Fig. 200 include some low molecular mass van der Waals glassformers (like OTP, isopropylbenzene, and toluene), some hydrogen-bonding system (like glucose and n-propanol), the mixture benzyl chloride in toluene, and the polymer PMMA. Additionally the figure displays also the JG β-relaxation of some epoxy compounds having multiple relaxation scenario (only the slowest secondary relaxation is the genuine JG) [242, 1052], the data of the JG β-relaxation for the polymer PET [1210], and the recently discovered JG β-relaxation of polyisoprene [978].

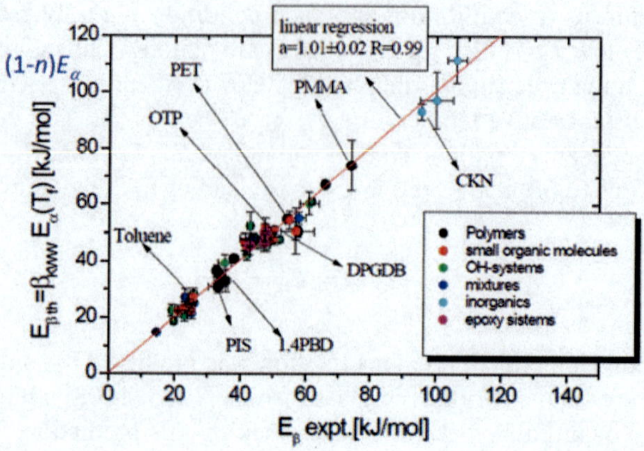

Fig. 200 Linear correlation between the experimental activation energy in the glassy state for the JG β-relaxation process (*abscissa*) and the activation energy predicted by the CM for the primitive relaxation (*ordinate*). *Symbols* for simple van der Waals molecules, H-bonded systems, polymers, chlorobenzene/toluene mixture, and epoxy oligomers are shown in the figure. The *solid line* is a linear regression of data (linear coefficient 0.99±0.01). Reproduced from [669] by permission

Application: Using the JG β-Relaxation to Predict the α-Relaxation Time Deep in the Glassy State

The Arrhenius T-dependence of τ_{JG} below T_g usually can be determined and can be extrapolated to lower temperatures. Using this relation between E_α and E_{JG} which is completely determined by $(1-n)$, one can estimate the magnitude of τ_α at any temperature in the glassy state. This has practical application in materials engineering in estimating τ_α, which is far too long to be determined directly by any means.

2.3.2.25 Aging of the JG β-Relaxation Used to Probe Structural Relaxation in the Glassy State

In Section 2.3.2.21 experimental evidences have been given to show that the JG relaxation is responsible for structural change deep in the glassy state on aging. Also in Section 2.3.2.24 the ratio of the activation energies of τ_α and τ_{JG} in the glassy state, E_{JG}/E_α, has been shown to be approximately equal to $(1-n)$. These results are indications of the JG β-relaxation playing a fundamental role in physical aging, and also there is still a connection between the JG β-relaxation and the α-relaxation in the glassy state. In particular, this connection should show up in some aspects of physical aging of non-equilibrium glass. Densification of the glass by physical aging should have concurrent effects on the JG relaxation and the α-relaxation, although the magnitudes of the effects are different. Because of this, physical aging can be studied through either the structural α-relaxation or the JG relaxation. Conventional study of aging is on the α-relaxation, which is a difficult task because it becomes inaccessible when its relaxation time τ_α increases on aging. Study of aging via the JG relaxation, although unconventional, nevertheless has the advantage that its relaxation time τ_{JG} is much shorter, and the change of the characteristics of the JG relaxation on aging can be monitored with ease and precision. This advantage was exploited by Casalini and Roland [1211] in their study of aging of 1,2-polybutadiene or poly(vinylethylene) (PVE) by measuring the variation the JG dielectric loss $\varepsilon''(\tilde{f}, t_{ag})$ with aging time t_{ag} at a fixed frequency $\tilde{f}$ in the vicinity of loss maximum, and at different aging temperatures T well below T_g. The time variation of $\varepsilon''(\tilde{f}, t_{ag})$ clearly deviates from single exponential decay, but is well described by the Kohlrausch stretched exponential function, $\exp[-(t_{ag}/\tau_{ag})^{\beta_{ag}}]$, with the stretching exponent $\beta_{ag} = 0.4$ for all aging temperatures (see Fig. 200, left panel). Although fixing β_{ag} equal to 0.4 for all aging temperatures is an assumption, this value is the same as the stretch exponent $\beta(T_g)$ of the Kohlrausch function whose Fourier transform fits well the dielectric α-loss peak of PVE close to T_g [981]. This coincidence between β_{ag} and β suggests that the variation of $\varepsilon''(\tilde{f}, t_{ag})$ with t_{ag} results from change in structure by aging, which is ultimately caused by the α-relaxation in this study, and τ_{ag} is somehow related to the α-relaxation.

The values of τ_{ag} determined in this manner at various aging temperatures are shown in the right panel of Fig. 200 together with the JG relaxation times τ_{JG} as a function of T in the glass state of PVE. Casalini et al. calculated τ_α from the CM, assuming the Kohlrausch exponent β in the glassy state retains the value of $\beta(T_g) = 0.40$ determined at T_g, consistent with the assumption that β_{ag} has the same value for all aging temperatures. It can be seen from the figure that both the absolute values and the temperature dependence of τ_α calculated from the CM for $T < T_g$, τ_α^{cm}, are in good agreement with τ_{ag}. These results suggest that although the aging time τ_{ag} is ultimately determined by the structural α-relaxation, it is still tied to τ_{JG} in the glassy state by the CM relation. Like the results shown in Section 2.3.2.24 above, it is possible to use the JG relaxation as a probe of the structural dynamics in the glassy state, which otherwise is inaccessible because of its immeasurably long relaxation time.

With the experimental protocol used to reach the various aging temperatures well below T_g, Casalini et al. assume they are studying a glass with a fixed structure, and the determined τ_{ag} are isostructural values. This assumption seems to be inconsistent with the definition of aging which must involve structural changes, and the isostructural hypothesis means no change in structure and hence no aging. Discussed in previous subsections, some glassformers are more sensitive to density as indicated by larger exponent γ in the TV^γ-dependence of τ_α in the liquid state. A weaker TV^γ-dependence with the same γ holds for τ_{JG}, to be shown in Section 2.3.2.29 later. Densification on aging shifts τ_{JG} in concert with τ_α to longer times on aging. Glassformers less sensitive to density (smaller γ) including PVE here show no such shift of τ_{JG} on aging. In this case, the increase of τ_α is suggested to be caused by increase of the dispersion of the structural α-relaxation or n, which reflects the structural change necessary for aging. Increase of n or decrease of β_{ag} with t_{ag} is a possibility, and may be consistent with the aging data of $\varepsilon''(\tilde{f}, t_{ag})$ when implemented in a more sophisticated fitting procedure than that used by Casalini et al.

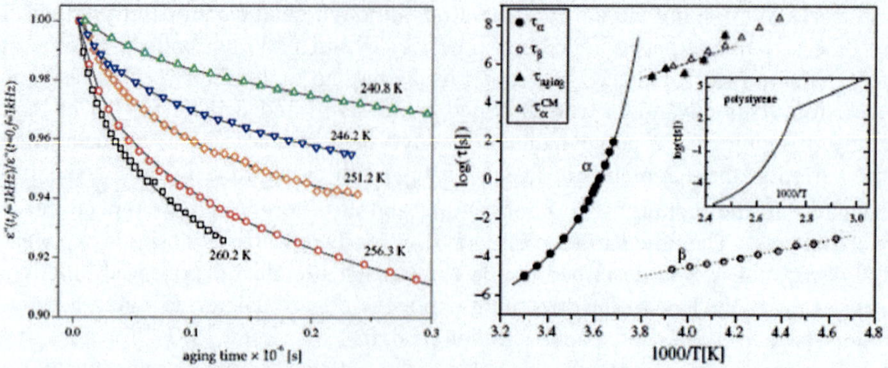

Fig. 201 (*Left*) Dielectric loss at $f = 1$ kHz vs. aging time t_{ag} normalized to the value at $t_{ag} = 0$ for PVE at five temperatures. The *solid lines* are fits to an expression describing the dependence of the normalized loss on t_{ag} by a stretched exponential function (see text). The characteristic time τ_{ag} of the stretched exponential increases with decreasing aging temperature indicated in the figure. The values of τ_{ag} are shown by *filled triangles* in the *right panel*. (*Right*) Relaxation times for the α and JG processes of PVE, together with the aging decay time τ_{ag} and the τ_α calculated from CM and labeled by τ_α^{cm}. τ_{ag} has the same behavior as τ_α^{cm}. The *solid line* is a VFTH fit to τ_α above T_g. The *dotted lines* are Arrhenius fits. The *inset* shows τ_α for polystyrene in the equilibrium liquid and glassy states, the latter determined from the intensity of second harmonic generation from chromophores within the polystyrene from [1212]. For more details see Casalini and Roland [1211]. Reproduced from [1211] by permission

Increase in n on aging beyond $n = 1 - \beta(T_g) = 0.6$ has the consequence of changing the temperature dependence of τ_{ag} from that obtained by Casalini et al. assuming $\beta(T_g) = 0.4$. This change may account for the deviations of τ_{ag} from τ_α^{cm} in detail which can be seen in Fig. 201, right panel.

Similar result was obtained from the JG β-relaxation of the pharmaceutical, telmisartan, on aging at 353.15 K by Adrjanowicz et al. [1204]. The isothermal dielectric loss spectra of telmisartan have been presented before in Fig. 199.

2.3.2.26 The Carbohydrates, Monosaccharides, Disaccharides, and Polysaccharides

Carbohydrates or saccharides, from the Greek word meaning sugar, constitute a large class of biomolecules that play many roles in biological processes, biochemical reactions, pharmaceutical processing as well as are a source of energy in living organisms [1213, 1214]. They are important in food science as one of the major ingredients, cryobiology as cryoprotectant [1215–1226], and pharmaceutical industry as matrix for amorphous drugs [1227, 1228]. The basic carbohydrate units are called monosaccharides. Among them, fructose and sorbose have six carbon atoms. Four of them and one oxygen form a five-membered ring, and hydroxyl groups are attached to three of the carbons and the resulting ring is a *furanose*. Each of the two remaining carbons and a hydroxyl group are attached to one of the two carbons adjacent to the oxygen. For glucose and galactose, five carbon atoms and one oxygen

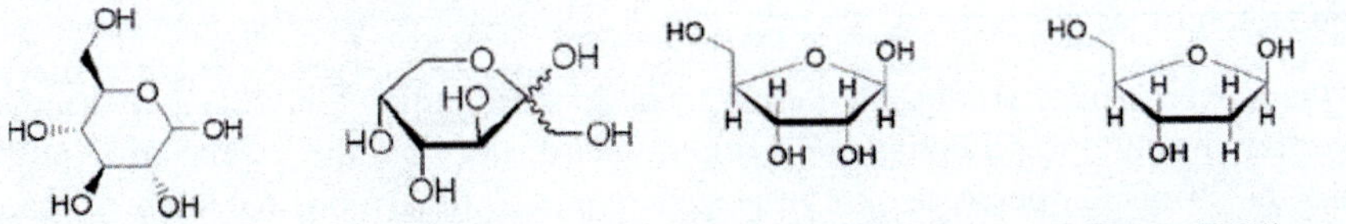

Glucose Fructose D-Ribose 2-Deoxy-D-Ribose

Fig. 202 Chemical structures of the monosaccharides: glucose, fructose, D-ribose, and 2-deoxy-D-ribose

form a six-membered ring, and hydroxyl groups are attached to four of the carbon, and the resulting ring is a *pyranose*. The remaining carbon and a hydroxyl group are attached to one of the carbon atoms adjacent to the oxygen. The structures of these two monosaccharides are depicted in Fig. 202, more can be found in [708].

D-Ribose ($C_5H_{10}O_5$) is a monosaccharide containing a total of five carbon atoms forming a pyranose ring. Closely related to D-ribose is 2-deoxy-D-ribose ($C_5H_{10}O_4$). The main difference between the two is that, in the case of 2-deoxy-D-ribose, the hydroxyl group attached to the carbon (#2) in D-ribose is reduced to hydrogen (see Fig. 202). D-Ribose and 2-deoxy-D-ribose are, respectively, the building blocks of the backbone chains in the nucleic acids, RNA (ribonucleic acid) and DNA (deoxyribonucleic acid). Monosaccharides can be linked together into disaccharides and polysaccharides (e.g., starch, cellulose, glycogen, chitosan, dextran) in a variety of ways by different glycosidic bonds. For instance, the disaccharide sucrose is formed by linking D-glucose and D-fructose together by a glycosidic bond as shown in Fig. 203. Notable disaccharides maltose and trehalose ($C_{12}H_{22}O_{11}$) are formed by linking two glucopyranose rings together by two different glycosidic linkages (see Fig. 203 here and more in [1012(c)]). While the pyranose and furanose rings of monosaccharide units are fairly rigid, the glycosidic bonds confer flexibility to disaccharide and polysaccharide molecules.

Carbohydrates in general are glass-forming substances. In view of their importance in science and applications, their molecular dynamics has been amply investigated using various kinds of spectroscopies in monosaccharides [1012(a), 1229–1238], disaccharides [1012(b), 1012(c), 1239–1242], and polysaccharides [1012(d), 1243–1252].

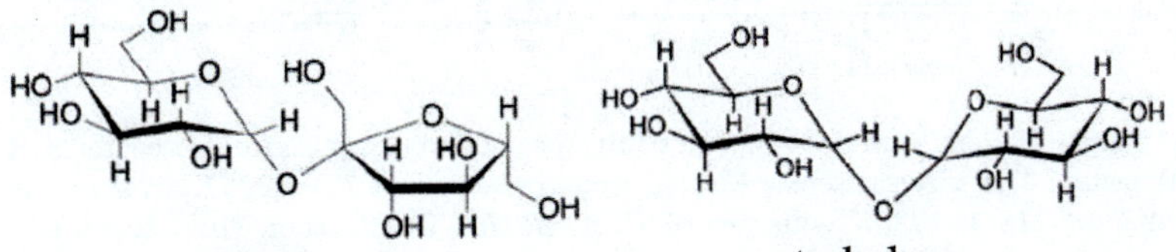

sucrose trehalose

Fig. 203 Chemical structures of the disaccharides: sucrose and trehalose

Monosaccharides

Earlier experimental studies of monosaccharides including glucose and fructose have found only one secondary relaxation much faster than the α-relaxation at the glass transition temperature. An important advance of a more recent broadband dielectric study over wide temperature ranges down deep into the glassy states is the discovery of the presence of a slower secondary relaxation in these two monosaccharides as well as in galactose, sorbose, and ribose [1012(a)]. The isothermal loss data of fructose and galactose are shown in Fig. 204. Naturally, this new secondary relaxation is called the β-relaxation and the previously known secondary relaxation the γ-relaxation. Dielectric relaxation measurements of fructose and ribose at elevated pressures show that the β-relaxation shifts to lower frequency with increasing pressure [1012(a)], mimicking the pressure dependence of the α-relaxation. The experimental facts together with other ancillary considerations lead us to conclude that the β-relaxation is the fundamental Johari–Goldstein β-relaxation of the monosaccharides. On the contrary, the relaxation time of the γ-relaxation of fructose shifts by 0.2 decade compared with 1.1 decade of the α-relaxation on applying pressure of 4800 bar.

The γ-relaxation possibly originates from the hydrogen-bonding scheme in these substances, but the data are not sufficient to make any definite conclusion. NMR study of the motion of the methylol in glucose [1233] has shown that the glucose ring and the methylol (CH_2OH) group mobility are strongly correlated and that the rotation of the methylol should not be used to explain the faster γ-relaxation process of glucose. Since the monosaccharides are hydrophilic, the presence of water in normally called dried samples and its motion (JG relaxation of water component) may give rise to the γ-relaxation, which is another possibility. This is supported by

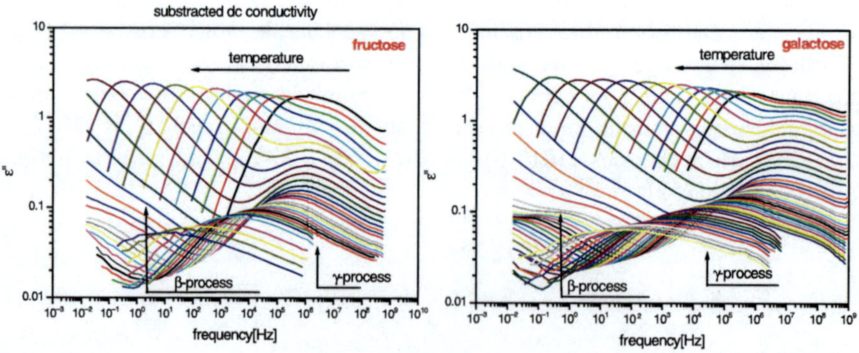

Fig. 204 (*Left*) The dielectric loss spectra of fructose at ambient pressure. Dc conductivity has been subtracted. The spectra were measured from $T = 333$ to $T = 225$ K with step of 4 K and from $T = 218$ to 178 K with step of 10 K. (*Right*) The dielectric loss spectra of galactose at ambient pressure. Dc conductivity has been subtracted. The spectra were measured from $T = 353$ K to $T = 305$ K with step of 4 K and from $T = 299$ K to $T = 175$ K with step of 4 K. Reproduced from [1012(a)] by permission

the γ-relaxation times, and activation energies of the monosaccharides are about the same as the JG relaxation of water confined in vermiculite clay, molecular sieves, and silica gels, as well in some aqueous mixtures [1093].

The JG β-relaxation of galactose can be seen as a resolved loss peak at temperatures much below T_g, but only as a shoulder near and below T_g. It only shows up as a shoulder at all temperatures in fructose, glucose, sucrose, and ribose [1012(a)]. The shoulders are sufficient evidence for the presence of the β-relaxation. However, its relaxation times were determined in [1012(a)] by modeling its contribution to the loss with a Cole–Cole function and fitting the loss spectrum with the additive contributions of the α- and γ-relaxations. As pointed out near the end of Section 2.3.2.1, this procedure is incorrect because it violates the fact that the JG β-relaxation is correlated with and is a precursor of the α-relaxation. Likely the procedure also overestimates τ_{JG} because the shoulder is assumed to be the high-frequency flank of the Cole–Cole distribution. This should be borne in mind when considering the τ_{JG} deduced in this way as shown by Fig. 7 in [1012(a)], especially when comparing it with the primitive relaxation time τ_0 calculated from the coupling model. Also, at temperatures above T_g, the α-loss peaks of fructose, glucose, galactose, and sorbose are broadened by the JG β-relaxation as well as the γ-relaxation, both of them having high dielectric strengths, and thus the true coupling parameter n of these monosaccharides would be smaller than that deduced from the width of the α-loss peaks. The situation in D-ribose and 2-deoxy-D-ribose is better in that the dielectric strengths of the γ-relaxations are much smaller than the α-relaxation. The Kohlrausch fit of the α-loss peak gives $n = 0.45$ [1012(e)]. The actual n of D-ribose and 2-deoxy-D-ribose could be somewhat smaller than 0.45 because residual broadening by the JG β-relaxation and presence of water in trace amount cannot be excluded. Nevertheless, the location of the primitive frequency ν_0 calculated from $n = 0.45$ is located as the shoulder found on the high-frequency flank of the α-loss peak.

Disaccharides

Dielectric measurements of disaccharides have revealed the presence of two secondary relaxations in their glassy states [1012(b), 1239, 1240]. Here we show the faster γ- and the slower β-relaxations by the dielectric loss data of seven disaccharides, sucrose, maltose, trehalose, lactulose, leucrose, lactose, and cellobiose, all in the glassy state in Fig. 205 (left panel). These disaccharides are each composed of two monosaccharides chosen from fructose, glucose, or galactose, and connected through different glycosidic bonds. Trehalose has the highest glass transition temperature T_g of the disaccharides. The faster γ-relaxations of the disaccharides have comparable relaxation times and activation enthalpies as the γ-relaxation in monosaccharides, and seems to be a common feature of the entire family of the carbohydrates (see the relaxation map in Fig. 206). Elevated pressures were applied to one of the disaccharides, leucrose, in dielectric measurements [1253], and the spectra in the right panel of Fig. 205 shows that at constant T the γ-relaxation does not shift with pressure P. Hence it is intramolecular in origin and has been shown

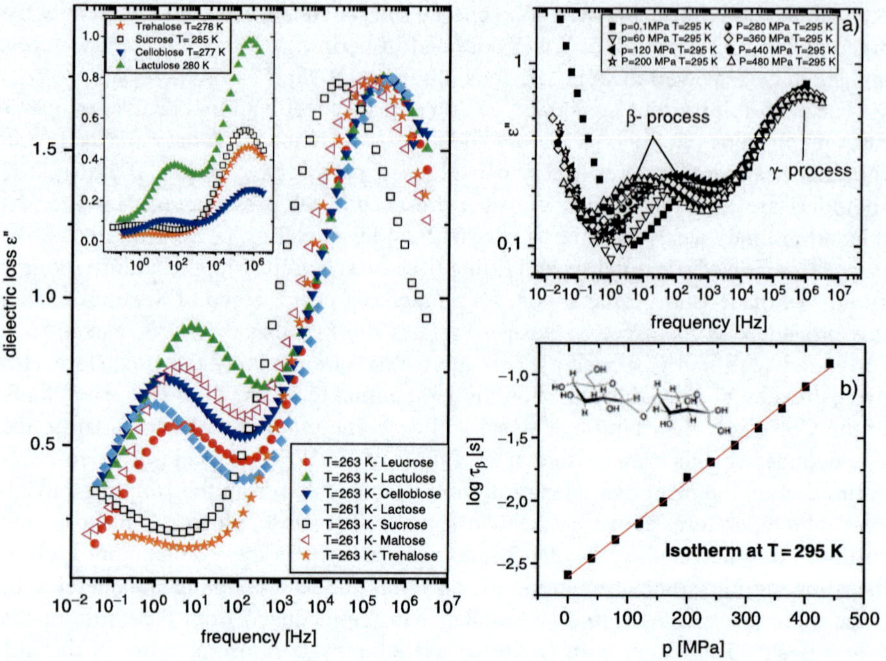

Fig. 205 (*Left*) Comparison of the loss spectra measured at almost the same temperature for all six disaccharides. In the *inset* are presented dielectric loss spectra measured for trehalose $T = 278$ K, lactulose $T = 280$ K, cellobiose $T = 277$ K, and sucrose $T = 285$ K, spectra of sucrose and maltose measured at the same γ-relaxation time. All spectra were vertically shifted to the height of γ-peak of maltose. Data from [1012(b)] replotted here. (*Right*) (**a**) Dielectric loss spectra of leucrose measured at $T = 295$ K and $P = 0.1$, 60, 120, 200, 280, 360, 440, and 480 MPa, showing the shift of the β-relaxation on elevating pressure, but not the γ-relaxation. (**b**) Pressure dependence of the isothermal relaxation times τ_β at 295 K. The activation volume $\Delta V_\beta = 21.5$ ml/mol. Data from [1253] replotted here

to be the intramolecular motion within the gluco- or fructopyranosyl rings of leucrose [1253]. From the similarity of the γ-relaxations in di- and monosaccharides, their motions are of the same type. On the other hand, the slower (β) relaxation moves to lower frequencies with increasing P and hence it is the JG β-relaxation of the disaccharide. These interpretations are supported by the insensitivity of the γ-relaxation time and sensitivity of the β-relaxation time to changes in density of the glass effected by different thermodynamic histories to arrive at the same final P and T [1253].

The origin of the slower (β) relaxation in sugars was a subject of our very recent studies. It was shown by Kaminski et al. that in sucrose this relaxation may originate from twisting rotation of the two monosugar rings (glucose and fructose) around the glycosidic bond [1012(b)]. The same point of view was presented by Meissner et al. [1250]. The properties of the β-relaxation of the other disaccharides including the relaxation time and activation energy are similar to that of leucrose. From this, we conclude that they all are JG β-relaxations.

Large conductivity contributions in the sucrose, lactulose, trehalose, maltose and leucrose studied only allow observation of the α-relaxation and determination of its relaxation time over a limited range of temperatures in the liquid state. It is impossible to obtain the true frequency dispersion of the α-relaxation and hence the coupling parameter n. In the case of lactose and cellobiose, the α-relaxation is totally masked by the conductivity in the whole studied temperature range. The relaxation times of the three processes, τ_α, τ_β, and τ_γ, determined are plotted against reciprocal temperature in Fig. 206.

De Gusseme et al. [1240] were able to probe the α-relaxation of trehalose ($C_{12}H_{22}O_{11}$) closer to T_g not possible with dielectric relaxation by means of specific heat spectroscopy performed by temperature-modulated differential scanning calorimetry (TMDSC). They also performed dielectric relaxation measurements and found the two secondary relaxations (β and γ) in addition to the α-relaxation. Their relaxation map is reproduced in Fig. 207. The T_g, at which the TMDSC relaxation time is 100 s, was 390 K, consistent with that determined by standard calorimetry. The measured τ_α by dielectric relaxation and TMDSC above T_g were combined to give rough estimate of the "fragility" index m. The value of m calculated for trehalose is 107, a rather high value for small molecular glassformers (130 according to [1012(b)]). The stretching exponent $\beta \equiv (1 - n) = 0.30$ was estimated from

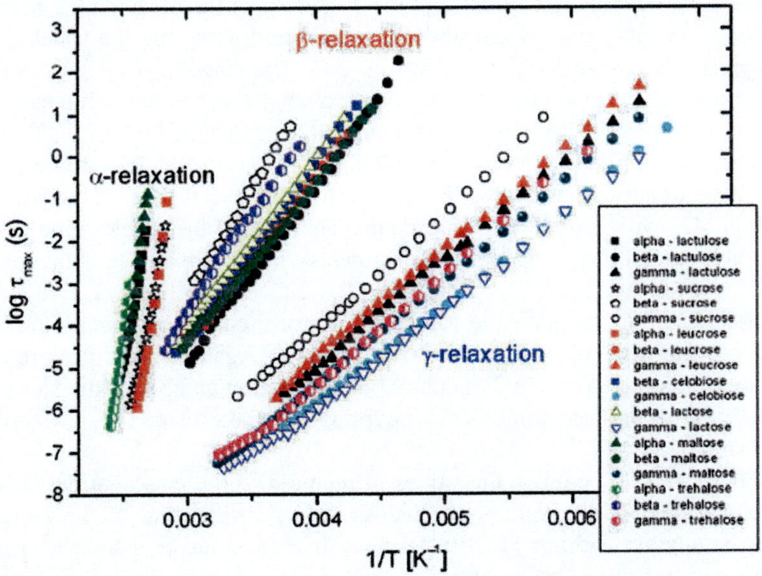

Fig. 206 Arrhenius plot of the relaxation times, τ_α, τ_β, and τ_γ, of three relaxation processes observed in the six disaccharides. Reproduced from [1012(b)] by permission

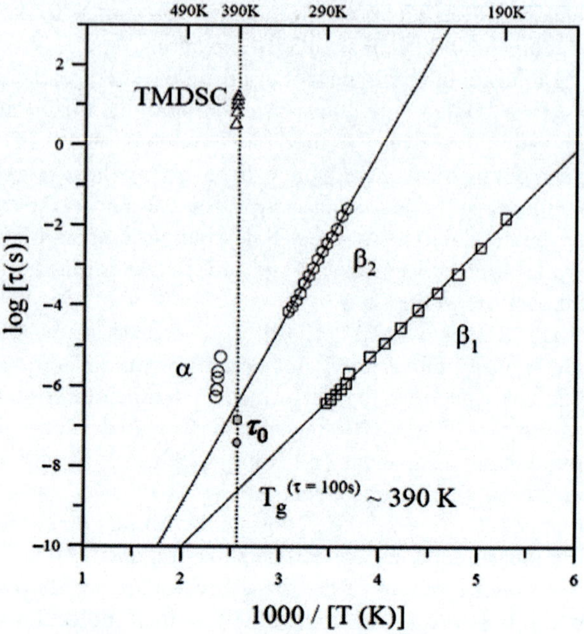

Fig. 207 Arrhenius relaxation map of trehalose, showing the α process and two distinct sub-T_g relaxation processes. Experimental data are from dielectric and specific heat spectroscopy (TMDSC) measurements. The *square* and the *circle* on the dotted line at $1000/T_g$ are the values of $\log[\tau_0(s)]$ calculated by the coupling model equation with $\beta_{KWW} \equiv (1-n)$ equal to 0.35 and 0.30, respectively. The value of $\beta_{KWW} = 0.30$ was deduced from TMDSC data by De Gusseme et al. [1240]. Reproduced from [1240] by permission

the isochronal specific heat spectroscopy data, according to the method proposed by Böhmer et al. [1254]. The uncertainty in determining this low value of β was not given, but for sure it is there because of the crudeness of the method. The primitive relaxation time $\tau_0(T_g)$ calculated with $\beta = 0.30$ and assuming the usual $t_c \approx 2$ ps is represented by the filled circle in Fig. 207. The calculated $\tau_0(T_g)$ is about a decade shorter than $\tau_\beta(T_g)$ obtained by extrapolating the Arrhenius T-dependence of τ_β measured below T_g. On the other hand, $\tau_0(T_g)$ calculated with $\beta = 0.35$ and represented by the filled square in Fig. 207 is closer to $\tau_\beta(T_g)$. Dielectric relaxation measurement of anhydrous maltose, an isomer of trehalose, was made by Noel et al. in 1996 [1241] with limited dynamic range. They found the α-relaxation and the γ-relaxation but not the β-relaxation. Nevertheless, they were able to obtain the frequency dependence of the α-loss peak. This allowed De Gusseme et al. to fit it to the Kohlrausch function and gave the value of 0.32 for the stretch exponent β, which is similar to the value of 0.30 deduced for trehalose. Again the loss peak reported by Noel et al. is broadened by conductivity contribution and hence the value of β corresponding to the true value of the coupling parameter $n \equiv (1-\beta)$ could be larger than 0.32.

It is interesting to compare the values of m and n of the disaccharides with the monosaccharides. The m value of trehalose is larger than that of the monosaccharides having six carbons [1012(a)], which in turn is larger than D-ribose, the monosaccharide having five carbons. The values of $n \approx 0.70$ or 0.65 of trehalose and $n \approx 0.68$ of maltose are larger than $n = 0.53$ (i.e., $\beta = 0.47$) of the monosaccharide glucose deduced from dielectric data by Chan et al. [1236] and 0.45 of

D-ribose by Kaminski et al. [1012(e)]. Thus, n increases with increasing number of carbon atoms or the molecular complexity in these sugars. This trend is expected by the coupling model as due to increase of intermolecular coupling with the number of carbon atoms of glassformers in the same family. An analogue of this is the series glycerol, threitol, xylitol, and sorbitol discussed before [841].

The disaccharides are effective as bioprotectant in the preservation of life, and trehalose appears to be the most efficient. In fact in nature, trehalose is widely found in various organisms that can survive adverse environmental conditions such as extreme drought and high or low temperatures in the state of almost complete dehydration and all metabolic processes are blocked [1255–1257]. The organisms enter into a state called *anhydrobiosis*, where almost all biological activities are suspended to allow them to survive the extreme conditions. These anhydrobiotic organisms include yeast, fungi, mushroom, and vegetation and life in desert areas. When dehydrated, all metabolic processes are blocked and life cycle in these organisms apparently ceases, but life is restarted upon rehydration. The molecular mechanisms that distinguish trehalose for its superior capabilities of biopreservation still remain unclear, despite various experimental and theoretical works [1258, 1259].

Polysaccharides

Polysaccharides including cellulose, starch, curdlan, and dextran belong to the group of polyglycanes that have the same repeating unit, namely the anhydroglucose unit. Their chemical structures differ mainly by the form of the glycosidic linkage in their polymer backbone. Cellulose has the $\beta(1-4)$ linkage, starch has the $\alpha(1-4)$ linkage, curdlan has $\beta(1-3)$ linkage, and dextran has the $\alpha(1-6)$ linkage. For each repeat unit, cellulose, starch, and curdlan have two hydroxyl side groups and one hydroxymethyl side group, while dextran has three hydroxyl groups and no hydroxymethyl group. The repeat units of three polysaccharides are shown in Fig. 208. The disaccharides, maltose and cellobiose, respectively, have the $\alpha(1\text{–}4)$ and $\beta(1-4)$ glycosidic linkages of two glucose units.

The JG β-relaxation has been found in the mono- and disaccharides as discussed before. It should be present in oligomers of saccharides and polysaccharides if it lives up to the expectation of being a universal and fundamental feature in all glassformers. We have seen that the γ-relaxation in the mono- and disaccharides invariably has larger relaxation strength than the JG β-relaxation, making the detection of the latter difficult. This is usually the case when a faster non-JG secondary relaxation with large dielectric strength is present such as the dialkyl phthalates discussed before. The oligomeric and polymeric carbohydrates are no exception. A dominant faster γ-relaxation with large strength found in the dielectric spectra of many polysaccharides has led some researchers to put emphasis on it instead of the slower β-secondary relaxation. On the other hand, dynamic mechanical spectroscopy has seen both relaxations with nearly equivalent strength. A critical review of data of polysaccharides and interpretations can be found in [1012(d)].

It turns out that the β- and γ-relaxations of cellulose and its oligomers (cellotribiose, cellotetrabiose, and cellopentabiose) have almost the same relaxation

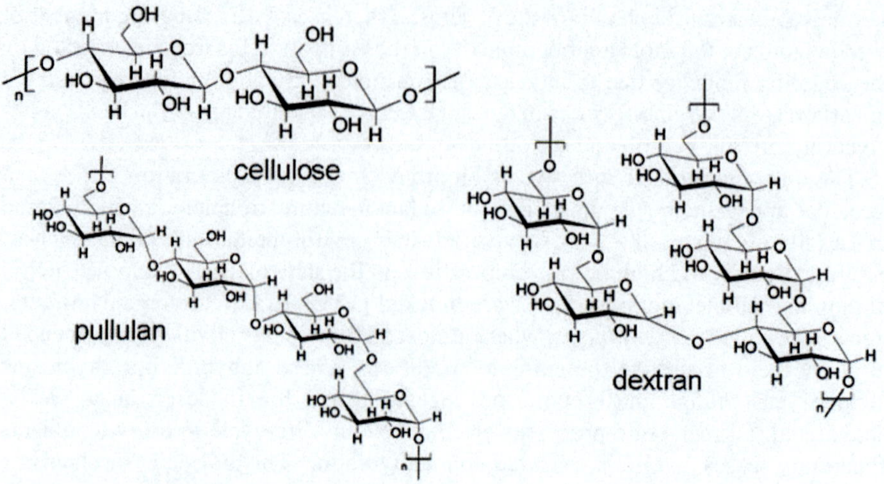

Fig. 208 Chemical structures of the repeat units of cellulose, pullulan, and dextran. Dextran is cross-linked

times and activation energies as those of its dimer, namely the disaccharide, cellobiose. These are shown by the comparison of their isothermal spectra at similar temperatures in Fig. 209 (left) and the T-dependences of their relaxation times in Fig. 209 (right). Since the β-relaxation of disaccharides has been identified as the JG β-relaxation, from the concurrence it follows that the β-relaxation of the polysaccharides are their JG β-relaxation. The molecular interpretation of the γ-relaxation of polysaccharides is as uncertain as that for the monosaccharides and disaccharides mentioned earlier. One possibility is reorientation of bound water [1248, 1260] or confined water in the polysaccharides [1261], which can been reinterpreted as the JG relaxation of water in the polysaccharides [1093].

2.3.2.27 JG β-Relaxation (or Primitive Relaxation) of Water

Water is the most abundant substance in our planet Earth, present in significant proportion in all living organisms, and without which life is impossible. Despite the fact that water molecule is simple compared with other glass-forming substances, its thermodynamic and kinetic properties are intriguing and their interpretation controversial. The study of glass transition of bulk water is hampered by crystallization. Water can be supercooled from the melting point at 273 K down to 235 K, but no further because it crystallizes [1263–1267]. Glass transition was reported to occur at a temperature T_g of about 136 K in vapor-deposited amorphous solid water (ASW) [1268, 1269]. Rapidly quenching water to temperatures below 100 K resulted in hyperquenched glassy water (HGW), which exhibits a glass to liquid like calorimetric transition at $T_g \approx 136$ K [1270–1272], and in low-density amorphous water (LDA) at $T_g \approx 129$ K[1273]. Analysis of dielectric data of ASW and HGW also

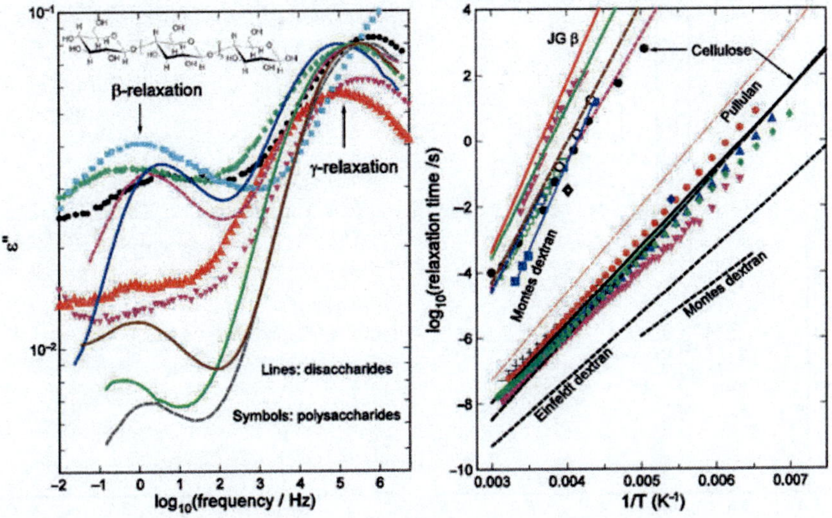

Fig. 209 (*Left*) Isothermal dielectric loss spectra showing the lower frequency β-relaxation and the higher frequency γ-relaxation of two polysaccharides, cellulose and dextran, and the pentamer, cellopentaose, obtained by Meissner et al. [1250(a)]. Cellulose (*green diamonds*, at 243 K), cellulose (*black circles*, at 258 K), dextran (*light blue squares*, at 258 K), cellopentaose (*red triangles*, at 243 K), and cellopentaose (*magenta inverted triangles*, at 258 K). The chemical structure of the pentamer cellopentaose is shown on *top*. These data are compared with the lower frequency JG β-relaxation and the higher frequency γ-relaxation of four disaccharides obtained by Kaminski et al. [1012(c)], and all shown by *lines*. Cellobiose (*magenta dashed line* at 263 K), maltose (*blue solid line* at 261 K), sucrose (*brown short dashed-long dashed line* at 285 K), and at the *bottom* trehalose (*green dotted line* at 268 K, and *grey dashed line* at 273 K). The ε″ of these four disaccharides have been scaled to have the same loss at the maximum of the γ-relaxation as that of the γ-relaxation of cellulose at 258 K. The actual data points of the disaccharides are not shown to avoid crowding. (*Right*) Relaxation map of relaxation times. On the *top* are the JG β-relaxation times of some disaccharides represented by *lines* (data not shown for the sake of clarity except for leucrose): sucrose (*red*), trehalose (*green*), cellobiose (*brown*), maltose (*magenta*) from Kaminski et al. [1012(c)], trehalose from De Gusseme et al. (*blue*) [1240]. JG β-relaxation times of the disaccharide, leucrose, are shown by *open green squares*. The *filled black circles* are the relaxation time of the so-called "δ-relaxation" (equivalent to β-relaxation in this chapter) of cellulose obtained by Einfeldt et al. [1249]. The three *black open circles* nearby are dynamic mechanical data of cellulose by Montes et al. [1247]. The *lone open diamond* is from the mechanical data of cellulose by Kubat et al. [1262]. The *blue line* represents the Arrhenius fit to the relaxation times of the so-called β-process of anhydrous dextran obtained by Montes et al. [1252] by a combination of dielectric and mechanical relaxation data represented by *filled squares* near it. The data of Scandola et al. [1245] consistent with that of Montes et al. are not shown. On the *bottom* are the γ-relaxation times of monosaccharides, fructose (+), glucose (*green diamonds*), and galactose (*magenta inverted triangles*), and disaccharides, maltose (*filled red circles*) and cellobiose (*blue triangles*). The *black solid and long dashed lines* represent the Arrhenius temperature dependences of the relaxation times of the β_E-relaxation of two kinds of cellulose, linters pulp and avicell, respectively, from Einfeldt et al. The β_E-relaxation (equivalent to the γ-relaxation in this chapter) times of the Merck starch are indistinguishable from that of linter pulp cellulose, and are not shown. The *long and short dashed red line* is the Arrhenius *T*-dependence of dried pullulan from Einfeldt et al. [1249] and the *lone filled diamond* is the isochronal dielectric data of pullulan at 10 Hz from Nishinari et al. [1244]. At the very *bottom* are the relaxation times of the β_E-relaxation (equivalent to the γ-relaxation in this chapter) obtained by Einfeldt et al. and the γ-relaxation time obtained by Montes et al. of dextran and combined with data of Scandola et al. and Nishinari et al. Reproduced from [1012(d)] by permission

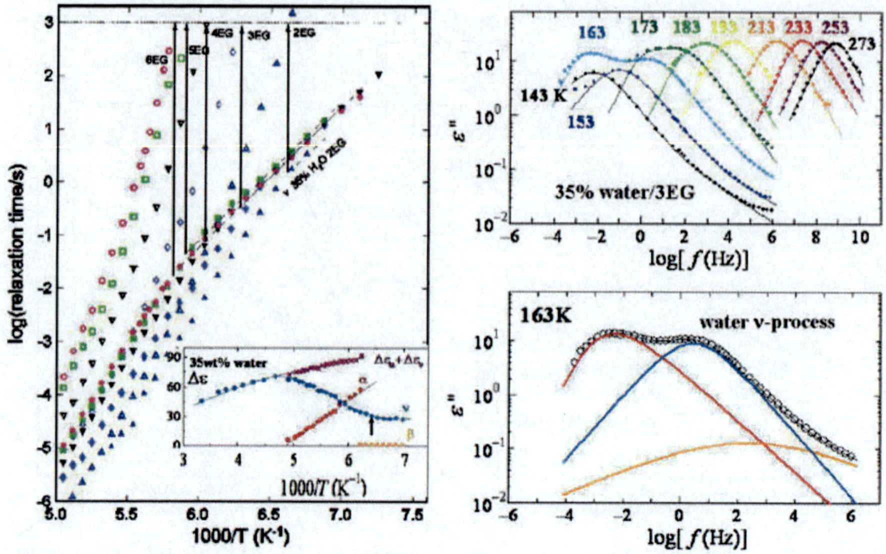

Fig. 210 (*Right*) Dielectric loss data of aqueous mixture of tri-ethylene glycol (3EG) with 35 wt% water at various temperatures in the *upper panel* from [1093, 1295]. The *lower panel* shows the water relaxation in the mixture (called the ν-process in [1295]) resolved as a peak at 163 K near T_g where τ_α reaches 10^3 s as shown in the *left panel*. Reproduced from [1295] by permission. (*Left*) Temperature dependence of the dielectric relaxation time τ_α (*open symbols*) and τ_ν (corresponding *closed symbols*) of mixtures of 35 wt% of water with various ethylene glycol oligomers. *Circles* for 6EG. *Squares* for 5EG. *Downward pointing triangles* for 4EG. *Diamond* for 3EG. *Upward-pointing triangles* for 2EG. Some of the data of the relaxation time of the ν-process, τ_ν (*closed symbols*), overlap and cannot be easily seen. For this reason, we use the *dashed lines* to indicate the Arrhenius temperature dependences assumed by τ_ν of the mixtures starting approximately at temperatures below T_g of the mixtures defined by $\tau_\alpha(T_g) = 10^3$ s. The *vertical arrow* locates $1000/T_g$ for each mixture. The change of temperature dependence of τ_ν at T_g is similar to that seen of the JG β-relaxation of a component in non-aqueous mixtures. The *inset* presents the dielectric strength $\Delta\varepsilon_\alpha$ of the α-relaxation and $\Delta\varepsilon_\nu$ of the ν-process in the mixture of 3EG with 35 wt% of water, as well as their sum. The *vertical arrow* indicates the location of $1000/T_g$. It can be seen that $\Delta\varepsilon_\alpha$ increases with decreasing temperature typical of α-relaxation, and $\Delta\varepsilon_\nu$ increases with increasing temperature and it assumes a stronger T-dependence above T_g, characteristic of secondary relaxation of glassformers

indicated a T_g of 136 K [1274]. However, crystallization to cubic ice sets in at 150 K and thus bulk supercooled water is inaccessible to experimentation in the temperature range between 150 and 235 K. Consequently, controversies abound in the studies of water. For example, Angell and coworkers [1275, 1276] proposed water's T_g may be 165 K based on the heat-release behavior of HGW, but this was contradicted by Johari [1277]. The possibility of T_g in the range of 160–165 K was given by computer simulation [1278], by the consideration of thermodynamic data [1279], and by thermal desorption experiments of layered nanometer-scale films of labeled ASW [1280]. None of these can be verified directly because supercooled bulk water is inaccessible in the relevant temperature range above 150 K and below 235 K.

Adiabatic calorimetric measurements of uncrystallized water confined in 1.1 nm pores of silica gel made by Oguni and coworkers found two glass transitions at 115 and 160 K [1281(a), 1282]. They interpreted the data by dividing the water into two parts: "interfacial water" and "internal water." The "interfacial water" molecules at the silica interface form hydrogen bonds with the oxygen of the silica wall of the nanopore and Oguni and coworkers suggested that they are responsible for the glass transition at 115 K. The "internal water" molecules were considered to be located in the central part of the pore, interacting with each other like in bulk water, and thus the observed $T_g = 160\,\mathrm{K}$ seems to support the proposed T_g of bulk water at 165 K by Angell and coworkers [1275, 1276]. However, Mallamace et al. calculated the number of water molecules that can fit across the diameter of 1.1 nm size pore [1284(a)]. Given that the diameter of a water molecule is 0.38 nm, they showed only three (=1.1/0.38) water molecules can fit across the 1.1 nm pore diameter. Out of these, two would be interfacial water molecules bonded to the silica wall, leaving only one water molecule in the "internal water" in the center of the pore not hydrogen bonded to silica. This state of "internal water" molecules with only one molecule layer not hydrogen bonded to silica cannot be considered equivalent to bulk water. Thus the observed T_g of 160 K of water in the 1.1 nm pores cannot be due to the cooperative α-relaxation of bulk water, and the data do not support the proposed $T_g = 160\,\mathrm{K}$ of bulk water by Angell and coworkers. The same reason was given by Mallamace et al. to refute the use of the 227 K peak in the heat capacity C_p of the "internal water" component of water confined in the 1.1 nm size pore of Oguni and coworkers by Angell in [1285] to support his proposed order–disorder transition of supercooled water at that temperature.

Adding to the confusion and controversy is the extension of the studies of water in several different forms of nanoconfinement using adiabatic calorimetry by Oguni and coworkers [1281(a), 1281(b), 1282, 1283]. As mentioned already, they reported the specific heat C_p of water confined in 1.1 nm pores of nanoporous silica gel has a peak at about 225 K [1281(a), 1282]. The revision of pore size from 3 in the original article [1281(a)] to 1.1 nm is explained by the paper by Johari et al. [1284(b)]. Angell [1285] have used the data of Oguni and coworkers to propose supercooled *bulk* water would show a broad λ-type transition at 225 K, resembling the λ-type disorder–order transition of binary metal alloys and the orientational transition of C60. In this proposal, the assumption was made that the heat capacity C_p of bulk water is the same as that of water confined in the SiO_2 gel. This assumption was shown by Johari et al. [1284(b)] to be invalid. They argued that the average volume of a water molecule is $\sim 30\,\text{Å}^3 (= 18/6.03 \times 10^{23})$, and therefore, the effective diameter of its circumscribed sphere is 0.38 nm. Accordingly, only about three water molecules (= 1.1/0.38) can fit across the diameter in a 1.1 nm pore, ignoring the excluded volume effect. Two of these would be interfacial molecules at the silica wall forming a nanoshell. This leaves only one H_2O molecule near the center of the 1.1 nm diameter cross section which would be hydrogen bonded exclusively to four other H_2O molecules. This state is not the same as that of a H_2O molecule in supercooled bulk water and therefore the data of Oguni and coworkers [1281(a), 1282] may not be used for a thermodynamic discussion of bulk water. It is possible that instead the interfacial water bonded to the silica wall is responsible for the C_p peak of water confined in the SiO_2 gel observed at 225 K by Oguni and coworkers.

Adiabatic calorimetry measurements were made of water confined to 1.2, 1.6, and 1.8 nm diameter linear pores of molecular sieves, MCM-41 (silica), by Oguni and coworkers [1283]. Glass transitions were found at 115 K, and near 165 K in 1.2 nm pores. On increasing the pore size, a new feature appears at higher temperatures in the relevant range of 160–220 K. The C_p of confined water in 1.8 nm diameter pores of MCM-41 appears like a glass softening endotherm, from which Oguni et al. [1283] suggested anew that glass transition of nanoconfined as well as bulk water occurs at 210 K, and not at 136 or 165 K. However, Johari [1284] showed that the exothermic and endothermic features observed on heating rapidly precooled and slowly precooled states of water in 1.8 nm pores of MCM-41 and the unusually large increase in the specific heat in the 210–230 K range are inconsistent with kinetic unfreezing of a disordered solid, or glass softening. The exotherm is attributable to the melt's gradual conversion to distorted ice-like structures and the endotherm to the reverse process until their fractional amounts reach a reversible equilibrium on heating. As argued by Johari [1284], his analysis also puts into question a conclusion based on neutron scattering studies by Liu et al. [1289] that supercooled water undergoes a structural and kinetic transition at ~225 K while remaining a liquid.

Another controversial subject is the proposed fragile-to-strong transition in supercooled water [1286–1288] also by Angell and coworkers. The term is used to describe an abrupt change of the structural relaxation time on supercooling water from a stronger to a weaker T-dependence, which was inferred from comparison of data of the hyperquenched glassy water at low temperatures and supercooled water above 235 K. Chen and coworkers claimed that they have observed directly the fragile-to-strong transition for both relaxation time and diffusivity in uncrystallized water confined in molecular sieves with 1.5 nm diameter [1289, 1290]. The break in temperature dependence of the relaxation time from neutron scattering, and diffusivity from NMR, occurs at 220 K. This is near 225 K the specific heat peak temperature of water confined in 1.1 nm silica gel nanopores, but was now reinterpreted by Angell and coworkers as order–disorder transition [1285]. Stanley and co-workers have attributed this change observed by Chen et al. to crossing a Widom or supercritical fluctuations line [1291].

However, the claim of observation of the fragile-to-strong transition by Chen and coworkers was contested by Swenson and coworkers [1292, 1293(a)]. These authors ruled out that the relaxation observed by Chen and coworkers is the α-relaxation of supercooled water. Cerveny et al. [1294] invoked dielectric data of water confined in molecular sieves with pore sizes 5 and 10 Å by Jansson and Swenson [1102] to show that the crossover of the structural relaxation time of supercooling water from VFTH temperature dependence to Arrhenius dependence actually occurs at lower temperature than that observed by Chen and coworkers, and is not due to the fragile-to-strong mechanism. Instead, they explained the crossover on supercooling water as due to finite size effect imposed by confinement within the pore, which occurs when the length scale of the structural relaxation can no longer extend further than the pore size. This finite size effect leads to a crossover from liquid-like behavior (structural α-relaxation time having the VFTH dependence) toward localized motions (Arrhenius dependence), the model proposed by Lorthioir et al. for binary polymer blends [1070]. The same explanation was given by Cerveny et al. for the crossover of the T-dependence of an observed relaxation of water in aqueous mixtures [1107, 1108]. As shown before in Sections 2.3.2.10 and 2.3.2.15 for several different binary mixtures, this explanation along the line of Lorthioir et al. is untenable because the structural α-relaxation was observed separately and together with the faster relaxation that changes T-dependence at T_g in the same temperature range. Hence the relaxation undergoing the crossover of the T-dependence cannot be the structural α-relaxation. More evidence against the interpretation of Cerveny et al. in aqueous mixtures will be found below. As for the fragile–strong transition observed by Chen and coworkers, several studies by others have failed to confirm it. These studies will be discussed later in appropriate subsections.

No attempt is made here to review in detail the conflicting views on dynamics of supercooled water and to cite the vast literature on this subject here. Instead the focus is on experimental evidences showing the existence of the JG relaxation of water in aqueous mixture and hydrated proteins, in uncrystallized water under nanoconfinement, and in other situations. As we shall see, this upshot enriches the understanding of the dynamics of water because JG relaxation is an integral part of the dynamics leading to glass transition, as shown by the multiple correlations and relations it has with the structural α-relaxation. The paper that has definitively identified the JG relaxation of water from experiments is by Capaccioli et al. [1093]. They show that the relaxation of the water component present in many different aqueous mixtures [1295] has the same properties as the JG β-relaxation of a component in binary mixtures of van der Waals liquids (discussed before in Section entitled "Well-Resolved JG Relaxation of a Component in Mixtures of van der Waals Liquids" of Section 2.3.2.8). The knowledge of the α- and JG β-relaxations of a component gained from the study of the van der Waals mixtures was transferred to elucidate and interpret the experimental data on the dynamics of water in aqueous mixtures, in nanoconfinement [1093], and in hydrated proteins [1094]. The exercise illustrates the benefit of "cross-field fertilization" in scientific research, i.e., the application of the clear understanding achieved in the study of one class of glass-forming materials to the class of aqueous mixtures where hydrogen bondings make the interpretation more difficult.

Before proceeding to the specifics in the following subsections, also it should be mentioned that Bergman, Swenson, and coworkers found evidence of relaxation process in mixtures of water with other glassformers [1261] and in water with one dimension or more dimensions reduced to nanometer scale [1102, 1296]. However, these authors interpreted the observed relaxation in aqueous mixtures [1261] as well as the observed process of a "two-water-layer phase" confined between the platelets of sodium vermiculite as the α-relaxation of bulk water [1296] not as the JG relaxation of water. For the "two-water-layer phase" [1296] Johari has shown that the interpretation of α-relaxation of bulk water is untenable [1297]. One reason given by Johari is that the dielectric strength decreases on decreasing the temperature, which is the opposite of that expected from the Curie law for α-relaxation. This fact can also be used to reject the interpretation by Cerveny et al. [1107, 1108] that it is the α-relaxation of water in aqueous mixtures. On the other hand, this observed T-dependence of the dielectric strength is consistent with the behavior of secondary relaxation and with the interpretation of the observed relaxation originating from water in aqueous mixtures as the JG β-relaxation of water. Under extreme confinement, cooperative α-relaxation becomes impossible because of the small size, but the local JG β-relaxation survives and is observed. I think this is also the current view of Bergman, Swenson, and coworkers. Actually, like water confined in sodium vermiculite, extreme confinement of the polymer poly(methylphenylsiloxane) (PMPS) of layers less than 2 nm thick in galleys of nanocomposites has produced the same result [495, 496], as discussed before in Section 2.2.5.8. That is, the α-relaxation of bulk PMPS is no longer observed, and the JG β-relaxation that survives is nearly the same as that found in bulk PMPS. The effect is the same for OTP, PDMS, and PMPS confined in 2.5 nm silanized glass pores, also discussed before in Section 2.2.5.8.

JG β-Relaxation of Water in Aqueous Mixtures

Small in size and having strong affinity for hydrogen bonding, water does not behave in exactly the same way as other glassformers in binary mixtures with another component. In aqueous mixtures the other component is usually hydrophilic and susceptible to hydrogen bonding with water. Examples include 1-propanol, ethylene glycol (EG), ethylene glycol oligomers (EGO), tri-propylene glycol, glycerol, xylitol, sorbitol, poly(ethylene glycol) (PEG), poly(vinylpyrro1idone) (PVP), and poly(vinylmethylether) (PVME). Broadband dielectric relaxation spectroscopy was used together with differential scanning calorimetry (DSC) to characterize the dynamics of these aqueous mixtures by several groups [131, 1069, 1071, 1093, 1106–1111, 1140, 1298, 1300–1310].

Representative results from some aqueous mixtures [1093, 1295] are shown in Figs. 210 and 211. The two figures on the right show as an example the dielectric loss spectra of aqueous mixture of 3EG with 35% water at various temperatures from [1295, 1093]. The results of τ_α (open symbols) and τ_ν (closed symbols) for mixtures of 35 wt% water with five ethylene glycol oligomers, 2EG, 3EG, 4EG, 5EG, and 6EG, are shown in the left panel. At lower temperatures, the water relaxation in the mixture (called the ν-process in [1295], which is the JG β-relaxation of water in the mixtures) is resolved as a peak at higher frequencies than the α-loss peak, particularly clear at 163 K near T_g where τ_α reaches 10^3 s. For aqueous mixtures, we shall use ν and JG β to denote the same secondary relaxation of the water component. Their relaxation times, τ_α and τ_ν, are shown as functions of reciprocal temperature altogether in Fig. 210. For ease of identification of the data of τ_α and τ_ν from the same mixture, the same shapes of the open (τ_α) and closed symbols (τ_ν) are used. Similar plots are given for 20, 30, and 40 wt% water mixtures with glycerol in Fig. 211. The dielectric loss spectra of mixtures with 30 and 40 wt% of water are shown in the right panel, and the relaxation map of τ_α and τ_ν in the left

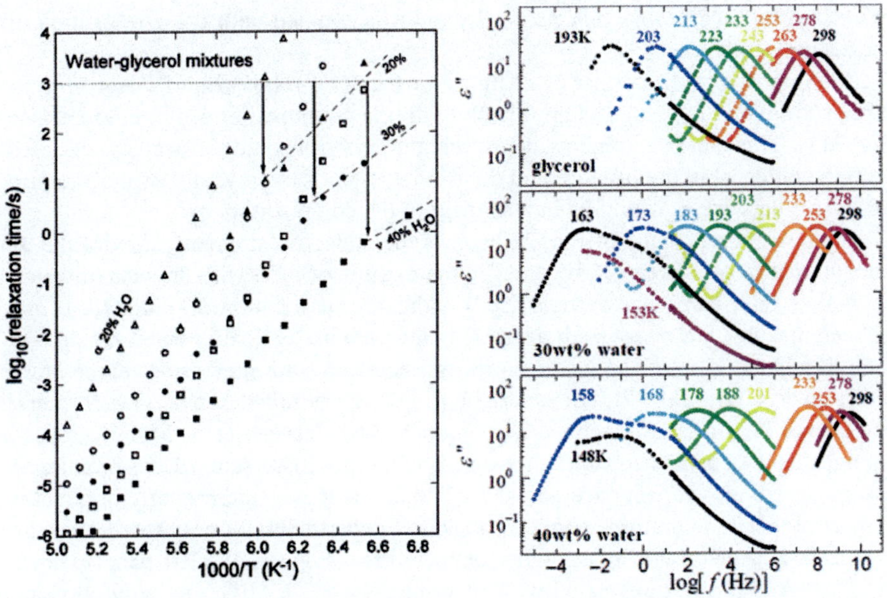

Fig. 211 (*Left*) Temperature dependence of the dielectric relaxation time τ_α (*open symbols*) and τ_ν (corresponding *closed symbols*) of mixtures of various wt% of water with glycerol. *Triangles* for 20 wt% water. *Circles* for 30 wt% water. *Squares* for 40 wt% water. Each *dashed line* indicates the Arrhenius temperature dependence assumed by τ_ν of the labeled mixture starting approximately at temperatures below T_g of the mixture defined by $\tau_\alpha(T_g) = 10^3$ s. The *vertical arrow* locates $1000/T_g$ for each mixture. The change of temperature dependence of τ_ν at T_g is similar to that seen of the JG β-relaxation of a component in non-aqueous mixtures. Reproduced from [1093] by permission. (*Right*) Dielectric loss spectra of neat glycerol, and mixtures of glycerol with 30 and 40 wt% water showing the α-process of the mixture and the water ν-process. Reproduced from [1295] by permission

panel. In all cases, τ_α has VFTH-like temperature dependence down to the longest times measured or down to temperatures near the T_g of the mixture indicated by heat capacity measurements wherever it is available. On the other hand, the temperature dependence of τ_ν is Arrhenius at lower temperatures but it crosses over to a stronger temperature dependence when temperature is increased above T_g. In these two figures, the dielectric glass transition temperature T_g, defined by $\log\left[\tau_\alpha(T_g)/s\right] = 3$, is indicated by the location of the vertical arrow for each mixture. It can be seen by inspection of the figures that the change of the temperature dependence of τ_ν occurs near the dielectric T_g in all cases. More examples from other aqueous mixtures showing the same behavior can be found in [131, 1069, 1071, 1093, 1106–1111, 1140, 1298–1310]. This general crossover property of the T-dependence of τ_ν or τ_{JG} of water in aqueous mixtures is exactly analogous to that of TBP, QN, and picoline in mixtures with tri-styrene shown in Figs. 147 and 149, and in neat glass-formers including xylitol and sorbitol (see Fig. 162). Hence the same interpretation holds, and we can conclude that the observed faster ν-process is the JG secondary relaxation of water *above and below* T_g, and it is coupled with the α-relaxation of the aqueous mixture.

It is important to note that τ_ν of the ν- or the JG β-relaxation of water in mixtures with glycerol changes significantly with the composition of the mixture (see Fig. 211). This fact is another evidence that the JG β-relaxation of water is coupled with the glycerol in the mixture and the α-relaxation of the glycerol–water mixture because otherwise τ_ν would be independent of the composition.

The inset of left panel of Fig. 210 shows the dielectric strengths, $\Delta\varepsilon_\alpha$ of the α-relaxation and $\Delta\varepsilon_\nu$ of the ν-processes in the mixture of 3EG with 35 wt% of water, as well as their sum. The vertical arrow indicates the location of $1000/T_g$. It can be seen that $\Delta\varepsilon_\alpha$ increases with decreasing temperature typical of α-relaxation. On the other hand, $\Delta\varepsilon_\nu$ of the water ν-process increases with increasing temperature *above as well as below* T_g, characteristic of secondary relaxation of glassformers, and cannot be α-relaxation above T_g as proposed by Cerveny et al. Moreover, $\Delta\varepsilon_\nu$ changes its T-dependence when crossing T_g from the glass state of the mixture to assume a stronger dependence above T_g. The fact that $\Delta\varepsilon_\nu$ increasingly dominates $\Delta\varepsilon_\alpha$ at higher temperatures above T_g makes it experimentally easy to resolve the ν-process and determine τ_ν at higher temperatures. Figure 212 (left panel) shows $\Delta\varepsilon_\alpha$ and $\Delta\varepsilon_\nu$ of mixtures of 35 wt% of water with 3EG, 5EG, and poly(ethylene glycol) of molecular mass of 600 (PEG600). The change of T-dependence of $\Delta\varepsilon_\nu$ on crossing T_g in all three aqueous mixtures are demonstrated in this figure. The right panel serves as a reminder of the same behavior exhibited by the JG β-relaxation of 16 wt% *tert*-butylpyridine (TBP) in mixture with tri-styrene. These figures are repeats of Figs. 183 and 184. They are shown again to heighten the awareness of the same effect found in aqueous mixtures. The similar behavior of $\Delta\varepsilon_\nu$ found in mixture of water with fructose also has been shown previously in Fig. 185, and in

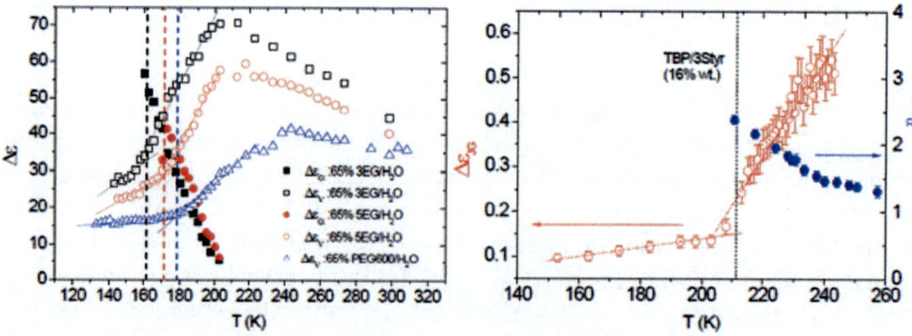

Fig. 212 (*Left*) Dielectric strengths $\Delta\varepsilon_\alpha$ and $\Delta\varepsilon_\beta$ of the mixtures of 35 wt% water and 65 wt% of 3EG, 5EG, and PEG600. *Full symbols* are for $\Delta\varepsilon_\alpha$ of the α-relaxation and *open symbols* for $\Delta\varepsilon_\beta$ of the JG β-relaxation. *Full lines* are linear fits of $\Delta\varepsilon_\beta$ below and above T_g. Each *dashed vertical line* indicates the temperature T_g near which the elbow-shape crossover of temperature dependence of $\Delta\varepsilon_\beta$ occurs. Here T_g is defined as the temperature at which $\log_{10}\left[\tau_\alpha(T_g)/\mathrm{s}\right] \sim 3$. (*Right*) The same behavior exhibited by the JG β-relaxation of 16 wt% *tert*-butylpyridine (TBP) in mixture with tri-styrene. Reproduced from [1093] by permission

mixture of water with glucose was reported by Tyagi and Murthy [1306]. Whenever data are available, the dielectric strength, $\Delta\varepsilon_\nu$ or $\Delta\varepsilon_{JG}$, of the ν- or JG-process of water in aqueous mixtures increases monotonically with temperature with an elbow shape that indicates a change of slope at T_g. This property is typical of secondary JG relaxation of a component in non-aqueous mixtures, and also in neat glassformers [1136, 1137], which serves as another indication that the structural α-relaxation and the JG β-relaxation are coupled. Hence, it can be considered as another indication that the process is the secondary JG relaxation of the water component coupled to the α-relaxation in the aqueous mixtures.

Despite the evidence given above for identifying the faster process in aqueous mixtures as the JG relaxation of water, other workers differ in their opinion. They have given different interpretations of the change from a stronger temperature dependence of its relaxation time above T_g to a weaker and Arrhenius dependence below T_g. Swenson and coworkers suggested that the observed faster process is a merged α- and β-process of water at temperatures above T_g [1106, 1261, 1292]. Cerveny and coworkers [1108, 1111, 1294] interpreted the relaxation at temperatures above T_g of the aqueous mixture not as the JG relaxation but as the structural α-relaxation of water, and the observed crossover of its relaxation time from stronger dependence of α-relaxation to a weaker Arrhenius T-dependence on supercooling the aqueous mixture past T_g as due to finite size effect of "confinement" by a frozen matrix. These two interpretations have in common of identifying the faster relaxation as the α-relaxation of the water component above T_g. They are incorrect because as shown for one example in Fig. 212, the dielectric strength of the faster process $\Delta\varepsilon_\nu$ decreases on decreasing the temperature at temperatures above T_g, opposite to the Curie law for $\Delta\varepsilon_\alpha$ of the α-relaxation. Moreover, the interpretation of the faster process in aqueous mixtures by Cerveny et al. is a repeat of that previously applied by Lorthior et al. to explain similar crossover in T-dependence of a relaxation of the more mobile component in binary polymer blends [1070] and van der Waals liquid mixtures [1095]. As discussed before in Section 2.3.2.15, this explanation has been ruled out experimentally at temperatures above T_g by the presence of a slower α-relaxation in parallel with the faster relaxation (considered also as the α-relaxation by Lorthior et al.) originating from the same component in several binary blends and mixtures, as shown before in Figs. 147 and 149. This simple fact rules out the interpretation by Lorthior et al. because this interpretation has to contend with the coexistence of two different α-relaxations of the more mobile component, which is impossible unless the mixture is heterogeneous. Also the dielectric strength of the observed relaxation of the more mobile component in mixtures of van der Waals liquids increases with increasing temperature above T_g (see Fig. 212, right panel for an example), which is another experimental fact that rules out its interpretation as the α-relaxation.

Emphasis has been made on the observed increase with increasing temperature of the dielectric strength $\Delta\varepsilon_\nu$ of the faster ν-process of water in aqueous mixture above T_g (see Fig. 212), opposite to the behavior of the dielectric strength $\Delta\varepsilon_\alpha$ of α-relaxation, which always increases with decreasing temperature. One may think that this is sufficient evidence to deter any interpretation of the ν-process as the α-relaxation of water above T_g of the mixture. So far this has not happened. Cerveny and coworkers [1107, 1108, 1111, 1294] still hold on to their interpretation of the faster relaxation in aqueous mixture above T_g as the α-relaxation of water. This is possibly due to them not having the chance of confronting the contradiction of the observed T-dependence of $\Delta\varepsilon_\nu$ to their interpretation. In fact in all their publications before the one in 2009 [1111], they never show their own data of $\Delta\varepsilon_\nu$ above T_g in any aqueous mixtures. The 2009 paper [1111] on dielectric relaxation of supercooled mixtures of ribose and deoxyribose with water reports isothermal spectra of the ν-process (called process II in the paper) at temperature *above* and *below* T_g of the mixtures with 30 wt% of water. These spectra taken at temperatures ranging

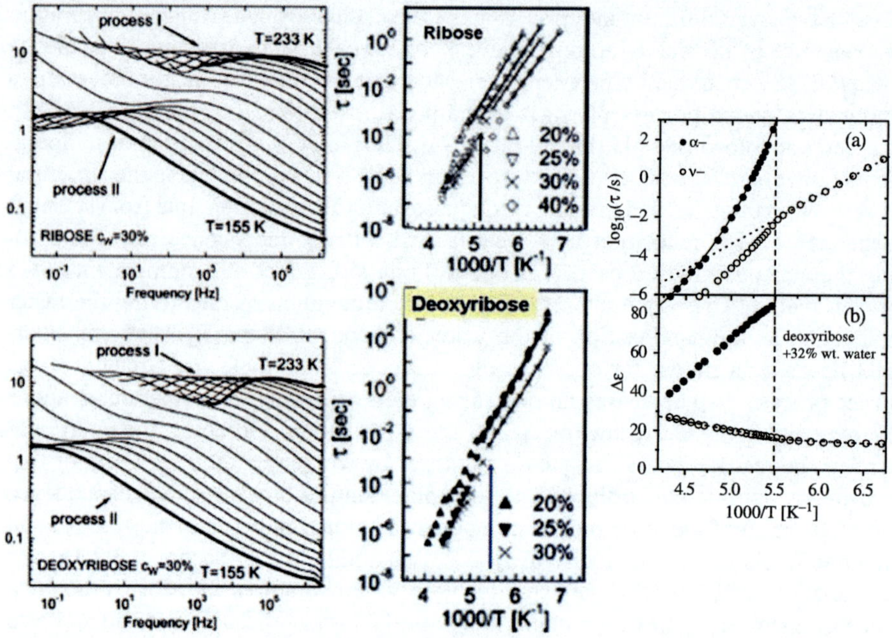

Fig. 213 (*Left*) Dielectric loss spectra for ribose (*upper*) and deoxyribose (*lower*) solutions with 30 wt% water in the temperature range of 155–233 K obtained by Pagnotta et al. [1312]. The calorimetric glass transition temperature T_g is about 190 K, slightly higher and lower for the mixture of ribose and deoxyribose, respectively. The solutions of the two sugars show two relaxation processes. The faster relaxation (process II) originating from water appears already at temperatures below the T_g and continues to be observed above T_g up to 233 K. Note that the dielectric strength of the faster relaxation monotonically increases with temperature above and below T_g. The slower structural α relaxation (process I) appears in the spectra at temperatures only when near and above T_g, and its relaxation time τ_α has the VFTH temperature dependence. (*Middle*) Temperature dependence of the faster water relaxation time τ_β (of process II) for mixtures of ribose with 20, 25, 30, and 40 wt% water (*upper*), and mixtures of deoxyribose with 20, 30, and 40 wt% water (*lower*). There is a change of T-dependence of τ_β at T_g for all these mixtures. This is indicated for 30 wt% water mixtures (×) by the *vertical arrow*, which is located at $1000/T_g$. Reproduced from [1312] by permission. (*Right*) The same result as the *middle panel* was found by Capaccioli, Ancherbak et al. (to be published) in deoxyribose + 30 wt% of water. The *lower panel* clearly demonstrates the monotonic increase of the dielectric strength of the faster β- or ν-relaxation originating from water above T_g, proving that it cannot be α-relaxation as suggested in [1312]

from 155 to 233 K are shown in Fig. 213 (left panel). The calorimetric glass transition temperature T_g has been determined to be about 190 K, slightly higher and lower for the mixture of ribose and deoxyribose, respectively. It can be seen by inspection of the spectra that the intensity or the dielectric strength of process II (i.e., $\Delta\varepsilon_\nu$) is increasing with increasing temperature *below* and *above* T_g. The very broad loss peak at the highest temperature of 233 K can be used as evidence of the loss contributed by process II having increased to the same level as the α-relaxation of the mixture. All these have been confirmed by repeat of the measurements and analysis of the data at the Univ. of Pisa by Capaccioli and coworkers.

The middle panel of Fig. 213 shows the temperature dependence of the faster water relaxation time τ_ν (of process II) for mixtures of ribose with 20, 25, 30, and 40 wt% water (upper), and mixtures of deoxyribose with 20, 30, and 40 wt% water (lower) from [1312]. There is a change of T-dependence of τ_β at T_g for all these mixtures, like that found in other aqueous mixtures. This is indicated for the 30 wt% water mixtures (×) by the vertical arrow, which is located on the x-axis at $1000/T_g$, with $T_g \approx 190$ K. Since Cerveny and coworkers have found the monotonic increase of $\Delta\varepsilon_\nu$ with increasing temperature *above* T_g from their own data of aqueous solutions of ribose and deoxyribose [1312], now they cannot ignore addressing the apparent contradiction of this experimental fact with their interpretation involving the α-relaxation of water *above* T_g.

Trend of τ_ν or τ_{JG} on Increasing Concentration of Water in Aqueous Mixtures: Used to Obtain the Upper Bound for the τ_{JG} of Bulk Water

The study of glass transition of bulk water is hampered by crystallization at lower temperatures before vitrification can be reached. Up to certain concentration depending on the hydrophilic solute, crystallization of water does not occur and the relaxation of dynamics of water in the mixture can be studied over an immense time range from picoseconds to thousands of seconds. Examples have been given before in Figs. 210–213. This is re-emphasized by the relaxation map of 35 wt% water mixture with hexaethylene glycol (6EG) in Fig. 214, showing the large dynamic range $10^{-2} < f < 10^{10}$ Hz of the ν- or JG process of water measured. The inset shows schematically the symmetrical ν-loss peak and the asymmetric α-loss peak at

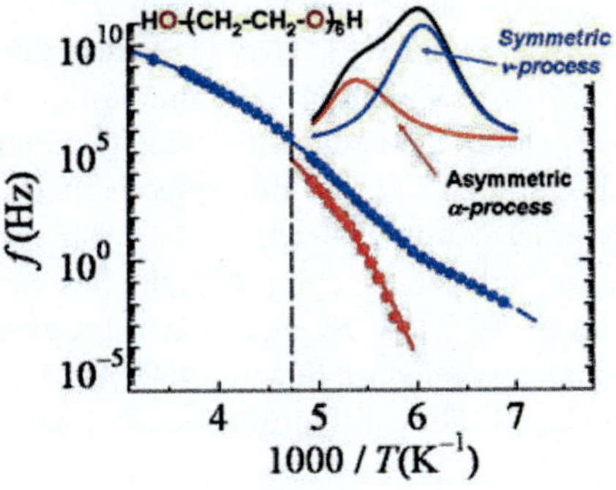

Fig. 214 Relaxation map of mixture of 35 wt% water with 6EG intended to demonstrate that τ_ν can be determined over a wide frequency range. The *inset* shows the more intense symmetric ν-loss peak and the asymmetric α-loss peak. Courtesy of N. Shinyashiki

temperatures above T_g. It also shows $\Delta\varepsilon_\nu$ is larger than $\Delta\varepsilon_\alpha$ above T_g, which can be seen in the left panel of Fig. 212. The ease of obtaining rich information of the ν-process in mixtures suggests the following strategy to deduce the dynamics of bulk water. On increasing the concentration of water in the mixture, the trend exhibited by τ_ν gives an upper bound of τ_{JG} of bulk water for all temperatures where τ_ν is determined, without being pre-empted by crystallization.

The general trend exhibited by τ_ν on increasing concentration of water in aqueous mixtures is elucidated by the data of 20, 30, and 40 wt% of water mixtures with glycerol shown before in Fig. 211. On increasing water content, τ_ν decreases for all measurement temperatures above and below T_g. In the glassy state where τ_ν has Arrhenius T-dependence, the activation energy also decreases. Mixtures at significantly higher wt% of water cannot be studied because water crystallizes in it. However, for the same weight fraction, higher molar concentration of water can be achieved by using solute of higher molecular weight such as the polymers, PVME [1108], poly(vinylpyrrolidone) (PVP) [1298, 1310], and hydrogel of poly (2-hydroxyethyl methacrylate) (PHEMA) [1313]. The τ_ν determined by dielectric relaxation measurements on mixtures of 35 wt% water with 3EG and poly(ethylene glycol) 600 (PEG600) [1093, 1299], 50 wt% of water with PVME [1108], 50 wt% of water with PVP [1298], and 38.6 wt% of water with PHEMA [1313] are shown in Fig. 215 (left panel). All these mixtures have a glass transition effected by the α-relaxation; none of the α-relaxation times is shown in the figure because the emphasis here is on their τ_ν. Included also are $\tau_\nu = 10^3$ s of water confined in MCM-41, silica gels, and in the hydration shell of bovine serum albumin (BSA) determined by adiabatic calorimetry [1281, 1282, 1285]. The two vertical arrows indicate the location of $1000/T_g$ for the mixtures with 3EG and with PEG600, and it is where the Arrhenius T-dependence of τ_ν below T_g (dashed line) changes to a stronger dependence above T_g. This is the case for all aqueous mixtures including those with PVME, PVP, and PHEMA, although not illustrated as such. It is clear by inspection of the figure that at all temperatures τ_ν decreases on increasing molar concentration of water in the mixtures in the order of 3EG, PEG600, PVME, PVP, and PEHMA, and its activation energy below T_g is within the range from 45 to 50 kJ/mol. The right panel shows the relaxation times at the high-temperature/short relaxation time region more clearly. There one can see the dielectric relaxation time of few layers of water in the hydration shell of lysozyme [1314, 1315] and bovine serum albumin (BSA) [1316, 1317], and the relaxation time, τ_c, of reorientational motion of heavy water (D_2O) in D_2O-hydrated myoglobin (0.35 g/g) from deuteron NMR [1318, 1319]. They are nearly the same as τ_ν of the mixtures.

From this trend of τ_ν on increasing molar concentration of water in the mixtures, we [1320] can deduce that the JG relaxation time of bulk water, $\tau_{JG,bulk}$, has to be shorter than τ_ν of all the aqueous mixtures shown in the figure at all temperatures. In other words, *the shortest* τ_ν *of the mixtures is an upper bound of* $\tau_{JG,bulk}$.

In both panels of Fig. 215, the relaxation time of water nanoconfined in molecular sieves, MCM-41 by dielectric relaxation [1281, 1282, 1287, 1093] and in graphite oxide by dielectric relaxation [1321(a)] and by neutron scattering [1321(b), (c)] are included. They differ by a factor of about 5. Any cooperativity that water relaxation

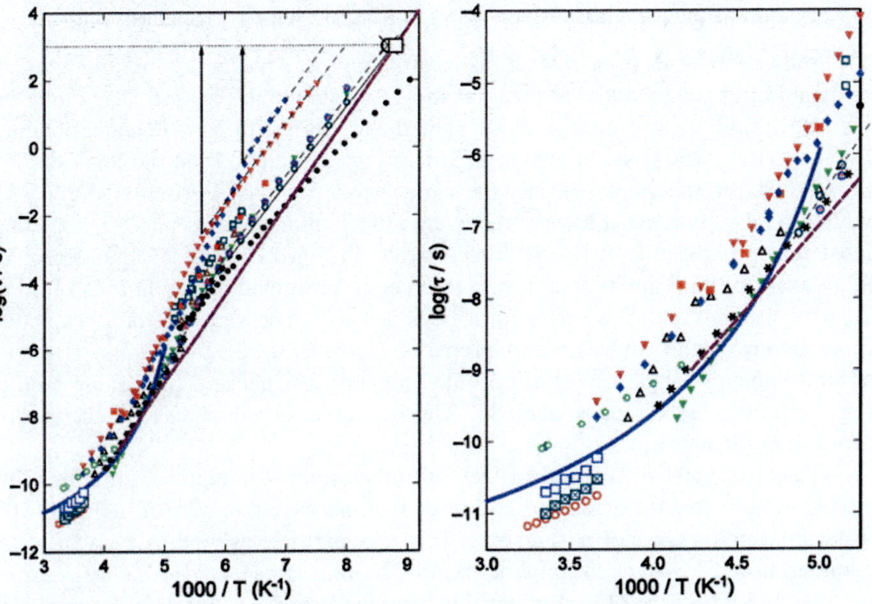

Fig. 215 (*Left*) $\log_{10}(\tau_\nu)$ vs. 1000/*T*, of the ν-relaxation of water component in 35 wt% water mixture with 3EG (*blue closed diamond*) and PEG600 (*purple closed inverted triangle*), 50 wt% water mixture with PVME (*red squares*), 50 wt% water with PVP (*pale blue open square*), 38.6 wt% of water with PHEMA (*magenta closed triangles*), 90 wt% water with polyvinyl alcohol at short times (*pale blue squares* with × inside), 80 wt% water with PVP at short times (*blue open squares*) [1310], and bulk water at short times (*brown open circles* [1311]). Dielectric relaxation time of water in the hydration shell of lysozyme (★) and bovine serum albumin (BSA) (*open black triangle*), and the relaxation time, τ_c, of reorientational motion of heavy water (D_2O) in D_2O-hydrated myoglobin (0.35 g/g) from deuteron NMR (*thick blue line*). The relaxation times of 20% water nanoconfined in graphite oxide from dielectric relaxation (*purple Arrhenius line* that fits the data, actual data not shown [1321(a)]), and from neutron scattering (*green open circles* [1321(b)]). Dielectric data of relaxation time of water confined in MCM-41 with pore diameter 2.14 nm at hydration levels 12 wt% (*open circles*), 22 wt% (*green* ▼), 55 wt% (*open blue diamonds*). The relaxation times of water in 20 wt% water with 1-propanol (*black* •) are related to the discussion of Fig. 216. (*Right*) The data of the *left panel* are magnified by restricting to the higher temperature range. Not shown are relaxation times of pure bulk water at higher temperatures in the picoseconds range determined by neutron scattering [1327], synchrotron radiation based inelastic ultraviolet scattering [1328], and heterodyne-detected optical Kerr effect [1329]. All of them is much shorter than the shortest dielectric relaxation times [1311] shown here

may have would be suppressed by the finite size effect caused by nanoconfinement. If the observed relaxation is local water motion, then its relaxation time τ_{conf} should be close to $\tau_{\mathrm{JG,bulk}}$ of bulk water. It can be seen in Fig. 215 that the shortest τ_ν of the mixtures is almost indistinguishable from τ_{conf} of water confined in MCM-41, and this further supports that *the shortest* τ_ν *of the mixtures as well as* τ_{conf} of water confined in MCM-41 is an upper bound of $\tau_{JG,bulk}$.

Observation of the α-Relaxation of Water in Mixture with 1-Propanol

The relaxation time of a relaxation originating from water in mixture with 1-propanol has been included in the relaxation map shown in Fig. 215. The dielectric loss spectra of pure 1-propanol (PrOH) and the mixture with 20 wt% water obtained by Sudo et al. [1307] are shown at 123 K in Fig. 216 a and b on the left side. For pure PrOH, the spectra are exactly the same as that found by Hansen et al. [348(b)] with the Debye process at lowest frequency and the unresolved α- and β-relaxations, labeled as processes I, II, and III, respectively, in Fig. 216. The T-dependences of all relaxation times are as found by Hansen et al. presented before in Fig. 115. For 20 wt% water mixture, all these processes are still observed and their relaxation times are nearly the same as pure 1-propanol. However, a new loss peak is observed at the frequency range lower than all the propanol loss peaks. This can be seen in Fig. 216b from the loss peak labeled I′. This new process originates from the motion of water in the mixture.

What distinguishes 1-propanol from all other solutes in aqueous mixtures discussed so far is that the α-relaxation time of 1-propanol, either pure or in the 20 wt% water mixture, is far shorter than even τ_ν of the mixtures as well as τ_{conf} of water confined in MCM-41 or in graphite oxide. The reader can verify this by comparing Fig. 115 with Fig. 215 or gather this from the low T_g of 1-propanol, which lies below 100 K. Due to this unusual circumstance, the relaxation process of water (I′)

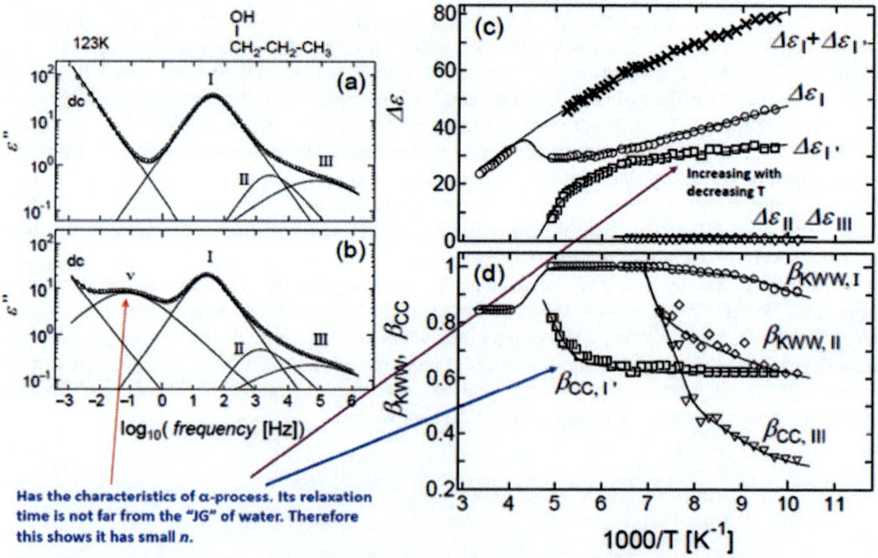

Fig. 216 (*Left*) Isothermal dielectric loss spectra at 123 K of pure 1-propanol (*a*), 20 wt% water mixture with 1-propanol. (*Right*) Temperature dependences of the relaxation strengths (*upper panel*) and frequency Cole–Cole and KWW dispersion parameters (*lower panel*) of all observed relaxation processes observed in the 20 wt% water mixture with 1-propanol. The interest here is mainly on the parameters of process I′ originating from water. Data replotted from [1307] with additional illustrations

found in the mixture is slower than all processes associated with 1-propanol (see the spectrum at 123 K in Fig. 216 (left)). Moreover, its properties are not like those of the local ν-process of water observed in all aqueous mixtures discussed before. This can be seen from the monotonic decrease of its relaxation strength, $\Delta\varepsilon_{I'}$, with increasing temperature as shown in Fig. 216 (right upper panel), and nearly constant dispersion width parameter $\beta_{CCI'}$ over a considerable range of temperature above T_g. These properties actually reflect that process I′ is the α-relaxation of water in the mixture. However, this a rather unusual α-relaxation because the temperature dependence of $\tau_{I'}$ is nearly Arrhenius, and not far from the local water relaxation times τ_ν in the mixtures as well as τ_{conf} of water confined in MCM-41 and graphite oxide as shown in Fig. 215 (left panel). These characteristics of the α-relaxation of 20 wt% water in 1-propanol suggest it has negligible intermolecular cooperativity and small coupling parameter. From what we have discussed on the component dynamics in binary mixtures in Sections 2.2.5.7 and 2.3.2.12, this can be readily understood because the α-relaxation of water in the mixture is facilitated by the presence of the much more mobile 1-propanol. From preliminary dielectric relaxation study of water/methanol mixtures by Li-Min Wang and coworkers, we have found similar dynamics of the water component as in the water/1-propanol mixture. This is not surprising because methanol like 1-propanol is more mobile than water.

Thus the α-relaxation of 20 wt% water in 1-propanol effectively is no different from the local secondary relaxation of water, and $\tau_{I'}$ in Fig. 215 offers another estimate of the JG β-relaxation time of bulk water $\tau_{JG,bulk}$, consistent with the estimate given before from the shortest τ_ν of the mixtures as well as τ_{conf} of water confined in MCM-41 and graphite oxide. However, $\tau_{I'}$ of water in more mobile 1-propanol can be faster than $\tau_{JG,bulk}$, and hence $\tau_{I'}$ is its lower bound. These facts inspire the construction of the shaded rectangle placed diagonally in Fig. 217. The vertical span of the rectangle is only half a decade, but it seems to contain the τ_ν of aqueous mixtures expected in the high water concentration limit, the τ_{conf} of water confined in MCM-41 and graphite oxide, and part of the $\tau_{I'}$ of water in mixture with 1-propanol. Furthermore, it covers the thousand seconds relaxation time of water confined in MCM-41 and hydration water in hydrated bovine serum albumin (BSA) determined by adiabatic calorimetry; the much shorter dielectric relaxation times of hydration water in hydrated myoglobin, lysozyme, and BSA; the very short dielectric relaxation time of water in 90 wt% water mixtures with PVA, 80 wt% water mixture with PVP [1310], and pure bulk water [1311]. Not included in all figures are the relaxation time of water in the hydration shell of collagen and elastin with Arrhenius T-dependence in the reciprocal temperature range of $3.6 < 1000/T < 7.3$ determined dielectrically by Gainaru et al. [1322(a)], and the nuclear spin-lattice relaxation data from ms to 10 ps range of Lusceac et al. [1322(b)]. The relaxation times are about the same as $\tau_{JG,bulk}$ of aqueous mixtures of PVP and other hydrated proteins in Figs. 215 and 217 with similar activation energy, and are not plotted to avoid overcrowding.

All these relaxation times mentioned above are good estimates of the JG β-relaxation time of bulk water $\tau_{JG,bulk}$, and thus the shaded rectangular box in Fig. 217 approximately locates within it the Arrhenius or near Arrhenius

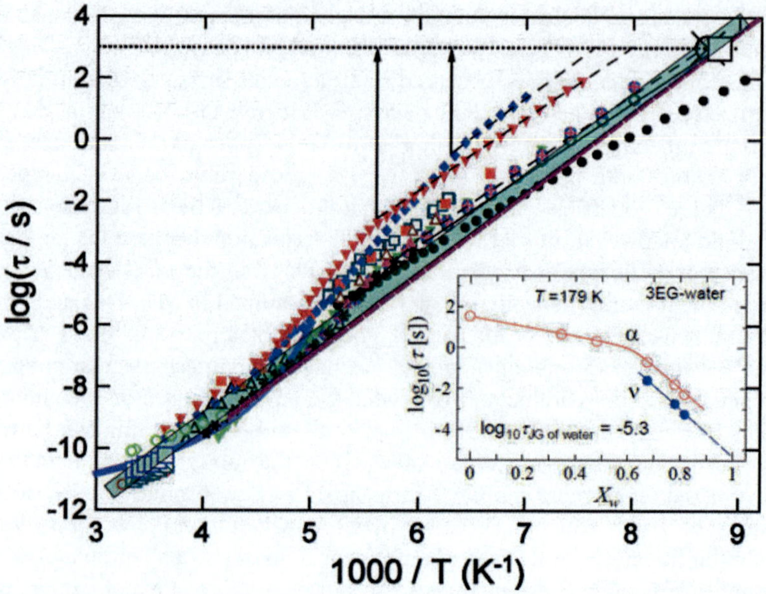

Fig. 217 Data same as in Fig. 215 (*left panel*). The *shaded rectangular box* is added to show all estimates of the secondary relaxation time τ_{JG} of bulk water seemed to be located within the box, and hence its relaxation time has approximately the Arrhenius *T*-dependence over 14 decades with activation energy of about 50 kJ/mol. The *inset* is a plot of the α- and ν-relaxation times of mixtures of 3EG with water against the molar fraction of water, X_W, at 179 K taken from [1295]. *Open circles* and *filled circles* represent the relaxation times of the α- and ν-relaxations, respectively. The *dashed line* is an extrapolation of log τ_ν to $X_W = 1$, i.e., bulk water. The extrapolation suggests log τ_{JG}(179 K) of bulk water is equal to –5.3, and it falls within the *shaded box* in the main figure

temperature dependence of $\tau_{JG,bulk}$ over 14 orders of magnitude. It has activation energy of about 46 kJ/mol according to the slope of the box. Since the strength of a hydrogen bond is approximately 20 kJ/mol, this corresponds to the energy required to break two hydrogen bonds in order for local rotation and translation of the water molecule. This is confirmed by the diffusion of water in channels of crystalline dihydrate of trehalose at temperatures from 25 to 65°C (1000/*T* from 3.36 to 2.96 and corresponds to the bottom end region of the box in Fig. 217) that has an activation energy of 49.4±3.9 kJ/mol [1323(a)]. There water moves between two discrete crystal positions by breaking two of the hydrogen bonds. The free energy of a hydrogen bond is around 23 kJ mol^{-1} from dielectric constant data [1323(b)].

The inset in Fig. 217 is a plot of the α- and ν-relaxation times of mixtures of 3EG with water against the molar fraction of water, X_W, at 179 K taken from [1295]. Open circles and filled circles represent the relaxation times of the α- and ν-relaxations, respectively. The dashed line is an extrapolation of log τ_ν to $X_W = 1$, i.e., bulk water. The extrapolation suggests log τ_{JG}(179 K) of bulk water is equal to –5.3. The reader can verify that it falls within the shaded box in the main figure. This is another indication that τ_{JG} of bulk water has been located in Fig. 217.

Deducing the Dynamics of Bulk Water from the α-Relaxation and ν-Relaxation in Aqueous Mixtures with Oligomers of Ethylene Glycol

If the aqueous mixture has large enough molar fraction of water, the α-relaxation necessarily involves water in cooperative motion with the solute. The CM equation (2.69) between the α-relaxation time τ_α and the JG β-relaxation time τ_{JG} of water in the mixture can be used to determine the coupling parameter n and hence the degree of cooperativity of the α-relaxation involving the water. Again taking $t_c = 2$ ps, n can be obtained from Eq. (2.69) when rewritten as

$$n = (\log \tau_\alpha - \log \tau_\nu)/(\log \tau_\alpha + 11.7), \tag{2.84}$$

since the ν-relaxation time τ_ν is the JG β-relaxation time of water in the mixture. Furthermore, on changing the solute systematically in such higher water content mixtures and observing the trend of the corresponding change of n, some insight into the dynamics of bulk water possibly may be obtained. This can be done by utilizing the 35–40 wt% water mixtures with the oligomers of ethylene glycol. Figure 218 presents the T-dependence of the α- and ν-relaxation times of the 35 wt% water mixtures with tri-ethylene glycol (3EG) and with di-ethylene glycol (2EG), and 40 wt% of water mixture with ethylene glycol (EG). From τ_α and τ_ν available at the lowest

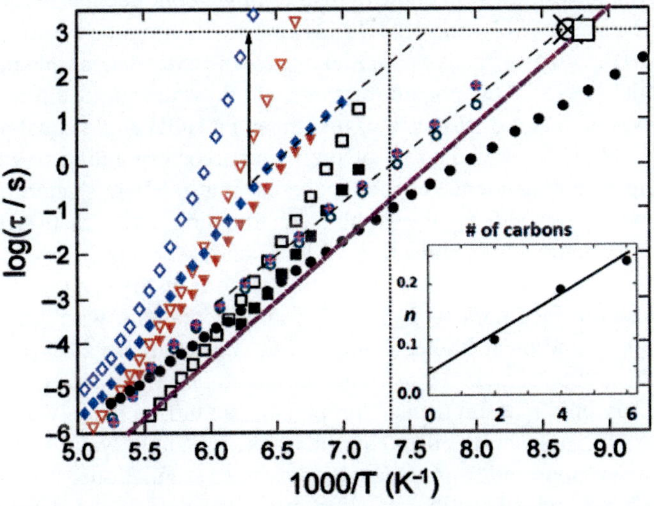

Fig. 218 Same data as in Figs. 215 and 217 except shown here in a restricted temperature range and with the addition of the α-relaxation times (*open symbols*) and ν-relaxation times (*closed symbols*) of mixture of 35 wt% water with di-ethylene glycol (*inverted triangles*), and 40 wt% of water with ethylene glycol (*squares*). The data of 35 wt% water with tri-ethylene glycol are still represented by *diamonds* like before. The *inset* is a plot of the coupling parameter n of the α-relaxation of aqueous solutions of tri-ethylene glycol, di-ethylene, and ethylene glycol against the number of carbon atoms of these glycols, and the dependence is linearly extrapolated to zero number of carbon atoms to deduce the coupling parameter of bulk water

temperature for each mixture, the values of coupling parameter n calculated by Eq. (2.74) are 0.24, 0.19, and 0.10 for mixtures with 3EG, 2EG, and EG, respectively. These values of n are plotted against the number of carbon atoms N_C of the solute in the inset of Fig. 218. A linear extrapolation of the data as shown suggests that n is nearly zero when there is no more carbon atoms. This extrapolated value of n cannot be strictly identified with that for bulk water. Nevertheless, with or without extrapolation, the n vs. N_C plot indicates that n of bulk water at low temperatures say below $T_g \approx 162$ K of tri-propylene glycol is small. At high temperatures between 0 and 35°C, precision complex permittivity data of pure water in the frequency range $0.2 \leq f \leq 410$ GHz by Buchner et al. [1311] also find Debye relaxation with $n = 0$. These short relaxation times data have been shown before in Fig. 215. The dielectric relaxation of water in 90 wt% water mixtures with PVA and 80 wt% water mixture with PVP [1310], at temperatures ranging from 0 to 25°C and relaxations within the range $10^{-11} < \tau < 10^{-10}$ s, show narrow symmetric loss peak. The fits by the Cole–Cole function with exponents $\beta = 0.9$ indicates that the coupling parameter n is less than 0.1. The reader may recognize that all the experimental data are obtained by dielectric relaxation except those from adiabatic calorimetry and one by NMR. These collections of dielectric relaxation data show that the coupling parameter of the α-relaxation of water is small for all relaxation times that range over 13 orders of magnitude.

Such small n deduced from dielectric data at long times is consistent with a molecular dynamics simulation of the dynamics of water molecules in deeply supercooled liquid states by Sciortino et al. [1324] in the time regime starting from 10^{-3} ps and up to 10^4 ps. The effective potential used is the extended simple point charge (SPC/E) model [1325]. This potential treats a single water molecule as a rigid set of point masses with an OH distance of 0.1 nm and a HOH angle equal to the tetrahedral angle 109.47°. After the end of the plateau corresponding to caged water dynamics, the time dependence of the incoherent intermediate scattering function, $F_s(Q,t)$, is well described by the Kohlrausch function with fractional exponent $\beta_K \equiv (1-n)$ being Q-dependent. At 206 K and $Q = Q_{max} = 18\ \text{nm}^{-1}$ or 1.8 Å^{-1}, where Q_{max} is the location of the maximum of the oxygen–oxygen partial static structure factor, $\beta_K(Q_{max}) \approx 0.7$ and increases toward unity when Q falls below Q_{max} and/or at higher temperatures. For example at $Q = 0.33$ Å^{-1}, $\beta_K(Q) \approx 0.85$ at 206 K and larger at higher temperatures. Since dielectric relaxation data are macroscopic measurements, comparison should be made with simulation data for $Q < Q_{max}$. Thus, the deduction of small n of bulk water from dielectric measurements is supported from molecular dynamics simulations.

Before leaving this subsection, mention must be made of the measurement of bulk water using techniques other than dielectric relaxation. All these other techniques have the limitations of only measuring the relaxation or diffusion of water at times shorter than 10 ps.

One is incoherent quasielastic neutron scattering investigations of diffusive motions in supercooled water by Chen et al. [1326] and Teixeira et al. [1327] over the temperature range $-20°\text{C} \leq T \leq 20°\text{C}$. They identified two relaxation times, both much shorter than all relaxation times shown before in Fig. 215

were interpreted as jump diffusion of protons and fluctuations of hydrogen bonds, respectively.

Masciovecchio et al. [1328] used the novel synchrotron radiation based inelastic ultraviolet scattering technique to measure the dynamic structure factor of normal and supercooled liquid at a momentum transfer $Q = 0.1$ nm^{-1} or 0.01 Å^{-1}, in the temperature range 260–340 K. The relaxation times are close to those determined by dielectric relaxation [1311] shown before in both panels of Fig. 215. This is expected because of the smaller Q used in the experiment than neutron scattering. These authors reported the Kohlrausch exponent β of about 0.6 for the correlation function used to fit the dynamic structure factor $S(Q, \omega)$. Although error estimates were given, it is not clear that this value of β can be taken with confidence because there are several contributions besides the structural relaxation to the observed $S(Q, \omega)$, and convolution with experimental resolution. This is to be contrasted with dielectric relaxation of water in the same temperature range [1311], where the $\beta \approx 1$ can be ascertained directly either by inspection of the observed loss peak or by fitting it by any of the empirical functions.

Torre et al. [1329] used heterodyne-detected optical Kerr effect experiments (HD-OKE) to study water in the time regime of $0.3 < t < 30$ ps and temperature from $254 \leq T \leq 314$ K. The data at times longer than 0.5 ps can be fitted by the Kohlrausch function with $\beta_K = 0.6$ for all temperatures, even smaller than the β_K values of Sciortino et al. at 206 K. The relaxation time obtained by Torre et al. is shorter by a factor of more than 5 than that of Sciortino, and shorter by about 25–100 times than the various relaxation times in the same temperature range of $254 \leq T \leq 314$ K as can be seen in Fig. 215. At present, the data of structural relaxation from Torre et al. cannot be reconciled with those of other spectroscopies, particularly dielectric relaxation. Also, β-relaxation of water was found in LiBr-6H_2O [1329], with τ_β much shorter than nanoconfined water.

On comparing the results obtained by the non-dielectric techniques in Fig. 215, it is clear that there is large variations of relaxation times. Moreover these techniques cannot provide data of relaxation time longer than 10^{-11} s, pertinent to the glass transition problem. Therefore, if the purpose is to understand the dynamics of water in relation to glass transition, it is wise to restrict consideration of experimental data coming solely from dielectric relaxation and calorimetry.

Deduction from T_g of Glassy Water

If cooled sufficiently fast as to avoid crystallization, water can be vitrified. The cooling rate required is about 10^6 K s^{-1} and the resulting material is called hyperquenched glassy water (HGW). However, at least two distinct forms of glassy water can form. The two forms most extensively studied are known as low-density amorphous ice (LDA) and high-density amorphous ice (HDA) [130]. There is also the amorphous solid water (ASW) films formed by molecular beam deposition onto Au(111) and Ru(001) single-crystal substrates at deposition rates between 0.1 and 7 μm h^{-1} and temperatures between 10 and 120 K. After annealing in vacuum from 77 to about 113 K [1269], ASW relaxes to a material that is indistinguishable from HGW. The glass transition temperatures of LDA, HGW, and ASW determined by differential scanning calorimetry (DSC) with heating rate of 30 K/min are all 136 K, and 124 K for LDA at 10 K/min [1269, 1271, 1330–1333] (see the review of Debenedetti [130]). The location of the vertical dotted line on the x-axis in Fig. 218 corresponds to 136 K. If these various forms of glassy water truly represent bulk water, and their $T_g = 136$ K is that of bulk water, then the coupling parameter of bulk water can be estimated from the JG β-relaxation time, $\tau_{JG,bulk}$, by using the CM equation (2.84). Taking $\tau_{JG,bulk}$ to lie within the shaded box in Fig. 217, and assuming τ_α from DSC is 10^2 s at 136 K, we find $n = 0.18$. If $\tau_\alpha = 10^3$ s at 136 K, then $n = 0.23$. If the true T_g of bulk water is lower than 136 K, or if t_c of water is shorter than 2 ps, then the coupling parameter n is even smaller. These small values of n of bulk water estimated from the DSC data of LDA, HGW, and ASW are consistent with similar conclusion deduced from aqueous mixtures and from simulations. At temperatures above $T_g = 136$ K, the T-dependence of $\tau_{JG,bulk}$ may not be perfectly Arrhenius as suggested by the shaded box in Fig. 217, but the deviation may not be large because n is already small at T_g and decreases as temperature rises, as found by molecular dynamics simulations [1324].

The Mystery of the Dynamics of Bulk Water Resolved?

The analyses of various experimental data mainly from a single technique, namely dielectric relaxation, with the help of adiabatic calorimetry, in the above have led to the following conclusions on the dynamic properties of the α- and the JG β-relaxation of bulk water if crystallization does not occur. The T-dependence of its JG β-relaxation time, $\tau_{JE,bulk}$, is nearly Arrhenius over 14 decades from 10^{-11} to 10^3 s. After breaking two hydrogen bonds, as suggested by the activation energy $E_a \approx 50$ kJ/mol of the JG relaxation time $\tau_\beta = \tau_\infty \exp(E_a/kT)$ of water, the water molecule is free to rotate and translate. The change from hydrogen-bonded to free rotation-translation engenders entropy change, which can explain the prefactor τ_∞ is of the order of 10^{-19} s deduced from the Arrhenius T-dependence of τ_β in Fig. 217. The coupling parameter n of the α-relaxation is small at T_g and smaller at higher temperatures, implying that the structural relaxation of water has low degree of intermolecular cooperativity and its relaxation time τ_α is not much longer than $\tau_{JG,bulk}$ at all temperatures above T_g. The last conclusion may not be farfetched because water is the smallest molecule in nature and the strength of hydrogen bond is only about 20 kJ/mol. If this is a correct description of the properties and behavior of bulk water, the mystery of the dynamics of bulk water is resolved.

The remarks given above applies only to *bulk* water, which has limited applicability because bulk water crystallizes at lower temperatures. In practice, water is studied not in the pure bulk state, but in situations where it does not crystallize. Examples include mixtures with hydrophilic substances, in hydration shell of biomolecules, confined in silica gels or molecular sieves discussed before, and other possibilities. Chemical and/or physical interaction of water in these cases can drastically modify the frequency dispersion of the α-relaxation of the water-coupled system, and change its relaxation time, leading to a system-specific glass transition temperature. The modification is less so for the local JG β-relaxation of bulk water. Nevertheless, because of the existence of the connection between the α- and the JG β-relaxation for all glassformers, the Arrhenius T-dependence of the JG β-relaxation time no longer holds above T_g of the system.

In view of the modifications of the α-relaxation in these systems, it is dangerous to identify the modified characteristics observed with that of bulk water. Furthermore, since the faster JG β-relaxation is also present and if observed, there is another danger for it to be misinterpreted as the α-relaxation of water either of the system itself or of bulk water. As discussed before, this happens in the interpretation of relaxation processes in aqueous mixtures by others. Swenson and coworkers suggested that the observed faster process is a merged α- and β-process of water at temperatures above T_g [1106, 1261, 1292]. Cerveny and coworkers [1107, 1108, 1111, 1294, 1312] interpreted the faster relaxation above T_g as the α-relaxation of water. The relaxation processes observed in water confined in nanometer space have been mentioned in the introduction and will be further discussed in the following section. The faster relaxation observed in water confined in molecular sieves (MCM-41) has been interpreted by Swenson and coworkers as the α-relaxation of water as well [1293].

The JG β-Relaxation of Water in Nanoconfinement

Mention of the dynamics of water confined in nanometer space has been made in the previous section, and used in deducing the JG β-relaxation time, $\tau_{JG,bulk}$, of bulk water. In view of the importance of these data on dynamics of water per se and interest in change of dynamics by nanoconfinement in general, a more detailed account of the experimental data is worthwhile to be given below.

The Dynamics of Water Nanoconfined in Molecular Sieves

The dynamics of water confined in molecular sieves had been studied by means of dielectric spectroscopy by Jansson and Swenson [1102]. In a cylindrical pore with diameter of 10 Å, two major relaxation processes were observed dielectrically by Jansson and Swenson as shown in the loss spectrum at 180 K in the lower part on the left side of Fig. 219. According to them, the much more intense slower process (open squares) is contributed by most of the water molecules in 10 Å pores, and is likely due to strong interactions with the hydrophilic inner surfaces of molecular sieves. Its relaxation time, τ_{sMS}, increases with decreasing temperature to attain 100 s at about 167–170 K (see relaxation map on the right side of Fig. 219). The relaxation time of the weaker and faster process, τ_{fMS} (closed squares), has Arrhenius temperature dependence at low temperatures below about 175 K. The magnitude of τ_{fMS} is nearly the same as the relaxation time $\tau_{6\,Å}$ (closed circles) of the 6 Å ultrathin water layer (two molecular layers) confined in fully hydrated Na-vermiculite clay measured by dielectric spectroscopy also shown in the figure [1296], and the JG relaxation time, $\tau_{JG}(T)$, of the water component in 50% water mixtures with PVME (not shown). Like τ_{JG} of this aqueous mixture, the Arrhenius temperature dependence of τ_{fMS} does not persist at higher temperatures, but changes to

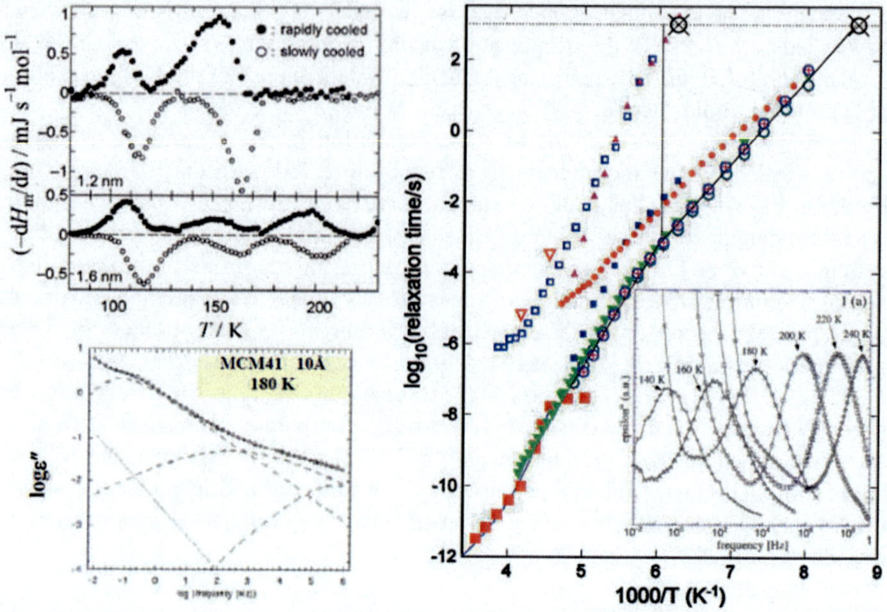

Fig. 219 (*Right*) Relaxation map of various relaxation times. Shown are the relaxation times, τ_{sMS} and τ_{fMS}, of a slower (*open squares*) and faster (*closed squares*) relaxation of water confined in molecular sieves of cylindrical pore with diameter of 10 Å [1102]. The Arrhenius temperature dependence of τ_{fMS} changes to stronger temperature dependence at approximately $T_g = 167$ K (indicated by the *vertical line*) where τ_{sMS} will reach about 10^3 s. The values of τ_{sMS} calculated by the CM equation are shown by *closed triangles* (see text). Shown also are more recent dielectric data [1293] of τ_{fMS} of water confined in molecular sieves MCM-41 C10 with pore diameter 2.14 nm at hydration levels of H=12 wt% (*open circles*), 22 wt% (*closed inverted triangles*), 55 wt% (*open circles* with + inside). The change of *T*-dependence of τ_{fMS} is made clear by the two *lines* with different slopes. The two *open inverted triangle* are τ_{sMS} deduced from the dielectric loss spectrum at 220 and 240 K (see *inset*) after subtracting off the conductivity. (*Left*) The dielectric loss spectrum at 180 K is shown by the figure on the *left* at the *bottom*. For comparison, we also include the dielectric relaxation time (*closed circles*) of 6 Å thick layer of water in fully hydrated Na-vermiculite clay [1296]. The two *larger circles* (⊗) are the relaxation times of τ_{sMS} and τ_{fMS} of water confined within 1.2, 1.6, and 1.8 nm nanopores of silica MCM-41 obtained by adiabatic calorimetry having the value of 10^3 s at *T*=115 and 165 K, respectively [1283]; data of 1.2 and 1.6 nm are shown on *top of the left panel*. The *red closed squares* are the average translational relaxation time τ_T determined by quasielastic neutron scattering of water confined in nanoporous silica matrices MCM-41-S with pore diameters of 14 Å [1289]. Dielectric data reproduced from [1102, 1293] by permission. Calorimetric data are reconstructed from data of [1283]

stronger temperature dependence at temperatures above 180 K. This crossover of temperature dependence of τ_{fMS} is the same as what we have seen in τ_{JG} of the JG-process in various non-aqueous and aqueous mixtures (see Fig. 215), and is a general property of JG β-relaxation. The striking similarity of this feature of τ_{fMS} and that shown by τ_{JG} leads us to interpret the observed faster process in a molecular sieve as the JG β-relaxation associated with slower process, which is the α-process of water strongly interacting with the hydrophilic inner surfaces of the molecular sieve. Such a relation between the two processes is further supported by applying the coupling model (CM) relation between τ_{JG} and τ_α to the present case of τ_{fMS} and τ_{sMS} for water confined in 10 Å pores. Using the CM relation adapted for the present case, $\tau_{sMS}(T) = [t_c^{-n}\tau_{fMS}(T)]^{1/(1-n)}$, τ_{sMS} are calculated from the data of τ_{fMS} (closed triangles) by assuming $\tau_c \approx 2$ ps as other molecular glassformers and $(1-n) \equiv \beta_K = 0.69$. With this choice of the Kohlrausch exponent, $\beta_K = 0.69$, the calculated τ_{sMS} (closed triangles) are in approximate agreement with the experimental values (open squares) in Fig. 219. The value of $\beta_K = 0.69$ chosen is not far from the value of 0.65 reported by Swenson et al. [1104] as the Kohlrausch exponent obtained by means of quasielastic neutron scattering after averaging the result from various scattering vectors.

More recent dielectric loss data were reported by Hedström et al. [1293] of water confined in molecular sieves MCM-41 C10 with pore diameter 2.14 nm at hydration levels of 12, 22, and 55 wt%, and are also shown in Fig. 219 (right panel). The isothermal loss data are shown in the inset. The resolved loss peaks correspond to the fast process of Jansson and Swenson [1102] with relaxation times comparable in magnitude to τ_{fMS}. There is also the change of T-dependence of τ_{fMS}, made clear by the two lines drawn with different slopes. The temperature of about 166 K at which the change occurs is indicated by the vertical solid line.

Adiabatic calorimetry measurements of the enthalpy relaxation of water confined within MCM-41 nanopores was made by Oguni and coworkers [1283]. Again a faster and a slower relaxation of water confined within 1.2 and 1.6 nm nanopores of silica MCM-41 have been found. Their adiabatic calorimetric relaxation times, τ_{sMS} and τ_{fMS}, have the characteristic value of 10^3 s at T=115 and 165 K, respectively, and are shown in Fig. 219 by the two larger circles (⊗). The temperature location of $\tau_{sMS} = 10^3$s from adiabatic calorimetry in Fig. 1 is 165 K. Remarkably, extrapolation of the Arrhenius T-dependence of the dielectric relaxation time τ_{fMS} of water confined in molecular sieves with pore diameter of 10 Å from Jansson and Swenson [1102] to lower temperatures shows it is consistent with the calorimetric $\tau_{sMS} = 10^3$s at 115 K. This temperature also is nearly the same as that obtained by extrapolating the Arrhenius dependence of the dielectric τ_{fMS} from Hedström et al. [1293] to 10^3 s, as demonstrated in Fig. 219. The agreement shows that τ_{fMS} from dielectric relaxation and calorimetry belong to the same process. Moreover, the dielectric τ_{sMS} of the slower process from Jansson and Swenson [1102] reaches 10^3 s at a temperature close to 165 K, the calorimetric temperature for $\tau_{sMS} = 10^3$s. Thus the adiabatic calorimetry data confirm the presence of the slower process and support the τ_{sMS} determined from the dielectric data by Jansson and Swenson.

One may ask why no τ_{sMS} was reported by Hedström et al. This is due to the large conductivity contributions at low frequencies masking the slower process at most temperatures (see inset of Fig. 219, right panel). However, at the two temperatures 220 and 240 K, the loss shows an excess wing on the high-frequency side of the conductivity contribution to ε'', indicating the presence of the unresolved slow process. After subtracting off the conductivity contribution, the slow process has been resolved, and its relaxation time determined at these two temperatures. The open inverted triangles are τ_{sMS} deduced in this way from the dielectric loss spectra at 220 and 240 K (see inset). They are close to τ_{sMS} of the resolved slow relaxation found by Jansson and Swenson (open squares), and this supports that the slow process also is present in the water confined in molecular sieves MCM-41 studied by Hedström et al. [1293].

It is important here to emphasize the change of T-dependence of τ_{fMS} found by dielectric relaxation [1102, 1293] occurs at about 166 K when τ_{sMS} from both the dielectric and the adiabatic calorimetry measurements reaches 10^3 s, just like the case of aqueous mixtures. This clearly shows that τ_{fMS} and τ_{sMS} are related in physically the same manner as in aqueous mixtures. Possibly the hydrogen-bonding interactions of water with the interface of the molecular sieves replace the role played by the hydrophilic solute in aqueous mixtures, and give rise to the slow process and the relation of its τ_{sMS} to $\tau_{fMS,}$ the primitive or JG β-relaxation of water. This scenario of water confined in solid materials like MCM-41 is found in water confined in solid hydrogels and silica gels to be discussed below. Again, there is the experimental evidence of the change of T-dependence of the water relaxation time near T_g of the slow process found by both dielectric spectroscopy and calorimetry (see Fig. 220). The scenario even can be extended to hydrated

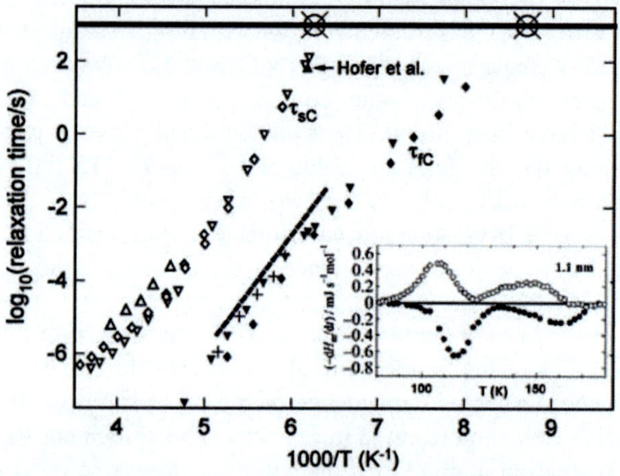

Fig. 220 τ_{sC} (*open diamonds*, and *open inverted triangles* for dry samples) and τ_{fC} (*closed diamonds*, and *closed inverted triangles* for dry samples), of slower and faster relaxation of water confined in aged silica hydrogels [1334]. τ_{fC} has Arrhenius T-dependence in the lower temperature range (activation enthalpy ≈50 kJ/mol.). τ_α or τ_{sPJ} (*open triangles*) and τ_{JG} or τ_{fPJ} (*dashed line*) of mixtures of 38.6 wt% of water with PHEMA [1335]; τ_α (*closed triangle*) of 34 and 42% of water with PHEMA (DSC from [1336]); τ_{JG} or τ_{fPJ} (+) of mixtures of 50 wt% of water with PHEMA [1335]. The two ⊗ are relaxation times of two processes both equal to 10^3 s detected by adiabatic calorimetry at T=115 and 160 K for water in the pores of silica gel with average diameter of 1.1 nm [1282]. *Inset*: Temperature dependence of the rates of spontaneous heat release and absorption, observed in the heat-capacity measurements of ordinary water (H_2O) by an intermittent heating method. Average pore diameter = 1.1 nm. The systematic heat-evolution and -absorption effects for the rapidly and slowly cooled samples are characteristics of a glass transition, and there are two transitions found in the 1.1 nm pores. Reproduced from [1282] by permission

proteins, a subject of extensive discussion in Section 2.3.2.28. There are about two layers of water surrounding the hydrated protein, and present are the slow process of the protein coupled to the hydration water by hydrogen bondings and the fast water relaxation. Remarkably, the hydration water relaxation time changes its T-dependence near T_g of the slow process found by both dielectric spectroscopy and calorimetry [131, 1071, 1093]. It is unfortunate that in publishing their most recent paper in 2010 [1112], Jansson and Swenson did not have the benefit of the detailed discussion here on the analogy between water confined in solid materials and water in biological and other soft systems, although our paper published in 2007 [1093] has already indicated that the dynamics of water confined in solid materials like MCM-41, hydrogels, and silica gels are similar to water in aqueous mixtures. Due to unawareness, they made the erroneous statement to the effect that the dynamics of water confined in solid materials is drastically different from that in hydrated proteins and other soft systems. Based on this wrong conception, they [1112] went on to suggest the way out of the conundrum they fabricated is to adopt the explanation of the dynamic crossover in soft systems by Cerveny et al. [1107(a)] as the change from cooperative α-relaxation of water above T_g to restricted motion of water in confined spaces below T_g. We have given before ample experimental evidence to show the explanation of Cerveny et al. is untenable not only in aqueous mixtures but also in mixtures of van der Waals liquids. Here I just remind the reader of the monotonic increase of the dielectric strength of the water in aqueous mixtures with increasing temperature above T_g (see Figs. 212 and 213 for examples), which is the signature of secondary relaxation and the opposite behavior of the cooperative α-relaxation of water above T_g proposed by Cerveny et al. The loss spectra of hydrated myoglobin at two hydration levels of $h = 0.33$ and 0.50 presented by Jansson and Swenson [1112] also contradict the interpretation by Cerveny et al. This can be seen from the monotonic increase of the dielectric strength of the water relaxation with temperature above the dynamic crossover temperature ~160 K (see panels (A) and (B) of Fig. 2 in [1112]). More examples of the same from other hydrated proteins can be found in [1317, 1358(a), 1358(c)].

The dynamics of water confined in nanoporous silica matrices MCM-41-S with pore diameters of 18 and 14 Å has been studied down to 200 K by quasielastic neutron scattering. The data were analyzed by using a relaxing cage model by Faraone et al. [1289(a)] and by Liu et al. [1289(b)]. These authors determined the temperature variation of the average translational relaxation time $\langle \tau_T \rangle$ and found it exhibits an abrupt change from VFTH-like at higher temperatures to Arrhenius law at $T \approx 225$ K with activation energies of 5.28 and 6.72 kcal/mol for 18 and 1.4 Å size pores, respectively. They interpreted this behavior as the fragile-to-strong liquid–liquid transition of water suggested by Angell and coworkers [563], now involving $\langle \tau_T \rangle$ of the structural α-relaxation of nanoconfined water. Their data (red closed squares) are plotted in Fig. 219. Similar behavior was observed in the T-dependence of the reciprocal of the self-diffusion coefficient of water confined in cylindrical nanoporous matrices (MCM-41-S) with 14 Å pores [1290].

As can be seen in Fig. 219, there is good agreement between these neutron scattering $\langle \tau_T \rangle$ of Chen and coworkers and the dielectric τ_{fMS} of Hedström et al. in the overlap region, except that the purported "fragile-to-strong" transition exhibited by the neutron data of $\langle \tau_T \rangle$ at about 225 K was not observed by the dielectric experiment. The neutron $\langle \tau_T \rangle$ with the VFTH dependence seems to be continuation of the dielectric τ_{fMS} of water confined in molecular sieves MCM-41 C10 with pore diameter 2.14 nm at hydration levels of 22 wt% (closed inverted green triangles).

As shown before in Figs. 217 and 218, the dielectric relaxation time τ_{fMS} of the faster process of water in molecular sieves has about the same activation enthalpy as τ_ν of the 50% water mixtures with PVME, PVP, and other aqueous mixtures shown in these figures as well as $\tau_{6\,Å}$ of the 6 Å water layer confined in fully hydrated Na-vermiculite clay. Moreover, the τ_{fMS} from adiabatic calorimetry located at T=115 K is nearly the same as value obtained by extrapolation of the Arrhenius dependence of τ_{JG} of the 50% water mixtures with PVME or with PVP to lower temperatures [1093]. These coincidences indicate that τ_{fMS} from dielectric relaxation and adiabatic calorimetry is an intrinsic feature of water, i.e., its primitive relaxation or JG β-relaxation, present in mixtures and in nanoconfinement.

Before closing this subsection, we revisit the adiabatic calorimetry data of water confined in the 1.8 nm diameter cylindrical pores of MCM-41 of Oguni et al. [1283]. They found on heating from 80 K the rapidly precooled sample of the water confined to 1.8 nm pores, an exothermic peak of enthalpy release rate at 210 K appeared in addition to the two peaks at 110 and 160 K found before in water confined in 1.2 and 1.6 nm pores (see Fig. 219). This observation led them to the conclusion that the glass-softening temperature T_g of *bulk* water is 210 K, in conflict with the other proposed values of T_g of supercooled water at 136 or 165 K. The conclusion also is at odds with the order–disorder transition at 227 K of supercooled water proposed by Angell [1285], as pointed out by Mallamace et al. [1284(a)]. Johari [1284(b)] offered a different interpretation of the exothermic peak observed in the 180–230 K range of water confined to 1.8 nm pores. He argued from other evidences it is inconsistent with glass relaxation and is attributed to growth and decay of ice-like distorted unit cells or of stacking faults that remain at equilibrium with the parent structure. The peak centered near 110 K was attributed by him to reorientation of H_2O in the nanoshell with little change in their center-of-mass position, which is not inconsistent with our interpretation of the JG β-relaxation of water. The intermediate peak in the range of 125–175 K was attributed by him to the change in the population of bonds between water and silica surface. Whatever the final outcome of the true interpretation of the intermediate exothermic peak and the T_g of nanoconfined and bulk water, the identification of the primitive or JG β-relaxation time of water as given above seems well supported by experiments.

The Dynamics of Water Nanoconfined in Hydrogels and Silica Gels

Since the primitive relaxation or JG β-relaxation of water is an intrinsic process, its characteristics should not be sensitive to other modes of nanoconfinement including silica hydrogels [1334], hydrogel of poly(2-hydroxyethyl methacrylate) (PHEMA) [1335], and in pores of silica gels [1282]. Two relaxation processes of pure water confined in aged silica hydrogels were also observed dielectrically [1334]. Their relaxation times are shown in Fig. 220 as a function of reciprocal temperature. The relaxation time, τ_{fC}, of the faster relaxation has Arrhenius temperature dependence in the lower temperature range with an activation enthalpy of about 50 kJ/mol. The resemblance of these properties of the faster relaxation to that of some of the JG-processes of water in aqueous mixtures suggests that it is the JG β-relaxation of water confined in hydrogels. The slower relaxation has VFTH temperature dependence for its relaxation time, τ_{sC}, like that of typical α-process. It was suggested [1334] that this relaxation originates from water molecules within the pores that do not interact strongly with the matrix and behave collectively. Another possibility is interaction of water with silica and silica being modified by hydrolysis. Whichever is the ultimate interpretation of the slower relaxation, it is an α-relaxation, and the faster relaxation with Arrhenius temperature dependence may be its JG β-relaxation. Their relaxation times, τ_{sC} and τ_{fC}, may be related by the CM equation just like the relation between τ_{sMS} and τ_{fMS} of water confined in a molecular sieve shown in Fig. 219. In fact, when we compare τ_{sC} with τ_{sMS}, and τ_{fC} with τ_{fMS} (not shown), the values of each pair are similar, indicating the origins of the fast process as well as the slow process are similar in the two cases.

We consider here also the relaxation of water confined in a hydrogel of PHEMA studied by Pathmanathan and Johari (PJ) [1335]. Within the experimental frequency range from 10 to 10^5 Hz, again there are two relaxation processes. The faster relaxation of the 38.6 wt% water has Arrhenius temperature dependence for its relaxation time, τ_{fPJ}, with activation energy of 60.8 kJ/mol, as shown in Fig. 220 by the Arrhenius fit to the data (dashed line) but without the data themselves to avoid crowding. It can be seen from this figure that the relaxation times τ_{fPJ} are near τ_{fC} of the faster process found in water confined in aged silica hydrogels as well as τ_{fMS} of water confined in a molecular sieve. This coincidence suggests that τ_{fPJ} is the JG relaxation time of water in the PHEMA hydrogel. The slower process of water in PHEMA hydrogels can only be observed at higher temperatures dielectrically. At lower temperatures, the dc conductivity dominates and preempts the observation of the slow process. Its relaxation time, τ_{sPJ}, determined from the dielectric loss spectra at several temperatures is shown as open triangles in Fig. 220. Again, at the same temperature, τ_{sPJ} is about the same as τ_{sC} or τ_{sMS}.

Calorimetric measurements of H_2O and D_2O in hydrogels of PHEMA by Hofer et al. [1336] show water has a very weak endothermic step at 132 ± 4 K for H_2O and another endothermic step at 162 ± 2 K (H_2O) and 165 ± 2 K (D_2O), all for a heating rate of 30 K/min. The lower temperature, 132 ± 4 K, is nearly the same as T_g of hyperquenched glassy water [1271, 1333, 1337]. The activation energy of the α-relaxation of hyperquenched glassy water is about 55 kJ/mol, which is also not far from 60.8 kJ/mol for τ_{fPJ}. Moreover, from the calorimetric data, PJ deduced that $\tau_{fPJ} = 53$ s at 135 K. The larger endothermic step at 162 K at the heating rate of 30 K/min indicates the relaxation time of the slower process, τ_{sPJ}, is also 53 s at 162 K. This result, shown by a lone closed triangle located at $1000/T = 6.17$ in Fig. 220 and indicated there that it comes from Hofer et al., seems to be the continuation of τ_{sPJ} obtained by dielectric measurements at higher temperatures, as could be suggested by drawing a line to connect them.

It can be seen by inspection of Fig. 220 that τ_{sPJ}, τ_{sC}, and τ_{sMS} (not shown) all have comparable values over the same temperature region where they increase toward long timescale of vitrification. The activation enthalpy of τ_{sPJ} from calorimetric relaxation near 162 K is about 120 kJ/mol, twice the activation energy of τ_{fPJ}. From the contrasting properties of τ_{sPJ} and τ_{fPJ} discussed above, we suggest that the slower process in PHEMA hydrogels is the relaxation of water interacting with PHEMA, like water interacting with the inner surface of molecular sieves [1102] or with the silica in silica hydrogels [1334]. The faster process could be considered as the JG β-relaxation of water in PHEMA hydrogels. Again τ_{sPJ} and τ_{fPJ} are related by the CM equation, and there is a change of the T-dependence of τ_{fPJ} at the temperature where τ_{sPJ} reaches the long time of 10^3 s like found for aqueous mixtures.

New study of water confined in the pores of silica gel with average diameter of 1.1 nm by adiabatic calorimetry reported again the presence of a faster and a slower process [1282]. We do not discuss the data of water confined in 6, 12, and 52 nm pores of silica gel in this reference because of the reported large fraction of ice formed in the central portion of these pores, which could complicate the observed results. The relaxation times of the slow and fast processes, τ_{scal} and τ_{fcal}, both equal to 10^3 s were detected by adiabatic calorimetry at T=115 and 160 K. They are shown by the two larger circles (⊗) all having relaxation time of 10^3 s in Fig. 220. When the data of τ_{sC} and τ_{sPJ} are extrapolated to longer times, the temperature at which they reach 10^3 s is close to 160 K. Therefore, the dielectric τ_{sPJ} and τ_{sC} are in approximate agreement with the calorimetric τ_{scal}, albeit the confining media are different in the three cases. Similarly, extrapolation of τ_{fPJ} and τ_{fC} to longer times also shows rough agreement with τ_{fcal}. Thus, the two processes of water confined in 1.1 nm pores of silica gel observed by adiabatic calorimetry have to be interpreted in the same way. In particular, the faster one is an intrinsic process of water, not very sensitive to the form of confinement or mixtures. These facts are at odds with the interpretation by Oguni et al. [1282] that the fast process, which they observed to have relaxation time of 10^3 s at its $T_g = 115$ K, is caused by the “interfacial water” molecules bonded to the silanol groups of the wall of the pores. The other relaxation they observed having relaxation time of 10^3 s at its $T_g = 160$ K is attributed to “internal water” molecules located in the central part of the pore, interacting with each other like in bulk water. As mentioned before, only one water molecule is in the “internal water” of the 1.1 nm diameter pores. This state of “internal water” molecules with only one molecule layer not hydrogen bonded to silica cannot give rise to cooperative α-relaxation of bulk water. The relaxation that has relaxation time 10^3 s at $T_g = 115$ K should be identified with the JG relaxation because of good agreement with the extrapolation of the dielectric JG relaxation time of water in some aqueous mixtures down to lower temperatures by its Arrhenius T-dependence. This is at odds with the interpretation of the relaxation of 10^3 s at 115 K as caused by the “interfacial water” molecules bonded to the silanol groups of the wall of the pores by Oguni et al. Bonding to the immobile silanol group should make the water molecule have longer relaxation time than the JG relaxation of water, and thus it is more reasonable to identify it with the relaxation observed at the higher temperature of 160 K.

The Dynamics of Water Confined in Cement Paste

The dynamics of supercool water confined into the calcium silicate hydrate (cement paste) was studied by quasielastic neutron scattering (QENS) and differential scanning calorimetry (DSC) by Zhang et al. [1289(e)]. The measured QENS spectrum was analyzed with the relaxing cage model. The translational relaxation time $<\tau>$ of the confined water plotted against 1000/*T* exhibits Arrhenius *T*-dependence for *T* below about 210 K with activation energy ~38 kJ/mol, and when extrapolated to lower temperatures $<\tau>$ reaches 100 s in the neighborhood of 110 K. Moreover, $<\tau>$ crosses over broadly in the range from 210 to 225 K to assume a stronger *T*-dependence at higher temperatures, which is accompanied by the observation of enthalpy relaxation by DSC in this temperature range. The latter indicates the presence of a slower relaxation with much longer relaxation time than $<\tau>$ undergoing glass transition in the temperature range. All these features of the dynamics of water confined in cement paste (see Fig. 220-1) are similar to those found for water confined in molecular sieves and silica gels shown before in Figs. 219 and 220, suggesting $<\tau>$ from QENS is the JG β-relaxation of water confined in cement paste. The slower relaxation detected by DSC at long times of the order of 100 s is water interacting with the cement paste, analogue of water interacting with the walls of molecular sieves and silica gels. The only difference is the higher T_g in the present case perhaps due to stronger interaction of water with the confining cement paste. This natural and analogous interpretation of $<\tau>$ from QENS differs from that of Zhang et al. who interpreted their observation to the so-called *dynamic crossover* or *fragile-to-strong transition* of water at 225 K. The difference of nearly 10 orders of magnitude in timescale between $<\tau>$ of the process observed by QENS and the relaxation time of the process detected by DSC means the two are very

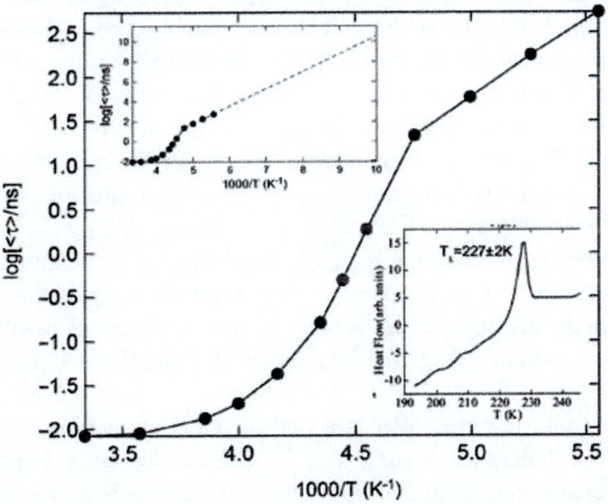

Fig. 220-1 Arrhenius plot of the translational relaxation time $<\tau>$ for water confined in aged cement paste. The *upper inset* shows extrapolation of the Arrhenius T-dependence down to 10 s. The *lower inset* shows the DSC curve of water confined in hydrated cement paste. Data from [1289(e)] are digitized and replotted here

different processes, and yet bear some relation to each other because of the change in T-dependence near T_g. This can be explained by our interpretation because of the connection between the α-relaxation and JG β-relaxation. However, it is not clear how it can be explained by the idea of *dynamic crossover* or *fragile-to-strong transition*.

Chen et al. [1289(g)] revisited the neutron scattering and DSC data of water confined in aged cement paste published in [1289(e)] summarily shown in Fig. 220-1. To support their claim that confined water exhibits the fragile-to-strong crossover in the neutron data at a temperature T_L coincident with the heat flow peak temperature of the DSC *cooling* scan, they claim this is a general phenomenon found in other glassforming liquids. They present viscosity, diffusion, and DSC data from the literature of several small molecular glassforming liquids to show their viscosity η and diffusion coefficient D also exhibit the fragile-to-strong crossover phenomena (FSC) at the peak temperature of heat capacity curve obtained by DSC at a *heating* rate of 10°C/min. These effects are considered by Chen et al. to be intrinsic properties of bulk liquid systems, and the FSC temperature T_L, suggested by them to be about 20% higher than the calorimetric glass transition temperature T_g, is an important transition temperature (or crossover temperature) for the dynamic behavior of liquids in general. The viscosity and DSC data of *o*-terphenyl, salol, alpha-phenyl-o-cresol used by Chen et al. are from the publication of Laughlin and Uhlmann [375]. The viscosity data of *tris*-naphthylbenzene (TNB) are not from Laughlin and Uhlmann, although incorrectly stated as such by Chen et al., but likely from Plazek and Magill [347, 743, 744] presented before in Fig. 11. The semilog plots of the viscosity η data against reciprocal temperature do give the impression that the T-dependence of η changes from a VFTH law to an Arrhenius law at some temperature T_L and the latter is maintained down to T_g. However, the use of rather large size symbols for the data may obscure any deviations. The data of η for values higher than 10^8 poise obtained by Laughlin and Uhlmann using a beam-bending technique lack precision, which can be seen from the scatter of the data in the plots. The most accurate data of η at lower temperatures where Arrhenius law is claimed to hold by Chen et al. are those of TNB from the creep compliance measurement of Plazek and Magill. Their TNB data and the good fit by the VFTH equation given by them originally have been shown before in Fig. 11. The appearance of satisfactory fit of the data to an Arrhenius law by Chen et al. is made possible by the large size symbols they use to represent the data, ignoring the deviation at the lowest temperature (see Fig. 3 in [1289(e)]). The crudeness of fitting the viscosity by the Arrhenius law can be seen from the discrepancy in its onset in two publications involving the same principal workers, which is $\log(\eta/\text{poise}) > 6$ in [1289(g)] but $\log(\eta/\text{poise}) > 4$ in [1289(h)].

If the fragile-to-strong crossover is a real physical happening and an universal behavior of liquid dynamics as suggested by Chen et al., then it should be found not only in viscosity and diffusion data and self-intermediate scattering function from simulations [1289(f)] but also by dielectric relaxation and photon correlation spectroscopy. Broadband dielectric spectroscopic data of many glassforming liquids including those considered by Chen et al. are available. The frequency range of dielectric measurements, typically from a few gigahertz down to 10^2 Hz, is broad enough to test the claim of Chen et al. However, the experimental data of the dielectric α-relaxation time τ_α of most if not all the glassforming liquids do not show change of T-dependence from the VFTH or other super-Arrhenius T-dependence to strictly Arrhenius law at any temperature above T_g down to T_g [236, 348, 424, 441, 455, 456, 621, 660, 661, 682, 971, 972, 1014, 1121, and many others]. Dynamic light scattering together with photon correlation spectroscopy also covers time range from nanoseconds down to 10^3 s and has not found evidence of the crossover to Arrhenius T-dependence either [348, 719].

Another experimental result that baffles understanding is the report [1289(i)] of breakdown of Stokes–Einstein relation of confined water in fully hydrated MCM-41-S with pore diameters of 14 and 18 Å when the average translational relaxation time $<\tau_T>$ determined by neutron scattering is longer than 2 ps but shorter than 100 ps, where it is found that $D \sim (<\tau_T>)^{0.74}$. This is different from the case of conventional glassformers. At these short relaxation times, the SE relation holds for conventional glassformers, and breakdown becomes evident only when the α-relaxation time becomes much longer than nanoseconds. The authors of [1289(i)] are unaware that heterogeneous dynamics cannot explain the breakdown of the Stokes-Einstein relation in OTP, TNB, sucrose benzoate and ROY as evident from their statement: "The SER breakdown and the dynamical heterogeneities are thus intimately connected".

The Dynamics of Water Nanoconfined in Clay

By confining water within a layered Na-vermiculite clay, the effect of nanoconfinement and the presence of intercalated sodium ions suppress crystallization of water, Bergman and Swenson found a relaxation process with an Arrhenius temperature dependence for its relaxation time in the 6 Å thick layer of water over a wide temperature range, including the usually inaccessible temperature region of bulk water [1296]. The data of this process have been shown in Fig. 219, and used as supporting evidence of the JG relaxation of water. However, Bergman and Swenson interpreted the process otherwise as evidence of the strong part of the structural α-relaxation of deeply supercooled bulk water to support the existence of a fragile-to-strong transition of supercooled water (i.e., the presumed change from a Vogel–Fulcher–Tammann–Hesse type temperature dependence of the structural relaxation time to an Arrhenius type) proposed to occur around 228 K by Angell and coworkers [563]. A critique of this interpretation was given by Johari [1297]. He pointed out that it is inappropriate on fundamental grounds to compare the dielectric properties of structural relaxation of bulk water with those of the two-molecule thick water layer between the platelets of sodium vermiculite clay. Moreover, he noticed that the change in dielectric strength with temperature can be determined from the area under the dielectric loss peaks published in [1296] at different temperatures. The estimate of the area seems to decrease on decreasing the temperature, which is the opposite of that expected from Curie law for the structural relaxation of liquids including water. On the other hand, this observed temperature dependence of the dielectric strength is in accord with secondary or JG relaxation.

The behavior of the water accommodated between the silicate layers of a clay $Na_8[Si_{32}O_{64}(OH)_8] \cdot 32H_2O$ was investigated by Oguni and coworkers using adiabatic calorimetry [1338]. Each sodium ion is coordinated octahedrally by six water molecules, half of which form hydrogen bonds with silanol-oxygen atoms, and the other half are located far from silanol oxygen and not forming a hydrogen bond with it. The water molecules, without strong hydrogen bonds with silicate layers, were found to show signature of glass transition at a temperature T_g of 106 K due to freezing of the water molecule in their orientations. From its relaxation time of 10^3 s at the T_g of 106 K and the assumption that the pre-exponential factor of its Arrhenius T-dependence is 10^{-14} s, the potential energy barrier of the reorientation relaxation was assessed to be about 34 kJ/mol, which is slightly smaller than the energy required to break two hydrogen bonds, and comparable to the observed activation energy of the JG relaxation of water in some aqueous mixtures.

The Dynamics of Water in Nanochannel of Crystal

Water confined in nanochannel of crystal was studied by adiabatic calorimetry by Oguni and coworkers [1339]. The crystalline material is $[Co(H_2bim)_3](TMA) \cdot 20H_2O$, where H_2bim is 2,2′-biimidazole, TMA is 1,3,5-benzene tricarboxylic acid, and $20H_2O$ represents the water forming nanochannel in the crystal. A glass transition was observed at a T_g of 107 K. It was discussed in the 2010 paper [1339] as glass transition of a small number of water molecules around the center of the water nanotube. It may be considered as another manifestation of the JG or primitive relaxation of water as suggested by its T_g of 107 K at 10^3 s, nearly the same as those found for the JG relaxation of water in other systems.

The Dynamics of Water Nanoconfined in Graphite Oxide

Dynamics of water intercalated in graphite oxide (GO) was investigated by Cerveny et al. [1321(a)] using dielectric spectroscopy over the frequency range $10^{-2} - 10^9$ Hz, temperature range 140–300 K, and water concentration from 0 wt% (anhydrous GO) up to 25 wt% without crystallization occurring. According to these authors, the interlayer distance during hydration increases from 5.67 to 8 Å which corresponds to the uptake of a water monolayer in the interlayer space of GO. The rotational water relaxation observed is slightly dependent on the hydration level. For water concentrations higher than 15 wt%, the authors claimed they have observed a change in the temperature dependence of the relaxation times between 10^{-7} and 10^{-6} s for 25 wt%. However, the change is slight and the authors also fitted the entire range by Arrhenius T-dependence, which was shown in the case of the 25 wt% sample together with the neutron scattering data by others [1321(b)] in Figs. 215, 217, and 218.

2.3.2.28 Hydrated Proteins

Proteins are biological macromolecules essential for function of living systems. The research in proteins and other biomolecules such as enzymes involves the study of their structure, dynamics, and function. By now, the problem of the structure of

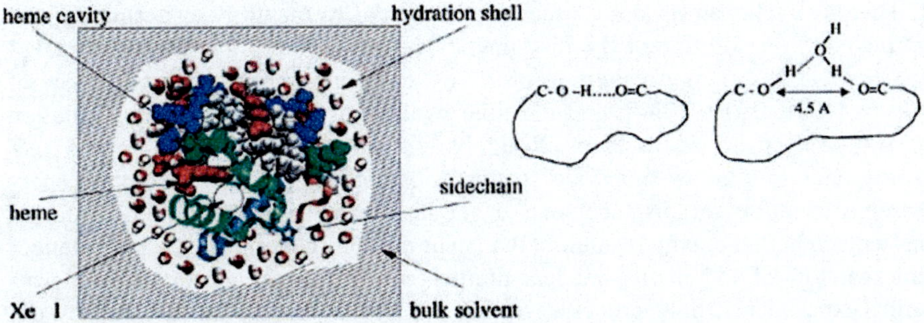

Fig. 221 (*Left*) A stylized look into Mb displays the parts of the protein that are involved in protein dynamics and function. The lower half of the structure shows three α-helices, part of the protein backbone, with one side chain as an example. The upper half provides a space-filling view. Between the two halves is a heme group (*in red*). Hydration waters are shown as *red-and-white spheres*. Two cavities that can hold ligands are also depicted. The hydration shell and the bulk solvent envelop the protein and dominate protein dynamics. Reproduced from [1341] by permission. (*Right*) Protein–water interaction: (*Left*, without water) hydrophilic interactions have to be saturated internally; (*Right*, with water) water allows the structure to expand; the optimal hydrogen bond configuration for the coupling of two functional sites is shown [1318]. Reproduced from [1318, 1341] by permission

many proteins including hemoglobin and myoglobin has been solved and the functions of numerous proteins are also known at least qualitatively. The structure of myoglobin is shown in Fig. 221 (left) together with the hydration shell and bulk solvent. Water penetrates the disordered fringes and activates liquid-like translational displacements of side chains, which allows the diffusion of small ligand molecules across the protein. Figure 221 (right) illustrates how in the dry system hydrogen bonds have to be saturated internally, while water molecules offer alternative binding sites which open the structure. Water thus introduces fluctuations, facilitates protein motions, and enables function.

However, the dynamics of proteins is still an unsolved problem. This is unsurprising because proteins are complex systems with structures more complex than ordinary supercooled liquids and glasses. Water is important for biological systems to function. Dehydrated proteins cannot function, but about two layers of water (called the hydration shell) surrounding the folded protein fully activates the protein functionality [133, 1340, 1341]. The dynamics of proteins is coupled to that of the solvents surrounding the protein molecules. However, in the literature, different views of this coupling and explanation of the dynamics have been offered [133, 134, 1318, 1319, 1344–1358].

Dynamic Transition from Neutron Scattering Experiments

Insight into motions of hydrated proteins apparently came first from the mean-squared displacement (msd), $<r^2>$, of fast processes in hydrated proteins measured by Mössbauer effect [1342, 1343] and by neutron scattering [133, 134, 1318, 1319,

1341–1347]. The timescales of the msd measured by neutron scattering depend on the energy resolution of the instrument. The backscattering spectrometer IN13 at Grenoble has energy resolution of 7 μeV corresponding to a time window of 100 ps [1318, 1319, 1344], and the time window of IN6 with energy resolution of 100 μeV covers about 15 ps. Roughly, instrument energy resolution 10 μeV corresponds to window timescale 100 ps. Mössbauer resonance absorption spectroscopy monitors the displacements of the heme iron inserted in the protein that has a much higher energy resolution than neutron backscattering and a much longer time window of 141 ns, the nuclear lifetime of ^{57}Fe. In the elastic neutron scattering experiments, only processes which are fast enough to be resolved by the spectrometer contribute to the loss and are probed. On increasing temperature, the measured msd exhibits a change from a weaker temperature dependence to a significantly stronger dependence at some temperature, T_D, traditionally referred to as the dynamic transition temperature. The temperature T_D is dependent on the protein, the solvent, and its weight fraction. For hydrated myoglobin, values of 200 K and up to 240 K [1319] have been reported, and 220 K for hydrated lysozyme [1347]. Neutron scattering measurements of myoglobin hydrated by fully deuterated water show the same characteristic temperature dependence of the msd, and this result indicates that the non-exchangeable hydrogens of the hydrated protein participate in the motions from which the observed msd originates. An example is taken from the paper by Tsai et al. [1345(a)] of four samples prepared to maximize the scattering from the protein relative to the other components (see Fig. 222, right panel). The samples are dry D-exchanged lysozyme, lysozyme powder hydrated with 30% D_2O, and dehydrated lysozyme-deuterated glycerol samples prepared with 20 and 50% deuterated glycerol. The T_D of the so-called "dynamic transition" is 210 K for the 30% D_2O sample, 270 K for the 50% deuterated glycerol sample, 330 K for the 20% deuterated glycerol sample, and not detected in the dehydrated lysozyme.

The phenomenon was first interpreted by Doster and coworkers [1344, 1354] as a "dynamic transition" of protein motions, which are coupled to the kinetic glass transition of the solvent near the protein surface. This interpretation was made more explicit by Doster [1319] by saying that "water near the protein surface can perform long range diffusion, which is arrested near 200 K [1318], it shows liquid behavior above 200 K and glassy behavior below. The slow α-process can be associated with the elementary step of diffusion. Hydration- and bulk water thus differ essentially in the magnitude of the diffusion coefficient and the rate of crystallization."

A different interpretation of such a rapid increase in the conformational msd was given by Frauenfelder and coworkers [134, 1341, 1348, 1355, 1356]. These authors suggested from their energy landscape model of hydrated proteins with hierarchical tiers that it is due to the β-fluctuations (in the hydration shell) in the lowest tier, which has Arrhenius temperature dependence for its relaxation time.

Zanotte et al. [1347] believed that the rapid increase of msd is not associated with a thermal event as detected by calorimetry and suggested that it is more appropriate to describe the phenomenon as "dynamic crossover" instead of "dynamic transition."

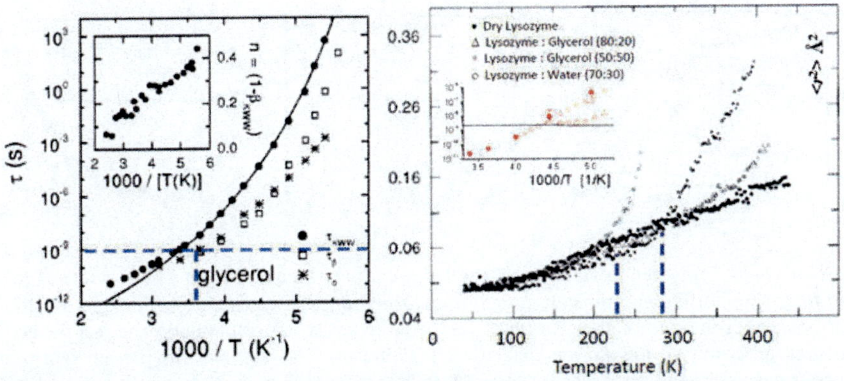

Fig. 222 (*Left*) Dielectric α-, JG β-, and primitive relaxation times of glycerol represented by *circles*, *squares*, and *asterisks*, respectively, taken from [424]. The *horizontal dashed line* indicates 1 ns. The *short vertical dashed line* indicates the temperature ~277 K above which τ_β of glycerol becomes shorter than 1 ns, the timescale corresponding to the neutron backscattering instrument with an energy resolution of 1 μeV. (*Right*) Mean-squared displacement as a function of temperature for lysozyme samples from Tsai et al. obtained by the neutron backscattering instrument with an energy resolution of 1 μeV [1345(a)]. The timescale corresponding to the neutron backscattering instrument at NIST with an energy resolution of 1 μeV is 1 ns. These authors determine T_D by visual inspection to be the temperature at which the slope of these curves changes: 210 K for 30% D_2O, 270 K for 50% glycerol, and 330 K for 20% glycerol. The temperature, $T_{1ns} = 277$ K, determined from the *left panel* and indicated by the *longer vertical dashed line* is consistent with $T_D \sim 270$ K of lysozyme-glycerol (50–50) given by Tsai et al. [1345(a)]. Going back to Fig. 217, there it has been shown that τ_β of water becomes shorter than 1 ns at temperatures above 238 K. The *shorter vertical dashed line* indicates this temperature, which is slightly higher than 210 K given by Tsai et al. Alternatively, the τ_β of water in hydrated lysozyme at hydration level of $h = 0.4$ obtained by Khodadadi et al. [1358(a)] shown in the *inset* extracted from [1358(a)] can be used to determine $T_{1\,ns} = 235$ K for 1 ns resolution and $T_{2ns} = 228$ K for 2 ns resolution as shown in the *inset* by the *horizontal line*. Data from [424, 1345(a), 1358(a)] are replotted in the redesigned figures

A most recent Raman and Brillouin scattering measurement was done to estimate the glass transition temperature, T_g, of hydrated lysozyme at hydration level of $h = 0.4$ [1357]. The measurements found glass transition in the hydrated lysozyme over a broad range of $T_g \sim 180 \pm 15$ K. However, the "dynamic transition" exhibited as a sharp rise in msd $< r^2 >$ occurs at temperatures around T_D~200–230 K [1358], significantly higher than T_g. The Brillouin scattering data determining T_g are presented in the left panel of Fig. 223. The fast dielectric relaxation process (▲) changes its T-dependence at T_g. The protein relaxation process measured by neutron spectroscopy (O) in lysozyme hydrated by D_2O at hydration level $h \sim 0.4$ [1357, 1358] exhibits the dynamic transition of its msd $< r^2 >$ at $T_D \sim 218$ K indicated by the arrows. The neutron scattering data of $< r^2 >$ were obtained by the backscattering spectrometer instrument with an energy resolution of 1 μeV with corresponding time window of 1 ns. The slower dielectric relaxation process (△) may be due to ice (see [1317]). The right panel presents the msd in dry and hydrated lysozyme and

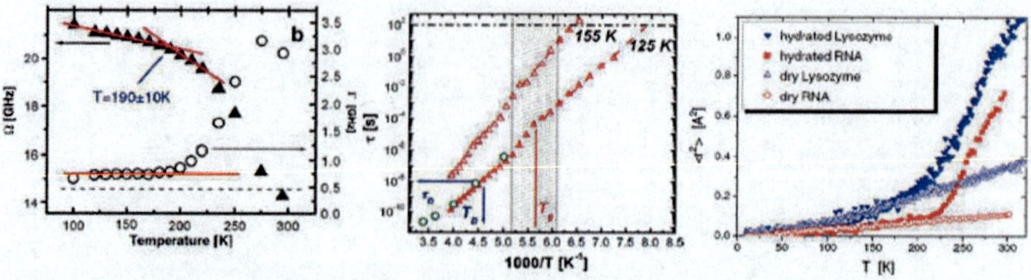

Fig. 223 (*Left*) Temperature variation of the Brillouin peak frequency Ω (▲) and width Γ (○). *Solid lines* show different temperature regimes. *Dashed line* indicates the width of the Brillouin spectrometer resolution function. (*Middle*) Temperature variations of the main and fast dielectric relaxation process (▲), the slower dielectric relaxation process (△), and the protein relaxation process measured by neutron spectroscopy (O) in lysozyme with hydration level $h \sim 0.4$ [1357]. The *arrows* indicate estimated T_D and averaged T_g of the hydrated lysozyme. The *shadowed area* marks the broad glass transition range. The *dashed-dotted line* marks relaxation time $\tau = 100$ s that usually corresponds to τ of the main relaxation process at T_g. *Numbers* show the temperatures at which these relaxation processes reach $\tau = 100$ s. (*Right*) Mean-squared atomic displacements in dry and wet lysozyme and RNA [1315, 1358(b)] obtained by using the neutron backscattering instrument at NIST with an energy resolution of 1 μeV. The sharp rise in $< r^2 >$ of hydrated biomolecules at T above 200 K is the dynamic transition. Reproduced from [1315, 1357, 1358(b)] by permission

RNA, obtained by the backscattering instrument with an energy resolution of 1 μeV, to show the dynamics transition occurring at about 220 K for both.

Thus, the "dynamic transition" has a different physical origin than glass transition of the hydrated protein. Actually the sharp increase of $< r^2 >$ at temperatures above T_D is caused by a hydration-induced fast relaxation process that enters the high-frequency window of the neutron spectrometer. As we shall see later on, high-frequency susceptibility from dielectric and neutron scattering measurements shows that the properties of hydration water and hydrated protein change smoothly when crossing T_D, confirming the "dynamic transition" is just the manifestation of the hydration water relaxation process entering the frequency window of the neutron spectrometers, and therefore the term "dynamic crossover" suggested by Zanotti et al. is more appropriate. The relaxation process that causes the sharp rise in $< r^2 >$ has characteristic relaxation time $\tau \sim 10^{-8} - 10^{-11}$ s at T_D depending on the spectral resolution of the instrument (1 μeV resolution corresponds to 240 MHz or about 1 ns). This is demonstrated in the next subsection by the $< r^2 >$ data of the 30% D_2O hydrated lysozyme, and the 50% dehydrated lysozyme–50% deuterated glycerol obtained by Tsai et al. using the neutron backscattering instrument with an energy resolution of 1 μeV.

Coupling Between the Protein Dynamics and the Solvent Dynamics

The coupling between protein dynamics and hydration-water dynamics is of fundamental importance for understanding protein function. Although some correlation between solvent and protein dynamics has been proposed based on experimental and simulation evidences [131, 134, 355, 1071, 1105, 1341, 1317–1319, 1345–1371], the interpretations of the origin differ greatly. In this subsection, relevant experiment and simulation data are discussed to elucidate the exact nature of the coupling.

The left panel of Fig. 222 shows the dielectric α- and the JG β-relaxation times of neat glycerol as well as the primitive relaxation times calculated from the experimental values of the stretch exponent of the Kohlrausch function that fits the spectrum as given in [424]. The horizontal dashed line indicates 1 ns. The short vertical dashed line indicates the temperature $T_{1ns} \sim 277$ K above which τ_{JG} of glycerol becomes shorter than 1 ns, the timescale corresponding to the neutron backscattering instrument with an energy resolution of 1 μeV. The value of $T_{1ns} \sim 277$ K is near the dynamic transition temperature $T_D \sim 270$ K of the lysozyme given by Tsai et al., as indicated in the plot of $< r^2 >$ vs. T by the longer vertical dashed line located at 277 K in the right panel of Fig. 222. The good correspondence between $T_{1\ ns}$ deduced from the dynamics of pure glycerol and T_D from msd of lysozyme (with 50% deuterated glycerol) is remarkable. It shows that lysozyme and glycerol are dynamically coupled, and the fast relaxation responsible for the dynamic transition is the JG β-relaxation of the solvent, glycerol.

For the 30% D_2O hydrated lysozyme, the temperature $T_{1\ ns}$ above which τ_{JG} of water becomes shorter than 1 ns has been determined before from the data of τ_{JG} of water in Fig. 217 to be within the range, $T_{1ns} \sim 208 - 238$ K. Measurement of τ_{JG} over the range from 10^{-3} to 10^{-11} s of water in hydrated lysozyme with hydration level $h = 0.4$ is available from the combined dielectric relaxation and nuclear scattering study by Khodadadi et al. [1358(a)] using the same nuclear spectrometer at NIST as Tsai et al. with resolution of 1 μeV. From Fig. 3 of this reference (shown partly as inset in Fig. 222, right panel), one can see $T_{1ns} \sim 235$ K for 1 ns resolution, and $T_{2ns} \sim 228$ K for 2 ns for the resolution used by Khodadadi et al. All of the above values of $T_{1\ ns}$ can be considered to be consistent with the dynamic transition temperature $T_D \sim 210$ K for 70% lysozyme–30% D_2O from Tsai et al. Again the rough agreement between $T_{1\ ns}$ or $T_{2\ ns}$ deduced from the dynamics of water and T_D from msd of lysozyme (with 30% D_2O) can be considered as evidence of dynamical coupling between the lysozyme and water, and the fast relaxation exhibiting the dynamic transition is the JG β-relaxation of water in the hydrated lysozyme.

To further show the relation of the secondary relaxation of the solvent to the dynamic transition, the neutron scattering msd data of dry trehalose-coated carbon monoxy myoglobin (CO-myoglobin) obtained by Cordone et al. using the IN16 spectrometer [1372] is compared with hydrated myoglobin using the IN13 spectrometer by Doster et al. [1344] in the left panel of Fig. 224. Also shown on the right panel for comparison are the MSD data of protein C-phycocyanin (C-PC) hydrated by D_2O with $h = 0.5$, and by D_2O with various deuterated trehalose included by Köper et al. [1373, 1374] in order to observe individual motions of the non-exchangeable protons of the protein. These include the C-PC/trahalose/D_2O samples with compositions 1.0/0.3/0.7, 1.0/0.75/0.75, and 1.0/1.0/0.0. The last one is entirely deuterated trehalose. It can be seen from both figures that no dynamic transition was observed for trehalose, and the msd and the density of state function are those of a harmonic solid, up to room temperature in the case of CO-myoglobin,

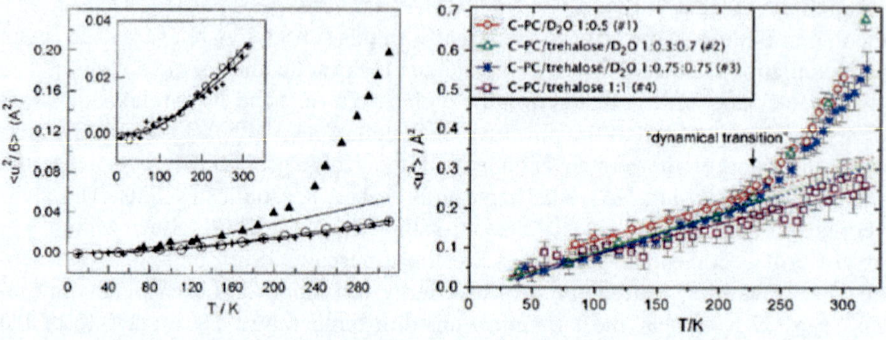

Fig. 224 (*Left*) Mean-squared displacements calculated from elastic neutron scattering measurements (IN16 with resolution of 1 μeV) and plotted as $<u^2>/6$ (Å^2) vs. temperature T (K), for comparison with data: (O) trehalose-coated CO-myoglobin and (+) trehalose, from Cordone et al. [1372]; (▲) D_2O hydrated myoglobin from Doster et al. using IN13 with resolution of 10 μeV [1344]. The *continuous lines* represent the fits of data points by harmonic vibrational contribution, including data from Doster et al. at low temperature (50–170 K). *Inset:* Data relative to trehalose on an expanded scale. As can be seen, the harmonic model is obeyed in the whole temperature range only for the trehalose sample, whereas non-harmonic contributions (protein-specific motions) are evident in the hydrated myoglobin sample at high temperature. (*Right*) MSDs $<u^2>$ of protonated C-PC samples embedded by deuterated trehalose/D_2O at different ratios as a function of temperature (40–318 K). The *solid* and *dotted lines* represent the harmonic behavior of each sample, respectively, before and after the "dynamical transition." Data from Köper et al. [1373, 1374] using IN13 with energy resolution (FWHM) of 8 μeV, which gives access to dynamics on a timescale above 150 ps and up to 400 ps. Reproduced from [1372, 1373] by permission. The temperature of the hydration water corresponding to 400 ps according to the middle panel of Fig. 223 is 235 K, which is in approximate agreement with the dynamic transition temperature indicated by Köper et al. using the *vertical arrow* in the right panel of Fig. 224

and up to above 320 K in the case of C-CP. No dynamical transition was observed, in contrast to proteins hydrated by D_2O. This can be explained by the γ-relaxation times of trehalose, τ_γ, being still longer than 1 ns and the JG β-relaxation time, τ_β, being much longer than 1 ns at 320 K, compared with the time window of 15 ps for IN6 and 400 ps for IN13. These can be seen by inspection of the relaxation map of trehalose given before in Fig. 206 in Section 2.3.2.26. Since the motion associated with the intramolecular γ-relaxation and the intermolecular JG β-relaxation do not occur in the neutron scattering temperature range, it is clear no dynamic transition from protein fluctuations can be observed in the protein/trehalose samples even when the protein is coupled to the trehalose via hydrogen bonding. This also explains on a molecular dynamics basis the role of trehalose in stabilizing protein by inhibiting some of its fluctuations which cause protein unfolding and denaturation. It is consistent with the relatively higher stability of hydrogen bonds with the protein [1375], and more water molecules can be substituted [1376] in the case of trehalose, as compared to other bioprotecting agents and sugars. Our explanation based on the secondary relaxations of trehalose is more direct than those based on the "high viscosity" hypothesis, according to which the denaturation processes are hindered by the higher glass transition temperature of trehalose solutions [1220, 1221, 1377–1382], at least as far as connection with the dynamic transition is concerned. This is because viscosity is related to the structural α-relaxation of trehalose, which definitely is not involved in causing the dynamic transition.

The identification of this fast process detected from the dynamic transition behavior of $< r^2 >$ by neutron scattering is a major problem in dynamics of hydrated biomolecules. It was considered as originating from the hydration water by some researchers, particularly those engaged in dielectric spectroscopy [1105, 1110, 1360–1362]. However, by replacing ordinary water with heavy water D_2O in hydrating the protein or deuterated glycerol in the study by various workers such as Bellissent-Funel [1374] and Tsai et al. [1345(a)], neutron scattering from the water is suppressed, and the observed scattering is from the hydrogen in the hydrated protein. This kind of neutron scattering data, such as that shown in the right panel of Fig. 222 by Tsai et al., in the right panel of Fig. 224 by Köper et al. [1373, 1374], in the left panel of Fig. 225 by Doster and coworkers [1319], and in the right panel of Fig. 225 by Wood et al. [1363], as well as by Sokolov and coworkers in the right panel of Fig. 223 [1358], definitely shows that this relaxation also can be detected by neutron scattering from hydrogen in the hydrated protein. Most interesting is the change in slope of the temperature dependence of $< r^2 >$ in the protein takes place at almost the same temperature as $< r^2 >$ of its solvent including hydration water This can be seen directly in Fig. 223 for hydrated myoglobin [1319] and for hydrated maltose binding protein (MBP) by Wood et al. [1363],

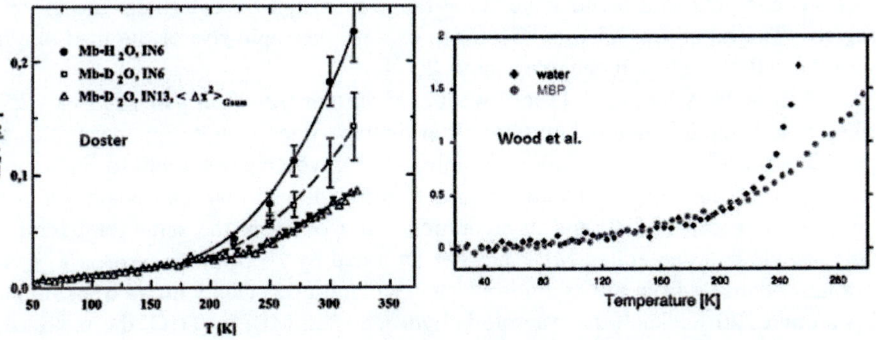

Fig. 225 (*Left*) Mean-squared displacements, IN6: D_2O-hydrated myoglobin (*closed circles*) and hydration water (H_2O-hydrated myoglobin) (*open squares*). Mean-squared displacements, IN13: D_2O-hydrated myoglobin (*open triangles*) and H_2O-hydrated myoglobin (*closed triangles*). Gaussian component with contribution of methyl groups removed [1319]. (*Right*) Mean-squared displacements of ns–ps motions in maltose binding protein (H-MBP-D_2O sample; *gray circles*) and in its hydration water (D-MBP-H_2O sample; *black diamonds*). Dynamical transitions (changes in slope of temperature-dependent mean-squared displacements) in the protein and in its hydration water take place at similar temperatures (~220 K). Data from Wood et al. [1363] using the backscattering spectrometer IN16 at the ILL. The instrumental energy resolution of 0.9 μeV (FWHM) implies that only motions in the ps–ns timescale are probed. Reproduced from [1319, 1363] by permission

and suggests dynamical coupling between protein and solvent motions at least on the ns–ps timescale of neutron scattering experiments. More discussions of these measurements and computer simulations [1363] are given below.

The connection between the relaxation of hydration water and the relaxation of the hydrated protein is revealed by comparing the msd observed in H_2O-hydrated myoglobin with that in D_2O-hydrated myoglobin, both at the same hydration level of 0.35 g/g. The data from the IN6 spectrometer with time window of 15 ps from Doster [1319] are shown in the left panel of Fig. 225. In both systems the msd increases in the same manner with a common onset at 180–200 K. In the H_2O-hydrated system, the msd is mainly contributed by water, because of the large cross-section and the large motional amplitude of water relative to protein structural fluctuations. In the D_2O-hydrated myoglobin, the msd comes from protein fluctuations with smaller amplitudes. The same two samples with H_2O and D_2O hydration were investigated on a 50 ps timescale using backscattering IN13 spectrometer in which the methyl group rotation contributes significantly. After the contribution of the methyl groups was subtracted, the resulting displacements are shown by open and closed triangles in Fig. 225. Both protein and water displacements are similar in magnitude and exhibit the same temperature dependence. The similar temperature dependence of the msd of protein and water suggests a common cause. Doster also explained the onset at 180 K to originate from fluctuations of hydrogen bonds, previously proposed to explain the temperature dependence of the IR amide bands of myoglobin in mixed solvents [1358]. The enthalpy and entropy differences are $\Delta H \approx 14(\pm 2)$ kJ/mol and $\Delta S/R \approx 3.5(\pm 0.3)$, respectively. He suggested the hydrogen-bond fluctuations to be the β-process which involves localized librational motions and acts as a precursor of long-range diffusion of water near the protein surface. The latter was used by Doster to explain the rapid rise of the msd due to what he called the liquid behavior above 200 K.

The study by Wood et al. [1363] went even further (see right panel of Fig. 225). They performed incoherent neutron scattering experiments on designed samples of a soluble protein (maltose binding protein, MBP), which allow them to separately probe protein and solvent dynamics, and directly showed that the dynamic transitions of the protein MBP and its hydration water occur at the same temperature. One sample is a deuterated MBP powder hydrated by H_2O; the measured $< r^2 >$ primarily coming from the hydration water dynamics revealed a broad transition at T_D around 220 K. The other sample is hydrogenated MBP in D_2O, the measured $< r^2 >$ originating from the protein dynamics exhibited a dynamical transition at the same temperature. The backscattering spectrometer IN16 at the ILL they used has instrumental energy resolution of 0.9 μeV (FWHM) implying that only motions in the ps–ns timescale are probed. As mentioned before in connection with Figs. 217 and 222, the temperature $T_{1\,\mathrm{ns}}$ above which τ_{JG} of water becomes shorter than 1 ns, the timescale corresponding to the neutron backscattering instrument with an energy resolution of 1 μeV, lies within the range, $T_{1\,\mathrm{ns}} \sim 208 - 238$ K. This range of $T_{1\,\mathrm{ns}}$ is consistent with the broad dynamic transition at T_D around 220 K.

Khodadadi et al. [1358(a)] found similar susceptibility spectra and the same relaxation times from neutron scattering and dielectric relaxation in lysozyme hydrated by D_2O (also see inset of the right panel of Fig. 222, and the middle panel of Fig. 223). Since neutron scattering probes exclusively the lysozyme and not the deuterated hydration water while dielectric relaxation probes also the deuterated hydration water, it follows the same fast relaxation was observed by both techniques, and the dynamics of lysozyme and water is coupled.

In Section 2.3.2.27 on dynamics of aqueous mixtures, we have already mentioned several properties of the dynamics that the two components, hydrophilic solute and water, are coupled together because of hydrogen bonding of water to the solute. One outstanding property is the change of T-dependence of the primitive relaxation or JG relaxation time, τ_0, τ_ν, or τ_{JG}, of water at T_g of the mixture, which depends strongly on the solute (see Figs. 210, 211, 212, 213, 214, 215, 216, 217, and 218). This property already signals the water relaxation is coupled to the solute. Here we underscore this coupling by the fact that τ_ν or τ_{JG} of water depends sensitively on the composition of the aqueous mixture shown by the example of mixtures of water with glycerol in the right panel of Fig. 226. Obviously, if the two components are not coupled at the local level, τ_0 or τ_{JG} of water should be independent of composition, which is not the case. In the left panel of Fig. 226 are dielectric relaxation time of the fast water dielectric relaxation process in hydrated lysozyme from Khodadadi et al. [1357, 1358], and in Fig. 227 the same process in hydrated collagen and elastin by Gainaru et al. [1322(a)], Luceac et al. [1322(b)], and Vogel et al. [1385] at different hydration levels h. The dependence of its relaxation time on hydration level

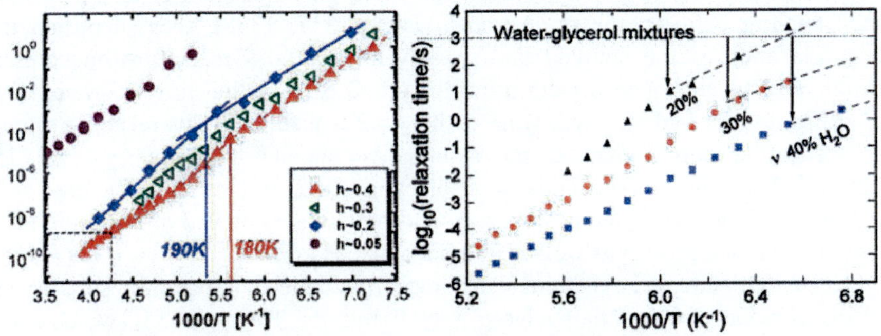

Fig. 226 (*Left*) Temperature variations of the relaxation time of the fast dielectric relaxation process in hydrated lysozyme at different hydration levels from Khodadadi et al. [1357, 1358], which is identified here as the JG β-relaxation time of water. *Arrows* indicate change in T-dependence at $T_g = 190$ for $h = 0.2$ and 180 for $h = 0.4$. Reproduced from [1357, 1358] by permission. (*Right*) The JG β-relaxation time τ_{JG} of water in 20, 30, and 40% mixtures with glycerol, the *arrows* indicate the change of T-dependence of the relaxation time at T_g where the α-relaxation time $\tau_\alpha = 10^3$ s (*dotted line on top*). The data of τ_α are not shown (but see Fig. 210) because the emphasis here is on the analogy between the hydrated lysozyme and water–glycerol mixtures. The glass transition temperatures T_g are indicated by the *vertical arrows* at which τ_α attains the value of 10^3 s. Note that τ_{JG} depends on the hydration level in the case of hydrated lysozyme, and composition in the case of the glycerol–water mixture, and also it changes T-dependence when crossing T_g. Both are indications that the α- and JG β-relaxation are coupled

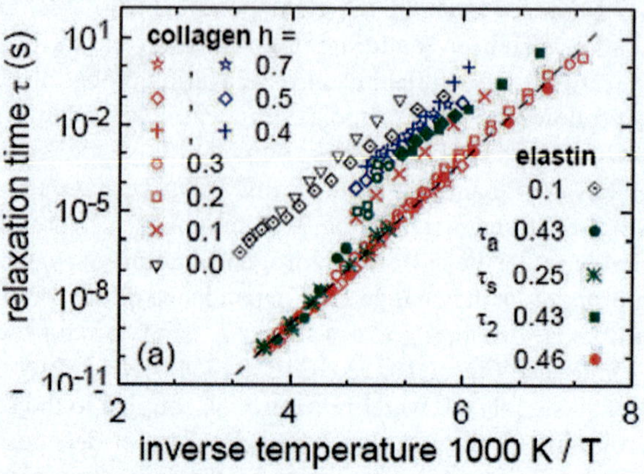

Fig. 227 The relaxation time of water in hydrated collagen and elastin. All are from dielectric data from Gainaru et al. [1322(a)] except τ_s and τ_a that are from NMR experiments based on the analysis of spin-lattice relaxation by assuming symmetric or asymmetric τ distributions, respectively, and τ_2 from stimulated-echo experiments by Vogel and coworkers [1322(b), 1385]. The *dashed line* represents an activation energy of 6330 K or 52 kJ/mol. Reproduced from [1322(b)] by permission

is made amply clear by inspection of these figures, and also the coupling between the biomolecules with water.

It is worthwhile to point out here that although Khodadadi et al. recognized the coupling between the protein and water [1357, 1358], they identified the fast relaxation as the coupled protein–solvent *structural* relaxation, in contrast to our identification here as the primitive or JG β-relaxation time of water. Our identification for hydrated lysozyme with $h \geq 0.2$ is justified by its relaxation times and activation energies being comparable to other aqueous mixtures (see Figs. 215, 226, and 227), and its T-dependence also changes at T_g (indicated by the *arrows* in the figure). Thus, the dynamics of water and protein are coupled down to the local level in hydrated protein as well as in other aqueous mixtures.

Evidence of coupled dynamics also comes from molecular dynamics simulations of the globular protein Ribonuclease A by Tarek and Tobias [1353], which show that the anharmonic and diffusive motions involved in the protein structural relaxation are correlated with the protein–water hydrogen bond dynamics, and also the complete structural relaxation of the protein requires relaxation of the hydrogen bond network via water translational displacement. They also found that arresting the protein–water hydrogen bond dynamics by inhibiting the solvent translational mobility, the protein atomic fluctuations are reduced throughout the protein, both in the backbone and side chains. The extent of the reduction is similar to that of dehydration of the system. This effect (see Fig. 3 in [1353]) was found by performing simulation in the 0.1–50 ps range at 300 K. At this temperature and short time range, there is no difference between the primitive or JG relaxation time and the α-relaxation of water as discussed before in Section 2.3.2.27 and suggested by the relaxation map of Fig. 215 or Fig. 217. The coupling parameter is effectively zero as shown by the relaxation time of water in various hydrated proteins obtained by neutron scattering, high-frequency dielectric relaxation, and deuteron NMR being close to τ_0 or τ_{JG} of water. Some examples have been shown in Figs. 215 and 217, and more to be shown later. Thus, there is no difference between saying the primitive/JG relaxation of water by us and the water translational displacement by Tarek and Tobias as being instrumental for protein atomic fluctuations to occur.

The dynamics of protein and the solvent are coupled, resulting in the protein relaxation and the solvent secondary relaxation in hydrated proteins to involve some motions of both the hydration water and the protein. Consequently, similar effects are seen in data obtained by techniques that can exclusively probe the motion of the protein or the water. The neutron scattering experiment of Wood et al. [1363] is a good example, where motions in maltose binding protein and hydration water were measured separately by using the H-MBP-D_2O sample, and the D-MBP-H_2O sample, respectively. To complement their neutron scattering data, Wood et al. carried out molecular dynamics simulations on the same system and reproduced the principal result observed by experiment: the dynamical transitions in both the protein and the hydration water occurring at the same temperature. The results of simulations are presented in Fig. 228a and b. The temperature dependence of the msd from the simulations for protein non-exchangeable H atoms (i.e., those probed by incoherent neutron scattering) exhibits a dynamical transition at 240 K (Fig. 228a), qualitatively consistent with the neutron data. Also plotted in Fig. 228a is the temperature dependence of the protein–water hydrogen-bond relaxation rates determined from a hydrogen-bond correlation function. The H-bond relaxation rate increases rapidly

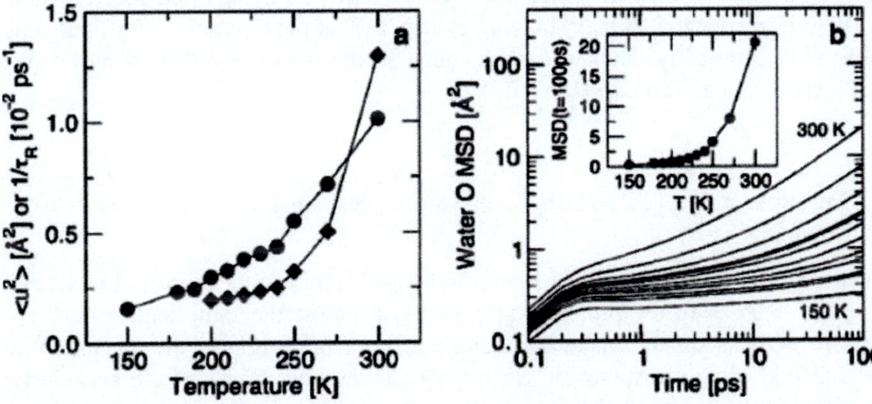

Fig. 228 (**a**) Temperature dependence of mean-squared displacements of non-exchangeable H atoms in MBP computed as 1 ns averages over the MD simulation trajectories (•), and protein–water H-bond network relaxation rates (♦). (**b**) Time evolution of mean-squared displacements of water O atoms over a range of temperature from 150 to 300 K (*bottom to top*: 150, 180, 190, 200, 210, 220, 230, 240, 250, 270, 300 K; the curve at 240 K, the temperature of the dynamical transition in the simulations, is indicated by a *heavy line*). (*Inset*) Temperature dependence of the water msds at 100 ps. Reproduced from [1363] by permission

with temperature at the dynamical transition of the protein, consistent with the previous conclusion by Tarek and Tobias [1353] that the protein transition is associated with relaxation of the network of protein–water H-bonds. The motion of the water molecules can be inferred from the time dependence of the msd of the O atoms of the water molecules from 0.1 to 100 ps plotted in Fig. 228b for several temperatures. At temperatures well below the protein dynamic transition $T_D \sim 240$ K in the MD simulations, after ballistic motion ends at about 0.3 ps, the msd increases very slowly with time corresponding to the nearly constant loss (NCL) if the data are shown as frequency dependence of the susceptibility, indicating that the water molecules are caged. The msd at 3 ps increases merely by a factor of about 2.5 in response to increase of T from 150 to 240 K. This weak T-dependence of the msd in the NCL regime of the hydrated protein is analogous to that observed in ordinary glassformers, discussed before in Section 2.3.2.6 and will be detailed in Sections 2.3.2.33, 2.3.2.34, 2.3.2.35, and 2.3.2.36. NCL is a universal feature of dynamics of caged molecules that precedes the primitive or JG β-relaxation. The latter is the first true relaxation process causing dissolution of caging. This action by the JG β-relaxation can be seen at higher temperatures in the msd rising after a few picoseconds. At the highest temperature of 300 K, the msd has not risen sufficiently to assume the $t^{1.0}$-dependence of diffusion even at 100 ps, the longest time of the simulation. At $T_D \sim 240$ K, the msd has time dependence of about $t^{0.56}$ at 100 ps. This time dependence indicates that the relaxation time of the α-relaxation of water for translation is longer than 100 ps at temperatures lower than 300 K and certainly so at $T_D \sim 240$ K, and therefore the rise of the msd is due to the presence of the primitive or JG β-relaxation of water. This deduction from the simulation data supports our interpretation that the dynamic transition originates from the primitive and JG β-relaxation of water entering the time window of the neutron spectrometer.

Other experimental studies showing evidences that the dynamics of protein and solvent are physically coupled can be found in the review by Ringe and Petsko [133]. Here we cite just two [1351, 1352].

The Primitive or the JG β-Relaxation of Solvent Responsible for the Dynamic Transition in $< r^2 >$

The neutron scattering observations of the change in slope of the temperature dependence of $< r^2 >$ in the protein taking place at almost the same temperature T_D as $< r^2 >$ of its solvent make the identification of the origin of the relaxation in the hydrated protein responsible for it very challenging. Some of the previously proposed explanations of the dynamic transition either cannot explain it or have no explicitly clear explanation of it. For example, Sokolov and coworkers [1358] argued, from their dielectric and neutron data of lysozyme at hydration level of $h = 0.4$, that this process is a *coupled protein–solvent structural relaxation*. But they honestly admitted having difficulty to rationalize why at the $T_g \sim 180$ K, the relaxation time of this fast hydrated protein relaxation is $\sim 10^{-4} - 10^{-5}$s and not 10^2 or 10^3 s as expected for structural relaxation. In fact, by extrapolation of the

Arrhenius T-dependence established at higher temperatures down to lower temperatures, the relaxation time reaches $\sim 10^2$ s only at a much lower temperature of $T \sim 125$ K (Fig. 223, middle panel). On the other hand, the fast relaxation responsible for $< r^2 >$ and its dynamic transition should be identified as the primitive or JG β-relaxation of water. This is because, first, the JG β-relaxation times τ_{JG} are short enough to match the 15 ps to 1 ns time windows of the neutron scattering measurements (see Fig. 215). Second, the relaxation times of the observed fast relaxation are nearly coincident with that of the JG β-relaxation of water in nanoconfinement or in various aqueous mixtures in the same temperature range (see Figs. 215 and 217 for a few examples). Third, the JG β-relaxation time of water in some aqueous mixtures is $\sim 10^{-4} - 10^{-5}$ s at the $T_g \sim 180$ K of hydrated lysozyme, similar in magnitude as that found for the fast relaxation. Fourth, the relaxation time of the fast relaxation changes its T-dependence at T_g of the hydrated protein, like JG β-relaxation time of water in various aqueous mixtures. All these can be verified by inspection of the figures presented before. Finally but not the least in importance, this identification is consistent with the simulations by Tarek and Tobias [1353] and by Wood et al. [1363], who found the protein structural relaxation are correlated with the protein–water hydrogen bond dynamics, and protein atomic fluctuations cannot occur without water translation in short timescale of the neutron scattering. As discussed before, the simulation by Wood et al. also found at $T_D \sim 240$ K that the msd is contributed by the JG β-relaxation.

It is worthwhile to reiterate that, when the JG β-relaxation of hydration water appears inside the short time window of neutron scattering, its τ_{JG} is not expected to be much shorter than τ_α, because of the small coupling parameter of water deduced before in Section 2.3.2.27. Hence the rapid rise of the msd above 250 K in H_2O-hydrated myoglobin in the left panel of Fig. 225 obtained by IN6 spectrometer with resolution of 15 ps can be reconciled with what Doster called the *liquid* behavior above 250 K. The explanation by Doster of the onset of the rise in msd at 180 K to originate from fluctuations of hydrogen bonds can be retained, because these are faster processes than the primitive or JG β-relaxation of water. However, another candidate is the caged water dynamics showing up in the msd with time dependence t^{c} with c a small positive number, or equivalently as the nearly constant loss (NCL) in the susceptibility. The major difference between our and Doster's explanations is the pivotal role played by the JG β-relaxation from which originates the rise of msd above 250 K. The hydrogen-bond fluctuations or the NCL precedes the JG β-relaxation. However, it is the JG β-relaxation but not the hydrogen-bond fluctuations or NCL which acts as the real precursor of diffusion of water near the protein surface. This is supported by the much smaller activation enthalpy 14 kJ/mol of hydrogen-bond fluctuations than 46–50 kJ/mol of the JG β-relaxation of water.

Here and there we have used the terms primitive or JG β-relaxation of hydration water, and structural α-relaxation of the hydrated protein. Although these terms seem to imply only one component is involved in each process, influence or participation of the other component must be borne in mind because of coupling of hydration water to protein by hydrogen bonding.

The primitive or JG β-relaxation times of water in various aqueous mixtures, in nanoconfinement, and in hydrated biomolecule are not exactly the same but not very different when it becomes shorter than 1 ns. The actual values depend on the water content, the hydrophilic solute, or the hydrated biomolecule. This can be seen from the relaxation maps in Figs. 215 and 217. Due to the mutual coupling of water and the hydrophilic solute or the biomolecule in dynamics, not only τ_{JG} but also the dynamics of the solute or hydrated biomolecule depends on the water content and on the chemical structure of the solute or the hydrated biomolecule. This dependence was the main point of contention in the 2010 paper of Khodadadi et al. [1358(d)], where they show results contradicting the traditional view that the dynamics of biological macromolecules simply follow the dynamics of hydration water (i.e., slaved dynamics). We have no disagreement with the results of Khodadadi et al. [1358(d)] since repeated emphasis on the coupled dynamics of biomolecules and hydration water has been given in previous subsections, and the relaxation maps have demonstrated the dependence. Nevertheless at short times probed by neutron scattering and for sufficiently high hydrated levels, the dependence of the dynamics of either the hydration water or the hydrated biomolecule on the biomolecule is not large. This explains why the dynamic transitions of various highly hydrated biomolecules occur at temperatures T_D that are not very different when spectrometers with the same time window are used. Evidence for this expected behavior can be seen from the D_2O-hydrated RNA and lysozyme data of Sokolov and coworker [1315, 1358(b)] shown before in the right panel of Fig. 223, and hydrated lysozyme data of Tsai et al. [1345(a)] in right panel of Fig. 222. Neutron scattering measurements of hydrated RNA and lysozyme were performed by using the high-flux backscattering instrument (resolution ∼1 μeV) at NIST, and $T_D \sim 220$ K. Zanotte et al. [1347] used the IN16 spectrometer at ILL with energy resolution 1 μeV, same as that in NIST, to measure the msd of 0.4 g D_2O g^{-1} hydrated lysozyme. They also found $T_D \sim 220$ K. The same IN16 spectrometer was used by Wood et al. [1363] to measure the msd of the D_2O-hydrated maltose binding protein (H-MBP-D_2O), and they found $T_D \sim 220$ K (see right panel of Fig. 225). Using the neutron backscattering spectrometer at NIST with a resolution of 0.8 eV, Chu et al. also found $T_D \sim 220$ K in the hydration water of the biopolymer RNA [1289(d)].

Another example is approximately the same T_D of D_2O-hydrated myoglobin in the study by Doster (see Fig. 225, left panel) and C-phycocyanin (C-PC) in various deuterated trehalose/water environments (see Fig. 224, right panel), both using the same IN13 spectrometer in ILL with energy resolution (FWHM) of 8 μeV. The fact that T_D is nearly the same in spite of the large variations in the chemical structure and backbone chemistry (amino acids or nucleic acids) of the protein is an indication that the dynamic transition does not reflect an intrinsic property of the protein. Instead, the dynamic transition originates from the primitive or JG β-relaxation of water, the relaxation time of which is insensitive to the protein at short times probed by neutron scattering. Of course, the role of coupling of water to the protein by hydrogen bonding is crucial to see the dynamic transition by neutron scattering from the hydrogen atoms in the protein.

Function of protein in biological and biochemical processes depends on the internal dynamics of the protein. From the above discussions, it is clear that the JG relaxation of hydration water plays an important role in enabling protein function.

Dried protein of course does not function, but then the problem in protein biopharmaceuticals industry (80 billion dollars in 2007) is how to stabilize the protein to avoid degradation on storage for timely use and application. A most recent review is given in [1383(a)]. The method commonly used is storage in a sugar-based glass, and the sugars are the disaccharides including sucrose and trehalose. There are collections of experimental data from several laboratories showing that the storage stability of dried proteins is determined by the fast local mobility of processes including the secondary relaxation and caged dynamic process of the sugar, rather than its structural relaxation or T_g [1383(b)–(e), 1384(a)–(c)]. Further discussion of this subject is deferred to a separate section at the end of this chapter.

Dielectric Relaxation and Enthalpy Relaxation Experiments

The knowledge acquired from the dynamics of aqueous mixtures in Section 2.3.2.26 can be put to use to understand the dynamics of hydration water and the role it has for protein dynamics. Dielectric relaxation data of a sample containing an equal weight fraction of myoglobin, water, and glycerol (i.e., 50 wt% water in the solvent, and a total solvent content $h = 2$) were published by Swenson and coworkers [1110]. Several relaxation processes were found in the dielectric spectra. The relaxation times of four such processes are shown in Fig. 229 together with some of the aqueous mixtures, 35 wt% water/PEG400, 50 wt% water/PVME, and 50 wt% water/PVP, shown before in Figs. 210 and 215. One of the two slower processes, with VFTH T-dependence of its relaxation time, τ_{Mb}, is likely originating from some cooperative protein conformational fluctuations coupled with water, giving rise to glass transition of the hydrated protein. By inspection of Fig. 229, at approximately 200 K, τ_{Mb} is of the order of 10^3 s. Thus, $T_{g\mathrm{Mb}}$ of the hydrated myoglobin is near or slightly higher than 200 K.

At temperatures below some T_x which is near $T_{g\mathrm{Mb}} \approx 200$ K, the relaxation time of the fastest relaxation process in the hydrated myoglobin, $\tau_{f\mathrm{Mb}}$ (black closed circles), is nearly the same as τ_{JG} of water in the aqueous mixtures of water with the polymers PVME and PVP (see Fig. 229). Moreover, as a function of temperature, $\tau_{f\mathrm{Mb}}$ exhibits a crossover from stronger temperature dependence at high temperatures to approximately Arrhenius dependence at lower temperatures, again like the behavior of τ_{JG} of water in the aqueous mixtures. The temperature at which the crossover takes place is at $T_x \approx T_{g\mathrm{Mb}} \approx 200$ K. Therefore, by analogy, the fastest relaxation in the mixture of myoglobin with 50 wt% water–glycerol solvent is naturally identified as the JG relaxation of the hydration water [131]. The second fastest process has been attributed to motions of polar side groups of the protein [1110], but it could be the α-relaxation of the glycerol–water mixture.

The result, $T_x \approx T_{g\mathrm{Mb}}$, found in hydrated myoglobin is exactly the same as we have seen before in aqueous mixtures. This is best brought out by the T-dependence of τ_{JG} of 50 wt% water/PVME and the location of T_{gm} of this aqueous polymer

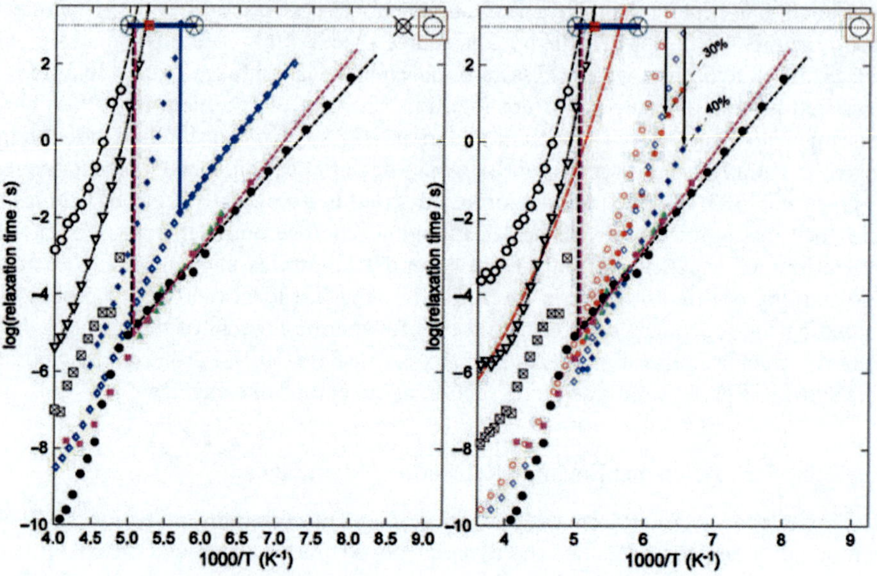

Fig. 229 (*Left*) Dielectric relaxation times of four processes observed by Swenson et al. in a sample containing an equal weight fraction of myoglobin, water, and glycerol [1110]. The relaxation times of the β-relaxation process of water (*black closed circles*), motions of polar side groups of the protein (*black squares with crosses inside*), and conformational protein fluctuations (*black open circles and inverted triangle*) all identified as such by the authors of [1110]. The *light black curves* are VF fits. The *broken black vertical line* indicates the temperature T_g of the hydrated myoglobin where one of the VF fits reaches 10^3 s. T_g is nearly coincident with T_x, the temperature above which the JG relaxation time assumes a stronger T-dependence than the Arrhenius dependence below it. The JG β-relaxation time, $\tau_\beta(T)$, of the water component in (**a**) 50 wt% water/PVME are indicated by *magenta closed squares*, and (**b**) in 50 wt% water/PVP by *green closed triangles*. The change of T-dependence of τ_β of 50 wt% water/PVME occurs at T_{gm} determined by DSC (located by the *vertical magenta line*) which is accidentally almost the same as T_x. *Open* and *closed blue diamonds* are $\tau_\beta(T)$ and $\tau_\alpha(T)$, respectively, for mixture of 35 wt% water with PEG400 [1295]. The *vertical line* indicates the temperature at which $\tau_\alpha(T) = 10^3$ s. The *large symbols on top* all having relaxation times of 10^3 s are from adiabatic calorimetric measurements of a 20% (w/w) aqueous solution of bovine serum albumin [1316]. The *two larger circles* (⊗) connected by the thick line indicate the range from 170 to 200 K of a slower relaxation found in the quenched sample. The *red square* in between them is for the annealed sample. The *open square with circle* inside is the relaxation occurring at T=110 K found for the quenched and annealed samples. The *other circle* (⊗) next to it represents the relaxation time of the secondary relaxation of water confined within 1.1, 1.6, and 1.8 nm nanopores of molecular sieves (silica MCM-41) obtained by adiabatic calorimetry having the value of 10^3 s at T=115 as found by adiabatic calorimetry [1283]. (*Right*) Same as on the left except for the additions of (1) the relaxation time of conformation fluctuations in Zn-substituted Mb (ZnPP-Mb) dissolved in a water–glycerol mixture with a volume ratio of 1:3 with a concentration of about 0.3 mM (ZnPP-Mb-W:3G) [1386] and shown here as the *red dashed line*; and (2) the dielectric relaxation time $\tau_\alpha(T)$ (*open symbols*) and $\tau_\beta(T)$ (corresponding *closed symbols*) of mixtures of 30 and 40 wt% of water with glycerol. *Circles* are for 30 wt% water/glycerol, and *squares* for 40 wt% water/glycerol. Each *dashed line* indicates the Arrhenius temperature dependences assumed by $\tau_\beta(T)$ of the labeled mixture starting approximately at T_g of the water/glycerol mixture defined by $\tau_\alpha(T_g) = 10^3$ s, located at the *vertical arrow* drawn. The change of temperature dependence of $\tau_\beta(T)$ at T_g is similar to that seen of the β-relaxation of the faster component in other aqueous and non-aqueous mixtures [1295] as well as in the hydrated Mb. Reproduced from [131] by permission

mixture, presented before in Fig. 215 and reproduced in Fig. 229, adopted from [131]. Not only is the τ_{JG} of the aqueous PVME mixture about the same as τ_{fMb} of the hydrated myoglobin, but also it exhibits the same kind of change of the T-dependence at the glass transition temperature, T_{gm} for the aqueous polymer mixture and $T_{gMb} \approx 200$ K for the hydrated myoglobin. Incidentally, the crossover temperatures of the two cases are almost the same. Therefore, the relaxation dynamic of hydrated myoglobin has a nearly exact analogy to that of the 50 wt% water solution of PVME, and is similar to that of other aqueous mixtures (Fig. 215), and non-aqueous mixtures (Fig. 147) [1093]. This general property found in hydrated myoglobin and the other systems naturally implies similar physical mechanism and analogous interpretation as the other simpler and better understood systems [131]. Specifically for hydrated myoglobin or other proteins, the secondary relaxation of water in the hydration shell and the conformation fluctuations of the protein are inseparable or symbiotic processes. The secondary relaxation of water hydrogen bonded to the protein is the first step in the evolution of processes with increasing complexity that eventually leads to cooperative motion involving the local conformations of the protein and the hydrogen-bonded water molecules altogether, and glass transition at T_{gMb} when the cooperative relaxation time becomes too long. The feed-back of the effect of this glass transition to the precursory secondary relaxation of water in the mixture is the cause of the change of temperature dependence of its relaxation time at T_{gMb} [131]. The literature on dynamics of hydrated proteins and other biomolecules is vast. Many published studies show evidences that the dynamics of protein and solvent are physically coupled [133]. In short, the JG β-relaxation of the water in the 50/50 wt% of water and glycerol solvent in hydrated myoglobin is the precursor and governor of both the local protein process and the cooperative conformational fluctuations in the protein responsible for the glass transition-like behavior of the protein.

Our interpretation of the fastest relaxation process of the hydrated myoglobin in Fig. 229 with relaxation time, τ_{fMb}, as the JG relaxation of the hydration water differs from that offered by Swenson et al. [1110]. They attributed the cause of this Arrhenius process in hydrated myoglobin at low temperatures (below $T_x \approx T_{gMb} \approx$ 200 K) to confinement of water resulting in a local β-process. At temperatures above $T_x \approx T_{gMb} \approx 200$ K, they interpreted the observed hydration water solvent process with a stronger temperature dependence of τ_{fMb} as a merged α- and β-process. This interpretation is debatable because the behavior of τ_{fMb} of the hydration water process is analogous in every respect to the JG β-relaxation process of water in aqueous and van der Waals liquid in non-aqueous mixtures [1093], and evidences from these other systems have shown that the process is entirely one and the same JG relaxation above and below the crossover temperature (see Fig. 147 for an example). Our interpretation also implies that the cooperative α-relaxation of the 50/50 wt% of the water/glycerol solvent of the myoglobin may also have been observed in the dielectric spectra of [1110]. Its relaxation times could be located in Fig. 228 at or near the data points represented by the squares (with crosses inside) and interpreted in [224] as due to the motion of polar side groups of myoglobin.

Calorimetric measurements have been widely used in studying the molecular motions in hydrated proteins. Examples from publications of earlier years are cited as [1351, 1387–1390]. Here we discuss the results of a recent study using adiabatic calorimetry by Kawai et al. [1316]. The relaxation processes of another hydrated protein, a 20% (w/w) aqueous solution of bovine serum albumin (BSA), were obtained by adiabatic calorimetry at temperatures ranging from 80 to 300 K [1316]. Measurements were made on sample after quenching it from 300 down to 80 K and on another sample after it had been annealed at temperatures in the range of 200–240 K. Shown in the left panel of Fig. 230 is the temperature dependence of the enthalpy relaxation rates, –d*H*/d*t*, for the quenched sample indicating the presence of two enthalpy relaxation. One is centered at around 110 K and the other over a wide temperature range from 120 to 190 K. The annealed sample showed three separate relaxation processes having $T_g = 110$ K, 135 K, and a broad range of temperatures higher than 180 K. Kawai et al. suggested these processes in order of increasing temperature to originate from the rearrangement motions of (1) primary hydrate water forming a direct hydrogen bond with the protein, (2) part of the internal water localized in the opening of a protein structure, and (3) the disordered region in the protein. We shall see that dielectric relaxation measurements on the same hydrated BSA offer a different interpretation of these processes.

Dynamics of processes probed by adiabatic calorimetric data is limited to long times of the order of 10^3 s. A better understanding of the processes can be gained by knowledge of their dynamics over a wider time/frequency range. Toward this goal, broadband dielectric measurements were performed by Shinyashiki et al. on the 20 and 40% (w/w) aqueous solution of BSA [1317] to complement the adiabatic calorimetry data of the same system [1316]. The measurements were made over frequency range of 12 decades from 2 mHz to 20 GHz isothermally at temperatures

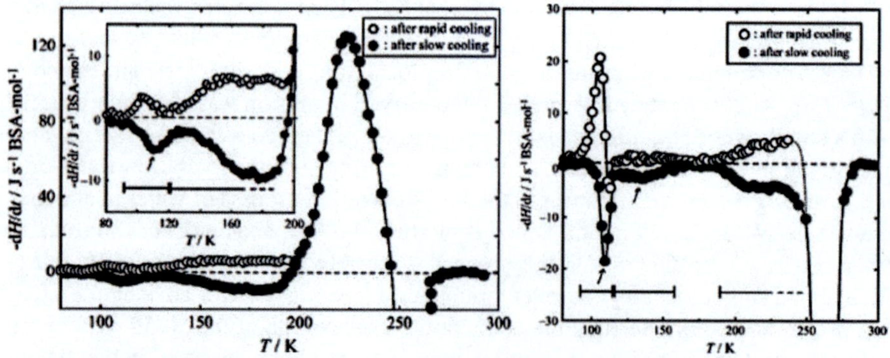

Fig. 230 (*Left*) Temperature dependence of -d*H*/d*t* for the quenched sample. *Horizontal bar* and *arrow* indicate the glass transition region and the T_g, respectively. *Inset* shows enlargement of $T = 80 - 200$ K region. (*Right*) Temperature dependence of -d*H*/d*t* for the annealed sample. *Horizontal bar* and *arrow* indicate the glass transition region and the T_g, respectively. Reproduced from [1316] by permission

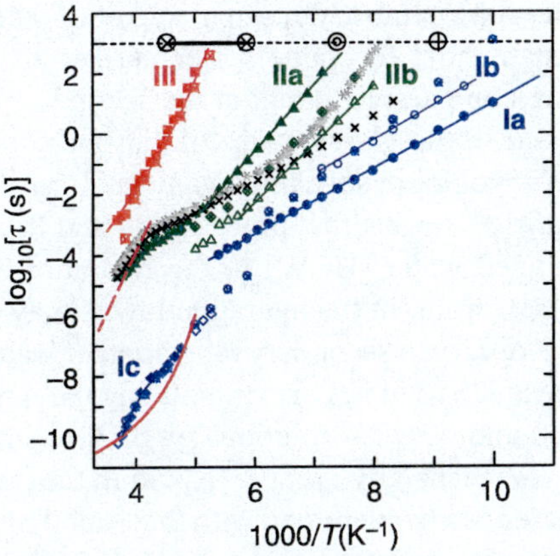

Fig. 231 Plots of relaxation times, τ, for 20 wt% BSA–water mixture against reciprocal temperatures of processes Ia (*solid circles*), Ib (*open circles*), Ic (*solid* and *open diamonds*), IIa (*solid triangles*), IIb (*open triangles*), III (*solid squares*). The *open symbols* of the processes IIb (*open triangles*), Ib (*open circles*), and parts of Ic (*open diamonds*) indicate that the loss peaks of them are not observed definitely (see text). The plots of the relaxation times of process Ia (*open circles with cross*), IIa (*open triangles with cross*), and III (*open squares with cross*) are obtained for 40 wt% BSA–water mixtures. *Crosses* and *asterisks* are the plots of pure ice by our measurement and that quoted from [28], respectively. Enthalpy relaxations with relaxation time of 10^3 s originating from different processes found by Kawai et al. [1316] are indicated by the *thick line* bounded on both sides by *circles with cross* in the broad range from 170 to 220 K, *double circle* at 135 K, and *circle with plus* at 110 K. The *magenta solid curve* represents the relaxation time τ_c of reorientational motion of heavy water (D_2O) near the surface of myoglobin in D_2O-hydrated myoglobin (0.35 g/g) determined by deuteron NMR [1318, 1319]. The *magenta straight dashed line* represents the relaxation time τ_{HN} obtained for the β-fluctuations in the hydration shell of myoglobin embedded in solid polyvinyl alcohol reported in [1389, 1390]. Reproduced from [1317] by permission

from 80 to 300 K. Figure 231 displays the temperature dependences of the relaxation times $\tau_p(T)$ of the various processes in 20% (w/w) aqueous solution of BSA. Subscripts are used to label the processes Ia, Ib, Ic, IIa, IIb, and III observed.

For comparison we include the relaxation time of pure ice, $\tau_{ice}(T)$, obtained by Shinyashiki et al. [1317] and Johari and Whalley [1391], and the data of the 40% (w/w) aqueous solution. It can be seen in the figure that $\tau_p(T)$ of processes p=IIa and IIb straddle $\tau_{ice}(T)$, and the three have similar activation energies in both the 20 and 40% solutions. This observation leads to the conclusion that processes IIa and p=IIb originate from crystallized bulk water in the aqueous solutions of BSA. Kawai et al. [1316] found in their 20% (w/w) aqueous solution of BSA a broad enthalpy relaxation by adiabatic calorimetry with relaxation time of 10^3 s centered at about 135 K in a sample after it was annealed at temperatures in the range of 200–240 K (see Fig. 230, right panel). Incidentally, τ_{IIa} of process IIa and τ_{ice} are near this enthalpy relaxation time at 135 K (shown by circle within circle in Fig. 230). This coincidence suggests processes IIa and IIb may be identified with the enthalpy relaxation at 135 K. The relaxation times, τ_{Ia} and τ_{Ib} of processes Ia and Ib, are comparable, have almost the same activation energies, and hence can be considered to originate from the same molecular mechanism. Moreover, their relaxation times also are comparable to that of the secondary (or Johari–Goldstein) relaxation of water found in the glassy states of various aqueous mixtures [1295] and in the glass state of myoglobin solvated by an equal fraction of water and glycerol [1110]. The similarities suggest that processes Ia and Ib are also the secondary relaxation of uncrystallized water (UCW) in the hydration water. The difference in various environments of water in the hydration layers may explain the broad frequency dispersion of the observed secondary relaxation of water, which requires two Cole–Cole processes Ia and Ib to fit it. On extrapolating the Arrhenius T-dependence of τ_{Ia} and τ_{Ib} of the 20 and 40 wt% solutions of BSA to lower temperatures, the relaxation time of 10^2 s is reached by τ_{Ia} and τ_{Ib} within the range from 110 to 95 K. These relaxation processes can be identified with that found by adiabatic calorimetry from the peak of the enthalpy relaxation rate at 110 K in the same system. This is because the enthalpy relaxation time of 10^3 s occurs in a neighborhood of 110 K (shown by circle at 110 K in Figs. 229 and 231), which is comparable to that found for our dielectric relaxation times τ_{Ia} and τ_{Ib}. Thus, the identification of the secondary relaxation of water in the hydrated BSA from dielectric measurements is substantiated by the same from adiabatic calorimetry measurements of Kawai et al. [1316] and vice versa. This interpretation also is supported by adiabatic calorimetry measurements of water confined in the pores of silica gel with average diameter of 1.1 nm [1282] (see Fig. 220). The study reported the presence of a faster and a slower process. The relaxation times of the faster process of confined water is 10^3 s at $T = 115$ K, comparable to what we found for τ_{Ia} and τ_{Ib}.

The relaxation times τ_{Ia} and τ_{Ic} of processes Ic and Ia are presented in two non-overlapping temperature regions, giving an impression that these are distinctly different processes. This is a false impression originating from unavoidable complications and uncertainties in data analysis caused by contribution to ε^* from the processes of the crystallized water overlapping that from process Ia. However, processes Ia and Ic originate from the same molecular motions detected through the same dipole moments. One is replaced by the other when crossing T_{gBSA}. The apparent mismatch of τ_{Ia} and τ_{Ic} in the neighborhood of T_{gBSA} is an artifact resulting from the procedure used to fit the dielectric spectra, and the assumption that they are two separately different relaxations in the first place. Interpreted this way as a single process and the secondary relaxation of water, the stronger T-dependence of its relaxation time for $T > T_{gBSA}$ than the Arrhenius T-dependence for $T < T_{gBSA}$ is to be noted. Such change of the T-dependence of the secondary relaxation of water in 20 and 40 wt% aqueous solutions of BSA has been found in many aqueous mixtures, myoglobin solvated by an equal fraction of water and glycerol, and even in the secondary relaxation of the faster component of binary van der Waals liquids. In the other systems, the change of T-dependence occurs at the glass transition temperature of the other component of the mixture. We shall see the same holds for our present system after identifying the molecular mechanism of process III in the following paragraph, and explanation given before in the other systems applies here.

Process III is the slowest among processes originating from molecular motions in the aqueous solutions of BSA. Compared with τ_{Ic} and τ_{II}, τ_{III} has much stronger T-dependence of the Vogel–Fulcher–Tammann–Hesse form, typically shown by structural α-relaxation of glass-forming liquids. As can be seen in Fig. 231, τ_{III} reaches 100 s at $T_{\mathrm{gBSA}} \approx 200$ K and 10^3 s at about 192 K. This means that this process is related to the enthalpy relaxation with relaxation time of 10^3 s seen in the region from 170 to 220 K by Kawai et al. (indicated in Fig. 231 by the thick line bounded on both sides by circles with cross in the broad range from 170 to 220 K). Thus, process III is responsible for the glass transition at T_{gBSA} of the hydrated BSA as seen by dielectric relaxation. The glass transition is driven by the cooperative motion of BSA and the water entering into the protein. The temperature at which τ_{III} attains 100 s agrees well with the broad range above 180 K of glass transition temperatures of protein observed by calorimetric measurements. This agreement suggests that process III corresponds to the molecular motion responsible for the glass transition of protein. The relation of processes Ia and Ic combined (secondary relaxation of water in the hydration shell) to process III in hydrated BSA resembles that found in myoglobin solvated by water/glycerol mixtures, as well as in many aqueous mixtures, in the dynamics of a component in binary mixtures of van der Waals glass-forming liquids, and in neat glassformers. The commonality is the change of the T-dependence of the secondary relaxation time from Arrhenius dependence at temperatures below T_g (τ_{Ia} and T_{gBSA} in the present case) to a stronger T-dependence above T_g (τ_{Ic} and T_{gBSA} in the present case). This indicates that the secondary relaxation of water in the hydration shell and the structural relaxation of the hydrated BSA are connected to each other as inseparable processes.

Excluding the relaxation of ice formed in the bulk water discussed in the above that is irrelevant for the present discussion, two major relaxation processes originating from water in the hydration shell and the hydrated myoglobin or BSA are observed. Their relaxation times determined from fits to the isothermal data of $\varepsilon^*(\nu)$ (not shown) are plotted against reciprocal temperature in Fig. 232 (left).

The loss peak of water in the hydration shell of the 20 wt% BSA–water mixture at lower temperatures seems to be composed of two processes close to each other, and their relaxation times are represented by open and closed blue squares. A single process is obtained for the 40 wt% BSA–water mixture and shown by green closed diamonds. All these relaxation times of water in the hydration shell, collectively denoted by τ_{W}, have Arrhenius T-dependence from 100 K and up to about 200 K. Above 200 K, the relaxation times of water in the hydration shell are represented by blue open circles and closed diamonds for the 20 and 40% (w/w) BSA–water mixtures, respectively. It can be seen that the Arrhenius T-dependence of τ_{W} at temperatures below about 200 K is replaced by a stronger and non-Arrhenius T-dependence after crossing some temperature in the neighborhood of 200 K. The gap between τ_{W} in this neighborhood is the artifact of fitting isothermal data by Cole–Cole functions, and should not be there. The relaxation times of the hydrated BSA, τ_{BSA}, are shown by the blue closed triangles and green open triangles for the 20 and 40% (w/w) BSA–water mixtures, respectively. With the much stronger

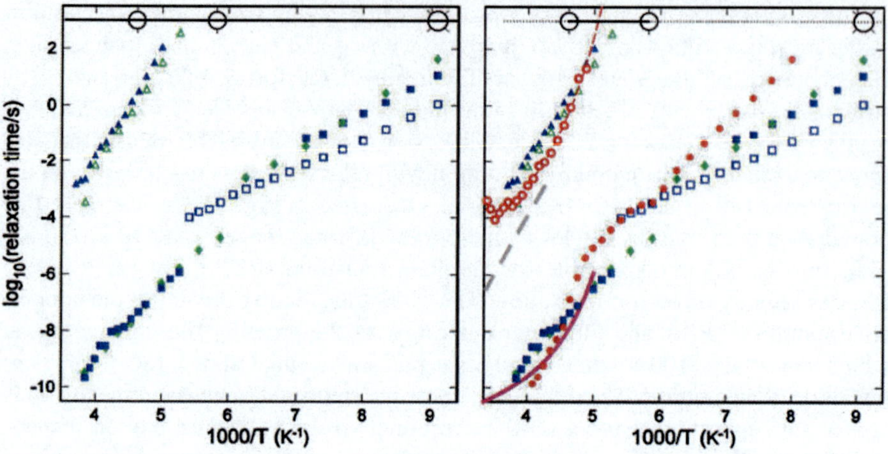

Fig. 232 (*Left*) Dielectric relaxation times of 20 and 40% (w/w) solution of bovine serum albumin (BSA) in water [1317]: The faster relaxation, with Arrhenius T-dependence below about 190 K and changing to a stronger T-dependence above it, is the β-relaxation of hydration water τ_W (*squares and diamonds*). The slower α-relaxation with relaxation times τ_{BSA} (*triangles*) reaches about 10^3 s at about 190 K. The relaxation of ice formed in the bulk water, dc conductivity, and other processes observed in the dielectric experiment are omitted to highlight these two relevant processes. The *two large circles joined by a line* indicate the temperature range of the α-process of hydrated BSA and the *large circle with four spokes* is the temperature of the water β-relaxation observed by adiabatic calorimetry at 10^3 s [1316]. (*Right*) Dielectric data of hydrated BSA same as in *left panel*. Added are the following. Dielectric relaxation times, τ_W (*red closed circles*) and τ_{Mb} (*red open circles*) in water/glycerol solution of myoglobin [1110]. The *magenta solid line* is VFTH fit to the relaxation time, τ_c, of reorientational motion of heavy water (D_2O) in D_2O-hydrated myoglobin (0.35 g/g) from deuteron NMR [1318, 1319]. The *grey straight dashed line* represents the relaxation time τ_{HN} obtained for the β-fluctuations in the hydration shell of myoglobin embedded in solid polyvinyl alcohol reported in [1392, 1393]

T-dependence, τ_{BSA} reaches 10^2–10^3 s near 200 K, suggesting it is the structural α-relaxation, and the glass transition of the hydrated BSA occurs at some temperature $T_{gBSA} \approx 200$ K. This happening is supported by heat capacity and enthalpy relaxation rate measurements of a 20 % (w/w) BSA–water mixture by Kawai et al. [1316]. These authors have found several enthalpy relaxations with relaxation time of 10^3 s originating from different processes. The one found in the broad range from 170 to 220 K is indicated by the thick line at 10^3 s bounded on both sides by ⊗ in Fig. 231. It can be seen from the figure that the dielectric τ_{BSA} falls within the temperature range of enthalpy relaxation. Kawai et al. found another enthalpy relaxation with relaxation time of 10^3 s in the temperature range from 100 to 120 and centered at 110 K (indicated by ⊕). This temperature is in rough agreement with the dielectric τ_W when extrapolated to 10^3 s, proving that adiabatic calorimetry and dielectric relaxation are observing the same water relaxation. Almost the same enthalpy relaxation was found for water confined within 1.1, 1.6, and 1.8 nm nanopores of silica MCM-41 [1283] and of silica gels [1282]. The τ_W of water in the hydration shell at temperature below T_{gBSA} is comparable in magnitude and activation energy to the dielectric JG relaxation of water in hydrophilic mixtures and nanoconfined water [1093]. This suggests that τ_W is truly the primitive or JG relaxation of water in the hydration shell. The change of the T-dependence of τ_W at T_{gBSA} is the same as that found in the JG relaxation of water in other aqueous mixtures, nanoconfined water, and a component in mixtures of two van der Waals glassformers [1093]. This general relation between the JG β-relaxation and the α-relaxation has been explained by the coupling model. The relaxation times, τ_{fMb} (equivalent to τ_W) and τ_{Mb}, of the JG relaxation of water and the α-relaxation of hydrated myoglobin from Swenson et al. [1110] are reproduced (red open and closed circles) in Fig. 231 (right) to compare with those of the hydrated BSA. Again, the JG relaxation time τ_W of water in the water/glycerol solvent of myoglobin is similar to what we find in hydrated BSA, and there is change of the T-dependence of its relaxation time at the glass transition temperature of the hydrated myoglobin where τ_{Mb} reaches 1000 s, which is also near 200 K. In the glassy state the relaxation times of JG relaxation of water in the water/glycerol solvent of myoglobin are longer and have larger activation energy than pure water in hydrated BSA. This is understandable because glycerol is less mobile than water. The JG relaxation of water with similar relaxation time and activation energy also was found by Svanberg et al. in hydrated palmitoyloleoyl phosphatidylcholine, an asymmetric lipid belonging to the phosphatidylcholine family with 16 and 18 carbon atoms in the tails [1394]. Also observed is a change of temperature dependence of the relaxation time at 210 K from one with activation energy of 56.3 kJ/mol at lower temperatures, and 31 kJ/mol above 210 K. This change to a weaker temperature dependence of water relaxation above 210 K obviously cannot be due to merging of α- and β-relaxation of water suggested for hydrated myoglobin [1110].

Also shown in Fig. 231 (right) is the relaxation time, τ_c, of reorientational motion of heavy water (D_2O) near the surface of myoglobin in D_2O-hydrated myoglobin (0.35 g/g) determined by deuteron NMR [1318, 1319]. The method probes the reorientational relaxation of the O–D vector, and τ_c obtained from about 200 to 300 K has been fitted by the VFTH T-dependence, $\tau_c(T) = (0.96)\exp[460/(T - 167)]$ ps, given in [1319] and shown here in Fig. 231 (right) by the solid magenta line. Good agreement between τ_c from NMR and τ_W from dielectric measurements can be seen. According to our interpretation like in other non-aqueous and aqueous mixtures, and nanoconfined water, τ_W as well as τ_c is the JG relaxation time of hydration water which is not undergoing glass transition at 200 K itself but is the precursor of the glass transition of the hydrated protein at near 200 K. In fact, the dielectric τ_W reaches 10^3 s near 110 K in the case of hydrated myoglobin, and at about 100 K for the hydrated BSA, which is near 110 K, the glass transition temperature of hydration water from calorimetry. As mentioned before, the coupling parameter of hydration water is small, and hence there is no big difference between the JG β-relaxation time, τ_W or τ_c, and the structural relaxation time τ_α of hydration water. Bearing this in mind, the interpretation here and in [131, 1317] can be reconciled with that offered by Doster and Settles on their τ_c from NMR in their statement: "The structural relaxation time related to translation of hydration water varies however in a super-Arrhenius manner with temperature as the neutron scattering and the deuteron relaxation data demonstrate." This is obvious because as said before, $\tau_\alpha \approx \tau_W$ or τ_c. However, their statement following the one cited above "These results provide evidence that hydration water turns into a glassy state near 200 K" cannot be reconciled. Only the hydrated myoglobin (to be distinguished from hydration water) turns into a laboratory-timescale glassy state (with relaxation time > 10^3 s) near 200 K. The deuteron NMR and the neutron scattering measure the JG β-relaxation which is not directly responsible for glass transition. The same interpretation by Doster, kinetic glass transition of the solvent near the protein surface coupled to the protein motions, was used earlier in the interpretation of the "dynamic transition," i.e., rapid increase of the msd of hydrated protein motion past $T_d \approx 200$ K seen by Mössbauer effect [1343] and neutron scattering [1344]. As discussed in the previous subsection, the JG β-relaxation in hydrated proteins is accountable for rapid increase of the msd of hydrated protein motion past $T_d \approx 200$ K.

A different interpretation of the rapid increase in the conformational MSD was given by Frauenfelder and coworkers [1392, 1393] based on earlier work by Fenimore et al. [135]. These authors stated that there is no "dynamical transition" in proteins near 200 K proposed by Doster and coworkers [1318, 1319, 1344]. Frauenfelder and coworkers suggested the rapid increase in the mean-squared displacement with temperature in many neutron scattering experiments can be quantitatively predicted by the β-fluctuations in the hydration shell. In their energy landscape model of hydrated proteins with hierarchical tiers, the β-fluctuation in the hydration shell is the lowest tier, which has Arrhenius temperature dependence for its relaxation time. Doster remarked that these suggestions by Frauenfelder et al. are misleading because diffusion within the hydration shell is ignored, and a well-defined energy landscape controlling protein dynamics does not account for the solvent. Although the term β-fluctuations used by Frauenfelder and coworkers sounds similar to ours, their interpretation of β-fluctuations is different from the JG β-relaxation of hydration water in hydrated protein that we advocate. Actually, in their dielectric relaxation study [1392, 1393] of hydrated myoglobin embedded in solid polyvinyl alcohol from 40 Hz to 110 MHz, they emphasized that the β-fluctuations in the hydration shell are different from the JG relaxations in glassformers. Their isothermal dielectric loss functions $\varepsilon''(\omega)$ of the β-fluctuations measured from 40 Hz to 110 MHz at 235 K and up to 295 K are fitted to the Havriliak–Negami (HN) functions. The HN relaxation times, τ_{HN}, have the Arrhenius T-dependence given by $\tau_{HN}/\mathrm{s} = 10^{-20.5}\exp[(79.2\,\mathrm{kJ/mol})/RT]$. As shown by the dashed grey line in Fig. 231 (right), τ_{HN} are about four decades longer than τ_W, the JG relaxation time of water in the hydration shell of BSA as well as of myoglobin. The fact that τ_{HN} of the β-fluctuations deduced is not much shorter than the α-relaxation time of the hydrated BSA and myoglobin suggests they may be the same process. Nevertheless, Frauenfelder and coworkers were able to explain the rapid increase in the mean-squared displacement measured by Doster and Settles [1318] using the IN13 neutron spectrometer at ILL with a time resolution $\log(\tau_n/\mathrm{s}) = -\log(\omega_n/\mathrm{s}^{-1}) = -10$ in the temperature range from 235 to 295 K by calculating the part of the dielectric loss spectrum with angular frequencies ω

that exceed $\omega_n = 10^{10}$ radians/s. This was carried out by extrapolating the HN fits to practically infinite frequencies. As can be seen by inspection of Fig. 231 (right), there are many decades of frequency separating ω_n and $\omega_{HN} = 1/\tau_{HN}$, and thus the extrapolation by the HN function can involve large uncertainties.

Using quasielastic neutron scattering S.-H. Chen et al. [1359] measured the msd of the hydrogen atom of hydrated water in hydrated lysozyme at the hydration level of $h = 0.3$ (i.e., 0.3 g H_2O per g dry lysozyme). At this hydration level, almost a monolayer of water covers the protein surface. An identical second sample was prepared using D_2O and the measurements were used to subtract out the incoherent scattering contribution from the protein hydrogen atoms in the H_2O sample. The results indicate that a "dynamic transition" of the msd at a temperature T_d between 200 and 220 K. Analyzing the neutron data using the relaxing cage model, they determined the average translational relaxation time, τ_T, for the hydration water in lysozyme. They found $\langle\tau_T\rangle$ follows a VFTH T-dependence above $T_L = 220$ K, but changes abruptly to have the Arrhenius T-dependence with activation energy of 3.13 kcal/mol, as shown by brown open circles in the lower left part of Fig. 233. Like for similar observation of nanoconfined water by the same group [1289], this behavior of $\langle\tau_T\rangle$ is interpreted as fragile-to-strong liquid–liquid transition involving the structural α-relaxation of the hydration water. The near coincidence of T_d and T_L has led Chen et al. to claim that the "dynamic transition" of hydrated proteins is effected by the fragile-to-strong dynamic crossover of the hydration water. The "dynamic transition" has been seen by neutron scattering in glycerol and lysozyme mixtures [1345(a), 1346]. In these systems, water and its purported fragile-to-strong dynamic

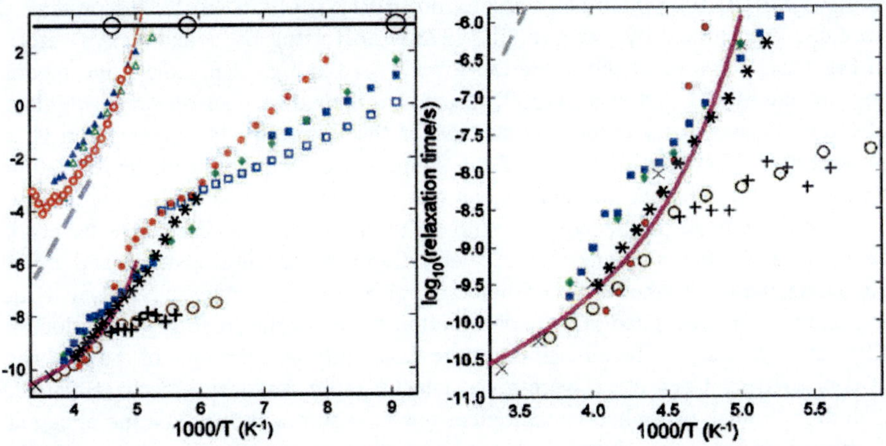

Fig. 233-1 (*Left*) Same as the right panel of Fig. 232 except for the additions of: the following. (1) Translational relaxation time, $\langle\tau_T\rangle$, for the hydration water in lysozyme from S.-H. Chen et al. [1359] (*large brown open circles*) show change from VFTH T-dependence above $T_L = 220$ K to Arrhenius T-dependence below T_L. (2) The slower and faster dielectric relaxation times of hydrated lysozyme by Khodadadi et al. [1358], $\tau_{1\,K}$ (∗) and $\tau_{2\,K}$ (+). Neutron relaxation time (×). Note the good agreement of $\tau_{2\,K}$ with $\langle\tau_T\rangle$ below T_L, and $\tau_{1\,K}$ with $\langle\tau_T\rangle$ above T_L. (*Right*) The high temperature part is magnified

crossover obviously are absent, and the "dynamic transition" observed cannot be explained by that proposed by Chen et al. The combined neutron scattering and dielectric relaxation times of D_2O-hydrated lysozyme by Khodadadi et al. [1357, 1358] also did not find the fragile-to-strong dynamic crossover, nor the dielectric data of hydrated collagen and elastin [1383].

It can be seen by inspection of Fig. 233-1, above $T_L = 220$ K where $\langle \tau_T \rangle$ follows the VFTH T-dependence of the so-called "fragile" water, there is good agreement between $\langle \tau_T \rangle$ of hydration water in lysozyme with the dielectric JG relaxation time, τ_W, of hydration water in myoglobin [1110] or in BSA [1317], and also with the reorientational time, τ_c, of heavy water (D_2O) near the surface of myoglobin in D_2O-hydrated myoglobin determined by deuteron NMR [1318, 1319]. Such agreement suggests, above $T_L = 220$ K, $\langle \tau_T \rangle$ is actually the primitive or JG relaxation time of hydration water. However, neither τ_W nor τ_c exhibits the abrupt change to the Arrhenius T-dependence at 220 K as Chen et al. found for $\langle \tau_T \rangle$. Although the dielectric data of Khodadadi et al. [1358] found the water relaxation time changes its T-dependence at 180 K, the T_g of the hydrated lysozyme [1357], as JG β-relaxation behaves in general. Extrapolating the Arrhenius T-dependence of $\langle \tau_T \rangle$ below $T_L = 220$ K given by Chen et al. to low temperatures, we find roughly 47 K would be the glass transition temperature for the "strong" water defined by $\langle \tau_T \rangle = 10^3$ s. This low value of T_g for hydration water is unrealistic, and shows once more that Chen et al. [1359] did not observe the α-relaxation of hydration water below $T_L = 220$ K as they claimed.

Most recently Doster et al. [1395(a)] made more sensitive measurements of water on perdeuterated C-phycocyanin (C-PC) using the new neutron backscattering spectrometer SPHERES, which has a resolution (FWHM) of 0.62 μeV. The protein is the same one studied by Köper et al. [1373, 1374] using IN13 and shown before in Fig. 224, but now at the hydration level $h = 0.3$ g/g. Using the same model and procedure as Chen et al. [1359] to analyze their own neutron scattering data of C-PC, Doster et al. found similar sign of the fragile-to-strong transition near $T_L = 220$ K. However, they showed the temperature dependence of the relaxation times is highly sensitive to data evaluation, and demonstrated that the procedure used by Chen et al. is inadequate. With an improved data analysis, Doster et al. obtained results that show no sign of such a fragile-to-strong transition, and are in good agreement with the results of dielectric relaxation and deuteron NMR, without any anomaly. They argued that the observation of the fragile-to-strong transition by Chen et al. is likely to be an artifact of the data analysis. Problems of the analysis of data used by Chen et al. were given later in more detail by Doster [1395(b)], including the use of a phenomenological power law model to adjust the apparent Q-dependence without giving a plausible justification, and finding the fragile-to-strong crossover at a temperature, where the line width of the spectrum falls below the instrumental resolution.

The results of a recently published dielectric relaxation study of lysozyme hydrated to a level of $h = 0.4$ g water per gram of protein by Khodadadi et al. [1358(c)] may offer another possibility to unravel the mystery of the identity of the "strong" water having Arrhenius T-dependence of $\langle \tau_T \rangle$ below $T_L = 220$ K. In the isothermal spectrum at temperatures between 203 and 223 K, one major loss peak shows up but in addition there is a faster process on its high-frequency flank with much weaker relaxation strength. The spectra in this temperatures range were fitted by the sum of two Cole–Cole distribution functions, and the relaxation times of the slower and major process, $\tau_{1\,\mathrm{K}}$, and the faster and minor process, $\tau_{2\,\mathrm{K}}$, determined. At temperatures above 223 K, the spectra show only the major loss peak and were fitted with a single Cole–Cole distribution function. The results presented in Fig. 233-1 show that $\tau_{1\,\mathrm{K}}$ (black asterisks) of hydration water in lysozyme is not too different from τ_W of hydration water in BSA or in myoglobin, or τ_c of heavy water (D_2O) in myoglobin from deuteron NMR for temperatures down to about 167 K. Moreover, $\tau_{1\,\mathrm{K}}$ is approximately the same as $\langle\tau_\mathrm{T}\rangle$ of Chen et al. above $T_\mathrm{L} = 220$ K, but unlike $\langle\tau_\mathrm{T}\rangle$ it continues to be present at temperatures below 220 K. Remarkably, $\tau_{2\,\mathrm{K}}$ of the faster minor relaxation (shown by + sign) has about the same value and small activation energy as $\langle\tau_\mathrm{T}\rangle$ of Chen et al. [1359] below $T_\mathrm{L} = 220$ K. This correspondence was also pointed out by Khodadadi et al. in [1358]. The comparison with the dielectric data suggests that quasielastic neutron scattering method and analysis used by Chen et al. could be more sensitive to the faster minor process having relaxation time $\langle\tau_\mathrm{T}\rangle \approx \tau_{2\mathrm{K}}$ at lower temperatures. Because of this, Chen et al. possibly missed detecting the slower major process below $T_\mathrm{L} = 220$ K, resulting in the abrupt change of temperature dependence of $\langle\tau_\mathrm{T}\rangle$ from the non-Arrhenius form of $\tau_{1\,\mathrm{K}}$ to the Arrhenius one of $\tau_{2\,\mathrm{K}}$. Hence, there is no fragile-to-strong dynamic crossover of hydration water at $T_\mathrm{L} = 220$ K.

In addition to dielectric relaxation, Khodadadi et al. [1358] studied the dynamics of lysozyme hydrated in D_2O using quasielastic neutron scattering. The neutron scattering data were cast in terms of the imaginary part of the susceptibility, $\chi''(Q,\nu)$, where Q is the scattering vector and ν the frequency, and the characteristic relaxation times from its peak frequency. There is only one peak in $\chi''(Q,\nu)$ and its relaxation time, $\tau_{n\mathrm{K}}$, is almost the same as $\tau_{1\,\mathrm{K}}$ over the temperature range from 200 to 295 K, and also comparable to τ_W of hydration water in BSA or in myoglobin, or τ_c of heavy water (D_2O) in myoglobin. Since neutrons scatter mostly from hydrogen atoms, the neutron scattering spectra of the lysozyme/D_2O sample reflect the motion of the protein and not the deuterated water of hydration. However, this observed protein motion is *not* the structural α-relaxation of the hydrated lysozyme. This is because we have learned from the dielectric data of hydrated myoglobin [1110], hydrated elastin and collagen [1383], and hydrated BSA [1316, 1317] that the structural α-relaxation of hydrated proteins is a separate entity with relaxation time different from τ_W or τ_c, and has a T_g of its own determined at 10^3 s by adiabatic calorimetry (see Figs. 230 and 233-1). Furthermore, $\tau_{1\,\mathrm{K}}$ and $\tau_{n\mathrm{K}}$ are comparable to τ_W or τ_c, indicating the former like the latter are the JG β-relaxation times and not the structural relation times of the hydrated protein. Altogether, these facts cannot be reconciled with the assignment of the motion with relaxation time $\tau_{1\,\mathrm{K}}$ or $\tau_{n\mathrm{K}}$ to "structural relaxation of the protein" or "protein's structural relaxation coupled to hydration water" by Khodadadi et al.

From the discussions above of the experimental data on dynamics of hydrated proteins, we conclude that the adiabatic calorimetric measurements [1316] are pivotal in proving that the two major relaxation processes, the primitive or JG β-relaxation of hydration water and the structural α-relaxation, are distinct but yet connected. They are distinct because their relaxation times reach 10^3 s at vastly different temperatures, $T_{g\beta}$ for the β-relaxation and $T_{g\alpha}$ for the α-relaxation, determined by adiabatic calorimetry. They are connected because of the observed change in the temperature dependence of the dielectric β-relaxation time when $T_{g\alpha}$ is crossed [1110, 1317, 1357, 1358]. These features are analogous to the dynamics found in many aqueous mixtures. Moreover, in the Arrhenius temperature regime, the β-relaxation time of hydration water in protein is comparable in magnitude and has similar activation enthalpy as the primitive or JG β-relaxation of many aqueous mixtures. There is also analogy with binary mixtures of van der Waals liquids, although the analogy is not total because in the absence of hydrogen bonding each component has its own α- and JG β-relaxations. There are other connections between the primitive or JG β-relaxation and the α-relaxation in mixtures and in neat glassformers beyond the one discussed here. These connections occur because the primitive relaxation is the precursor of the many-body relaxation dynamics evolving with time until the α-relaxation is reached. The explanation has been given by the coupling model, and the same explanation applies to the dynamics of hydrated proteins.

Without the benefit of considering the full spectrum of long-time adiabatic calorimetric and low-frequency dielectric data, wrong conclusions and interpretations were made by researchers that were based on short-time neutron scattering data or deuteron NMR data alone. Chen et al. [1389] and Khodadadi et al. [1358] concluded that they have observed the structural α-relaxation of the hydrated protein. Doster and Settles [1318, 1319] interpreted their deuteron NMR data as the structural relaxation related to translation of hydration water of myoglobin. However, as clearly demonstrated in several figures shown before, the neutron $\langle\tau_T\rangle$ above $T_L = 220$ K of Chen et al., the neutron τ_{nK} and dielectric $\tau_{1\,K}$ of Khodadadi et al., as well as τ_c from deuteron NMR data of Vogel [1322(b)] cited by Doster and Settles are the JG β-relaxation times of hydration water at temperatures above $T_{g\alpha}$ unequivocally determined by adiabatic calorimetry in the case of hydrated BSA. The structural relaxation time of the hydrated BSA and myoglobin determined by dielectric relaxation spectroscopy has much longer times than $\langle\tau_T\rangle$, τ_{nK}, $\tau_{1\,K}$, and τ_c obtained by the other authors.

Studies of dynamics of the hydration shell in hydrated collagen and elastin by Vogel and coworkers using ^{2}H-NMR [1322(b), 1385(a)] and by Gainaru et al. using dielectric relaxation [1322(a)] were reported. In the first ^{2}H-NMR study [1385(a)], a change in T-dependence of τ_β at $T \approx 200$ K was reported, but it is no longer present in the second report of another study [1322(b), 1322(c)]. In fact, Lusceac et al. stated that the temperature dependence can roughly be described by an Arrhenius law, with the possibility of a weak crossover in the vicinity of 220 K. These authors were careful to say that this is only a possibility because the crossover is so weak that it may not be real. Thus it can be concluded that none of these investigations of hydrated collagen and elastin found evidence of the existence of a sharp fragile-to-strong transition reported for lysozyme at about 220 K by Chen and coworkers

[1359] (see Fig. 227). The dielectric data of elastin at hydration level $h = 0.40$ (g water/1 g protein) found a secondary relaxation with Arrhenius T-dependence for its relaxation time τ_β at all temperatures below $T_g = 275$ K with activation energy of 53 kJ/mol corresponding to the energy needed to break two hydrogen bonds to free water molecules to rotate and translate. The characteristics of τ_β is similar in every respect to what we found and called the primitive or JG relaxation in hydration water of proteins [131, 1071, 1317], various aqueous mixtures, and nanoconfined water [1093]. The change of T-dependence of τ_β at T_g was not observed in hydrated collagen and elastin by dielectric relaxation because T_g is much higher than other hydrated proteins, and the data were taken at temperatures below T_g (see Fig. 227).

Although we often use the term primitive or JG relaxation in general for secondary relaxations which are related to the primary relaxation, not all JG relaxations in different systems have the same characteristics. The amplitudes of rotation and/or translation of the JG relaxation depend on the chemical structure and bonding of the glassformer. Water is the smallest molecule in nature that involves hydrogen bonding. After breaking two hydrogen bonds, as suggested by the activation energy $E_a \approx 45$–53 kJ/mol of the JG relaxation time $\tau_\beta = \tau_\infty \exp(E_a/kT)$ of hydration water, the water molecule is free to rotate and translate. The change from hydrogen-bonded to free rotation–translation engenders entropy change, which nicely explains the prefactor τ_∞ is of the order of 10^{-19} s (see Fig. 217). This can take place and water molecules can translate to longer distance by repeated breaking and remaking hydrogen bonds even at temperatures much below the glass transition temperature of the hydrated protein [1385(b)]. On the other hand, the JG β-relaxation of a polymer such as polybutadiene, which has bonded interaction along the chain, cannot translate, and the rotation angle is restricted by the conformations [1322(d)]. Even motion in toluene is more restricted by dense packing and steric constraints [1322(d)] compared with water. Most conventional JG β-processes in van der Waals glassformers involve only spatially localized molecular motions. The combined rotational and translational freedom of the JG relaxation of water can rationalize the observation that its ^{2}H-NMR line-shape is Lorentzian at higher temperatures down to about 210 K ($\tau_\beta \sim 10^{-7}$ s) in hydrated myoglobin, and at all temperatures investigated in elastin and collagen as found by Lusceac et al. [1322(b)]. The anisotropy of the NMR interaction is averaged out by the presence of the translational motion, rendering the rotational motion of water isotropic, rather than anisotropic.

For some idiosyncratic reason, Lusceac et al. made negative comment on our use of the term "JG β-relaxation" for the observed secondary relaxation of water in [1093, 1094, 1140(a)]. Specifically they stated: "Moreover, the isotropy of the high temperature motion is strong evidence against the conjecture that the water dynamics can be identified with the JG β-process...." From our usage of this term, Lusceac et al. forcefully asserted that necessarily we are saying that the water-specific secondary relaxation must have the same characteristics as JG β-process of conventional glassformers. Nowhere in [1093, 1094, 1140(a), 1071] we have said that, and in fact sometimes we have alternatively called it the ν-process or the primitive relaxation of water. Thus, the assertion of Lusceac et al. is a concoction coming from their own idiosyncrasy. We are experts on secondary relaxations and are fully aware of the fact that not all "JG β-relaxations" have the same characteristics. As discussed before in various subsections, secondary relaxation that we classify as JG β-relaxation has to satisfy some criteria indicating that it is connected to the α-relaxation [120, 685]. It is found not only in polymers and van der Waals glassformers, but also in other lesser known glassformers. These include metallic glasses, ionic liquids, inorganic glasses with covalent bonding, and molten salts. The difference in physical structure and chemical bonding of these glassformers necessarily makes the characteristics of the JG β-relaxation different. Metallic glasses such as $Pd_{77}Cu_6Si_{17}$ and $Zr_{65}Al_{7.5}Cu_{27.5}$ have secondary relaxation which is referred to in the literature as JG β-relaxation [960(b)] not only by us. So is one of the secondary relaxations in ionic liquids such as 1-butyl-3-methylimidazolium *bis*(trifluoromethylsulfonyl)imide (BMIM-BMSF) shown in Fig. 111-2, the secondary relaxation of which involves rotation and translation of the cation and/or anion molecules. These secondary relaxations satisfy some criteria for us to call it JG β-relaxation. It is obvious that the JG β-relaxation in metallic glass and ionic liquid must involve translation motion of atoms and molecules, respectively, and different from the conventional JG β-relaxation in polymers and van der Waals glassformer having only spatially localized molecular motions. The atoms in metallic glasses do not even rotate when translating. If anyone follows the logic of Lusceac et al., he or she would make the following objection: "The fact that secondary relaxation in metallic glasses and ionic liquids involves translational motion of atoms and molecules, respectively, is strong evidence against the conjecture that it can be identified with the JG β-process." This hypothetical objection is bunk and no more than splitting hair in semantics. The situation in real life is like persons who were only familiar with snakes in their own country that move smoothly along its own contour (reptate), but had never been in the southwestern part of USA and Mexico to see a sidewinder. When finally the person came to USA and saw a sidewinder, and was told that it is just another snake, the person objected to the use of the term "snake" for the sidewinder. This is because, according to his or her experience, conventionally snakes do not move by lateral and looping bodily motion as the sidewinder does. This objection is without merit, because the sidewinder is indeed a snake according to zoology. The triviality of the objection of calling the water-specific relaxation as JG β-relaxation of water can be debunked in the same way as the objection of calling *sidewinder* as *snake* as follows. Water in secondary relaxation of aqueous mixtures and hydrated proteins (*sidewinder*) moves differently than secondary relaxation of toluene or polybutadiene (*Python in Greek mythology, or any common snake you can find in Northern Europe*), but it is legitimate to call them all by the same name, JG β-relaxations (*snakes*), according to the classification by Ngai and Paluch [120] (*by zoologists*).

Furthermore, even within the same family of glassformers, the angle of rotation of JG relaxation in the glassy state also can vary greatly. The ^{2}H-NMR study of JG β-relaxation of neat polyalcohols, including glycerol, threitol, xylitol, and sorbitol [239], established an empirical relation, $\sin\Theta \approx (\Delta\varepsilon_\beta/\Delta\varepsilon_\alpha)^{1/2}$ between the semi-angle Θ of reorientation of the JG relaxation and the ratio of its dielectric strength to that of the α-relaxation. For sorbitol, $(\Delta\varepsilon_\beta/\Delta\varepsilon_\alpha)$ is about 2.5×10^{-2}, and the semi-angle is about 10° at $T = T_g/1.3$ which is about three times larger than that of threitol. For some of the aqueous mixtures [1093], the ratio $\Delta\varepsilon_\beta/\Delta\varepsilon_\alpha$ is much larger than sorbitol. For example, in a mixture of 35 wt% of water with trimer of ethylene glycol, $\Delta\varepsilon_\beta/\Delta\varepsilon_\alpha \approx 1$. Hence the empirical relation suggests the amplitude of the motion of the water JG relaxation in hydrated proteins can be of the order of the tetrahedral angle as observed by ^{2}H-NMR. By clarifying the special nature of the JG secondary relaxation of water in the above, its properties are fully consistent with those observed by Lusceac et al. using ^{2}H-NMR. Lusceac et al. prefer to use "water-specific β process" to describe the process and not the term JG β-relaxation we use, although they have no idea of its function like we do, not only for hydrated proteins, aqueous mixtures, but also for glass-forming materials in general. This is clear from the statement made by Lusceac et al. themselves: "... suggesting the existence of a water-specific β process. However, exact origin and possible universality of this β process are open questions."

We have no problem in replacing the term "JG β-relaxation of water" by "water-specific β process" if anyone prefers. This is because nothing is changed or lost in our explanation of the hydration water dynamics in hydrated proteins by the change in terminology. The main purpose of our use of the term "JG β-relaxation" is to distinguish secondary relaxations that involve the motion of the entire molecule, have properties connected to the α-relaxation and relaxation time in approximate agreement with the primitive relaxation time of the CM [685], and inseparable from the α-relaxation [998, 999]. For the secondary relaxation of water in aqueous mixtures or hydrated proteins, one such property is the change in T-dependence of its relaxation time at T_g, which is found for JG β-relaxation in conventional glassformers. We could have simply called them secondary relaxations that belong to the special class, but this is cumbersome. It is more expedient when referring to these relaxations by giving them a name. I do not have the audacity to call them the Ngai secondary relaxation, which arguably can be justified because I am the first one to distinguish them from other secondary relaxations, and point out their symbiotic relation with the structural α-relaxation and its fundamental role it plays in glass transition. But instead I chose to call them "JG β-relaxations." This term serves also the ancillary purpose of honoring Johari and Goldstein for their seminal work that shows even totally rigid glassformers can have a secondary relaxation. However, our ideas on the "JG β-relaxation" differ from some of that issued by Johari as well as by Goldstein. It does not make sense for anyone to identify the view of these two illustrious colleagues on what we call the "JG β-relaxation" with ours. Under any circumstance, when others say that it is not a "JG β-relaxation" when referencing us like Lusceac et al. did for the "water-specific β process," they should justify it by experimental facts to counter our very own definition of "JG β-relaxation," and not what were written by Johari and Goldstein, jointly or separately. The experimental facts of the "water-specific β process" in hydrated proteins have been amply discussed, and clarification of the nature of secondary relaxation of water has been given above. There is no contradiction of the facts with our own definition of the "JG β-relaxation," and hence the "water-specific β process" is no more or no less than the JG β-relaxation of water. The secondary relaxation of glycerol has been shown to cause the dynamic transition in glycerol-coated lysozyme (see Fig. 222, left and right, and discussion). This secondary relaxation of glycerol satisfies all our criteria for it to be the "JG β-relaxation" according to our definition of this term [120, 195, 239, 424], and its motion is not the same as that of water.

The deuteron NMR experiment of Lusceac et al. has confirmed the process is a "water-specific β process" at temperatures *above* and *below* T_g. This fact contradicts the interpretations of others in [1105–1108, 1111]. One interpretation [1107, 1108, 1111] proposed that at high temperatures above T_g, water reorientation is an α-relaxation. Another proposed that it is a merged α- and β-process above T_g [1105, 1106]. These interpretations also contradict the monotonic increase of the dielectric strength of the water relaxation with temperature above and below T_g, which is the signature of secondary relaxation. On the other hand, it is well known that the dielectric strength of α-relaxation has the opposite temperature dependence. There are ample experimental evidence for the monotonic increase of the dielectric strength of the "water-specific β process" with temperature above T_g in aqueous mixtures (see Figs. 212 and 213 for example), and also in hydrated proteins [1112, 1317, 1358(a), 1358(c)].

Additional Proofs of the Primitive or JG β-Relaxation of Water Is Responsible for the Dynamic Transition Observed by Neutron Scattering and Mössbauer Spectroscopy in Hydrated Proteins and Enzymes

Combining Data from Hydrated Biomolecules, Aqueous Mixtures, and Nanoconfined Water

The most convincing way we can demonstrate that the fast water-specific relaxation observed in hydrated biomolecules is the primitive or JG β-relaxation of water is to compare their relaxation times τ_W shown before in Fig. 233-1 with the JG β-relaxation times of the water component in several aqueous mixtures and water nanoconfined. Some of the data of τ_W are shown again in the left panel of Fig. 233-2 with large symbols or thick lines to distinguish them from those of τ_{JG} of water in various mixtures and in confinement. Added are the data of τ_W found in hydrated collagen and elastin [1322(a)], shown here by the thick line with Arrhenius temperature dependence in the measurement temperature range faithfully reproduced. The adiabatic calorimetry data of water confined in molecular sieves and silica gels are added to that of the hydrated BSA on the line with $\tau = 10^3$ s. The small open and closed pale blue diamonds are τ_α and τ_{JG} of 35 wt% of water mixture with triethylene glycol, and are shown here to reemphasize the generally observed change of T-dependence of τ_{JG} at T_g of aqueous mixtures as well as the hydrated BSA [1316, 1317], myoglobin [1318, 1319], and lysozyme [1358]. The solutes in the aqueous mixtures include PVP, PEHMA, PVME, PVA, PEG, PEI, and 1-propanol, and water is confined in MCM-41, and graphite oxide. High-frequency dielectric data of bulk water is also included in the plot. All data shown here have been discussed before.

We can see from the figure that, when τ_W of the water-specific process in hydrated biomolecules becomes short ($<10^{-4}$ s for the examples here), it is nearly indistinguishable from τ_{JG} of aqueous mixtures and confined water. At low

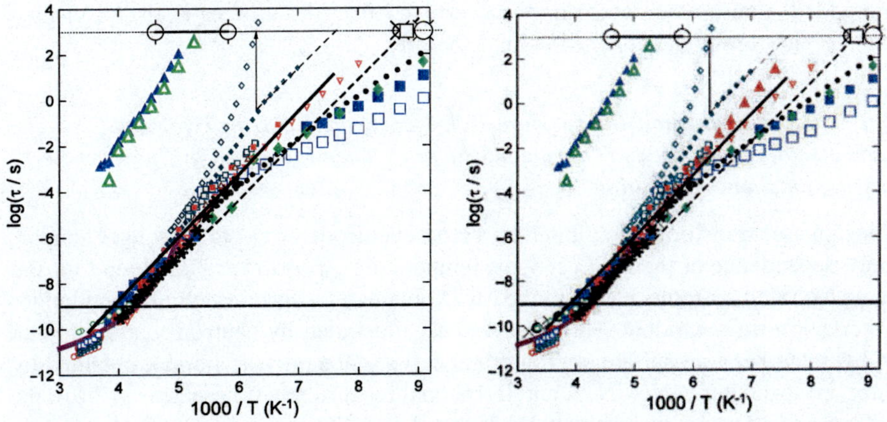

Fig. 233-2 $\log_{10}(\tau)$ vs. 1000/*T*. (*Left*) The *larger symbols* are τ_α and τ_W of the hydrated BSA and lysozyme as in Fig. 233-1. Data of Swenson et al. and Chen et al. therein are omitted here. The *thick magenta solid line* is VFTH fit to the relaxation time, τ_c, of reorientational motion of heavy water (D_2O) in D_2O-hydrated myoglobin (0.35 g/g) from deuteron NMR. The relaxation times from neutron scattering of hydrated lysozyme are now represented by *red closed circles*. The *thick line* is τ_W of hydrated collagen and elastin. All *smaller symbols* and *thinner lines* are for τ_{JG} or τ_ν of aqueous mixtures and confined water. (1) 35 wt% water mixture with 3EG (*faint blue open diamonds* for τ_α and *closed diamond* for τ_{JG}), 50 wt% water mixture with PVME (*red closed squares*), 50 wt% water with PVP (*blue open square*), 38.6 wt% of water with PHEMA (*brown open triangles*), 20 wt% water with 1-propanol (*black* •). At times shorter than 10^{-10}s are τ_{JG} of 90 wt% water with polyvinyl alcohol (*pale blue squares* with × inside), 80 wt% water with PVP at short times (*blue open squares*), and bulk water at short times (*brown open circles*). The *long thinner dashed line* is the Arrhenius fit to τ_{JG} of 20% of water nanoconfined in graphite oxide from dielectric relaxation, actual data not shown, and *green open circles* with $\tau_{JG} < 10^{-9}$ s are from neutron scattering. Dielectric data of relaxation time of water confined in MCM-41 with pore diameter 2.14 nm at hydration levels of 22 wt% (*green* ▼). The *horizontal dashed line* indicates $\tau_W = 10^{-9}$ s, the upper bound of the time window of neutron scattering instrument having energy resolution of 1 μeV. (*Right*) All data shown in the *left panel* together with the following additions. Water JG or water-specific relaxation times τ_{JG} from the neutron scattering from [1354(c)], dielectric relaxation [1105], Mössbauer spectroscopy [1354(c)], and NMR of hydrated myoglobin [1322(c)], and neutron scattering from hydration water on C-phycocyanin (C-PC) [1395(a)]. For τ_{JG} shorter than 10^{-6} s, the new additions overlap the previous ones and cannot be resolved, indicating that τ_{JG} are approximately the same for different proteins and hydrophilic solutes. Thus, the water JG relaxation is responsible for the dynamic transition observed by neutron scattering and Mössbauer spectroscopy

temperatures in the glassy state, the deviations of the dielectric τ_W of water in hydrated BSA from τ_{JG} are noteworthy but the deviations are probably due to large uncertainty in determining τ_W from the very broad loss peaks in the spectra. Nevertheless, the adiabatic τ_W of hydrated BSA is in good agreement with the averaged τ_{JG} from various aqueous mixtures and confined water. In the right panel of Fig. 233-1, more data of the relaxation time of the process responsible for the dynamic transition observed by neutron scattering and Mössbauer spectroscopy of hydrated myoglobin and neutron scattering of C-phycocyanin (C-PC) are added. These new data cannot be resolved from each other in the short-time regime, supporting the conclusion made before.

The Dynamic Transition Temperature T_d is Nearly the Same for Hydrated Biomolecule with Different Chemical Structures when Measured by Spectrometers Having the Same Resolution

The dynamic transition is exhibited by a crossover from a weak to a stronger temperature dependence of the MSD at some temperature T_{d}. A lower T_{d} is found for the same hydrated biomolecule from the MSD obtained by spectrometer having longer observation time window. This trend is best illustrated by neutron scattering data of hydrated heparan sulfate, an important cell surface polysaccharide, obtained by three spectrometers, IN6, IN13, and IN16, on three different timescales t_{exp} of 10 ps, 100 ps, and a few ns, respectively [1370(b)]. The MSD of the dry HS, 0.15 g D_2O/g HS (HS-0.15 g/g) and 0.43 g D_2O/g HS (HS-0.43 g/g) samples are shown here in Fig. 233-3 from (A) to (C). T_{d} of the dynamic transition observed in the 0.43 g

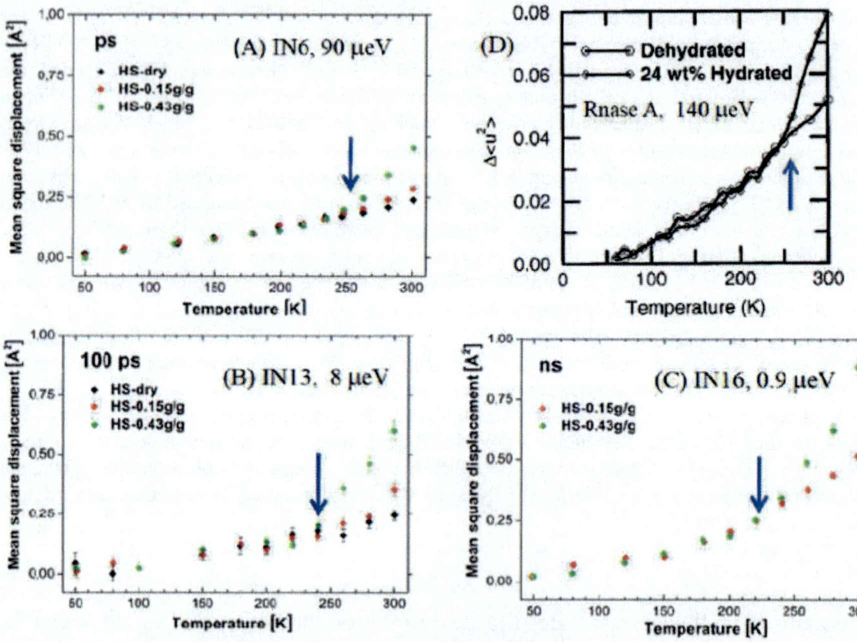

Fig. 233-3 Counter-clockwise. **(A)**–**(C)** Mean-squared displacements of hydrated heparan sulfate (HS) extracted for the dry HS, 0.15 g D_2O/g HS (HS-0.15 g/g) and 0.43 g D_2O/g HS (HS-0.43 g/g) samples from measurements using, respectively, IN6 (~10 ps), IN13 (~100 ps), and IN16 (few ns). Reproduced from [1370(b)] by permission. **(D)** Mean-squared displacement of 24 wt% hydrated ribonuclease (RNase) measured by the Fermi Chopper spectrometer at NIST with resolution of 140 μeV comparable to IN6 with resolution of 90 μeV. Reproduced from [1345(b)] by permission. The *arrow* indicates the location of T_{d} for each case

D_2O/g HS (HS-0.43 g/g) sample are suggested by the locations of the arrows in panels (A)–(C). The decrease of T_d on increasing the long time limit of the observation time window is clear. This trend has been explained by the so-called "frequency-window model" as an instrumental energy resolution effect [1364(b)]. That is, the increase of the MSD on increasing T past T_d is caused by a relaxation process entering the time window of the instrument. The relaxation process according to us is the primitive or JG β-relaxation of hydration water. As discussed before, this relaxation is coupled by hydrogen bonding to the biomolecules, and that is why the dynamics transition can be observed in the 0.43 g D_2O/g HS (HS-0.43 g/g) samples.

Panel (D) of Fig. 233-3 shows the MSD of 24 wt% hydrated ribonuclease (RNase) measured by the Fermi Chopper spectrometer at NIST with resolution of 140 μeV comparable to IN6 with resolution of 90 μeV [1345(b)]. Despite the difference of the two biomolecules, T_d is about 260 K, the same for both HS-0.43 g/g and the hydrated RNase. It can be seen from Fig. 233-2 at $T = 260$ K, the JG β-relaxation time of the hydration water τ_{JG} starts to become comparable to the 10 ps timescale t_{exp} of these spectrometers (see Fig. 233-2). The same can be said for $T_d \sim 240$ K and $T_d \sim 225 - 230$ K from the MSD obtained by IN13 and IN16, respectively, on HS-0.43 g/g. Again from Fig. 233-2, the most probable JG β-relaxation time starts to become comparable to 100 ps and ns at 240 and 225 K, respectively. Hence from these data, it seems that the JG β-relaxation causes the dynamic transition.

The effect of the instrumental energy resolution on the MSD obtained by elastic-incoherent neutron scattering has also been demonstrated in partially deuterated lysozymes hydrated by D_2O, and H_2O at $h = 0.4$ [1380(b)]. Data were obtained for the same sample both by the IN13 and the IN10 spectrometers. The elastic energy resolution (FWHM) of IN13 was 8 μeV, which corresponds to an elastic time resolution of 500 ps, and the MSD of the hydrated lysozyme shows the dynamic transition at $T_d \approx 240$ K. On the other hand, the elastic energy resolution (FWHM) of the IN10 spectrometer was 1 μeV, which corresponds to an elastic time resolution of 4000 ps, and the MSD of the same hydrated lysozyme shows the dynamic transition at a lower $T_d \approx 220$ K.

Next we compare the MSD of three different hydrated biomolecules and disaccharide aqueous mixtures all obtained from neutron scattering using IN13 in Fig. 233-4. Panel A is the MSD of protonated C-PC samples embedded by deuterated trehalose/D_2O at different ratios as a function of temperature [1373] shown before in Fig. 224. Panel B presents the MSD of trehalose–, maltose–, and sucrose–$19H_2O$ and $-19D_2O$ mixtures [1361(a) and (b)]. Panel C gives the total MSD of single-stranded DNA (ssDNA) at 0 and 0.55 h, and double-stranded DNA (dsDNA) at 0 and 0.55 h. Panel D gives the MSD of ferric horse myoglobin (met–Mb) hydrated with D_2O at 20% level, and also the MSD when it was encapsulated in silica gel. Although no uniform criterion can be used to determine T_d, its values are about the same (230–240 K) for all four systems. There is true dynamic transition at $T_d \approx 230 - 240$ K for the three hydrated biomolecules (Panels A, C, and D). However, the observed change of T-dependence of MSD of the disaccharide aqueous mixtures (Panel B) should not be considered as dynamic transition equivalent

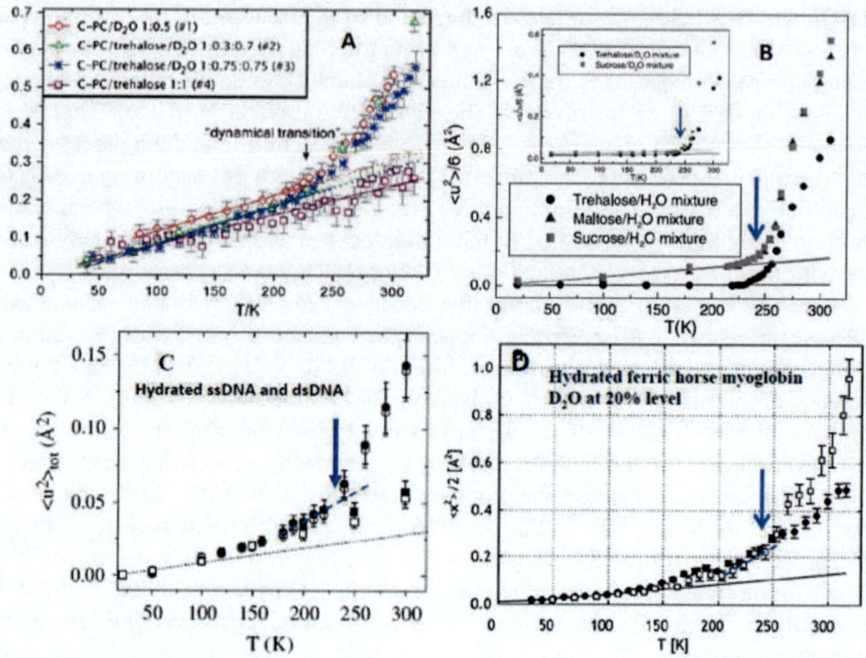

Fig. 233-4 MSD of different hydrated biomolecules and aqueous mixtures of three disaccharides obtained by IN13 with energy resolution (FWHM) of 8 μeV, which gives access to dynamics on a timescale above 150 ps and up to 400 ps. (**a**) MSD of protonated C-PC samples embedded by deuterated trehalose/D_2O at different ratios as a function of temperature (40–318 K). Reproduced from [1373] by permission. (**b**) Temperature dependence of the MSD for trehalose–, maltose–, and sucrose–H_2O and –D_2O mixtures. Reproduced from [1361] by permission. (**c**) Temperature dependence of the total MSD for single-stranded DNA (ssDNA) at 0 h (*empty squares*), ssDNA at 0.55 h (*empty circles*), double-stranded DNA (dsDNA) at 0 h (*full squares*) and dsDNA at 0.55 h (*full circles*). The *lower dashed line* is the Gaussian contribution $< u^2 >_G$ to the total MSD of dsDNA at 0 h. Reproduced from [1346(c)] by permission. (**d**) Temperature dependence of the MSD of ferric horse myoglobin (met–Mb) hydrated with D_2O at 20% level. *Filled circles*: met–Mb encapsulated in silica gel; *open squares*: met–Mb hydrated powder. MSD obtained using the dynamical heterogeneity model, for methyl hydrogens. Reproduced from [1362] by permission

to that observed in the hydrated biomolecules where T_d is higher than T_g. This is because Magazu and coworkers [1363(a) and (b)] found in the disaccharide aqueous mixtures that the observed T_d is the same as $T_g \approx 238$ K for the trehalose mixture, ≈ 235 K for the maltose mixture, and ≈ 233 K for the sucrose mixture, as previously determined by experiment [1221]. In many other glassformers, starting from low temperatures, on increasing temperature the MSD measured by neutron scattering exhibits a marked increase after crossing T_g [114,115(a),507,509,510], and is explained by the sensitivity to change from glassy state to the liquid state at T_g of either the nearly constant loss (NCL) from caged dynamics [507, see Section 2.3.2.34] or very fast secondary relaxation (see Section (2.3.2.15)). This is exactly the origin of the change in T-dependence of the MSD found by Magazu and coworkers [1373] in the disaccharide aqueous mixtures. In the disaccharide aqueous mixtures, the change is *not* due to the water JG relaxation entering the time window of the IN13 spectrometer at 230–240 K, while it is the case for the JG relaxation of the hydration water in hydrated proteins. In fact dielectric relaxation data of aqueous mixture of trehalose [1140(c)] at the same concentration as Magazú et al. show the water JG relaxation times τ_{JG} at $T_g = 238$ K is two decades longer than the timescale $t_{exp} \sim 100$ ps of the IN13 spectrometer used. Dielectric data of aqueous mixture of sucrose at slight lower concentration also indicate the same situation [1103]. Thus the water JG relaxation cannot explain the change of T-dependence at T_g of the MSD observed in the aqueous mixtures of trehalose and sucrose, which nevertheless has an explanation from the NCL. Trehalose and sucrose have a fast γ-relaxation with short relaxation time τ_γ but not short enough at 230–240 K to contribute to the MSD [1012(c)]. Only if τ_γ becomes much shorter at 230–240 K to become comparable to $t_{exp} \sim 100$ ps when trehalose and sucrose are dissolved in water will the γ-relaxation contribute to the observed change of the MSD T-dependence at T_g of these aqueous mixtures, but at this time there is no evidence for this.

The change of T-dependence of the MSD of glassformers at T_g is usually mild [114,115(a),507,509,510]. On the other hand, the MSD of the aqueous disaccharide mixtures exhibits much stronger increase at temperatures above T_g (see Panel B). This feature may be due to the translation motion of the water JG relaxation in the mixtures when the relaxation times enter the time window of the IN13 spectrometer at some temperature T_d higher than T_g, and it is the true analogue of the dynamic transition seen in hydrated biomolecules [1361(d)]. Thus the MSD data actually show two crossovers of T-dependence, one at T_g due to NCL and another one at a higher temperature T_d due to the water JG relaxation in the mixtures. The MSD data of aqueous glucose mixtures from Di Bari et al. [1346(e)] obtained also by IN13 support this interpretation as well as ^{1}H-NMR measurements [1346(f)] by Van den Dries et al. Their data of aqueous glucose mixtures shown in Fig. 233-4′(a) clearly show the larger change of T-dependence of the MSD occurs at 20 or more degrees higher than T_g, and has to be associated with the dynamic transition at T_d. Additional confirmation that this is indeed the dynamic transition can be taken from the IN13 neutron scattering data of lysozyme embedded in glucose–water matrices published by Cornicchi et al. in [1346(b)] as a function of the water content [Fig. 233-4′(b)]. Although there are some uncertainties in the values of T_d for lysozyme embedded in glucose–water reported in Fig. 233-4′(a), on increasing water content, the observed trend of decreasing T_d is consistent with the expected decrease of the water JG relaxation time. The dynamic transition temperatures T_d given by Cornicchi et al. are nearly the same as that of the aqueous glucose mixtures found by Di Bari et al. using the same IN13 spectrometer, and by Van den Dries et al. from ^{1}H-NMR measurements [see Fig. 233-4′(a)]. The finding that T_d for lysozyme embedded in glucose–water is nearly the same as T_d follows from two established facts: (1) the coupling of lysozyme to the glucose–water solvent and (2) the water JG relaxation time of the solvent when shorter than 150 ps resolution of IN13 is insensitive to whether with or without lysozyme at sufficiently high concentration of water. The

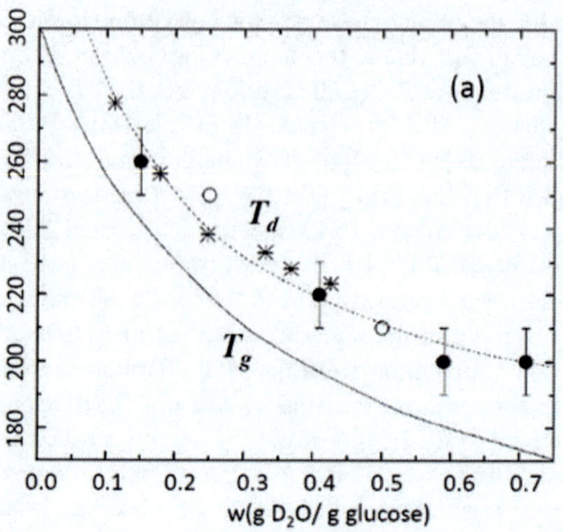

Fig. 233-4′a Dynamic transition temperatures T_d for lysozyme embedded in glucose–water matrices from [1346(b)] by Cornicchi et al. (•) and for glucose–water matrices (o) from [1346(e)] by Di Bari et al. as a function of the water content obtained from IN13 neutron scattering spectroscopy with energy resolution of 9 μeV (FWHM) make accessible motions faster than ~150 ps. For comparison, the critical temperatures T_c of glucose–water mixtures (*) obtained from ^{1}H-NMR measurements [1346(f)] by Van den Dries et al. are also reported. The *continuous line* represents the glass transition temperatures T_g of glucose–water systems determined by differential scanning calorimetry [1220] and dielectric relaxation technique [1241]. The *dashed line* is guide for the eye

Fig. 233-4′b Total MSD vs. *T* for the samples measured by Cornicchi et al. using the IN13 neutron scattering spectrometer with water content ranging from 0 to 0.7*w* (g D_2O/g glucose)

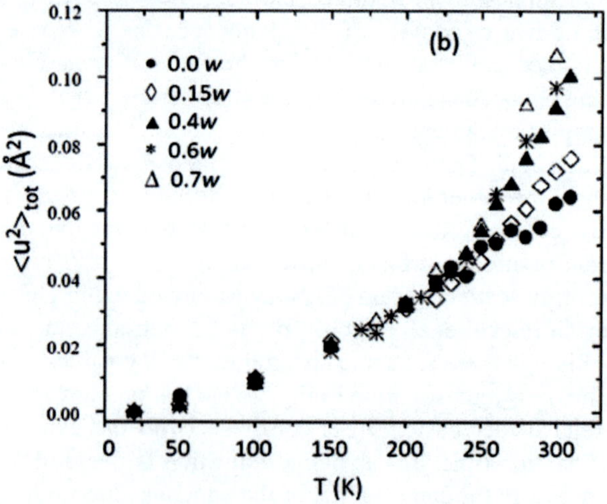

latter has been shown before in Fig. 233-2 for hydrated proteins and for aqueous hydrophilic solute mixtures in Fig. 217.

According to our interpretation, the dynamic transition of proteins and biomolecules embedded in a solvent containing water originates from the water

JG relaxation time entering the time window of the nuclear scattering spectrometer when temperature is increased above T_d. This explanation implies that dynamic transition is also observed in the water-containing solvent even without the presence of the protein or biomolecule. Also, at sufficiently high concentration of water in the solvent, the dynamic transition temperature T_d should not be too different for the two cases. These natural consequences of the explanation are consistent with the observation of dynamic transition with similar values of T_d in the glucose–water by Di Bari et al. and in the lysozyme embedded in glucose–water by Cornicchi et al. The same should apply to the case of pure water as the solvent. Although experiment cannot be performed on bulk water due to intervention by crystallization, the dynamics of water confined in a hydrophobic mesoporous carbon material has been studied by incoherent quasielastic neutron scattering [1289(f)]. This study by Chu et al. found change of T-dependence of the MSD-like dynamic transition of hydrated biomolecules in the neighborhood of 200-210 K using the HFBS spectrometer at NIST with an energy resolution of 0.8 μeV (FWHM). The temperature 200–210 K is comparable to $T_d \sim 210 - 220$ K found for lysozyme:water (70:30) by Tsai et al. [1345(a)] using the same HFBS spectrometer at NIST (see Fig. 222, right), and $T_d \sim 220$ K found for hydration water of maltose binding protein (D-MBP-H_2O) by Wood et al. using IN16 with resolution of 0.9 μeV (FWHM) [1363] (see Fig. 225, right).

On the left side of Fig. 233-5 in panel (b) is the classic MSD of the non-exchangeable hydrogens in D_2O-hydrated myoglobin (0.35 g/g) measured with the backscattering spectrometer IN13 by Doster. Only the collective Gaussian component is shown, the methyl group contribution was subtracted [1354(c)]. According to Doster, only processes faster than 50 ps contribute to the displacements, and this is consistent with $T_d \sim 240$ K observed. Mössbauer spectroscopy data of myoglobin heme iron in myoglobin covered by two different solvents shown in panel (a) will be discussed in the next subsection.

The top panel on the right side of Fig. 233-5 show the T-dependence of MSD of maltose binding protein (MBP) hydration water in the D-MBP-H_2O sample from measurements using IN5 (filled circles) by Paciaroni et al. [1346(d)], and compare with the MSD of the same sample obtained by using IN16 (empty circles) and from molecular dynamics simulations (stars) by Wood et al. [1363], shown before in Fig. 225. The MSD of D-MBP-H_2O coming from the picoseconds time-window of IN5 and the nanosecond time-window of IN16 appeared to Paciaroni et al. to have the dynamic transition occurring at nearly the same temperature. The following is the statement made by them in [1346(d)]: "Quite interestingly, if one looks at the MSD, as estimated in both of the spectrometers, the dynamical onset of the protein hydration water motions takes place at around the same temperature." If this conclusion is correct, then it is the first example contradicting many other experiments showing that T_d decreases with the larger time-window of the experiments such as in Fig. 233-3 for hydrated heparan sulfate (HS) [1370(b)]. In the same panel of Fig. 233-5, Paciaroni et al. present the MSD of MBP(H) + D_2O (filled squares) measured on IN5, and the dynamic transition occurs at higher temperature with T_d equal to or large than 250 K. The large difference between D-MBP-H_2O and MBP(H) +

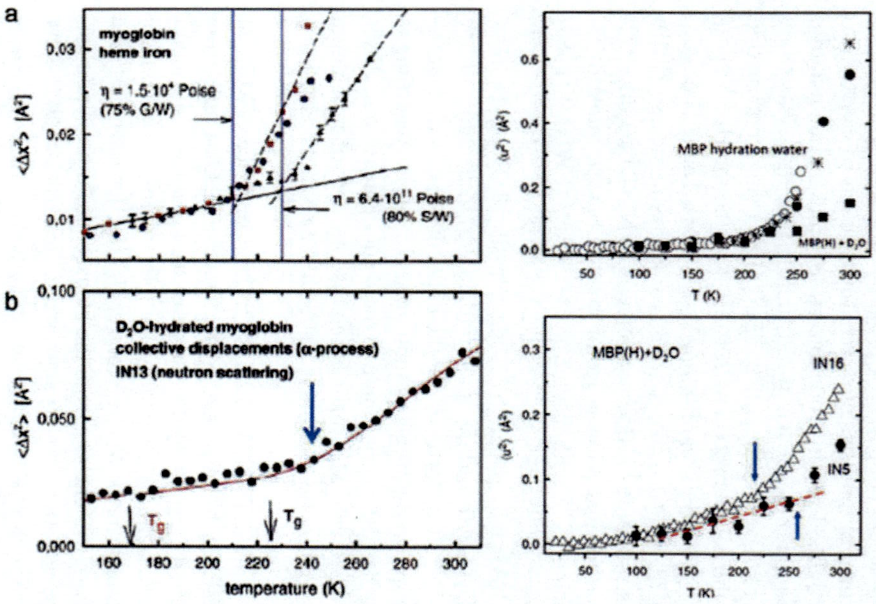

Fig. 233-5 (*Left*) (**a**) Mean-squared displacements of the heme iron of myoglobin measured with Mössbauer spectroscopy in various solvents: *full circles*: water in myoglobin crystals, *squares*: 75% glycerol–water, and *triangles*: 80% sucrose–water solution. The viscosity at the onset temperatures is indicated. The experimental timescale is 140 ns. (**b**) Mean-squared displacements of the non-exchangeable hydrogens in D_2O-hydrated myoglobin (0.35 g/g) measured with the backscattering spectrometer IN13 (ILL). Only the collective Gaussian component is shown, the methyl group contribution was subtracted. The *full line* is a fit assuming that the α-relaxation water crosses the experimental time window of IN13. Instrumental time window. Reproduced from [1354(c)] by permission. (*Right, upper panel*) Mean-squared displacements vs. *T* of MBP hydration water, as calculated from measurements on IN5 (*full circles*), on IN16 [1363] (*empty circles*), and from molecular dynamics simulations [1363] (*stars*), together with MSD of MBP(H) + D_2O (*full squares*) measured on IN5.The experimental MSD of hydration water were corrected for the contribution of exchangeable protein protons. The MSD reported by Wood et al. [1363] had been recalculated and rescaled for comparison. (*Right, lower panel*) MSD of MBP(H)+D_2O measured on IN5 (*full circles*) and on IN16 (*empty triangles*). Data from [1346(d)] are replotted

D_2O in their values of T_d from MSD measured by IN5 also contradicts that T_d is the same from MSD measured by IN16 [1363], as well as the belief that the dynamics of the protein and the hydration water are coupled as far as the dynamic transition phenomenon is concerned. Before accepting this exception proposed by Paciaroni et al. as a matter of fact, one should notice that there is no MSD data of D-MBP-H_2O obtained by IN5 (filled circles) in the critical temperature range from 225 to 250 K. Without data filling in this critical T-range, the actual T_d of D-MBP-H_2O from IN5 experiment is still unknown and the exception to the rule may not be real. On the other hand for MBP(H) + D_2O, the T_d obtained from IN16 is evidently lower than that from IN5. This normal behavior can be seen in the lower panel on the right side of Fig. 233-5.

The MSD of three different hydrated biomolecules from ns – ps motions obtained by IN16 are presented in Fig. 233-6. Panel A are the MSD extracted by linear regression from the Guinier plots in the low Q-range (0.19 Å^2 $< Q^2 <$ 1.13 Å^2) for samples of human serum butyrylcholinesterase (HuBChE) including HuBChE-H_2O, HuBChE-D_2O, and HuBChE-dry from [1364(c)]. Human serum butyrylcholinesterase is a homotetrameric sialoglycoprotein synthesized in the liver. Panel B are the MSD of wild-type (WT), the AA, and AA+Y RC mutants from *Rhodobacter sphaeroides*, compared to the buffer (D_2O, 0.1% Triton X-100 detergent, HEPES 10 mM, pD 7.93 (pH 7.5)) from [1365]. Panel C are MSD in maltose binding protein samples, H-MBP-D_2O and D-MBP-H_2O shown before in Fig. 225. Panel D are MSD of 0.4 g D_2O g^{-1} hydrated lysozyme taken from the work of Zanotti, Gibrat, and Bellissent-Funel [1347]. Nearly the same $T_d \sim 220 - 225$ K is found for the four different hydrated biomolecules. At this temperature, the relaxation time of the hydration water τ_{JG} matches the timescale of the IN16 spectrometer t_{exp} of the order of ns, as can be verified by inspection of Fig. 233-2.

Nearly the same T_d is found in widely different hydrated biomolecules by using the same spectrometer or different spectrometers with the same timescale. This fact is good indication of the relaxation process responsible for the dynamic transition being insensitive to the biomolecule. Since these hydrated biomolecules have different T_g and hence very different τ_α at the same temperature, the fact rules out the involvement of the α-relaxation in the dynamic transition. The primitive or JG β-relaxation time of the hydration water is a strong candidate for the relaxation process because not only are its relaxation times τ_{JG} nearly independent of the biomolecules when shorter than 10 ns (as demonstrated before in Fig. 233-2 for various hydrated biomolecules, aqueous mixtures, and confined water), but also at T_d they match the timescale t_{exp} of the spectrometer used for all the hydrated biomolecules considered here. However, even at timescales shorter than 10 ns, the relaxation times τ_{JG} do depend on the chemical structure of the hydrated biomolecule because of the coupling between the hydrated water with it. We have seen ample examples of this dependence from the data of τ_{JG} in aqueous mixtures, albeit usually the data were taken at longer times than 10 ns. For hydrated biomolecules this dependence is in accord with the spread of τ_{JG} from the collection of experimental data shown in Fig. 233-7 as well as the data published by Khodadadi et al. [1358(d)].

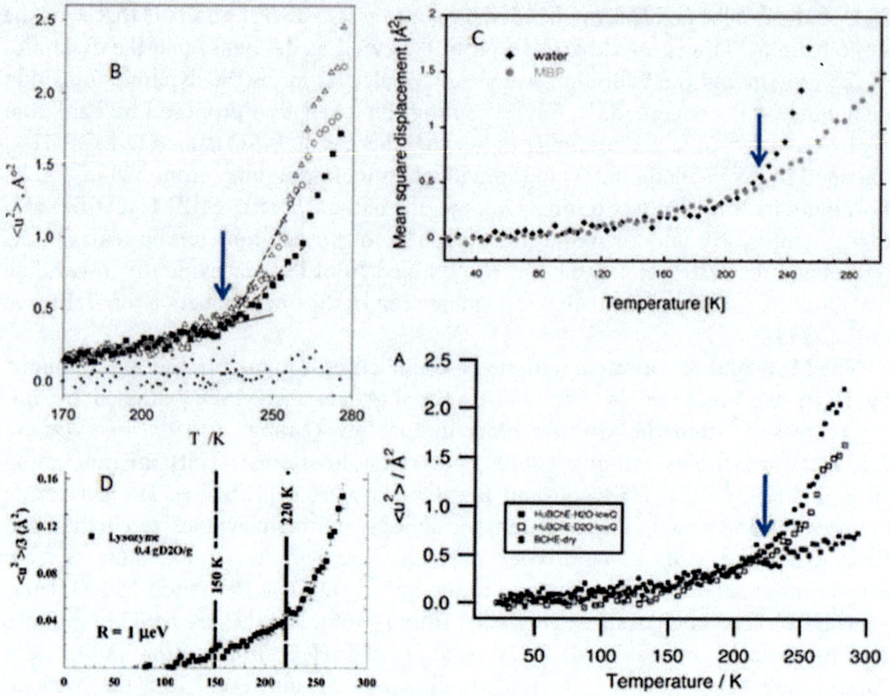

Fig. 233-6 MSD from IN16. (**a**) MSD of HuBChE-H_2O, HuBChE-D_2O, and HuBChE-dry. Reproduced by permission from [1364(c)]. (**b**) MSD of wild-type (WT) (*closed squares*) and the AA (*open circles*) and AA+Y (*open triangles*) RC mutants from *R. sphaeroides*, compared to the buffer (D_2O, 0.1% Triton X-100 detergent, HEPES 10 mM, pD 7.93 (pH 7.5)). *Error bars* are only shown for the WT RC but are of similar magnitude for the mutant proteins. The *dots* correspond to the solvent and detergent alone measured and analyzed under the same conditions. The *solid line* is a linear fit of the lower temperature portion of MSD common to all proteins, and *dashed lines* indicate the differing slopes of MSD for the mutants and the WT above T_d, the dynamical transition temperature (approximately 220–230 K for all three proteins). This change is attributed to the onset of anharmonicity and diffusive dynamics. Above T_d, until the melting temperature (277 K in D_2O), larger amplitude motions, corresponding to increased flexibility, are found for the mutated proteins. Reproduced from [1365] by permission. (**c**) MSD of ns – ps motions in maltose binding protein (H-MBP-D_2O sample; *gray circles*) and in its hydration water (D-MBP-H_2O sample; *black diamonds*). Dynamical transitions (changes in slope of temperature-dependent mean-squared displacements) in the protein and in its hydration water take place at similar temperatures (~220 K). Reproduced from [1363] by permission. (**d**) Dynamic transition in 0.4 g D_2O g^{-1} hydrated lysozyme. $T_d = 220$ K was chosen by the authors of [1347]. Reproduced from [1347] by permission

Primitive or JG β-Relaxation of Hydration Water Mistaken as α-Relaxation of Hydrated Protein

The role of the primitive or JG β-relaxation (or water-specific secondary relaxation) of hydration water in causing the dynamic transition as well as the connection it

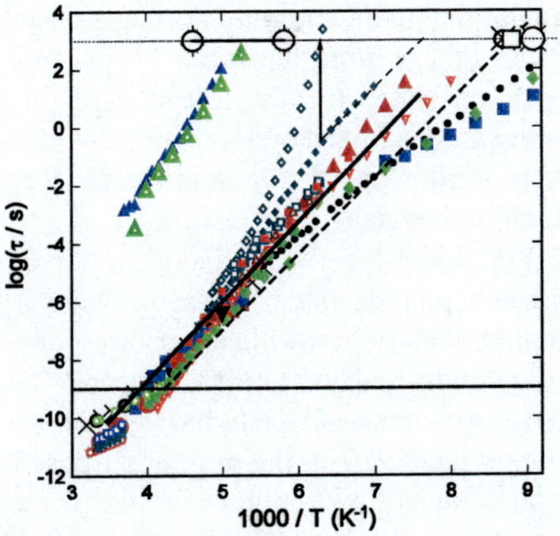

Fig. 233-7 Plots of $\log_{10}(\tau)$ vs. 1000/*T* for various systems: for 20 wt% water–BSA mixture, τ_α (*large blue closed triangles*) and τ_υ (*large closed blue squares*); for 40 wt% water–BSA mixture, τ_α (*large open green triangles*), and τ_υ not shown. The *large symbols on top* all having relaxation times of 10^3 s are from adiabatic calorimetric measurements of a 20% (w/w) aqueous solution of bovine serum albumin. The *two larger circles connected by the thick line* indicate the range of a slower relaxation found in the quenched sample. The *open square* is the relaxation occurring at $T = 110$ K found for the quenched and annealed samples. The *other circle* (⊗) next to it represents the relaxation time of the secondary relaxation of water confined within 1.1, 1.6, and 1.8 nm nanopores of molecular sieves (silica MCM-41) obtained by adiabatic calorimetry having the value of 10^3 s at $T = 115$. For hydrated lysozyme, τ_υ (*green* ★). The relaxation times from neutron scattering of hydrated lysozyme are now represented by *large red closed circles*. The *thick straight line* is τ_υ of hydrated collagen and elastin (Arrhenius fit to data only). Water-specific relaxation times τ_υ of hydrated myoglobin from neutron scattering (*medium black closed circles*, look like open circles because of overlaps with other data), Mössbauer spectroscopy (*single large inverted black triangle*), dielectric relaxation (*large closed red triangles*), and NMR (*large* ×). Neutron scattering data of hydration water on C-phycocyanin (C-PC) (*large closed squares*). For τ_υ shorter than 10^{-6} s, the new additions overlap the previous ones and cannot be resolved, indicating that τ_υ are approximately the same for different proteins and hydrophilic solutes. Thus, the water JG relaxation is responsible for the dynamic transition observed by neutron scattering and Mössbauer spectroscopy. The *horizontal dashed line* indicates $\tau = 10^{-9}$ s, roughly the upper bound of the time window of neutron scattering instrument having energy resolution of 1 μeV. All *smaller symbols* and *thinner lines* are for τ_{JG} or τ_υ of aqueous mixtures and confined water. (1) 50 wt% water mixture with 3EG (*faint blue open diamonds* for τ_α and *closed diamond* for τ_υ), 50 wt% water mixture with PVME (*red closed squares*), 50 wt% water with PVP (*blue open squares*), 38.6 wt% of water with PHEMA (*brown open triangles*), 20 wt% water with 1-propanol (*black* •). At times shorter than 10^{-10} s are τ_υ of 90 wt% water with poly(vinyl alcohol) (*pale blue squares* with × inside), 80 wt% water with PVP at short times (*blue open squares*), and bulk water at short times (*brown open circles*). The *long thinner dashed line* is the Arrhenius fit to τ_{JG} or τ_υ of 20% of water nanoconfined in graphite oxide from dielectric relaxation, actual data not shown, and *green open circles* with $\tau_\upsilon < 10^{-9}$ s are from neutron scattering. Dielectric data of relaxation time of water confined in MCM-41 with pore diameter 2.14 nm at hydration levels of 22 wt% (*red inverted triangles*)

has with the α-relaxation of the hydrated protein is made clear by the various experimental facts presented in this section. However, at this time, the identity of the JG β-relaxation of hydration water and the role it plays in the dynamics of the hydrated biomolecule are not recognized or appreciated by the research community. An example is given here to illustrate the common misconception of the primitive or JG β-relaxation of hydration water.

The right panel of Fig. 233-8 is a reproduction of a figure in [1354(c)] by Doster. It is the Arrhenius plot of the α-relaxation rate $k_S \equiv 1/\tau_\alpha$ deduced from viscosity measurements and specific heat spectroscopy of various solvents: 80% sucrose–water (diamonds magenta), 90% glycerol–water (blue triangles), 75% glycerol–water (green squares), 60% ethylene glycol–water (red circles), and hydration water of myoglobin (HW). For the present purpose in this subsection, we consider only the $k_S \equiv 1/\tau_\alpha$ data of hydration water of myoglobin (HW) in this figure. The black closed circles are k_S determined by Doster with neutron scattering, open triangles are from dielectric relaxation by Swenson et al. of hydrated myoglobin with 0.8 g water per g protein [1105], the lone inverted closed triangle is from Mössbauer spectroscopy [1354(b), 1318], the two open diamonds at the bottom are taken from calorimetry of hydrated myoglobin with $T_g = 170$ K [1354(c)]. Doster also referenced calorimetry data of myoglobin crystal [1351(b)] and hydrated bovine serum albumin (BSA) [1281(b)] in his figure caption, but I do not understand the connection of the figure to these data. The lone right pointing triangle of HW was not described by Doster [1354(c)]. The fits of these data of HW to a VFTH equation are also shown (full lines). Doster considered all of these data as the reciprocal of the α-relaxation time τ_α of the myoglobin hydration water. The VFTH fit exhibits prominent curvature, and this has led Doster [1354(c)] to proclaim that "Water is a highly fragile liquid." This statement is incorrect because Doster mixed up the cooperative α-relaxation of the hydrated myoglobin responsible for glass transition with the JG β-relaxation of the hydration water (or the water-specific secondary relaxation). This point becomes clear when the complete set of dielectric relaxation data are included as shown in the left panel of Fig. 233-8.

The same symbols are used in the left panel of Fig. 233-8 for all the relaxation time data of HW presented by Doster in the right panel. All of them are black in color and large in size to distinguish them from added data from other sources. Most cogent are the large red closed triangles, which are the water-specific secondary relaxation times of hydrated myoglobin from dielectric measurements by Swenson et al. [1105]. The red closed triangles should be exactly the same as the large black open triangles (corresponding to those shown by Doster in the right panel) at the same temperature. The slight differences between the two are due to digitizing the same data from the figure of Doster [1354(c)] and from the publication of the original data by Swenson et al. [1105]. It is clear from the left panel of Fig. 233-8 that Doster took only the higher temperature part of the complete set of the dielectric relaxation time of the hydration water relaxation (large black open triangles) and considered them to be τ_α of the myoglobin hydration water. The data continuing with Arrhenius T-dependence down to lower temperatures, which has the signature of secondary relaxation, are not considered. If there is a reason for Doster to do

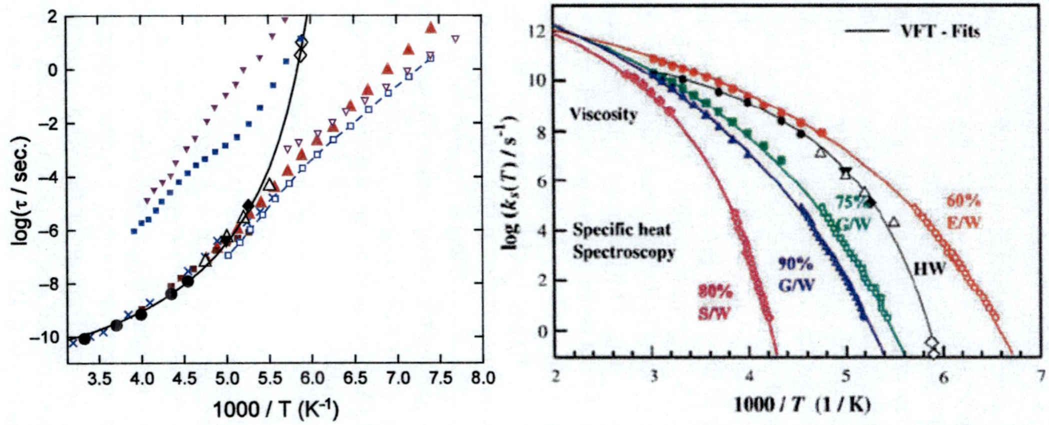

Fig. 233-8 (*Right*) This is a reproduction of Fig. 3 of [1354(c)] by W. Doster. It is the Arrhenius plot of the α-relaxation rate $k_S \equiv 1/\tau_\alpha$ from viscosity and specific heat spectroscopy of various solvents: 80% sucrose–water (*diamonds magenta*), 90% glycerol–water (*blue triangles*), 75% glycerol–water (*green squares*), 60% ethylene glycol–water (*red circles*), and hydration water of myoglobin (HW). For the hydrated myoglobin, the *black closed circles* are from neutron scattering, *open triangles* are from dielectric relaxation, the *lone inverted closed triangle* is from Mössbauer spectroscopy, the *two open diamonds* at the *bottom* are from calorimetry. The *lone right pointing triangle* of HW was not described by Doster [1354(c)].The fits to a VFTH equation are also shown (*full lines*). For the source of the data and the parameters of the VFTH fits, see [1354(a), (c)]. (*Left*) The same data of HW together with the VFTH fit in the *left panel* are reproduced in Arrhenius plot of $\tau_\alpha \equiv 1/k_S$ shown in the main figure. Same symbols all *black* in color and with *large size* are used for ease in distinguishing them from the other data. The latter are described as follows. *Large size red closed triangles* are the full set of dielectric data obtained by Swenson et al. [1105], only part of which were used by Doster (*large open black triangles*). The *smaller size red open triangles* are water JG relaxation or water-specific secondary relaxation from myoglobin at hydration level of $h = 0.33$ [1112]. The *inverted open* and *closed triangles* are, respectively, the water JG relaxation time τ_{JG} and the presumably τ_α of 0.8 g of water/g of myoglobin [1109]. The NMR relaxation times of hydrated myoglobin are represented by the multiplication signs [1322(c)]. The *closed squares* are from neutron scattering from hydration water on C-phycocyanin (C-PC) [1395(a)]

this, it is perhaps he took the suggestion of Swenson and coworkers [1112] that the observed dielectric relaxation is a β-relaxation below T_g, and a merged α- and β-relaxation of water above T_g. This suggestion of Swenson and coworkers has been discussed before and shown to be untenable. To reinforce this in the present case, the relaxation times of two relaxation processes observed by Swenson and coworkers in 0.8 g of water/g of myoglobin [1109] are shown in the left panel of Fig. 233-8 by the inverted purple open and inverted closed triangles. The hydration level of the myoglobin in this publication [1109] is the same as the sample in [1105], and naturally the relaxation times of the faster relaxation (open inverted triangles) are practically the same as the secondary relaxation (large closed red triangles) from [1105]. The relaxation time of the slower process (inverted closed triangles) reaches 100 s at about 175 K consistent with the calorimetry data (open diamonds), and hence it is the cooperative α-relaxation of the hydrated myoglobin. In the latest publication of dielectric study of hydrated myoglobin by Swenson and coworkers, they found again the slow and the fast processes at hydration level of $h = 0.33$ [1112]. The relaxation times of the two processes are shown by open and closed blue squares in the left panel of Fig. 233-8. The relaxation time of the slower process (closed blue squares) reaches 100 s at about 170 K, and thus is the cooperative α-relaxation of the hydrated myoglobin. Having identified the slow process to be the cooperative α-relaxation leading to glass transition at $T_g \sim 170$ K, consistent with the calorimetry data (large open diamonds), the association of the selected faster water-specific relaxation time data (large open triangles) with the calorimetry data by Doster is questionable. The conclusion by Doster that "Water is a highly fragile liquid" is mistaken because it is obtained by mixing up the JG β-relaxation of hydration water with the cooperative α-relaxation of the hydrated myoglobin. The simultaneous presence of the water-specific relaxation responsible for the dynamic transitions and a slower relaxation responsible for the calorimetric glass transition is not unique to hydrated myoglobin. The hydrated bovine serum albumin (BSA) discussed before together with the data presented in Figs. 230–232 is such an example.

In the left panel of Fig. 233-8, the data represented by the multiplication signs are the relaxation times of the water-specific process in hydrated myoglobin obtained from NMR spectroscopy by Lusceac et al. [1322(c)]. They are in good agreement with the neutron scattering and dielectric relaxation data, and indicate they are the primitive or JG β-relaxation of the hydration water. Justification has been given before that the intermolecular coupling between water molecules themselves is small and practically zero at short relaxation times probed by neutron scattering and by the NMR spectroscopy. Hence the α-relaxation or the JG β-relaxation of the hydration water is close to each other, and NMR spectroscopy cannot tell them apart. Also, JG β-relaxation of water molecule itself can translate once the hydrogen bonds are broken to make its rotational motion isotropic like the characteristic of α-relaxation. Therefore, for the relaxation observed at short times above T_g by NMR, it makes hardly any difference by calling it the α-relaxation instead of the β-relaxation of the hydration water. We see no value in insisting that the relaxation has to be interpreted exclusively as the α-process of hydration water above T_g by Lusceac et al.

Added to the left panel are relaxation times of hydration water on C-phycocyanin (C-PC) determined by neutron scattering [1395(a)]. Represented by closed brown squares, they are in agreement with the relaxation times of hydration water on myoglobin. This is another example of the insensitivity of the primitive or JG β-relaxation time of the hydration water to the biomolecule at high frequencies.

Glycerol–Water and Sucrose–Water as Solvents

The dynamic transition also was observed in myoglobin covered by solvents that are aqueous mixtures instead of pure water [1354(a)–(c)]. This has been shown before in the left panel (a) of Fig. 233-5 by the mean-squared displacements of the heme iron of CO-myoglobin (MbCO) measured with Mössbauer spectroscopy in 75% glycerol–water (75% G/W) and 80% sucrose–water (80% S/W) solution. The experimental timescale t_{exp} of Mössbauer spectroscopy is 140 ns. The viscosity η of the solvent at the onset temperatures T_d indicated in the figure shows that it has the very high value of 6.4×10^{11} poise and corresponding very long τ_α of order of 1 s for 80% S/W. This is clearly telling us that the dynamic transition at 140 ns timescale has nothing to do with the α-relaxation or viscosity of the solvent in this case. On the other hand, Doster tried to overcome this difficulty of relating the dynamic transition to viscosity of the solvent by making the assumption: "These results suggest instead, that the viscosity near the protein surface differs from its bulk value."

Before we proceed to discuss the dynamic transition of myoglobin with 75% G/W and 80% S/W as solvents, I hasten to remind the reader that pure water as the solvent is a special case. As discussed in the above, there is hardly any difference between the α-relaxation and the β-relaxation of the hydration water at the short timescale of neutron scattering. Hence, for H_2O- or D_2O-hydrated proteins, it is legitimate for Doster [1354(c)] to say that "the onset of the dynamical transition depends on the solvent viscosity near the protein surface." However, this statement does not apply generally for all solvents, an example is 80% S/W. According to us, the generally valid statement is "the onset of the dynamical transition depends on the primitive or JG β-relaxation of the solvent near the protein surface." To support our statement, τ_α and the β-relaxation time τ_{JG} of the solvent are considered in relation to the dynamics transition of myoglobin found by Mössbauer spectroscopy when the solvent is either 75% G/W or 80% S/W. The relation can be gleaned from the Arrhenius plot of τ_α and τ_{JG} of the two solvents in the left panel of Fig. 233-9. The τ_α of 80% S/W was taken from the VFTH equation given in [1354(a)]. The β-relaxation time τ_{JG} of 80% S/W is not available, and instead τ_{JG} of 70% S/W is shown by the thick dotted line obtained by dielectric relaxation spectroscopy [1103]. The actual data of τ_{JG} of 70% S/W having this Arrhenius T-dependence were determined experimentally no shorter than about $10^{-6.5}$ s. The horizontal dashed line locates t_{exp} of Mössbauer spectroscopy at 140 ns. The dynamic crossover temperature $T_d \approx 240$ K of myoglobin covered with 80% S/W [1354(b)] is indicated by the vertically pointing arrow on the left. It can be seen from the figure that at T_d, τ_α is about 10^{-2} s which is five orders of magnitude longer than t_{exp}, and hence cannot

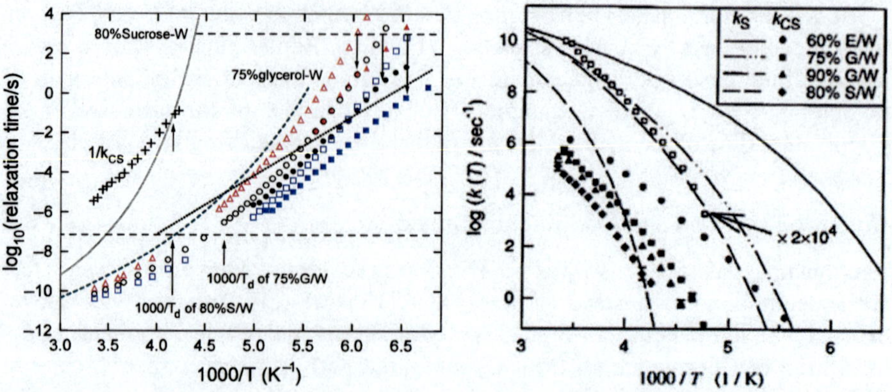

Fig. 233-9 (*Left*) Temperature dependence of the dielectric relaxation time τ_α (*open symbols*) and the corresponding τ_{JG} of water (*closed symbols* of the same kind) of three mixtures of various wt% of water with glycerol. *Triangles* for 20 wt% water. *Circles* for 30 wt% water. *Squares* for 40 wt% water. Each *down pointing vertical arrow* is used to indicate the change of T-dependence of τ_{JG} of water at T_g of the mixture with T_g defined by $\tau_\alpha(T_g) = 10^3$ s. Note that the α-relaxation of the mixture and the water JG relaxation merge together at short times and τ_α is practically the same as τ_{JG}. The *dashed* and *solid lines* are the VFTH dependences of τ_α of the mixtures of glycerol with 25 wt% water and sucrose with 20 wt% of water, respectively, from [1354(a)]. These are labeled here as 75% glycerol–water and 80% sucrose–water, and as 75% G/W and 80% S/W in [1354(a), (c)]. The *horizontal dashed line* corresponds to 140 ns, the long time limit of Mössbauer spectroscopy. From *right* to *left*, the *two arrows* pointing at the 140 ns line indicate $1000/T_d$, where T_d is the temperature of the dynamic transition determined from the mean-squared displacements of the heme iron of myoglobin measured with Mössbauer spectroscopy in 75% glycerol–water (T_d = 215 K) and 80% sucrose–water solution (T_d = 240 K) shown before in Fig. 233-5a. The *upper arrow* pointing at the τ_α line of 80% sucrose–water indicates that $\tau_\alpha(T_d)$ is about 0.1 s. The *dotted line* is the Arrhenius T-dependence of τ_{JG} of water in 70% sucrose–water with activation energy of 0.59 eV from [1103]. The *plus signs* are the reciprocals of the entry rates of CO across the surface of myoglobin, $1/k_{SB}$, as function of reciprocal temperature. (*Right*) Arrhenius plot of ligand escape rates k_{CS} (*closed symbols*) and solvent relaxation rate k_S (*lines*) for horse myoglobin in different solvents, 60% ethylene glycol/water, 75% and 90% glycerol/water, and 80% sucrose/water. For a comparison of the temperature dependence, additionally shown is the ligand escape rate of 75% glycerol/water scaled by a factor of 2×10^4 (*open squares*) to the corresponding solvent relaxation rate. Reproduced from [1354(a)] by permission

have anything to do with the dynamics transition. On the other hand, τ_{JG} of 70% S/W comes close to t_{exp} at a temperature close to T_d. This indicates that indeed the primitive or JG β-relaxation of 80% S/W is responsible for the dynamics transition observed by Mössbauer spectroscopy.

Shown also in Fig. 233-9 are τ_α and τ_{JG} of 80, 70, and 60% of glycerol–water mixtures presented before in Fig. 211. The α- and β-relaxations of these aqueous mixtures have merged together at t_{exp} = 140 ns of Mössbauer spectroscopy, and $\tau_\alpha \approx \tau_{JG}$. These two, effectively the same, relaxation times reach t_{exp} = 140 ns at a temperature close to the dynamic crossover temperature T_d = 215 K [1354(b)] of myoglobin covered by 75% G/W (indicated by the vertically pointing arrow). Like the case of proteins hydrated by pure water, in the present case with 75% G/W as the solvent, its α- and β-relaxations are practically the same. Thus, one can say either the α-relaxation or the β-relaxation of the 75% G/W solvent is responsible for the dynamic transition. Lichtenegger et al. [1354(b)] gave viscoelastic relaxation times of 75% G/W in the temperature interval between 220 and 230 K to decrease from 2 ms to 200 ns in the range of the ^{57}Fe-nuclear lifetime. Their viscoelastic relaxation times are longer than the dielectric relaxation times. The latter are more consistent with $T_d \sim 215$ K for myoglobin in 75% glycerol/water given by Lichtenegger et al. themselves [1354(b)]. See also Fig. 233-5a.

If the solvent is 80% S/W, then it is definitely not the α-relaxation but the β-relaxation of the 80% S/W solvent which is responsible for the dynamic transition. This is evident from the viscosity of 80% S/W being of the order of 10^{12} poise, many orders of magnitude higher than 75% G/W at the respective T_d of myoglobin mixed with these two solvents (see panel (a) of Fig. 233-5). The viscosity of the solvent coupled to the protein is not less than the neat solvent.

The statement by Doster [1354(c)], "... dynamical transition depends on the solvent viscosity near the protein surface," is valid for solvents including pure water, 75% G/W, 90% G/W, and 60% ethylene glycol–water (60% E/W), where the times of the α-relaxation and the JG β-relaxation are practically the same at the timescale of neutron scattering experiments, i.e., $\tau_\alpha(T \geq T_d) \approx \tau_{JG}(T \geq T_d)$. However, this statement does not hold straightforwardly if the solvent is 80% S/W, which has $\tau_\alpha(T_d) >> \tau_{JG}(T_d)$, unless the assumption is made by Lichtenegger et al. [1354(b)] that the microviscosity of the solvent at protein–solvent interface is much lower than in the bulk. A possible way to justify this assumption is partial demixing of water and sucrose in the protein solvation shell [1354(b)], but that has to be so severe for the relaxation times to be reduced by many orders of magnitude. Also this assumption may have contradicted the observation of the dynamics of the 80% sucrose/water solvent by Mössbauer spectroscopy by adding $K_4{}^{57}Fe(CN)_6$ (potassium ferrocyanide, PFC) to the solvent. The sample has a concentration of 3.3% PFC and 96.7% sucrose/water solution (concentration of sucrose/water, 78.7:21.3, w/w). They found the temperature dependence of the Mössbauer parameters derived for this sample is remarkably similar to those of myoglobin with 80% S/W as solvents. The mean-squared displacements of the ^{57}Fe nucleus of the two samples as a function of temperature show dynamic transition at practically the same temperature in the range of 230–240 K. The data of the two samples are reproduced here in Fig. 233-10, and comparison also is made with that obtained with deoxymyoglobin crystals by Parak et al. [1343]. Unless by happenstance there is the same partial demixing of water and sucrose in the PFC sample, the same dynamic transition temperature found in the two samples casts doubt on the explanation by Lichtenegger et al.

Validity of the Doster's statement can be restored if it is modified as follows: "The dynamic transition depends on the relaxation time of the primitive or JG β-relaxation of the solvent near the protein surface." The view from this modified statement is not congruent either with the suggestion of Frauenfelder and coworkers that the viscosity-independent β-processes in the hydration shell controlling the onset of the

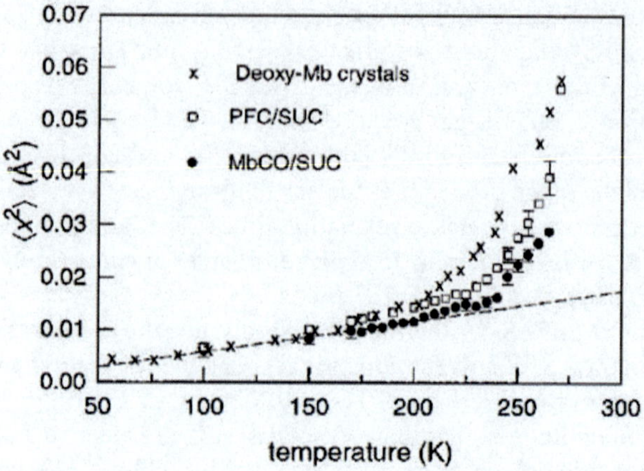

Fig. 233-10 Mean-squared displacements, $< x^2 >$, of ^{57}Fe vs. temperature: potassium ferrocyanide in 80% sucrose/water (□), MbCO in 80% sucrose/water (•), and deoxymyoglobin crystals (×) for comparison. The *dashed line* represents a linear extrapolation of $< x^2 >$ to $T = 0$ K. Reproduced from [1354(b)] by permission

dynamics transition [1355, 1356]. The difference can be bought out by comparing in detail the nature of the viscosity-independent β-processes in the hydration shell of these authors [1355, 1356] with our JG β-relaxation of our explanation.

Ligand Escape Rate

The relation between the structural α-relaxation rate $k_S \equiv 1/\tau_\alpha$ of the solvent and the ligand escape rate, k_{CS}, in myoglobin with 80% S/W as the solvent was found to be different from that common to samples when the solvent is either 75% G/W, 90% G/W, or 60% E/W. This difference was found by Kleinert et al. [1349(a)], reiterated by Doster [1354(c)], and reproduced here in the right panel of Fig. 233-8. The solvent relaxation rate k_S was determined by specific heat spectroscopy data as shown in the right panel of Fig. 233-8. The ligand escape rates k_{CS} in myoglobin with 75% G/W, 90% G/W, or 60% E/W as solvent have about the same temperature dependence as k_S, explicitly demonstrated in the figure for 75% G/W by multiplying k_{CS} by a factor of 2×10^4 to match the corresponding k_S. However, k_{CS} in myoblobin with 80% S/W as solvent has a much weaker T-dependence than k_S of the solvent, which can be seen in both the right and the left panels of Fig. 233-9. Moreover, the comparison of the T-dependence of $1/k_{CS}$ in myoblobin with 80% S/W as solvent with other relaxation times suggests it is comparable to that of $\tau_{JG}(T)$ of the same solvent. Although $\tau_{JG}(T)$ of 80% S/W has not been determined in the same temperature range of $1/k_{CS}$, its T-dependence is expected to be stronger than the Arrhenius T-dependence below T_g as shown for the 70% S/W, and hence similar to that of $1/k_{CS}$. This suggests that the ligand escape rate, k_{CS}, in myoblobin with 80% S/W as solvent is controlled by the JG β-relaxation of the solvent, and not its viscosity. The same conclusion applies to k_{CS} in myoglobin with 75% G/W, 90% G/W, or 60% E/W as solvent because $\tau_\alpha(T) \approx \tau_{JG}(T)$ in the temperature range of the k_{CS} data presented in the figure. Hence, the result showing that $k_{CS}(T)$ and $k_S(T)$ have the same T-dependence in the figure can be restated as $k_{CS}(T)$ and $k_{JG}(T)$ have the same T-dependence, where $k_{JG}(T) \equiv 1/\tau_{JG}(T)$.

There is yet another difference between the ligand CO escape rate k_{CS} of MbCO with 80% S/W as solvent than the others. k_{CS} are generally much smaller than the structural relaxation rate k_S of the solvents. The 80% S/W sample is an exception. Its k_{CS} approaches k_S and even exceeds it at 240 K. As Doster stated [1354(c)], this result makes sense only if he assumes that the viscosity near the protein surface is lower than in the bulk. However, this assumption need not be made if one accepts that k_{CS} is controlled by the primitive or JG β-relaxation of the solvent in general. It is worthwhile to mention here that there are experimental evidences showing that diffusion of gas molecules such as methane and CO_2 in polystyrene with activation energy of 49 kJ/mol or less [1354(d)] in both the liquid and glassy states is controlled by the secondary relaxation. This is also supported by diffusion of methane in poly(ethylene terephthalate) with activation energy of 12.22 kcal/mol [1354(e)], and diffusion of helium in BPA-polycarbonate having activation energy of 3.9 kcal/mol from experiment and 3.7 kcal/mol from simulation [1354(f)].

Applicability of the Mode Coupling Theory?

Mode coupling theory (MCT) was applied by Doster and coworkers as early as 1990 [1344] to describe the dynamics of hydrated proteins. Doster maintained in recent papers [1354(c), 1395(b)] that the theory should be applicable to density fluctuations of the protein hydration shell. To support this assertion, he showed the intermediate scattering function $I(q,t)$ of myoglobin hydration water derived by Fourier transforming H_2O/D_2O difference spectra (IN6, ILL) to the time domain. His results are reproduced here in Fig. 233-11.

The two-steps decay of $I(q,t)$ found are considered as the β-process and α-relaxation of MCT, and hence as evidence of the applicability of MCT. According to Doster [1354(c)], on decreasing temperature, the α-process slows down, and the timescale of the fast MCT β-process is nearly independent of the temperature, only the amplitude decreases at low temperatures. The timescale of the MCT β-process is shorter than ~2 ps. This, together with the insensitivity of its timescale to temperature, indicates that the β-process is due to vibrational density of states or to hydrogen bond fluctuations accountable by an asymmetric two-state model [1395(b)]. An analogue of it and identification with vibrational density of states can be found in the neutron scattering study of glycerol [1396]. Thus, the data of $I(q,t)$ from IN6 alone do not tell much about the dynamics of hydration water in the temperature range of $320 \geq T \geq 180$ K, even though it is consistent with MCT in the two-steps decay and the excess dynamic structure factor was in agreement with MCT prediction. The susceptibility $\chi''(\omega)$ obtained from measurements using IN13 over energy range, corresponding to times within $0.1 < t < 30$ ps, exhibits a minimum and its frequency dependence was well fitted by the MCT equation [1344]. However, it is

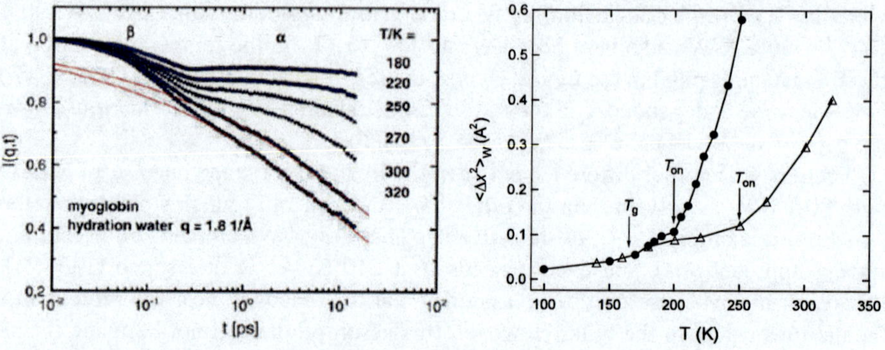

Fig. 233-11 (*Left*) Intermediate scattering function $I(q,t)$ of myoglobin hydration water derived by Fourier transforming H_2O/D_2O difference spectra (IN6, ILL) to the time domain. The time range of the α- and β-processes of MCT are indicated in the figure by Doster. With decreasing temperature the α-process slows down. The timescale of the fast β-process is nearly independent of the temperature, only the amplitude decreases at low temperatures, according to Doster [1354(c), 1395(b)]. Reproduced from [1354(c)] by permission. (*Right*) Proton mean-squared displacements of hydration water adsorbed to C-PC obtained by neutron scattering taken with spectrometer having resolution of 2 ns (*full circles*) and 15 ps (*open triangles*, myoglobin). The *arrows* indicate the glass temperature T_g and the two resolution-dependent onset temperatures T_{on} of the dynamic transition. Redrawn from a figure in [1395(b)]

noteworthy that caged dynamics in terms of the susceptibility minimum as described by MCT for glassformers in general is restricted to temperatures higher than T_c, which is higher than T_g. Starting at some temperature near or below T_c, the minimum becomes so flat or changes to a power law, $\chi''(\omega) \propto \omega^{-c}$ with $c << 1$, which cannot be described by either the idealized MCT prediction considered in [1344], or the other versions of MCT [33–35]. More discussions of the limited applicability of various versions of MCT will be given in Sections 2.3.2.33–2.3.2.36, and 2.3.2.37. Actually, we have come across an example from hydrated maltose binding protein (MBP) and hydration water, H-MBP-D_2O and D-MBP-H_2O by Wood et al. [1363]. Wood et al. carried out molecular dynamics simulations on the same system and reproduced the principal result observed by experiment. Here we revisit the evolution of the MSD of the O atoms of the water molecules plotted in Fig. 228b for each temperature from 150 to 300 K. Well below the protein dynamical transition temperature T_d(~ 240 K) in the MD simulations, following the initial, subpicosecond rise, corresponding to ballistic motion, the MSD are essentially flat, and the corresponding susceptibility $\chi''(\omega)$ will have flat minimum, which is better described as nearly constant loss. The condition persists from 150 K up to at least 220 K, indicating none of these data can be fitted by the idealized MCT. Interestingly, the $\chi''(\omega)$ data of hydrated myoglobin from IN13 that exhibit the minimum were taken at $T \geq 250$ K, which is above $T_d \sim 240$ K (see Fig. 233-5b). Data of $\chi''(\omega)$ at 220 K were limited to times less than 3 ps. If the hydrated myoglobin data were like the simulations of Wood et al. then $\chi''(\omega)$ will exhibit a flat minimum that cannot be described by idealized MCT. Thus, MCT fails to describe the fast dynamics at temperatures below T_d, for sure in hydrated MBP and maybe in other hydrated proteins, and questionable is the proposal that it can explain the dynamics transition, and the long time dynamics of hydrated proteins at lower temperatures down to T_g.

The right panel of Fig. 233-11 is a plot of the proton mean-squared displacements (MSD) of hydration water adsorbed to protein C-phycocyanin (C-PC) obtained by neutron scattering taken with spectrometer having resolution of 2 ns (full circles) [1395(a), (b)]. Also shown for comparison is the MSD of hydrated myoglobin taken by IN6 with resolution of 15 ps (open triangles). The arrows indicate the glass temperature T_g and the two resolution-dependent onset temperatures T_{on} of the dynamic transition. For hydrated C-PC, below $T_g \sim 170$ K the displacements are dominated by harmonic vibrational motion with linear increase of the MSD with temperature. After crossing $T_g \sim 170$ K, the MSD shows a stronger increase with temperature, and then even stronger increase at the dynamics transition with onset at $T_{on} \approx 220$. The same changes of T-dependence of the MSD at T_g and T_{on} are found in hydrated myoglobin except $T_{on} \approx 250$ K because of the 15 ps timescale of the IN6 (see Fig. 233-3 for MSD of other hydrated proteins obtained by IN6 or spectrometer with comparable resolution that have $T_{on} \approx 250$ K). The changes in the T-dependence of the MSD on crossing T_g found in these two hydrated proteins by Doster, as well as in hydration water of hydrated maltose binding protein (MBP) by Wood et al. (see Fig. 233-6), are just examples of a general phenomenon shown by all glassformers [114, 115(a), 507–510, 1062]. This phenomenon will be discussed in Sections 2.3.2.33–2.3.2.36, and an explanation given before will be reiterated over there. To the best of my knowledge, MCT has no explanation for this phenomenon, at least so far.

Returning to the intermediate scattering function data in the left panel of Fig. 233-11, actually the relaxation time estimated from $I(q,t)$ at 320 K is of the order of 10 ps, and at the range of 270–250 K is of the order of 100 ps. These estimates are consistent with the primitive or JG β-relaxation time of hydration water (see Fig. 233-2). No distinction between secondary and primary relaxation of water can be made at such short times and high temperatures because the two processes have merged together. Also, water molecules have weak intermolecular coupling and the separation between α- and β-relaxations is not large even at lower temperatures as discussed before.

Dynamic Transition due to ν-Relaxation: Evidence from $<x^2>^{\gamma}$ of the Heme Iron

Parak and coworkers in 2007 [132] considered the new modes of motions causing the dynamic transition observed at T_d in hydrated myoglobin crystal (Mbmet) by Mössbauer spectroscopy to be associated with what they called the flexible state of the myoglobin. They analyzed the mean-squared displacements, $<x^2>^{\gamma}$, of the heme iron, which is the sum of vibration part, $<x^2_{vib}>^{\gamma}$, and the contribution from the flex state $<x^2_{flex}>^{\gamma}$. Below T_d the Mössbauer spectra can be fitted by Lorentzians with common line widths, and can be explained by $<x^2_{vib}>^{\gamma}$ alone. With rising temperatures, new modes of motions of the flex state contribute and become measurable above T_d giving rise to $<x^2_{flex}>^{\gamma}$ and the broad lines observed in the Mössbauer spectra. The motions in the flexible state were described by a Brownian oscillator using the Langevin equation plus a restoring force, together with the Gibbs free energy needed to arrive at the free state given by $\Delta G_{\text{flex}} = \Delta H_{\text{flex}} - T\Delta S_{\text{flex}}$. Using this model, Parak et al. fitted the $<x^2>^{\gamma}$ data in the temperature range from 80 to 340 K, yielding $\Delta H_{\text{flex}} = 35$ kJ/mol and $\Delta S_{\text{flex}} = 121.3$ J/K/mol. These parameters suggest the relaxation rate of the process corresponding to the motions in the flex state is given by $\tau = \tau_{\text{at}} \exp[(\Delta H_{\text{flex}} - T\Delta S_{\text{flex}})/RT]$. Its activation energy of $\Delta H_{\text{flex}} = 35$ kJ/mol is remarkably close to activation energy of the relaxation time, τ_ν, of the hydration water in the range of 40–50 kJ/mol. The entropy factor, $\exp\left[-\frac{\Delta S_{\text{flex}}}{R}\right]$, evaluates to $10^{-6.3}$. Thus, from the attempt frequency, $(\tau_{\text{at}})^{-1} \sim 10^{-13}$ or 10^{-14} s, the result of Parak et al. translates to the pre-exponential factor, $\tau_{\infty,\text{f}} \equiv \tau_{\text{at}} \exp\left[-\frac{\Delta S_{\text{flex}}}{R}\right]$, having order of magnitude of 10^{-19} or 10^{-20} s. Remarkably, this is about the same as the preexponential factor of τ_ν of the hydration water. In fact the average of the relaxation times τ_ν of the hydrated proteins shown in Fig. 233-2 when fitted by Arrhenius T-dependence, $\tau_\nu = \tau_\infty \exp\left[H/RT\right]$, has $H \sim 48$ kJ/mol and $\tau_\infty \sim 10^{-19.2}$ s. The approximate agreement of H and τ_∞ from τ_ν with ΔH_{flex} and $\tau_{\infty,\text{f}}$ from the analysis by Parak et al. is likely not an accident. This is because in myoglobin crystal (Mbmet), the heme iron is coordinated by a water molecule. Also, as found by replacement of H_2O by D_2O in experiments [1363], the same dynamic transition was observed from motion of water or from the protein, indicating the motion of the hydration water is coupled to the protein. Thus, the results from Parak et al. provide another support that the primitive or ν-relaxation of hydration water coupled to the flexible state is causing the dynamic transition.

2.3.2.29 TV^{γ}-Dependence of τ_{JG}

If intermolecular potential $V(r)$ for liquids is a repulsive inverse power potential (IPP), $U(r) = \varepsilon(\sigma/r)^q$ where r is the intermolecular distance, q is a constant, and ε and σ have respective dimensions of energy and length, it was shown by Hoover et al. [1397] that the canonical partition function and hence also all thermodynamic properties depend on a single density-temperature variable, $\rho(\varepsilon/kT)^{1/\gamma}$ with $\gamma = q/3$, rather on T and density ρ (or volume V) separately. If distance is scaled by $\rho^{-1/3}$, and time scaled by the characteristic time $\rho^{-1/3}(m/kT)^{1/2}$, Hoover and Ross pointed out that the scaled equations of motion of the particles are simplified to show that the forces depend only on the density–temperature product variable $\rho(\varepsilon/kT)^{1/\gamma}$ with $\gamma = q/3$. Hence, for a fixed value of $\rho(\varepsilon/kT)^{1/\gamma}$ with $\gamma = q/3$, the dynamic variables of the system when expressed in terms of the scale distance and time at any density or temperature are the same. In other words, all dynamical quantities can be cast in forms that depend on the single combined variable ρ^{γ}/T or alternatively $T^{-1}V^{-\gamma}$ with $\gamma = q/3$ [1401]. Experimentally, the dependence on ρ^{γ}/T was demonstrated by the dynamic structure factor of OTP from quasielastic neutron scattering (QENS) being invariant when measured over several combinations of T and P, subject to the condition that the quantity TV^{γ} with $\gamma = 4$ remains constant by Tölle and coworkers [1402, 1403]. Dynamic light-scattering α-relaxation times for OTP, measured at various T and P, were shown to superpose when plotted as a function of ρ^4/T or $1/TV^4$ by Dreyfus et al. [1404, 1405]. The results of these pioneering experimental works are shown together in Fig. 234a.

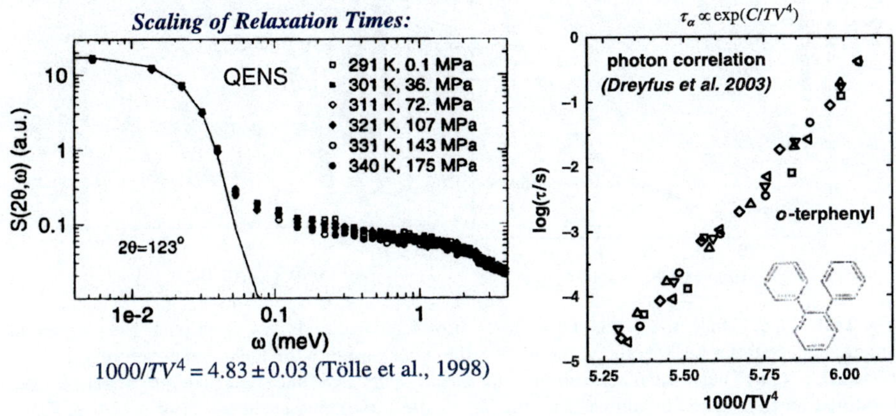

Fig. 234a $1/TV^4$ scaling of QENS dynamics structure factor and α-relaxation time from photon correlation spectroscopy demonstrated by Tölle and coworkers [1402, 1403] and Dreyfus et al. [1404, 1405], respectively, for *o*-terphenyl. Reproduced from [1402, 1404] by permission

Recent models of energy landscapes, used to describe glass transition, are also based on this inverse power law potential but with the addition of a density-dependent constant to account for the long-range attractive forces [1399–1401].

Superposition of viscosity data for OTP with the same exponent, $\gamma = 4$, has also been demonstrated [1406]. Suggested by these results for OTP indicating that for local properties the repulsive forces are paramount and the power q of the repulsive inverse power law depends on the glassformer, Casalini and Roland investigated possible scaling of other glassformers mainly by broadband dielectric spectroscopy [1407–1409]. Treating the exponent γ as a material-specific constant, α-relaxation times τ_{α} from different combinations of T and P for various molecular and polymeric glassformers (see Fig. 234b) can be expressed as a unique function of ρ^{γ}/T, with the exponent varying in the range $0.13 \le \gamma \le 8.5$ for the glassformers investigated to date [119]. Here we show by experimental results that τ_{JG} also depends on the variable $T^{-1}V^{-\gamma}$ with the same γ as that of τ_{α}.

Let us first consider the small molecular glassformers that have narrow α-loss peak and an excess wing on the high-frequency flank (but otherwise no other resolved secondary relaxation in their dielectric spectra), and recall the experimental evidences indicating that the excess wing is an unresolved JG β-relaxation. The glassformers, salol, PC, PDE, and BMMPC, all have τ_{α} as functions of $T^{-1}V^{-\gamma}$

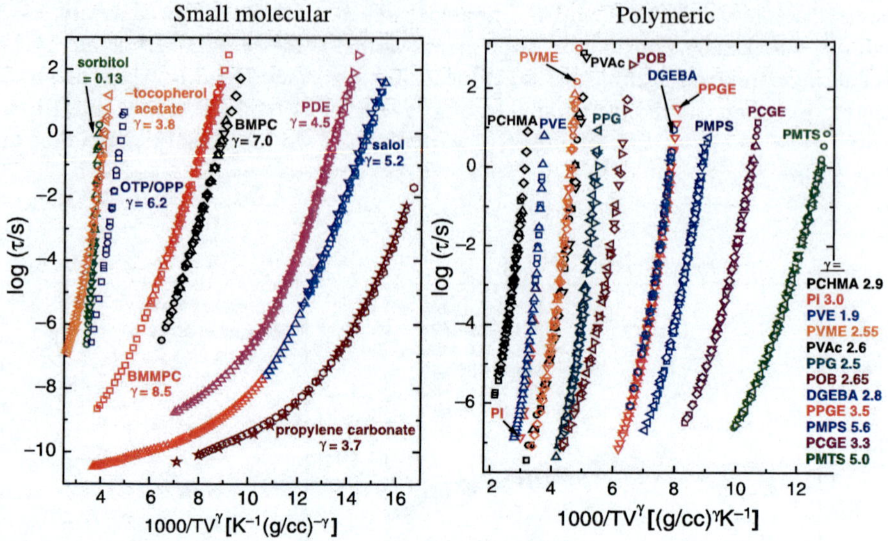

Fig. 234b (*Left*) Dielectric α-relaxation times of molecular liquids as a function of the reciprocal of temperature times the specific volume, with the latter raised to the indicated power of γ [1407, 1408(a)]. (*Right*) Dielectric α-relaxation times of polymers as a function of the reciprocal of temperature times the specific volume, with the latter raised to the indicated power of γ [1409]. Data from [1408(a), 1409] are collected together and replotted in the two figures here

with different γ, as shown in Fig. 234b. We have shown before in Section 2.2.4 in the liquid state of these glassformers that the entire dispersion, the α-loss peak plus the excess wing or the unresolved JG β-relaxation, is invariant to various combinations of temperature and pressure at constant $\tau_\alpha(T^{-1}V^{-\gamma})$. This property immediately implies the invariance of τ_{JG} to changes in T and P while maintaining $\tau_\alpha(T^{-1}V^{-\gamma})$ constant, and it follows that τ_{JG} must also be a function of $T^{-1}V^{-\gamma}$ with the same γ.

Next we consider glassformers that have a resolved JG β-relaxation in the liquid state. Experimental data given in Section 2.3.2.8 have shown that τ_{JG} is invariant to changes in T and P while maintaining $\tau_\alpha(T^{-1}V^{-\gamma})$ constant in benzoin isobutylether (BIBE), dipropyleneglycol dibenzoate (DPGDB), and oligomer of PMMA. Hence, for these glassformers, τ_{JG} must also be a function of $T^{-1}V^{-\gamma}$ with the same γ.

Finally we consider the α- and JG β-relaxation of a component in binary mixtures including picoline [676] and quinaldine [677] dissolved in tri-styrene. On varying the combinations of P and T while keeping τ_α constant, the ratio τ_α/τ_{JG} and the frequency dispersion of the α-relaxation are both invariant (see "Well-Resolved JG Relaxation of a Component in Mixtures of van der Waals Liquids" in Section 2.3.2.8). These two properties were observed at several values of τ_α above T_g. From the invariance of the ratio τ_α/τ_{JG} at constant τ_α and the $T^{-1}V^{-\gamma}$-dependence of τ_α, it follows that τ_{JG} also is a function of $T^{-1}V^{-\gamma}$ albeit its functional form, $f_{JG}(T^{-1}V^{-\gamma})$, is different from $f_\alpha(T^{-1}V^{-\gamma})$ of τ_α. Moreover, $f_{JG}(T^{-1}V^{-\gamma})$ is a weaker function of $T^{-1}V^{-\gamma}$ than $f_\alpha(T^{-1}V^{-\gamma})$ which can be inferred from the well-known experimental fact that the ratio τ_α/τ_{JG} increases with increasing τ_α applicable also to neat glassformers.

The importance of the general results discussed above is that the $T^{-1}V^{-\gamma}$-dependence of the relaxation rate is inherent to the JG β-relaxation, which has transpired long before the commencement of the α-relaxation. It may be inferred from this that the origin of the $T^{-1}V^{-\gamma}$-dependence of molecular mobility may be attributed to the JG β-relaxation. In the view of the coupling model (CM), the stronger $T^{-1}V^{-\gamma}$-dependence of τ_α than τ_{JG} is due to its many-body nature, which magnifies all effects [196]. Here it is relevant to mention the work of Rault [766] in which he considered the VFTH temperature dependence of the cooperative α-relaxation to originate from participation of increasing number of non-cooperative β-relaxations. The conclusion derived solely from experimental data is anticipated by the CM from the relation between τ_α and τ_0, and $\tau_0 \approx \tau_{JG}$, which is combined to become

$$\tau_\alpha(T,P,V,S) = [t_c^{-n}\tau_0(T,P,V,S)]^{1/(1-n)} \approx [t_c^{-n}\tau_{JG}(T,P,V,S)]^{1/(1-n)}. \quad (2.85)$$

Here we write out explicitly all the possible dependences of the relaxation times on the thermodynamic variables P, V, and T and entropy S, which are related by $(\partial S/\partial P)_T = -(\partial V/\partial T)_P$ [742]. From this it follows that if τ_α can be written as a function of $T^{-1}V^{-\gamma}$, then τ_{JG} must also be a function of $T^{-1}V^{-\gamma}$. The relation between τ_α and τ_{JG} becomes

$$\tau_\alpha(T^{-1}V^{-\gamma}) = [t_c^{-n}\tau_0(T^{-1}V^{-\gamma})]^{1/(1-n)} \quad \text{or} \quad [t_c^{-n}\tau_{JG}(T^{-1}V^{-\gamma})]^{1/(1-n)}. \quad (2.86)$$

This is a CM prediction that both τ_α and τ_{JG} (or τ_0) depend on the same variable $T^{-1}V^{-\gamma}$, but their dependences are different and related by the equation above.

Causality Implies that $T^{-1}V^{-\gamma}$-Dependence Originating from the Primitive or JG β-Relaxation

The experimental evidences given above have clearly shown that τ_{JG} (or τ_0) depends on the same variable $T^{-1}V^{-\gamma}$ as τ_α. From this and the fact that the JG β-relaxation or the primitive relaxation has transpired long before the α-relaxation, it follows from the principle of causality that the dependence of τ_α on the variable $T^{-1}V^{-\gamma}$ originates from the same of τ_{JG} (or τ_0).

Molecular Dynamics Simulation: Indication of $T^{-1}V^{-\gamma}$-Dependence Originating from the Primitive or JG β-Relaxation

Before ending this section, we cite the results from other studies that support the $T^{-1}V^{-\gamma}$-dependence of dynamic quantities actually originate from that of the primitive relaxation of the CM or the JG β-relaxation. Earlier it was found that the static structure factor and the local structure in non-associated liquids are sensitive to the repulsive part of the potential at short distance [69, 1410] but not the nature of the attractive potential extending to longer distance. Recently, Coslovich and Roland [1411(a)] simulated binary Lennard-Jones liquids with intermolecular potential $U(r) \propto [(\sigma/r)^m - (\sigma/r)^6]$, with the attractive exponent fixed at the value of 6, and the repulsive exponent m varied over the values of 8, 12, 24, and 36. In agreement with experimental results of various glassformers, the diffusion coefficients for the simulated Lennard-Jones liquids in normal and moderately supercooled states are a unique function of the variable ρ^γ/T as shown in Fig. 235.

The magnitude of the exponent γ is always larger than $m/3$ due to the contributions of the attractive term, but not far from it. For $m = 36$ ($\gamma = 13.4$, and $m/3 = 12$), $m = 24$ ($\gamma = 9.1$, and $m/3 = 8$), $m = 12$ ($\gamma = 5.0$, and $m/3 = 4$), and $m = 8$ ($\gamma = 3.5$, and $m/3 = 2.7$). Instead γ is determined by the steepness of the repulsive part of $U(r)$ in the region bounded by r_c and $r_{1/2}$ which are *closest* approach distance between the particles and separation corresponding to half-height of the radial distribution functions between large particles $g_{11}(r)$, respectively, in the supercooled regime. This is illustrated in Fig. 236-1.

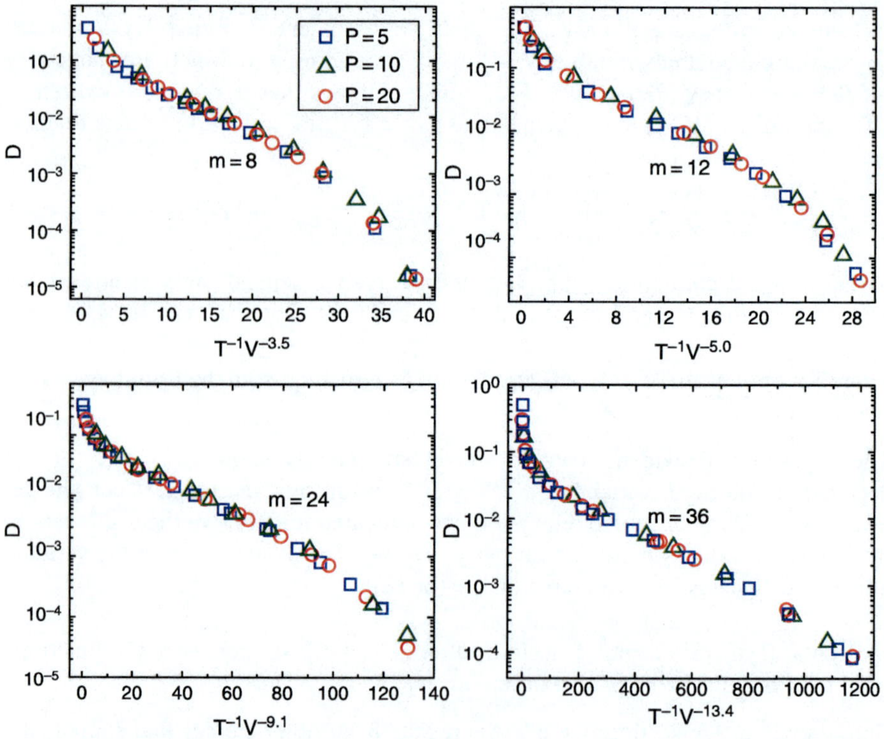

Fig. 235 Reduced diffusion coefficients D as a function of $1/(TV^\gamma)$ for different values of the repulsive exponent m at different pressures: P=5 (*squares*), P=10 (*circles*), and P=20 (*triangles*). For $m = 36$ ($\gamma = 13.4$), $m = 24$ ($\gamma = 9.1$), $m = 12$ ($\gamma = 5.0$), and $m = 8$ ($\gamma = 3.5$). The estimated uncertainty on γ is ±0.1 (±0.2 for $m = 36$). Reproduced from [1411(a)] by permission

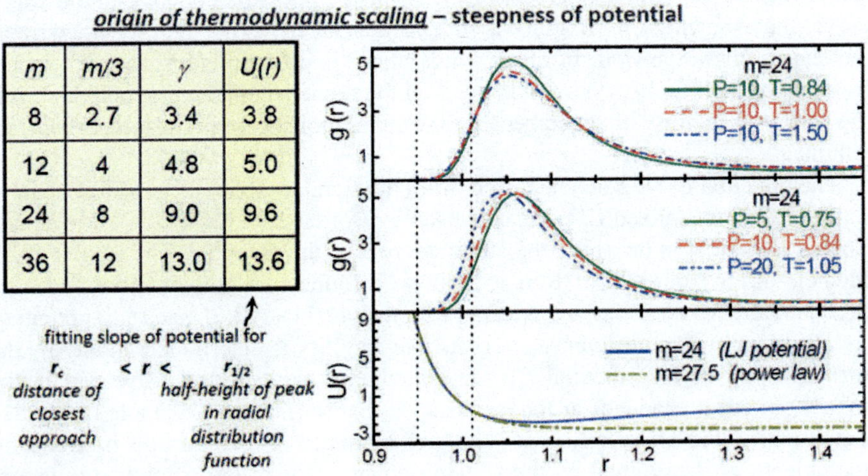

Fig. 236-1 *Top panel*: Radial distribution functions between large particles $g_{11}(r)$ at $P = 10$ for $T = 1.50$ (*dotted line*), $T = 1.00$ (*dashed line*), and $T = 0.84$ (*solid line*). *Middle panel*: $g_{11}(r)$ at the lowest equilibrated T: $T = 0.75$ at $P = 5$ (*dotted line*), $T = 0.84$ at $P = 10$ (*dashed line*), and $T = 1.05$ at $P = 20$ (*solid line*). *Bottom panel*: pair potential $U_{11}(r)$ (*solid line*) and the intermediate power law, r^q (*dotted line*) that fits $U_{11}(r)$ in the range from $r_c = 0.95$ to $r_{1/2} = 1.01$. This range is bounded by the two *vertical dotted lines* in all panels. After Coslovich and Roland [1411(a)]. The extreme right column of the table lists the values of q of the fits. The exponent γ that scales the diffusion coefficient in Fig. 235 given in the third column is close to q. Redrawn from data of [1411(a)]

The distance within the range $r_c < r < r_{1/2}$ is smaller than the average separation between the large particles corresponding to the first peak of the pair distribution function $g_{11}(r)$. At such small distances I would think that the primitive or the local JG β-relaxation is more directly related to the repulsive part of $V(r)$ than the cooperative α-relaxation which involves more particles and longer length scale. Thus, I take this as direct evidence that the dependence of dynamic quantities on ρ^γ/T originates in the primitive relaxation of the CM or the local JG β-relaxation. This original dependence of τ_{JG} (or τ_0) on $T^{-1}V^{-\gamma}$ is passed onto τ_α and magnified in τ_α of the structural α-relaxation at later times through action of the many-body dynamics. The stronger dependence of τ_α on $T^{-1}V^{-\gamma}$ than τ_{JG} (or τ_0) also follows directly from the CM equation (2.86) by raising the dependence to the superlinear power of $1/(1-n)$.

Scaling in Soft Spheres: Fragility Invariance on the Repulsive Potential Softness

The purely repulsive inverse power law potential $U(r) = \varepsilon(\sigma/r)^q$ has been utilized in recent simulation studies by De Michele et al. [1412(a)] of several binary mixtures of soft particles with $q = 6$, 8, 12, and 18, which change the "softness" of the interaction and value of $\gamma = q/3$ [69]. For different power q of the inter-particle repulsive potential, the results show that the diffusion coefficients as a function of T

for various values of the exponent of the repulsive potential collapse onto a single curve, thus showing that fragility is independent of the exponent q characterizing the short-range repulsion. In other words, there is no correlation between γ and fragility, or the fragility does not depend on the repulsive interaction potential. It is the isochoric fragility obtained because the simulations were performed at constant volume.

The potential of De Michele differs from the standard ($q = 12$, $p = 6$) Lennard-Jones or the generalized LJ potentials used by Bordat et al [264] or Coslovich and Roland [1411(a)] in dropping the attractive part of the potential. The results of De Michele differ from that of Bordat et al. who found that both fragility and non-exponentiality of the binary LJ systems with the (12,11), (12,6), and (8,5) potentials are all different. The difference suggests that fragility is determined mainly by the attractive part of the potential $U(r)$ in conjunction with the repulsive part in the anharmonicity it produces at longer distance than r_c. This is palatable to the fact that fragility, the characteristic of the cooperative α-relaxation and its transport coefficients, is determined by the slowing down effect of the many-body dynamics involving motion of multiple particles and longer length scales than presented by the repulsive interaction potential, in addition to change of volume and entropy with temperature.

Another way to understand the result of De Michele et al. is to recall that for the purely repulsive inverse power law potential they used, all thermodynamic and dynamics properties depend on the product variable $TV^{q/3}$. Hence it is sufficient to obtain either the T-dependence or the V-dependence of any quantity. One consequence of this is that the fragility does not change upon changing the density of the soft binary mixture of De Michele et al. at fixed q. Since intermolecular coupling strength depends on density, this property implies that intermolecular coupling is not an important factor in determining the fragility for all binary mixtures of De Michele et al. Based on experimental data as well as the CM, we have already shown before that, in addition to change in entropy and density, intermolecular coupling and the resulting many-body effects on τ_α determines fragility (see Section 2.2.3). From this and the fact that intermolecular coupling does not enter into consideration in determining the fragility of all the binary mixtures of De Michele, the observed invariance of fragility to the density change is hence explained.

The simulations of De Michele et al. and the explanation given here are consistent with the $T^{-1}V^{-\gamma}$-dependence of dynamic quantities of glassformers determined by the repulsive potential at short distance, and $T^{-1}V^{-\gamma}$-dependence of τ_α actually stems from that of the primitive relaxation of the CM or the JG β-relaxation. In real non-hydrogen-bonded glassformers, the presence of the attractive part of the intermolecular potential $U(r)$ can change the actual value of γ from $\gamma = q/3$ determined by the repulsive part of $U(r)$ acting alone (see table in Fig. 236-1). Except for the modification of γ-value, the scaling properties of thermodynamic and dynamic real glassformers are passed on from that of the inverse power law (IPL) from Hoover and Ross [1397, 1398], and others [1401]. The short-range repulsive part of $U(r)$ imparts the ρ^γ/T-dependence onto the primitive relaxation time τ_0 or the JG β-relaxation time τ_{JG}. The longer range attractive part of $U(r)$ enters into many-molecule dynamics, which involves more particles and grows in length scale with increase of time, until finally the terminal α-relaxation is reached. The essential effect of the many-molecule dynamics is to make τ_α much longer than τ_0 or τ_{JG} but it still depends on density and temperature via the same product variable ρ^γ/T. However, since the effect of the many-body dynamics on slowing down the primitive relaxation is non-linear, the functional form of the dependence of τ_α on ρ^γ/T is different from that of τ_0 or τ_{JG}, although they have the same γ. A stronger dependence of τ_α on ρ^γ/T can be expected intuitively from the participation of more molecules.

Lack of Correlation Between γ and the Characteristics (n, m_P, ξ_{het}) of the α-Relaxation: Another Support of Its ρ^γ/T-Dependence Originating from That of τ_0 or τ_{JG}

The evidences given in the above suggest that the ρ^γ/T-dependence originates from that of τ_0 or τ_{JG}, and the stronger dependence of τ_α on ρ^γ/T is due to the build-up of many-body dynamics. In the context of the CM, $\tau_\alpha(\rho^\gamma/T)$ is obtained from $\tau_0(\rho^\gamma/T)$ by Eq. (2.86) by raising the dependence to the superlinear power of $1/(1-n)$, i.e., $\tau_\alpha(\rho^\gamma/T) \propto [\tau_0(\rho^\gamma/T)]^{1/(1-n)}$. From this relation, it is clear that the non-exponentiality index n of the α-relaxation and γ are *independent* parameters and they bear no obvious correlation with each other. This relation also indicates that the isobaric fragility index m_P (say at ambient pressure) of the α-relaxation is determined by both n and γ, and therefore like n, m_P bears no correlation with γ.

The values of γ and m_P of many glassformers have been given in Tables 2 and 3 in [119]. In some glassformers, different values from more than one source are given. We [1412(b)] use exclusively the data from dielectric relaxation from the works published by Casalini, Paluch, and Roland cited in [119], and by Alba-Simionesco et al. [547(c)], and whenever there is a choice between the two we take the data from the works of Casalini, Paluch, and Roland for the sake of consistency with the results from the same group of collaborators. These are exactly the same data used by Casalini and Roland in [557] to support their claim that the degree of fragility can be deduced solely from the ρ^γ/T-dependence of τ_α.

Added are new dielectric data of vitamin E ($\gamma = 3.9$, $n = 0.35$, $m_P = 63$, $m_V = 41.6$) [1413(b)], diphenyl-vinylene carbonate (DPVC) ($\gamma = 3.3$, $n = 0.29$, $m_P = 92$, $m_V = 61$) [697], heptapropylene glycol dimethylether (7PGDE) ($\gamma = 3.1$, $n = 0.52$) [652(c)], and benzoin isobutylether (BIBE) ($\gamma = 3.8$, $n = 0.39$, $m_P = 78.4$, $m_V = 44.3$) [687]. In Fig. 236-2, we plot n against γ, and m_P against γ to test if there is any correlation between γ and n or m_P, the characteristics of the α-relaxation. It is clear from the plots that there is neither any correlation between γ and n nor between γ and m_P. Thus, the observed ρ^γ/T-dependence of τ_α bears no relation to fragility m_P and non-exponentiality of the structural α-relaxation even if glycerol and sorbitol are excluded.

Our conclusion contradicts that of Casalini and Roland (CR) [557]. Instead, they concluded that fragility is a consequence of the relative interplay of temperature and density effects indicated by the ρ^γ/T-dependence near T_g, and from the observed

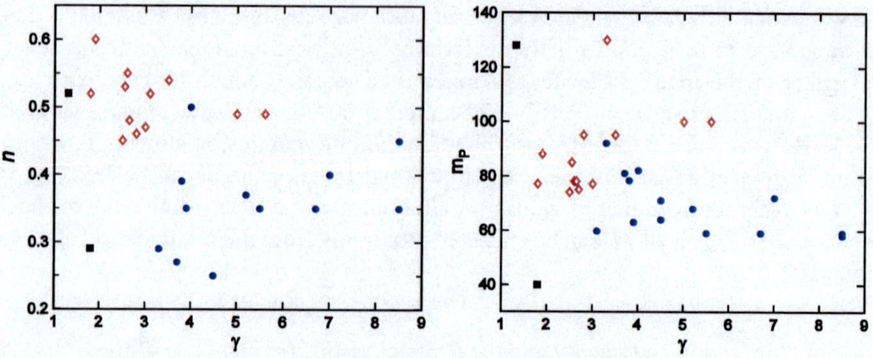

Fig. 236-2 (*Left*) Plot of n against γ, where the *closed circles* are data of small molecular van der Waals glassformers, the *open diamonds* are polymeric glassformers, and the *two closed squares* are data of glycerol (smaller value) and sorbitol (larger value). (*Right*) Plot of isobaric fragility index, m_P, against γ. The *symbols* stand for the same glassformers as in the *left panel*. There is neither any correlation between n and γ, nor between m_P and γ

ρ^γ/T-dependence of τ_α they can infer whether a glassformer is strong or fragile, i.e., its fragility. Specifically, strong behavior reflects a substantial contribution from density (larger γ), while the relaxation of fragile liquids is more thermally activated (smaller γ). This they illustrated in a figure, which is reproduced here in the

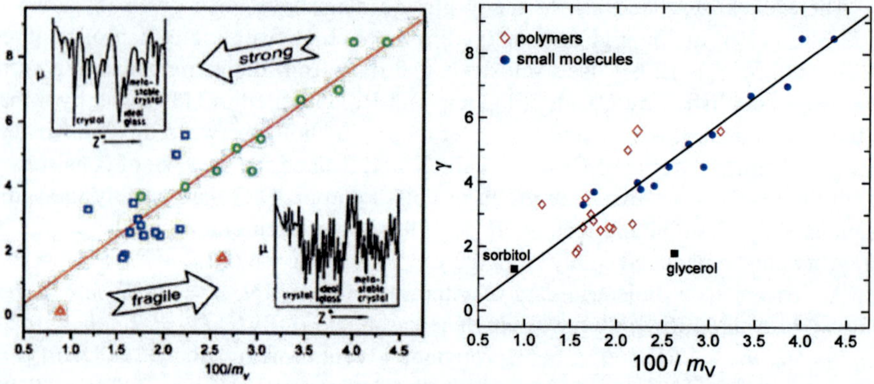

Fig. 236-3 (*Left*) Parameter γ vs. the inverse isochoric fragility m_V for 26 materials. The data (in order of increasing m_V) are: PCB62, BMMPC, BMPC, PCB54, PCB42, KDE, salol, glycerol, PDE, PMPS, OTP, PECH, PMTS, PVME, PVAc, PPG, PC, DGEBA, PI, PPGE, PVAc(2), PVE, PB, PCGE, and sorbitol. The *solid line* is a linear fit (correlation coeff.=0.92) to all data (polymeric, *open squares*; small molecular, *open circles*) except the H-bonded materials (*open triangles*). Representations of the potential energy hypersurface taken from Angell. Reproduced from [557] by permission. (*Right*) Reproduction of the anticorrelation between γ and m_V by using the data of the same 26 materials given in Tables 2 and 3 in [119] and references cited in [557], and new data of vitamin E, DPVC, and BIBE. The *closed circles* stand for data of small molecular van der Waals glassformers, the *open diamonds* are for polymeric glassformers, and the two *closed squares* are data of glycerol and sorbitol

left panel of Fig. 236-3. In this figure they have plotted γ against the reciprocal of isochoric fragility, $100/m_V$, for 26 materials, demonstrating the anticorrelation between the two quantities, that is larger γ (strong dependence on density ρ) corresponding to smaller isochoric fragility m_V. We independently reproduced the same anticorrelation with the values of γ and m_V of the 26 glassformers considered by CR taken from the tables in [119] together with the new data of vitamin E, DPVC, and BIBE (see right panel of Fig. 236-3). Actually this anticorrelation between γ and m_V is natural or unsurprising, because stronger dependence of τ_α on ρ (larger γ) implies that its T_g-scaled T-dependence at constant ρ (isochoric condition) will be weaker (smaller m_V). Thus, its manifestation when considering real glassformers does not indicate anything deeper. This is also the opinion of CR which can be inferred from their own statement: "As discussed above, γ is a measure of the relative contribution of T and v, which means that γ should also be related to m_V." So, what is the crux of the argument CR used to arrive at their conclusion?

The crux of the argument offered by CR in their own words is as follows. "For a given material, a dominant short-range repulsive potential gives rise to stronger (less fragile) dynamics. Larger γ implies steeper potential wells as depicted in Fig. 3 with the sketch taken from Angell, and hence a liquid structure more resistant to changes in T. Relaxation is facilitated by changes of the energy barriers (from changes in intermolecular distances); thus the effect of V becomes more important for strong liquids. For fragile liquids, the potential energy surface is characterized by flatter minima (illustrated in Fig. 3), so that thermally activated motion can proceed. Evidently the shape of the potential affects its anharmonicity, a steeper potential (larger γ) being more harmonic. According to this interpretation, the fragility of liquids increases with the anharmonicity of the potential ...." It is clear from these statements that CR considered only the short-range repulsive part of the intermolecular potential and deemed it solely determines the depth and anharmonicity of the potential well. This bold hypothesis may not work for the real glassformers in Fig. 236-3 having a wide range of chemical structures and sizes because the attractive part of the potential can vary from one glassformer to another in its relation to the repulsive part, and the variation can change the overall shape and anharmonicity of the resulting intermolecular potential.

Whether the argument of CR is valid or not, the fact remains that from experiment there is no correlation between γ and fragility or non-exponentiality, the two principal characteristics of the structural α-relaxation, as shown in Fig. 236-2. Here it is worthwhile to emphasize that the term fragility usually means isobaric fragility, and it is measured by the steepness index m_P. Furthermore, the representations of the potential energy hypersurface in Fig. 236-3 by CR were taken from Angell's publication [216], where Angell related the potential energy hypersurface to isobaric fragility m_P at ambient pressure, and not m_V.

In dismissing the lack of correlation between fragility and γ found by De Michele et al. [1412(a)], CR made the statement: "Of course, our observed correlation between fragility and γ is an experimental fact, notwithstanding any connection of the latter to the intermolecular potential." If by fragility CR means isochoric fragility, this statement cannot make connection with De Michele et al. because they studied soft particles having only the repulsive potential, while CR considered real glassformers with both the repulsive and the attractive potentials. If by fragility CR means isobaric fragility, then their statement has no support because they did not examine the correlation between m_P and γ themselves. No one has checked this out either until now, 5 years after the publication of [557] by CR. We find no correlation between m_P and γ as indicated by the plot in Fig. 236-2.

The dynamics of colloidal suspensions of soft spherical particles with the softness of the particles ranging from very soft, intermediate soft, to the conventional hard spheres have been shown before in Fig. 34c for fragility and Fig. 34d. A less steep repulsive potential is expected for softer spheres. However, the softer spheres have lower fragility, opposite to the trend suggested by CR that less steep repulsive potential engenders higher fragility.

Important Role of the Attractive Part of the Intermolecular Potential in Determining the α-Relaxation: Support from Molecular Dynamics Simulations

Another way to show that the short-range repulsive part of the intermolecular potential does not solely determine the characteristics of the α-relaxation is the results of molecular dynamics simulation of Berthier and Tarjus (BT) on binary Lennard-Jones systems of Kob and Andersen [1412(c)]. Similar in spirit to the work of Bordat et al. [264], BT modify the standard LJ potential of Eq. (2.4) with $q = 12$ and $p = 6$ by truncating its range and reducing it to the purely repulsive part of the pair potentials proposed by Weeks-Chandler-Andersen (WCA) [1410]. Otherwise the other parameters of the LJ potential are the same. The results of BT reproduced are here in Fig. 236-4.

Characterized by the pair correlation functions $g(r)$ obtained by summing over the partial $g_{AA}(r)$, $g_{AB}(r)$, and $g_{BB}(r)$ weighed by the concentrations of the species A and B, the static structures of the LJ and the WCA liquids are about the same, especially at higher temperatures as shown in the left panel for two temperatures of $T = 0.45$ and 1.5 at the typical liquid density ρ of 1.2 (all quantities measured in LJ units). This is consistent with the studies by Hoover and Ross [1396, 1397], Hiwatari

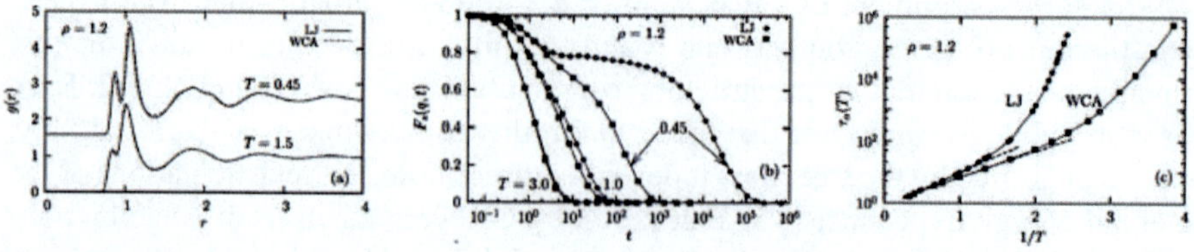

Fig. 236-4 Comparison between the LJ and its WCA description at the typical liquid density $\rho = 1.2$ for several temperatures. (*Left*) Static pair correlation functions $g(r)$ as a function of r; (*middle*) time dependence of the self-intermediate scattering function for the majority component A; (*right*) Arrhenius plot of the relaxation time τ_α defined from $F_s(q, t = \tau_\alpha) = 1/e$ (the lowest T is 0.43 for LJ and 0.26 for WCA), with high-T Arrhenius fits shown as *dashed lines*. Despite similar pair structure, both systems display dramatically different α-dynamics, but note the faster secondary relaxation is the same. Reproduced from [1412(c)] by permission

et al. [1401], and Chandler et al. [1410] showing that the structure of dense non-associated liquids, the pair correlation functions, and the thermodynamics all can be predicted from the short-ranged repulsive part of the intermolecular potentials.

However, the dynamics of the two liquids is very different at lower temperatures. This is shown in the middle panel by the self intermediate scattering function, $F_s(q,t)$, with $q_{AA} = 7.2$, which corresponds to the position of the peak of the total static structure factor at $\rho = 1.2$. At high temperatures, there is good agreement in $F_s(q,t)$, but the difference between the two systems rapidly increases at lower temperatures and becomes very large at the lowest accessible temperatures. The relaxation time τ_α determined by $F_s(q, t = \tau_\alpha) = 1/e$ of the WCA model is then more than three orders of magnitude faster than that of the LJ model (see middle and right panels). Since the static pair correlation function is practically the same while the dynamics changes a lot, BT pointed out that their observation contradicts the mode coupling theory, which predicts the dynamics of the liquids is uniquely determined by the static pair correlation functions. The results of BT also unambiguously show that the attractive part of the intermolecular potential in the LJ liquid plays the pivotal role of considerably slowing down the α-relaxation on cooling the liquid to approach glass transition, consistent with our point of view.

There is yet another difference in the dynamics of the α-relaxation between the LJ and WCA models. The density scaling of the relaxation time τ_α determined that is empirically found in real glass-forming liquids and polymers [1402–1409], and in the LJ model as well (see left panel of Fig. 236-5 and [1412(a)–(d)]), is absent when attractive forces are truncated. There is large change of fragility with density for $\rho \leq 1.4$ of the WCA liquid with the short-ranged repulsive potential. These findings show in another way the crucial role played by the attractive component of the pair potential in determining the characteristics of the α-relaxation.

An interesting feature of the $F_s(q,t)$ in Fig. 236-4 not mentioned by BT is that the faster secondary relaxation resolved at lower temperatures is the same for the LJ and WCA model potentials (see the example at $T = 0.45$), in contrast to the

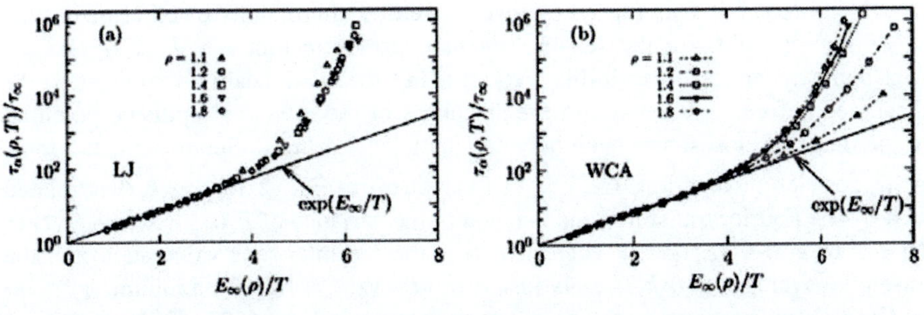

Fig. 236-5 Rescaling of the relaxation data for the LJ and WCA models over a wide range of densities and temperatures. BT used the activation energy scale $E_\infty(\rho)$ obtained by fitting the high-T data to an Arrhenius formula. (*Left*) Arrhenius plot of the relaxation time for a scaled temperature $T/E_\infty(\rho)$ for LJ. (*Right*) Same plot for WCA. Note the large change of fragility with density for $\rho \leq 1.4$, not seen in the LJ data. Reproduced from [1412(c)] by permission

large difference in the α-relaxation. This feature follows as a consequence of the proposal I made before that the short-ranged repulsive potential dictates the density and temperature dependence of the local primitive or JG β-relaxation. This local relaxation is the same for the LJ and the WCA liquids because the repulsive part of the potential in the LJ model is identical to that in the WCA model. Moreover, at higher temperatures, the α-relaxation merges with the local relaxation in the WCA model as can be seen from $F_s(q,t)$. This explains why density scaling works for τ_α in the WCA model at higher temperatures (see the Arrhenius part in Fig. 236-5, right panel).

From the importance of the attractive part of the intermolecular potential demonstrated by BT, they also pointed out the danger of theoretical treatments of glass transition in which the structure and the dynamics of viscous glass-forming liquids are controlled by the short-ranged repulsive forces, and the attractive part of the interactions is treated as a mere cohesive background amenable to perturbative treatment. Examples they mentioned are studies based on the mode coupling theory of glasses [1412(d), (e)], the self-consistent phonon theory [582], model potential energy landscapes [1399], as well as recent work focusing on the correlations between pressure and energy fluctuations [1411(c), (d)] and on the density scaling of the relaxation time [1411(a)]. They commented that the short-ranged repulsive potential is also the predominant factor in the "jamming scenario," which postulates that the physics of glasses and glass-forming liquids is controlled by a zero-temperature critical point characteristic of the jamming [1412(f), (g)], or in the glass transition of spheres with finite-ranged repulsive interactions [1412(h)]. In both cases, the longer-ranged attraction was considered as a perturbation.

$T^{-1}V^{-\gamma}$ or ρ^{γ}/T Scaling Holds for the Full-Time Dependence of Both $F_s(k,t)$ and $\chi_4(t)$ Using the Same Scaling Exponent γ

Coslovich and Roland published another simulation [1411(b)] of the Kob–Andersen binary mixtures of particles interacting with the Lennard-Jones potential, $U(r) \propto [(\sigma/r)^{12} - (\sigma/r)^6]$. This time they investigated by simulations whether or not the ρ^γ/T scaling holds for the self-intermediate scattering function, $F_s(k,t)$, and the four-point dynamic susceptibility $\chi_4(k,t)$ using the *same* scaling exponent γ. As discussed before, γ is related to the steepness of the effective repulsive potential in the range of closest approach between particles. Earlier in Section 2.2.2.7, four-point correlation function, $G_4(\vec{r},t)$, was defined by Eq. (2.38). $\chi_4(k,t)$ is related to it by the Fourier transform and is equal to the variance of $F_s(k,t)$, i.e., $\chi_4(k,t) = N[< f_s^2(k,t) > -F_s^2(k,t)]$, where $f_s(k,t)$ is the instantaneous value such that the ensemble average $< f_s(k,t) >$ is just $F_s(k, t)$. $\chi_4(k,t)$ has a maximum, $\chi_4^{max} = \chi_4(k, t_{max})$, proportional to the dynamic correlation volume [523, 527] at $t = t_{max}$, and it turns out that t_{max} is approximately the same as the structural relaxation time τ for k that matches the position of the first peak in the static structure factor [527].

Actually an answer can be given in the affirmative that the ρ^γ/T scaling holds for the self-intermediate scattering function, $F_s(k,t)$, and the four-point dynamic susceptibility $\chi_4(k,t)$ using the *same* scaling exponent γ, before carrying out the simulation by making use of previously known experimental facts. The data presented in Section 2.2.4 have amply shown the co-invariance of the α-dispersion (or n) and τ_α to various combinations of T and P [618], and thus n is also a function of ρ^γ/T. Moreover, in Section 2.2, various experimental facts have been invoked to support that n is a surrogate for the extent or intensity of the many-body α-relaxation, and naturally n correlates with the length scale or the dynamic correlation volume of the many-body α-relaxation. As a matter of fact, for the few glassformers that have the length scale measured by multidimensional NMR, the length scale of dynamic heterogeneity ξ_{het} from NMR has been shown to correlate with n, as shown before in Fig. 33-2. On combining these two properties, it can be expected that the ρ^γ/T scaling holds for $F_s(k,t)$ and $\chi_4(k,t)$ using the *same* scaling exponent γ.

The simulations were carried out in the *NVT* ensemble for five isochoric paths in the density range of $1.150 \leq \rho \leq 1.1350$, at a fixed reduced wave vector $k^* = k(\rho^{1/3}) = 7.44$, which matches the position of the first peak in the static structure factor ($k = 7.0$) for the well-studied density $\rho = 1.2$. In the beginning of Section 2.3.2.29, mention has been made of the general result by Hoover and Ross [1396, 1397] for the dynamic behavior of systems having an inverse power potential (IPP), $U(r) = \varepsilon(\sigma/r)^q$. When distance is scaled by $\rho^{-1/3}$ and time scaled by $\rho^{-1/3}T^{-1/2}$, the dynamics properties of the IPP system at any density or temperature are the same for a fixed value of $\rho(\varepsilon/kT)^{1/\gamma}$ with $\gamma = q/3$. However, the presence of the attractive term in the (12,6) LJ potential modifies the slope of the repulsive term and the steepness of the effective repulsive potential in the range of closest approach between particles. Consequently γ is not exactly 4.

With time t scaled to become the reduced time $t^* = t(\rho^{1/3}T^{1/2})$, the reduced relaxation times τ^*, defined as the time for $F_s(k,t^*)$ to decay by a factor of e, are shown in the left panel of Fig. 236-6 vs. the scaling variable ρ^γ/T. The material constant $\gamma = 5.1 \pm 0.1$ provides the optimal collapse of τ^* onto a single curve; this value is in accord with the scaling behavior found previously for the diffusion coefficient of this system (see Fig. 235, $m = 12$, $\gamma = 5.0$). The intermediate scattering functions $F_s(k,t^*)$ were obtained for each of the five densities at the respective temperatures corresponding to a fixed value of $\rho^\gamma/T = 5.07$.

In the inset of the left panel of Fig. 236-6, these correlation functions are plotted as a function of reduced time, and, $F_s(k^*,t^*)$ has essentially the same shape for state points for which ρ^γ/T is constant. Thus, not only do the relaxation times τ^* superpose as a function of ρ^γ/T, but the entire t-dependence of the correlation functions is invariant for state points for which ρ^γ/T is constant. This result has the fundamental implication that the characteristic times of the faster processes, including the primitive relaxation time τ_0 or the JG β-relaxation time τ_{JG}, are function of ρ^γ/T with the *same* scaling exponent γ as the structural relaxation time τ_α. One can recognize that this is consistent with the experimental findings that τ_0 (or τ_{JG}), τ_α, and n (or the shape of the α-relaxation) are co-invariants to changes in P and T (see Section 2.3.2.8). Since the primitive or the JG β-relaxation has transpired

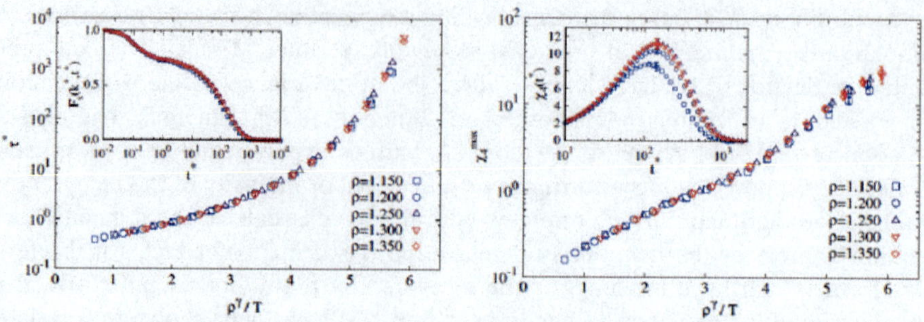

Fig. 236-6 (*Left*) Reduced relaxation times $\tau^* = \tau(\rho^{1/3}T^{1/2})$ as a function of ρ^γ/T with $\gamma = 5.1$ for all studied densities. *Inset*: Self-intermediate scattering functions as a function of reduced time $\tau^* = t(\rho^{1/3}T^{1/2})$ for state points at which $\rho^\gamma/T = 5.07$: $T = 0.402$ (at $\rho = 1.15$), $T = 0.50$ (at $\rho = 1.20$), $T = 616$ (at $\rho = 1.30$), and $T = 0.912$ (at $\rho = 1.350$). A constant reduced wave vector $k = k^*(\rho^{1/3}) = 7.44$ is considered. (*Right*) Maximum of four-point susceptibility as a function of ρ^γ/T with $\gamma = 5.1$ for all studied densities. *Inset*: Four-point dynamic susceptibility as a function of reduced time $\tau^* = t(\rho^{1/3}T^{1/2})$ for state points at which $\rho^\gamma/T = 5.07$ (same state points as inset of the figure on the left). Reproduced from [1411(b)] by permission

before the structural α-relaxation, the result of $F_s(k^*, t^*)$ from the simulation provides another support for the ρ^γ/T or $T^{-1}V^{-\gamma}$-dependence of the α-relaxation time to originate from that of τ_0 (or τ_{JG}). The stronger ρ^γ/T-dependence of τ_α with the same γ is merely a magnification from that of τ_0 (or τ_{JG}) by the many-body effect.

Correlation of Heterogeneous Dynamics Length Scale, ξ_{het}, with n

Since χ_4^{max} is proportional to the dynamic correlation volume, N_c, of the α-relaxation (or the corresponding length scale, ξ_{het}, of heterogeneous dynamics discussed before in Section 2.2.2), from the results of Fig. 236-6 we can conclude that N_c for a given material (defined in the simulations by the Kob–Andersen model) depends only on the magnitude of τ^* or τ_α, independent of T or ρ. The α-dispersion or n also depends only on the magnitude of τ^* or τ_α, independent of T or ρ. An example is shown in the inset in the left panel of Fig. 236-6 for one τ^*. Not shown are the same that hold for the self intermediate scattering functions for other values of τ^* or τ_α. Moreover, from unpublished results of Coslovich and Roland, the values of n increase monotonically with ρ^γ/T just like χ_4^{max} and N_c (or ξ_{het}) shown in the right panel of Fig. 236-6. Thus the simulations demonstrate the existence of a correlation between N_c (or ξ_{het}) and n, which has support from multidimensional NMR experiments on real glassformers [422] discussed before in Section 2.2.2. N_c from χ_4^{max} (or ξ_{het} from NMR measurement) offers valuable information on the heterogeneous dynamics of the α-relaxation. However, this quantity is not easy to acquire and not available for most glassformers. In contrast, the width of the α-dispersion or n is readily obtained by direct measurements using various spectroscopic techniques. Although the quantitative relation between N_c (or ξ_{het}) and n is not obvious, both quantities reflect different measures of the extent of the many-body relaxation dynamics. Hence for understanding the effects of many-body relaxation dynamics on physical quantities relevant to glass transition, it is much more expedient to use n than N_c (or ξ_{het}).

2.3.2.30 Invariance of the Primitive Relaxation Time, τ_0, to Variations of P and T While Keeping τ_α Constant Deduced from the Same Observed on the Normal Mode Relaxation Time, τ_n, of Type-A Polymers

Dielectric relaxation data were obtained for various temperatures T and pressures P on the polymers, poly(oxybutylene) (POB, $M_n = 4800$ and $M_w/M_n = 1.10$ [1413], polypropylene glycol (PPG, $M = 400$ Da) [701, 1414], and 1,4-polyisoprene (PI) [637, 638, 1414]. These polymers have dipole moment along the backbone of the polymer which enables motions of longer length scale of the chain, called normal modes, to be detected in the dielectric spectra [1415]. The local segmental α-relaxation is made visible dielectrically as usual by the normal component of the dipole moment. Thus, dielectric relaxation spectra of these so-called type-A polymers show not only the α-relaxation but also the longer time normal or Rouse modes of the polymer chains. Dielectric measurements were carried out not only at ambient pressure but also at elevated pressures on these polymers and the results provide additional information of the relation between the α-relaxation and the normal modes. The loss spectra of POB obtained at ambient pressure and elevated pressure in the left panel of Fig. 237 show the shape (or frequency dispersion) of the α-loss peak is invariant to changes in the combination of P and T that maintains τ_α constant at $\sim 10^{-3}$ s. This is a general property of the α-relaxation of glassformers discussed before in Section 2.2.4.

What is new in the left panel of Fig. 237 is that the shape of the normal mode loss peak also does not change, and the separation of the two loss peaks remains the same. The latter means that the ratio, τ_n/τ_α, is invariant to variations of P and T while maintaining τ_α constant. This is shown in another manner by the plot of $\log \tau_\alpha$ vs. $\log \tau_n$ at three constant temperatures and varying pressure. The two other type-A polymers, PI and PPG, behave like POB, and this is shown in Fig. 238. The reader may recall the similar result for the JG β-relaxation that the ratio τ_α/τ_{JG} (or τ_α/τ_0) is the same for different combinations of P and T having the same τ_α, shown before in Section 2.3.2.8. This apparently general relation of the normal modes to the α-relaxation is interesting as well as challenging to explain.

In Section 2.2.5.9 we have seen another general property of polymer dynamics, which is the breakdown of thermorheological simplicity of relaxation mechanisms of different time/length scales. Specifically, the relaxation times of the normal modes have weaker temperature dependence than that of the local segmental relaxation, and this causes viscoelastic anomalies observed in polymers. These have been shown before in Figs. 99, 102, and 103 for unentangled PPG, PS, and PMPS, respectively. Explanation has been given by the CM in the literature [173, 360, 837, 904] as well as conveyed by Eqs. (2.63)–(2.65). The basic difference between normal modes and local segmental α-relaxation is their coupling parameters. The coupling

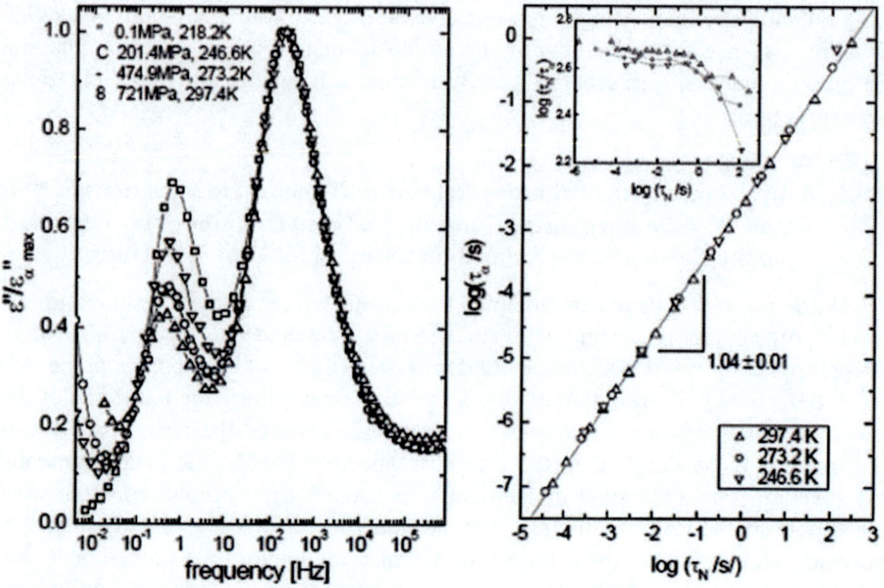

Fig. 237 (*Left*) Comparison of the peaks for the normal and segmental modes at an approximately constant value of $\tau_\alpha \sim 10^{-3}$ s. The *ordinate* has been normalized by the peak intensity of the segmental mode. The relative intensity of the normal mode increases with increase of either pressure or temperature. Note that there is negligible change in either the peak shapes or the separation of the two peaks [1413]. (*Right*) Segmental relaxation times as a function of normal mode relaxation times for three isotherms. The pressure variation was in the range from 11 to 952 MPa. The slope of the fitted line reveals the near equivalence of the pressure dependences for the two relaxation modes. The fact that the data falls on the master line means, for any fixed τ_α, that τ_n is the same independent of P and T. The *inset* shows the logarithm of the ratio of the two relaxation times, plotted vs. the normal mode relaxation time (*symbols* same as in main figure). Reproduced from [1413] by permission

parameter n_α is from the intermolecular potential between repeat units. On the other hand, n_n is from entanglement coupling of longer length scale than the molecular weight of entanglement M_e. For unentangled polymers, the Rouse type of normal modes, n_n is zero. For fully entangled linear polymers, n_n is about 0.41 deduced from the experimentally observed $M^{3.4}$-dependence of the terminal relaxation time or viscosity and compared with the CM predicted $M^{2/(1-n_n)}$ dependence. The traditional approach to polymer dynamics [29] assumes that relaxation mechanisms of all length scales have one and the same monomeric friction coefficient $\zeta_0(T,P)$ governing the temperature and pressure dependence of their relaxation times. The consequence of this assumption is that all viscoelastic mechanisms shift uniformly with temperature, and the ideal behavior is called thermorheological simplicity. In reality, the α-relaxation time has stronger T-dependence [165–168, 170–172, 206, 867–869, 909, 910] and P-dependence [637–639, 1413, 1414] than the chain modes of longer length scale (see Section 2.2.5.9). According to the CM it is due to the difference in the coupling parameters n_α and n_n, and the fact that many polymers have

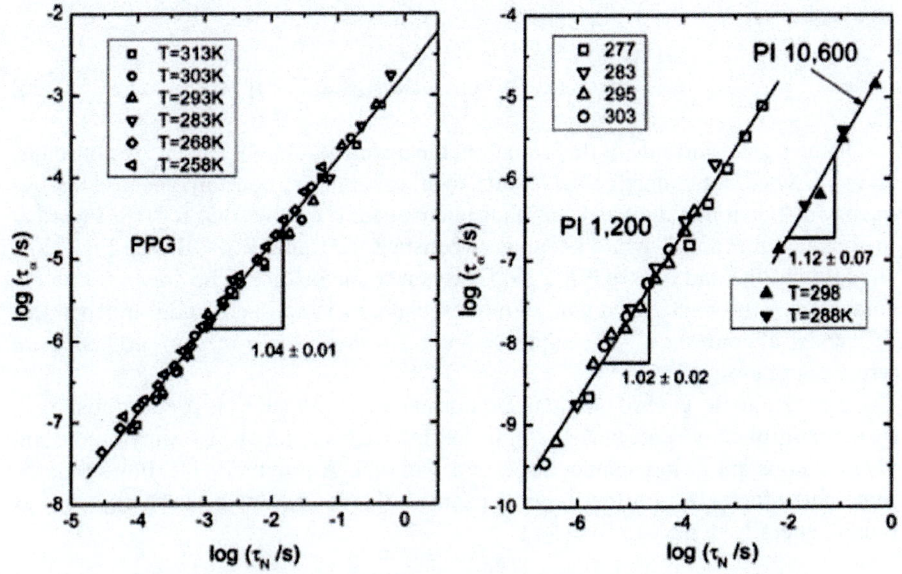

Fig. 238 (*Left*) Segmental relaxation times as a function of normal mode relaxation times for six isotherms measure for PPG (M = 4000 Da); data from [701] and shown as such in [1413]. The slope was obtained by simultaneously fitting all data. (*Right*) Segmental relaxation times as a function of normal mode relaxation times for PI of respective molecular weight equal to 1200 Da (*open symbols*) from [638], and 10,600 Da (*filled symbols*) from [637], and plotted as such in [1413]. The temperatures (K) are as indicated. Reproduced from [1413] by permission

n_α larger than n_n. The primitive relaxation times of all viscoelastic mechanisms have the same primitive friction coefficient $\zeta_0(T,P)$, but after many-body dynamics due to coupling is taken into account by the CM equation, the dependence is modified to $[\zeta_0(T,P)]^{1/(1-n_\alpha)}$ for the α-relaxation and $[\zeta_0(T,P)]^{1/(1-n_\mathrm{n})}$ for the normal mode. With the primitive relaxation times of the α-relaxation and the longest normal mode written as $\tau_0 = A_0\zeta_0(T,P)$, and $\tau_{0\mathrm{n}} = A_\mathrm{n}\zeta_0(T,P)$, the resultant relaxation times after many-body effects have been taken into account are given by

$$\tau_\alpha(T,P) = [t_\mathrm{c}^{-n_\alpha} A_0\zeta_0(T,P)]^{1/(1-n_\alpha)} \tag{2.87}$$

and

$$\tau_\mathrm{n}(T,P) = [\hat{t}_\mathrm{c}^{-n_\mathrm{n}} A_\mathrm{n}\zeta_0(T,P)]^{1/(1-n_\mathrm{n})}, \tag{2.88}$$

where $\hat{t}_\mathrm{c}$ for entangled normal modes has to be distinguished from t_c for α-relaxation because the former is longer than the latter. For the simplest case of unentangled and Rouse-like normal modes where $n_\mathrm{n} = 0$, Eq. (2.88) reduces to

$$\tau_\mathrm{n}(T,P) \equiv \tau_{0\mathrm{n}} = A_\mathrm{n}\zeta_0(T,P), \tag{2.89}$$

and we have

$$\tau_n(T,P)/\tau_\alpha(T,P) = (A_n/A_0)[t_c/\tau_\alpha(T,P)]^{n_\alpha}. \tag{2.90}$$

On the right-hand side of this equation, the quantities A_n, A_0, and t_c are constants, and n_α is invariant to changes in P and T while keeping τ_α constant because the frequency dispersion of the α-relaxation is unchanged. It follows that τ_n/τ_α is invariant to changes in P and T while keeping τ_α constant, as found in POB (see Fig. 237), and also in PPG and PI (see Fig. 238). The same conclusion can be drawn for barely entangled or fully entangled normal modes where n_n is non-zero and if independent of P and T at constant τ_α. The expression used to show this is messy, and hence not written out here.

Returning to the case of unentangled normal modes, since τ_n is proportional to τ_0 by the multiplicative constant A_n/A_0 from the experimental observation of constant τ_n/τ_α at constant τ_α independent of P and T in type-A polymers, it follows that the same holds for τ_0/τ_α. Furthermore, the same holds for τ_{JG}/τ_α because $\tau_{JG} \approx \tau_0$ as found generally in glassformers.

2.3.2.31 *TV^γ*-Dependence of the Primitive Relaxation Time, τ_0, Same as that of the Normal Mode Relaxation Time, τ_n, of Type-A Polymers

Like other polymeric and non-polymeric glassformers, the logarithm of the α-relaxation time, $\log(\tau_\alpha)$, of the type-A polymers, POB, PPG, and PI, obtained by dielectric measurements for various temperatures T and pressures P yields a master curve when plotted vs. $T^{-1}V^{-\gamma}$ [119, 1402–1409]. The analysis was extended to the dielectric normal mode in these type-A polymers [1413, 1414]. The normal mode relaxation times (strictly speaking, the longest normal mode relaxation times, τ_n) superpose to a single master curve when plotted against $T^{-1}V^{-\gamma}$, using the *same* value of γ as for the local segmental α-relaxation times, τ_α. This concurrent property of the α-relaxation and the normal mode in POB and PPG are displayed in Fig. 239, and can be expressed ins term of two different functions of the same variable $T^{-1}V^{-\gamma}$ by $\tau_\alpha = f_\alpha(T^{-1}V^{-\gamma})$, and $\tau_n = f_n(T^{-1}V^{-\gamma})$. As can seen by the comparison in the figure, $f_n(T^{-1}V^{-\gamma})$ for τ_n has a weaker dependence on $T^{-1}V^{-\gamma}$ than $f_\alpha(T^{-1}V^{-\gamma})$ for τ_α.

This result is can be derived from Eqs. (2.87) and (2.88). Experiments have shown that $\tau_\alpha = f_\alpha(T^{-1}V^{-\gamma})$, and mean $\zeta_0(T,P)$ in Eq. (2.87) necessarily has the dependence on the product variable $T^{-1}V^{-\gamma}$, and it should be rewritten as $\zeta_0(T^{-1}V^{-\gamma})$), and

$$\tau_\alpha(T^{-1}V^{-\gamma}) = [t_c^{-n_\alpha} A_0 \zeta_0(T^{-1}V^{-\gamma})]^{1/(1-n_\alpha)}. \tag{2.91}$$

Substituting $\zeta_0(T^{-1}V^{-\gamma})$ into Eq. (2.88), we arrive at

$$\tau_n(T^{-1}V^{-\gamma}) = [\hat{t}_c^{-n_n} A_n \zeta_0(T^{-1}V^{-\gamma})]^{1/(1-n_n)}. \tag{2.92}$$

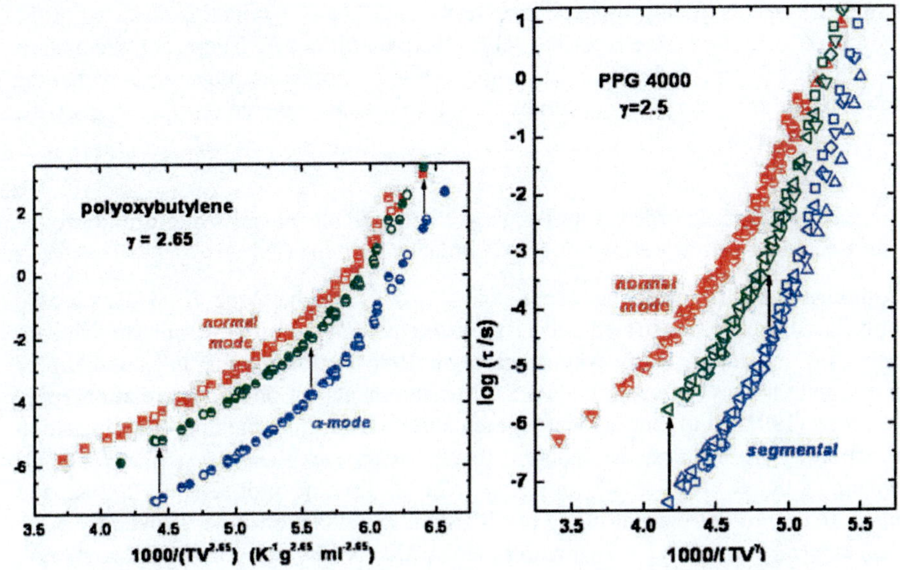

Fig. 239 (*Left*) Dielectric normal mode (*squares*) and local segmental (*circles*) relaxation times of POB measured as a function of temperature at $P = 0.1$ MPa (■, •) and as a function of pressure at $T = 246.6$ K (*lower half filled* □, *lower half filled* O), 273.2 K (□, O), and 297.4 K (*upper half filled* □, *upper half filled* O). The data were superposed using an exponent $\gamma = 2.65$. The *arrows* indicate the constant vertical shift has been applied to all data points of log τ_α. This operation clearly shows that τ_α has a stronger TV^γ-dependence than τ_n for the same γ. (*Right*) Normal mode and local segmental relaxation times as a function of the quantity $T^{-1}V^{-\gamma}$, where $\gamma = 2.5$ for PPG400, $T = 258$ (O), 268 (◁), 283 (△), 293 (▽), 303 (◇), and 313 K (□). Reproduced from [1413, 1414] by permission

Thus, τ_α and τ_n are both functions of $T^{-1}V^{-\gamma}$, but their dependences on $T^{-1}V^{-\gamma}$ are different. Usually, n_α is larger than n_n, and certainly in the case of the Rouse normal modes of unentangled polymers because $n_n = 0$. Thus, a stronger dependence of τ_α on $T^{-1}V^{-\gamma}$ than τ_n is predicted by the two equations given above. The prediction can be tested quantitatively after the values of n_α and n_n have been determined by analysis of the dielectric spectra performed in [1415]. The experimental findings were explained by the CM equations (2.91) and (2.92). The values of n_α and n_n were determined from the dielectric spectra, and the stronger dependence of τ_α on $T^{-1}V^{-\gamma}$ than τ_n was explained quantitatively. For details, see [1415].

Combining the results of this subsection and the previous one, for polymers it can be said that all three processes, the JG β-relaxation, the local segmental α-relaxation, and the normal chain mode relaxation, are functions of the same $T^{-1}V^{-\gamma}$ variable, albeit the function for the α-relaxation is different from those of the other two relaxations, and they can all be different if entanglement coupling enters into the relaxation of the normal modes. In the explanation of the relation between τ_α and τ_n on their $T^{-1}V^{-\gamma}$-dependences given by the CM via the equations above, playing an important role is the primitive friction factor $\zeta_0(T^{-1}V^{-\gamma})$, which is also that of the primitive relaxation time τ_0 or the JG β-relaxation time τ_{JG}. Therefore, the experimental fact that τ_α and τ_n are functions of $T^{-1}V^{-\gamma}$ offers another support that the dependence of τ_α on $T^{-1}V^{-\gamma}$ originates from the dependence of τ_0 or τ_{JG} on the same variable.

An Earlier Example of Relation Between *P*- and *T*-Dependences of Normal Mode and Local Segmental Relaxation in Accordance with the CM

Although the different dependences of τ_α and τ_n on the same $T^{-1}V^{-\gamma}$ variable only has been demonstrated recently, different temperature and pressure dependences of τ_α and τ_n were found before in 1986 in amorphous and unentangled poly(methylphenyl siloxane) (PMPS). Measurements by photon correlation spectroscopy [199] found for the local segmental relaxation from the fit to the correlation function by the Kohlrausch function that $\tau_\alpha = \tau_{\alpha\infty} \exp[(B_\alpha + a_\alpha P)/R(T - T_0)]$ where $B_n/R = 1659 \pm 43$ K, $a_\alpha = 67.6 \pm 1.5\ \mathrm{cm^3/mol}$, $T_0 = 207$ K, and $n_\alpha = 0.56$. For the normal mode of diffusion, its relaxation time is given by $\tau_n = \tau_{n\infty} \exp[(B_n + a_n P)/R(T - T_0)]$ with $B_n/R = 732 \pm 18$ K, $a_n = 32.4 \pm 0.8\ \mathrm{cm^3/mol}$, $T_0 = 207$ K, and $n_n = 0$ because the polymer is unentangled. The CM equations (2.87) and (2.88) leading to the following predictions are $(1 - n_\alpha)B_\alpha = B_n$, and $(1 - n_\alpha)a_\alpha = a_n$. Both predictions have been verified [201].

2.3.2.32 Calorimetric Detection of JG Relaxation

One of the conventional methods to detect glass transition is by heat capacity measurement. Commercial instruments for thermal analysis such as DSC usually observe the response of the structural α-relaxation at $T_{g\alpha}$, but do not have the sensitivity to pick up that of the JG β-relaxation. This is obviously because the local primitive or JG β-relaxation involves the motion of lesser molecules and with smaller amplitudes than the α-relaxation, and consequently has smaller magnitudes in all thermodynamic and dynamic quantities. Nevertheless, use of the precision adiabatic calorimetry was able to detect heat capacity changes at another temperature $T_{g\beta}$ below $T_{g\alpha}$ in isopropylbenzene, propylene carbonate, and 1,3-diphenyl-1,1,3,3-tetramethyldisiloxane, and *o*-terphenyl by Oguni and coworkers [1048, 1075, 1416]. This heat capacity response arises from the JG β-relaxation, and $T_{g\beta}$ is called the (calorimetric) JG β-glass transition temperature at which $\tau_{JG}(T_{g\beta})$ is typically of the order of 10^3 s. Similarly, as discussed before in Sections 2.3.2.27 and 2.3.2.28, the JG β-relaxation of water in nanoconfinement and in hydrated proteins have been found by adiabatic calorimetry. Mentioned before in Section 2.3.2.16 is the detection by adiabatic calorimetry of the JG β-relaxation of cyclohexanol in its plastic crystalline phase [1048]. Thus adiabatic calorimetry is a very useful tool to detect JG relaxation particularly when it cannot be resolved by other techniques especially at long relaxation times of the order of 10^3 s.

Relation to the Correlation Index c of Fujimori–Oguni

Fujimori and Oguni (FO) [1075] have proposed to plot both $\log[\tau_\alpha(T)]$ and $\log[\tau_\beta(T)]$ against the scaled temperature, $T_{g\beta}/T$, where $T_{g\beta}$ is the β-glass temperature at which $\tau_\beta(T_{g\beta})$ is equal to 10^3 s. This plot is shown in Fig. 240 for four glassformers, propylene carbonate, 1,3-diphenyl-1,1,3,3-tetramethyldisiloxane (PMS), *o*-terphenyl, and isopropylbenzene. The conventional glass temperature, denoted by $T_{g\alpha}$, is defined as the temperature at which $\tau_\alpha(T_{g\alpha})$ is also equal to 10^3 s. FO defined a "correlation index," $c \equiv (1 - T_{g\beta}/T_{g\alpha})$, which they believe to be related to the size of the correlated region of molecules executing the α-relaxation at $T_{g\alpha}$. Small or zero value of c according to FO means the absence of correlation, and larger value of c indicates that the molecules form correlated region. Although this proposal by FO is interesting, a theoretical theory or model that takes FO from c to the size of correlation is absent.

It turns out that the "correlation index" c of FO and the coupling parameter n of the CM are linearly proportional to each other and given by $c = 14.7n/(3-\log\tau_{\beta\infty})$. This was shown in [251]. The belief of FO that their quantity c has something to do with cooperativity (or correlation) of molecular motions has a theoretical basis provided by the coupling model.

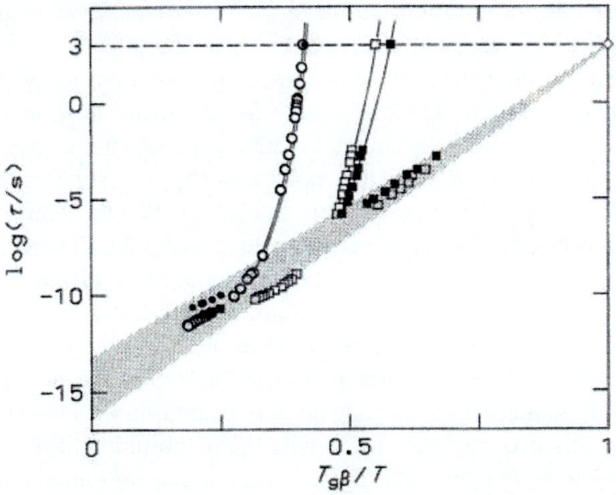

Fig. 240 Arrhenius plots of the α- and β-relaxation times τ against the inverse of temperature scaled by $T_{g\beta}$, the respective β-glass transition temperatures at $\tau = 1$ ks: (O) propylene carbonate ($c = 0.59$); (•) PMS ($c = 0.59$); (□) *o*-terphenyl ($c = 0.45$); (■) isopropylbenzene ($c = 0.43$). The *shaded area* represents the probable region of β-relaxation times. From [1075] of Fujimori and Oguni. They pointed out that all the relaxation times of the α-processes at high temperatures are well in the shaded region, indicating validity of their proposition that the β-process could be elementary for the α-process in what they called the correlated region of molecules. Reproduced from [1075] by permission

2.3.2.33 JG β-Relaxation Causes Cage Decay and Terminates the NCL

We have seen from many of the previous subsections that the primitive or JG β-relaxation is well connected to the α-relaxation. Without the JG β-relaxation as the precursor, the α-relaxation would not have taken place. There is yet another important role played by the JG β-relaxation in controlling the dynamics at earlier times when all molecules are confined within cages, each of which is formed by neighboring molecules acting by the intermolecular potential. Except at very high temperatures, the caged dynamics starts early at time of the order of ps for molecular glassformers after appearance of vibrations, and the loss is due to motion within the anharmonic potential of the cage and the fluctuations of the cages themselves. Unlike rotational and translational motions involved in secondary and primary relaxations, the loss $\varepsilon''(\nu)$ or $\chi''(\nu)$ observed in this caged regime has no characteristic relaxation time and thus its frequency dependence is a power law, $p\nu^{-c}$, with c positive and usually small. From the relation of susceptibility to intermediate scattering function and mean-squared displacement in Chapter 1, this loss corresponds to a mean-squared displacement with time dependence of $p't^c$. It continues indefinitely until the cages start to decay by the onset of a secondary relaxation involving the motion of the entire molecule, which is the primitive or JG β-relaxation. Non-JG secondary relaxation involving rotation of only a part of the molecule is not as effective as JG β-relaxation in this respect. At lower temperatures where τ_{JG} is long compared with ps, the loss within the cage, $p\nu^{-c}$, extends over many decades of frequency before ν_{JG} is reached when cage decay has already commenced. This necessarily means that the exponent c has to become smaller at lower temperature as experimentally observed (and longer τ_{JG}) for otherwise at times before reaching τ_{JG} the mean-squared displacement, whether translational or rotational, would be too large to be consistent with the molecules that are still caged. The caged dynamics observed as $p\nu^{-c}$ with $c << 1$ is naturally called the nearly constant loss (NCL), which was briefly discussed before in Section 2.3.2.6. It is also observed by quasielastic neutron scattering in terms of the mean-squared displacement $<u^2(t)>$ with t-dependence $\sim t^c$ with $c << 1$ at low enough temperatures. An example from $<u^2(t)>$ of water O atoms of hydrated protein over a range of temperature was given before in Fig. 228b. Anharmonicity of the intermolecular potential as well as the amplitudes of fluctuations of cages generally decreases with decreasing temperature, and hence this is one of the reason why the magnitude of the NCL decreases mildly with decreasing temperature. There is yet another reason which will become clear below.

The time when the cages start to decay marks the termination of the NCL regime [195, 508, 685], and τ_{JG} or τ_0 is the time of dissolution of cages through local relaxation by rotation and/or translation of the entire molecular unit. Actually we have seen this before from the discussion in Section 2.3.2.28 of the dynamic transition of biomolecules covered by water or other solvents. The $<u^2>$ measured by Mössbauer spectroscopy and neutron scattering at temperature below the dynamic transition temperature T_d comes from vibrations and NCL of caged dynamics (in some cases also the rotation of the methyl group), and exhibits mild increase with temperature. This NCL regime is terminated at T_d, beyond which $<u^2>$ rises rapidly with temperature. Via Figs. 222, 224, 233-3 to 233-8, we have verified that the primitive or JG β-relaxation of the solvent at T_d matches the timescale of the spectrometer used, and hence these experimental findings provide clear evidence of the termination of NCL by the primitive or JG β-relaxation.

This expected relation between the NCL and the primitive or the JG β-relaxation has been verified by dielectric relaxation data of ionic conductors and molecular glassformers [149, 1057–1059, 1061, 1417], which show that the primitive frequency $\nu_0 \equiv 1/(2\pi\tau_0)$ or the JG β-relaxation frequency $\nu_{JG} \equiv 1/(2\pi\tau_{JG})$ is located below but not far from $\nu_{NCL} \equiv 1/(2\pi t_{x1})$, the lower bound of the NCL frequency regime where $\varepsilon''(\nu) = p\nu^{-c}$ with $c << 1$. The loss in between ν_{NCL} and ν_0 or ν_{JG} represents the frequencies of transitory and local relaxation processes that transpire during course of cage decay. The dielectric loss data of NMEC, glycerol, propylene carbonate, PIB, PI in Figs. 133–136 and other glassformers [508] show the relation between ν_{NCL} and ν_{JG} with the latter estimated by the primitive relaxation frequency $\nu_0 \equiv 1/(2\pi\tau_0)$ of the CM calculated by Eq. (2.68). The termination of the NCL by the primitive or JG relaxation frequency is further illustrated for NMEC, threitol, and xylitol in Fig. 241 here. From this relation between NCL and the primitive or JG relaxation, argument has been advanced to show the level of NCL of the same glassformer is inversely proportional to $\log(t_{x1}) - \log(t_{ons})$, where t_{ons} is the onset time of the NCL [507].

Onset Time of NCL

It is worthwhile to mention that the time regime of NCL does not extend indefinitely to shorter times. The onset time of NCL, t_{ons}, in van der Waals liquids is about 2 ps,

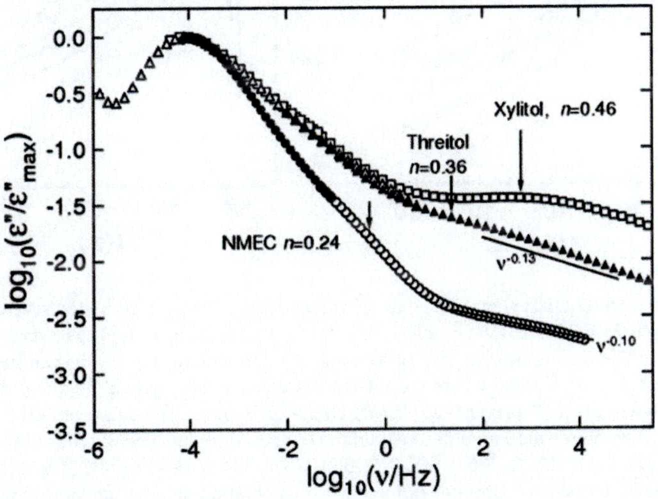

Fig. 241 The dielectric loss data scaled by the maximum loss of three molecular glassformers by NMEC ($n = 0.24$), threitol ($n = 0.36$), and xylitol ($n = 0.46$) that have approximately the same primary relaxation time τ. The *vertical arrows* indicate the location of the calculated ν_0. Reproduced from [149] by permission

which can be gathered by inspection of the signal $S(t)$ obtained by optical heterodyne detection (OHD) of the optical Kerr effect (OKE) in salol [330a] and OTP [330b] shown in Fig. 242. With the appropriate selection of polarization conditions, the imaginary part of the Fourier transform of the OHD-OKE signal, $S(t)$, is proportional to susceptibility $\chi''(\omega)$ data obtained from depolarized light scattering, i.e., $\chi''(\omega) \propto \mathrm{Im\,FT}\{S(t)\}$. Thus, if $S(t) \propto t^{-1+c}$, then $\chi''(\omega) \propto \omega^{-c}$. At 247 K for salol and 284 K for OTP, $S(t) \propto t^{-1+c}$ and $\chi''(\omega) \propto \omega^{-c}$, with $c << 1$ is the NCL, and it can be seen that it starts at $t \sim 2$ ps. Therefore, $t_{\mathrm{ons}} \sim 2$ ps. The OHD-OKE was applied to other supercooled molecular liquids including benzophenone and 2-biphenylmethanol [332]. Like salol and OTP, the data show the onset of the intermediate power law (IPL), $S(t) \propto pt^{-1+c}$ with $c << 1$, at $t_{\mathrm{ons}} \sim 1-2$ ps. In all cases, this NCL feature in the dynamics is followed in the course of time by the merged JG relaxation and the α-relaxation which decays exponentially at long times, suggesting that the caged dynamics manifested by the NCL is terminated by the relaxation. The so-called intermediate power law (IPL) of $S(t) \propto t^{-1+c}$ with $c << 1$ is the time-domain equivalent of the near constant loss (NCL) observed in susceptibility spectrum [1056, 1062].

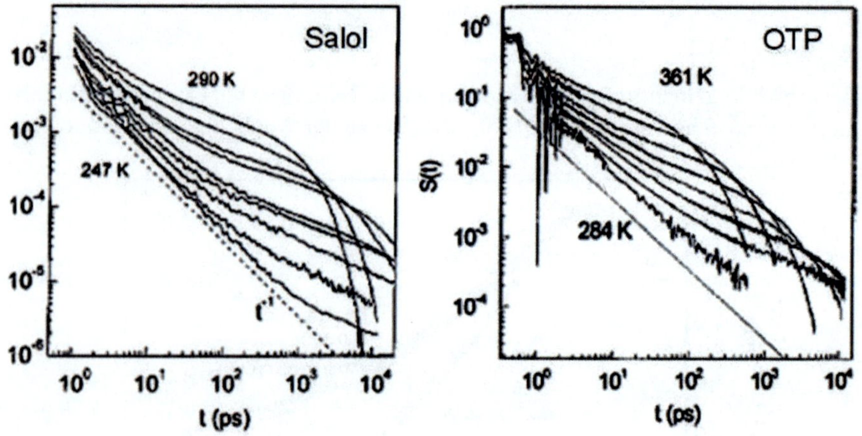

Fig. 242 (*Left*) OHD-OKE signal $S(t)$ data of salol from ~1 ps to ~10 ns for several temperatures: 247, 250, 253, 257, 261, 266, 270, 280, and 290 K (*bottom to top*). The data are scaled, $S(t = 0) = 1$. The data appear to be a power law at intermediate times. This can be seen most clearly in the $T = 247$ K data, where the intermediate power law span is $2 < t < 500$ ps. The *dashed line* represents a t^{-1} power law. (*Right*) OHD-OKE signal $S(t)$ data from OTP plotted for some of the temperatures measured to illustrate the temperature-dependent trends in the data. The temperatures are 284.0, 292.0, 296.0, 303.9, 309.8, 319.7, 334.5, and 361.0 K. The data were normalized at 300 fs. The slowest portions of the lowest temperature data sets are not shown. The data sets at each temperature have the same basic shape, but as the temperature is lowered, the time span of the data expands. The *dashed line* represents a t^{-1} power law. Reproduced from [330a, 330b] by permission

Molecular dynamics simulations of OTP by Wahnström and Lewis also show the onset of NCL is at about 2 ps at temperatures equal to 205 K and below it (see Fig. 6) [315].

The coincidence of t_{ons} and t_c, the crossover time from primitive relaxation to Kohlrausch α-relaxation appearing in the CM equation (2.67), may not be accidental because both are indicators of the time when intermolecular interaction starts to exert its influence manifested by either caging or α-relaxation. In terms of the theoretical basis of the CM discussed in Section 2.2.1.3, whether it is caging or α-relaxation, the physics involves onset of chaos in non-linear Hamiltonian dynamics, and their onset times are naturally the same.

NCL in CKN

The dielectric relaxation measurements of $0.4Ca(NO_3)_2$–$0.6KNO_3$ (CKN) by Lunkenheimer and coworkers [1034, 1418, 1419] over a wide frequency range from 4×10^{11} Hz to 10^{-2} Hz and consistent with those obtained published earlier by Howell et al. at lower frequencies and down to the glassy state [55] are most revealing on all the relaxation mechanisms. Although conductivity relaxation is measured by dielectric relaxation, its conductivity relaxation time is not much different from the structural relaxation time at temperatures higher than 373 K, indicating the two relaxations are coupled together. For frequencies higher than 0.1 GHz and temperatures above the putative MCT critical temperature T_c=375 K, the dielectric loss spectra exhibit minima consistent with neutron scattering [1420] and depolarized light scattering (DLS) [1421] in the same spectral range. However, the "knee" found in DLS spectra at lower temperatures [1421], and supposedly verifying a prediction of the idealized MCT, turned out to be an experimental artifact [1422].

In reality, at lower temperatures, the NCL appears in the dielectric loss spectra, and its frequency range increases with decreasing temperature [195, 1057, 1058]. Shown in the inset of Fig. 243 are the electric loss modulus, $M''(\nu) \equiv \mathrm{Im}M^*(\nu)$, data of $0.4Ca(NO_3)_2$–$0.6KNO_3$ (CKN) at 342 K from the Lunkenheimer et al. For definition of the complex electric modulus $M^*(\nu)$, see Section 1.2.2. The glass transition temperature T_g is 333 K for CKN. The dashed line is the fit to the $M^*(\nu)$ data by the imaginary part of $M_n^*(\nu)$, obtained from the Fourier transform of the Kohlrausch function, $\exp[-(t/\tau_K)^{1-n}]$, with $n = 0.34$ or $\beta \equiv (1-n) = 0.66$ and $\tau_K = 6.6 \times 10^{-4}$ s according to Eq. (1.56). From this fit, the corresponding $\varepsilon_n^*(\nu)$ is obtained from the identity $\varepsilon_n^*(\nu) = 1/M_n^*(\nu)$. The main figure shows the data represented by $\varepsilon''(\nu)$ and the dashed line is the imaginary part of $\varepsilon_n^*(\nu)$. The ε'' data at 342 K demonstrate the presence of the NCL over about three decades, and there is a crossover to a power law $\varepsilon''(\nu) \propto \nu^{-\beta}$ at lower frequencies, with $\beta = 1-n$, and eventually $\varepsilon'' \propto \nu^{-1}$ at even lower frequencies in the dc conductivity regime. The data in the M'' representation show also the crossover from NCL to the power law $M'' \propto (\nu)^{-\beta}$. The data $\varepsilon''(\nu)$ deviate from Im $\varepsilon_n^*(\nu)$ at higher frequencies and we determine ν_{x2} as the frequency above which the deviations are more than 10%. Similarly, we determine ν_{x1} as the frequency below which the data is more than 10% larger than the ν^{-c}-dependence ($c << 1$) chosen to represent

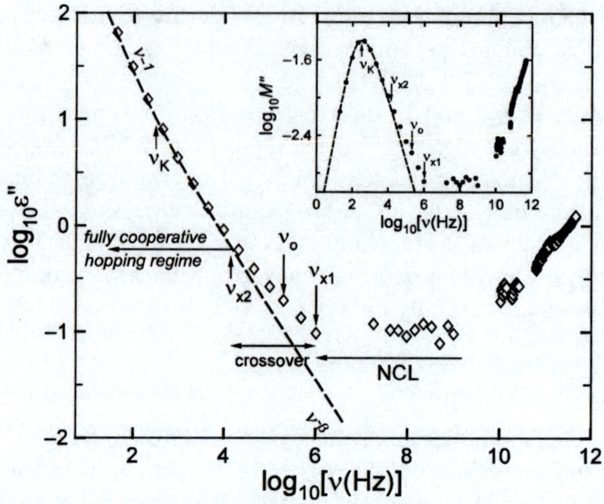

Fig. 243 Dielectric loss ε'' as a function of frequency of CKN at 342.5 K showing the existence of the NCL over three decades in frequency. The *dashed line* is Im $\varepsilon_n^*(\nu)$ from the fully cooperative ion conductivity relaxation calculated from $M_n^*(\nu)$, the fit by the Kohlrausch function with $n = 0.34$ to the complex electric loss modulus data. The *dashed line* in the *inset* shows this fit to the imaginary part of M^*, the experimental data. The deviation of the data from the Kohlrausch fit at higher frequencies is marked by one crossover frequency, ν_{x2}. The deviation of the data from the NCL at lower frequencies is marked by the other crossover frequency, ν_{x1}. The locations of ν_K and the calculated independent relaxation frequency, ν_o, are also indicated. Data after [1034, 1418, 1419]. Reproduced from [195] by permission

the NCL. When data are not sufficient as in the present case, the ν_{x1} shown in the figure is somewhat arbitrary. Either from $\varepsilon''(\nu)$ or $M''(\nu)$, we can discern three spectral regimes discussed in the previous section: (1) an NCL regime with $\alpha \approx 0$ for $\nu > \nu_{x1}$ or $t < t_{x1} \equiv 1/(2\pi\nu_{x1})$, (2) an intermediate crossover regime defined by $\nu_{x1} > \nu > \nu_{x2}$ or $t_{x1} < t < t_{x2} \equiv 1/(2\pi\nu_{x2})$, and (3) the fully cooperative regime for $\nu < \nu_{x2}$ or $t > t_{x2}$, where the Fourier transform of the Kohlrausch function fits well the $M''(\nu)$ data. The location of the primitive relaxation frequency, ν_0, is calculated by the CM equation from $n = 0.34$ and $\tau_K = 6.6 \times 10^{-4}$ s with $t_c = 2$ ps, previously deduced for CKN from high-frequency/high-temperature measurements (see Fig. 8). Note that a factor of two uncertainty in determining t_c or ν_c introduces via the CM equation uncertainty of only 2^n in ν_0, which is negligible for the present purpose. The most important point to make is that ν_0 lies well inside the intermediate time regime and satisfies the inequality $\nu_{x1} >> \nu_0 >> \nu_{x2} >> \nu_K$ or $t_{x1} << \tau_0 << t_{x2} << \tau_K$ which is necessary in order to justify the dual roles of the primitive relaxation, i.e., the terminator of the caged or the NCL regime and the precursor of the fully cooperative relaxation.

Having explained the procedure of determining the crossover frequencies by an example, the isothermal $\varepsilon''(\nu)$ data for many temperatures obtained by Lunkenheimer and coworkers had been analyzed in the same manner [1057,1058].

Some representative data at $T = 342$, 350, 356, and 361 K are reproduced here in the left panel of Fig. 244. At 361 K, $\varepsilon''(\nu)$ already exhibits a very flat minimum from 10^{10} to 10^8 Hz. The NCL is even more evident in $\varepsilon''(\nu)$ at lower temperatures. Isothermal data similar to that shown for 342 K were produced by Lunkenheimer et al. at 325.0, 350.3, 356.4, 361.0, 365.4, 370.0, 385.0, 393.0, 417.0, 440.0, and 468.0 K. Kohlrausch fits to these data in the electric modulus representation had previously been done by Lunkenheimer [1034], and from his work the parameters, n and ν_K, are obtained for all temperatures. From these parameters, ν_0 are calculated. Analyses of these data at other temperatures similar to that shown for 342 K give the parameters, ν_{x1} and ν_{x2}, which define the frequency regimes for each temperature. The primitive relaxation frequency, $\nu_o \equiv 1/(2\pi\tau_o)$, is found to be located between ν_{x1} and ν_{x2}.

In the right panel of Fig. 244-1, ν_{x1}, ν_0, ν_{x2}, and ν_K are plotted as a function of temperature. At all temperatures where these four frequencies can be determined, the inequality $\nu_{x1} >> \nu_0 >> \nu_{x2} >> \nu_K$ is satisfied, and it shows once again that the primitive relaxation terminates the caged dynamics in the NCL regime and heralds the onset of the cooperative processes leading finally to the

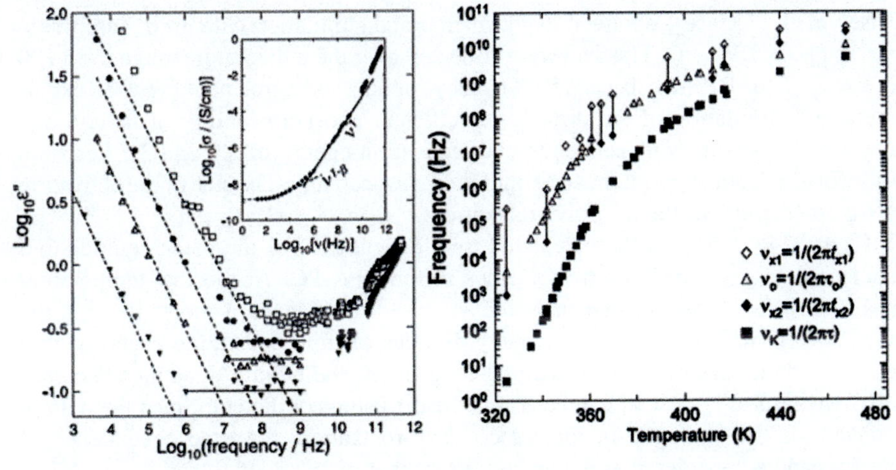

Fig. 244-1 (*Left*) The *inset* shows the CKN data of $\log_{10}\sigma'(\nu)$ vs. log(frequency/Hz) of CKN at T=342 K to show the near linear frequency dependence of the NCL in Fig. 1. The *dashed line* that includes the dc conductivity and the high-frequency power law $\nu^{1-\beta}$ or ν^n corresponds to the Kohlrausch fit to the electric modulus (shown by the dashed line in the inset of Fig. 243) with $n = 0.34$. In the main figure, the *symbols* are dielectric loss data of CKN by Lunkenheimer et al. at T=342, 350, 356, and 361 K (going from *bottom* to *top*). The *dashed lines* are the loss due to cooperative ion hopping calculated from the Kohlrausch fits to the electric modulus data and converted to a dielectric loss function of frequency. The crossover frequencies ν_{x1} and ν_{x2} (not shown) are determined from these data in the same manner as explained in the text (and in Fig. 242). (*Right*) *Solid squares* are the Kohlrausch conductivity relaxation frequency ν_K obtained from the fits to the data of CKN [1034, 1418, 1419] at various temperatures in the electric modulus representation. *Open triangles* are the primitive frequencies ν_o calculated from the fits. The *open* and *closed diamonds* are the crossover frequencies, ν_{x1} and ν_{x2}. Reproduced from [195] by permission

Kohlrausch relaxation. The crossover region is narrower at higher temperatures. Although broader at lower temperatures, the region is only about two and a half decades wide at 324 K, the lowest measurement temperature below T_g. The separation between ν_0 and ν_K becomes smaller at higher temperatures. This trend is expected from the CM relation, $(\log \nu_0 - \log \nu_K) = n[\log \nu_c - \log \nu_K]$, where $\nu_c \equiv 1/2\pi t_c$, because of the increase of ν_K and also the possible decrease of n on increasing temperature.

The interpretation of the evolution of dynamics from NCL to primitive or JG relaxation and to the cooperative relaxation given above for CKN which is both an ionic conductor and a glassformer applies verbatim to glassy as well as crystalline ionic conductors, the dielectric spectra of which show all the common features [1057, 1058]. There are plenty of such examples, and the common features will be discussed in Chapter 3 on universal properties of many-body relaxation. The same remark applies to experimental data and interpretation of polymeric and non-polymeric molecular glassformers, inorganic and metallic glassformers and colloidal suspensions.

Evidence from Helium Transport in Glassy BPA-PC

Gusev et al. [1354(e)] obtained experimental and simulation data on helium transport in glassy BPA-PC. The data were collected over the temperature range from 100 to 300 K (T_g of BPA-PC is ca. 450 K). They found good agreement between experiment and simulation of the diffusion coefficient and permeability of helium and they have Arrhenius T-dependence with activation energy of 3.9 and 3.7 kcal/mol for diffusion from experiment and simulation, respectively. The size of the activation energy is like that of the primitive relaxation.

Figure 244-2 depicts the calculated time dependence of the mean-squared displacement $< r^2 >$ of helium molecules in the BPA-PC. At the low temperature of 110 K, after a few picoseconds the $< r^2 >$ of the helium molecules has the $< r^2 > \propto t^c$ with $c \sim 0.14$ over many decades of time, indicating caged helium molecule dynamics or NCL. Finally at long times, the helium dynamics becomes diffusive with $< r^2 > = 6Dt$, where D is the diffusion coefficient. Since the helium molecule is the analogue of local secondary relaxation, the time dependence of $< r^2 >$ can be considered as evidence for termination of caged dynamics (or NCL) by onset of secondary relaxation.

NCL Is Not an Additive Contribution

Although the presence of NCL is clear from the dielectric (Fig. 240) and mechanical (Figs. 169 and 170) spectra, and the optical Kerr effect (OHD-OKE, Fig. 241), none of these spectra can rule out that the NCL is an additive contribution to the secondary and primary relaxation that appears at longer times. The data cannot rule out that the NCL can exist at even longer times than these relaxations. In fact, the high-frequency dielectric spectra of glycerol has fitted successfully by assuming $\varepsilon''(\nu) = c_1\nu^{-b} + c_2\nu + c_3$, where the last term represents the NCL [682(b)]. The

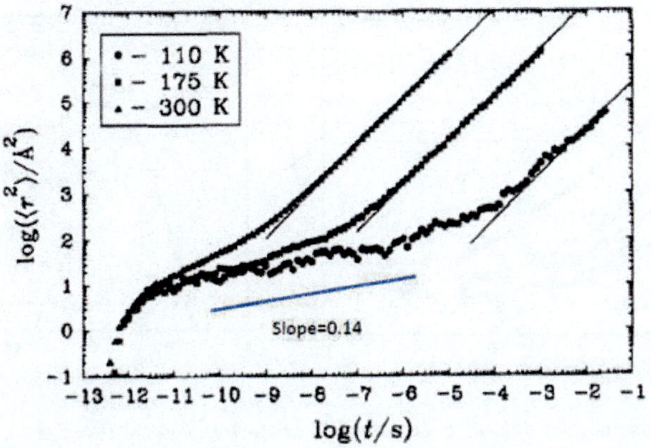

Fig. 244-2 Dynamics of helium molecules in BPA-PC. *Symbols* at 300 and 175 K show averages over 1500 random walks (500 walks in each of the three microstructures) and *symbols* at 110 K over 150 random walks. *Solid lines* with slope 1 exactly represent the Einstein diffusive mode $< r^2 >= 6Dt$. The onset of Einstein diffusion occurs at several dozen angstroms. Reproduced from [1354(e)] by permission

OHD-OKE signal, $S(t)$, is fitted by the function $S(t) = [pt^{-1+c} + dt^{b-1}]\exp(-t/\tau_\alpha)$, where the first term with $c << 1$ corresponds to the NCL and is assumed to be an additive term [332]. These assumptions are at odds with our idea that NCL is terminated at by primitive or JG relaxation. Our idea can be substantiated by comparing dielectric relaxation and dynamic light-scattering experimental data of CKN in the same temperature and frequency ranges. The comparison is made in Fig. 245 [1451]. While the dielectric data admit the possibility that the NCL is an additive contribution to the observed $\varepsilon''(\nu)$, the dynamic light-scattering data of $\chi''(\nu)$ at 468 K clearly rule this out, because if otherwise $\chi''(\nu)$ would show a flat loss at low frequencies which is not observed.

NCL in Ionic Liquid

The orientational dynamics of the supercooled ionic organic liquids *N*-propyl-3-methylpyridinium *bis*(trifluoromethylsulfonyl)imide (PMPIm) were studied by optically heterodyne-detected optical Kerr effect (OHD-OKE) in the short time range from 1 to about 3×10^5 ps [333]. The shape of the decay curve is very similar to those that have been observed earlier for supercooled molecular liquids by the same technique, including the onset of the intermediate power law (IPL), pt^{-1+c} with $c << 1$, at about 2 ps, to be followed by the merged JG relaxation and the α-relaxation which eventually decays exponentially. At 233 K, the IPL appears in the range from 2 to 30 ps [333]. Thus this is another example of NCL terminated by the JG relaxation now merged with the α-relaxation. As an aside, the authors of the OHD-OKE study interpreted their data in terms of the two-correlators schematic MCT model with the orientation coupled to density fluctuations by Götze and Sperl

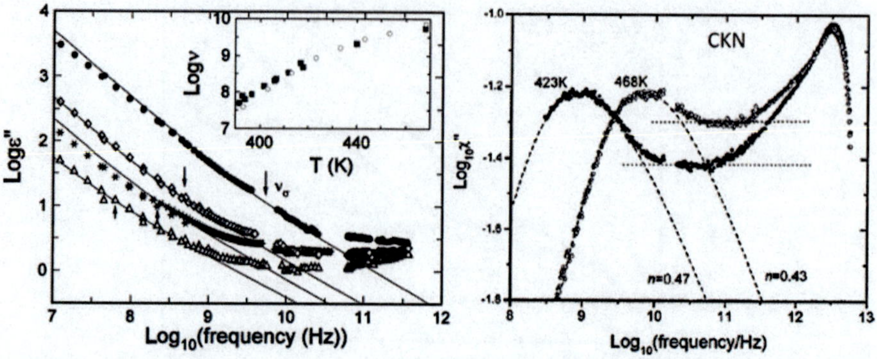

Fig. 245 (*Left*) Isothermal data of $\varepsilon''(\nu)$ of CKN taken by Lunkenheimer et al. at higher temperatures of 393, 412, 417, and 468 K (from *bottom* to *top*). Each *line* is obtained by fitting the isothermal electric modulus data $M^*(\nu)$ by the Kohlrausch function and transforming the obtained fit to $\varepsilon''(\nu)$. The *inset* compares the temperature dependence of the frequency of the structural relaxation peak in $\chi''(\nu)$ from the dynamic light-scattering measurement of Li et al. [1421] on CKN (*open circles*) with the frequency of the $M''(\nu)$ peak (*filled squares*). (*Right*) $\chi''(\nu)$ data from the dynamic light scattering [1421] of CKN at 423 and 462 K. The peak at lower frequencies is the structural relaxation. The *dashed lines* are the fit by the imaginary part of the Fourier transform of the Kohlrausch functions with parameters given by Li et al. Reproduced from [1451] by permission

[34]. However, as acknowledged by the authors themselves [333], the empirical equation they used to analyze the data is not compatible with MCT. For more discussion of IPL, pt^{-1+c} with $c << 1$, and relation to NCL and MCT, see Section 2.3.2.37.

NCL in Metallic Glasses

Another example is the isochronal mechanical loss G'' spectrum of metallic glasses presented in Fig. 169 by the 1 Hz shear loss modulus data of $Pd_{43}Ni_{10}Cu_{27}P_{20}$ [956], and in Fig. 170 by the 5.4 kHz shear loss modulus data of $Zr_{65}Al_{7.5}Cu_{27.5}$ [957]. They clearly show the presence of both the NCL and the JG β-relaxation either resolved or in the form of the excess wing. The NCL is identified in the semilog plot of G'' vs. T by its characteristic weak T-dependence, approximately proportional to $\exp(T/T_0)$ in these figures, also found in several molecular glassformers by dielectric, mechanical, and spin-lattice relaxation spectroscopies (see Fig. 6 in [971]). The presence of both the excess wing (or IPL with $c \approx 0.2$ or other similar values) and the NCL (IPL with $c \approx 0$, or logarithmic decay of orientation correlation function) in metallic glass is interesting because there is no rotation. Hence rotational–translational coupling is absent in metallic glass, and the recent versions of MCT by Götze and Sperl [34, 35] and Cummins [1119] that incorporate rotational–translational coupling cannot explain the IPL with $c \approx 0$ or NCL in metallic glasses. The isochronal dynamic shear modulus data, $G^*(T,\nu) = G'(\nu) + iG''(\nu)$, of the metallic glass $Pd_{43}Ni_{10}Cu_{27}P_{20}$ at $\nu = 1$ Hz by Pelletier et al. were able to resolve the JG relaxation as a shoulder instead of showing up as an excess wing [956] (see Fig. 169). At temperatures lower than about 150°C, the temperature dependence of

$G''(\nu)$ becomes weak and indicates it is the contribution from the NCL. The data demonstrate once more that it is the JG relaxation terminating the NCL.

Extending the Frequency Range of NCL in Glycerol to 12 Decades

It is interesting to put together the observation of NCL in $\chi''(\nu)$ of glycerol ($T_g = 185$ K) at 160 K by dynamic light scattering in the 1–40 GHz range by Kisluik et al. [1423] (see Fig. 1b in this reference), and also in $\varepsilon''(\nu) = p\,\nu^{-c}$ with $c \approx 0.07 \pm 0.02$ at nearly the same temperatures by dielectric spectroscopy in the $10^{-2} - 10^5$ Hz range by Kudlik et al. [971] (see Fig. 3 and Eq. (7) in this reference). The NCL observed in these two widely different frequency ranges at the same temperature has the same property and is likely exemplification of the same physical process. Although no data of either $\chi''(\nu)$ or $\varepsilon''(\nu)$ were obtained in the intermediate frequency range from 10^5 to 10^9 Hz, if measurable at 160 K, in all likelihood they are also NCL interpolating the low-frequency $\varepsilon''(\nu)$ data and the high-frequency $\chi''(\nu)$ data. This is because there is no other relaxation process in glycerol that will appear within the intermediate frequency region at 160 K that one can think of. The excess wing hiding the JG relaxation and the α-loss peak of glycerol at 160 K are located at lower frequencies than 10^{-2} Hz. The result indicates that NCL pervades over the entire frequency range from 10^{-2} Hz to 40 GHz (over 12 decades) at 160 K and lower temperatures, and is terminated by the JG relaxation residing in the excess wing. Moreover, the dielectric loss data of glycerol [971] show that between 160 K and $T_g = 185$, the NCL is 2–3 orders of magnitude smaller than the maximum loss of the α-relaxation. The small NCL relative to the α-relaxation in glycerol is consistent with the small value of the coupling parameter, $n(T_g) = 0.29$ (obtained from the Kohlrausch fit of dielectric dispersion of the α-relaxation of glycerol [195]), and the correlation between the size of NCL and n [507] to be discussed in a subsection after the following one.

2.3.2.34 Change of T-Dependence of NCL at T_g in Analogy to the JG β-Relaxation Strength

The connection that NCL has with the primitive or JG β-relaxation discussed in the above has other consequences. One consequence manifests itself by the change of temperature dependence of NCL from a weaker increase with temperature in the glassy state ($T < T_g$) to a stronger one $T > T_g$ in the liquid state [115, 149, 507, 1056, 1424]. Examples include CKN shown here in Fig. 246 and PIB and PI shown before in Figs. 135 and 136. The sensitivity of the NCL to glass transition at T_g parallels that of the JG β-relaxation which has its dielectric strength as well as its relaxation time changing temperature dependence when crossing T_g as discussed before in Sections 2.3.2.15 and 2.3.2.17. This parallel behavior implies the existence of a connection between NCL and JG β-relaxation. The connection comes from the fact that the primitive relaxation or the JG β-relaxation is the terminator of the NCL [507]. For a derivation of the observed change in T-dependence of the NCL when crossing T_g from the latter fact, see [507].

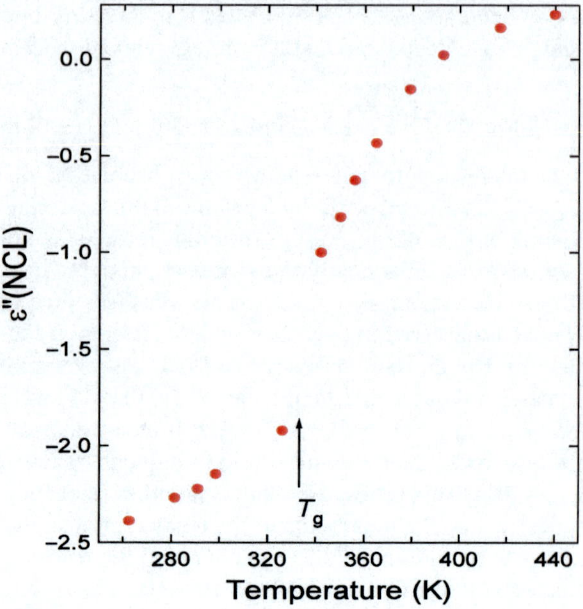

Fig. 246 The dielectric NCL intensity of CKN as a function of temperature suggesting a change in slope near $T_g = 333$ K

Support from Colloidal Particles Suspension by Confocal Microscopy Experiment

To support the interpretation of NCL as the manifestation of caged molecular dynamics and the caged decay and termination of NCL as caused by the onset of primitive or JG β-relaxation given above, we cite the experimental results obtained on colloidal particles (hard spheres) suspension by confocal microscopy experiment by Weeks et al. [141(a)] shown before in Fig. 1, and by Weeks and Weitz [141(b)]. Here we select the data in the case of volume fraction $\phi = 0.56$ and reproduce them in Fig. 246. This spectroscopy has the advantage that motions of all colloidal particles can be observed continuously as a function of time, and hence the change in dynamics is seen directly. In the right panel, Weeks et al. [141(a)] show from the trajectories of particles that they spent most of their time confined in cages formed by their neighbors and moved significant distances only during quick and rare cage rearrangements. A typical particle takes average time τ_c of 500 s to shift position and leaves the cage. This time, identifiable with the primitive relaxation of the CM, is indicated by the vertical arrow in the log–log plot of mean-squared displacement, $< \Delta x^2(\Delta t) >$, against Δt in the left panel of Fig. 247 and in Fig. 1 (further discussion of the other issues in the figure and the insets are deferred till Chapter 3). It can be seen that $< \Delta x^2(\Delta t) > \sim (\Delta t)^c$ with $c \approx 0.13$ at times shorter than 200 s, during which effectively the particles are all confined in cages because τ_c of 500 s has not yet been reached. The $< \Delta x^2(\Delta t) >$ for $\Delta t < 200$ s corresponds to susceptibility with frequency dependence $\chi''(\nu) \sim (\nu)^{-c}$ with $c \approx 0.13$, which is the NCL in the caged regime. The lower-right inset is a plot of the cage correlation function $C_{cage}(\Delta t)$ against Δt for three systems with $\phi = 0.56$, 0.52, and 0.46 (from top

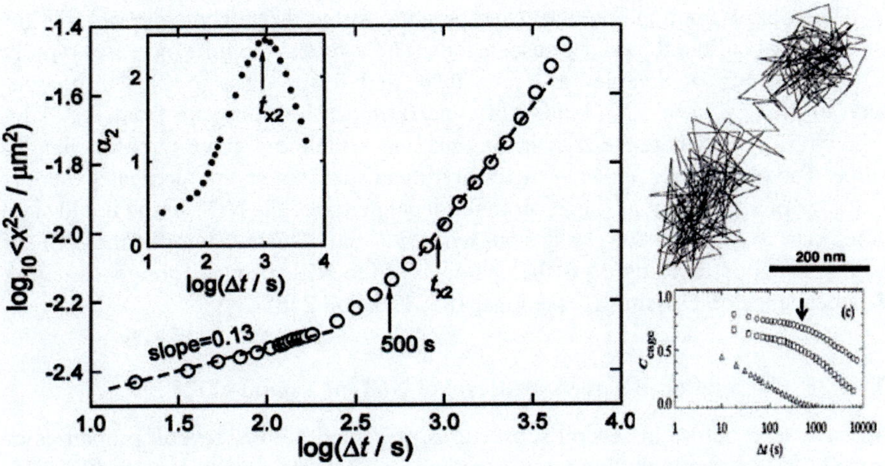

Fig. 247 (*Left*) Mean-squared displacement $<\Delta x^2(\Delta t)>$ for volume fractions $\phi = 0.56$ from Weeks et al. [141]. One *vertical arrow* indicates $\Delta t = 500$ s, the time when a typical particle shifts position and leaves the cage determined by confocal microscopy experiment [141]. The other *vertical arrow* indicates the time t_{x2} when the non-Gaussian parameter $\alpha_2(\Delta t)$ assumes its maximum as shown in the *inset* on the *upper-left corner*. The *dashed line* has slope 0.13 that indicates the NCL regime. In the *lower-right corner* is a plot of the cage correlation function $C_{cage}(\Delta t)$ against Δt for three systems with $\phi = 0.56$, 0.52, and 0.46 (from *top* to *bottom*), and the *vertical arrow* indicates $\Delta t = 500$ s. (*Right*) A 2D representation of a typical trajectory in 3D for 100 min for $\phi = 0.56$ from Weeks et al. [141] to illustrate that particles spent most of their time confined in cages formed by their neighbors and moved significant distances only during quick rare cage rearrangements. The particles shown took ~500 s to shift position. Reproduced from [141(a), 141(b)] by permission

to bottom), and the vertical arrow indicates $\Delta t = 500$ s. It can be seen that for $\phi = 0.56$, $C_{cage}(\Delta t)$ undergoes significant decay starting at $\Delta t = 500$ s.

At $\phi = 0.60$, the colloidal particles are in the glassy state, and the change from $\phi = 0.56$ to $\phi = 0.60$ in Fig. 1 is the analogue of cooling a glassformer from the supercooled liquid state to the glassy state. The caged regime where $<\Delta x^2(\Delta t)>\sim (\Delta t)^c$ extends to longer times because the average time τ_c of particles leaving the cage is longer. There is concomitant decrease in the magnitude of $<\Delta x^2(\Delta t)>$ due to decrease in size of cage with increase in ϕ, and the exponent c becomes smaller. All these properties have analogues in the behavior of the NCL of glassformers on cooling, as can be seen in the data of NMEC, CKN, glycerol, propylene carbonate, etc., as well as the inter-comparison (see Fig. 241). In particular, the analogue of τ_c is τ_0 or τ_{JG} in glassformers. The clear picture of motions of colloidal particles from confocal microscopy leads to unequivocal understanding of the origin of the NCL, which is in accord with our interpretation as the loss due to motions within the cages defined by the anharmonic intermolecular potentials and fluctuations of the cages themselves. The cages decay and the NCL regime is terminated starting at times of the order of the primitive or JG relaxation time. This relation between the NCL and the primitive or JG relaxation explains why their properties are interrelated.

The volume fraction ϕ replaces temperature as the thermodynamic variable for the athermal colloidal particles suspensions. The α-relaxation time increases rapidly with increasing ϕ, diverges at $\phi_g \approx 0.58$, and forms glass [142]. The magnitude of NCL at $\phi = 0.53$, 0.56, 0.60, and 0.61 can be estimated from Fig. 1 by $< \Delta x^2(\Delta t_{x1}) >$, where the Δt_{x1} is the time near which $\log(< \Delta x^2(\Delta t) >)$ changes slope. The plot of $< \Delta x^2(\Delta t_{x1}) >$ linearly against ϕ (not shown) indicates there is a change of slope near ϕ_g, i.e., a stronger dependence of the NCL in the liquid state when $\phi < \phi_g$ than in the glassy state when $\phi > \phi_g$. This behavior of the NCL of hard spheres on crossing ϕ_g is the analogue of the NCL of glassformers changing T-dependence on crossing T_g (see Figs. 135, 136, and 246).

2.3.2.35 Correlation Between the Level of NCL at T_g and $n(T_g)$

As have seen before in several subsections presented before, several properties of the JG β-relaxation including relaxation time and strength are correlated with $n(T_g)$, the frequency/time dispersion parameter of the α-relaxation. Combining this with the connection between the NCL with the JG relaxation shown in Section 2.3.2.33, it is not surprising that some correlation between the NCL and $n(T_g)$ can be expected and shown to be true in the following.

From Quasielastic Neutron Scattering

The NCL at high frequencies and low temperatures near and below T_g can be probed by quasielastic neutron scattering (QENS). From the incoherent scattering function, $S(Q, \omega, T)$, the elastic part of the scattering, $S_{el}(Q, \Delta, \omega, T)$, is operationally defined by the integral of $S(Q, \omega, T)$ over ω within $-\Delta\omega < \omega < \Delta\omega$, where Q is the momentum transfer and $\Delta\omega$ is the resolution frequency width. After normalizing $S_{el}(Q, \Delta\omega, T)$ measured at temperature T by its value at $T = 0$, $S_{el}(Q, \Delta\omega, T = 0)$, one defines a Debye–Waller factor, $W(Q, \Delta\omega, T)$ and a mean-squared displacement $< u^2(T) >$ of the fast relaxation by

$$\begin{aligned} S_{el}(Q, \Delta\omega, T)/S_{el}(Q, \Delta\omega, T = 0) &= \exp[-2\,W(Q, \Delta\omega, T)] \\ &= \exp[- < u^2(T) > Q^2/3]. \end{aligned} \tag{2.93}$$

Thus, $< u^2(T) >$ is a measure of intensity of the fast relaxation after the vibrational contribution has been removed. If temperature is near and below T_g, this fast relaxation has to be the relaxation in the caged regime or the NCL. This is because at these low temperatures and short times corresponding to the resolution width of neutron scattering spectrometers, neither the α-relaxation nor the primitive or JG relaxation can contribute in many glassformers. Even water in aqueous mixtures or in nanoconfinement, its primitive or JG relaxation contributes to quasielastic neutron scattering only above T_g. The spectrometers IN6, IN10, IN13, and IN16 of the Institut Laue-Langevin have energy resolutions (FWHM) of 80, 1, 8, and 0.9 μeV, respectively. The elastic intensities measured by IN6, IN10, IN13, and

IN16 include fast relaxation with frequencies higher than 19, 0.24, 1.9, and 0.22 GHz, respectively. The T-dependences of the fast relaxation intensity from neutron scattering data of different glassformers obtained using the same neutron spectrometer expressed as a mean-squared displacement, $< u^2(T) >$, are shown in Fig. 248.

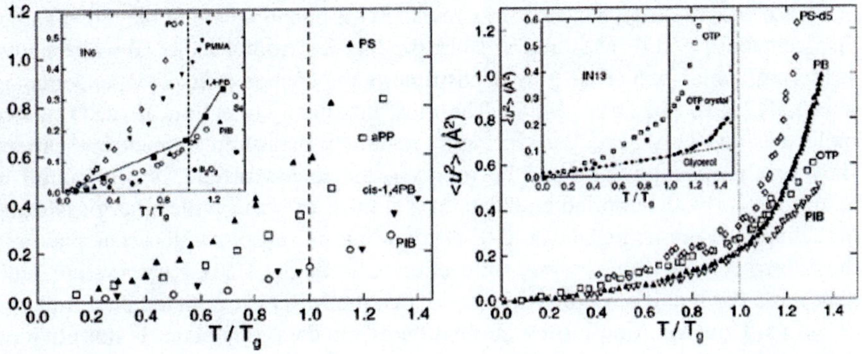

Fig. 248 (*Left*) The *inset* shows the mean-squared displacement $<u^2>$ as a function of T/T_g of several glassformers (B_2O_3, Se, 1,4-polybutadiene (1,4-PB), PMMA, and bisphenol A polycarbonate (PC)) obtained from incoherent neutron scattering by the IN6 spectrometer of the Institut Laue–Langevin together with the data on PIB taken with the LAM-40 spectrometer by Kanaya et al. The main figure shows $< u^2 >$ of a number of amorphous polymers (PIB, *cis*-1,4-polybutadiene (*cis*-1,4-PB), atactic polypropylene (aPP), and polystyrene (PS)) obtained using the LAM-80ET spectrometer with 0.02 meV energy resolution by Kanaya et al. (*Right*) The mean-squared displacement $<u^2>$ as a function of T/T_g for several polymeric glassformers (deuterated polystyrene (PS-d5), polybutadiene (PB), and PIB) together with *o*-terphenyl (OTP) from incoherent neutron scattering obtained with the IN10 spectrometer, except for PB, which was obtained with the IN16 spectrometer. The *inset* shows data on OTP, crystalline OTP, and glycerol obtained with the IN13 spectrometer. Reproduced from [115(a)] by permission

From these figures it can be seen that $< u^2(T) >$, identified here as the NCL, exhibits a change in slope around T_g. For the same class of glassformers, their NCL can be compared after temperature has been scaled by the respective T_g as already been done in Fig. 248 where $< u^2(T/T_g)>$ is plotted against T/T_g. As shown before using experimental data in [115(a), 1424] glassformers having larger $n(T_g)$ have stronger increase of $< u^2(T/T_g) >$ or NCL with increasing T/T_g in the glassy state. The correlation is particularly strong when glassformers of the same class are compared. From this it follows at once that, at $T = T_g, < u^2(T_g) >$ is larger for glassformer with larger $n(T_g)$. Also found above T_g is that glassformer with larger $n(T_g)$ has more rapid increase of $< u^2(T/T_g) >$ as a function of T/T_g. These correlations have been rationalized in [507] by using two facts: (1) the NCL is terminated by the primitive or JG β-relaxation and (2) both the primitive and JG β-relaxation times, $\tau_0(T/T_g)$ and $\tau_{JG}(T/T_g)$, are shorter for glassformers having larger $n(T_g)$ shown before in Section 2.3.2.1. The correlations are analogues of correlation between $n(T_g)$ and the T_g-scaled T-dependence of $\tau_\alpha(T/T_g)$ quantified by the fragility index m (see Section 2.2.3). This remarkable analogy between this property of the fast caged relaxation or NCL and the slow structural relaxation suggests that the mechanisms giving rise to them bear some relation to each other. Qualitative arguments given in [507] have led to the result that the level of NCL measured by $< u^2(T/T_g) >$ is proportional to $1/\log \tau_0(T/T_g)$ or $1/\log \tau_{JG}(T/T_g)$. From this proportionality, the correlation between $< u^2(T/T_g) >$ and $n(T_g)$ follows because $\tau_{JG}(T/T_g)$ is shorter for glassformer with larger n (see Section 2.3.2.1).

Niss et al. [1422(d)] have recently revisited the phenomenon of the mean-squared displacement (MSD) measured at short time scales from 4 ns to 10 ps by neutron scattering, which can sense glass transition by the change of its T-dependence near $T = T_g$ [115(a), 507, 510, 1424]. The most complete collection of MSD of small molecular and polymeric glassformers previously measured by neutron scattering showing this phenomenon can be found in the review article [115(a)] published in 2000. Niss et al. supplied new data of a number of small molecular glassformers including DBP, cumene, DHIQ, TPP, *m*-toluidine, and sorbitol. Niss et al. claimed in their abstract there is one-to-one connection between the MSD's temperature dependence and the liquid's fragility index m when the MSD is evaluated on a time scale of ~4 ns. However, on closer examination of the data, this claim is not obviously supported by the plot of $< u^2(T) >$ or $< u^2(T) >/a^2$ against T/T_g. Here a is the intermolecular distance assumed to scale with density ρ at T_g according to $a^2 \propto \rho^{-2/3}$. There is better correlation between m and the values of $d \ln < u^2(T) >/d \ln T$ calculated at $T = T_g$ from data in the temperature range from T_g to $\sim 1.1T_g$, as shown in Fig. 4 of Niss et al. [1422(d)]. This correlation is used by Niss et al. to support the elastic model [583] for glass transition, which relates a larger MSD to a shorter α-relaxation time, and the temperature dependence of the α-relaxation time is governed by the temperature dependence of the MSD. The elastic model regards the observed MSD $< u^2(T) >$ is totally vibrational in origin for all temperatures below and above T_g and hence is very different from our interpretation especially at higher temperatures where at least part of the $< u^2(T) >$ corresponds to the nearly constant loss of motion of molecules confined in the anharmonic cage.

The pivotal support of the correlation in Fig. 4 of Niss et al. is the one point from DHIQ which has much larger values of both $d\ln < u^2(T) >/d\ln T$ and m. Without DHIQ, the results of the rest of the glassformers, except for glycerol, cluster together and cannot be used to support the correlation (see inset of Fig. 248-1). Glycerol is also special in not having any resolved secondary relaxation like the others. In interpreting the $< u^2(T) >$ of DHIQ as totally vibrational by the elastic model, there is a problem not anticipated by Niss et al. Apparently they are not aware of the discovery of the exceptionally fast JG β-relaxation in the glassy state of DHIQ by Paluch et al. [667]. By either extrapolating the Arrhenius T-dependence of the JG β-relaxation time τ_{JG} determined in the glassy state to T_g or using the calculated primitive relaxation time τ_0 at temperatures above T_g as shown before in [667] and reproduced here in Fig. 248-1, we can see that τ_{JG} likely will enter into the time window shorter than ~4 ns of the IN10 neutron scattering spectrometer at temperatures in the range from T_g to $1.1T_g$. This is illustrated in Fig. 248-1 where the horizontal lines from top to bottom indicate times of 100 s, 4 ns, and 0.4 ns, and the vertical lines correspond to $1000/T_g$ and $1000/(1.1T_g)$ for both sets of data obtained at ambient pressure and at 0.5 GPa. Without subtracting off the probable significant contribution of the JG β-relaxation to the observed $< u^2(T) >$ and $d\ln < u^2(T) >/d\ln T$ of DHIQ, the data from DHIQ cannot be used in Fig. 4 of Niss et al. to support $< u^2(T) >$ is entirely vibrational. Hence it is premature to claim that the linear relation between $d\ln < u^2(T) >/d\ln T$ and m supports the elastic model from $< u^2(T) >$ obtained by IN10 at time scales shorter than 4 ns. Moreover, by comparing the time scale of 0.4 ns of IN13 spectrometer with $\tau_{JG}(T)$ estimated by τ_0 in Fig. 248-1, it can be verified that $\tau_{JG}(T)$ of DHIQ is longer than 0.4 ns in the range from T_g to $1.1T_g$, and therefore the JG β-relaxation does not contribute to $< u^2(T) >$ obtained by IN13. Consequently, $d\ln < u^2(T) >/d\ln T$ of DHIQ from IN13 is much reduced than the same quantity from IN10, as found experimentally by Niss et al. (see inset of Fig. 248-1). This exceedingly large reduction can be considered as proof of significant contribution to the $< u^2(T) >$ and $d\ln < u^2(T) >/d\ln T$ of DHIQ observed by IN10.

The $< u^2(T) >$ observed by neutron scattering at temperatures below $1.1T_g$ when transformed to susceptibility will appear as nearly constant loss (NCL), implying it has the same origin as the NCL observed by broadband dielectric spectroscopy over frequencies higher than the relaxation frequency f_s of the fastest secondary relaxation present. At sufficiently low temperatures when f_s is low, say 10 Hz, the NCL can be found to extend from say 10^3 Hz up to the highest dielectric measurement frequency, typically 10^9 Hz. Examples of the latter have been given in

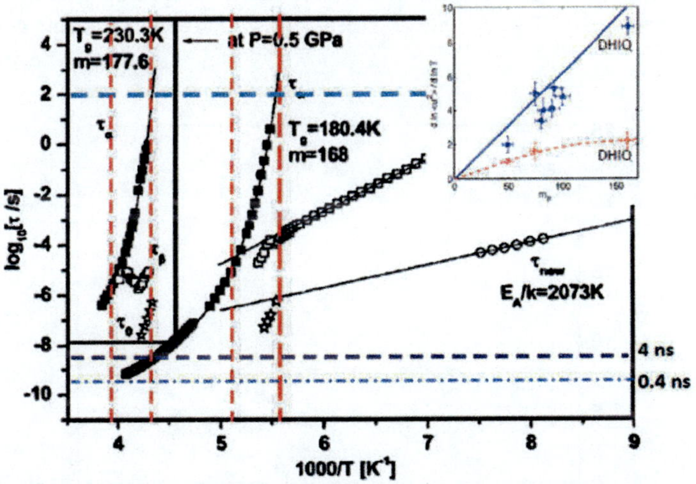

Fig. 248-1 Relaxation map of DHIQ showing the α- and γ-relaxation times at ambient and elevated pressure of 0.5 GPa. In the figure, the γ-relaxation time is labeled by τ_β. The relaxation time of the JG β-relaxation found at ambient pressure is denoted by τ_{new}. The stars are the primitive relaxation times τ_0 calculated by the CM relation. The three horizontal lines are located at 100 s, 4 ns, and 0.4 ns. Data taken from [667]. The vertical lines indicate $1000/T_g$ and $1000/(1.1T_g)$. The inset is a plot of $d\ln <u^2(T)>/d\ln T$ against fragility index m_P reproduced from [1422(d)] with permission. *Full symbols* in *blue* are data obtained from IN10 on 4 ns time scale, and the *open symbols* in *red* are data obtained from IN13 on a 10 times shorter time scale of 0.4 ns

Figs. 241, 244-1, and 246. Various correlations of the observed NCL with different characteristics of the α-relaxation have been found, including the degree of non-exponentiality and fragility in Fig. 241 and change of T-dependence near T_g in Fig. 246. The NCL observed at such low frequencies cannot be related to vibration and the elastic model, and yet the fast process corresponding to the NCL correlates with the α-relaxation.

From Susceptibility Spectra

Susceptibility spectra from dielectric relaxation and dynamic light scattering of many glassformers have been analyzed to obtain the constant loss $\Delta\chi''(T)$ for different glassformers. The NCL data were compared in a plot of $\Delta\chi''(T)/\Delta\chi''(T_g)$ against T/T_g [1424]. When compared in a plot against T_g-scaled temperature, the NCL of different glassformers exhibited a pattern that correlates with the $n(T_g)$. The rapidity of increase of $\Delta\chi''(T)/\Delta\chi''(T_g)$ with T/T_g is quantified by the steepness index, R, defined in a similar way as the other steepness or "fragility" index, m, by

$$R \equiv \mathrm{d}\left\{\log\left[\Delta\chi''(T)/\ \Delta\chi''(T_g)\right]\right\}/\mathrm{d}\left[\log\left(T/T_g\right)\right]\Big|_{T_g} \tag{2.94}$$

R is a meaningful parameter for differentiating the temperature dependences of the constant loss near T_g of various glassformers. Its values for a number of glassformers are plotted against $\beta_\alpha(T_g) \equiv 1 - n(T_g)$ in Fig. 249, which shows a strong correlation

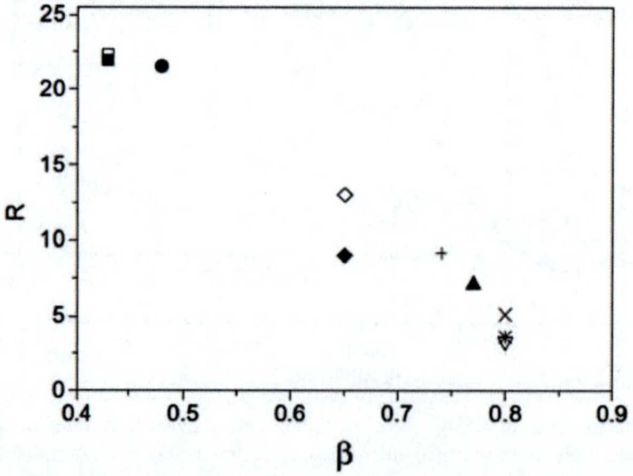

Fig. 249 Steepness index of the scaled constant loss, R (defined in Eq. (2.94)), vs. the Kohlrausch exponent of the α-relaxation at T_g, $\beta_\alpha(T_g)$, for light-scattering measurements on CKN(□), OTP(•), glycerol (◇), and $ZnCl_2$ (∇); dielectric measurements on CKN (■), glycerol (♦), propylene carbonate (+), and cyclo-octanol(▲); and neutron scattering measurements on $Na_{0.5}Li_{0.5}PO_3$(*) and $Zr_{46.8}Ti_{8.2}Cu_{7.5}Ni_{10}Be_{27.5}$ (×). Reproduced from [1424] by permission

between R and $n(T_g)$. This correlation found for the fast NCL is remarkably similar to that between the fragility m of the slow α-relaxation and $n(T_g)$, suggesting once more that the mechanisms giving rise to them bear some relation to each other.

An objective method to compare the NCL of different glassformers from their susceptibility spectra having the same peak frequency ν_{max} is to normalize χ'' by χ''_{max} in each case. This procedure has been followed in comparing the dielectric susceptibility spectra of three small molecular glassformers (NMEC, threitol, and xylitol) as shown before in Fig. 241 where ε'' is normalized by ε''_{max}. Here the comparison is made on the dynamic light-scattering spectra of the inorganic glassformers CKN and $ZnCl_2$ in the left panel, and CKN and B_2O_3 in the right panel of Fig. 250-1.

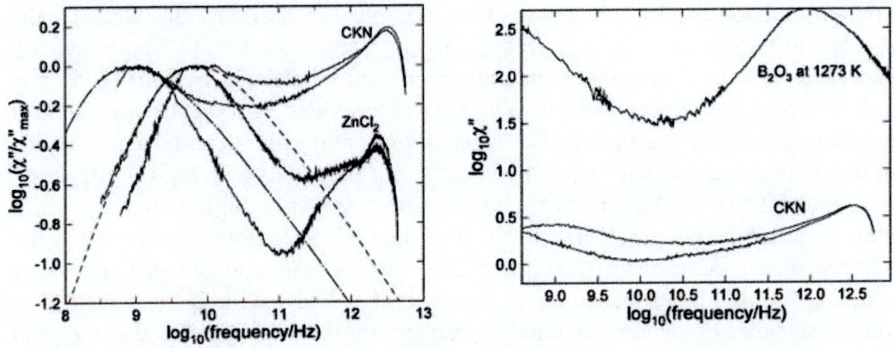

Fig. 250-1 (*Left*) Susceptibility minima from dynamic light scattering for CKN at 468 and 423 K [1421]. The *dashed* and *dashed-dotted curves* are the fits by the Kohlrausch functions with $n \equiv (1-\beta)$ equal to 0.43 and 0.47 for 468 and 423 K, respectively. Also shown are data of $ZnCl_2$ with n nearly equal to zero from Lebon et al. [1426(a)]. (*Right*) Depolarized light-scattering susceptibility spectrum of B_2O_3 at 1273 K [1426(c)] compared with CKN data at 468 and 423 K [1421]

These figures demonstrate, after normalizing ε'' by ε''_{max} and χ'' by χ''_{max}, the NCL increases with the width of the α-dispersion or the value n at the same value of τ_α for the different glassformers. Here χ''_{max} denotes the maximum loss of the α-relaxation. The susceptibility minima from dynamic light scattering for CKN at 468 and 423 K obtained by Li et al. [1421] are much higher than that of $ZnCl_2$ from the paper by Lebon et al. [1426(a)]. This indicates that the NCL contribution to χ''/χ''_{max} of CKN is larger than that of $ZnCl_2$, which correlates with the larger n of CKN equal to 0.43 and 0.47 at 468 and 423 K, respectively, as compared with the nearly zero value of $ZnCl_2$ [1426(a)]. The fragility or steepness index m of CKN (~93) is also much larger than that of $ZnCl_2$ (~30). The same remarks apply to the comparison of CKN with B_2O_3. At 1273 K, the high-frequency slope of $\log \chi''(\nu)$. $\log \nu$ is -0.8, which suggests n of B_2O_3 at this temperature is about 0.20. Also, $m = 32$ for B_2O_3 [112].

While the much deeper susceptibility minima of the "strong" glassformers $ZnCl_2$ and B_2O_3 compared with CKN can be rationalized in terms of NCL by the correlation of NCL with n or m [507, 1426(b)], they violate the predictions of MCT as pointed out before by Lebon et al. and Brodin et al.

NCL and the Non-ergodicity Parameter or Factor

The intermediate self scattering function $F_s(Q,t)$ obtained by neutron scattering or by molecular dynamics simulations exhibits a two-step decay separated by a plateau with height $f_s(Q,t)$ often referred to as the non-ergodicity parameter. At the end of plateau, the decay of $F_s(Q,t)$ down to zero is often approximately described by $F_s(Q,t) = f_s(Q,t)\exp[-(t/\tau_\alpha)^{1-n}]$. The plateau, where $F_s(Q,t)$ decreases very slowly, comes from motions of caged molecules by the anharmonic intermolecular potentials. When $F_s(Q,t)$ has been Fourier transformed to frequency space, the plateau becomes the NCL in the susceptibility spectrum $\chi''(\omega)$. From this interpretation of the non-ergodicity parameter, we can expect that more anharmonic intermolecular potential gives rise to larger NCL and lower plateau value $f_s(Q,t)$, provided τ_α is kept constant. This expectation has been verified by Bordat et al. from molecular dynamics simulations of three binary LJ particle systems with different potentials (see Eq. (2.68)) [264]. Also found is that n is larger and the fragility index m of τ_α is higher for system with more anharmonic potential. Hence, we have the anticorrelation (correlation) of the non-ergodicity parameter $f_s(Q,t)$ (the NCL) with the non-exponentiality parameter n and the fragility index m. Moreover, Bordat et al. calculated the T_g-scaled temperature dependence of the non-ergodicity parameter for the three systems above T_g, and found that this quantity anticorrelates with n and m. The result is in accordance with the anticorrelation of the non-ergodicity parameter with fragility found by Scopigno et al. [571(a)] from their analysis of inelastic X-ray scattering data of small molecular glasses at temperatures below T_g. Both results found that $[f_s(Q,t)]^{-1}$ plotted against T/T_g increases linearly from unity at $T/T_g = 0$ with a slope that correlates directly to the fragility of the corresponding system or liquid, although the simulation results were obtained above T_g and the X-ray data below T_g.

However, there is breakdown of the correlation in the inelastic X-ray scattering data when the polymers, PVC, BPA-PC, PMMA, PET, PIB, PVAc, and PS, are added to the original data of mainly small molecular glassformers in Scopigno et al. by Buchenau and Wischnewski [572]. Except for PET, all the m values of all the polymers used by Buchenau et al. are taken from Böhmer et al. [112], which are determined by mechanical creep compliance and considered to be most accurate. The m value of PET equal to 80 used by Buchenau et al. was taken from dielectric relaxation data of not totally amorphous PET, and is much lower than the value of 124 of [1089, 1090]. Recently Scopigno et al. [571(b)] reconsidered the inelastic X-ray data of some of polymers from Buchenau et al. data. For the polymer glasses, BPA-PC, PMMA, PET, PIB, PVAc, and PS, they believed that the deviations from the correlation are due to contributions of the secondary relaxation process from internal degrees of freedom. They suggested a way to remove the effect of the secondary relaxation, and restore the correlation between their parameter non-ergodicity factor, α, defined by $[f_s(Q,t)]^{-1} = 1 + \alpha T/T_g$ and m of the polymers but still not for PIB and PS [571]. Notwithstanding, their m values for polymers are taken from dielectric relaxation measurements, which are not the same as the mechanical m values used by Buchenau. If the mechanical m values of the polymers were used, the correlation, $m \propto \alpha$, is not as ideal as that shown in Fig. 250-2.

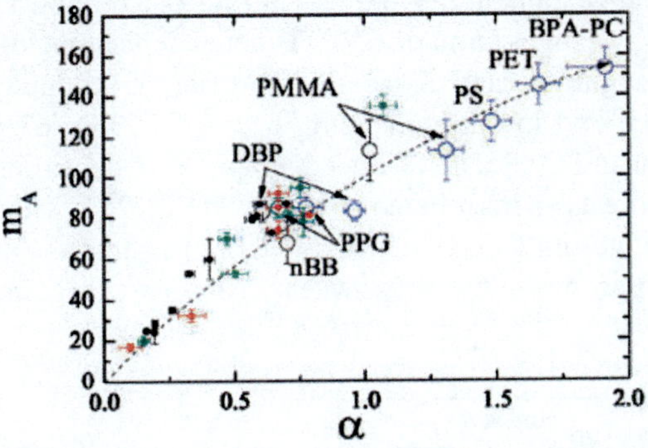

Fig. 250-2 Correlation between the temperature dependencies of viscosity (fragility, m) and non-ergodicity factor (α) accounting for the presence of secondary relaxations for all polymers possessing secondary relaxation processes plus dibutylphthalate (DBP), n-butylbenzene (n-BB). IXS (*black symbols*, in order of increasing fragility: NiZr, BeF2, silica, cyanoadamantane, glycerol, 1,4-polybutadiene (1,4 PB), normal-butylbenzene, salol, m-toluidine, o-terphenyl (OTP), selenium, and m-tricresylphosphate (mTCP). Compressibility measurements from Brillouin spectroscopy (*blue symbols*, in order of increasing fragility: dibutylpthalate, poly(propylene glycol) (PPG), PMMA, PS, PET, BPA-PC), photon correlation spectroscopy (*red symbols*, in order of fragility: As_2O_3, B_2O_3, 0.5Li–0.5$NaPO_3$, OTP, $NaPO_3$, $LiPO_3$). Sound velocity jump at T_g (*green symbols*, in order of increasing fragility: germanium dioxide (GeO_2); propylene glycol (PG); poly(butyl acrylate) (PBA); poly(propylene glycol)-diglycidylether (PPGDE); poly(propylene glycol) (PPG); diglycidylether of bisphenol A (DGEBA); 1,2-polybutadiene (1,2 PB)). *Large symbols*, explicitly labeled, are used for those systems having non-zero residual entropy, i.e., subject to a correction. For details and source of data, see [571(b)]. Reproduced from [571(b)] by permission

Sidebottom et al. [1427] used photon correlation spectroscopy to measure $F_s(Q,t)$ in a series of glass-forming liquids with widely different fragility and demonstrated a correlation between the non-ergodicity parameter $f_s(Q,t)$ of the liquid and its fragility index m at temperatures above T_g. There is also the correlation of $f_s(Q,t)$ with the non-exponentiality parameter n, although the correlation is not as good as with m. Their data have been incorporated into Fig. 250-2.

2.3.2.36 Fast Relaxation (NCL) Senses the Hole Volume from PALS

Positron annihilation lifetime spectroscopy (PALS) is a technique to deduce the unoccupied volume, or the so-called "free volume," of glassformers. The positronium lifetime, τ_3, is measured, and usually the dimension of the hole modeled by a cavity with radius R is *nearly* linearly proportionality to τ_3. Positron annihilation measurements of τ_3 on four widely studied small molecular glassformers, *o*-terphenyl, glycerol, propylene glycol (PG), and propylene carbonate (PC), over a wide temperature range both below and above the glass transition [387] have been discussed before in Section 2.2.5.1 and shown in Fig. 64. There, we have demonstrated from the data of all four glassformers that the change of the T-dependence of $\tau_3(T)$ at T_g and T_B mimics that shown by the α-relaxation in the T-dependence of $\tau_\alpha(T)$, its derivative or generalized "fragility" $m(T) \equiv \frac{d \log \tau_\alpha}{d(T/T_g)}$, and the Kohlrausch stretch exponent $\beta(T) \equiv [1-n(T)]$. As a reminder, T_B is the temperature at which $\tau_\alpha(T)$ changes from one VFTH dependence to another, accompanied by the change of magnitude and T-dependence of $n(T)$. Examples of these can be seen in Figs. 59–63, and in particular, Fig. 61 for PC, Fig. 62 for OTP and PG, and Fig. 63 for salol and OTP. The T_g-scaled temperature dependence of $\tau_3(T/T_g)$ is reproduced once more as an inset in the left panel of Fig. 251, where the temperature T_r, where $\tau_3(T)$ changes its T-dependence, is nearly the same as T_B for all four glassformers. The main part of the figure shows $m(T)$ and its change in T-dependence at $T_B \approx T_r$.

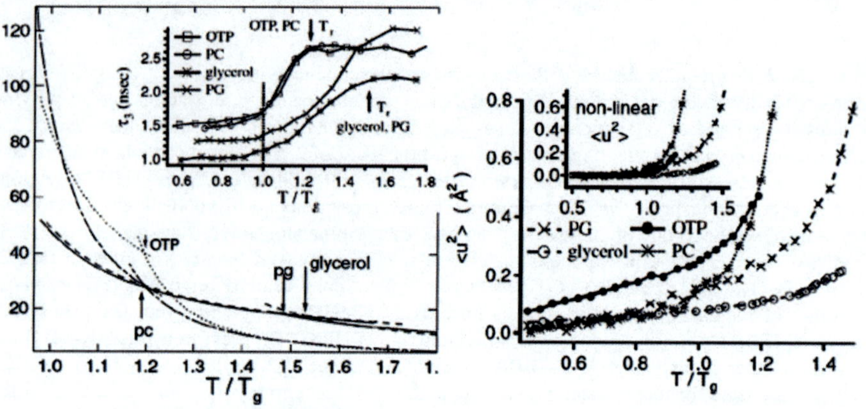

Fig. 251 (*Left*) τ_3 as a function of the T_g-scaled temperature for OTP, PC, PG, and glycerol. A crossover in the thermal dependence of τ_3 is observed both at T_g and an upper temperature T_r, indicated by the *vertical arrows*. Typical standard uncertainties in τ_3 are ±0.025 ns. (*Right*) $< u^2 >$ as function of T_g-scaled temperature for OTP, PC, PG, and glycerol. The *inset* shows the deviations from a linear fit to the data established in the region $T/T_g < 1$. Standard uncertainties in $<u^2>$ are typically less than the size of the data markers. Reproduced from [387] by permission

Yet another connection of $\tau_3(T/T_g)$ has been made to the fast relaxations of these same four glassformers obtained by quasielastic neutron scattering in terms of the mean-squared displacement, $< u^2(T/T_g) >$ [387]. The data were obtained from the use of different neutron scattering spectrometers, but all having comparable resolutions $\approx 1\mu$eV, and only motions with frequencies higher than 0.24 GHz, or times shorter than 0.66 ns can contribute to $<u^2>$. These timescales are comparable to that of PALS experiment. Shown in the right panel of Fig. 251, the T-dependences of $<u^2>$ are approximately but not exactly linear below T_g. If the region $T/T_g < 1$

is fit to a straight line, the inset of the right panel of Fig. 251 emphasizes the deviations from linearity in this region. The T-dependence of this generally non-linear $<u^2>$ bears a striking resemblance to the trend in τ_3 and m. Strong increases in the non-linear $<u^2>$ occur near $T/T_g \sim 1.2$ in OTP and PC and $T/T_g \sim 1.5$ in glycerol and PG. The T-dependences of τ_3, m, and $<u^2>$ all undergo changes at T_B and T_g.

It is remarkable that the fast relaxations (0.24 GHz or higher) reflected in $<u^2>$ sense the change of hole volume at T_g, which is determined by macroscopic variables, like enthalpy or specific volume, at timescales closer to a hundred or a thousand seconds. Even the dielectric relaxation data occur on a timescale much slower than the neutron scattering $<u^2>$. Stickel and coworkers [621] reported that the dielectric α-process has a frequency $\nu_\alpha \sim 10^{-1}$Hz at T_g in PC, and the unresolved JG β-relaxation is close by. This intriguing behavior can be understood if $<u^2>$ represents the NCL of molecules moving inside their cages of unoccupied volume defined by the anharmonic intermolecular potential of its neighbors, as well as from fluctuations of the cages due to motion of the neighbors. In a glass this cage is relatively rigid, but heating through T_g introduces an abrupt increase of the temperature dependence of the cage size. This would naturally lead to an increase in the rate of change of $<u^2>$ with temperature after crossing T_g. At T_B, the JG β-relaxation has merged with the α-relaxation, and their common relaxation frequency ν_α is $\sim 10^7$ Hz for PC (see Fig. 61), only one decade slower than the $\approx 10^8$ Hz motions reflected in $<u^2>$. For T less than T_B, the fast relaxation in $<u^2>$ has little or no direct contribution from the α-relaxation or the JG β-relaxation. However, these relaxations make contribution to $<u^2>$ at temperatures exceeding T_B, and this explains the more rapid rise of $<u^2>$.

For a harmonic solid, $<u^2>$ increases linearly with T. The highly non-linear variations of $<u^2>$ in Fig. 251 thereby indicate anharmonic motions of the caged particles. From their similar temperature variations for T less than T_B, the picture arises that both τ_3 and $<u^2>$ reflect the level of anharmonicity in the intermolecular caging. Glycerol appears to form a strongly caged glass with harmonic-like motions (linear in $<u^2>$) persisting nearly to T_g. Very strong caging and a low degree of anharmonicity are consistent with the small nanopores (smallest τ_3 values) in glycerol. This is also consistent with the smaller coupling parameter n and the smaller separation between τ_{JG} and τ_α in glycerol, which has been proposed as the cause of lower level of the NCL in the above and in [507]. Conversely, the $<u^2>$ of OTP deviates from linearity well below T_g, indicating an onset of anharmonicity deep in the glassy state. This implies relatively weaker caging, supported by the large τ_3 values in OTP, more anharmonic and stronger intermolecular potential, and the larger coupling parameter n than glycerol. Thus, τ_3 and its temperature dependence provide a measure of the anharmonicity of the caging potential in these small molecule organic glassformers.

There are also studies of the fast relaxation of caged molecules by neutron and Raman scattering, and combined with measurements of unoccupied volume fraction with PALS in a few polymeric systems, *cis*-1,4-PB, PIB, and aPP, by Kanaya and coworkers [1428–1430]. The data show that $<u^2>$ and the hole volume have remarkably similar temperature dependence. In particular, like $< u^2(T) >$, the hole volume as a function of temperature exhibits a change in slope at T_g.

Experimental Evidence of $<u^2(T)> \propto V_h(T)$, and Explanation

Most interesting is the study by both quasielastic light scattering (QELS) and quasielastic neutron scattering (QENS) by Duval and coworkers [1431–1433]. They subtracted from the measured $< u^2(T) >$ of PMMA a linear temperature-dependent term contributed by harmonic vibrations. The result, $\Delta<u^2(T)>$, has nearly the same temperature dependence as the dynamic hole volume fraction, $\Delta F_h(T)$, obtained from PALS by Yee, Gidley, and coworkers [1434] for a wide range of temperatures at and below T_g. The data of $\Delta<u^2(T)>$ from QELS and QENS are nearly identical but only those from QENS are shown in Fig. 252. The result immediately points out that the fast relaxation or NCL senses the hole volume, which is consistent with other properties of the NCL discussed earlier including the change of T-dependence when T_g is crossed. It is also unsurprising on recalling that NCL is connected to the JG β-relaxation, and we already know from Section 2.3.2.29 that the primitive or JG relaxation time has the TV^γ-dependence and the primitive or JG relaxation strength changes T-dependence on crossing T_g. Furthermore, if the temperature dependence of the primitive or the JG relaxation time, τ_0 or τ_{JG}, is written in the form suggested by Doolittle [26, 29], $\tau_0 \propto \exp[V_0/V_h(T)]$, where V_h is the hole volume, then it together with the relation, $<u^2(T)>_{NCL} \propto 1/\log \tau_0$, derived in [507], leads to $<u^2(T)>_{NCL} \propto V_h(T)$. The result explains the relation between free

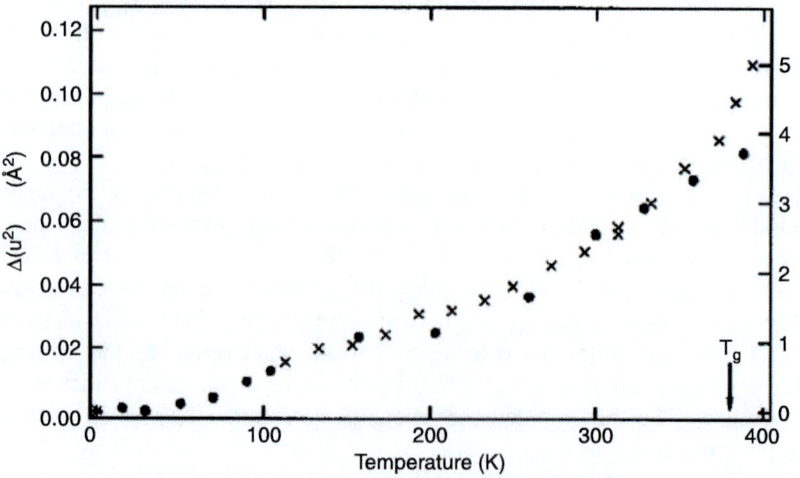

Fig. 252 Comparison between the mean-squared displacement $\Delta< u^2 >$ represented by *closed circles*, and the dynamic hole volume fraction $\Delta F_h(\times)$. Data from Duval and coworkers [1431–1433], and Hristov et al. [1434]. Reproduced from [1431] by permission

volume and the NCL fast relaxation found from experimental studies of Bartos et al. [1430], Duval et al. [1431, 1432], and Ngai et al. [387].

2.3.2.37 Comparison of the MCT Description of Caged Dynamics with NCL

Revision of the Idealized MCT (2002) in Response to the Observation of the Logarithmic Decay of the Orientational Correlation Function and Its Equivalent, the NCL in the Susceptibility

We have heuristically described the dynamics of molecules or ions at times when all molecules are mutually caged and before the onset of the primitive relaxation (i.e., caged regime), and identified them with the intermediate power (IPL), pt^{-1+c} observed by OHD-OKE, the NCL in susceptibility, $\chi''(\omega) \propto \omega^{-c}$, or the MSD, $<u^2(t)>$, all with $c << 1$. However, no rigorous theory has been given to support it at this time. This is in contrast to the mode coupling theory (MCT) which gives mathematically rigorous results to describe the caged regime. The MCT predictions on the caged dynamics are compared here with our heuristic description and experimental data. There are several versions of MCT introduced in the past two decades. In earlier years the one version most considered and compared with experiments is the idealized or schematic MCT [31, 32, 1435, 1436], which is limited to temperatures above the critical temperature T_c high above T_g. Its predictions were often used to compare with short-time dynamics observed by neutron scattering [309, 1437–1441] and light-scattering experiments [32, 1421, 1442–1445]. Idealized MCT predicts at temperatures above T_c a susceptibility minimum produced by a crossover in the frequency dependence of the susceptibility $\chi''(\nu)$ from ν^a for higher frequencies to ν^{-b} for lower frequencies, where the exponents a and b are related and determined by the control parameter λ, and some scaling relations. For $T < T_c$, it predicts a low-frequency crossover from ν^1 to ν^a, which produces a "knee," and was found in the original depolarized light-scattering spectra of $Ca_{0.4}K_{0.6}(NO_3)_{1.4}$ at low temperatures [1421], but it turns out to be an experimental artifact [1422]. Despite the idealized MCT having been favorably compared with experiments and MD simulations of fast processes in a number of glassformers at earlier years (often the "fragile" ones, but not the "strong" ones such as $ZnCl_2$ [1426] and B_2O_3 [1446]), some new experiments started at the year 2000 have shown features of the fast processes in "fragile" glassformers that cannot be explained. The first came from optical heterodyne-detected optical Kerr effect (OHD-OKE) experiments on salol (phenyl salicylate) and OTP [330]. OHD-OKE experiment measures an impulse response function $S(t)$, the imaginary part of the Fourier transform of which is directly proportional to the dynamic susceptibility $\chi''(\nu)$ from depolarized light scattering, i.e., $\chi''(\nu) \propto \text{Im FT}\{S(t)\}$. The entire data extending from 2 ps up to about 3×10^4 ps were well fitted by the empirical function [332]

$$S(t) = [pt^{-1+c} + dt^{-1+b}]\exp(-t/\tau_\alpha). \tag{2.95}$$

In a range of time longer than 2 ps and extending to some longer times that increase with decreasing temperature, a portion of the OHD-OKE signal, $S(t)$, exhibits a power law decay, pt^{-1+c} (the first term inside the bracket in equation above). Here p is the amplitude, and the positive fractional exponent c has significant temperature dependence and decreases to a value close to 0 at temperatures somewhat below the critical temperature T_c of MCT. Since this new feature appears on timescales longer than intramolecular vibrations, and before the von Schweidler power law of MCT (the second term inside the bracket in the equation), it is called the intermediate power law (IPL). The IPL with $c \approx 0$ is not predicted by the idealized MCT. Neither can the extended mode coupling theory [1440, 1443, 1444] that introduces, in addition, a full range of hopping times explain the IPL and its temperature dependence near and below T_c. If c were exactly zero, then the IPL would correspond to a logarithmic time dependence of the polarization–polarization correlation function. Perhaps it is for this reason that this part of the OHD-OKE data was sometimes referred to as logarithmic decay of the orientational correlation function [33, 34, 332]. Logarithmic decay of correlation function was found also by photon correlation spectroscopy in dense colloidal suspensions [254]. The existence of the IPL with exponent $c \approx 0$ or equivalently the logarithmic decay had been well recognized by Götze and Sperl [33]. In 2002, a special solution of MCT for states near higher-order glass-transition singularities that have correlation function with logarithmic decay was proposed by them to explain the OHD-OKE data [33].

The OHD-OKE signal $S(t)$ measures the time derivative of the polarizability–polarizability correlation function, which is equivalent to the time derivative of the orientational correlation function $\phi(t)$ of the molecules. Therefore, the imaginary part of the Fourier transform of $S(t)$ is the susceptibility spectrum $\chi''(\nu)$ that can be obtained directly by dynamic light-scattering experiment [1062, 1053]. In particular, the time dependence t^{-1+c} of the IPL corresponds to the frequency dependence ν^{-c} of $\chi''(\nu)$. Since c becomes small at lower temperatures, the t^{-1+c} decay of $S(t)$ corresponds to ν^{-c} of $\chi''(\nu)$ with $c << 1$, which is appropriately called the nearly constant loss (NCL) in the susceptibility $\chi''(\nu)$ or the dielectric loss $\chi''(\nu)$ spectrum. Immediately following the OHD-OKE experiment on salol [330] is the publication of $\chi''(\nu)$ data of polyisobutylene (PIB) in the $5 < \nu < 40$ GHz range from dynamic light-scattering experiment by Sokolov et al. [1062]. This publication reported the observation of susceptibility spectra that have the power law dependence, $\chi''(\nu) = p'\nu^{-c}$ with $c << 1$, an exact correspondent of the IPL, pt^{-1+c} with $c << 1$. The measured $\chi''(\nu)$ of PIB is nearly frequency independent over a broad temperature range from below T_g to higher than $1.3T_g$. The same was found in another polymer, 1,4-polyisoprene (PI) [1063]. The amplitude p' has an approximately linear dependence on temperature, increasing by about a factor of 4 from $T_g \approx 200$ K to nearly 290 K [507] (see Figs. 135 and 136). The temperature dependence of p' of PIB and PI is weaker below T_g than above T_g and exhibits a change of slope at T_g. The sensitivity of the NCL in these two polymers to glass transition is remarkable [114, 510] when considering that it was observed in the GHz range, corresponding to times of order of 10^{-10} s, whereas glass transition at T_g occurs at timescale of thousands of seconds. If NCL originates from excursions of caged molecule within anharmonic potential and fluctuations of cages, then the following possible explanation can be considered. The amplitudes of fluctuations and excursions depend on temperature through density and entropy. The change in the temperature dependences of density and entropy when crossing T_g is well known, and this change leads to corresponding change of the T-dependence of amplitudes of fluctuations and excursions, and hence the intensity of the NCL. No further change of T-dependence of p' was observed above T_g, up to temperatures as high as $1.3T_g$.

Similar observations of the presence of the NCL were made from dynamic light-scattering data of glycerol and poly(methylmethacrylate) (PMMA) [1423]. It is worthwhile to point out that while dynamic light scattering can measure the NCL of glycerol in the $5 < \nu < 40$ GHz range at 160 K (i.e., below T_g) and 200 K (i.e., above T_g), the best broadband dielectric relaxation measurement of glycerol by Lunkenheimer and coworkers [972] lacks the sensitivity to detect the low dielectric loss in the same frequency and temperature ranges. Consequently, no dielectric loss $\varepsilon''(\nu)$ data of glycerol were reported in the GHz to 40 GHz frequency range at temperatures below 213 K. The $\varepsilon''(\nu)$ data obtained above 100 GHz and below 1 GHz at 195 K or below 10^{-1} GHz at 184 K (see Fig. 4 in [972]), for example, were interpolated by Brodin et al. [1047] by a smooth curve to suggest the presence of a minimum (see Fig. 1a in [1047]) instead of the NCL found directly in the dynamic light-scattering data [1423]. Dielectric measurements in other glassformers that have no resolved JG relaxation as glycerol suffer the same problem of not having the sensitivity to see the weak NCL in the same high-frequency region as in the case of glycerol. These include propylene carbonate (at $T = 163$ K, $T_g = 159$ K) [972], 2-picoline, propylene glycol, and 4-*tert*-butylpyridine [1447, 1448]. All these glassformers have smaller n, and their weak NCL follows from the correlation of the NCL intensity as well as the JG relaxation with n [149, 507].

A Digression: NCL Should Not Be Confused with Excess Wing or IPL with $c \approx 0.2$

This warning has been issued before in Section 2.3.2.6, but repeated here because more experimental facts have been given here in the present subsection, and it is important to remove any confusion introduced in identifying the excess wing of types $\mathcal{A}, \mathcal{B}$ and $\mathcal{C}$ (see Section 2.3.2.6 for definitions) observed by dielectric relaxation with the IPL observed by OHD-OKE in the interpretation of the dynamics by others [1047, 1053, 1054].

Returning to the dielectric data of glycerol at 195 and 184 K [972], the loss measured at frequencies lower than 10^{-1} GHz shows power law decrease of $\varepsilon''(\nu) \sim \nu^{-c}$ with $c \approx 0.23$ and 0.17, respectively, in excess on the high-frequency side of the fit of the α-relaxation loss peak by either the Cole–Davidson empirical function or the one-sided Fourier transform of the Kohlrausch stretched exponential function. Similar excess loss, $\varepsilon''(\nu) \sim \nu^{-c}$ with $c \approx 0.2 - 0.3$, was found in the susceptibility spectrum of the other glassformers [972, 1447, 1448], and is often called the "excess wing," which corresponds in the time domain to the IPL, $\varepsilon(t)$

or $\chi(t) \sim t^{-1+c}$, with $c \approx 0.2 - 0.3$. Perhaps it is the inability of dielectric spectroscopy to reveal the presence of the NCL at high frequencies (>1 GHz) that has led Brodin et al. to emphasize exclusively the excess wing with $c \approx 0.2$ or about as a fundamental feature of the dynamics at all temperatures [1047, 1053, 1054], but not the NCL although it was actually seen by experiments. NCL was found by dynamic light scattering [1062, 1063, 1423] as $\chi''(\nu) = p'\nu^{-c}$, and in OHD-OKE [330, 332] as the IPL, t^{-1+c}, with c nearly zero in both cases. Even in the high-frequency range ($10^{-3} < \nu < 10^{2}$ GHz), dielectric relaxation experiment has found the NCL in $0.4Ca(NO_3)_2$–$0.6KNO_3$ (CKN, $T_g = 335$ K) as $\varepsilon''(\nu) = p'\nu^{-c}$ with $c \approx 0$ (see Fig. 244). This is made possible by the sizeable dielectric loss higher than 0.1 of CKN at 342 K and higher temperatures [1034, 1057] (at these temperatures the structural α-relaxation and the ionic conductivity relaxation are practically coupled together [55]). Thus from the experimental evidences given above, the NCL found at higher frequencies and temperatures is also an important and fundamental feature of the dynamics, and should not be forgotten for the sake of emphasizing the excess wing in dielectric relaxation (at frequencies below 1 GHz) and dynamic light scattering or IPL in OHD-OKE all with $c \approx 0.2$ or about as done by Brodin et al. [1053, 1054]. This is also clear from the fact that the NCL is needed to explain the flat susceptibility instead of a minimum at high frequencies from dynamic light scattering [1062, 1063, 1423] and some IPL with c nearly zero from OHD-OKE [330, 332]. Later on, we shall return to discuss the origin of this so-called excess wing and its relation to the NCL, which is actually the precursor of the excess wing.

Even the state-of-the-art commercial dielectric relaxation cannot measure the complex dielectric permittivity $\varepsilon^*(\nu)$ at frequencies higher than 2 GHz at temperatures near T_g to capture the NCL in glassformers including glycerol, propylene carbonate [972], 2-picoline, propylene glycol and 4-*tert*-butylpyridine [1447, 1448], and benzophenone (BZP) [1120, 1121]. However, a special experimental method using a circular parallel-plate capacitor was able to measure the $\varepsilon^*(\nu)$ at the frequencies from 100 MHz to 20 GHz for five glassy polymers, PMMA, polystyrene (PS), polyethylene terephthalate (PET), polyethylene naphthalate (PEN), and polyvinylchloride (PVC), at room temperature by Nozaki and coworkers [1149a]. Power law dependence of the loss factor with very small exponent c was found in these polymers in the glassy state, which is definitely identifiable with the NCL. Combining this high-frequency dielectric technique with conventional lower frequency measurements, Nozaki and coworkers [1149b] were able to find the NCL with $c \approx 0.1$ of sorbitol and xylitol in the glassy states that essentially pervades the entire frequency range from 2 GHz down to 100 Hz. For sorbitol, such a vast expanse of the NCL was found at 183, 163, and 143 K. The β-relaxation time of sorbitol has been determined in the glassy state [615], and it has the Arrhenius temperature dependence given by $\tau_\beta/\mathrm{s} = 1.3\times10^{-17}\exp(E_\beta/RT)$ with $E_\beta = 61$ kJ/mol. At 143 K, $\log_{10}(\tau_\beta/\mathrm{s})$ calculated by the Arrhenius relation is equal to 5.4, and the frequency of the β-loss maximum ν_β would be located at about 10^{-6} Hz. If the β-relaxation represented by a symmetric Cole–Cole function were assumed to be responsible for the observed NCL, its half-width at half-maximum is at least 15 decades. This immense width is not easy to comprehend or justify. Published isochronal dielectric loss data of xylitol at 10 kHz also by Nozaki and coworkers (see Fig. 3a in this [1450]) capture the α-relaxation peak and the β-relaxation, and $\log(\varepsilon'')$ exhibits the $\exp(T/T_0)$-dependence at lower temperatures down to the lowest temperature of measurement of about 70 K. The β-relaxation time of xylitol has been determined in the glassy state [615], and it has the Arrhenius temperature dependence given by $\tau_\beta/\mathrm{s} = 10^{-15}\exp(E_\beta/RT)$ with $E_\beta = 52$ kJ/mol. At 150, 100, and 70 K, $\log_{10}(\tau_\beta/\mathrm{s})$ calculated by the Arrhenius relation is equal to 3.1, 12.2, and 23.8, respectively. If the β-relaxation were responsible for the observed loss at 10 kHz (equivalent to $\log_{10}(\tau_\beta/\mathrm{s})$) and assuming it is represented by a symmetric Cole–Cole function, its loss maximum would be 7.9, 17, and 28.6 decades of frequency below 10 kHz at temperatures 150, 100, and 70 K, respectively. The corresponding complete span of the symmetric Cole–Cole function would be twice as large. Thus, unthinkably huge breadths of the β-relaxation are needed to account for the observed loss data of xylitol at these lower temperatures. This is highly improbable, and hence also is the hypothesis that the high-frequency (excess) wing of the β-relaxation is responsible for the observed loss having the $\exp(T/T_0)$ temperature dependence at temperatures below 150 K. Instead, the results are additional proofs that the NCL is a separate and fundamental process of glass-forming substances that should be distinguished from the loss in excess of the KWW α-relaxation coming from processes involving cooperative motion of increasing number of molecules and initiated by the Johari–Goldstein β-relaxation [973].

A New MCT (2004) That Includes Rotational–Translational Coupling

In 2003, additional observations of IPL, pt^{-1+c} with $c << 1$, by OHD-OKE in the supercooled liquids, benzophenone (BZP), and 2-biphenylmethanol (BPM) were presented [332], together with refurbished IPL data that were previously obtained on *ortho*-terphenyl (OTP), salol [330], and dibutyl phthalate (DBP) [331]. In more recent past, IPL was found by OHD-OKE in aspirin [1195] as mentioned before in Section 2.3.2.23, and in the orientational dynamics of the supercooled ionic organic liquids *N*-propyl-3-methylpyridinium *bis*(trifluoromethylsulfonyl)imide (PMPIm) [333]. Depolarized light-scattering (DLS) experiments on BZP reported good agreement with the OHD-OKE results [1047]. The observed IPL may be consistent with the logarithmic decay of orientation correlation function obtained by a special solution of MCT near a high-order singularity proposed in 2002 to replace the idealized MCT [33]. However, this way to explain the IPL was abandoned by Fayer and coworkers [332] in 2003 because this solution by MCT may apply only in very rare situations for systems that have special inter-particle interaction potentials such as colloids, while the logarithmic decay was found in the five glassformers with different chemical structures and interaction potentials. This may not be obvious from the fact that Fayer and coworkers were able to fit the *T*-dependence of *p* by the $(T - T_c)^{1/2}$-dependence suggested by idealized MCT. The amplitude *p* of IPL in all the five glassformers increases weakly with temperature. It turns out that a

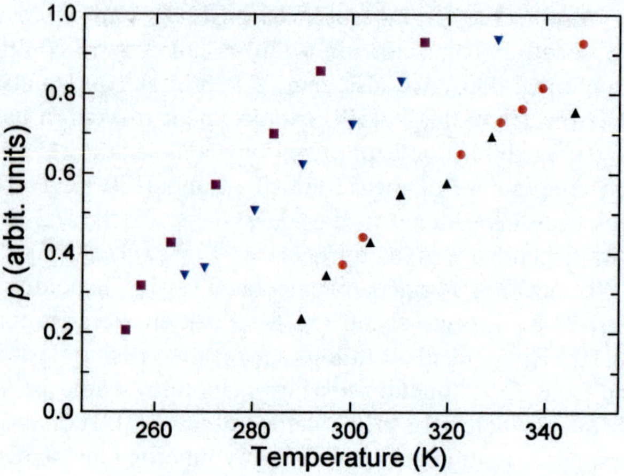

Fig. 253 The amplitudes *p* of the intermediate power laws of four liquids obtained by OHD-OKE are replotted vs. temperature. OTP (*up triangles*), salol (*down triangles*), BZP (*squares*), and BPM (*circles*). The vertical scale for each liquid is given by arbitrary units. The plots demonstrate that *p* has approximately linear *T*-dependence of the nearly constant loss

nearly linear dependence on T is also an acceptable description of the same data as shown in Fig. 253, similar to the nearly linear T-dependence of the NCL found in PIB, PI, glycerol, and PMMA by dynamic light scattering above 1 GHz [1062, 1063, 1423], and as shown in Figs. 135 and 136 for PIB and PI, respectively. Thus, for OHD-OKE data, the $(T - T_c)^{1/2}$-dependence of p is not unique, and the nearly linear T-dependence is equally acceptable. The latter or other comparable forms of T-dependence are the characteristic of NCL in glassformers with different chemical structures that include CKN in Figs. 244 and 246 from dielectric loss, and various glassformers in Figs. 248 and 251 from $< u^2(T) >$ obtained by neutron scattering, and PIB, PI, glycerol, and PMMA by dynamic light scattering above 1 GHz.

In 2004, by including rotational–translational coupling in a schematic model, Götze and Sperl [34], and later on Sperl [35] showed that the IPL data of salol and benzophenone (BZP) from OHD-OKE could be explained by the nearly logarithmic decay of correlation function obtained from the two-correlators model of MCT. One correlator is for density fluctuations and the other for reorientational fluctuations of the molecules. By assuming the presence of strong rotational–translational coupling, the model yields time dependence of response functions consistent with those measured for BZP and salol by OHD-OKE. They demonstrated that the fast β-relaxation of MCT can lead to a symmetrical Cole–Cole (CC) loss peak [217] with its high-frequency flank having the frequency dependence of ν^{-c}. The conventional susceptibility loss minimum scenario of the idealized MCT results if the CC loss peak frequency is larger than the microscopic excitations, and the data can be fit by the idealized MCT as usual. According to the new two-correlators model of MCT, the CC loss peak with frequency dependence given by the imaginary part of $1/[1 + (i2\pi\nu\tau_{CC})^c]$ also can be located at much lower frequencies. In this case, the high-frequency flank of the CC peak gives rise to a weaker power law ν^{-c} (i.e., either an excess wing or logarithmic decay if $c \approx 0$) on the high-frequency side of the α-loss peak. These two scenarios are best illustrated in Fig. 254 taken from

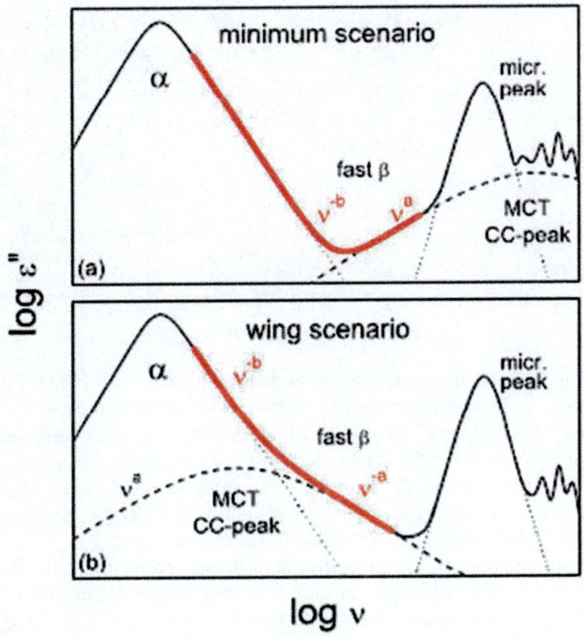

Fig. 254 Schematic loss spectra for two scenarios of two-correlators version of MCT [34, 35]. Depending on the frequency of the CC peak, the conventional case of a fast β minimum (**a**) or a wing at the right flank of the α-peak (**b**) can arise. Figure taken from [1120, 1121] and reproduced by permission

[1120, 1121]. It was shown by Sperl [35] that this scenario can fit the IPL data of BZP from OHD-OKE. The OHD-OKE signal *S*(*t*) of BZP and BMP are shown in the right panel of Fig. 255. Depolarized light-scattering (DLS) experiments on BZP reported good agreement with OHD-OKE in showing the power law ν^{-c} indicating the excess wing [1054].

Test by Broadband Dielectric Relaxation Measurements of BZP

Broadband dielectric relaxation measurement were made on BZP by the Augsburg group [1120, 1121] to test the two-correlators model of MCT. Their dielectric loss data are reproduced in the left panel of Fig. 255. These data have been shown before in Fig. 176 for a different purpose than here, and it is worthwhile to show the data again here for comparison with the OHD-OKE data. Indeed the Augsburg group found a symmetric CC peak at low temperatures (200 K and below) that is located at frequencies higher than the α-loss peak. At higher temperature it tends to merge with the α-loss peak and was seen as an excess wing from 210 to 225 K. However, the characteristics of the dielectric CC relaxation departs from that of the MCT CC relaxation. As pointed out by the Augsburg group, the departures are the following [1120, 1121]. (1) The relaxation time of the MCT CC peak from the MCT analysis of the OKE data at 251 K [35] is about two orders of magnitude shorter than that of the dielectric CC peak obtained by extrapolation to the same temperature (see Fig. 256). (2) The frequency and width of the MCT CC peak should be temperature

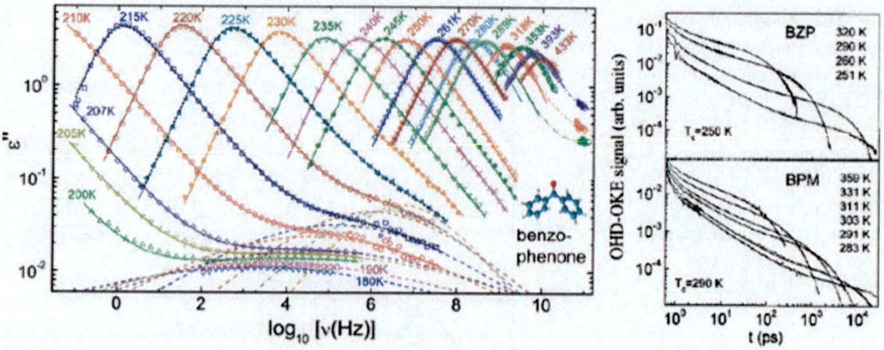

Fig. 255 (*Left*) Frequency-dependent dielectric loss of BZP for various temperatures. The *solid lines* are fits with the sum of a Cole–Davidson (CD) and a Cole–Cole (CC) function (at $T \leq 190$ K the CD amplitude and at $T \geq 235$ K the CC amplitude were set to zero). The *dashed lines* show the CC part of the fits accounting for the observed secondary relaxation. The *dotted lines* represent fits of the 215 and 225 K curves performed with the convolution ansatz. The *dashed-dotted lines* at high frequencies are drawn to guide the eyes. Reproduced from [1121] by permission. (*Right*) OHD-OKE signal of BZP and BPM data and fits to $S(t) = (pt^{-1+c} + dt^{b})\exp(-t/\tau_{\alpha})$ for some representative temperatures from 2 ps to the ends of the data. This empirical equation fits the data well over the full ranges of time and temperature. Figure from [332] reproduced by permission. The chemical structure of BZP is shown

independent, while that of the dielectric CC peak in BZP are strongly temperature dependent, and the width parameter c reaches values outside the range allowed by MCT. Other problems of the two-correlators MCT in explaining the excess wing were pointed out by Brodin et al. in [1047, 1053, 1054]. Although the intermediate power law or the excess wing can be successfully reproduced by the theory at temperatures above the MCT critical temperature T_c, the excess wing continues to be observed in BZP and other glassformers at lower temperatures down to T_g. According to Brodin et al. [1047, 1053, 1054], this poses another problem for this 2004 version of MCT, because still it cannot explain the characteristics of dynamic susceptibility at longer timescales and lower temperatures down to T_g.

Let us compare the BZP dielectric data of the Augsburg group at 250 K with the OHD-OKE signal at 251 K. The latter was well fit by the empirical equation [332]

$$S(t) = (pt^{-1+c} + dt^{-1+b})\exp(-t/\tau_{\alpha}). \tag{2.96}$$

The highest frequency in the dielectric measurements at 250 K is about 2 GHz (corresponding to ~100 ps). The IPL, pt^{-1+c} with $c = 0.13$, found by OHD-OKE at 251 K occurs in the time regime shorter than about 10 ps. Thus dielectric relaxation measurements were not able to see the IPL in the form of the NCL, which would be $\varepsilon''(\nu) = p'\nu^{-c}$ with $c = 0.13$, the equivalent of IPL, pt^{-1+c} with $c = 0.13$, found by OHD-OKE. Dielectric measurement is not able to see the excess wing that follows after the NCL in time either. This is because the former occurs in the intermediate time regime longer than 10 ps and shorter than 10^2 ps (see Fig. 255).

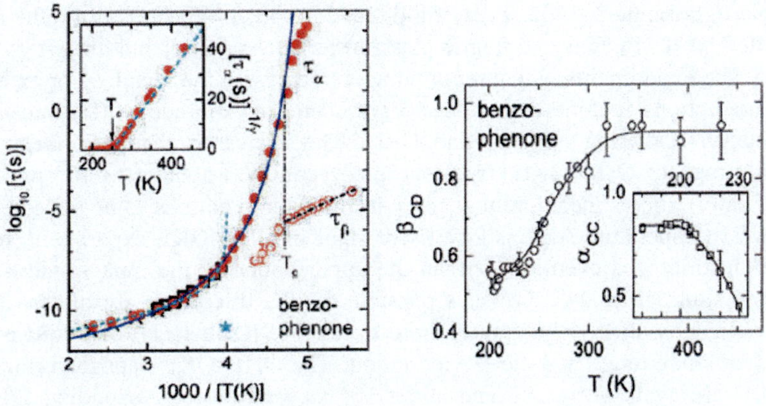

Fig. 256 (*Left*) Relaxation times determined by Lunkenheimer et al. [1121] from the fits shown in Fig. 255 (*full and open circles*). In addition, literature data from OKE and DLS experiments are shown (*open squares* and ×, respectively). The *solid line* is a crude fit of the dielectric data with a VFTH function, $\tau_\alpha = (4.2 \times 10^{-13}\ \mathrm{s})\exp[718.2/(T - 189)]$. The *dashed line* is a fit of the dielectric data for $T \geq 255$ K with the MCT equation, $\tau_\alpha \sim (T - T_c)^{-\gamma}$ using the parameters $T_c = 250$ K and $\gamma = 1.92$ as reported by Cang et al. [332]. The *dashed-dotted line* is a fit of the secondary relaxation time at $T < T_g$ assuming an Arrhenius law with pre-exponential factor $= 8.8 \times 10^{-15}$ s, and activation energy $E = 0.36$ eV. The *star* shows the CC relaxation time as determined by Sperl [35] from a 2004 version of MCT analysis of the OKE data. The *inset* shows the dielectric data and MCT fit, using a representation that should linearize according to the MCT prediction, $\tau_\alpha \sim (T - T_c)^{-\gamma}$ with $\gamma = 1.92$ [35]. (*Right*) Width parameters β_{CD} and α_{CC} of the α-relaxation and the secondary relaxation (*inset*), respectively. They were determined from the fits of the experimental loss spectra shown as solid lines in Fig. 255. The *lines* are drawn to guide the eyes. Reproduced from [1121] by permission

In the case of OHD-OKE experiments carried out at higher temperatures, the JG β- and α-relaxations have already merged together, and it is the merged relaxation that terminates the IPL having exponent c, which decreases toward the value $c \approx 0$ with decreasing temperature. The merged relaxation is given by the term $t^{-1+b}\exp(t/\tau_\alpha)$ in Eq. (2.96), which in the frequency domain corresponds to imaginary part of the Cole–Davidson function, $1/(1 - i\omega\tau_\alpha)^b$. Such action of the merged JG β- and α-relaxation in terminating the IPL can be seen by the other product term, $pt^{-1+c}\exp(-t/\tau_\alpha)$, needed in the function given by Eq. (2.96) to fit OHD-OKE data. The factor $\exp(-t/\tau_\alpha)$ in the product effectively ends the contribution of the IPL pt^{-1+c} at time longer than τ_α. The action ensures that the term, pt^{-1+c}, is not an additive contribution to the merged β- and α-relaxation, a fact that has been shown before in the context of NCL (i.e., when $c \approx 0$) from a combined consideration of depolarized light-scattering and dielectric relaxation data of the glass-forming molten salt, $0.4Ca(NO_3)_2 - 0.6KNO_3$(CKN). See Fig. 245 and [1451].

The following is a legitimate question to be asked: why did OHD-OKE produce data that deviate from the traditional and idealized MCT while before 2000 other techniques like quasielastic neutron scattering [32, 1437, 1438, 1441] and dynamic light scattering [32, 1421, 1440, 1442, 1443, 1445] have found the idealized MCT works in the same fragile glassformer such as salol and *ortho*-terphenyl (OTP)? The experiments which reported agreement with the idealized or extended MCT predictions include depolarized light scattering on salol, OTP, and CKN, and neutron scattering on OTP and CKN. The answer is that the much wider spectral range of OHD-OKE (from sub-picoseconds to about 5×10^5 ps in the case of salol) allows meaningful spectra at lower temperatures to be collected. At the lowest temperature such as 247 K for salol in OHD-OKE experiment, the α-relaxation time τ_α is even longer than the upper bound of the time window. This favorable situation at 247 K makes possible for the true caged dynamics, exemplified as IPL with $c \approx 0$ (logarithmic decay or NCL), to be observed over a broad frequency range and clearly identified. The OHD-OKE data for τ_α of OTP at 284 K are again very long and about 10^5 ps, and the corresponding IPL has $c = 0.15 \pm 0.03$. On the other hand, the windows of experiments cited above are for times shorter than 200 ps in neutron scattering, and for frequencies higher than 0.1 GHz in depolarized light scattering. Therefore they are limited to much shorter times or higher frequencies compared with OHD-OKE signal. Moreover, the data collected were usually performed at higher temperature whereby the merged JG β- and α-relaxations or its high-frequency flank appears in the susceptibility spectrum $\chi''(\nu) \propto \mathrm{Im}\{1/(\mathrm{i} - \omega\tau_\alpha)^b\}$, and is seen as the von Schweidler law with $\chi''(\nu) \propto \nu^{-b}$, of the idealized MCT. Encroachment of the high-frequency flank of the α-relaxation from the low-frequency side into the already limited experimental frequency window leaves no room for the NCL to be observed in the neutron and dynamic light-scattering experiments in the past. The consequence is necessarily that the observation of a susceptibility minimum and its characteristics, including the exponents b and a of the power laws defining it, had been compared favorably with the predictions of the idealized MCT in the years before the advent of the OHD-OKE data in 2000 [330], and the dynamic light-scattering data of PIB in 2001 [1062].

Returning to the discussion of BZP dielectric data of Lunkenheimer et al. [1121] shown in the right panel of Fig. 257, the α-relaxation strength, $\Delta\varepsilon_{\mathrm{CD}}$, of BZP obtained as a function of temperature from fits of the α-relaxation by the Cole–Davidson function shows a bump in a neighborhood of temperature around 240 K which is the MCT T_c and also T_B of BZP. Here T_B is the temperature at which $\tau_\alpha(T)$ changes its T-dependence from one VFTH law to another one as revealed by the Stickel plot in the left panel of Fig. 257, which determines $T_B = 240$ K. Incidentally, T_B coincides with T_c. The product, $T\Delta\varepsilon_\alpha$, plotted against T in the upper inset of the right panel exhibits a cusp anomaly near 240 K like that predicted by the idealized MCT for the non-ergodicity parameter at T_c. This product $T\Delta\varepsilon_\alpha$ also was considered in earlier works by Schönhals et al. for various glassformers including dibutyl phthalate, salol, propylene carbonate, glycerol, and propylene glycol [1452]. Their results are reproduced in the left panel of Fig. 258.

There was either no cusp at T_c like in glycerol and propylene glycol, or a pseudo-cusp with the wrong shape was observed, like in salol, and disagreements with MCT predictions were noted by Schönhals et al. Thus, the MCT-like cusp of $T\Delta\varepsilon_\alpha$ found

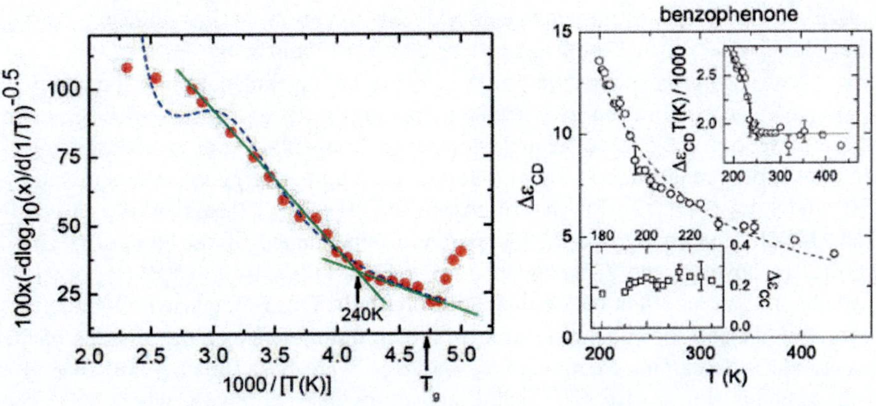

Fig. 257 (*Left*) Stickel plot of dielectric data of $\tau_\alpha(T)$ showing the change from one VFTH law to another one at $T_B = 240$ K. The scatter at higher temperature is due to no sufficiently finely spaced data points in performing the derivative. Courtesy of P. Lunkenheimer. (*Right*) Relaxation strengths as obtained from the fits of the experimental loss spectra. The *line* in the main frame is a fit of the α-relaxation strength, $\Delta\varepsilon_{CD}$, using the Curie–Weiss law. In the *upper inset*, $\Delta\varepsilon_{CD}$ multiplied by temperature is shown. The *line* is a fit with the MCT prediction, leading to $T_c = 240$ K. The *lower inset* shows the relaxation strength of the secondary process. Reproduced from [1121] by permission

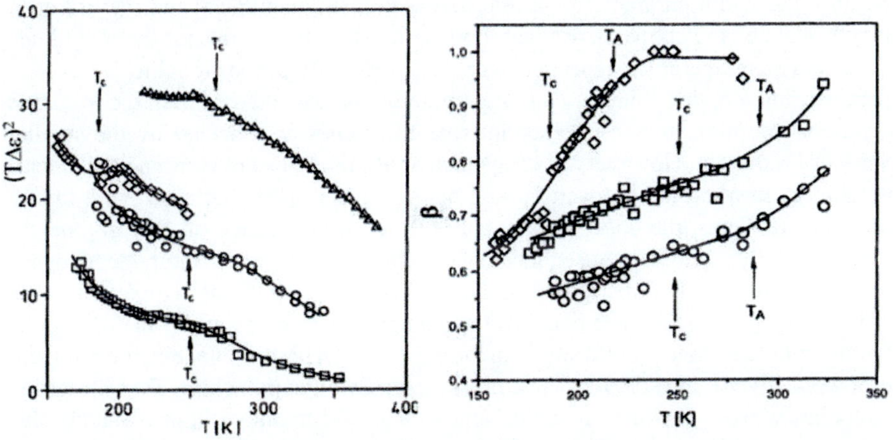

Fig. 258 (*Left*) $(T\Delta\varepsilon)^2$ vs. T. Glycerol (O); propylene glycol (□); propylene carbonate (◇); salol (Δ). *Lines* are guides for the eye. T_c is the critical temperature of MCT. (*Right*) Width parameters β_{CD} vs. T. Glycerol (◯); propylene glycol (□); propylene carbonate (◇). *Lines* are guides for the eye. T_c is the critical temperature of MCT and T_A is the temperature above which the temperature dependence of $\tau_\alpha(T)$ becomes Arrhenius. Reproduced from [1452] by permission

in BZP is an exception as far as dielectric relaxation is concerned, possibly caused by the mode of change of the T-dependence of the width parameter β_{CD} when crossing $T_B \approx T_c$ in BZP unlike that of the other glassformers. The difference is that β_{CD}

of BZP in Fig. 256 does not change after T falls below $T_B \approx T_c$, while β_{CD} of the other glassformers shows the largest decrease after T falls below $T_B \approx T_c$.

For this reason and others, we propose an alternative explanation of the anomaly of $\Delta\varepsilon_\alpha$ from consideration of what is happening at T_B as follows. As discussed before in Section 2.2.5.1, correlated changes of T-dependence of τ_α and the dielectric relaxation strength $\Delta\varepsilon$ at T_B were found in a number of glassformers including DBP and salol (Fig. 72), PC and oligomers of PG (Fig. 73), and PVAc (Fig. 74) [758, 1452]. The change of the temperature dependence of the dynamics causing the crossover of the T-dependence of τ_α from $(\text{VFTH})_1$ to $(\text{VFTH})_2$ at T_B is expected to have an effect on the relaxation strength. This is because both the relaxation strength and the relaxation time τ_α are complimentary characteristics of the dynamics. Concomitant changes of τ_α and $\Delta\varepsilon_\alpha$ were best shown [758] in a plot of $\Delta\varepsilon_\alpha$ against log ν_α for DBP, salol, and propylene carbonate, where ν_α is the α-loss peak frequency and is approximately the same as $1/(2\pi\tau_\alpha)$. These plots indicate two different dependences of $\Delta\varepsilon_\alpha$ on log ν_α, which when extrapolated intersect at a crossover frequency ν_B. It turns out that for all the glassformers ν_B is nearly the same as $1/[2\pi\tau_\alpha(T_B)]$. Similar results were found also for the polymer, poly(vinyl acetate) [238, 759], and most recently in an amorphous pharmaceutical, ibuprofen [1004]. Therefore, the change of the dependence of $\Delta\varepsilon_\alpha$ on log ν_α also occurs at the temperature T_B. This phenomenon can also be explained by the more rapid increase of n with decreasing temperature (frequency) after crossing T_B (ν_B), because the width parameter or n reflects the non-exponentiality and the extent or length scale of the cooperative many-body α-relaxation, and in turn the magnitude of $\Delta\varepsilon_\alpha$. The temperature dependence of $\Delta\varepsilon_\alpha$ is roughly proportional to $1/T$ at temperature much higher than T_B consistent with the Kirkwood–Fröhlich theory based on the assumption of non-interacting isolated dipoles as reflected by the smaller values of n therein. However, this dependence of $\Delta\varepsilon_\alpha$ does not continue when temperature is lowered to approach T_B and below T_B because n increases with falling temperature. Thus, the observed anomaly of the T-dependence of $\Delta\varepsilon_\alpha$ in a neighborhood of T_B is another consequence of the change of the temperature dependence of n or the degree of cooperativity of the α-relaxation with falling temperature after crossing T_B [237, 253]. The detail of the change of n in crossing T_B in BZP is different from the other glassformers including salol, DBP, propylene carbonate, and propylene glycol. In BZP, n or $(1 - \beta_{CD})$ increases on approaching T_B with falling temperature from above to reach a value at T_B, and maintain approximately this value when falling below T_B (see Fig. 4 in [44]). On the other hand, in the other glassformers, n is slowly varying and small when $T > T_B$ and rapidly increases with decreasing temperature after crossing T_B [237, 253]. This difference in the change of temperature dependence of n when crossing T_B of BZP from the other glassformers may explain the cusp anomaly of $T\Delta\varepsilon_\alpha$ found in BZP but not in the others.

Before we close, it is worthwhile to remind the reader of the discussion given before in Section 2.3.2.16 on the nature of the resolved secondary relaxation of BZP found by the Augsburg group. There we have demonstrated that it is not the JG β-relaxation, which is not resolved. This conclusion is consistent with the narrow α-loss peak that is well fit by the Fourier transform of the Kohlrausch function with relatively large exponent $\beta_{KWW} \equiv (1-n) = 0.70$. The JG β-relaxation was resolved after BZP was mixed with tri-styrene to enhance its coupling parameter, as shown in Fig. 177.

Tests with Pressure and Temperature Dependences

Combined pressure and temperature dependences of the high-frequency depolarized light-scattering susceptibility of *ortho*-terphenyl (OTP) were obtained by Patkowski et al. [644] with the purpose of comparing with idealized MCT predictions and the conclusions of a previous analysis of incoherent neutron scattering data by Tölle and coworkers [1403]. Patkowski et al. found that the time–temperature–pressure superposition does not work for OTP in the frequency range of the susceptibility minimum and both a and b exponents of the MCT are temperature and pressure dependent. At constant temperature, the exponent a increases with increasing pressure, being in most cases outside the MCT range ($a < 0.39$) and approaching 1 at high pressure, while the exponent b decreases with increasing pressure, reaching unreasonable values (< 0.2) at low temperatures. At a constant pressure, the exponent a decreases while the exponent b increases with increasing temperature. This kind of behavior is not consistent with the idealized MCT, in which both exponents a and b are interrelated by $[\Gamma^2(1+a)/\Gamma(1+2a)] = [\Gamma^2(1-b)/\Gamma(1-2b)]$, and they must be constant. When reinterpreted in terms of NCL, these behaviors of the susceptibility minimum can be rephrased as the susceptibility $\chi''(\nu)$ showing up as a power law, $p'\nu^{-c}$, with c decreasing with increasing τ_α isothermally at higher pressures or isobarically at lower temperatures. This behavior is no different from the behavior of IPL, pt^{-1+c}, on decreasing temperature observed by OHD-OKE. As described before, the idealized MCT cannot account for the IPL. If one insists on using MCT to explain the data of OTP, then the choice has to be the two-correlators MCT model. However, OTP has a resolved JG relaxation, which merges with the α-relaxation at high frequencies of the depolarized light-scattering study, but bifurcates from the α-relaxation at lower frequencies and temperatures. The Cole–Cole relaxation of the two-correlators MCT model is non-descript and cannot be identified with the JG relaxation of OTP, which has specific properties including various correlations with the α-relaxation discussed in various subsections before.

Patkowski et al. also found the pressure-induced shift of the position of the minimum is much weaker than that of the α-peak. Thus there is no universal scaling of the entire susceptibility including both the α-peak and the MCT susceptibility minimum, contradicting the previously reported conclusions obtained from the analysis of incoherent neutron scattering data by Tölle et al. [1403]. Actually, this is obvious from the fact that caged dynamics (NCL) does not involve either rotational or translational relaxation, and has no characteristic relaxation time as opposed to genuine relaxation processes such as the JG relaxation or α-relaxation.

Change of Caged Dynamics for Different Combinations of T and P at Constant Density and Static Structure Factor $S(Q)$

Frick et al. [647] performed neutron scattering spectroscopy on protonated polybutadiene to investigate the inelastic incoherent scattering law, $S(Q,\omega)$. These experiments give information on the self-correlation function $G_S(r,t)$ for the protons mainly. $G_S(r,t)$ is related by a space-time Fourier transformation to the $S(Q,\omega)$ as shown before in Eq. (1.119). The dynamic susceptibility $\chi''(Q,\omega)$ is obtained by the ratio $S(Q,\omega)/n(\omega,T)$. Here $n(\omega,T)$ is the Bose occupation factor used to correct for temperature changes due to phonon contributions in order to stress the relaxation contributions. Frick et al. also found that the static structure factor $S(Q)$ of deuterated polybutadiene is unchanged when P and T are varied along constant density lines. In MCT, the static structure factor controls the dynamics of the density correlation function. Thus, for different combinations of P and T having the same density, MCT predicts that the same dynamic susceptibility, $\chi''(Q,\omega)$, should be observed. This prediction is contradicted by the experimental findings of Frick et al. in that both the fast relaxation and α-relaxation change on varying T and P while maintaining the density constant. This is demonstrated for the fast relaxation at one constant density in Fig. 259 for constant $Q = 1.6\,\text{Å}^{-1}$. Frick et al. stated that even given the fact that $S(Q)$ is composed of different unknown partial structure factors, their data still contradict idealized MCT prediction, with the exception of the unlikely case that different partial structure factors cancel each other.

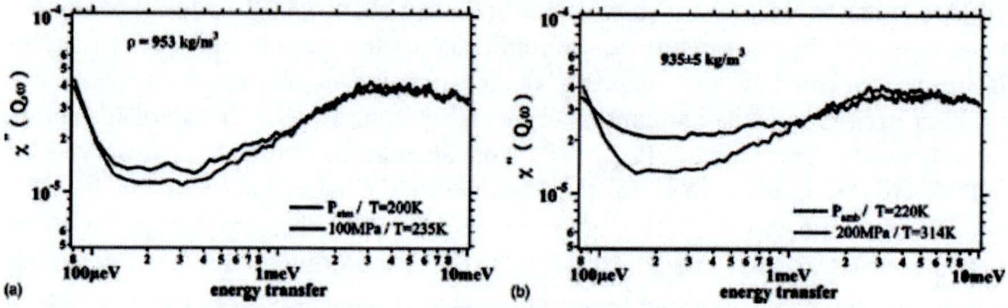

Fig. 259 (**b**) The change of the dynamic susceptibility of protonated polybutadiene (h6-PB) at $Q = 1.6\,\text{Å}^{-1}$ in the fast relaxation regime for two different combinations of P and T measured by inelastic incoherent neutron scattering along an *isochore*. The two sets of data at ($P = 0.1\,\text{MPa}$, $T = 220\,\text{K}$) and ($P = 200\,\text{MPa}$, $T = 314\,\text{K}$) are from an isochore with density=935±5 kg/m^3. The slight uncertainty explains some deviations in the vibration regime. (**a**) The same behavior was found along an isochore with density = 953 kg/m^3 for dynamic susceptibility data at ($P = 0.1\,\text{MPa}$, $T = 200\,\text{K}$) and ($P = 100\,\text{MPa}$, $T = 235\,\text{K}$). Figure from Frick et al. [647], reproduced by permission

The change of the α-relaxation also was reported by Frick et al. For two (T,P) combinations at the same density, the one with a higher T has shorter α-relaxation time and higher intensity of the fast relaxation. The fast relaxation shows up like a susceptibility minimum, but the minimum is so flat, particularly for the (T,P) combination having a higher temperature (see Fig. 259b), that it resembles the dynamic light-scattering data of PIB at 260 K in Fig. 135 and PI at 230 K in Fig. 136. Thus the flat minimum suggests it may actually be the NCL, which is definitely the case of PIB because of availability of evidence in Fig. 135 at lower temperatures near and below T_g where the α-relaxation time, τ_α, becomes very long. The flat susceptibility minimum observed in polybutadiene by Frick et al. cannot be fitted by the idealized MCT because of serious deviations from the MCT relation between the critical exponents a and b. An example of such deviation was reported earlier by Zorn et al. in neutron scattering studies of polybutadiene at ambient pressure [310].

On the other hand, if the data are interpreted as NCL instead of idealized MCT, the experimental findings of Frick et al. that $\chi''(Q,\omega)$ changes with different combinations of P and T having the same static structure factor can be easily understood. In particular, the observation that the (T,P) combination having a longer α-relaxation time τ_α has a lower NCL intensity and hence a less shallow susceptibility minimum is rationalized as follows. First, this finding is in accord with the empirical fact that the intensity of NCL increases on decreasing τ_α. There are many examples of this shown before in previous sections such as Fig. 133 for NMEC and *tert*-butylpyridine, Figs. 135 and 135 for PIB and PI, Fig. 244 for CKN, and Figs. 1 and 79 for colloidal particles. Second, longer τ_α corresponds to lower temperature or higher pressure, and increase in density. The latter means that for tighter cages, more restricted motion of caged molecule, and smaller amplitude of cage fluctuation, there is lower intensity of NCL.

2.3.2.38 Conversion of α-Relaxation to the JG β-Relaxation or Primitive Relaxation by Suppression of Cooperativity

The properties of the JG β-relaxation or primitive relaxation discussed in the subsections in Section 2.3 all point to the existence of its relation to the structural α-relaxation. At any temperature or pressure, the relation arises from the local JG β-relaxation or primitive relaxation being the basic motion generating the many-molecule dynamics, which increases in length scale with time to become the terminal α-relaxation with the maximal possible length scale, $\xi(T)$, and correlation function given by the Kohlrausch correlation function with exponent, $n(T)$. If one can reduce or suppress the involvement of many bodies in the α-relaxation, the α-relaxation would have lesser or even no difference from the local JG β-relaxation or primitive relaxation. One way to accomplish this feat is to reduce the spatial dimension of the glassformer to make it less than $\xi(T)$. Since one can calculate the primitive relaxation time and make a prediction of the resultant $\tau_\alpha(T)$ in the event that many-body cooperative dynamics can be removed totally. Since $\xi(T_g)$ of bulk glassformer having n not small is of the order of nanometers near and above T_g (see Section 2.2.2), confinement of the supercooled liquid in nanometer-size glass pores is an effective method to reduce many-body cooperative dynamics. The proviso is that the liquid does not chemically bond to or physically interact with the walls, which can be met if the surfaces of the confining walls have been rendered passive by treatment such as with silane for silica-related glass pores. This condition may not be achieved in many experiments reviewed in [1453, 1454]. Nevertheless, there are several studies of the relaxation of the nanoconfined liquids in silanized glass pores that show large reduction of intermolecular cooperativity. Examples include the liquids 1,2-diphenylbenzene (i.e., *ortho*-terphenyl (OTP)) [492–494], salol [490], and the polymers, poly(dimethyl siloxane) and poly(methylphenyl siloxane) which are hydrophobic and form no hydrogen bonds with the glass pore surface [498–500]. Under this ideal condition, molecules nearer the smooth wall experience reduction of intermolecular coupling due to the finite size and the absence of other molecules on the side of the wall. Intermolecular coupling is reduced for all molecules when the pore size becomes comparable or smaller than the many-molecule dynamics length scale. Experimentally, the density of the confined liquid may be lower than the bulk [493], and this happening also facilitates the reduction of intermolecular coupling because at lower density the molecules are further apart.

Reduction of intermolecular coupling in the CM means that the coupling parameter $n(T)$ becomes smaller than the bulk value $n_b(T)$. Had this effect been the only factor, the dispersion of the liquid in the pore will be narrower than in the bulk. However, there is an additional complication. The cooperative dynamics of molecules situated at various distances from the wall is necessarily affected differently and thus n and τ_α are position dependent. The overall observed relaxation dispersions of confined liquids are additionally broadened by the spatial distribution of the molecules in the pores. This is a dominant effect on the observed dispersion because all molecules inside the pore are sampled by experiment, and it could eclipse the effect of the reduction of $n(T)$ of the molecules particularly those located nearer the smooth wall. Hence, we cannot obtain any of the reduced values of $n(T)$ of molecules from the experimentally observed width of the dispersion of the liquid confined in pores. Nevertheless, we expect that $n(T)$ to decrease with the decreasing size of the pores. The CM equation can be rewritten as $\tau_\alpha = \tau_0(\tau_0/t_c)^{n/(1-n)}$ or $\tau_\alpha = \tau_{JG}(\tau_{JG}/t_c)^{n/(1-n)}$. Since τ_0 or τ_{JG} is usually much longer than $t_c = 2$ ps and $n/(1-n)$ is a monotonic decreasing function with decreasing n, it is clear that $\tau_\alpha(T)$ of nanoconfined liquids decreases on decreasing the size of the pores. Consequently, the difference between τ_α and τ_0 or τ_{JG} becomes smaller [494, 506, 900]. If, in sufficiently small size pores, $n \to 0$, then $\tau_\alpha \to \tau_0$ or τ_{JG}, and the characteristics of the α-relaxation will not be very different from that of the JG relaxation. These expected changes are seen experimentally, and described in the following subsections.

PDMS and PMPS Confined in Silanized Glass Pores

The trend expected from the above discussion has been shown before in Figs. 28 and 29 by the dielectric relaxation data of low molecular weight poly(dimethyl siloxane) (PDMS) and poly(methylphenyl siloxane) (PMPS) confined in silanized glass pores of various sizes down to 2.5 nm obtained by Schönhals and coworkers [498–500]. Change from random coils to more stretched conformation of these polymers was also observed in confinement, and the induced orientations are an additional cause of enhancement of the mobility of molecules like non-linear deformation discussed in Section 2.2.5.10. The extremely broad loss peaks of the confined polymers originating from distribution of environments of the confined polymer can be seen from the isothermal spectra given in these figures. In the case of bulk PDMS ($M_w \approx 1400\,\mathrm{g\,mol}^{-1}$, $T_g = 152\,\mathrm{K}$), the location of the primitive frequency ν_0 corresponding to τ_0 calculated from τ_α and $n = 0.48$ values of bulk PDMS at one temperature near T_g by the CM equation is indicated by the short horizontal line in Fig. 260. Note that τ_0 is shorter than the bulk τ_α by about five orders of magnitude. On the other hand, τ_α of PDMS in 5 nm pores at the same temperature is only a factor of 4 longer than τ_0, and its temperature dependence is weaker than τ_α of bulk PDMS, having an apparent activation enthalpy of 48 kJ/mol commonly found for JG relaxation of polymeric glassformers. Moreover, the dielectric strength $\Delta\varepsilon$ (see Figs. 28 and 29) as well as the thermal capacity ΔC_p of the α-relaxation (see Fig. 30) undergoes marked decrease from the bulk value on decreasing the pore size. At 5 nm, the dielectric strength and ΔC_p of the α-relaxation are significantly smaller than the same quantities of bulk PDMS. They become similar to that observed for the JG relaxation of bulk glassformers by adiabatic calorimetry [1048, 1075]. Moreover, the temperature dependence of $\Delta\varepsilon$ changes from decreasing with increasing temperature like the behavior of α-relaxation to increasing with increasing temperature as exhibited by secondary relaxation (Figs. 28 and 29). All these changes on decreasing the size indicate that in nanometer size pores down to 5 nm or less, the α-relaxation is not much different from the JG relaxation.

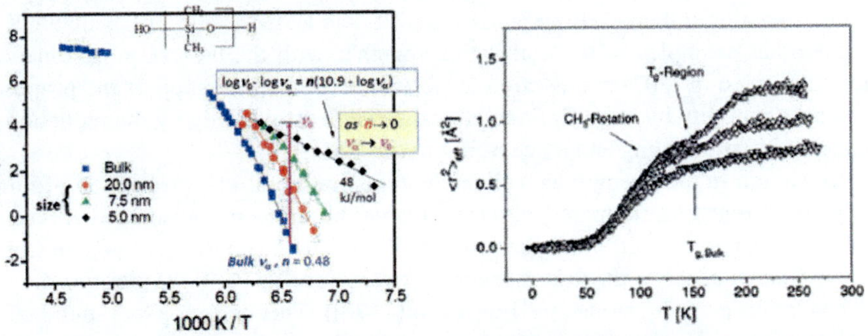

Fig. 260 (*Left*) Loss peak frequency, f_p, of PDMS in bulk or in nanopores of various sizes. Shown also is the primitive frequency ν_0 calculated by the CM equation using known values of $n = 0.48$ of bulk PDMS and ν_α at one temperature indicated by the *vertical arrow*. (*Right*) Temperature dependence of the effective mean-squared displacement $< r^2 >_{\mathrm{eff}}$ for PDMS: from *top* to *bottom*, (Δ) 7.5 nm, (◇) 5.0 nm, (∇) 2.5 nm. The step at about 100 K is ascribed to local methyl group rotations and the second step to the increase of the displacements near the glass transition. For the 2.5 nm pores no increase is observed around T_g. Reproduced from [500] by permission

It is worthwhile here to also consider the temperature dependences of the mean-squared displacement, $< r^2 >_{\mathrm{eff}}$, of PDMS measured by neutron scattering in bulk PDMS and PDMS confined in the different pore sizes of 7.5, 5, and 2.5 nm. The neutron spectrometer used is IN16 in ILL at Grenoble with energy resolution

1 μeV, corresponding to the observation times of the order of $t \sim \hbar/E \sim 4$ ns. Based on the relaxation map in Fig. 260, the JG β-relaxation time near T_g suggested by the calculated τ_0 is about 10^{-5} s and at 175 K it is about 10^{-7} s. Both are much longer than 4 ns. Hence the caged molecular dynamics is responsible for the measured $< r^2 >_{\rm eff}$ of PDMS confined in the 7.5 nm pore in the left panel of Fig. 260 up to 175 K, and the change to a stronger T-dependence after crossing T_g, estimated to be about 144 K from $\tau_\alpha = 10^2$ s. This behavior of $< r^2 >_{\rm eff}$ is a particular case of the general property of caged molecular dynamics found in other glassformers (see Figs. 248 and 251). The same behavior is found in the NCL of caged molecular dynamics, which is the equivalent of $< r^2 >_{\rm eff}$ in the susceptibility, $\chi''(Q,\omega)$. Most interesting in the right panel of Fig. 260 is the trend exhibited by $< r^2 >_{\rm eff}$ on decreasing the pore size. The magnitude of $< r^2 >_{\rm eff}$ not only decreases over the entire temperature range, but also increases above T_g. The same said in the above about PDMS confined in nanopores was found for PMPS. The data of $< r^2 >_{\rm eff}$ have been presented before in Fig. 31. The plot of the slope $d< r^2 >_{\rm eff} /dT$ for $T > T_{g,\rm bulk}$ vs. inverse pore size given in the inset of Fig. 31 shows this trend clearly for PMPS. We have concluded before that intermolecular coupling and n are further reduced on decreasing pore size. Hence the trend of $< r^2 >_{\rm eff}$ observed correlates with the decrease in n. The reader may recall that the same correlation is observed when comparing $< r^2 >_{\rm eff}$ of OTP with glycerol in Fig. 248 (right) and Fig. 251, $< r^2 >_{\rm eff}$ of PIB with PS or PMMA in Fig. 248, and the NCL in the susceptibility spectra of $ZnCl_2$ with CKN. In all these other cases, the correlation of the trend in caged $< r^2 >_{\rm eff}$ or NCL with n is observed and can be verified. That is glycerol has smaller n than OTP, so is PIB than PS and PMMA, and $ZnCl_2$ than CKN. These other examples of the correlation together with the theoretical rationalization, discussed in [507] and Section 2.3.2.35, can be used to support the proposed enhanced mobility by nanoconfinement caused at least partially by the reduction in intermolecular coupling and n according to the CM.

Reduction in the strength as well as the temperature of glass transition of ultrathin polystyrene film compared with the bulk was observed by Ellison and Torkelson [1170] from reduced difference between the liquid-state and the glassy-state slopes of the T-dependence of the fluorescence of pyrene-labeled PS in the ultrathin single-layer film on glass substrate (see Fig. 1a in [1170]). This behavior was seen before by Kawana and Jones [1455] in thin polystyrene films supported on silicon from the decrease of the difference between the expansivities on the glass and liquid sides of the transition with decreasing film thickness. The decrease of T_g with decreasing thickness h of the PS film can be explained by the decrease of intermolecular coupling and the averaged n. Since n is proportional to the extent and hence the strength of the many-molecule α-relaxation dynamics (see Section 2.2), the reduction in the α-relaxation strength in glass transition with decreasing either film thickness or confining pore size is also explained by the decrease of the averaged n. It is worthwhile to point out for polymers of high molecular weights, induced chain orientations in ultrathin films are an additional cause of reduction of intermolecular coupling [1456].

OTP Confined in Silanized Glass Pores

The dynamics of OTP confined in silanized nanometer pores were studied by photon correlation spectroscopy [492, 493], and the data were shown before in Fig. 92 and discussed before in Section 2.2.5.8, and in [494]. The α-relaxation times τ_α determined from the non-exponential correlation function of OTP confined in the smallest 2.5 nm pores at two temperatures are shown by closed squares in Fig. 92. Shown also are τ_α (from PCS) and τ_{JG} from dielectric data of bulk OTP, and τ_0 calculated from the bulk OTP parameters of τ_α and n obtained from PCS and from dielectric measurements [492, 493]. There are six orders of magnitude reduction of the bulk τ_α at 243.15 K caused by reduction of intermolecular cooperativity in OTP when it is confined in 2.5 nm pores (with possibly lower density), and this change is correctly predicted by the primitive relaxation time τ_0 of the CM. In other words, the measured α-relaxation of OTP confined in 2.5 nm pores is almost a realization of the primitive relaxation process in bulk OTP. Reference [493] pointed out the possibility that the density of OTP in the nanopores is lesser than in the bulk. This is just another cause of reduction of intermolecular coupling and n in addition to finite size effect, and its presence does not alter our calculation of τ_0 and comparison with the observed τ_α of OTP confined in 2.5 nm pores.

Dynamic mechanical response of liquid salol confined in mesoporous silica with coated inner surfaces has been measured by Kuppensteiner et al. [490] as a function of temperature and frequency (1–100 Hz) for various pore sizes (2.4–7.3 nm). The α-relaxation time τ_α together with T_g is reduced on decreasing pore size, which can be attributed to the reduction of length scale of the many-body relaxation and hence n. Bulk salol has n=0.37, and the largest reduction of τ_α due to extreme confinement would be τ_0 that can be calculated from τ_α of bulk salol by the CM equation. This is done in Fig. 260-1 for τ_α=100 s at T=T_g=220 K. The calculated τ_0 is 10^{-3} s. The observed τ_α in coated 2.4 nm pores is about 2×10^{-2} s, which is

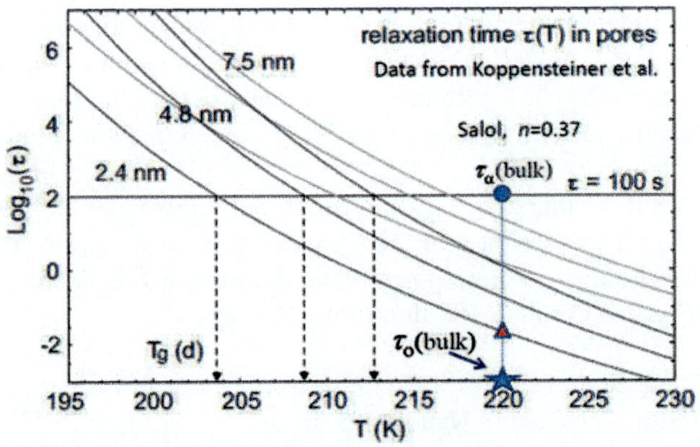

Fig. 260-1 Relaxation time in treated pore centers (*black lines*). *Horizontal line* shows τ=100 s. *Gray lines* are relaxation times in untreated pores [490]. Shown also is the primitive relaxation time of bulk salol calculated with n=0.37 determined in [596]

about 20 time longer than the largest reduction possible. This indicates that many-body effect is not totally suppressed, and the α-relaxation has not been converted to the primitive or the JG β-relaxation, consistent with the VFTH T-dependence of the observed relaxation time. The reduction of τ_α of salol in coated 2.4 nm pores is smaller than τ_α of OTP in the same pores [492–494(a)]. This is consistent with the larger $n \sim 0.50$ of OTP than 0.37 of salol.

PMPS Ultrathin Films in Nanocomposites

Similar effects were found in thin poly(methylphenylsiloxane) (PMPS) films with thickness of the order of 1.5–2.0 nm intercalated into galleries of organically modified layered silicates by Anastasiadis et al. [495]. The modified layered silicates were prepared by a cation exchange reaction between the Na^+ in the silicate hosts and dioctadecyl-dimethyl-ammonium bromide at exchange capacity in order to render the originally hydrophilic silicate surface organophilic. Hybrids were prepared by mixing dry organosilicate and PMPS ($M_w = 2600$, $T_g = 223$ K) with various PMPS concentrations from 15 up to 30%. The observed α-relaxation time τ_α in the PMPS film is much shorter than in the bulk at the same temperature (see left panel of Fig. 261). The root-mean-squared end-to-end distance of the PMPS chains is estimated to be of the order of 3 nm, which is about twice the thickness of the films. It can be expected that there are significant induced orientations in the chains due to severe chain confinement. This, together with the extremely small thickness of the film, suggests that we may now have large reduction of intermolecular coupling of the local segmental relaxation. Hence, in the thin film, the coupling parameters n may be reduced to approach zero value, and the observed α-relaxation time $\tau_\alpha(T)$ becomes nearly equal to the primitive relaxation time $\tau_0(T)$ calculated from the parameters, $\tau_\alpha(T)$ and n, of bulk PMPS obtained by fitting the dielectric α-loss peak of PMPS [496]. The coupling parameters n in the Kohlrausch functions used in the fits have the values 0.50, 0.48, 0.46, and 0.40, at 228, 233, 238, and 243 K, respectively. This expected change of $\tau_\alpha(T)$ from the bulk values to somewhere in the neighborhood of $\tau_0(T)$ when confined to 1.5 nm thin film is supported in the comparison made in Fig. 261 with the experimental data at the four temperatures, 228, 233, 238, and 243 K. The reduction of $\tau_\alpha(T)$ is comparable with, albeit more severe than, that seen in PMPS confined in 2.5 nm glass pores discussed before (see Fig. 29). Another similar case is thin supported polystyrene films [1457]. The changes of $\tau_\alpha(T)$ on decreasing film thickness from 90 to 6 nm are similar to that found in PDMS confined in glass pores when the pore size is decreased from the bulk to 5 nm (Fig. 260) or in PMPS thin films (Fig. 29).

PEO Ultrathin Film Confined in Nanocomposites

More recent results have been reported on the effects of confinement on the segmental relaxation of ~1 nm poly(ethylene oxide) (PEO) films confined within the galleries of hydrophilic silicates, sodium-montmorillonite (Na^+MMT) obtained by

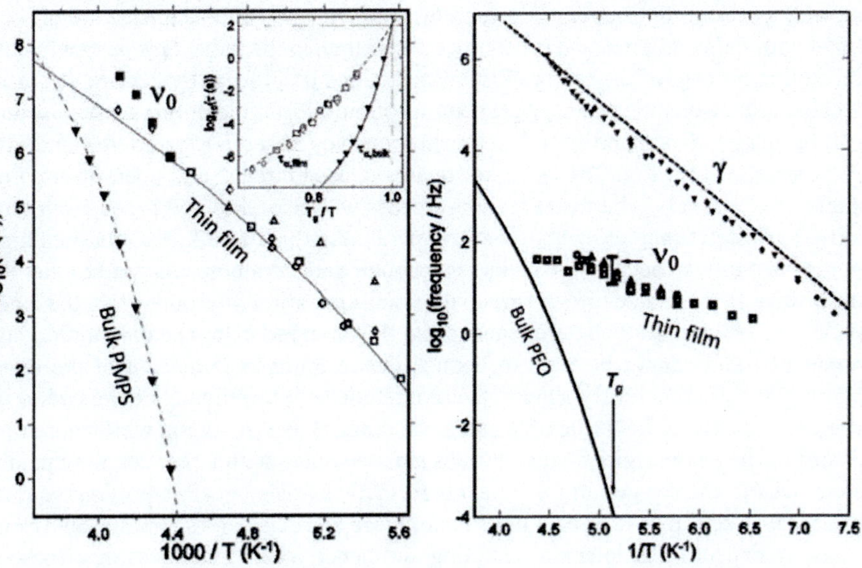

Fig. 261 (*Left*) Temperature dependencies of the various relaxation times of bulk and thin film of PMPS obtained by dielectric spectroscopy. α-Relaxation rates, $-\log\,\tau_{\alpha,\mathrm{bulk}}$, of bulk PMPS (▼). Relaxation rates, $-\log\,\tau_{\alpha,\mathrm{film}}$, of 1.5 nm thin films of PMPS prepared by mixing dry organosilicate and PMPS with various PMPS concentrations from 15 up to 30% (△ 15%, ○ 20%, □ 25%, 30% ◇). Primitive relaxation rate, ν_0, of bulk PMPS calculated by the CM from bulk PMPS data at temperatures above T_g (■). The *lines* are VFTH fits to $-\log\,\tau_{\alpha,\mathrm{bulk}}$ and $-\log\,\tau_{\alpha,\mathrm{film}}$. The *inset* compares the T_g-scaled temperature dependence of the two relaxation times. (*Right*) Comparison of the observed segmental relaxation frequency of PEO confined in galleys of Na+MMT (from [1458]) and the calculated primitive frequency from ν_α of the bulk PEO studied in [1458] and the two values $n = 0.34$ and 0.38 of the coupling parameters n. Reproduced from [1096] by permission

Elmahdy et al. using dielectric spectroscopy [1458]. It was found that confinement results in an acceleration of the PEO segmental dynamics and changes the VFTH temperature dependence of τ_α found in bulk PEO to an Arrhenius temperature dependence with low activation energy (see right panel of Fig. 261). In bulk and pure PEO, a fast γ-relaxation has been observed together with the segmental α-relaxation. This γ-relaxation, called the β-relaxation by Elmahdy et al., is not the JG β-relaxation of PEO as can be judged by a comparison with the relaxation map of PEO in Fig. 157. The α-relaxation times have the VFTH temperature dependence of $\tau_\alpha = \tau_\infty \exp[B/(T - T_0)]$, with $\tau_\infty = 1.0 \times 10^{-11}$s, $B = 2700 \pm 300$ K, and $T_0 = 112 \pm 10$ K. The relaxation times for the γ-relaxation of the pure PEO follow an Arrhenius dependence, $\tau_\gamma = \tau_{\gamma\infty} \exp(E_\gamma/RT)$, with a single activation energy $E_\gamma = 35 \pm 0.5$ kJ/mol and $\tau_{\gamma\infty} = 6 \times 10^{-15}$ s, characteristic of a local process.

The temperature dependences of the relaxation frequencies $f_\alpha \equiv (1/2\pi\tau_\alpha)$ and $f_\gamma \equiv (1/2\pi\tau_\gamma)$ of bulk PEO are shown in Fig. 261. For the 50 wt% PEO hybrid as well as for all hybrids with lower PEO content, the PEO segments are all confined within the galleries and relax much faster than in the bulk. This *new* segmental relaxation process of the ~1 nm PEO films (called α'-process by Elmahdy et al.) is located in between the α- and γ-relaxations of bulk PEO. It exhibits Arrhenius temperature dependence and its relaxation frequencies observed for 50, 70, and 80% PEO are shown in Fig. 261 by open squares, closed circles, and open inverted triangles, respectively. The faster local γ-process was also observed (open circles and closed inverted triangles) with a rate comparable to that of bulk PEO (dashed line), understandably so because of its local character and intramolecular nature that it is insensitive to finite size effect. Due to the semi-crystallinity of bulk PEO, the amorphous region is spatially heterogeneous and the observed α-loss peak is additionally broadened and cannot be used to deduce the coupling parameter n of the amorphous PEO. As previously shown, the magnitude of n of polymers is related to the chemical structure [244]. Flexible chain backbones have smaller n, while more rigid chains having more rigid pendant groups are associated with larger coupling parameters. Due to the oxygen in the repeat unit, PEO is a very flexible polymer and the chemical structure of its repeat units is not more complicated than polyisobutylene. Thus we expect intermolecular coupling and hence n of PEO is less than those of PIB. The value of n of PIB is 0.45 [208–210], which can be considered as an upper bound of the actual value of n of PEO. Another indication that n of amorphous PEO is smaller comes from the small steepness index $m \equiv d[\log \tau_\alpha(T)]/d(T_g/T)$ at $\tau_\alpha(T = T_g) = 100$ s. From the VFTH dependence given for $\tau_\alpha(T)$ of bulk PEO [1458], m is 29 compared with 45 of PIB and much larger values of other amorphous polymers including PMPS. In Fig. 261 (right), the two short horizontal lines terminating the vertical arrows indicate the primitive relaxation frequencies, ν_0, of *bulk* PEO calculated by the CM equation with two assumed but reasonable values of $n = 0.34$ and 0.38 at the temperature at which τ is about 1000 s [1096]. These calculated ν_0 are about the same as the observed segmental relaxation frequencies of PEO confined in the galleys of sodium-montmorillonite.

POB Confined in Glass Pores

The repeat unit of poly(oxybutylene) (POB) has dipole moment with components normal and parallel to the backbone so that both the segmental α-relaxation and the normal modes (global chain motion) can be probed by broadband dielectric spectroscopy. Kyritsis and coworkers reported [503] changes of dynamics of the α-relaxation and the normal modes of both hydroxyl- and methyl-ended poly(oxybutylene) (POB) confined in Vycor glass with mean pore diameter of 4 nm. They found, when confined, methyl-ended POB ($M_n = 970$ g/mol) has segmental α-mobility significantly higher than in the bulk state, leading to a decrease of T_g. In contrast, such effect is not observed for the normal (Rouse) modes, the relaxation rate of which is the same whether confined or not (see Fig. 262). This interesting observation has an immediate explanation from the CM based on previous explanations [837] of other different behaviors of segmental α-relaxation and normal modes of bulk polymers reviewed in [165, 87]. As discussed in Sections

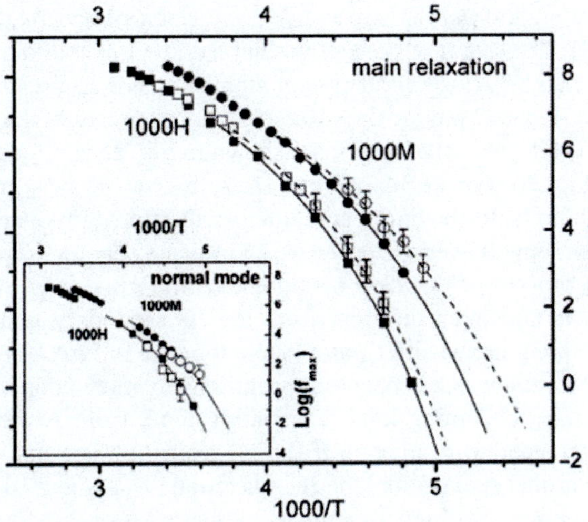

Fig. 262 Arrhenius plots of the peak frequencies of the α-relaxation for OH-ended POB with molecular weight $M_n = 970$ (1000 H, *squares*) and methyl group-ended POB with $M_n = 970$ (1000 M, *circles*) in bulk (*closed symbols*) and in confinement (*open symbols*). In the *inset* are the Arrhenius plots for the corresponding normal mode processes using the same symbols. Reproduced from [503] by permission

2.3.2.30, 2.3.2.31, and 2.2.5.9, the normal Rouse mode relaxation time, τ_n, has weaker temperature and pressure dependences compared with the α-relaxation time, τ_α. This difference is explained by the CM based on the fact that the normal (Rouse) modes are entropic and not intermolecularly coupled, and hence has zero coupling parameter. Hence $\tau_n(T,P) \propto \zeta_0(T,P)$, the primitive monomeric friction coefficient, whether in the bulk or nanoconfined. On the other hand, for the intermolecularly coupled α-relaxation, $\tau_\alpha(T,P) \propto [\zeta_0(T,P)]^{1/(1-n_\alpha)}$, and hence the stronger T- and P-dependences of τ_α than those of τ_n. The different effects that nanoconfinement in Vycor glass have on τ_α and τ_n observed by Kyritsis can be explained following the same line as used before in [1096]. When nanoconfined, the α-relaxation coupling parameter, n_α, of POB is reduced. Then it follows from $\tau_\alpha = \tau_0(\tau_0/t_c)^{n_\alpha/(1-n_\alpha)}$ that τ_α of the nanoconfined POB is significantly reduced (since $\tau_0 \gg t_c$ under the experimental condition). On the other hand, the normal (Rouse) modes are entropic in nature and intermolecularly uncoupled to start with, and remain so when POB is nanoconfined, and hence τ_n of POB in nanoconfinement is the same as in bulk POB.

The size effect on POB is not sufficiently small enough to reduce n to near zero to reveal the JG β-relaxation. Nevertheless, together with the fact that the same primitive monomeric friction coefficient $\zeta_0(T)$ governing both τ_n of the normal mode and τ_{JG} of the JG relaxation (see Section 2.3.2.30), the null effect on τ_n by nanoconfinement supports that τ_{JG} of the unresolved JG β-relaxation in POB is insensitive nanoconfinement.

Polyethylene-Like Glass Transition of Alkyl Nanodomains in Side-Chain Polymers

In Section 2.2.2.5 we have discussed the polyethylene-like glass transition driven by a process called α_{PE} occurs within the alkyl nanodomains formed by nanophase separation of incompatible main and side chain parts in higher member of the poly(n-alkyl acrylates) (PnAA), and poly(n-alkyl methacrylates) (PnAMA) [297, 298]. The size d of the self-assembled alkyl nanodomains is mainly determined by the number, C, of alkyl carbon atoms per side chain and increases from 0.5 nm for $C = 4$ up to 2 nm for $C = 10$. The T-dependences of the α-relaxation time, $\tau_{C,\mathrm{PE}}$, for $C = 4$, 8, and 10 of PnAA are shown in Fig. 25 and for $C = 4$, 7, and 10 of PnAMA in Fig. 26. For PnBA or PnBMA with $C = 4$, $\tau_{4,\mathrm{PE}}$ has Arrhenius T-dependence with prefactor having magnitude of vibrational time, and fragility index m near the minimum allowed value. These all indicate near total absence of cooperativity in the α_{PE} process of the alkyl nanodomains in PnBA or PnBMA with $C = 4$, and hence α_{PE} is not much different from the JG secondary relaxation β_{PE} of polyethylene [481]. On increasing C, alkyl nanodomains in PnAA and PnAMA have larger size, and the α_{PE} process becomes progressively more cooperative. This trend is reflected by $\tau_{C,\mathrm{PE}}$ becoming longer, deviating more from Arrhenius T-dependence to have larger steepness index m. If the JG relaxation is a universal feature accompanying any primary relaxation, the JG relaxation β_{PE} associated with the α_{PE} process should exist also in the higher members of the PnAA or PnAMA series, and their relaxation times $\tau_{C,\mathrm{JG}}$ should be nearly the same as the observed values of $\tau_{4,\mathrm{P}}$ [481]. JG relaxation has not been resolved from the dielectric or mechanical spectra of PnAA or PnAMA with $C > 4$ at normal cooling rates. Possibly this is due to the overlap of the JG relaxation process β_{PE} with the α_{PE} process if $\tau_{C,\mathrm{JG}}$ is not much shorter than $\tau_{C,\mathrm{PE}}$. However, rapid quenching introduces additional free volumes which increase the amplitude of the localized JG relaxation in the alkyl nanodomains and can help to resolve it, as done before in other glassformers to reveal the unresolved JG relaxation. In fact, by rapidly quenching $C = 10$ poly(n-decyl methacrylate) and $C = 7$ poly(n-heptyl methacrylate) [482], the JG relaxation in alkyl nanodomains has been resolved, as shown before for C10 in Fig. 27. Here in Fig. 263, the data of quenched C7 poly(n-heptyl methacrylate) are reproduced to show the results are general. The relaxation peak has some structure and two values of the relaxation frequencies are given for both cases of $C = 10$ and 7. It can be seen that these relaxation frequencies are not far from $1/\tau_{4,\mathrm{PE}}$ as we have expected. The additional free volume in the quenched samples disappears as the sample densifies during physical aging at temperatures slightly below T_g of the α_{PE} process. Consequently, the JG relaxation process can no longer be resolved.

Having proven the existence of the JG relaxation and located its relaxation time $\tau_{C,\mathrm{JG}}$ of polyethylene-like glass transition in nanodomains in PnAA and PnAMA, we can use the relation between $\tau_{C,\mathrm{JG}}$ and $\tau_{C,\mathrm{PE}}$ given by Eq. (2.69) to calculate the coupling parameter n_C when $\tau_{C,\mathrm{PE}} = 1$ s. The results have been shown before in Figs. 25 and 26 with $n_C = 0.35$ and 0.24 for the $C = 10$ and 8 members of PnAA, respectively, and $n_C = 0.33$ and 0.20 for the $C = 10$ and 7 members of PnAMA, respectively.

Recently, Arbe et al. [482(c)] have joined in the study of the dynamics in poly(n-butyl methacrylate), PnBMA ($C = 4$), and poly(n-hexyl methacrylate), PnHMA($C = 6$). In addition to calorimetry and dielectric relaxation, they introduced neutron diffraction and neutron spin echo spectroscopy in the study. Using isotopic labeling, neutron scattering probed the collective dynamics at the main

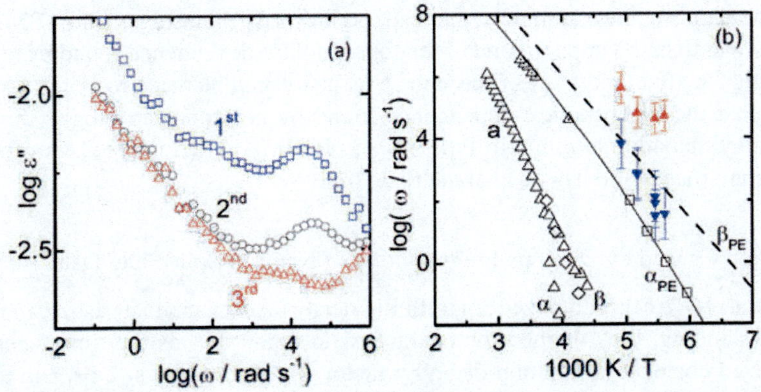

Fig. 263 (**a**) Dielectric loss ε'' vs. logarithm angular frequency ω measured for poly(*n*-heptyl methacrylate) at –90°C. The curves are measured on quenched samples (first sweep: t_e = 00 min) and after further annealing of the same sample for t_e = 20 min (second sweep) or t_e = 40 min (third sweep) at a temperature T_e slightly below the conventional glass temperature T_g of α_{PE}. (**b**) Arrhenius plots for poly(*n*-heptyl methacrylate), *C7*. The peak maxima frequencies from dielectric measurements on quenched samples (first sweeps as shown in previous figure) are indicated by *closed triangles*. The relaxation frequencies for the conventional glass transition α or a (*open symbols*) and for the polyethylene-like glass transition α_{PE} (*symbols with dot at center*) in these samples are shown for comparison. Data from dielectric spectroscopy (*triangles*), shear (*diamonds*), and heat capacity spectroscopy (*squares*) are included. The *thick dashed line* corresponds to the β_{PE} process in P*n*BMA and also to α_{PE} for $C = 4$ having relaxation angular frequencies equal to $1/\tau_{4,PE}$. Reproduced from [482] by permission

chain and the side group for both polymers, and the self-motions of hydrogen atoms in the side groups of P*n*HMA. The intent of Arbe et al. in making the neutron scattering measurements, a selective technique providing space/time resolution at a molecular level, is to check the validity of the interpretation of the dynamics of the alkyl nanodomain in P*n*AA and P*n*AMA by Beiner and coworkers [297, 481, 482(a), 482(b)] as discussed above. The correlation functions probed by neutron scattering are in the short time range, $2 \times 10^{-12} < t < 10^{-8}$ s. For the side groups in the nanodomains, Arbe et al. found that the correlation function decays nearly as a logarithmic function of time. For P*n*BMA, the logarithmic *t*-dependence persists over the range $4 \times 10^{-12} < t < 10^{-8}$ s at temperatures in the range $330 \leq T \leq 420$ K, and over the range $4 \times 10^{-12} < t < 10^{-9}$ s for $T = 450$ and 480 K. If this nearly logarithmic time dependence is transformed to susceptibility $\chi(\omega)$, the latter would correspond to a nearly constant loss (NCL) reminiscent of caged dynamics, rather than the structural relaxation process (α_{PE}) and not even the β_{PE} process. As discussed before in Section 2.3.2.33, the JG β-relaxation is to be distinguished from the NCL because it causes cage decay and terminates caged dynamics. Thus, the NCL precedes the JG β-relaxation in time. This can be checked from the dielectric relaxation times, $\tau_\beta{}'$, of the β'^{DS} process in the nanodomains obtained by Arbe et al. By extrapolating the nearly Arrhenius *T*-dependence of $\tau_\beta{}'$ at higher temperatures of P*n*BMA and P*n*HMA in Figs. 3 and 4 of [482(c)] to higher temperatures of neutron scattering measurements, the readers can verify themselves for P*n*BMA that τ_β' is longer than the upper bound of the logarithmic t-dependence found by neutron scattering for all temperatures. The same conclusion can be made for P*n*HMA. This proves that the logarithmic t-dependence found by neutron scattering is the NCL of the alkyl nanodomain in both P*n*BMA and P*n*HMA. NCL is not a conventional relaxation process and has no characteristic time.

Confusion Caused by Dubious Interpretation of Neutron Spin-Echo Data

Arbe et al. [482(c)] recognized logarithmic decays lack a characteristic timescale. Notwithstanding, they defined the characteristic time, τ_s^{SG}, as the time where the normalized correlation function decays a factor $1/e$ of its value at 2 ps, and took it to be the structural relaxation time of the side groups in the alkyl nanodomains. The reader can find evidence of the identification of τ_s^{SG} as structural relaxation time of side groups from various statements one can find in [482(c)]. Examples include the following two statements: "... the structural relaxation of the side groups in PBMA and PHMA exhibits extremely stretched (logarithmic-like) decays (Fig. 10)," and "At least for the highest temperatures, both structural times τ_s^{MC} and τ_s^{SG} approximately follow a power-law dependence on n_C." From the proof given in the previous paragraph that the logarithmic t-dependence is the NCL, it is clear that it cannot be identified as the structural relaxation time of the side chains in the alkyl nanodomains as done by Arbe et al., and the τ_s^{SG} obtained has no meaning.

Moreover, the values of τ_s^{SG} obtained in a dubious way and its mistaken identity as structural relaxation have undesirable consequences. An example is the conclusion made by Arbe et al. on comparing the relation between τ_s^{SG} and τ_s^{MC} (where the superscript MC stands for main chain) of PBMA and PHMA in their Fig. 9: "With increasing side-chain length the dynamics within the nanodomains becomes increasingly faster than at the main-chain level; a dynamic asymmetry clearly develops between main chain system and PE-like nanodomains for high-order PnMAs." This conclusion from neutron scattering contradicts the trend of the relaxation time of the structural relaxation, α_{PE}, becoming increasingly closer to the relaxation time, α, of the main chain with increasing side chain length or C [297]. This trend was well established by dielectric and mechanical shear data of poly(n-alkyl acrylates) from $C = 4$ up to $C = 10$, poly(n-alkyl methacylates) from $C = 4$ up to $C = 12$, poly(di-n-alkyl itaconates) from $C = 7$ up to $C = 11$, and hairy rod polyimides from $C = 12$ up to $C = 16$ in addition to amorphous polyethylene. The reader can find this general trend of these polymers clearly in Fig. 2c and b of the seminal paper by Beiner and Huth in 2003 [297]. Thus, the mistaken identity of τ_s^{SG} and its artificially manufactured values from the logarithmic decay of the correlation function obtained by neutron scattering offer no help, and instead the conclusion derived from it has contradicted the real physics already firmly established earlier [297]. By the way and for the same reason, it is meaningless to compare the logarithmic t-dependence of the normalized dynamic structure factor of P*n*BMA and P*n*HMA with the truly structural relaxation of PEMA having the Kohlrausch function time dependence with stretch exponent $\beta_K = 0.5$ and of PE in Fig. 12 of

[482(c)]. However, it is meaningful to compare logarithmic t-dependence of the normalized dynamic structure factor of PnBMA and PnHMA in the context that they are caged dynamics corresponding to NCL in the susceptibility $\chi(\omega)$. The observation that the logarithmic t-dependence of PnHMA decays at shorter times than PnBMA at the same temperature is equivalent to PnHMA having a higher level of NCL than PnBMA. This can be explained from the correlation between the level of NCL at T_g and the coupling parameter, $n(T_g)$, discussed before in Section 2.3.2.35, and the fact that $n(T_g)$ decreases with decrease in side chain length or C established before in [481] and supported by [482(a)].

Discussed so far are the collective dynamics of side groups probed by neutron spin echo (NSE) on PnBMA and PnHMA samples that are fully deuterated. Arbe et al. also performed NSE on the PnHMA-d5 sample to examine incoherent scattering from side group hydrogens. The PnHMA-d5 sample is a partially deuterated sample where the side group is protonated and the main chain (including the α-methyl group) is deuterated. They found a logarithmic-like decay for this correlation function as well. On comparing self-dynamics and collective dynamics of side groups, Arbe et al. found large difference in τ_s^{SG} of the two kinds of dynamics. For comparable conditions of scattering vector, the collective relaxation observed at 330 K decays to $1/e$ at τ_s^{SG} of about 40 ns, while the self-correlation function of protons at 310 K relaxes at $\tau_s^{SG} \approx 70$ ps, more than two orders of magnitude faster. Correctly interpreting the self-dynamics and collective dynamics of side groups as the caged dynamics, this difference can be explained as follows. It is intuitively clear that the mean-squared displacement from the self and individual motion of the hydrogen atoms in the side group is larger than the mean-squared displacement of the collection motion of the atoms in the side chain. This difference corresponds to faster logarithmic decay of the correlation function of the incoherent scattering from side group hydrogens as observed.

The emphasis on the NSE data together with the mistaken identification as the structural relaxation of the alkyl nanodomains has led Arbe et al. to interpret the dynamics of alkyl nanodomains in ways different from that of [297, 481, 482(a)] discussed in the above. Instead they related it to the dynamics of the fast component dynamically confined by the slower component in miscible systems such as dynamically asymmetric polymer blends, which is the same idea shown before to be untenable in Section 2.3.2.15. They also suggested "In the case of dynamically asymmetric polymer blends, the anomalous dynamic features were interpreted in the framework of the MCT, invoking high-order MCT transition." The latter is the 2002 version of MCT proposed by Götze and Sperl [33], which has been rejected by Fayer and coworkers [332] on the grounds that the current MCT higher order singularity case should occur only in very rare situations for systems like colloids that have very specific inter-particle interaction potentials, and thus it is not capable of explaining the universality the logarithmic decay found in different glassformers and now in the alkyl nanodomains of PnBMA and PnHMA. That is the reason why Götze and Sperl proposed a new version of MCT to explain it by invoking translational–rotational coupling in 2004 [34, 35], which also has encountered problem of its own as discussed in Section 2.3.2.37. Mention is also made by Arbe et al.

that the high-order MCT transition has also been proposed to explain the logarithmic relaxation in hydrated proteins found by Lagi et al. [1289(c)] in simulation of the self-dynamics of protein amino acids of hydrated lysozyme powder. The self-intermediate scattering functions of the amino acid residue center of mass show a logarithmic decay over three decades of time, from 2 ps to 2 ns, followed by an exponential α-relaxation. This corresponds exactly to the OHD-OKE signal of small molecular glassformers with time dependence given by Eq. (2.96) found by Fayer and coworkers [330–333], and also to the time dependence of the mean-squared displacement of water O atoms in hydrated maltose-binding protein showing the plateau obtained by molecular dynamics simulation by Wood et al.[1363] and shown in Fig. 228b. These are nearly constant loss from caged dynamics as shown before.

Finally, a comment is necessary on the following statements made by Arbe et al.: "However, it would be difficult to identify the dielectric process β'^{DS} with such dynamic glass transition (α_{PE}) in PHMA. The main reason is that the β'^{DS} relaxations in PHMA and PBMA are nearly indistinguishable, and we have shown above that for PBMA this process is different from the structural SG-relaxation. In this sense, this process could be interpreted as a kind of Johari–Goldstein relaxation of the alkyl nanodomains also in higher order PnMAs, as suggested by Ngai and Beiner." The statement that the relaxation times of the β'^{DS} relaxations in P*n*HMA and P*n*BMA are close together is correct as found before by Beiner and coworkers in the same polymers where α_{PE} is used instead of β'^{DS}. However, the rest of the statements could give a wrong message to confuse the real physics already at hand. Beiner and coworkers have studied poly(*n*-alkyl acrylates) from $C = 4$ up to $C = 10$, poly(*n*-alkyl methacylates) from $C = 4$ up to $C = 12$, in contrast to only two, P*n*BMA ($C = 4$) and P*n*HMA ($C = 6$), by Arbe et al. As shown before by Beiner and coworkers [297, 481, 482(a)], the observed relaxation time of the observed process (called α_{PE} by Beiner and coworkers, and β'^{DS} by Arbe et al.) in alkyl nanodomain of poly(*n*-alkyl acrylates) and poly(*n*-alkyl methacylates) having longer side chains (larger *C*) occurs at increasingly higher temperatures and becomes increasingly super-Arrhenius (i.e., larger fragility index *m*). It is true that the difference of the α_{PE} relaxation times as well as *m* between P*n*BMA and P*n*HMA is small, but the differences of these quantities between P*n*BMA and higher members of poly(*n*-alkyl methacylates) with *C* ranging from 7 to 12 become increasingly larger on increasing *C*. Therefore, the statement "the β'^{DS} relaxations in PHMA and PBMA are nearly indistinguishable," by Arbe et al. is correct but cannot be generalized to poly(*n*-alkyl methacylates) with $C \geq 7$. The α_{PE} found by Beiner et al. in poly(*n*-alkyl methacylates) with $C \geq 7$ have VFTH temperature dependence for its relaxation time and are genuine structural relaxation of the alkyl nanodomains. It is the finite size effect on reducing the length of the side chain that leads to nearly total suppression of intermolecular coupling (cooperativity) when $C=3$, the case of P*n*BMA, and makes the structural relaxation time no different from that of the primitive or JG β-relaxation. This is clear from the CM equations (2.67) and (3.69), which reduce to $\tau_\alpha = \tau_0 = \tau_\beta$ when $n \to 0$. The nearly Arrhenius *T*-dependence of the α_{PE} or β'^{DS} process in P*n*BMA is consistent with the small fragility index expected for structural relaxation that has vanishing intermolecular coupling or small coupling parameter n [112]. Additional strong evidence that the observed α_{PE} process in poly(n-alkyl methacylates) for $C = 7$ and 10 is the genuine structural relaxation of the alkyl nanodomain can be drawn from the discovery of the faster JG β-relaxation in poly(n-decyl methacrylate), $C = 10$, and poly(n-heptyl methacrylate), $C = 7$, after rapid quenching (see Figs. 27 and 263) [482(a)]. Rapid quenching is a known technique to find the JG β-relaxation, which is otherwise too weak to be resolved in the equilibrium liquid state. The relaxation times of the JG β-relaxation found by rapid quenching of $C = 10$ and 7 turn out to be nearly the same as the Arrhenius relaxation time of the α_{PE} or β'^{DS} process in PnBMA, and thus reconfirming that the latter is the JG β-relaxation of the alkyl nanodomains for all PnAMA from $C = 4$ to 12

Thin Polymer Films from Computer Simulations

Employing the bond-fluctuation lattice model, Monte Carlo simulations [1459–1461] of low molecular weight thin polymer films provide information on Rouse modes which involve standing wave-like motion of the entire chain in thin polymer films. In one simulation [1459], the diffusion coefficient of unentangled chains parallel to the wall confining the thin film is almost the same as that of the bulk in the whole temperature region studied, indicating the same for the Rouse modes. Reference [1460] showed that the relaxation time of the incoherent intermediate scattering function or the coherent polymer scattering function for the wave vector $q = 2.94$ (the value at the peak of the static structure factor and correspond to the length scale of a bond) is significantly reduced with decreasing film thickness. However, the reductions of the relaxation times of the same functions are much weaker for $q = 1.05$. Since at $q = 1.05$ the correlation functions probe density fluctuation on the length scale of a chain, this result is consistent with the lack of change of the relaxation time of the long wavelength Rouse modes [1459]. Another simulation [1461] reported the thickness dependence of the fifth Rouse mode, which has the same length scale as the incoherent scattering function with wave vector q at the maximum of the static structure factor. The relaxation time, τ_q^s, of the incoherent scattering function decreases with decreasing film thickness D from bulk to 30 ($\approx 7.5R_g$) and down to 6 ($\approx 1.5R_g$) and the effect becomes larger with decreasing temperature (see Fig. 264, left). Here R_g is the radius of gyration of the polymer. On the other hand, the relaxation time of the fifth Rouse mode, τ_5, instead increases with decreasing film thickness, although the increase from the bulk value down to $D = 6$ is not large (a factor ~2) and not sensitive to temperature compared with the decrease of τ_q^s (see Fig. 264, right). The T-dependence of τ_5 in the thin film is parallel to that of τ_q^s in the bulk. In the thin film with $D = 6$ ($\approx 1.5R_g$), the presence of induced orientation of the chains parallel to the surfaces can decrease the entropy of the Rouse modes and their relaxation rates and explain the increase of τ_5. These contrasting behaviors of the local segmental relaxation and the Rouse mode on decreasing film thickness from simulations are analogous to that found by dielectric relaxation in POB confined in Vycor glass with mean pore diameter of 4 nm

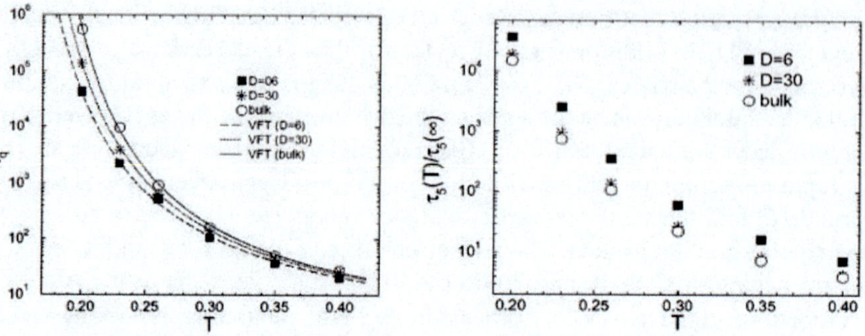

Fig. 264 (*Left*) The temperature dependence of the relaxation times τ_q^s for two film thicknesses, $D = 6$ ($\approx 1.5R_g$) and $D = 30$ ($\approx 7.5R_g$), and the bulk (unconfined system of linear dimension $L = 60$). (*Right*) The temperature dependence of the relaxation times, $\tau_5(T)$, of the fifth Rouse mode. The ordinate is divided by the infinite temperature result $\tau_5(\infty)$, to illustrate the influence of film thickness with decreasing temperature. Two films, $D = 6$ and $D = 30$, are compared with the bulk. Reproduced from [1461] by permission

and shown in Fig. 262. Conventional theories of modified Rouse modes for undiluted polymers [29] have the relaxation times of the Rouse modes proportional to the same monomeric friction coefficient that also governs the local segmental relaxation time (τ_q^s in this case). Accordingly, the Rouse relaxation times and the local segmental relaxation times should see the same change on decreasing film thickness and maintain to have the same temperature dependence. Thus the observed different changes in the magnitude as well as the temperature dependence on decreasing film thickness of τ_q^s of the segmental relaxation compared to τ_5 of the Rouse modes are truly anomalous and at variance with conventional theories.

On comparing the variation of τ_q^s with that of τ_5 of the bulk polymer in the same entire range of temperature from 0.4 to 0.2 in the two panels of Fig. 264, one can check that τ_q^s has a stronger T-dependence than τ_5. This result is consistent with the data from mechanical compliance and dielectric relaxation [165, 201, 206, 866–869] measurements of low molecular weight polymers, which showed breakdown of thermorheological simplicity of the viscoelastic spectrum and the local segmental relaxation time to have a stronger temperature dependence than the relaxation times of the Rouse modes. An explanation of the mechanical data has been given by the CM [837] by Eqs. (2.59) and (2.61), and the coupling parameter n_R of Rouse modes has zero values as compared with non-zero value of n_α, the coupling parameter of the local segmental relaxation. The same explanation applies to the stronger T-dependence of τ_q^s than τ_5. Furthermore, in the thin film with thickness $D = 6$, the temperature dependence of τ_q^s now becomes weaker than in the bulk and is nearly the same as that of τ_5. This can be explained by the decrease of n_α in the thin film with $D = 6$.

Only the fifth Rouse mode of chain length equal to 10 was reported in [1461]. Wavelength of this fifth Rouse mode is comparable to the length scale corresponding to the maximum of the structure factor and hence is rather short. For this reason, the fifth Rouse mode may not have exactly the properties predicted by the Rouse model. An example is the deviation of the time correlation function of the fifth Rouse mode from a linear exponential function of time found in another simulation by the same group [1462]. Viscoelastic measurements in polyisobutylene were able to solve the Rouse modes with short wavelengths and showed that they deviate from the Rouse model predictions (see Figs. 94 and 95). For this reason, they were called sub-Rouse modes. The coupling parameter of the sub-Rouse modes is intermediate between the local segmental motion and the Rouse modes [56]. A more appropriate comparison between the relaxation times of the longer wavelength Rouse modes and τ_q^s from the Monte Carlo simulations was given in [1459, 1460].

Reversal of Effect due to Physical or Chemical Interaction with Substrate or Confining Wall

In all the cases described above, the enhanced molecular mobility and concomitant reduction in T_g of the nanoconfined liquid are observed provided strong physical or chemical interaction with the wall or the substrate is absent. If present, the constraints imposed by the chemical bonds or the strong intermolecular constraint/interaction with the immobile substrate or wall tend to slow down the molecules at the surface. The less mobile surface molecules at coordinate $z = 0$ impose stronger intermolecular constraints on the neighboring inner layer ($z = 1$) of molecules and consequently slow down their α-relaxation than that in the bulk liquid. The effect propagates inward away from the surface from one layer ($z = j$) to another ($z = j + 1$), albeit with decreasing magnitude. In the CM, the effect on the α-relaxation, layer by layer, is described by $n(z = 0)$ for the surface layer significantly larger than n_{bulk}, and $n(z = j)$ for inner layer $z = j$ still larger than n_{bulk} but decreasing monotonically with increasing j, and n_{bulk} is its lower bound. The postulated increase of $n(z = j)$ here is analogous to the increase of n of the more mobile component in binary mixtures discussed before in Sections 2.2.5.7 and 2.3.2.10, and justifiable formally through a mathematical procedure [182, 183]. The correlation function of α-relaxation of molecules in layer j is described by the Kohlrausch function

$$\phi(z,t) = \exp\{-[t/\tau(z)]^{1-n(z)}\}, \tag{2.97}$$

where $\tau(z)$ is given by CM relation

$$\tau(z) = [t_c^{-n(z)}\tau_0]^{1/(1-n(z))}. \tag{2.98}$$

The primitive relaxation time τ_0 should be independent of z if density is also independent of z.

This CM description has support from molecular dynamics simulations by Scheidler et al. [322, 1463] of a binary Lennard-Jones (LJ) liquid thin film confined on two sides by walls made of frozen configurations of the same system and remain interacting with the liquid via the LJ potential. The film thickness is 15.0 with the unit of length given by σ_{AA}, the length parameter in the LJ potential between the A-particles. The center of the film corresponds to a distance $z = 7.5$ from the walls. Scheidler et al. calculated the spatially resolved self-part of the intermediate scattering function, $F_s(q, z, t)$, which takes into account only the particles at a distance z from the wall. The results are shown in the left panel of Fig. 265. For each z, they fit the time dependence of τ_c of the Kohlrausch functions,

$$F_s(q, z, t) = A(z) \exp -[t/\tau(z)]^{\beta(z)}. \tag{2.99}$$

Since $F_s(q, z, t)$ for $z = 7.5$ at the center of the film is identical to the one in the bulk as found by Scheidler et al., the fit parameters for this highest z-value correspond to the bulk values. On approaching the wall or as z is decreased, β decreases and τ increases monotonically as shown in the inset and the main part of Fig. 265 (right), respectively, for $T = 0.55$ and $q = 7.2$ (at the maximum in the static structure factor for A–A correlation). Thus the CM expected increase of $n(z) = 1 - \beta(z)$ with decreasing distance from the wall is verified by the z-dependence of β obtained by Scheidler et al. The next step is to calculate $\tau(z)$ via Eq. (2.98). The value of t_c is about 2 in the units of LJ time $(m\sigma_{AA}^2/48\varepsilon_{AA})^{1/2}$, of the simulation. First, Eq. (2.98) was used to deduce $\tau_o(z)$ at $z = 7.5$ from the "experimental" values of

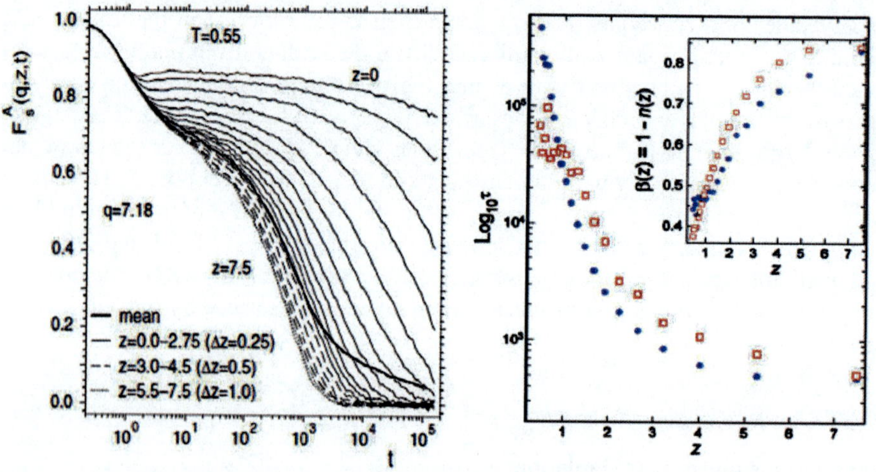

Fig. 265 (*Left*) Time dependence of $F_s(q, z, t)$ for the A particles, for various values of z at $T = 0.55$, and $q = 7.18$, the location of the main peak in the static structure factor. The different curves correspond to different distances z from the wall. Δz gives the difference in z for neighboring curves. Reproduced from [1463] by permission. (*Right*) $\mathrm{Log}_{10}\tau(z)$ vs. z at $T = 0.55$ and $q = 7.2$. *Filled circles* are from the molecular dynamics simulation results of Scheidler et al. [1463]. *Open squares* are calculated by the CM using the Kohlrausch exponents, $\beta(z)$, obtained by Scheidler et al. by fitting their self-intermediate scattering function $F_s(q, z, t)$ to the Kohlrausch functions (*left panel*). These values of $\beta(z)$ are shown as a function of z in the inset (*filled circles*). *Open squares* in the inset are the values of $\beta(z)$ required to get exact agreement of the calculated $\tau(z)$ with that obtained from simulations. Data from [900] are replotted in figure here

$\tau(z)$ and $\beta(z)$ at the same z. The value obtained $\tau_o(z) = 200$ is assumed to be constant throughout the film, i.e., independent of z, although it is possible that $\tau_o(z)$ may have some slight dependence on z if particles density of the film is not uniform particularly near the wall. With $\tau_o(z)$ fixed at 200, $t_c = 2$, and $(1\text{–}n(z))$ taken from the data of Scheidler et al. (filled triangles in inset of Fig. 265, right), the CM relation between $\tau(z)$ and $\tau_o(z)$ enables $\tau(z)$ to be calculated, and the results are shown in the figure by open circles. The calculated $\tau(z)$ reproduce the data quite well within a factor of 2 for $z \geq 0.618$. The calculated $\tau(z)$ for the three smaller $z = 0.561, 0.510$, and 0.463 are smaller by a factor of about 7 than the data. At these z values, $\tau(z)$ is outside the time window of the simulation. Consequently there is large uncertainty in the determination of $\beta(z)$ from fitting the $F_s(q, z, t)$ to the Kohlrausch function by Scheidler et al. Exact agreement between data and calculated $\tau(z)$ can be achieved by letting $\beta(z)$ be adjustable in the CM equation. The values of $\beta(z)$ required are shown as open diamonds in the inset of Fig. 265. One can see by inspection that, for exact agreement at $z = 0.561$, 0.510, and 0.463 between data and calculation, $\beta(z)$ has to be equal, respectively, to 0.396, 0.392, and 0.373 instead of 0.453, 0.465, and 0.442 given by Scheidler et al. The differences are not large and can be considered within the errors in determining $\beta(z)$ by fitting $F_s(q, z, t)$ data.

Similar effects were found by Weeks and coworkers [1464] in a colloidal suspension confined between two quasiparallel rigid walls by confocal microscopy to directly observe the motion of colloidal particles. The suspension is a mixture of two particle sizes to prevent wall-induced crystallization. The total volume fraction ϕ of the particles is ≈ 0.42, and the motion if unconfined is nearly diffusive, with the MSD growing almost with slope 1 on the log–log plot. This behavior is similar to monodisperse samples with a volume fraction of $\phi \leq 0.4$. Thus, this sample is liquid like and far below the glass transition $\phi_g \approx 0.6$. In thinner regions, the motion parallel to the walls slows, as seen in the sequence of solid curves in Fig. 266-1(a). This slowing starts at a thickness of $H = 16.28$ μm (second curve from the top) and slows dramatically for thinner samples. Similarly slowing of motion perpendicular to the walls is shown in Fig. 266-1(b). These results suggest that motion is slower in confinement, and glassy behavior is produced in sufficiently thin region for a

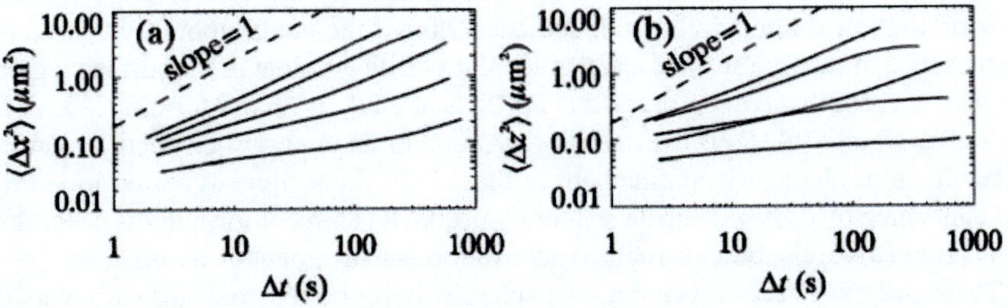

Fig. 266-1 Mean-squared displacements. **(a)** Data for sample A, showing motion parallel to the walls, for thickness H = bulk, 16.28, 11.06, 9.41, and 6.92 μm (from *top* to *bottom*). The *dashed line* has a slope of 1 **(b)** similar to **(a)** but for motion perpendicular to the walls. Data are ordered by thickness at long Δt as in **(a)**. Reproduced from [1464] by permission

sample which is a liquid in an unconfined geometry. One can see in Fig. 266-1(a) on decreasing the thickness H that the MSD does not simply shift to longer times but its time dependence changes to become increasingly flatter or more dispersive. Interpreted in terms of the CM, this can be taken as evidence of increasing coupling parameter n of the particles confined in thinner region, and the observed slowing down explained. Unfortunately, due to contributions by particles of two sizes, the time dependence of the MSD in Fig. 266-1 does not allow the coupling parameter of either one of the two types of particles to be determined.

Polymer Surface Mobile Layer

From the simulation data of Scheidler et al. [1463] discussed above and the explanation given by the CM analysis [900], it can be generalized to effects on the structural relaxation of thin polymer films layer by layer by the free (polymer–air) surface or the interface with a substrate or another polymer. The effect can propagate a long way into the thin film. This is consistent with the results found experimentally by Torkelson and coworkers for thin PMMA films [1143] that both the substrate interface and the free surface perturb the relaxation rate at least 100 nm into the film interior, although the effect of free surface is opposite to that of the substrate. At the free surface, cooperative segmental mobility is enhanced and T_g is reduced. On the other hand, at the substrate interface hydrogen bonds suppress cooperative segmental mobility and lead to an increased T_g. The competition of different effects from free surface, interface, and substrate on structural relaxation has been illustrated by experiments by Ellison and Torkelson [1170], and the results rationalized by the CM [506]. Tanaka and coworkers [1465(a)] used dynamic mechanical measurement and analysis to probe the local segmental dynamics of thin and ultrathin polystyrene (PS) films supported on substrates, which is either polyimide (PI) or SiO_x. PI has higher T_g than PS. For the 1.46×10^6 molecular weight PS film with thickness equal to 226 nm on PI substrate, the temperature dependence of the relaxation time deduced from isochronal mechanical loss modulus E'' is essentially the same as that of bulk PS. However, as the film thickness is decreased, the E'' relaxation peak becomes increasingly broader on both the lower and higher temperature sides, as shown by the data of 50 and 37 nm thick PS films. The broadening of E'' over a range of about 40 K on the lower temperature side indicates higher molecular mobility caused by the presence of the free surface and a gradual mobility gradient in the surface region, as observed by Ellison and Torkelson [1170]. Since $2R_g$ is 66 nm for the 1.46×10^6 molecular weight PS (R_g is the radius of gyration of an unperturbed chain), induced orientations of chains by confinement in the 37 nm thick film suggested in [1456] and confirmed by diffuse neutron scattering probe of chain conformations of PS thin films [1466(b)] are another possible cause of the enhancement of mobility.

There are other techniques that can selectively probe the free surface layers of polymer films on substrate. Kajiyama, Tanaka, and coworkers measured the dynamical response of the surface of ~200 nm thick monodisperse PS films with molecular weights ranging from 4.9 to 1460 kg/mol by lateral force spectroscopy, a technique using atomic force microscope (AFM). This AFM technique most sensitive to the surface layers gives a good measure of the segmental relaxation time and hence the T_g of the surface layers. The segmental relaxation time of the surface τ_s so obtained is much shorter than in the bulk as can be seen by the comparison in the left panel of Fig. 266-2, and the corresponding T_gs of the PS films are markedly lower than the bulk. According to the CM, the surface layer is much more mobile than the bulk because of reduced intermolecular coupling and hence smaller coupling parameter n_s by the absence of molecules on one side. The smaller n_s of the surface layer can be deduced from the observed τ_s and the primitive relaxation time τ_0 by the CM equation. Since τ_0 is the same for the surface layer as well as for the bulk, it can be calculated at any temperature from the bulk relaxation time τ_b and the known coupling parameter $n_b = 0.64$ for mechanical relaxation [165], again using the CM equation. This procedure was first used to obtain the reduced coupling parameter of freely standing thin PS films (M_w = 767 kg/mol, R_g = 57 nm) in [1456]; the data and the results of the CM analysis are reproduced in the middle and right panels of Fig. 266-2. The coupling parameter of the freely standing film with thickness = 23.4 is reduced to $n_{\text{film}} = 0.40$ compared with $n_b = 0.64$.

Tanaka et al. [1465(c)] followed the same procedure to deduce the reduced value of the coupling parameter of the surface layer. From the surface τ_s at

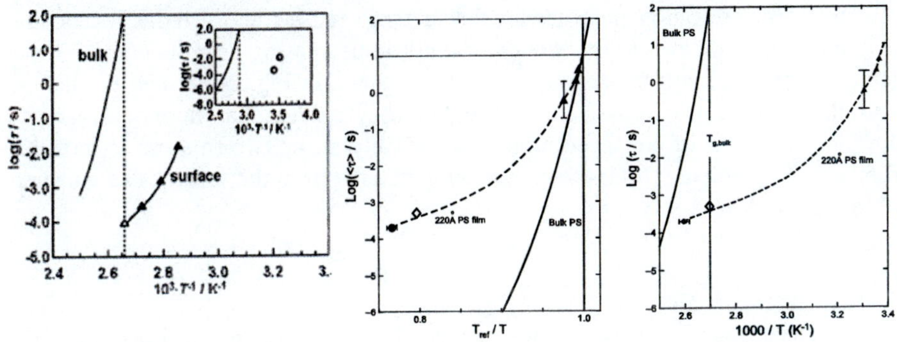

Fig. 266-2 (*Left*) Semilogarithmic plot of segmental relaxation time against reciprocal absolute temperature for the PS film with M_n of 140 K. The *solid line* denotes the log τ vs. 1000/T relation for the corresponding bulk PS calculated from Vogel–Fulcher equation. The *vertical dashed line* indicates the glass transition temperature of bulk PS, T_{gb}. The *closed* and *open triangles* are the experimental relaxation times at the surface and the one calculated at T_{gb} by extrapolating experimental surface τ, respectively. The *inset* is the same plot for the PS film with M_n of 4.9 K, with the experimental surface τ values represented by *filled circles*. Data from [1465(b)] are replotted here. (*middle*). The *three filled triangles* and the *circle* are the relaxation times of the freely standing thin PS film found experimentally. The *dashed curve* is an extrapolation of the possible temperature dependencies of the measured relaxation time of the film. The *open diamond* represents the value of $\tau_{\text{film}} = 10^{-3.3}$ s calculated by the CM equation at $T = T_{gb} = 370$ K with $n_{\text{film}} = 0.40$. The *solid curve* is the local segmental motion retardation time τ_b of high molecular weight bulk PS from creep compliance data. (*Right*) A comparison of the T_{ref}-scaled temperature dependences of τ_b and τ_{film}. The legends of *symbols* and *lines* are the same as given in the *middle panel*. The *two dotted lines* indicate that both relaxation times are equal to 10 s at $T_{\text{ref}}/T = 1$. Reproduced from [1456] by permission

$T_{gb} = 9.03 \times 10^{-5}$ s, they deduced the coupling parameter at the PS surface n_s to be approximately 0.2, which is significantly smaller than the bulk value of 0.64. It is not totally reduced to zero value because there are still polymer molecules inside the surface.

Tanaka et al. [1465(c)] reported that the apparent activation energy of τ_s is significantly smaller than τ_b, consistent with the findings of [1456] in the freely standing PS film. In fact the right panel of Fig. 226-2 shows the T_{ref}-scaled temperature dependencies of τ_b and τ_s, with T_{ref} defined as the temperatures at which the relaxation times are equal to 10 s. The much weaker T_{ref}-scaled temperature dependencies of τ_s compared with τ_s is consistent with the smaller coupling parameter of the surface layer than the bulk.

Forrest and coworkers [1465(d), 1465(e)] measured the surface relaxation time τ_s of PS and iPMMA by monitoring the temporary recovery of nanoholes fabricated on the free surface of the films. Their findings are consistent with those of Kajiyama and Tanaka [1465(b), 1465(c)].

Napolitano and Wübbenhorst [1465(f)] have fabricated freely standing films of atactic polystyrene (M_w=932,000 g/mol, M_w/M_n=1.2, R_g=28 nm) and investigated the dynamics by dielectric spectroscopy. They found large reduction in the glass transition temperature as the film thickness is decreased from bulk down to 40 nm. They emphasize the finding of a temperature-dependent asymmetric broadening of the structural relaxation peak toward lower temperatures and a reduction of the dynamic fragility down to the monomer limit on decreasing film thickness. These features are similar to those found by others shown in Fig. 266-2 and consistent with the removal of intermolecular coupling. They verified that these observations are manifestations of enhanced molecular mobility at the two free surfaces of the film and the gradient of glass transition temperatures from the surface to the center of the thin film.

Concluding Remarks on Nanoconfinement

The results given above for various nanoconfined glassformers in Figs. 92, 260, 2.261, and 2.263 demonstrate the veracity of the primitive relaxation and its relaxation time τ_0. They are actually observed as the α-relaxation and its relaxation time τ_α when the size of the glassformer is scaled down to a few nanometers. This transformation of the α-relaxation of the bulk glassformer to the primitive of the JG relaxation occurs because the many-molecule dynamics in the bulk are suppressed when the size is reduced to the nanometer scale. We recall the observation of the JG relaxation of water confined in molecular sieves and silica gels discussed before in Section 2.3.2.27, but no comparison with τ_0 was given because the parameters of the α-relaxation of bulk water needed for the calculation are unknown. Some studies of the dynamics of thin polymer films go beyond the local segmental relaxation to include the modes of longer length scales, such as the Rouse modes, and the terminal modes if the polymer is entangled. In bulk amorphous polymers, it is well established experimentally that the temperature and pressure dependences of the more global viscoelastic modes are different from that of the local segmental relaxation [165]. In considering the viscoelastic data of thin polymer films, one should be mindful of these facts in bulk polymers. When analyzing or interpreting the data of thin films, it is logical to require the theory or model used to be able to account for, or at least consistent with, these known bulk polymer properties. Otherwise, the theory constructed for thin films will not be worthwhile. The CM satisfies this requirement. Extended to consider polymer thin films [496, 505, 506], the CM predicts an increase of the mobility of the local segmental motions as discussed in the preceding section, but the lack of such a change for the mobility of the Rouse modes (see Figs. 262 and 264) and the diffusion of entire polymer chains [1466]. For the latter it is because chains in a thin film formed by a high molecular weight entangled polymer are still entangled. Therefore the coupling parameter of diffusing entangled polymer chains n_D is either unchanged from its value in the bulk polymer [496, 505, 506] or may be even slightly increased from the bulk value due to reduction in entropy of the chains by confinement. These varied predictions of the CM for the segmental relaxation and the chain modes of unentangled and entangled polymer are in accord with experiments and computer simulations.

Since the discovery of more than a decade ago of the deviation from bulk values of the glass transition temperature due to confinement of glassformers in nanometer spaces [485, 1171, 1467], many experiments have shown there can be significant changes of T_g as well as other dynamic properties in the nanoconfined materials. However, the changes vary greatly depending on the nature of the interfaces, the chemical structure of the nanoconfined glassformer, the experimental methods used, and in the case of polymers the length scale of the dynamics probed [496, 1453, 1468, 1469]. Just for T_g alone, it can decrease, increase, or remain the same depending upon the experimental or simulation conditions. As concluded in the most recent and comprehensive review [1453], "... the existing theories of T_g are unable to explain the range of behaviours seen at the nanometre size scale, in part because the glass transition phenomenon itself is not fully understood." We fully agree with the above quoted remark by Alcoutlabi and McKenna, and furthermore we can identify the reason why the existing theories of T_g fail to explain the range of behaviors of nanoconfinement. Most theories of dynamics of nanoconfinement are based on conventional theories of glass transition for bulk materials. But, as has been shown in [165] and in Section 2.2.5.9, conventional theories and models of glass transition are not able to account for some general dynamic properties of bulk glassformers. What causes the conventional theories to be deficient already for bulk glassformers is the lack of adequate treatment of the many-molecule dynamics or at least taking into account the dispersion of the structural α-relaxation as a fundamental element. Thus, it is unsurprising to find that, when applied to nanoconfinement, the conventional theories have limited success in explaining the range of behaviors of nanoconfinement.

Analogue in a Secondary Relaxation of BPA-PC

Some special secondary relaxation in aromatic backbone polymers such as the γ-relaxation in bisphenol A polycarbonate (BPA-PC) is intermolecularly coupled. It involves the π-flips of the two phenyl rings in the backbone of a repeat unit, which can be regarded as a primitive motion. Because the phenylene rings are bulky, the primitive motion is coupled to the same in other repeat units resulting in the correlation function describable by a Kohlrausch function with $n \approx 0.66$ and relaxation time τ_γ having activation energy E_γ of about 38 kJ/mol in the glass state of BPA-PC [884]. By dissolving BPA-PC at low concentration in a solvent, NMR measurements showed motion of the phenyl rings becomes much faster, and its relaxation time, τ_{sol}, has a smaller activation energy E_{sol} of about 10–13 kJ/mol [1470, 1471], comparable to the value of the energy barrier obtained by molecular dynamics computation [883, 1472, 1473]. We can view τ_{sol} to be the primitive relaxation time of the γ-relaxation because intermolecular coupling is drastically reduced in dilute solution. The ratio E_γ/E_{sol} is consistent with the CM prediction that it should be equal to $1/(1-n)$. The larger observed value of the ratio than predicted may be due to difference in the environment between bulk BPA-PC and its dilute solution.

David et al. [250] have reached the same conclusion of cooperative dynamics from their mechanical relaxation studies of secondary relaxation in other aromatic backbone polymers including poly(aryl ether ether ketone), poly(phenylene sulfide), and poly(aryl ether sulfone). These authors have also used the coupling model predictions to interpret their experimental data.

2.3.2.39 Connection Between the Fast Primitive Relaxation and the Slow Structural Relaxation of Aging Colloidal Suspension of Laponite

Discussed before in Section 2.2.5.5 of Section 2.2.5 is the effect of shear on the non-equilibrium structural dynamics of an aging colloidal suspension of 3 wt% Laponite from light-scattering experiment by Di Leonardo et al. [343]. With or without the application of shear and for any scattering vector $\boldsymbol{q}$ and aging time t_w, the intermediate scattering function $F(\boldsymbol{q}, t, t_w)$ conforms to $\exp(-t/\tau_f)$ form at short times, with $\tau_f \propto q^{-2}$ of normal Brownian motion [338], and thus it corresponds to the primitive diffusion of the CM. At longer times, $F(\boldsymbol{q}, t, t_w)$ follows the time dependence of $\exp[-(t/\tau_s)^{1-n}]$. For practical reason, $F(\boldsymbol{q}, t, t_w)$ was fitted by the sum $A\exp(-t/\tau_f) + (1-A)\exp[-(t/\tau_s)^{1-n}]$. Increase of the average structural relaxation time $< \tau_s >$ by aging proceeds unaffected by the applied shear at shorter aging times t_w such that τ_s is less than $\dot{\gamma}^{-1}$, the reciprocal of the shear rate $\dot{\gamma}$. However, the shear flow significantly reduces aging when $< \tau_s(t_w) >$ is greater than $\dot{\gamma}^{-1}$ as shown before in Fig. 81 for four different shear rates. The onset of reduction of both $< \tau_s >$ and n occurs at t_w for which $\langle \tau_s(t_w) \rangle \dot{\gamma} \approx 1$, and the sizes of the reductions increase with the shear rate $\dot{\gamma}$. Di Leonardo et al. also study the primitive relaxation time τ_f on aging and the effect by shear. The results of τ_f vs. t_w for different values of the shear rate, including $\dot{\gamma} = 0$, are shown in the left panel of Fig. 267.

In the presence of shear flow, τ_f seems to follow the behavior of the $\tau_f(\dot{\gamma} = 0)$ until the *slow* structural relaxation time τ_s becomes comparable to $\dot{\gamma}^{-1}$, i.e., when $\langle \tau_s(t_w) \rangle \dot{\gamma} \approx 1$ is satisfied. Beyond this time, τ_f increases slower than in the absence of shear. The behavior of τ_f is similar to τ_s, shown before in Fig. 81. Thus, τ_f exhibits the crossover to the shear-dominated aging dynamics at exactly the same

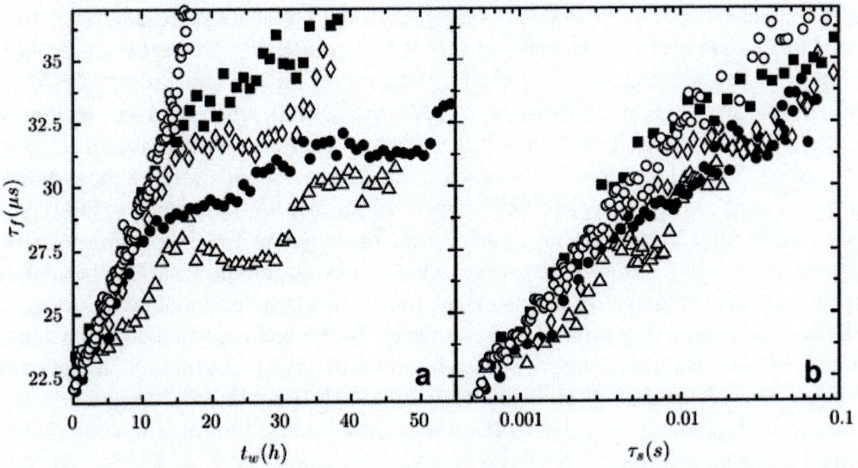

Fig. 267 (**a**) Crossover to shear-dominated aging as observed in the timescale of fast dynamics. τ_f is reported as a function of waiting time t_w during aging under different shear rates $\dot{\gamma}$: (Δ) 446, (•) 223, (◇) 67, (■) 22 s^{-1}. (O) refer to aging without shear. (**b**) All curves collapse when plotting τ_f against the respective slow τ_s and the crossover disappears. *Symbol sizes* are representative of error bars as obtained from the fitting procedure. Reproduced from [343] by permission

time as τ_s. In the right panel of Fig. 267, τ_f is plotted against τ_s for the four different shear rates. All curves tend to collapse together. All the above indicate a close link between the fast and the slow dynamics in Laponite solutions. Since the fast or primitive diffusion transpires before in time than the slow diffusion, the crossover to the shear-dominated aging dynamics property must originate from the former from causality.

2.3.2.40 Which Criteria Are Most Critical for Identification of the Johari–Goldstein β-Relaxation?

In the previous subsections, various properties have been used to distinguish secondary relaxation which is true precursor of the structural α-relaxation in the sense that the latter cannot occur without the former taking place at shorter times. These secondary relaxations are called the Johari–Goldstein (JG) β-relaxations in order to distinguish from other secondary relaxations. I hasten to point out that my usage of the term JG β-relaxation differs from other workers not only in the way to specify secondary relaxation but also in the theoretical interpretation of it, including past and recent works by Johari and Goldstein together or separately. In my understanding, the existence of fundamental relation between the α- and the JG β-relaxation implies that some properties of the two relaxations are well connected or inseparable from each other. One may expect that the JG β-relaxation mimics α-relaxation in some observed properties. Examples include change of the secondary relaxation time, τ_β, on elevating pressure, by aging, and by mixing with another glassformer, as well as change of T-dependence of τ_β (dielectric relaxation strength $\Delta\varepsilon$) from Arrhenius (weaker) to a stronger (stronger) T-dependence when crossing T_g from below. We may add the ratio E_β/RT_g used by others for identification of JG relaxation if it falls in the neighborhood of 24 [971, 995, 1007], where E_β is the activation energy of τ_β in the glassy state, albeit the ratio can vary over a range for genuine JG relaxations, and values close to 24 can be assumed by secondary relaxations which do not belong to the JG class [996]. However, mimicking properties exhibited by a secondary relaxation may not be necessary or sufficient for identifying it as the JG β-relaxation. For example, a few secondary relaxations show shift in relaxation time with pressure such as *m*-fluoroaniline [665] and phenolphthalein dimethylether (PDE) [1065] and aging [1042], and change in T-dependence when crossing relaxation T_g of the relaxation time such as *m*-fluoroaniline [971, 1026] and the dielectric strength $\Delta\varepsilon$ such as diethyl phthalate (DEP) [1474]. But, after all, these secondary relaxations studied are not JG β-relaxation as shown for *m*-fluoroaniline in [671] by Hensel-Bilowka et al.

On the other hand, τ_β of some genuine JG β-relaxation does not change at all or at least some changes to mimic the observed significant change of τ_α in cases including the following: (1) on applying pressure in hydrogen-bonded glass-formers such as sorbitol [659(b)] and even in poly(ethylene methacrylate) [1475]; (2) on aging of sorbitol and chlorobenzene in *cis*-decalin [1148] by Johari, and poly(vinylethylene) by Casalini and Roland [1157]; and (3) on mixing PEMA with poly(4-vinylphenol)/poly(ethyl methacrylate) [1072] and mixing *tert*-butylpyridine (TBP) with tri-styrene or quinaldine (QN) with tristyrene [672b]. The latter are shown in Fig. 268, where one can see by inspection that not only the change with

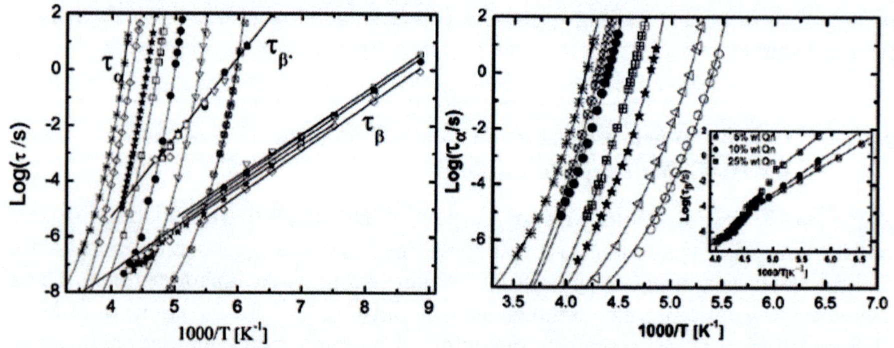

Fig. 268 (*Left*) Arrhenius plot for mixtures of TBP with tri-styrene (3Styr): (∗) 100% 3Styr, (◇) 5 wt% TBP/3Styr, (★)16 wt% TBP/3Styr, (□) 25 wt% TBP/3Styr, (•) 40 wt% TBP/3Styr, (▽) 60 wt% TBP/3Styr, (⊗) 100 % TBP. *Lines* are fits with the VFTH equation for the structural α-relaxation and Arrhenius equation for β′ and β-processes. The β-process is the JG process. (*Right*) Arrhenius plot for mixtures of quinaldine with tri-styrene: (∗) 100% 3Styr, (⊗) 5 wt% Qn/3Styr, (•)10 wt% Qn/3Styr, (⊞) 25 wt% Qn/3Styr, (★) 40 wt% Qn/3Styr, (◁) 80 wt% Qn, (O) 100% Qn. *Lines* in the main figure are fits with the VFTH equation for the structural α-relaxation and with the Arrhenius equation for the β-process in the *inset*. Reproduced from thesis of K. Kessairi, [672b], by permission

composition of the mixture of τ_β is small compared with τ_α but also they are in opposite directions. However, the β-relaxations in these two mixtures are genuine JG β-relaxation or primitive relaxation of the CM according to the most critical criteria that the two relaxations are connected as shown before from invariance of the ratio τ_β/τ_α to various combinations of P and T at constant τ_α, and agreement with the CM equation for some composition of mixture of QN with tri-styrene (see Figs. 148 and 149), and for mixture of picoline with tri-styrene (see Fig. 146). Basically, τ_β of genuine JG β-relaxation of TBP and QN does not need to shift or shift in the same direction as τ_α of some genuine JG β-relaxation on mixing with tri-styrene. This is because τ_β of TBP and QN depends on how they pack with tri-styrene, while τ_α depends sensitively on the intermolecular coupling and variation with composition of the mixture and its change is much larger and predictable when mixed with tri-styrene which is much less mobile. Even without much change of τ_β or change in opposite direction to that of τ_α on increasing tri-styrene content, the increasing separation between τ_β and τ_α on increasing tri-styrene content in the mixture is fully consistent with the CM prediction by Eq. (2.69) because of the corresponding increase of n due to enhanced intermolecular constraint/interaction. The case of mixing PEMA with poly(4-vinylphenol)/poly(ethyl methacrylate) [1072] is similarly explained by the increase of n of the α-relaxation of the PEMA component on increasing the higher T_g component.

A similar situation arises on elevating pressure in sorbitol [659b]. There τ_β of the JG β-relaxation of these hydrogen-bonded glassformers changes little compared with the significant increase of τ_α of pressure. The activation energy of τ_β in the glassy state of sorbitol is 61 kJ/mol, not much larger than the activation energy of uncrystallized water in nanoconfinement, in some aqueous mixtures and hydrated proteins (see Fig. 217), and comparable to the energy to break two hydrogen bonds. From this, it is not that surprising that τ_β is not very sensitive to pressure. Keeping τ_α fixed, the separation between τ_β and τ_α increases on increasing pressure. Notwithstanding, this is accompanied by the corresponding increase in n as evidenced by the broadening of the α-loss peak on elevating pressure as shown in the left panel of Fig. 269 for sorbitol [659(b)], and this feature enables the increase in separation between τ_β and τ_α on increasing pressure to be explained by the CM equation (2.69). Broadening of the α-loss peak by pressure is found in all the polyalcohols with fewer number of carbon atoms including glycerol, threitol (see Fig. 46), and xylitol in the right panel of Fig. 269 [556].

The cases of aging of sorbitol and chlorobenzene in *cis*-decalin [1148] by Johari and poly(vinylethylene) by Casalini and Roland [1157] can be similarly explained by the increase of n of the α-relaxation with aging time, although this cannot be checked because the spectra fall in very low frequencies and cannot be measured.

From the above, it is clear that properties of the α-relaxation mimicked by the secondary relaxation may be helpful to identify the JG β-relaxation, but there are exceptions, and hence they are by no means crucial. I may also mention that usually if there are two secondary relaxations present, the slower one is the JG β-relaxation, but this is violated in DHIQ where the faster secondary relaxation turns out to be the

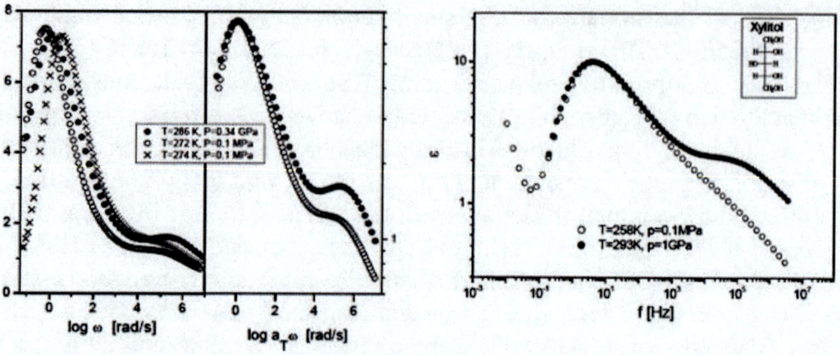

Fig. 269 (*Left*) The dielectric loss of sorbitol (■) at elevated pressure (340 MPa) at 286.4 K and at atmospheric pressure at (O) 272 and (×) 274 K. The *right panel* shows in a double logarithmic plot the attempted superposition obtained by shifting the 272 K spectrum by 0.8 decades. The conductivity contribution, assumed proportional to $1/\omega$, has been subtracted from the spectra. Data from [659(b)] replotted. (*Right*) Dielectric loss spectra of xylitol at ambient pressure and elevated pressure of 1 GPa for the same peak frequency. Data from [556] replotted

one [667]. The crucial experimental evidences are those that show the β-relaxation is directly connected to the α-relaxation such as via the CM equation. An example of direct connection comes from recent NMR studies [998, 999], where modification of the secondary relaxation by a procedure has the consequence of changing the α-relaxation. Another example comes from measurements of dielectric relaxation over broad frequency range at elevated pressure up to about 2 GPa. These studies of several neat glassformers, BIBE, PPGE, DGEBA, and PVAc, and binary mixtures discussed in Section 2.3.2.8, have shown that the distance of separation between the two relaxations measured by the difference, $(\log \tau_\alpha - \log \tau_\beta)$, is invariant to widely different combinations of P and T while holding τ_α constant. Large changes of volume and entropy are effected by elevating pressure up to four orders of magnitude even compensated by increase in temperature to hold τ_α constant, hence the invariance of $(\log \tau_\alpha - \log \tau_\beta)$ is a strong indication that the two relaxations are rigorously related to each other. Another property that connects the β-relaxation to the α-relaxation is τ_β (or τ_0) having the same TV^γ-dependence as τ_α discussed in Section 2.3.2.29. These are the critical experimental evidences showing the α- and β-relaxations are not independent of each other and can be considered as the foremost criteria for JG β-relaxation. These criteria can stand on their own strength without support by theoretical model, although this relation between the two relaxations is also a prediction of the coupling model (CM). First, as shown repeatedly for many glassformers, there is the approximate agreement between the observed secondary relaxation time, τ_β, and the primitive relaxation time, τ_0, of the CM, i.e., $\tau_\beta \approx \tau_0$. Second, there is a generally observed invariance of the dispersion or n of the α-relaxation to different combinations of T and P while maintaining τ_α constant [120, 618]. On combining these two with the CM relation, $\log \tau_\alpha - \log \tau_0 = n \log (\tau_\alpha / t_c)$, one is led to the conclusion that ($\log \tau_\alpha - \log \tau_\beta$) remains unchanged independent of the combinations of T and P if τ_α is maintained constant.

The Case of PEMA

Given in the preceding paragraphs are examples of genuine JG β-relaxation, which do not mimic some properties of the α-relaxation, such as increase of relaxation time on elevating pressure or by aging. Therefore, if the relaxation time τ_β of a secondary relaxation in a glassformer is found not to change significantly with pressure or aging, this observation alone cannot rule out that it is a genuine JG β-relaxation. Due to the specific mode of motion of the JG β-relaxation and the chemical structure of the glassformer, it is possible that the motion is insensitive to changes in specific volume or entropy. A case can be taken from the dielectric study of the dynamics of low molecular weight poly(ethyl methacrylate) (PEMA) ($M_w = 2.0 \times 10^3$ g/mol) by Mpoukouvalas, Floudas, and Williams (MFW) [1475]. They reported that the β-relaxation time changes little with pressure. Presumably from this behavior, they concluded the observed β-relaxation of PEMA is *intramolecular* in origin, which is tantamount to saying that it is not the JG β-relaxation having fundamental connection to the α-relaxation. The conclusion of MFW contradicts the interpretation of the secondary relaxation as the JG β-relaxation of PEMA and likewise in other poly(*n*-alkyl methacrylates) given in various places in this book and in [252]. The connection between β-relaxation and the α-relaxation in PEMA is also suggested by the fact that the β-process is not an independent motion, but its rotation around the C–C backbone bond is coupled with a rocking motion of the backbone, previously demonstrated by NMR studies [753–755, 1084, 1476]. The involvement of motion of the backbone in the β-relaxation of PEMA distinguishes itself from other secondary relaxations of some other polymers, which the side group or part of it moves independently of the chain backbone. Motion of the epoxide group in DGEBA and PPGE forming the γ-relaxation discussed before is an example.

When dielectric data are available at elevated pressures, the critical identification of JG β-relaxation is not whether τ_β having significant dependence on pressure P, but is the invariance of ($\log \tau_\alpha - \log \tau_\beta$) to widely different combinations of P and T while keeping τ_α constant. This criterion is a far more demanding to be satisfied by the data than the lesser criteria based on properties of the secondary relaxation mimicking the α-relaxation including pressure dependence of τ_β. The published data of MFW are detailed enough for us to perform a test of using this criterion; especially the data were made available to us generously by George Floudas. In Fig. 270 we plot the α- and β-relaxation times of PEMA against reciprocal temperature at four different pressures $P_i = 0.1$ MPa (squares), 300 MPa (diamonds), 600 MPa (circles), and 1200 MPa (triangles). Closed symbols stand for the α-relaxation and the same symbols but now open stand for the corresponding β-relaxation. We take a constant value of $-\log(\tau_\alpha(T_{gi}; P_i)) = -2.5$ at $T = T_{gi}$ for each pressure P_i and obtain the corresponding value of $-\log(\tau_\beta(T_{gi}; P_i))$. The operation is indicated by any one of the vertical arrows. The tip of each arrow locates $-\log(\tau_\beta(T_{gi}; P_i))$ from the data for each applied pressure. It can be seen by inspection of Fig. 270

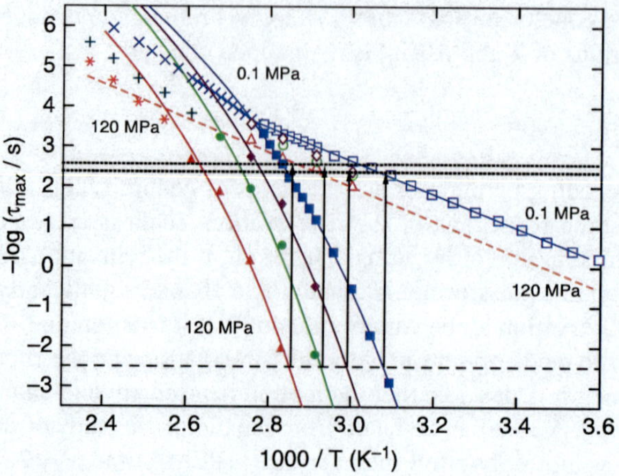

Fig. 270 Plot of α- and β-relaxation times of PEMA ($M_w = 2.0\times10^3$ g/mol) at 0.1 MPa (*squares*), 300 MPa (*diamonds*), 600 MPa (*circles*), and 1200 MPa (*triangles*). The tip of each *arrow* indicates $-\log(\tau_\beta(T_g)/s)$ at the glass transition temperature, T_g, defined by $-\log(\tau_\alpha(T_g)/s) = -2.5$ for each applied pressure. The spread of the values of $-\log(\tau_\beta(T_g)/s)$ for the different pressures is less than 0.2 of a decade, well within the errors and uncertainties in determining τ_α and τ_β by fitting the loss spectra by superposition of two HN functions. The *solid lines* are fits to the temperature dependence of $-\log(\tau_\alpha)$ by the Vogel–Fulcher–Tammann–Hesse equation

that $-\log(\tau_\beta(T_{gi};\ P_i))$ is nearly independent of pressure, while T_{gi} changes with pressure by about 25 K. The spread of the values of $-\log(\tau_\beta(T_{gi};\ P_i))$ bounded by the two horizontal lines is less than 0.2 of a decade, well within the errors and uncertainties in determining τ_α and τ_β by fitting the loss spectra by superposition of two HN functions or by the Williams ansatz in [1475] by MFW. The same is true for a different choice of $-\log(\tau_\alpha(T_{gi};\ P_i)) = -0.5$. Thus, the β-relaxation of PEMA satisfies the foremost criterion for JG β-relaxation. Truly it is the JG β-relaxation of PEMA and not intramolecular in origin as concluded by MFW, because otherwise τ_β would bear no such rigorous relation to τ_α at different combinations of T_{gi} and P_i.

The same data of $-\log(\tau_\alpha)$ and $-\log(\tau_\beta)$ in Fig. 270 have been plotted by MFW [1475] themselves as a function of pressure at constant temperature, reproduced here in the right panel of Fig. 271. As appeared and without more detailed description, the pressure insensitivity of τ_β is clear. As mentioned before, this property has led MFW to conclude the observed β-relaxation of PEMA is *intramolecular* in origin, and it is not the JG β-relaxation having fundamental connection to the α-relaxation. Although this conclusion has been dismissed by the strong connection between τ_β and τ_α at different combinations of T_{gi} and P_i shown in Fig. 270, it is worthwhile to point out much of the data of τ_β plotted by MFW in the right panel of Fig. 271 are in the glassy state of PEMA, where it is well known that the pressure dependence is understandably weak. Thus, the inclusion of τ_β is the glassy state in Fig. 271 (right) risks giving a false impression that τ_β has almost no pressure dependence at all even above T_g.

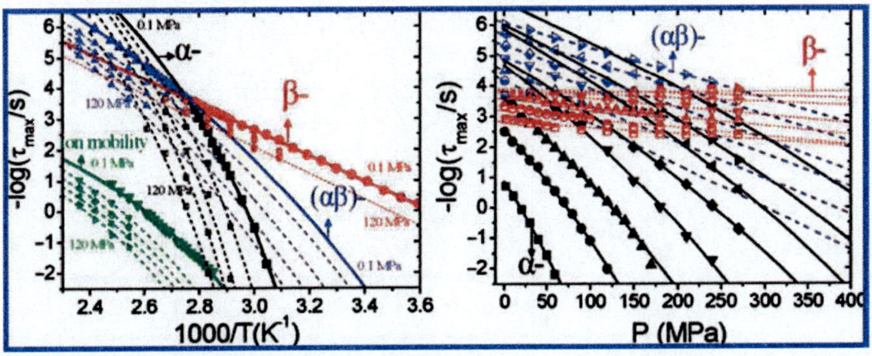

Fig. 271 (*Left*) Arrhenius representation of the relaxation times at maximum loss for the processes (α, β, (αβ), and ion mobility as indicated) for PEMA ($M_w = 2.0 \times 10^3$ g/mol) and for pressures in the range of $0.1 < P < 120.0$ MPa plotted in 30.0 MPa steps. The *lines* are fits to the VFTH equation for the slow, (αβ)-, and α-processes and to the Arrhenius equation for the β-process. (*Right*) Pressure dependence of the relaxation times at maximum loss for the (αβ)-, α-, and the β-process of PEMA ($M_w = 2.0 \times 10^3$ g/mol) for temperatures in the range of $353.15 < T < 423.15$ K, in steps of 10 K. The lines are fits to the pressure-modified VFTH equation for the α-process whereas linear fits are made for the β- and (αβ)-processes. Reproduced from [1475] by permission

To remove this false impression possibly given to the reader by this plot of MFW, the same data for four temperatures $T_j = 343.15$ (squares), 353.15 (circles), 363.15 (up triangles), and 373.15 K (down triangles) are reproduced here in Fig. 272. The × sign is used to mark the symbols that represent data of $-\log(\tau_\beta)$ in the glassy state of PEMA at each T_j. Pressure insensitivity of $-\log(\tau_\beta)$ shown by these marked data is unsurprising because the glass is not as compressible as the liquid. However, without this distinction, all the data of $-\log(\tau_\beta)$ when viewed together can create a false impression of spectacular pressure insensitivity of $-\log(\tau_\beta)$. The solid lines in the figure are fits to the pressure dependence of $-\log(\tau_\alpha)$ by the pressure-modified VFT equation for the α-process [1475]. The tip of each arrow indicates $-\log(\tau_\beta(P_{gj};\, T_j))$ at the glass transition pressure, P_{gj}, as when $-\log(\tau_\alpha(P_{gj};\, T_j)) = -2.5$ for each temperature T_j. The spread of the values of $-\log(\tau_\beta(P_{gj};\, T_j))$ is 0.2 of a decade if not included the highest temperature and is 0.4 of a decade for all. Again, the foremost criterion for JG β-relaxation is satisfied within the errors and uncertainties in determining τ_α and τ_β by fitting the loss spectra.

Besides interpreting the β-relaxation of PEMA as intramolecular in origin, MFW [1475] also revived the Williams ansatz [77, 968, 1477], which essentially says the β-relaxation relaxes with correlation function $\varphi_\beta(t)$ a portion, (1–A), of the total relaxation strength of the dipole, and the α-relaxation with correlation function $\varphi_\alpha(t)$ relaxes the remaining strength A. The combined relaxation function is given by $\phi(t) = A\phi_\alpha(t) + (1 - A)\phi_\beta(t)\phi_\alpha(t)$. The two correlation functions $\varphi_\beta(t)$ and $\varphi_\alpha(t)$ are *independent* of each other, and hence no relation exists between their relaxation times. Basically, the Williams ansatz assumes independent contributions from the α- and β-processes in the time domain, which has been rejected

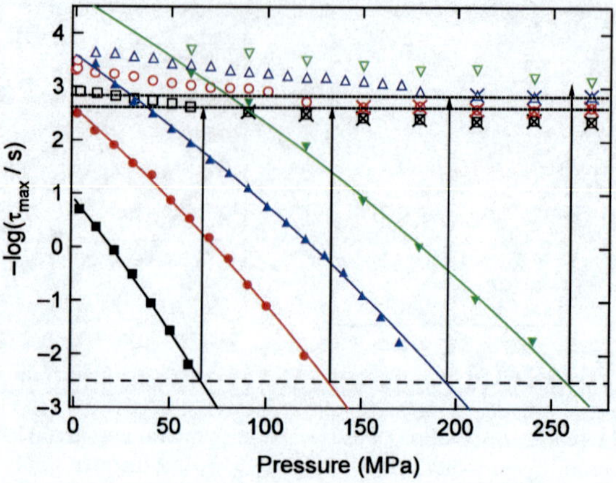

Fig. 272 Pressure dependence of the relaxation times at maximum loss for the α- and the β-processes of PEMA ($M_w = 2.0 \times 10^3$ g/mol) for four temperatures, 343.15 (*squares*), 353.15 (*circles*), 363.15 (*up triangles*), and 373.15 K (*down triangles*). The × sign is used to mark the *symbols* that represent data already in the glassy state of PEMA. The *solid lines* are fits to the pressure dependence of $-\log(\tau_\alpha)$ by the pressure-modified VFT equation for the α-process. The tip of each *arrow* indicates $-\log(\tau_\beta(P_g)/s)$ at the glass transition pressure, P_g, as when $-\log(\tau_\alpha(P_g)/s) = -2.5$ for each temperature. The spread of the values of $-\log(\tau_\alpha(P_g)/s)$ is 0.4 of a decade, well within the errors and uncertainties in determining τ_α and τ_β by fitting the loss spectra by superposition of two HN functions

by Böhmer et al. [998, 999] using their NMR data in non-polymeric glassformers. The invariance of $(\log \tau_\alpha - \log \tau_\beta)$ to variations in combinations of P and T shown here for PEMA (and elsewhere for non-polymeric glassformers) contradicts the assumption of independent contributions from the α- and β-processes in PEMA in the Williams ansatz. The reader may recall in Section 2.3.2.15 the exercise of using the Williams ansatz to analyze the relaxation data of the aqueous mixture of 26% wt. fraction of water in PPG400 at ambient and elevated pressures. The results from the Williams ansatz fail to explain the experimental observations, as shown in Fig. 164-2.

A possible rescue of the Williams ansatz is to impose a relation between the two relaxation times of $\varphi_\beta(t)$ and $\varphi_\alpha(t)$ like that given by the CM relation. However, even after this modification, $\phi(t) = A\phi_\alpha(t) + (1 - A)\phi_\beta(t)\phi_\alpha(t)$ is still an ansatz. There is no guarantee that it gives accurate description of the evolution of the many-body relaxation dynamics starting from the JG β-relaxation to end up with the α-relaxation, which is an unsolved problem. In PEMA, MFW have observed the marked decrease in β-relaxation strength $\Delta\varepsilon_\beta$ with increasing pressure is accompanied by a complementary increase in the α-relaxation strength $\Delta\varepsilon_\alpha$. It is correct that this is consistent by the concept of partial and total relaxations by the α- and β-relaxations in the Williams ansatz, but it does not prove validity of the ansatz. It is also consistent with the fact that the two relaxations are not independent but connected.

2.3.2.41 Broadening of α-Relaxation and Concomitant *Increase* of Separation from the JG β-Relaxation in Glycerol and Threitol at Elevated Pressure

If the JG β-relaxation or the primitive relaxation, its counterpart in the CM, is fundamental for glass transition and performs the indispensable role of generating the structural α-relaxation by many-body relaxation, then it better shows its presence in all glassformers. Glycerol is perhaps one of the most studied glassformer, and yet no experiment has ever been able to fully resolve its JG β-relaxation and to settle once and for all that it exists. At ambient pressure, an excess wing shows up on the high-frequency flank of the α-loss peak in the liquid state as well as the glassy state, giving no direct evidence of the presence of the JG β-relaxation [682, 972, 1023, 1034]. As shown before by Fig. 121a, physical aging of glycerol over many days did find the excess wing change its shape slightly to reveal a broad shoulder, which has been taken as evidence of the presence of JG β-relaxation. Nevertheless, possibly there are skeptics who still doubt the universal presence of the JG β-relaxation using the example of glycerol.

The CM equation (2.69) demonstrates that the separation between the JG β-relaxation from the α-relaxation measured by ($\log \tau_\alpha - \log \tau_{JG}$) is small for glycerol and other glassformers that have small value of the coupling parameter n, and this helps to rationalize why it is difficult to resolve the JG β-relaxation in glycerol. Following this idea from the CM, the JG β-relaxation will be resolved if somehow its coupling parameter can be increased. One method demonstrated before in Sections 2.3.2.10 and 2.3.2.16 is by mixing the glassformer with another less mobile glassformer. The increase in n can be seen by broadening of the frequency dispersion of the α-relaxation, and the JG β-relaxation becomes fully resolved due to larger separation distance ($\log \tau_\alpha - \log \tau_{JG}$) in the mixture. Many such examples can be found in Figs. 147, 152, 171, 172, 175, and 177. These are examples of van der Waals glassformers. Unfortunately, this method does not apply to glycerol with hydrogen bonds. Mixture of glycerol and sorbitol discussed in Section 2.3.2.12 is the closest analogue but the JG β-relaxation of the glycerol component cannot be resolved either.

On elevating pressure accompanied by higher temperature on glycerol, we have seen from Fig. 46 that the α-loss peak broadens. This indicates an increase of the coupling parameter n caused by breaking hydrogen bonds under high pressure and at high temperature. The past experiment [659] was carried out up to the highest pressure of 1.8 GPa and at 255 K, but despite the broadening of the α-loss peak, the excess wing remains unchanged except for higher intensity. It is possible that higher pressure than 1.8 GPa and higher temperature than 255 K are needed in order to break more hydrogen bonds and further increase the intermolecular coupling before the JG β-relaxation in glycerol can be resolved. This feat has been accomplished in a recent dielectric relaxation experiment by Pronin et al. [1478], where super high pressure up to 6 GPa was applied to glycerol. The $\varepsilon'(\nu)$ and $\varepsilon''(\nu)$ spectra at 294 K and different pressures from 1.7 and up to 4.5 GPa are shown in the left panels of Fig. 273-1, and at 4.5 GPa and various temperatures from 363 down to 293 K

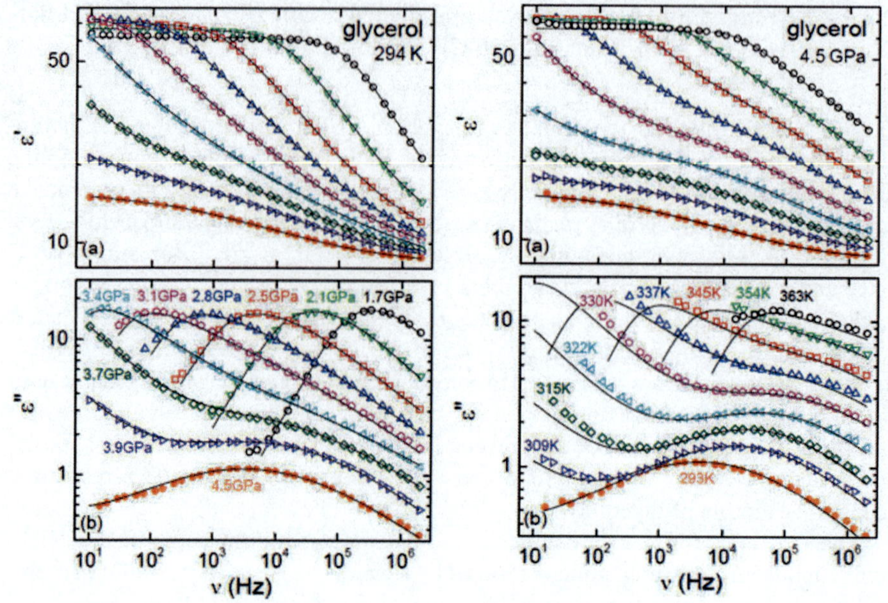

Fig. 273-1 (*Left*) Glycerol dielectric loss spectra at different pressures and 294 K. The lines are fits with the sum of the Cole–Davidson function model for the α-peak and the Cole–Cole function for the β-peak. (*Right*) Glycerol spectra under 4.5 GPa at various temperatures. The lines are fits with the sum of the Cole–Davidson function model for the α-peak and the Cole–Cole function for the β-peak. The significant conductivity contributions showing up at low frequencies have been neglected in the fits. Data from Pronin et al. [1478] and reproduced by permission

are shown in the right panels. Pronin et al. were able to fully resolve a secondary relaxation of glycerol in the liquid state as well as in the glassy state starting at pressure higher than 3 GPa, never seen before in the dielectric loss spectra. The resolved secondary relaxation is well separated by many decades in frequency from the α-relaxation. Johari pointed out the need for Pronin et al. to correct the data for the stray capacitance error. This correction has been done in the data shown in Fig. 273-1.

Concurrent with the appearance of the resolved secondary relaxation is the significant broadening of the dispersion of the α-relaxation, consistent with the increase in coupling constant n at higher pressure and temperature, resulting in the secondary relaxation being more separated from the α-relaxation. The increase of the isobaric fragility or steepness index m_P with pressure P shown in the right panel of Fig. 273-2 provides another support for the increase of n, from the correlation between n and m (see Section 2.2.3) which should hold since glycerol is considered throughout. It is clear from both panels of the figure that relaxation time τ_β of the secondary relaxation is pressure sensitive and its relaxation strength increasing with temperature. However, as discussed before, these properties of the secondary relaxation are necessary but not sufficient to identify the secondary relaxation to the JG β-relaxation. To do so, Pronin et al. applied the criterion based on CM prediction by

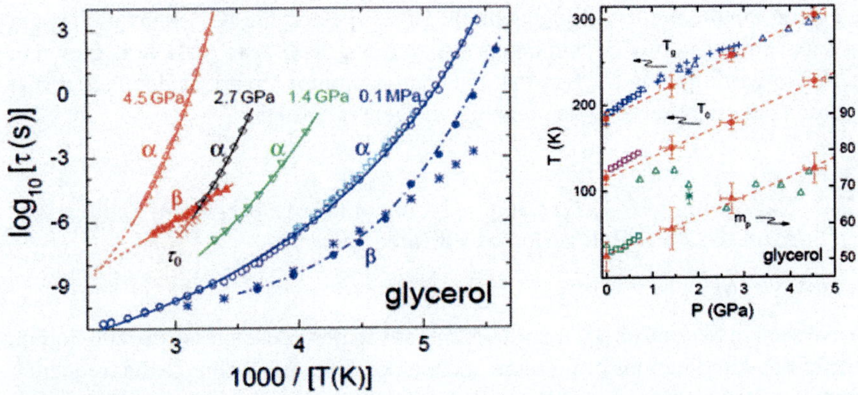

Fig. 273-2 (*Left*) Temperature dependences of relaxation times at four different pressures. The *open symbols* denote the α-relaxation times τ_{CD}. For 0.1 MPa and 4.5 GPa, β-relaxation times τ_{CC} are provided (*closed circles* and *triangles*, respectively). The predictions for $\tau_\beta(T)$ from the primitive relaxation time $\tau_0(T)$ calculated by the CM equation (2.69) are shown as *stars* (for 0.1 MPa) and *crosses* (for 4.5 GPa). The *solid lines* are fits using VFTH law for $\tau_{CD}(T)$ and the Arrhenius law for $\tau_{CC}(T)$ at 4.5 GPa. The *dash-dotted line* is drawn to guide the eyes. After Pronin et al. [1478]. Reproduced by permission. (*Right*) Pressure dependences of the VFTH temperature T_0 and T_g (*left axis*) and of the isobaric fragility m_P (*right axis*). The *closed squares*, *circles*, and *triangles* represent data of T_g, T_0, and m_P at four pressures, 0.1, 1.4, 2.7, and 4.5 GPa from Pronin et al. [1478]. The *open squares* at 0.1–0.7 GPa were taken from [1479]. The *crosses* denote T_g from viscosity measurements of [1480]. The *open triangles* show T_g and m_P from [556]. The single points at 1.8 GPa (*stars*) are T_g and m_P from [1481(a)]

Eq. (2.69). If the observed secondary relaxation is the JG β-relaxation, its most probable relaxation time τ_β should be equal to the primitive relaxation time τ_0 at least in order of magnitude. Calculation of τ_0 can be carried only after the coupling parameter n has been determined by fitting the α-relaxation by the Fourier transform of the Kohlrausch function. The reader can see in the right panels of Fig. 273-1 from the isothermal data at 4.5 GPa that the peak frequency, $\nu_\beta \equiv 1/(2\pi\tau_\beta)$, only can be determined without using any fitting procedure at $T = 330$ K and below, and marginally at 337 K. Above 337 K, the two relaxations overlap over limited frequency range available to the super high-pressure experiment, and there is no way to determine ν_β without large uncertainty. Pronin et al. obtained τ_α and τ_β by fitting the data with the sum of the Cole–Davidson (CD) function for the α-peak and the Cole–Cole function for the β-peak, and also the exponent (1–n) of the Kohlrausch function from the exponent of the CD function. With the parameters τ_α and n determined for $T \leq 337$ K, the primitive relaxation times τ_0 were calculated by the CM equation (2.68) and are shown in the left panel of Fig. 273-2. Considering the uncertainties in determining τ_α and n from the data, the agreement between τ_β and τ_0 in order of magnitude is fulfilled and allowed Pronin et al. to identify the β-relaxation to be the JG β-relaxation of glycerol.

In close relation to the results described above for glycerol is the data of threitol under applied pressure shown earlier in Fig. 46. Threitol has one more carbon atom than glycerol in the family of polyalcohols. Like glycerol, the JG β-relaxation is not resolved at ambient pressure, but it was resolved at 0.54 GPa and 241 K as shown in Fig. 46. Notwithstanding the fact that the data of threitol are not as detailed as that of Pronin et al., they serve to demonstrate the general property.

2.3.2.42 Narrowing of α-Relaxation and Concomitant *Decrease* of Separation from the JG β-Relaxation at Elevated Pressure

Tri(dimethylsiloxy)phenylsilane

We have seen in Sections 2.3.2.1 and 2.3.2.8 for van der Waals glass-forming liquids that different combinations of pressure and temperature do not change the frequency dispersion of the α-relaxation if the most probable α-relaxation time τ_α is kept constant. Moreover, the JG β-relaxation time τ_{JG} remains the same. This remarkable property was called co-invariance of *n*, τ_α, and τ_β to *P* and *T* variations and is consistent with the CM equation from which follows the co-invariance of *n*, τ_α, and τ_0 to *P* and *T* variations. The hydrogen-bonded glassformers behave differently, with broadening of the α-relaxation dispersion at elevated *P* and *T* and shorter τ_β while τ_α is kept constant. The cases include glycerol (Figs. 46 and 273-1), threitol (Fig. 46), xylitol (Fig. 269), and sorbitol (Figs. 162, 190, and 269). This behavior of hydrogen-bonded glassformers correspond to increase of *n* and increase of the ratio τ_α/τ_β at elevated *P* and *T* at constant τ_α. This behavior is again consistent with the CM equation because it predicts the increase of the ratio τ_α/τ_0 to follow as a consequence of the increase of *n*. The latter reflects the increase of intermolecular coupling by breaking hydrogen bonds.

Now, what about the possibility of having glassformers that behave differently from the two classes discussed above. Specifically, on elevating *P* and *T* while keeping τ_α constant, the frequency dispersion of the α-relaxation narrows and *n* decreases. If there are such glassformers, will the ratio τ_α/τ_β decrease as predicted by the CM equation from the observed decrease of *n*? The answer is in the affirmative, although examples from this third class of glassformers are few. So far there are only two glassformers behaving this way. One example is from the dielectric relaxation study of tri(dimethylsiloxy)phenylsilane (TDMSPS) [1481(b)]. As shown in this reference and here in the left panel of Fig. 273-3, for nearly the same peak frequency, the α-loss peak becomes narrower on elevating pressure up to 1.7 GPa. The fit to the Fourier transform of the Kohlrausch function requires $n = 0.43$ at 1.7 GPa and $n = 0.49$ at ambient pressure. In both cases the primitive relaxation frequencies calculated from the CM equation are in good agreement with the locations of the JG relaxation frequency. Narrowing of the α-loss peak with applied pressure of 1.7 GPa is accompanied by smaller separation distance between the α- and the β-relaxations, $\log \nu_\beta - \log \nu_\alpha$. This effect is also shown by the relaxation map in the right panel of Fig. 273-3, where the inset demonstrates weaker T_g-scaled *T*-dependence and smaller isobaric fragility index m_P at 1.7 GPa than at ambient pressure. Thus, the decrease of *n* is accompanied by corresponding decrease of m_P, in accordance with the correlation between *n* and *m*.

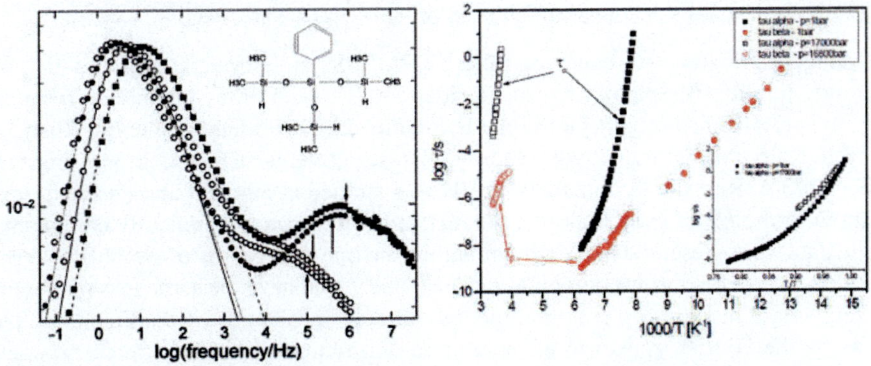

Fig. 273-3 (*Left*) *Solid symbols* stand for loss data of TDMSPS at ambient pressure, while *open symbols* are for loss data taken at 1.8 GPa and higher temperatures. The *lines* are fits to the Kohlrausch functions with $n = 0.43$ at 1.7 GPa and $n = 0.49$ at ambient pressure. In both cases the primitive relaxation frequencies calculated from the CM equation indicated by the *vertical arrows* are in good agreement with the location of the JG loss peak frequencies. Narrowing of the α-loss peak with applied pressure of 1.7 GPa is accompanied by smaller separation distance between the α- and the β-relaxations, $\log \nu_\beta - \log \nu_\alpha$, in TDMSPS. This unique behavior of TDMSPS is contrasted by the changes in opposite directions of hydrogen-bonded glassformers, where the α-loss peak broadens and the separation distance increases with applied pressure. (*Right*) The relaxation map showing the smaller separation distance between the α- and the β-relaxations, $\log \nu_\beta - \log \nu_\alpha$, as well as weaker T_g-scaled T-dependence and fragility index m in TDMSPS at 1.7 GPa than at ambient pressure. Reproduced from [1481(b)] by permission

The chemical structure of TDMSPS can be seen in the left panel of Fig. 273-3. Likely what makes TDMSPS so unique is the bulky but flexible Si-based part of the molecule joined to the otherwise rigid phenyl ring. High pressure can alter the relation of the phenyl rings to each other through shielding by the compressed Si-based parts of the molecules. Intermolecular coupling is thereby reduced and the α-dispersion is narrowed as observed. Supports of this explanation come from the concomitant decrease of the separation between the α-relaxation and the JG β-relaxation and from the decrease in the fragility index m.

At a constant structural relaxation time, dielectric relaxation data of a vinyl-terminated carboranylenesiloxane (VCS) taken at ambient and elevated pressures also show that the structural relaxation dispersion function of VCS narrows with increasing pressure and temperature [1481(e)], and hence this is one of the few cases of non-hydrogen-bonded glassformers where the temperature–pressure superposition breaks down. This can be explained by a reduction of the intermolecular interactions due to the altered spatial relations and weakened interactions of its carborane cages to one another in this vinyl-terminated carboranylenesiloxane. The mechanism is the increased shielding of the carborane cages by the flexible arms of the molecules on elevating pressure. This explanation of narrowing of the α-loss peak by reduction of intermolecular coupling is supported by concomitant reduction of the fragility parameter m with increasing pressure and the correlation between n and m expected to hold when restricted to the same glassformer.

Protic Ionic Liquid-Verapamil Hydrochloride ($C_{27}H_{38}N_2O_4{\cdot}HCl$)

Another glassformer behaving like TDMSPS at elevated pressure is the ionic liquid, verapamil hydrochloride (VH), with the chemical formula: (±)-5-((3,4-dimethoxyphenethyl)methylamino)-2-(3,4-dimethoxyphenyl)-2-iso-propylvaleronitrile monohydrochloride. Its structure is depicted in the inset of Fig. 273-4. Relaxation dynamics of VH was studied at ambient and elevated pressures by dielectric spectroscopy [1481(c) and (d)]. Since it is ionically conducting, the data were presented as electric modulus; the imaginary parts of which are shown in Fig. 273-4 for various combinations of P and T that have the same loss maximum frequency. On elevating pressure, the loss modulus peak narrows continuously. The fits by the Fourier transform of the Kohlrausch function requires stretch exponent $\beta_K = (1 - n) = 0.61$ at ambient pressure, and $\beta_K = (1 - n) = 0.68$ at the highest pressure applied. The excess wing on the high-frequency side of the loss peak is the unresolved JG β-relaxation with frequency, f_{JG}, suggested by the primitive frequency, f_0, calculated by the CM equation. The excess wing as well as f_{JG} (suggested by the calculated f_0) shifts to lower frequency in conjunction with the narrowing of the α-loss peak on both the high- and low-frequency sides. This effect is consistent with the decrease of separation between the α- and the primitive or JG β-relaxation concomitant with the narrowing of the dispersion of the α-relaxation at elevated pressure.

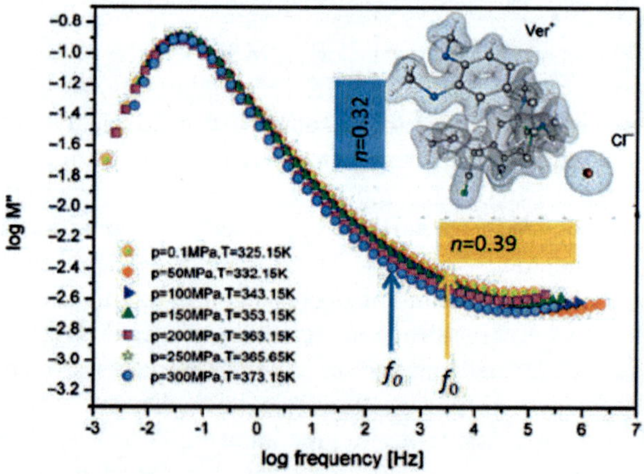

Fig. 273-4 The main panel in the figure shows the loss modulus for several combinations of P and T that have the same loss peak frequency. The *two vertical arrows* indicate the location of the calculated primitive frequency, f_0, or the unresolved JG relaxation frequency, f_{JG}, at $P = 0.1$ and 300 MPa with $\beta_K = (1 - n) = 0.61$ and 0.68, respectively

After obtaining the effect of pressure on the temperature dependences $\tau_\alpha(T)$, it was found that the isobaric fragility, m_P, also decreases with increasing pressure. Like TDMSPS, the decrease of n is accompanied by corresponding decrease of m_P. Hence, the correlation between m_P and n holds on compressing VH by applied pressure.

2.3.2.43 JG β-Relaxation of Aqueous Mixture Under High Pressure: Water–Propylene Glycol Oligomer Mixtures

In various places of Sections 2.2 and 2.3, the introduction of elevated pressure in conjunction with temperature to dynamics of glassformers in both the liquid and glassy states has yielded many interesting and general properties that provide insight into the mechanism of the dynamics. So far, none of the glassformers studied in this way involves water and aqueous mixtures. This is understandable in the case of pure water because of the limitations imposed including crystallization even at ambient pressure. Many aqueous mixtures have no such problem and are amenable to the study of the dynamics at elevated pressure. Nevertheless, there is hardly any published experimental data of aqueous mixture under applied high pressure that show the dynamics over wide time/frequency range.

Only very recently has the data become available by the collaborative effort of researchers in the University of Silesia, Poland, and University of Pisa, Italy. Given in a paper by Grzybowska et al. [1304(b)], the results show properties of the dynamics at elevated pressure that have bearing on those of binary mixtures of van der Waals glassformers and of neat glassformers with or without the involvement of hydrogen bonds. Grzybowska et al. performed broadband dielectric spectroscopy measurements of aqueous mixtures of polypropylene glycol of molecular weight $M_w = 400$ g/mol (PPG400) at several concentrations up to 36 wt% of water at ambient pressure from room temperature down to 100 K. The water-rich mixture, PPG400+26 wt% H_2O, which corresponds to mole fraction of water of $x_w = 0.89$, was measured under applied pressure up to 1.8 GPa over temperatures ranging from 198 to 295 K.

At ambient pressure and low temperatures in the glassy state, anhydrous PPG400 has two resolved secondary relaxation, β and γ, shown in Fig. 273-5 (upper left). The faster γ-relaxation is not the JG β-relaxation because in the liquid state its relaxation frequency is much higher than the primitive relaxation frequency (indicated by the bold arrow in Fig. 273-5 (middle right)) calculated with n=0.37 from the Kohlrausch function that fits the α-loss peak. Similar data and conclusion have been shown before in Fig. 125.

Addition of water significantly changes the relaxation dynamics of propylene glycol by the appearance of a new relaxation originating from water, called ν-relaxation in previous works [1093, 1295]. This is clear from the evolution of the spectra in the glassy state and in the liquid state in Fig. 273-5 (upper left and upper right), and also from the rapid increase of the dielectric strength $\Delta\varepsilon_\nu$ of the ν-relaxation on increasing of water wt% in Fig. 273-5 (upper right). Like other aqueous mixtures there is a threshold concentration (mole fraction of water is 0.67 in the present system) above which dielectric strength of ν-process is rapidly increasing with water concentration and the relaxation time does not change much with the concentration. In this concentration regime the ν-process is believed to reflect the true dynamics of water molecules predominantly surrounded by other water molecules. The spectra of the aqueous mixtures show two well-resolved relaxations, the slow α-relaxation and the fast ν-process, just like other aqueous mixtures shown

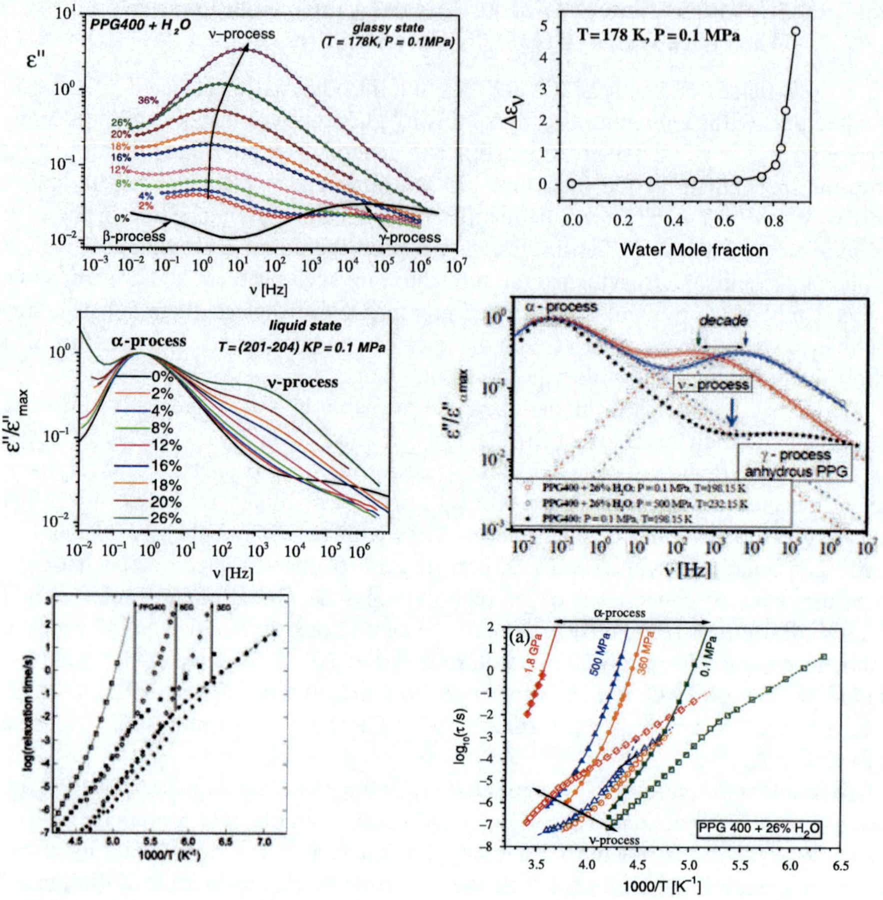

Fig. 273-5 (*Upper left*) Dielectric loss spectra of aqueous mixtures of PPG400 with various concentrations of water in the glassy state at $T = 178$ K and $P = 0.1$ MPa. (*Upper right*) Figure shows the dependences of dielectric strength of the secondary relaxation, originating from water in the aqueous mixtures, called the ν-process, against mole fraction of water. (*Middle left*) Figure presents spectra of the water mixtures with the same α relaxation times in the liquid state. (*Middle right*) Comparison of dielectric spectra with the same α relaxation times, obtained at different P and T combinations for 26 wt% water mixture of PPG400 and anhydrous PPG400. *Solid lines* denote fits of entire spectra as a superposition of the two HN functions (*dotted lines*). (*Lower left*) Comparison of the loss spectra of the aqueous mixture of PPG400 containing 26 wt% water at $P = 500$ MPa and $T = 232.15$ K with that of two 35 wt% water mixtures with ethylene glycol oligomers, 3EG and 6EG. Data from [1093, 1295] are replotted here. (*Lower right*) Temperature dependences of the α and ν relaxation times at various pressures for PPG400+26% H_2O. All figures are redrawn in different styles and representations than those published in [1304(b)]

before in Section 2.3.2.27. Their relaxation times, τ_α and τ_ν, are compared in Fig. 273-5 (lower left) with those of two other aqueous mixtures, 35 wt% water mixtures with ethylene glycol oligomers 3EG and 6EG [1093, 1295]. Except for PPG400+26 wt% H_2O having a higher T_g, the properties found for τ_α and τ_ν are similar to the aqueous mixtures of 3EG and 6EG shown before in Figs. 214 and 215. There is a change of temperature dependence of τ_ν from Arrhenius below T_g to a stronger T-dependence above T_g. The location of T_g for each of the aqueous mixtures in the lower left panel of Fig. 273-5 is indicated by the vertically pointing arrow. In the glassy state the Arrhenius activation energy of τ_ν is about the same as that of the other two mixtures, which reflects the nature of the local relaxation of water. Although τ_ν is longer in PPG400+26 wt% H_2O than the other two mixtures, the change is small compared with the many more decades of increase of τ_α coming from the significant increase of T_g. Here, the reader is reminded that the same crossover of T-dependence at T_g is found in the JG β-relaxation time of a component in binary mixtures of van der Waals liquids discussed in Section 2.3.2.27.

Crossover of τ_ν at Elevated Pressure

Dielectric measurements of PPG400+26 wt% H_2O obtained at elevated pressure of 360 and 500 MPa exhibit two well-resolved relaxation peaks coming from the slower α-process and the faster ν-process, just like at ambient pressure. This is shown by the loss spectrum obtained at $P = 500$ MPa and $T = 232.15$ K in Fig. 273-5 (middle right). There, comparison is made with the loss spectrum taken at $P = 0.1$ MPa and $T = 198.15$ K of the same mixture having the same α-relaxation frequency (time), ν_α (τ_α). It can be seen that the ν-relaxation frequency (time), ν_ν (τ_ν), is not the same for the two cases, with ν_ν (τ_ν) being higher (shorter) by one decade at 500 MPa than at 0.1 MPa. This situation is different from that seen for the relation between the α- and the JG β-relaxation times of a component in binary mixtures of van der Waals liquids (see Figs. 144, 146, and 148), where there is no change in both the dispersion of the α-relaxation (or n) and ν_{JG} (τ_{JG}) for various combinations of P and T that keeps ν_α (τ_α) constant. This is due to the change in structure of hydrogen-bonded systems when subjected to elevated pressure and higher temperature by disrupting the hydrogen-bonding network. The result is the change of intermolecular coupling. The consequence is change in the relation between τ_α and the primitive relaxation time τ_0 or τ_{JG} according to the CM, as seen before in Section 2.3.2.41 for glycerol, threitol, and propylene glycol in Fig. 46, xylitol in Fig. 269, and sorbitol in Figs. 48, 190, and 269. The preponderance of hydrogen bonding in PPG400+26 wt% H_2O makes the structure susceptible to change at elevated pressure accompanied by higher temperature. The situation is just like glycerol, threitol, xylitol, and sorbitol. Thus, the increase in separation between ν_α and ν_{JG} or ν_0 found in PPG400+26 wt% H_2O at $P = 500$ MPa and $T = 232.15$ K than at $P = 0.1$ MPa and $T = 198.15$ K in Fig. 273-5 (middle right) follows the same behavior established for the polyalcohols and receives the same explanation.

From the dielectric spectra of PPG400+26 wt% H_2O taken under isobaric condition with $P = 0.1$, 360, 500, and 1800 MPa, and isothermal condition with $T = 213$ and 233.8 K, the relaxation times, τ_α and τ_ν, are obtained and plotted against $1000/T$ in Fig. 273-5 (lower right). For all four different pressures, the temperature dependence of τ_α is well described by the VFTH equation and reaches 100 s at the dielectric T_g which increases with pressure, and τ_ν undergoes the change from

Arrhenius T-dependence to a stronger T-dependence when crossing some temperature, T_{cross}, which is nearly the same as T_g. The value of τ_ν at T_g, $\tau_\nu(T_g)$, is shorter at higher pressure. The crossover of T-dependence of τ_ν at $T_{cross} \sim T_g$ and the trend of decrease of $\tau_\nu(T_g)$ with pressure are the same as that exhibited by τ_{JG} at ambient and elevated pressures of xylitol and sorbitol [1088] and shown before in Fig. 162. The explanation of these observations in xylitol and sorbitol by the CM applies verbatim to PPG400+26 wt% H_2O. The crux of the CM explanation is the inseparable connection between the primitive relaxation and the structural relaxation leading to the crossover at T_g. The increase of intermolecular coupling (and coupling parameter n) due to suppression of hydrogen bonding by applied pressure and higher temperature results in increased separation between τ_α and τ_0. The decrease of $\tau_{\nu cross}$ on increasing pressure when fitted by the exponential function in the inset (b) indicates that $\tau_{\nu cross}(P \to \infty) = 3$ μs, not much shorter than $\tau_{\nu cross}(P = 1.8\ \text{GPa})$.

Found in the glassy states of PPG400+26 wt% H_2O at various pressures are slight differences of the magnitude of τ_ν and its activation energy. The spectra measured at the highest pressure of 1.8 GPa have the α-relaxation peak hidden by the large contribution of dc-conductivity, and its relaxation times in Fig. 273-5 (middle right) were obtained indirectly from the real part of permittivity. The data of τ_α over limited temperature range and the VFTH fit cannot yield accurate determination of the isobaric fragility index, m_P, and conclusion made in [1304(b)] based on it may not be valid.

2.3.2.44 JG β-Relaxation of Aqueous Mixture Under High Pressure: Water–Fructose Mixtures

Similar results on applying pressure to aqueous solution of the saccharide, fructose, at various concentrations were obtained by Ancherbak et al. [1140b]. The relaxation times, τ_α and τ_ν, obtained behave in the same way as those found PPG400+26 wt% H_2O. In a plot of τ_α and τ_ν against $1000/T$, there is a change of the T-dependence of τ_ν from Arrhenius below T_g to a stronger dependence above T_g, where T_g is the temperature at which $\tau_\alpha \sim 10^3$ s (see Fig. 273-6). The water or ν-relaxation is clearly resolved in the fructose aqueous solutions, and its dielectric strength $\Delta\varepsilon_\nu$ shows monotonic increase with temperature in a range of temperature above and below T_g (see Fig. 273-6). This temperature dependence of $\Delta\varepsilon_\nu$ provides firm evidence that the observed water or ν-process is a secondary relaxation for all temperatures above and below T_g and is not an α-relaxation of water above T_g that becomes a secondary relaxation below T_g, as proposed by other workers in aqueous mixtures [1106, 1261].

2.3.2.45 Evidences of Primitive or β-Relaxation and Faster Relaxation Are Responsible for the Stabilization of Dried Protein in Sugar-Based Glass

Biopharmaceuticals including proteins, drugs, and vaccines are important parts of modern medicine with sales of products close to 100 billion dollars per year.

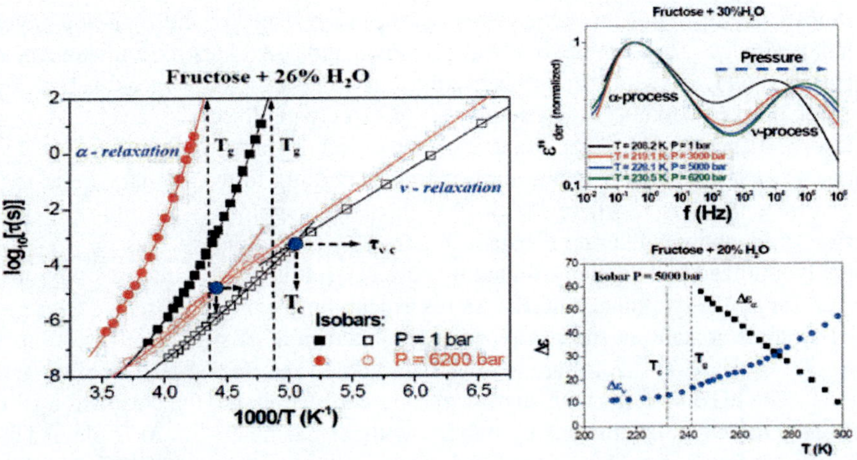

Fig. 273-6 (*Left*) Relaxation times, τ_α and τ_ν, of the structural relaxation of the aqueous mixture of fructose + 26 wt% water at ambient pressure and elevated pressure of 6200 bar, showing the change of T-dependence of τ_ν at some temperature T_c near T_g. Note that T_g increases with pressure, and $\tau_\nu(T_g)$ becomes shorter on elevating pressure. (*Right on top*) The latter is illustrated by the spectra taken at several pressures and temperatures all having the α-loss peak frequency. The ν-loss peak moves to higher frequencies on increasing pressure. (*Right at bottom*) The dielectric strength of the α- and ν-relaxation showing the $\Delta\varepsilon_\nu$ is a monotonic increasing function of temperature above and below T_g, the hallmark of secondary relaxation at all temperatures. Figures courtesy of K. Grzybowska and coworkers

The problem of the biopharmaceuticals is inadequate stability in aqueous solution to allow storage, transport, and distribution. Thus, such products are commonly dried to increase stability, but even the dry product is usually not stable during storage. Saccharides and polyols are commonly added to form a glass to protect proteins against degradation during storage [1383(a)–(i), 1384(a)–(d)]. Low molecular weight non-reducing saccharides stabilize effectively, and the disaccharides sucrose and trehalose are commonly used for commercial products. This process to achieve stabilization is like that in nature where living organisms manage to survive unfavorable conditions of extreme drought and high temperatures by producing trehalose to form glassy state to block metabolism. The water substitute hypothesis and the vitrification hypothesis have been commonly used to rationalize stabilization of the biopharmaceuticals by the added sugars to form a glass. The water substitute hypothesis suggests that the excipient sugar operates by replacing water lost to form hydrogen bonds to the protein, and hence thermodynamically stabilizes the native conformation.

On the other hand, the vitrification hypothesis explains increased stability by the much reduced molecular mobility for all degradation processes to proceed when the system is transformed into the glassy state, resulting in significantly lower degradation rate. This hypothesis implies that the difference between the storage temperature T and T_g, $(T - T_g)$, is a reliable measure for stability. Recent works found that $(T - T_g)$ is not a critical parameter for stability by comparing formulations with different sugars of the same protein at the same sugar/protein ratio and water content. This was shown in several ways by Pikal and coworkers for human growth hormone (hGH) in lyophilized disaccharide formulations [1383(h)] as well as by Yoshioka and coworkers in β-galactosidase (β-GA) lyophilized with stachyose, trehalose, and sucrose [1384(a)]. Pikal et al. measured degradation by the loss of the purity, P, that varies with square root of time according to the simple expression, $P = P(t = 0) - kt^{1/2}$, where $P(t = 0)$ is the percentage purity at time zero (i.e., initial value) and k is the rate constant. Figure 274-1a is a plot of k vs. $T - T_g$ for many lyophilized hGH systems including pure hGH. It is divided into two sections, the left for all the lyophilized hGH systems except those using sucrose and the right hGH systems containing sucrose. For the non-sucrose hGH systems all with large negative values of $T - T_g$, there is obviously no correlation between stability and $T - T_g$. The hGH systems with sucrose show a correlation, but the correlation is in the wrong direction, i.e., increasing stability with increasing $T - T_g$. As explained by Pikal et al. there is reduction of the T_g of a protein system (T_g, dry≈136°C) by addition of sucrose (T_g, dry ≈75°C), but stabilization is enhanced by increasing levels of sucrose. From this one can see for the comparison of stability between different formulations. It is important to have control of water content, otherwise one can be led to the wrong conclusion. Figure 274-1(b) will be discussed later. It is interesting that there is correlation between stability (reciprocal of rate constant k) and

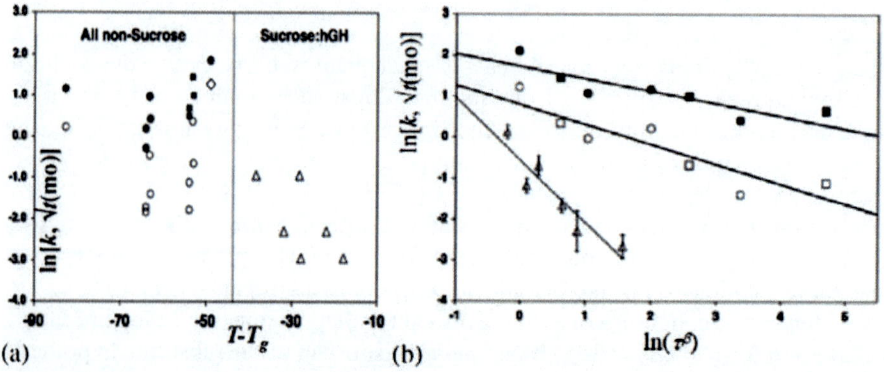

Fig. 274-1 (**a**) Correlation of stability of hGH with the difference between storage temperature, T, and the glass transition temperature of the formulation, T_g. *Open symbols* are aggregation, *filled symbols* are chemical degradation, *diamonds* are pure protein, and *circles* are disaccharide systems other than sucrose. Sucrose data are *triangles*. Reproduced from [1383(h)] by permission. (**b**) Correlation between stability (reciprocal of rate constant k) and relaxation time constant, τ^{β}, where τ is the structural relaxation time and β is the Kohlrausch exponent. *Circles* and *squares* are pure human growth hormone (hGH), stachyose (1:1), and trehalose (1:1, 1:3, 1:6 weight ratios of protein/saccharide). The *triangles* represent aggregation data in sucrose formulations. The rate constant is from square root of time (months) kinetics and τ is in hours. All input data for τ are for mean moisture samples. *Error bars* (standard errors) for the non-sucrose formulations are smaller than the symbol size. Note that sucrose formulations are more stable than expected based upon the relaxation time constant τ^{β}. Reproduced from [1383(h)] by permission

relaxation time constant, τ^{β}, where τ is the structural relaxation time and β is the Kohlrausch exponent [1383(h)] as shown in Fig. 274-1b. There, circles and squares are pure human growth hormone (hGH), stachyose (1:1), and trehalose (1:1, 1:3, 1:6) weight ratios of protein/saccharide. The triangles represent aggregation data in sucrose formulations. The rate constant is from square root of time (months) kinetics, and τ is in hours. All input data for τ are for mean moisture samples. Note that sucrose formulations are more stable than expected based upon the relaxation time constant τ^{β}. This aspect will be further discussed later.

The left panel of Fig. 274-2 compares the rate constants k of degradation for different formulations obtained by Pikal and coworkers [1383(h)]. Here we compare only the rate constants of pure hGH, and hGH lyophilized with stachyose, trehalose, and sucrose all at the same storage temperature $T = 40°C$. Light shading represents chemical degradation, while the dark bars represent degradation by aggregation, which is primarily non-covalent dimer formation. All formulations are nominally "dry." The numeral inside brackets in the labels of the sugars indicates the ratio of sugar/hGH. From this figure, it can be seen that at the same ratio of sugar/hGH, stability increases in the order of stachyose, trehalose, and sucrose, while T_g decreases in the same order [1383(h)]. Thus stability anticorrelates instead of correlates with the absolute value of $(T - T_g)$. Yoshioka and coworkers [1384(a)] also found for the β-GA lyophilized with stachyose, trehalose, and sucrose at 12% relative humidity; the aggregation rate at temperatures below and above T_g was also lowest for sucrose, intermediate for trehalose, and largest for stachyose, suggesting again that β-GA aggregation rate is not related to $(T - T_g)$. The data of Yoshioka et al. showing this trend are presented in the right panel of Fig. 274-2, where the ordinate is the time required for 10% degradation (t_{90}) calculated from the apparent first-order rate constant measured.

Failure to correlate protein stability with global mobility indicated by $(T - T_g)$ has led others to suggest and investigate the possibility that the local motion or local mobility may play an important role in stability using neutron scattering [1383(b), (d), (e), (h)] and carbon (^{13}C) T_1 and $T_{1\rho}$ of nuclear magnetic resonance relaxation times by Pikal and coworkers [1383(g)] and Yoshioka and coworkers [1384(a)–(e)]. Yoshioka et al. [1384(a)] investigated the local mobility of β-GA, as determined by spin-lattice relaxation in the rotating frame at 25°C and 12% RH. Spin-lattice relaxation was significantly retarded by the addition of excipient. The relaxation time $T_{1\rho}$ of β-GA carbonyl carbon was more markedly increased by the addition of sucrose than by the addition of stachyose. The effect of trehalose on $T_{1\rho}$ was intermediate when compared to those for sucrose and stachyose. The same conclusion was obtained from ^{13}C NMR measurements of T_1 at higher frequencies than $T_{1\rho}$. Thus, the aggregation rate of β-GA lyophilized with stachyose, trehalose, and sucrose correlates with the local mobility of β-GA, as indicated by $T_{1\rho}$, rather than with $(T - T_g)$. This indicates that stabilization during storage is effected by reduction of local protein mobility, and reduction is most effective by sucrose and less effective by trehalose and stachyose in this order. From the T/T_g-dependence of t_{90} for aggregation of β-GA lyophilized with sucrose, trehalose, and stachyose in the right panel of Fig. 274-2, it can be seen that the temperature dependence of t_{90} all

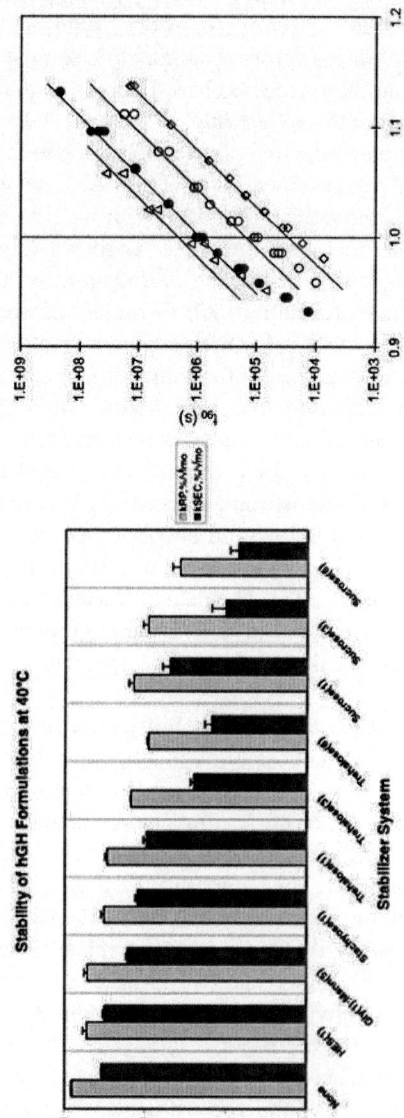

Fig. 274-2 (*Left*) Comparison of stability of freeze-dried hGH formulations at 408°C. The *light shaded bars* are rate constants, based on square root of time kinetics, for chemical degradation as measured by reverse phase HPLC assay, and the *dark shaded bars* represent rate constants, based upon square root of time kinetics, for aggregation as measured by size exclusion HPLC assay. All formulations are nominally "dry." *Error bars* are standard errors as given by the regression analysis. The *numeral inside brackets* in the labels of the sugars indicate the ratio of sugar/hGH. Reproduced from [1383(h)] by permission. (*Right*) T/T_g-dependence of t_{90} for aggregation of β-galactosidase (β-GA) lyophilized with sucrose (Δ), trehalose (○,●) or stachyose (◇). The weight fraction of excipient: 0.33 (Δ,○,◇) and 0.5 (●). Reproduced from [1384(a)] by permission

show a change in slope around T_g. This behavior comes as a surprise to Yoshioka and coworkers because they had already concluded that local mobility is mainly responsible for β-GA aggregation, and they thought that the temperature dependence of local mobility and hence t_{90} should not show a change in slope near T_g. To resolve this conundrum, they assume that local mobility of protein is coupled with structural relaxation. In support of this assumption they cited their NMR spin-lattice relaxation data of bovine serum γ-globulin lyophilized with dextran [1384(d)], where the local mobility of both the protein and dextran exhibited change in the slope of temperature dependence at the same temperature. This finding suggested that the local mobility of β-GA is coupled with the structural relaxation of β-GA lyophilized with sucrose, trehalose, or stachyose. To us and from the previous subsections showing the connection between the primitive or JG β-relaxation which determines the local mobility with the structural relaxation, the change of T-dependence of the local mobility from Arrhenius law below T_g to a stronger dependence above T_g becomes natural. Hydrogen bonds of the disaccharide to the protein are analogous to water to protein (see Section 2.3.2.28 or water to hydrophilic solutes such as PPG400 discussed in the previous subsection and in Section 2.3.2.27. In these water-related systems, multiple examples have demonstrated that the primitive or JG β-relaxation time of water changes its T-dependence at the glass transition temperature of the hydrated protein or the aqueous solution. In analogy to hydrated proteins, the primitive or JG β-relaxation of the disaccharide is coupled to the structural relaxation of the β-GA lyophilized by the disaccharide, and consequently there is the similar change of T-dependence of the primitive or JG β-relaxation time of the disaccharide at T_g, as experimentally found for the polysaccharide dextran in the case of bovine serum γ-globulin lyophilized with dextran [1384(d)]. The local mobility of the protein is dictated by the local primitive or JG β-relaxation of the disaccharide or polysaccharide. Therefore, from the experimental deduction that degradation time t_{90} is controlled by local mobility, naturally it follows that t_{90} exhibits change of T-dependence of the primitive or JG β-relaxation time at T_g of the disaccharide or polysaccharide. From Fig. 274-2, the Arrhenius activation energy of t_{90} for $T < T_g$ is about 120 kJ/mol. This is similar to the activation energy of JG β-relaxation time τ_{JG} of trehalose of about 100 kJ/mol for the T-range higher than 322 K corresponding to the same range as measured by Yoshioka and coworkers. The similar magnitudes of the activation energies of t_{90} and τ_{JG} can be considered as additional support of the explanation.

Cicerone and coworkers [1383(b), (d), (e), (h)] employed extensively elastic and inelastic neutron scattering of the fast dynamics on nanosecond timescale of the disaccharide/protein formulations [1383(d), (e), (h)] as well as the disaccharides [1383(b)]. Here we show the mean-squared displacement, $< u^2(T) >$, for the 6:1 sucrose/hGH formulation and the 6:1 trehalose/hGH formulation. Results for $< u^2(T) >$ show significantly less amplitude of motion for the 6:1 sucrose/hGH formulation than for the 6:1 trehalose/hGH formulation. Thus, stability does correlate with the fast dynamics measured by neutron scattering in the glass, as found in other lyophilized proteins. However, $< u^2(T) >$ corresponds to dynamics on a nanosecond timescale while stability is determined on the much longer timescale of months, and this observed correlation presents another conundrum in these and other protein formulations. The authors of [1383(h)] were puzzled by the following disparity between the degradation rate and $< u^2(T) >$ despite the correlation between them. First, the activation energies for both the sucrose and trehalose systems (6:1 disaccharide/hGH) are ≈18 kcal/mol or 75 kJ/mol for aggregation is much larger than ≈1.4 kcal/mol for $< u^2(T) >$. Furthermore, the ratio of $< u^2(T) >$ for trehalose/sucrose is 1.2 at 40°C while the corresponding ratio for rate constants is 2.1. They appropriately concluded that obviously the relationship between fast dynamics and stability is more complex than a direct proportion. Their puzzle can be removed by recalling the fact that, for T below T_g, the measured $< u^2(T) >$ originates from caged dynamics. Its magnitude is determined by the primitive relaxation time τ_0 or JG β-relaxation time τ_{JG}, as discussed before in Sections 2.3.2.33 and 2.3.2.35. In Section 2.3.2.35 is shown the correlation between $< u^2(T_g) >$ or NCL in susceptibility at T_g and $n(T_g)$. Also we recall from Sections 2.3.2.10, 2.3.2.11, 2.3.2.12, 2.3.2.13, and 2.3.2.14 that smaller $n(T_g)$ implies longer τ_0 or τ_{JG}. Combining these two, we can deduce that smaller $< u^2(T_g) >$ corresponds to longer $\tau_0(T_g)$ or $\tau_{JG}(T_g)$. The T_g values for the sucrose and trehalose formulations are 348 and 386 K, respectively. From Fig. 274-3 we find that the value of $< u^2(T_g) >= 1.65\text{Å}^2$ for the sucrose formulation is smaller than $< u^2(T_g) >= 2.65\text{Å}^2$ for the trehalose formulations. The final deduction from these is that $\tau_0(T_g)$ or $\tau_{JG}(T_g)$ is longer, and hence the local mobility is lower (and smaller degradation rate) for the sucrose formulation than the trehalose formulation. This is possibly the origin of the observed anticorrelation between $< u^2(T) >$ and stability of the protein formulation. By the way, dielectric data have shown that $\tau_{JG}(T_g)$ of pure sucrose is several orders of magnitude longer than $\tau_{JG}(T_g)$ of pure trehalose (see Fig. 206 and [1012(c)]). Also

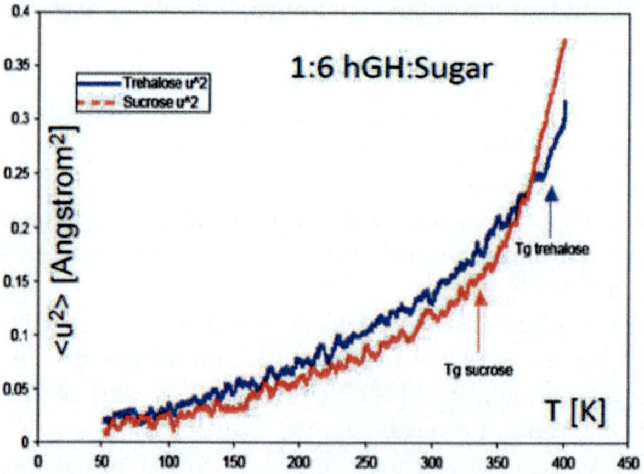

Fig. 274-3 Mean-squared displacement, $< u^2(T) >$, measured by neutron scattering for 1:6 hGH/disaccharide trehalose and sucrose formulations of hGH. The mean-squared amplitude of hydrogen motion (Å^2) is plotted against temperature. The T_g values for the sucrose and trehalose formulations are 348 and 386 K, respectively. Uncertainties (standard error) are roughly the size of the plotting symbol. Reproduced from [1383(h)] by permission

the fragility index m of pure sucrose is 100, which is less than 130 for trehalose [1012(c)], which also indicates smaller $n(T_g)$ of sucrose than trehalose from the correlation between m and $n(T_g)$ for glassformers of the same family. The larger m of trehalose than sucrose found by dielectric relaxation in [1012(c)] is consistent with that deduced from heat capacity data of Shamblin et al. [1185] and the smaller D value of 5.1 for trehalose than the value of 7.3 for sucrose. Dranca et al. [992(d)] calculated the thermodynamic fragility index using the empirical equation proposed by Wang and Angell [573], given before in Section 2.2.3.5, and the experimentally obtained mid-point T_g values of 346 and 392 K, respectively, for sucrose and trehalose, $\Delta C_p(T_g)$ value of 0.55 J K^{-1} g^{-1} (188.3 J K^{-1} mol^{-1}) for both sucrose and trehalose, and the heat of fusion, ΔH_m, for sucrose and anhydrous trehalose from literature values of 41.4 and 53.4 kJ mol^{-1}, respectively. Using these values, the thermodynamic fragility index m was found to be 88 for sucrose, larger than 77 found for trehalose. This is opposite to the trend found in the experimentally determined kinetic fragility index m, with trehalose having a much larger value than that of sucrose [1012(c)]. This is another example of non-equivalence of the kinetic and thermodynamic fragility discussed amply in Section 2.2.3.5. Dranca et al. correctly found from their calorimetric method the JG β-relaxation of trehalose. The value of the activation energy obtained by them at the lowest annealing temperature is 75 kJ mol^{-1}, comparable to the value of 87 kJ mol^{-1} from direct dielectric measurements [1012(c)]. However, for sucrose, they gave 42 kJ mol^{-1} for the activation energy of β-relaxation, much smaller than 98 kJ mol^{-1} for the JG β-relaxation from dielectric relaxation [1012(c)]. On the other hand, their value of 42 kJ mol^{-1} is not far from 52 kJ mol^{-1} for the γ-relaxation obtained by dielectric measurement. It is important to correct this wrong identification of the JG β-relaxation of sucrose here because of the connection between the JG β-relaxation with local mobility for stabilization of proteins lyophilized by disaccharides and the comparison between sucrose and trehalose systems.

Another quantity that Pikal and coworkers use to correlate kinetic processes in pharmaceuticals and biopharmaceuticals is τ^{β}, where τ is the structural relaxation time determined in the glassy state of the protein formulation and $\beta \equiv (1 - n)$ is the exponent of the Kohlrausch function that fits the time dependence of the relaxation. They used τ^{β} instead of τ for good reasons [1383(i), (j), (k)] and since it is derived from the structural relaxation time, they continue to call it the structural relaxation time and relate it to global mobility [1383(c), (g), (h)]. In several proteins lyophilized by the disaccharides they find good correlation between stability (reciprocal of rate constant k) and τ^{β} as shown in Fig. 274-1b for several disaccharide formulations. Again in this plot of ln$k$ vs. lnτ^{β}, they found that sucrose formulations are more stable, but not expected based upon comparison of τ^{β}. This is because at equivalent stabilizer/hGH ratios, the sucrose systems have much smaller τ^{β} and yet more stable (smaller k) than the other stabilizer systems as can be seen in Fig. 274-1b. Since Pikal and coworkers looked upon τ^{β} as surrogate of the structural relaxation time, the results led them to the conclusion that the sucrose systems have *higher* molecular mobility and yet they are more stable than the other stabilizer systems. This conclusion creates an apparent contradiction with the other conclusion that local mobility determines stability and the superior stability of sucrose formulations is due to them having lower local mobility [1383(g), 1384(a)]. The contradiction can be resolved by noting that τ^{β} is more related to the primitive relaxation time than the structural relaxation time, pointed out before in Section 2.3.2.22 and Eq. (2.82). Identifying τ_{α} with τ here, the CM equation (2.68) can be rewritten in the form $\tau^{\beta} = t_c^{-n}\tau_0 \approx t_c^{-n}\tau_{\mathrm{JG}}$, recognizing that t_c is a constant with value $\approx$2 ps for organic glassformers, and $\beta \equiv (1-n)$. In the papers by Pikal and coworkers, τ was measured in hours. Changing the unit of time from second to hour, the equation can be rewritten as

$$\ln \tau_0 \approx \ln \tau_{JG} = \ln \tau^{\beta} - 35n. \tag{2.100}$$

Since $\ln\tau_0$ or $\ln\tau_{\mathrm{JG}}$ is no different from $\ln\tau^{\beta}$ except for the added constant term, it follows from Fig. 274-1b that stability correlates with $\ln\tau_0$ or $\ln\tau_{\mathrm{JG}}$ and hence local mobility when restricting individually to any of the other disaccharide systems. Stability now correlates with $\ln\tau_0$ or $\ln\tau_{\mathrm{JG}}$ and hence stability also correlates with local mobility; the findings by NMR of Pikal and coworkers and Yoshioka and workers are thus restored.

Furthermore, Eq. (2.100) confers a bonus when comparing stability vs. $\ln\tau_0$ or $\ln\tau_{\mathrm{JG}}$ for all the disaccharides systems in Fig. 274-1b combined. The last term, –35n, in Eq. (2.100) is a considerable correction to $\ln\tau^{\beta}$ that can make the τ_0 or τ_{JG} of the sucrose formulation longer than that of the trehalose formulation, if n is larger or β is smaller for the trehalose formulation than the sucrose formulation. If so, this means that $\ln\tau_0$ or $\ln\tau_{\mathrm{JG}}$ of the sucrose formulation can be *longer* and hence its local mobility *lower* than that of the trehalose formulation, and this brings back consistency with the conclusion based on direct local mobility measurements by NMR spin-lattice relaxation times and neutron scattering of mean-squared displacement. The crux to arrive at this from Eq. (2.100) is based on the assumption that $n(\beta)$ of sucrose formulation is smaller (larger) than $n(\beta)$ of trehalose. This is a possibility in view of the correlation between the fragility index m and n for glassformers of the same family, and the fact that m of pure sucrose is smaller than pure trehalose. Also the ratio $\tau_{\alpha}/\tau_{\mathrm{JG}}$ at T_g is larger for trehalose than sucrose (see Fig. 206 and [1012(c)]). Both characteristics are indicators of larger n for trehalose than sucrose. Notwithstanding, this assumption may or may not hold against rigorous experiment test in protein lyophilized by sucrose and trehalose. The research field of stabilization of biopharmaceuticals can benefit by more experimental investigations using other techniques to gain more information on the dynamics beyond the ones employed. Although the connection I have made of the existing data with primitive or JG β-relaxation is plausible, more experimental confirmation is needed. This is the reason why I put the discussion of the research on stabilization of biopharmaceuticals last in this chapter. Lest one forgets, this is such a very important field for the populace and the economy worldwide and, therefore, the problem of stabilization of biopharmaceuticals deserves much more studies in the future.

New Evidence of Dependence of Stability on β-Relaxation

In the 6th International Discussion Meeting on Relaxation in Complex Systems, August 2009, Rome, Italy (paper under preparation), Cicerone reported empirical evidence that enzyme stability is clearly linked to the β-relaxation in many formulations by presenting data of more than 100 glasses containing wide range of glassformers. The peroxidase (HRP) was type II from horse radish, and the alcohol dehydrogenase (ADH) was from bakers' yeast. The collection of data has the β-relaxation time, τ_β, varying over four decades from 10^{-4} to 10^{-8} s, and Cicerone found strong positive correlation between the logarithm of the enzyme destabilization time, τ_{deact}, and log τ_β. The two are well approximated by linear relation, $\log(\tau_{\text{deact}}/h) \approx \log(\tau_\beta)+\text{const}$. From this relation, Cicerone can predict the absolute stability within a factor of 3 just by measuring the β-relaxation time. More details of this spectacular finding by Cicerone will be published together with the data by him in the near future. I thank M. Cicerone for letting me to show his results in Fig. 274-4 here before submitting to publication in *PNAS*. Further study on this subject is worthwhile to identify the nature of the secondary relaxation considered by Cicerone and coworkers. For the trehalose–glycerol mixture used, it seems that the secondary relaxation considered is the γ-relaxation rather than the JG β-relaxation of trehalose in the mixture for mass fraction of glycerol equal to 0.15, 0.24, and 0.36 [1383(o)].

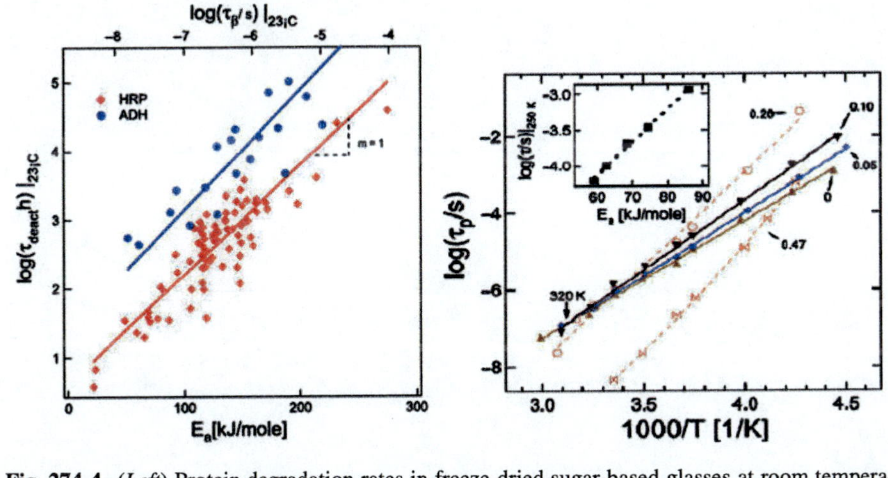

Fig. 274-4 (*Left*) Protein degradation rates in freeze-dried sugar-based glasses at room temperature. E_a is the apparent activation energy for protein degradation process. τ_β is obtained from E_a of the protein degradation process from the mapping between E_a for peak dielectric relaxation and τ_β in the inset to the *right panel*. This mapping is made to test the hypothesis that the protein degradation processes are linked to β-relaxation. The slope of 1 between the ordinate and top abscissa supports this hypothesis. (*Right*) Arrhenius plot of peak dielectric relaxation time. *Inset* shows the plot of log τ at 250 K vs. E_a. Courtesy of Marcus Cicerone

It is worthwhile to mention earlier studies by neutron scattering of fast dynamics [1383(b), 1383(l)] that are consistent with protein stability being linked to the β-relaxation recently found by Cicerone and coworkers. The consistency comes from the fact that shorter the β-relaxation the more intense are the shorter time processes including the caged dynamics, as shown before in Section 2.3.2.35 and [507]. All the observations contradict the traditional point of view that formulations using higher T_g solvents should lead to stronger suppression of protein dynamics and activity [1388, 1383(m), 1383(n)]. A basic assumption underlying the traditional point of view is that the structural α-relaxation of the solvent is the most important process for influencing the protein dynamics. This turns out to be wrong, and instead it is the primitive or JG β-relaxation of the solvent.

A Procedure for the Enhancement in the Stability of Formulation?

Whichever is the sugar used to stabilize biopharmaceuticals including sucrose and trehalose, there is an experimental procedure I can suggest to suppress further the local mobility of the JG β-relaxation, and hence increase the stability of the formulation. The procedure is to apply elevated pressure to the formulation in the equilibrium liquid state first before subsequent cooling to reach the target glassy state, and then pressure is released. This procedure may result in a glassy state denser than that obtained by cooling alone with the effect of increasing τ_0 or τ_{JG} (suppressing the local mobility) and enhancement of the stability of the formulation. Various pressure and temperature paths can be explored with the goal of obtaining the condition for optimal stability.

Chapter 3
Universal Properties of Relaxation and Diffusion in Interacting Complex Systems

3.1 Introduction

This book is principally concerned with the dynamics of glass forming in relation to the glass transition problem. Glass-forming substances and systems are usually constituted of densely packed structural units interacting with anharmonic potentials. Consequently, the view of the author is that relaxation and diffusion of glassformers has to be treated with emphasis on the many-body nature of the irreversible process. At any fixed temperature T and pressure P, the dynamics and properties originating from the many-body relaxation is already a challenging problem to be solved first, before the changes observed on varying T and P are to be accounted for. Examples taken from some of the properties of neat and homogeneous glassformers at fixed T and P discussed in Chapter 2 include the following: (i) the Kohlrausch α-correlation function $\exp\left[-(t/\tau_\alpha)^{1-n}\right]$ for the structural α-relaxation when $\tau_\alpha >> 1\text{ps}$; (ii) the Kohlrausch α-correlation function replaced by the linear exponential function at times shorter than t_c (≈ 2 ps for molecular glassformers, and much longer for colloidal particle suspensions); (iii) the $Q^{-2/(1-n)}$ scattering vector dependence of τ_α; (iv) the appearance of the universal Johari-Goldstein (JG) relaxation strongly connected to the α-relaxation as seen from NMR study and by the ratio of its relaxation time τ_{JG} to τ_α correlated with n and approximately given by the relation $\log(\tau_\alpha/\tau_{JG}) = n(\log\ \tau_\alpha + 11.7)$ for molecular glassformers; and (iv) caged relaxation at shorter times exemplified by the nearly constant loss (NCL) which is terminated at longer times of the order of τ_{JG} by the universal JG relaxation. These properties at fixed T and P, plus more discussed in Chapter 2 apparently all controlled by n, are indications of the manifestation of many-body relaxation, and most challenging to be solved.

Glass transition compounds the difficulty of the problem by requiring further consideration of the change of the many-body relaxation with changes of T and P. The latter bring in changes of their conjugate thermodynamic variables including entropy S and specific volume V. All these changes of T, P, V, and S force modification of the many-body relaxation and spawn more conspicuous properties such as (i) decrease of V (or free volume) and increase of S (or configuration entropy S_c); (ii) VFTH temperature dependence (including its inadequacy) of τ_α; (iii) T_g-scaled

K.L. Ngai, *Relaxation and Diffusion in Complex Systems*, Partially Ordered Systems,
DOI 10.1007/978-1-4419-7649-9_3, © Springer Science+Business Media, LLC 2011

T dependence of τ_α or "fragility," (iv) the breakdown of the Stokes-Einstein (SE) relation and the Debye-Stokes-Einstein (DSE) relation (onset and T dependence), and (v) T dependence of n. These are traditionally the properties of glassformers that researchers have exclusively focused on to construct theories of glass transition such as the free volume theory, the configuration entropy theory of Adam-Gibbs, their generalizations, and some modern approaches. These theories can explain the rapid increase of τ_α on cooling and compressing leading to glass transition but leave out the rich dynamic properties of the α-relaxation untouched, not even those at fixed T and P that are worth consideration (see Sections 2.2.1–2.2.5 on α-relaxation). The multiple subsections in Sections 2.3 have presented many different experimental facts demonstrating that the JG β-relaxation is universal, fundamental, and inseparable from any consideration of the structural α-relaxation. None of the traditional theories of glass transition have considered the JG β-relaxation and its connection to the α-relaxation. Neither are the more recent theories including those bringing into consideration the dynamic heterogeneous nature of the α-relaxation from the breakdown of SE and DSE relations, nor are the theory focusing on the fast relaxation while molecules are still caged. Here, I remind the readers of one of the many examples of the connection between the JG β-relaxation and the α-relaxation discussed in Section 2.3. This example is the invariance of the α-dispersion (or n) as well as the ratio τ_α/τ_{JG} to various combinations of T and P while keeping τ_α constant. Also, both τ_α and τ_{JG} are functions of the *same* product variable TV^γ. These new experimental findings provide further evidence for the many-body nature of the α-relaxation and the universal JG β-relaxation is its precursor and originator. However, none of the past and recent theories mentioned above has provided explanation for these new and important findings. As far as the author knows, the coupling model (CM) is the only model having the equivalent of the JG β-relaxation in the primitive relaxation, the device via the CM equation (2.68) to capture and explain many of the effects of many-body relaxation on the structural α-relaxation, and the capability of addressing all the dynamic processes in order of increasing time from caged relaxation (NCL), JG secondary to the structural α-relaxation, and their inter-relations.

If the above discussion restricted to glass-forming materials is not sufficient to support the call for incorporating many-body relaxation into the theory of glass transition, I shall bring in some dynamic properties of other interacting systems which are not glass forming or in which the interest of the process in the system is not in glass transition per se. If the physics of many-body relaxation and diffusion is fundamental, the properties of the dynamics manifested will show up analogously in these various non-glass-forming interacting systems. It is therefore a worthwhile undertaking to broaden the study of glass-forming materials by bringing in these other processes in different systems. For the past three decades, the author has actively engaged in the study of relaxation and diffusion in several fields which are unrelated to the glass transition problem. This is partly motivated by the CM, which is supposedly applicable to relaxation in many-body interacting systems in general. But, the higher motivation is to explore the possibility that *relaxation in interacting many-body systems exhibits universal properties.* If indeed the properties of many-body relaxation are universal, it would be an important discovery that has immediate impact on various scientific disciplines. To accomplish this goal, we can take advantage of the properties of the α-relaxation and JG β-relaxation of glass-formers matching the concepts and predictions of CM discussed in Chapter 2 by using them as guides to look for parallels in other interacting many-body systems.

First is the crossover of the correlation function from linear exponential relaxation $\exp(-t/\tau_0)$ to Kohlrausch stretched exponential relaxation $\exp\left[-(t/\tau)^{1-n}\right]$ in a neighborhood of some time t_c. Second is to find the equivalent of the primitive relaxation (or the analogue of the JG relaxation in glassformers) either directly or by identification with a feature in the experiment data or molecular dynamics simulation. Third is to validate the relation between the primitive relaxation time τ_0 and the observed many-body relaxation time τ given by the equation

$$\tau = [t_c^{-n}\tau_0]^{\frac{1}{1-n}}, \tag{3.1}$$

and the relation becomes quantitative if t_c is known. Even without knowing the exact value of t_c, the predicted dependence of τ on various variables U given by

$$\tau(U) \propto [\tau_0(U)]^{1/(1-n)} \tag{3.2}$$

can be tested against experiment. In some cases, dependences of τ on more than one variable U are predicted by the same n, thus making the test by experiment even more stringent and satisfying. The dependence of τ_0 on U is not anomalous because it is the relaxation time of simple (not many-body) relaxation and either is known from basic theory or can be independently determined by experiment or simulations in situations where many-body effects are eliminated. If successful, Eq. (3.2) serves to explain the observed dependence of τ on U, which is often found to be anomalous in other interacting systems similar to those we have seen in glassformers in Chapter 2. The success can be viewed as evidence of a strong connection between τ and τ_0, like that found between τ_α and τ_{JG} or τ_0 (since $\tau_{JG} \approx \tau_0$) in glassformers.

Because the anomalous dependence of $\tau(U)$ comes from raising $\tau_0(U)$ to the superlinear power $1/(1-n)$ in Eqs. (3.1) and (3.2), the degree of the abnormality increases with increasing n. This offers multiple checks of Eqs. (3.1) and (3.2) by experiment by increasing or decreasing the interaction even in the same system. Depending on the system, it can be accomplished in different ways such as increasing or decreasing concentration, temperature, association, chemical cross-links, changing the other component in binary mixtures, and densification.

Indeed, my collaborators and I as well as other researchers independently have found anomalous dynamic properties in interacting systems which are analogues to those of glassformers discussed in Chapter 2. In particular, the dispersion of the relaxation or the diffusion governs or correlates with the anomalous properties. The collection of experimental data in various systems of different fields shows analogous dynamic properties and hence is a strong evidence supporting the presence of the many-body relaxation in interacting systems and the dynamics manifested is universal and fundamental. Moreover, these analogous anomalous properties can be explained by the CM.

Complete and detailed description of this large collection of data cannot be made in this book because this would make the size of this book unduly large. Therefore selected sets of data are presented, and most of them are just briefly described and the corresponding explanations by the CM given here are also abbreviated. References are given for the interested reader to look for the details. A few sets of data have been presented before in Chapter 2 for specific purpose, not the same as in this chapter, which is to demonstrate the universal relaxation properties of different interacting systems.

Attempt is made to present the similar properties in other systems as much as possible in about the same order as introduced in Chapter 2 for the relaxation processes in glass-forming materials and systems. This way makes it easier for the reader to find and collate the properties to correspond to the glass-forming systems, all of which are the vestige of many-body dynamics in relaxation of interacting systems with glass-forming substances as a special class.

3.2 Universal Properties

3.2.1 The Kohlrausch Stretched Exponential Correlation Function $\exp[-(t/\tau)^{1-n}]$

Materials of interest in research on relaxation and diffusion these days vary greatly in chemical structure and physical type. However, many of them share one common characteristic and that is the relaxing or diffusing units have mutual interactions with each other. The simplest class consists of relaxation and diffusion processes in bulk materials that involve identical interacting units with no other complications, such as identical molecules in glass-forming viscous liquid, colloidal particles nearly all of the same size in colloidal suspensions [141–143], identical mobile ions in crystalline or polycrystalline [283, 1060, 1482–1486], vitreous, or molten ionic conductors [54–56, 115(b), 147, 232, 279–295, 1056–1059], and a monodisperse entangled polymer melts where all polymer chains have the same repeat unit and are of the same length [359, 360, 837]. In these ideal cases, the time correlation function of the many-body relaxation is well described by the Kohlrausch function as in the case of bulk and neat glassformers. Liquid or glassy state is not required to have the Kohlrausch correlation function as exemplified by the Li ion motion in polycrystalline $LiAlSi_2O_6$ (β-spodumene) [283, 1485, 1060] and in the crystalline ion conductors such as Na β-alumina [1486], $Li_{0.18}La_{0.61}TiO_3$ (LLTO) [1483], and yttria-stabilized zirconia (YSZ), an oxygen ion conductor of fluorite structure used commercially as electrolytes in solid oxide fuel cells [1482, 1484]. These crystalline cases elucidate that ion-ion interaction rather than randomness is principally responsible for the slowing and stretching of the relaxation.

However, one must be aware that some materials and systems have built-in disorder, randomness, or fluctuation. Examples include concentration fluctuations in miscible blends of two different liquids or polymers, random copolymers, ionic conductors containing two different kinds of mobile ions, and polydisperse polymer melts with a broad distribution of chain lengths. Spatial heterogeneity of various kinds includes liquid or polymer nanoconfined by wall, free-standing, or supported ultrathin polymer films, repeat units of a cross-linked (semicrystalline) polymer located at different distances from the junction points (the crystalline-amorphous boundaries), and density fluctuations present in the non-equilibrium glassy state. In these more complicated cases, the frequency dispersion of the observed relaxation is broadened by these factors in addition to many-body relaxation originating from interactions. The resulting dispersions often deviate from those of the Kohlrausch function. In the event when it can be forced fitted by a Kohlrausch function, the value of n appearing in the exponent $\beta \equiv (1-n)$ of the Kohlrausch function does not reflect that coming exclusively from many-body relaxation and should not be used in Eqs. (3.1) and (3.2) to make predictions. Bearing this caution in mind, we can also consider the more complicated materials where there is mutual interaction between the units. We are not interested in relaxation of systems where the dispersions are solely due to disorder, randomness, or fluctuation, because no new physics is involved.

3.2.1.1 Mean-Square Displacement of Diffusion in Interacting Systems

Glass-Forming Systems

For systems exhibiting diffusion in space, many-body effects are more explicitly shown by the mean-square displacement (MSD) $<r^2(t)>$ as a function of time. Normal diffusion will have $<r^2(t)>= 6Dt$ for all times, a result that follows from exponential correlation function for diffusion or random walk. The actual $<r^2(t)>$ of interacting systems deviates from this simple time dependence. Previously in Section 2.2.1.1, the subdiffusion having a fractional power law dependence of $<r^2(t)>\propto t^{1-n}$ has been shown to be present at times before $<r^2(t)>$ crosses over to the terminal steady-state diffusion at long times and the dependence $<r^2(t)>=6Dt$ takes over. Such is in the case of the results of solvent in polymer solutions from dynamic lattice liquid simulation [327, 328] (see Fig. 7) and also of the dynamic light-scattering data of colloidal particle suspensions with $\phi = 0.465$ (see Fig. 9). Preceding the subdiffusion is the primitive diffusion with $<r^2(t)>= 6D_0t$ in these two systems. From the transition from primitive to subdiffusion, one can determine the crossover time t_c of the coupling model (CM). In terms of correlation times τ and τ_0 and the elementary diffusion length R, we have $D = R^2/\tau$ and $D_0 = R^2/\tau_0$, respectively. The CM equations (2.12) or (2.14) between τ and τ_0 provide a relation between D and D_0.

Here we add the results from the molecular dynamics simulation of glass-forming binary Lennard-Jones particles by Donati et al. [321]. The $<r^2(t)>$ data from this paper are shown in Fig. 275. At the highest temperature of 0.5495 as well as the lowest temperature of 0.4510 (in L-J unit), subdiffusion having the fractional power law dependence of $<r^2(t)>\propto t^{1-n}$ with $(1-n) = 0.77$ appears at earlier times before the terminal free diffusion with $<r^2(t)>= 6Dt$ takes over. The onset time of

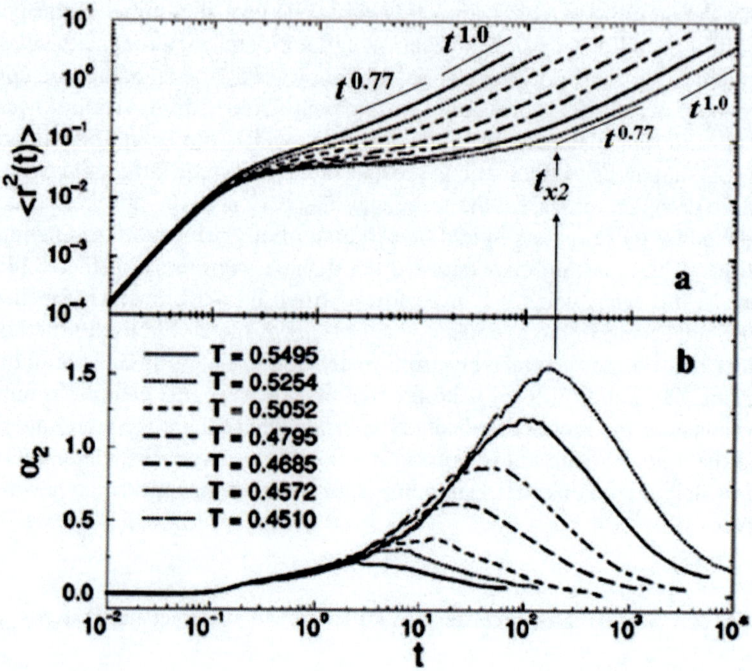

Fig. 275 (**a**) Mean-square displacement $< r^2(t) >$ of the A particles of binary L-J liquid vs. time for several values of T. (**b**) Non-Gaussian parameter $\alpha_2(t)$ vs. time for the same values of T as in (**a**). From Donati et al. [321], and reproduced by permission

the subdiffusion is denoted by t_{x2} and its location for $T = 0.451$ is indicated approximately by the arrow in Fig. 275. At lower temperatures, the particles are initially caged, and the caged dynamics contributes a power law $< r^2(t) > \propto t^\gamma$ to the MSD with $\gamma << 1$ or a nearly constant loss (NCL) with frequency dependence $\omega^{-\gamma}$ to the susceptibility $\chi''(\omega)$. In order of time, the caged dynamics term $< r^2(t) > \propto t^\gamma$ is followed by the subdiffusion term $< r^2(t) > \propto t^{1-n}$ and in turn by the terminal diffusion term $< r^2(t) >= 6Dt$. The value of the fractional power (1-n) of the subdiffusion is comparable to the stretch exponent β in the Kohlrausch function used to fit the intermediate scattering function $F_s(q,t)$ for $q = q_{max}$ [321]. On decreasing temperature, the term $< r^2(t) > \propto t^\gamma$ persists to longer times, accompanied by decrease in both $< r^2(t) >$ and the exponent γ reflecting increasing degree of caging. At $T = 0.451$, γ is about 0.1. This description of dynamics from the simulations is along the line as in Chapter 2. It needs to be pointed out that the same results from simulations of the binary LJ mixtures were interpreted by others [265, 321] in terms of the idealized mode coupling theory (MCT). In fact these results were considered as verification of the predictions of the original version of MCT, although it cannot explain the dynamic lattice liquid simulation results and the dynamic light-scattering data of colloidal particle suspensions with $\phi = 0.465$.

For diffusion in d-dimension, $F_s(q,t)$ was sometimes approximated by

$$F_s(q,t) \approx \exp\left[-\frac{q^2}{2}\langle r^2(t)\rangle/2d + \frac{q^4}{2}\left(\frac{\langle r^2(t)\rangle}{2d}\right)^2 \alpha_2(t)\right], \tag{3.3}$$

where the non-Gaussian parameter $\alpha_2(t)$, given before by Eq. (1.125), is a measure of the deviation from the Gaussian dynamics, because it vanishes when the dynamics is Gaussian. Shown in Fig. 275(b) is $\alpha_2(t)$ obtained by Donati et al. from their simulations of binary L-J mixture. It exhibits a peak and the maximum of $\alpha_2(t)$ occurs close to t_{x2}, the onset time of subdiffusion, which is demonstrated for $T = 0.451$ by the vertical arrow in Fig. 275. We have seen before from several examples in Section 2.3.2.33 that t_{x2} is a few times longer than the primitive relaxation time τ_0. This, together with the location of t_{x2} for $T = 0.451$, suggests possibly that $\tau_0 \approx 100$ (L-J units). From the discussion in Section 2.2.1.1, t_c of the binary L-J particles is about 1. Donati et al. fitted $F_s(q_{max}, t)$ to the Kohlrausch function having stretched exponent $(1-n) = 0.75$ and $\tau = 655$. The approximate values of these parameters are consistently related by the CM equation (2.14) $\tau = \left[(t_c)^{-n}\tau_0\right]^{1/(1-n)}$.

The MSD $< \Delta x^2(\Delta t) >$ and the non-Gaussian parameter $\alpha_2(\Delta t)$ of colloidal particle suspensions determined by confocal microscopy at various volume fractions ϕ [141] show the same behavior as the L-J system. These quantities for $\phi = 0.56$ have been shown before in Fig. 247 in connection with the nearly constant loss (i.e., the initial plateau in the MSD) from caged dynamics and the primitive relaxation time indicated by Δt when a typical particle changes position and leaves the cage, which is about 500 s for $\phi = 0.56$. The subdiffusion with $< \Delta x^2(\Delta t) > \propto (\Delta t)^{1-n}$ and its onset at t_{x2} are illustrated in the figure. Also shown is the initial plateau having $< \Delta x^2(\Delta t) > \propto (\Delta t)^{0.13}$, contributed by motions of particles trapped in anharmonic cages, which gives rise to the susceptibility $\chi''(\omega)$ having the frequency dependence of $\omega^{-0.13}$ (i.e., the NCL). For the sake of easy comparison of binary L-J mixtures and Li^+ ions in Li_2SiO_3 with the MSD at different temperatures to be discussed below, the $< \Delta x^2(\Delta t) >$ for $\phi = 0.46$ and 0.52 are shown together with $\phi = 0.56$ in Fig. 276. Returning to Fig. 247, the inset therein presents the time dependence of the non-Gaussian parameter $\alpha_2(\Delta t)$, which also peaks at t_{x2} as in the case of the binary L-J mixture.

Ion Dynamics in Glassy Ionic Conductor

All the examples given above are glass-forming systems, and the similarity of the MSD of some of them may not be surprising. Now we present the MSD of other interacting systems unrelated to glass transition, and the remarkable thing is that they are also similar. The similarity has repercussion on theoretical interpretation of the dynamics in some of the systems, which will become clear later.

Historically, Kohlrausch studied the electrical relaxation of a Leyden jar (an alkali silicate glassy ionic conductor) in 1854 at Göttingen [2, 3] in which he found that the time decay of ionic conductivity relaxation has the stretched exponential

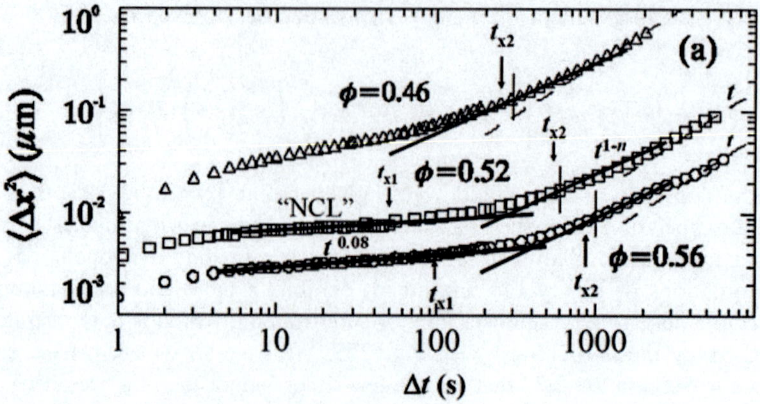

Fig. 276 Mean-square displacements of colloidal particles in suspension from Weeks and Weitz [800]. The *diagonal dashed straight lines* indicate estimates for $<\Delta x^2(\Delta t)>\sim 2Dt$. The *solid lines* indicate $<\Delta x^2(\Delta t)>\propto (\Delta t)^{1-n}$ starting at about t_{x2}. The NCL regime is depicted by the nearly flat line representing $<\Delta x^2(\Delta t)>\propto (\Delta t)^c$ with c small in the range from 0.08 to 0.13. This power law terminates at t_{x1}. *Vertical bars* indicate estimate of the cage rearrangement timescale Δt^* by finding the maximum of the non-Gaussian parameter. The proximity of Δt^* to t_{x2} is clear from the figure. The primitive or independent relaxation time τ_0 calculated by the CM equation is near t_{x2} [1058]

time dependence given by Eq. (1.1). If scientific activity in ionic conductivity relaxation and diffusive transport in glasses is considered to have started by Kohlrausch, then by now the research field is about 160 years old. Many research efforts in this area since then will not be mentioned here, except those having results on the ion dynamics that brings out the universal dynamic properties. The first example is the dynamics of mobile ions in glassy ionic conductors. Habasaki and coworkers have carried out molecular dynamics simulations of Li ion dynamics in the metasilicate glass Li_2SiO_3 since 1997, and many properties have been uncovered [468–472, 475, 802, 1058]. The $<r^2(t)>$ and $\alpha_2(t)$ of Li ion motions at different temperatures is shown in Fig. 277.

There is hardly any difference in their behaviors as functions of time and change with temperature as the binary L-J mixture and the colloidal particles (on changing volume fraction). In particular, prior to the free diffusion is the presence of the subdiffusion term $<r^2(t)>\propto t^{1-n}$ which is preceded by the caged dynamics term $<r^2(t)>\propto t^c$. The data at 700 K in the main figure show the subdiffusion $<r^2(t)>\propto t^{0.64}$ and its onset time at t_{x2} in between 20 and 30 ps. This is about the same time at which $\alpha_2(t)$ at 700 K peaks as indicated in the figure. The same is true for other temperatures. The exponent c of caged dynamics term $< r^2(t) >\propto t^c$ is 0.11 or 0.13, which is about the same as that of the binary L-J mixture (Fig. 275) and the colloidal particles with $\phi = 0.56$ (Fig. 247 and 276).

The self-part of the van Hove function, discussed in Chapter 1 and defined before in Eq. (1.112) by $G_s(\vec{r},t) = (1/N)\sum_{i=1}^{N} \langle \delta(\vec{r}_i(t) - \vec{r}_i(0) - \vec{r})\rangle$, of the Li ions obtained

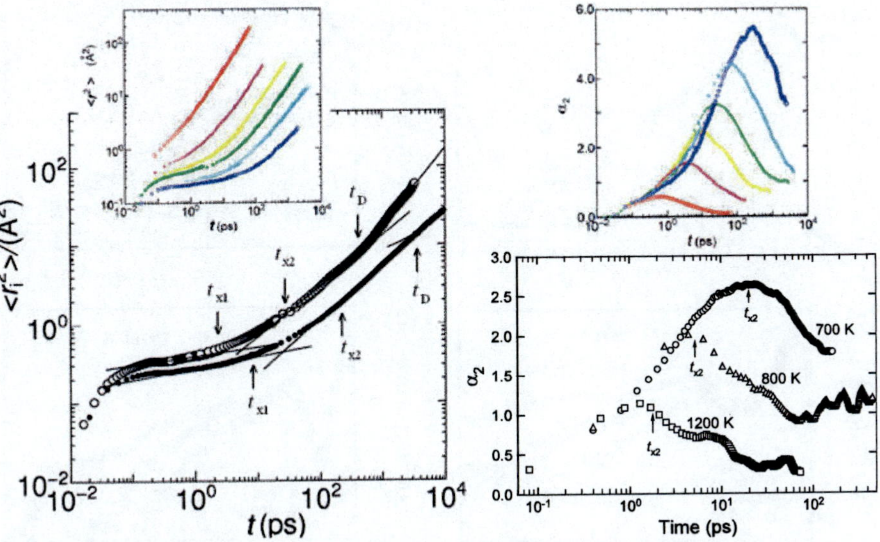

Fig. 277 (*Left*) MSD of Li^+ ions in Li_2SiO_3 at 600 K (*lower*) and 700 K (*upper*). The t_{x1}, t_{x2}, and t_D are crossover times that separate out the four time regimes explained in the text. $t_{x1} \approx 2$ ps at 700 K and $t_{x1} \approx 10$ ps at 600 K. The *inset* shows the MSD of Li ions at 1673, 1000, 800, 700, 600, and 500 K from top to bottom. (*Right*) The non-Gaussian parameter $a_2(t)$ for Li^+ ions in Li_2SiO_3 calculated from their time-dependent displacement distribution function at 700, 800, and 1200 K (*lower panel*) and at 1673, 1000, 800, 700, 600, and 500 K from *bottom* to *top* (*upper panel*). At higher temperatures, the maximum of $a_2(t)$ for the Li^+ ions moves to shorter times and at all temperatures the maximum is located near t_{x2}

by the molecular dynamic simulations [1487–1489] is shown for different times within four time regimes in Figs. 278(a)–(d). The times in the four time regimes are (1) 0.40, 0.8, and 1.6 ps ($t < t_{x1}$); (2) 3.2, 6.4, 12.8, and 25.6 ps ($t_{x1} < t \approx t_{x2}$); (3) 40, 80, 160, and 200 ps ($t_{x2} < t < \tau$); and (4) 400, 800, and 1600 ps ($t > \tau$). Here $\tau \approx 229$ ps is the stretched exponential relaxation time of the incoherent scattering function $F_s(k, t)$ for $k = 2\pi/3\,(\text{Å}^{-1})$. Such k is chosen because it corresponds to $\sqrt{\langle r^2 \rangle} \approx 3$ Å, the distance between Li sites. This together with $(1 - n) \equiv \beta = 0.64$, $t_c = 1$ ps, and the CM equation (3.1) enables us to calculate τ_0 and the result is 32 ps, which is an order of magnitude longer than t_{x1} but is nearly the same as $t_{x2} \approx 27$ ps. The onset of terminal diffusion with $< r^2(t) > \propto t^{1.0}$ starts at $t_D \approx 400$ ps, which is slightly longer than $\tau \approx 229$ ps.

$G_s(r, t)$ of Li^+ ions has similar dependences on space and time as the same function of the binary L-J mixture [321] and the analogous distribution function $P(\Delta x)$ of colloidal particles [141]. Furthermore, the Li ion dynamics is heterogeneous with the presence of faster and slower ions, like that found in the other two systems as well as in glass-forming liquids.

In regime (1), $t < t_{x1}$, in the framework of the CM, the fact that τ_0 is significantly longer than t_{x1} explains the small probability of independent jump of the ions out

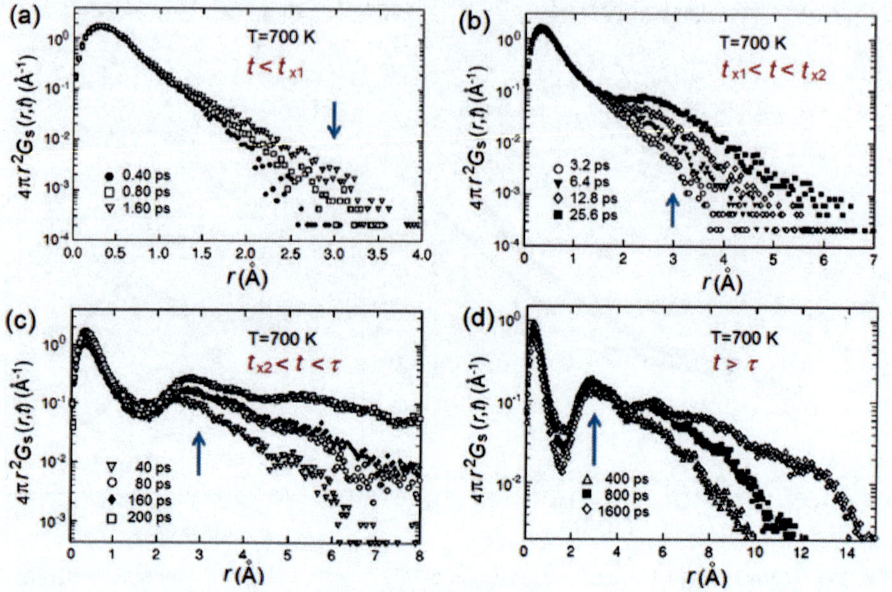

Fig. 278 $4\pi r^2 G_s(r,t)$ of Li^+ ions in Li_2SiO_3 at 700 K as a function of r at different times. $G_s(r,t)$ is the self-part of the van Hove function for the Li ion. (**a**) 0.40, 0.8, and 1.6 ps ($t < t_{x1}$). (**b**) 3.2, 6.4, 12.8, and 25.6 ps $t_{x1} < t \approx t_{x2}$. (**c**) 40, 80, 160, and 200 ps ($t_{x2} < t < \tau$); $\tau \approx 229$ ps is calculated (see text). (**d**) 400, 800, and 1600 ps. The *vertical arrows* indicate the distance of 3 Å, roughly the separation between neighboring Li^+ sites. Data from [468] are replotted together with additional illustrations in all figures

of their cages during the caged ion regime. It also explains the very slow decay of the cage indicated by either the plateau-like MSD or practically no decrease of the normalized area, $A_1(t)$, of the first peak of the self-part of the Li ion van Hove function, used to gauge the cage decay (see left panel of Fig. 279). $A_1(t)$, obtained by integrating $4\pi r^2 G_s(r,t)$ over the first peak domain of r <2 Å, can be considered as the analogue of the cage decay function of Weeks and Weitz for colloidal particles from confocal microscopy shown in Fig. 247. Alternatively, caged dynamics can be seen from $A_2(t)$ obtained by integrating $4\pi r^2 G_s(r,t)$ over the first neighboring shell region with r between $r_1 = 2$ and $r_2 = 4$ Å. $A_2(t)$ remains very small for $t < t_{x1}$, indicating negligible number of ions hopping over to the neighboring sites.

In the time regime (2), $t_{x1} < t < t_{x2}$, there is non-negligible probability for independent ion jumps to neighboring sites, particularly when time increases towards t_{x2} because t_{x2} is comparable to τ_0. This can be seen by the development of a shoulder and then a peak in $4\pi r^2 G_s(r,t)$ as t increases from 3.2 up to 25.6 ps. Therefore the cages start to decay in regime (2), a property corroborated by the departure of the MSD from the relation $< r^2(t) > \propto t^{0.11}$ in Fig. 277 (left). It is also corroborated by the initially small decrease and increase of the normalized area $A_1(t)$ and $A_2(t)$, respectively, in the regime $t_{x1} < t < t_{x2}$.

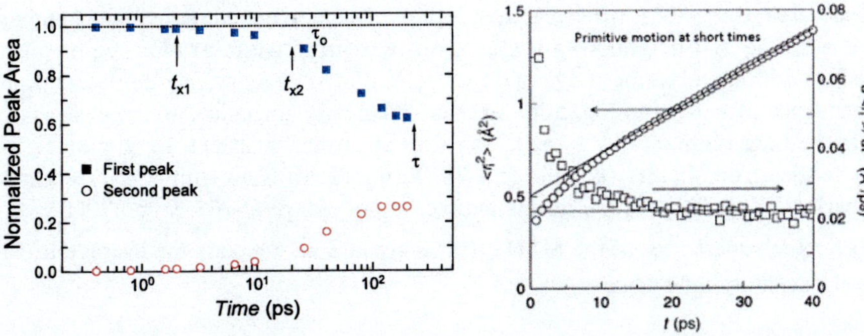

Fig. 279 (*Left*) Normalized areas of the first peak $A_1(t)$ and the second peak $A_2(t)$ of the self-part of the Li ion van Hove functions at 700 K at different times shown up to 200 ps. The positions of t_{x1}, t_{x2}, τ_o, and τ are indicated by the *vertical arrows*. Data from [468] replotted. (*Right*) Plot of $< r^2 >$ at 700 K against time, showing the linear relation within the approximate range of $15 < t < 40$ ps, which is the primitive relaxation regime. Plot of $d < r^2 > /dt$ at 700 K against time, showing that $d < r^2 > /dt$ is nearly constant within the range 15<t<40 ps. Data from [802] replotted together with illustration in the figure

In the time regime (3), $t_{x2} < t < \tau$, t is longer than τ_0 for practically all times. Therefore, the Li ions are capable of executing the independent jump having the primitive relaxation time τ_0 with significant probability. But, instead of independent jumps with no correlation with each other, they participate in some cooperative or correlated hopping process. This is evidenced by the observed stretched exponential correlation function for the intermediate scattering function $F_s(k,t)$ in regime (3), instead of linear exponential decay for independent hops [1488]. This is like $F_s(k,t)$ found by dynamic light scattering in colloidal particle suspension by Segre and Pusey [143]. Thus t_{x2} marks the change from apparent free jump to the cooperative/correlated jump process. Again this description is corroborated by even more rapid decrease of A_1 in regime (3) than in regime (2), as can be seen by the steeper rise of the MSD with the power law dependence of $< r^2(t) > \propto t^{1-n}$, with $(1-n) \equiv \beta \approx 0.64$ in Fig. 277 (left), and population in the second peak of $4\pi r^2 G_s(r,t)$ given by A_2. This is exactly like the colloidal supercooled liquid with $\phi = 0.56$ from the data of confocal microscopy shown in Figs. 247 and 276 in the corresponding time regime $t_{x1} < t < t_{x2}$, and likewise the cage correlation function of Weeks and Weitz (see inset in Fig. 247) show much faster decay when t>t_{x2} than in regime 2.

In the right panel of Fig. 279 is the linear plot of $< r^2(t) >$ against linear time in the abridged time region, $15 < t < 40$ ps, within which lies the calculated primitive relaxation time τ_0. The plot shows the linear relation $< r^2(t) > \propto t^{1.0}$ within the approximate range of $15 < t < 40$ ps, supported by d $< r^2 > /dt$ being nearly constant within the range $15 < t < 40$ ps. This is evidence for the existence of the primitive relaxation because it gives rise to independent ion jumps, and their contribution to the MSD should be linearly proportional to time. Existence of the primitive relaxation is additionally supported by the time development of A_2, the area of the second peak of $4\pi r^2 G_s(r,t)$. In the 15–40 ps region, $A_2(t)$ increases linearly with time [802], where the area is proportional to net increase in the number of ions in the first neighboring shell region. The primitive relaxation regime also exists at 600 K within the range of $150 < t < 250$ ps as shown by the linear time dependence of $< r^2 >$ in Fig. 5 in [802] and the nearly constant d $< r^2 > /dt$.

In the time regime (4), $t > \tau$, Li ion sites at further distances are populated as can be seen from $4\pi r^2 G_s(r,t)$ in Fig. 278. Terminal diffusion with $< r^2(t) > \propto t^{1.0}$ is reached. At the onset time of this regime (about 600 ps at 700 K and 2000 ps at 600 K), the square root of the MSD $\sqrt{\langle r^2 \rangle}$ is about 3 Å, which is the average Li-Li ion site separation distance.

3.2.1.2 Space-Time Pictures of Motions of Li+ Ions Equivalent to Those of Motions of Colloidal Particles by Confocal Microscopy

Habasaki and I [802] have generated space-time pictures to elucidate the motion of Li ions in Li_2SiO_3 at 700 K at different times, which is equivalent to the real-time pictures of motion of colloidal particles provided by confocal microscopy [141, 800]. These pictures of Li ions at progressively longer times of 4, 8, 16, 24, 48, 200, 400, and 1000 ps give clear information of the motions of Li ions in the different dynamic regimes.

Before presenting these space-time pictures, let us remind the reader that from the time dependences of $< r^2(t) >$ and $G_s(r,t)$, several dynamic regimes of Li^+ ions at 700 K have been delineated by the three times, t_{x1}, t_{x2}, and t_D shown in Fig. 277. The space-time pictures of the motion of Li^+ ions are discussed in the regimes defined by these three times.

Figure 280(a)-(d) shows the displacements of the Li ions at four chosen times 4, 8, 16, and 24 ps, respectively, at 700 K. The displacements of the Li ions are indicated by the vectors originating from the positions of the ions at an arbitrary chosen initial time t_0 in three dimensions for a part of the basic cell of the simulation. The amplitudes of the vectors are coded according to the color scheme shown. These results are supposed to represent the behavior of the ions within the time regime (2) bounded approximately by $t_{x1} \sim 2\,\text{ps} < t < t_{x2} \sim 30\,\text{s}$. According to the color code of these figures, there are few isolated jumps to nearest neighboring Li sites with distance of about 3 Å, and the number of such jumps increases with time. These isolated jumps are the primitive relaxation of the CM that we are looking for and have been identified before in the right panel of Fig. 279 by their contribution to the MSD that is linearly proportional to time within the approximate range $15 < t < 40$ ps, within which lies $\tau_0 \approx 32$ ps.

The primitive relaxation observed in MSD with $< r^2(t) > \propto t^{1.0}$ in the range of $15 < t < 40$ ps ends at longer times past τ_0 or t_{x2} when more ions jump to neighboring sites and these jumps are no longer isolated events. This can be seen starting at 48 ps in the panel (a) of Fig. 280 (lower four). With further increase in time, more ions have displacements equal to or larger than the distance between neighboring Li sites. The ions undergoing these large amplitude motions form clusters and the vectors of their motions tend to be linked together, and the length scale of the vectors linked together increases with time. Conversely the ions limited to local motions

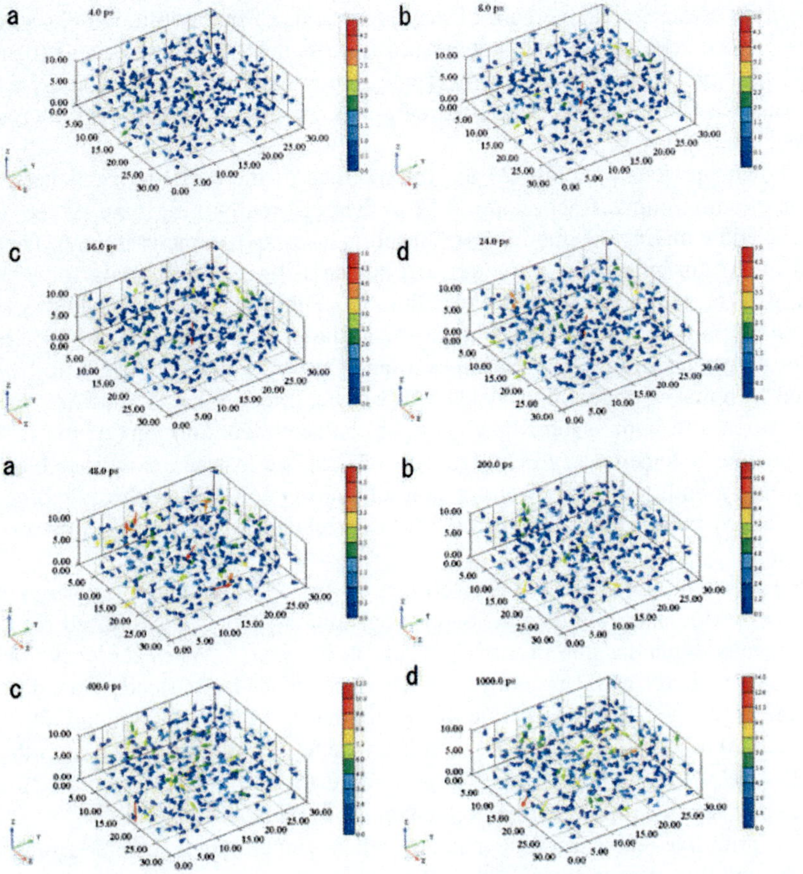

Fig. 280 (*Upper four*) Motion of Li ions in Li_2SiO_3 at 700 K at four different times: (**a**) 4 ps, (**b**) 8 ps, (**c**) 16 ps, and (**d**) 24 ps. The positions of the Li ions at any of the indicated chosen times are represented by the vectors originating from the positions of the ions at an initial time in three dimensions for a part of the basic cell of the simulation. The values of axes are in Å. The *colors* are used to indicate the lengths of the vectors (the values shown in the legend are also in Å). Note that the color codes of the scales for 4, 8, 16, and 24 ps are the same. The rare maximum displacements of the Li ions are 2.5, 3.5, 3.5, and 4.0 Å for 4, 8, 16, and 24 ps, respectively. But note that most of the displacements have magnitudes smaller than the maximum value. (*Lower four*) Motion of Li ions in Li_2SiO_3 at 700 K at four times: (**a**) 48 ps, (**b**) 200 ps, (**c**) 400 ps, and (**d**) 1000 ps. The positions of the Li ions at any of the indicated chosen times are represented by the vectors originating from the positions of the ions at an initial time in three dimensions for a part of the basic cell of the simulation. The values of axes are in Å. The *colors* are used to indicate the lengths of the vectors (the values shown in the legend are also in Å). Note that the code of the color scales for 48 ps is different from that of 200 and 400 ps, and from that of 1000 ps. The maximum displacements of the Li ions are 4.5, 7.8, 9.0, and 11.8 Å for 48, 200, 400, and 1000 ps, respectively. But note that most of the displacements have magnitudes smaller than the maximum value. Reproduced from [802] with permission

also form clusters. The increase of the length scale of the mobile ion clusters with time can be seen in going in succession from panels (a), (b), (c), and (d) at 48, 200, 400, and 1000 ps respectively. These features of Li^+ ions motion are exactly the same as those found in colloidal particle suspensions by confocal microscopy [141].

Shown previously in Fig. 23 are the motions of the colloidal particles in suspension with volume fraction $\phi = 0.52$ at the cage rearranging time $\Delta t^* = 600$ s, estimated by finding the maximum of the non-Gaussian parameter $\alpha_2(\Delta t)$. The location of Δt^* is indicated by the vertical bar crossing the MSD data for $\phi = 0.52$ in Fig. 276, and it can be seen from this figure that $\Delta t^* \approx t_{x2}$. Therefore, since $t_{x2} \approx 30$ ps from the Li metasilicate glass, the pattern of motion of Li^+ ions at 24 or 48 ps shown in Fig. 280 (upper d) and (lower a), respectively, is to be compared with that of colloidal particles with volume fraction $\phi = 0.52$ at $\Delta t^* = 600$ s. The patterns in both systems have similar character. The motions of the Li^+ ions are obviously dynamically heterogeneous. There are fast and slow ions classified as such by both temporal (waiting time of jumps) and spatial (back-and/or-forth correlation) behaviors [1488, 1489]. Back-correlated jumps return to the original positions.

(4) At even longer times, the two types of ions, fast and slow, exchange roles. This feature is also found in glass-forming liquids [226–228], referred to as heterogeneous dynamics and discussed before in Chapter 1. When these motions are coarse grained, the MSD assumes the linear dependence t, and steady-state diffusion regime is reached. The onset time of this regime t_D is about 600 ps at 700 K and about 2000 ps at 600 K. As can be seen from the left panel of Fig. 277, at t_D the root MSD $\sqrt{\langle r^2 \rangle}$ is about 3 Å, which is the average Li-Li ion site separation distance, although even in a diffusive regime, dynamics is still heterogeneous.

The primitive relaxation time increases when temperature is lowered from 700 to 600 K, and this change explains the lengthening of the time/frequency range of the NCL, as observed by experiment. The primitive relaxation also marks the onset of many-molecule dynamics. As time increases, increasing number of ions participate in cooperative motion until the maximum number, dictated by the ion-ion interaction and the matrix, is reached and steady-state diffusion begins.

In total, the processes found in the mean-square displacement of Li^+ ions in Li metasilicate include the caged dynamics, the primitive relaxation, the heterogeneous jump dynamics, and steady-state diffusion. The primitive relaxation terminates the caged ion dynamics, and it also marks the onset of many-molecule dynamics. These processes are isomorphic to those found in molecular glass-forming liquids and colloidal particles. Previously in ionic conductors the isomorphism was incomplete because of the absence of direct experimental evidence for the primitive relaxation, whereas in molecular glassformers, the evidence is provided by the universal Johari-Goldstein relaxation. By finding the primitive relaxation by molecular dynamics simulation presented in this study, we show that the underlying physics of the dynamics of ionic systems is exactly parallel to that of molecular glass-forming liquids, in spite of the fact that the Li^+ ions are moving in the metasilicate glassy matrix and have nothing in common with supercooled liquids.

From the above discussion, it is clear that time dependences of the MSD, the non-Gaussian parameter, and the heterogeneous dynamics of Li ions in Li_2SiO_3 are isomorphic to those of the binary L-J mixture and the colloidal particle suspension. The isomorphism indicates that the dynamics of diffusion in interacting systems has universal properties and suggests a common explanation for all. In the next subsection to follow, conductivity relaxation data of various crystalline and glassy ionic conductors [195, 1057, 1058] are introduced to support this isomorphism. Details of the experimental data are presented later in Sections 3.2.6.3, 3.2.6.4, and 3.2.7. The implication of universal properties found entirely in different classes of interacting systems poses problem for theory/model that is applicable to explain one system but not the others. Such is the case of MCT designed specifically for glass-forming liquids. It is a fluid- and density-based theory and does not seem natural to apply to dynamics of glassy ionic conductors such as lithium metasilicate glass where the Li ions are diffusing into the glassy matrix [1057, 1058]. Crystalline ionic conductors such as Na β-alumina [1486], $Li_{0.18}La_{0.61}TiO_3$ (LLTO) [1482, 1484, 1491], yttria-stabilized zirconia (YSZ), $(Y_2O_3)_x(ZrO_2)_{1-x}$ [149, 1492], and the pyrochlores $Gd_2Ti_{2-y}Zr_yO_7$ with $0.5 \le y \le 2$ [150–153, 1493] show exactly the same ion dynamics as glassy ones, and there the applicability of MCT is hard to believe to be possible. The main support of the idealized MCT to glass-forming liquids comes principally from good fits by its prediction to the frequency dependence of the observed susceptibility spectrum $\chi''(\nu)$ and the predicted various power law dependences on $(T - T_c)$ of the susceptibility minimum value $\chi''_{\min}$ and frequency $\nu_{\min}$ as well as the relaxation time of the α-relaxation. Success of these fits seems not sufficient to prove applicability of MCT to glassformers because equally successful good fits were obtained for the dielectric loss $\varepsilon''(\nu)$ of silver ions in silver iodide-silver selenate glassy ionic conductor [1494] to which MCT does not apply.

3.2.1.3 Support from Conductivity Relaxation Data of Crystalline, Glassy, and Molten Ionic Conductors

There is a relation between MSD and the complex conductivity [65]:

$$\sigma^*(\omega) = \omega^2 \frac{N_{\text{ion}} q^2}{6 H_R k T} \int_0^\infty < r^2(t) > e^{-i\omega t} dt, \qquad (3.4)$$

where N_{ion} is the number density of mobile ions, q the ion charge, k the Boltzmann constant, and T the temperature. H_R, the Haven ratio, is the ratio of the self (tracer)-diffusion coefficient D^* and the conductivity diffusion coefficient D_σ. For ionic conductors with one kind of ions, D_σ is calculated from the measured dc conductivity σ_{dc} via the Nernst-Einstein equation [1490]. D^* is larger than D_σ and $H_R \le 1$ because of the many-ion diffusion process necessitated by ion-ion interaction. In the absence of interaction, $H_R = 1$.

For $t < t_{x1}$, it follows from Eq. (3.4) that the MSD of caged dynamics $<r^2(t)> \propto t^c$ with $c << 1$ correspond to $\sigma'(\omega) \propto \omega^{1-c}$ for the real part of the ac conductivity $\sigma'(\omega)$ and $\varepsilon''(\omega) \propto \omega^{-c}$ for the dielectric loss at frequencies $\omega > (1/t_{x1})$.

For $t_{x2} < t < \tau$, the subdiffusion term $\langle r^2(t)\rangle \propto t^{1-n}$ gives rise to the power law frequency dependences $\sigma'(\omega) \propto \omega^n$ for the real part of the ac conductivity and $\varepsilon''(\omega) \propto \omega^{-(1-n)}$ for the dielectric loss in the frequency range $(1/\tau) < \omega < (t_{x2})^{-1}$.

For $t > \tau$, after these motions of fast and slow ions have been coarse grained, the MSD assumes the $t^{1.0}$ dependence of steady-state diffusion regime and $\sigma'(\omega)$ becomes a constant, which is the dc conductivity σ_{dc} and $\varepsilon''(\omega) \propto \omega^{-1}$.

From the parameters n and τ, the primitive relaxation time τ_0 can be calculated by the CM equation (3.1). Once τ_0 has been calculated, the relation $t_{x2} \approx \tau_0$ can be checked by the experimental data. Good agreement was obtained for many ionic conductors [1057, 1058], some of which are presented in Sections 3.2.6.3, 3.2.6.4, and 3.2.7.

Polymer Chain Dynamics

More than 10 years ago, MSD of the center of mass of polymer chains in both unentangled and entangled polymer liquids was found to display a subdiffusive diffusion behavior at shorter times, which crosses over to terminal free diffusive dynamics at a characteristic decorrelation time τ_{dec} [1462, 1495–1502]. This subdiffusive behavior is characterized by an exponent that depends on the degree of polymerization N as well as temperature T and density of the sample. It is considered anomalous because the conventional theories of polymer dynamics, such as the Rouse model for unentangled polymer dynamics and the reptation model for entangled dynamics, predict no such subdiffusive center-of-mass motion at short times, in disagreement with simulations and experiments. The disagreement is not unexpected in our view because both models are mean-field theories of single-chain dynamics and many-body effects on diffusion leading to cooperative motion and dynamic heterogeneity in the liquid are neglected. The previous observations of subdiffusive dynamics have been mainly by simulations and some in a small range of times. Recently new data of neutron spin echo (NSE) performed by Zamponi et al. [1503] have extended the experimental time range up to about 200 ns to fully observe the subdiffusive behavior at shorter times and its crossover to the long-time normal diffusive regime. The four polyethylene samples investigated have molecular weights between 0.5 and 5.5 kg/mol covering the range from the low degree of polymerization regime of unentangled polymers to the entangled regime. All samples exhibit anomalous subdiffusive dynamics at short time and the transition to normal (Brownian-like) diffusion at long times. The deviation of the observed subdiffusion from normal diffusion increases with molecular weights.

The experimental work by Zamponi et al. is accompanied by the predictions of the theoretical approach of Guenza and coworkers [1504, 1505]. In this approach, specific intermolecular interactions are considered in the construction of a generalized Langevin equation for cooperative dynamics of a group of polymer chains. An analytical expression for the potential acting between the centers of mass of a pair of polymer chains was derived which allows for a formal solution of the set of coupled equations of motion through normal-mode transformation. The theory predicts anomalous subdiffusive dynamics of the center-of-mass motion and crossover to normal diffusion after some long time. The subdiffusive dynamics is heterogeneous and the distribution is not Gaussian. All the aforementioned properties of the dynamics of polymer chains obtained by the theory of Guenza are analogous to the dynamics of Li^+ ions in Li metasilicate glass, binary LJ liquids, and colloidal particle suspension, discussed in Section "Ion Dynamics in Glassy Ionic Conductor."

Essentially, the theory of Guenza relates this anomalous behavior of chain diffusion to intermolecular correlation and the cooperative motion of a group of polymer chains in a dynamically heterogeneous liquid. Thus, the theory takes care of the many-chain diffusion in interacting chains systems in the same spirit as the coupling model for many-body relaxation for interacting systems. It is opposite in spirit to the single-chain, mean-field models of Rouse and reptation. Applications of the CM to polymer diffusion and viscoelasticity will be given later in this chapter.

Figure 281 presents the center-of-mass MSD data from neutron spin echo experiments (symbols) and the theoretical predictions from Guenza's theory (shorter dashed lines) for three samples with number of segments of the protonated diffusion PE chains $N = 106$, 192, and 377. Long dashed lines with slope exactly equal to 1 represent normal Brownian-like diffusion. Experimental data show the presence of subdiffusion with t^{1-n} dependence of the MSD over a period of time as indicated by the solid lines. The characteristic exponent $(1-n)$ decreases with increasing degree

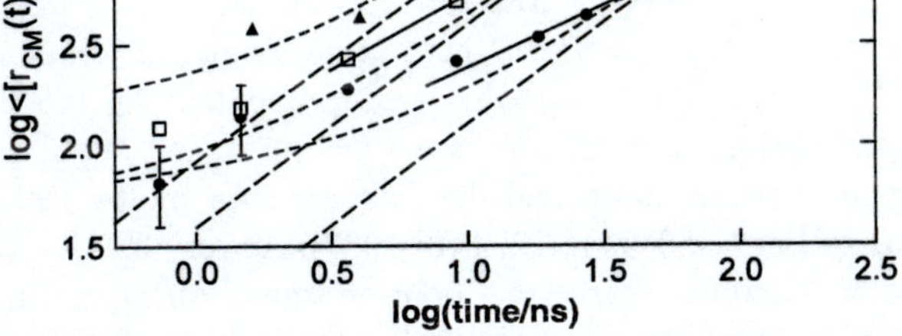

Fig. 281 The center-of-mass mean-square displacement extracted from the data at $q = 0.3\ \mathrm{nm}^{-1}$. *Symbols*: experimental data for $N = 106$ (*up triangles*), 192 (*open squares*), and 377 (*filled circles*). In the log–log plot, the *long dashed lines* have slope equal to 1, indicating terminal free diffusion at long times. *Short dashed lines* are calculated result from cooperative dynamics-generalized Langevin equation. *Solid lines* have slope equal to $(1-n)$ and are drawn to indicate the presence of subdiffusion. Data at shorter times have large uncertainties (errors not shown) and are not considered for the present purpose of showing the existence of the subdiffusion. Redrawn from the data of Zamponi et al. [1503]

of polymerization. Even in the unentangled polymers with $N=36$ and 106, the subdiffusion appears, and thus the Rouse model of a single chain is not fully accurate when applied to undiluted polymers, although the deviation is not large because the fractional exponent $(1-n)$ is close to 1. Intermolecular interaction is enhanced by entanglements in polymer with larger N, and this is reflected by the decrease in $(1-n)$. For $N=198$ and 377, the solid lines have slope or $(1-n)$ equal to 0.72 and 0.63, respectively. The MSD seems to level off with decreasing time past the subdiffusive regime. This behavior is reminiscent of the plateau in the MSD of caged Li ions, L–J, and colloidal particles shown in Figs. 275–277.

The presence of the subdiffusion in polymer chain diffusion with similar properties as those found in the dynamics of diffusion of Li ions, L–J, and colloidal particles further substantiates the universal behavior of diffusion in many-body interacting systems. These systems have widely different interaction potentials and yet the dynamics is similar. Specific to polymer dynamics, it also serves to show the inadequacy of conventional approaches to polymer melt dynamics such as the Rouse and the reptation models.

3.2.2 *Stronger Interaction/Constraints Lead to Larger n*

In Section 2.2.1 on glassformers, molecular structure with enhanced intermolecular interaction/constraint has increased stretching (or n) in the Kohlrausch correlation function of the structural relaxation. Analogues of this property abound in other interacting systems. Some examples are given below.

3.2.2.1 Ionically Conducting Systems

The concentration of mobile ions in many glassy ionic conductors can be changed by orders of magnitude. Naturally the interaction or the correlation between the motion of the ions is weak at very low concentration when the ions are far apart and increases with increasing concentration. Naturally interionic interaction vanishes and n is practically zero at very low ion concentrations and increases monotonically with concentration, provided there is no change in the structure of the glassy matrix. This expectation is realized experimentally in the broadening of the electric loss modulus peak $M''(\omega)$ as a function of the angular frequency ω with increasing ion concentration and the increase in n of the Kohlrausch function fit according to Eq. (1.56) [115(b), 148, 1057]. At very low ion concentration, $M''(\omega)$ narrows to become nearly the Fourier transform of a linear exponential function of time. The classic example is mobile Na^+ ions in a Vycor glass $x\text{NaO} - (1-x)[0.04B_2O_3 - 0.96SiO_2]$ at 313°C with $x = 0.00044$ and containing very few Na^+ ions by Simmons and coworkers [1506]. The family of Ag ion conductors $(AgI)_x(AgPO_3)_{1-x}$ for $0 \leq x \leq 0.6$ shows monotonic narrowing of $M''(\omega)$ with decreasing x and the number of mobile Ag ions [1507]. In fact, fits to $M''(\omega)$ by the Fourier transform of the derivative of the Kohlrausch function for $\Phi(t)$ in Eq. (1.56) show decrease of n with the decrease in x [1508]. Here the reader is warned that there are publications by other workers where they concluded that the same data when plotted as $\log_{10}\sigma(\omega)$vs. $\log_{10}\omega$ can be reduced to a universal curve when subjected to some scaling procedure. Here $\sigma(\omega)$ or $\sigma'(\omega)$ is the real part of the complex conductivity. From these results, these workers further conclude that the dynamics of migrating ions is the same independent of chemical structure and ionic concentration [1509–1511], in contradiction to what is indicated by the *same* data plotted as the electric loss modulus $M''(\omega)$. The width of the $M''(\omega)$ loss peak varies over a wide range when many ionic conductors are considered [115(b), 149]. In order to maintain their assertion that $\sigma(\omega)$ is universal for all ionic conductors, the proponents claimed that the electric loss modulus $M''(\omega)$ is not an appropriate representation of the data, although the complex $M^*(\omega)$ is no more than the reciprocal of the complex $\varepsilon^*(\omega)$, and $\sigma(\omega)$ is the real part of $i\omega\varepsilon_0\varepsilon^*(\omega)$. The attack on the electric modulus representation of data was shown to be invalid and refuted in [707]. The reasons why scaling of $\log_{10}\omega$ to an "universal" curve is fallacious were given in [115(b)]. They include (1) the differences in $\sigma(\omega)$ of different ionic conductors that are obscured in a log–log plot together with horizontal shift of data and (2) the presence of the near constant loss (NCL) contribution with $\sigma(\omega) = A\omega^{\mu}$ with $\mu \sim 1$ at higher frequencies irrespective of ion concentration and chemical structure. Demonstration of the absence of universal scaled $\log_{10}\sigma(\omega)$ vs. $\log_{10}\omega$ was given by comparing data from $Na_2O - 3SiO_2$ glass at $-0.5°C$ with a high concentration of mobile Na^+ ions with those of the Vycor glass at 313°C with $x = 0.00044$ and containing very few Na^+ ions. The same conclusion was made in the comparison of $\log_{10}\sigma(\omega)$ data for two $xK_2O - (1-x)GeO_2$ glasses with $x = 0.20$ and 0.0023 by Jain [1512]. The claim, that the dynamics of migrating ions is the same independent of chemical structure and ionic concentration, not only is bold but also was proffered without any theoretical support. As time tells, it serves no useful purpose, and only confusion of the real issue.

3.2.2.2 Entangled Polymer Chains

The simplest example is the comparison of an entangled monodisperse linear chain polymer with an entangled monodisperse multiple-arm, star-branched molecule with the same number of repeat units in each arm. Here, the change is in the architecture of the polymer and not in the chemical structure of the repeat unit. Obviously the multiple-arm stars have larger constraints on the movement of an individual macromolecule compared to the linear polymer. The CM predicts that n of the stars (or other architectures like H-polymers) is larger than that of the linear molecules. Experimentally this prediction is exemplified by a broader dispersion of the stars than the linear chains (see, for example, [1513]). As we shall see later, there are other predictions that accompany the predicted change of n, which have been verified by experiments.

3.2.2.3 Semidilute Polymer Solutions and Associating Polymer Solutions

More detailed description of these systems is given in a section to follow. The polymer chains in these solutions are still interacting via either entanglement or association. Naturally, increase of concentration of the polymer chains enhances interchain coupling and constraints, and the expected increase of n in the Kohlrausch correlation function together with the consequences it has on other properties including concentration dependence and scattering angle dependence was found by experiments [198, 767, 778–780, 1514]. These aspects are subjects of further discussion to be given later. Here we cite the dielectric loss data of chain normal modes of poly(2,6-dichloro-1,4-phenylene oxide) (PDCPO) solutions from [1514]. With increasing concentration ϕ of the polymer such that the product ϕM approaches M_e, the molecular weight between entanglements, not only does the observed loss maximum frequency shifts to the low frequencies but also the loss curve simultaneously broadens on both sides of the peak maximum. The analysis of the spectra using the CM [925] relates the broadening to increase in coupling parameter n between chains with increasing concentration of PDCPO.

3.2.2.4 Junction Dynamics of Cross-Linked Polymers

A specific chemical moiety can link up a fixed number of polymer chains at a junction point. Multiple junction points convert the polymer chain to a network. The dynamics of the junctions has been studied experimentally [1515–1517]. The constraint on the motion of the junctions and hence $n \equiv (1 - \beta)$ is expected to decrease by lowering the density of junctions or the addition of a diluent. These expected changes by the CM were indeed observed by NMR measurements of the junction dynamics by Shi et al. [1515]. There are consequences of the dependence of n on junction density in the dynamic properties, and these are given in a later section.

3.2.3 *Crossover from* $\exp(-t/\tau_0)$ *to* $\exp[-(t/\tau)^{1-n}]$ *at* t_c

Like glassformers, there are experimental evidences in other interacting systems supporting the crossover of the correlation function from $\exp(-t/\tau_0)$ to $\exp[-(t/\tau)^{1-n}]$ in a neighborhood of a temperature-insensitive time t_c. We have mentioned before in Section 2.2.1.1 for polymers and small molecular van der Waals liquids that $t_c \approx 2$ ps. As we see below, the magnitude of t_c differs from one class of systems to another, increasing with decreasing strength of interaction and/or constraints. This trend is understandable, because in the limit of zero strength of interaction or no constraints, many-body relaxation is reduced to one-body relaxation, and its correlation function $\exp(-t/\tau_0)$ should hold for all times and hence $t_c \to \infty$.

3.2.3.1 Ionically Conducting Systems

Fast Glassy Ionic Conductors

There are many fast glassy ionic conductors where the mobility of the ions is so high that the conductivity relaxation time τ_σ is 10 orders of magnitude or more lesser than the structural relaxation time τ_α at T_g. In contrast, in $0.4Ca(NO_3)_2 - 0.6KNO_3$

(CKN) discussed in Section 2.2.1.1, τ_σ is only about 2–3 orders of magnitude lesser than τ_α at T_g [55, 1418]. These fast glassy ionic conductors can have τ_σ becoming very short by raising temperature but still staying within the glassy state. This is an example of relaxation of an interacting system which does not involve glass transition because T here is always significantly below T_g. At lower temperatures where $\tau_\sigma >> t_c$ of the order of 1 ps, the conductivity relaxation correlation of glassy and crystalline ionic conductors is well described by the Kohlrausch function $\exp[-(t/\tau_\sigma)^{1-n}]$, as shown by fits of the data in the electric modulus $M^*(\omega)$ representation by the Fourier transform of the derivative of the Kohlrausch function (see Section 1.2.2 and [55, 362, 707, 1518, 1519]). The conductivity $\sigma(\omega)$ corresponding to the $M^*(\omega)$ obtained with $\exp[-(t/\tau_\sigma)^{1-n}]$ has the ω^n dependence at high frequencies when $\omega\tau_\sigma >> 1$ and approaches the dc conductivity σ_{dc} in the low-frequency limit when $\omega\tau_\sigma << 1$ [1057, 1058]. The crossover of the correlation function from $\exp[-(t/\tau_\sigma)^{1-n}]$ to $\exp(t/\tau_0)$ with decreasing time past t_c is observed via the corresponding $\sigma(\omega)$ crossing over from the ω^n dependence to a frequency-independent value with increasing frequency past $\omega_c \equiv (t_c)^{-1}$. Here, two examples of the crossover are chosen from the work of Cramer et al. on two fast glassy ionic conductors $0.44LiBr-0.56Li_2O-B_2O_3$ [1520], and $0.5Ag_2S-0.5GeS_2$ glass [1521]. In inorganic glasses, the vibrational contribution to $\sigma(\omega)$ extends from high frequencies down to low frequencies with an ω^2 dependence [336], which is also found in these glassy ionic conductors as shown in Figs. 282 and 283. The Li

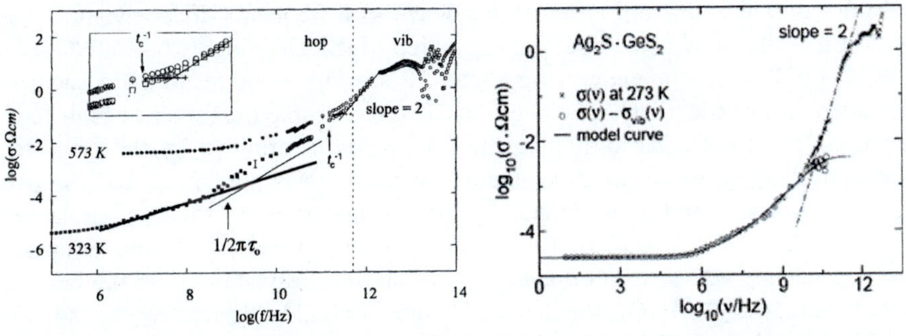

Fig. 282 (*Left*) The ac conductivity data of $0.44LiBr-0.56Li_2O-B_2O_3$ reconstructed from the data of Cramer et al. [1520]. Shown also is the high-frequency vibrational contribution to $\sigma(\omega)$ extending down to low frequencies with an ω^2 dependence. Subtracting off this vibrational contribution from the data at 573 K leaves a frequency-independent σ_0 (+). The inset shows this more clearly as well as the crossover of $\sigma(\omega)$ at a frequency of about 10^{11} Hz from frequency-independent σ_0 to the ω^n dependence. At lower temperatures such as 323 K, there is a contribution with near-$\omega^{1.0}$ dependence to $\sigma(\omega)$, the counterpart of the near constant loss in $\varepsilon''(\omega)$ at high frequencies, which obscures the crossover of diffusing ions at 10^{11} Hz. The nearly constant loss is due to loss while the mobile ions are still caged. (*Right*) Frequency-dependent conductivity $\sigma(\nu)$ of $0.5Ag_2S \cdot 0.5GeS_2$ glass at 273 K. Below 10 GHz, $\sigma(\nu)$is essentially caused by the hopping motion of the silver ions, while above 20 GHz it is essentially due to the low-frequency flank of the vibrational component. The ion contribution $\sigma_{ion}(\nu) = [\sigma(\nu) - \sigma_{vib}(\nu)]$ exhibits a crossover to primitive ionic conductivity relaxation at $10^{10.5}$ Hz. Figure reproduced from [1521] by permission

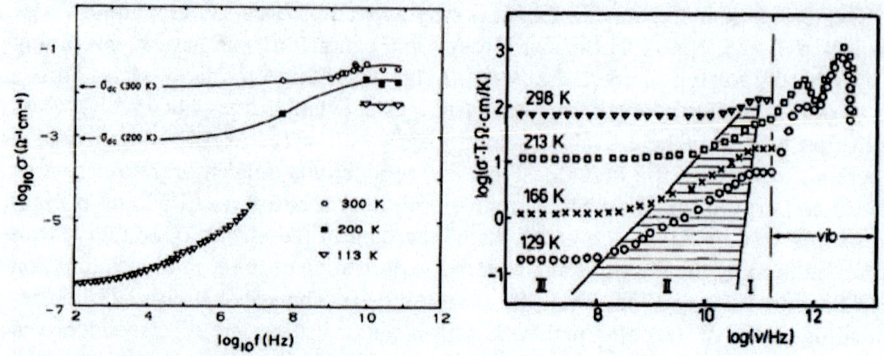

Fig. 283 (*Left*) Frequency dependence of $\sigma(\nu)$ of Na β-alumina at three temperatures showing crossover after 10^{10} Hz. Reproduced from [1522(a)] by permission. (*Right*) The product $\sigma(\nu)T$ of $RbAg_4I_5$ crystals at four different temperatures. The crossover is seen for 166 and 129 K. Reproduced from [1525, 1526] by permission

ion conductivity contribution is obtained after the tail of the vibrational contribution with an ω^2 -dependence has been subtracted off.

The results are shown for $0.44LiBr{-}0.56Li_2O{-}B_2O_3$ in the left panel of Fig. 282 at the high temperature of 573 K where ionic diffusion is the dominant contribution to $\sigma(\omega)$; there is a crossover of $\sigma(\omega)$ at a frequency $\omega_c \equiv 1/t_c$ of the order of 10^{11} Hz from an ω^n dependence to a frequency-independent value σ_0, just like that found for structural relaxation of the glassformer $0.4Ca(NO_3)_2{-}0.6KNO_3$ (CKN) at temperatures far above T_g (see Fig. 8 and recall that structural relaxation and conductivity relaxation in CKN are coupled together at high temperatures [55, 1418]). According to the CM, σ_0 is determined by the primitive relaxation time τ_0, while the dc conductivity σ_{dc} is governed by τ_σ, and the two relaxation times τ_σ and τ_0 are related by the CM equation (3.1) with τ_σ replacing τ. In the glassy state, both σ_{dc} (or τ_σ) and σ_0 (or τ_0) have Arrhenius temperature dependence with activation energies E_{dc} and E_0, respectively. In the case of $0.44LiBr{-}0.56Li_2O{-}B_2O_3$, the data of Cramer et al. [1520] yield $E_{dc} = 0.49\,\text{eV}$ or 5670 K, and $E_0 = 0.22\,\text{eV}$ or 2553 K. The exponent $(1{-}n)$ of the Kohlrausch function is estimated to be within the range of $0.40 < (1-n) < 0.48$. The other CM equation (3.2) predicts that

$$E_0 = (1-n)E_{dc}. \tag{3.5}$$

The product on the right side of this equation lies within the bounds $2268 < (1-n)E_{dc} < 2721$ K, and is consistent with $E_0 = 2553$ K [148].

The right panel of Fig. 282 shows $\sigma(\nu)$ having the same crossover from the ω^n dependence to a constant value of primitive relaxation in another fast glassy Ag ion conductor $0.5Ag_2S{-}0.5GeS_2$ at 273 K after subtracting the vibrational ω^2 dependence from the data.

Fast Crystalline Ionic Conductors

Many-body relaxation dynamics requires interaction between the units. Although disorder fosters interaction, by no means it is essential to observe the vestige of many-body relaxation dynamics. Thus, we can expect to see the crossover to primitive conductivity relaxation in fast crystalline ionic conductors. In fact there are experimental evidences for the crossover in crystalline Na β-alumina [1486, 1522, 1523], Na β''-alumina [1524, 1525], and $RbAg_4I_5$ [1525, 1526]. The crossover of conductivity of Na β-alumina at $\nu_c \approx 10^{10}$ Hz and $RbAg_4I_5$ at about 10^{11} Hz can be seen by inspection of the data presented in the left and right panels of Fig. 283, respectively. For Na β-alumina, the activation energy E_0 of $\sigma(\nu)$ for $\nu > \nu_c$ in the range of $200 < T < 300$ K is ≈ 810 K. On the other hand, the activation energy of dc conductivity E_{dc} is ≈ 1600 K at 300 K and ≈ 1700 K at 200 K (see Fig. 287). The coupling parameter n is $\approx$0.50 at 300 K and $\approx$0.53 at 200 K [1522(a)]. From these values, one can verify that the relation $E_0 = (1-n)E_{dc}$ of Eq. (3.5) is satisfied.

The crossover behavior is thus found in the dynamics of ions in fast ionic conductors in the glassy or crystalline state, in the same manner as that observed on the structural relaxation of glassformers including molten CKN by dielectric relaxation (Fig. 8), small molecular and polymeric glassformers by quasielastic neutron scattering (Figs. 4 and 5) and computer simulations (Figs. 6 and 7), and colloidal particle suspension by light scattering (Fig. 9). This universal crossover occurs for all many-body relaxation processes in systems with interactions. If the CM has captured the general physics of complex many-body correlation in relaxation of interacting systems, then it is not surprising that the crossover and other properties are universal, as we have seen in various ionic conductors and in other systems that are neither ionic conductors nor glassformers, which we discuss later on in the chapter.

Susceptibility Minimum of Glassy Ionic Conductors Well Fit to Mode Coupling Theory (MCT)

At lower temperatures where the conductivity relaxation time τ_σ is not as high to see the crossover to primitive relaxation, the $\varepsilon''(\nu)$ spectra of the fast solid-state ionic conductors show up as minima at high frequencies just like the susceptibility minima of glass-forming substances. The latter, when well fit by the predictions of the idealized MCT, often has been taken as confirmation of the theory for glass transition, as has been claimed [32] for susceptibility data from neutron and light scattering as well as dielectric susceptibility $\varepsilon''(\nu)$ spectra of CKN and CRN. In brief, the MCT prediction [31, 32] for the minimum in the dynamic susceptibility $\chi''(\nu)$ or $\varepsilon''(\nu)$ at high frequencies ν and temperatures higher than T_c is

$$\chi''(\nu) = \chi''_{\min}\left[a(\nu/\nu_{\min})^{-b} + b(\nu/\nu_{\min})^{a}\right]/(a+b), \tag{3.6}$$

where $\chi''_{\min}$ and $\nu_{\min}$ are the height and the location of the minimum, respectively. The exponents a and b are related to each other via the control parameter λ by the two equations $\lambda = \Gamma^2(1-a)/\Gamma(1-2a) = \Gamma^2(1+b)/\Gamma(1+2b)$, where Γ is the Gamma function. The first equation implies that $a < 0.395$ and hence a pronounced sublinear increase of $\chi''(\nu)$ at $\nu > \nu_{\min}$ is predicted. For $T > T_c$, MCT predicts the following critical temperature dependences:

$$\nu_{\min} \propto (T - T_c)^{1/2a}, \; \chi''_{\min} \propto (T - T_c)^{1/2}, \text{ and } \tau_\alpha \propto \eta/T \propto (T - T_c)^{-\gamma}, \quad (3.7)$$

where τ_α is the characteristic time of the α-relaxation and $\gamma = (1/2a) + (1/2b)$. It has become a common practice of researchers to check the validity of the MCT equations (3.6) and (3.7) for the $\chi''(\nu)$ data and determine the critical temperature T_c in the process.

The similarity of the $\varepsilon''(\nu)$ spectra of glassy ionic conductors to the susceptibility spectra of glass-forming liquids invites fitting the former to the predictions of MCT. This has been done [1494] on the dielectric loss data of the fast glassy ionic conductor $0.48(AgI)_2{-}0.52Ag_2SeO_4$ obtained by Cramer and Buscher [1528] and is shown in Fig. 284 (left). The shapes of the minima of $\varepsilon''(\nu)$ are well fit by Eq.

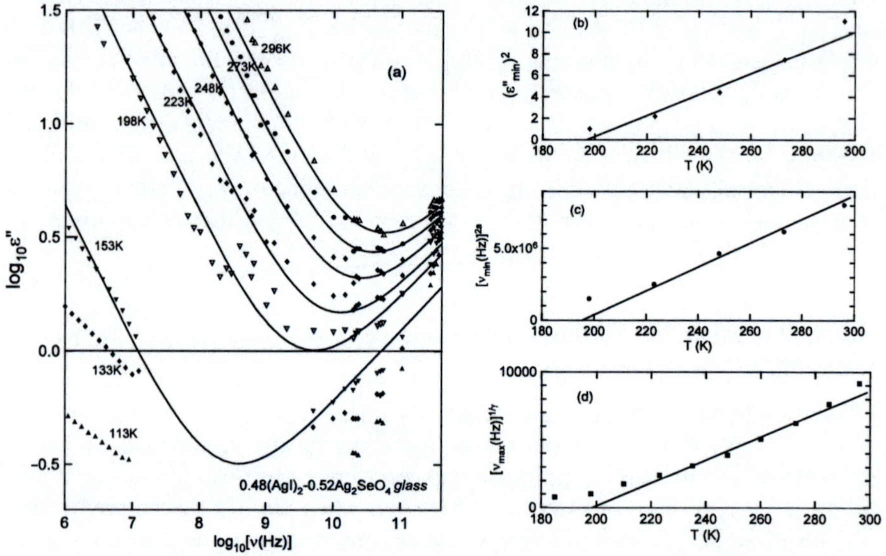

Fig. 284 (**a**) Frequency dependence of the dielectric loss of the glassy ionic conductor $0.48(AgI)_2{-}0.52Ag_2SeO_4$ at 296, 273, 248, 223, 198, 153, 133, and 113 K (*from top to bottom*). *Symbols* stand for experimental data from [1528] and are replotted here. The *solid lines* are fits using the MCT expression, Eq. (3.6), with $\lambda = 0.176$ ($a = 0.32$ and $b = 0.614$). Temperature dependences of $(\varepsilon''_{\min})^2$ is shown in (**b**), $(\nu_{\min})^{2a}$ in (**c**), and $(\nu_{\max})^{1/\gamma} \equiv (1/2\pi\tau_\alpha)^{1/\gamma}$ in (**d**). *Straight lines* in (**b**), (**c**), and (**d**) are fits to linear dependence on $(T{-}T_c)$ of these quantities according to the predictions of MCT by Eq. (3.7). Extrapolation of the linear fits determines T_c. It lies within the range of $196 < T_c < 210$ K from the extrapolations of the three quantities. All figures are redrawn from the data in [1494]

(3.5) from the MCT β-process, and the critical laws of MCT equation (3.7) are also obeyed as shown by the other plots in Fig. 284 and like that found for CKN shown in Fig. 285 [1418]. Notwithstanding the success, it does not seem possible that a fluid theory based on density fluctuations like MCT is applicable to motions of atomic size ions in the matrix of solid-state ionic conductors, glassy or crystalline. This exercise demonstrates that susceptibility minimum and the associated two-step decay of the intermediate scattering function are generally found in relaxation of other interacting systems beyond the glass-forming liquids and can have similar shapes. Moreover, the good fit of the susceptibility by the MCT predictions of the minimum by Eq. (3.6) and the critical laws Eq. (3.6) found in a case where the theory should not apply is an admonition of the danger of validating the MCT as the theory of fast relaxation in glass-forming liquids from merely the good fits to their susceptibility data such as that shown in Fig. 285.

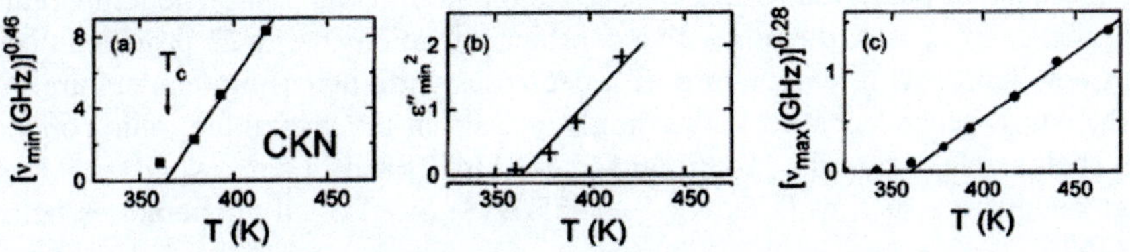

Fig. 285 Temperature dependence of various parameters obtained from the analysis of the high-frequency data of CKN in terms of MCT. The *solid lines* are in accord with the laws given by Eq. (3.7) with a critical temperature of 360 K. Reproduced from [1418] by permission

Susceptibility Minimum of Glassy Ionic Conductors Is Actually Caused by the NCL of Caged Ions

In the case of fast ionic conductors, glassy or crystalline, the nearly constant loss (NCL) is a well-known and well-observed contribution to the dielectric loss that goes back to as early as 1946 [1529–1531], is discussed in the book by Wong and Angell in 1976 [1055], extended to a gigahertz range for the first time by Robert Cole and coworkers in 1989 [1532], and followed by many works [124, 1056–1061, 1533–1535]. Hence, it is more or less certain that the minimum in the $\varepsilon''(\omega)$ of fast ionic conductors is due to the NCL lying in between the vibration absorption at high frequencies and the ω^{-n} dependence at lower frequencies contributed by the ion conductivity. This occurs when the conductivity relaxation time becomes short to limit the range of the NCL and the $\varepsilon''(\omega)$ minimum is observed [1494].

Crossover of Temperature Dependence of σ_{dc} at High Temperatures

We have shown before in Section 2.2.1.2 for glass-forming liquids that another way to see the crossover to primitive relaxation is by the change in T dependence of the viscosity at very low values corresponding to the structural relaxation time τ_α of the order of picoseconds. The analogy in the case of ion dynamics in fast glassy or crystalline ionic conductors is the change in T dependence of the dc conductivity σ_{dc}, which reflects the corresponding change of the conductivity relaxation time τ_σ according to the Maxwell relation $\sigma_{dc} = \epsilon_0\varepsilon_\infty / < \tau_\sigma >$. Here $\epsilon_0 = 8.854 \times 10^{-14}$ F/cm is the permittivity of free space, ε_∞ is the high-frequency dielectric constant, and $< \tau_\sigma >$ is the mean relaxation time. At lower temperatures where $\tau_\sigma << \tau_c$ and the correlation function is $\exp\left[-(t/\tau_\sigma)^{1-n}\right]$, τ_σ has the Arrhenius T dependence

$$\tau_\sigma(T) = \tau_\infty^* \exp(E_a^*/kT), \tag{3.8}$$

and $< \tau_\sigma >= \left[\Gamma(1/\beta)/\beta\right] \tau_\sigma$, where $\beta \equiv (1-n)$, and Γ stands for the gamma function [55]. The activation energy E_a^* and the prefactor τ_∞^* are usually anomalous because of many-ion relaxation due to ion–ion interactions. The temperature dependence of τ_σ is Arrhenius with a constant activation energy E_a^* provided that n does not change with temperature. If n decreases with increasing temperature, the Arrhenius T dependence no longer holds as seen in the crystalline ionic conductors such as Na β-alumina [1486] and LLTO [1482] and in some very fast glassy ionic conductors such as $0.525Ag_2S$+$0.475(B_2S_3$:$SiS_2)$ at all temperatures below T_g [1536].

On raising the temperature, τ_σ becomes shorter and when it is of the order of picoseconds or shorter, t_c is crossed and primitive motion of ions takes over with correlation function given by $\exp(-t/\tau_0)$ with

$$\tau_0(T) = \tau_\infty \exp\left(E_a/kT\right) \tag{3.9}$$

and now $<\tau_\sigma>= \tau_\sigma$. The primitive activation energy E_a can be identified with a realistic energy barrier of an independent ion hop. The reciprocal of the prefactor τ_∞ is the attempt angular frequency, which should correspond to some peak frequency of the infrared or Raman spectrum. It follows from Eq. (3.2) that

$$E_a = (1-n)E_a^* \quad \text{and} \quad \tau_\infty = (t_c)^n(\tau_\infty^*)^{1-n}. \tag{3.10}$$

The temperature dependence of τ_0 is mirrored by that of the measured σ_{dc}. Therefore, on increasing T, the T dependence of σ_{dc} will eventually cross over to the Arrhenius T dependence having the same activation energy E_a, albeit it may be slightly modified at the higher temperatures when $1/\tau_0$ approaches the vibration frequencies. These properties associated with the expected crossover of σ_{dc} at high temperatures have been found in many ionic conductors [1538] including the molten salt CKN, fast glassy ionic conductors such as $0.48(AgI)_2$–$0.52Ag_2SeO_4$ [1528] and $0.525Ag_2S$+$0.475(B_2S_3$:$SiS_2)$ [1536, 1537], and crystalline ionic conductors such as yttria-stabilized zirconia, $(ZrO_2)_{1-x}(Y_2O_3)$ [1539–1543]. As an example, σ_{dc} data of yttria-stabilized zirconia (YSZ) are shown in Fig. 286 [1539]. In the lower temperature regime, where σ_{dc} has the Arrhenius T dependence,

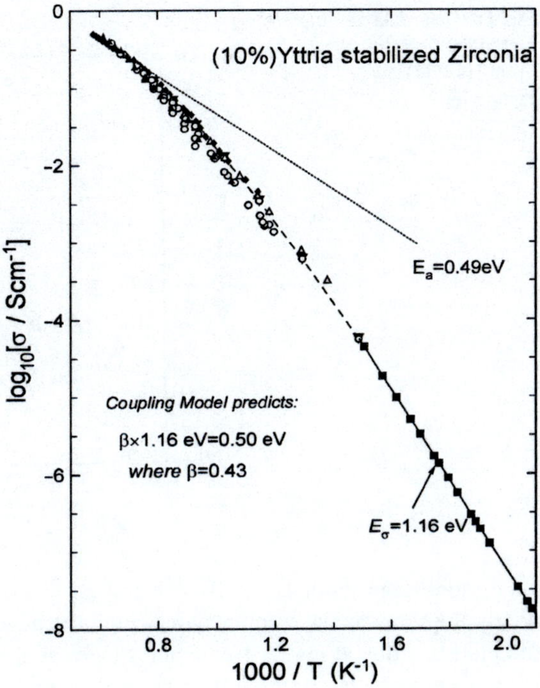

Fig. 286 σ_{dc} data of yttria-stabilized zirconia. Reproduced from [1539] by permission

$\sigma_{dc}(T) = \sigma^*_\infty \exp(-1.14\ \text{eV}/kT)\ \Omega^{-1}\,\text{cm}^{-1}$; from experimentally measured frequency dispersion, the oxygen ion hopping correlation function was found to have the Kohlrausch function with $n = 0.56$.

As temperature increases, $\sigma_{dc}(T)$ becomes non-Arrhenius, but at the highest temperatures, on approaching 1 $\Omega^{-1}\,\text{cm}^{-1}$, it returns to another Arrhenius dependence described by $\sigma_{dc}(T) = 10^{3.6} \times \exp(-0.49\,\text{eV}/kT)\,\Omega^{-1}\text{cm}^{-1}$ Such behavior is evidence of crossover from primitive conductivity relaxation to coupled conductivity relaxation because at the high conductivity levels approaching 1 $\Omega^{-1}\,\text{cm}^{-1}$, the conductivity relaxation times are likely to be comparable with $t_c \approx 2$ ps. For YSZ, the value of ϵ_∞ is 28 [1541], and hence from the Maxwell equation $\sigma_{dc} = \varepsilon_0\varepsilon_\infty / <\tau_\sigma>$, $\sigma_{dc} = 1\ \Omega^{-1}\,\text{cm}^{-1}$ corresponds to $<\tau_\sigma>= 2.5 \times 10^{-12}$ s. Oxygen–oxygen ion interaction slows down the conductivity at lower temperatures when $\tau_\sigma >> t_c$, and its higher activation energy $E^*_a = 1.16\,\text{eV}$ is correctly predicted from the smaller (primitive) activation energy ($E_a = 0.49$ eV) at high temperatures by Eq. (3.5). This can be verified by the equation $0.49\ \text{eV} = (1 - n)\,1.16\,\text{eV}$ being satisfied if $n = 0.57$, which is close to the value of 0.56 independently determined by fitting the frequency dependence of the conductivity relaxation data by the Kohlrausch function [1492, 1542].

Figure 287 shows that the dc conductivity of Na β-alumina also crosses over to a weaker T dependence when σ_{dc} exceeds about 0.4 $\Omega^{-1}\,\text{cm}^{-1}$, which corresponds to $< \tau_\sigma >= 1.0 \times 10^{-11}$s because ε_∞ is 50 [1522a]. This indicates that t_c for

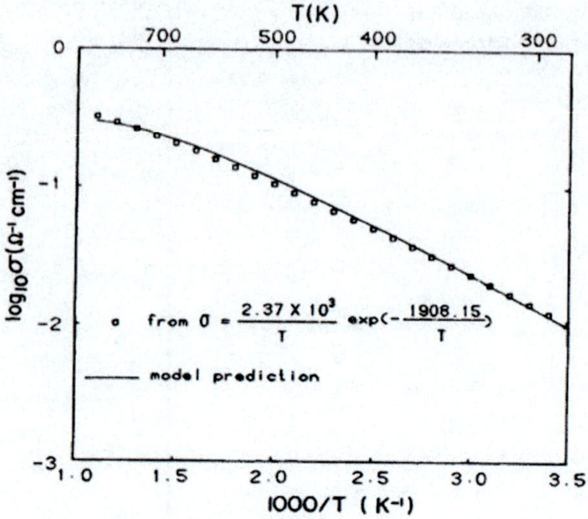

Fig. 287 σ_{dc} data of Na β-alumina (*open circles*). From [1522(a)] by permission

Na β-alumina is about 1.0×10^{-11}s and $\nu_c = 1.6 \times 10^{10}$ Hz. Interestingly, this is nearly the same as the frequency at which the ac conductivity $\sigma(\nu)$ crosses over to the plateau value at higher frequencies (see the left panel of Fig. 283). Thus, there is consistency in determining the crossover to primitive relaxation from temperature dependence of σ_{dc} and from frequency dependence of isothermal $\sigma(\nu)$. Earlier in connection with the left panel of Fig. 283, the relation $E_0 = (1 - n)E_{dc}$, has been verified for Na β-alumina. Except for the change in notation, this relation is the same as $E_a = (1 - n)E_a^*$ in Eq. (3.10).

The values of E_a calculated in the same way by Eq. (3.10) from the values of E_a^* and n deduced from the data at lower temperatures of many ionic conductors, glassy or crystalline, are given in Table 3.1. Each is about the same as or slightly larger than (but by no more than 20%) the value of E_a obtained from the high-temperature σ_{dc} data which have τ_σ shorter than t_c as described above. Furthermore, the reciprocal of the primitive attempt time τ_∞ calculated by Eq. (3.10) from the anomalous τ_∞^* is also in rough agreement with the value deduced from the high-temperature σ_{dc} data after the crossover. It corresponds well to the peak angular frequency of vibrational spectrum. These good correspondences between the calculated τ_∞ and E_a from the experimentally determined parameters τ_∞^*, E_a^*, and n in the glassy state with their counterparts at high temperature are expected because both sets of parameters are for independent diffusion of the ion, and the difference between them is caused only by the difference in T and density. These changes should not have a large effect on the primitive attempt frequency and the activation energy of local and independent hopping of the ion. There are cases in which n decreases with increasing temperature and approaches zero value while τ_σ is still longer than t_c. This causes the crossover of T dependence of σ_{dc} data to occur at temperature for which τ_σ is still longer than t_c, but the relations in Eq. (3.10) are still valid.

Table 3.1 Ionic transport and conductivity relaxation parameters. For the glass-forming melts, $\beta \equiv (1-n)$ and E_a^* were obtained in the glassy state from conductivity relaxation measurements, while the activation energies E_a were obtained from the high-temperature melt dc conductivity data. For some glassy ionic conductors, the activation energies E_a^{glass} were obtained from neutron scattering or high-frequency microwave and far-infrared conductivity data at temperatures all within the glassy state. All activation energies are in units of kJ/mol

	β	E_a^*	βE_a^*	E_a	E_a^{glass}	$\beta E_a^*/E_a$
Glass-forming ionic conductors						
$LiCl–7H_2O$	0.46	34	15.6	14.4		1.08
$CdF_2–LiF–AlF_3–PbF_2$	0.77	109	83.9	68.4		1.23
ZBLAN20	0.68	85	57.8	50		1.16
ZBLAN10	0.66	79	52.1	46		1.13
ZBLA	0.61	72	43.9	36		1.22
$(Li_2O)–3(B_2O_3)$	0.52	84	43.7	40		1.09
$(Na_2O)–3(SiO_2)$	0.55	64	35.2	33.5		1.05
$0.56Li_2O–0.45LiBr–B_2O_3$	0.44	47.1	20.7		21.1	0.98[a]
$AgPO_3$	0.66	49.5	32.7	28.5		1.14
$(AgI)_{0.1}–(AgPO_3)_{0.9}$	0.59	43	25.4	22.5		1.13
$(AgI)_{0.2}–(AgPO_3)_{0.8}$	0.57	39.5	22.5	19.8		1.14
$(AgI)_{0.3}–(AgPO_3)_{0.7}$	0.54	32.9	17.8	15.6		1.14
$(AgI)_{0.4}–(AgPO_3)_{0.6}$	0.51	32.0	16.3	13.3		1.23
$(AgI)_{0.5}–(AgPO_3)_{0.5}$	0.48	26.9	12.9	10.1	8.7	1.27
$(AgI)_{0.6}–(AgPO_3)_{0.4}$	0.48	26.9	12.9	7.9		1.29
$(AgI)_{0.7}–(Ag_2MoO_4)_{0.3}$	0.44	19.3	8.5	9.0		0.95
$0.48(AgI)_2–0.52Ag_2SeO_4$	0.51	25.1	12.8	13.6		0.94
$(Ag_2S)_{0.5}(GeS_2)_{0.5}$	0.45	32.8	14.8		14.5	1.02[a]
Crystalline conductors						
Na β-Al_2O_3	0.5	13.4[b]	6.7	6.8[c]	6.74	0.99
$RbAg_4I_5$	0.47[d]	9.8[c]	4.6		4.2[c]	1.09[a]
$(Y_2O_3)_{0.095}(ZrO_2)_{0.905}$	0.43	111.9	48.2	48.2		1

[a]Calculated from the ratio $\beta E_\sigma/E_a^{glass}$.
[b]At 300 K.
[c]Obtained by plotting log σ against $1/T$.
[d]Estimated from log σ vs. log f data.

Anomalously Short Prefactor τ_∞^*

In Fig. 287, we show the low-temperature conductivity relaxation data of (9.5%)YSZ from León et al. [1542, 1539, 1540] in terms of $\tau_\sigma(T)$. Its temperature dependence is Arrhenius as indicated by the straight line through the data points in the figure, which corresponds to

$$\tau_\sigma(T) \equiv \tau_\infty^* \exp\left(E_a^*/kT\right) = 10^{-16.44} \exp\left(1.16\ eV/kT\right)\ \text{s}. \tag{3.11}$$

The prefactor τ_∞^* of the experimentally observed $\tau_\sigma(T)$ is very short and its reciprocal corresponds to an unphysically high attempt frequency of the oxygen ions. On the other hand, the independent hopping relaxation times $\tau_0(T)$ calculated from $\tau_\sigma(T)$ via Eq. (3.5) with $t_c = 2$ ps are given by

$$\tau_0 (T) = \tau_\infty \exp \left(E_a/kT\right) = 10^{-13.74} \exp \left(0.50\ eV/kT\right)\ \text{s}, \tag{3.12}$$

and it is represented by the thicker dashed line in Fig. 288. Remarkably, this calculated $\tau_0(T)$ has nearly the same value as well as temperature dependence as $\tau_\sigma(T)$ at high temperatures. This is no surprise because $\tau_\sigma(T)$ is reaching $t_c = 2$ ps at the high temperatures, and hence it crosses over to $\tau_0(T)$. The reciprocal of the prefactor τ_∞ of the calculated τ_0 now has the magnitude of an attempt frequency, as it should.

The actual attempt frequency of oxygen ions in YSZ was determined experimentally from the hyper-Raman spectrum obtained by Shin and Ishigame [1544]. The

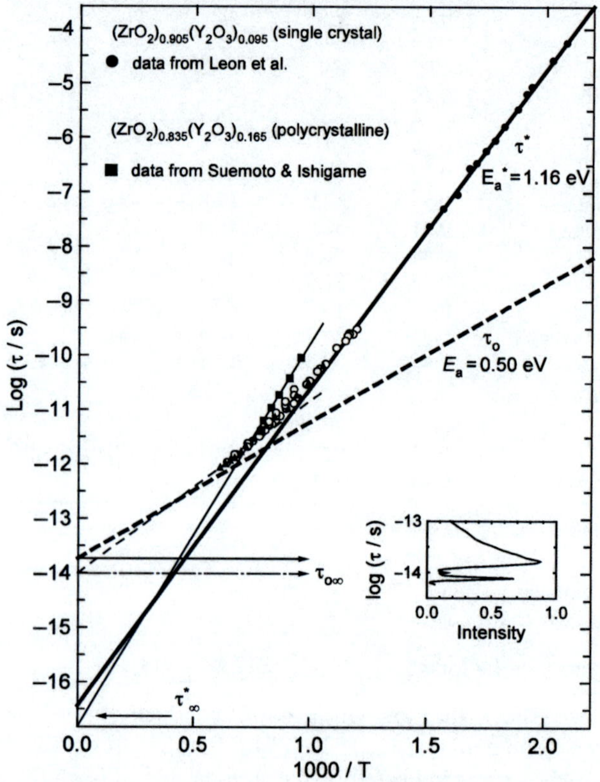

Fig. 288 *Closed circles* and *filled squares* are the relaxation times τ obtained from the oxygen ion conductivity relaxation data (by León et al. [1542]) and quasielastic light-scattering data (by Suemoto and Ishigami [1545]) of YSZ. The *inset* shows the hyper-Raman spectrum from Shin and Ishigami [1544]. *Open circles* and *closed triangles* are τ deduced from dc conductivity data. The *thick solid* and *thin solid lines* through the data points are Arrhenius fits. The *thick* and *thin broken lines* represent the independent relaxation time τ_0 calculated from τ of conductivity relaxation (*thick solid line through circles*) and from quasielastic light scattering (*thin solid line through closed squares*), respectively. The intercepts of the *broken lines* give prefactors $\tau_{0\infty}$ (values indicated by the two horizontal arrows) in good agreement with the frequencies in the narrow band at 690 cm^{-1} of the hyper-Raman spectra shown in the inset. Reproduced from [1482] by permission

vibrational modes shown have displacement of the oxygen ion toward the vacancy and are candidates for the vibrational frequency. Shin and Ishigame argued that the highest frequency narrow band located at 690 cm^{-1}, corresponding to time of $(\omega_\infty)^{-1} = 10^{-14.1}$ s, contributes most effectively to the hopping of the oxygen ions and can be identified by their attempt frequency. Having determined the reciprocal of the true attempt (angular) frequency ω_∞ of the oxygen ions from experiment to be $10^{-14.1}$ s, comparison of it can be made in Fig. 288 with the prefactors τ^*_∞ and τ_∞. The figure has used the notation $\tau_{0\infty}$ instead of τ_∞. The true angular attempt frequency $(\tau_\infty)^{-1}$ deduced from the CM is only 2.4 times smaller than the experimentally determined ω_∞. This small discrepancy between $(\tau_\infty)^{-1}$ and ω_∞ is well within the uncertainty in determining $\tau_0(T)$ from Eq. (3.10) due to experimental errors in n and $\tau_\sigma(T)$ obtained by León et al. The upper horizontal arrow in Fig. 288 indicates the good correspondence between τ_∞, determined by the intercept with the y-axis at $1/T = 0$, and $(\omega_\infty)^{-1}$ from the hyper-Raman spectrum depicted by the inset. On the other hand, as mentioned earlier, the angular frequency $1/\tau^*_\infty = 10^{16.4}s^{-1}$ is unphysical because its value is more than 200 times higher than the vibrational frequency ω_∞.

Quasielastic light scattering (QELS) in YSZ by tandem Fabry–Perot interferometry was measured by Suemoto and Ishigame [1545] using the same samples as in the hyper-Raman scattering experiment. Light scattering is due to fluctuation of the polarizability caused by ionic motion. They found that the shape of the scattered light intensity peak as a function of temperature taken at constant frequency f, in the range from 1.8 to 24 GHz is non-Lorentzian and in good agreement with that coming from a correlation function that has the Kohlrausch form. The dependence of the temperature of the intensity maximum on f is converted to a dependence on $(2\pi f)^{-1} \equiv \tau^*_{QELS}$ and the data are shown in Fig. 287 by closed squares. The thinner straight line through the data point is the best fit to an Arrhenius temperature dependence:

$$\tau^*_{QELS} \equiv \tau^*_{QELS,\infty}\exp\left(E^*_{a,QELS}/kT\right) = 10^{-16.82}\exp\left(1.40\ eV/kT\right). \quad (3.13)$$

The unphysically high apparent attempt frequency $\tau^*_{QELS,\infty}$ is again evident from the prefactor. The QELS correlation time τ^*_{QELS} has a slightly larger activation energy than the conductivity relaxation time $\tau_\sigma(T)$ obtained by León et al. [1542]. This difference arises because the concentration of yttria is higher in the sample studied by QELS than that by conductivity relaxation (16.5 vs. 9.5 mol %) and it is known from conductivity measurement that E^*_a increases with yttria content when above roughly 8 mol%. The spectral shape is well fitted by an expression proportional to $\chi'(\omega)/\omega$, where $\chi'(\omega)$ is the imaginary part of the susceptibility function calculated [1544] by a Fourier transform of the time derivative of the Kohlrausch function. In the process, the coupling parameter n_{QELS} was determined to have the value of 0.55. The independent ion hopping correlation time $\tau_{QELS,o}$ calculated from τ^*_{QELS} has the Arrhenius dependence

$$\tau_{QELS,o}(T) \equiv \tau_{QELS,\infty}\exp\left(E_{a,QELS}/kT\right) = 10^{-14.0}\exp\left(0.63\ eV/kT\right)\ s \quad (3.14)$$

and is plotted as a function of temperature in Fig. 288 (the thinner dashed line).

The situation in QELS is similar to conductivity relaxation in that the experimentally determined attempt frequency $1/\tau^*_{QELS,\infty}$ is too high to be real. However, the attempt frequency $(\tau_{\mathrm{QELS},\infty})^{-1}$ of the independent ionic hopping motion deduced by the CM nearly coincides with the measured frequency of the vibrational mode (690 cm^{-1}). The near coincidence is indicated by the lower horizontal arrow located at the intercept of $\tau_{\mathrm{QELS,o}}$ (T) with the y-axis at $(1000/T) = 0$. This horizontal arrow points almost at the 690-cm^{-1} peak position of the observed vibrational band. Thus, the QELS data reaffirm the interpretation of the CM that $\tau_{\mathrm{QELS,o}}$ and τ^*_{QELS} are the ion hopping correlation time without and with the effects of many-body interactions between the ions, respectively.

The Meyer–Neldel Rule or Compensation Law

The many-body effects simultaneously make the prefactor τ^*_∞ unphysically short and the activation energy E^*_a larger than the true activation energy E_a. This dual effect is sometimes referred to as the Meyer–Neldel rule [1546] or the compensation law. It makes repeated appearance in other interacting systems, some of which are to be shown later.

Quasielastic Neutron Scattering Studies of Glassy Ionic Conductors

Quasielastic neutron scattering (time of flight) measurements at short times (in the picosecond range) of the ionic diffusion coefficient as a function of temperature in superionic glasses and short-time ion diffusion in fast ion glasses including the systems AgI–$AgPO_3$ and Ag_2S–GeS_2 with and without AgI have been made by two groups [1547–1550]. It was found that the activation enthalpy of the short-time diffusion coefficient E_a is smaller than that of the dc conductivity and approximately equal to βE^*_a [1522(b), 1523, 1058, 1551]. Here β is the Kohlrausch exponent and E^*_a is the activation enthalpy of conductivity relaxation observed at much lower temperatures and frequencies. These neutron scattering experiments measure the ion diffusion with correlation times of the order of picoseconds. The short-time diffusion E_a from neutron scattering is compared with βE^*_a calculated from the long-time conductivity relaxation data in Table 3.2. The good agreement between E_a and βE^*_a suggests the crossover of dynamics near 1 ps.

3.2.3.2 Entangled Polymer Chains

Crossover from Primitive (Rouse) Dynamics to Many-Chain Cooperative Dynamics

The length scale of entanglement interaction (i.e., the entanglement distance) depends on the polymer but usually is of the order of a few nanometers. The radius

Table 3.2 Activation enthalpies E_a of 0.5AgI–0.5$AgPO_3$ [1547, 1548] and Ag_2S–GeS_2 [1549, 1550] from short-time ionic diffusion obtained by neutron time-of-flight measurements. The Kohlrausch exponent $\beta \equiv (1-n)$ and the activation enthalpies E_σ of conductivity relaxation observed in the same glasses at lower temperatures and frequencies in the many-particle hopping region are also included. The near equality between E_a and βE_σ is found, consistent with the change to independent relaxation at times shorter than $t_c \approx 2$ ps and probed by quasielastic neutron scattering (QENS)

Glass	E_a^*	$\beta \equiv (1-n)$	βE_a^*	E_a (QENS)
AgI–$AgPO_3$	21 (kJ/mol)	0.44	9.2 (kJ/mol)	8.7 (kJ/mol)
Ag_2S–GeS_2	0.34 eV	0.45	0.153 eV	0.15 eV

of gyration of the near-Gaussian chain depends on the chain length and can be even larger. The length scale of interaction between entangled chains in polymer melts is much larger than that between repeat units of the same polymer. Consequently, the strength of interaction between entangled chains is weaker than that between repeat units and we expect that the crossover time t_c from many-chain complex dynamics to primitive single-chain dynamics would be much longer than 1 or 2 ps, the crossover time found for local segmental motion in polymers and for structural relaxation of small molecular liquids and ionic conductivity relaxation. The single-chain dynamics without considering chain–chain entanglement interactions is given by the time-honored Rouse model modified for undiluted polymer [29], which has proven to give excellent description of the viscoelastic properties of low molecular weight unentangled polymers except at temperatures close to T_g [29, 1552].

Evidence of crossover from cooperative many-chain dynamics to Rouse dynamics at t_c of the order of nanoseconds can be found from the neutron spin echo spectroscopic data of coherent scattering from poly(ethylene propylene) copolymer and poly(dimethylsiloxane) chains by Richter et al. [1553]. They found the Rouse dynamics at short times from the experimental intermediate scattering function $S(Q,t)$ but it ceases to hold at times longer than a few nanoseconds, after which the dynamics slows down as expected for cooperative many-chain dynamics as shown in Fig. 289.

3.2.3.3 Colloidal Suspensions

Hard-Sphere Colloidal Particles

The diffusion of colloidal particles with a mean radius of about 10^2 nm [141, 143] suspended at high concentrations in a liquid constitutes another problem of relaxations in a many-body system with hard-sphere interaction. This subject has been discussed before in Section 2.2.1.1 and Fig. 9 shows the mean-square displacement $< r^2(t) >$ obtained from the light-scattering dynamic structure factor $S(Q,t)$ by the relation $S(Q,t) = \exp(-Q^2\langle r^2(t)\rangle/6)$ at particles volume fraction $\phi = 0.465$. $<r^2(t)>$ has $t^{1.0}$ dependence for $t<2$ ms and crosses over to assume a fractional power $t^{0.65}$ dependence for $t>2$ ms, which eventually gives way to another $t^{1.0}$ dependence for $t>0.2$ s. The first crossover in the neighborhood defined by

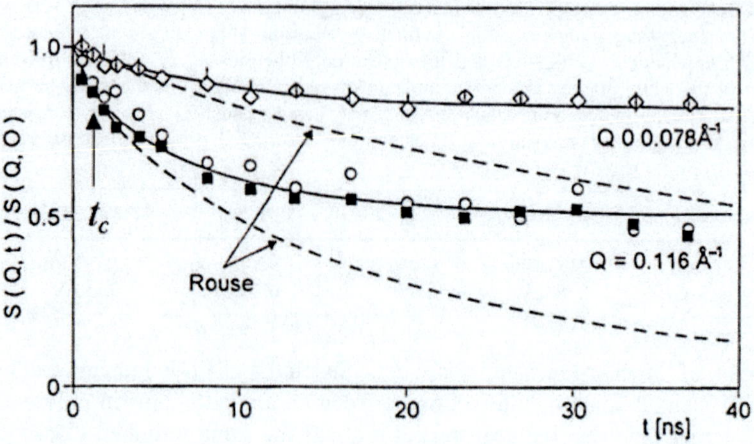

Fig. 289 Evidence of such a t_c can be found from the neutron spin echo spectroscopic data of coherent scattering from poly(ethylene propylene) copolymer and poly(dimethylsiloxane) chains by Richter et al. [1553]. The intermediate scattering function $S(Q,t)$ follows that of the Rouse dynamics (*dashed line*) but ceases to hold at times longer than a few nanoseconds, after which the dynamics slows down. Same data of polymers in dilute solution (not shown) follows the Rouse dynamics. Reproduced from [1553] by permission

2ms $< t < 8$ ms is due to the change of the particle diffusion correlation function from $\exp(-t/\tau_0)$ to $\exp\left[-(t/\tau)^{1-n}\right]$, with $n = 0.35$. Thus, t_c can be taken to be 4 ms. As shown in [1554], the diffusion correlation function $\exp\left[-(t/\tau)^{1-n}\right]$ for t>8 ms is responsible not only for the $t^{0.65}$ dependence for t>2 ms but also for its crossover to steady-state diffusion near 0.2 s. An analysis of the data for $\phi = 0.57$ was given in [1554], but in retrospect the results from this analysis are doubtful. This is because at higher volume fraction such as $\phi = 0.57$ and above, $S(Q, t)$ or $<r^2(t)>$ is the complicated by the particles being caged for a long period of time. This caged colloidal particle dynamics is equivalent of the "intermediate power law" (IPL) observed in small molecular and polymeric glassformers by optical Kerr effect, and as nearly constant loss (NCL) by and dynamics light scattering and dielectric relaxation discussed before in Sections 2.3.2.6, 2.3.2.23, and 2.3.2.33. Its appearance in $S(Q, t)$ or $< r^2(t) >$ obscures the crossover and makes it difficult to ascertain the relaxation times τ_0.

Aqueous Suspension of Laponite

Also discussed before in Section 2.2.1.1, from dynamic light scattering in aqueous suspension of Laponite, the intermediate scattering function $f(\boldsymbol{q}, t)$ at short times has the $\exp(-t/\tau_f)$ dependence, with τ_f having the q^{-2}-dependent normal Brownian motion [338], and thus it is the primitive diffusion in the CM. At longer times, $f(\boldsymbol{q}, t)$ follows the time dependence of $\exp[-(t/\tau_s)^{1-n}]$ [343, 344]. In practice, $f(\boldsymbol{q}, t)$ was fitted by the sum

$$f(q,t) = A\exp(-t/\tau_f) + (1-A)\exp[-(t/\tau_s)^{1-n}]. \tag{3.15}$$

An analysis of the data in terms of the CM by Zulian et al. gave values of the crossover time t_c for samples with different Laponite concentrations [344]. The results of the analysis show that t_c becomes shorter on increasing Laponite weight concentration C_W, following a linear decrease of $\log_{10}(t_c)$ with C_W. The value of t_c is ~1 ms at low C_W of 0.4 wt% and decreases to ~20 μs at high C_W of 3%. This trend is consistent with the CM which predicts that t_c is determined by the interaction. Increase of C_W certainly enhances the interaction between the rigid discs in the Laponite solutions. From the very original CM [47] reviewed in Section 2.2.1.2, it is clear that stronger the interaction, larger the ω_c and shorter the $t_c \equiv (\omega_c)^{-1}$. Also found is that the fast and slow relaxation times τ_f and τ_s of aqueous suspension of Laponite are related by the CM equation (3.3).

3.2.3.4 Semidilute Polymer Solutions

Adding a solvent to an entangled polymer separates the chains further apart, but the chains are still entangled if the length of the chains is long and the concentration is not too low. Such systems are called semidilute polymer solutions. If the crossover from cooperative many-chain dynamics to primitive Rouse dynamics exists, its crossover time t_c would be longer than undiluted entangled polymers, which is a few nanoseconds (see Fig. 289), because entanglement interaction is weakened by the presence of the solvent. Figure 290 shows as an example from light-scattering experiment performed on aqueous gelatin solutions a semidilute entangled polymer solution by Ren et al. [776]. The light-scattering dynamic structure factor $S(Q,t)$ shows clear evidence of sharp crossover from exponential decay to Kohlrausch decay at $t_c \leq 10^{-4}$ s for 3 and 7% aqueous gelatin (linear polymer) solutions at $T = 45°C$ (above gel point) and 27°C (below gel point).

The same was found in pre-gelling silica solution by Martin and Wilcoxon [1555, 1556] using quasielastic light scattering. The initial decay rate is proportional to Q^2, where Q is the scattering vector, indicating it is the primitive relaxation. The slower relaxation that follows has the Kohlrausch stretched exponential time dependence. In the pre-gel silica solution, t_c lies between 10^{-5} and 10^{-4} s, and the Kohlrausch fractional exponent $(1-n)$ is equal to 0.65.

3.2.3.5 Polymeric Cluster Solutions

Similar light-scattering experiments on polymeric cluster solution by Adam et al. [780] and Delsanti et al. [1557] also have the crossover from exponential decay to Kohlrausch stretched exponential decay at t_c with similar order of magnitude as semidilute polymer solutions. In all these cases, the Kohlrausch exponent $\beta \equiv (1-n)$ decreases with increasing concentration of the polymer clusters, naturally caused by increasing coupling between the relaxing units and hence the coupling parameter of the CM.

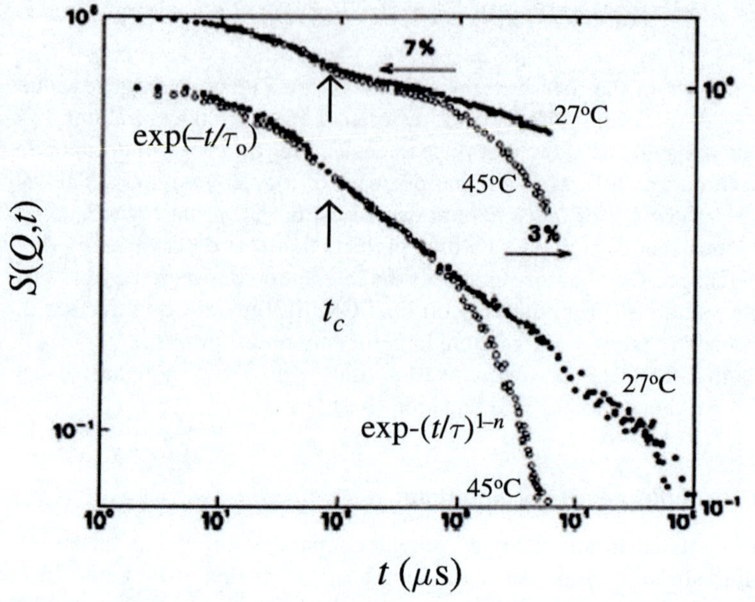

Fig. 290 Light-scattering dynamic structure factor $S(Q,t)$ for 3% (*bottom* and *right* y-axis) and 7% (*top* and *left* y-axis) aqueous gelatin (linear polymer) solutions at $T = 45°C$ (above gel point) and 27°C (below gel point), showing the crossover time $t_c \leq 10^{-4}$ s. The x-axis is logarithm of time in micro seconds ranging from $1\mu s$ to $10^5 \mu s$. Reproduced from [776] by permission

3.2.3.6 Associating or Aggregating Polymer Solutions

The time correlation data obtained from the dynamic light-scattering (DLS) experiments on polymer solutions of various kinds by Nyström and coworkers [778, 1558–1560] and earlier by others [1561] have revealed the existence of two relaxation modes, one single exponential at short times followed by a Kohlrausch stretched exponential at longer times. The first-order electric field correlation function $g^1(q,t)$ obtained from DLS (see Section 1.3.2) in all these reports was fitted by Eq. (3.15) to deduce the parameters τ_f, τ_s, n, and A, and their q dependences. Here are some examples of associating polymer:

(1) Aqueous solutions of a hydrophobically end-capped poly(oxyethylene) (POE) urethane (HPOEU) by Nyström et al. [778].
(2) Aqueous solutions of associating diblock and triblock poly(oxyethylene)-containing copolymers of the same type by Thuresson et al. [1558].
(3) Semidilute aqueous mixtures of various compositions of oppositely charged and hydrophobically modified polyelectrolytes. In the system, the cationic polymer is a N,N-dimethyl-N-dodecylammonium derivative of hydroxyethylcellulose and the hydrophobically modified polyacrylate, which was chosen as the negatively charged polyelectrolyte [1559].

(4) An example of aggregating polymer solutions is aqueous solutions of low-methoxyl pectins. Pectins are important water-soluble, anionic heteropolysaccharides of plant origin. Structurally they are biopolymers consisting of a linear backbone of randomly connected (1→4)-linked α-D-galactosyluronic acid residues partially esterified with methanol and interrupted by (1→2)-linked α-L-rhamnopyranosyl residues. Low temperatures and increased pectin concentrations promote the formation of multichain aggregates. The time correlation data obtained from dynamic light-scattering experiments revealed, for all solutions with polymer from 0.5 up to 1.5 wt%, the existence of two relaxation modes, one single exponential at short times followed by a Kohlrausch stretched exponential at longer times. In the semidilute regime at 1.5 wt%, on lowering the temperature, chain associations are enhanced in the solutions resulting in increasingly longer values of the Kohlrausch relaxation time of the slow mode, and it has a stronger wave vector dependence. These features have been rationalized in the framework of the CM by Kjøniksen et al. [1560].

(5) Similar results were obtained by Narayanan et al. also by DLS in aqueous solutions of pectin by the addition of calcium chloride to induce gelation [1562].

(6) The crossover was found in two other systems: aqueous alginate modified by association and gelation via reaction by Bu et al. [1563] and semidilute aqueous hyaluronic acid solutions by Maleki et al. [1564(a)].

(7) Similar result was found in rod-like micelles formed at 3 M NaCl, which become increasingly entangled with increasing concentration of the copolymer PEO–PPO–PEO triblock copolymers [1565]. Length of the micellar rods increases with increase in NaCl concentration at constant copolymer concentration. Another system is a thermosensitive methoxy-poly(ethylene glycol)-*block*-poly(*N*-isopropylacrylamide)-*block*-poly(4-styrenesulfonic acid sodium) triblock copolymer having the following composition: $MPEG_{45}$-*b*-$P(NIPAAM)_n$-*b*-P(SSS) with $n = 17$, 48, and 66 [1564(b)]. The fast relaxation mode always is diffusive with $D = 1/\tau_f q^2$.

3.2.4 Anomalous $Q^{-2/(1-n)}$ Dependence of τ

Quasielastic neutron scattering data of molecular glassformers presented in Section 2.2.5.2 have shown that the slow structural α-relaxation time τ_α does not have the normal Q^{-2} dependence on the scattering vector Q, while the fast primitive relaxation time τ_0 does. Instead τ_α has the stronger but anomalous $Q^{-2/(1-n)}$ dependence, where $(1-n)$ is the fractional exponent of the Kohlrausch correlation function of the α-relaxation. This is also the case for non-molecular glassformers including colloidal particle suspension [143] and Laponite [344] from dynamic light-scattering measurements. This is one of several properties of the α-relaxation indicating that they are governed by the dispersion or n, and the anomalies associated with them originate from many-body dynamics. The anomalous $Q^{-2/(1-n)}$ dependence of τ_α is derivable from the CM equations (3.1) or (3.2) simply by substituting the known or found Q^{-2} dependence of τ_0 into them. If this anomaly of structural relaxation of glassformers is indeed originating from many-body relaxation as proposed in Section 2.2, then it should show up in relaxation of interacting systems which are not glassformers and the relaxation process is entirely different from structural relaxation of glassformers, as long as they are relaxations involving units that are interacting with each other.

Almost all the systems discussed above in Sections 3.2.3.3–3.2.3.6 show that the slow relaxation time τ_s of the Kohlrausch stretched exponential decay in Eq. (3.15) has the anomalous $Q^{-2/(1-n)}$ dependence, while τ_f has the normal Q^{-2} dependence. All relaxations of the systems in Sections 3.2.3.4–3.2.3.6 involve motions of polymer chains totally unrelated to the structural α-relaxation and the glass transition problem. Because these relaxations also exhibit the crossover from exponential to Kohlrausch decay in some neighborhood of t_c, the CM equations (3.1) and (3.2) hold and explain the $Q^{-2/(1-n)}$ dependence of τ_s. For other related applications of the CM, see the review in [1567]. To be precise, τ_s and $(1 - n)$ are two separate parameters in the Kohlrausch function. While τ_s has the experimentally determined Q dependence given by $\tau_s \propto Q^{-\alpha_s}$, the exponent (1–$n$) is independent of Q, and it turns out that α_s is in reasonably good agreement with $2/(1 - n)$. The systems showing the $Q^{-2/(1-n)}$ dependence of τ_s vary greatly. They include (i) aqueous solutions of linear random coil molecules of swine gelatin, (ii) synthetic polymer (polyurethane) cluster solutions and pre-gelling silica solution, (iii) aqueous solutions of a hydrophobically end-capped poly(oxyethylene) urethane, (iv) aqueous solutions of associating diblock and triblock poly(oxyethylene)-containing copolymers, (v) semidilute aqueous mixtures of various compositions of oppositely charged and hydrophobically modified polyelectrolytes, (vi) aqueous solutions of the biopolymer pectin, alginate, and hyaluronic acid, (vii) rod-like micelles formed by the copolymer PEO–PPO–PEO triblock copolymers in the presence of NaCl, and (viii) aqueous solutions of hydroxypropylcellulose (HPC) by probe diffusion measurements [198, 777, 1566].

Not mentioned in the above is the three-component microemulsion system containing bis(2-ethylhexylsulfosuccinate) (AOT) (surfactant), D_2O, and decane (oil), and the measurements by dynamic light-scattering, small-angle neutron scattering, and neutron spin echo (NSE) spectroscopic measurements by Sheu et al. [1568]. The time-dependent density correlation function exhibits the Kohlrausch form with stretch exponent (1–n), and its relaxation time has the anomalous $Q^{-2/(1-n)}$ dependence.

The $Q^{-2/(1-n)}$ -dependence of τ_s is thus shared by these diverse systems besides the colloidal suspension and Laponite. This suggests that the same $Q^{-2/(1-n)}$ dependence of the structural α-relaxation time of glassformers is a universal manifestation of many-body relaxation.

Before leaving this topic, the data from three systems are shown. All illustrate the increase of n or decrease of the Kohlrausch stretch exponent $\beta \equiv (1 - n)$ when interaction is enhanced by one way or the other, as well as corresponding increase of the exponent α_s of the experimentally determined Q dependence $\tau_s \propto Q^{-\alpha_s}$ and the good agreement of α_s with $2/\beta$.

The first example is taken from light-scattering experiment performed on aqueous gelatin solutions, the semidilute entangled linear polymer solutions by Ren et al. [776] shown in Fig. 291. The dynamic structure factor $S(Q,t)$ has been presented in Fig. 290.

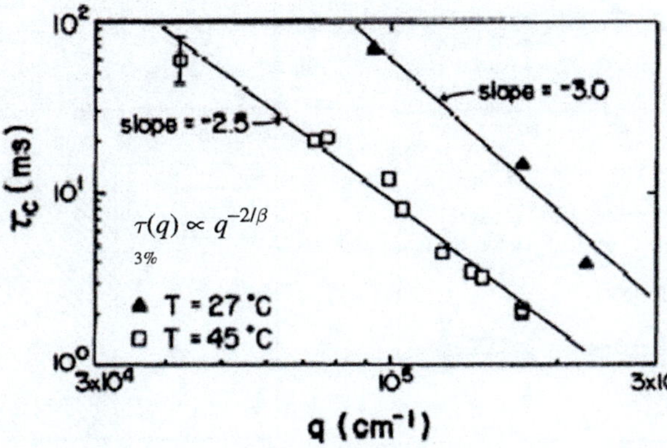

Fig. 291 Data of the slow-chain diffusion relaxation time τ_c (symbols) in aqueous gelatin semidilute solutions (*symbols*) showing its q dependence, which is well described by the $q^{-2/(1-n)}$ dependence (*lines*) predicted by the CM. Reproduced from [776] by permission

The fast-chain diffusion process with exponential decay has the normal q^{-2} dependence for its relaxation time τ_f. The q dependence of the slow polymer chain diffusion time τ_s from the dynamic structure factor $S(q,t)$ is shown by symbols in Fig. 291 for 3% aqueous gelatin solutions (τ_c in the figure). The data at 45°C (open squares) were taken above the gel point (~30°C) and the Kohlrausch stretched exponential fit to the time dependence of $S(q,t)$ has the stretch exponent $(1-n) = 0.81$. Also shown in the same figure are the data of τ_c at 27°C, which is below the gel point and the stretch exponent $(1-n) = 0.67$. The larger n at 27°C is due to enhanced intermolecular coupling in the gel by the presence of cross-links. The lines in the figure have the predicted $q^{-2/(1-n)}$ dependence using exactly the values of 0.81 and 0.67 for $(1–n)$ obtained from the Kohlrausch fits. It can be seen from the figure that these lines well describe the q dependence of τ_c.

The second example come from 0.1, 0.5, and 1.5 wt% of pectin solutions by Kjøniksen et al. [1560]. Part (a) of the left panel of Fig. 292-1 presents the temperature dependence of the stretched exponent β of the Kohlrausch relaxation function of the slow mode for the pectin solutions. Part (b) shows the effects of temperature and polymer concentration on α_s and $2/\beta$, as well as good agreement between these two quantities.

The third example shown in the right panel of Fig. 292-1 comes from the semidilute aqueous mixtures of various compositions of anionic hydrophobically modified polyacrylate (HM-P-) and cationic hydrophobically modified hydroxyethylcellulose (HM-P+) with total polymer concentration of 1 wt% by Tsianou et al. [1559]. Part (a) of the figure presents some examples of the wave vector q dependences of the fast (τ_f^{-1}) and the slow (τ_s^{-1}) inverse relaxation times for several mixture ratios

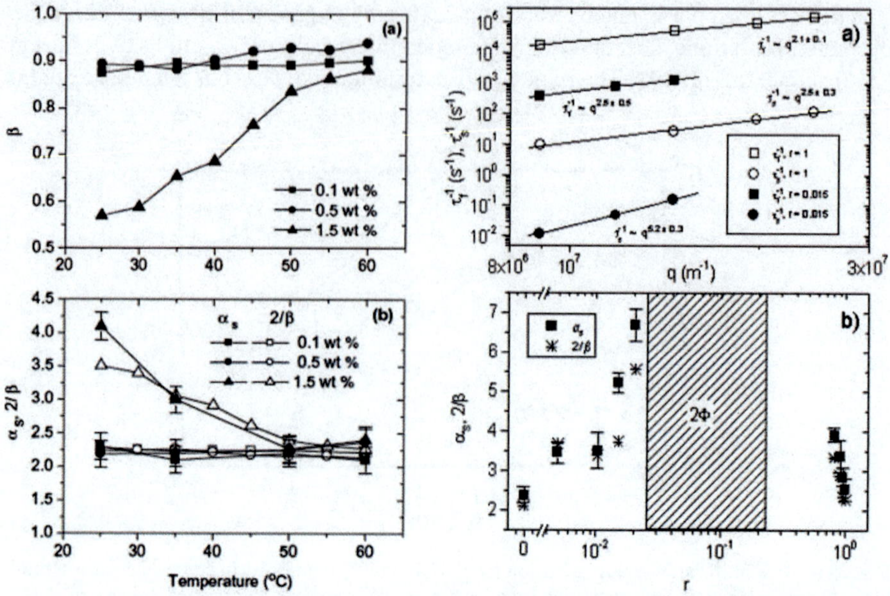

Fig. 292-1 (*Left*) (**a**) Temperature dependence of the stretched exponent β for pectin solutions at the concentrations shown. (**b**) Effects of temperature and polymer concentration on the quantities $2/\beta$ and α_S, illustrating the q dependence of the slow inverse relaxation mode. Reproduced from [1560] by permission. (*Right*) (**a**) Illustration of the wave vector dependences of the fast (τ_f^{-1}) and the slow (τ_s^{-1}) inverse relaxation times for semidilute aqueous mixtures of various compositions of anionic hydrophobically modified polyacrylate (HM-P-) and cationic hydrophobically modified hydroxyethylcellulose (HM-P+) with total polymer concentration of 1 wt% by Tsianou et al. [1559]. The mixture ratio r is indicated. (**b**) Effects of the mixture ratio on the quantities $2/\beta$ (see the main text for explanation) and α_S, illustrating the q dependence of the slow inverse relaxation time. Reproduced from [1559] by permission

indicated. Part (b) shows the effects of the mixture ratio outside the two-phase region (shaded area labeled by 2Φ) on the quantities $2/\beta$ and α_S, and the good correspondence between them. It is understood that the intermolecular interactions are strengthened when approaching the two-phase region because of polymer clustering. Thus the accompanying observed decrease of β and the good agreement between $2/\beta$ and α_s are in accord with the CM predictions [1567].

The fourth example is given by the family of thermosensitive methoxy-poly (ethylene glycol)-*block*-poly(*N*-isopropylacrylamide)-*block*-poly(4-styrenesulfonic acid sodium) triblock copolymers having the following composition: $MPEG_{45}$-*b*-P(NIPAAM)$_n$-*b*-P(SSS) with $n = 17$, 48, and 66 [1564(b)]. The structure of the triblocks is shown by the chemical formula and the illustration on the right side of Fig. 292-2. Overall, there is good agreement between α_S and $2/\beta_S$, where α_s is the power in the observed dependence $\tau_S \propto Q^{-\alpha_S}$ and $\beta_S \equiv (1 - n_s)$ is the Kohlrausch exponent of the stretched exponential correlation function for the slow mode. Remarkably, even the temperature dependences of α_S and $2/\beta_S$ are about the same.

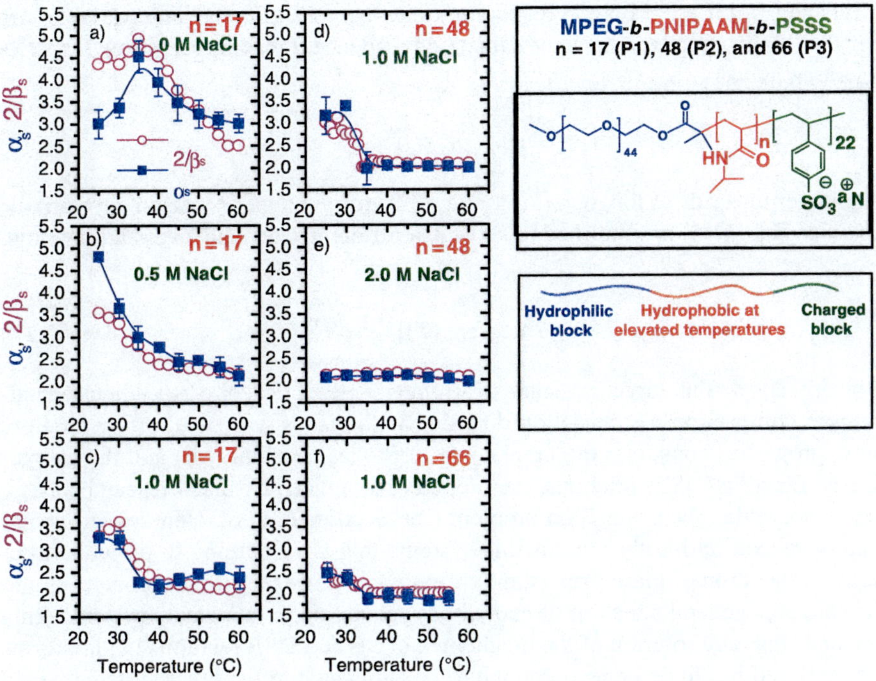

Fig. 292-2 (*Left*) Effects of temperature and salt addition on the quantities $2/\beta$ (see text for explanation) and α_s, expressing the q dependence of the slow relaxation time, for 1 wt% solutions of M(PEG)$_{45}$-*b*-P(NIPAAM)$_n$-*b*-P(SSS)$_{22}$. Courtesy of Bo Nyström. The data shown are to be published as [1564(b)] in the future. (*Right*) Chemical structure and illustration

3.2.5 Different Correlation Functions of the Same Relaxation Can Have Different Kohlrausch Exponents (1– n), Relaxation Times τ, and T-Dependences

Previously in Section 2.2.5.6, experimental data of the structural α-relaxation in the same glassformers were presented to show that the Kohlrausch correlation functions $\langle\mu(0)\mu(t)\rangle = \exp[-(t/\tau_\mu)^{1-n_\mu}]$ of different dynamic variables μ can differ in the stretch exponent $(1 - n_\mu)$ and the value of the relaxation time τ_μ as well as its T dependence. This situation occurs when comparing data obtained from different techniques such as translational diffusion, rotational diffusion, dielectric relaxation, dynamic light scattering, mechanical relaxation, and NMR, which involve various dynamic variables. This property has been interpreted as caused by different dynamic variables weighing the effect of many-body relaxation differently, resulting in different degrees of stretching of the α-relaxation to different longer times. Correlation function for dynamic variable μ that is more stretched (i.e., the exponent n_μ appearing in the Kohlrausch function is larger) has longer relaxation time τ_μ and stronger T dependence of τ_μ. These properties are predicted by the CM equations

(3.1) and (3.2) if the primitive relaxation times $\tau_{0\mu}$ of all the dynamic variables are equal or about equal and have the same T dependence. It can be seen from Eq. (3.1) rewritten as

$$\tau_\mu = \tau_{0\mu}(\tau_{0\mu}/t_c)^{n_\mu/(1-n_\mu)} \tag{3.16}$$

that larger n_μ leads to longer τ_μ. Like Eq. (3.2) for the dependence on any variable U, when Eq. (3.16) is simplified to show just different T dependences and rewritten as

$$\tau_\mu(T) \propto [\tau_{0\mu}(T)]^{1/(1-n_\mu)}, \tag{3.17}$$

it clearly shows that larger n_μ leads to stronger T dependence of τ_μ. When specializing to comparison of translation diffusion with viscosity or with probe rotation, these properties constitute the breakdown of Stokes-Einstein (SE) and the Debye-Stokes-Einstein (DSE) relations, the property recognized by many researchers as a key to solve the glass transition problem (see Section 2.2.5.6). Here, experimental data of relaxation in other interacting systems that have nothing to do with structural α-relaxation of glassformers are presented to show the similar properties, and they are thus general behavior of many-body relaxation in interacting systems. This demands that any solution of the breakdown of SE and DSE relations in glassformers proffered has to be general enough to be applicable to the other systems. Since the CM is designed for relaxation and diffusion of any interacting systems, naturally its explanation for the breakdown of SE and DSE relations in glassformers given in Section 2.2.5.6 is general enough to satisfy this requirement.

3.2.5.1 Glassy Ionic Conductors: Conductivity vs. NMR

This subject has been touched before in Section 2.2.1 to support the CM explanation for the breakdown of SE and DSE relations in glassformers. More experimental data and details are given here to stress the universal behavior of relaxation and diffusion in interacting systems.

Experimental Data

The fact that the dynamics of ion diffusion as probed by nuclear spin relaxation (NSR) and electrical conductivity relaxation (ECR) in glassy ionic conductors is significantly different was discovered by Tatsumisago et al. [147, 279] and by Kanert and coworkers [280–282] at almost the same time in 1992 on different ionic conductors. Subsequently, other workers have repeatedly confirmed this anomalous property [283–289] that resembles the breakdown of the DE and DSE laws. Experimentally both the electrical conductivity and NSR correlation functions have the Kohlrausch forms and Arrhenius T dependences for their relaxation times:

$$\phi_M(t) = \exp[-(t/\tau_M)^{1-n_M}], \quad \text{where} \quad \tau_M = \tau_{M,\infty}\exp(E_M/kT) \tag{3.18}$$

and

$$\phi_s(t) = \exp[-(t/\tau_s)^{1-n_s}], \quad \text{where} \quad \tau_s = \tau_{s,\infty}\exp(E_s/kT). \tag{3.19}$$

Notwithstanding, τ_s of the relaxation of the nuclear spin of the diffusing ion is considerably longer than the electrical conductivity relaxation time τ_M. The activation energies E_s of τ_s is larger than E_M of τ_M. At this level, the relations $\tau_s >> \tau_M$ and $E_s > E_M$ from the dynamics of ions are analogues of the relations between the viscosity (or rotation relaxation) and self (or probe)-diffusion, which constitute the breakdown of the SE and DSE relations. Moreover, it was found that $n_s > n_M$, the analogue of $n_\eta > n_D$ or $n_r > n_D$, which is the key to the explanation of the breakdown of the SE and DSE relations by the CM in Section 2.2.5.6. The effect was also found by a Monte Carlo simulation experiment of a disordered Coulomb lattice gas model of the ionic conductor [467, 1569, 1570]. It is also present in the NMR and conductivity relaxation data of the non-glassy fast ionic conductor sodium β-alumina, Na-βAl_2O_3, found by Bjorkstam and Villa [290] as early as 1980 and brought back into attention in 1993 by [291]. These experiments have proven that the decoupling between NSR and conductivity relaxation is a general phenomenon of interacting many-ion dynamics, glassy or crystalline.

Before proceeding further, I hasten to point out that electrical conductivity relaxation (ECR) is a macroscopic probe involving the measurement of capacitance $C(\omega)$ and conductance $G(\omega)$, from which the parallel quantities $\varepsilon^*(\omega)$, $\sigma^*(\omega)$ or $M^*(\omega)$ are obtained. Therefore, the macroscopic electrical conductivity correlation function in Eq. (3.18) fitting the $M^*(\omega)$ is not the microscopic conductivity correlation function $\phi_\sigma(t)$ given by

$$\phi_\sigma(t) = \exp[-(t/\tau_\sigma)^{1-n_\sigma}], \quad \text{where} \quad \tau_\sigma = \tau_{\sigma,\infty}\exp(E_\sigma/kT). \tag{3.20}$$

The Kohlrausch exponents of $\phi_M(t)$ and $\phi_\sigma(t)$ are the same and both denoted by n, because the time dependence or frequency dispersion is faithfully reproduced by the macroscopic measurement. The difference is between the macroscopic τ_M and the microscopic τ_σ, which will be given later, although they have the same activation energy E_σ and hence the prefactors $\tau_{\sigma,\infty}$ and $\tau_{M,\infty}$ differ. To avoid confusion with notations, the relations between the parameters of electrical and microscopic conductivity relaxation are written explicitly as follows:

$$n_M = n_\sigma, \; E_M = E_\sigma, \; \tau_{M,\infty} \neq \tau_{\sigma,\infty}. \tag{3.21}$$

On the other hand, nuclear spin-lattice relaxation is a microscopic probe of the motion of the mobile ions, and the SLR, $T_1^{-1}(\omega_L, T)$, corresponds to the microscopic correlation function.

Here we show the three examples, all of which involve macroscopic conductivity relaxation measurements and hence τ_M. The first one is from the 1992 work of Kanert and coworkers on a heavy metal fluorozirconate glass with the following composition (in mol%): 27.4ZrF_4, 27.4HfF_4, 19.8BaF_2, 3LaF_3, 3.2AlF_3, 20NaF

(ZBLAN glass; $T_g = 553$ K). The F^- ions are responsible for ionic conductivity, which was measured over ranges of frequencies and temperatures. The conductivity relaxation time τ_M, its activation energy E_σ, and the coupling parameter n_σ determined are shown in Fig. 293 and Table 3.3. The ^{19}F spin-lattice relaxation (SLR) times τ_s has been measured at different Larmor frequencies. The τ_s of ZBLAN was obtained from the ^{19}F spin-lattice relaxation rate $T_{1\rho}^{-1}(\omega_L, T)$ in the rotating frame at Larmor frequencies of 28, 42, and 62 kHz matching the high frequencies used to measure the conductivity relaxation time τ_M [282]. The results of τ_s are shown in Fig. 293, and n_s and E_s determined from fits to temperature dependence of $T_{1\rho}^{-1}(\omega_L, T)$ are given in Table 3.3. All measurements were in the glassy state where the temperature dependences of τ_s and τ_M are undoubtedly Arrhenius, and their difference is brought out by the comparison of the $T_{1\rho}^{-1}(\omega_L, T)$ with the product σT at the *same* frequency of about 25 kHz. The fact that the characteristic temperatures $T_s \approx 525$ K and $T_M \approx 425$ K of τ_s and τ_M, respectively, are so different makes perfectly clear SLR and conductivity relaxation are not the same, although both

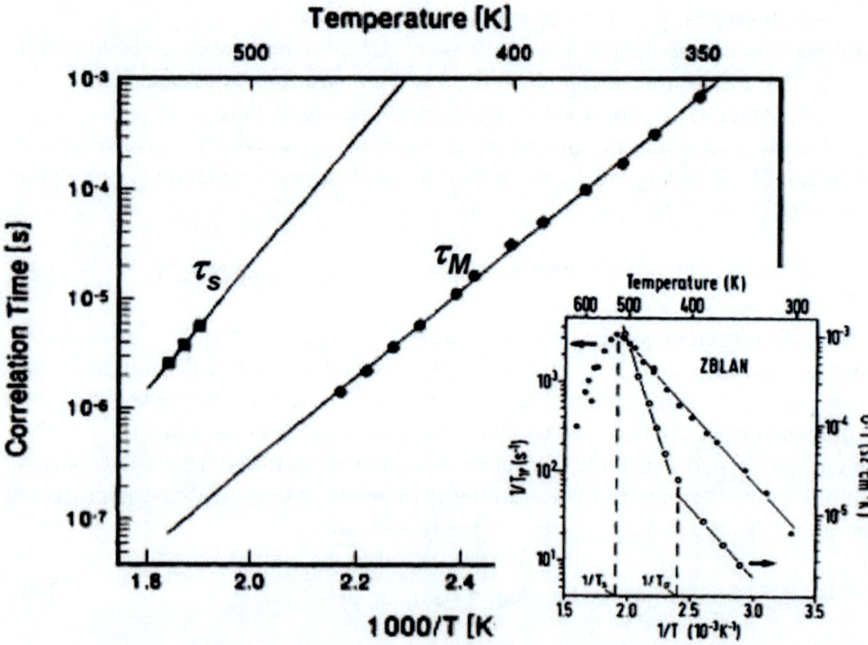

Fig. 293 ZBLAN conductivity correlation time τ_M from data analysis of the electric modulus and NSR correlation time τ_s from frequency dependence of $1/T_{1\rho}$ maxima plotted against $1000/T$. The inset is a comparison of the T dependence of the ionic motion-induced part of the ^{19}F spin-lattice relaxation (SLR) rate $1/T_{1\rho}$ and of σT observed at the *same* frequency of about 25 kHz. These isochronal data clearly show the different temperatures $T_s \approx 525$ K and $T_M \approx 425$ K of the SLR maximum and of the crossover point of σT from one T dependence to another. Reproduced from [282] by permission

Table 3.3 The relaxation time τ_s, activation energy E_s, and coupling parameters n_s for SLR, and the corresponding quantities τ_M, E_σ, and n_s for ECR for three different glassy ionic conductors

Ionic conductor	Dynamic variable *U*	$\beta_U \equiv (1-n_U)$	E^*_U (*K*)	$(1-n_U)E^*_U$ (*K*)
$(Li_2S)_{0.56}(SiS_2)_{0.44}$	*Spin* (SLR)	0.35	5845	2046
$(Li_2S)_{0.56}(SiS_2)_{0.44}$	*M* (ECR)	0.52	3911 (from τ_M)	2034
			4000 (from σ_{dc})	2080
$(LiCl)_{0.6}(Li_2O)_{0.7}(B_2O_3)$	*Spin* (SLR)	0.35	7400	2590
$(LiCl)_{0.6}(Li_2O)_{0.7}(B_2O_3)$	σ (ECR)	0.50	5500	2750
ZBLAN[a]	*spin* (SLR)	0.30	13346	3830
ZBLAN[a]	σ (ECR)	0.44	9284	4085

[a]27.4ZrF, 27.4HfF, 19.8BaF, 3LaF, 3.2AlF$_3$, 20NaF (in mol%)

measure the motion of the F$^-$ ions. For details of the procedure used to obtain the parameters, see [282].

The second example is the 1992 study by Tatsumisago et al. [147, 279] on Li$^+$ ion motion in the fast glassy ionic conductor 0.6LiCl–0.7Li$_2$O–B$_2$O$_3$. The right panel of Fig. 294 is the Arrhenius plot of the correlation time of the mobile Li$^+$ ion τ_s deduced from ^{7}Li nuclear spin-lattice relaxation measurement and the macroscopic electrical conductivity relaxation time τ_M from dielectric measurements. In this work, the Larmor frequencies of the SLR in the 10–100-MHz range are much higher than the conductivity relaxation measurements, unlike the rotating frame SLR measurements in the tens of kilohertz range of Kanert and coworkers. Nevertheless, the fact that τ_s is much longer than τ_M and has a larger activation energy E_s than E_σ is clear

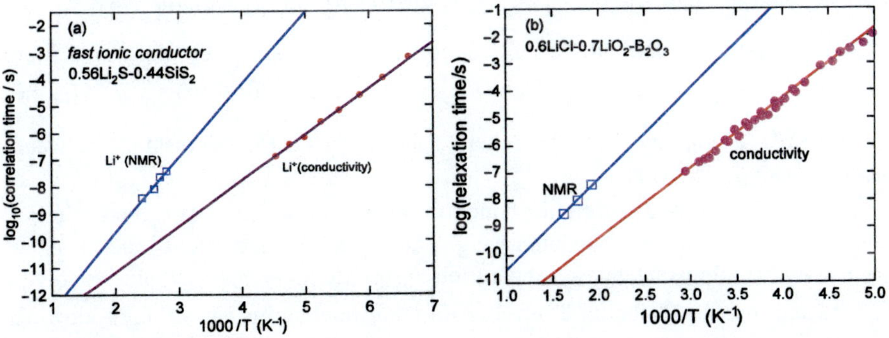

Fig. 294 (**a**) Arrhenius plot of the correlation time of the mobile ion τ_s deduced from ^{7}Li nuclear spin-lattice relaxation measurement (*open squares*) and the macroscopic electrical conductivity relaxation time τ_σ (•) from measurements of τ_σ. Data from [286] and [292] are replotted here. (**b**) Arrhenius plot of the correlation time of the mobile ion τ_s deduced from ^{7}Li nuclear spin-lattice relaxation measurement (*open squares*) and the macroscopic electrical conductivity relaxation time τ_σ or τ_{EM} (*closed circles*) from measurements of 0.6LiCl–0.7Li$_2$O–B$_2$O$_3$. Data from [279] and [291] are replotted here. The *upper full line* is τ_s calculated by the CM (see text)

from the figure. Furthermore, the coupling parameters n_s for SLR and n_σ for conductivity relaxation have been determined from the data. All parameters are given in Table 3.4.

The third example is taken from glassy Li^+ conductor $(Li_2S)_{0.56}(SiS_2)_{0.44}$ by Borsa et al. [286]. The quality of the data is similar to that of Tatsumisago et al., and the Arrhenius plot of τ_s and τ_M shown in the left panel of Fig. 294 shows the same features. All parameters deduced from the data are entered into Table 3.3.

The CM Explanation

The frequency-dependent conductivity $\sigma(\omega)$ is given in Eq. (1.130) by the Fourier transform of the current–current correlation function, or equivalently the velocity–velocity correlation function $\phi_\sigma(t) = N^{-1}\sum_{ij}\langle v_i(0)v_j(t)\rangle$. An expression for $\sigma(\omega)$ in terms of the mean-square displacement is given by Eq. (3.4). In this expression, the Haven ratio [1490] though less than unity is not very small for glassy ionic conductors, which indicates that the tracer diffusion correlation function $\phi_D(t) = \langle r(0)r(t)\rangle$ is only slightly different from $\phi_\sigma(t)$ and the $i{\neq}j$ cross terms is not important.

If spin-lattice relaxation (SLR) measurements are made using the mobile ion nucleus, and the ion SLR mechanism is caused by time-dependent fluctuations of the nuclear spin coupling energy ω_{ij} between interacting ion pairs (i,j), the resulting SLR rate $T_1^{-1}(\omega_L, T)$ as a function of temperature at the Larmor frequency ω_L is given by the expression $\kappa[J(\omega_L, T) + 4J(2\omega_L, T)]$ [92]. Here κ is the coupling constant and $J(\omega,T)$, the spectral density function, is the real part of the Fourier transform $J(\omega, T) \equiv \mathrm{Re}\int_0^\infty \phi_s(t/\tau_s)\exp(-i\omega t)dt$ of the pair–pair correlation function $\varphi_s(t) = \sum_{i,j} < \omega_{ij}(0)\omega_{ij}(t) >$. If the ion NSR is via magnetic dipole or quadrupolar interactions, it is governed by the correlation function $\phi_s(t) = (1/N)\sum_{i\neq j}(1/N)\sum_{i\neq j}\left\langle F_{ij}^{(q)}(t)\, F_{ij}^{(q)}(0)\right\rangle$, where $F_{ij}^{(q)}(t) = q\sqrt{8\pi/15Y_2(q)}\,\Omega_{ij}\Big/r_{ij}^3$, Y_2 is the spherical harmonics, r_{ij} is the distance between two ions, and $q = 1, 2$ [92].

$\phi_\sigma(t)$ is effectively a correlation function for an ion, while $\phi_s(t)$ is a correlation function for a pair of ions. The latter weighs more heavily on the contributions from ion pairs at smaller distance of separation r_{ij}, particularly for $\phi_s(t)$ that involves $F_{ij}^{(q)}(t)$ and the $(r_{ij})^{-3}$ factor. Therefore, ion–ion interaction has stronger slowing and stretching effect on $\phi_s(t)$ than on $\phi_\sigma(t)$ and this translates in the context of the CM to a larger n_s in the SLR correlation function in Eq. (3.19) than n_σ in the electrical conductivity correlation function in Eq. (3.18) or the macroscopic conductivity relaxation function in Eq. (3.20), e.g., $n_s > n_M$. The same argument has been used to justify the coupling parameter for center–of-mass diffusion is smaller than that for viscosity in the CM explanation of the breakdown of the Stokes–Einstein relation in Section 2.2.5.6. The independent relaxation of the ion and its primitive relaxation time $\tau_0 = \tau_{0,\infty}\exp(E_a/kT)$ as well as t_c are the same for both NSR and $\sigma(\omega)$. On applying Eq. (3.1) separately to NSR and $\sigma(\omega)$, we have [282, 291–294, 1571]

$$\tau_s = [(t_c)^{-n_s}\tau_0]^{1/(1-n_s)} = \tau_0[\tau_0/t_c]^{n_s/(1-n_s)}, \tau_\sigma = [(t_c)^{-n_\sigma}\tau_0]^{1/(1-n_\sigma)} = \tau_0[\tau_0/t_c]^{n_\sigma/(1-n_\sigma)}. \tag{3.22}$$

Since $n_s > n_\sigma$ and the fact that in the experimental investigations all relaxation times including τ_0 are much longer than $t_c \approx 1-2$ ps, these equations lead us to the result that $\tau_s >> \tau_\sigma$, as observed. In particular, we have the following relations between the activation energies and prefactors:

$$(1-n_s)E_s = (1-n_\sigma)E_\sigma = E_a,\ \tau_{s,\infty} = \tau_\infty(\tau_\infty/t_c)^{n_s/(1-n_s)},\ \tau_{\sigma,\infty} = \tau_\infty(\tau_\infty/t_c)^{n_\sigma/(1-n_\sigma)}. \tag{3.23}$$

The first relation is well obeyed by the experimental data from electrical conductivity relaxation (ECR), in view of the relations between the parameters in $\phi_M(t)$ of ECR and $\phi_\sigma(t)$ given in Eq. (3.21), within experimental errors as demonstrated in Table 3.3. The prefactor τ_∞ of the primitive relaxation time τ_0 is shorter than $t_c \approx 1-2$ ps because it is the reciprocal of vibrational frequency. From this and $n_s > n_\sigma$, it can be deduced from the second and the third relations under Eq. (3.23) that $\tau_{s,\infty} << \tau_{\sigma,\infty} << \tau_\infty$, which is in accord with experiments. This can be seen if the Arrhenius T dependences of τ_s and τ_M in Figs. 293 and 294 are extrapolated to infinite temperature. From the ZBLAN data in Fig. 293, $\tau_{s,\infty} = 6.6\times10^{-17}$ and $\tau_M, \infty = 4.1\times10^{-15}$. The experimental value of the prefactor $\tau_{s,\infty}$ is also in agreement with that calculated from the second CM equations in Eq. (3.23), assuming that the prefactor $\tau_{0,\infty}$ of τ_0 is 2×10^{-13} s [282]. The comparison of the experimental prefactor $\tau_{s,\infty}$ with calculation has to wait later after the microscopic prefactor $\tau_{\sigma,\infty}$ has been deduced from the experimental $\tau_{M,\infty}$ by a relation in the next subsection.

An Analogue of Breakdown of SE and DSE Relations and Decoupling of Dynamic Variables in the Same Glassformer

The decoupling of nuclear SLR from electrical conductivity relaxation is an exact analogy of the breakdown of SE and DSE relations in supercooled liquids. The SLR is the counterpart of viscosity or rotational relaxation, and electrical conductivity relaxation is the counterpart of self-diffusion or probe translational diffusion. It is also the analogue of the observation of different magnitudes and temperature dependence of the structural α-relaxation times for dielectric, mechanical, light-scattering, and enthalpy relaxation in a variety of glassformers by experiments and by molecular dynamics simulations [266–275], discussed before in Section 2.2.1(v). The relations $\tau_s >> \tau_\sigma$, $E_s > E_\sigma$, and $n_s > n_\sigma$ have analogues in the breakdown of SE and DSE relations (see Section 2.2.5.6 and Fig. 247). In the present case, the coupling parameters n_s and n_σ are known from experiment to provide a critical test of the prediction of the CM, which turned out to be valid in general for glassy ionic conductors.

Quantitative Relation Between τ_s and τ_σ *Beyond* $(1 - n_s)E_s = (1 - n_\sigma)E_\sigma = E_a$

As discussed before, electrical conductivity relaxation (ECR) measurements give macroscopic information about ion motion dynamics. The capacitance $C(\omega)$ and conductance $G(\omega)$ measured, as well as the quantities $\varepsilon^*(\omega)$, $\sigma^*(\omega)$, or $M^*(\omega)$ obtained, give no clue to ion concentration, charge, and jump length, and thus clearly it is not microscopic. Notwithstanding, the measured electrical conductivity relaxation (ECR) time τ_M and the microscopic conductivity relaxation time τ_σ have the same activation energy E_σ, emphasized in Eq. (3.21), but their prefactors $\tau_{\sigma,\infty}$ and $\tau_{M,\infty}$ differ. There is a way to calculate τ_σ from τ_M given by a comparison between microscopic and electric conductivity relaxation in [56(b)].

Starting from the velocity–velocity correlation function, an explicit expression for microscopic complex conductivity $\sigma^*(\omega\tau_\sigma)$ is derived via a stochastic transport theory which is given as

$$\sigma^*(\omega\tau_\sigma) = (Nq^2/kT\tau_\sigma)(r_{\text{rms}}^2/6)\left\{\frac{1}{\widetilde{\phi}_\sigma(\omega\tau_\sigma)} - i\omega\tau_\sigma\right\}, \tag{3.24}$$

where N is the density of the mobile ions, q the ion charge, k the Boltzmann constant, and T the temperature. The quantity r_{rms}^2 represents the mean-squared displacement of an ion in hopping to nearest neighboring sites, and $\widetilde{\phi}_\sigma(\omega\tau_\sigma)$ is the Laplace–Fourier transform of $\phi_\sigma(t) = \exp[-(t/\tau_\sigma)^{1-n_\sigma}]$. It has also been shown in [56(b)] that the complex conductivity from ECR can be rewritten in the form

$$\sigma_M^*(\omega\tau_M) = (\varepsilon_o/M_\infty\tau_M)\left\{\frac{1}{\widetilde{\phi}_M(\omega\tau_M)} - i\omega\tau_M\right\}, \tag{3.25}$$

which is formally the same as $\sigma^*(\omega\tau_\sigma)$ in the previous equation. From this, τ_σ is related to τ_M by

$$\tau_\sigma/\tau_{EM} = (Nq^2 r_{\text{rms}}^2)/(6kT\varepsilon_o\varepsilon_\infty). \tag{3.26}$$

This relation enables the microscopic τ_σ to be calculated from the experimental macroscopic τ_M and the quantities N, q, and r_{rms}^2 which can be deduced from the chemical and physical structure of the ionic conductor. This has been carried out for the three glassy and one crystalline ionic conductors [1572], and three of these Li^+ ion conductors $0.56Li_2S$–$0.44SiS_2$, $0.45Li_2S$+$0.55GeS_2$, and $Li_{0.5}La_{0.5}TiO_3$(LLTO) are shown in Fig. 295.

After obtaining the microscopic τ_σ, the primitive relaxation time τ_0 can be deduced from the second part of Eq. (3.22) and then the SLR time τ_s can be calculated using this τ_0 by the first part of Eq. (3.22). The three steps outlined can be combined into one equation:

$$\tau_s = [t_c^{(\beta_s-\beta_\sigma)}(\tau_M N q^2 r_{\text{rms}}^2/6kT\varepsilon_o\varepsilon_\infty)^{\beta_\sigma}]^{1/\beta_s}, \tag{3.27}$$

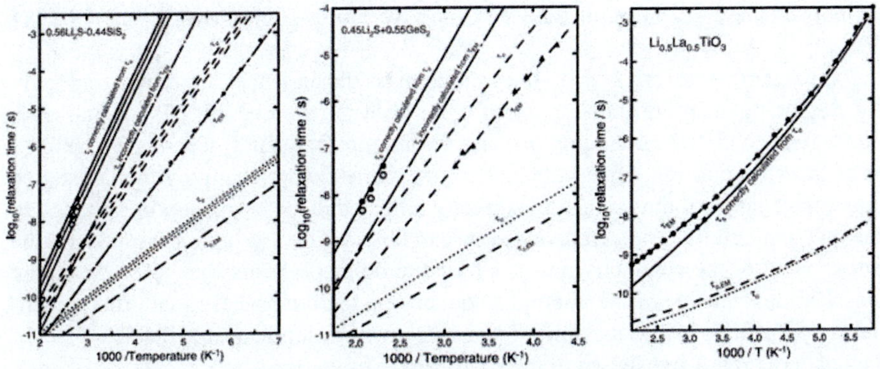

Fig. 295 (*Left*) Arrhenius plot of the correlation time of the mobile ion τ_s deduced from ^{7}Li nuclear spin-lattice relaxation measurement and the macroscopic electrical conductivity relaxation time τ_M (labeled here as τ_{EM}) for $0.56Li_2S$–$0.44SiS_2$. Experimental data of τ_s (*filled diamonds* and *open diamonds* determined by two different methods) and τ_{EM} (*closed circles*). The three *dashed lines* in descending order are the corresponding microscopic ion hopping relaxation time τ_σ calculated from Eq. (3.26) for r_{rms} = 3, 2.5, and 2 Å and other known parameters given in [1572]. The most probable value of r_{rms} is 2.5 Å. The three *dotted lines* in descending order are the corresponding microscopic ion hopping relaxation time τ_0 calculated from τ_σ by solving the second part of Eq. (3.22) for the three cases of r_{rms} = 3, 2.5, and 2 Å. After obtaining τ_0, the three full lines in descending order are τ_s calculated from the first part of Eq. (3.22) with r_{rms} = 3, 2.5, and 2 Å, respectively, and other parameters given in [1572]. There is good agreement with the experimental τ_s. Also shown is the primitive relaxation time $\tau_{0,EM}$ (*dashed line at the bottom*) calculated from τ_{EM} by solving an equation similar to Eq. (3.22), with τ_{EM} replacing τ_σ. Note that τ_s calculated from $\tau_{0,EM}$ (*dashed–dotted line*) cannot explain the spin-relaxation time data. (*Middle*) Arrhenius plot of the correlation time of the mobile ion τ_s deduced from ^{7}Li nuclear spin-lattice relaxation measurement and the macroscopic electrical conductivity relaxation time τ_{EM} for $0.45Li_2S$+$0.55GeS_2$. Experimental data of τ_s (*filled and open circles*) and τ_{EM} (*triangles*). Other lines are the same as in the caption of the *left panel*, except here only results for r_{rms} = 2.5 Å are shown. (*Right*) Same as in the caption of Fig. 1 for $Li_{0.5}La_{0.5}TiO_3$ (LLTO), a crystalline ionic conductor. The value of r_{rms} = 3.87 Å is known. Experimental data of τ_s (*filled diamonds*) and τ_{EM} (*filled circles*). Reproduced from [1572] by permission

where $\beta_s \equiv (1 - n_s)$ and $\beta_\sigma \equiv (1 - n_\sigma)$. This relation now enables the CM to quantitatively account for the difference between the experimental values of τ_s and τ_M. Figure 295 demonstrates excellent agreement with the data. A corollary of Eq. (3.27) is the relation between E_s and E_σ:

$$E_s/E_\sigma = (1 - n_\sigma)/(1 - n_s), \tag{3.28}$$

a prediction that has previously been verified (see Table 3.3).

Isotope Mass Dependence of σ

Before leaving the subject of dynamics of ions in ionic conductors, we mention an interesting experimental finding of anomalous isotope mass dependence of conductivity by Jain and Peterson in 1982 [1573] that has been shown to provide strong support of the application of the CM to ion dynamics from the very start in 1984 [362].

The experimental study is on the isotope mass dependence of the ion conductivity. Replacing the more abundant isotope ^{7}Li by ^{6}Li in the Li_2O:2.88B_2O_3 glasses is one way to change the prefactor, because the attempt frequency depends on the mass m of the vibrating ion. Given the rest are the same except changing the Li isotope mass, the change of the prefactor is directly reflected in the corresponding change of the dc conductivity. The dc conductivity can be measured to high accuracy and the ratio $^6\sigma/^7\sigma$ of the conductivities was found to be much larger than $\sqrt{7/6}$ from the classical dependence of the attempt frequency on the reciprocal square root dependence of the mass m or other similar values from more sophisticated theories [1573]. The data of $^6\sigma/^7\sigma$ are shown in the right panel of Fig. 296.

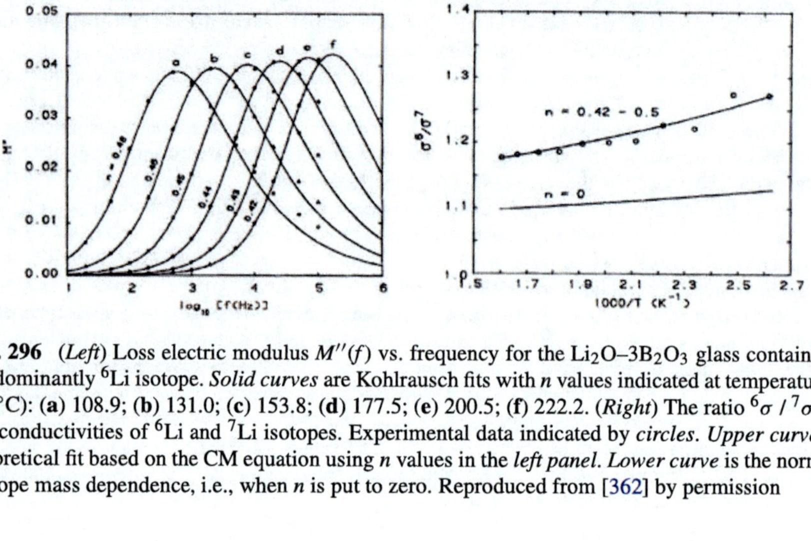

Fig. 296 (*Left*) Loss electric modulus $M''(f)$ vs. frequency for the Li_2O–3B_2O_3 glass containing predominantly ^{6}Li isotope. *Solid curves* are Kohlrausch fits with n values indicated at temperatures (in °C): (**a**) 108.9; (**b**) 131.0; (**c**) 153.8; (**d**) 177.5; (**e**) 200.5; (**f**) 222.2. (*Right*) The ratio $^6\sigma\,/\,^7\sigma$ of the conductivities of ^{6}Li and ^{7}Li isotopes. Experimental data indicated by *circles*. *Upper curve* is theoretical fit based on the CM equation using n values in the *left panel*. *Lower curve* is the normal isotope mass dependence, i.e., when n is put to zero. Reproduced from [362] by permission

The observed anomalous isotope mass m dependence of dc conductivity has been explained quantitatively from the CM relation

$$\tau_\infty^*(m) = [t_c^{-n}\tau_\infty(m)]^{1/(1-n)} \tag{3.29}$$

between the prefactor $\tau_\infty^*(m)$ of the observed conductivity relaxation time τ_σ and the primitive prefactor $\tau_\infty(m)$ using the actual value of n obtained by fitting the frequency dispersion of the data in an electric modulus representation by Fourier transform of the Kohlrausch function. The values of the coupling parameter n are obtained from the fits to the loss electric modulus data shown in the left panel of Fig. 296. The calculated $^6\sigma/^7\sigma$ are in good agreement with the experimental data.

So far, no other theory except the CM has successfully explained this isotope mass dependence.

3.2.5.2 Entangled Polymer Chains: Self-Diffusion vs. Viscosity

There are direct experimental data of self-diffusion and viscosity of entangled polymers showing that they have different activation energies and molecular weight dependences. This property is a close analogue of the breakdown of Stokes–Einstein relation found for small molecular glassformers and the difference between SLR and conductivity of mobile ions in glassy and crystalline ionic conductors. However, note that this anomaly is observed from the motion of polymer chains that has nothing to do with glass transition, nor with ions in glasses or crystals. The discussion of the dynamics of entangled polymer melts below in parallel with other different systems serves to illustrate several other general properties of relaxing/diffusing interacting systems.

Activation Enthalpy and Molecular Weight Dependences of Viscosity of Entangled Polymers

Some high molecular weight entangled polymers such as polyethylene (PE), hydrogenated polybutadiene (HPB), and poly(dimethylsiloxane) (PDMS) have low glass temperatures and the terminal relaxation can be measured at temperatures high above T_g of the polymers. At the high temperatures the terminal relaxation time as well as the viscosity exhibited Arrhenius behavior. The activation energies E_η, determined from the zero-shear viscosity or terminal zone shift factors, were often found to be larger than the conformational transition barrier energy E_a, which should govern the temperature dependence of the mobility of the polymer chains high above T_g had they been able to relax independently of each other. Examples of PE and HPB in Table 3.4 show E_η for shear viscosity of entangled linear chains, which is larger than the known conformation energy barrier of about 3.5 kcal/mol for PE and 4.2 kcal/mol for HPB [359, 360, 833–836, 1575]. The difference between the activation energies of shear viscosity η and self-diffusion coefficient D will be the subject of discussion in a later section.

Linear Entangled Polymers

Earlier attempt [360] and more sophisticated approach by taking into account of constraint mitigation later [837] have applied the CM to the terminal relaxation of entangled linear polymers. The coupling parameter can be obtained from fits to the dielectric spectrum [837] or the mechanical spectrum [360] of monodisperse polymers. The value of n_η obtained falls within the range of $0.40 \leq n_\eta \leq 0.43$. An alternative method to determine n_η is from the experimentally observed ratio J_S/J_N, which is a measure of the terminal dispersion [29]. Here J_N is the plateau compliance and J_S is the steady-state compliance [165, 171, 248]. If the Kohlrausch function is used to fit the terminal dispersion, then

$$\frac{J_S}{J_N} = \beta\left[\Gamma\left(\frac{2}{\beta}\right)/\Gamma\left(\frac{1}{\beta}\right)\right]^2, \tag{3.30}$$

Table 3.4 Parallel explanation of the molecular weight and temperature dependences of zero shear viscosity of polyethylene (PE), hydrogenated polybutadiene (HPB), and polydimethylsiloxane (PDMS). The actual conformational energy barrier of PE is taken from measurement on alkanes [360] and that of HPD is not known, but it is expected to be similar to PE. For references to the sources of data on PE and HPB, see Table 2 in [1577]. For PDMS, the experimental activation energy for viscosity E_η is 3.56 kcal/mol [1578] compared with E_a=2 kcal/mol from neutron scattering experiment [1579, 1580]

Linear polymer	Measured shear viscosity activation enthalpy, E_η (kcal/mol)	Observed M dependence of shear viscosity	Coupling parameter for shear viscosity (n_η)	Predicted M dependence of shear viscosity, from $\tau_0 \propto M^{2.0}$	Predicted conformational energy barrier, E_a (kcal/mol)	Actual conformational energy barrier (kcal/mol)
PE	6.35	$M^{3.5}$	0.43	$M^{3.5}$	3.6	3.5
HPD	7.2	$M^{3.4}$	0.41	$M^{3.4}$	4.2	
PDMS	3.56	$M^{3.5}$	0.43	$M^{3.5}$	2.03	2.0

where Γ is the gamma function and $\beta = (1 - n_\eta)$. The right-hand side of Eq. (3.30) is a monotonic increasing function of n_η that reaches the value of 2 when $n_\eta = 0.40$ [360, 904]. Experimental value of J_S/J_N of monodisperse polymer melts is about 2 or slightly larger [165, 171, 248], and thus it substantiates that n_η falls within the range of $0.40 \le n_\eta \le 0.43$.

Incidentally, the coupling parameter n_η does not depend on the length of the chain as long as the molecular weight exceeds the molecular weight M_e for entanglement because the situation is self-similar. The terminal relaxation time τ_η and η are related by the Maxwell equation $\eta = G_N \tau_\eta$, where G_N is the plateau modulus. The values of n_η in the range of $0.40 \le n_\eta \le 0.43$ can immediately explain the $M^{3.4}$ or the $M^{3.5}$ dependence of η and τ_η [905] based on Eq. (3.4) and the well-known $M^{2.0}$ dependence of the primitive relaxation time τ_0 of uncoupled motion of a single chain from the Rouse model modified for undiluted polymers [29]. The observed anomalous dependence of τ_η on M and T

$$\tau_\eta (M, T) \propto M^{3.4} \exp\left(E_\eta / RT\right) \tag{3.31}$$

have been explained simultaneously by the generalization of Eq. (3.4)

$$\tau_\eta (M, T) \propto [\tau_0 (M, T)]^{1/(1-n_\eta)} = M^{2/(1-n_\eta)} \exp\left(E_a / (1 - n_\eta) RT\right), \tag{3.32}$$

with a *single* value of n_η. The results can be seen from Table 3.4 for PE and HPD. The companion relation for activation enthalpies is $E_a = (1 - n_\eta) E_\eta$. The enhanced molecular weight dependence of τ_η and η is analogous to the enhanced isotope mass dependence of Li^+ ion conductivity in Li borate glasses [362, 1573], discussed before in this section.

For PE, from the experimental value of $E_\eta = 6.35$ kcal/mol [833], $E_\mathrm{a} = (1 - n_\eta)E_\eta$ yields 3.6 kcal/mol for the primitive activation energy. This is in agreement with the value of 3.5 kcal/mol (14.8 kJ/mol) for the conformational energy barrier of bulk PE from molecular dynamics simulations by Boyd and coworkers [1581] and also consistent with the estimation of the potential energy for rotating the carbon–carbon bonds in alkanes in the range from 3.1 to 3.7 kcal/mol by Flory [1575]. There is additional support from ^{13}C-NMR measurements of local segmental relaxation in two unentangled low molecular weight polyethylene $C_{44}H_{90}$ and one with $M_\mathrm{w} = 2150$ g/mol in the nanoseconds range by Qiu and Ediger [1582]. At this short time range, many-body dynamics is not that important. This is supported by the ^{13}C-NMR experimental correlation function well described by an exponential correlation function $\exp(-t/\tau_\alpha)$, and τ_α has an Arrhenius temperature dependence with activation enthalpy equal to 16.7 kJ/mol (or 4 kcal/mol), which is close to the primitive activation energy E_a deduced from $E_\eta = 6.35$ kcal/mol by the product $(1 - n_\eta)E_\eta$, which evaluates to 3.6 kcal/mol (or 15.1 kJ/mol).

For PDMS with $n_\eta = 0.43$, Eq. 3.32) yields a molecular weight dependence of $M^{3.5}$ and $E_\eta = 3.5$ kcal/mol based on $E_\mathrm{a} = 2$ kcal/mol from neutron scattering data of Allen and coworkers [1579, 1580], which is in good agreement with the experimental value of $E_\eta = 3.56$ kcal/mol (14.9 kJ/mol) from rheological measurements of entangled PDMS with $M_\mathrm{w} = 10^5$ g/mol ($M_e = 12{,}000$ g/mol) [1578, 1583]. Conversely from the experimental value of $E_\eta = 3.56$ kcal/mol, $E_\mathrm{a} = (1 - n_\eta)E_\eta$ yields 2.03 kcal/mol, the primitive activation energy from neutron scattering (Table 3.4). Nevertheless, viscosity measurement on PDMS with very low molecular weight of 1250 and 550 g/mol by Ding et al. [1584] reported $E_\eta = 14.8$ and 13.5 kJ/mol, almost the same E_η of entangled PDMS. These values of E_η are larger than $E_\mathrm{a} = 2$ kcal/mol (8.4 kJ/mol) of local segmental relaxation of high molecular weight PDMS from neutron scattering. The E_η of low M_w PDMS from Ding et al. may be that of the sub-Rouse modes, which is cooperative in nature [872].

The viscosity η of entangled linear atactic polypropylene (aPP) does not have Arrhenius temperature dependence [361, 902]. Effort has been made to measure η at very high temperatures [361] where the local segmental relaxation time is of the order of picoseconds [875]. Because t_c is about 2 ps for polymers, the local segmental relaxation is essentially in the primitive relaxation mode, and we can expect its relaxation rate at these high temperatures governed by the conformational transition barrier energy of aPP, which is about 4 kcal/mol [1585]. The mobility of the terminal relaxation of the chains and the viscosity should have the same activation energy at the same temperatures. However, from the measurement of viscosity, the apparent activation energy of the viscosity E_η of entangled aPP is estimated to be 7.2 kcal/mol at the very high temperatures [361]. This larger activation energy can be explained again by Eq. (3.32), which predicts $E_\eta = 7.0$ kcal/mol from $n_\eta = 0.43$ and $E_a = 4$ kcal/mol. By the way, the reptation theory [41–44] cannot predict E_η as done by the CM.

The above shows that the CM equation(3.32) can explain both the molecular weight and temperature dependence of the viscosity of entangled polymers. Rendell et al. [1577] have shown that the same CM can explain additionally the concentration dependence of the terminal relaxation and viscosity of entangled polymer solutions. Explicitly, if ϕ is the polymer volume fraction, the predicted combined M and ϕ dependences of viscosity is $\eta \propto M^{3.4}\phi^{3.4}$, in agreement with experiment [905].

Branched Entangled Polymers

Changing the architecture of polymers from linear chains by branching to multiple-arm star will enhance the molecular constraints overall and hence the effective coupling parameter if the arms of the star are sufficiently long for them to fully entangle. Larger coupling parameter n_η for such a star than the linear value results in stronger molecular weight dependence and higher activation energy of the viscosity as predicted by Eq. (3.32). The relaxation times of the Rouse–Ham modes for a star-branched polymer molecule still have the M^2 dependence like the Rouse modes for a linear molecule [29], and Eq. (3.32) still holds. The same conformation energy barrier is deduced [359, 1577]. Such predictions are borne out by data of star-branched PE and HPB as shown in Table 3.5.

Table 3.5 Shear viscosity data of star-branched PE

Star-branched polymer	Measured shear viscosity activation enthalpy, E_η (kcal/mol)	Observed M dependence of shear viscosity	Coupling parameter for shear viscosity n_η	Predicted M dependence of shear viscosity	Predicted conformational energy barrier, E_a (kcal/mol)	Actual conformational energy barrier (kcal/mol)
PE	11.66	$M^{6.56}$	0.70	$M^{6.67}$	3.5	3.5
HPD	14.29		0.71	$M^{6.9}$	4.15	n.a.

From [833]. The data of HPB stars are taken from the Ph.D. thesis of V. R. Raju [1586]. For more details, see [359, 1577]

Again, a larger coupling parameter can explain simultaneously the molecular weight dependence and the activation energy for the stars. The larger values of the coupling parameter are also consistent with the observed broader dispersions of the terminal relaxation of the monodisperse stars [1577, 1586].

Different Activation Energies and M Dependence of Self-Diffusion than Viscosity in Some Entangled Polymers

Self-diffusion of entangled polymer chains was measured for PE and HPB [1587]. For the self-diffusion coefficient of HPB, only data cited here and in [1577] are considered and not all the data combined like the way discussed by Lodge in [1588]. From experiments, the self-diffusion coefficient D of these two polymers has the M^{-2} dependence and Arrhenius T dependence with activation enthalpy E_D listed

Table 3.6 Parallel explanation of the molecular weight and temperature dependences of self-diffusion of polyethylene (PE) and hydrogenated polybutadiene (HPB). The actual conformational energy barrier of PE is taken from measurement on alkanes [360]. HPD is not known, but it is expected to be similar to PE. For references to the sources of data, see Table 2 in [1577]

Linear polymer	Measured self-diffusion activation enthalpy, E_D (kcal/mol)	Observed M dependence of self-diffusion coefficient	Coupling parameter for self-diffusion, n_D	Predicted M-dependence of self-diffusion from $\tau_0 \propto M^{2.0}$	Predicted conformational energy barrier, E_a (kcal/mol)	Actual conformational energy barrier (kcal/mol)
PE	5.47	M^{-2}	1/3	M^{-2}	3.7	3.5
HPB	7.2	M^{-2}	1/3	M^{-2}	4.1	n.a.

in Table 3.6. Theoretically, D is the ratio $(R_g)^2/\tau_D$, where $R_g(\sim M)$ is the radius of gyration of the linear chain and τ_D is the center-of-mass diffusion time of the entangled chain. The latter is related to the primitive diffusion time τ_{0D} by the CM equation

$$\tau_D(M,T) = [t_c^{-n_D}\tau_{0D}(M,T)]^{1/(1-n_D)}. \tag{3.33}$$

Since τ_{0D} is the uncoupled center-of-mass diffusion of a single chain, it can be identified with that of the Rouse model, which has the $M^{2.0}$ dependence. With the $M^{2.0}$ dependence of τ_{0D}, Eq. (3.33) yields the $M^{3.0}$ dependence of τ_D if $n_D = 1/3$. From the $M^{3.0}$ dependence of τ_D and $(R_g)^2 \propto M$, it follows from $D = (R_g)^2/\tau_D$ that $D \propto M^{-2.0}$ as found by experiment. Support of $n_D = 1/3$ comes from the fact that the product $(1 - n_D)E_D$ is almost the same as $(1 - n_\eta)E_\eta$, and also E_a, the conformational energy barrier of these two polymers. These correspondences can be seen in Table 3.6.

Confirmation by the Neutron Spin Echo Experiment of Zamponi et al.

Here we recall the center-of-mass mean-square displacement (MSD) data from neutron spin echo experiment of Zamponi et al. [1503] shown in Fig. 281. For the entangled polyethylene chains with $N = 192$ and 377, the subdiffusion term t^{1-n} has exponent (1–n) values of 0.72 and 0.63, respectively. The molecular weights of the polymers with $N = 192$ and 377 are about $1.5M_c$ and $3M_c$, respectively, where M_c is the molecular weight for entanglement. Substituting these values of (1–n) for (1–n_D) in Eq. (3.33), one gets the molecular weight dependences

$$D \propto M^{-1.8},\ \text{for}\, N = 192,\ \text{and}\ D \propto M^{-2.2},\ \text{for}\, N = 377 \tag{3.34}$$

which are consistent with the neutron spin echo data within the uncertainty of determining the exponent of the subdiffusion.

Analogue of the breakdown of SE relation in supercooled liquids, and decoupling of SLR and conductivity relaxation in glassy and crystalline ionic conductors

The results above from comparison of self-diffusion and viscosity of entangled polymer chains, namely $n_D < n_\eta$ and $E_D < E_\eta$, have almost exact analogues in the breakdown of Stokes–Einstein relation of molecular glassformers [438, 449–457] and ionic liquids [158], and the CM explanations given here for entangled polymers follow the same parent CM equation as those given for the analogues [159, 268] discussed in Sections 2.2.5.7. Furthermore, the result

$$(1 - n_D)E_D \approx (1 - n_\eta)E_\eta = E_a \tag{3.35}$$

for entangled polymers is the analogue of Eq. (3.23), $(1-n_s)E_s \approx (1-n_\sigma)E_\sigma = E_a$, between nuclear spin relaxation and conductivity relaxation in the case of ionic conductors (see also Table 3.3). Both cases show that the activation energy of the primitive relaxation E_a can be recovered from that of the slow many-body relaxation by multiplying it by the fractional Kohlrausch exponent, and the result is the same independent of the dynamic variable chosen.

3.2.5.3 Semidilute Polymer Solutions

Polymer dynamics in semidilute and concentrated solutions has been a subject of considerable interest in the past decades. Some works in this area have been referenced before [776–779, 1555–1567] in connection with some universal properties of relaxation and diffusion in interaction systems discussed before. Compared to polymer melts, introduction of the polymer concentration c as an additional variable has made the problems in polymer solutions richer. Examples of other studies on various properties of semidilute polymer solutions related to c dependence are given in [834, 1590–1606] and reviews in [1607, 1608]. There are various problems of polymer diffusion in semidilute solutions of high molecular weight polymers as well as in concentrated solutions. In the concentrated solutions, the problem is the observed $M^{2.5}$ dependence of the polymer diffusion coefficient D, where M is the molecular weight [1609, 1610]. In semidilute solutions, there are a number of observed properties that need to be explained. Here we take the observed properties from two systems.

Probe Diffusion vs. Viscosity in Semidilute Polymer Solutions

The first one is the hydroxypropylcellulose (HPC):water semidilute aqueous solutions. The dynamics of several length scale of this semidilute polymer solution was studied by measurement of optical probe diffusion in it by light scattering. The optical probes carboxylate modified polystyrene sphere in water solutions of nominal 300 kDa (M_W=415 kDa) HPC at 25°C. The size of probes employed R_{probe} are 760,

189, 102, and 20 nm [1566, 1611, 198]. The optical probe diffusion D and shear viscosity η experimental data of hydroxypropylcellulose/water solutions obtained by Phillies and coworkers are summarized briefly as follows. All these were obtained from the field correlation function $g^{(1)}(t)$ from light scattering of optical probes suspended in solution. The parameters characterizing the quantities above depend on the probe size and the polymer molecular weight.

(a) The field correlation function has the stretched exponential time dependence

$$g^{(1)}(t) = A_o \exp[-(t/\tau_{\mathrm{D}})^{1-n_{\mathrm{D}}}]. \tag{3.36}$$

(b) The concentration dependence of D is given by the stretched exponential form

$$D = D_o \exp(-\alpha_{\mathrm{D}} c^{u_{\mathrm{D}}}). \tag{3.37}$$

(c) The τ_{D} has the anomalous dependence on the scattering vector q given by

$$\tau_{\mathrm{D}} \propto q^{-2/(1-n_{\mathrm{D}})}. \tag{3.38}$$

(d) The solution viscosity η is found to be described well by another stretched exponential c dependence

$$\eta = \eta_o \exp(\alpha_\eta c^{u_\eta}), \quad \text{with} \quad \alpha_\eta \propto M^{\gamma_\eta} \tag{3.39}$$

valid up to very large M and c. The parameters α_η, u_η, and γ_η for viscosity in general have values different from the corresponding α_{D}, u_{D}, and γ_{D} for probe diffusion [198, 777, 1606]. Probes in 300-kDa hydroxypropylcellulose show Stokes–Einstein behavior, $D = kT/6\pi\eta R_{\mathrm{probe}}$, for large ($R_{\mathrm{probe}} > 55$ nm) spheres but increasingly non-Stokes–Einstein behavior for small probes ranging from 33 down to 0.5 nm in radius.

Some of these observed dynamic properties of semidilute polymer solution at different length scales determined by the probe diameter are obviously analogues of glassformers. These include the Kohlrausch stretched exponential correlation function in Eq. (3.28) and the superlinear q dependence of τ_{D} in Eq. (3.37). Here is another example of the breakdown of Stokes–Einstein relation $D = kT/6\pi\eta R_{\mathrm{probe}}$, which has nothing to do with glass transition. In optical probe experiments, motions of chain segments with length scales comparable to R_{probe} determine the probe diffusion coefficient D. On the other hand, η is determined by the motions of entire polymer chains, which experience stronger dynamic constraints and larger coupling parameter than do chain segments having length scale of R_{probe} if R_{probe} is much smaller than R_{g} of the polymer. Thus, we expect $n_\eta(c) > n_D(c)$ to hold when the probe size is small. If $n_\eta(c) > n_{\mathrm{D}}(c)$, it follows immediately from the CM-derived expressions [198] that η will have a stronger c dependence than D. The reader may see that the CM explanation given here for the breakdown of SE relation of probe diffusion in semidilute solution of entangled polymer chains is isomorphic to that given for the same phenomenon in glass-forming liquids.

A theoretical treatment of the dynamics of semidilute polymer solutions has been proposed by using the CM of polymer chain dynamics together with scaling arguments. The theory has many predictions, which are remarkably consistent with all known experimental facts of HPC/water system. The formalism involves some cumbersome equations and will not be reproduced here. The interested reader can consult [198, 779]. The upshot of scaling is to allow extraction of values for the coupling parameter n_D by three separate paths from the concentration, time, and scattering vector dependencies of $g^{(1)}(t)$. Values of n_D from these three distinct physical approaches are shown to be mutually consistent, especially in the higher concentration, large-probe-particle regime in which the scaling arguments are most likely to be valid. In other words, as a function of c and M, the same n_D explains simultaneously the concentration and molecular weight dependences of D and τ_D, time dependence of $g^{(1)}(t)$, and scattering vector dependences of τ_D. By the way, it is well known that reptation model cannot give satisfactory explanation of the dynamics of semidilute polymer solutions, not even the dependence of D on c and M.

Linear and Star-Branched Polystyrene Tracer Diffusion in Semidilute Polymer Solutions

A large body of tracer diffusion coefficient measurements of linear and star-branched polymers in dilute, semidilute and concentrated solutions have accumulated since 1986 [1594–1598]. Naturally, attempts [1595, 1596] have been made in the past to examine these data in the light of the reptation model. Unfortunately, it was found that these data could not be described by the reptation mechanism even with constraint release being included [1594–1598, 1607]. In 1989, an explanation based on combining concentration scaling theory [834, 1590, 1612] and the coupling model was proposed and part of the results was published in an abbreviated format by Ngai and Lodge [1613]. Since then, publication of the entire analysis of the experimental data was put on hold, partly because at that time there is no *direct* evidence to support the application of the coupling model to diffusion in semidilute solutions. However, in the intervening years from 1989 till 1996, new experimental investigations [776–778, 780, 1557, 1566, 1602, 1603, 1614, 1615] have provided the most direct evidence of the applicability of the coupling model. These include the crossover from primitive linear exponential relaxation $\exp(-t/\tau_0)$ to Kohlrausch stretched exponential relaxation $\exp\left[-(t/\tau)^{1-n}\right]$ and the $Q^{-2/(1-n)}$ dependence of τ, discussed before in Sections 3.2.3 and 3.2.4. Encouraged by these developments uniquely in favor of the CM, the scaling–coupling theory of dynamics of semidilute polymer solutions was given in full in 1996 by Ngai and Phillies [198] and applied to probe diffusion data of Phillies and coworkers addressed in the previous subsection. In the same year, the scaling–coupling theory was applied [779] to the analysis of the linear, 3-arm star and 12-arm star polystyrene tracer diffusion in solutions of linear poly(vinyl methyl ether) of various molecular weights and concentrations by dynamic light-scattering measurements by Lodge and coworkers [1595–1598] and forced Rayleigh scattering measurement of linear polystyrene tracer in polystyrene matrix by Yu and coworkers [1594]. The scaling–coupling theory [779] explained the empirical stretched exponential form $D = D_o \exp(-\alpha c^u)$ for the concentration dependence of the diffusion coefficient originates from the monotonic increase of the coupling parameter n with c/c^*. Conversely, the measured stretched exponential concentration dependence of D has been used to calculate n as a function of c, tracer molecular weight M, and matrix molecular weight P. At constant c and fixed P, these n values, $n(M)|_{c,P}$, have been used to calculate the tracer molecular weight dependence of the diffusion coefficient $D(M)|_{c,P}$ at constant c and fixed P. The calculated values of $D(M)|_{c,P}$ are in good agreement with measurements. In particular, tracer molecular weight dependence of $D \propto M^a$, with a larger than 2 and as large as approximately 3, observed in matrix solutions at higher concentrations can be explained. Full dynamic constraints are imposed on the tracer by the matrix solution if $P >> M$. However, when the latter condition is not satisfied, some of the dynamic constraints are mitigated in the same matrix solution, leading to a decrease of the coupling parameter. The effects of mitigation of dynamic constraints under the condition of $M > P$ on the diffusion of linear and branched PS in PVME solutions are demonstrated by using directly the experimental data or alternatively by the decrease of the calculated values of n. Thus, the dependences of D on the three variables c, M, and P can be explained from the corresponding dependences of the coupling parameter n through the physics contained in the coupling model with the assist of scaling. The temperature dependence of $D(c, M, P)$ has also been predicted to depend sensitively on $n(c/c^*, M, P)$. This prediction should be a critical test of the coupling model, but till now there is little data of temperature dependence of tracer diffusion to compare with. Details of the analysis by the model and comparison of predictions with experiments can be found in [779].

3.2.6 Recovering or Discovering the Primitive Relaxation

In Chapter 2 where we addressed principally glass-forming substances, the universal secondary relaxation of the Johari–Goldstein (JG) kind found by various spectroscopies is a manifestation of the primitive relaxation in view of their similar character and behavior, and the good correspondence between the JG relaxation time τ_{JG} from experiment and the calculated τ_0. Analogues of the JG relaxation in glassformers actually are already found in the other interacting systems discussed in Sections 3.2.3 and 3.2.4. Dynamic light-scattering experiments have shown the presence of the primitive relaxation at short times with $\exp(-t/\tau_f)$ decay and τ_f has the normal q^{-2} dependence. The experimental correlation function is usually well fitted by Eq. (3.6) in the form of the sum $f(q,t) = A\exp(-t/\tau_f) + (1-A)\exp[-(t/\tau_s)^{1-n}]$. To make obvious the equivalence to JG β-relaxation as often found in dielectric loss spectra, $f(q,t)$ is Fourier transformed to obtain the susceptibility $\chi''(\omega)$ in Fig. 297.

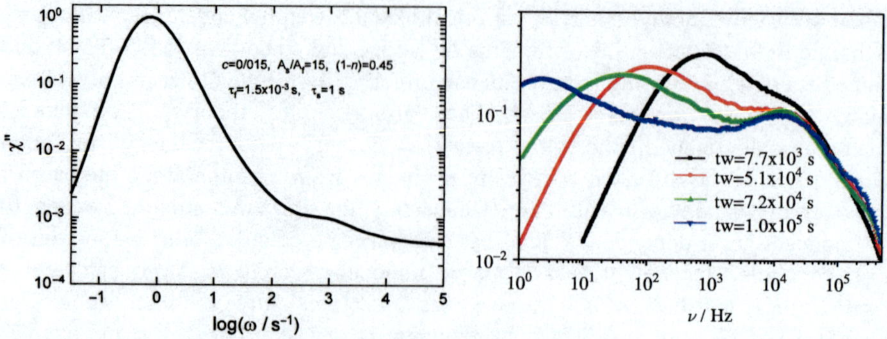

Fig. 297 (*Left*) Susceptibility obtained by Fourier transform of $f(q,t) = A\exp(-t/\tau_f) + (1-A)\exp\left[-(t/\tau_s)^{1-n}\right]$ from fit to dynamic light-scattering data of an aqueous solution of a hydrophobically end-capped poly(oxyethylene) urethane (HPOEU) at concentration $c = 0.015$ g/ml by Nyström et al. [778]. The parameters A, τ_f, τ_s, and (1–n) are shown in the figure. (*Right*) Susceptibility data obtained by numerical Fourier transform of PCS correlation function of the Laponite colloidal suspension at different aging times t_w performed by S. Capaccioli (Laponite concentration $C_w = 2.5\%$, salt concentration $C_s = 2 \times 10^{-3}$ M). Data are from Fig. 2-D of [342] and provided digitally by L. Zulian

The parameters A, τ_f, τ_s, and (1–n) are taken from dynamic light-scattering study of an aqueous solution of a hydrophobically end-capped poly(oxyethylene) urethane (HPOEU) at concentration $c = 0.015$ g/ml by Nyström et al. [778]. The bump at higher frequencies is the analogue of resolved JG β-relaxation of glassformers.

The time-domain photon correlation spectroscopy data of colloidal suspension of Laponite [340–342] discussed before have also been fitted by $f(q,t) = A\exp(-t/\tau_f) + (1-A)\exp\left[-(t/\tau_s)^{1-n}\right]$. Shown in the right panel of Fig. 297 is the susceptibility $\chi''(\omega)$ obtained by Fourier transform of the data at four different aging times t_w. The JG β-relaxation appearing at higher frequencies is like that commonly found in other glassformers by dielectric spectroscopy.

We have seen in the many examples given before that the relaxation time τ of many-body interacting systems is anomalous in magnitude as well as dependences on variables. In the context of the coupling model (CM), these anomalies originate from the modification of the normal primitive relaxation time τ_0 according to Eqs. (3.1) and (3.2), principally by raising τ_0 to the superlinear power $1/(1-n)$. Larger n leads to greater modification of τ_0 and higher degrees of the anomaly. All these are consistent with many-body effects as the cause of the anomalies because n is a measure of the amount of stretching in the Kohlrausch correlation function, which is an obvious indicator of many-body effects. By examples, it has been shown how the CM explains many anomalies in different systems. Now we turn this around and use the experimentally observed anomalies to support verity of the CM. This is done by calculating τ_0 from the experimental τ and n according to Eqs. (3.1) and (3.2), and verifying that the calculated τ_0 is in agreement with either the expected value or the actual value obtained by another source. Actually, this way of recovering τ_0 has been demonstrated before at several places in this chapter. The recovered prefactor

τ_∞ of τ_0 from conductivity relaxation or dynamic light scattering in YSZ has been shown to agree well with the attempt frequency of the oxygen ion independently determined by hyper Raman spectroscopy (see Fig. 286). The recovered activation energy E_a of τ_0 of Ag ions in AgI–$AgPO_3$ or Ag_2S–GeS_2 is in agreement with that directly determined by neutron scattering (see Table 3.2). The recovered activation energy E_a of τ_0 of PE and PDMS chains is consistent with the conformation energy barriers independently determined (see Tables 3.4 and 3.5). In this section, we add more examples from diverse interacting systems to reaffirm the physical reality of the primitive relaxation and in turn the validity of the CM.

3.2.6.1 Influence of Mesophase Structures on the β-Relaxation in Side-Chain Liquid Crystal Polymers (SCLCPs)

This particular problem gives us another occasion to consider the anomalously large activation energy accompanied by unphysically high attempt frequency in many-body relaxations with Arrhenius T-dependent relaxation times. Such dual anomalies have been encountered before in ionic conductors in this chapter and are referred to in the literature as the "compensation law" or the "Meyer–Neldel" rule. The system is a side-chain liquid crystal polymer (SCLCP) and the chemical structure of its repeat unit of the 10 samples studied by Schönhals and coworkers [1616–1618] are shown schematically below.

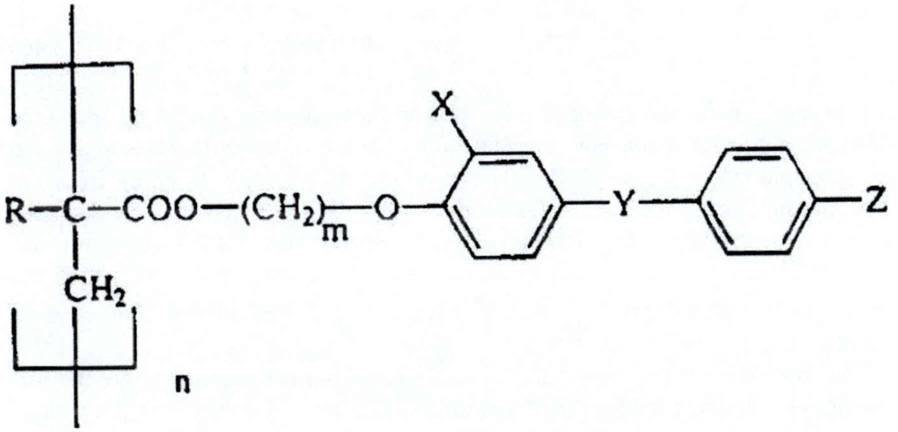

The backbone or the main chain is either poly(methacrylate) or poly(acrylate), and the side chain consists of mesogenic groups separated from the main chain by a spacer consisting of a chain of $-CH_2-$ units. The mesogenic groups are derivatives of (p-alkoxy-phenyl)-benzoate that have optical activity [1616–1618]. The process studied by Schönhals and coworkers is the rotational fluctuations of the mesogenic group about its long axis called the β-relaxation. Although called such, this β-relaxation is cooperative because of its size, like the γ-relaxation in BPA-polycarbonate [883–890].

Table 3.7 Experimental data of the most probable $f^*_{\beta\infty}$ and E^*_β, appearing in the Arrhenius equation $f^*_{\max,\beta}(T) = f^*_{\beta\infty}\exp(-E^*_\beta/kT)$ for 10 polymers with various mesophase structures from dielectric relaxation measurements. The most probable coupling parameter n and the primitive pre-exponential factor $f_{\beta\infty}$ have been obtained from the predictions of the coupling model

Polymer	Mesophase	$\log(f^*_{\beta\infty}/Hz)$	E^*_β (kJ/mol)	$\log(f_{\beta\infty}/Hz)$	n
P1	Isotropic	12.8	45.7	11.44	0.716
P2	Nematic	13.8	46.5	11.71	0.720
P3	Nematic	14.8	53.3	11.85	0.756
P4	Smectic A	16.5	57.5	12.17	0.774
P5	Smectic A	16.4	56.4	12.17	0.77
P6	Smectic A	17.1	61.0	12.22	0.787
P7	Smectic B	17.5	62.8	12.27	0.793
P8	Smectic B	18.2	63.2	12.40	0.794
P9	Smectic B	18.5	64.9	12.42	0.80
P10	Smectic B	19.4	68.9	12.51	0.81

Different mesophases were obtained by a variation of the number of methylene units in the spacer and in the tail groups Z. The mesophases of 10 SCLCPs are shown in Table 3.7. It was found by dielectric spectroscopy that the most probable relaxation frequency $f^*_{\max,\beta}$ of the β-relaxation has the Arrhenius temperature dependence

$$f^*_{\max,\beta}(T) = f^*_{\beta\infty}\exp\left(-E^*_\beta \big/ kT\right), \tag{3.40}$$

but the pre-exponential factor $f^*_{\beta\infty}$ as well as the activation energy E^*_β increases (obeying a compensation law) significantly with the order of the mesophase. In the isotropic state, $f^*_{\beta\infty} = 10^{12.8}$ Hz and $E^*_\beta = 45.7$ kJ/mol. In going from the isotropic state to the nematic, the smectic A, and the smectic B mesophases (i.e., with increasing order of the mesophase and decreasing mean lateral mesogenic distance as obtained by X-ray measurement), both $f^*_{\beta\infty}$ and E^*_β increase to reach the anomalously high values of $f^*_{\beta\infty} = 10^{19.4}$ Hz and $E^*_\beta = 68.9$ kJ/mol. These results are collected in Table 3.7.

The interesting experimental facts found by varying the mesophase are the following: (1) the simultaneous increases of $\log f^*_{\beta\infty}$ and E^*_β with decreasing mean lateral mesogenic distance that obey the compensation law and (2) the unphysically large pre-exponential factors found in the smectic mesophases. These are analogues of the same properties found in other interacting systems including the crystalline and glassy ionic conductors. The CM has been applied to explain them [1619]. Concentration fluctuations occur in the environments of the mesogenic group because of the internal blending of the mesogenic groups with the long methylene spacer chains, which broadens the β-relaxation dispersion, making it impossible to determine the coupling parameter by fitting the frequency dispersion to the Fourier transform of a Kohlrausch function. Notwithstanding, the reduction of the data to primitive relaxation time offers a way to understand the anomalies. The most probable $\tau_\beta^*(T)$ is related to $f^*_{\max,\beta}(T)$ by the approximate relation $1/\left[2\pi\tau_\beta^*(T)\right] = f^*_{\max,\beta}(T)$. Since the observed most probable frequency of the β-relaxation has an Arrhenius dependence, it follows from Eq. (3.3) that the primitive relaxation time $\tau_\beta(T)$ also has another Arrhenius T dependence $\tau_\beta = \tau_{\beta\infty}\exp\left(E_{a\beta}/RT\right)$. Here $E_{a\beta}$ and $\tau_{\beta\infty}$ are respectively the activation enthalpy and prefactor of the primitive relaxation time of rotational motion of the mesogenic group about its long axis. Due to the close resemblance in chemical structures of the mesogenic group to the monomer unit of BPA-PC and the strong similarity of rotational motion of the mesogenic group (β-relaxation) in the SCLCPs studied by Schönhals et al. and the rotations of the phenyl rings in the γ-relaxation in BPA-PC (see Section 2.3.2.38), we can take $E_{a\beta}$ and $\tau_{\gamma\infty}$ to have approximately the same values as the corresponding quantities $E_{a\gamma}$ and $\tau_{\gamma\infty}$ deduced earlier for the γ-relaxation. The values $E_{a\gamma} = 13$ kJ/mol and $\tau_{\gamma\infty} = 6.5 \times 10^{-13}$ s have been obtained from NMR measurement in dilute solution of BPA-PC [1470] and from theoretical calculations [1472, 1473]. Therefore the value $E_{a\beta} = 13$ kJ/mol is used for the analysis of the experimental data of Schönhals et al.

The concentration fluctuations give rise to a distribution of environments seen by the mesogenic groups. The different environments introduce a distribution of primitive relaxation times about the most probable value $\tau_\beta(T) = \tau_{\beta\infty}\exp\left(E_{a\beta}/RT\right)$ as well as possibly a distribution of coupling parameter with most probable value n. From the CM equation (3.3) the corresponding most probable effective relaxation time $\tau_\beta^*(T)$ is given by $\tau_\beta^*(T) \equiv \tau^*_{\beta\infty}\exp\left(E^*_{a\beta}/RT\right)$ with

$$\tau^*_{\beta\infty} = \left[t_c^{-n}\tau_{\beta\infty}\right]^{1/(1-n)} \quad \text{and} \quad E^*_{a\beta} = E_{a\beta}/(1-n). \tag{3.41}$$

X-ray measurement [1618] shows that an increase in the order of the mesophase is accompanied by a decrease in the mean lateral mesogenic distance. From this result, we expect that the increase in order of the mesophase will pack the mesogenic groups closer together, enhance their mutual constraints, and cause the coupling parameter n to increase. In fact, the values of n calculated from the expression $\left(1 - E_{a\beta}/E^*_{a\beta}\right)$ for different mesophases are in agreement with this expected trend, as shown in Table 3.7 as well as in Fig. 298.

With the values of n for the SCLCPs with various mesophases determined, the critical stage is reached to test if these values of n will be able to explain the compensation law followed by $f^*_{\beta\infty} \equiv 1/(2\pi\tau^*_{\beta\infty})$ and the large spread of its observed values ranging from $10^{12.8}$ Hz to $10^{19.4}$ Hz, which is exhibited in the right panel of Fig. 298 in a plot of $f^*_{\beta\infty}$ against n (solid triangles) of the various mesophases. This test is made by solving $f^*_{\beta\infty} = (2\pi)^{-1}\left[t_c^{-n}\tau_{\beta\infty}\right]^{-1/(1-n)}$ for the primitive prefactor $\tau_{\beta\infty}$ for each SCLCP with its experimentally determined $f^*_{\beta\infty}$ and n from Table 3.7 and $t_c = 2$ ps for polymers. The results for $\tau_{\beta\infty}$ give immediately $f_{\beta\infty} = 1/\left(2\pi\tau_{\beta\infty}\right)$. The values of $f_{\beta\infty}$ obtained for all 10 SCLCPs are also listed in Table 3.7 and shown in the right panel of Fig. 298. It is important to recognize that

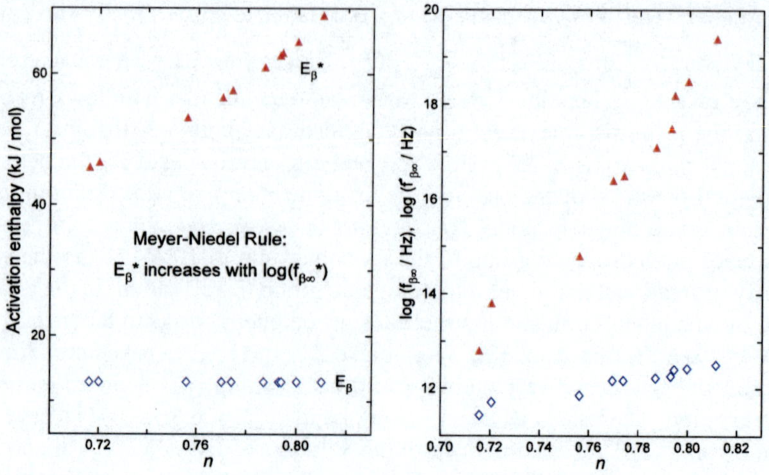

Fig. 298 (*Left*) The *x*-axis is the coupling parameter *n* of 10 SCLCPs with various mesophase structures (see Table 3.7). Each of the 10 *open diamonds* stands for the point (n, E_β). Each of the 10 *solid triangles* stands for the point (n, $E^*_\beta E^*_\beta$). (*Right*) The *x*-axis is the coupling parameter *n* of 10 SCLCPs with various mesophase structures. Each of the 10 *solid triangles* stands for the point (n, $f^*_{\beta\infty}$). Each of the 10 *open diamonds* stands for the point (n, $f^*_{\beta\infty}$), where $f_{\beta\infty}$ is calculated with $t_c = 2$ ps and $f^*_{\beta\infty}$ (from experiment, see Table 3.7). Reproduced from [1619] by permission

these results have been obtained in effect without any adjustable parameters because *n* has been independently determined by the relation between the experimental value of the activation enthalpy $E^*_{a\beta}$ and the known value of $E_{a\beta}$.

In the right panel of Fig. 298, $f^*_{\beta\infty}$ and the corresponding $f_{\beta\infty}$ (open diamonds) are plotted simultaneously against *n* of the SCLCPs. In contrast to the large spread of almost seven decades shown by $f^*_{\beta\infty}$, the primitive prefactor $f_{\beta\infty}$ has a narrow spread of one decade. The trend that $f_{\beta\infty}$ increases moderately with *n* is apparent from the figure. Since *n* increases with decreasing lateral mesogenic distance, the observed trend can be restated as a moderate increase of $f_{\beta\infty}$ with decreasing lateral mesogenic distance. The latter is physically possible and reasonable because $f_{\beta\infty}$ is an attempt frequency which, like a vibration frequency, is expected to increase with decreasing inter-mesogenic distance because the latter causes the potential energy well to become steeper. All values of $f_{\beta\infty}$ lie within the narrow range of physically reasonable values, $10^{11.5} \leq f_{\beta\infty} \leq 10^{12.5}$, and can be readily interpreted as the pre-exponential factor of the rotational frequency of the mesogenic group. Moreover, the values are close to the primitive prefactor of the γ-relaxation in BPA-PC $f_{\gamma\infty}$ defined as $f_{\gamma\infty} = 1/(2\pi\tau_{\gamma\infty})$ and have been previously determined to have the value of $10^{11.48}$ Hz. In fact, this value of $f_{\gamma\infty}$ obtained from an earlier work [1620] is nearly identical to $f_{\beta\infty}$ of the SCLCP with isotropic mesophase (P1 in Table 3.7). The good correspondence between the order of magnitude of the primitive prefactor of the β-relaxation of the SCLCPs and that of the closely related γ-relaxation of BPA-PC is not an accident but the result of similar physics, i.e., cooperative dynamics of densely packed mesogenic groups and phenyl rings of similar chemical structures.

The observed dramatic increases of the pre-exponential factor $f^*_{\beta\infty}$ and the activation energy $E^*_{a\beta}$ of the β-relaxation rate with the mesophase structure can now be understood in the framework of the coupling model. The increase in the order of the mesophase enables the mesogenic groups to pack closer together and causes intermolecular cooperativity or the coupling parameter n to increase. The increase of n is responsible for the large observed variations of $f^*_{\beta\infty}$, and $E^*_{a\beta}$ and the compensation law obeyed by their variations. The prefactor and the activation energy of the primitive relaxation time recovered from the anomalous experimental data are physically reasonable. This supports reality of the primitive relaxation as well as its relation to the many-body cooperative relaxation as described by the CM equations.

3.2.6.2 Dynamics of Cross-Linked Junction of a Polymer Network

The classical approach to rubber elasticity considers only the chain entropy and internal energy contributions to the elastic energy of a network. Any effect of intermolecular interactions is ignored, based on the idea that such interactions are independent of the chain configurations and thus do not modify the stress. The deviation of experimental results from the predictions of classical theory, as exemplified by the strain dependency of the elastic modulus (defined as the ratio $\sigma/(\lambda - \lambda^{-2})$, where σ is the stress and λ the extension ratio), has led to refinements which take into account the effect of interactions among the network chains. In the Flory description [1621–1623], the intermolecular interactions reduce the configurations available to the network chains by constraining the network junctions. The reduced effectiveness of these constraints upon deformation gives rise to a strain-dependent elastic modulus.

Rubber elasticity theories restrict their concerns to the properties of the network in mechanical equilibrium. However, experimental studies reveal analogies in the behavior of elastomers and polymer melts [1624], which bespeak a connection between elastic and dynamic properties, suggesting that unifying concepts may underlie thermodynamic theories and dynamic models. By providing information on the dynamics of networks, recent experimental advances reinforce the idea that the microscopic motions and the elastic properties can be usefully interrelated. For example, quasielastic neutron scattering measurements probe the motions of the network cross-links [1625, 1626]. In end-linked polydimethylsiloxane, it was observed that the junctions diffuse on the same timescale as the network chains, although the spatial extent of the cross-link motion is reduced by intermolecular constraints [1626].

^{31}P-NMR spin-lattice relaxation measurements on a series of polytetrahydrofuran networks with tris(4-isocyanatophenyl)thiophosphate junctions have been carried out [1515] to study junction dynamics and their dependencies on cross-link density and diluent concentration. These less conventional techniques (as far as the study of rubber elasticity is concerned) provide information that is consistent with existing rubber elasticity models, yet demonstrates that a void exists. There is a conspicuous absence of theories which specifically describe the junction dynamics. The new experimental findings invite a new approach from a totally dynamic point of view; mechanical equilibrium properties can be considered as only a special or a limiting case (albeit a very important one). The CM has provided such a new approach [1516, 1517]. The adequacy of this application of the CM to address dynamic response of networks is demonstrated here by analyzing the informative ^{31}P-NMR spin-lattice relaxation data [1515] and by revisiting the change of mechanical relaxation on deforming glassy polymers presented before in Section (2.2.5.9 (I)). Before addressing the experimental data, it is instructive to show the connection between the constrained junction model of Flory for rubber elasticity and the CM approach for dynamics of junctions.

Connections Between Flory's Constrained Junction Model and the Coupling Model

The restrictions on the configurations available to the cross-link junctions of a network, arising from its presence in dense phase, govern the mechanical properties of elastomers. These constraints also give rise to intermolecular correlations and cooperativity of the junction motions, and hence can be described by the CM. The close connection between the elasticity of networks and junction dynamics can be illustrated through a comparison of the predictions of the CM for constraint dynamics with the constrained junction model of Flory. This comparison given in Table 3.8 is enlightening because it shows that the gist of the CM for interacting or constrained dynamics is compatible with the idea of Flory in his constrained junction model.

Flory modeled the effects of restrictions imposed by neighboring chains on the fluctuations of network junctions, deriving an expression for the modification of the elastic stress for a network of phantom chains. Analogously, the CM (when applied to junctions in polymer networks) uses the dynamics of the constraints on junctions to model the slowing down of the motions of the network junctions caused by interactions with neighboring chains and thus obtain the modification of the (primitive) correlation function of relaxation of junctions $C_0(t)$. From the idea that phantom chains are able to freely pass through one another, we expect that $C_0(t)$ has an exponential time dependence $\exp\left[-(t/\tau_0)\right]$. At temperatures sufficiently above T_g, τ_0 can be well approximated by the Arrhenius temperature dependence $\tau = \tau_\infty \exp(E_a/RT)$ with the activation energy E_a given by the true microscopic conformational energy barrier to motion of junctions in the phantom network model. Following the general physical principle behind the CM, the junction constraints will modify $C_0(t)$ to $C_c(t) = \exp[-(t/\tau^*)^{1-n}]$ with $\tau^*(T) \equiv \tau^*_\infty \exp\left(E^*_a/RT\right)$, and will give rise to the two relations $\tau^*_\infty = (t_c^{-n}\tau_\infty)^{1/(1-n)} = \tau_\infty(\tau_\infty/t_c)^{n/(1-n)}$ and $E^*_a = E_a/(1-n)$. These relations tells us that the degrees of modification of the pre-exponential and the activation energy from the phantom network values τ_∞ and E_a to the constrained values τ^*_∞ and E^*_a increase with n. Typically τ_∞ is of the order of 10^{-13}–10^{-14} s. This, together with $t_c \geq 2$ ps, indicates that the ratio τ_∞/t_c is much less than unity. Hence, both $\log(\tau_\infty/\tau^*_\infty)$ and the ratio E^*_a/E_a increase with n or the severity of constraints. Higher cross-link densities and cross-link functionality will

Table 3.8 Close connection between the elasticity of networks and junction dynamics illustrated through a comparison of the predictions of the CM for constraint dynamics with the constrained junction model of Flory. The comparison is borne out by NMR results on networks

Network feature	Anticipated effect	Flory model, f_c	CM n	CM $\frac{E_a^*}{E_a} - 1$	CM $\log(\tau_\infty/\tau_\infty^*)$	^{31}P-NMR[a] n	^{31}P-NMR[a] $\frac{E_a^*}{E_a} - 1$	^{31}P-NMR[a] $\log(\tau_\infty/\tau_\infty^*)$
Higher cross-link density	More firmly imbedded junction	Higher[b]	Higher	Higher	Higher	Higher[g]	Higher[h]	Higher[i]
Diluent	Reduced severity of constraints	Lower[c]	Lower	Lower	Lower	Lower[g]	Lower[h]	Lower[i]
Higher cross-link density	More constrained junctions	Higher[d]	Higher	Higher	Higher			
Extension	Alleviation of constraints	Lower[e]	Lower[f]	Lower[f]	Lower[f]			

[a]Reference [1515]; [b]Reference [1627]; [c]Reference [1623, 1628]; [d]Reference [1627, 1629]; [e]Reference [1623, 1630–1633]; [f]Tentative (see [940]); [g]See Fig. 299 (left); [h]See Fig. 299 (right); [i]See Fig. 300

enhance the strength of the intermolecular constraints and consequently n. On the other hand, with the addition of diluent, the junctions become increasingly isolated from the neighboring chain segments so that the strength of the intermolecular constraints and hence n decreases. Such a decrease of n with diluent concentration has been seen in the case of the local segmental motion in amorphous polymer solutions (see Section 2.2.5.7) and the terminal motion of barely entangled polymer solutions (see Section 3.22). The data of these other systems have been explained by a reduction of intermolecular constraints and the coupling parameter. In fact, in these cases the experimental data gave direct evidence of the decrease of n with diluent concentration [849, 1634].

Comparison with Experimental Results of ^{31}P-NMR Spectroscopy

New insight has been provided by the solid-state ^{31}P-NMR spin-lattice relaxation measurements of Shi et.al. [1515] on a series of polytetrahydrofuran networks. The relaxation properties of the tris(4-isocyanatophenyl) thiophosphate junctions were characterized in networks with strand molecular weights ranging from 250 to 2900. The dominant mechanism for ^{31}P nuclear spin relaxation was identified to be chemical shift anisotropy. The spin-lattice relaxation times measured over a wide range of temperatures were fitted by spectral density functions derived from the appropriate Fourier transforms of the stretched exponential correlation function, with τ^* assumed to have the Arrhenius temperature dependence. From these fits, Shi et al. obtained the coupling parameter n, the apparent pre-exponential factor τ^*_∞, and the apparent activation energy E^*_a for the networks of different cross-link densities, both neat and diluted. The coupling parameter was found to increase with decreasing molecular weight between cross-links and, at constant cross-link density, decrease with the addition of diluent. Averages of the values of n determined for each sample from ^{31}P spin-lattice relaxation data by two parallel methods (cross polarization and direct polarization) are plotted in Fig. 299. These results are in accord with expectations based on the CM [1516, 1517].

The apparent activation energy E^*_a is also found to increase significantly with higher cross-link density and, at constant cross-link density, decrease with the addition of diluent. In the right panel of Fig. 299 we have plotted the averages of the E^*_a values determined by the two NMR methods. The products $(1-n)E^*_\mathrm{a}$ obtained by using the experimentally determined values of the coupling parameter (left panel) and the apparent activation energy E^*_a for all samples have essentially the same value of 26.2 kJ/mol. This constancy is predicted by the coupling model, since the product, being the true microscopic energy barrier E_a to motion of junctions in a phantom network, is independent of intermolecular interactions or constraints and hence should be independent of cross-link density and dilution. It is worthwhile to point out that 26.2 kJ/mol is of the correct magnitude for the intramolecular conformational energy barrier of polymers with structure similar to the PTHF network junctions. Alternatively, we have calculated E^*_a by the relation $E^*_\mathrm{a} = E_\mathrm{a}/(1-n)$ using the experimentally determined coupling parameters (right panel of Fig. 299) and a constant value of $E_\mathrm{a} = 26.2\,\mathrm{kJ/mol}$. The results in the right panel of Fig. 299

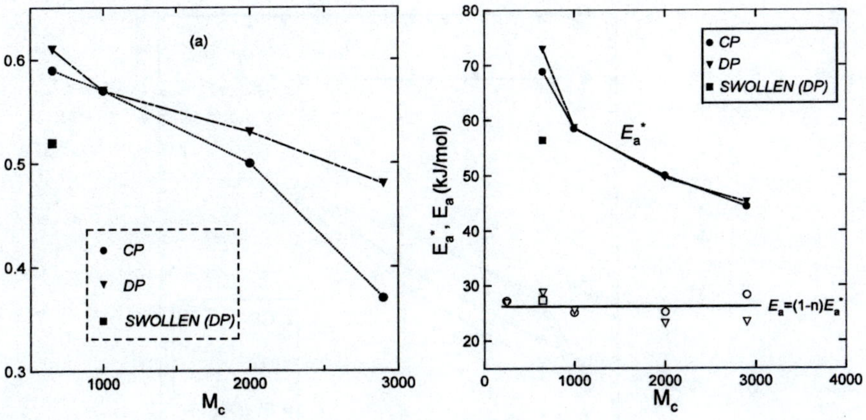

Fig. 299 (*Left*) The coupling parameter of junction dynamics n (*filled circles*) obtained by taking the average of the two values determined by Shi et al. from their ^{31}P-NMR data taken using cross and direct polarizations [1515] for four polymer networks with different molecular weights between cross-links M_c. The *filled square* is the corresponding coupling parameter obtained in a similar manner for a swollen sample with $M_c = 650$. The *smooth curve* drawn through the data points is a visual aid only. (*Right*) *Filled triangles* represent the apparent activation enthalpy E_a^*, obtained by an average of the two values determined by Shi et al. from their ^{31}P-NMR data using direct and cross polarizations for four polymer networks. *Filled square* is E_a^* for the swollen polymer network having $M_c = 650$. The *corresponding unfilled symbols* are the theoretical predicted values of the coupling model calculated by the equation $E_a^* = E_a/(1-n)$ using a constant $E_a = 26.2$ kJ/mol and the coupling parameters n for the same five network polymers given in the *left panel*. A smooth curve has been drawn through the calculated points to guide the eyes. Data from [1516] are replotted in all figures here

suggest good agreement with the experimental values of E_a^* for five samples of different cross-link densities and diluent concentrations.

The raw ^{31}P-NMR data give unphysically short times of the pre-exponential factor τ_∞^* (10^{-16} to 10^{-19} s) and the trend that τ_∞^* decreases dramatically with higher cross-link density (see Fig. 300). The experimental values for τ_∞^* in this figure are the geometric mean of the values obtained by the two NMR methods used. These anomalous properties of τ_∞^* can be explained by the other equation of the CM $\tau_\infty^* = (t_c^{-n}\tau_\infty)^{1/(1-n)} = \tau_\infty(\tau_\infty/t_c)^{n/(1-n)}$. The prefactor τ_∞ corresponds to relaxation of junctions of a phantom network. Thus, its value should be independent of intermolecular interaction and hence of cross-link density and diluent concentration. Since τ_∞ has not been determined directly by experiment, it is taken to be an adjustable (albeit constant) parameter. With the experimentally determined n and a value for the crossover time, $t_c = 4 \times 10^{-12}$ s and $\tau_\infty = 1.5 \times 10^{-14}$ s, the calculated values of τ_∞^* are found to be in good agreement with the experimental values obtained for all five samples, as shown in Fig. 300. The trends of a rapid decrease of τ_∞^* with cross-link density and an increase upon dilution are reproduced.

Universal Features

The physical aspects of junction dynamics of networks discussed above are isomorphic to the dynamics of β-relaxation of the mesogenic group in side-chain liquid

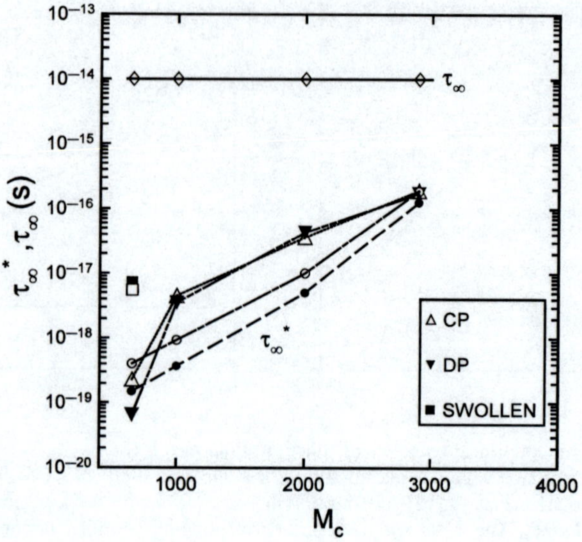

Fig. 300 *Filled diamonds* are the apparent pre-exponential factor τ_∞^* determined by Shi et al. from their experimental data [1515] for the four polymer networks by cross and direct polarizations. *Filled square* is τ_∞^* for the swollen polymer network with $M_c = 650$. The corresponding *open symbols* represent the theoretical predictions of the coupling model calculated by the equation indicated in the figure using constant values of $\tau_\infty = 1.5\times10^{-14}$ s and $t_c = 4\times10^{-12}$ s independent of M_c and diluent concentration. The coupling parameters used here are taken from the data in Fig. 299. For clarity a smooth curve is drawn through the calculated points. Data from [1516] are replotted here

crystalline polymers discussed in Section 3.2.6.1 and ion dynamics in crystalline and glassy ionic conductors (Section 3.2.5). These are entirely different systems and relaxation processes, and yet the anomalies found are analogous.

Effect of Deformation

In Flory's constrained junction model of elasticity, elongation alleviates the restrictions of junctions from neighboring segments [1623], in that the "domain" of the constraints extends along the stretch direction. The elastic behavior of the network thus becomes more phantom-like at higher elongations, in agreement with experiments [1630–1633]. We anticipate that this alleviation of the constraints on the junctions by elongation will be manifested in the junction dynamics. The junctions have higher mobility at higher elongations. If this is the case, the coupling model predicts a decrease of n with extension, along with corresponding changes in relaxation time τ^* and its E_a^* and τ_∞^*. Such experiment has not been carried out to reveal the expected changes in the network dynamics associated with elongation. An analogue of this is the non-linear deformation of glassy polymers including bisphenol A polycarbonate and polystyrene. As discussed in Section 2.2.5.9(I), decrease of n and shorter τ^* by large deformation were found experimentally in glassy polymers.

3.2.6.3 Cooperative Oxygen Ion Dynamics in $Gd_2Ti_{2-y}Zr_yO_7$

The oxygen ion conductors with pyrochlore structure $A_2B_2O_7$ are promising materials for use in fuel cells [1635–1637]. The pyrochlores $Gd_2Ti_{2-y}Zr_yO_7$ are particularly interesting since the concentration of mobile oxygen vacancies can be increased by substitution of Zr for Ti, and oxygen ion conductivity shows the highest value found among materials with pyrochlore structure. For $y < 1.8$ the conductivity is comparable to that of YSZ (10^{-2} S/cm at 700°C). Molecular dynamics [151, 152] and static lattice energy minimization simulations [153] have suggested that oxygen diffusion in $Gd_2Ti_{2-y}Zr_yO_7$ occurs by hopping from 48*f* site to 48*f* site. This result has been later confirmed by XPS measurements [1637]. The oxygen occupancy of 48*f* sites is 1 (or very close to 1) for Zr contents below $y = 0.6$ but decreases progressively as Zr content is further increased [1638]. These vacancies in 48*f* sites are responsible for oxygen hopping and diffusion, and explain the increase of more than two orders of magnitude in dc conductivity at 600°C observed when increasing Zr content from $Gd_2Ti_{1.4}Zr_{0.6}O_7$ to $Gd_2Zr_2O_7$ [1635]. Although the activation energies E_σ of the dc conductivity increases monotonically with y from $E_\sigma = 0.70$ eV for $y = 0.5$ to $E_\sigma = 1$ eV for $y = 2$ [1639, 1640].

More recent measurements of E_σ by Moreno et al. [150] are shown in the left panel of Fig. 301 for six samples with Zr content of $y = 0.5, 0.7, 0.9, 1.3, 1.8$, and 2. However, the energy barrier for oxygen hopping from 48*f* to 48*f* sites E_a has been previously calculated. From molecular dynamics simulations [151, 152], E_a ranges from 0.57 to 0.64 eV, and from static lattice energy minimization simulations [153], $E_a = 0.58$ eV. Two problems or anomalies present themselves. One is

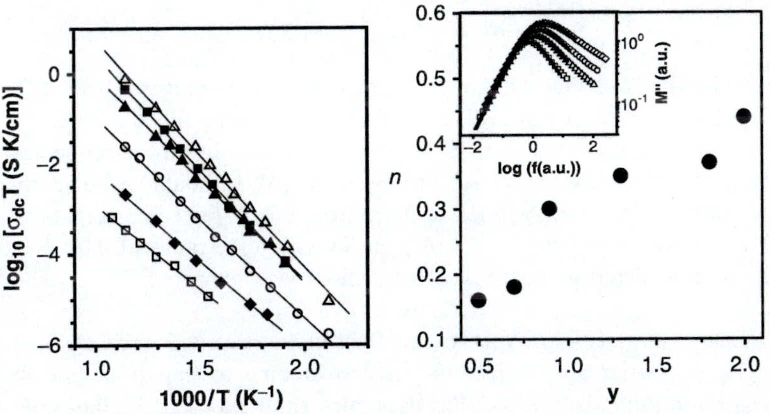

Fig. 301 (*Left*) Arrhenius plot of dc conductivity in the series $Gd_2Ti_{2-y}Zr_yO_7$ for different Zr contents: $y = 0.5$ (*open squares*), 0.7 (*closed diamonds*), 0.9 (*open circles*), 1.3 (*closed triangles*), 1.8 (*open triangles*), and 2 (*closed squares*). (*Right*) Dependence of the exponent n obtained from fits of electric modulus data to Fourier transform of the Kohlrausch function, as a function of Zr content in $Gd_2Ti_{2-y}Zr_yO_7$. The *inset* shows the imaginary part of the electric modulus for samples (*from top to bottom*) with $y = 2, 1.3, 0.9$, and 0.5. Note that experimental data in the *inset* have been horizontally and vertically shifted for clarity. Reproduced from [150] by permission

that the actual energy barrier for oxygen ion hopping E_a is significantly smaller than the observed E_σ. The other is the monotonic increase of E_σ with y. These anomalies are analogous to those found in other interacting systems. These include the conductivity relaxation in other ionic conductors, β-relaxation in side-chain liquid crystalline polymers, and motions of polymer chains, where the anomalies originate from many-body relaxation dynamics and have been explained by the CM through the coupling parameter n.

To resolve the anomalies, the key step was taken by Moreno et al. in determining n for the conductivity relaxation as a function of frequency in $Gd_2Ti_{2-y}Zr_yO_7$ using impedance spectroscopy. The imaginary part of the electric modulus $M''(f)$ from the impedance spectroscopy measurements of $Gd_2Ti_{2-y}Zr_yO_7$ is shown in the inset of the right panel of Fig. 301. It can be seen that the width of the $M''(f)$ loss peak increases with y. $M''(f)$ calculated by using the Kohlrausch function for $\Phi(t)$ in Eq. (1.56) was used to fit the data, and the coupling parameter n was determined for all six samples from the good fits found. The results shown in the right panel of Fig. 301 show systematic and significant increase in n as Zr content is increased from $y = 0.5$ to 2.0. It is known that increasing Zr content above $y = 0.5$ results in creating vacant 48f sites which are responsible for oxygen hopping motion [1638], resulting in higher charge carrier (oxygen vacancies at 48f sites) concentration. The increase also creates higher disorder in both the cationic and anionic sublattices [1637]. Both effects act in concert to enhance mutual interactions between the carriers, which lead to the increase of the coupling parameter n with y as expected by the CM, as observed by experiment. From this, the observed increase of the dc conductivity activation energy E_σ with y follows as a consequence of the CM equation (3.23), now written as

$$E_\sigma(y) = E_a / \left[1 - n(y)\right], \tag{3.42}$$

together with the facts that the energy barrier for oxygen hopping from 48f to 48f sites E_a is independent of y and n is a monotonic increasing function of y.

Furthermore, from the values of $E_\sigma(y)$ and $n(y)$ obtained from experiment, we can form the product $[1-n(y)]E_\sigma(y)$. According to the CM, it should be the microscopic energy barrier E_a for an oxygen ion to jump from one 48f site to a neighboring 48f sites. The products from the six samples all lie within the range of 0.60 ± 0.03 eV and can be considered as nearly independent of y as expected. This is shown in Fig. 302.

The mean value of 0.60 eV is in excellent agreement with recent calculations of the energy barrier an oxygen ion must overcome to hop from one 48f site to a neighboring one from molecular dynamics simulations (0.57–0.64 eV) [151, 152] and from static lattice energy minimization simulations (0.58 eV) [153]. This result offers proof of the verity of the primitive relaxation time. The observed slow relaxation time and its properties are the consequence of the many-ion cooperative dynamics originating from Coulomb interaction between the ions. Cooperative oxygen ion dynamics is not only a subject of fundamental interest but also a key factor in determining the dc ionic conductivity value for practical applications.

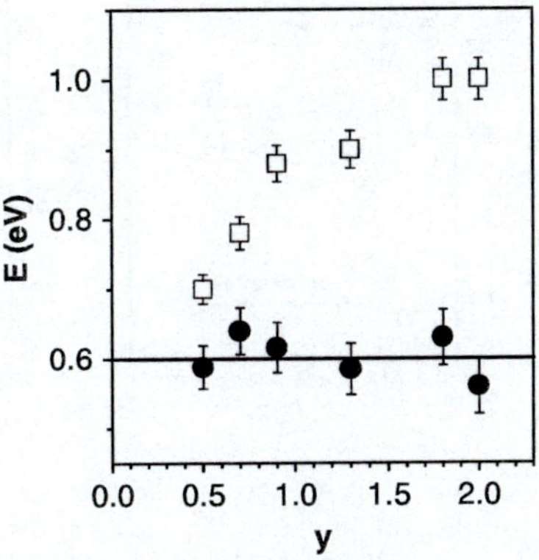

Fig. 302 Activation energies E_{σ} (□) and $E_a = (1-n)E_{\sigma}$ (•) as a function of Zr content in $Gd_2Ti_{2-y}Zr_yO_7$. *Solid line* represents the average value $E_a = 0.60$ eV obtained for the primitive energy barrier for oxygen hopping. Reproduced from [150] by permission

In the discussion above for $Gd_2Ti_{2-y}Zr_yO_7$, increase in activation energy E_{σ} when increasing Zr content is explained by the increase in the degree of correlations among mobile oxygen ions in the oxygen ion diffusion process as the structure becomes more disordered. This explanation is supported by the concomitant increase in n, which is a measure of the many-ion conductivity relaxation dynamics. If the effect and the interpretation are general, one would expect that replacing Gd in $Gd_2Ti_{2-y}Zr_yO_7$ by other lanthanides such as Y and Dy, both the exponent n and the activation energy E_{σ} also will increase with Zr content in $Y_2Ti_{2-y}Zr_yO_7$ and in $Dy_2Ti_{2-y}Zr_yO_7$. This is in fact observed by Díaz-Guillén et al. [1645]. It is noted that the increase in E_{σ} occurs despite increasing cell volume, indicating that it is caused by the enhancement of ion–ion correlations and the increase in n, in accordance with the coupling model predictions.

As for the primitive energy barrier $E_a(y)$, its value is given by the product $[1 - n(y)]E_{\sigma}(y)$, according to the CM. We have seen from Fig. 302 for the $Gd_2Ti_{2-y}Zr_yO_7$ series that the calculated values of $E_a(y)$ are approximately constant and in agreement with the values obtained by molecular dynamics simulations and by static lattice energy minimization simulations. The same was found for the two other series $Y_2Ti_{2-y}Zr_yO_7$ and $Dy_2Ti_{2-y}Zr_yO_7$, as shown in Fig. 303(a)–(c), which are like Figs. 301 and 302 for the $Gd_2Ti_{2-y}Zr_yO_7$ series.

Moreover, increasing order and concomitant decrease in E_{σ} as well as n was found by annealing at a higher temperature of $Dy_2Ti_{0.45}Zr_{0,56}O_7$ [1646]. On the other hand, E_a calculated by the product $[1 - n]E_{\sigma}$ remains nearly independent of the annealing temperature. All these supplementary data support once more the verity of the primitive relaxation through its primitive activation energy.

3.2.6.4 The Crystalline Lithium Ionic Conductor $Li_{3x}La_{2/3-x}TiO_3$ (LLTO)

For some ionic conductors that have the dc conductivity σ_{dc} measured over an extended range of temperature to reach high σ_{dc} levels, the complete temperature dependence of σ_{dc} is non-Arrhenius. Examples can be taken from some ionic conductors in Table 3.1, and they can be glassy, molten, or even crystalline. Specific

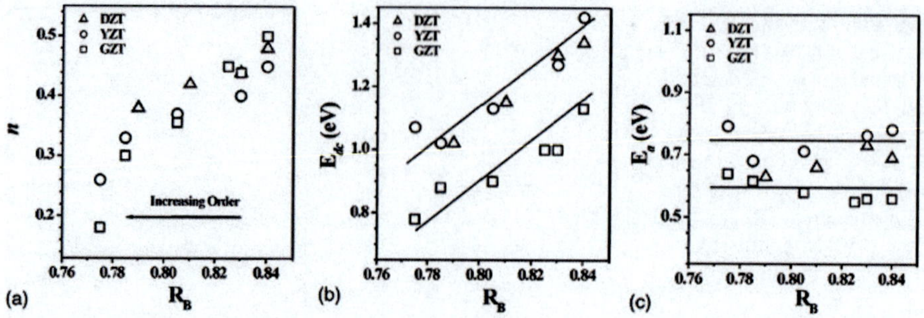

Fig. 303 Dependences on the average cation size R_B of (**a**) the exponent n, (**b**) the dc conductivity activation energy E_{dc}, and (**c**) the primitive activation energy E_a for samples in the series of $Dy_2Zr_{2\text{-}y}Ti_yO_7$ (DZT), $Y_2Zr_{2\text{-}y}Ti_yO_7$ (YZT), and $Gd_2Zr_{2\text{-}y}Ti_yO_7$ (GZT). Note that the absolute scales in (**b**) and (**c**) are the same, and the *solid lines* are shown only to emphasize the trends. Reproduced from [1645] by permission

examples include the fast silver ion-conducting glasses by Kincs and Martin [1536] and the crystalline lithium ionic conductors $Li_{3x}La_{2/3\text{-}x}TiO_3$ (LLTO); a special case is $Li_{0.18}La_{0.61}TiO_3$ by León et al. [1484]. This non-Arrhenius behavior, seemingly a universal feature in fast ionic conductors as dc conductivity approaches values about 10^{-1} S/cm (see Table 3.1), may not be too surprising. This is because when $\sigma_{dc} \approx 1$ S/cm, the conductivity relaxation time τ_σ would be shorter than the crossover (to primitive relaxation) time $t_c \approx 1$ ps as discussed in Section 3.2.3 and demonstrated in [1537, 1538]. However, the crystalline ionic conductor LLTO exhibits non-Arrhenius behavior at levels of σ_{dc} much lower than 1 S/cm, as shown in the left panel of Fig. 304.

For crystalline ionic conductors such as LLTO and glassy ionic conductors that remain deep in the glassy state throughout the entire temperature range, the observed non-Arrhenius T dependence of σ_{dc} or τ_σ offers a challenge for an explanation. This is because the activation energy should be constant since the structure and hence the energy barrier opposing ion hopping remains unchanged. By the same token, the primitive activation energy E_a in the CM should also be constant. Notwithstanding, the explanation from the CM model is the change of ion–ion correlation effects caused by decrease of coupling parameter n_σ as τ_σ becomes shorter on increasing temperature. There is an analogue of this behavior in the structural relaxation of many glass-forming substances (see Chapter 2), where the coupling parameter n_α of the structural α-relaxation shows monotonic decrease with decrease in τ_α by raising temperature. It can be readily seen from the CM equation $\tau_\sigma(T) = [t_c^{-n_\sigma(T)}\tau_0(T)]^{1/(1-n_\sigma(T))}$ that $\tau_\sigma(T)$ is non-Arrhenius even though $\tau_0(T) = \tau_{0\infty}\exp(E_a/kT)$ is strictly Arrhenius if n_σ is temperature dependent. The same applies to σ_{dc} because $\sigma_{dc}(T) = \varepsilon_0\varepsilon_\infty / <\tau_\sigma(T)>$, where $<\tau_\sigma(T)> = [\Gamma(1/\beta_\sigma)/\beta_\sigma]\tau_\sigma(T)$ and $\beta_\sigma = (1 - n_\sigma)$. This CM explanation also offers an opportunity to recover the primitive relaxation time and the primitive activation energy from the experimental data.

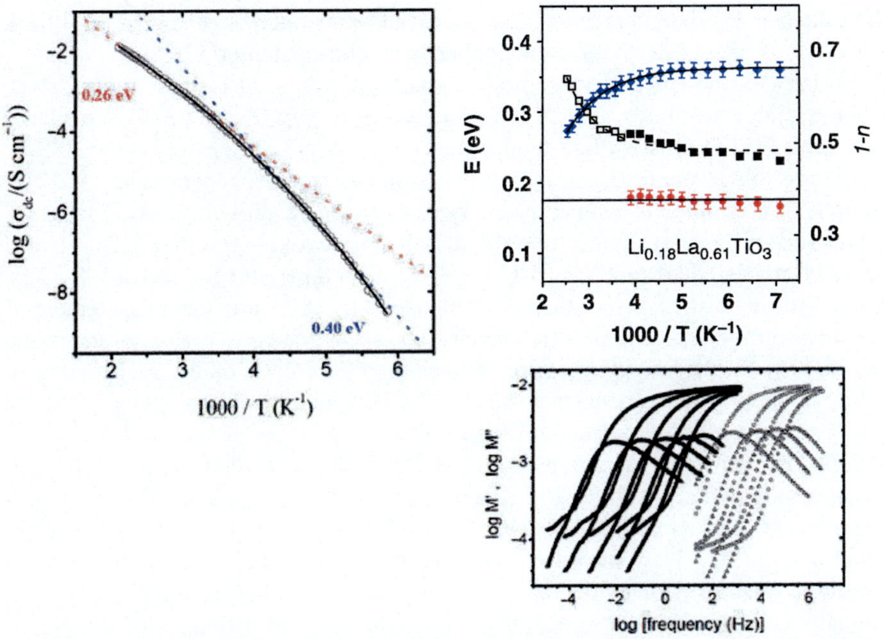

Fig. 304 (*Left*) dc conductivity vs. 1000/T of LLTO, showing a strongly non-Arrhenius behavior. *Dashed lines* indicate the low- and high-temperature activation energies of the Li ion spin relaxation rate, which are 0.4 and 0.26 eV, respectively. *Solid line* is a fitting to VFTH law. Data from León et al. [1484] is replotted here. (*Right*) (*Upper*) Temperature dependence of the dc activation energy E_σ (*diamonds*) and of the Kohlrausch exponent $[1 - n(T)]$ (*solid squares*) describing the dynamic response of ions in $Li_{0.18}La_{0.61}TiO_3$. E_σ has been determined up to 385 K, but $[1 - n(T)]$ only up to 256 K. *Circles* represent the microscopic activation energy E_a estimated by using the CM equation (3.23). It has the constant value of 0.175 eV (*horizontal line*). *Open squares* are the predicted values of the exponent $[1 - n(T)]$ for $256 < T < 385$ K in order that the same constant value, $E_a = 0.175$ eV, is maintained at the higher temperatures according to the relation $E_a = [1 - n(T)]E_\sigma$. Data from [1482] is replotted here. (*Lower*) Spectra of the real (*open circles*) and imaginary parts (*open squares*) of the electric modulus of $Li_{0.18}La_{0.61}TiO_3$ obtained from 121, 131, 141, 150, and 160 K from *left to right*. *Solid symbols* represent spectra of real and imaginary parts of the electric modulus at higher temperatures, 191, 212, 232, and 256 K from the *left to right*. The narrowing of the loss peak on increasing temperature is clear. Data from [1647(a)] is replotted

Combination of time domain and dynamic frequency measurements of the complex electric modulus $M^*(\nu)$ of LLTO has been performed to investigate ion dynamics in the wide frequency range of 11 decades from 10^{-5} to 10^6 Hz and temperatures up to 256 K [1647(a)]. The spectra are shown in the right (lower) panel of Fig. 304, where the narrowing of the loss modulus peak on increasing temperature can be clearly seen. Indeed, this study have found that the exponent $[1 - n_\sigma(T)]$ of the Kohlrausch function used to fit the spectra increases with increasing temperature as shown by the solid squares in the right (upper) panel of Fig. 304. The primitive activation energy $E_a(T)$ calculated via Eq. (3.42) by the product $[1 - n_\sigma(T)E_\sigma(T)]$ are shown by the open circles in the same figure. It can be seen that within errors the values of $E_a(T)$ obtained are independent of temperature as it should be since it corresponds to the microscopic energy barrier of the crystalline LLTO.

So far, the published electric modulus data give $[1 - n(T)]$ for T up to 256 K, while sizeable decrease of $E_\sigma(T)$ continues up to 385 K. If the CM explanation holds, $E_a(T)$ must still be temperature independent all the way up to 385 K. Assuming this is the case and taking the temperature-independent value of $E_a = 0.175$ eV determined at temperatures lower than 256 K, $[1 - n(T)]$ must have the values shown by open squares in order that the product $[1 - n_\sigma(T)]\, E_\sigma(T)$ remains constant and is equal to $E_a = 0.175$ eV for all temperatures up to 385 K. This offers a further critical test of the CM explanation by extending the measurement of the electric modulus at higher frequencies up to the gigahertz region to determine $[1 - n_\sigma(T)]$ directly at temperatures higher than 256 K. Actually, such measurements of $M^*(\nu)$ for frequencies as high as 1 GHz and temperature reaching 385 K have been made but so far unpublished [1647(b)]. The coupling parameter $n_\sigma(T)$ obtained by fitting the electric modulus by Eq. (1.56) with $\Phi(t)$ as the Kohlrausch function at temperature above 256 K and up to 385 K is in agreement with the predicted values of the exponent $[1 - n_\sigma(T)]$ for $256 < T < 385$ K shown in the right panel of Fig. 304. New measurements confirming the same will be published in the future [1647(c)]. Using the experimentally obtained values of $E_\sigma(T)$ and $n_\sigma(T)$, the calculated products $(1 - n(T))E_\sigma(T)$ should be good estimates of the primitive energy barrier E_a. An approximately constant value of $E_a = 175 \pm 10$ meV is obtained over the whole temperature range (line in the right upper panel of Fig. 304), which is remarkably similar to the value of 170 mV previously determined from electrical conductivity relaxation and NMR spin-lattice relaxation for the microscopic energy barrier [1484]. This result strongly supports the interpretation of a decrease in ion–ion correlation with increasing temperature as the origin of the observed decrease in the exponent n, which in turn explains the parallel decrease of E_σ. For the present purpose, the primitive activation energy E_a is not only recovered from the data but also indeed independent of temperature as it should be in a crystalline ionic conductor.

3.2.6.5 The Crystalline Lithium Ion Conductor $Li_{1.2}Ti_{1.8}Al_{0.2}(PO_4)_3$

Rivera et al. [1647(c)] have also analyzed in a similar way the recent experimental conductivity data of the Nasicon-type ionic conductor $Li_{1.2}Ti_{1.8}Al_{0.2}(PO_4)_3$ obtained by Arbi et al. [1647(d)]. It also shows a non-Arrhenius temperature dependence of the dc conductivity and a decrease of the exponent n with increasing temperature, although it has been measured in a smaller temperature range (from 173 to 323 K). Figure 305 shows the temperature dependence of the dc activation energy E_σ obtained from [1647(d)] (open squares), which changes from 350 meV at low temperatures to 220 meV at the highest temperatures. Using the values of the exponent n determined by those authors (solid squares in the figure), Rivera et al. have calculated the energy barrier E_a by the product $(1 - n(T))E_\sigma(T)$ and found that again an approximately constant value of E_a (open circles in the figure) is obtained over the whole temperature range. This similar finding in a different

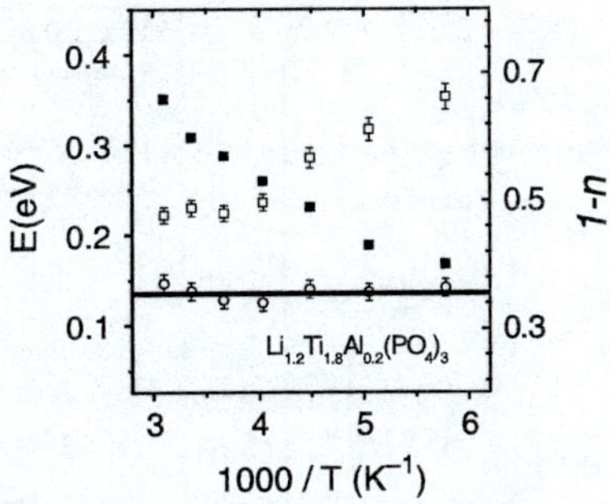

Fig. 305 Temperature dependence of the dc activation energy E_{dc} (*open squares*) and of the exponent n (*solid squares*) describing the dynamic response of ions in $Li_{1.2}Ti_{1.8}Al_{0.2}(PO_4)_3$ after Arbi et al. [1647(d)]. *Open circles* represent the microscopic activation energy estimated by using the coupling model. Reproduced from [1647(c)] by permission

ionic conductor further supports our conclusions derived from the CM analysis of temperature-dependent dynamics in LLTO.

3.2.6.6 Ionic Conductivity of Nanometer Thin Films of Yttria-Stabilized Zirconia

Discussions of the change in dynamics of various glassformers on reducing one or more dimensions to nanometer scale were made Section 2.3.2.33. There experimental data have shown the suppression of the many-body relaxation, obviously due to the reduction in its length scale, and much faster structural α-relaxation than in the bulk. Under extreme confinement or restricting to consideration of the surface layer of thin films, the suppression can be so severe that the observed α-relaxation time τ_α approaches or even assumes the primitive relaxation time τ_0 calculated from the CM or the JG β-relaxation time τ_{JG} of the bulk glassformer. The effect has been demonstrated in ultrathin poly(methylphenylsiloxane) (PMPS) and poly(ethylene oxide) (PEO) films with thickness of the order of 1.5–2.0 nm intercalated into galleries of organically modified layered silicates in Fig. 261 and PDMS confined in 2.5-nm glass pores in Fig. 260. Here we present experimental data of the oxygen ion conductivity relaxation in thin films of yttria-stabilized zirconia (YSZ) to demonstrate that the same effect is observed. This is another example of the universal dynamics manifested in relaxation of interacting systems.

Before we proceed to discuss the oxygen ion conductivity in thin films of YSZ, mention has to be made on the earlier studies by Maier and coworkers in which they found substantial increase of the dc ionic conductivity in superlattices of CaF_2 and BaF_2 when the thickness of the individual layers was decreased down to 16 nm [1648]. The first case involving YSZ is the large enhancement of the in-plane electrical conductivity σ of highly textured thin film of yttria-stabilized zirconia (YSZ) deposited onto an MgO substrate found by Kosacki et al. [1649]. The enhancement started to be observed for films with thickness d less than 60 nm, and σ increases

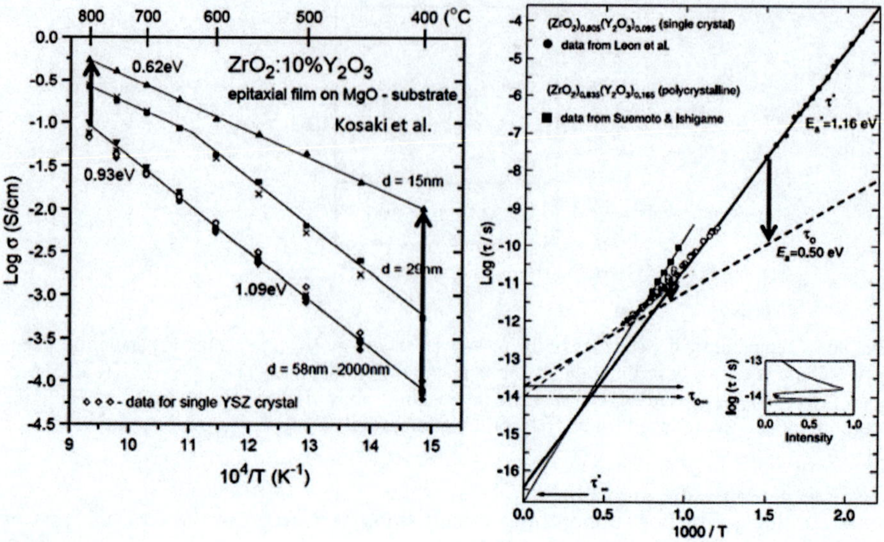

Fig. 306-1 (*Left*) Temperature dependence of the electrical conductivity determined for epitaxial YSZ thin films of different thicknesses. The *thick vertical arrows* indicate the order of magnitude increase of the conductivity at $T \sim 400$ and 800°C. Reproduced from [1649] by permission. (*Right*) Same as Fig. 288 except for the two *thick arrows* at $T \sim 400$ and 800°C to indicate that the difference between the measured conductivity relaxation time of bulk YSZ and the calculated primitive relaxation time is of nearly same order of magnitude as the increase of conductivity of the 15-nm YSZ thin film at the same temperature in the *left panel*. Note that the compositions of the samples in the two figures are not exactly the same

by orders of magnitude with decreasing d down to 15 nm as shown in the left panel of Fig. 306-1. The thick arrows at $T{\sim}400$ and 800°C indicate enhancement of σ by about 2 and 0.75 decades, respectively.

The observed increase in conductivity σ was attributed by Kosacki et al. to a significant and dominating contribution from the oxygen transport at the interface, which becomes increasingly important with decrease in the film thickness. From an analysis of the data by these authors, it was estimated that the thickness of interface layer is about 1.6 nm, and interfacial conductivity can be orders of magnitude greater than the bulk-like, lattice-related conductivity depending on the temperature. In the deposited YSZ thin film, there is reduction of inter-ion coupling because of the absence of oxygen ions on one side of the air-YSZ surface as well as in the 1.6-nm interfacial YSZ layer on top of the MgO substrate. The analogy with supported polymer thin films on substrate is evident, where we have seen the surface layer has much higher segmental mobility than bulk from AFM probe of the surface layer of polystyrene thin films by Tanaka et al. [1465(c)] and Forrest and coworker [1465(d), (e)]. The reduction of coupling parameter and the relation to the primitive relaxation time have been demonstrated by Tanaka et al. This is also the case in the observed enhancement of conductivity in the 15-nm YSZ film. The two arrows in the right panel of Fig. 306-1 indicate the difference in order of magnitude between the bulk conductivity relaxation time τ_σ and the primitive relaxation time τ_0 at 400 and 800°C. They are nearly the same as the enhancement of σ observed in the 15-nm epitaxial YSZ thin film at the same temperatures in the left panel of Fig. 306-1. Also the activation energy of σ in the 15-nm epitaxial YSZ thin film is 0.62 eV, which is not far from the 0.50 eV of τ_0, considering the fact that the YSZ in the two cases does not have the same mole percentage of yttria.

Motivated by the observation of Kosacki et al. that enhancement of conductivity increases with decreasing YSZ film thickness and the suggestion that it comes from oxygen ions in the interfacial layer, Garcia-Barriocanal et al. fabricated heterostructures where YSZ layers (with 8 mol% nominal yttria content) in the thickness range from 62 nm down to 1 nm were sandwiched between two 10-nm-thick layers of insulating $SrTiO_3$ (STO) [1650(a), 1650(b)]. Also, superlattices were grown, alternating 10-nm-thick STO films with YSZ layers of thickness between 62 and 1 nm. The thickness of 1 nm corresponds to the dimension of two unit cells of YSZ. The ultrathin layer of YSZ grows rotated by 45° around the *c*-axis and strains to match the STO lattice. Because the bulk lattice constants of STO and YSZ are 0.3905 and 0.514 nm, respectively, the epitaxial growth of the YSZ on top of the STO engenders a large, expansive strain in the thin YSZ layers of 7% in the *ab* plane. The real part of the lateral electrical conductivity σ' of the thinnest 1-nm YSZ film between two 10-nm-thick STO films plotted against frequency *f* is shown in the left panel of Fig. 306-2. The dc ionic conductivity σ_{dc} of this [$STO_{10\,nm}/YSZ_{1\,nm}/STO_{10\,nm}$] trilayer was determined by the plateau level of σ' (*f*) and the values are identified by the stars in the figure. The electronic contribution to the ac measurements was negligible, and thus the measured ac transport is attributable to oxygen ion diffusion process [1650(a), 1650(b)]. In the right panel of Fig. 306-2, the temperature dependence of σ_{dc} of the [$STO_{10\,nm}/YSZ_{1\,nm}/STO_{10\,nm}$] trilayers is shown together with data from a single crystal and the 700-nm film. Whereas the bulk and the thick

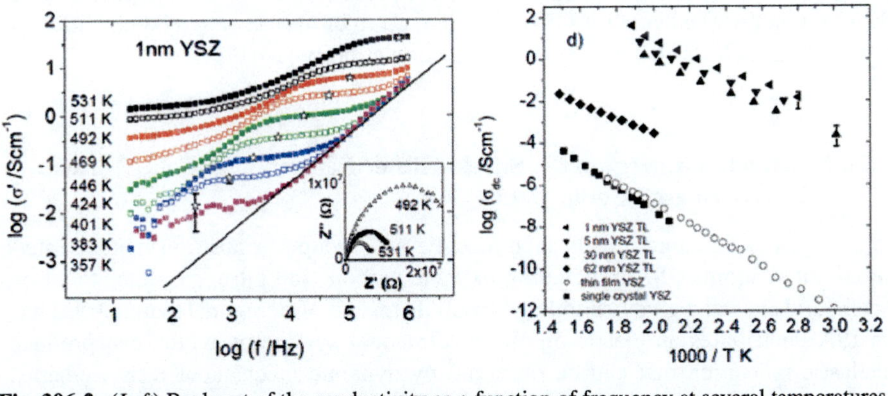

Fig. 306-2 (*Left*) Real part of the conductivity as a function of frequency at several temperatures for a 1-nm YSZ (~8 mol.% yttria content) trilayer with 10-nm STO *top* and *bottom* layers grown on a (100) STO substrate. Isotherms were measured in the range of 357–531 K. The *solid line* represents likely stray capacitance from the STO and not an NCL contribution as originally suggested in [1650(a)]. *Stars* identify the value of σ_{dc}. (*Inset*) Imaginary vs. real part of the impedance (Nyquist) plots at 492, 511, and 531 K. The high-frequency contribution is a Debye-like process characterized by a conductivity exponent $n = 0$

film of YSZ show the well-known Arrhenius behavior with an activation energy $E_\sigma \sim 1.1$ eV, the trilayers show much larger conductivity values and smaller values of the activation energy $E_{\sigma,\text{film}}$ but compensated by much larger pre-exponential factor $\sim \sigma_{\infty,\text{film}}$. The thickest trilayer (62-nm YSZ) already shows an increase of about two orders of magnitude in the high-temperature dc conductivity, and $E_{\sigma,\text{film}}$ decreases to 0.72 eV. When decreasing the thickness of the YSZ layer to 30 nm and then to 1 nm, the dc conductivity increases another three or more orders of magnitude, and $E_{\sigma,\text{film}}$ decreases further to 0.6 eV, nearly the same as 0.62 eV for the 15-nm-thin film of YSZ deposited on MgO (see left panel of Fig. 306-1) and 0.50 eV, the activation energy of the primitive oxygen ion hop. The values of the pre-exponential factor $\sim \sigma_{\infty,\text{film}}$ are as high as $\sim 10^7$ (ohm cm)$^{-1}$. Garcia-Barriocanal et al. concluded that the high conductance observed comes from some interface conduction mechanism, which they supported by conductance measurements of superlattices with [$YSZ_{1\,\text{nm}}/STO_{10\,\text{nm}}$] as the repeat unit. From the good epitaxial quality of the heterostructures, they believe that the large in-plane expansive strain on the YSZ interface plane, together with the high concentration of vacant oxygen positions and probable positional disorder, leads to uncorrelated ion diffusion at the interface, resulting in the reduction of the activation energy and the huge enhancement of the ionic conductivity. Here, to the cause of the effect, we add the reduction of inter-ion coupling due to the absence of ions on both sides of the YSZ thin films. Like Garcia-Barriocanal et al. have done before, the reduction in coupling parameter n from the bulk value of 0.56 [1542] accounts for the concomitant decrease of activation energy from the bulk value $E_\sigma \sim 1.1$ eV to 0.60 eV according to the CM equation (3.42). Garcia-Barriocanal et al. suggested that the very large pre-exponential factor $\sim \sigma_{\infty,\text{film}}$ is possibly due to an increase in the concentration of oxygen vacancies and a larger entropy term [$\exp(S/k_B)$] is due to an increased number of available positions for the oxygen ions (enhanced positional disorder for the oxygen vacancies) [1650(b)]. Pennycook et al. reported density-functional calculations that trace the origin of the effect to a combination of lattice-mismatch strain and O sublattice incompatibility [1650(c)].

3.2.6.7 Activation Energy of the Snoek–Köster Relaxation in Cold-Worked, Body-Centered Cubic Metals

In body-centered cubic metals such as α-Fe and group-V transition metals, interstitial solute atoms (ISAs) including oxygen, carbon, and nitrogen prefer to sit on octahedral sites and give rise to local strain distortion of tetragonal symmetry. Thus an ISA contributes an elastic dipole of tetragonal symmetry, which can produce anelastic relaxation that can be observed by dynamic mechanical measurements. Recognizing this possibility, Snoek found these anelastic relaxations as peaks by internal friction measurements of Q^{-1} as a function of temperature [1651]. If the ISA concentration is low, the Snoek peak is well described by an exponential Debye relaxation with a single relaxation time τ_S, which is related to the reciprocal of the jump transition rate of the ISA to neighboring sites. It was found that τ_S has an Arrhenius temperature dependence $\tau_S = \tau_{S\infty}\exp(H_S/kT)$, where H_S is the activation enthalpy of diffusion of the interstitial atoms. After the ISA-containing metal has been cold worked, an additional internal friction peak, called the Snoek–Köster (SK) peak, appears at a higher temperature [1652]. Both the peak position and the peak height depend on the degree of cold work and the ISA concentration [1653–1655]. The activation enthalpy H_{SK} and the width of the SK peak are usually substantially larger than the corresponding quantities of the Snoek peak [1653–1655].

It is generally agreed that the SK relaxation is caused by the bowing out of a dislocation segment under applied stress, dragging along with it the surrounding Cottrell atmosphere of rather high density of ISAs, as suggested by Schoeck [1656]. However, Schoeck's own model did not take into account of the effect that interactions and correlations between the diffusing ISAs have on the SK relaxation. In an application of the CM to the SK relaxation [1653–1655], this effect was incorporated into Schoeck's model. The primitive jump relaxation time of a single ISA without the influence of others is given by the Snoek expression. The interaction/correlation between the diffusing ISAs slows down the diffusion coefficient, resulting in the correlation function of the SK relaxation to the Kohlrausch stretched exponential form:

$$C_{SK}(t) = \exp[-(t/\tau_{SK})^{1-n}], \quad \text{with} \quad \tau_{SK} = \tau_{\infty}^{SK}\exp[H_{SK}/kT]. \tag{3.43}$$

The CM equation relates its activation energy H_{SK} to the activation enthalpy of Snoek H_S by

$$H_{SK} = H_S/(1-n). \tag{3.44}$$

Since the Snoek relaxation activation enthalpies H_S are available from the literature, the SK relaxation offers an excellent opportunity to critically test this CM relation and also to recover the primitive activation energy of the SK relaxation. To verify this relation, literature data of the internal friction peaks from SK relaxation were analyzed and fitted to those calculated from $C_{SK}(t)$. The fits yield the coupling parameters n and the activation energy H_{SK}. These quantities together with H_S are shown in Table 3.9 for α-Fe, Ta, and Nb containing nitrogen (N), carbon (C) and oxygen (O) as ISAs. The product $(1-n)H_{SK}$ in the last column is to be compared with H_S in the neighboring column. It can be seen that H_{SK} is larger than H_S, but the product $(1-n)H_{SK}$ is approximately the same as H_S.

These results of getting the primitive activation energy H_S from H_{SK} of SK relaxation in metals are analogues of the procedure used in the cases of (1) β-relaxation in side-chain liquid crystalline polymers discussed in Section 3.2.6.1 and (2) junction dynamics in Section 3.2.6.2 and Li ion dynamics in Sections 3.2.6.4 and 3.2.6.5.

Reduction of n in Snoek–Köster Relaxation by Decreasing Carbon Concentration

One way of reducing the interaction between ISAs is to decrease their concentration. This can be accomplished in ultra-high purity iron where the carbon concentration

Table 3.9 Values of the coupling parameter n and activation enthalpy H_{SK} of the Snoek–Köster (SK) relaxation together with the activation enthalpy of the Snoek relaxation H_S for α-Fe, Ta, and Nb containing nitrogen (N), carbon (C), and oxygen (O) as ISAs

High-purity specimen	Doped ISA	H_{SK}(eV)	n	H_S(eV)	$(1-n)H_{SK}$ [eV]
α-Fe	N	1.40	0.38	0.796	0.87
α-Fe	C	1.84	0.60	0.87	0.74
Ta (*A*)	O	1.45	0.30	1.10	1.02
Ta (*B*)	O	1.68	0.48	1.10	0.88
Ta monocrystal	O	2.24	0.46	1.10	1.21
Nb	O	1.49	0.40	1.15	0.90
Nb monocrystal	O	1.68	0.50	1.15	0.84

is smaller. One can expect from the CM a reduction of n and a smaller activation energy H_{SK} of the SK relaxation in the ultra pure iron from Eq. (3.44) [1655]. This expectation is fulfilled by the results of an experiment on SK relaxation in ultra-high purity iron carried out by Magalas [1655]. The data of the relaxation parameters H_{SK}, τ_∞^{SK}, and T_{max}^{SK}, of the Snoek–Köster relaxation are presented in Table 3.10. Here T_{max}^{SK} is the temperature of the SK peak at 1 Hz. It can be seen from the table that H_{SK} of the ultra pure iron is reduced to 0.95 eV, which is slightly higher than the value of $H_S = 0.87$ eV (see Table 3.9) and much smaller than 2.1 eV for the H_{SK} of the less pure iron.

The prefactor τ_∞^{SK} of the ultra-high purity iron is 6.26×10^{-12} s compared with 3.71×10^{-19} s of less pure iron. The latter has not only the anomalously short τ_∞^{SK} but also the large $H_{SK} = 2.12$ eV, an example of the Meyer–Neldel rule or compensation law of relaxation time in interacting systems. These have analogues in the side-chain liquid crystalline polymers, the junction dynamics, and Li ion conductors. Let us take $n = 0.60$ from the SK relaxation of carbon ISA in Fe given in Table 3.9 from another experiment and assume that this is the same as that of 1000 ppm carbon sample of Magalas, and also assume $t_c = 10^{-11}$ s. With

Table 3.10 Relaxation parameters of stable SK peak in deformed ultra-high purity iron with carbon as ISA at concentration of 25 ppm, compared with another sample with carbon at concentration of 1000 ppm

Carbon concentration					
Total	In solid solution (at ppm)	Deformation at RT (%)	H_{SK}(eV)	τ_∞^{SK} (s)	T_{max}^{SK} (K) at 1 Hz
1000	110	24*	2.12 ± 0.04	3.71×10^{-19}	546.8 ± 0.6
25	5/10	13**	0.95 ± 0.01	6.26×10^{-12}	461.3 ± 0.4

* Successive deformation at RT: 13% ($T_a = 673$ K) + 6% ($T_a = 673$ K) + 6% ($T_a = 673$ K)
**First run-up to $T_a = 673$ K, deformation 13%

these assumptions, near physical value of the primitive prefactor $\tau_\infty = 10^{-14}$ s is obtained from the CM equation

$$\tau_\infty^{SK} = \left[(t_c)^{-n}\tau_\infty\right]^{1/(1-n)} \tag{3.45}$$

and $\tau_\infty^{SK} = 3.71 \times 10^{-19}$ of less pure iron.

3.2.6.8 Precipitates in Al–Ag Alloys, Ta–H, and Ti–H Systems

The thermal-activated relaxation originating from motion in solute clusters of precipitate in a number of alloys was studied by internal friction measurement at a fixed frequency. The energy loss Q^{-1} measured by internal friction plotted as a function of reciprocal temperature shows up as a broad and asymmetric peak. The shape deviates strongly from the Debye-type relaxation peak with exponential relaxation function but is well accounted for by the Lévy function, which is basically the Laplace transform of the Kohlrausch function [1657(a)]. From the best fit of the position and shape of each internal friction peak, Wang and coworkers [1657(a)] obtained the Kohlrausch exponent (1–n) and the activation energy E_a^* (fit). The results of E_a^* (fit) and n shown together with the experimentally observed value E_a^* (obs.) are given in Table 3.11 for five alloys Al–Ag, Ti–H and Ta–H. The compositions of the alloys are as follows: Al–30 wt% Ag; A1–20 wt% Ag; A1–20.1 wt% Ag; Ti–5.4 at.% H; and Ta–5 at.% H.

Because of solute clustering in the precipitates, Wang and coworkers argued the presence of interaction between the solute atoms and many-body relaxation in the motions of solute atoms, and hence the application of the CM to deduce the true activation energy E_a. Clustering of H in hydrides has been suggested in the light of neutron scattering experiment [1658]. This approach differs from the interpretations proposed by others for the internal friction peak associated with precipitates, all of which neglect the many-body relaxation aspect of the mechanism [1659, 1660]. From the results of Wang and coworkers using the CM, it is easy to understand why the internal friction peak due to the Ag clusters in Al–Ag alloys has the same peak

Table 3.11 The observed apparent activation energy E_a^* (obs.), the calculated apparent activation energy E_a^* (fit), the true activation energy E_a, and the coupling parameter n

Alloy	Structure of alloy	Structure of precipitate	E_a^* (obs.) (eV)	E_a^* (fit) (eV)	E_a (eV)	n
Al–30 wt% Ag	FCC	HCP	1.2 ± 0.1	1.3	0.91	0.3
Al–20 wt% Ag	FCC	HCP	1.2 ± 0.1	1.3	0.91	0.3
Al–20.1 wt% Ag	FCC	HCP	1.09 ± 0.02	1.3	0.91	0.3
Ta–5 at.% H	BCC	BCT	0.38 ± 0.05	0.40	0.10	0.75
Ti–5.4 at.% H	HCP	FCT	0.47 ± 0.01	0.47	0.25	0.44

temperature as that due to γ' precipitates. The latter is due to the stress-induced change in the particle shape (i.e., precipitation and dissolution) for which the rate-limiting process is atom migration around the matrix–precipitate interface. For the Ta–H system, it is puzzling why E_a^* (obs.) (=0.38 eV) in the β_2 phase is much larger than the activation energy (0.12 eV) of hydrogen in α-Ta phase (see Table 3.11). However, it becomes understandable if we take into account of the fact that the true activation energy E_a is 0.10 eV, which is comparable with 0.12 eV.

3.2.6.9 Grain Boundary Relaxation

Kong and coworkers [1657(b)–(e)] made internal friction measurements on various <111> tilt and twist grain boundaries in high-purity Al bicrystals. The temperature dependence of the grain boundary internal friction peak originating from grain boundary relaxation was determined. The activation enthalpy and the pre-exponential factor of the relaxation time are related by the compensation law, just like that discussed for the relaxation parameters of the SK peak in Table 3.10. The large values of the activation parameters measured for high-angle tilt and twist grain boundaries are attributed to a cooperative motion of atoms (group mechanism) during boundary relaxation and are explained by application of the CM.

3.2.6.10 Conformational Transition Energy Barrier of Polymers

This subject already has been touched upon in Section 3.2.5.2 and results are shown in Tables 3.4–3.6. For polymers with low T_g such as PE and PDMS, the primitive activation energy E_a, which is the conformational transition energy barrier in this case, can be obtained from the measured activation energies E_η and E_D of viscosity η and self-diffusion coefficient D by the relations $E_a = (1 - n_\eta)E_\eta$ and $E_a = (1 - n_D)E_D$, respectively The success has been demonstrated before. The value of E_a obtained is supported by measurements of the local segmental relaxation in (1) PDMS by neutron scattering [1579, 1580], (2) PE by ^{13}C-NMR measurements [1582] and molecular dynamics simulations [1581], and (3) aPP by molecular dynamics simulations [1585].

3.2.7 Changes Effected by Mixing or Interfacing

Shown in Section 2.2.5.7 are many examples from experiments and molecular dynamics simulations of the many-body structural relaxation of glassformers being modified by mixing with another glassformer, and in Section 2.3.2.38 by the interfacing with another glassformer or confining immobile solid wall. In this section, analogues of the changes in properties found for structural relaxation in glassformers on mixing or interfacing found in entirely different processes and/or systems are presented. These are other examples of universal dynamics of relaxation and diffusion in interacting systems.

3.2.7.1 Global Chain Dynamics of Each Component in Binary Polymer Blends

In the binary mixtures of glassformers, each component has its own structural α-relaxation and temperature dependence of its relaxation time. Both the relaxation time and the width of the dispersion of the more mobile or faster component increase with increasing concentration of the less mobile or slower component. These changes have been explained by enhancement of intermolecular coupling of the faster component when it is replaced in part by the slower component. In the context of the CM, enhancement of intermolecular coupling is equivalent to increase in the coupling parameter n_α, from which several predictions in agreement with experiments follow including the increased separation between the α-relaxation and the JG β-relaxation. On the other hand, the relaxation time of the slower component decreases on addition of the faster component. For polymeric glassformers, examples discussed before are local segmental α-relaxation in blends of PVME with PS (see Fig. 166), PEO with PMMA (see Fig. 157), and LB-PBD with CR-PBD (see Fig. 170). Here we show analogues of these properties of local segmental α-relaxation from polymer *chain* relaxation in binary miscible polymer blends.

A new rheo-optical technique to measure dynamic infrared dichroism was invented by Fuller and applied to binary polymer blends [1661]. The technique was able to measure not only the total relaxation modulus from global chain motion but also the contribution of each component in a bidisperse entangled polymer melt. The bulk relaxation is measured simultaneously by using birefringence. The first family of blends studied is nearly monodisperse pairs of hydrogenated and deuterated polyisoprenes of molecular weights 53,000 and 370,000, both several times the entanglement molecular weight [1661]. This study was repeated with binary blends of nearly monodisperse poly(ethylene propylene) (PEP) samples of molecular weights 53 K, 125 K, and 370 K, all above the critical molecular weight for entanglement. The results of both studies are similar, and shown in Fig. 306-3 is the relaxation dynamics of each of the two components of the blends of 53 K PEP with 370 K PEP together with the components in their pure state. The blends have compositions of $\Phi_L = 75$, 50, 30, 20, and 10% by volume of the longer chains.

It can be seen by comparing the left panel for the longer chains and right panel for the shorter chains that the global chain dynamics of the two components is different in all blends. The long-chain relaxation is significantly altered by the addition of short chains to the surrounding matrix, as had been inferred from dynamic mechanical relaxation measurement of bulk polymer [1663, 1664]. The dichroism relaxation of the long chains presented in the left panel of Fig. 306 directly shows the change to faster relaxation and narrower dispersion monotonically on increasing the volume fraction of the short chains. Another observation is that increasing presence of long chains in the blend produces a significant retardation of the relaxation of the short chains, accompanied by broadening of the dispersion (see right panel of Fig. 306). All these changes of the *global* chain relaxation properties individually of the long and short chains are analogous to those found in structural relaxation of glassformers. This serves as another example of universal properties of interacting

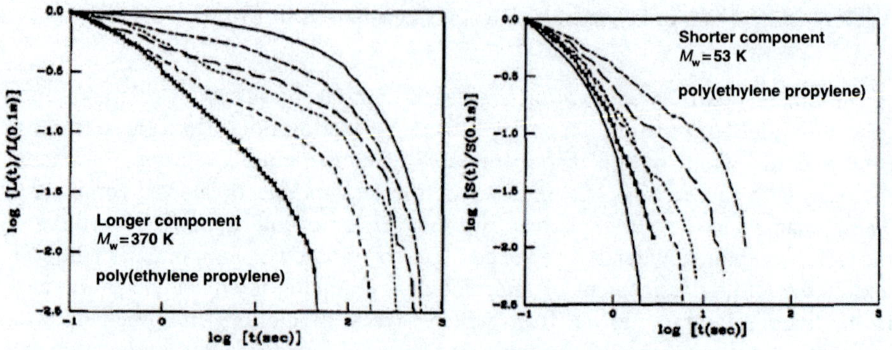

Fig. 306-3 (*Left*) Long-chain dichroism relaxation following a step strain for the S/L blends, where S and L stand for shorter and longer nearly monodisperse poly(ethylene propylene) chains with molecular weights of 53 K and 370 K, respectively. From right to left are data of long chain in the order of 100, 75, 50, 30, 20, and 10% by volume of the longer chains. (*Right*) Short-chain dichroism relaxation following a step strain for the S/L blends. From left to right are data of short chain in the order of 100, 75, 50, 30, 20, and 10% by volume of the shorter chains. Reproduced from [1662] by permission

systems. For the present case, the interaction between polymer chains of the two components comes from the entanglements as well as the short-range or nematic-like orientational coupling between the polymer chains of the two components. The nematic-like orientational coupling was known to many authors [1665–1675], and existence of such interaction has been established in the study of stress optical coefficients for rubbers [1676, 1677] before the dichroism relaxation studies of the blends published in 1989 [1661]. The effect of the "nematic interaction" is to force neighboring polymer segments to orient in the same direction. One of clearest evidence of nematic-like orientational coupling comes from deuterium nuclear magnetic resonance study [1674] of short probe polymer chains dissolved in a strained elastomer. The data showed the presence of imposed orientational anisotropy of the probe chain at very short length scales of the bond level.

Explanation from the CM

The CM applied to global chain dynamics of entangled polymers [202, 203, 359, 360, 837, 870, 872, 904] readily explains the observed changes of global chain dynamics in the blends from the decrease and increase in the entanglement coupling parameters n_L and n_S of the long and short chains respectively as follows. Short chains cannot entangle the long chains as well as long chains themselves, and hence entanglement coupling is decreased by the addition of short chains. This, together with faster motion of the short chains, should reduce the entanglement coupling and its parameter $n_L(\Phi_L)$ of the long chains in the blends by adding short chains. The CM equation, the counterpart of Eq. (3.33), for the long chains is

$$\tau_L(\Phi_L, T) = [t_c^{-n_L(\Phi_L)}\tau_{0L}(\Phi_L, T)]^{1/(1-n_L(\Phi_L))}, \tag{3.46}$$

with $t_c \approx 1$ ns for entanglement coupling (see Fig. 289). This equation together with the monotonic decrease of $n_L(\Phi_L)$ with decrease in Φ_L readily explains the changes of the dynamics of the long chains shown in Fig. 306. On the other hand, the long chains replacing short chains have much longer relaxation time. Hence, the unrelaxed modes of the long chains impose stronger constraint on the short chains, effectively enhancing the entanglement coupling parameter $n_S(\Phi_L)$. In more technical terms, the unrelaxed long chains in the blend enforce sustained orientation anisotropy on the level of Rouse segments of the short chains at times past $\tau_s(\Phi_L = 0, T)$, the global chain relaxation time of the short chains without the addition of long chains. The sustained orientation anisotropy result in prolonged relaxation of the short chains in the blend and relaxation time $\tau_S(\Phi_L, T)$ longer than $\tau_S\,(\Phi_L = 0, T)$. This effect of the long chains on the short chains originates from the short-range or nematic-like orientational coupling between polymer chains in the melt [1665–1675]. This intermolecular orientational coupling interaction between the unrelaxed long chains and the short chains is the cause of the increase of $n_S(\Phi_L)$ with increase of Φ_L in the CM description. A CM expression for $\tau_S(\Phi_L, T)$ similar to Eq. (3.46) predicts the increase of $\tau_S(\Phi_L, T)$ with increasing Φ_L.

Explanation from the Reptation-Based Model After Incorporating Nematic-Like Coupling

The component global chain dynamics found by dichroism relaxation studies of the entangled blends [1661, 1662] cannot be explained by various models which incorporate the effects of constraint release into a reptation-based model [43, 1678–1681]. An example is the strong retardation of the relaxation of the short chains with increasing Φ_L. These reptation-based models predict that the time required for a short chain to completely diffuse out from the initial "tube" of constraints imposed by the mean field just after the strain is applied should be the same for all Φ_L. This prediction is at odds with experiment as can be seen from the left panel of Fig. 306 that the measurable relaxation of the short chains in the blend with 75 vol.% long chains takes place at times more than an order of magnitude longer than those of the monodisperse short chains. Another experimental observation at variance with the reptation-based models is the terminal long-chain relaxation appearing to become single exponential with a long relaxation time for long-chain volume fractions Φ_L less than 0.5, which decreases as Φ_L decreases [1661]. In order to reconcile the dichroism relaxation of the entangled blends [1661, 1662] with the reptation-based model, Doi et al. [1682] pointed out that nematic interaction has to be incorporated into the model. This was implemented by Ylitalo et al. on top of the reptation with constraint release model of Rubinstein–Helfand–Pearson (RHP) [1678]. The RHP model describes the stress relaxation of a blend as the result of two independent mechanisms: reptation, by which the stressed chain disengages from the deformed tube, and constraint release, by which the deformed tube relaxes. With the addition of the short-range nematic interaction, the decorated reptation model can explain the data.

Difference in Spirit of the Explanations by the CM and the Reptation–Constraint Release–Nematic Interaction

It is instructive to point out the difference in spirit of the explanation of the global chain dynamics of entangled blends [1661, 1662] by the CM and the reptation–constraint release–nematic interaction model [1662]. The CM emphasizes that the global chain dynamics of entangled blends is analogous to similar results of component structural relaxation of mixtures of two glass-forming liquids. For the global chain dynamics of entangled blends, the interactions, including long-range entanglement and short-range intermolecular orientation coupling, are different from the Lennard–Jones-like interaction between repeat units in structural relaxation of polymers. Nevertheless, both are genuine many-body relaxation in interacting systems that exhibit similar dynamic properties governed by the same physics and laws. Although the CM equations are used in the explanation given here, these can be replaced by those from a more sophisticated and rigorous theory of universal relaxation of interacting systems when it appears in the future. This spirit of the CM explanation is in stark contrast to the explanation given by the reptation–constraint release–nematic interaction model. The reptation model itself, without the other mechanisms added, considered that the entangled polymer chain dynamics observed including the $M^{3.4}$ dependence of viscosity is the consequence of the topology of long linear chains and the fact that neighboring chains cannot cross. The model views each polymer chain confined by neighboring chains in a tube-like region with diameter equal to the average distance between entanglements. This tube limits the lateral motion of the chain but allows its curvilinear diffusion along its mean contour or its reptation [41–44]. Each linear chain moves like a slithering snake, basically independent of other chains. Thus, the reptation model emphasizes that the particular geometry of entangled thread-like polymer chains is the principal cause of the observed dynamics. Obviously it does not apply to other many-body relaxation processes in other interacting systems, and not even the structural relaxation in the *same* polymer. Cyclic polymers have no ends and cannot reptate, but yet the viscoelastic and diffusion properties of entangled cyclic polymers are essentially the same as linear polymers including similar molecular weight dependence. Actually the reptation model has used the geometric argument to reduce the many-chain relaxation problem to a one-chain relaxation (reptation motion) problem, which provides immediate answers to be compared with experiments [42, 44] including the time dependence of the end-to-end vector correlation function $\phi_e(t)$ given by

$$\phi_e(t) =< R_e(t) \cdot \boldsymbol{R}_e(0) > / < \boldsymbol{R}_e(0)^2 \geq \sum\nolimits_{p=1,odd}^{\infty} \left(\frac{8}{p^2\pi^2}\right) \exp\left(-\frac{p^2 t}{\tau_d}\right). \quad (3.47)$$

where τ_d is the disentanglement time of the reptation-tube model. Remarkably, this expression for $\phi_e(t)$ is the same for the Rouse model, except that τ_d is to be replaced by the Rouse time τ_R. The dispersion predicted by Eq. (3.47) of the pure reptation model is too narrow and cannot explain the terminal frequency dispersion of monodisperse linear polymers in the dynamic modulus or creep experiment measurements. This, together with other problems including $\tau_d \propto M^{3.0}$ and not the observed $M^{3.4}$ dependence of viscosity, has led to the addition of the new mechanism of constraint release to the reptation model [43, 1678–1681] in order to bring consistency with experiment. The constraint release mechanism is due to the motion of neighboring chains that form the fixed tube of the original reptation model and has the flavor of many-chain dynamics. However, it is implemented as a correction of the chain reptative motion independent of the other chains. The reptation–constraint release model still is leaning heavily on the premise that each linear chain moves like a slithering snake, and therefore it is far from solving the problem by treating it as relaxation of many chains interacting by entanglements. Basically the problem of all reptation-based models is that they have accepted from the start that the chain motion is reptation and its predictions, and corrections are made by adding new mechanisms such as constraint release or tube renewal to bring consistency of the model with experiment in monodisperse polymers [43, 1678–1681]. The component chain dynamics of bidisperse entangled polymer blends discussed here [1661, 1662] requires another correction of including nematic interaction, leading ultimately to the reptation–constraint release–nematic interaction model. The deviation of the center-of-mass MSD data from neutron spin echo experiments by Zamponi et al. [1503] from the reptation prediction (see Fig. 281) would require another correction. Instead of correction after correction on the reptation motion, it may be worthwhile to restart anew to solve the problem as a many-chain relaxation problem, without jumping to reptation but taking into account of the limited lateral motion of thread-like molecules. Although the CM does not describe the lateral constraints as elaborately as done in the tube model [42, 44], lateral constraints and their mitigation have been considered in an attempt to rationalize the experimental results of monodisperse polymers in [837], and the component dynamics of bidisperse entangled blends here.

3.2.7.2 Other Examples of Change of Global Chain Dynamics of Entangled Polymers by Mixing

Plasticized Poly(Cyclohexy1 Methacrylate)

Polarized photon correlation measurements of high molecular weight, entangled neat and plasticized poly(cyclohexyl methacrylate) (PCHMA) by dioctyl phthalate (DOP) were able to observe both the density fluctuations caused by the local segmental relaxation and concentration fluctuations due to cooperative diffusion of chain diffusion [203]. The plasticized PCHMA contain $\Phi_{DOP} = 5, 10$, and 15 wt% DOP. Over a common temperature range, the data show that the local segmental relaxation time exhibits a stronger temperature dependence than does chain diffusion, a result which is consistent with the breakdown of thermorheological simplicity of polymers observed before by mechanical spectroscopy [165–171]. The other observation relevant in the present context is the time correlation function of entangled chain diffusion narrows on increasing the content of DOP in the mixture.

The trend indicates a decrease of the entanglement coupling parameter. The narrowing of the dispersion on increasing weight percentage of DOP in the plasticized PCHMA is accompanied by the observation of a weaker $(T - T_g)$ dependence of the chain diffusion relaxation time. Following the CM relation similar to Eq. (3.46), this trend corroborates the decrease of entanglement coupling parameter, which in turn is caused by the decrease in chain entanglement interaction by the presence of the plasticizer to separate the chains further apart.

Dilute Polyisoprene Probes in Networks

Adachi and coworkers [1685] studied the dielectric normal mode relaxation of nearly monodisperse PI trapped in cross-linked natural rubber networks [1684] and in PB networks. The experimental data were reviewed by Adachi and Kotaka [1686, 1687]. The PB networks used were loosely cross-linked with the molecular weight between cross-links $M_x > M_e$, where M_e is the molecular weight between entanglements. The dielectric relaxation loss curve $\varepsilon''(\omega)$ is broader than that of pure bulk monodisperse entangled PI. The experimental data of 14 K-PI trapped in a PB network labeled BR(0.1:20) are shown in Fig. 307 (filled circles). In the same figure the solid curve represents the $\varepsilon''(\omega)$ calculated by the CM with the entanglement coupling parameter n equal to 0.45 [837]. This value of n is determined by Eq. (3.32) from the observed $(M_{PI})^{3.6}$ dependence of the dielectric relaxation time of the PI chains. The dielectric loss peak in Fig. 307 is significantly broader than the loss peak of pure monodisperse PI melt, and the relaxation time of the former is much longer than the latter. These changes of PI when introduced into PB network are similar to those experienced by the short chains when long chains are present in the bidisperse entangled polymer blends studied by Fuller and coworkers [1661, 1662].

3.2.7.3 Mixed Alkali Effect in Ionic Conductors

The "mixed-alkali (MA) effect" refers to deviation from linear additivity in isotherms of several dynamic properties of ions in alkali oxide glasses as one kind of alkali ion is replaced by another kind [1689–1692]. For example, it is

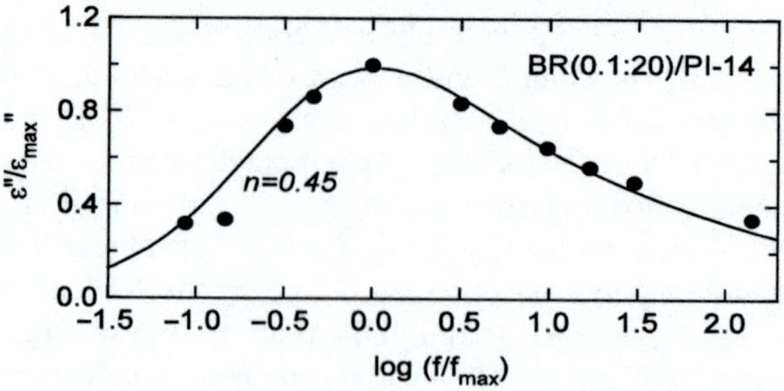

Fig. 307 Normalized $\varepsilon''(\omega)$ data of free PI chains with $M_{PI} = 13,500$ trapped in loosely cross-linked polybutadiene network with $M_x = 11,000$ (*solid circles*). Normalized $\varepsilon''(\omega)$ curve calculated with $n = 0.45$ by the CM. Reproduced from [837] by permission

frequently observed that the electrical conductivity σ of a single alkali ion glass rapidly decreases when a small amount of the alkali A is replaced by alkali B. A similar effect occurs on the other composition extreme with the roles of A and B interchanged, giving rise in many cases to a deep minimum in σ at some intermediate composition. The left panel of Fig. 308 shows the MA effect on the dc conductivity of the $0.2[x\mathrm{Na_2O}–(1–x)\mathrm{Rb_2O}]–0.8\mathrm{B_2O_3}$ mixed-alkali borate glasses with x= 0.0, 0.2, 0.4, 0.6, 0.8. The right panel shows tracer diffusion coefficients of ^{22}Na and ^{86}Rb at 653 K in these glasses. The data come from the work of Imre et al. [1693], and similar results had been obtained by Jain et al. [1694].

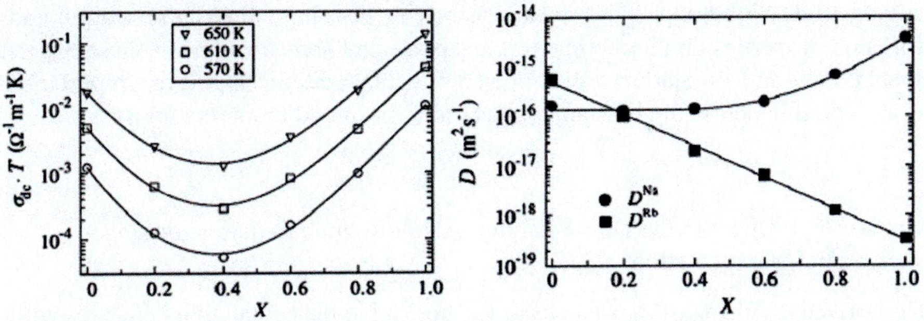

Fig. 308 *(Left)* Isotherms of the product $\sigma_{dc}T$ of $0.2[x\mathrm{Na_2O}–(1–x)\mathrm{Rb_2O}]–0.8\mathrm{B_2O_3}$ glasses for three temperatures. *(Right)* Composition dependence of the tracer diffusion coefficients of ^{22}Na and ^{86}Rb at 653 K in these glasses at 653 K. Reproduced from [1693] by permission

The effect turns out to be more general and has been found in glasses with other mobile non-alkali monovalent cations and anions (such as F^-) and in crystalline ionic conductors (such as alkali β-alumina). Various experimental techniques have shown that the local environments of the two different alkali ions in the MA glasses are different [1695–1700]. The alkalis retain more or less the same local environment as in the respective single alkali (SA) glasses, and they are randomly distributed in the glass structure. Since there is a large difference between the sites for the two alkali ions with large differences in their site energies [1701, 1702], jumps of an alkali ion to the sites of a different alkali ion cannot occur. This implies that ions of one kind tend to block the pathways for the other kind of ions and vice versa, as found out also by reverse Monte Carlo method [1703]. It follows that there are fewer sites for ionic motion of the more mobile alkali species when some of them are replaced by the less mobile alkali species, and these fewer sites are more separated in space. This is one of the main causes of the MA effect. The other is the immobilization or the reduction of mobility of ions of one kind by the presence of relatively immobile ions of the other kind [1704, 1705]. Both effects lead to decrease in the number of mobile ions and increase in their average distance of separation, resulting in diminishing ion–ion interaction and correlation. The situation can be likened to the effect found on the reduction of ion concentration in single alkali glasses, where it is well known that reduction of ion concentration increases the activation energy of conductivity due to lower level of cooperativity between ions [115, 1706]. Thus the MA effect, when considered in terms of the change in activation enthalpy and isothermal conductivity, can be qualitatively understood by its similarity to the effect of decreasing the alkali ion concentration of an SA glass, a suggestion that was made earlier by Greaves and Ngai [1706(b)]. The reduction of cooperativity or the degree of many-ion dynamics of ions by the presence of immobile ions of another kind is observed by molecular dynamics simulation [1707]. More recent results that enhance understanding of the MA effect come from molecular dynamics simulations [1707–1711] and experiments [1693, 1711], some of which will be further discussed.

Consistent with the purpose of this chapter of showing universal relaxation and diffusion properties on interacting systems presented here are some of those experimental facts and simulation results of MA effect that either support the importance of ion–ion interaction or show analogous properties in other interacting systems.

Mixed Alkali Effect in the Dilute Foreign Alkali Regime Requires Ion–Ion Interaction for Explanation

The derivative $(\partial \ln \sigma_{dc}/\partial x)$ is a measure of how rapid the conductivity changes with replacement of the majority ions. Here x is the fraction of the majority ions and (1–x) the fraction of the minority ions. The most rapid change measured by the largest value of $(\partial \log \sigma_{dc}/\partial x)$ occurs at $x = 1$, i.e., the dilute limit of the foreign alkali region. This very rapid decrease in σ_{dc} by replacement of the host alkali ions by a small amount of foreign alkali ions indicates that many host ions suffer large decreases of mobility (or are immobilized) per single foreign alkali ion introduced. Such a very large number Z of host alkali ions immobilized by a foreign alkali ion was deduced from the experimental values of the limiting rate of change of σ_{dc} with the host ion fraction x, $\lim_{x\to 1}(\partial \ln \sigma_{dc}/\partial x)$, by Moynihan et al. [1693], assuming that the single alkali (SA) silicate glass they studied was a strong electrolyte. Z increases with decreasing temperature and reaches values of order of around 50 at the lowest temperature studied. However, at that time it was difficult for Moynihan et al. to understand how a single foreign alkali can immobilize so many host alkali ions. In fact, this puzzling result led Moynihan et al. to conclude that it was unacceptable and that the faulty result originated from the assumption that the SA glass is a strong electrolyte. However, very recent NMR chemical shift measurements of MA glasses by Eckert [1712] have confirmed that indeed many host alkali ions are affected by every single foreign ion. Thus, it appears to be true that a single foreign alkali ion can immobilize a large number of host alkali ions. Furthermore, we have confirmation from molecular dynamics simulations of diffusion of Li ions in Li_2SiO_3, where a small proportion of Li ions randomly chosen were frozen, and also of Li in $(Li_{1-x}K_x)2SiO_3$ with small x [1707]. This is an important experimental fact, because it can be used to rule out immediately some current models of the MA effect by showing that in the dilute foreign alkali region the models do not predict a large number of host ions being immobilized by a single foreign ion.

An explanation of largest value of $(\partial \log \sigma_{dc}/\partial x)$ occurs in the dilute limit of the foreign alkali region was provided [1705] by drawing the analogy to the simulation results of mobile particles confined by walls formed by immobile particles [322, 1463]. All immobile and mobile particles interact with each other by the same Lennard–Jones potential, and the effects of immobile particles on the mobile particles are obtained from the simulation. The immobile or frozen particles were made to form a "wall," and the effects that the immobile particles have on the dynamics of the mobile particles at any distance z away from the wall were obtained from the spatially resolved self-part of the intermediate scattering function $F_s(z, t)$ with wave vector at the maximum in the static structure factor. The results have been discussed before in Section 2.3.2.38 and results shown in Fig. 265. Particles in the layer closest to the wall (smallest z) have their relaxation time increased by orders of magnitude. The immense slowing down of the particles by the immobile particles does not stop at the first layer at the interface. From Fig. 265 one can see that the effect propagates to the next layer and so on, although the increase of relaxation time becomes less with increasing distance away from the wall. The immobilizing effect of the immobile wall particles on the initially (before the wall is introduced) mobile particles is not limited to the immediate layer but propagates further away to affect more layers.

In the MA glass in the dilute foreign alkali limit, the foreign alkali has mobility orders of magnitude smaller than the host ions (see right panel of Fig. 308 for the case that Rb is the foreign ion in $0.2\,[x\mathrm{Na_2O} - (1-x)\mathrm{Rb_2O}] - 0.8\mathrm{B_2O_3}$, and there is interaction between the Rb ion and the host Na ions. Thus, if this result from Lennard–Jones liquid can be used as an indication of the effect that an immobile foreign alkali ion can have on the dynamics of the initially mobile host ions, we are led to conclude that not just the five or six nearest neighbor host alkali ions will be severely immobilized by the foreign alkali ion but also second nearest neighbor host alkali ions will be immobilized, although to a lesser extent, and so on. The immobilizing effect propagates and dies off with distance away from the foreign alkali ion. It is now conceivable in the glasses of composition $0.242\,[x\mathrm{K_2O} + (1-x)\mathrm{Na_2O}] - 0.758\mathrm{SiO_2}$ studied by Moynihan et al. [1693] in the dilute Na^+ ion composition range that an Na^+ ion can effectively immobilize 48 K^+ ions, as the data at 25°C of the value of $\lim_{x\to 1}(\partial \ln\ \sigma/\partial x)$ require, under the assumption that the alkali silicate glass is a strong electrolyte. The early work [1693] did not have the benefit of recent progress in the knowledge of the change in dynamics of interacting systems, but only the crude concept of immobile complexes of alkali composition NaK_Z. This large number Z of K^+ ions immobilized by an Na^+ ion required by the data at 25°C was considered inconceivable in the framework of this crude concept and led Moynihan et al. to fault the assumption that the alkali silicate glasses are strong electrolytes [1693]. However, in the intervening years and at the present time, the structural and dynamic studies of alkali oxide glasses seem to show that they are indeed strong electrolytes. Now, with the justification that the large Z is reasonable when ion–ion interactions are taken into account, the data at 25°C can be rationalized without rejecting the presumption that alkali silicate glasses are strong electrolytes.

The dynamics of L–J particles confined by walls made of immobile L–J particles show interesting temperature dependence. The large increase of relaxation time compared to the bulk was shown before at a low temperature in Fig. 265, but the increase diminishes rapidly with increasing temperature. In the inset of Fig. 309, the ratio of the relaxation times τ of particles at $z=0.8$ close to the wall to that at $z=7.5$ at the center of the confined liquid is plotted as a function of temperature. The data are taken again from [1463]. It can be seen that the ratio $\tau(z = 0.8)/\tau(z = 7.5)$ decreases rapidly with temperature. Thus, the degree of immobilization caused by the frozen particles in the wall decreases with increasing temperature. When this temperature dependence of immobilization is applied to the mixed alkali effect, the observed decrease of $\lim_{x\to 1}(\partial \ln \sigma/\partial x)$ or the corresponding Z with temperature [1693] as shown in Fig. 309 can be rationalized by the same explanation as given before for the confined L–J liquid [900].

The discussion of the results above serves the purpose of highlighting the importance of ion–ion correlation/interaction in considering the MA effect. Maass [1713]

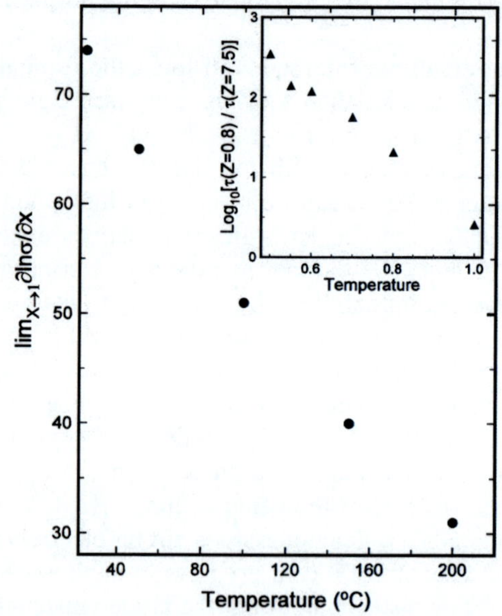

Fig. 309 Temperature dependence of the limiting slopes $\lim_{x\to 1}(\partial \ln \sigma/\partial x)$ of $\ln \sigma$ vs. x isotherms obtained for $0.242[xK_2O + (1 - x)Na_2O] - 0.758SiO_2$ glasses by Moynihan et al. [1693] in the dilute Na^+ ion composition range. This quantity, an indication of how many host K^+ ions are immobilized by a single foreign Na^+ ion, is a decreasing function of temperature. The *inset* shows ratio of the relaxation time $\tau(z = 0.8)$ of particles in the L–J liquid located at $z=0.8$ close to the frozen L–J glass forming the wall to $\tau(z = 7.5)$ far from the wall. The ratio, an indication of the extent of the immobilization of particles in the L–J liquid, is also a decreasing function of temperature like $\lim_{x\to 1}(\partial \ln \sigma/\partial x)$ in the main figure. Results in the *inset* are obtained from data of molecular dynamics simulation result of Schneider et al. [322, 1463] on $F_s(z,t)$ and shown in Fig. 265. Reproduced from [1705] by permission

had called attention to interaction between the moving ions in considering the MA effect, although for a different reason than we discussed here. Swenson et al. also mention this point made by Maass in their work [1714]. Ion–ion correlation certainly plays an important role in determining the ionic conductivity and its frequency dispersion in SA glasses, as indicated by various experimental facts discussed in Sections 3.2.2–3.2.6 and incorporated into the coupling model.

Immobilization of Li Ions by Frozen Li Ions in the Confining Walls

The immobilization of more mobile ions by less mobile ions in the MA effect has been rationalized by analogy to similar effect found in L–J particles confined by frozen L–J particles forming the walls. To demonstrate the physics of two systems closer, a molecular dynamics simulation was made on the dynamics of Li ions in Li_2SiO_3 glass confined by parallel walls formed by the same glass except that the Li ions therein were frozen but still interacting with the Li ion in the confined Li_2SiO_3 glass [1707]. The geometry of the systems is illustrated in the left panel of Fig. 310. The self-part of the density–density correlation function $F_s(k,t;z)$ was obtained at any distance z from the wall for Li ions located at $z \le 0$. The unit of z is $L/10$, where L is the side length of the basic cell of the simulation. The $F_s(k,t;z)$ obtained has the Kohlrausch stretched exponential time dependence $\exp[-(t/\tau(k;z))]^{\beta(k,z)}$. The frozen Li ions in the walls cause slowing down of Li ion as seen by the increase of $\tau(k;z)$ in the middle panel of Fig. 310 and the concomitant increase in stretching of $F_s(k,t;\ z)$ as seen by the decrease of $\beta(k;z)$ in the right panel for two chosen values of $k = 2\pi/10$ and $2\pi/3$. The effect is largest for the Li ions closest to the wall and decreases monotonically with distance from the wall. The values of $\tau(k,z)$ for Li ions residing in the innermost $z=5$ region are close to those for the unmodified bulk Li_2SiO_3 glass, which are $\tau(k) = 1292$ ps for $k = 2\pi/10$ and 340 ps for $k = 2\pi/3$. Values of $\beta(k;z)$ in the innermost $z=5$ region are also close to those

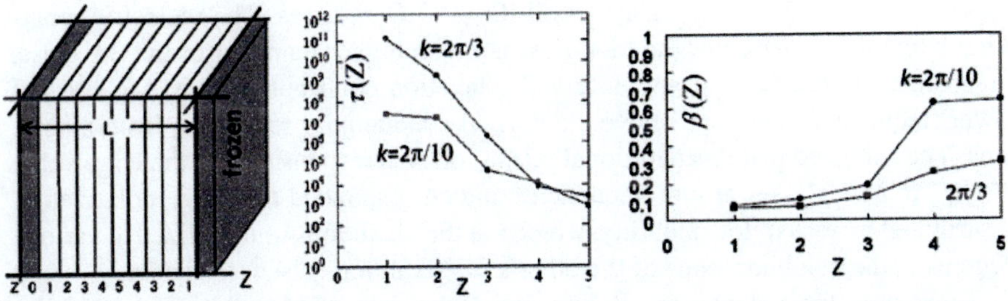

Fig. 310 (*Left*) Li metasilicate glass confined by two walls of the same Li metasilicate glass except that the Li ions therein are frozen. (*Middle*) Plot of $\tau(z)$ against z for the stretched exponential region of $F_s(k,t)$ of Li ions in Li_2SiO_3 glass (in ps). *Filled circle*: $k = 2\pi/10$. *Filled square*: $k = 2\pi/3$. (*Right*) $\beta(z)$ against z for the stretched exponential region of $F_s(k,t)$ of Li ions. *Filled circle*: $k = 2\pi/10$. *Filled square*: $k = 2\pi/3$. The values of $\beta(z)$ in the z=5 region are close to the values for the bulk Li_2SiO_3, which are $\beta = 0.67$ and 0.32 for $k = 2\pi/10$ and $2\pi/3$, respectively. Data from [1707] are replotted in all figures here

for the unmodified bulk system, which are $\beta(k) = 0.67$ for $k = 2\pi/10$ and 0.32 for $k = 2\pi/3$. The changes in the dynamics of Li ions interacting with the frozen Li ions in the confining walls are exact analogues of the L–J particles confined by frozen L–J particles in the confining walls obtained by Scheidler et al. [322, 1463]. This becomes obvious when comparing the dependence of $\tau(k; z)$ and $\beta(k; z)$ on distance from the wall in Fig. 310 with that of the same quantities for the confined L–J particles in Fig. 265. Thus, this is another example of universal behavior of relaxation in interacting systems.

Additional Evidence of Immobilization of Ions and Reduction of Ion–Ion Interaction of Mobile Ions in MA Effect

In MA glasses there are two kinds of ions and even the more mobile kind has some immobilized, less mobile, and more mobile ions. The electric modulus representation is known to suppress low-frequency contributions, not only from the electrode polarization but also from the less mobile and the immobile ions in the MA glasses. The story is different, however, when the same MA data are shown in the $\varepsilon^*(f)$ or the complex impedance Z^* vs. log f representation. In these representations of the data, the low-frequency contributions from less mobile and immobile ions are emphasized and show up prominently as large deviations from the prediction of the Kohlrausch relaxation at low frequencies well below the frequency maximum in the M'' vs. log f plot. The large deviations from Kohlrausch relaxation were first seen in the complex impedance plots of Z'' vs. Z' of MA glasses and later at lower frequencies in $\varepsilon'(f)$, the real part of $\varepsilon^*(f)$. Whereas Z'' vs. Z' plots of actual data for the single alkali glasses or Z'' vs. Z' curves predicted by the Kohlrausch fit to the SA glass electric modulus data display approximately the shape of a submerged semicircle, the data for MA glasses show large deviations from this shape. This deviation was pointed out by Moynihan and Boesch [1716] and by Tomozawa and coworkers [1717, 1718]. Tomozawa and coworkers further showed enormous differences between SA and MA glasses in $\varepsilon'(f)$ and $Z^*(f)$ at low frequencies by comparing the data of Li_2O–$2SiO_2$ and $[0.5Na_2O+0.5LiO_2]$–$2SiO_2$ glasses. The dielectric relaxation strength $(\varepsilon_s - \varepsilon_\infty)$ of the MA glass is much higher than that of the SA glass, suggesting the presence of an additional relaxation contribution in MA glasses at lower frequencies, over and above the normal conductivity relaxation from mobile ions. They showed that this additional relaxation was responsible for the large value of $\varepsilon'(f)$ in MA glasses at low frequencies and corresponded to a large polarization contributed by ions of low mobility, which are the alkali ions immobilized to various degrees in the neighborhoods of the other kind alkali ions discussed above.

Therefore, when shown as $\varepsilon'(f)$ vs. log f, the data of MA glasses support the presence of contributions to polarization from immobilized or partially immobilized alkali ions that are not accounted for by the electric modulus $M^*(f)$ due to its suppression of the low-frequency polarizations from the immobilized or partially immobilized ions. In effect, by showing the data as $M^*(f)$ one has selected only the faster dielectric response of the mobile ions in the MA glass. Kohlrausch fit to the $M^*(f)$ data of MA glasses is often found to be successful, but the Kohlrausch exponent β of the fit merely reflects the width of the dispersion because of contributions from ions with different mobility. The frequency dispersion of $M^*(f)$ data of MA glasses is narrower than the parent single alkali glasses as found in experiments by many workers [1704, 1717, 1719–1725]. This can be considered as evidence for the reduced ion–ion interaction due to fewer mobile ions and wider separations between them in the MA glasses mentioned before. An example is shown in the left and middle panels of Fig. 311 by the variation of the Kohlrausch exponent β with x from fits using the Kohlrausch function to the electric modulus of the alkali silicate glasses $[xK_2O + (1-x)Na_2O] - 3SiO_2$. Here β for all x are obtained from electric modulus data that have nearly constant values of τ (within a factor 2 of 3×10^{-3} s). The data show that β increases or the dispersion narrows with increasing amounts of foreign alkali in the dilute foreign-alkali region, leading to a maximum in β at an intermediate composition close to the minimum of dc conductivity. The same was found by Karlsson et al. in mixed alkali $Li_xRb_{1-x}PO_3$ glasses [1725], and they also consider this as indicating that the mixed alkali glasses behave as single alkali glasses of effectively lower concentrations. Another example in the right panel of Fig. 311 shows the narrower loss modulus $M''(f)$ data of the MA glass, $0.242[xK_2O+(1-x)Na_2O]-0.758SiO_2$ with $x = 0.950$, in the dilute Na^+ ion composition range compared with the parent SA with $x = 1.00$ studied by Moynihan et al. [1693].

Decrease of the Strength of MA Effect with Increase in Ion–Ion Distance

If ion–ion interaction/correlation is an important element in determining the MA effect, then a natural consequence is that the MA effect should go away on

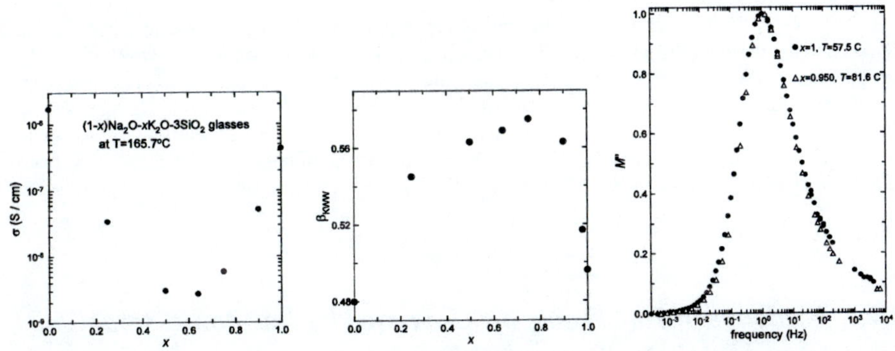

Fig. 311 (*Left*) The variation of the dc conductivity with x for the alkali silicate glasses $[xK_2O + (1-x)Na_2O] - 3SiO_2$. Data taken from [1726] and replotted. (*Middle*) The variation of the Kohlrausch exponent β with x from fits using the Kohlrausch function to the electric modulus. Here β for all x is obtained from electric modulus data that have nearly constant values of τ (within a factor 2 of 3×10^{-3} s). The data show that β increases or the dispersion narrows with increasing amounts of foreign Na in the dilute foreign Na region, leading to a maximum in β at an intermediate composition. Data taken from [1726] and replotted. (*Right*) Electric modulus spectra of the MA glass, $0.242[xK_2O + (1-x)Na_2O] - 0.758SiO_2$ with $x = 0.950$, in the dilute Na^+ ion composition range compared with the parent SA with $x = 1.00$ studied by Moynihan et al. [1693]

increasing the mean ion–ion separation distance. Tomozawa et al. [1722, 1727] measured the MA effect in alkali germanate glasses of molar percentage composition $0.019[xK_2O + (1-x)Na_2O] - 99.981GeO_2$ with an extremely low (0.019 mol%) total alkali content for x values of 0, 0.44 and 0.71. Their results at a temperature of 350°C show that the log σ vs. x plot did not exhibit the minimum associated with the MA effect. In glasses with low alkali content there is a negligible effect of ion–ion interaction on the dynamics of ion transport in the SA or the MA cases. After mixing the alkalis, the ions remain far apart, and there is no immobilization of one kind of alkali ions by another kind as supported by the normal semicircular trace of the data in a plot of Z″ vs. Z′. This trend of decreasing magnitude of the MA effect on increasing separation of the ions was verified by Voss et al. [1711] by measurements of a number of Na–Rb alumino-germanate glasses and Na–Rb borate glasses that permit wide range of the values of $< d_{\mathrm{ion}} > / < d_{\mathrm{network}} >$, the average Na–Na distance to the network distance. The left panel of Fig. 312 shows the strength of the mixed-alkali effect Δ_{MAE} as a function of $< d_{\mathrm{ion}} > / < d_{\mathrm{network}} >$.

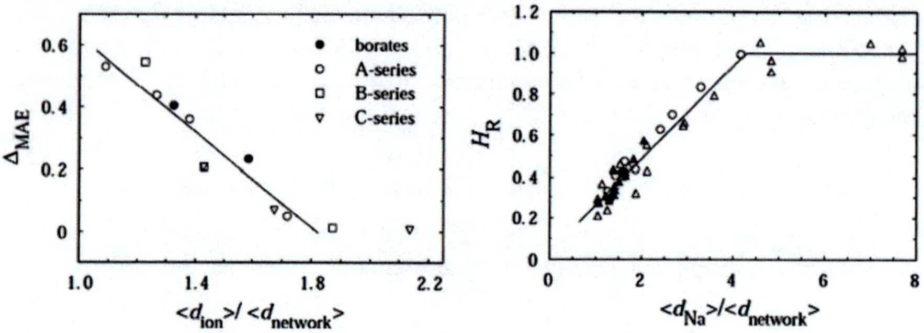

Fig. 312 (*Left*) Strength of the mixed-alkali effect Δ_{MAE} as a function of the average Na–Na distance to the network distance $< d_{\mathrm{ion}} > / < d_{\mathrm{network}} >$. *Open symbols* represent Na–Rb alumino-germanate glasses, whereas the *two filled circles* refer to Na–Rb borate glasses. Reproduced from [1711] by permission. (*Right*) Haven ratio H_R as a function of the ratio of the average Na–Na distance to the network distance $< d_{\mathrm{Na}} > / < d_{\mathrm{network}} >$. Values for Na-borate glasses: •, Voss et al. [1715] ($T = 380$°C); o, Kelly et al. [1728] ($T = 300$°C). Na-alumino-germanate glasses: ▲, from [1711] ($T = 380$°C), Δ, Kelly et al. [1728] ($T = 300$°C). Reproduced from [1711] by permission

3.2.8 Evidence of Ion Transport Governed by Ion–Ion Interaction from Molecular Dynamics Simulation

To further support ion–ion interaction and the many-ion relaxation governing ion transport in glassy ionic conductors, we cite an important observation by Binder [1732(a)] and Horbach et al. [1732(b)] from the results of molecular dynamics simulations on the Na_2O–SiO_2 melt.

From the simulations, Horbach et al. obtained the probability function $P_{\mathrm{Na-Na}}(t)$ that a "bond" that exists between a Na ion and another Na ion at $t=0$ is also

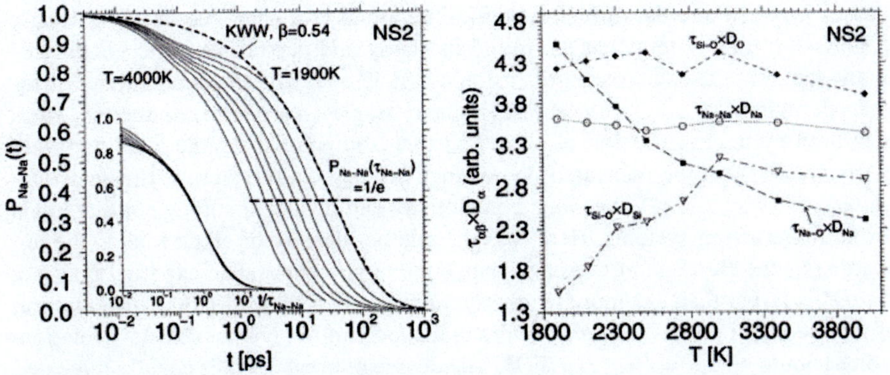

Fig. 313 (*Left*) Time dependence of $P_{\mathrm{Na-Na}}$, the probability that a bond between two sodium atoms which exists at time zero is also present at time t, for all temperatures investigated. *Inset*: plot of the same data vs. the scaled time $t/\tau_{\mathrm{Na-Na}}$, as a function of temperature. (*Right*) Temperature dependence of the products $\tau_{\alpha\beta}D_\alpha$ shows whether or not the diffusion constant D_α is correlated with the lifetime of a bond $\tau_{\alpha\beta}$. Reproduced from [1732(b)] by permission

present at time t. The time dependence of $P_{\mathrm{Na-Na}}(t)$ are shown in the left panel of Fig. 313, together with the fits to the Kohlrausch function having exponent $\beta_{\mathrm{KWW}} = (1-n) = 0.54$ and relaxation time $\tau_{\mathrm{Na-Na}}$. The two Na ions are considered as "bonded" if their distance is less than r_{min}, the location of the first minimum of the pair distribution function $g_{\mathrm{Na-Na}}(r)$. They also obtained the probability functions $P_{\alpha\beta}(t)$ for other atoms (ions), where α and β range over Na, Si, and O, and the corresponding lifetimes $\tau_{\alpha\beta}$ of the α–β bonds for different temperatures. The lifetimes $\tau_{\alpha\beta}$ can now be correlated with the diffusion constants by plotting different products $\tau_{\alpha\beta}D_\alpha$ vs. temperature, which are shown in the right panel of Fig. 313. The product $\tau_{\mathrm{Na-Na}}D_{\mathrm{Na}}$ is essentially constant over the whole temperature range. Therefore, the sodium diffusion constant seems to be linked to Na–Na "bonds," and the role of Na-Na interaction/correlation in governing Na ion transport is clear.

On the other hand $\tau_{\mathrm{Na-O}}D_{\mathrm{Na}}$ increases with decreasing temperature. This means that the elementary diffusion step for the sodium diffusion is not related to that of a Na–O bond, although the nearest neighbor distance is smaller for Na – O($r_{\mathrm{Na-O}} = 2.2$ Å) than for Na – Na ($r_{\mathrm{Na-O}} = 3.3$ Å). In other words, Na ion diffusion is decoupled from the silicate matrix.

3.2.9 *Haven Ratio, Breakdown of Nernst–Einstein Relation: Analogue of Breakdown of Stokes–Einstein Relation*

The breakdown of the Stoke–Einstein relation in molecular glassformers discussed in Section 2.2.5.6 has been considered to be one of the most important and general characteristics by the glass transition research community. It occurs not only in molecular glassformers such as OTP and tri-naphthyl benzene but also in metallic glasses [812], where the diffusing entities are atoms. An analogue of this anomaly in glass-forming systems can be found in glassy and crystalline ionic conductors by the fact that the self (tracer)-diffusion coefficient D^* and the conductivity diffusion coefficient D_σ are not the same. In glassy single-alkali ionic conductors, D^* is measured by tracer diffusion of an radioactive isotope of the same alkali, and D_σ is calculated from the measured dc conductivity σ_{dc} via the Nernst–Einstein relation $\sigma_{dc} = (Nq^2/k_B T)D_\sigma$ between conductivity and diffusion coefficient that holds for non-interacting systems. Here N is the number density of alkali ions, q the ion charge, k_B the Boltzmann constant, and T the temperature. The experimental fact that D^* is larger than D_σ in many glassy ionic conductors with large concentration of mobile alkali ions is evidence of the breakdown of the Nernst–Einstein relation, which should be $\sigma_{dc} = (Nq^2/k_B T)D^*$. The degree of breakdown is usually indicated by the Haven ratio $H_R = D^*/D_\sigma$, which has values less than equal to one [1490]. The smaller the H_R is, the larger is the breakdown. Since ion–ion interaction and many-ion dynamics have been identified as important in glassy and crystalline ionic conductors containing high concentration of mobile ions, it is natural to consider it as the cause of the breakdown. Simulation of Li ion motion in lithium silicate glasses by Heuer et al. [1728] has also concluded that the inverse of the Haven ratio can be considered as a measure of the degree of "collectivity" in ionic motion.

One way to support this is consistency of the prediction that H_R should increase with decreasing ion–ion interaction strength as can be realized by increasing the average separation between ions. The limit of $H_R = 1$ is reached and Nernst–Einstein relation holds when ion–ion interaction becomes negligible. The equivalent of this in the CM description is the expected increase of H_R with decrease of the coupling parameter n with decreasing ion concentration, where n can be obtained from the fit to the electric modulus data by the Kohlrausch function $\exp\left[-(t/\tau)^{1-n}\right]$. The decrease of n with decreasing ion concentration in the same family of glassy ionic conductors has been verified before in [1706(a), 1721] and in crystalline ionic conductors in [150]. The reader may recall that there is an analogous example in the breakdown of the Stokes–Einstein relation. As mentioned before in Section 2.2.5.6, silica having smaller n than OTP exhibits no breakdown of the Stokes–Einstein relation [822]. For glassy ionic conductors, the support of the ion–ion interaction as the cause of the breakdown of the Nernst–Einstein relation can be drawn from the collection of experimental data of the dependence of H_R on total alkali content of alkali borate, germinate, and silicate glasses by Kelly et al. [1729] and Na alumino-germanate and Na borate glasses by Voss et al. [1711]. Kelly et al. have reported for each family of alkali oxide glass that H_R increases with decreasing total alkali content for alkali borate, germanate, and silicate glasses. Voss et al. combined their Haven ratio data of Na borate and Na alumino-germanate glasses with the data of Kelly et al. and plotted all data of H_R altogether against the ratio of the average Na–Na distance to the network distance $< d_{Na} > / < d_{network} >$. This plot shown here in the right panel of Fig. 312 verifies the expected increase of H_R on increasing ion–ion distance, and the limit $H_R = 1$ is attained at low alkali concentration.

The degree of breakdown of the Stokes–Einstein relation of glassformers increases with decreasing temperature toward T_g (see Section 2.2.5.6). Similar T

dependence of the degree of breakdown of the Nernst–Einstein relation was found by simulation and experiment. Decrease of the Haven ratio with decreasing temperature was found by Knödler et al. [1730] by simulation of a stochastic lattice gas model with ion–ion interaction included into the Hamiltonian and by experimental measurements in a Rb borate glass $0.2Rb_2O$–$0.8B_2O_3$ [1731] However, H_R of the corresponding Na borate glass $0.2Na_2O$–$0.8B_2O_3$ shows no temperature dependence.

Explanation of the breakdown of the Stokes–Einstein relation in glassformers given by the CM was based on the correlation functions for center-of-mass diffusion and viscosity which weigh differently the effects of the many-body relaxation, resulting in different coupling parameters and relaxation times. A similar reasoning was given to explain the difference in relaxation times between nuclear spin relaxation and conductivity relaxation of glassy ionic conductors (see Section 3.2.5.1). The origin of the Haven ratio being less than unity in glassy ionics may also be explained by the slight difference between the tracer diffusion correlation function $C_D(t) =< \boldsymbol{r}_i(0)\boldsymbol{r}_i(t) >$ and the conductivity correlation function $C_\sigma(t) = (1/N)\Sigma_{ij} < \boldsymbol{v}_i(0)\boldsymbol{v}_j(t) >$, where $\boldsymbol{r}_i$ and $\boldsymbol{v}_i$ are the position and the velocity of the diffusing ions, respectively. The $i \neq j$ cross-correlation terms in $C_\sigma(t)$ do not appear in the $C_D(t)$.

3.2.9.1 The Haven Ratio for Mixed Alkali Glass

Having shown for single alkali glasses that the Haven ratio H_R anti-correlates with ion–ion interaction and many-ion cooperative dynamics, here the "common" Haven ratios of the mixed alkali glasses $0.2\,[xNa_2O - (1-x)Rb_2O] - 0.8B_2O_3$ obtained by Imre et al. [1693] are used to further support the decrease of ion–ion interaction and many-ion dynamics in the mixed alkali effect.

The common $H_R(x)$ of the mixed alkali glass with composition x is defined by

$$H(x) = \left(\frac{Nq^2}{k_B T \sigma_{dc}(x)}\right)\left[xD^{Na}(x) + (1-x)D^{Rb}(x)\right], \tag{3.48}$$

where D^{Na} and D^{Rb} are the tracer diffusion coefficients of Na and Rb, respectively. The right panel of Fig. 314 shows that H_R increases on introducing foreign alkali into the host, and it peaks at the same composition ($x = 0.4$) as the activation enthalpy ΔH_σ of dc conductivity (see left panel). These properties of H_R and ΔH_σ corroborate in indicating the decrease of ion–ion interaction and many-ion dynamics as one of the manifestations of the mixed alkali effect.

3.2.10 Caged Dynamics, Nearly Constant Loss, and Termination by the Primitive Relaxation

For glassformers, we have discussed in Sections 2.3.2.6 and 2.3.2.33 the short time regime when the molecules are still caged and the local JG β-relaxation has not yet

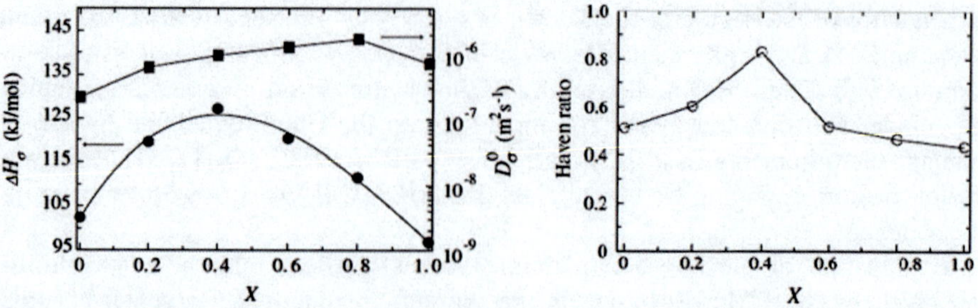

Fig. 314 (*Left*) Activation enthalpy ΔH_σ and pre-exponential factor D_σ^0 of the conductivity diffusion coefficient as a function of composition below the glass transition temperature for $0.2\,[xNa_2O - (1-x)Rb_2O] - 0.8B_2O_3$ glasses. (*Right*) Composition dependence of the common Haven ratio of the same glass. Reproduced from [1693] by permission

appeared. In this caged regime, the dynamics shows up in susceptibility generally as the nearly constant loss (NCL) over an extended range of frequencies at lower temperatures. The lower bound of the frequency range of NCL is higher but not by much in frequency than the JG β-relaxation frequency ν_{JG} or the primitive frequency ν_0. This relation between the lower bound of NCL and ν_{JG} or ν_0 found in general for various glassformers supports the interpretation of the JG β-relaxation or the primitive relaxation terminates the caged dynamics regime and the NCL. Examples can be found in Figs. 1, 247, and 276 for colloidal particle suspensions; Fig. 119 for propylene carbonate; Fig. 135 for PIB; Fig. 241 for threitol and NMEC; and Figs. 243 and 244 for $0.4Ca(NO_3)_2 - 0.6KNO_3$ (CKN). More examples from metallic glasses (see Fig. 170 for $Zr_{65}Al_{7.5}Cu_{27.5}$) and ionic liquids can be found in Section 2.3.2.33. On increasing temperature, both ν_α and ν_{JG} increase and they tend to merge together, and this also increases the lower bound of the frequency regime in which the NCL can be detected. At high enough temperatures, the NCL regime is shrunk to the extent that it cannot be seen and is replaced by a susceptibility minimum. In time-domain experiments such as optical Kerr effect (OHD-OKE), the counterpart of the NCL is the term pt^{-1+c} with $c{\approx}0$, called the intermediate power law (IPL), that usually appears at temperature below the critical temperature T_c of MCT and at times before the $dt^{b-1}\exp(-t/\tau_\alpha)$ term, which accounts for the α-relaxation at high temperatures (see Section 2.3.2.33).

In glass-forming systems involving long-range diffusion such as binary mixtures of L–J particles and Li ion in Li metasilicate glass at low temperatures by simulations and colloidal particle suspensions at high concentrations by confocal microscopy, the counterpart of the NCL is the mean-square displacement $<r^2(t)>$ having the t^c dependence with $c{\approx}0$ shown in Figs. 275 and 276. For the glass-forming ionic molten salt CKN, the NCL of caged dynamics was shown before by data of dielectric loss $\varepsilon''(\omega) \propto \omega^{-c}$ with $c{\approx}0$ in Figs. 243 and 244, consistent with $< r^2(t) > \propto t^c$. This is because the relation between $< r^2(t) >$ and the complex conductivity $\sigma^*(\omega)$ shown in Eq. (3.3). The t^c dependence of $< r^2(t) >$ gives rise to the ω^{1-c} dependence of $\sigma'(\omega)$, the real part of $\sigma^*(\omega)$, from which the ω^{-c}-dependence of $\varepsilon''(\omega)$ follows from the Maxwell relation $\sigma^*(\omega) = i\varepsilon_0\varepsilon^*(\omega)$. For

CKN, also shown in the figures is the relation of the lower bound of the NCL regime to the primitive relaxation frequency. The latter is slightly lower than the former. For colloidal particle suspension at volume fraction $\phi = 0.56$, the primitive relaxation time can be identified with the average time τ_c of 500 s for a typical particle to shift position and leaves the cage given by Weeks et al. from their experiment [141]. This time, indicated in the log–log plot of MSD in Fig. 247, clearly terminates the $t^{0.13}$ dependence of the MSD, which ends at about 250 s. The cage correlation function obtained by Weeks and Weitz [141(b)] also starts its decay at about 500 s as shown in the inset of the figure. The characteristics of the NCL and its termination by some primitive relaxation are common to relaxation in different families of interacting systems, the majority of examples come from the glassformers.

In this section, experimental conductivity relaxation data of NCL in glassy, molten, and crystalline ionic conductors are presented to show exactly the same behavior as the other interacting systems. For the benefit of making clear comparison of the characteristics of the NCL in ionic conductors with other systems, the rich dielectric loss data of the glassformer *N*-methyl-ε-caprolactam (NMEC) [508, 1733] are presented once more in Fig. 315. The data show clearly the NCL, the excess wing (i.e., the unresolved JG β-relaxation), and the structural α-relaxation.

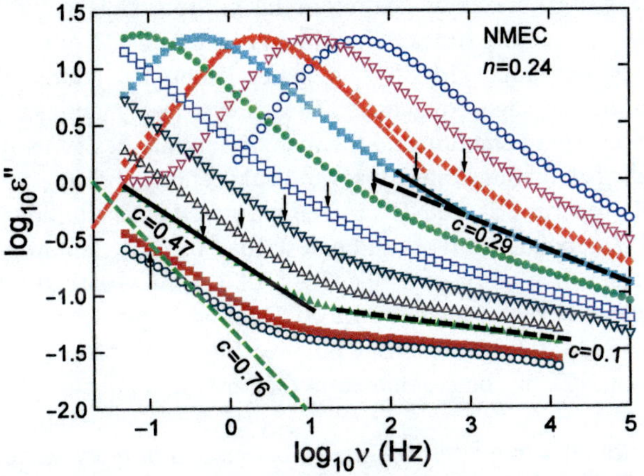

Fig. 315 Dielectric loss spectra of NMEC at temperatures above and slightly below T_g (*from right to left*: 186, 184, 182, 180, 178, 176, 174, 172, 170, 168, 164, 162 K) showing the following three distinct relaxation mechanisms. (1) The α-loss peaks and the fits (*red and pale blue dashed lines*) at two temperatures by the Fourier transform of the KWW function $\varphi(t) = \exp[-(t/\tau_\alpha)^{1-n}]$, with $n = 0.24$. (2) The excess loss $\varepsilon''(\nu) \sim \nu^{-c}$ with $c = 0.29$ (*long dashed line*) at 178 K and with $c = 0.47$ (*solid line*) at 168 K. The very different slopes of the excess loss at the same levels at the two temperatures rule out frequency–temperature superposition of the loss data of the excess loss in this region. (3) The emergence of the NCL $\varepsilon''(\nu) \sim \nu^{-c}$ with $c \approx 0.1$ (*black short dashed line*) with lower intensity at lower temperatures. The *green dashed line* with the label $c \approx 0.76$ shows the frequency dependence of the high-frequency flank of the KWW fit to the (unobserved) loss peak at 168 K. The *vertical arrows* indicate the relaxation time of the unresolved Johari–Goldstein β-relaxation suggested by the coupling model. Data are supplied by R. Richert and replotted here

Indicated by arrows in the figure, the locations of the primitive relaxation frequency calculated from the CM equation (3.1) are consistent with the interpretation that the NCL is terminated by the primitive relaxation.

Actually the best systems to show the NCL of caged dynamics are the glassy and crystalline ionic conductors. In contrast to glass-forming liquids, the existence of NCL in glassy ionic conductors has been known for more than half a century. Evidences for its existence were suggested repeatedly over the span of several decades starting from 1946 by Garton [1529] and many to follow [56(a), 115(b), 146, 149, 467, 1055–1061, 1424, 1522, 1523, 1530–1535, 1730, 1734–1751]. Not only there is an abundance of experimental data available but also the measurements are made over wide range of frequency and temperature in which the NCL has been detected. In many cases the spectral information is also complete to allow the primitive relaxation frequency/time to be calculated from the relaxation time and stretch exponent of the Kohlrausch function used to fit the electric modulus data and show the primitive relaxation terminates the NCL regime. Examples demonstrating this can be found in [1057–1059] and will be further discussed here. The existence of NCL in crystalline or polycrystalline ionic conductor including Na β-alumina [1486], YSZ [1484, 1755], LLTO [1059, 1755] and $LiAlSi_2O_6$ [1060] is particularly instructive to demonstrate that they exhibit the same caged dynamics as found in glass-forming systems. Some examples of the universal properties of caged dynamics will be presented later on in this section. Thus, any theory of caged dynamics, with applicability restricted to liquids such as MCT or to glasses such as random energy barrier model [1478], is not general enough to address the apparently universal properties of caged dynamics. To address the dynamics of ions beyond caged regime, ion–ion interaction and the many-ion dynamics generated have to be included, as has been demonstrated in several subsections above.

One general properties of the NCL in glassy as well as crystalline ionic conductors is its weak T dependence that can be described by either a power law, T^{α} with α not much greater than unity, or $\exp(T/T_0)$ [1056, 1058, 1059, 1060, 1752], similar to the temperature dependence of the NCL found in glassformers by dielectric relaxation [508, 971, 1026] and OHD-OKE (see Fig. 253). More examples of general properties of NCL in ionic conductors that are analogous to glassformers are presented below.

NCL exists not only in glassy $LiAlSi_2O_6$ but also in polycrystalline $LiAlSi_2O_6$. The magnitude of the NCL in polycrystalline $LiAlSi_2O_6$ is only about a factor between 2 and 3 times smaller than that in the glassy $LiAlSi_2O_6$ at the same temperature [1060]. It is not necessary to have structural disorder as in a glass or a liquid to generate the NCL. The comparable magnitudes of the NCL in polycrystalline and glassy $LiAlSi_2O_6$ suggest that the mechanism that gives rise to the NCL is somehow related to the local motion of the ion confined within the cage formed by the other ions. The cage defined by the anharmonic potential confining the ion is likely the only characteristic shared by both forms of $LiAlSi_2O_6$ as far as the NCL is concerned. Anharmonicity of the potential as well as mean-square displacement of the caged ion is expected to be larger in the glassy state than in the polycrystals, and this may explain the larger NCL observed in the former rather than in the latter.

The discussion and the analysis of conductivity relaxation data of crystalline and glassy ionic conductors follow the same line as those for CKN in Section 2.3.2.33. There two characteristic times t_{x1} and t_{x2} and the corresponding frequencies ν_{x1} and ν_{x2} are used to delineate three dynamic regimes. These two times are also involved in the description of the motions of ions in the space–time pictures generated by molecular dynamics simulations of Li metasilicate glass. The pictures presented in Fig. 280 are helpful when reading the following description. The very first dynamic regime is located at frequencies ν higher than ν_{x1} or at times t shorter than $t_{x1} \equiv 1/(2\pi\nu_{x1})$, where $\varepsilon''(\nu) = p\nu^{-c}$ or $\sigma'(\nu) \propto \nu^{1-c}$, and c is small. This is the caged ion regime or the NCL regime. The cage decays after t_{x1} by the onset of the primitive relaxation, which is an independent ion hop to neighboring sites in the case of ionic conductors. Naturally, we have $\tau_0 > t_{x1}$ or $\nu_{x1} > \nu_0$. The NCL regime is followed by the transition region defined by $\nu_{x1} > \nu > \nu_{x2}$ or $t_{x1} < t < t_{x2} \equiv 1/(2\pi\nu_{x2})$, in which more ions are hopping to neighboring sites and the motions become increasingly more cooperative with time. Here t_{x2} is the onset time of the "fully cooperative regime," where the Fourier transform of the derivative of the Kohlrausch function $\exp\left[-(t/\tau_K)^{1-n}\right]$ fits well the data in the complex modulus $M^*(\nu)$ representation for all $\nu < \nu_{x2}$. Perhaps it is appropriate to call this the "fully cooperative regime" because the Kohlrausch function is the characteristic function of the stable Levy distribution, a generalization of the Gaussian distribution [223]. Although τ_0 and ν_0 have been mentioned earlier, after all they can be calculated by Eq. (3.1) using the parameters τ_K and n of the Kohlrausch function, together with $t_c \approx 2\,\text{ps}$ or $\nu_c \approx 10^{11}$ Hz from high-frequency/high-temperature measurements shown in Section 3.2.3.1. If conductivity relaxation data are presented in terms of $\varepsilon''(\nu)$ or $\sigma'(\nu)$, instead of $M''(\nu)$, then ν_{x2} is defined as the frequency above which the power laws $\varepsilon''(\nu) \propto \nu^{-1+n}$ and $\sigma'(\nu) \propto \nu^n$ no longer hold. Demonstration of the evolution of the dynamics in the order of the frequencies ν_{x1}, ν_0, ν_{x2}, and $\nu_K = 1/\tau_K$ has been given before for CKN in several figures of Section 2.3.2.33. For data presented as the MSD, $<r^2(t)>$, t_{x2} is the onset time of the t^n dependence of $<r^2(t)>$. It has been shown together with t_{x1} and τ_0 in the plot of $<r^2(t)>$ vs. time for Li ions in metasilicate glass (Fig. 277), binary mixtures of L–J particles (Fig. 275), and colloidal particle suspensions (Fig. 276) in Section 3.2.1 before.

Conductive relaxation data of some crystalline and glassy ionic conductors are now shown to have the same time evolution of dynamics as glassformers in support of the universal dynamics in interacting systems. To familiarize the reader with such data given in terms of $\sigma'(\nu)$, the same data of the molten salt, CKN, presented before in terms of $\varepsilon''(\nu)$ and $M''(\nu)$ in Figs. 243 and 244 are recast in this form in the left panel of Fig. 316. Shown also are the locations of ν_{x1}, ν_0, ν_{x2}, and ν_K. The line with slope equal to 1 is drawn to indicate $\sigma'(\nu) \propto \nu^{1-c}$ with $c << 1$, corresponding to the NCL. The dashed line represents $\sigma'(\nu)$ exactly calculated from the Kohlrausch fit to $M^*(\nu)$, and it gives the dc conductivity at low frequencies and the power law $\sigma'(\nu) \propto \nu^n$ at high frequencies. Actually, $\sigma'(\nu)$ data of glassy ionic conductors 0.44LiBr–0.56Li_2O–B_2O_3 at 323 and 573 K and 0.5Ag_2S–0.5GeS_2 glass at 273 K in Fig. 282 have shown some of the same features already. Here we show in more detail the $\sigma'(\nu)$ data of other glassy ionic conductor 0.80LiF–0.20$Al(PO_3)_3$ from

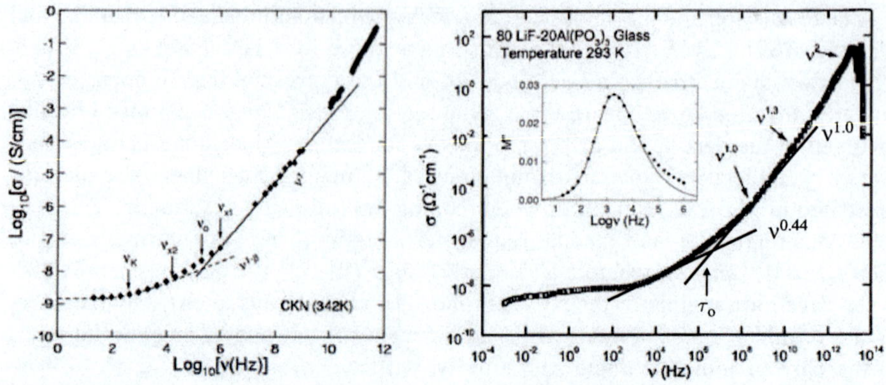

Fig. 316 (*Left*) Replotting the CKN data of Fig. 243 as log $\sigma(\nu)$ vs. log(frequency/Hz) to show the nearly linear frequency dependence of the NCL. The *dashed line* that includes the dc conductivity and the high-frequency power law $\nu^{1-\beta}$ or ν^n corresponds to the Kohlrausch fit to the electric modulus with n=0.34. Reproduced from [1057] by permission. (*Right*) Log $\sigma(\nu)$ at several temperatures of a glassy ionic conductor 0.80LiF–0.20Al$(PO_3)_3$ vs. log(frequency/Hz). The data show a near linear frequency ($\nu^{1.0}$)-dependent conductivity and a broad crossover to the $\nu^{0.44}$ dependence (obtained from the Kohlrausch function fit to the electric modulus in the *inset*) and finally to the dc conductivity. ν_0 (erroneously replaced by τ_0) is the independent relaxation frequency of an ion hopping out of its cage, which is calculated from the CM. Redrawn from data of Kulkarni et al. [1756]. Reproduced from [1057] by permission

the work of Kulkarni et al. [1756] in the right panel of Fig. 316, the data of glassy 0.5Ag_2S–0.5GeS_2 from the work of Ribes et al. [1757] in the left panel of Fig. 317, and the data of 0.48$(AgI)_2$–0.52Ag_2SeO_4 from the work of Cramer and Buscher [1528] in the right panel of Fig. 317. In all cases, $n = (1-\beta)$ was determined by the fit to the *same* data in the electric modulus representation by the Fourier transform of the Kohlrausch function $\Phi(t)$ in Eq. (1.56), and τ_0 is calculated from the CM equation.

The reader may recall that 0.5Ag_2S–0.5GeS_2 was discussed before in Section 3.2.3 on the comparison between activation energies of short-time silver ion diffusion measured by quasielastic neutron scattering and long-time conductivity relaxation (see Table 3.2). The glass 0.48$(AgI)_2$–0.52Ag_2SeO_4 was also discussed in Section 3.2.3 in a similar context. As an aside, the dielectric loss data of this glassy ionic conductor was shown to fit the MCT predictions (see Fig. 284), even though MCT does not apply to ionic conductivity relaxation in glasses. This warns the possibility that good fits to susceptibility minima to MCT predictions in supercooled liquids do not necessarily validate MCT as the uniquely correct description of caged dynamics of glassformers.

In all the ionic conductors shown in Figs. 316 and 317, the caged dynamics (NCL) regime ending at ν_{x1} is terminated by the primitive relaxation as evidenced by the location of its relaxation frequency ν_0 not much longer than ν_{x1}. At frequencies lower than ν_0, there is a gradual transition of $\sigma'(\nu)$ before it finally assumes the frequency dependence of the cooperative Kohlrausch ion hopping starting at ν_{x2}.

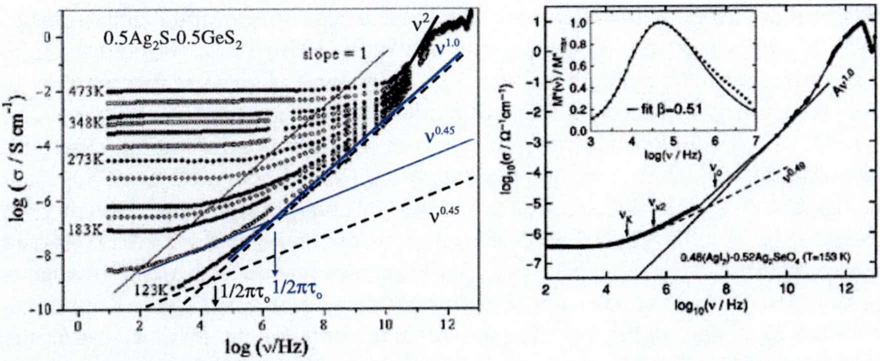

Fig. 317 (*Left*) Log $\sigma(\nu)$at several temperatures of a glassy ionic conductor, $0.5Ag_2S$–$0.5GeS_2$, from the work of Ribes et al. [1757] vs. log(frequency/Hz). The data show a near-linear frequency ($\nu^{1.0}$)-dependent conductivity and a broad crossover to the $\nu^{0.45}$ dependence (obtained from the Kohlrausch function fit to the electric modulus) and finally to the dc conductivity. $\nu_0(= 1/2\pi\tau_0)$ is the independent relaxation frequency of an ion hopping out of its cage, which is calculated from the CM. Reproduced from [1057] by permission. (*Right*) Log $\sigma(\nu)$vs. log(frequency/Hz) at 153 K of a glassy ionic conductor $0.48(AgI)_2$–$0.52Ag_2SeO_4$ from the work of Cramer and Buscher [1528]. The data show a near-linear frequency ($\nu^{1.0}$)-dependent conductivity and a broad crossover to the $\nu^{0.49}$ dependence (obtained from the Kohlrausch function fit to the electric modulus) and finally to the dc conductivity. $\nu_0(= 1/2\pi\tau_0)$ is the independent relaxation frequency of an ion hopping out of its cage, which is calculated from the CM. The *dashed line* is the cooperative ion conductivity relaxation. Reproduced from [1057] by permission

The latter begins with the fractional power $\sigma'(\nu) \propto \nu^{-n}$ and on decreasing frequency it flattens out to assume the frequency-independent dc conductivity σ_{dc} at low frequencies. This behavior is clearly demonstrated in Figs. 316 and 317.

The examples of glassy ionic conductors discussed above all have sizeable n in the exponent (1–n) of the Kohlrausch function, $n = 0.34, 0.44, 0.45$, and 0.49 for CKN, $0.80LiF$–$0.20Al(PO_3)_3$, $0.5Ag_2S$–$0.5GeS_2$, and $0.48(AgI)_2$–$0.52Ag_2SeO_4$, respectively. The sizeable n values are expected from significant many-ion cooperative relaxation in these ionic conductors containing high concentration of mobile ions. It is instructive to compare them with the dynamics of ions in conductors with lower ion concentrations. In Section 3.2.2.1, the decrease of n reflecting diminishing many-ion cooperative dynamics with decreasing ion concentration has been discussed. Here, at constant ν_K, the dependence of the magnitude of the caged dynamics (NCL) on ion concentration (or n) is discussed, as well as the dependences of the characteristic frequencies ν_{x1}, ν_0, and ν_{x2}. This comparison is analogous to the comparison made before for the glassformers, NMEC ($n = 0.24$), threitol ($n = 0.36$), and xylitol ($n = 0.46$) in Fig. 241, and for CKN vs. $ZnCl_2$, and CKN vs. B_2O_3 in Fig. 250 at constant or approximate constant structural relaxation frequency ν_α. Although ν_{x1}, ν_0, and ν_{x2} are not illustrated in these figures, it can be gathered by inspection of the spectra that glassformer with smaller n has lower values of ν_{x1}, ν_0, and ν_{x2} as well as lower NCL (relative to the α-relaxation). In addition, the NCL is also terminated by the primitive or JG relaxation frequency in each of these glassformers. Although similar trend can be deduced from several comparisons of NCL in ionic conductors at the same temperature [1056, 1060, 1061], the cogent comparison has to be made at the same ν_K. An example is given as follows.

A comparison of the entire range of dynamics including the NCL of two glassy oxide Na ion conductors with very different ion concentrations is shown in Fig. 318. Plotted in the left panel of the figure are $\log[\sigma'(\nu)]$ vs. $\log \nu$ of a Na_2O–$3SiO_2$ glass at –0.5°C with a high concentration of mobile Na^+ ions [1518] and of a Vycor glass $xNa_2O - (1-x)[0.04B_2O_3 - 0.96SiO_2]$ at 313°C with small $x = 0.00044$ and hence containing very few Na^+ ions [1506]. The conductivity relaxation frequency ν_K of the two glasses at these temperatures has been determined by the Kohlrausch fit to the *same* data in the electric modulus representation, which is shown in the right panel of Fig. 318. ν_K is practically the same as the peak frequency ν_{max} at the maximum of $M''(\nu)$. The ν_K of the two glasses differ only by a factor of about 2, and thus this near-isoconductivity relaxation time condition is satisfied for making objective comparison of the NCL in these two ionic conductors.

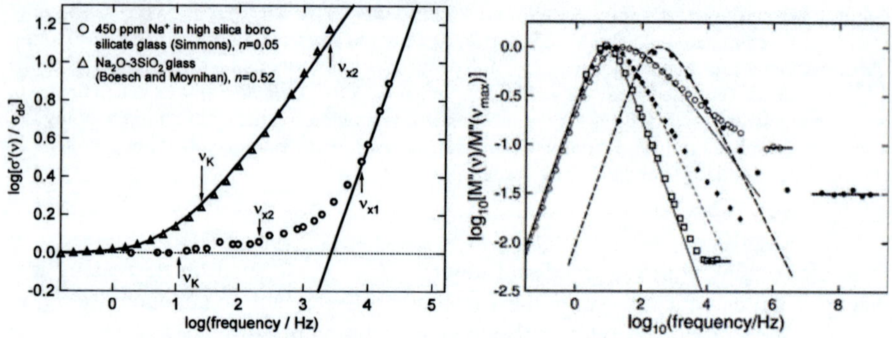

Fig. 318 (*Left*) Isothermal electrical relaxation data shown as plots of $\log\left[\sigma'(f)/\sigma_{dc}\right]$ vs. $\log f$ for a Na_2O–$3SiO_2$ glass at –0.5°C and for a Vycor glass $xNa_2O-(1-x)[0.04B_2O_3-0.96SiO_2]$ at 313°C with $x = 0.00044$. Reproduced from [1057] by permission. (*Right*) Scaled $M''(\nu)$ data of several ionic conductors. The Na trisilicate glass (*open circles*) and the Kohlrausch fit with $n = 0.50$ (*dashed-dotted line*). Vycor glass $xNa_2O - (1-x)[0.04\,B_2O_3 - 0.96\,SiO_2]$ with $x = 0.00044$ (*open squares*) and the Kohlrausch fit with $n = 0.05$ (*solid line*). The CKN data at 333 K (*closed diamonds*) and at 342 K (*closed circles*), together with the Kohlrausch fits with $n = 0.28$ (*thinner dashed line*) and 0.33 (*thicker dashed line*), respectively. The *horizontal lines* indicate the levels of NCL. Reproduced from [1482] by permission

In the left panel of the figure, the $M''(\nu)$ data of the Na trisilicate glass, the Vycor glass, and CKN at two temperatures 333 and 342 K are compared after normalizing $M''(\nu)$ by $M''(\nu_{max})$, the maximum value at the loss peak. The NCL appearing at the high-frequency end, $A_M\nu^{-c}$ with $c << 1$, is indicated for each case by the horizontal line. It can be seen from the comparison that ionic conductor with broader dispersion or large n in $M''(\nu)$ has higher level of NCL relative to the $M''(\nu_{max})$. In the order of increasing ratio NCL/$M''(\nu_{max})$, Vycor glass has $n = 0.05$, CKN has n=0.28 and 0.33, and the Na trisilicate glass has $n = 0.50$. The reader may recognize the similarity between the trend of the NCL in ionic conductors shown in the left panel of Fig. 318 and that of the NCL in glassformers in Figs. 241 and 250 at constant or approximate constant structural relaxation frequency ν_α. The correlation between NCL and n found in the dynamics of the two systems is another evidence for universal properties of interacting systems.

In the plot of $[\sigma'(\nu)/\sigma_{dc}]$ against log ν in the left panel of Fig. 318, it can be seen that the Vycor glass has ν_{x2} much closer to ν_K than the Na_2O–$3SiO_2$ glass, which is another indication of much narrower dispersion of the former compared with the latter. The same is true for ν_{x1}; the location of it for the Vycor glass is well inside the range of frequencies shown, while that for the Na_2O–$3SiO_2$ glass is outside and not shown (data at higher frequencies are truncated in this plot). The NCL appears in this plot as $\sigma'(\nu)/\sigma_{dc} \approx A_\sigma \nu^{1-c}$, with $c << 1$. The larger NCL of Na trisilicate glass than Vycor glass is reflected in this plot by larger value of A_σ of the former compared with the latter. A similar situation was found by Jain et al. in the comparison of $\sigma'(\nu)$ for two $xK_2O - (1-x)GeO_2$ glasses with $x = 0.20$ and 0.0023 [1512]. Whether considered by $A_M \nu^{-c}$ in $M''(\nu)/M''(\nu_{max})$ at constant ν_K or by $A_\sigma \nu^{1-c}$ in $\sigma'(\nu)/\sigma_{dc}$, NCL is smaller in ionic conductor having smaller n. This trend is the same as that found for the glassformer such as the pair NMEC and xylitol in Fig. 241, and the pair CKN and $ZnCl_2$ in Fig. 250 (at the same ν_α). For ionic conductors, the trend has been rationalized in [1058, 1417] by the fact that ν_0 terminates the NCL and ν_0 is located at higher frequency (at constant ν_K). A similar explanation for glassformer is by the fact that ν_{JG} or ν_0 terminates the NCL and ν_{JG} or ν_0 is located at higher frequency (at constant ν_α) for glassformer with smaller n [507]. Incidentally, this trend in glassformers readily explains the deep and sharp susceptibility minimum found in "strong" glassformers having small n such as $ZnCl_2$ or B_2O_3 (Fig. 250–1). This is because the depth of the minimum is determined by the NCL which is low in these glassformers. That is also the reason why the idealized MCT fails to fit the susceptibility minimum $\chi''(\nu)$ of these "strong" glassformers because the predicted spectral shape of the minimum is given by $\chi''(\nu) = \chi''_{min}\left[a(\nu/\nu_{min})^{-b} + b(\nu/\nu_{min})^{a}\right]/(a+b)$. Since a has to be less than 0.395, MCT cannot explain the sharp minimum seen in the "strong" glassformers. Conversely, the higher level of NCL of "fragile" glassformers with larger n can produce a shallower minimum satisfying the condition $a < 0.395$, and MCT fit can be successful. The main point is that NCL is responsible for the susceptibility minimum, which can be shallow or deep depending on the NCL level controlled by n. This is another example of the connection of the faster dynamics to the ultimate many-body relaxation dynamics characterized by its dispersion or the coupling parameter n.

The examples from CKN, $0.80LiF$–$0.20Al(PO_3)_3$, $0.5Ag_2S$–$0.5GeS_2$, and $0.48(AgI)_2$–$0.52Ag_2SeO_4$ have been chosen to elucidate the NCL in isothermal spectra because the measurements were made over enormous broad frequency range for us to see the relation between the characteristic frequencies ν_{x1}, ν_0, ν_{x2}, and ν_K. If measurement is made over a limited range of frequency typically from 10 Hz to 1 MHz, fast and slow processes only can be seen together in isochronal spectra of $\sigma'_\nu(T)$ at constant ν. The faster caged dynamics appears in $\sigma'_\nu(T)$ at lower temperatures, and it is unmistakably the NCL because $\sigma'_\nu(T)$ has the frequency dependence of ν^{1-c} with $c << 1$ at constant T [1059, 1758]. The crossover from NCL to primitive ion relaxation and subsequently to cooperative ion dynamics can be seen from the progressive change in T dependence of $\sigma'_\nu(T)$. The best example is the analysis of $\sigma'_\nu(T)$ measured over finely spaced temperatures down to 8 K in the frequency range 10 Hz–1 MHz reported by León and coworkers on two different Li ionic conductors, one is the crystalline $Li_{0.18}La_{0.61}TiO_3$ (LLTO) and the other is the glassy $61SiO_2$–$35Li_2O$–$3Al_2O_3$–$1P_2O_5$ [1059], and by Rivera et al. in $(Y_2O_3)_{0.16}(ZrO_2)_{0.84}$ and glassy Li, Na, K, and Rb triborate glasses [1755]. At lower temperatures, an NCL corresponding to linear frequency-dependent ac conductivity is the dominant contribution. As temperature is increased, a crossover from the NCL to a fractional power law frequency dependence of the ac conductivity ν^n is observed at fixed T (see left panel of Fig. 319 for LLTO).

The data in the left panel of the figure also show the weak dependence of the NCL that is proportional to $\exp(T/T_o)$. For each isochronal at a fixed frequency ν, the open square indicates the crossover point (T_x, $\sigma'_\nu(T_x)$) from the weak T dependence of NCL to the stronger activated T dependence of the ac conductivity ν^n with a criterion explained in [1758]. Naturally, the crossover can be identified with termination of the NCL regime. The coordinates of the crossover points are shown again as open squares in the Arrhenius plot of conductivity in the right panel of Fig. 319. The closed circles are additional crossover points at higher frequencies determined by a method discussed in [1758]. Shown also is the dc conductivity as a function

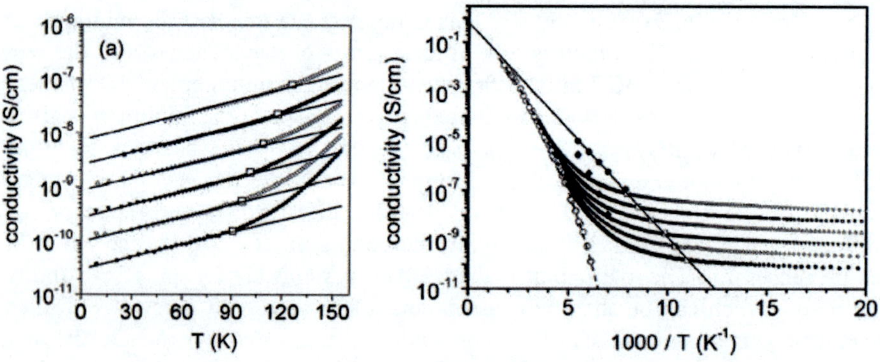

Fig. 319 (*Left*) Conductivity vs. temperature plots for LLTO. Conductivity data are shown for different frequencies (300 Hz, 1, 3, 10, 30, and 100 kHz, from *bottom* to *top*) in a linear temperature scale. *Open square symbols* represent the crossover between the linear frequency-dependent conductivity (NCL regime) and the power law ν^n regime, which is due to ionic hopping. *Solid lines* are fits to exponential temperature dependence of conductivity data in the NCL regime. (*Right*) Arrhenius plots of conductivity data of LLTO at different frequencies (300 Hz, 1, 3, 10, 30, and 100 kHz, from *bottom* to *top*). *Open square symbols* represent the crossover points obtained from the *left panel*, and *solid circles* are the crossover points at higher frequencies determined by method discussed in [1758]. The *solid line* is a fit to an Arrhenius law for the crossover temperature dependence. *Open circles* are dc conductivity data. *Closed diamonds* are the crossover points calculated by assuming the validity of the augmented Jonscher equation, which shows an activation energy close to that observed for the dc conductivity. Reproduced from [1758] by permission

of temperature (open circles), which is non-Arrhenius as the reader may recall from Section 3.2.6.4.

The solid line is a fit to an Arrhenius law for the crossover points, showing the relation $\sigma'_\nu(T_x) = \sigma_\infty \exp(-E_m/kT_x)$, where $E_m = 0.17 \pm 0.03$ eV, and $\sigma_\infty \approx 1$ S/cm, which corresponds to a conductivity relaxation time τ of the order of magnitude of 10^{-11} s from the Maxwell relation $\sigma = \varepsilon_0\varepsilon_\infty/\tau$. It turns out that E_m is nearly the same as the temperature-independent primitive activation energy $E_\mathrm{a} \approx 0.175$ eV of the independent ion hopping relaxation time τ_0 that was calculated by the products $[1 - n_\sigma(T)]\,E_\sigma(T)$ over a range of temperatures for LLTO (see Section 3.2.6.4 and Fig. 304, upper right). Hence E_m is identified with the energy of the barrier preventing the Li^+ ions to abandon their cages, and the relaxation time corresponding to $\sigma'_\nu(T_x)$ is the primitive ion relaxation time τ_0. The result is another indication that the NCL occurs while most of the ions are still caged and is terminated by the primitive ion relaxation. Naturally an estimate in order of magnitude of the time for the ions to leave their respective cages is given by the primitive relaxation time of the CM. At a longer time t_{x2} after which essentially all ions have high probability of successfully jumping out of their cages, they start the cooperative ion conductivity relaxation process well described by the Kohlrausch correlation function.

3.2.10.1 Caution for Those Who Prefer Data Represented by $\sigma'(\nu)$ than $M^*(\nu)$

Before leaving Fig. 316, it is instructive to consider a popular expression of Jonscher [124], often used to interpret conductivity relaxation of ionic conductors. Jonscher's expression for the real part of the conductivity is

$$\sigma(\nu) = \sigma_0[1 + (\nu/\nu_\mathrm{p})^n], \tag{3.49}$$

where σ_0 is the dc conductivity, ν_p the characteristic relaxation frequency, and n the fractional exponent. At higher frequencies the ν dependence given by Eq. (3.49) no longer works and is replaced by ν^{1-c} with $c << 1$, the NCL term. To include the NCL, some workers add a term $A\nu$ to the right-hand side of Eq. (3.49):

$$\sigma'(\nu) = \sigma_0\left[1 + \left(\frac{\nu}{\nu_p}\right)^n\right] + A_\sigma \nu. \tag{3.50}$$

Recalling Fig. 245, where it is shown the NCL in glassformers is not an additive contribution to the susceptibility, one would be skeptical of this approach in ionic conductors. Since σ_0 and ν_p are thermally activated with the dc activation energy E_σ, meanwhile A_σ shows a much milder temperature dependence, it can be shown from Eq. (3.50) that the crossover between the two terms should be thermally activated with the energy E_σ. The crossover point is readily obtained by equating the NCL term to the real part of the ac conductivity given by the Jonscher's expression at the crossover frequency. The locations of the crossovers shown by the closed diamonds in Fig. 319 indeed have the T dependence of σ_dc and do not correspond in any way to the NCL.

Jonscher's expression for the $\sigma'(\nu)$ in Eq. (3.49) gives no characteristic times explicitly. Users of this expression usually define a characteristic time via the characteristic frequency ν_c by the equation $\sigma'(\nu_c) = 2\sigma_{dc}$. The ν_c so defined may be close to ν_K in ionic conductors having higher concentrations of ions and larger n like Na_2O–$3SiO_2$ glass shown in the left panel of Fig. 318, and has activation energy nearly the same as σ_{dc}. However, when $\sigma'(\nu_c) = 2\sigma_{dc}$ is applied to ionic conductors with low concentration of ions and small n such as the Vycor glass $x Na_2O-(1-x)[0.04B_2O_3-0.96SiO_2]$, ν_c is more than two decades higher than ν_K and it has much weaker T dependence than σ_{dc}. Thus, $\sigma'(\nu_c) = 2\sigma_{dc}$ yields ν_c associated with different dynamics for ions with different concentration and/or width of dispersion and n. Imre et al. [1693] applied this to determine ν_c, which they called ν_{on}, to the series of the mixed alkali glasses $0.2[xNa_2O-(1-x)Rb_2O]-0.8B_2O_3$ with $x = 0.0$, 0.2, 0.4, 0.6, and 0.8. The mixed alkali (MA) effects were shown before in Figs. 308 and 311. The activation enthalpies ΔH_σ of dc conductivity show a maximum at some $X = 0.4$ in Fig. 311 and also in Fig. 320, but the activation enthalpies $\Delta H_{\nu_{on}}$ of ν_{on} do not, which is almost independent of composition. Thus ΔH_σ shows the mixed alkali effect as usual, but $\Delta H_{\nu_{on}}$ do not. Imre et al. were taken by surprise. But this is no surprise in view of the discussion above. The number of mobile ions in the mixed alkali glasses is fewer than that in the single alkali glasses, and ν_{on} in the MA glasses no longer corresponds to dc conductivity. Consequently, the activation enthalpy $\Delta H_{\nu_{on}}$ of MA glasses is smaller than ΔH_σ, and it shows no MA effect or at least smaller effect than ΔH_σ. The danger of using the Jonscher's expression to interpret deeper properties of ion dynamics is made clear by this example.

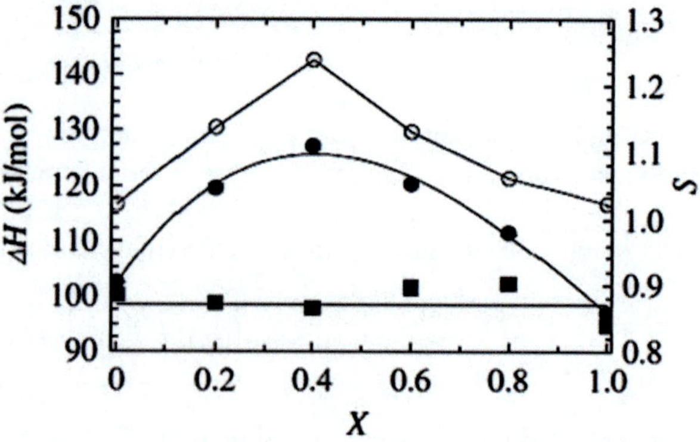

Fig. 320 Composition dependence of activation enthalpies ΔH_σ (•) and $\Delta H_{\nu_{on}}$ (■) for the $0.2[xNa_2O-(1-x)Rb_2O]-0.8B_2O_3$ mixed alkali glasses – *left axis*. The quantity S (○) – *right axis* is of no interest here and not discussed in the text. Reproduced from [1693] by permission

3.2.10.2 Rationalization of the Observed Properties of NCL by Its Relation to the Primitive Relaxation

Crystalline and Ionic Conductors

From the close relation between $t_{x1}(T)$ and $\tau_0(T)$, $t_{x1}(T)$ has the same activation energy as E_a of $\tau_0(T)$. This has been demonstrated in Fig. 319 for Li ion dynamics in LLTO and other crystalline and glassy ionic conductors [195, 1058, 1059, 1755, 1758], and hence we have $t_{x1}(T) = t_\infty \exp(E_a/kT)$. Given the fact that the NCL is terminated at $t_{x1}(T)$) by the same process (i.e., the primitive ion hop), $< r^2(t_{x1}(T)) >$ is expected to have the same value independent of temperature or $t_{x1}(T)$. From its onset at t_{on} and up to $t_{x1}(T)$, $< r^2(t_{x1}(T)) >$ or the NCL increases by the same amount for all temperatures. However, because t_{x1} is thermally activated, this same increase of $< r^2(t_{x1}(T)) >$ is spread over a number of decades of time given by $\{\log_e[t_{x1}(T)] - \log_e(t_{on})\}/2.303$. Therefore $< r^2(t_{x1}(T)) >$ is inversely proportional to $\log_e\left[t_{x1}(T)/t_{on}\right]$. From this proportionality and the relation between conductivity and mean-square displacement (Eq. (3.4)), the magnitude A of the NCL, $\varepsilon''(\nu) = A\nu^{1-c}$ with $c << 1$, is given by the proportionality relation [1058]

$$A \propto \frac{1}{E_a}[1 - (kT/E_a)\log_e(t_{on}/t_\infty)]^{-1}. \tag{3.51}$$

This expression is well approximated by

$$A \propto (E_a)^{-1}\exp(T/T_0), \quad \text{for} \quad T << T_0 \tag{3.52}$$

with

$$T_o \approx E_a/k\,\log_e(t_{on}/t_\infty). \tag{3.53}$$

Note that T_0 is a positive number because $\log_e(t_{on}/t_\infty)$ is a positive number from the fact that $t_{x1} < \tau_0$, and hence t_∞ is even shorter than τ_∞, the prefactor of τ_0, which is the reciprocal of a vibrational attempt frequency. Thus the weak temperature dependence of the NCL is captured by the interpretation of the origin of the NCL within the framework of the CM. Since the result given by Eqs. (3.51) and (3.52) is obtained from a qualitative argument, we do not expect that it will accurately describe the temperature dependence of the NCL, particularly if considered over a very extended temperature range where another source of contribution to NCL may come into play at very low temperatures and high frequencies, such as transitions between the asymmetric double-well potentials (ADWPs) [1744, 1755, 1759]. Of course, the ADWPs constitute a different source of the NCL than the slow cage decay process discussed here.

Note that according to Eq. (3.51), at constant temperature, A should decrease with increasing E_a, had other factors like ion concentration that determines the absolute value of A are the same for the ionic conductors to be compared. If this condition holds for many ionic conductors, then an approximate anticorrelation between A and E_a may exist at constant T. This is an important prediction and can be tested by experiment. Previously, Rivera et al. [1061] measured the intensity of the NCL A in four alkali triborate glasses M_2O–$3B_2O_3$, with M = Li, Na, K, and Rb. They observed that A approximately has the $m^{-1/3}$-dependence and this dependence led them to speculate vibrational relaxation as the origin of the NCL. The new interpretation of the NCL given in this work was not available at the time of writing

[1061]. Now it replaces the earlier speculation. Its prediction, $A \propto 1/E_a$, replaces the less general observation that $A \propto m^{-1/3}$. The former can be rigorously tested by the alkali triborate glasses data of Rivera et al. [1061]. This test has been done by calculating E_a from the experimental determined activation energy E_a^* of τ_K by using the relation $E_a = (1 - n)E_a^*$ and $(1-n)$ from the Kohlrausch fit of the electric modulus loss peak [1752]. The results of E_a in fact is inversely proportional to A in the family of alkali triborate glasses investigated by Rivera et al., and this is shown in the left panel of Fig. 321 among other ionic conductors.

The predicted approximate correlation between the magnitude A of the NCL and the reciprocal of primitive activation energy E_a by Eq. (3.51) can be tested beyond the family of alkali borate glasses. This has been carried out by León et al. at $T = 100$ K for many ionic conductors [1752]. The results in the left panel of Fig. 321 show the remarkable agreement between experimental data for a variety of ionic conductors and the correlation predicted. The solid line in the figure represents an exact inversely proportional relation. It is indeed noteworthy that such a correlation holds considering there may be other factors which affect the magnitude of the

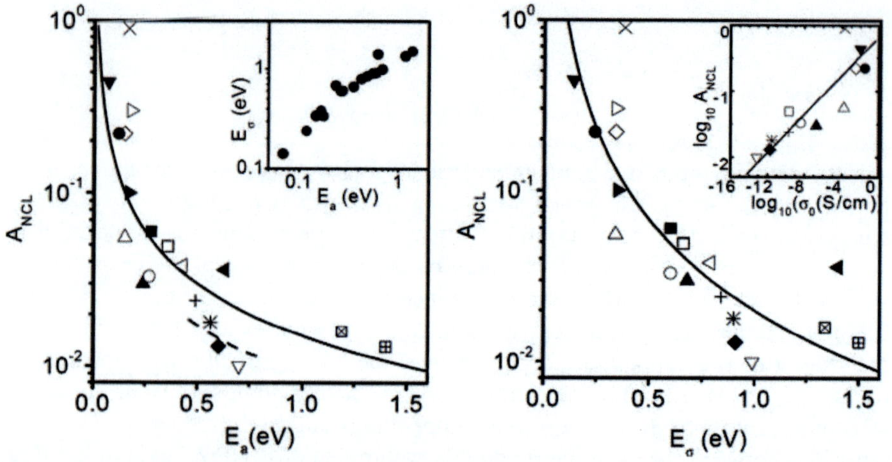

Fig. 321 (*Left*) A remarkable correlation is observed when plotting the magnitudes A of the NCL at $T = 100$ K vs. the activation energy E_a for a variety of ionic conductors. $Li_2O-3B_2O_3$ (+), $Na_2O-3B_2O_3$ (∗), $K_2O-3B_2O_3$ (◆), $Rb_2O-3B_2O_3$ (▽), $LiPO_3$ (▲), $35Li_2O-61SiO_2-3Al_2O_3-P_2O_5$ (○), $Li_{0.18}La_{0.61}TiO_3$ (×), $(ZrO_2)_{0.84}(Y_2O_3)_{0.16}$ (◀), $Na_2O-3SiO_2$ (□), $(AgI)_{0.3}-(AgPO_3)_{0.7}$ (▷), $(AgI_2)_{0.48}-(Ag_2SeO_4)_{0.52}$ (◇), Na−βAl_2O_3 (▼), $(Li_2S)_{0.56}-(SiS_2)_{044}$ (▶), $(Ag_2S)_{0.5}-(GeS_2)_{0.5}$ (Δ), Ag_7GeSe_5I (•), $(LiF)_{0.8}(Al(PO_3)_3)_{0.2}$ (■), $xK_2O-(1-x)GeO_2 (x = 0.2)$ at T = 381 K (◁), $xK_2O-(1-x)GeO_2$ $(x = 0.02)$ at T = 367 K (⊠), $xK_2O-(1-x)GeO_2$ $(x = 0.0023)$ at $T = 374$ K (⊞). *Solid line* represents an inversely proportional relation as derived from the model for the NCL (see Eq. (3.51)). *Dashed line* represents a fit to Eq. (3.51) for the family of alkali triborate glasses. The *inset* shows that the smaller the value of E_a, the smaller the activation energy E_σ for the dc conductivity for the set of ionic conductors shown in the main panel. (*Right*) Main figure shows the correlation between the magnitude A of the NCL at $T = 100$ K and the activation energy, E_σ, of the dc conductivity. *Symbols* are the same as those used in the *left panel*. The *inset* shows the same NCL data vs. the dc conductivity at room temperature. *Lines* are guides for the eye. Reproduced from [1752] by permission

NCL and have not been taken into consideration. The question arises if the correlation still holds when considering the activation energy E_σ for the dc conductivity instead of E_a. It has been shown [1706(a)] for different families of ionic conductors that the one with smaller activation energy E_σ has larger coupling parameter n and hence correspondingly smaller E_a, and rationalized therein by enhanced inter-ionic coupling in ionic conductors with smaller primitive energy barrier E_a. This correlation between E_a and E_σ is obeyed by the ionic conductors considered here and is shown in the inset of the left panel of Fig. 321. From this empirical relation between E_a and E_σ, and the correlation that holds between A and E_a, we also expect a correlation between A and E_σ. The magnitude A of the NCL is plotted in the right panel of Fig. 321 vs. the activation energy E_σ, and a strong correlation indeed exists between them. This correlation is noteworthy because very different time and spatial scales are involved in dc conductivity and the NCL. While the dc conductivity is related to charge transport over long range at long times, the NCL observed at high frequencies in the ac conductivity originates from motion while the ions are still caged. The inset shows that a correlation is also observed between A and the logarithm of the dc conductivities at room temperature σ_o. Since $\sigma_0(T) = \sigma_\infty \exp(-E_\sigma/kT)$ and the prefactor σ_∞ takes not too different values in all ionic conductors, the rule that higher the E_σ, the smaller the dc conductivity at room temperature is usually obeyed. Therefore, the correlation observed between A and the dc conductivities at room temperature can also be considered simply to follow from the correlation between A and E_a deduced from Eq. (3.51).

Even though NCL originates in the caged ions regime, yet its magnitude correlates with the dc conductivity either via its activation enthalpy or its value at room temperature. This happens because NCL is connected to the primitive ion hopping motion, which in turn is connected to the many-ion cooperative hopping motions and finally the dc conductivity. Without this insight, the correlation of the NCL with dc conductivity may have tempted others to jump to the conclusion that the NCL is associated with ion hopping motions of the mobile ions. In fact this erroneous conclusion was proffered by Murugavel and Roling [1760] from their own findings of a correlation between the magnitude of the NCL at T=173 K and the dc conductivity at room temperature of Li, K, and Na alumino-silicate and alumino-germanate glasses. These authors also found that the magnitude of the NCL is higher for the sodium ion conducting glass as compared to the lithium ion conducting glass. This observation clearly is opposite to alkali mass dependence of the NCL found by Rivera et al. [1061] for the alkali triborate glasses. Interestingly, in the case of the alumino-silicate glasses, the dc conductivity of the sodium ion conducting glass is approximately twice as high as the dc conductivity of the lithium ion conducting glass. This is opposite for the alkali borate glasses, where lighter alkali has larger dc conductivity at room temperature (see inset in the right panel of Fig. 321). The reversal of trend in the alkali alumino-silicate glasses is caused by the replacement of silicon by the trivalent aluminum, which changes the structure of the oxide glass. Thus, the results of the alkali alumino-silicate glasses from Murugavel and Roling [1760], as well as the alkali borate glasses from Rivera et al., are in accord with the prediction stemming from Eq. (3.51). Here, the NCL is loss due to ions while still caged due to the anharmonic potential and fluctuations of the cages.

Glassformers

Experimentally, the NCL found in non-ionic glassformers in the equilibrium liquids state or the glassy state is no different from the NCL of crystalline, molten, or glassy ionic conductors. Conceptually they have similar origin due to caging by anharmonic potential and fluctuations of cages, and are terminated by the primitive relaxation. Therefore the same steps leading to the magnitude and temperature dependence of the NCL in ionic conductors given above can be applied to deduce the properties of the NCL in non-ionic glassformers.

Let us now consider two non-ionic glassformers of the same family but having two different coupling parameters. For the same structural relaxation time τ_α, it follows from Eq. (3.1) that the glassformer with the larger coupling parameter has a shorter primitive relaxation time τ_0, and hence also a shorter t_{x1}. Since $< u^2 >_{NCL}$ increases by the same amount in the period $t_{on} < t < t_{x1}$ in the two glassformers, the magnitude of $< u^2 >_{NCL}$ is inversely proportional to $[\log_e(t_{x1}) - \log_e(t_{on})]$. Thus, under the condition of isochronal α-relaxation, the glassformer having larger n is predicted to have larger $< u^2 >_{NCL}$. If the glass transition temperature T_g of all glassformers is defined uniformly to be the temperature at which τ_α has reached the same long time, say 10^3 s, then the prediction is larger $< u^2(T_g)_{NCL}$ for glassformer having larger n. This prediction is supported by experimental data of mean-square displacement in various glassformers obtained from neutron scattering by spectrometers of the same frequency resolution shown in the left panel of Fig. 248 and by other spectroscopies in Figs. 249 and 250. Although the vibrational contribution has not been subtracted off from $< u^2(T) >$, it can be inferred from the data of polymers in the main part of Fig. 248 that $< u^2(T_g) >_{NCL}$ increases in ascending order from PIB, *cis*-PB, aPP, and PS. This trend correlates with the corresponding increase of n at T_g from 0.45 for PIB, 0.50 for *cis*-PB, 0.60 for aPP, and 0.64 for PS [112, 165]. The same holds for the data from IN6 in the inset of Fig. 248, with B_2O_3 having the smallest n equal to 0.40 [112] and polycarbonate the largest $n = 0.65$. The IN10 and IN16 data in the right panel of Fig. 248 as well as the IN13 data in the inset tell the same story. The inset compares $< u^2(T) >$ data of glycerol, which has $n = 0.29$ [195], with that of OTP, which has $n = 0.50$ (from dielectric relaxation) or 0.45 (from photon correlation spectroscopy), and it can be seen that $< u^2(T_g) >$ of OTP is significantly larger than that of glycerol.

3.2.11 A Problem Related to Glass Transition: Breakdown of Thermorheological Simplicity and Associated Viscoelastic Anomalies in Polymers

Any theory of glass transition should be applicable to amorphous polymers because the phenomena are no different from other classes of glassformers as shown here and there in Chapter 2. In addition to structural relaxation and JG secondary relaxation, polymers have slower relaxation (viscoelastic) mechanisms involving longer length scales up to the entire chain length. The relationship between the structural relaxation and these other relaxation modes is a basic problem in polymer viscoelasticity. This is because of the presence of several viscoelastic anomalies throughout the viscoelastic spectrum, which become more prominent when temperature is lowered towards T_g, suggesting relation to glass transition. Some of these anomalies have been discussed in Section 2.2.5.9. There, the purpose is to show that the degree in their manifestation is governed by or dependent on the characteristic of the structural (local segmental) relaxation, including its width of the dispersion (or n), which in turn depends on the chemical structure of the repeat unit. The reader is just reminded in here of these viscoelastic anomalies and the fact that they are caused by the structural (local segmental) relaxation of the polymer, and hence are related to the glass transition problem. We have already said that any viable theory of glass transition must be general enough to be applicable to polymers. Moreover, it should be able to explain the breakdown of thermorheological simplicity and associated viscoelastic anomalies near T_g, because these are caused by the rapid increase of the relaxation time of the structural relaxation with decreasing temperature toward vitrification, which is the core of the glass transition problem.

Before closing this section, we discuss an apparent conundrum encountered [1582] and also an invalid explanation [1761, 1762] offered by other researchers in recent consideration of the difference in the temperature dependences of the local segmental relaxation time τ_α and the viscosity η, or the temperature dependence of their shift factor $a_{T,\alpha}$ and $a_{T,\eta}$ of polymers.

3.2.11.1 A Conundrum

Polyethylene (PE) is a very flexible carbon backbone polymer, and its viscosity and $a_{T,\eta}$ have Arrhenius temperature dependence instead of the VFTH dependence. The viscosity of high molecular weight polyethylene (PE) was discussed in Section 3.2.5.2 and results given in Tables 3.4 and 3.5. The Arrhenius temperature dependence of η has activation enthalpy $E_A = 6.4$ kcal/mol or 26.8 kJ/mol [833]. On the other hand, ^{13}C-NMR measurements of unentangled polyethylene, $C_{44}H_{90}$, and another one with $M_w = 2150$ in the nanosecond range by Qiu and Ediger [1582] found that the local segmental relaxation is best described by an exponential correlation function $\exp(-t/\tau_\alpha)$, and τ_α has an Arrhenius temperature dependence with activation enthalpy $E_{A,\alpha}$ equal to 4 kcal mol or 16.7 kJ/mol. Similar value of 14.8 kJ/mol for activation energy of conformational transitions in PE was obtained by Boyd et al. [1581]. Evidently, $E_{A,\alpha}$ is less than $E_{A,\eta}$, showing once more that the temperature dependences of $a_{T,\alpha}$ and $a_{T,\eta}$ are different. Their own result from PE posed a problem for Qiu and Ediger [1582], who were led to make the drastic conclusion: "The activation energy for the conformational dynamics was found to be 4 kcal/mol, which is significantly less than the flow activation energy. Therefore, in contrast to some other well-studied examples, conformational transitions are not the fundamental motions for flow in polyethylene." The result also creates a conundrum for us because it is opposite to the stronger T dependence of $a_{T,\alpha}$ than that of $a_{T,\eta}$ found by mechanical (shear creep) measurements in high molecular weight PS,

PVAc, and atactic poly(propylene) and explained by the CM equations (2.59) and (2.65) in Section 2.2.5.9.

The conundrum can also be resolved by the CM as follows. In contrast to the mechanical measurements at times longer than 0.1 s and temperatures near T_g, the ^{13}C-NMR experiment was carried out at short times of the order of nanosecond. From the exponential correlation function $\exp(-t/\tau_\alpha)$ observed by ^{13}C-NMR, it is clear that $n_\alpha \approx 0$ for the local segmental relaxation at the short time range. Therefore from Eq. (1.59) now with $n_\alpha \approx 0$, the measured activation energy of 4 kcal/mol by the ^{13}C-NMR experiment should be close to the primitive activation energy $E_a \approx 3.6$ kcal/mol deduced in Section 3.2.5.2, or the potential energy barrier of 3.5 kcal/mol (14.8 kJ/mol) for conformational transition in bulk polyethylene obtained by molecular dynamics simulation [1581]. Moreover, 4 kcal/mol can be explained by the CM with $E_a = 3.6$ kcal/mol and $n_\alpha = 0.1$. On the other hand, the entanglement coupling parameter $n_\eta \approx 0.40$ does not change with temperature because the chains are always entangled, and hence from Eq. (1.65), $E_{A,\eta} = E_a/(1 - n_\eta) = 6.4$ kcal/mol. This is the reason why $E_{A,\eta}$ is *larger* than the activation energy for the conformational dynamics determined by the ^{13}C-NMR experiment. The conundrum is resolved. The conformational transition is still the fundamental motions for flow in polyethylene, albeit many-chain relaxation caused by entanglement interaction slows down the flow and enhances its activation energy $E_{A,\eta}$ over and above E_a of the conformational transition.

3.2.11.2 Problems Encountered in an Explanation of the Breakdown of Thermorheological Simplicity

Although the breakdown of thermorheological simplicity discovered by Plazek and coworkers and confirmed by others to be general phenomenon, solution of the problem was largely put aside by the polymer physics research community. Since the phenomenon occurs when temperature is approaching T_g and it involves the structural relaxation, there is every reason to believe that it has strong connection to the glass transition problem. This connection is innate in the CM explanation [206, 209, 837, 868–872]. Despite the fundamental nature of the problem, the CM is the one and only theoretical attempt that both addresses the glass transition problem and has given an explanation of the various breakdowns of thermorheological simplicity and associated viscoelastic anomalies of entangled and unentangled polymers. In contrast, the reptation theory of viscoelasticity of polymers neither touches the glass transition problem nor offers explanation for the viscoelastic anomalies. Other theories and models of glass transition made no effort to solve the viscoelastic anomalies either. This challenge is largely ignored by the majority of the research community. However there are still many colleagues who recognized the importance of this fundamental problem in polymer viscoelasticity. One notable person is the late Prof. John D. Ferry, who discussed this problem in his textbook *Viscoelastic Properties of Polymers*, 3rd ed. [29]. He gave me much needed encouragement to tackle the problem during my visit to the University of Wisconsin in 1992, and thereafter. After Don Plazek and I finished writing a draft of the review on the viscoelastic anomalies of polymers [165], we sent it to Prof. Ferry for his comments before submitting it for publication. He sent us back valuable comments to improve the review, and the most memorable one is his suggestion of the title of the review *Identification of Different Modes of Molecular Motion in Polymers That Cause Thermorheological Complexity* [165]. It is most appropriate for the review and we used it exactly as he suggested.

Among other researchers who recognized the fundamental importance of the viscoelastic anomalies to both viscoelasticity of polymers and glass transition are Sokolov and Schweizer. I thank them for reviving interest of this fundamental problem to the research community by offering their own explanation [1761]. They recognized the connection of the breakdown of thermorheological simplicity to the glass transition problem and proposed the weaker temperature dependence of relaxation times of chain modes compared with local segmental relaxation, in their own words, as follows: ". . . . decoupling of the temperature dependence of the local and chain scale relaxation times in polymer liquids is the combined consequence of heterogeneous dynamics on the nanometer segmental scale and the diffusive nature of macromolecular relaxation for which heterogeneities are temporally and spatially averaged out. This idea provides an explanation for the long-standing puzzle of the breakdown of time-temperature superposition in polymer melts." "The underlying qualitative idea is that chain relaxation is a mean-field process in the sense that heterogeneous dynamics and local chain stiffness effects are (largely) averaged over, in contrast to nanometer-scale segmental relaxation."

Unfortunately, the explanation proffered by SS is not viable for several reasons [1762]. The temperature dependence of the chain dynamics coming from temporal and spatial averaging of the heterogeneous segmental dynamics, suggested by Sokolov and Schweizer (SS), is no different from the effective friction factor of the averaged segmental dynamics. In other words, the observed segmental dynamics is also an average over its temporal and spatial heterogeneities, and thus the friction factor of its averaged relaxation time $< \tau_\alpha >$ has the same temperature dependence as that obtained by the averaging performed for the chain modes suggested by Sokolov and Schweizer. The anomaly remains unexplained.

There are other problems in the explanation given by SS. One is posed by the viscosity η and self-diffusion coefficient D data of high molecular weight PE and HPB discussed in Section 3.2.5.2. According to the averaging proposed by SS, they should have the *same* chain friction obtained by averaging out the spatially heterogeneous dynamics of the local structural relaxation, but this is contradicted by experiments showing that η has higher activation energy than D (see Table 3.4). The explanation of SS rests on the assumption that the decoupling of translational and reorientational motions in *non-polymeric* liquids is due to heterogeneity of the local dynamics. However, for three well-known cases of such decoupling, *o*-terphenyl [456], tris(naphthyl)benzene [455, 820, 821], and sucrose benzoate [457], the distribution of relaxation times is unchanged with variation in temperature over the range of T/T_g from about 1.02 to 1.2. As acknowledged by Mapes et al. [454]: "The dielectric relaxation and photon correlation spectroscopy results imply a relaxation time distribution whose shape is too temperature independent to cause [translational-rotational decoupling], at least within the framework of current approaches." Thus dynamic heterogeneity fails to explain even the decoupling of viscosity from self-diffusion in non-polymeric glassformers. Lest the reader forgets, a viable explanation of this has been given by the CM [797]. The point herein is that it is premature for SS to explain a second phenomenon using an explanation that does not yet satisfactorily account for the first phenomenon. Next, SS support their ansatz by its consistency with "the almost universal behavior of the temperature dependence of chain relaxation" [1762]. This universal behavior is the putative equivalence of the T_g-normalized temperature dependence (fragility) of the viscosity η or global relaxation time τ_c for all polymers. However, with the exception of polystyrenes of different molecular weights (wherein temperature dependences of low molecular weight samples reflect contributions from a changing compliance and the sub-Rouse modes), in addition to the usual friction factor [832], the fragility of the chain dynamics of polymers differs substantially as shown in [1763]. Thus, while coarse averaging of T_g-normalized temperature dependences can yield single curves with large error bars (Fig. 3b in [1761]), the data points per se exhibit substantial scatter. For example, the ordinate value of a fragility plot of τ_c poly(vinyl ethylene) is more than twice that of poly(methylphenylsiloxane) at $T_g/T \sim 0.92$; there is a similar difference between polyisobutylene and polystyrene at $T_g/T \sim 0.95$ [1763]. Other examples can be cited; there is no universal behavior.

In spite of the shortcomings of the explanation from SS, they are commended for their recognition of the importance of this fundamental and valiant attempt to solve it.

3.2.12 *Looking Out for Universal Dynamics in Other Complex Interacting Systems*

Demonstrated in the preceding subsections of this chapter, similar relaxation and diffusion dynamics are shared by diverse interacting systems. Among the systems exhibiting similar dynamics, the class of glassformers is the most intensively and extensively studied, and has by far the most examples to support universal properties. But even the lesser systems discussed before, there are sufficient experimental studies to provide evidence of similar properties. All the results suggest universality of relaxation and diffusion in interacting systems. Therefore, it is a worthwhile pursuit to look for universal dynamic properties in other interacting systems that have not been either explored or attracted enough attention. This pursuit can bring tremendous benefits to any of these systems because the dynamic properties particularly the anomalous ones can be immediately understood or explained by the knowledge gained from the better known systems. Caution must be exercised when considering complex interacting systems possessing extraneous factors to complicate the observed dynamics and obscure the vestige of the universal many-body relaxation. The contributions from the extraneous factors such as multicomponents and spatial heterogeneity have to be separated or removed before the universal dynamics can be revealed. Before closing this chapter, we give two examples of such complex interacting systems that show features in the dynamics resembling those of glassformers.

3.2.12.1 Charge Density Wave Systems

The system is the modulated electronic superstructure called charge-density wave (CDW) that appears in some quasi-one-dimensional materials such as $K_{0.3}MoO_3$ and orthorhombic TaS_3 at low temperatures [1764–1770]. The first evidence of glass state in CDW systems have been found in the heat capacity C_p measurements at low temperature [1765] showing up as the excess contribution above the Debye contribution from vibrations, $C_p \sim T^3$, like in glasses related to the Boson peak in susceptibility found at high frequencies.

CDW is electronic in origin and far flung from any of the systems discussed before. The first indication of similarity of dynamics in CDW to glassformers was put forth by Kriza and Mihály in 1986 [1766] including the stretched exponential dielectric relaxation they observed in $K_{0.3}MoO_3$ and the connection with the CM. More striking resemblance of dynamics with glassformer was found by Starešinić et al. from dielectric relaxation measurements of semiconducting CDW systems in orthorhombic TaS_3 [1768] and $K_{0.3}MoO_3$ [1769]. They found two relaxations α and β in the dielectric loss spectra (see the example for TaS_3 in the left panel of Fig. 322) and also in thermally stimulated current spectra of o-TaS_3 and $K_{0.3}MoO_3$ (right panel of Fig. 322).

The relaxation time τ_α of the α-process increases rapidly with decreasing temperature and reaches very long time at some temperature T_g (see left panel of Fig. 323) like structural relaxation of glassformers undergoing glass transition. The relaxation time τ_β of the faster β-relaxation changes T dependence near T_g to assume an Arrhenius dependence at temperatures below T_g, resembling the behavior of the JG β-relaxation of glassformers. Starešinić and Biljaković [1771] explained their results by the model of cooperatively relaxing CDW domains of coherent phase

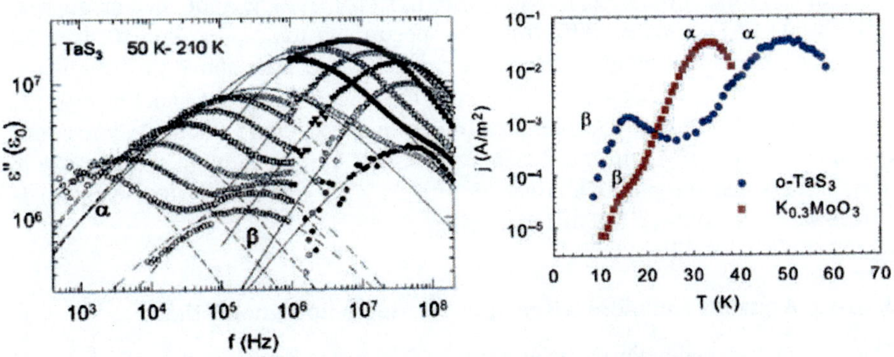

Fig. 322 (*Left*) Imaginary part ϵ'' of the dielectric function of orthorhombic TaS_3 as a function of frequency at several temperatures. The *lines* represent single-process Cole–Cole fits of the dielectric loss data for $T > 50$ K. The *solid lines* correspond to the α-process above 1 MHz, the *dashed lines* to the α-process below 1 MHz, and the *dash-dotted lines* correspond to the β-process seen below 60 K. Reproduced from [1768] by permission. (*Right*) Thermally stimulated current spectra of o-TaS_3 and $K_{0.3}MoO_3$. The sample is cooled in the external electric field and heated in short-circuit configuration and the depolarization current is recorded. Maxima correspond to the "melting" of frozen relaxation processes. Reproduced from [1770] by permission

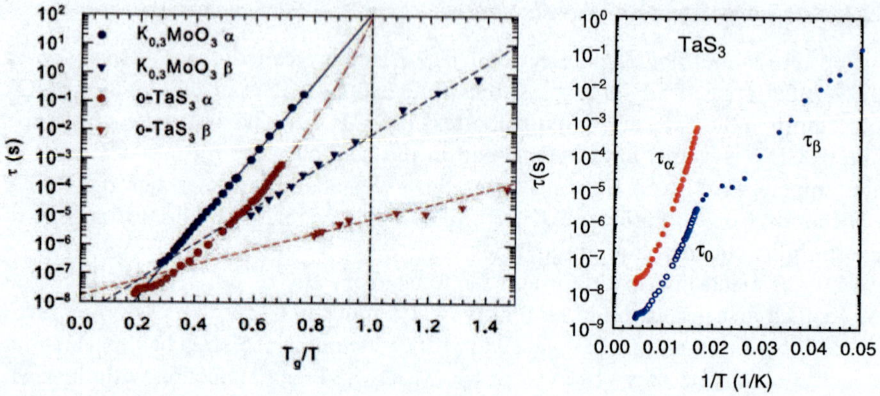

Fig. 323 (*Left*) Characteristic relaxation times τ_α and τ_β of the α and β processes in CDW systems *o*-TaS_3 and $K_{0.3}MoO_3$ as a function of the inverse temperature normalized to the glass transition temperature T_g. T_g is 42 K for *o*-TaS_3 and 23 K for $K_{0.3}MoO_3$. *Full line* represents the activated increase of τ_α in $K_{0.3}MoO_3$ and *dash-dotted line* the Vogel–Fulcher increase of τ_α in *o*-TaS_3. *Dashed lines* represent the activated increase of τ_β in both systems. Reproduced from [1770] with permission. (*Right*) Data of τ_α and τ_β in *o*-TaS_3 (*filled circles*, same as in the *left panel*). The *open circles* are the primitive relaxation time τ_0 in *o*-TaS_3 calculated by Starešinić and Biljaković (unpublished) using the CM relation with the parameters given in the text

coupled through electrostatic interaction. At lower temperatures, in the absence of free carrier screening, this causes slowing down and subsequent freezing of the elastic degrees of freedom, i.e., the α-process. The β-process represents the dynamics of topological defects developed on the domain boundaries. They used the CM relation to calculate the primitive relaxation time τ_0 from the experimental values of τ_α of TaS_3 by using the CM equation $\tau_0 = (t_c)^n(\tau_\alpha)^{1-n}$. They obtained an approximate value of $(1 - n) = 0.67$ from the spectra and took $t_c = 2 \times 10^{-11}$ s, the reciprocal of the real pinning resonance frequency, because above this frequency, CDW is depinned. The calculated values of τ_0 shown in the right panel of Fig. 323 are in agreement with τ_β. Despite the progress made by Starešinić, Biljaković, and coworkers [1768–1770] in the dynamics of CDW system through the analogy to glassformers and the possible connection to the CM, more study has to be made before we can arrive at a full understanding.

3.2.12.2 Aqueous Colloidal Dispersions of Magnetic Nanoparticles

The nanometric magnetic particles dispersed in water have many applications in technology [1772, 1773] and are also systems of fundamental interest as interacting hard spheres in which the inter-particle interactions can be tuned by varying external parameters. One class of these colloidal dispersions is constituted of chemically stable nanocrystals of γ-Fe_2O_3 oxidized in maghemite with diameter of about 10 nm or less dispersed in water and stabilized by electrostatic repulsion [1773]. The last step was achieved by coating the surface of the nanoparticles with citrate molecules, which ensures a negative charge at neutral pH, compensated by sodium counterions. The inter-particle interactions can be continuously tuned by varying the osmotic pressure π through the ionic strength [1774]. With the ionic strength held constant, the viscosity increases with the volume fraction Φ until the samples no longer flow, and the state was referred to as the "macroscopic solid."

Each particle bears a dipole moment $\mu(\sim 10^4\,\mu\text{B})$ and an uniaxial optical anisotropy linked to μ through the magnetic energy of anisotropy of the nanocrystal. In a fluid dispersion, without field, the crystal axes are oriented at random and no optical birefringence is observed. The orientation of the particles via their magnetic moment under an applied field induces a magnetobirefringence Δn in the solution due to the mechanical orientation of the nanoparticle axes. The rotational relaxation of the nanoparticles was studied by the relaxation of the magneto-induced birefringence signal. Indeed, the birefringence response of the suspension to a small external field pulse is a transient orientation followed by a relaxation to a randomly orientated state. If the sample is monodisperse and dilute (independent particles), the relaxation of the birefringence is well described by a linear exponential decay according to Perrin [12] with the relaxation time $\tau_R = 3\eta V_H/kT$, where η is the solvent viscosity and V_H the hydrodynamic volume of the particle. The samples studied by Mériguet et al. [1775, 1776] were polydisperse, and polydispersity induces a distribution of the τ_R, and hence departure from the ideal single exponential decay is observed even for dilute dispersion. For a dilute dispersion of particles with a volume fraction close to $\Phi \sim 0.01$ with a mean diameter of 7 nm, a relaxation time of $\tau = 4\,\mu\text{s}$ was measured. Because of polydispersity ($\sigma = 0.35$) of the sample, the relaxation function is not a single exponential. This can be seen in the right panel of Fig. 324 at low Φ. The data for all Φ are shown in Figs. 324 and 325. On the left panel of Fig. 324 are the normalized optical intensity $I(t)/I_0$ for citrate concentrations of 0.03 mol l^{-1} at increasing volume fraction of nanoparticles. The right panel of Fig. 324 shows the decrease of the Kohlrausch stretch exponent $\alpha \equiv (1-n)$ vs. normalized volume fraction Φ/Φ^*, where $\Phi^* = 0.18$.

On increasing Φ, the relaxation is slowed down. Up to a volume fraction of 11%, the relaxation function and the relaxation time depend weakly on Φ. On further increase of Φ, a dramatic slowing down was observed by the stretching of the relaxation function to longer times (see Fig. 324) and the rapid increase of the relaxation time τ_2 (see left panel of Fig. 325). This occurs before the macroscopic solidification of the sample. The decay of the optical intensity was well fitted by a sum of a single exponential accounting for the short-time relaxation and a stretched exponential of Kohlrausch for the long-time behavior:

$$I(t) = I_1 \exp\left(-\frac{t}{\tau_1}\right) + I_2 \exp\left[-\left(\frac{t}{\tau_2}\right)^{1-n}\right],$$

where $I_1 + I_2 = I_0 = I(t=0)$. The reader may recall this time dependence of the correlation function is the same as in other systems including the aqueous colloidal

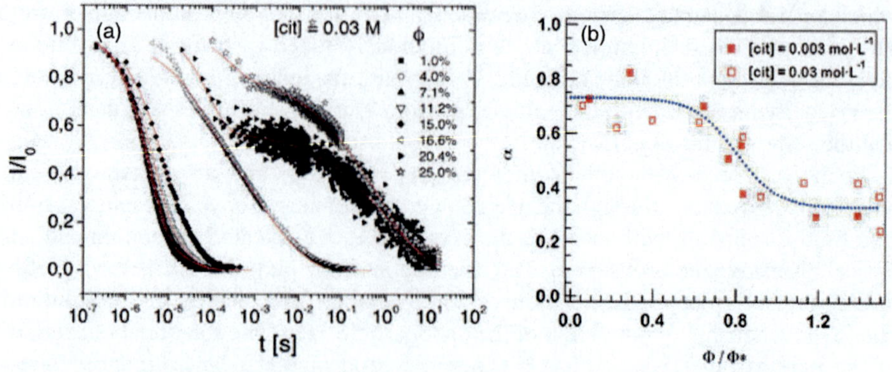

Fig. 324 (*Left*) Normalized optical intensity $I(t)/I_0$ for citrate concentrations of 0.03 mol/l at increasing volume fraction of nanoparticles. Here ϕ is Φ expressed as percentage. (*Right*) Variation of the Kohlrausch stretch exponent $\alpha \equiv (1-n)$ vs. normalized volume fraction Φ/Φ^*, where $\Phi^* = 0.18$. The *dotted line* is guide to the eyes. Reproduced from [1775] by permission

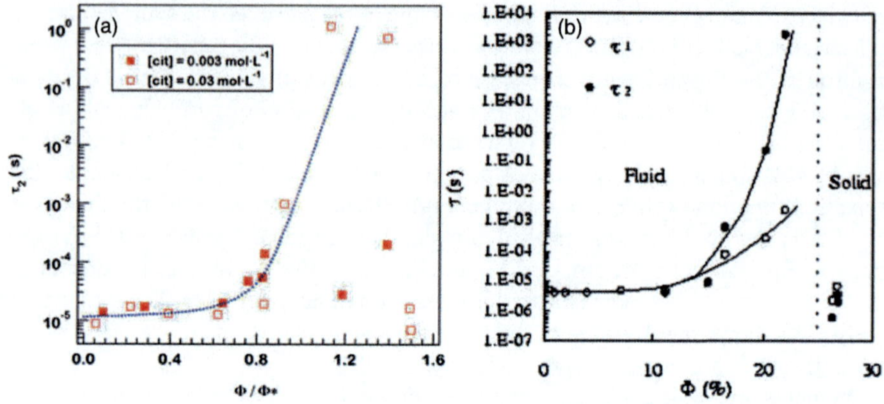

Fig. 325 (*Left*) Variation of the relaxation time τ_2 with the normalized volume fraction Φ/Φ^*, where $\Phi^* = 0.18$. Reproduced from [1775] by permission. (*Right*) Variation of the relaxation times τ_1 and τ_2 with the volume fraction Φ. Reproduced from [1776] by permission

suspensions of a synthetic clay, Laponite (see Section 2.3.2.39), and in associating or aggregating polymers solutions in Section 3.2.3.6.

The exponent $\alpha \neq (1-n)$ of the stretched exponential exhibits a significant decrease starting near the same $\Phi/\Phi^* = 1$ (Fig. 324), where the relaxation time τ_2 also shows the rapid increase (left panel of Fig. 325). This is another example of the correlation between the width of the dispersion (or n) and the relaxation time found in many cases discussed before including the aqueous colloidal suspensions of Laponite and the associating or aggregating polymers solutions. The right panel of Fig. 325 shows variation of both the relaxation times τ_1 and τ_2 with the volume fraction Φ. It can be seen that the separation between the two relaxation times increases as n increases with Φ. Again this trend is found in many other interacting systems.

A more recent paper by Wandersman et al. of the same group [1777] studied the freezing of the orientational degrees of freedom of strongly interacting magnetic and charged nanoparticles by magneto-induced birefringence measurement, as the colloidal glass transition is approached. Using a magneto-induced birefringence technique, they show that the rotational dynamics drastically slows down following a modified Vogel–Fulcher law with the volume fraction Φ replacing T, for Φ above a threshold, the value of which depends on the range of electrostatic repulsion between nanoparticles. The birefringence measurement provides the time correlation of the orientational degrees of freedom of the the nanoparticles, which is given by $C_2(t) =< P_2(\vec{e}_i(t)\cdot \vec{e}_i(0)) >$, where P_2 is the second Legendre polynomial and $\vec{e}_i$ is the direction of the optical axis of the ith nanoparticle. For dilute ferrofluids, the magneto-induced birefringence relaxations are nearly exponential. The relaxation becomes increasingly more non-exponential and stretched to longer times on increasing volume fraction, but at short times it remains exponential component. For this reason, the time dependence of the relaxation was fitted by the function $A\exp(-t/\tau_1) + (1-A)\exp[-(t/\tau_2)^{1-n}]$, like the earlier study by Mériguet et al. [1775, 1776]. The data are shown in Fig. 326 (left). In the fitting process, A, τ_2, and $(1-n)$ are treated as fitting parameters, while the short characteristic time τ_1 is assumed to be constant and set equal to its value at low volume fraction. This latter assumption is different from that of [1776], where τ_1 increases with volume fraction as shown in Fig. 325 (right). The changes of the fitting parameters τ_2 and $(1-n)$ with the volume fraction are reproduced from [1777] in Fig. 326 (right).

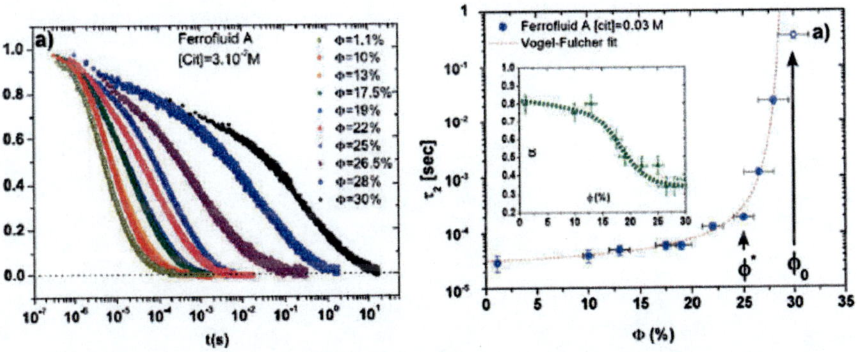

Fig. 326 (*Left*). Normalized birefringence relaxations of ferrofluids A at different volume fractions Φ increasing from left to right. The initial time $t = 0$ is defined when the magnetic field is cut off. (*Right*) Rotational characteristic time τ_2 extracted from the fit of birefringence relaxations as a function of the volume fraction Φ. The *open symbol* is the sample at volume fraction $\Phi = 30\%$ for which no stationary birefringence can be reached after a pulse of magnetic field. The *dashed line* is a Vogel-Fulcher fit of the data. *Inset*: Evolution of the stretched exponent $\alpha \equiv (1-n)$ as a function of the volume fraction. The *dotted line* is a guide for the eye. Reproduced from Ref. [1777] by permission.

Despite the assumption in fitting the data in [1777] that is different from that in [1775, 1776], both experiments essentially show the same universal features of relaxation in other interacting systems. There are more interacting systems exhibiting the universal properties but are not discussed in this volume.

Chapter 4
Afterword

It is hard to believe that a problem in physics introduced at the beginning of the scientific age still remains unsolved till now, the electronic and information age. The glass transition problem is such the case. The long history of the glass transition problem can be appreciated by citing just two among many general phenomena, namely the fractional exponential time dependence of relaxation in glass-forming systems and the super-Arrhenius temperature dependence of the relaxation time. One hundred and forty six years have gone by since the discoveries of the stretched exponential time relaxation function $\phi(t) = \exp[-(t/\tau)^{1-n}]$ by Kohlrausch in 1854 and almost a century since the super-Arrhenius temperature dependence of the relaxation time published by Vogel in 1921. However, till 2010, researchers are still engaged in various attempts to explain these two phenomena [24, 25, 26(a)–26(z)], let alone many other critical experimental facts found in diverse types of glassformers that have accumulated over the intervening years. Evidently at this time there is no universally accepted theoretical or even phenomenological solution of the glass transition problem. The importance of the glass transition problem in condensed matter physics and physical chemistry and the challenge for solution are recognized by Nobel laureates, many eminent scientists, and even recently the *NY Times* in a 2008 editorial for the benefit of the general public. The large number of papers related to the glass transition published in recent years is astounding and is evidence of the awareness of the importance of the problem by the research community. The problem is not only fundamental in physics and chemistry but also occurs in many practical applications involving a broad range of materials and processes. Examples from everyday commonly encountered substances and materials in modern life are water, aqueous mixtures, solvents, polymers, biomolecules, biopolymers, colloids, carbohydrates, pharmaceuticals, biopharmaceuticals, glasses of all kinds (inorganic, organic, polymeric, and metallic), and the more exotic or technologically advanced materials encountered in nanotechnology, biotechnology, energy generation and storage, electronics, and aerospace industries. Solution of the glass transition problem will benefit a broad range of research area as well as technological applications.

So, what is the difficulty that prevents solution of the glass transition problem all these years? The glass transition occurs by changing temperature or pressure, with

K.L. Ngai, *Relaxation and Diffusion in Complex Systems*, Partially Ordered Systems,
DOI 10.1007/978-1-4419-7649-9_4,

concomitant changes in thermodynamic variables including density and entropy. The latter are major factors in causing the change of structural relaxation and transport coefficients, and vitrification follows as one of the consequences. Some glassformers show change of relaxation time or viscosity over 15 orders of magnitude over some modest range of temperature and change in density and entropy. These spectacular effects may have caused many researchers to focus solely on the thermodynamics factors and their changes in constructing theories and models to solve the glass transition problem. This distraction by the change of the structural relaxation with temperature has led many not recognizing that glass transition in part is a many-body relaxation problem. This is because in most glass-forming liquids, the basic structural units have non-trivial or anharmonic interactions with each other, and therefore motion of fundamental units cannot be independent of each other. Holding the temperature (pressure) of a viscous liquid constant before vitrification occurs, the problem is reduced primarily to the many-body relaxation and diffusion. The effects of the many-body relaxation on the structural relaxation must be accounted for before considering the change in temperature (pressure) and combining the two effects on glass transition. From this discussion, it is clear that the glass transition problem involves two major issues and problems. Viable solution of the problem requires adequate account of both issues and appropriately putting them together to render predictions consistent with all experimental facts. Attacking the glass transition problem head on, without the awareness that actually it consists of two problems, will not end up with success. It is imperative to find some way to handle the many-body relaxation problem. But this is easier said than done because many-body relaxation and diffusion is an unsolved problem in statistical mechanics.

Before being exposed to the problem of the relaxation in complex classical systems in 1979 [47, 48, 178], my research activities were on quantum electronic physics. I knew nothing about the glass transition problem. Perhaps this is a blessing in disguise because the situation allows me to focus my attention on the many-body relaxation problem from the start in constructing the coupling model (CM). Although I have not solved the problem, some gains have been made by the concepts introduced and the predictions produced by the CM. The success can be traced back to the construct of the primitive relaxation and the relation between the primitive relaxation time τ_0 and the many-body relaxation time τ. This important CM relation comes from the recognition that the origin of the many-body relaxation can be traced back to phase space dynamics in classical chaos (i.e., non-linear Hamiltonian dynamics that is universal for interacting systems with anharmonic potential), and the fact that classical chaos has an onset time t_c, the magnitude of which depends on the strength of interaction. Also entering this CM relation between τ_0 and τ is the coupling parameter n that appears in the fractional exponent of the Kohlrausch correlation function $\phi(t) = \exp[-(t/\tau)^{1-n}]$. The involvement of n in this CM relation is most fitting because n, or equivalently some measure of the width of the dispersion of the relaxation, is a direct measure of the extent of the many-body involvement in the relaxation. This is intuitively obvious since in the limit of $n \to 0$, the Kohlrausch correlation function becomes the exponential Debye relaxation, and $\tau \to \tau_0$ follows from the CM equation. The CM equation spawns many predictions which have been successfully tested against observed properties of glass-forming liquids. But this is not the only major thrust of this work. Instead, throughout this work, the emphasis is also on the volume of diverse experimental data and facts to convince the reader that many-body dynamics is ubiquitous in glass-forming liquids and cannot be ignored or dealt with lightly in constructing theories or models of glass transition. This mission is accomplished by demonstrating in many different ways and cases that n (or the width of the dispersion) either governs or correlates with the observed properties of the structural relaxation. In most if not all current theories and models of glass transition, the dispersion of the structural relaxation either is not considered at all or if considered bears no significant relation to the structural relaxation time. Thus, they are definitely incomplete as solution of the glass transition problem and risk the danger of contradicting the experimental facts and being proven incorrect. This is one of important message that this work delivers. It should be considered as an objective deduction, entirely from experiments, and not biased by the view of the author and the CM. Notwithstanding, it must be said that the predictions of the CM are fully consistent with the experiments.

The evidences presented are convincing that I do not see how future attempts to solve the glass transition problem can ignore taking into account of the many-body dynamics or at least having incorporated either n or the width of the dispersion to play the important role. So far in the published literature, no theory of glass transition, except the CM, has given n such an important role as required by experimental findings.

Another major thrust of the work is the demonstration from experimental data and facts of the connections that the faster dynamic processes have with the structural relaxation. The secondary relaxations belonging to a special class show this connection through many different ways. Secondary relaxation involving the motion of the entire basic structural unit of the glassformer belongs to this class. To the best of my knowledge, the possibility of the existence of secondary relaxation in the special class strongly connected to the primary structural relaxation was first brought out by the 1998 paper [251]. There it was established in general for glassformers of different types and at any fixed value of the primary α-relaxation time τ_α that the separation of the secondary β-relaxation time τ_β from τ_α in logarithmic scale is proportional to n or the width of the dispersion of the α-relaxation. The search for this correlation was motivated by the CM in which the primitive relaxation is strongly connected to the structural relaxation because it is the precursor of the evolution in the many-body dynamics with time, leading ultimately to the structural α-relaxation. Moreover, the primitive relaxation is a local independent relaxation like secondary relaxation that involves the motion of the entire unit, and $\log(\tau_\alpha/\tau_0)$ is rigorously proportional to n according to the CM. Since 1998, many other experimental results have found from various properties the existence of inseparable connection between secondary relaxation of a special kind and the structural α-relaxation, and approximate agreement of the relaxation times $\tau_0 \approx \tau_{JG}$ in diverse kinds of glassformers (see the multiple items in Section 2.3).

Since not all observed secondary relaxations belong to a special class, there is a need to distinguish them from the intramolecular secondary relaxations which involve the motion of a part of the basic structural unit and do not have any such found connections with the α-relaxation. This was not done in the 1998 paper where they are simply called β-relaxation [251], but in later publications I have chosen the term "Johari–Goldstein β-relaxation" for the secondary relaxations belonging to this special class. There is a good reason for this choice because of the important discovery by these two colleagues, Johari and Goldstein, in establishing the existence of secondary relaxation in totally rigid small molecular glassformers and the fact that the secondary relaxation observed must involve the motion of the entire basic structural unit like the primitive relaxation of the CM. Some polymers with no side group like 1,2- and 1,4-polybutadienes and polyvinylchloride also have secondary relaxation that must involve the motion of the entire repeat unit, and they send effectively the same message as the findings of Johari–Goldstein (JG) in small molecular glassformers. This is perhaps the reason why some colleagues object to the usage of the term JG β-relaxation by me and my collaborators. Our use of the term JG β-relaxation may have caused some workers to identify the interpretation of secondary relaxations by Johari or Goldstein with our own interpretation. This should not be done, because their interpretations are not the same as ours. To avoid confusion, I have used both the primitive relaxation and the JG β-relaxation at the same time in various places. Currently, the term JG β-relaxation is widely used by other workers for other purposes or interpretations. One example is the use to distinguish the caged fast process of mode coupling theory also called β-relaxation from normal secondary β-relaxation. With due respect to others for their own use of the same term JG β-relaxation, it is important not to mix up the meaning of the same term by others with ours [685].

The importance of the primitive or the JG β-relaxation is not only in its connection to the structural relaxation but also that it has properties similar to those of the structural relaxation. There are a number of such similar properties. Here we mention only one, namely that the JG β-relaxation time is also density and entropy dependent and more importantly its dependence is on exactly the same product variable TV^{γ} as the α-relaxation time. Recent experiments and simulations employing combination of temperature and elevated pressure have brought this result out clearly. Since the primitive or the JG β-relaxation transpires before the α-relaxation, from causality it is possible that the origin of the TV^{γ} dependence of τ_{α} may have originated from the TV^{γ} dependence of τ_0 or τ_{JG}. The many-body nature of the α-relaxation involving participation of more structural units naturally makes τ_{α} to have a stronger TV^{γ} dependence compared to τ_0 or τ_{JG}. Conventional theories and models of glass transition take into account of the dependence on density and entropy exclusively for the structural relaxation and its relaxation time τ_{α}. The finding that τ_0 or τ_{JG} is also dependent on exactly the same TV^{γ} variable as τ_{α} is an important one. Theories and models of glass transition published in the literature either do not consider secondary relaxation at all or if considered bear no relation to the α-relaxation.

In addition to be fundamental for consideration of glass transition, the primitive or JG β-relaxation has impact on processes in various systems of practical importance such as avoidance of crystallization of amorphous drugs in pharmaceutical industry, function of hydrated biomolecules, bioprotection of living organism, stabilization of lyophilized biopharmaceuticals, enhancement of shelf life of products in food industry, efficiency of portable energy storage materials and fuel cell materials for green energy production, improved performance of various complex materials and devices in engineering, etc. The basic features of dynamics of the systems in these applied fields are isomorphic to glassformers, which is unsurprising because they are all interacting systems and the same physics of many-body relaxation and diffusion basically governs their behavior or properties. The bonus this confers is that the improved understanding of the relaxation and diffusion properties of prototypical glass-forming systems and substances gained by the presentations in this book can help to make progress in these applied research fields, which are important for the populace and the economy worldwide. I do expect that this cross-field fertilization will bring more benefits to applied research areas in the future.

At times short compared with the primitive or the JG β-relaxation times, naturally motion of the entire basic structural unit has not yet transpired, and all structural units of the glassformer are localized within "cages" defined by the intermolecular potential. The cages start to decay with the onset of local relaxation, and the caged dynamics regime ends at times comparable to the primitive time τ_0 or the JG β-relaxation time τ_{JG}. Thus, the primitive or the JG β-relaxation performs yet another important role of terminating the shorter time regime of caged dynamics. From this connection, the susceptibility of the caged dynamics correlates with τ_0 or τ_{JG}. This, together with the correlation of τ_0 or τ_{JG} with n of the structural α-relaxation, leads to the correlation between the susceptibility of the caged dynamics and n or other characteristics of the α-relaxation that correlates with n such as "fragility." There are plenty of experimental supports of these correlations. No mention of the Boson peak is made throughout the text because this vibrational feature of the glassy state is unrelated to glass transition. Any link between the properties of the Boson peak and structural α-relaxation comes naturally from the fact that both depend on the interatomic or the intermolecular potential.

Many-body relaxation and diffusion should be occurring generally in all interacting systems and share the same basic characteristics. The glass-forming liquids amply discussed in Chapter 2 belong to a special class of interacting systems. The immense number of glassformers studied by a large variety of experimental techniques in the past century provides the richest collection of data that allow us to extract the fundamental characteristics of relaxation and diffusion in this class of interacting systems. Notwithstanding, if the physics of relaxation and diffusion in interacting systems were general, analogues of the fundamental characteristics of glass-forming liquids and glasses should be found in relaxation and diffusion processes of other interacting systems that have nothing to do with glass transition. The main thrust of the last chapter is to present from experiments exact analogues of the fundamental characteristics of glass-forming liquids and glasses in other interacting systems. Although in some complex interacting systems, extraneous factors may obscure direct observation of the universal characteristics, they should be present. This is one of the important points made in the book because it establishes that the manifestations of many-body relaxation and diffusion are fundamental and universal, independent of the chemical and physical structures of the interacting system. It points out that many-body relaxation and diffusion is a fundamentally important problem in physics that impacts many branches of science. The problem is still unsolved and deserves much attention. Considering many of the substances and systems discussed in this volume are so complex and widely different in chemical composition and physical structure, it is remarkable that universal properties of relaxation and diffusion exist and are shared by all of them. It seems to be the consequence of another unexplored fundamental physical law of Nature, in this case of irreversible processes in interacting systems.

Furthermore, the precepts and predictive relations of the CM have been found applicable to relaxation and diffusion processes of the other interacting systems. They also have the primitive relaxation $\phi(t) = \exp(-(t/\tau_0))$ at shorter times and the Kohlrausch stretched exponential relaxation $\phi(t) = \exp[-(t/\tau)^{1-n}]$ at longer times, with the former crossing over to the latter at some time t_c determined by the strength of the interaction. The parameter n appearing in the exponent of the Kohlrausch function measures the extent of the many-body dynamics. The CM equation $\tau = [t_c^{-n}\tau_0]^{1/(1-n)}$ holds as well. From this relation, it follows that n either governs or correlates with the properties of τ, as found experimentally in the relaxation and diffusion processes of other interacting systems. In view of the success the CM has in correlating and explaining experimental facts, its constructs can be used as foundation as well as guide for anyone in attempting to solve the many-body relaxation problem rigorously in the future.

The principal purpose of this work is to convince the research community especially those engaged in glass transition that many-body relaxation and diffusion in interacting systems is an indispensable part of the problem. Without solution for it, or at least some practical way to account for the effects, the glass transition problem will never be completely settled. To make this convincing, numerous results from experimental investigations and computer simulations are presented to show the ubiquitous effects originating from many-body relaxation. The fabric of this work is a large collection of experimental facts presented together in a coherent manner, and the message it conveys can be judged for objectivity by the reader. This has not been done before in this field of research. Perhaps the reason why I was able to do so is the help given by the CM. It anticipates universal behavior of many-body relaxation and diffusion in all interacting systems in general, and it also has the precepts and predictions to look for them in experimental data. Ultimately, credit goes to all the experimentalists who labored long and hard to obtain these valuable clues on the characteristics of many-body relaxation and diffusion in various interacting systems. In relation to experiments, my contribution is merely putting their results together to show unity, simplicity, and beauty coming out from variety, disorder, and complexity. Newton (1676) acknowledged the importance of works of scientists before him by the statement: "If I have seen a little further, it is by standing on the shoulders of Giants." If I am allowed to modify this famous statement of Newton to express my appreciation of the works of the experimentalists that have made possible my small contribution to science, I would offer the following:

If I have seen a little further, it is by constantly looking behind the shoulders of the experimentalists.

I certainly agree that the premise of the CM has not yet been developed to become a full theory. However, contribution to progress in natural science is not limited to construction of a theory and also includes experimental or empirical discovery, keen observation, and basic ideas on which theory with mathematical rigor can be built upon. These contributions are vital in the field of glass transition alone because at present no existing theory can fully explain all the experimental facts and is universally accepted as the solution. Lest one forgets, the field of glass transition is experimentally driven by the discovery of properties and phenomena in myriad of materials and systems of diverse structures over centuries and continued in the future. As for the dismissal of the CM for serving no purpose except fitting data by some individuals with exquisite style of doing research, I shall leave the readers of this book to make their own judgment after having acquainted themselves with the many predictions of the CM with confirmations by experiments, as well as applications to a variety of research areas, described in the chapters. The approach of the CM is in accord with that recommended by Richard Feynman, as he explained in his book *The Character of Physical Law*: "In general, we look for a new law by the following process. First we guess it. Then we compute the consequences of the guess to see what would be implied if this law that we guessed is right. Then we compare the result of the computation to nature, with experiment or experience, compare it directly with observation, to see if it works. If it disagrees with experiment it is wrong. In that simple statement is the key to science." Thus far, all the predictions of the CM work suggesting a new physical law have been found. Naturally the construction of rigorous mathematical theory (not necessarily by me) to firmly establish the physical law is the next step. The CM also satisfies the criterion of scientific research given by Karl R. Popper: "One can sum up all this by saying that the criterion of *the scientific status of a theory is its falsifiability, or refutability, or testability*," originally published in his book *Conjectures and Refutations* (1963). Out of seven points he put forth to arrive at this criterion, two are most relevant to the CM. The following are excerpts of these two points: "Confirmations should count only if they are the result of *risky predictions*; ..." and "Every genuine *test* of a theory is an attempt to falsify it, or to refute it. Testability is falsifiability; but there are degrees of testability: some theories are more testable, more exposed to refutation, than others; they take, as it were, greater risks." From the numerous comparisons with various experiments discussed in the chapters, I submit that it is justified to claim that the CM has taken greater risks. Most risky are the predictions of the CM made before the experimentalists went out to test them. Some experimentalists can confirm that certain experimental results indeed verify the CM predictions made before the experiments were carried out.

References

1. R. Brown, Edinb. Phil. J. **5**, 358 (1828); Philos. Mag. **4**, 161 (1828); See also A. Fick, Poggendorff's Annalen **94**, 59 (1855), and (in English) Philos. Mag. S.4, **10**, 30–39 (1855).
2. R. Kohlrausch, *Theorie des elektrischen Rückstandes in der Leidener Flasche*; von R. Kohlrausch, Pogg. Ann. Phys. Chem. **91**, 179–214 (1854). For experimental data, see R. Kohlrausch, Pogg. Ann. Phys. Chem. **91**, 56 (1854).
3. F. Kohlrausch, *Ueber die elastische Nachwirkung bei der Torsion*, Pogg. Ann. Phys. Chem. **119**, 337–368 (1863). *Beiträge zur Kenntniss der elastischen Nachwirkung*, Pogg. Ann. Phys. Chem.**128**, 1–20, 207–228, 399–419 (1866).
4. F. Kohlrausch, *Beiträge zur Kenntniss der elastischen Nachwirkung*, Pogg. Ann. Phys. Chem. **128**, 1–20, 207–228, 399–419 (1866).
5. J.C. Maxwell, Philos. Trans. R. Soc. (London) **157**, 49 (1867).
6. J.P. Joule, Mem. Manchr Literary Philos. Soc., 3rd ser., **3**, 292 (1867); The Scientific Papers of J. P. Joule, Vol. **1**, Physical Society, London, p. 558 (1884).
7. S.V. Nemilov, G.P. Johari, Philos. Mag. **83**, 3117 (2003).
8. A. Einstein, *Investigations on the Theory of Brownian Movement*, Dover, New York, NY (1956).
9. J. Stachel, Ed., *The Collected Papers of Albert Einstein*, Vol. **2**, pp. 170–182, 206–222, Princeton University Press (1987).
10. G.G. Stokes, *On the Effect of the Internal Friction of Fluid on the Motion of Pendulums*, Proc. Camb. Philos. Soc. **9**, 8–106 (1851).
11. A. Einstein, Ann. d. Physik **17**, 549 (1905).
12. J. Perrin, *Le Mouvement Brownien et la Réalité Moleculaire*, Ann. Chim. Phys. (8me Serie), 5–114 (1909); J. Phys. Radium **5**, 497 (1934); *Atoms*, Constable, London (1916). See also R. Newburgh, J. Peidle, W. Rueckner, *Einstein, Perrin, and the Reality of Atoms:1905 Revisited*, Am. J. Phys. **74**, 478 (2006).
13. M.V. Smoluchowski, Ann. d. Physik. **21**, 756 (1906).
14. P. Langevin, *Sur la Theorie du Mouvement Brownien*, C. R. Acad. Sci. (Paris) **146**, 530 (1908). In English, Am. J. Phys. **65**(11), 1079 (1997).
15. R. Kubo, M. Toda, N. Hashitsumi, *Statistical Physics II, Nonequilibrium Statistical Mechanics*, Springer, Berlin (1985).
16. F. Perrin, J. Phys. Radium **5**, 497 (1934).
17. A. Einstein, Ann. d. Physik **19**, 371 (1906).
18. P. Debye, Physik. Z. **13**n, 97 (1912).
19. P. Debye, Ber. Deut. Phys. Ges. **55**, 777 (1913).
20. P. Debye, *Polar Molecules*, Dover, New York, NY (1929).
21. S. Chandrasekhar, Rev. Modern Phys. **15**, 1 (1943).
22. J.K.G. Dhont, *An Introduction to Dynamics of Colloids*, Elsevier, Amsterdam (1996).
23. B.J. Berne, R. Percora, *Dynamic Light Scattering*, Wiley, New York, NY (1976).

K.L. Ngai, *Relaxation and Diffusion in Complex Systems*, Partially Ordered Systems, DOI 10.1007/978-1-4419-7649-9, © Springer Science+Business Media, LLC 2011

24. K. Chang, *The Nature of Glass Remains Anything but Clear* (The New York Times, July 29, 2008). See a technical response to this New York Times article entitled *The Nature of Glass: Somethings are Clear*, by K.L. Ngai, S. Capaccioli, D. Prevosto, M. Paluch, published in *Metastable Systems under Pressure*, NATO Science for Peace and Security Series-A: Chemistry and Biology, S. Rzoska, A. Drozd-Rzoska, V. Mazur, Eds., Springer, pp. 3–30 (2010).
25. (a) P.W. Anderson, *Through the Glass Lightly*, Science **267**, 1616 (1995); (b) R.G. Palmer, D.L. Stein, E. Abrahams, P.W. Anderson, Phys. Rev. Lett. **53**, 958 (1984); (c) M. Cardona, R.V. Chamberlin, W. Marx Ann. Phys. (Leipzig) **16**(12), 842 (2007). These authors assert that the original paper of Kohlrausch was forgotten until it was cited by Palmer et al. in 1984 [25(b)]. They went on to say that unfortunately Palmer et al. [25(b)] did not refer to the correct 1854 paper by R. Kohlrausch but to a wrong one his published in 1847 [Pogg. Ann. Phys. Chem. **72**, 353 (1847)]. Actually I and my coworkers have to share the blame for incorrectly citing the 1847 paper of Kohlrausch in our 1983 paper [K.L. Ngai et al., Phys. Rev. B **28**, 6073 (1983).]. This paper published one year earlier than Palmer et al. [25(b)], as well as our interaction with Palmer and Stein at the *Workshop on Relaxation in Disordered Systems* that I organized in 1983, might have misled Palmer et al. to follow suit in citing the wrong 1847 paper of Kohlrausch. Incidentally in their paper [25(b)], Palmer et al. did cite the Proceedings of the 1983 Workshop I organized as well as my first publication of the Coupling Model in 1979 [K.L. Ngai, Comment Solid State Phys. **9**, 127 (1979).].
26. (a) P.-G. de Gennes, *A Simple Picture for Structural Glasses*, C. R. Physique **3**, 1263 (2002); (b) G.H. Frederickson, H.C. Anderson, J. Chem. Phys. **83**, 5822 (1985); (c) T.R. Kirkpatrick, D. Thirumalai, P.G. Wolynes, Phys. Rev. A **40**, 1045 (1989); (d) J.Y. Cavaille, J. Perez, G.P. Johari, Phys. Rev. B **39**, 2411 (1989); (e) W. Götze, L. Sjögren, Rep. Prog. Phys. **55**, 241–376 (1992); (f) J.C. Philips, Rep. Prog. Phys. **59**, 1133 (1996); (g) D. Kivelson, G. Tajus, J. Non-Cryst. Solids **235–237**, 86 (1998); (h) F. Stillinger, Phys. Rev. E **59**, 48 (1999); (i) M. Mézard, G. Parisi, J. Phys. Condens. Matter. **12**, 6655 (2000); (j) S. Franz, M. Mézard, G. Parisi, J. Stat. Phys. **97**, 459 (1999); (k) J. Grigera, A. Cavagna, I. Giardina, G. Parisi, Phys. Rev. Lett. **88**, 55502 (2002); (l) J. Bendler, J. Fontanella, M.F. Shlesinger, Chem. Phys. **284**, 311 (2002); (m) F. Ritort, P. Sollich, Adv. Phys. **52**, 219 (2003); (n) C. Toninelli, G. Biroli, D.S. Fisher, Phys. Rev. Lett. **92**, 185504 (2004); (o) A.V. Granato, V.A. Khonik, Phys. Rev. Lett. **93**, 155502 (2004); (p) J.P. Garrahan, D. Chandler, Proc. Natl. Acad. Sci. U.S.A. **100**, 9710 (2003); L. Berthier, D. Chandler, J. Garrahan, Europhys. Lett. **69**, 320 (2005); (q) M. Tokuyama, I. Oppenheim, Phys. Rev. E **50**, R16 (1994); Physica A **216**, 85 (1995); (r) J. Dudowicz, K.F. Freed, J.F. Douglas, J. Phys. Chem. B **109**, 21350 (2005); Adv. Chem. Phys. **137**, 125 (2008); (s) A. Widmer-Cooper, P. Harrowell, Phys. Rev. Lett. **96**, 185701 (2006); (t) P. Mayer, K. Miyazaki, D.R. Reichman, Phys. Rev. Lett. **97**, 095702 (2006); (u) V. Lubchenko, P.G. Wolynes, Annu. Rev. Phys. Chem. **58**, 235 (2007); J.D. Stevenson, P.G. Wolynes, Nat. Phys. **6**, 62 (2010); (v) E.J. Saltzman, K.S. Schweizer, J. Phys. Condens. Matter **19**, 205123 (2007); (w) G. Biroli, J.-P. Bouchaud, J. Phys. Condens. Matter **19**, 205101 (2007); (x) J.S. Langer, Phys. Today, February 8 (2007), Phys. Rev. E **78**, 051115 (2008); J.S. Langer, S. Mukhopadhyay, Phys. Rev. E **77**, 061505 (2008); (y) N. Xu, M. Wyart, A.J. Liu, S.R. Nagel, Phys. Rev. Lett. **98**, 175502 (2007); (z) U. Buchenau, J. Chem. Phys. **131**, 074501 (2009), Phys. Rev. B **80**, 172201 (2009).
27. A.K. Doolittle, D.B. Doolittle, J. Appl. Phys. **28**, 901 (1957).
28. (a) M.H. Cohen, D. Turnbull, J. Chem. Phys. **31**, 1164 (1959); (b) M.H. Cohen, G.S. Grest, Phys. Rev. B **20**, 1077 (1979).
29. J.D. Ferry, *Viscoelastic Properties of Polymers*, 3rd ed., Wiley, New York, NY (1980).
30. G. Adam, J.H. Gibbs, J. Chem. Phys. **43**, 139 (1965).
31. W. Götze, in *Liquids, Freezing and the Glass Transition*, Proceedings of the Les Houches Summer School of Theoretical Physics, Session LI (1989), J.-P. Hansen, D. Levesque, J. Zinn-Justin, Eds., North-Holland, Amsterdam, p. 287 (1991).

32. W. Götze, J. Phys. Condens. Matter **11**, A1–A45 (1999); W. Götze, *Complex Dynamics of Glass-Forming Liquids*, International Series of Monographs on Physics 143, Oxford Science Publications, Oxford (2009).
33. W. Götze, M. Sperl, Phys. Rev. E. **66**, 011405 (2002).
34. W. Götze, M. Sperl, Phys. Rev. Lett. **92**, 105701 (2004).
35. M. Sperl, Phys. Rev. E **74**, 011503 (2006).
36. M. Goldstein, J. Chem. Phys. **51**, 3728 (1969).
37. C.A. Angell, Science **267**, 1924 (1995).
38. S. Sastry, P.G. Debenedetti, F.H. Stillinger, Nature **393**, 554 (1998).
39. P.G. Debenedetti, F.H. Stillinger, Nature **410**, 259 (2001).
40. D.J. Wales, *Energy Landscapes*, Cambridge University Press, Cambridge (2003).
41. P.G. de Gennes, J. Chem. Phys. **55**, 572 (1971).
42. M. Doi, S.F. Edwards, J. Chem. Soc. Faraday Trans. II **14**, 1789, 1802, 1818 (1978).
43. W.W. Graessley, Adv. Polym. Sci. **47**, 67 (1982).
44. M. Doi, S.F. Edwards, *The Theory of Polymer Dynamics*, Clarendon Press, Oxford (1986).
45. T.C.B. McLeish, Adv. Phys. **51**, 1379 (2002).
46. J.C. Dyre, T.B. Schrøder, Rev. Mod. Phys. **72**, 873 (2000).
47. K.L. Ngai, Comment Solid State Phys. **9**, 127 (1979).
48. K.L. Ngai, Comment Solid State Phys. **9**, 141 (1979).
49. L. Boltzmann, Pogg. Ann. Physik **7**, 108 (1876).
50. R. Balescu, *Equilibrium and Nonequilibrium Statistical Mechanics*, Wiley, New York, NY (1975).
51. H.B. Callen, T.A. Welton, Phys. Rev. **83**, 34 (1951).
52. I.L. Hopkins, R.W. Hamming, J. Appl. Phys. **28**, 906 (1957).
53. J.E. McKinney, H.V. Belcher, J. Res. Nat. Bur. Stand. **A67**, 43 (1963).
54. P.B. Macedo, C.T. Moynihan, R. Bose, Phys. Chem. Glasses **13**, 171 (1972).
55. F.S. Howell, R.A. Bose, P.B. Macedo, C.T. Moynihan, J. Phys. Chem. **78**, 639 (1974).
56. (a) For a review of electric modulus, see I.M. Hodge, K.L. Ngai, C.T. Moynihan, J. Non-Cryst. Solids **351**, 104 (2005); (b) K.L. Ngai, C. León, Phys. Rev. B, **60**, 9396 (1999); (c) K.L. Ngai, R.W. Rendell, Phys. Rev. B, **61**, 9393 (2000).
57. R. Richert, H. Wagner, Solid State Ion. **105**, 167 (1998).
58. N.O. Birge, S.R. Nagel, Phys. Rev. Lett. **54**, 2674 (1985).
59. M. Beiner, J. Korus, H. Lockwenz, K. Schröter, E. Donth, Macromolecules **29**, 5183 (1996).
60. M. Beiner, S. Kahle, E. Hempel, K. Schröter, E. Donth, Europhys. Lett. **44**, 321 (1998).
61. R. Greiner, F.R. Schwarzl, Rheol. Acta, **23**, 378 (1984).
62. A.J. Kovacs, Fortschr. Hochpolym.-Forsch. **3**, 394 (1964).
63. R. Richert, J. Chem. Phys., **113**, 8404 (2000).
64. C. Bauer, R. Böhmer, S. Moreno-Flores, R. Richert, H. Sillescu, D. Neher, Phys. Rev. E **61**, 1755 (2000).
65. R. Kubo, J. Phys. Soc. Jpn, **12**, 570 (1957).
66. N.H. March, M.P. Tosi, *Atomic Dynamics in Liquids*, McMillan, London (1976).
67. S.W. Lovesey, *Condensed Matter Physics: Dynamic Correlations*, Benjamin, Reading, MA (1980).
68. J.P. Boon, S. Yip, *Molecular Hydrodynamics*, McGraw Hill, New York, NY (1980).
69. J.-P. Hansen, I.R. McDonald, *Theory of Simple Liquids*, Academic, London (1986).
70. S. Dattagupta, *Relaxation Phenomena in Condensed Matter Physics*, Academic, Orlando, FL (1987).
71. G.D.J. Phillies, *Elementary Lectures in Statistical Mechanics*, Springer, New York, NY (2000).
72. R. Kubo, Rep. Prog. Phys. **29**, 255 (1966).
73. R. Zwanzig, Ann. Rev. Phys. Chem. **16**, 67 (1965).
74. S.H. Glarum, J. Chem. Phys. **33**, 639 (1960).

75. R.H. Cole, Ann. Rev. Phys. Chem. **40**, 1 (1989).
76. G. Williams, Adv. Polym. Sci. **33**, 60 (1979).
77. G. Williams, D.C. Watts, in *NMR Basic Principles and Progress*, P. Diehl, E. Flick, E. Kosfeld, Eds., Springer, Berlin, Vol. **4**, p. 271 (1971).
78. D.M.F. Edwards, P.A. Madden, I.R. McDonald, Mol. Phys. **51**, 1141 (1984).
79. H. Takeuchi, R. Roe, J. Chem. Phys. **94**, 7446 (1991).
80. D. Bedrov, G.D. Smith, Macromolecules **38**, 10314 (2005).
81. B. Chu, *Laser Light Scattering*, Academic, New York, NY (1974).
82. B.J. Berne, R. Pecora, *Dynamic Light Scattering*, Wiley, New York, NY (1976).
83. C.H. Wang, *Spectroscopy of Condensed Media*, Academic, New York, NY (1985).
84. R. Pecora, Ed., *Dynamic Light Scattering*, Academic, New York, NY (1986).
85. (a) G.D. Patterson, in Ref. [84], Chapter 6, p. 260; (b) C.P. Lindsey, G.D. Pattersn, J.R. Stevens, J. Polym. Sci. Part B. Polym. Phys. **17**, 1547 (1979); (c) C.P. Lindsey, G.D. Patterson, J. Chem. Phys.**73**, 3348 (1980).
86. W. Brown, *Light Scattering: Principles and Development*, Clarendon Press, Oxford (1996).
87. K.L. Ngai, G. Floudas, D.J. Plazek, A.K. Rizos, in *Encyclopedia of Polymer Science and Technology*, Wiley, New York, NY (2002).
88. C.H. Wang, Mol. Phys. **41**, 541 (1980).
89. K.M. Zero, R. Pecora, Macromolecules, **15**, 87 (1982).
90. A.K. Rizos, K.L. Ngai, Phys. Rev. B **46**, 8126 (1992).
91. G. Petekidis, D. Vlassopoulos, G. Fytas, R. Ruelkens, G. Wegner, Macromolecules **31**, 6129 (1998).
92. A. Abragam, *The Principles of Nuclear Magnetism*, Clarendon Press, Oxford (1962).
93. K. Schmidt-Rohr, H.W. Spiess, *Multidimensional Solid State NMR and Polymers*, Academic, London (1994).
94. R. Böhmer, G. Diezemann, G. Hinze, E. Rössler, Prog. Nucl. Mag. Reson. Spectrosc. **39**, 191 (2001).
95. S.W. Lovesey, *Theory of Neutron Scattering from Condensed Matter*, Oxford University Press, Oxford (1994).
96. (a) T. Springer, *Quasielastic Neutron Scattering for the Investigation of Diffusive Motions in Solids and Liquids*, Springer Tracts in Modern Physics, Vol. **64**, Springer, Berlin (1972); (b) F. Mezei, in *Liquids, Freezing and Glass Transition*, J.P. Hansen, D. Levesque, J. Zinn-Justin, J., Eds., North Holland, Amsterdam, p. 629 (1991).
97. L. Van Hove, Phys. Rev. **95**, 249 (1954).
98. A. Rahman, Phys. Rev. A **136**, 405 (1964).
99. M.S. Green, J. Chem. Phys. **19**, 1036 (1951).
100. H. Nyquist, Phys. Rev. **32**, 110 (1928).
101. R. Kubo, Rep. Progr. Phys. **29**, 255 (1966). Cited before as [72].
102. H.K. Onnes, Commun. Phys. Lab. Univ. Leiden, Nos. **119, 120, 122** (1911).
103. J. Bardeen, L.N. Cooper, J.R. Schrieffer, Phys. Rev. **106**, 162 (1957).
104. See articles in Ann. N.Y. Acad. Sci. **279** (1976), **371** (1981), **484** (1986).
105. G.P. Johari, M. Goldstein, J. Chem. Phys. **53**, 2372 (1970).
106. N.G. McCrum, B.E. Read, G. Williams, *Anelastic and Dielectric Effects in Polymeric Solids*, Wiley, New York, NY (1991).
107. See articles in *Relaxation in Complex Systems*, K.L. Ngai, G.B. Wright, Eds., National Technical Information Service, US Department Commerce, Springfield, VA (1984).
108. G.P. Johari, in *Plastic Deformation of Amorphous and Semicrystalline Materials*, Les Houches Lectures (Les Editions de Physiques, France, 1982), p. 109 (1982).
109. See articles in *Proceedings of the 1st, 2nd , 3rd, 4th and 5th International Discussion Meetings on Relaxations in Complex Systems*, J. Non-Cryst. Solids **131–133** (1991), **172–174** (1994), **235–237** (1998), **307–310** (2002), **352**, issues 42–49 (2006).
110. See articles in *Disorder Effects on Relaxational Processes*, R. Richert, A. Blumen, Eds., Springer, Berlin (1994), in particular pp. 89–150 of the article by K.L. Ngai entitled *Universal Patterns of Relaxations in Complex Correlated Systems.*

111. K.L. Ngai, Evidences for Universal Behavior of Condensed Mater at Low Frequencies/Long Times, in *Non-Debye Relaxation in Condensed Matter*, T.V. Ramakrishnan, M. Raj Lakshmi, Eds., World Scientific, Singapore, pp. 23–193 (1987).
112. R. Böhmer, K.L. Ngai, C.A. Angell, D.J. Plazek, J. Chem. Phys. **99**, 4201 (1993).
113. See papers in *Supercooled Liquids, Advances and Novel Applications*; *ACS Symposium Series*, Vol. **676**, J.T. Fourkas, D. Kivelson, U. Mohanty, K. Nelson, Eds., American Chemical Society, Washington, DC (1997).
114. C.A. Angell, K.L. Ngai, G.B. McKenna, P.F. McMillan, S.W. Martin, J. Appl. Phys. **88**, 3113 (2000).
115. (a) K.L. Ngai, J. Non-Cryst. Solids **275**, 7 (2000); (b) K.L. Ngai, C.T. Moynihan, *The Dynamics of Mobile Ions in Ionically Conducting Glasses and Other Materials*, Bull. Mater. Res. Soc. **23** (11), 51 (1998); (c) K.L. Ngai, IEEE Trans. Dielectrics Electrical Insulations **8**, 329 (2001).
116. P.G. Debenedetti, F.H. Stillinger, Nature **410**, 259 (2001).
117. See articles in *Adv. Chem. Phys. Part B, Fractals, Diffusion and Relaxation in Disordered Complex Systems*, Y.P. Kalmykov, W.T. Coffey, S.A. Rice, Eds., Vol. **133**, Wiley, New York, NY (2006).
118. P.G. Debenedetti, *Metastable Liquids: Concepts and Principles*, Princeton University Press, Princeton, NJ.
119. C.M. Roland, S. Hensel-Bielowka, M. Paluch, R. Casalini, Rep. Prog. Phys. **68**, 1405 (2005).
120. K.L. Ngai, R. Casalini, S. Capaccioli, M. Paluch, C.M. Roland, *Adv. Chem. Phys. Part B, Fractals, Diffusion and Relaxation in Disordered Complex Systems*, Y.P. Kalmykov, W.T. Coffey, S.A. Rice, Eds., Chapter 10, Wiley, New York, NY, Vol. **133**, pp. 497–585 (2006).
121. E. Donth, *The Glass Transition*, Springer Series in Material Science, Vol. **48**, Springer, Berlin, Heidelberg (2001).
122. See articles in *Structure, Dynamics and Properties of Silicate Melts*, Reviews in Mineralogy, J.F. Stebbins, P.F. McMillan, D.B. Dingwell, Eds., Mineralogical Society of America, Washington, DC, Vol. **32** (1995).
123. S.V. Nemilov, *Thermodynamics and Kinetic Aspects of the Vitreous State*, CRC Press, Ann Arbor, MI (1995).
124. A.K. Jonscher, *Dielectric Relaxation in Solids*, Chelsea Dielectric Press, London (1983).
125. A. Inoue, *Bulk Amorphous Alloys*, Trans Tech Publications, Zürich (1998).
126. W.L. Johnson, Mater. Res. Bull. **11**, 104 (1999).
127. F. Faupel et al., Rev. Mod. Phys. **75**, 237 (2003).
128. W.H. Wang, C. Dong, C.H. Shek, Mater. Sci. Eng. R **44**, 45–89 (2004).
129. See articles in *Amorphous Metallic Alloys*, F.E. Luborsky, Ed., Butterworth, London (1983).
130. P.G. Debenedetti, J. Phys. Condens. Matter **15**, R1669–R1726 (2003).
131. K.L. Ngai, S. Capaccioli, N. Shinyashiki, J. Phys. Chem. B **112**, 3826 (2008).
132. F. Parak, G.U. Nienhaus, J. Non-Cryst. Solids **131–133**, 362 (1991); F.G. Parak, K. Achterhold, S. Croci, M. Schmidt, J. Biol. Phys. **33**, 371 (2007).
133. D. Ringe, G.A. Petsko, Biophys. Chem. **105**, 667–680 (2003).
134. P. Fenimore, H. Frauenfelder, B. McMahon, R.D. Young, Proc. Natl. Acad. Sci. **101**, 14408 (2005).
135. G. Cottone, G. Ciccotti, L. Cordone, J. Chem. Phys. **117**, 9862 (2002).
136. B.C. Hancock, G. Zografi, J. Pharm. Sci. **86**, 1 (1997).
137. C. Bhugra, R. Shmeis, S.L. Krill, M.J. Pikal, J. Pharm. Sci. **97**, 455 (2008).
138. J. Einfeldt, D. Meissner, A. Kwasniewski, Prog. Polym. Sci. **26**, 1419 (2001).
139. K. Kaminski, E. Kaminska, S. Hensel-Bielowka, E. Chelmecka, M. Paluch, J. Ziolo, P. Wlodarczyk, K.L. Ngai, J. Phys. Chem. B **112**, 7662 (2008).
140. A. Robert, E. Wandersman, E. Dubois, V. Dupuis, R. Perzynski, Europhys. Lett. **75**, 764 (2006).

141. (a) E.R. Weeks, J.C. Crocker, A.C. Levitt, A. Schofield, D.A. Weitz, Science **287**, 627 (2000); (b) E.R. Weeks, D.A. Weitz, Phys. Rev. Lett. **89**, 095704 (2002).
142. W. van Megan, S.M. Underwood, Phys. Rev. E **49**, 4206 (1994); For new results showing deviations from MCT prediction see G. Brambilla et al., Phys. Rev. Lett. **102**, 085703 (2009); ibid. **104**, 169602 (2010); and D. El Masri et al., J. Stat. Mech. (2009) P07015.
143. P.N. Segre, P.N. Pusey, Phys. Rev. Lett. **77**, 771 (1996). Equilibrium dynamics of concentration suspensions of colloidal hard spheres have been extended to times $\sim 10^5$ s and volume fraction up to 0.5970 by G. Brambilla et al., Phys. Rev. Lett. **102**, 085703 (2009); ibid. **104**, 169602 (2010); D. El Masri et al., J. Stat. Mech. P07015 (2009). See comments by J. Reinhardt et al., and by van Megan et al., Phys. Rev. Lett. **105**, 199604, and **104**, 169601 (2010), and replies by G. Brambilla et al., Phys. Rev. Lett. **105**, 199605, and **104**, 169602 (2010).
144. D.L. Leslie-Pelecky, N.O. Birge, Phys. Rev. Lett. **72**, 1232 (1994).
145. R. Brand, P. Lunkenheimer, A. Loidl, J. Chem. Phys. **116**, 10386 (2002); Th. Bauer, M. Köhler, P. Lunkenheimer, A. Loidl, C.A. Angell, J. Chem. Phys. **133**, 144509 (2010).
146. C.A. Angell, Chem. Rev. **90**, 523 (1990).
147. C.A. Angell, Annu. Rev. Phys. Chem. **172**, 1 (1992).
148. K.L. Ngai, J. Non-Cryst. Solids **203**, 232 (1996).
149. K.L. Ngai, J. Habasaki, C. Leon, A. Rivera, Z. Phys. Chem. **219**, 47 (2005).
150. K.J. Moreno, G. Mendoza-Suárez, A.F. Fuentes, J. García-Barriocanal, C. León, J. Santamaria, Phys. Rev. B **71**, 132301 (2005).
151. P.J. Wilde, C.R.A. Catlow, Solid State Ionics **112**, 173 (1998).
152. P.J. Wilde, C.R.A. Catlow, Solid State Ionics **112**, 185 (1998).
153. M. Pirzada, R.W. Grimes, L. Minervini, J.F. Maguire, K.E. Sickafus, Solid State Ionics **140**, 201 (2001).
154. T. Welton, Chem. Rev. **99**, 2071 (1999).
155. M.J. Earle, K.R. Seddon, Pure Appl. Chem. **72**, 1398 (2000).
156. P. Wasserscheid, W. Keim, Angew. Chem. Int. Ed. **39**, 3772 (2000).
157. S.H. Chung, R. Lopato, S.G. Greenbaum, H. Shirota, E.W. Castner Jr., J.F. Wishart, J. Phys. Chem. B **111**, 4885 (2007).
158. N. Ito, R. Richert, J. Phys. Chem. B **111**, 5016 (2007).
159. K.L. Ngai, J. Phys. Chem. B **110**, 26211 (2006).
160. P.W. Anderson, Phys. Today, **43**, 9 (1990).
161. F. Crick, *What Mad Pursuit*, Basic Books, New York, NY (1988).
162. C.A. Reynolds, B. Serin, W.H. Wright, L.B. Nesbitt, Phys. Rev. **78**, 487 (1950); E. Maxwell, Phys. Rev. **78**, 477 (1950).
163. H. Fröhlich, Phys. Rev. **79**, 845 (1950).
164. J. Bardeen, Rev. Mod. Phys. **23**, 261 (1951).
165. K.L. Ngai, D.J. Plazek, Rubber Chem. Tech. Rubber Rev. **68**, 376 (1995).
166. D.J. Plazek, J. Phys. Chem. **69**, 3480 (1965).
167. D.J. Plazek, J. Polym. Sci. Part A-2 **6**, 621 (1968).
168. D.J. Plazek, V.M. O'Rourke, J. Polym. Sci. Part A-2 **9**, 209 (1971).
169. (a) D.J. Plazek, E. Riande, H. Markovitz, N. Raghupathi, J. Polym. Sci. Polym. Phys. Ed. **17**, 2189 (1979); (b) D.J. Plazek, Polymer J. **12**, 43 (1980).
170. D.J. Plazek, J. Polym. Sci. Polym. Phys. Ed. **20**, 729 (1982).
171. D.J. Plazek, J. Rheol. **40**, 987 (1996); J. Non-Cryst. Solids **353**, 3783 (2007).
172. K.L. Ngai, D.J. Plazek, S.S. Deo, Macromolecules **20**, 3047 (1987).
173. K.L. Ngai, in *Physical Properties of Polymers*, 3rd ed., Chapter 2, Cambridge University Press, Cambridge (2003).
174. K. Popper, *The Logic of Scientific Discovery*, but see also the Duhem-Quine thesis in *Can Theories Be Refuted? Essays on the Duhem-Quine Thesis*, S.G. Harding ed., D. Reidel Publishing, Dordrecht, Holland (1975).
175. K.L. Ngai, Comment Solid State Phys. **9**, 127 (1979).

176. K.L. Ngai, Comment Solid State Phys. **9**, 141 (1979).
177. K.L. Ngai, C.T. White, Phys. Rev. B **20**, 2475 (1979).
178. K.L. Ngai, A.K. Jonscher, C.T. White, Nature **277**, 185 (1979).
179. A.K. Rajagopal, K.L. Ngai, S. Teitler, J. Phys. C **17**, 6611 (1984).
180. (a) K.L. Ngai, A.K. Rajagopal, S. Teitler, Physica **133A**, 213 (1985); (b) K.L. Ngai, A.K. Rajagopal, R.W. Rendell, S. Teitler, Phys. Rev. B **28**, 6073 (1983); (c) A.A. Jones, J.F. O'Gara, P.T. Inglefield, J.T. Bendler, A.F. Yee, K.L. Ngai, Macromolecules **16**, 685 (1983).
181. K.L. Ngai, A.K. Rajagopal, S. Teitler, J. Chem. Phys. **88**, 6088 (1988).
182. A.K. Rajagopal, K.L. Ngai, S. Teitler, Nucl. Phys. B (Proc. Suppl.) **5A**, 97 (1988).
183. K.L. Ngai, A.K. Rajagopal, S. Teitler, Nucl. Phys. B (Proc. Suppl.) **5A**, 103 (1988).
184. K.L. Ngai, R.W. Rendell, J. Non-Cryst. Solids **131–133**, 942 (1991).
185. A.K. Rajagopal, K.L. Ngai, S. Teitler, J. Non-Cryst. Solids **131–133**, 282 (1991); J. Chem. Phys. **92**, 243 (1990).
186. K.L. Ngai, S.L. Peng, K.Y. Tsang, Physica A **191**, 523 (1992).
187. K.L. Ngai, C.M. Roland, G.N. Greaves, J. Non-Cryst. Solids **182**, 172 (1995).
188. K.Y. Tsang, K.L. Ngai, Phys. Rev. E **54**, R3067 (1996).
189. K.Y. Tsang, K.L. Ngai, Phys. Rev. E **56**, R17 (1997).
190. K.L. Ngai, R.W. Rendell, in *Supercooled Liquids, Advances and Novel Applications*, J.T. Fourkas, D. Kivelson, U. Mohanty, K. Nelson, Eds., Chapter 4, ACS Symposium Series Vol. **676**, American Chemical Society, Washington, DC, pp. 45–66 (1997).
191. R.W. Rendell, Phys. Rev. E **48**, R17 (1993).
192. K.L. Ngai, K.Y. Tsang, Phys. Rev. E **60**, 4511 (1999).
193. K.L. Ngai, J. Phys. C **12**, 6437 (2000).
194. K.L. Ngai, IEEE Trans. Dielectr. Electr. Insulation **8**, 329 (2001).
195. K.L. Ngai, J. Phys. Condens. Matter **15**, S1107 (2003).
196. K.L. Ngai, J. Non-Cryst. Solids **351**, 2635 (2005).
197. K.L. Ngai, Adv. Colloid Interface. Sci. **64**, 1 (1996).
198. K.L. Ngai, G.D.J. Phillies, J. Chem. Phys. **105**, 8385 (1996); K.L. Ngai, Macromol. Symp. **146**, 117 (1999).
199. G. Fytas, Th. Dorfmuller, B. Chu, J. Polym. Sci. B Polym. Phys. **22**, 1471 (1984).
200. G. Fytas, Th. Dorfmuller, Y.H. Lin, B. Chu, Macromolecules, **14**, 1088 (1981).
201. K.L. Ngai, G. Fytas, J. Polym. Sci. B Polym. Phys. **24**, 1683 (1986).
202. G. Fytas, K.L. Ngai, Macromolecules **21**, 804 (1988).
203. G. Fytas, G. Floudas, K.L. Ngai, Macromolecules **23**, 1104 (1990).
204. A. Patkowski, W. Steffen, G. Meier, E.W. Fischer, J. Non-Cryst. Solids **172–174**, 52 (1994).
205. W. Steffen, A. Patkowski, H. Gläser, G. Meier, E.W. Fischer, Phys. Rev. E **49**, 2992 (1994).
206. D.J. Plazek, C. Bero, S. Neumeister, G. Floudas, G. Fytas, K.L. Ngai, Colloid Polym. Sci. **272**, 1430 (1994).
207. A.K. Rizos, K.L. Ngai, Macromolecules **31**, 6217 (1998).
208. A.K. Rizos, T. Jian, K.L. Ngai, Macromolecules **28**, 517 (1995).
209. K.L. Ngai, D.J. Plazek, A.K. Rizos, J. Polym. Sci. B Polym. Phys. **35**, 5992 (1997).
210. A.K. Rizos, K.L. Ngai, D.J. Plazek, Polymer **38**, 6103 (1997).
211. S. Kahle, J. Gapinski, G. Hinze, A. Patkowski, G. Meier, J. Chem. Phys. **122**, 074506 (2005).
212. G. Williams, D.C. Watts, Trans. Faraday Soc. **66**, 80 (1970).
213. G. Williams, D.C. Watts, S.B. Dev, A.M. North, Trans. Faraday Soc. **67**, 1323 (1971).
214. P.K. Dixon, Phys. Rev. B **42**, 8179 (1990).
215. C.A. Angell, in *Relaxation in Complex Systems*, K.L. Ngai, G.B. Wright, Eds., US Department Commerce, Springfield, MA, p. 3 (1985).
216. C.A. Angell, J. Non-Cryst. Solids **131–133**, 13 (1991).
217. K.S. Cole, R.H. Cole, J. Chem. Phys. **9**, 341 (1941).
218. D.W. Davidson, R.H. Cole, J. Chem. Phys. **19**, 1484 (1951).
219. S. Havriliak, S. Negami, J. Poly. Sci. C **14**, 99 (1966).

220. P.K. Dixon, L. Wu, S. Nagel, B.D. Williams, J.P. Carini, Phys. Rev. Lett. **65**, 1108 (1990).
221. R. Bergman, J. Appl. Phys. **88**, 1356 (2000).
222. T. Blochowicz, C. Gainaru, P. Medick, C. Tsichirwitz, E.A. Rössler, J. Chem. Phys. **124**, 134503 (2006).
223. W. Feller, *An Introduction to Probability Theory and Its Applications*, Vols. **1** and **2**, Wiley, New York, NY (1967).
224. M.F. Shlesinger, G.M. Zaslavsky, J. Klafter, Nature (London) **363**, 31 (1993).
225. J. Klafter, A. Blumen, M.F. Shlesinger, Phys. Rev. A **35**, 3081 (1987).
226. (a) K. Schmidt-Rohr, H.W. Spiess, Phys. Rev. Lett. **66**, 3020 (1991); (b) J. Leisen, K. Schmidt-Rohr, H.W. Spiess, Physica A **201**, 79 (1993).
227. A. Heuer, M. Wilhelm, H. Zimmermann, H.W. Spiess, Phys. Rev. Lett. **75**, 2851 (1997).
228. For review of other works, see R. Böhmer et al., J. Non-Cryst. Solids **235–237**, 1 (1998); R. Böhmer, Curr. Opin. Solid State Mater. Sci. **3**, 378 (1998).
229. C.T. Moynihan, J. Schroeder, J. Non-Cryst. Solids **160**, 52 (1993).
230. B. Schiener, R.V. Chamberlin, G. Diezemann, R. Böhmer, J. Chem. Phys. **107**, 7746 (1997).
231. R. Richert, J. Phys. Chem. B **101**, 6323 (1997).
232. C.T. Moynihan, J. Non-Cryst. Solids **235–237**, 781 (1998).
233. H. Sillescu, J. Non-Cryst. Solids **243**, 81 (1999).
234. K.L. Ngai, R.W. Rendell, in *The Symmetric and Fully Distributed Solution to a Generalized Dining Philosophers Problem, Relaxation in Complex Systems and Related Topics*, I.A. Campbell C. Giovannella, Eds., Plenum Press, New York, NY, pp. 309–316 (1990).
235. H. Sillescu, R. Böhmer, G. Diezemann, G. Hinze, J. Non-Cryst. Solids **307–310**, 16 (2002).
236. F. Stickel, *Untersuchung der Dynamik in niedermolekularen Flüssigkeiten mit Dielektrischer Spektroskopie*, Shaker, Aachen (1995).
237. C. León, K.L. Ngai, J. Phys. Chem. B **103**, 4045 (1999).
238. K.L. Ngai, C.M. Roland, Polymer **43**, 567 (2002).
239. A. Döß, M. Paluch, H. Sillescu, G. Hinze, Phys. Rev. Lett. **88**, 095701 (2002).
240. R. Nozaki, D. Suzuki, S. Ozawa, Y. Shiozaki, J. Non-Cryst. Solids **235–237**, 393 (1998).
241. C. León, K.L. Ngai, C.M. Roland, J. Chem. Phys. **110**, 11585 (1999).
242. S. Corezzi, M. Beiner, H. Huth, K. Schröter, S. Capaccioli, R. Casalini, D. Fioretto, E. Donth, J. Chem. Phys. **117**, 2435 (2002).
243. M. Paluch, C.M. Roland, J. Gapinski, A. Patkowski, J. Chem. Phys. **118**, 3177 (2003).
244. K.L. Ngai, C.M. Roland, Macromolecules **26**, 6824 (1993).
245. A.K. Rizos, R.M. Johnsen, W. Brown, K.L. Ngai, Macromolecules **28**, 5450 (1995).
246. G. Floudas, A.K. Rizos, W. Brown, K.L. Ngai, Macromolecules **27**, 2719 (1994).
247. K.L. Ngai, D.J. Plazek, V.M. O'Rourke, Macromolecules **30**, 5450 (1997).
248. K.L. Ngai, I. Echeverria, D.J. Plazek, Macromolecules **29**, 7937 (1997).
249. L. David, A. Sekkat, S. Etienne, J. Non-Cryst. Solids **172–174**, 214 (1994).
250. L. David, C. Girard, R. Dolmazon, M. Albrand, S. Etienne, Macromolecules **29**, 8343 (1996).
251. K.L. Ngai, J. Chem. Phys. **109**, 6982 (1998).
252. K.L. Ngai, T.R. Gopalkrishnan, M. Beiner, Polymer **47**, 7222 (2006).
253. K.L. Ngai, J. Chem. Phys. **111**, 3639 (1999).
254. E. Bartsch, M. Antonietti, W. Schupp, H. Sillescu, J. Chem. Phys. **97**, 3950 (1992).
255. W. van Megen, S.M. Underwood, J. Chem. Phys. **88**, 7841 (1988).
256. M.P. Allen, D.J. Tildesley, Eds., *Computer Simulation of Liquids*, Oxford Science, Oxford (1987).
257. J.E. Lennard-Jones, Proc. R. Soc. A **106**, 463 (1924).
258. L.J. Lewis, G. Wahnström, Phys. Rev. E **50**, 3865 (1994).
259. B.W.H. van Beest, G.J. Kramer, R.A. van Santen, Phys. Rev. Lett. **64**, 1955 (1990).
260. J. Horbach, W. Kob, K. Binder, Philos. Mag. B **235** and **237**, 320 (1998).
261. F. Sciortino, W. Kob, Phys. Rev. Lett. **86**, 648 (2001).
262. G.J. Kramer, A.J.M. de Man, R.A. van Santen, J. Am. Chem. Soc. **64**, 6435 (1991).

263. J. Horbach, W. Kob, K. Binder, J. Phys. Condens. Matter **15**, S903 (2003).
264. P. Bordat, F. Affouard, M. Descamps, K.L. Ngai, Phys. Rev. Lett. **93**, 105502 (2004); J. Non-Cryst. Solids **352**, 4630 (2006).
265. W. Kob, H.C. Andersen, Phys. Rev. E **51**, 4626 (1995).
266. C.T. Moynihan, P.B. Macedo, C.J. Montrose, P.K. Gupta, M.A. DeBolt, J.F. Dill, B.E. Dom, P.W. Drake, A.J. Esteal, P.B. Elterman, R.P. Moeller, H. Sasabe, J.A. Wilder, Ann. N.Y. Acad. Sci. **279**, 15 (1976).
267. K.L. Ngai, S. Mashimo, G. Fytas, Macromolecules **21**, 3030 (1988).
268. K.L. Ngai, J. Phys. Chem. B **103**, 10684 (1999).
269. L. Comez, D. Fioretto, L. Palmieri, L. Verdini, P.A. Rolla, J. Gapinski, A. Patkowski, W. Steffen, E.W. Fischer, Phys. Rev. E **60**, 3086 (1999).
270. K.L. Ngai, Philos. Mag. B **79**, 1783 (1999).
271. H. Takeuchi, R.-J. Roe, J. Chem. Phys. **94**, 7446 (1991).
272. K.L. Ngai, J. Chem. Phys. **98**, 7588 (1993).
273. R.-J. Roe, J. Chem. Phys. **94**, 7446 (1992), **100**, 1610 (1994); J. Non-Cryst. Solids **235–238**, 308 (1998).
274. D. Chakrabarti, B. Bagchi, Phys. Rev. Lett. **96**, 187801 (2006).
275. K.L. Ngai, J. Non-Cryst. Solids **353**, 709 (2007).
276. K.L. Ngai, A.K. Rajagopal, T.P. Lodge, J. Polym. Sci. B Polym. Phys. **28**, 1367 (1990).
277. P.A.M. Dirac, *Lectures on Quantum Mechanics*, Belfer Graduate School of Science Monograph Ser. No. 2, Yeshiva University, New York, NY (1964).
278. D. Bonn, W.K. Kegel, J. Chem. Phys. **118**, 2005 (2003).
279. M. Tatsumisago, C.A. Angell, S.W. Martin, J. Chem. Phys. **97**, 6968 (1992).
280. S. Estalji, O. Kanert, J. Steinert, H. Jain, K. L. Ngai, Phys. Rev. B **43**, 7481 (1991); S. Estalji, R. Küchler, O. Kanert, R. Bolter, H. Jain, K.L. Ngai, J. Phys. (Paris) Colloq. C2 **2**, 159 (1992).
281. O. Kanert, R. Küchler, S. Estalji, K.L. Ngai, H. Jain, in *The Physics of Non-Crystalline Solids*, L.D. Pye et al., Eds., Taylor and Francis, New York, NY, p. 178 (1992).
282. O. Kanert, R. Küchler, K.L. Ngai, H. Jain, Phys. Rev. B **49**, 76 (1994).
283. B. Munro, M. Schrader, P. Heitjans, Ber. Bunsenges. Phys. Chem. **96**, 1718 (1992).
284. M. Trunnel, D.R. Torgeson, S.W. Martin, F. Borsa, J. Non-Cryst. Solids **139**, 257 (1992).
285. W. Franke, P. Heitjans, Ber. Bunsenges. Phys. Chem. **96**, 1674 (1992); W. Franke, P. Heitjans, B. Munro, M. Schrader, in *Proceedings of XII International Conference on Defects in Insulating Materials*, Schloss Nordkirchen, Germany, O. Kanert, J.-M. Spaeth, Eds., World Scientific, Singapore, p. 1009 (1993).
286. F. Borsa, D.R. Torgeson, S.W. Martin, H.K. Patel, Phys. Rev. B **46**, 795 (1992).
287. F. Borsa, D.R. Torgeson, A. Pradel, M. Ribes, J. Non-Cryst. Solids **172–174**, 1315 (1994).
288. K.H. Kim, D.R. Torgeson, F. Borsa, S.W. Martin, Solid State Ionics **90**, 29 (1995).
289. V. Blache, J. Förster, H. Jain, O. Kanert, R. Küchler, K.L. Ngai, Solid State Ionics **113–115**, 723 (1998).
290. J. Bjorkstam, M. Villa, Phys. Rev. B **22**, 5033 (1980).
291. K.L. Ngai, J. Chem. Phys. **98**, 6424 (1993).
292. K.L. Ngai, Phys. Rev. B **48**, 13481 (1993).
293. K.L. Ngai, Solid State Ionics **61**, 345 (1993).
294. K.L. Ngai, J. Non-Cryst. Solids **162** (1993) 268.
295. M. Meyer, P. Maass, A. Bunde, Phys. Rev. Lett. **71**, 573 (1993).
296. M. Beiner, Macromol. Rapid Commun. **22**, 869 (2001).
297. M. Beiner, H. Huth, Nat Mater. **2**, 595 (2003); S. Pankaj, M. Beiner, J. Phys. Chem. B **114**, 15459 (2010).
298. S. Hiller, O. Pascui, H. Budde, O. Kabisch, D. Reichert, M. Beiner, New J. Phys. **6**, 10 (2004).
299. K.L. Ngai, E. Kamińska, M. Sekuła, M. Paluch, J. Chem. Phys. **123**, 204507 (2005).
300. J.H. Simmons, P.B. Macedo, J. Res. Natl. Bur. Stand. **75A**, 175 (1971).
301. R.E. Robertson, J. Polym. Sci. Polym. Sym. **68**, 173 (1978); J. Polym. Sci. Polym. Phys. **17**, 597 (1979).

302. E. Donth, J. Non-Cryst. Solids **53**, 325 (1982).
303. A.J. Kovacs, J.J. Aklonis, J.M. Hutchinson, A.R. Ramos, Polym. Sci. Polym. Phys. Ed. **17**, 1097 (1979).
304. For a review, see G.B. McKenna, in *Comprehensive Polymer Science*, Polymer Properties, C. Booth, C. Price, Eds., Pergamon, Oxford, Vol. **2**, pp. 311–362 (1989).
305. S.H. Glarum, J. Chem. Phys. **33**, 1371 (1960).
306. (a) M.F. Shlesinger, E.W. Montroll, Proc. Natl. Acad. Sci. USA. **81**, 1280 (1984); (b) M.F. Shlesinger, Annu. Rev. Phys. Chem. **39**, 269 (1988); (c) J. Bendler, J. Fontanella, M. Shlesinger, Chem. Phys. **284**, 311 (2002).
307. J. Colmenero, A. Alegria, A. Arbe, B. Frick, Phys. Rev. Lett. **69**, 478 (1992); J. Colmenero, A. Arbe, A. Alegria, Phys. Rev. Lett. **71**, 2603 (1993); J. Colmenero, A. Arbe, A. Alegria, J. Non-Cryst. Solids **172–174**, 126, (1994).
308. J. Colmenero, A. Arbe, G. Coddens, B. Frick, C. Mijangos, H. Reinecke, Phys. Rev. Lett. **78**, 1928 (1997).
309. M. Kiebel, E. Bartsch, O. Debus, F. Fujara, W. Petry, H. Sillescu, Phys. Rev. B **45**, 10310 (1992).
310. (a) R. Zorn, A. Arbe, J. Colmenero, B. Frick, D. Richter, U. Buchenau, Phys. Rev. E **52**, 781 (1995); (b) R. Zorn, D. Richter, B. Frick, B. Farago, Physica (Amsterdam) **201A**, 52 (1993).
311. A. Triolo, R.E. Lechner, A. Desmedt, T.F. Telling, V. Arrighi, Macromolecules **35**, 7039 (2002); A. Triolo, O. Russina, V. Arrighi, F. Juranyi, S. Janssen, C.M. Gordon, J. Chem. Phys. **119**, 8549 (2003).
312. (a) V.G. Sakai, J.K. Maranas, Z. Chowdhuri, I. Peral, J.R.D. Copley, J. Polym. Sci. B Polym. Phys. **43**, 2914 (2005); (b) V.G. Sakai, J.K. Maranas, I. Peral, J.R.D. Copley, Macromolecules **41**, 3701 (2008). In this paper, the fast process observed at times before $t_c=1$ ps is broader than a linear exponential function. This is possible for polymers, because of chain connectivity.
313. K.L. Ngai, J. Colmenero, A. Arbe, A. Alegria, Macromolecules **25**, 6727 (1992).
314. J. Colmenero, A. Arbe, A. Alegria, K.L. Ngai, J. Non-Cryst. Solids **172–174**, 229 (1994).
315. G. Wahnström, L.J. Lewis, Physica A **201**, 150 (1993).
316. C.M. Roland, K.L. Ngai, L.J. Lewis, J. Chem. Phys. **103**, 4632 (1995).
317. C.M. Roland, K.L. Ngai, Mech. Time-Dependent Mater. **66**, 109 (1997).
318. P. Sindzingre, M. Klein, J. Chem. Phys. **96**, 4681 (1992).
319. K.L. Ngai, C.M. Roland, J. Phys. Chem. B **101**, 4437 (1997).
320. S. Kämmerer, W. Kob, R. Schilling, Phys. Rev. E **56**, 5450 (1997).
321. C. Donati, S.C. Glotzer, P.H. Poole, W. Kob, S.J. Plimpton, Phys. Rev. E **60**, 3107 (1999).
322. P. Scheidler, W. Kob, K. Binder, Eur. Phys. J. E **12**, 5 (2003).
323. C. De Michele, D. Leporini, Phys. Rev. E. **63**, 036701 (2001).
324. J. Colmenero, F. Alvarez, A. Arbe, Phys. Rev. E **65**, 041804 (2002).
325. A. Neelakantan, J.K. Maranas, J. Chem. Phys. **120**, 465 (2004).
326. F. Affouard, E. Cochin, F. Danède, R. Decressain, M. Descamps, J. Chem. Phys. **123**, 084501 (2005).
327. T. Pakula, J. Teichmann, Mater. Res. Soc. Symp. Proc. **455**, 211 (1997).
328. P. Polanowski, T. Pakula, J. Chem. Phys. **117**, 4022 (2002).
329. K.L. Ngai, F.S. Liu, Phys. Rev. B **24**, 1049 (1981). K.L. Ngai, R.W. Rendell, Philos. Mag. **77**, 621 (1998).
330. (a) G. Hinze, D.D. Brace, S.D. Gottke, M.D. Fayer, J. Chem. Phys. **113**, 3723 (2000); (b) S.D. Gottke, D.D. Brace, G. Hinze, M.D. Fayer, J. Phys. Chem. **105**, 238 (2001).
331. D.D. Brace, S.D. Gottke, H. Cang, M.D. Fayer, J. Chem. Phys. **116**, 1598 (2002).
332. H. Cang, V.N. Novikov, M.D. Fayer, J. Chem. Phys. **118**, 2800 (2003).
333. J. Li, I. Wang, K. Fruchey, M.D. Fayer, J. Phys. Chem. A **110**, 10384 (2006).
334. C. Cramer, K. Funke, T. Saatkamp, Philos. Mag. B **71**, 701 (1995); K.L. Ngai, C. Cramer, T. Saatkamp, K. Funke, in *Non-Equilibrium Phenomena in Supercooled Fluids, Glasses*

and Amorphous Materials, M. Giordano, D. Leporini, M.P. Tosi, Eds., World Scientific, Singapore, pp. 3–24 (1996).
335. K. Funke, B. Heimann, M. Vering, D. Wilmer, J. Electrochem. Soc. **148**, A395 (2001).
336. U. Strom, J.R. Hendrickson, R.J. Wagner, P.C. Taylor, Solid State Commun. **15**, 1871 (1974); U. Strom, P.C. Taylor, Phys. Rev. B **16**, 5512 (1977).
337. K.N. Pham, S.U. Egelhaaf, P.N. Pusey, W.C.K. Poon, Phys. Rev. E **69**, 011503 (2004).
338. B. Abou, D. Bonn, J. Meunier, Phys. Rev. E **64**, 021510 (2001).
339. D. Bonn, H. Tanaka, G. Wegdam, H. Kellay, J. Meunier, Europhys. Lett. **45**, 52 (1999).
340. B. Ruzicka, L. Zulian, G. Ruocco, Phys. Rev. Lett. **93**, 258301 (2004).
341. B. Ruzicka, L. Zulian, G. Ruocco, J. Phys. Condens. Matter **16**, S4993 (2004).
342. B. Ruzicka, L. Zulian, G. Ruocco, Langmuir **22**, 1106 (2006).
343. R. Di Leonardo, S. Gentilini, F. Ianni, G. Ruocco, J. Non-Cryst. Solids **352**, 4928 (2006).
344. L. Zulian, B. Ruzicka, G. Ruocco, S. Capaccioli, J. Non-Cryst. Solids **353**, 3885 (2007).
345. S. Jabbari-Farouji, G.H. Wegdam, D. Bonn, Phys. Rev. Lett. **99**, 065701 (2007).
346. K.L. Ngai, G. Floudas, A.K. Rizos, J. Chem. Phys. **106**, 6957 (1997).
347. K.L. Ngai, J.H. Magill, D.J. Plazek, J. Chem. Phys. **112**, 1887 (2000).
348. (a) C. Hansen, F. Stickel, R. Richert, E.W. Fischer, J. Chem. Phys. **108**, 6408 (1998); (b) C. Hansen, F. Stickel, T. Berger, R. Richert, E.W. Fischer, J. Chem. Phys. **107**, 1086 (1997).
349. E.P. Wigner, *Statistical Properties of Real Symmetric Matrices with Many Dimensions*, in Proceedings of the 4th Canadian Mathematical Congress, University of Toronto Press, Toronto, Canada, pp. 174–184 (1957).
350. A collection of original papers by E.P. Wigner and others on random matrices and applications can be found in *Statistical Theories of Spectra: Fluctuations*, Charles E. Porter, Academic, New York, NY (1965).
351. M.L. Mehta, *Random Matrices*, 2nd ed., Academic, New York, NY (1991).
352. J. Dyson, J. Math. Phys. **3**, 140, 157, 166, 1191, 1199 (1962).
353. O. Bohigas, Random Matrix Theories and Chaotic Dynamics, in *Chaos and Quantum Physics*, Les Houches Summer School Session, Elsevier, Amsterdam, Vol. **52**, p. 87 (1991).
354. L.P. Gorkov, G.M. Eliashberg, Soviet Phys. JETP **21**, 940 (1965).
355. K.L. Ngai, S. Capaccioli, J. Phys. Condens. Matter **19**, 205114 (2007).
356. (a) D. Bedrov, G.D. Smith, Macromolecules **39**, 8526 (2006); (b) G.D. Smith, D. Bedrov, J. Polym. Sci. B Polym. Phys. **45**, 627 (2007); (c) D. Bedrov, G.D. Smith, reported at the 6th International Discussion Meeting on Relaxation in Complex Systems, August 30–September 4th 2009, and to be published in Special Issue of J. Non-Cryst. Solids (2011).
357. The Dining Philosopher Problem, a classic multi-process synchronization problem in computer science first set forth by E. Dijkstra in 1965 and reformulated by Sir Charles Anthony Richard Hoare. G. Kolata, Science **223**, 917 (1984).
358. A.A. Jones, J.F. O'Gara, P.T. Inglefield, J.T. Bendler, A.F. Yee, K.L. Ngai, Macromolecules **16**, 685 (1983).
359. G.B. McKenna, K.L. Ngai, D.J. Plazek, Polymer **26**, 1651 (1985).
360. K.L. Ngai, D.J. Plazek, J. Polym. Sci. B Polym. Phys. Ed. **24**, 619 (1986).
361. K.L. Ngai, C.H. Wang, G. Fytas, D.L. Plazek, D.J. Plazek, J. Chem. Phys. **86**, 4768 (1987).
362. K.L. Ngai, R.W. Rendell, H. Jain, Phys. Rev. B **30**, 2133 (1984).
363. R.S. McKay, J.D. Meiss, Eds., *Hamiltonian Dynamic Systems*, Adam Hilger, Bristol (1987).
364. S.W. McDonald, A.N. Kaufman, Phys. Rev. Lett. **42**, 1189 (1979).
365. O. Bohigas, M.J. Giannoni, C. Schmit, Phys. Rev. Lett. **52**, 1 (1984).
366. G. Casati, F. Valz-Gris, I. Guarneri, Lett. Nuovo Cimento **28**, 279 (1980).
367. M.V. Berry, Proc. R. Soc. Lond. A **413** 183 (1987).
368. M.C. Gutzwiller, *Chaos in Classical and Quantum Mechanics*, Springer, New York, NY (1990).
369. This subject is still actively being pursued at recent times, see J.P. Keating, S. Müller, Proc. R. Soc. A **463**, 3241 (2007).

370. H.-J. Stöckmann, *Quantum Chaos: An Introduction*, Cambridge University Press, Cambridge (1999).
371. C. Jaffe, W.P. Reinhardt, J. Chem. Phys. **77**, 5191 (1982).
372. R.B. Shirts, W.P. Reinhardt, J. Chem. Phys. **77**, 5204 (1982).
373. S.H. Strogatz, R. Mirollo, P.C. Mathews, Phys. Rev. Lett. **68**, 2730 (1992).
374. Von W. Oldekop, Glasstechnische Berichte **30**, 8 (1957).
375. W.T. Laughlin, D.R. Uhlmann, J. Phys. Chem. **76**, 2317 (1972).
376. E. Fermi, Phys. Rev. **75**, 1169 (1949).
377. K.Y. Tsang, M.A. Lieberman, Physica D **11**, 147 (1984); Phys. Lett. **103A**, 175 (1984); Physica D **21**, 401 (1986).
378. M.A. Lieberman, K.Y. Tsang, Phys. Rev. Lett. **55**, 908 (1985).
379. M.A. Lieberman, A.J. Lichtenberg, Phys. Rev. A **5**, 1852 (1972).
380. E. Fermi, J. Pasta, S. Ulam, in *Collected Papers of E. Fermi*, E. Segré, Ed., University of Chicago Press, Chicago, Vol. **2**, p. 978 (1965).
381. G.P. Tsironis, S. Aubry, Phys. Rev. Lett. **77**, 5225 (1996).
382. A. Bikaki, N.K. Voulgarakis, S. Aubry, G.P. Tsironis **59**, 1234 (1999).
383. S. Flach, A.V. Gorbach, Phys. Rep. **467**, 1–116 (2008).
384. G. Kalosakas, K.L. Ngai, S. Flach, Phys. Rev. E **71**, 061901 (2005).
385. M.H. Cohen, D. Turnbull, J. Chem. Phys. **31**, 1164 (1959).
386. R.A. Pethrick, F.M. Jacobsen, O.E. Morgensen, M. Eldrup, J. Chem. Soc. Faraday Trans. II **76**, 225 (1980).
387. K.L. Ngai, L-R Bao, A.F. Yee, C.L. Soles, Phys. Rev. Lett. **87**, 215901 (2001).
388. J. Bartos, J. Kristiak, J. Phys. Chem. B **104**, 5666 (2000).
389. J. Bartoš, O. Šauša, J. Krištiak, T. Blochowicz, E. Rössler, J. Phys. Condens. Matter **13**, 11473 (2001).
390. G. Dlubek, M.Q. Shaikh, K. Raetzke, F. Faupel, J. Pionteck, M. Paluch, J. Chem. Phys. **130**, 144906 (2009).
391. W. Kauzmann, Chem. Rev. **43**, 219 (1948).
392. J.H. Gibbs, E.A. DiMarzio, J. Chem. Phys. **28**, 373 (1958).
393. H. Vogel, Phys. Z. **22**, 645 (1921).
394. G.S. Fulcher, J. Am. Ceram. Soc. **8**, 339 (1925).
395. G. Tammann, W. Hesse, Z. Anorg. Allg. Chem. **156**, 245 (1926).
396. J.H. Magill, J. Chem. Phys. **47**, 2802 (1967). It turns out that the tri(naphthyl) benzene used in this work was not the tri-α isomer but the 1,3-bis(1-naphthyl)-5-(2-naphthyl) benzene, ααβ-TNB (C.M. Whitaker, R.J. McMahon, J. Phys. Chem. **100**, 1081 (1996)).
397. Z. Wojnarowska, K. Adrjanowicz, P. Wlodarczyk, E. Kaminska, K. Kaminski, K. Grzybowska, R. Wrzalik, M. Paluch, K.L. Ngai, J. Phys. Chem. B **113**, 12536 (2009).
398. R. Richert, C.A. Angell, J. Chem. Phys. **108**, 9016 (1998).
399. D. Prevosto, M. Lucchesi, S. Capaccioli, R. Casalini, P.A. Rolla, Phys. Rev. B **67**, 174202 (2003).
400. G.P. Johari, J. Chem. Phys. **112**, 8958 (2000).
401. K.L. Ngai, J. Phys. Chem. B **103**, 5895 (1999); K.L. Ngai, R.W. Rendell, D.J. Plazek, J. Chem. Phys. **94**, 3018 (1991).
402. C.M. Roland, S. Capaccioli, M. Lucchesi, R. Casalini, J. Chem. Phys **120**, 10640 (2004).
403. S. Corezzi, S. Capaccioli, R. Casalini, D. Fioretto, M. Paluch, P.A. Rolla, Chem. Phys. Lett. **320**, 113 (2000).
404. M. Goldstein, J. Chem. Phys. **64**, 4767 (1976).
405. G.P. Johari, J. Chem. Phys. **112**, 7518 (2000).
406. G.P. Johari, J. Chem. Phys. **113**, 751 (2000).
407. G.P. Johari, Philos. Mag. B **81**, 1935 (2001).
408. G.P. Johari, J. Chem. Phys. **116**, 1744 (2002).
409. G.P. Johari, J. Chem. Phys. **116**, 2043 (2002).
410. G.P. Johari, Chem. Phys. **265**, 217 (2001).

411. G.P. Johari, J. Phys. Chem. B **105**, 3600 (2001).
412. G.P. Johari, J. Non-Cryst. Solids **288**, 148 (2001).
413. G.P. Johari, J. Non-Cryst. Solids **307–310**, 387 (2002).
414. C.A. Angell, S. Borick, J. Non-Cryst. Solids **307–310**, 393 (2002).
415. G.P. Johari, J. Phys. Chem. B **107**, 5048 (2003).
416. K.L. Ngai, O. Yamamuro, J. Chem. Phys. **111**, 10403 (1999).
417. C.M. Roland, P.G. Santangelo, K.L. Ngai, J. Chem. Phys. **111**, 5593 (1999).
418. C.M. Roland, P.G. Santangelo, D.J. Plazek, K.M. Bernatz, J. Chem. Phys. **111**, 9337 (1999).
419. G.P. Johari, M. Beiner, C. Macdonald, J.J. Wang, Non-Cryst. Solids **278**, 58 (2001).
420. O. Yamamuro, I. Tsukushi, A. Lindqvist, S. Takahara, M. Ishikawa, T. Matsuo, J. Phys. Chem. B. **102**, 1605 (1998); M. Ishikawa, O. Yamamuro, T. Matsuo, H. Takakura, N. Achiwa, J. Korean Phys. Soc. **32**, S842 (1998); S. Takahara, O. Yamamuro, T. Matsuo, J. Phys. Chem. **99**, 9589 (1995).
421. U. Tracht, M. Wilhelm, A. Heuer, H. Feng, K. Schmidt-Rohr, H.W. Spiess, Phys. Rev. Lett. **81**, 2727 (1998).
422. S.A. Reinsberg, A. Heuer, B. Doliwa, H. Zimmermann, H.W. Spiess, J. Non-Cryst. Solids **307–310**, 208 (2002).
423. R. Casalini, K.L. Ngai, C.M. Roland, Phys. Rev. B **68**, 014201 (2003).
424. K.L. Ngai, P. Lunkenheimer, C. León, U. Schneider, R. Brand, A. Loidl, J. Chem. Phys. **115**, 1405 (2001).
425. E. Hempel, G. Hempel, A. Hensel, C. Schick, E. Donth, J. Phys. Chem. B **104**, 2460 (2000).
426. K. Schröter, J. Non-Cryst. Solids **352**, 3249 (2006).
427. G. Kolata, Science **223**, 917 (1984).
428. D. Lehman, M. Rabin, in *Conference Record of the 8th Annual ACM Symposium on Principles of Programming Language*, Association for Computing Machinery, Williamsburg, VA, January 26–28, pp. 133–138 (1981).
429. M. Ediger, Annu. Rev. Phys. Chem. **51**, 99 (2000).
430. H. Sillescu, J. Chem. Phys. **104**, 4877 (1996).
431. R. Böhmer, G. Hinze, G. Diezemann, B. Geil, H. Sillescu, Europhys. Lett. **36**, 55 (1996).
432. A. Heuer, Phys. Rev. E **56**, 730 (1997).
433. G. Hinze, Phys. Rev. E **57**, 2010 (1998).
434. S.C. Kuebler, A. Heuer, H.W. Spiess, Phys. Rev. E **56**, 741 (1997).
435. U. Tracht, A. Heuer, S.A. Reinsberg, H.W. Spiess, Appl. Magn. Reson. **17**, 227 (1999).
436. G. Hinze, G. Diezemann, H. Sillescu, Europhys. Lett. **44**, 565 (1998).
437. S.Y. Grebenkin, J. Phys. Chem. B **112**, 15369 (2008).
438. M.T. Cicerone, M.D. Ediger, J. Chem. Phys. **103**, 5684 (1995).
439. H. Wendt, R. Richert, Phys. Rev. E **61**, 1722 (2000).
440. E.V. Russell, N.E. Israeloff, Nature (London) **408**, 695 (2000).
441. F. Qi, T. El Goresy, R. Böhmer, A. Doss, G. Diezemann, G. Hinze, H. Sillescu, T. Blochowicz, C. Gainaru, E.A. Rössler, H. Zimmermann, J. Chem. Phys. **118**, 7431 (2003).
442. C.Y. Wang, M.D. Ediger, J. Chem. Phys. **112**, 6933 (2000).
443. D. Bingemann, N. Wirth, J. Gmeiner, E.A. Rössler, Macromolecule **40**, 5379 (2007).
444. B. Schiener, R. Böhmer, A. Loidl, R.V. Chamberlin, Science **274**, 752 (1996).
445. B. Schiener, R.V. Chamberlin, G. Diezemann, R. Böhmer, J. Chem. Phys. **107**, 7746 (1997).
446. S.C. Glotzer, J. Non-Cryst. Solids **274**, 342 (2000), and references therein.
447. L. Berthier, Phys. Rev. E **69**, 020201 (2004).
448. T. Pakula, J. Mol. Liquids **86**, 109 (2000).
449. M.T. Cicerone, F.R. Blackburn, M.D. Ediger, J. Chem. Phys. **102**, 471 (1995).
450. M.T. Cicerone, P.A. Wagner, M.D. Ediger, J. Phys. Chem. B **101**, 8727 (1997).
451. D. Bainbridge, M.D. Ediger, Rheol. Acta **36**, 209 (1997).
452. M.D. Ediger, J. Non-Cryst. Solids **235–237**, 10 (1998).

453. S.F. Swallen, P.A. Bonvallet, R.J. McMahon, M.D. Ediger, Phys. Rev. Lett. **90**, 15901 (2003); S.F. Swallen et al., J. Phys. Chem. B **113**, 4600 (2009).
454. M.K. Mapes, S.F. Swallen, M.D. Ediger, J. Phys. Chem. B **110**, 507 (2006).
455. R. Richert, K. Duvvuri, L.J. Duong, J. Chem. Phys. **118** 1828 (2003).
456. R. Richert, J. Chem. Phys. **123**, 154502 (2005).
457. J.R. Rajian, W. Huang, R. Richert, E.L. Quitevis, J. Chem. Phys. **124**, 014510 (2006).
458. L. Comez, D. Fioretto, L. Palmieri, L. Verdini, P.A. Rolla, J. Gapinski, A. Patkowski, W. Steffen, E.W. Fischer, Phys. Rev. E **60**, 3086 (1999).
459. A. Arbe, J. Colmenero, M. Monkenbusch, D. Richter, Phys. Rev. Lett. **81**, 590 (1998).
460. A. Arbe, A. Alegría, J. Colmenero, M. Monkenbusch, D. Richter, J. Phys. Condens. Matter **11**, A363 (1999).
461. B. Frick, G. Dosseh, A. Cailliaux, C. Alba-Simionesco, Chem. Phys. **292**, 311 (2003).
462. A. Heuer, H.W. Spiess, Phys. Rev. Lett. **82**, 1335 (1999).
463. A. Arbe, J. Colmenero, M. Monkenbusch, D. Richter, Phys. Rev. Lett. **82**, 1336 (1999).
464. J. Colmenero, A. Arbe, F. Alvarez, M. Monkenbusch, D. Richter, B. Farago, B. Frick, J. Phys. Condens. Matter **15**, S1127 (2003).
465. M. Zamponi, A. Wischnewski, M. Monkenbusch, L. Willner, D. Richter, P. Falus, B. Farago, M.G. Guenza, J. Phys. Chem. B **112**, 16220 (2008).
466. P. Maass, N. Petersen, A. Bunde, W. Dieterich, H.E. Roman, Phys. Rev. Lett. **66**, 52 (1991).
467. W. Dieterich, D. Knödler, P. Pendzig, J. Non-Cryst. Solids **172–174**, 1237 (1994); , J. Mol. D. Knödler, P. Pendzig, W. Dieterich, Solid State Ionics **70–71**, 356 (1994); D. Knödler, P. Pendzig, W. Dieterich, Solid State Ionics **86–88**, 29 (1996).
468. J. Habasaki, K.L. Ngai, Y. Hiwatari, Phys. Rev. E **66**, 021205 (2002).
469. J. Habasaki, K.L. Ngai, J. Chem. Phys. **122**, 214725 (2005).
470. J. Habasaki, J. Non-Cryst. Solids **353**, 3956 (2007).
471. J. Habasaki, K.L. Ngai, Phys. Chem. Chem. Phys. **9**, 4673 (2007).
472. J. Habasaki, K.L. Ngai, J. Chem. Phys. **129**, 034503 (2008).
473. A. Arbe, J. Colmenero, F. Alvarez, M. Monkenbusch, D. Richter, B. Farago, B. Frick, Phys. Rev. Lett. **89**, 245701 (2002).
474. B. Farago, A. Arbe, J. Colmenero, R. Faust, U. Buchenau, D. Richter, Phys. Rev. E **65**, 051803 (2002).
475. J. Habasaki, K.L. Ngai, Y. Hiwatari, J. Chem. Phys. **120**, 8195 (2004).
476. X.H. Qiu, M.D. Ediger, J. Phys. Chem. B **107**, 459 (2003).
477. K.L. Ngai, S. Capaccioli, J. Phys. Condens. Matter **19**, 205114 (2007).
478. R.W. Rendell, K.L. Ngai, S. Mashimo, J. Chem. Phys. **87**, 2359 (1987).
479. D.J. Plazek, K.L. Ngai, Macromolecules **24**, 1222 (1991).
480. R. Böhmer, C.A. Angell, Phys. Rev. B **45**, 10091 (1992).
481. K.L. Ngai, M. Beiner, Macromolecules **37**, 8123 (2004).
482. (a) T.R. Gopalakrishnan, M. Beiner, J. Phys. Conf. Ser. **40** (2006); (b) M. Beiner, AIP Conf. Proc. **832**, 134 (2006) *American Institute of Physics*, Melville, NY; (c) A. Arbe, A.-C. Genix, S. Arrese-Igor, J. Colmenero, D. Richter, Macromolecules **43**, 3107 (2010).
483. D. Kivelson, S.A. Kivelson, X. Zhao, Z. Nussinov, G. Tarjus, Physica A **219**, 27 (1995).
484. E.W. Fischer, E. Donth, W. Steffen, Phys. Rev. Lett. **68**, 2344 (1992).
485. C.L. Jackson, G.B. McKenna, J. Non-Cryst. Solids **131–133**, 221 (1991).
486. For a review, see M. Alcoutlabi, G.B. McKenna, J. Phys. Condens. Matter **17**, 461 (2005).
487. See papers published in *Dynamics in Confinement*, Eur. Phys. J. E **12** (2003).
488. (a) P. Pissis, D. Daoukaki-Diamanti, L. Apekis, C. Christodoulides, J. Phys. C **6**, L32 (1994); (b) G. Barut, P. Pissis, R. Pelster, G. Nimitz, Phys. Rev. Lett. **80**, 3543 (1998); (c) P. Pissis, A. Kyritsis, G. Barut, R. Pelster, G. Nimtz, J. Non-Cryst. Solids **235–237**, 444 (1998).
489. F. Kremer, A. Huwe, M. Arndt, P. Behrens, W. Schwieger, J. Phys. Condens. Matter **11**, A175 (1999).

490. M. Arndt, R. Stannarius, H. Groothues, E. Hempel, F. Kremer, Phys. Rev. Lett. **79**, 2077 (1997); M. Arndt, R. Stannarius, W. Gorbatschow, F. Kremer, Phys. Rev. E **54**, 5377 (1996); J. Koppensteiner, W. Schranz, M.A. Carpenter, Phys. Rev. B **81**, 024202 (2010).
491. A. Huwe, F. Kremer, P. Behrens, W. Schwieger, Phys. Rev. Lett. **82**, 2338 (1999).
492. T. Ruths, Ph.D. Thesis, Johannes Gutenberg University, Mainz (1997).
493. A. Patkowski, T. Ruths, E.W. Fischer, Phys. Rev. E **67**, 021501 (2003).
494. (a) K.L. Ngai, J. Phys. Condens. Matter **11**, A119 (1999); (b) Eur. Phys. J. E **12**, 93 (2003).
495. S.H. Anastasiadis, K. Karatasos, G. Vlachos, E. Manias, E.P. Giannelis, Phys. Rev. Lett. **84**, 915 (2000).
496. K.L. Ngai, Eur. Phys. J. E **8**, 225 (2002).
497. D. Morineau, Y. Xia, Ch. Alba-Simionesco, J. Chem. Phys. **117**, 8966 (2002).
498. A. Schönhals, H. Goering, Ch. Schick, B. Frick, R. Zorn, Eur. J. Phys. E Soft Matter **12**, 173 (2003).
499. A. Schönhals, H. Goering, Ch. Schick, B. Frick, R. Zorn, Colloid Polym. Sci. **282**, 882 (2004).
500. A. Schönhals, H. Goering, Ch. Schick, B. Frick, R. Zorn, J. Non-Cryst. Solids **351**, 2668 (2005).
501. R. Richert et al., J. Phys. Chem. B **107**, 895 (2003).
502. C. Alba-Simionesco et al., J. Phys. Condens. Matter **18**, R15–R68 (2006).
503. A. Kyritsis, J. Non-Cryst. Solids, **353**, 4318 (2007).
504. J. Mijović, H. Lee, J. Kenny, J. Mays, Macromolecules **39**, 2172 (2006).
505. K.L. Ngai, Eur. Phys. J. E **12**, 93 (2003).
506. K.L. Ngai, J. Polym. Sci. B Polym. Phys. **44**, 2980 (2006).
507. K.L. Ngai, Philos. Mag. **84**, 1341 (2004); K.L. Ngai, Johari-Goldstein or primitive relaxation: Terminator of caged dynamics and precursor of the α-relaxation. AIP Conf. Proc. **708**, 515–523 (2004).
508. S. Capaccioli, M.S. Thayyil, K.L. Ngai, J. Phys. Chem. B **112**, 16035 (2008).
509. U. Buchenau, R. Zorn, Europhys. Lett. **18**, 523 (1992).
510. C.A. Angell, J. Phys. Condens. Matter **12**, 6463 (2000).
511. S. Franz, G. Parisi, J. Phys. Condens. Matter **12**, 6335 (2000).
512. C. Donati, S. Franz, G. Parisi, S.C. Glotzer, J. Non-Cryst. Solids **307**, 215 (2002).
513. G. Biroli, J.-P. Bouchaud, Europhys. Lett. **67**, 21 (2004).
514. L. Berthier, Phys. Rev. Lett. **91**, 055701 (2003).
515. Y. Hiwatari, T. Muranaka, J. Non-Cryst. Solids **235–237**, 19 (1998).
516. B. Doliwa, A. Heuer, Phys. Rev. E **61**, 6898 (2000).
517. C. Bennemann, C. Donati, J. Baschnagel, S.C. Glotzer, Nature **399**, 246 (1999).
518. G. Biroli, J.P. Bouchaud, Europhys. Lett. **67**, 21 (2004).
519. S.C. Glotzer, V.V. Novikov, T.B. Schrøder, J. Chem. Phys. **112**, 509 (2000).
520. N. Lacevic, F.W. Starr, T.B. Schrøder, V.N. Novikov, S.C. Glotzer, Phys. Rev. E **66**, 030101R (2002).
521. N. Lacevic, S.C. Glotzer, J. Phys. Condens. Matter **15**, S2437 (2003).
522. N. Lacevic, F.W. Starr, T.B. Schrøder, S.C. Glotzer, J. Chem. Phys. **119**, 7372 (2003).
523. L. Berthier, G. Biroli, J.-P. Bouchaud, L. Cipelletti, D. El Masri, D. L'Hôte, F. Ladieu, M. Pierno, Science **310**, 1797 (2005).
524. L. Berthier, G. Biroli, J.-P. Bouchaud, W. Kob, K. Miyazaki, D.R. Reichman, J. Chem. Phys. **126**, 184503 (2007).
525. C. Dalle-Ferrier, C. Thibierge, C. Alba-Simionesco, L. Berthier, G. Biroli, J.-P. Bouchaud, F. Ladieu, D. L'Hôte, G. Tarjus, Phys. Rev. E **76**, 041510 (2007).
526. S. Capaccioli, G. Ruocco, F. Zamponi, J. Phys. Chem. B **112**, 10652 (2008).
527. C. Toninelli, M. Wyart, L. Berthier, G. Biroli, J.-P. Bouchaud, Phys. Rev. E **71**, 041505 (2005); L. Berthier, G. Biroli, arXiv:1011.2578, but unaware dynamic heterogeneity fails to explain breakdown of SE relation in TNB, OTP and SB.
528. (a) C. Toninelli, M. Wyart, L. Berthier, G. Biroli, J.-P. Bouchaud, Phys. Rev. E **71**, 041505 (2005); (b) C.M. Roland, D. Fragiadakis, D. Coslovich, S. Capaccioli, K.L. Ngai, J. Chem. Phys. **133**, 124507 (2010).

529. A.K. Rizos, K.L. Ngai, Phys. Rev. E **59**, 612 (1999).
530. N.A. Bokov, N.S. Andreev, Sov. J. Glass Phys. Chem. **15**, 243 (1989).
531. C.T. Moynihan, J. Schroeder, J. Non-Cryst. Solids **160**, 52 (1993).
532. C.A. Angell, L.-M. Wang, S. Mossa, Y. Yue, J.R.D. Copley, in *Slow Dynamics in Complex Systems*, 3rd Intern'al Symposium, M. Tokuyama, I. Oppenheim, Eds., AIP Conference Proceedings, Vol. **708**, p. 473 (2004).
533. J.P. Johari, J. Chem. Phys. **126**, 114901 (2007).
534. C.A. Angell, L. Monnerie, L.M. Torrel, MRS Symposium Proceedings, Material Research Society, Pittsburgh, PA, Vol. **215**, p. 3 (1991).
535. K.L. Ngai, J. Non-Cryst. Solids **95** and **96**, 969 (1987). See also I. Hodge, J. Non-Cryst. Solids **202**, 164 (1996) and comment of this paper by C.M. Roland, K.L. Ngai, J. Non-Cryst. Solids **212**, 74 (1997).
536. J.C. Phillips, J. Non Cryst. Solids **34**, 153 (1979), **43**, 37 (1982), Solid State Phys. **37**, 93 (1981), Rep. Prog. Phys. **59**, 1133 (1996).
537. M.F. Thorpe, J. Non Cryst. Solids **57**, 355 (1983); H. He, M.F. Thorpe, Phys. Rev. Lett. **54**, 2107 (1985).
538. W. Bresser, P. Boolchand, P. Suranyi, Phys. Rev. Lett. **56**, 2493 (1986).
539. P. Boolchand et al., Solid State Ionics **39**, 81 (1990).
540. M. Tatsumisago, B.L. Halfpap, J.L. Green, S.M. Lindsay, C.A. Angell, Phys. Rev. Lett. **64** 1549 (1990).
541. W.A. Kamitakahara et al., Phys. Rev. B **44**, 94 (1991).
542. R. Fabian, Jr., D.L. Sidebottom, Phys. Rev.B **80**, 064201 (2009).
543. X.W. Feng, W.J. Bresser, P. Boolchand, Phys. Rev. Lett., **78**, 4422 (1997).
544. D. Selvanathan, W.J. Bresser, P. Boolchand, Phys. Rev. B, **61**, 15061 (2000).
545. D.G. Georgiev, P. Boolchand, M. Micoulaut, Phys. Rev. B, **62**, R9228 (2000).
546. Y. Wang, P. Boolchand, M. Micoulaut, Europhys. Lett. **52**, 633 (2000).
547. (a) K.U. Schug, H.E. King, Jr., R. Bohmer, J. Chem. Phys. **109**, 1472 (1998); (b) R. Casalini, C.M. Roland, Phys. Rev. Lett. **92**, 245702 (2004); (c) C. Alba-Simionesco, A. Cailliaux, A. Alegria, G. Tarjus, Europhys. Lett. **68**, 58 (2004).
548. M. Paluch, S. Hensel-Bielowka, T. Psurek, J. Chem. Phys. **113**, 4374 (2000).
549. M. Paluch, C.M. Roland, S. Pawlus, J. Chem. Phys. **116**, 10932 (2002).
550. M. Paluch, C.M. Roland, J. Non-Cryst. Solids **316**, 413 (2003).
551. R. Casalini, C.M. Roland, Phys. Rev. B **71**, 073501 (2005).
552. M. Paluch, S. Hensel-Bielowka, J. Ziolo, J. Chem. Phys. **110**, 10978 (1999).
553. M. Paluch, S. Hensel-Bielowka, J. Ziolo, Phys. Rev. E **61**, 526 (2000).
554. D.M. Colucci, G.B. McKenna, J.J. Filliben, A. Lee, D.B. Bowman, J.D. Russell, J. Polym. Polym. Sci., Phys. Ed. **35**, 1561 (1997).
555. D. Huang, D.M. Colucci, G.B. McKenna, J. Chem. Phys. **116**, 3925 (2002).
556. M. Paluch, R. Casalini, S. Hensel-Bielowka, C.M. Roland, J. Chem. Phys. **116**, 9839 (2002).
557. R. Casalini, C.M. Roland, Phys. Rev. B **72**, 031503 (2005).
558. K. Niss, C. Dalle-Ferrier, G. Tarjus, C. Alba-Simionesco, J. Phys. Condens. Matter **19**, 076102 (2007).
559. C.M. Roland, K.L. Ngai, Macromolecules **29**, 5747 (1996).
560. P.G. Santangelo, C.M. Roland, Macromolecules **31**, 4581 (1998).
561. D. Huang, G.B. McKenna, J. Chem. Phys. **114**, 5621 (2001).
562. L.-M. Wang, V. Velikov, C.A. Angell, J. Chem. Phys. **117**, 10184 (2002).
563. K. Ito, C.T. Moynihan, C.A. Angell, Nature **398**, 492 (1999).
564. L.-M. Martinez, C.A. Angell, Nature (London) **410**, 663 (2001).
565. R.J. Speedy, Mol. Phys. **95**, 169 (1998).
566. A. Scala, F. Starr, E. La Nave, F. Sciortino, H.E. Stanley, Nature **406**, 166 (2000).
567. S. Sastry, Nature **409**, 164 (2001).
568. F.H. Stillinger, P.G. Debenedetti, J. Chem. Phys. **116**, 3353 (2002).

569. D.J. Wales, J.P.K. Doye, Phys. Rev. B **63**, 214204 (2001).
570. G. Ruocco, F. Sciortino, F. Zamponi, C. De Michele, T. Scopigno, J. Chem. Phys. **120**, 10666 (2004).
571. (a) T. Scopigno, G. Ruocco, F. Sette, G. Monaco, Science **302**, 849 (2003). There is breakdown of the correlation when experimental data of polymers including PVC, BPA-PC, PMMA, PET, PIB, PVAc, and PS from Buchenau and Wischnewski [572] are added to the original data of mainly small molecular glassformers in Scopigno et al. Correlation was restored after accounting for the effect of the secondary relaxation processes by T. Scopigno, D. Cangiolosi, G. Ruocco, Phys. Rev. B **81**, 100202 (2010), but not restored for PIB and PS as found by C. Dalle-Ferrier et al., Macromolecules **43**, 8977 (2010).
572. U. Buchenau, A. Wischnewski, Phys. Rev. B **70**, 092201 (2004).
573. L.-M. Wang, C.A. Angell, J. Chem. Phys. **118**, 10353 (2003).
574. (a) L.-M. Wang, C. A. Angell, R. Richert, J. Chem. Phys. **125**, 074505 (2006); (b) L.-M. Wang, J. Phys. Chem. B **113**, 5168 (2009); (c) L.-M. Wang, Z. Chen, Y. Zhao, R. Liu, Y. Tian, J. Alloys Compounds **504S**, S201 (2010).
575. V. Lubchenko, P.G. Wolynes, J. Chem. Phys. **119**, 9088 (2003).
576. J.D. Stevenson, P.G. Wolynes, J. Phys. Chem. B **109**, 15093 (2005).
577. T.R. Kirkpatrick, D. Thirumalai, Phys. Rev. A **40**, 1045 (1989).
578. X. Xia, P.G. Wolynes, Proc. Natl. Acad. Sci. U.S.A. **97**, 2990 (2000).
579. G. Parisi, G. Ruocco, F. Zamponi, Phys. Rev. E **69**, 061505 (2004).
580. A. Sokolov, A. Kisliuk, D. Quitmann, A. Kudlik, E. Rössler, J. Non-Cryst. Solids **172–174**, 138 (1994).
581. R.W. Hall, P.G. Wolynes, J. Chem. Phys. **86**, 2943 (1987).
582. R.W. Hall, P.G. Wolynes, J. Phys. Chem. B **112**, 301 (2008).
583. J. Dyre, Rev. Mod. Phys. **78**, 953 (2006).
584. V.N. Novikov, A.P. Sokolov, Nature (London) **431**, 961 (2004).
585. V.N. Novikov, Y. Ding, A.P. Sokolov, Phys. Rev. E **71**, 061501 (2005).
586. G.P. Johari, Philos. Mag. **86**, 1567 (2006).
587. S.N. Yannopoulos, G.P. Johari, Nature **442**, E7 (2006).
588. (a) C.M. Roland, S. Hensel-Bielowka, M. Paluch, R. Casalini, Rep. Prog. Phys. **68**, 1405 (2005); (b) C.M. Roland, R. Casalini, J. Chem. Phys. **122**, 134505 (2005); (c) R. Casalini, C.M. Roland, Phys. Rev. B **71**, 014210 (2005).
589. P. Santangelo, C.M. Roland, Macromolecules **31**, 4581 (1998).
590. A.K. Rizos, K.L. Ngai, Macromolecules **31**, 6217 (1998).
591. Y. Ding, V.N. Novikov, A.P. Sokolov, Macromolecules **37**, 9264 (2004).
592. K. Kunal, M. Paluch, C.M. Roland, J.E. Puskas, Y. Chen, A.P. Sokolov, J. Polym. Sci. B Polym. Phys. **46**, 1390 (2008).
593. C.M. Roland, K.L. Ngai, Macromolecules **29**, 5747 (1996).
594. S. Shahriari, A. Mandanici, L.-M. Wang, R. Richert, J. Chem. Phys. **121**, 8960 (2004).
595. K.L. Ngai, Phys. Rev. B **71**, 214201 (2005).
596. C.M. Roland, R. Casalini, M. Paluch, Chem. Phys. Lett. **367**, 259 (2003).
597. M. Paluch, K.L. Ngai, S. Hensel-Bielowka, J. Chem. Phys. **114**, 10872 (2001).
598. L. Torell, L. Borjesson, M. Elmroth, J. Phys. Condens. Matter **2**, SA207 (1990).
599. K. Schröter, E. Donth, J. Non-Cryst. Solids, **307–310**, 270 (2002).
600. R. Zorn, G.B. McKenna, L. Willner, D. Richter, Macromolecules **28**, 8552 (1996).
601. M. Connolly, F. Karasz, M. Trimmer, Macromolecules **28**, 1872 (1995).
602. A. Patkowski, M. Paluch, H. Kriegs, J. Chem. Phys. **117**, 2192 (2002).
603. S. Capaccioli, M. Paluch, unpublished.
604. K. Adrjanowicz, K. Kaminski, M. Paluch, K.L. Ngai, J. Phys. Condens. Matter **22**, 125902 (2010).
605. J.N. Sherwood, *The Plastically Crystalline State*, Wiley, New York, NY (1979).
606. K. Adachi, H. Suga, S. Seki, Bull. Chem. Soc. Jpn. **41**, 1073 (1968).
607. A. Srinivasan, F.J. Bermejo, A. de Andres, J. Dawidowski, J. Zuniga, A. Criado, Phys. Rev. B **53**, 8172 (1996).

608. M.A. Ramos, S. Vieira, F.J. Bermejo, J. Dawidowski, H.E. Fischer, H. Schober, M.A. Gonzalez, C.K. Loong, D.L. Price, Phys. Rev. Lett. **78**, 82 (1997).
609. M.A. Miller, J. Ruiz, F.J. Bermejo, N.O. Birge, Phys. Rev. B **57**, R13977 (1998).
610. B. Kuchta, M. Descamps, F. Affouard, J. Chem. Phys. **109**, 6753 (1998).
611. R. Brand, P. Lunkenheimer, U. Schneider, A. Loidl, Phys. Rev. B **62**, 8878 (2000).
612. S. Benkhof, A. Kudlik, T. Blochowicz, E. Rössler, J. Phys. Condens. Matter **10**, 8155 (1998).
613. R. Brand, P. Lunkenheimer, A. Loidl, Phys. Rev. B **56**, R5713 (1997).
614. K. Kessairi, S. Capaccioli, V. Castelvetro, M. Lucchesi, K.L. Ngai, to be published.
615. A. Döss, M. Paluch, H. Sillescu, G. Hinze, J. Chem. Phys. **117**, 6582 (2002).
616. S. Capaccioli, D. Prevosto, K. Kessairi, M. Lucchesi, P. Rolla, J. Non-Cryst. Solids **353**, 3984 (2007); **357**, 251 (2011).
617. (a) D. Weitz, J. Mattsson, H. Wyss, A. Fernandes-Nieves, K. Miyusaka, D. Reichman, "Soft colloids make stronger glasses", talk given at the 6th IDMRCS, Rome, Italy, 8/31–9/4 (2009); (b) J. Mattsson, H.M. Wyss, A. Fernandez-Nieves, K. Miyazaki, Z. Hu, D.R. Reichman, D.A. Weitz, Nature **462**, 83 (2009); (c) E. Bartsch, M. Antonietti, W. Schupp, H. Sillescu, J. Chem. Phys. **97**, 3950 (1992).
618. K.L. Ngai, R. Casalini, S. Capaccioli, M. Paluch, C.M. Roland, J. Phys. Chem. B **109**, 17356 (2005).
619. K. Akabori, H. Atarashi, M. Ozawa, T. Kondo, T. Nagamura, K. Tanaka, Polymer **50**, 4868 (2009).
620. D.U. Ahn, S.Y. Kwak, Macromol. Mater. Eng. **286**, 17 (2001).
621. F. Stickel, E.W. Fischer, R. Richert, J. Chem. Phys. **104**, 2043 (1996).
622. J. Colmenero, A. Alegria, P.G. Santangelo, K.L. Ngai, C.M. Roland, Macromolecules **27**, 407 (1994).
623. A. Gilchrist, J.E. Earley, R.H. Cole, J. Chem. Phys. **26**, 196 (1957).
624. G. Williams, Trans. Faraday Soc. **60**, 1548 (1964), **61**, 1564 (1965), **62**, 2091(1966).
625. H. Sasabe, S. Saito, J. Polym. Sci. Polym. Phys. Ed. **6**, 1401 (1968); Polym. J. (Tokyo) **3**, 624 (1972).
626. H. Sasabe, S. Saito, M. Asahina, H. Kakutani, J. Polym. Sci. Polym. Phys. Ed. **7**, 1405 (1969).
627. J.P. Johari, E. Whalley, Faraday Symp. Chem. Soc. **6**, 23 (1972).
628. T. Atake, C.A. Angell, J. Phys. Chem. **83**, 3218 (1979).
629. G. Fytas, A. Patkowski, G. Meier, Th. Dorfmüller, Macromolecules **15**, 21 (1982).
630. G. Fytas G, Th. Dorfmüller, C.H. Wang, J. Phys. Chem. **87**, 5041 (1983).
631. W. Heinrich, B. Stoll, Colloid Polym. Sci. **263**, 873 (1985).
632. M. Naoki, K. Matsumoto, M. Matsushita, J. Phys. Chem. **90**, 4423 (1986); M. Naoki, H. Endou, K. Matsumoto, J. Phys. Chem. **91**, 4169 (1987); M. Naoki, S. Katahira, J. Phys. Chem. **95**, 431 (1991).
633. (a) W.F. Oliver, C.A. Herbst, S.M. Lindsay, G.H. Wolf, Phys. Rev. Lett. **67**, 2795 (1991); (b) R.L. Cook, H.E. King, Jr., C.A. Herbst, D.R. Herschbach, J. Chem. Phys. **100**, 5178 (1994).
634. O. Yamamuro, S. Takahara, H. Suga, J. Non-Cryst. Solids **183**, 144 (1995); S. Takahara, M. Ishikawa, O. Yamamuro, T. Matsuo, J. Phys. Chem. B **103**, 792 (1999).
635. K.U. Schug, H.E. King, Jr., R. Böhmer, J. Chem. Phys. **109**, 1472 (1998).
636. S.P. Andersson, O. Andersson, Macromolecules **31**, 2999 (1998).
637. G. Floudas, T. Reisinger, J. Chem. Phys. **111**, 5201(1999).
638. G. Floudas, C. Gravalides, T. Reisinger, G. Wegner, J. Chem. Phys. **111**, 9847 (1999).
639. G. Floudas, G. Fytas, T. Reisinger, G. Wegner J. Chem. Phys. **111**, 9129 (1999).
640. S. Corezzi, P.A. Rolla, M. Paluch, J. Ziolo, D. Fioretto, Phys. Rev. E 60, 4444 (1999).
641. J. Köplinger, G. Kasper, S. Hunklinger, J. Chem. Phys. **113**, 4701 (2000).
642. A. Tölle, Rep. Prog. Phys. **64**, 1473 (2001).
643. J. Gapinski, M. Paluch, A. Patkowski, Phys. Rev. E **66**, 011501 (2002).

644. A. Patkowski, M. Matos Lopes, E.W. Fischer, J. Chem. Phys. **119**, 1579 (2003).
645. A. Alegría, D. Gómez, J. Colmenero, Macromolecules **35**, 2030 (2002).
646. C. Dreyfus, A. Aouadi, J. Gapinski, M. Matos-Lopes, W. Steffen, A. Patkowski, R.M. Pick, Phys. Rev. E **68**, 011204 (2003); C. Dreyfus, A. Le Grand, J. Gapinski, W. Steffen, A. Patkowski, Eur. J. Phys. **42**, 309 (2004).
647. B. Frick, C. Alba-Simionesco, K.H. Andersen, L.P. Willner, Phys. Rev. E **67**, 051801 (2003).
648. A. Cailliaux, C. Alba-Simonesco, B. Frick, L. Willner, I. Goncharenko, Phys. Rev. E **67**, 010802 (2003).
649. B. Frick, G. Dosseh, A. Cailliaux, C. Alba-Simionesco, Chem. Phys. **292**, 311 (2003).
650. T. Psurek, S. Hensel-Bielowka, J. Ziolo, M. Paluch, J. Chem. Phys. **116**, 9882 (2002).
651. S. Hensel-Bielowka, T. Psurek, J. Ziolo, M. Paluch, Phys. Rev. E **63**, 62301 (2001).
652. (a) R. Casalini, C.M. Roland, J. Chem. Phys. **119**, 4052 (2003); (b) R. Casalini, C.M. Roland, J. Chem. Phys. **119**, 11951 (2003); (c) C.M. Roland, R. Casalini, R. Bergman, J. Mattsson, Phys. Rev. B **77**, 012201 (2008); (d) S.P. Andersson, O. Andersson, Macromolecules **31**, 2999 (1998).
653. R. Casalini, M. Paluch, C.M. Roland, J. Chem. Phys. **118**, 5701 (2003).
654. S. Pawlus, R. Casalini, C.M. Roland, M. Paluch, S.J. Rzoska, J. Ziolo, Phys. Rev. E **70**, 061501 (2004).
655. R. Casalini, M. Paluch, C.M. Roland, J. Phys. Chem. A **107**, 2369 (2003).
656. R. Casalini, M. Paluch, C.M. Roland, Phys. Rev. E **67**, 031505 (2003).
657. R. Casalini, C.M. Roland, Phys. Rev. Lett. **91**, 015702 (2003).
658. R. Casalini, C.M. Roland, Phys. Rev. B **69**, 094202 (2004).
659. (a) S. Hensel-Bielowka, J. Ziolo, M. Paluch, C.M. Roland, J. Chem. Phys. **117**, 2317 (2002); (b) S. Hensel-Bielowka, M. Paluch, J. Ziolo, C.M. Roland, J. Phys. Chem. B **106**, 12459 (2002).
660. S. Pawlus, M. Paluch, M. Sekula, K.L. Ngai, S.J. Rzoska, J. Ziolo, Phys. Rev. E **68**, 021503 (2003).
661. M. Sekula, S. Pawlus, S. Hensel-Bielowka, J. Ziolo, M. Paluch, C.M. Roland, J. Phys. Chem. B **108**, 4997 (2004).
662. G. Floudas, Prog. Poly. Sci. **29**, 1143 (2004).
663. M. Naoki, K. Ujita, S. Kashima, J. Phys. Chem. **97**, 12356 (1993).
664. T. Okuchi, G.D. Cody, Ho-kwang Mao, R.J. Hemley, J. Chem. Phys. **122**, 244509 (2005).
665. A. Reiser, G. Kasper, S. Hunklinger, Phys. Rev. Lett. **92**, 125701 (2004).
666. P. Papadopoulos, D. Peristeraki, G. Floudas, G. Koutalas, N. Hadjichristidis, Macromolecules **37**, 8116 (2004).
667. M. Paluch, S. Pawlus, S. Hensel-Bielowka, E. Kaminska, D. Prevosto, S. Capaccioli, P.A. Rolla, K.L. Ngai, J. Chem. Phys. **122**, 234506 (2005).
668. R. Casalini, C.M. Roland, Macromolecules **38**, 1779 (2005).
669. S. Capaccioli, D. Prevosto, M. Lucchesi, P.A. Rolla, R. Casalini, K.L. Ngai, J. Non-Cryst. Solids **351**, 2643 (2005).
670. C.M. Roland, R. Casalini, S. Hensel-Bielowka, M. Paluch, Rep. Prog. Phys. **68**, 1405 (2005).
671. S. Hensel-Bielowka, M. Paluch. K.L. Ngai, J. Chem. Phys. **123**, 014502 (2005).
672. (a) K. Kessairi, S. Capaccioli, D. Prevosto, S. Sharifi, P.A. Rolla, J. Non-Cryst. Solids **353**, 4273 (2007); (b) K. Kessairi, Ph.D. Thesis, Department of Physics, University of Pisa, Pisa, Italy (2007).
673. D. Prevosto, S. Capaccioli, S. Sharifi, K. Kessairi, M. Lucchesi, P.A. Rolla, J. Non-Cryst. Solids **353**, 4278 (2007).
674. S. Capaccioli, K. Kessairi, D. Prevosto, M. Lucchesi, P. Rolla, J. Non-Cryst. Solids **353**, 3984 (2007).
675. D. Prevosto, K. Kessairi, S. Capaccioli, M. Lucches, P.A. Rolla, Philos. Mag. **87**, 643 (2007).

676. M. Mierzwa, S. Pawlus, M. Paluch, E. Kaminska, K.L. Ngai, J. Chem. Phys. **128**, 044512 (2008).
677. K. Kessairi, S. Capaccioli, D. Prevosto, M. Lucchesi, S. Sharifi, P.A. Rolla, J. Phys. Chem. B, **112**, 4470 (2008).
678. D. Prevosto, S. Capaccioli, M. Lucchesi, P.A. Rolla, K.L. Ngai, J. Non-Cryst. Solids **355**, 705 (2009).
679. M.S. Thayyil, S. Capaccioli, P. Rolla, K.L. Ngai, Philos. Mag. **88**, 4047 (2008).
680. M.S. Thayyil, S. Capaccioli, D. Prevosto, K.L. Ngai, Philos. Mag. **88**, 4007 (2008).
681. (a) L.P. Singh, S.S.N. Murthy, Phys. Chem. Chem. Phys. **11**, 5110 (2009); (b) L.P. Singh, S.S.N. Murthy, G. Singh, Phys. Chem. Chem. Phys. **11**, 9278 (2009).
682. (a) U. Schneider, R. Brand, P. Lunkenheimer, A. Loidl, Phys. Rev. Lett. **84**, 5560 (2000); (b) P. Lunkenheimer et al., **77**, 318 (1996).
683. C. Svanberg, R. Bergman, P. Jacobsson, Europhys. Lett. **64**, 358 (2003).
684. E. Kaminska, K. Kaminski, S. Hensel-Bielowka, M. Paluch, K.L. Ngai, J. Non-Cryst. Solids 352, 4672 (2006).
685. K.L. Ngai, M. Paluch, J. Chem. Phys. **120**, 857 (2004).
686. M. Paluch, S. Pawlus, S. Hensel-Bielowka, E. Kaminska, D. Prevosto, S. Capaccioli, P.A. Rolla, K.L. Ngai, J. Chem. Phys. **122**, 234506 (2005).
687. S. Capaccioli, R. Casalini, D. Prevosto, P.A. Rolla, unpublished.
688. T. Psurek, J. Ziolo, M. Paluch, Physica A **331**, 353 (2004).
689. R. Casalini, T. Psurek, M. Paluch, C.M. Roland, J. Mol. Liq. **111**, 53 (2004).
690. T. Psurek, S. Hensel-Bielowka, J. Ziolo, M. Paluch, J. Chem. Phys. **116**, 9882 (2002).
691. S. Hensel-Bielowka, T. Psurek, J. Ziolo, M. Paluch, Phys. Rev. E **63**, 62301 (2001).
692. M. Naoki, M. Matsushita, Bull. Chem. Soc. Jpn. **56**, 2396 (1983).
693. M. Masuko, A. Suzuki, S. Hanai, H. Okabe, Jpn. J. Tribol. **42**, 455 (1997).
694. A. Suzuki, M. Masuko, T. Nakayama, H. Okabe, Jpn. J. Tribol. **42**, 467 (1997).
695. A. Suzuki, M. Masuko, T. Nikkuni, Tribol. Int. **33**, 107 (2000).
696. G. Williams, in *Dielectric Spectroscopy of Polymeric Materials*, J.P. Runt, J.J. Fitzgerald, Eds., American Chemical Society, Washington, DC, p. 3 (1997).
697. S. Capaccioli, D. Prevosto, K. Kessairi, M. Lucchesi, P.A. Roland, Philos. Mag. **87**, 681 (2007).
698. S.H. Zhang, R. Casalini, J. Runt, C.M. Roland, Macromolecules **36**, 9917 (2003).
699. C.M. Roland, R. Casalini, P. Santangelo, M. Sekula, J. Ziolo, M. Paluch, Macromolecules **36**, 4954 (2003).
700. C.M. Roland, R. Casalini, M. Paluch, J. Polym. Sci. Polym. Phys. Ed. **42**, 4313 (2004).
701. C.M. Roland, T. Psurek, S. Pawlus, M. Paluch, J. Polym. Sci. Polym. Phys. **41**, 3047 (2003).
702. M. Mierzwa, S. Pawlus, M. Paluch, J.R. Sangoro, F. Kremer, to be published.
703. K. Mpoukouvalas, G. Floudas, B. Verdonck, F.E. Du Prez, Phys. Rev. E **72**, 011802 (2005).
704. R.D. Rogers, K.R. Seddon, Eds., *Ionic Liquids IIIA, B:Fundamentals, Progress, Challenges, and Opportunities*; ACS Symposium Series, American Chemical Society, Washington, DC, Vol. **901** (2005).
705. P. Wasserscheid, T. Welton, Eds., *Ionic Liquids in Synthesis*, Wiley-VCH, Weinheim, Germany (2003).
706. A. Rivera-Calzada, K. Kaminski, C. Léon, M. Paluch, J. Phys. Chem. B **112**, 3110 (2008).
707. For a review of the use of electric modulus to characterize ionic conductivity relaxation, see, I.M. Hodge, K.L. Ngai, C.T. Moynihan, J. Non-Cryst. Solids **351**, 104 (2005).
708. For references, see K. Kaminski, E. Kaminska, P. Wlodarczyk, S. Pawlus, D. Kimla, A. Kasprzycka, M. Paluch, J. Ziolo,W. Szeja, K.L. Ngai, J. Phys. Chem. B **110**, 25045 (2006).
709. G. Fytas, G. Meier, Th. Dorfmuller, A. Patkowski, Macromolecules **15**, 214 (1982).
710. G. Fytas, A. Patkowski, G. Meier, Th. Dorfmuller, Macromolecules **15**, 874 (1982).
711. G. Fytas, A. Patkowski, G. Meier, Th. Dorfmuller, J. Chem. Phys. **80**, 2214 (1980).
712. G.D. Patterson, J.R. Stevens, P.J. Carroll, J. Chem. Phys. **77**, 622 (1980).

713. S.W. Smith, B.D. Freeman, C.K. Hall, Macromolecules **30**, 2052 (1997).
714. G. Fytas, Th. Dorfmuller, C.H. Wang, J. Phys. Chem. **87**, 5041 (1983).
715. Y.-H. Hwang, G.Q. Shen, J. Phys. Condens. Matter **11**, 1453 (1999).
716. H. Leyser, A. Schulte, W. Doster, W. Petry, Phys. Rev. E **51**, 5899 (1995).
717. A. Tölle, Rep. Prog. Phys. **64**, 1473 (2001).
718. L. Comez, D. Fioretto, H. Kriegs, W. Steffen, Phys. Rev. E **66**, 032501 (2002).
719. L. Comez, S. Corezzi, D. Fioretto, H. Kiegs, A. Best, W. Steffen, Phys. Rev. E **70**, 011504 (2004).
720. D.L. Sidebottom, C.M. Sorensen, Phys. Rev. B **40**, 461 (1988).
721. T. Berger, Ph.D. Thesis, Max Planck Institute for Polymer Research, Mainz, Germany (1996).
722. M. Paluch, C.M. Roland, J. Gapinski, A. Patkowski, J. Chem. Phys. **118**, 3177 (2003).
723. M. Paluch, A. Patkowski, E.W. Fischer, Phys. Rev. Lett. **85**, 2140 (2000).
724. M. Paluch, J. Gapinski, A. Patkowski, E.W. Fischer, J. Chem. Phys. **114**, 8048 (2001).
725. A. Patkowski, M. Paluch, H. Kriegs, J. Chem. Phys. **117**, 2192 (2002).
726. M. Paluch, R. Casalini, A. Best, A. Patkowski, J. Chem. Phys. **117**, 7624 (2002).
727. A. Patkowski, J. Gapinski, G. Meier, Colloid Polym. Sci. **282**, 874 (2004).
728. J. Gapinski, M. Paluch, A. Patkowski, Phys. Rev. E **66**, 011501 (2002).
729. C.M. Roland, K.J. McGrath, R. Casalini, Macromolecules **38**, 8729 (2005).
730. A. Alegrıa, D. Gomez, J. Colmenero, Macromolecules **35**, 2030 (2002).
731. A. Zetsche, E.W. Fischer, Acta Polym. **45**, 168 (1994).
732. A. Kamath, R.H. Colby, S.K. Kumar, Phys. Rev. E **67**, 010801 (2003).
733. T.P. Lodge, T.C.B. McLeish, Macromolecules **33**, 5278 (2002).
734. (a) C.M. Roland, K.L. Ngai, Macromolecules **24**, 2261 (1991); (b) Macromolecules **25**, 363 (1992); (c) Macromolecules **28**, 4033 (1995); (d) Macromolecules **33**, 3184 (2000); (e) J. Rheol. **36**, 1691 (1992).
735. A. Alegria, J. Colmenero, K.L. Ngai, C.M. Roland, Macromolecules **27**, 4486 (1994).
736. K.L. Ngai, C.M. Roland, Rubber Chem. Tech. **77**, 579 (2004).
737. K.L. Ngai, S. Capaccioli, J. Phys. Chem. B **108**, 11118 (2004).
738. K.L. Ngai, C.M. Roland, Macromolecules **37**, 2817 (2004).
739. S. Capaccioli, K.L. Ngai, J. Phys. Chem. B **109**, 9727 (2005).
740. C.M. Roland, R. Casalini, Macromolecules **40**, 3631 (2007).
741. M.S. Thayyil, S. Capaccioli (2009), and to be published.
742. G.N. Lewis, M. Randall, *Thermodynamics*, 2nd ed., McGraw-Hill, New York, NY (1961).
743. D.J. Plazek, J.H. Magill, J. Chem. Phys. **45**, 3038 (1996).
744. D.J. Plazek, J.H. Magill, J. Chem. Phys. **49**, 3678 (1968).
745. R.J. Greet, J.H. Magill, J. Phys. Chem. **71**, 1746 (1967).
746. C.M. Roland, K.L. Ngai, J. Chem. Phys. **106**, 1187 (1997).
747. (a) M. Mierzwa, S. Pawlus, M. Paluch, J.R. Sangoro, F. Kremer, to be published (2010); (b) A. Rivera, E. Rössler, Phys. Rev. B **73**, 212201 (2006); (c) M. Mierzwa, S. Pawlus, M. Paluch, A. Rivera-Calzada, presented at 6th Intern'al Discussion Meeting on Relaxation in Complex Systems, August 2009, Rome, Italy.
748. A. Pimenov, P. Lunkenheimer, H. Rall, R. Kohlhaas, A. Loidl, R. Böhmer, Phys. Rev. E **54**, 676 (1996).
749. C. Mai, S. Etienne, J. Perez, G.P. Johari, Philos. Mag. B **50**, 657 (1985).
750. C.M. Roland, Soft Matter. **4**, 2316 (2008).
751. The same can be seen from the temperature dependence of the full width at half maximum of the loss peak presented in Stickel's thesis, Ref. [236].
752. E. Eckstein, J. Qian, R. Hentschke, T. Thurn-Albrecht, W. Steffen, E.W. Fischer, J. Chem. Phys. **113**, 4751 (2000).
753. M. Wind, R. Graf, S. Renker, H.W. Spiess, W. Steffen, J. Chem. Phys. **122**, 014906 (2005).
754. K. Schmidt-Rohr, A.S. Kudlik, H.W. Beckham, A. Ohlemacher, U. Pawelzik, C. Boeffel, Macromolecules **27**, 4733 (1994).

755. A.S. Kudlik, H.W. Beckham, K. Schmidt-Rohr, D. Radloff, U. Pawelzik, C. Boeffel, Macromolecules **27**, 4746 (1994).
756. R. Casalini, M. Paluch, C.M. Roland, J. Phys. Condens. Matter **15**, S859 (2003).
757. R. Casalini, C.M. Roland, Phys. Rev. B **71**, 014210 (2005).
758. A. Schönhals, Europhys. Lett. **56**, 815 (2001).
759. M. Tyagi, A. Alegría, J. Colmenero, J. Chem. Phys. **122**, 244909 (2005).
760. A. Arbe, J. Colmenero, J. Gómez, D. Richter, B. Farago, Phys. Rev. Lett. **60**, 1103 (1999).
761. V.G. Sakai, C. Chen, J.K. Maranas, Z. Chowdhuri, Macromolecules **37**, 9975 (2004).
762. A. Arbe, D. Richter, J. Colmenero, B. Farago, Phys. Rev. E **54**, 3853 (1996).
763. C.K. Hall, E. Helfand, J. Chem. Phys. **77**, 3275 (1982).
764. K.L. Ngai, R.W. Rendell, J. Non-Cryst. Solids **131–133**, 942 (1991).
765. V.N. Novikov, A.P. Sokolov, Phys. Rev. E **67**, 031507 (2003).
766. J. Rault, J. Non-Cryst. Solids, **271**, 177 (2000).
767. E.A. DiMarzio, A.J.M. Yang, J. Res. Natl. Inst. Stand. Technol. **102**, 135 (1997).
768. P.A. O'Connell, G.B. McKenna, J. Chem. Phys. **110**, 11054 (1999).
769. C.T. Thurau, M.D. Ediger, J. Chem. Phys. **118**, 1996 (2003).
770. D. Richter, A. Arbe, J. Colmenero, M. Monkenbusch, B. Farago, R. Faust, Macromolecules **31**, 1133 (1998).
771. A. Arbe, D. Richter, J. Colmenero, B. Farago, Physica B **234–236**, 437 (1997).
772. D. Richter, M. Monkenbusch, A. Arbe, J. Colmenero, B. Farago, R. Faust, J. Phys. Condens. Matter **11**, A297 (1999).
773. G. Arialdi, J.-P. Ryckaert, D.N. Theodorou, Chem. Phys. **292**, 371 (2003). In this paper, the authors reported $Q^{-3.3}$-dependence at Q<1 Å^{-1} and $(1-n)=0.58$ at Q<0.375 Å^{-1}. For $(1-n)=0.58$, the value of the exponent $2/(1-n)=3.45$ is consistent with the observed value of 3.3.
774. J. Wuttke, W. Petry, G. Coddens, F. Fujara, Phys. Rev. E. **52**, 4026 (1995).
775. K.L. Ngai, C.M. Roland, J. Phys. Chem. B **101**, 4437 (1997).
776. S.Z. Ren, W.F. Shi, W.B. Zhang, C.M. Sorensen, Phys. Rev. A **45**, 2416 (1992).
777. G.D.J. Phillies, C. Richardson, C.A. Quinlan, S.Z. Ren, Macromolecules **26**, 6849 (1993).
778. B. Nyström, H. Walderhaug, F.N. Hansen, J. Phys. Chem. **97**, 7743 (1993).
779. K.L. Ngai, Adv. Colloid Interf. Sci. **64**, 1 (1996).
780. M. Adam, M. Delasanti, J.P. Munch, D. Durand, Phys. Rev. Lett. **61**, 706 (1988).
781. H.N. Ritland, J. Am. Ceram. Soc. **39**, 403 (1956).
782. A.Q. Tool, J. Am. Ceram. Soc. **29**, 240 (1946).
783. O.S. Narayanaswamy, J. Am. Cer. Soc. **54**, 491 (1971).
784. (a) C.T. Moynihan, P.B. Macedo, C.J. Montrose, P.K. Gupta, M.A. DeBolt, J.F. Dill, B.E. Dom, P.W. Drake, A.J. Esteal, P.B. Elterman, R.P. Moeller, H. Sasabe, J.A. Wilder, Ann. NY Acad. Sci. **279**, 15 (1976); (b) G.W. Scherer, J. Am. Ceram. Soc. **67**, 504 (1984).
785. I.M. Hodge, Macromolecules **20**, 2897 (1987).
786. I.M. Hodge, J. Non-Cryst. Solids **131–133**, 435 (1991), ibid. **169**, 211 (1994).
787. K.L. Ngai, in *The Physics of Non-Crystalline Solids*, L.D. Pye et al., Eds., Taylor & Francis, London, p. 178 (1992).
788. L.C.E. Struik, *Physical Aging in Amorphous Polymers and Other Materials*, Elsevier, New York, NY (1978).
789. S.L. Simon, D.J. Plazek, J.W. Sobieski, E.T. McGregor, J. Polym. Sci. B Polym. Phys. **35**, 929 (1997).
790. P.A. O'Connell, C.R. Schultheisz, G.B. McKenna, in *The Physics of Glassy Polymers*, A. Hill, M. Tant, Eds., ACS, New York, NY, pp. 199–217 (1998).
791. S.L. Simon, J.W. Sobieski, D.J. Plazek, Polymer **42**, 2555 (2001).
792. C.T. Thurau, M.D. Ediger, J. Chem. Phys. **116**, 9089 (2002).
793. M.T. Cicerone, M.D. Ediger, J. Phys. Chem. B **97**, 10489 (1993).
794. M.D. Ediger, T. Inoue, M.T. Cicerone, F.R. Blackburn, Macromol. Symp. **101**, 139 (1996).
795. F. Blackburn, C.-Y. Wang, M.D. Ediger, J. Phys. Chem. **100**, 18249 (1996).

796. M.T. Cicerone, M.D. Ediger, J. Chem. Phys. **104**, 7210 (1996).
797. K.L. Ngai, Philos. Mag. **87**, 357 (2007).
798. R.E. Courtland, E.R. Weeks, J. Phys. Condens. Matter **15**, S359 (2003).
799. G.C. Cianci, R.E. Courtland, E.R. Weeks, AIP Conf. Proc. **832**, 21 (2006).
800. E.R. Weeks, D.A. Weitz, Phys. Rev. Lett. **89**, 095704 (2002).
801. W.K. Kegel, A. van Blaaderen, Science **287** 290 (2000).
802. J. Habasaki, K.L. Ngai, J. Non-Cryst. Solids **352**, 5170 (2006).
803. D.J. Plazek, K.L. Ngai, R.W. Rendell, Polym. Eng. Sci. **24**, 1111 (1984).
804. D. Bonn, S. Tanase, B. Abou, H. Tanaka, J. Meunier, Phys. Rev. Lett. **89**, 015701 (2002).
805. J.H. Magill, D.J. Plazek, Nature. **209**, 70 (1966).
806. J.H. Magill, D.J. Plazek, J. Chem. Phys. **46**, 3757 (1967); D.J. Plazek, J.H. Magill, J. Chem. Phys. **49**, 3678 (1968).
807. J.H. Magill, H.-M. Li, J. Cryst. Growth **20**, 135 (1973).
808. F. Fujara, H. Geil, H. Sillescu, G. Fleischer, Z. Phys. B **88**, 195 (1992).
809. I. Chang, F. Fujara, B. Geil, G. Heuberger, T. Mangel, H. Sillescu, J. Non-Cryst. Solids **172–174**, 248 (1994).
810. T. Hikima, Y. Adachi, M. Hanaya, M. Oguni, Phys. Rev. B **52**, 3900 (1995).
811. M. Hatase, M. Hanaya, M. Oguni, J. Non-Cryst. Solids **333**, 129 (2004).
812. (a) U. Geyer, et al. Appl. Phys. Lett. **69**, 2492 (1996); (b) A. Bartsch, K. Rätzke, A. Meyer, F. Faupel, Phys. Rev. Lett. **104**, 195901 (2010); (c) A. Meyer, Phys. Rev. B **66**, 134205 (2002); (d) G. Biroli, J.P. Bouchaud, J. Phys. Condens. Matter **19**, 205101 (2007).
813. M.T. Cicerone, F.R. Blackburn, M.D. Ediger, Macromolecules **28**, 8224 (1995).
814. Y. Hwang, M.D. Ediger, J. Polym. Sci. Polym. Phys. Ed. **34**, 2853 (1996).
815. C.-Y. Wang, M.D. Ediger, J. Phys. Chem. B **104**, 1724 (2000).
816. D.D. Deppe, A. Dhinojwala, J.M. Torkelson, Macromolecules **29**, 3898 (1996).
817. G. Heuberger, H. Sillescu, J. Phys. Chem. **100**, 15255 (1996).
818. D.B. Hall, D.D. Deppe, K.E. Hamilton, A. Dhinojwala, J.M. Torkelson, J. Non-Cryst. Solids **235**, 48 (1998).
819. I. Chang, H. Sillescu, J. Phys. Chem. B **101**, 8794 (1997).
820. X.R. Zhu, C.H. Wang, J. Chem. Phys. **84**, 6086 (1986).
821. K. Zemke, K. Schmidt-Rohr, J.H. Magill, H. Sillescu, H.W. Spiess, Mol. Phys. **80**, 1317 (1993).
822. M.L.F. Nascimento, E.D. Zanotto, Phys. Rev. B **73**, 024209 (2006).
823. J. Horbach, W. Kob, Phys. Rev. B **60**, 3169 (1999).
824. M.D. Ediger, P. Harrowell, L. Yu, J. Chem. Phys. **128**, 034709 (2008).
825. F.H. Stillinger, J.A. Hodgdon, Phys. Rev. E, **50**, 2064 (1994).
826. G. Tarjus, D. Kivelson, J. Chem. Phys. **103**, 3071 (1995).
827. X.Y. Xia, P.G. Wolynes, Phys. Rev. Lett. **86**, 5526 (2001); J. Phys. Chem. B **105**, 6570 (2001).
828. S. Merabia, D. Long, Eur. Phys. J. E **9**, 195 (2002).
829. H. Wagner, R. Richert, J. Phys. Chem. B **103**, 4071 (1999).
830. T. Jorg, R. Bohmer, H. Sillescu, H. Zimmermann, Europhys. Lett. **49**, 748 (2000).
831. O. Urakawa, S.F. Swallen, M.D. Ediger, E.D. von Meerwall, Macromolecules **37**, 1558 (2004).
832. C.M. Roland, K.L. Ngai, D.J. Plazek, Macromolecules **37**, 7051 (2004).
833. R.A. Mendelsohn, W.A. Bowles, F.L. Finger, J. Polym. Sci. A2 **8**, 105 (1970).
834. M. Tirrell, M. Rubber, Chem. Technol. Rubber Rev. **57**, 523 (1984).
835. V.R. Raju, H. Rachapudy, W.W. Graessley, J. Polym. Sci. B Polym. Phys. **17**, 1223 (1979).
836. C.R. Bartels, B. Crist, W.W. Graessley, Macromolecules **17**, 2702 (1984).
837. K.L. Ngai, D.J. Plazek, R.W. Rendell, Rheol. Acta **36**, 307 (1997).
838. R.-J. Roe, D. Rigby, H. Furuya, H. Takeuchi, Comput. Polym. Sci. **2**, 32 (1992). R.-J. Roe, J. Chem. Phys. **100**, 1610 (1994).

839. C.J.F. Böttcher, P. Bordewijk, *Theory of Electric Polarization*, Elsevier, Amsterdam, Vol. **II** (1973).
840. G.F. Signorini, J.-L. Barrat, M.L. Klein, J. Chem. Phys. **92**, 1294 (1990).
841. A. Doss, G. Hinze, B. Schiener, J. Hemberger, R. Bohmer, J. Chem. Phys. **107**, 1740 (1997).
842. R. Casalini, D. Fioretto, A. Livi, M. Lucchesi, P.A. Rolla, Phys. Rev. B, **56**, 3016 (1997).
843. T. Blochowicz, E.A. Rössler, Phys. Rev. Lett. **92**, 225701 (2004).
844. T. Blochowicz, *Broadband Dielectric Spectroscopy in Neat and Binary Molecular Glass Formers*, Ph.D. Thesis, Logos Verlag, Berlin (2003).
845. A. Zetsche, F. Kremer, W. Jung, H. Schulze, Polymer **31**, 1883 (1990).
846. S. Kamath, R.H. Colby, S.K. Kumar, K. Karatasos, G. Floudas, G. Fytas, J.E.L. Roovers, J. Chem. Phys. **111**, 6121 (1999).
847. S. Kamath, R.H. Colby, S.K. Kumar, Macromolecules **36**, 8567 (2003).
848. K. Adachi, Y. Ishida, J. Poly. Sci. Polym. Phys. Ed. **14**, 2219 (1976).
849. K.L. Ngai, Macromolecules **24**, 4865 (1991).
850. M.S. Thayyil, S. Capaccioli, P.A. Rolla, K.L. Ngai, Philos. Mag. **88**, 4047 (2008).
851. D. Bingemann, N. Wirth, J. Gmeiner, E.A. Rossler, Macromolecules **40**, 5379 (2007).
852. L. Zhang, T.R. Lutz, J. Zhao, M.D. Ediger, J. Polym. Sci. B Polym. Phys. **45**, 2252 (2007).
853. K. Duvvuri, R. Richert, J. Phys. Chem. B **108**, 10451 (2004).
854. K.L. Ngai, S. Capaccioli, C.M. Roland, Macromolecules **39**, 8543 (2006).
855. T.R. Lutz, Y. He, M.D. Ediger, H. Cao, G. Lin, A.A. Jones, Macromolecules **36**, 1724 (2003).
856. B. Min, X. Qiu, M.D. Ediger, M. Pitsikalis, N. Hadjichristidis, Macromolecules **34**, 4466 (2001).
857. J. Zhao, L. Zhang, M.D. Ediger, Macromolecules **41**, 8030 (2008). Measurements were only made for 2%, 50% and 100% PEO.
858. A. Arbe, A. Alegria, J. Colmenero, S. Hoffmann, L. Willner, D. Richter, Macromolecules **32**, 7572 (1999).
859. G.-C. Chung, J.A. Kornfield, S.D. Smith, Macromolecules **27**, 5729 (1994).
860. J.C. Haley, T.P. Lodge, J. Chem. Phys. **122** (2005) 234914.
861. R.H. Colby, Polymer **30**, 1275 (1989).
862. J.C. Haley, T.P. Lodge, J. Rheol. **48** (2004) 463.
863. K.L. Ngai, D.J. Plazek, in *Physical Properties of Polymers Handbook*, 2nd ed., Chapter 26, J.E. Mark, Ed., Springer, Berlin, pp. 455–478 (2007).
864. J.Y. Cavaille, C. Jordan, J. Perez, L. Monnerie, G.P. Johari, J. Polym. Sci. B Polym. Phys. **25**, 1235 (1987).
865. L.I. Palade, V. Verney, P. Attené, Macromolecules **28**, 7051 (1995).
866. R.W. Gray, G. Harrison, J. Lamb, Proc. R. Soc. London Ser. A **356**, 77 (1977).
867. J. Cochrane, G. Harrison, J. Lamb, D.W. Phillips, Polymer **21**, 837 (1980).
868. K.L. Ngai, A. Schönhals, E. Schlosser, Macromolecules **25**, 4915 (1992).
869. D.J. Plazek, A. Schönhals, E. Schlosser, K.L. Ngai, J. Chem. Phys. **98**, 6488 (1993).
870. D.J. Plazek, X.D. Zheng, K.L. Ngai, Macromolecules **25**, 4920 (1992).
871. K.L. Ngai, D.J. Plazek, C. Bero, Macromolecules **26**, 1065 (1993).
872. K.L. Ngai, D.J. Plazek, A.K. Rizos, J. Polym. Sci. B Polym. Phys. **35**, 599 (1997).
873. R. Zorn, G.B. McKenna, L. Wilner, D. Richter, Macromolecules **28**, 8552 (1995).
874. C.G. Robertson, C.M. Rademacher, Macromolecules **37**, 10009 (2004); C.G. Robertson, L.-I. Palade, J. Non-Cryst. Solids **352**, 342 (2006).
875. C.M. Roland, K.L. Ngai, P.G. Santangelo, X.H. Qiu, M.D. Ediger, D.J. Plazek, Macromolecules **34**, 6159 (2001).
876. K.L. Ngai, J. Non-Cryst. Solids **353**, 709 (2007).
877. J.C. Haley, T.P. Lodge, Yiyong He, M.D. Ediger, E.D. von Meerwall, J. Mijovic, Macromolecules **36**, 6142 (2003).
878. C.M. Roland, P.G. Santangelo, K.L. Ngai, G. Meier, Macromolecules **26**, 6164 (1993).

879. P.G. Santangelo, C.M. Roland, K.L. Ngai, A.K. Rizos, H. Katerinopoulos, J. Non-Cryst. Solids **172–174**, 1084 (1994).
880. (a) A. Rizos, K.L. Ngai, Macromolecules **27**, 7076 (1994); (b) L.-M. Wang, Y. Tian, R. Liu, R. Richert, J. Phys. Chem. B **114**, 3618 (2010); (c) L.-M. Wang, Y. Tian, R. Liu, K.L. Ngai, J. Phys. Chem. B **115**, 719 (2011).
881. C.M. Roland, P.G. Santangelo, Z. Baram, J. Runt, Macromolecules **27**, 5382 (1994).
882. C.M. Roland, Macromolecules **28**, 3463 (1995).
883. A.A. Jones, J.F. O'Gara, P.T. Inglefield, J.T. Bendler, A.F. Yee, K.L. Ngai, Macromolecules **16**, 658 (1983).
884. K.L. Ngai, R.W. Rendell, A.F. Yee, D.J. Plazek, Macromolecules **24**, 61 (1991).
885. E.W. Fischer, G.P. Hellmann, H.W. Spiess, F.J. Horth, U. Ecarius, M. Wehrle, Macromol. Chem. Phys. Suppl. **12**, 189 (1985).
886. C. Schmidt, K.J. Kuhn, H.W. Spiess, Prog. Colloid Polym. Sci. **71**, 71 (1985).
887. J.Y. Jho, A.F. Yee, Macromolecules **24**, 1905 (1991).
888. C. Xiao, J. Wu, A.F. Yee, L. Xie, D. Gidley, K.L. Ngai, A.K. Rizos, Macromolecules **32**, 7913 (1999).
889. A.K. Rizos, L. Petihakis, K.L. Ngai, J. Wu, A.F. Yee, Macromolecules **32**, 7921 (1999).
890. B. Bergquist, Y. Zhu, A.A. Jones, P.T. Inglefield, Macromolecules **32**, 7925 (1999).
891. Y. He, T.R. Lutz, M.D. Ediger, J. Chem. Phys. **119**, 9956 (2003).
892. T.R. Lutz, Y. He, M.D. Ediger, Macromolecules **38**, 9826 (2005).
893. C.L. Jackson, G.B. McKenna, J. Chem. Phys. **93**, 9002 (1991).
894. C. Schick, E. Donth, Phys. Scr. **43**, 423 (1991).
895. G. Liu, Y. Li, J. Jonas, J. Chem. Phys. **95**, 6892 (1991).
896. J. Schüller, Y. Mel'nichenko, R. Richert, E.W. Fischer, Phys. Rev. Lett. **73**, 2224 (1994).
897. S. Stapf, R. Kimmich, R.O. Seitter, Phys. Rev. Lett. **75**, 2855 (1995).
898. K.L. Ngai, J. de Physique IV **10**, issue P7, 21 (2000).
899. K.L. Ngai, J. Thermal Anal. Calorimetry **57**, 745 (1999).
900. K.L. Ngai, Philos. Mag. B, **82**, 291 (2002).
901. J.D. Ferry, Macromolecules **24**, 5237 (1991).
902. D.J. Plazek, D.L. Plazek, Macromolecules **16**, 1469 (1983).
903. D.J. Plazek, M.J. Rosner, D.L. Plazek, J. Polym. Sci. B Polym. Phys. **26**, 473 (1988).
904. K.L. Ngai, D.J. Plazek, J. Polym. Sci. Polym. Phys. Ed. **23**, 2159 (1985).
905. (a) D.J. Plazek, G. Berry, in *Glass: Science and Technology, Vol. 3, Viscosity and Relaxation*, Chapter 7, D.R. Uhlmann, N. J. Kreidl, Eds., Academic Press, New York, (1986); (b) G. C. Berry, T.G. Fox, Adv. Polym. Sci. **5**, 261 (1968).
906. D.J. Plazek, I.C. Chay, K.L. Ngai, C.M. Roland, Macromolecules **28**, 6432 (1995). See also recent publications by (a) M. Paluch, S. Pawlus, A.P. Sokolov, K.L. Ngai, Macromolecules **43**, 3103 (2010); (b) X. Wu, X. Zhou, C. Liu, Z. Zhu, J. Appl. Phys. **106**, 013527 (2009); (c) J.R. Wu, G.S. Huang, Q.Y. Pan, J. Zheng, Y.C. Zhu, B. Wang, Polymer **48**, 7653 (2007); (d) X.B. Wu, Z.G. Zhu, J. Phys. Chem. B **113**, 11147 (2009); (e) X. Wu, H. Wang, C. Liu, Z. Zhu, Soft Matter, **7**, 579 (2011).
907. K.L. Ngai, D.J. Plazek, Macromolecules **35**, 9136 (2002).
908. R.W. Gray, G. Harrison, J. Lamb, J. Polym. Sci. Polym. Phys. Ed. **14**, 1361 (1976).
909. A. Schönhals, E. Schlosser, Phys. Scr. **T49A**, 233 (1993).
910. A. Schönhals, Macromolecules **26**, 1309 (1993).
911. K. Adachi, H. Hirano, Macromolecules **31**, 3958 (1998).
912. T. Nicolai, G. Floudas, Macromolecules **31**, 2578 (1998).
913. K.L. Ngai, R. Casalini, C.M. Roland, Macromolecules **38**, 4363 (2005).
914. P.G. Santangelo, K.L. Ngai, C.M. Roland, Macromolecules **26**, 2682 (1993).
915. E.R. Fitzgerald, L.D. Grandine, J.D. Ferry, J. Appl. Phys. **24**, 640 (1953). For a historical perspective, see J.D. Ferry, Macromolecules **24**, 5237 (1991).
916. E. Catsiff, A.V. Tobolsky, J. Colloid Sci. **10**, 375 (1955); A.V. Tobolsky, *Properties and Surture of Polymers*, Wiley, New York, NY (1960).
917. M.L. Williams, J.D. Ferry, J. Colloid Sci. **10**, 474 (1955).

918. D. Richter, M. Monkenbusch, J. Allgeier, A. Arbe, J. Colmenero, B. Farago, Y. Choel Bae, R. Faust, J. Chem. Phys. **111**, 6107 (1999).
919. A. Arbe, M. Monkenbusch, J. Stellbrink, D. Richter, B. Farago, M. Almdal, R. Faust, Macromolecules **34**, 1281 (2001).
920. S. Krushev, W. Paul, G.D. Smith, Macromolecules **35**, 4198 (2002).
921. E. Schlosser, A. Schönhals, Prog. Colloid Polym. Sci. **91**, 158 (1993).
922. A. Schönhals, E. Schlosser, Phys. Scr. **T49A**, 233 (1993).
923. H.W. Spiess, J. Non-Cryst. Solids **131–133**, 766 (1991).
924. S. Kaufmann, S. Wefing, D. Schaefer, H.W. Spiess, J. Chem. Phys. **93**,197 (1990).
925. K.L. Ngai, R.W. Rendell, Macromolecules **20**, 1066 (1987).
926. I.M. Ward, *Mechanical Properties of Solid Polymers*, 2nd ed., Wiley, New York, NY (1983).
927. H.E.H. Meijer, L.E. Govaert, Prog. Polym. Sci. **30**, 915 (2005).
928. H. Eyring, J. Chem. Phys. **4**, 283 (1936).
929. R.E. Robertson, J. Chem. Phys. **44**, 3950 (1966).
930. M.C. Boyce, D.M. Parks, A.S. Argon, Mech. Mater. **7**, 15 (1988).
931. C.P. Buckley, D.C. Jones, Polymer **36**, 3301 (1995).
932. J.M. Caruthers, D.B. Adolf, R.S. Chambers, P. Shrikhande, Polymer **45**, 4577 (2004).
933. K. Chen, K.S. Schweizer, Europhys. Lett. **79**, 26006 (2007).
934. K.L. Ngai, A.F. Yee, in *Relaxations in Complex Systems*, K.L. Ngai, G. Wright, Eds., Naval Research Laboratory, Washington, DC (1984).
935. R.W. Rendell, K.L. Ngai, G.R. Fong, A.F. Yee, R.J. Bankert, Polym. Eng. Sci. **27**, 2 (1987).
936. A.F. Yee, R.J. Bankert, K.L. Ngai, R.W. Rendell, J. Polym. Sci. B Polym. Phys. **26**, 2463 (1988).
937. G.B. McKenna, L.Z. Zapas, J. Polym. Sci. Polym. Phys. Ed. **23**, 1647 (1985); Polym. Eng. Sci. **26**, 725 (1986).
938. (a) G.B. McKenna, J. Phys. Condens. Matter **15**, S737 (2003); (b) A.L. Flory, G.B. McKenna, Polymer, **46**, 5211 (2005).
939. B. Bernstein, E.A. Kearsley, L.J. Zapas, Trans. Soc. Rheol. **9**, 27 (1965).
940. C.M. Roland, K.L. Ngai, A.F. Yee, Rubber Chem. Technol. **66**, 615 (1993).
941. H.-N. Lee, K. Paeng, S.F. Swallen, M.D. Ediger, Science **323**, 231 (2009).
942. S.S. Sternstein, in *Treatise on Materials Science and Technology*, J.M. Schultz, Ed., Academic, New York, NY, Vol. **10B** (1977).
943. D.L. Questad, Polym. Eng. Sci. **26**, 269 (1986).
944. R. Pixa, C. Goett, D. Froelich, Polym. Bull. **14**, 53 (1985). R. Pixa, B. Grisoni, T. Gay, D. Froelich, Polym. Bull. **16**, 381 (1986).
945. G.P. Johari, Ann. N.Y. Acad. Sci. **279** (1976) 117.
946. G.P. Johari, J. Chem. Phys. **58**, 1766 (1973).
947. K. Pathmanathan, G.P. Johari, J. Phys. C **18**, 6535 (1985).
948. R.K. Chan, K. Pathmanathan, G.P. Johari, J. Phys. Chem. **90**, 6358 (1986).
949. (a) M.G. Parthun, G.P. Johari, Macromolecules **29**, 3254 (1992); (b) M.G. Parthun, G.P. Johari, J. Chem. Phys. **103**, 7611 (1995).
950. K. Pathmanathan, G.P. Johari, J. Polym. Sci. B Polym. Phys. **25**, 379 (1987).
951. C. Mai, S. Etienne, J. Perez, G.P. Johari, Philos. Mag. B **50**, 657 (1985).
952. C. Mai, S. Etienne, J. Perez, G.P. Johari, J. Non-Cryst. Solids **74**, 119 (1985).
953. (a) K.L. Ngai, Phys. Rev. E **57**, 7346 (1998); (b) K.L. Ngai, Physica A **261**, 36 (1998).
954. H.S. Chen, J. Non-Cryst. Solids **46**, 289 (1981).
955. H. Okumura, H.S. Chen, A. Inoue, T. Masumoto, J. Non-Cryst. Solids **130**, 304 (1991), ibid. **150**, 401 (1992); L. Hu et al., App. Phys. Lett. **96**, 221908 (2010).
956. J.M. Pelletier, B. Van de Moortele, I.R. Lu, Mater. Sci. Eng. **A336**, 190 (2002).
957. (a) P. Rösner, K. Samwer, P. Lunkenheimer, Europhys. Lett. **68**, 226 (2004); (b) Z.F. Zhao, P. Wen, C.H. Shek, W.H. Wang, Phys. Rev. B **75**, 174201 (2007).

958. T. Ichitsubo, E. Matsubara, T. Yamamoto, H.S. Chen, N. Nishiyama, J. Saida, K. Anazawa, Phys. Rev. Lett. **95**, 245501 (2005).

959. (a) A.S. Argon, Acta Metall. **27**, 47 (1979); (b) A. Argon, L.T. Shi, Acta Metall. **31**, 499 (1983); (c) C.A. Schuh, A.C. Lund, Nat. Mater. **2**, 449 (2003); (d) F. Spaepen, Acta Metall. **25**, 407 (1977); (e) P.S. Steif, F. Spaepen, J.W. Hutchinson, Acta Metall. **30**, 447 (1982); (f) J.S. Langer, Phys. Rev. E **70**, 041502 (2004); (g) W.J. Johnson, K. Samwer, Phys. Rev. Lett. **95**, 195501 (2005); (h) M. Chen, Annu. Rev. Mater. Res. **38**, 445 (2008); (i) W.H. Wang, presentation made at "The Physics of Glass," KITP, UCSB, May 13th (2010). H.B. Yu, W.H. Wang, H.Y. Bai, Y. Wu, M.W. Chen, Phys. Rev. B **81**, 220201(R) (2010); (j) D. Pan, A. Inoue, T. Sakurai, M.W. Chen, Proc. Natl. Acad. Sci. USA. **105**, 14 769 (2008); (k) S.G. Mayr, Phys. Rev. Lett. **97**, 195501 (2006); (l) J.J. Lewandowski, W.H. Wang, A.L. Greer, Philos. Mag. Lett. **85**, 77 (2005), J. Appl. Phys. **99**, 093506 (2006); (m) D. Rodney, presented a talk entitled "Thermally-activated events in glasses from atomic-scale simulations: Relation between STZ and β-relaxation" in "The Physics of Glass," KITP, UCSB, May 21st (2010), and D. Rodney, C. Schuh, E. Homer, E. Bitzek, P. Derlet, to be published. In this presentation he showed by molecular dynamics simulations of $Cu_{50}Zr_{50}$ quenched metallic glass in 2D with and without stress applied. By sampling phase space locally in the two cases, they found that the local shear and its evolution with time in the microstructure as well as the distribution of energy barriers of local motions are the same in the two cases. Without stress, the results reflect the primitive or JG β-relaxation, and with stress, the results reflect the STZ. Hence, the similar results in the two cases support that the primitive or JG β-relaxation is responsible for STZ, as shown experimentally in (i) by Wang and coworkers; (n) H.E.H. Meijer, L.E. Govaert, Prog. Polym. Sci. **30**, 915 (2005); (o) N.T. Tsui et al., Polymer **49**, 4703 (2008); (p) L.P. Chen, A.F. Yee, E.J. Moskala, Macromolecules **32**, 5944 (1999); (q) H. Eyring, J. Chem. Phys. **4**, 283 (1936); (r) M.C. Boyce, D.M. Parks, A.S. Argon, Mech. Mater. **7**, 15 (1988); (s) J. Richeton et al., Philos. Mag. **87**, 3629 (2007).

960. (a) J. Hachenberg, K. Samwer, J. Non-Cryst. Solids **352**, 5110 (2006); (b) J. Hachenberg, D. Bedorf, K. Samwer, R. Richert, A. Kahl, M.D. Demetriou, W.L. Johnson, Appl. Phys. Lett. **92**, 131911 (2008); (c) D. Bedorf, K. Samwer, J. Non-Cryst. Solids **356**, 340 (2010).

961. G.P. Johari, M. Goldstein, J. Chem. Phys. **55**, 4245 (1971).

962. G.P. Johari, J. Chim. **82**, 283 (1985).

963. G.P. Johari, in *Lecture Notes in Physics*, Th. Dorfmüller, G. Williams, Eds., Springer, Berlin, Vol. **277**, p. 90 (1987).

964. M. Goldstein, J. Phys. Paris. Colloq. **C2**, 2 (1975), J. Chem. Phys. **132**, 041104 (2010); J. Non-Cryst. Solids **357**, 249 (2011), Special issue of papers presented in the 6th IDMRCS, Rome, Italy, August 30th–September 4th (2009).

965. G.P. Johari, in *Plastic Deformation of Amorphous and Semicrystalline Materials*, Les Houches Lectures 1982 (les Editions de Physique, France, 1982), p. 109.

966. J.P. Johari, J. Non-Cryst. Solids **307–310**, 317 (2002).

967. F. Stillinger, T.A. Weber, Phys. Rev. A **28**, 2408 (1983).

968. G. Williams, D.C. Watts, Trans. Faraday Soc. **67**, 1971 (1971).

969. F. Garwe, A. Schönhals, H. Lockwenz, M. Beiner, K. Schröter, E. Donth, Macromolecules **29**, 247 (1996).

970. (a) K.L. Ngai, Macromolecules **32**, 7140 (1999); (b) S. Kahle, J. Korus, E. Hempel, R. Unger, S. Höring, K. Schröter, E. Donth, Macromolecules **30**, 7214 (1997).

971. A. Kudlik, S. Benkhof, T. Blochowicz, C. Tschirwitz, E. Rössler, J. Mol. Struct. **479**, 201 (1999).

972. P. Lunkenheimer, U. Schneider, R. Brand, A. Loidl, Contemp. Phys. **41**, 15 (2000).

973. K.L. Ngai, D. Prevosto, S. Capaccioli, C.M. Roland, J. Phys. Condens. Matter **20**, 244125 (2008).

974. S. Capaccioli, K. Kessairi, D. Prevosto, M. Lucches, P.A. Rolla, J. Phys. Condens. Matter **19**, 205133 (2007).

975. R. Casalini, S. Capaccioli, to be published.
976. S. Capaccioli, K. Kessairi, D. Prevosto, M. Lucchesi, K.L. Ngai, J. Non-Cryst. Solids **352**, 4643 (2006).
977. D. Prevosto, S. Capaccioli, P.A. Rolla, M. Paluch, S. Pawlus, S. Hensel-Bielowka, E. Kaminska, J. Non-Cryst. Solids **352**, 4685 (2006).
978. C.M. Roland, M.J. Schroder, J.J. Fontanella, K.L. Ngai, Macromolecules **37**, 2630 (2004); R.B. Bogoslovov et al., Macromolecules **43**, 2904 (2010). In this 2010 paper, the authors fit the spectra of PBD assuming the α- and β-relaxations are additive, resulting in their value of $n = 0.57$ larger than those obtained before in the spirit of the CM [195,874,981,979], which are in the range of $0.50 \leq n \leq 0.54$.
979. M.J. Schroder, K.L. Ngai, C.M. Roland, J. Polym. Sci. B Polym. Phys. **45**, 342 (2007).
980. S.A. Lusceac, C. Gainaru, M. Vogel, C. Koplin, P. Medick, E.A. Roessler, Macromolecules **38**, 5625 (2005); R.D. Deegan, S.R. Nagel, Phys. Rev. B **52**, 5653 (1995).
981. R. Casalini, K.L. Ngai, C.G. Robertson, C.M. Roland, J. Polym. Sci. B Polym. Phys. **38**, 1841 (2000).
982. M. Beiner, S. Kahle, E. Hempel, K. Schröter, E. Donth, Macromolecules **31**, 8973 (1998).
983. V.A. Bershtein, L.M. Egorova, R.E. Prud'homme, J. Macromol. Sci. Phys. **B36**, 513 (1997).
984. V.A. Bershtein, L.M. Egorova, *Differential Scanning Calorimetry of Polymers, Physics, Chemistry, Analysis, Technology*, Ellis Horwood, New York, NY (1994).
985. J.M. Pelletier, J. Perez, L. Duffrene, A. Sekkat, J. Non-Cryst. Solids **258**, 119 (1999).
986. L. Hornbøll, Y. Yue, J. Non-Cryst. Solids **354**, 350 (2008).
987. Y.Z. Yue, J. Non-Cryst. Solids **354**, 1112 (2008).
988. H.S. Chen, T.T. Wang, J. Appl. Phys. **52**, 5898 (1981).
989. H.S. Chen, C.R. Kurkjian, J. Am. Ceram. Soc. **66**, 613 (1983).
990. H.S. Chen, A. Inoue, J. Non-Cryst. Solids **61–62**, 805 (1984).
991. A. Inoue, T. Masumoto, H.S. Chen, J. Mater. Sci. **19**, 3953 (1984).
992. (a) S. Vyazovkin, I. Dranca, Pharm. Res. **23**, 422 (2006); (b) S. Vyazovkin, I. Dranca, J. Phys. Chem. B **109**, 18637 (2005); (c) R. Surana, A. Pyne, R. Suryanarayanan, Pharm. Res. **21**, 867 (2004); (d) I. Dranca, S, Bhattacharya, S. Vyazovkin, R. Suryanarayanan, Pharm. Res. **26**, 1064 (2009).
993. A. Napolitano, P.B. Macedo, J. Res. Nat. Bur. Stand. **72A**, 425 (1968).
994. L. Hu, Y. Yue, J. Phys. Chem. B **112**, 9053 (2008). For a review, see LiNa Hu, ChunZhi Zhang, YuanZheng, Yue, XiuFang Bian, Chin. Sci. Bull. **55**, 457 (2010). R. Ruiz Santos, E. Prokhorov, J. González-Hernández, G. Luna-Bárcenas, Yu. Kovalenko, J. Non-Cryst. Solids **356**, 2541 (2010). The observed τ_β is in accord with CM prediction.
995. A. Kudlik, C. Tschirwitz, T. Blochowicz, S. Benkhof, E. Rössler, J. Non-Cryst. Solids **235–237**, 406 (1998).
996. K.L. Ngai, S. Capaccioli, Phys. Rev. E **69**, 031501 (2004).
997. K.L. Ngai, M. Paluch, J. Phys. Chem. B **107**, 6865 (2003).
998. R. Böhmer, G. Diezemann, B. Geil, G. Hinze, A. Nowaczyk, M. Winterlich, Phys. Rev. Lett. **97**, 135701 (2006).
999. A. Nowaczyk, B. Geil, G. Hinze, R. Böhmer, Phys. Rev. E **74**, 041505 (2006).
1000. (a) A. De Gusseme, L. Carpentier, J.F. Willart, M. Descamps, J. Phys. Chem. B **107**, 10879 (2003); (b) L. Carpentier, R. Decressain, S. Desprez, M. Descamps, J. Phys. Chem. B **110**, 457 (2006); (c) L. Carpentier, R. Decressain, A. De Gusseme, C. Neves, M. Descamps, Pharm. Res. **23**, 798 (2006).
1001. V. Yu Kramarenko, T.A. Ezquerra, V.P. Privalko, Phys. Rev. E **64**, 051802 (2001).
1002. L. Wu, Phys. Rev. B **43**, 9906 (1991).
1003. R. Nath, A. Nowaczyk, B. Geil, R. Böhmer, J. Non-Cryst. Solids **353**, 3788 (2007).
1004. A.R. Brás, J.P. Noronha, A.M.M. Antunes, M.M. Cardoso, A. Schönhals, F. Affouard, M. Dionísio, N.T. Correia, J. Phys. Chem. B **112**, 11087 (2008).
1005. R.H. Cole, D.W. Davidson, J. Chem. Phys. **20**, 1389 (1952).

1006. S.K. Garg, C.P. Smyth, J. Phys. Chem. **69**, 1294 (1965).
1007. A. Kudlik, C. Tschirwitz, S. Benkhof, T. Blochowicz, E. Rössler, Europhys. Lett. **40**, 649 (1997).
1008. O.E. Kalinovskaya, J.K. Vij, J. Chem. Phys. **112**, 3262 (2000).
1009. G.P. Johari, O.E. Kalinovskaya, J.K. Vij, J. Chem. Phys. **114**, 4634 (2001).
1010. (a) S.S.N. Murthy, M. Tyagi, J. Chem. Phys. **117**, 3837 (2002); (b) M. Mizukami, H. Fujimori, M. Oguni, Solid State Commun. **100**, 83 (1996).
1011. (a) L.-M. Wang, R. Richert, J. Phys. Chem. B **109**, 11091 (2005); (b) H. Huth, L.-M. Wang, C. Schick, R. Richert, J. Chem. Phys. **126**, 104503 (2007).
1012. (a) K. Kaminski, E. Kaminska, P. Wlodarczyk, S. Pawlus, D. Kimla, A. Kasprzycka, M. Paluch, J. Ziolo, W. Szeja, K.L. Ngai, J. Phys. Chem. B **110**, 25045 (2006); (b) K. Kaminski, E. Kaminska, S. Hensel-Bielowka, E. Chelmecka, M. Paluch, J. Ziolo, P. Wlodarczyk, K.L. Ngai, Phys. Chem. B **112**, 7662 (2008); (c) K. Kaminski, E. Kaminska, P. Wlodarczyk, S. Pawlus, D. Kimla, A. Kasprzycka, M. Paluch, J. Ziolo, W. Szeja, K.L. Ngai, J. Phys. Chem. B, **112**, 12816 (2008); (d) K. Kaminski, E. Kaminska, K.L. Ngai, M. Paluch, P. Wlodarczyk, A. Kasprzycka, W. Szeja, J. Phys. Chem. B, **113**, 10088 (2009); (e) K. Kaminski, E. Kaminska, P. Wlodarczyk, M. Paluch, J. Ziolo, K.L. Ngai, J. Phys. Condens. Matter. **20**, 335104 (2008).
1013. D. Prevosto, S. Capaccioli, M. Lucchesi, S. Sharifi, K. Kessairi, K.L. Ngai, Philos. Mag. **88**, 4063 (2008).
1014. D. Prevosto, S. Capaccioli, M. Lucchesi, P. Rolla, K.L. Ngai, J. Chem. Phys. **120**, 4808 (2004).
1015. G. Dominguez-Espinosa, R. Díaz-Calleja, E. Riande, Macromolecules **39**, 5043 (2006), ibid. **39**, 3071 (2006).
1016. (a) A. Mandanici, W. Huang, M. Cutroni, R. Richert, J. Chem. Phys. **128**, 124505 (2008).
1017. S. Capaccioli, M.S. Thayyil, K.L. Ngai (2009), unpublished.
1018. (a) A. Mandanici, W. Huang, M. Cutroni, R. Richert, Philos. Mag. **88**, 3961 (2008); (b) A. Mandanici, M. Cutroni, Mater. Sci. Eng. A **521–522**, 279 (2009).
1019. S. Maślanka, M. Paluch, W.W. Sułkowski, C.M. Roland, J. Chem. Phys. **122**, 084511 (2005).
1020. K.L. Ngai, E. Kamińska, M. Sekuła, M. Paluch, J. Chem. Phys. **123**, 204507 (2005).
1021. R.L. Leheny, S.R. Nagel, Europhys. Lett. **39**, 447 (1997).
1022. A. Schönhals, F. Kremer, F. Stickel, Phys. Rev. Lett. **71**, 4096 (1993).
1023. A. Hofmann, F. Kremer, E.W. Fischer, A. Schönhals, in *Disorder Effects on Relaxational Processes*, R. Richert, A. Blumen, Eds., Springer, Berlin, p. 309 (1994).
1024. C.F. Behrens, T. Christensen, T.G. Christiansen, J. Dyre, N.B. Olsen, Phys. Rev. Lett. **76**, 1553 (1996).
1025. S. Adichtchev, T. Blochowicz, C. Gainaru, V.N. Novikov, E.A. Rössler, C. Tschirwitz, J. Phys. Condens. Matter **15**, S835 (2003).
1026. J. Wiedersich, T. Blochowicz, S. Benkhof, A. Kudlik, N.V. Surovtsev, C. Tschirwitz, V.N. Novikov, E. Rössler, J. Phys. Condens. Matter **11**, A147 (1999).
1027. K. Pathmanathan, G.P. Johari, Polymer **29**, 303 (1988).
1028. C. Svanberg, R. Bergman, P. Jacobson, L. Börjesson, Phys. Rev. B **66**, 054304 (2002).
1029. J. Mattsson, R. Bergman; P. Jacobsson, L. Börjesson, Phys. Rev. Lett. **90**, 075702 (2003).
1030. K.L. Ngai, Philos. Mag. B **82**, 291 (2002).
1031. N.B. Olsen, J. Non-Cryst. Solids **235–237**, 399 (1998).
1032. M. Jimenez-Ruiz, M.A. Gonzalez, F.J. Bermejo, M.A. Miller, N.O. Birge, I. Cendoya, A. Alegria, Phys. Rev. B **59**, 9155 (1999).
1033. M. Wagner, R. Richert, J. Chem. Phys. **110**, 11660 (1999).
1034. P. Lunkenheimer, *Dielectric Spectroscopy of Glassy Dynamics*, Shaker, Aachen (1999).
1035. P. Lunkenheimer, A. Loidl, Chem. Phys. **284**, 205 (2002).
1036. P. Lunkenheimer, R. Wehn, Th. Riegger, A. Loidl, J. Non-Cryst. Solids. **307–310**, 336 (2002).

1037. S. Hensel-Bielowka, M. Paluch, Phys. Rev. Lett. **89**, 025704 (2002).
1038. S. Pawlus, S. Hensel-Bielowka, K. Grzybowska, J. Ziolo, M. Paluch, Phys. Rev. B **71**, 174107 (2005).
1039. K. Grzybowska, A. Grzybowski, M. Paluch, S. Capaccioli, J. Chem. Phys. **125**, 044904 (2006).
1040. K. Grzybowska, A. Grzybowski, J. Ziolo, S.J. Rzoska, M. Paluch, J. Phys. Condens. Matter **19**, 376105 (2007).
1041. K. Kaminska, E. Kaminski, M. Paluch, K.L. Ngai, J. Chem. Phys. **124**, 164511 (2006).
1042. S. Sharifi, D. Prevosto, S. Capaccioli, M. Lucchesi, M. Paluch, J. Non-Cryst. Solids, **353**, 4313 (2007).
1043. K.L. Ngai, K. Grzybowska, A. Grzybowski, E. Kaminska, K. Kaminski, M. Paluch, S. Capaccioli, J. Non-Cryst. Solids **354**, 5085 (2008).
1044. J. Dyre, B. Olsen, Phys. Rev. Lett. **91**, 155703 (2003).
1045. N.B. Olsen, T. Christensen, J.C. Dyre, Phys. Rev. Lett. **86**, 1271 (2001).
1046. A.I. Nielsen, T. Christensen, B. Jacobsen, K. Niss, B. Olsen, R. Richert, J.C. Dyre, arXiv:0712.2827v ; J. Chem. Phys. **130**, 154508 (2009).
1047. (a) A. Brodin, C. Gainaru, V. Porokhonskyy, E.A. Rössler, J. Phys. Condens. Matter **19**, 205104 (2007); (b) C. Gainaru, R. Kahlau, E.A. Rössler, R. Böhmer, J. Chem. Phys. **131**, 184510 (2009); (c) C. Gainaru, W. Hiller, R. Böhmer, Macromolecules **43**, 1907 (2010).
1048. M. Mizukami, H. Fujimori, M. Oguni, Solid State Commun. **100**, 83 (1996).
1049. G. Meier, B. Gerharz, D. Boese, E.W. Fischer, J. Chem. Phys. **94**, 3050 (1991).
1050. S. Kahle, J. Gapinski, G. Hinze, A. Patkowski, G. Meier, J. Chem. Phys. **122**, 074506 (2005).
1051. M. Beiner, K.L. Ngai, Macromolecules **38**, 7033 (2005).
1052. D. Pisignano, S. Capaccioli, R. Casalini, M. Lucchesi, P.A. Rolla, A. Justl, E. Rössler, J. Phys. Condens. Matter **13**, 4405 (2001).
1053. A. Brodin, E.A. Rössler, J. Chem. Phys. **125**, 114502 (2006).
1054. A. Brodin, E.A. Rössler, J. Chem. Phys. **126**, 244508 (2007).
1055. J. Wong, C.A. Angell, *Glass Structure by Spectroscopy*, Dekker, New York (1976).
1056. K.L. Ngai, J. Chem. Phys. **110**, 10576 (1999).
1057. K.L. Ngai, C. León, Phys. Rev. B **66**, 064308 (2002).
1058. K.L. Ngai, J. Habasaki, Y. Hiwatari, C. León, J. Phys. Condens. Matter **15**, S1607 (2003).
1059. C. León, A. Rivera, A. Várez, J. Sanz, J. Santamaria, K.L. Ngai, Phys. Rev. Lett. **86**, 1279 (2001).
1060. A.K. Rizos, J. Alifragis, K.L. Ngai, P. Heitjans, J. Chem. Phys. **114**, 931 (2001).
1061. A. Rivera, C. León, C.P.E. Varsamis, G.D. Chryssikos, K.L. Ngai, C.M. Roland, L.J. Buckley, Phys. Rev. Lett. **88**, 125902 (2002).
1062. A.P. Sokolov, A. Kisliuk, V.N. Novikov, K.L. Ngai, Phys. Rev. B, **63**, 172204 (2001).
1063. A.P. Sokolov, unpublished (2003).
1064. R. Casalini, C.M. Roland, S. Capaccioli, J. Chem. Phys. **126**, 184903 (2007), and to be published.
1065. D. Prevosto, S. Sharifi, S. Capaccioli, P. Rolla, S. Hensel-Bielowka, M. Paluch, J. Chem. Phys. **127**, 114507 (2007).
1066. D. Prevosto, S. Capaccioli, M. Lucchesi, S. Sharifi, K. Kessairi, K.L. Ngai, Philos. Mag. **88**, 4083 (2008).
1067. M.S. Thayyil, S. Capaccioli, paper presented at the 6th International Discussion Meeting on Relaxation in Complex Systems, Rome, Italy, August 30–September 4th, 2009, and to be published.
1068. D. Bedrov, G.D. Smith, Euro Polym. J. **42**, 3248 (2006).
1069. T. Psurek, S. Maslanka, M. Paluch, R. Nozaki, K.L. Ngai, Phys. Rev. E **70**, 011503 (2004).
1070. (a) C. Lorthioir, A. Alegrıa, J. Colmenero, Phys. Rev. E **68**, 031805 (2003); (b) C. Lorthioir, A. Alegrıa, J. Colmenero, Euro. Phys. J. E **12**, S127 (2003); (c) T. Blochowicz et al., J. Phys. Chem. B (2011) in press. The interpretation of α' coming from some MTHF by confinement is muddled by fact that the α-relaxation time of MTHF is identical to the reorientation time of 3S detected by PCS.

1071. (a) K.L. Ngai, S. Capaccioli, M. Shahin Thayyil, N. Shinyashiki, J. Therm. Anal. Calorimetry **99**, 123 (2010); (b) S. Capaccioli, K.L. Ngai, S. Ancherbak, P.A. Rolla, N. Shinyashiki, J. Non-Cryst. Solids, **357**, 641 (2011).
1072. S. Zhang, X. Jin, P.C. Painter, J. Runt, Macromolecules **35**, 3636 (2002).
1073. X. Jin, S. Zhang, J. Runt, Polymer **43**, 6247 (2002).
1074. X. Jin, S. Zhang, J. Runt, Macromolecules **37**, 8110 (2004).
1075. H. Fujimori, M. Oguni, Solid State Commun. **94**, 157 (1995); M. Oguni, J. Sekine, H. Mine, J. Non-Cryst. Solids **352**, 4665 (2006).
1076. I. Alig, G.P. Johari, J. Polym. Sci. B Polym. Phys. **31**, 299 (1993).
1077. M.G. Parthun, G.P. Johari, J. Chem. Phys. **102**, 6301 (1995), **103**, 440 (1995), **103**, 7611 (1995), Macromolecules **29**, 3254 (1992).
1078. S. Monserrat, F. Roman, P. Colomer, Polymer **44**, 101 (2003).
1079. D.A. Wasylyshyn, G.P. Johari, J. Chem. Phys. **104**, 5683 (1996); J. Polym. Sci. B Polym. Phys. **35**, 437 (1997).
1080. E. Tombari, C. Ferrari, G. Salvetti, G.P. Johari, J. Phys. Condens. Matter **9**, 7017 (1997); J. Phys. Chem. Phys. **1**, 1965 (1999).
1081. N.M. Alves, J.L. Gomez Ribelles, J.A. Gomez Tejedor, J.F. Mano, Macromolecules **37**, 3735 (2004).
1082. M.S. Sanchez, G.G. Ferrer, C.T. Cabanilles, J.M.M. Duenas, M.M. Pradas, J.L.G. Ribelles, Polymer **42**, 10071 (2001).
1083. S. Reissig, M. Beiner, S. Zeeb, S. Höring, E. Donth, Macromolecules **32**, 5701 (1999).
1084. M. Wind, R. Graf, S. Renker, H.W. Spiess, Macromol. Chem. Phys. **206**, 142 (2005).
1085. T. Fujima, H. Frusawa, K. Ito, Phys. Rev. E **66**, 031503 (2002).
1086. E. Donth, K. Schröter, S. Kahle, Phys. Rev. E **60**,1099 (1999).
1087. (a) C. Svanberg, R. Bergman, Philos. Mag. **87**, 383 (2007); (b) R. Bergman, F. Alvarez, A. Alegrıa, J. Colmenero, J. Chem. Phys. **109**, 7546 (1998); (c) R. Bergman, C. Svanberg, Phys. Rev. E **72**, 043501 (2005).
1088. M. Paluch, C.M. Roland, S. Pawlus, J. Ziolo, K.L. Ngai, Phys. Rev. Lett. **91**, 115701 (2003).
1089. A. Sanz, A. Nogales, T.A. Ezquerra, N. Lotti, L. Finelli, Phys. Rev. E **70**, 021502 (2004).
1090. A. Nogales, A. Sanz, T.A. Ezquerra, J. Non-Cryst. Solids **352**, 4649 (2006).
1091. E.L. Gjersing, S. Sen, P. Yu, B.G. Aitken, Phys. Rev. B **76**, 214202 (2007).
1092. F. Affouard, J.-F. Willart, M. Descamps, J. Non-Cryst. Solids **307–310**, 9 (2002).
1093. S. Capaccioli, K.L. Ngai, N. Shinyashiki, J. Phys. Chem. B **111**, 8197 (2007).
1094. K.L. Ngai, S. Capaccioli, N. Shinyashiki, J. Phys. Chem. B **112**, 3826 (2008).
1095. D. Cangialosi, A. Alegria, J. Colmenero, J. Chem. Phys. **128**, 224508 (2008).
1096. K.L. Ngai, J. Non-Cryst. Solids **353**, 4237 (2007).
1097. T.P. Lodge, E.R. Wood, J.C. Haley, J. Polym. Sci. B Polym. Phys. **44**, 756 (2006); J. Zhao, M.D. Ediger, Y. Sun, L. Yu, Macromolecules **42**, 6777 (2009); S. Arrese-Igor et al., Macromolecules **43**, 6406 (2010). Note that the β-relaxation of PI in 20 and 35%PI blends is misidentified by them as α-relaxation in Fig. 8.
1098. K. Adachi, I. Fujihara, Y. Ishida, J. Polym. Sci. Polym. Phys. Ed. **13**, 2155 (1975); Polym. J. **11**, 233 (1979).
1099. G. Floudas, W. Steffen, E.W. Fischer, W. Brown, J. Chem. Phys. **99**, 695 (1993).
1100. N. Taniguchi, O. Urakawa, K. Adachi, Macromolecules **37**, 7832 (2004).
1101. M.S. Thayyil, S. Capaccioli (2010) to be published.
1102. H. Jansson, J. Swenson, Eur. Phys. J. E **12**, S51 (2003).
1103. H. Jansson, R. Bergman, J. Swenson, J. Non-Cryst. Solids **351**, 2858 (2005).
1104. J. Swenson, H. Jansson, W.S. Howells, S. Longeville, J. Chem. Phys. **122**, 084505 (2005).
1105. J. Swenson, H. Jansson, R. Bergman, Phys. Rev. Lett. **96**, 247802 1–4 (2006).
1106. S. Cerveny, G.A. Schwartz, A. Alegría, R. Bergman, J. Swenson, J. Chem. Phys. **124**, 194501 (2006).
1107. (a) S. Cerveny, A. Alegría, J. Colmenero, Phys. Rev. E **77**, 031803 (2008); (b) J. Chem. Phys. **128**, 044901 (2008); (c) J. Chem. Phys. **124**, 194501 (2006).

1108. (a) S. Cerveny, J. Colmenero, A. Alegría, Macromolecules **38**, 7056 (2005); (b) J. Non-Cryst. Solids **353**, 4523 (2007).
1109. H. Jansson, R. Bergman, J. Swenson, J. Phys. Chem. B **109**, 24134 (2005).
1110. J. Swenson, H. Jansson, J. Hedström, R. Bergman, J. Phys. Condens. Matter **19**, 205109 (2007).
1111. S.E. Pagnotta, S. Cerveny, A. Alegría, J. Colmenero, J. Chem. Phys. **131**, 085102 (2009).
1112. H. Jansson, J. Swenson, Biochim. Biophys. Acta **1804**, 20 (2010).
1113. H.S. Chen, N. Morito, J. Non-Cryst. Solids **72**, 287 (1985).
1114. K.L. Ngai, J. Non-Cryst. Solids **352**, 404 (2006).
1115. Z.F. Zhao, P. Wen, C.H. Shek, W.H. Wang, Phys. Rev. B **75**, 174201 (2007).
1116. A. Brodin, R. Bergman, J. Mattsson, E.A. Rössler, Eur. Phys. J. B **36**, 349 (2003).
1117. S.K. Nayak, S.S.N. Murthy, J. Chem. Phys. **99**, 1607 (1993).
1118. A. Mermet, E. Duval, A. Polian, M. Krisch, Phys. Rev. E **66**, 31510 (2002).
1119. H.Z. Cummins, J. Phys. Condens. Matter **17**, 1457 (2005).
1120. L.C. Pardo, P. Lunkenheimer, A. Loidl, Phys. Rev. E **76**, 030502(R) (2007).
1121. P. Lunkenheimer, L.C. Pardo, M. Kohler, A. Loidl, Phys. Rev. E **77**, 031506 (2008).
1122. K.M. Gough, T.A. Wildman, J. Am. Chem. Soc. **112**, 9141 (1990).
1123. Z. Rappoport, S.E. Biali, M. Kaftory, J. Am. Chem. Soc. **112**, 7742 (1990).
1124. Md. Shahin, S.S.N. Murthy, J. Chem. Phys. **122**, 014507 (2005).
1125. M.S. Thayyil, S. Capaccioli, D. Prevosto, S. Ancherbak, P.A. Rolla, M. Lucchesi, K.L. Ngai, to be published (2010).
1126. M.A. Desando, D.L. Gourlay, S. Walker, J. Chem. Soc. Faraday Trans. II, **79**, 559 (1983).
1127. O. Yamamuro, M. Ishikawa, I. Kishimoto, J.J. Pinvidic, T. Matsuo, J. Phys. Soc. Jpn. **68**, 2969 (1999).
1128. R. Puertas, J. Salud, D.O. Lopez, M.A. Rute, S. Diez, J.L. Tamarit, M. Barrio, M.A. Perez-Jubindo, M.R. de la Fuente, L.C. Pardo, Chem. Phys. Lett. **401**, 368 (2005).
1129. L.P. Singh, S.S.N. Murthy, J. Chem. Phys. **129**, 094501 (2008).
1130. O. Haida, H. Suga, S. Seki, J. Chem. Thermodyn. **9**, 1133 (1977).
1131. F.X. Hassion, R.H. Cole, J. Chem. Phys. **23**, 1756 (1955).
1132. J. Barthel, K. Bachhuber, E. Buchner, H. Hetzenauer, Chem. Phys. Lett. **165**, 369 (1990).
1133. R. Brand, P. Lunkenheimer, U. Schneider, A. Loidl, Phys. Rev. B **62**, 8878 (2000).
1134. M. Jimenez-Ruiz, M.A. Gonzales, F.J. Bermejo, M.A. Miller, N.O. Birge, I. Cendoya, A. Alegria, Phys. Rev. B **59**, 9155 (1999).
1135. S. Benkhof, A. Kudlik, T. Blochowicz, E. Rössler, J. Phys. Condens. Matter **10**, 8155 (1998).
1136. G.P. Johari, G. Power, J.K. Vij, J. Chem. Phys. **116**, 5908 (2002).
1137. G.P. Johari, G. Power, J.K. Vij, J. Chem. Phys. **117**, 1714 (2002).
1138. N.B. Olsen, T. Christensen, J.C. Dyre, Phys. Rev. E **62**, 4435 (2000).
1139. G. Power, G.P. Johari, J.K. Vij, J. Chem. Phys. **119**, 435 (2003).
1140. (a) N. Shinyashiki, M. Shinohara, Y. Iwata, T. Goto, M. Oyama, S. Suzuki, W. Yamamoto, S. Yagihara, T. Inoue, S. Oyaizu, S. Yamamoto, K.L. Ngai, S. Capaccioli, J. Phys. Chem. B **112**, 15470 (2008). For pressure dependence, see (b) S. Ancherbak, S. Capaccioli, D. Prevosto, M.S. Thayyil, P.A. Rolla, N. Shinyashiki, J. Phys. Chem. B, to be submitted in (2010); (c) S. Ancherbak, S. Capaccioli, to be published.
1141. S. Yagihara, M. Yamada, M. Asano, Y. Kanai, N. Shinyashiki, S. Mashimo, K.L. Ngai, J. Non-Cryst. Solids **235–237**, 412 (1998).
1142. B. Geil, G. Diezemann, R. Böhmer, Phys. Rev. E **74**, 041504 (2006).
1143. (a) R.D. Priestley, C.J. Ellison, L.J. Broadbelt, J.M. Torkelson, Science **309**, 456, (2005); (b) Macromolecules **38**, 654 (2005).
1144. S. Hensel-Bielowka, S. Pawlus, M. Paluch, J. Ziolo, S.J. Rzoska, D. Prevosto, S. Sharifi, presented at 4th Workshop on Nonequilibrium Physics, Pisa, Italy (2006).
1145. (a) R. Casalini, C.M. Roland, J. Chem. Phys. **131**, 114501 (2009); (b) R. Casalini, C.M. Roland, J. Polym. Sci. B Polym. Phys. **48**, 582 (2010).
1146. D. Prevosto, S. Capaccioli, M. Lucchesi, P.A. Rolla, M. Paluch, S. Pawlus, Phys. Rev. B **73** 104205 (2006).

1147. M. Paluch, S. Pawlus, S. Hensel-Bielowka, K. Kaminski, T. Psurek, S.J. Rzoska, J. Ziolo, C.M. Roland, Phys. Rev. B **72**, 224205 (2005).
1148. G.P. Johari, J. Chem. Phys. **77**, 4619 (1982).
1149. K. Pathmanathan, J.Y. Cavaille, G.P. Johari, J. Polym. Sci. Polym. Phys. **27**, 1519 (1989).
1150. E. Muzeau, G.P. Johari, Chem. Phys. **149**, 173 (1990).
1151. (a) D. Prevosto, S. Capaccioli, S. Sharifi, K. Kessairi, M. Lucchesi, P.A. Rolla, J. Non-Cryst. Solids **353**, 4278 (2007); (b) T. Viciosa, G. Pires, J.J. Moura Ramos, J. Mol. Liq. **148**, 114 (2009); (c) H.P. Diogo, J.J. Moura Ramos, J. Polym. Sci. B Polym. Phys. **47**, 820 (2009).
1152. G. Power, J.K. Vij, G.P. Johari, J. Chem. Phys. **124**, 074509 (2006).
1153. M.B.M. Mangion, G.P. Johari, J. Polym. Sci. B Polym. Phys. **29**, 437 (1991).
1154. H.H. Song, R.J. Roe, Macromolecules **20**, 2723 (1987).
1155. P. Lunkenheimer, R. Wren, U. Schneider, A. Loidl, Phys. Rev. Lett. **95**, 055702 (2005).
1156. P. Lunkenheimer, R. Wehn, T. Riegger, A. Loidl, J. Non-Cryst. Solids **307**, 336 (2002).
1157. R. Casalini, C.M. Roland, Phys. Rev. Lett. **102**, 035701 (2009).
1158. F. Simon, Z. Anorg. Allgem. Chem. **203**, 219 (1931).
1159. E. Jenckel, Z. Electrochem. **45**, 202 (1939).
1160. J. Kubát, Nature **204**, 378 (1965).
1161. B.E. Read, G.D. Dean, P.E. Tomlins, J. Non-Cryst. Solids **172–174**, 562 (1994).
1162. S. Etienne, N. Hazeg, E. Duval, A. Mermet, A. Wypych, L. David, J. Non-Cryst. Solids **353**, 3871 (2007).
1163. S.V. Nemilov, Glass Phys. Chem. **26**, 511 (2000).
1164. M.R. Tant, G.L. Wilkes, Polym. Eng. Sci. **21**, 874 (1981).
1165. J.M. Hutchinson, Prog. Polym. Sci. **20**, 703 (1995).
1166. Y. Huang, D.R. Paul, Polymer **45**, 8377 (2004).
1167. J. Kurchan, Nature **433**, 222 (2005).
1168. R.D. Priestley, P. Rittigstein, L.J. Broadbelt, K. Fukao, J.M. Torkelson, J. Phys. Condens. Matter **19**, 205120 (2007).
1169. P. Rittigstein, J.M. Torkelson, unpublished.
1170. C.J. Ellison, J.M. Torkelson, Nat. Mater. **2**, 695 (2003).
1171. J.L. Keddie, R.A.L. Jones, R.A. Cory, Europhys. Lett. **27**, 59 (1994).
1172. P. Papon, J. Leblond, P.H.E. Meijer, *The Physics of Phase Transitions*, Springer, Berlin (2002).
1173. T. Hikima, Y. Adachi, M. Hanaya, M. Oguni, Phys. Rev. B **52**, 3900 (1995).
1174. (a) T. Hikima, M. Hanaya, M. Oguni, J. Non-Cryst. Solids **235–237**, 539 (1998). J Mol Struct **479**, 245 (1999); (b) M. Hatase, M. Hanaya, T. Hikima, M. Oguni, J. Non-Cryst. Solids **307–310**, 257 (2002); (c) F. Paladi, M. Oguni, Phys. Rev. B **65**, 144 (2002).
1175. D. Turnbull, J.C. Fisher, J. Chem. Phys. **17**, 71 (1949).
1176. H. Tanaka, Phys. Rev. E, **68**, 011505 (2003). T. Konishi, H. Tanaka, Phys. Rev. B **76**, 220201 (2007).
1177. (a) Y. Sun, H. Xi, S. Chen, M.D. Ediger, L. Yu, J. Phys. Chem. B **112**, 5594 (2008); (b) Y. Sun, H. Xi, M.D. Ediger, R. Richert, L. Yu, J. Chem. Phys. **131**, 074506 (2009); (c) H. Xi, Y. Sun, L. Yu, J. Chem. Phys. **130**, 094508 (2009); (d) S. Capaccioli, K.L. Ngai, M. Paluch, to be published.
1178. D.Q.M. Craig, P.G. Royall, V.L. Kett, M.L. Hopton, Int. J. Pharm. **179**, 179 (1999).
1179. L. Yu, Adv. Drug Deliv. Rev. **48**, 27 (2001).
1180. H. Imaizumi, N. Nambu, T. Nagai, Chem. Pharm. Bull. **28**, 2565 (1980).
1181. M. Otsuka, N. Kaneniwa, Chem. Pharm. Bull. **36**, 4026 (1988).
1182. M. Yoshioka, B.C. Hancock, G.J. Zografi, Pharm. Sci. **83**, 1700 (1994).
1183. V. Andronis, G.J. Zografi, J. Non-Cryst. Solids **271**, 236 (2000).
1184. Y. Aso, S. Yoshioka, S. Kojima, J. Pharm. Sci. **89**, 408 (2000); Y. Aso, S. Yoshioka, S. Kojima, J. Pharm. Sci. **93**, 384 (2004).
1185. S.L. Shamblin, X. Tang, L. Chang, B.C. Hancock, M.J. Pikal, J. Phys. Chem. B **103**, 4113 (1999).

1186. J. Alie, J. Menegotto, P. Cardon, H. Duplaa, A. Caron, C. Lacabanne, M. Bauer, J. Pharm. Sci. **93**, 218 (2004).
1187. G.P. Johari, S. Kim, R.M. Shanker, J. Pharm. Sci. **94**, 2207 (2005); **96**, 1159 (2007).
1188. R. Nath, T. El Goresy, B. Geil, H. Zimmermann, R. Böhmer, Phys. Rev. E. **74**, 021506 (2006).
1189. R. Nath, A. Nowaczyk, B. Geil, R. Böhmer, J. Non-Cryst. Solids **353**, 3788 (2007).
1190. C. Bhugra, S. Rambhatla, A. Bakri, S.P. Duddu, D.P. Miller, M.J. Pikal, D. Lechuga-Ballesteros, J. Pharm. Sci. **96**, 1258 (2007). For a review, see C. Bhugra, M.J. Pikal, J. Pharm. Sci. **97**, 1329 (2008).
1191. (a) C. Bhugra, R. Shmeis, S.L. Krill, M.J. Pikal, J. Pharm. Sci. **97**, 455 (2007); (b) C. Bhugra, R. Shmeis, S.L. Krill, M.J. Pikal, Pharm. Res. **23**, 2277 (2006); (c) O. Korhonen, C. Bhugra, M.J. Pikal, J. Pharm. Sci. **97**, 3830 (2008).
1192. S. Bhattacharya, R. Suryanarayanan, J. Pharm. Sci. **99**, 2935 (2009).
1193. E. Fukuoka, M. Makita, S. Yamamura, Chem. Pharm. Bull. (Tokyo) **37**, 1047 (1989); E. Fukuoka, M. Makita, Y. Nakamura, ibid. **39**, 2087 (1991).
1194. G.P. Johari, D. Pyke, Phys. Chem. Chem. Phys. **2**, 5479 (2000).
1195. H. Cang, J. Li, H.C. Andersen, M.D. Fayer, J. Chem. Phys. **123**, 064508 (2005).
1196. A. Nowaczyk, B. Geil, G. Hinze, R. Böhmer, Phys. Rev. E **74**, 041505 (2006).
1197. A. Detken, P. Focke, H. Zimmermann, U. Haeberlen, Z. Olejniczak, Z.T. Lalowicz, Z. Naturforsch. Teil A **50**, 95 (1995).
1198. N.T. Correia, J.J.M. Ramos, M. Descamps, G. Collins, Pharm. Res. **18**, 1767 (2001).
1199. Z. Wojnarowska, K. Adrjanowicz, P. Wlodarczyk, E. Kaminska, K, Kaminski, K. Grzybowska, R. Wrzalik, M. Paluch, K.L. Ngai, J. Phys. Chem. B **113**, 12536 (2009).
1200. G.P. Johari, S. Kim, R.M. Shanker, J. Pharm. Sci. **94**, 2207 (2005).
1201. T. El Goresy, R. Böhmer, J. Phys. Condens. Matter **19**, 205134 (2007).
1202. E. Tombari, S. Presto, G.P. Johari, R.M. Shanker, J. Pharm. Sci. **95**, 1006 (2006).
1203. K. Kaminski, E. Kaminska, K. Adrjanowicz, K. Grzybowska, P. Wlodarczyk, M. Paluch, A. Burian, J. Ziolo, P. Lepek, J. Mazgalski, W. Sawicki, J. Pharm. Sci. **99**, 94 (2009).
1204. (a) K. Adrjanowicz, K. Kaminski, M. Paluch, K.L. Ngai, J. Phys. Condens. Matter **22**, 125902 (2010); (b) K. Grzybowska, M. Paluch, A. Grzybowski, Z. Wojnarowska, L. Hawelek, K. Kolodziejczyk, K.L. Ngai, J. Phys. Chem. B **114**, 12792 (2010).
1205. J.Y. Cavaille, J. Perez, G.P. Johari, Phys. Rev. B **39**, 2411 (1989).
1206. K. Abbes, G. Vigier, J.Y. Cavaille, L. David, A. Faivre, J. Perez, J. Non-Cryst. Solids **235–237**, 286 (1998).
1207. A. Faivre, G. Niquet, M. Maglione, J. Fornazero, J.F. Jal, L. David, Eur. Phys. J. B, **10**, 277 (1999).
1208. B. Sixou, A. Faivre, L. David, G. Vigier, Mol. Phys. **99**, 1845 (2001).
1209. M.L. Cerrada, G.B. McKenna, Macromolecules **33**, 3065 (2000).
1210. S.P. Bravard, R. Boyd, Macromolecules **36**, 741 (2003).
1211. R. Casalini, C.M. Roland, Phys. Rev. Lett. **102**, 035701 (2009).
1212. A. Dhinojwala, G.K. Wong, J.M. Torkelson, J. Chem. Phys. **100**, 6046 (1994).
1213. H.J. Gabius, Naturwissenschaften **87**, 108 (2000).
1214. T.L. Lindhorst, Chem. Z **34**, 38 (2000).
1215. F. Amalfa, Biophys. J. **78**, 2452 (2000).
1216. N.J. Benaroudj, Biol. Chem. **276**(26), 24261 (2002); N. Benaroudj, H.H. Lee, A.L. Goldberg, J. Biol. Chem. **276**, 24261 (2001).
1217. J.H. Crowe, Annu. Rev. Physiol. **60**, 73 (1998).
1218. G.M. Wang, A.D.J. Haymet, J. Phys. Chem. B **102**, 5341 (1998).
1219. J.H. Crowe, J.F. Carpenter, L.M. Crowe, Annu. Rev. Physiol. **60**, 73 (1998).
1220. J.L Green, C.A. Angell, J. Phys. Chem. **93**, 2880 (1989).
1221. C. Branca, S. Magazu, G. Maisano, P. Migliardo, J. Chem. Phys. **111**, 281 (1999); C. Branca, A. Faraone, S. Magazu, G. Maisano, F. Migliardo, P. Migliardo, V. Villari, Rec. Res. Dev. Phys. Chem. **3**, 361 (1999).

1222. G. Cottone, G. Ciccotti, L. Cordone, J. Chem. Phys. **117**, 9862 (2002).
1223. G. Cottone, S. Giuffrida, G. Ciccotti, L. Cordone, Proteins **59** (2), 291 (2005).
1224. A. Hedoux, R. Ionov, J.F. Willart, A. Lerbret, F. Affouard, Y. Guinet, M. Descamps, D. Prevost, L. Paccou, F. Danede, J. Chem. Phys. **124**, 14703 (2006).
1225. A. Lerbret, P. Bordat, F. Affouard, A. Hedoux, Y. Guinet, M. Descamps, J. Phys. Chem. B **111**, 9410 (2007).
1226. A. Hedoux, J.-F. Willart, L. Paccou, Y. Guinet, F. Affouard, A. Lerbret, M. Descamps, J. Phys. Chem. B **113**, 6119 (2009).
1227. D. Champion, H. Hervet, G. Blond, M. Le Meste, D. Simatos, J. Phys. Chem. B **101**, 10674 (1997).
1228. (a) S.L. Shamblin, B.C. Hancock, M.J. Pikal, Pharm. Res. **23**, 2254 (2006); (b) S.A. Luthra, I.M. Hodge, M. Utz, M.J. Pikal, J. Pharm. Sci. **97**, 5240 (2008).
1229. L. Finegold, F. Franks, R.H.M. Hatley, J. Chem. Soc. Farad. Trans. **85**, 2945 (1989).
1230. D.J. Freedberg, Am. Chem. Soc. **124**, 2358 (2002).
1231. L. Poppe, H.J. van Halbeek, Am. Chem. Soc. **114**, 1092 (1992).
1232. C. Baraguey, D. Mertens, A. Dolle, J. Phys. Chem. B, **106**, 6331 (2002).
1233. D. van Dusschoten, U. Tracht, A. Heuer, H.W. Spiess J. Phys. Chem. A **103**, 8359 (1999).
1234. G.R. Moran, K.R. Jeffrey, J.M. Thomas, J.R. Stevens, Carb. Res. **328**, 573 (2000).
1235. Gangasharan, S.S.N. Murthy, J. Chem. Phys. **99**, 9865 (1993).
1236. R.K. Chan, K. Pathmanathan, G.P. Johari, J. Phys. Chem. **90**, 6358 (1986).
1237. E. Tombari, C. Cardelli, G. Salvetti, G.P. Johari, J. Mol. Struct. **559**, 45 (2001).
1238. S.S.N. Gangasharan, S.S.N. Murthy, J. Phys. Chem. **99**, 12349 (1995).
1239. T.R. Noel, R. Parker, S.G. Ring, Carbohydr. Res. **329**, 839 (2000).
1240. A. De Gusseme, L. Carpentier, J.F. Willart, M. Descamps, J. Phys. Chem B, **107**, 10879 (2003).
1241. T.R. Noel, R. Parker, S.G. Ring, Carbohydr. Res. **282**, 193 (1996).
1242. F. Amalfa, Biophys. J. **78**, 2452 (2000).
1243. K. Nishinari, D. Chatain, C. Lacabanne, J. Macromol. Sci. Phys. B **22**, 795 (1983–1984).
1244. K. Nishinari, N. Shibuya, K. Kainuma, Makromol. Chem. **186**, 433 (1985).
1245. M. Scandola, G. Ceccorulli, M. Pizzoli, Int. J. Biol. Macromol. **13**, 254 (1991).
1246. H. Montes, K. Mazeau, J.Y. Cavaille, J. Non-Cryst. Solids **172–174**, 990 (1994).
1247. H. Montes, K. Mazeau, J.Y. Cavaille, Macromolecules **30**, 6977 (1997).
1248. B. Schartel, J. Wendling, J.H. Wendorff, Macromolecules **29**, 1521 (1996).
1249. J. Einfeldt, D. Meissner, A. Kwasniewski, Prog. Polym. Sci. **26**, 1419 (2001).
1250. (a) D. Meissner, J. Einfeldt, A. Kwasniewski, J. Non-Cryst. Solids **275**, 199 (2000); (b) Macromol. Chem. Phys. **201**, 1969 (2000).
1251. M.F. Butler, R.E. Cameron, Polymer **41**, 2249 (2000).
1252. H. Montes, J.Y. Cavaille, Polymer **40**, 2649 (1999).
1253. K. Kaminski, E. Kaminska, S. Hensel-Bielowka, S. Pawlus, M. Paluch, J. Ziolo, J. Chem. Phys. **129**, 084501 (2008).
1254. R. Böhmer, E. Sanchez, C.A. Angell, J. Phys. Chem. **96**, 23, 9089 (1992).
1255. J.H. Crowe, L.M. Crowe, Science **223**, 701 (1984).
1256. G. Bianchi, A. Gamba, C. Murellie, F. Salamini, D. Bartels, Plant J. **1**, 355 (1991).
1257. D. Panek, Braz. J. Med. Biol. Res. **28**, 169 (1995).
1258. (a) L. Cordone, G. Cottone, S. Giuffrida, J. Phys. Condens. Matter **19**, 205110 (2007); (b) F. Francia, M. Dezi, A. Mallardi, G. Palazzo, L. Cordone, G. Venturoli, J. Am. Chem. Soc. **130**, 10240 (2008).
1259. A. Lerbret, P. Bordat, F. Affouard, M. Descamps, F. Migliardo, J. Phys. Chem. **109**, 11046 (2005).
1260. V.J. McBrierty, C.M. Keely, F.M. Coyle, X. Huan, J.K. Vij, Faraday Disc. **103**, 225 (1996).
1261. S. Cerveny, G.A. Schwartz, R. Bergman, J. Swenson, Phys. Rev. Lett. **93**, 245702 (2004).
1262. J. Kubát, C. Pattyranie, Nature **215**, 390 (1967).
1263. H. Tanaka, Nature **380**, 328 (1996).

1264. O. Mishima, H.E. Stanley, Nature **396**, 329 (1998).
1265. C.A. Angell, Nature **331**, 206 (1988).
1266. R.S. Smith, B.D. Kay, Nature **398**, 788 (1999).
1267. P.G. Debenedetti, J. Phys. Condens. Matter **15**, R1669 (2003).
1268. M. Sugisaki, H. Suga, S. Seki, Bull. Chem. Soc. Jpn **41**, 2591 (1968).
1269. A. Hallbrucker, E. Mayer, G.P. Johari, J. Chem. Phys. **93**, 4986 (1989).
1270. A. Hallbrucker, E. Mayer, J. Phys. Chem. **91**, 503 (1987).
1271. G.P. Johari, A. Hallbrucker, E. Mayer, Nature **330**, 552 (1987).
1272. I. Kohl, L. Backmann, A. Hallbrucker, E. Mayer, T. Loerting, Phys. Chem. Chem. Phys. **7**, 3210 (2005).
1273. A. Hallbrucker, E. Mayer, G.P. Johari, J. Phys. Chem. **93**, 7751 (1989).
1274. G.P. Johari, J. Chem. Phys. **122**, 144508 (2005).
1275. V. Velikov, S. Borick, C.A. Angell, Science **294**, 2335 (2001).
1276. Y. Yu, C.A. Angell, Nature **427**, 717 (2004).
1277. G.P. Johari, J. Chem. Phys. **116**, 8067 (2002).
1278. N. Giovambattista, C.A. Angell, F. Sciortino, H.E. Stanley, Phys. Rev. Lett., **93**, 047801 (2004).
1279. F.W. Starr, C.A. Angell, E.L. Nave, S. Sastry, A. Acala, F. Sciortino, H.E. Stanley, Biophys. Chem. **105**, 573 (2003).
1280. S.M. McClure, D.J. Safarik, T.M. Truskett, C.B. Mullins, J. Phys. Chem. B **110**, 11033 (2006).
1281. (a) S. Maruyama, K. Wakabayashi, M.A. Oguni, AIP Conf. Proc. **708**, 675 (2004); (b) K. Kawai, T. Suzuki, M. Oguni, Biophys. J. **90**, 3732 (2006).
1282. M. Oguni, S. Maruyama, K. Wakabayashi, A. Nagoe, Chem. Asian **2**, 514 (2007).
1283. M. Oguni, Y. Kanke, S. Namba, AIP Conf. Proc. **982**, 34 (2008).
1284. (a) F. Mallamace, E. Tombari, G. Salvetti, G.P. Johari, Condens Matter. arXiv: 0806.4775; (b) G.P. Johari, Thermochim. Acta **492**, 29 (2009).
1285. C.A. Angell, Science **319**, 582 (2008).
1286. C.A. Angell, J. Phys. Chem. **97**, 6339 (1993).
1287. K. Ito, C.T. Moynihan, C.A. Angell, Nature **398**, 492 (1999).
1288. F. Starr, C.A. Angell, H.E. Stanley, Physica A **323**, 51 (2003).
1289. (a) A. Faraone, L. Liu, C.Y. Mou, C.W. Yen, S.H. Chen, J. Chem. Phys. **121**, 10843 (2004); (b) L. Liu, S.-H. Chen, A. Faraone, C.-W. Yen, C.-Y. Mou, Phys. Rev. Lett. **95**, 117802 (2005); (c) M. Lagi, P. Baglioni, S.-H. Chen, Phys. Rev. Lett. **103**, 108102 (2009); (d) X. Chu, E. Fratini, P. Baglioni, A. Faraone, S.-H. Chen, Phys. Rev. E **77**, 011908 (2008); (e) Y. Zhang, M. Lagi, E. Fratini, P. Baglioni, E. Mamontov, S.-H. Chen, Phys. Rev. E **79**, 040201(R) (2009); (f) X.-Q. Chu, K.-H. Liu, M.S. Tyagi, C.-Y. Mou, S.-H. Chen, Phys. Rev. E **82**, 020501(R) (2010); (g) S.H. Chen, Y. Zhang, M. Lagi, S.H. Chong, P. Baglioni, F. Mallamace, J. Phys.: Condens. Matter **21**, 504102 (2009); (h) F. Mallamace, C. Branca, C. Corsaro, J. Spooren, S.-H. Chen, H.E Stanley, J. Non-Cryst. Solids doi:10.1016/j.jnoncrysol.2010.06.049; (i) F. Mallamace, C. Branca, C. Corsaro, N. Leone, J. Spooren, H.E. Stanley, S.-H. Chen, J. Phys. Chem. B **114**, 1870 (2010).
1290. F. Mallamace, M. Broccio, C. Corsaro, A. Faraone, U. Wanderlingh, L. Liu, C.-Y. Mou, S.-H. Chen, J. Chem. Phys. **124**, 161102 (2006).
1291. L. Xu, P. Kumar, S.V. Buldyrev, S.-H. Chen, P.H. Poole, F. Sciortino, H.E. Stanley, Proc. Natl. Acad. Sci. U.S.A. **102**, 16558 (2005).
1292. J. Swenson, Phys. Rev. Lett. **97**, 189801 (2006).
1293. J. Hedstrom, J. Swenson, R. Bergman, H. Jansson, S. Kittaka, Eur. Phys. J. Special Top. **141**, 53 (2007).
1294. S. Cerveny, J. Colmenero, A. Alegria, Phys. Rev. Lett. **97**, 189802 (2006).
1295. N. Shinyashiki, S. Sudo, S. Yagihara, A. Spanoudakis, A. Kyritsis, P. Pissis, J. Phys. Condens. Matter **19**, 205113–205112 (2007).
1296. R. Bergman, J. Swenson, Nature **403**, 283 (2000).

1297. G.P. Johari, Chem. Phys. **258**, 277 (2000).

1298. S.K. Jain, G.P. Johari, J. Phys. Chem. **92**, 5851 (1988).

1299. S. Sudo, M. Shimomura, T. Saito, T. Kashiwagi, N. Shinyashiki, S. Yagihara, J. Non-Cryst. Solids **305**, 197 (2002).

1300. S. Sudo, S. Tsubotani, M. Shimomura, N. Shinyashiki, S. Yagihara, J. Chem. Phys. **121**, 7332 (2004).

1301. S. Sudo, M. Shimomura, K. Kanari, N. Shinyashiki, S. Yagihara, J. Chem. Phys. **124**, 044901 (2006).

1302. S. Sudo, M. Shimomura, N. Shinyashiki, S. Yagihara, J. Non-Cryst. Solids **307–310**, 356 (2002). S. Sudo, N. Shinyashiki, S. Yagihara, J. Mol. Liq. **90**, 113 (2001).

1303. Y. Hayashi, A. Puzenko, I. Balin, Y.E. Ryabov, Y. Feldman, J. Phys. Chem. B, **109**, 917 (2005).

1304. (a) K. Grzybowska, A. Grzybowski, S. Pawlus, S. Hensel-Bielowka, M. Paluch, J. Chem. Phys. **123**, 204506 (2005); (b) K. Grzybowska, M. Paluch, A. Grzybowski, S. Pawlus, S. Ancherbak, D. Prevosto, S. Capaccioli, J. Phys. Chem. Lett. **1**, 1170 (2010).

1305. M. Shinohara, T. Goto, M. Oyama, S. Suzuki, N. Shinyashiki. S. Yagihara, T. Inoue, S. Oyaizu, S. Yamamoto, AIP Conf. Proc. **832**, 145 (2006).

1306. M. Tyagi, S.S.N. Murthy, Carbohydr. Res. **341**, 650 (2006).

1307. S. Sudo, S. Tobinai, N. Shinyashiki, S. Yagihara, AIP Conf. Proc. **832**, 149 (2006).

1308. S. Cerveny, A. Alegría, J. Colmenero, Phys. Rev. E **77**, 031803 (2008).

1309. H.-J. Kwon, J.-A. Seo, H.K. Kim, Y.-H. Hwang, AIP Conf. Proc. **982**, 728 (2008).

1310. N. Shinyashiki, M. Shimomura, T. Ushiyama, T. Miyagawa, S. Yagihara, J. Phys. Chem. B **111**, 10079 (2007).

1311. R. Buchner, J. Barthel, J. Stauber, Chem. Phys. Lett. **306**, 57 (1999).

1312. S.E. Pagnotta, S. Cerveny, A. Alegría, J. Colmenero, J. Chem. Phys. **131**, 085102 (2009).

1313. K. Pathmanathan, G.P. Johari, J. Polym. Sci. B Polym. Phys. **28**, 675 (1990); J. Chem. Soc. Faraday Trans. **90**, 1143 (1994).

1314. S. Khodadadi, S. Pawlus, J.H. Roh, V. Garcia Sakai, E. Mamontov, A.P. Sokolov, J. Chem. Phys. **128**, 195106 (2008).

1315. G. Caliskan, R.M. Briber, D. Thirumalai, V. Garcia-Sakai, S.A. Woodson, A.P. Sokolov, J. Am. Chem. Soc. **128**, 32 (2006).

1316. K. Kawai K., T. Suzuki, M. Oguni, Biophys. J. **90**, 3732 (2006).

1317. N. Shinyashiki, W. Yamamoto, A. Yokoyama, T. Yoshinari, S. Yagihara, R. Kita, K.L. Ngai, S. Capaccioli, J. Phys. Chem. B **113**, 14448 (2009).

1318. W. Doster, M. Settles, Biochim. Biophys. Acta **1749**, 173 (2005).

1319. W. Doster, Eur. Biophys. J. **37**, 591 (2008).

1320. K.L. Ngai, S. Capaccioli, N. Shinyashiki, S. Achebak, Philos. Mag. to be published.

1321. (a) S. Cerveny, F. Barroso-Bujans, A. Alegria, J. Colmenero, J. Phys. Chem. C **114**, 2604 (2010); (b) A. Buchsteiner, A. Lerf, J. Pieper, J. Phys. Chem. B. **110**, 22328 (2006); (c) A. Lerf, A. Buchsteiner, J. Pieper, S. Schottl, I. Dekany, T. Szabo, H.P. Boehm, J. Phys. Chem. Solids **67**, 106 (2006).

1322. (a) C. Gainaru, A. Fillmer, R. Böhmer, J. Phys. Chem. B **113**, 12628 (2009); (b) S.A. Lusceac, M.R. Vogel, C.R. Herbers, arXiv:0904.4424v1 [cond-mat.Soft] 28 Apr 2009; (c) Biochim. Biophys. Acta Proteins Proteom. **1804**, 41 (2010); (d) S. Lusceac, M. Rosenstihl, M. Vogel, C. Gainaru, A. Fillmer, R. Böhmer, J. Non-Cryst.Solids 357, 655 (2010); (e) M. Vogel, E. Rössler, J. Chem. Phys. **114**, 5802 (2001).

1323. (a) D. Kilburn, S. Townrow, V. Meunier, R. Richardson, A. Alam, J. Ubbink, Nat. Mater. **5**, 632 (2006); (b) S.J. Suresh, V.M. Naik, J. Chem. Phys. **113**, 9727 (2000).

1324. F. Sciortino, P. Gallo, P. Tartaglia, S.-H. Chen, Phys. Rev. A **54**, 6331 (1996).

1325. H.J.C. Berendsen, J.R. Grigera, T.P. Straatsma, J. Phys. Chem. **91**, 6269 (1987).

1326. H.-S. Chen, J. Teixeira, R. Nicklow, Phys. Rev. A **26**, 3477 (1982).

1327. J. Teixeira, M.-C. Bellissent-Funel, S.H. Chen, A.J. Dianoux, Phys. Rev. A **31**, 1913 (1985).

1328. C. Masciovecchio, S.C. Santucci, A. Gessini, S. Di Fonzo, G. Ruocco, F. Sette, Phys. Rev. Lett. **92**, 255507 (2004).
1329. R. Torre, P. Bartolini, R. Righini, Nature **424**, 296 (2004); L. Bove et al., J. Chem. Phys. **134**, 034514 (2011).
1330. G.P. Johari, A. Hallbrucker, E. Mayer, Science **273**, 90 (1996).
1331. Y.P. Handa, D.D. Klug, J. Phys. Chem. **92**, 3323 (1988).
1332. A. Hallbrucker, E. Mayer, G.P. Johari, Philos. Mag. B **60**, 179 (1989).
1333. G.P. Johari, A. Hallbrucker, E. Mayer, J. Chem. Phys. **92**, 6742 (1990).
1334. M. Cammarata, M. Levantino, A. Cupane, A. Longo, A. Martorana, F. Bruni, Eur. Phys. J. E **12**, S63 (2003).
1335. K. Pathmanathan, G.P. Johari, J. Polym. Sci. Polym. Phys. B **28**, 675 (1990); J. Chem. Soc. Faraday Trans. **90**, 1143 (1994).
1336. K. Hofer, E. Mayer, G.P. Johari, J. Phys. Chem. **94**, 2689 (1990); **95**, 7100 (1991).
1337. G.P. Johari, G. Astl, E. Mayer, J. Chem. Phys. **92**, 809 (1990).
1338. K. Watanabe, M. Oguni, Bull. Chem. Soc. Jpn **80**(9), 1758 (2007).
1339. K. Watanabe, M. Oguni, M. Tadokoro, Y. Oohata, R. Nakamura, J. Phys. Condens. Matter **18**, 8427 (2006); J. Phys. Chem. B **114**, 2091 (2010).
1340. R.B. Gregory, Ed., *Protein–Solvent Interactions*, Marcel Dekker, New York, NY (1995).
1341. P.W. Fenimore, H. Frauenfelder, B.H. McMahon, R.D. Young, Proc. Natl Acad. Sci. USA. **101**, 14408 (2004).
1342. F. Parak, H. Formanek, Acta Crystallogr. A **27**, 573 (1971); F. Parak, M. Fischer, G.U. Nienhaus, J. Mol. Liq. **42**,145 (1989).
1343. F. Parak, E.W. Knapp, D. Kucheida, J. Mol. Biol. **161**, 177 (1982).
1344. W. Doster, S. Cusak, W. Petry, Nature **337**, 754 (1989), Phys. Rev. Lett. **65**, 1083 (1990).
1345. (a) A.M. Tsai, D.A. Neumann, L.N. Bell, Biophys. J. **79**, 2728 (2000); (b) A.M. Tsai, T.J. Udovic, D.A. Neumann, Biophys. J. **81**, 2339 (2001).
1346. (a) A. Paciaroni, S. Cinellli, G. Onori, Biophys. J. **83**, 1157 (2002); (b) E. Cornicchi, M. Marconi, G. Onori, A. Paciaroni, Biophys. J. **91**, 289 (2006); (c) E. Cornicchi, S. Capponi, M. Marconi, G. Onori, A. Paciaroni, Philos. Mag. **87**, 509 (2007); (d) A. Paciaroni, A. Orecchini, E. Cornicchi, M. Marconi, C. Petrillo, M. Haertlein, M. Moulin, F. Sacchetti, Philos. Mag. **88**, 4071 (2008); (e) M. Di Bari, A. Deriu, G. Albanese, F. Cavatorta, Chem. Phys. **292**, 333 (2003); (f) I.J. Van den Dries, N.A.M. Besseling, D. Van Dusschoten, M.A. Hemminga, E. Van Der Linden, J. Phys. Chem. B. **104**, 9260 (2000).
1347. J.-M. Zanotti, G. Gibrat, M.-C. Bellissent-Funel, Phys. Chem. Chem. Phys. **10**, 4865 (2008).
1348. P.W. Fenimore, H. Frauenfelder, B.H. McMahon, F.G. Parak, Proc. Natl. Acad. Sci. USA. **99**, 16047 (2002).
1349. H. Frauenfelder, P.W. Fenimore, G. Chen, B.H. McMahon, Proc. Natl. Acad. Sci. USA **103**, 15469 (2006).
1350. V. Lubchenko, P.G. Wolynes, H. Frauenfelder, J. Phys. Chem. B **109**, 7488 (2005).
1351. (a) Y. Miyazaki, T. Matsuo, H. Suga, J. Phys. Chem. B **104**, 8044 (2000); (b) Y. Miyazaki, T. Matsuo, H. Suga, Chem. Phys. Lett. **213**, 303 (1993).
1352. M.M. Teeter, A. Yamano, B. Stec, U. Mohanty, Proc. Natl. Acad. Sci. USA. **98**, 11242 (2001).
1353. M. Tarek, D.J. Tobias, Phys. Rev. Lett. **88**, 138101 (2002).
1354. (a) T. Kleinert, W. Doster, H. Leyser, W. Petry, V. Schwarz, M. Settles, Biochemistry **37**, 717 (1998); (b) H. Lichtenegger, W. Doster, T. Kleinert, A. Birk, B. Sepiol, G. Vogl, Biophys. J. **76**, 414 (1999); (c) W. Doster, Biochim. Biophys. Acta Proteins Proteom. **1804**, 3 (2010); (d) J. Han, R.H. Boyd, Polymer **37**, 1797 (1995); (e) R.K. Bharadwaj, R.H. Boyd, Polymer **40**, 4229 (1999); (f) A.A. Gusev, U.W. Suter, D.J. Moll, Macromolecules **28**, 2582 (1995).
1355. G. Chen, P.W. Fenimore, H. Frauenfelder, F. Mezei, J. Swenson, R.D. Young, Philos. Mag. **88**, 3877 (2008); H. Frauenfeder, F. Mezei, Acta Cryst. D **66**, 1229 (2010).
1356. H. Frauenfelder, G. Chen, J. Berendzen, P.W. Fenimore, H. Jansson, B.H. McMahon, I.R. Stroe, J. Swenson, R.D. Young, Proc. Natl. Acad. Sci. USA. **106**, 5129 (2009).

1357. S. Khodadadi, A. Malkovskiy, A. Kisliuk, A.P. Sokolov, Biochim. Biophys. Acta Proteins Proteom. **1804**, 15 (2010).
1358. (a) S. Khodadadi, S. Pawlus, J.H. Roh, V.G. Sakai, E. Mamontov, A.P. Sokolov, J. Chem. Phys. **128**, 195106 (2008); (b) A.P. Sokolov, J.H. Roh, E. Mamontov, V.G. Sakai, Chem. Phys. **345**, 212 (2008); (c) S. Khodadadi, S. Pawlus, A.P. Sokolov, J. Phys. Chem. B **112**, 14273 (2008); (d) S. Khodadadi, J.H. Roh, A. Kisliuk, E. Mamontov, M. Tyagi, S.A. Woodson, R. M. Briber, A.P. Sokolov, Biophys. J. **98**, 1321 (2010).
1359. S.-H. Chen, L. Liu, E. Fratini, P. Baglioni, A. Faraone, E. Mamontov, Proc. Natl. Acad. Sci. USA. **103**, 9012 (2006).
1360. (a) N. Miura, Y. Hayashi, S. Mashimo, Biopolymers **39**, 183 (1996); (b) N. Nandi, K. Bhattacharyya, B. Bagchi, Chem. Rev. **100**, 2013 (2000); (c) A. Oleinikova, P. Sasisanker, H. Weingartner, J. Phys. Chem. B **108**, 8467 (2004).
1361. (a) S. Magazú, F. Migliardo, C. Mondelli, M. Vadala, Carbohydrate Res. **340**, 2796 (2005). (b) S. Magazú, G. Maisano, F. Migliardo, C. Mondelli, Biophys. J. **86**, 3241 (2004). (c) B. Varga, F. Migliardo, E. Takacs, B. Vertessy, S. Magazú, M.T.F. Telling, J. Biol. Phys. **36**, 207 (2010). (d) S. Capaccioli, K. L. Ngai, to be published. (e) See also related work by F. Gabel, M.-C. Bellissent-Funel, Biophys. J. **92**, 4054 (2007).
1362. G. Schiro, M. Sclafani, C. Caronna, F. Natali, M. Plazanet, A. Cupane, Chem. Phys. **345**, 259 (2008); J. Am. Chem. Soc. **132**, 1371 (2010).
1363. K. Wood, A. Frölich, A. Paciaroni, M. Moulin, M. Hartlein, G. Zaccai, D.J. Tobias, M. Weik, J. Am. Chem. Soc. **130**, 4586 (2008).
1364. (a) A.L. Tournier, J. Xu, J.C. Smith, Biophys. J. **85**, 1871 (2003); (b) T. Becker, J.A. Hayward, J.L. Finney, R.M. Daniel, J.C. Smith, Biophys. J. **87**, 1436 (2004); (c) F. Gabel, M. Weik, B.P. Doctor, A. Saxena, D. Fournier, L. Brochier, F. Renault, P. Masson, I. Silman, G. Zaccai, Biophys. J. **8**, 3152 (2004).
1365. S. Sacquin-Mora, P. Sebban, V. Derrien, B. Frick, R. Lavery, C. Alba-Simionesco, Biochemistry **46**, 14960 (2007).
1366. P. Kumar, Z. Yan, L. Xu, M.G. Mazza, S.V. Buldyrev, S.H. Chen, S. Sastry, H.E. Stanley, Phys. Rev. Lett. **97**, 177802 (2006).
1367. J.E. Curtis, T.E. Dirama, G.A. Carri, D.J. Tobias, J. Phys. Chem. B. **110**, 22953 (2006).
1368. T. Li, A.A. Hassanali, Y.T. Kao, D. Zhong, S.J. Singer, J. Am. Chem. Soc. **129**, 3376 (2007).
1369. F. Gabel, M.C. Bellissent-Funel, Biophys. J. **92**, 4054 (2007).
1370. (a) G. Zaccai, Science **288**, 1604 (2000); (b) M. Jasnin, L. van Eijck, M.M. Koza, J. Peters, C. Laguri, H. Lortat-Jacob, G. Zaccai, Phys. Chem. Chem. Phys. **12**, 3360 (2010).
1371. L. Zhang, L. Wang, Y.T. Kao, W. Qiu, Y. Yang, O. Okobiah, D. Zhong, Proc. Natl. Acad. Sci. USA. **104**, 18461 (2007).
1372. L. Cordone, M. Ferrand, E. Vitrano, G. Zaccai, Biophys. J. **76**, 1043 (1999).
1373. I. Köper, S. Combet, W. Petry, M.-C. Bellissent-Funel, Eur. Biophys. J. **37**, 739 (2008).
1374. M.-C. Bellissent-Funel, Pramana J. Phys. **63**, 91 (2004).
1375. E.C. Lopez-Dıez, S. Bone, Biochim. Biophys. Acta **1673**, 139 (2004).
1376. M. Sola-Penna, J.R. Meyer-Fernandes, Arch. Biochem. Biophys. **360**, 10 (1998).
1377. F. Affouard, P. Bordat, M. Descamps, A. Lerbret, S. Magazu, F. Migliardo, A.J. Ramirez-Cuesta, M.F.T. Telling, Chem. Phys. **317**, 258 (2005).
1378. E. Cornicchi, G. Onori, A. Paciaroni, Phys. Rev. Lett. **95**, 158104 (2005).
1379. S. Giuffrida, G. Cottone, L. Cordone, Biophys. J. **91**, 968 (2006).
1380. (a) S. Magazu, F. Migliardo, M.F.T. Telling, Eur. Biophys. J. **36**, 163 (2007). (b) S. Magazu, F. Migliardo, A. Benedetto, J. Phys. Chem. B **114**, 9268 (2010).
1381. M.F. Mazzobre, M.D. Buera, J. Chirife, Biotechnol. Prog. **13**, 195 (1997).
1382. K.D. Rector, J.W. Jiang, M.A. Berg, M.D. Fayer, J. Phys. Chem. B **105**, 1081 (2001).
1383. (a) M.C. Manning, D.K. Chou, B.M. Murphy, R.W. Payne, D.S. Katayama, Pharm. Res. (2010). doi: 10.1007/s11095-009-0045-6; (b) M.T. Cicerone, C.L. Soles, Biophys. J. **86**, 3836 (2004); (c) B. Wang, S. Tchessalov, M.T. Cicerone, N.W. Warne, M.J. Pikal, J. Pharm. Sci. doi: 10.1002/jps.21622 (2008); (d) M.T. Cicerone, A. Tellington, L. Trost,

A. Sokolov, BioProcess Int. January issue, 1, (2003); (e) M.T. Cicerone, C.L. Soles, Z. Chowdhuri, M.J. Pikal, L. Chang, Am. Pharm. Rev. **8**(22), 24 (2005); (f) A.M. Abdul-Fattah, V. Truong-Le, L. Yee, L. Nguyen, D.S. Kalonia, M.T. Cicerone, M.J. Pikal, J. Pharm. Sci. **96**, 1983 (2007); (g) S.A. Luthra, M.J. Pikal, M. Utz, J. Pharm. Sci. **97**, 4336 (2008); (h) M.J. Pikal, D. Rigsbee, M.L. Roy, D. Galreath, K.J. Koavach, Bingquan (Stuart) Wang, J.F. Carpenter, M.T. Cicerone, J. Pharm. Sci. **97**, 5106 (2008); (i) For a review see Liuquan (Lucy) Chang, M.J. Pikal, J. Pharm. Sci. **98**, 2886 (2009); (j) J. Liu, D.R. Rigsbee, M.J. Pikal, J. Pharm. Sci. **91**,1853 (2002); (k) K. Kawakami, M.J. Pikal, J. Pharm. Sci. **94**, 948 (2004); (l) G. Caliskan, D. Mechtani, J.H. Roh, A. Kisliuk, A.P. Sokolov, S. Azzam, M.T. Cicerone, S. Lin-Gibson, I. Peral, J. Chem. Phys. **121**, 1978 (2004); (m) D.P. Miller, R.E. Anderson, J.J. de Pablo, J. Pharm. Res. **15**, 1215 (1998); (n) R.H.M. Hatley, J.A. Blair, J. Mol. Catal. B Enzymatic **7**, 11 (1999); (o) A. Anopchenko, T. Psurek, D. VanderHart, J.F. Douglas, J. Obrzut, Phys. Rev. E **74**, 031501 (2006).

1384. (a) S. Yoshioka, T. Miyazaki, Y. Aso, T. Kawanishi, J. Pharm. Res. **24**, 1660 (2007); (b) S. Yoshioka, T. Miyazaki, Y. Aso, J. Pharm. Res. **23**, 961 (2006); (c) S. Yoshioka, Y. Aso, S. Kojima, J. Pharm. Res. **13**, 926 (1996); (d) S. Yoshioka, Y. Aso, S. Kojima, S. Sakurai , T. Fujiwara, H. Akutsu, J. Pharm Res. **16**, 1621 (1999); (e) S. Yoshioka, Y. Aso, S. Kojima, J. Pharm. Res. **13**, 926 (1996).
1385. (a) M. Vogel, Phys. Rev. Lett. **101**, 225701 (2008). The observation of "fragile-to-strong" transition claimed in this paper was withdrawn in the publication that followed it [1322(b)]; (b) R. Knauss, G. Fleischer, W. Gründer, J. Kärger, A. Werner, Magn. Reson. Med. **36**, 241 (1996).
1386. Y. Shibata, A. Kurita, T. Kushida, Biophys. J. **75**, 521 (1998).
1387. G. Sartor, E. Mayer, G.P. Johari, Biophys. J. **66**, 249 (1994).
1388. J.L. Green, J. Fan, C.A. Angell, J. Phys. Chem. **98**, 13780 (1994).
1389. C. Inoue, M. Ishikawa, J. Food Sci. **65**, 1187 (2000).
1390. A. Cupane, M. Leone, V. Militello, Biophys. Chem. **104**, 335 (2003).
1391. G.P. Johari, E. Whalley, J. Chem. Phys. **75**, 1333 (1981).
1392. G. Chen, P.W. Fenimore, H. Frauenfelder, F. Mezei, J. Swenson, R.D. Young, Philos. Mag. **88**, 3877 (2008).
1393. H. Frauenfelder, G. Chen, J. Berendzen, P.W. Fenimore, H. Jansson, B.H. McMahon, I.R. Stroe, J. Swenson, R.D. Young, Proc. Natl. Acad. Sci. USA. **106**, 5129 (2009).
1394. C. Svanberg, P. Berntsen, A. Johansson, T. Hedlund, E. Axén, J. Swenson, J. Chem. Phys. **130**, 035101 (2009).
1395. (a) W. Doster, S. Busch, A.M. Gaspar, M.-S. Appavou, J. Wuttke, H. Scheer, Phys. Rev. Lett. **104**, 098101 (2010); (b) W. Doster, J. Non-Cryst. Solids, **357**, 622 (2011).
1396. J. Wuttke, W. Petry, G. Coddens, F. Fujara, Phys. Rev. E **52**, 4026 (1995).
1397. W.G. Hoover, M. Ross, Contemp. Phys. **12**, 339 (1971); W.G. Hoover, M. Ross, K.W. Johnson, D. Henderson, J.A. Barker, B.C. Brown, J. Chem. Phys. **52**, 4931 (1970).
1398. P.G. Debenedetti, F.H. Stillinger, T.M. Truskett, C.J. Roberts, J. Phys. Chem. B **103**, 7390 (1999).
1399. M.S. Shell, P.G. Debenedetti, E. La Nave, F. Sciortino, J. Chem. Phys. **118**, 8821 (2003).
1400. R.J. Speedy, J. Phys. Condens. Matter **15**, S1243 (2003).
1401. Y. Hiwatari, H. Matsuda, T. Ogawa, N. Ogita, A. Ueda, Prog. Theor. Phys. **52**, 1105 (1974).
1402. A. Tölle, Rep. Prog. Phys. **64**, 1473 (2001).
1403. A. Tölle, H. Schober, J. Wuttke, O.G. Randl, F. Fujara, Phys. Rev. Lett. **80**, 2374 (1998).
1404. C. Dreyfus, A. Aouadi, J. Gapinski, M. Matos-Lopes, W. Steffen, A. Patkowski, R.M. Pick, Phys. Rev. E **68**, 011204 (2003).
1405. C. Dreyfus, A. Le Grand, J. Gapinski, W. Steffen, A. Patkowski, Eur. Phys. J. **42**, 309 (2004).
1406. G. Tarjus, D. Kivelson, S. Mossa, C. Alba-Simionesco, J. Chem. Phys. **120**, 6135 (2004).

1407. R. Casalini, C.M. Roland, Phys. Rev. E **69**, 062501 (2004).
1408. (a) C.M. Roland, R. Casalini, J. Chem. Phys. **121**, 11503 (2004); (b) R. Casalini, S. Capaccioli, C.M. Roland, J. Phys. Chem. B **110**, 11491 (2006).
1409. R. Casalini, C.M. Roland, Colloid Polym. Sci. **283**, 107 (2004).
1410. J.D. Weeks, D. Chandler, H.C. Andersen, J. Chem. Phys. **54**, 5237 (1971). D. Chandler, J.D. Weeks, H.C. Andersen, Science **220**, 787 (1983).
1411. (a) D. Coslovich, C.M. Roland, J. Phys. Chem. B **112**, 1329 (2008); (b) D. Coslovich, C.M. Roland, J. Chem. Phys. **131**, 151103 (2009); (c) U.R. Pedersen, N.P. Bailey, T.B. Schroder, J.C. Dyre, Phys. Rev. Lett. **100**, 015701 (2008); (d) N.P. Bailey, U.R. Pedersen, N. Gnan, T.B. Schroder, J.C. Dyre, J. Chem. Phys. **129**, 184507 (2008), J. Chem. Phys. **129**, 184508 (2008); (e) C.M. Roland, D. Fragiadakis, D. Coslovich, S. Capaccioli, K.L. Ngai, "Correlation of Non-exponentiality with Dynamic Heterogeneity from Four-point Dynamic Susceptibility $\chi_4(t)$ and its Approximation $\chi_T(t)$", J. Chem. Phys. **133**, 124507 (2010).
1412. (a) C. De Michele, F. Sciortino, A. Coniglio, J. Phys. Condens. Matter **16**, L489 (2004); (b) S. Capaccioli, K.L. Ngai, M. Paluch, unpublished (2010); (c) L. Berthier, G. Tarjus, Phys. Rev. Lett. **103**, 170601 (2009). arXiv:1005.0914 (May, 2010); (d) T. Voigtmann, W.C.K. Poon, J. Phys. Condens. Matter. **18**, L465 (2006); (e) T. Voigtmann, Phys. Rev. Lett. **101**, 095701 (2008); (f) C.S. O'Hern, S.A. Langer, A.J. Liu, S.R. Nagel, Phys. Rev. E **68**, 011306 (2003); (g) N. Xu, M. Wyart, A.J. Liu, S.R. Nagel, Phys. Rev. Lett. **98**, 175502 (2007); (h) L. Berthier, T.A. Witten, Europhys. Lett. **86**, 10001 (2009); Phys. Rev. E **80**, 021502 (2009).
1413. (a) C.M. Roland, R. Casalini, M. Paluch, J. Polym. Sci. Polym. Phys. Ed. **42**, 4313 (2004); (b) K. Kaminski, S. Maslanka, J. Ziolo, M. Paluch, K.J. McGrath, C.M. Roland, Phys. Rev. E **75**, 011903 (2007).
1414. R. Casalini, C.M. Roland, Macromolecules **38**, 1779 (2005).
1415. K.L. Ngai, R. Casalini, C.M. Roland, Macromolecules **38**, 4363 (2005).
1416. H. Fujimori, M. Oguni, J. Chem. Thermodyn. **26**, 367 (1994).
1417. A. Rivera, J. Santamaria, C. León, K.L. Ngai, J. Phys. Condens. Matter **15**, S1633 (2003).
1418. A. Pimenov, P. Lunkenheimer, H. Rall, R. Kohlhaas, A. Loidl, R. Böhmer, Phys. Rev. E **54**, 676 (1996).
1419. P. Lunkenheimer, A. Pimenov, A. Loidl, Phys. Rev. Lett. **78**, 2995 (1997).
1420. F. Mezei, W. Knaak, B. Farago, Phys. Rev. Lett. **58**, 571 (1987).
1421. G. Li, W.M. Du, X.K. Chen, H.Z. Cummins, N.J. Tao, Phys. Rev. A **45**, 3867 (1992).
1422. (a) N.V. Surovtsev, J.A.H. Wiedersich, V.N. Novikov, E. Rossler, A.P. Sokolov, Phys. Rev. B **58**, 14 888 (1999); (b) J. Gapinski, W. Steffen, A. Patkowski, A.P. Sokolov, A. Kisliuk, U. Buchenau, M. Russina, F. Mezei, H. Schober, J. Chem. Phys. **110**, 2312 (1999); (c) H.C. Barshilia, G. Li, G.Q. Shen, H.Z. Cummins, Phys. Rev. A **59**, 5625 (1999); (d) K. Niss, C. Dalle-Ferrier, B. Frick, D. Russo, J. Dyre, C. Alba-Simionesco, Phys. Rev. E, **82**, 021508 (2010).
1423. A. Kisliuk, V.N. Novikov, A.P. Sokolov, J. Polym. Sci. B Polym. Phys. **40**, 201 (2002).
1424. R. Casalini, K.L. Ngai, J. Non-Cryst. Solids **293–295**, 318 (2001).
1425. L. Larini, A. Ottochian, C. De Michele, D. Leporini, Nat. Phys. **4**, 42 (2008).
1426. (a) M.J. Lebon, C. Dreyfus, G. Li, A. Aouadi, H.Z. Cummins, R.M. Pick, Phys. Rev. E **51**, 4537 (1995); (b) K.L. Ngai, C.M. Roland, Phys. Rev. E **54**, 6969 (1996); (c) A. Brodin, L. Borjesson, D. Engberg, L.M. Torell, A.P. Sokolv, Phys. Rev. B **53**, 11511 (1996).
1427. D.L. Sidebottom, B.V. Rodenburg, J.R. Changstrom, Phys. Rev. B **75**, 132201 (2007).
1428. T. Kanaya, K. Kaji, J. Bartos, M. Klimova, Macromolecules **30**, 1107 (1997).
1429. T. Kanaya, T. Tsukushi, K. Kaji, J. Bartos, J. Kristiak, Phys. Rev. E **60**, 1906 (1999).
1430. J. Bartos, P. Bandzuch, O. Sausa, K. Kristiakova, J. Kristiak, T. Kanaya, W. Jenninger, Macromolecules **30**, 6906 (1997).
1431. E. Duval, A. Mermet, N. Surovtsev, A.J. Dianoux, J. Non-Cryst. Solids **235–237**, 203 (1998).

1432. E. Duval, A. Mermet, N. Surovtsev, J.F. Jal, A.J. Dianoux, Philos. Mag. B **77**, 457 (1997).
1433. A. Mermet, E. Duval, N. Surovtsev, J.F. Jala, A. Dianoux, A.F. Yee, Europhys. Lett. **38**, 515 (1997).
1434. H.A. Hristov, B. Bolan, A.F. Yee, L. Xie, D.W. Gidley, Macromolecules **29**, 8507 (1996).
1435. U. Bengtzelius, W. Götze, A. Sjölander, J. Phys. C **17**, 5915 (1984).
1436. E. Leutheuser, Phys. Rev. A **29**, 2765 (1984).
1437. W. Petry, E. Bartsch, F. Fujara, M. Kiebel, H. Sillescu, B. Farago, Z. Phys. B **83**, 175 (1991).
1438. J. Wuttke, M. Kiebel, E. Bartsch, F. Fujara, W. Petry, H. Sillescu, Z. Phys. B **91**, 357 (1993). M. Kiebel, E. Bartsch, O. Debus, F. Fujara, H. Sillescu, Phys. Rev. B **45**, 10301 (1992).
1439. F. Mezei, J. Non-Cryst. Solids **131–133**, 317 (1991), data interpreted in terms of MCT by authors in next reference.
1440. H.Z. Cummins, W.M. Du, M. Fuchs, W. Götze, S. Hildebrand, A. Latz, G. Li, N.J. Tao, Phys. Rev. E **47**, 4223 (1993).
1441. A. Tölle, H. Schober, J. Wuttke, F. Fujara, Phys. Rev. E **56**, 809 (1997).
1442. G. Li, M. Du, A. Sakai, H.Z. Cummins, Phys. Rev. A **46**, 3343 (1992).
1443. W.M. Du, G. Li, H.Z. Cummins, M. Fuchs, J. Toulouse, L.A. Knauss, Phys. Rev. E **49**, 2192 (1994).
1444. M. Fuchs, W. Götze, S. Hildebrand, A. Latz, J. Phys. Condens. Matter **4**, 7709 (1992).
1445. H.Z. Cummins, Y.H. Hwang, G. Li, W.M. Du, W. Losert, G.Q. Shen, J. Non-Cryst. Solids **235–237**, 254 (1998).
1446. D. Engberg, A. Wischnewski, U. Buchenau, L. Börjesson, A.J. Dianoux, A.P. Sokolov, L.M. Torell, Phys. Rev. B **58**, 9087 (1998).
1447. T. Blochowicz, C. Tschirwitz, S. Benkhof, E.A. Rössler, J. Chem. Phys. **118**, 7544 (2003).
1448. T. Blochowicz, C. Gainaru, P. Medick, C. Tschirwitz, E.A. Rössler, J. Chem. Phys. **124**, 134503 (2006).
1449. (a) K. Murata, A. Hanawa, R. Nozaki, J. Appl. Phys. **98**, 084107 (2005); (b) R. Nozaki, J. Non-Cryst. Solids, to be published (2010).
1450. E. Kubo, A. Minoguchi, H. Sotokawa, R. Nozaki, J. Non-Cryst. Solids **352**, 4724 (2006).
1451. K.L. Ngai, R. Casalini, Phys. Rev. B **66**, 132205 (2002).
1452. A. Schönhals, F. Kremer, A. Hofmann, E.W. Fischer, E. Schlosser, Phys. Rev. Lett. **70**, 3459 (1993).
1453. M. Alcoutlabi, G.B. McKenna, J. Phys. Condens. Matter. **17**, R461 (2005); S.F. Swallen et al., J. Phys. Chem. B **113**, 4600 (2009).
1454. C. Le Quellec, G. Dosseh, F. Audonnet, N. Brodie-Linder, C. Alba-Simionesco, W. Haeussler, B. Frick, Eur. Phys. J. Special Top. **141**, 11 (2007).
1455. S. Kawana, R.A.L. Jones, Phys. Rev. E **63**, 021501 (2001); Eur. Phys. J. E **10**, 223 (2003).
1456. K.L. Ngai, A.K. Rizos, D.J. Plazek, J. Non-Cryst. Solids **235–237**, 435 (1998).
1457. Z. Fakhraai, J.A. Forrest, Phys. Rev. Lett. **95**, 025701 (2005).
1458. M.M. Elmahdy, K. Chrissopoulou, A. Afratis, G. Floudas, S.H. Anastasiadis, Macromolecules **39**, 5170 (2006).
1459. J. Baschnagel, K. Binder, Mater. Res. Soc. Symp. Proc. **543**, 157 (1999).
1460. J. Baschnagel, C. Mischler, K. Binder, J. Phys. IV **10**, Pr7 (2000).
1461. C. Mischler, J. Baschnagel, K. Binder, Adv. Colloid Interf. Sci. **94**, 197 (2001). http://arxiv.org/abs/condmat / 0012277.
1462. T. Kreer, J. Baschnagel, M. Müller, K. Binder, Macromolecules **34**, 1105 (2001).
1463. P. Scheidler, W. Kob, K. Binder, G. Parisi, Philos. Mag. B **82**, 283 (2002).
1464. C.R. Nugent, K.V. Edmond, H.N. Patel, E.R. Weeks, Phys. Rev. Lett. **99**, 025702 (2007).
1465. (a) K.-I. Akabori, K. Tanaka, T. Nagamura, A. Takahara, T. Kajiyama, Macromolecules **38**, 9735 (2005); (b) T. Kajiyama, K. Tanaka, N. Satomi, A. Takahara, Macromolecules **31**, 5151 (1998); (c) K. Tanaka, A. Takahara, T. Kajiyama, Macromolecules **33**, 7588 (2000); (d) T.Z. Fakhraai, J.A. Forrest, Science **319**, 600 (2008); (e) D. Qi, Z. Fakhraai, J.A. Forrest, Phys. Rev. Lett. **101**, 096101 (2008); (f) S. Napolitano, M. Wübbenhorst, Polymer **51**, 5309 (2010).

1466. (a) T. Kuhlmann, J. Kraus, P. Müller-Buschbaum, J. Non-Cryst. Solids **235**, 457 (1998); (b) J. Kraus, P. Müller-Buschbaum, T. Kuhlmann, D.W. Schubert, M. Stamm, Europhys. Lett. **49**, 210 (2000).
1467. G. Reiter, Europhys. Lett. **23**, 579 (1993).
1468. J.A. Forrest, Eur. Phys. J. E **8**, 261 (2002).
1469. G.B. McKenna, Eur. Phys. J. E **12**, 191(2003).
1470. J.F. O'Gara, S.G. Dejardins, A.A. Jones, Macromolecules **14**, 62 (1981).
1471. J.J. Connolly, E. Gordon, A.A. Jones, Macromolecules **17**, 722 (1984).
1472. A.E. Tonelli, Macromolecules **5**, 558 (1972).
1473. J.T. Bendler, Ann. N.Y. Acad. Sci. **371**, 299 (1981).
1474. G. Power, J.K. Vij, G.P. Johari **124**, 044513 (2006).
1475. K. Mpoukouvalas, G. Floudas, G. Williams, Macromolecules **42**, 4690 (2009).
1476. M. Wind, R. Graf, A. Heuer, H.W. Spiess, Phys. Rev. Lett. **91**, 155702–1 (2003).
1477. G. Williams, Chem. Rev. **72**, 55 (1972).
1478. A.A. Pronin, M.V. Kondrin, A.G. Lyapin, V.V. Brazhkin, A.A. Volkov, P. Lunkenheimer, A. Loidl, Phys. Rev. B **81**, 041503 (2010).
1479. A. Reiser, G. Kasper, Europhys. Lett. **76**, 1137 (2006).
1480. C.A. Herbst, R.L. Cook, H.E. King, Nature **361**, 518 (1993).
1481. (a) S. Pawlus, M. Paluch, J. Zioło, C.M. Roland, J. Phys. Condens. Matter. **21**, 332101 (2009); (b) S. Pawlus, M. Paluch, E. Kaminska, K.L. Ngai, J. Phys. Chem. B **110**, 7678 (2006); (c) Z. Wojnarowska, M. Paluch, A. Grzybowski, K. Adrjanowicz, K. Grzybowska, K. Kaminski, P. Wlodarczyk, J. Pionteck, J. Chem. Phys. **131**, 104505 (2009); (d) K. Adrjanowicz, K. Kaminski, M. Paluch, P. Wlodarczyk, K. Grzybowska, Z. Wojnarowska, L. Hawelek, W. Sawicki, P. Lepek, R. Lunio, J. Pharm. Sci. **99**, 828 (2010); (e) S. Pawlus, M. Paluch, J. Ziolo, M.K. Kolel-Veetil, J. Phys.: Condens. Matter **22**, 415101 (2010).
1482. C. León, J. Habasaki, K.L. Ngai, Z. Phys. Chem. **223**, 1311 (2009).
1483. C. León, M.L. Lucía, J. Santamaría, M.A. París, J. Sanz, A. Várez, Phys. Rev. B **54**, 184 (1996).
1484. C. León, J. Santamaría, M.A. París, J. Sanz, J. Ibarra, L.M. Torres, Phys. Rev. B **56**, 5302 (1997).
1485. R. Böhmer, P. Lunkenheimer, M. Lotze, A. Loidl, Z. Phys. B Condens Matter **100**, 583 (1996).
1486. K.L. Ngai, U. Strom, Phys. Rev. B **38**, 10350 (1988).
1487. J. Habasaki, I. Okada, Y. Hiwatari, Phys. Rev. B **55**, 6309 (1997); **81**, 141503 (2010).
1488. J. Habasaki, K.L. Ngai, Y. Hiwatari, J. Chem. Phys. **122**, 054507 (2005).
1489. J. Habasaki, Y. Hiwatari, Phys. Rev. E **59**, 6962 (1999).
1490. Y. Haven, B. Verkerk, Phys. Chem. Glass **6**, 38 (1965).
1491. A. Rivera, J. Santamaría, C. León, T. Blochowicz, C. Gainaru, E.A. Roessler, Appl. Phys. Lett. **82**, 2425 (2003).
1492. K.L. Ngai, Philos. Mag. **77**, 187 (1998).
1493. M.R. Díaz-Guillén, K.J. Moreno, J.A. Díaz-Guillén, A.F. Fuentes, K.L. Ngai, J. Garcia-Barriocanal, J. Santamaria, C. León, Phys. Rev. B **78**, 104304 (2008).
1494. R. Casalini, K.L. Ngai, C.M. Roland, J. Chem. Phys. **112**, 5181 (2000).
1495. K. Kremer, G. Grest, J. Chem. Phys. **92**, 5057 (1990).
1496. W. Paul, G.D. Smith, D.Y. Yoon, B. Farago, S. Rathgeber, A. Zirkel, L. Wilner, D. Richter, Phys. Rev. Lett. **80**, 2346 (1998).
1497. G.D. Smith, W. Paul, M. Monkenbusch, D. Richter, Chem. Phys. **261**, 61 (2000).
1498. G.D. Smith, W. Paul, M. Monkenbusch, D. Richter, J. Chem. Phys. **114**, 4285 (2001).
1499. J.T. Padding, W.J. Briels, J. Chem. Phys. **114**, 8685 (2001).
1500. J.T. Padding, W.J. Briels, J. Chem. Phys. **115**, 2846 (2001).
1501. N.C. Karayiannis, V.G. Mavrantzas, D.N. Theodorou, Phys. Rev. Lett. **88**, 105503 (2002).
1502. N.C. Karayiannis, A.E. Giannousaki, V.G. Mavrantzas, D.N. Theodorou, J. Chem. Phys. **117**, 5465 (2002).

1503. M. Zamponi, A. Wischnewski, M. Monkenbusch, L. Willner, D. Richter, P. Falus, B. Farago, M.G. Guenza, J. Phys. Chem. B **112**, 16220 (2008).
1504. M.G. Guenza, Phys. Rev. Lett. **88**, 025901 (2002); Macromolecules **35**, 2714 (2002); J. Chem. Phys. **119**, 7568 (2003).
1505. E.J. Sambriski, G. Yatsenko, M.A. Nemirovskaya, M.G. Guenza, J. Phys. Condens. Matter **19**, 205115 (2007).
1506. J.H. Simmons, P.B. Elterman, C.J. Simmons, R.K. Mohr, J. Am. Ceram. Soc. **62**, 158 (1979).
1507. M.B.M. Mangion, G.P. Johari, Phys. Chem. Glasses **29**, 225 (1988).
1508. K.L. Ngai, Solid State Ionics **105**, 231 (1998).
1509. D.L. Sidebottom, P.F. Green, R.K. Brow, J. Non-Cryst. Solids **183**, 151 (1995).
1510. B. Roling, A. Happe, K. Funke, M.D. Ingram, Phys. Rev. Lett. **78**, 2160 (1997).
1511. B. Roling, Solid State Ionics **105**, 185 (1998).
1512. H. Jain, S. Krishnaswami, Solid State Ionics **105**, 129 (1998).
1513. V.R. Raju, H. Rachapudy, W.W. Graesssley, J. Polym. Sci. Polym. Phys. Ed. **17**, 1223 (1979).
1514. S.Z. Ren, W.F. Shi, W.B. Zhang, C.M. Sorensen, Phys. Rev. A **45**, 2416 (1992).
1515. J.-F. Shi, L.C. Dickinson, W.J. MacKnight, J.C.W. Chien, Macromolecules **26**, 5908 (1993).
1516. K.L. Ngai, C.M. Roland, Macromolecules **25**, 2454 (1994).
1517. K.L. Ngai, C.M. Roland, A.F. Yee, Rubber Chem. Tech. **66**, 817 (1993).
1518. C.T. Moynihan, L.P. Boesch, N.L. Laberge, Phys. Chem. Glasses **14**, 122 (1973).
1519. C.T. Moynihan, J. Non-Cryst. Solids **172–174**, 1395 (1994); ibid. **203**, 359 (1996); Solid State Ionics **105**, 75 (1998).
1520. C. Cramer, K. Funke, M. Buscher, A. Happe, Philos. Mag. B **71**, 713 (1995).
1521. C. Cramer, S. Brueckner, Y. Gao, K. Funke, R. Belin, G. Taillades, A. Pradel, J. Non-Cryst. Solids **307–310**, 905 (2002).
1522. (a) K.L. Ngai, U. Strom, Phys. Rev. B **38**, 10350 (1988); (b) K.L. Ngai, J. Phys. (Paris) Colloq. C2, **2**, 61 (1992).
1523. K.L. Ngai, O. Kanert, Solid State Ionics **53–55**, 936 (1992).
1524. C. Cramer, R. Graeber, M.D. Ingram, T. Saatkamp, D. Wilmer, K. Funke, Mater. Res. Soc. Symp. Proc. **369**, 233 (1995).
1525. K. Funke, Prog. Solid State Chem. **22**, 111 (1993).
1526. R. Hoppe, T. Kloidt, K. Funke, Ber. Bunsenges. Phys. Chem. **95**, 1025 (1991).
1527. K. Funke, J. Non-Cryst. Solids **172–174**, 1215 (1994).
1528. C. Cramer, M. Buscher, Solid State Ionics **105**, 109 (1998).
1529. G.C. Garton, Discuss. Farad. Soc. A **42**, 161 (1946).
1530. A.E. Owen, Prog. Ceram. Sci. **3**, 77 (1963).
1531. J.M. Stevels, Handbuch Phys. **20**, 372 (1957).
1532. A. Burns, G.D. Chryssikos, E. Tombari, R.H. Cole, W.M. Risen, Phys. Chem. Glasses **30**, 264 (1989).
1533. A.S. Nowick, A.V. Vaysleyb, W. Liu, Solid State Ionics **105**, 121 (1998).
1534. X. Lu, H. Jain, J. Phys. Chem. Solids **55**, 1433 (1994).
1535. A.S. Nowick, B.S. Lim, A.V. Vaysleb, J. Non-Cryst. Solids **172–174**, 1243 (1994).
1536. J. Kincs, S.W. Martin, Phys. Rev. Lett. **76**, 70 (1995).
1537. K.L. Ngai, A.K. Rizos, Phys. Rev. Lett. **76**, 1296 (1996).
1538. K.L. Ngai, G.N. Greaves, C.T. Moynihan, Phys. Rev. Lett. **80**, 1018 (1998).
1539. K.L. Ngai, Philos. Mag. B **77**, 187 (1998).
1540. K.L. Ngai, J. Non-Cryst. Solids **248**, 194 (1999).
1541. C. León, P. Lunkenheimer, K.L. Ngai, Phys. Rev. B **64**, 184304 (2001).
1542. C. León, L. Lucia, J. Santamaria, Phys. Rev. B **55**, 882 (1997), Philos. Mag. B **75**, 629 (1997).
1543. A. Pimenov, J. Ullrich, P. Lunkenheimer, A. Loidl, C.H. Ruescher, Solid State Ionics **109**, 111 (1998).

1544. S. Shin, M. Ishigame, Phys. Rev. B **34**, 8875 (1986).
1545. T. Suemoto, M. Ishigame, Phys. Rev. B **33**, 2757 (1986).
1546. W. Meyer, H. Neldel, Z. Tech. **18**, 588 (1937).
1547. A.J. Dianoux, M. Tachez, R. Mercier, J.P. Malugani, J. Non-Cryst. Solids **131–133**, 973 (1991).
1548. M. Tachez, R. Mercier, J.P. Malugani, A.J. Dianoux, Solid State Ionics **20**, 93 (1986).
1549. A.P. Owens, A. Pradel, M. Ribes, S.R. Elliott, J. Non-Cryst. Solids **131–133**, 1104 (1991).
1550. A.P. Owens, A. Pradel, M. Ribes, S.R. Elliott, Mater. Res. Soc. Symp. Proc. **210**, 621 (1991).
1551. G.N. Greaves, K.L. Ngai, Phys. Rev. B **52**, 6358 (1995).
1552. K.L. Ngai, D.J. Plazek, Viscoelastic Behavior of Rubber, in *Science and Technology of Rubber*, 3rd ed., Chapter 5, J.E. Mark, B. Erman, F.R. Eirich, Eds., Elsevier, Amsterdam (2005).
1553. D. Richter, R. Butera, L.J. Fetters, J.S. Huang, B. Ewen, C. Lartigue, Phys. Rev. Lett. **64**, 1389 (1992).
1554. K.L. Ngai, R.W. Rendell, Philos. Mag. B **77**, 621 (1998).
1555. J.E. Martin, J.P. Wilcoxon, Phys. Rev. Lett. **61**, 898 (1988).
1556. J.E. Martin, J.P. Wilcoxon, J. Odinek, Phys. Rev. A **43**, 858 (1991).
1557. M. Delsanti, J.P. Munch, M. Adam, D. Durand, in *Relaxation in Complex Systems and Related Topics*, I.A. Campbell, C. Giovannella, Eds., Plenum Press, New York, NY (1990).
1558. K. Thuresson, S. Nilsson, A.-L. Kjøniksen, H. Walderhaug, B. Lindman, B. Nyström, J. Phys. Chem. B **103**, 1425 (1999).
1559. M. Tsianou, A.-L. Kjøniksen, K. Thuresson, B. Nyström, Macromolecules **32**, 2974 (1999).
1560. A.L. Kjøniksen, M. Hiorth, B. Nyström, Eur. Polym. J. **40**, 2427 (2004).
1561. E. Raspaud, D. Lairez, M. Adam, J.-P. Carton, Macromolecules **27**, 2956 (1994).
1562. J. Narayanan, V.W. Deotare, R. Bandyopadhyay, A.K. Sood, J. Colloid Interf. Sci. **245**, 267 (2002).
1563. H. Bu, A.-L. Kjøniksen, B. Nyström, Eur. Polym. J. **41**, 1708 (2005).
1564. (a) A. Maleki, A.-L. Kjøniksen, B. Nyström, Carbohydr. Res. **342**, 2776 (2007); (b) A.-L. Kjøniksen, K. Zhu, M. Behrens, J. Skov Pedersen, B. Nyström, J. Phys. Chem. B (2010).
1565. R. Ganguly et al., J. Colloid Interf. Sci. **315**, 693, (2007).
1566. G.D.J. Phillies, C.A. Quinlan, Macromolecules **28**, 160 (1996).
1567. K.L. Ngai, *Dynamics of Semidilute Solutions of Polymers and Associating Polymers*, Adv. Colloid Interf. Sci. **64**, 1 (1996).
1568. E.Y. Sheu, S.-H. Chen, J.S. Huang, J.C. Sung, Phys. Rev. A **39**, 5867 (1989).
1569. P. Meyer, P. Maass, A. Bunde, J. Non-Cryst. Solids **172–174**, 1292 (1994).
1570. P. Maass, M. Meyer, A. Bunde, Phys. Rev. B **51**, 8164 (1995).
1571. K.L. Ngai, Solid State Ionics **105**, 225 (1998).
1572. K.L. Ngai, C. León, J. Non-Cryst. Solids **315**, 124 (2003).
1573. H. Jain, N.L. Peterson, Philos. Mag. A **46**, 351 (1982).
1574. G.H. Vineyard, J. Phys. Chem. Solids **3**, 121 (1957).
1575. P.J. Flory, *Statistical Mechanics of Chain Molecules*, Oxford University Press, New York, NY (1989).
1576. K.L. Ngai, C.M. Roland, J. Poly. Sci. Poly. Phys. Ed. **35**, 2503 (1997).
1577. K.L. Ngai, R.W. Rendell, A.K. Rajagopal, S. Teitler, Ann. NY Acad. Sci. **484**, 150 (1986); R.W. Rendell, K.L. Ngai, G.B. McKenna, Macromolecules **20**, 2250 (1987).
1578. C.M. Roland, P.G. Santangelo, J. Non-Cryst. Solids, **307–310**, 835 (2002).
1579. G. Allen, P.N. Brier, A.G. Goodyear, J.S. Higgins, Farad. Symp. Chem. Soc. **6**, 169 (1972).
1580. G. Allen, J.S. Higgins, A. Maconnachie, R.E. Ghosh, J. Chem. Soc. Farad. Trans. 2, **78**, 2117 (1982).
1581. R.H. Boyd, R.H. Gee, J. Han, Y.J. Jin, J. Chem. Phys. **101**, 788 (1994).
1582. X.H. Qiu, M.D. Ediger, Macromolecules **33**, 490 (2000).
1583. D.J. Plazek, J. Danhauser, J.D. Ferry, unpublished data.

1584. Y. Ding, V.N. Novikov, A.P. Sokolov, R. Casalini, C.M. Roland, Macromolecules **37**, 9273 (2004).
1585. D.N. Theodorou, private communication.
1586. V.R. Raju, Ph.D. Dissertation, Northwestern University (1980).
1587. For reviews of experimental data, see M. Tirrell, Ref. [834], and T.P. Lodge, N.A. Rotstein, S. Prager, Adv. Chem. Phys. **79**, 1 (1990).
1588. T.P. Lodge, Phys. Rev. Lett. **83**, 3218 (1999).
1589. K.L. Ngai, Macromol. Symp. **146**, 117 (1999).
1590. P.-G. DeGennes, *Scaling Concepts in Polymer Physics*, Cornell University Press, Ithaca, NY (1979).
1591. E.J. Amis, C.C. Han, Y. Matsushita, Polymer **25**, 650 (1984).
1592. T. Yang, A.M.J. Jamieson, J. Colloid Interf. Sci. **126**, 220 (1988).
1593. M. Mustafa, P.S. Russo, J. Colloid Interf. Sci. **129**, 240 (1989).
1594. H. Kim, T. Chang, J.M. Yohanan, L. Wang, H. Yu, Macromolecules **19**, 2737 (1986).
1595. L.M. Wheeler, T.P. Lodge, B. Hanley, M. Tirrell, Macromolecules **20**, 1120 (1987).
1596. L.M. Wheeler, T.P. Lodge, Macromolecules **19**, 2983 (1986); ibid. **22**, 3399 (1989).
1597. T.P. Lodge, P. Markland, Polymer **28**, 1377 (1987).
1598. T.P. Lodge, P. Markland, L.M. Wheeler, Macromolecules **22**, 3409 (1989).
1599. G.D.J. Phillies, Macromolecules **19**, 2367 (1986); J. Phys. Chem. **96**, 10061 (1992).
1600. A.N. Semenov, Physica A **166**, 263 (1990).
1601. W. Brown, K. Schillen, S. Hvidt, J. Phys. Chem. **96**, 6038 (1992).
1602. J. Bauer, W. Burchard, Phys. II France **2**, 1053 (1992).
1603. W. Burchard, P. Lang, L. Schulz, T. Coviello, Makromol. Chem. Macromol. Symp. **58**, 21 (1992).
1604. Z. Sun, C.H. Wang, Macromolecules **27**, 5667 (1994).
1605. M. Mustafa, D.L. Tipton, M.D. Barkley, P.S. Russo, F.D. Blum, Macromolecules **26**, 370 (1993).
1606. S. Bu, P.S. Russo, Macromolecules **27**, 1187 (1994).
1607. T.P. Lodge, N.A. Rotstein, S. Prager, Adv. Chem. Phys. **79**, 1 (1990).
1608. J. Skolnick, A. Kolinski, Adv. Chem. Phys. **78**, 1 (1989).
1609. N. Nemoto, T. Kojima, T. Inoue, K. Osaki, Macromolecules **22**, 3793 (1989).
1610. N. Nemoto, M. Kishine, T. Inoue, K. Osaki, Macromolecules **24**, 1648 (1991).
1611. G.D.J. Phillies, M. Lacroix, J. Phys. Chem. B **101**, 39 (1997).
1612. D.W. Schaefer, C.C. Han, in *Dynamic Light Scattering*, R. Pecora, Ed., Plenum Press, New York, NY, p. 181 (1985).
1613. K.L. Ngai, T.P. Lodge, ACS Polym. Preprint **30**, 67 (1989).
1614. B. Nyström, K. Thuresson, B. Lindmann, Langmuir **11**, 1994 (1995).
1615. B. Nyström, B. Lindmann, Macromolecules **28**, 967 (1995).
1616. A. Schönhals, D. Wolff, J. Springer, Macromolecules **28**, 6254 (1995).
1617. A. Schönhals, D. Wolff, J. Springer, Polym. Adv Technol. **7**, 853 (1996); SPIE Proc. **424**, 2779 (1996).
1618. A. Schönhals, H.-E. Carius, in *Dielectric Properties of Polymeric Liquid Crystals*, Chapter 3, Liquid Crystal Book Series, Vol. **4**, W. Brostow, A.A. Collyer, Eds., Chapman and Hall, New York, NY.
1619. K.L. Ngai, A. Schönhals, J. Polym. Sci. Polym. Phys. Ed. **36**, 1927 (1998).
1620. K.L. Ngai, R.W. Rendell, A.F. Yee, Macromolecules **21**, 3396 (1988).
1621. P.J. Flory, Polym. J. **17**, 1 (1985).
1622. P.J. Flory, Rub. Chem. Technol. **52**, 110 (1979).
1623. P.J. Flory, B. Erman, Macromolecules **15**, 800 (1982); **15**, 806 (1982).
1624. M.H. Wagner, J. Schaeffer, J. Rheol. **37**, 643 (1993).
1625. J.S. Higgins, K. Ma, R.H. Hall, J. Phys. C. **14**, 4995 (1981).
1626. R. Oeser, B. Ewen, D. Richter, B. Farago, Phys. Rev. Lett. **60**, 1041 (1988).
1627. J. Sanjuan, M.A. Llorente, J. Polym. Sci. Polym. Phys. Ed. **26**, 235 (1988).

1628. J.E. Mark, Rub. Chem. Technol. **48**, 495 (1975).
1629. P.J. Flory, B. Erman, J. Polym Sci. Polym. Phys. Ed. **22**, 1953 (1984).
1630. R.W. Brontzman, J.E. Mark, Macromolecules **19**, 667 (1986).
1631. B. Erman, A. Kloczkowski, J.E. Mark, Macromolecules **22**, 1432 (1989).
1632. D.B. Adolf, J.G. Curro, Macromolecules **20**, 1646 (1987).
1633. C.M. Roland, Rubber Chem. Technol. **62**, 880 (1989).
1634. K.L. Ngai, R.W. Rendell, Macromolecules **20**, 1066 (1987).
1635. P.K. Moon, H.L. Tuller, Solid State Ionics **28–30**, 470 (1988).
1636. H.L. Tuller, Solid State Ionics **52**, 135 (1992).
1637. J. Chen, J. Lian, L.M. Wang, R.C. Ewing, R.G. Wang, W. Pan, Phys. Rev. Lett. **88**, 105901 (2002).
1638. C. Heremans, B.J. Wuensch, J.K. Stalick, E. Prince, J. Solid State Chem. **117**, 108 (1995).
1639. M. Pirzada, R.W. Grimes, L. Minervini, J.F. Maguire, K.E. Sickafus, Solid State Ionics **140**, 201 (2001).
1640. A.J. Burggraaf, T. Van Dijk, M.J. Verkerk, Solid State Ionics **5**, 519 (1981).
1641. O. Porat, C. Heremans, H.L. Tuller, Solid State Ionics **94**, 75 (1997).
1642. E.V. Tsipis, V.V. Kharton, J.R. Frade, J. Eur. Ceram. Soc. **25**, 2623 (2005).
1643. S.A. Kramer, H.L. Tuller, Solid State Ionics **82**, 15 (1995).
1644. M.A. Subramanian, G. Aravamudan, G.V. Subba Rao, Prog. Solid State Chem. **15**, 55 (1983).
1645. M.R. Díaz-Guillén, K.J. Moreno, J.A. Díaz-Guillén, A.F. Fuentes, K.L. Ngai, J. Garcia-Barriocanal, J. Santamaria, C. León, Phys. Rev. B **78**, 104304 (2008).
1646. K.J. Moreno, A.F. Fuentes, U. Amador, J. Santamarıa, C. León, J. Non-Cryst. Solids **353**, 3947 (2007).
1647. (a) A. Rivera, J. Santamaría, C. León, T. Blochowicz, C. Gainaru, E.A. Roessler, Appl. Phys. Lett. **82**, 2425 (2003); (b) A. Rivera, J. Santamaría, C. León, T. Blochowicz, C. Gainaru, E.A. Roessler, unpublished data; (c) A. Rivera, J. Santamaría, C. León, to be published; (d) K. Arbi, M. Tabellout, M.G. Lazarraga, J.M. Rojo, J. Sanz, Phys. Rev. B **72**, 094302 (2005).
1648. J. Maier, Solid State Ionics **131**, 13 (2000); W. Puin, S. Rodewald, R. Ramlau, P. Heitjans, J. Maier, Solid State Ionics **131**, 159 (2000); N. Sata, K. Eberman, K. Eberl, J. Maier, Nature **408**, 946 (2000).
1649. I. Kosacki, T. Christopher, M. Rouleau, P.F. Becher, J. Bentley, D.H. Lowndes, Solid State Ionics **176**, 1319 (2005).
1650. (a) J. Garcia-Barriocanal, A. Rivera-Calzada, M. Varela, Z. Sefrioui, E. Iborra, C. Leon, S. Pennycook, J. Santamaria, Science **321**, 676 (2008); **324**, 465b (2009); (b) J. Garcia-Barriocanal, A. Rivera-Calzada, M. Varela, Z. Sefrioui, M.R. Díaz-Guillén, K.J. Moreno, J.A. Díaz-Guillén, E. Iborra, A.F. Fuentes, S.J. Pennycook, C. Leon, J. Santamaria, Chem. Phys. Chem. **10**, 1003 (2009); (c) T.J. Pennycook, M.J. Beck, K. Varga, M. Varela, S.J. Pennycook, S.T. Pantelides, Phys. Rev. Lett. in press (2010).
1651. J.L. Snoek, Physica (Utrecht) **8**, 711 (1941).
1652. A. Seeger, M. Weller, J. Diehl, Z.L. Pan, J.X. Zhang, T.S. Kê, Z. Metallkd. **73**, 1 (1982).
1653. Y. Wang, M. Gu, L. Sun, K.L. Ngai, Phys. Rev. B **50**, 3525 (1994).
1654. K.L. Ngai, Y.N. Wang, L.B. Magalas, J. Alloys Compounds **211/212**, 327 (1994).
1655. L.B. Magalas, K.L. Ngai, *M3DIII: Mechanics and Mechanisms of Material Damping*, Alan Wolfenden, V.K. Kinra, Eds., STP **1304**, 189 (1997).
1656. G. Schoeck, Acta Metall. **11**, 617 (1963).
1657. (a) Y.N. Wang, M. Gu, L.H. Sun, J. Phys.: Condens. Matter **1**, 10039 (1989). (b) Y. Shi, W.B. Jiang, Q.P. Kong, P. Cui, Q.F. Fang, M. Winning, Phys. Rev. B **73**, 174101 (2006). (c) Q.P. Kong, W.B. Jiang , Y. Shi, P. Cui, Q.F. Fang, M. Winning, Mater. Sci. Eng. A **521–522**, 128 (2009). (d) W.B. Jiang, Q.P. Kong, D.A. Molodov, G. Gottstein, Acta Mater. **57**, 3327 (2009). (e) W.B. Jiang, Q.P. Kong, P. Cui, F.Q. Fang, D.A. Molodov, G. Gottstein, Phil. Mag. **90**, 753 (2010).

1658. G. Cannelli, F.M. Mazzolai, Nuovo Cimento B **64**, 171 (1969).
1659. M. Koiwa, O. Yoshimari, J. Phys. Coll. **46**, C10–C99 (1985).
1660. A.S. Nowick, B.S. Berry, *Anelastic Relaxation in Crystalline Solids*, Academic, New York, NY, p. 490 (1972).
1661. J.A. Kornfield, G.G. Fuller, D.S. Pearson, Macromolecules **22**, 1334 (1989).
1662. C.M. Ylitalo, J.A. Kornfield, G.G. Fuller, D.S. Pearson, Macromolecules **24**, 749 (1991).
1663. J.P. Montfort, G. Marin, D. Monge, Macromolecules **17**, 1551 (1984).
1664. M.J. Struglinski, W.W. Graessley, Macromolecules **18**, 2630 (1985).
1665. J.-P. Jarry, L. Monnerie, Macromolecules **12**, 316 (1979).
1666. M. Fukuda, G.L. Wilkes, R.S. Stein, J. Polym. Sci. A-2 **9**, 1417 (1971).
1667. B. Deloche, E.T. Samulski, Macromolecules **14**, 575 (1981).
1668. P. Schmidt, B. Schneider, Makromol. Chem. **184**, 2075 (1983).
1669. M.M. Jacobi, R. Stadtler, W. Gronski, Macromolecules **19**, 2884 (1986).
1670. P. Sotta, B. Deloche, J. Herz, A. Lapp, D. Durand, J.-C. Rabadeux, Macromolecules **20**, 2769 (1987).
1671. B. Erman, J.-P. Queslel, L. Monnerie, Polymer **29**, 1823 (1988).
1672. J.F. Tassin, L. Monnerie, L.J. Fetters, Macromolecules **21**, 2404 (1988).
1673. F. Boue, B. Farnoux, J. Bastide, A. Lapp, J. Herz, Cl Picot, Europhys. Lett. **1**, 637 (1986).
1674. B. Deloche, A. Dubault, J. Herz, A. Lapp, Europhys. Lett. **1**, 629 (1986).
1675. B. Erman, J.-P. Jarry, L. Monnerie, Polymer **28**, 727 (1987); B. Erman, L. Monnerie, Macromolecules **22**, 3342 (1989).
1676. E.A. DiMarizio, J. Chem. Phys. **36**, 1563 (1962).
1677. T. Tanaka, G. Allen, Macromolecules **10**, 426 (1977).
1678. M. Rubinstein, E. Helfand, D.S. Pearson, Macromolecules **20**, 822 (1987).
1679. M.J. Marrucci, Polym. Sci. Polym. Phys. Ed. **23**, 159 (1985).
1680. J.L. Viovy, J. Phys. (Les Ulis., Fr.) **46**, 847 (1985).
1681. H. Watanabe, M. Tirrell, Macromolecules **22**, 927 (1988).
1682. M. Doi, D.S. Pearson, J. Kornfield, G.G. Fuller, Macromolecules **22**, 1488 (1989).
1683. G. Fytas, G. Floudas, K.L. Ngai, Macromolecules **23**, 1104 (1990).
1684. B.T. Poh, K. Adachi, T. Kotaka, Macromolecules **20**, 2563 (1987).
1685. K. Adachi, T. Nakamoto, T. Kotaka, Macromolecules **22**, 3111 (1989).
1686. K. Adachi, T. Kotaka, Polym. Yearbook **6**, 43 (1990).
1687. K. Adachi, T. Kotaka, Prog. Polym. Sci. **18**, 585 (1993).
1688. J.A. Zawada, C.M. Ylitalo, G.G. Fuller, R.H. Colby, T.E. Long, Macromolecules **25**, 2896 (1992).
1689. R.M. Hakim, D.R. Uhlman, Phys. Chem. Glasses **8**, 174 (1967).
1690. J.O. Isard, J. Non-Cryst. Solids **1**, 235 (1969).
1691. D.E. Day, J. Non-Cryst. Solids **21**, 343 (1976).
1692. J.R. Henrickson, P.J. Bray, Phys. Chem. Glasses **13**, 43 (1972).
1693. A.W. Imre, S. Voss, H. Mehrer, J. Non-Cryst. Solids **333**, 231 (2004).
1694. H. Jain, N.L. Peterson, H.L. Downing, J. Non-Cryst. Solids **55**, 283 (1983).
1695. G.N. Greaves, S.J. Gurman, C.R.A. Catlow, A.V. Chadwick, S. Houde-Walter, C.M.B. Henderson, B.R. Dobson, Philos. Mag. A **64**, 1059 (1991).
1696. F. Ali, A.V. Chadwick, G.N. Greaves, M.C. Jermy, K.L. Ngai, M.E. Smith, Solid State Nucl. Magn. Reson. **5**, 133 (1995).
1697. G.N. Greaves, A.V. Chadwick, M.C. Jermy, M.E. Smith, R. Zhu, J. Nono-Cryst. Solids **235–237**, 766 (1998).
1698. B. Gee, H. Eckert, J. Phys. Chem. **100**, 3705 (1996).
1699. J. Swenson, A. Matic, A. Brodin, L. Börjesson, W.S. Howells, Phys. Rev. B **58**, 11331 (1998).
1700. C. Rau, P. Armand, A. Pradel, C.P.E. Varsamis, E.I. Kamitsos, D. Granier, A. Ibanez, E. Philippot, Phys. Rev. B **63**, 184204 (2001).
1701. T. Uchino, T. Sakka, Y. Ogata, M. Iwasaki, J. Non-Cryst. Solids **146**, 26 (1992).

1702. J. Habasaki, I. Okada, Y. Hiwatari, J. Non-Cryst. Solids **208**, 181 (1996).
1703. J. Swenson, S. Adams, Phys. Rev. Lett. **90**, 155507 (2003).
1704. C.T. Moynihan, N.S. Saad, D.C. Tran, A.V. Lesikar, J. Am. Ceram. Soc. **63**, 458 (1980).
1705. K.L. Ngai, Y. Wang, C.T. Moynihan, J. Non-Cryst. Solids **307**, 999 (2002).
1706. (a) K.L. Ngai, S.W. Martin, Phys. Rev. B **40**, 10550 (1989); (b) G.N. Greaves, K.L. Ngai, Phys. Rev. B **52**, 6358 (1995).
1707. J. Habasaki, K.L. Ngai, Y. Hiwatari, J. Chem. Phys. **121**, 925 (2004).
1708. J. Habasaki, I. Okada, Y. Hiwatari, J. Non-Cryst. Solids **183**, 12 (1995); ibid. **208**, 181 (1996).
1709. J. Habasaki, Y. Hiwatari, J. Non-Cryst. Solids **307**, 930 (2002), J. Non-Cryst. Solids **320**, 281 (2003).
1710. H. Lammert, A. Heuer, Phys. Rev. B **72**, 214202 (2005).
1711. S. Voss, S.V. Divinski, A.W. Imre, H. Mehrer, J.N. Mundy, Solid State Ionics **176**, 1383 (2005).
1712. H. Eckert, Paper presented at the 4th Intern'al Discussion Meeting on Relaxations in Complex Systems, Hersonnisos, Crete, Greece, June (2001).
1713. P. Maass, J. Non-Cryst. Solids **255**, 35 (1999).
1714. J. Swenson, A. Matic, C. Karlsson, L. Börjesson, C. Meneghini, W.S. Howells, Phys. Rev. B **63** (2001) 132202.
1715. S. Voss, F. Berkemeier, A.W. Imre, H. Mehrer, Z. Phys. Chem. **218** (2004) 1353.
1716. C.T. Moynihan, L.P. Boesch, 79th Annual Meeting of the American Ceramic Society, Chicago, Paper 26-G-77.
1717. M. Tomozawa, J.M. Hyde, J.F. Cordaro, M. Yoshiyagawa, Phys. Chem. Glasses **33**, 69 (1992).
1718. J.M. Hyde, M. Tomozawa, M. Yoshiyagawa, Phys. Chem. Glasses **28**, 174 (1987).
1719. K.L. Ngai, H. Jain, Solid State Ionics **18** and **19**, 362 (1986).
1720. T.J. Higgins, L.P. Boesch, V. Voltera, C.T. Moynihan, P.B. Macedo, J. Am. Ceram. Soc. **56**, 334 (1973).
1721. K.L. Ngai, J. Mundy, H. Jain, O. Kanert, J. Balzer-Jollenbeck, Phys. Rev. B **39**, 6169 (1989).
1722. M. Tomozawa, J.M. Hyde, J.F. Cordaro, M. Yoshiyagawa, Phys. Chem. Glasses **33**, 69 (1992).
1723. W.C. Huang, H. Jain, J. Non-Cryst. Solids **212**, 117 (1997).
1724. A.R. Kulkarni, P. Lunkenheimer, A. Loidl, Mater. Chem. Phys. **63**, 93 (2000).
1725. C. Karlsson, A. Mandanici, A. Matic, J. Swenson, L. Börjesson, J. Non-Cryst. Solids **307**, 1012 (2002), Phys. Rev. B **68**, 064202 (2003).
1726. L.P. Boesch, Ph.D. Thesis, Catholic University of America, Washington, DC (1983).
1727. J.F. Cordaro, M. Tomozawa, J. Am. Ceram. Soc. **65**, C50 (1982).
1728. A. Heuer, M. Kunow, M. Vogel, R.D. Banhatti, Phys. Chem. Chem. Phys. **4**, 3185 (2002).
1729. J.E. Kelly, J.F. Cordaro, M. Tomozawa, J. Non-Cryst. Solids **41**, 47 (1980).
1730. D. Knödler, P. Pendzig, W. Dieterich, Solid State Ionics **86–88**, 29 (1996).
1731. S. Voss, A.W. Imre, H. Mehrer, Phys. Chem. Chem. Phys. **6**, 3669 (2004).
1732. (a) K. Binder, J. Non-Cryst. Solids **274**, 332 (2000); (b) J. Horbach, W. Kob, K. Binder, Chem. Geol. **174**, 87 (2001).
1733. Data of *N*-methyl-ϵ-caprolactam (NMEC) are kindly supplied by R. Richert.
1734. M. Pollak, G.E. Pike, Phys. Rev. Lett. **28**, 1449 (1972). This paper dealt with electronic systems.
1735. S.R. Elliott, Adv. Phys. **36**, 135 (1987).
1736. U. Strom, K.L. Ngai, Solid State Ionics **9** and **10**, 283 (1989).
1737. A. Hunt, Solid State Commun. **80**, 151 (1991).
1738. K.L. Ngai, U. Strom, O. Kanert, Phys. Chem. Glasses **33**, 109 (1992).
1739. O. Kanert, R. Küchler, K.L. Ngai, H. Jain, Phys. Rev. **49**, 76 (1994).
1740. H. Jain, J. Non-Cryst. Solids **131–133**, 961 (1991).

1741. W.K. Lee, J.F. Liu, A.S. Nowick, Phys. Rev. Lett. **67**, 1559 (1991).
1742. X. Lu, H. Jain, O. Kanert, J. Non-Cryst. Solids **172–174**, 1436 (1994).
1743. D.L. Sidebottom, P.F. Green, R.K. Brow, Phys. Rev. Lett. **74**, 5068 (1995).
1744. H. Jain, X. Lu, J. Non-Cryst. Solids **196**, 285 (1996).
1745. O. Kanert, R. Küchler, J. Dieckhofer, X. Lu, H. Jain, K.L. Ngai, J. Non-Cryst. Solids **172–174**, 1277 (1994).
1746. A.S. Nowick, Solid State Ionics **136** and **137**, 1307 (2000).
1747. K. Funke, D. Wilmer, Solid State Ionics **136** and **137**, 1329 (2000).
1748. J.C. Dyre, T.B. Schrøder, Rev. Mod. Phys. **72**, 873 (2000).
1749. B. Roling, Phys. Chem. Chem. Phys. **3**, 5093 (2001).
1750. B. Roling, C. Martiny, S. Murugavel, Phys. Rev. Lett. **87**, 085901 (2001).
1751. C. León, A. Rivera, J. Santamarıa, C.T. Moynihan, K.L. Ngai, Phys. Rev. Lett. **89**, 079601 (2002).
1752. C. León, K.L. Ngai, A. Rivera, Phys. Rev. B **69**, 134303 (2004).
1753. J.C. Dyre, P. Maass, B. Roling, D.L. Sidebottom, Rep. Prog. Phys. **72**, 046501 (2009).
1754. W. Dieterich, P. Maass, Solid State Ionics **180**, 446 (2009).
1755. A. Rivera, J. Santamarıa, C. León, J. Sanz, C.P.E. Varsamism, G.D. Chryssikos, K.L. Ngai, J. Non-Cryst. Solids **307–310**, 1024 (2002).
1756. A. Kulkarni, P. Lunkenheimer, A. Loidl, Ceramics Trans. **92**, 115 (1999).
1757. M. Ribes, E. Bychkov, A. Pradel, J. Optoelectronics Adv. Mater. **3**, 665 (2001).
1758. C. León, A. Rivera, A. Várez, J. Sanz, J. Santamaría, C.T. Moynihan, K.L. Ngai, J. Non-Cryst. Solids **305**, 88 (2002).
1759. K.S. Gilroy, W.A. Phillips, Philos. Mag. B **43**, 735 (1981).
1760. S. Murugavel, B. Roling, J. Non-Cryst. Solids **330**, 122 (2003).
1761. A.P. Sokolov, K.S. Schweizer, Phys. Rev. Lett. **102**, 248301 (2009).
1762. K.L. Ngai, D.J. Plazek, C.M. Roland, Phys. Rev. Lett. **103**, 159801 (2009).
1763. K.L. Ngai, D.J. Plazek, C.M. Roland, Macromolecules **41**, 3925 (2008).
1764. R.J. Cava, R.M. Fleming, P. Littlewood, E.A. Rietman, L.F. Schneemeyer, R.G. Dunn, Phys. Rev. B **30**, 3228 (1984).
1765. K. Biljakovic,´J.C. Lasjaunias, F. Zougmore, P. Monceau, F. Levy, L. Bernard, R. Currat, Phys. Rev. Lett. **57**, 1907 (1986).
1766. G. Kriza, G. Mihály, Phys. Rev. Lett. **56**, 2529 (1986).
1767. G. Grüner, Rev. Mod. Phys. **60**, 1129 (1988).
1768. D. Starešinić, K. Biljaković, W. Brütting, K. Hosseini, P. Monceau, H. Berger, F. Levy, Phys. Rev. B **65**, 165109 (2002).
1769. D. Starešinić, K. Hosseini, W. Brütting, K. Biljaković, E. Riedel, S. van Smaalen, Phys. Rev. B **69**, 113102 (2004).
1770. K. Biljakovic,´D. Starešinić, D. Dominko, J.C. Lasjaunias, Physica B **404**, 456 (2009).
1771. D. Starešinić, K. Biljaković, presented at the 5th Intern'al Discussion Meeting on Relaxation in Complex Systems, Lille, France, July (2005).
1772. R. Rosensweig, *Ferrohydrodynamics*, Cambridge University Press, Cambridge, (1985).
1773. B. Berkovsky, *Magnetic Fluids and Applications Handbook*, Begell House, New York, NY (1996).
1774. F. Cousin, E. Dubois, V. Cabuil, Phys. Rev. E, **68**, 021405 (2003).
1775. G. Mériguet, E. Dubois, V. Dupuis, R. Perzynski, J. Phys. Condens. Matter **18**, 10119 (2006).
1776. G. Mériguet, E. Dubois, V. Dupuis, R. Perzynski, AIP Conf. Proc. **708**, 124 (2004).
1777. E. Wandersman, V. Dupuis, E. Dubois, R. Perzynski, Phys. Rev. E **80**, 041504 (2009).

Index

C

D

E

F

T